ASM METALS Reference Book

Third Edition

Editor
Michael Bauccio

Acquisitions/Editorial
Veronica Flint

Technical Editor
Sunniva Collins

Production Project Manager
Suzanne Hampson

Production Project Specialist
Dawn Levicki

ASM INTERNATIONAL®

The Materials
Information Society

First printing, December 1993
Second printing, September 1994

This book is a collective effort involving hundreds of technical specialists. It brings together a wealth of information from worldwide sources to help scientists, engineers, and technicians solve current and long-range problems.

Comments, criticisms, and suggestions are invited, and should be forwarded to ASM International.

Library of Congress Cataloging in Publication Data

ASM International®
Materials Park, OH 44073-0002

ASM metals reference book/editor, Michael Bauccio.—3rd ed. p. cm.
ISBN: 0-87170-478-1
1. Metals—Handbooks, manuals, etc. 2. Metal-work—Handbooks, manuals, etc.
I. Bauccio, Michael. II. American Society for Metals.
III. Title: A.S.M. metals reference book.
IV. Title: Metals reference book. TA459.A78 1993
620.1'6—dc20
93-28716 CIP

Printed in the United States of America

Foreword

This two-volume handbook is a reference for engineering designers (materials engineers, metallurgists, etc.) involved in the complex and continuous process of materials selection.

Many types of materials are used for construction of commercial and military equipment. The primary materials used are metallics (for example, steel, aluminum, titanium, and nickel), nonmetallics (such as plastics and ceramics), and composites. Each of these classes of materials has a wide range of properties, presenting the designer with the formidable challenge of first choosing the proper class of material for a specific structure, and then determining the most appropriate material within the class selected. In making these decisions, the designer must review mechanical properties, corrosion resistance properties, and economic data for each material under consideration for completion of each structural engineering project.

Because the materials selection process depends to a large degree on mechanical and corrosion resistance properties, these handbooks emphasize the presentation of property data. Pertinent information on materials properties will assist the designer in selecting the most appropriate material for a specific application.

Michael L. Bauccio
Editor

About the Editor

Michael L. Bauccio has over 12 years of engineering experience and is an engineer with The Boeing Company, Seattle WA. Before joining Boeing he was an engineering specialist with Bell Helicopter Textron, Inc., Fort Worth TX. Mr. Bauccio has a BS in biological sciences from Loyola University, New Orleans LA; an MS in pharmacology and toxicology from St. John's University, Jamaica NY; and an MS in materials science and chemical engineering from the Sever Institute of Technology at Washington University, St. Louis MO.

Table of Contents

Glossary of Metallurgical and Metalworking Terms

A

abrasion. (1) A process in which hard particles or protuberances are forced against and moved along a solid surface. (2) A roughening or scratching of a surface due to *abrasive wear*. (3) The process of grinding or wearing away through the use of abrasives.

abrasive. (1) A hard substance used for *grinding, honing, lapping, superfinishing, polishing*, pressure blasting, or *barrel finishing*. Abrasives in common use are alumina, silicon carbide, boron carbide, diamond, cubic boron nitride, garnet, and quartz. (2) Hard particles, such as rocks, sand or fragments of certain hard metals, that wear away a surface when they move across it under pressure. See also *super-abrasives*.

abrasive belt. A coated abrasive product, in the form of a belt, used in production grinding and polishing.

abrasive blasting. A process for cleaning or finishing by means of an abrasive directed at high velocity against the workpiece.

abrasive disk. (1) A grinding wheel that is mounted on a steel plate, with the exposed flat side being used for grinding. (2) A disk-shaped, coated abrasive product.

abrasive erosion. Erosive wear caused by the relative motion of solid particles which are entrained in a fluid, moving nearly parallel to a solid surface. See also *erosion*.

abrasive flow machining. Removal of material by a viscous, abrasive media flowing under pressure through or across a workpiece.

abrasive jet machining. Material removal from a workpiece, by impingement of fine abrasive particles which are entrained in a focused, high-velocity gas stream.

abrasive machining. A machining process in which the points of abrasive particles are used as machining tools. Grinding is a typical abrasive machining process.

abrasive wear. The removal of material from a surface when hard particles slide or roll across the surface under pressure. The particles may be loose or may be part of another surface in contact with the surface being abraded. Compare with *adhesive wear*.

abrasive wheel. A grinding wheel composed of an abrasive grit and a bonding agent.

absolute density. See *density, absolute*.

Ac_{cm}, Ac_1, Ac_3, Ac_4. Defined under *transformation temperature*.

accelerated corrosion test. Method designed to approximate, in a short time, the deteriorating effect under normal long-term service conditions.

accelerated-life test. A method designed to approximate, in a short time, the deteriorating effect obtained under normal long-term service conditions. See also *artificial aging*.

accelerated testing. A test performed on materials or assemblies that is meant to produce failures caused by the same failure mechanism as expected in field operation but in significantly shorter time. The failure mechanism is accelerated by changing one or more of the controlling test parameters.

acicular ferrite. A highly substructured nonequiaxed *ferrite* formed upon continuous cooling by a mixed diffusion and shear mode of transformation that begins at a temperature slightly higher than the transformation temperature range for upper bainite. It is distinguished from *bainite* in that it has a limited amount of carbon available; thus, there is only a small amount of carbide present.

acicular ferrite steels. Ultralow carbon (%) steels having a microstructure consisting of either acicular ferrite (low-carbon bainite) or a mixture of acicular and equiaxed ferrite.

acid. A chemical substance that yields hydrogen ions (H^+) when dissolved in water. Compare with *base*. (2) A term applied to slags, refractories, and minerals containing a high percentage of silica.

acid bottom and lining. The inner bottom and lining of a melting furnace, consisting of materials like sand, siliceous rock, or silica brick that give an acid reaction at the operating temperature.

acid copper. (1) Copper electrodeposited from an acid solution of a copper salt, usually copper sulfate. (2) The solution referred to in (1).

acid embrittlement. A form of *hydrogen embrittlement* that may be induced in some metals by acid.

acid rain. Atmospheric precipitation with a pH below 5.6 to 5.7. Burning of fossil fuels for heat and power is the major factor in the generation of oxides of nitrogen and sulfur, which are converted into nitric and sulfuric acids washed down in the rain. See also *atmospheric corrosion*.

acid refractory. Siliceous ceramic materials of a high melting temperature, such as silica brick, used for metallurgical furnace linings. Compare with *basic refractories*.

acid steel. Steel melted in a furnace with an acid bottom and lining and under a slag containing an excess of an acid substance such as silica.

acoustic emission. A measure of integrity of a material, as determined by sound emission when a material is stressed. Ideally, emissions can be correlated with defects and/or incipient failure.

actinide metals. The group of radio-active elements of atomic numbers 89 through 103 of the periodic system—namely, thorium, protactinium, uranium, neptunium, plutonium, americium, curium, berkelium, californium, einsteinium, fermium, mendelevium, nobelium, and lawrencium.

activation. (1) The changing of a passive surface of a metal to a chemically active state. Contrast with *passivation*. (2) The (usually) chemical process of making a surface more receptive to bonding with a coating or an encapsulating material.

activation energy. The energy required for initiating a metallurgical reaction—for example, plastic flow, diffusion, chemical reaction. The activation energy may be calculated from the slope of the line obtained by plotting the natural log of the reaction rate versus the reciprocal of the absolute temperature.

active. The negative direction of *electrode potential*. Also used to describe corrosion and its associated potential range when an electrode potential is more negative than an adjacent depressed corrosion rate (passive) range.

active metal. A metal ready to corrode, or being corroded.

activity. A measure of the *chemical potential* of a substance, where the chemical potential is not equal to concentration, that allows mathematical relations equivalent to those for ideal systems to be used to correlate changes in an experimentally measured quantity with changes in chemical potential.

addition agent. (1) A substance added to a solution for the purpose of altering or controlling a process. Examples: wetting agents in acid pickles; brighteners or antipitting agents in plating solutions; inhibitors. (2) Any material added to a charge of molten metal in a bath or ladle to bring the alloy to specifications.

adhesion. (1) In frictional contacts, the attractive force between adjacent surfaces. In physical chemistry, adhesion denotes the attraction between a solid surface and a second (liquid or solid) phase. This definition is based on the assumption of a reversible equilibrium. In mechanical technology, adhesion is generally irreversible. In railway engineering, adhesion often means friction. (2) Force of attraction between the molecules (or atoms) of two different phases. Contrast with *cohesion*. (3) The state in which two surfaces are held together by interfacial forces, which may consist of valence forces, interlocking action, or both.

adhesive. A substance capable of holding materials together by surface attachment. Adhesive is a general term and includes, among others, cement, glue, mucilage, and paste.

adhesive bonding. A materials joining process in which an adhesive, placed between the faying surfaces (adherends) solidifies to produce an adhesive bond.

adhesive wear. (1) Wear by transference of material from one surface to another during relative motion due to a process of solid-phase welding. Particles that are removed from one surface are either permanently or temporarily attached to the other surface. (2) Wear due to localized bonding between contacting solid surfaces leading to material transfer between the two surfaces or loss from either surface. Compare with *abrasive wear*.

adjustable bed. Bed of a press designed so that the die space height can be varied conveniently.

Ae$_{cm}$, Ae$_1$, Ae$_3$, Ae$_4$. Defined under *transformation temperature*.

age hardening. Hardening by *aging* (heat treatment) usually after rapid cooling or cold working.

age softening. Spontaneous decrease of strength and hardness that takes place at room temperature in certain strain hardened alloys, especially those of aluminum.

aging. (1) The effect on materials of exposure to an environment for a prolonged interval of time. (2) The process of exposing materials to an environment for a prolonged interval of time in order to predict in-service lifetime.

aging (heat treatment). A change in the properties of certain metals and alloys that occurs at ambient or moderately elevated temperatures after hot working or a heat treatment (quench aging in ferrous alloys, natural or artificial aging in ferrous and nonferrous alloys) or after a cold working operation (strain aging). The change in properties is often, but not always, due to a phase change (precipitation), but never involves a change in chemical composition of the metal or alloy. See also *age hardening, artificial aging, interrupted aging, natural aging, overaging, precipitation hardening, precipitation heat treatment, progressive aging, quench aging, step aging*, and *strain aging*.

air acetylene welding (AAW). A fuel gas welding process in which coalescence is produced by heating with a gas flame or flames obtained from the combustion of acetylene with air, without the application of pressure, and with or without the use of filler metal.

air bend die. Angle forming dies in which the metal is formed without striking the bottom of the die. Metal contact is made at only three points in the cross section: the nose of the male die and the two edges of a V-shape die opening.

air bending. Bending in an *air bend die*.

air carbon arc cutting (AAC). An arc cutting process in which metals to be cut are melted by the heat of a carbon arc and the molten metal is removed by a blast of air.

air classification. The separation of metal powder into particle-size fractions by means of an air stream of controlled velocity; an application of the principle of *elutriation*.

air-hardening steel. A steel containing sufficient carbon and other alloying elements to harden fully during cooling in air or other gaseous media from a temperature above its transformation range. The term should be restricted to steels that are capable of being hardened by cooling in air in fairly large sections, about 2 in. (50 mm) or more in diameter. Same as self-hardening steel.

air-lift hammer. A type of gravity-drop hammer in which the ram is raised for each stroke by an air cylinder. Because length of stroke can be controlled, ram velocity and therefore the energy delivered to the workpiece can be varied. See also *drop hammer* and *gravity hammer*.

alclad. Composite wrought product comprised of an aluminum alloy core having one or both surfaces a metallurgically bonded aluminum or aluminum alloy coating that is anodic to the core and thus electrochemically protects the core against corrosion.

alkali metal. A metal in group IA of the periodic system—namely, lithium, sodium, potassium, rubidium, cesium, and francium. They form strongly alkaline hydroxides, hence the name.

alkaline cleaner. A material blended from alkali hydroxides and such alkaline salts as borates, carbonates, phosphates, or silicates. The cleaning action may be enhanced by the addition of surface-active agents and special solvents.

alkaline earth metal. A metal in group IIA of the period system—namely, beryllium, magnesium, calcium, strontium, barium, and radium—so called because the oxides or "earths" of calcium, strontium, and barium were found by the early chemists to be alkaline in reaction.

alligatoring. (1) Pronounced wide cracking over the entire surface of a coating having the appearance of alligator hide. (2) The longitudinal splitting of flat slabs in a plane parallel to the rolled surface. Also called fish-mouthing.

alligator skin. See *orange peel*.

allotriomorphic crystal. A crystal whose lattice structure is normal but whose external surfaces are not bounded by regular crystal faces; rather, the external surfaces are impressed by contact with other crystals or another surface such as a mold wall, or are irregularly shaped because of nonuniform growth. Compare with *idiomorphic crystal*.

allotropy. (1) A near synonym for *polymorphism*. Allotropy is generally restricted to describing polymorphic behavior in elements, terminal phases, and alloys whose behavior closely parallels that of the predominant constituent element. (2) The existence of a substance, especially an element, in two or more physical states (for example, crystals).

allowance. (1) The specified difference in limiting sizes (minimum clearance or maximum interference) between mating parts, as computed arithmetically from the specified dimensions and tolerances of each part. (2) In a foundry, the specified clearance. The difference in limiting sizes, such as minimum clearance or maximum interference between mating parts, as computed arithmetically. See also *tolerance*.

alloy. (1) A substance having metallic properties and being composed of two or more chemical elements of which at least one is a *metal*. (2) To make or melt an alloy.

alloy cast iron. Highly alloyed cast irons containing more than 3% alloy content. Alloy cast irons may be of a type of white iron, gray iron, or ductile iron.

alloying element. An element added to and remaining in a metal that changes structure and properties.

alloy plating. The codeposition of two or more metallic elements.

alloy powder, alloyed powder. A metal powder consisting of at least two constituents that are partially or completely alloyed with each other.

alloy steel. Steel containing specified quantities of alloying elements (other than carbon and the commonly accepted amounts of manganese, copper, silicon, sulfur, and phosphorus) within the limits recognized for constructional alloy steels, added to effect changes in mechanical or physical properties.

alloy system. A complete series of compositions produced by mixing in all proportions any group of two or more components, at least one of which is a metal.

all-position electrode. In arc welding, a filler-metal electrode for depositing weld metal in the flat, horizontal, overhead, and vertical positions.

all-weld-metal test specimen. A test specimen wherein the portion being tested is composed wholly of weld metal.

alpha brass. A solid-solution phase of one or more alloying elements in copper having the same crystal lattice as copper.

alpha ferrite. See *ferrite*.

alternate immersion test. A corrosion test in which the specimens are intermittently exposed to a liquid medium at definite time intervals.

Alumel. A nickel-base alloy containing about 2.5 Mn, 2 Al, and 1 Si used chiefly as a component of pyrometric thermocouples.

aluminizing. Forming of an aluminum or aluminum alloy coating on a metal by hot dipping, hot spraying, or diffusion.

amalgam. A dental alloy produced by combining mercury with alloy particles of silver, tin, copper, and sometimes zinc.

amorphous. Not having a crystal structure; noncrystalline.

amorphous solid. A rigid material whose structure lacks crystalline periodicity; that is, the pattern of its constituent atoms or molecules does not repeat periodically in three dimensions. See also *metallic glass*.

anelastic deformation. Any portion of the total deformation of a body that occurs as a function of time when load is applied and which disappears completely after a period of time when the load is removed.

anelasticity. The property of solids by virtue of which strain is not a single-value function of stress in the low-stress range where no permanent set occurs.

angle of bite. In the rolling of metals, the location where all of the force is transmitted through the rolls; the maximum attainable angle between the roll radius at the first contact and the line of roll centers. Operating angles less than the angle of bite are termed contact angles or rolling angles.

angle of nip. In rolling, the *angle of bite*. In roll, jaw, or gyratory crushing, the entrance angle formed by the tangents at the two points of contact between the working surfaces and the (assumed) spherical particles to be crushed.

anion. A negatively charged ion that migrates through the electrolyte toward the *anode* under the influence of a potential gradient. See also *cation* and *ion*.

anisotropy. The characteristic of exhibiting different values of a property in different directions with respect to a fixed reference system in the material.

annealing carbon. See *temper carbon*.

annealing. A generic term denoting a treatment consisting of heating to and holding at a suitable temperature followed by cooling at a suitable rate, used primarily to soften metallic materials, but also to simultaneously produce desired changes in other properties or in microstructure. The purpose of such changes may be, but is not confined to: improvement of machinability, facilitation of cold work, improvement of mechanical or electrical properties, and/or increase in stability of dimensions. When the term is used unqualifiedly, full annealing is implied. When applied only for the relief of stress, the process is properly called *stress relieving* or stress-relief annealing.

In ferrous alloys, annealing usually is done above the upper critical temperature, but the time-temperature cycles vary widely both in maximum temperature attained and in cooling rate employed, depending on composition, material condition, and results desired. When applicable, the following commercial process names should be used: *black annealing, blue annealing, box annealing, bright annealing, cycle annealing, flame annealing, full annealing, graphitizing*, in-process annealing, *isothermal annealing, malleabilizing*, orientation annealing, *process annealing, quench annealing, spheroidizing, subcritical annealing*.

In nonferrous alloys, annealing cycles are designed to: (a) remove part or all of the effects of cold working (recrystallization may or may not be involved); (b) cause substantially complete coalescence of precipitates from solid solution in relatively coarse form; or (c) both, depending on composition and material condition. Specific process names in commercial use are *final annealing, full annealing, intermediate annealing, partial annealing, recrystallization annealing*, stress-relief annealing, *anneal to temper*.

annealing twin. A *twin* formed in a crystal during recrystallization.

anneal to temper. A final partial anneal that softens a cold-worked nonferrous alloy to a specified level of hardness or tensile strength.

anode. (1) The electrode of an electrolyte cell at which oxidation occurs. Electrons flow away from the anode in the external circuit. It is usually at the electrode that corrosion occurs and metal ions enter solution. (2) The positive (electron-deficient) electrode in an electrochemical circuit. Contrast with *cathode*.

anode copper. Special-shaped copper slabs, resulting from the refinement of *blister copper* in a reverberatory furnace, used as anodes in electrolytic refinement.

anode effect. The effect produced by polarization of the *anode* in electrolysis. It is characterized by a sudden increase in voltage and a corresponding decrease in amperage due to the anode becoming virtually separated from the electrolyte by a gas film.

anode efficiency. Current efficiency at the *anode*.

anode film. (1) The portion of solution in immediate contact with the *anode*, especially if the concentration gradient is steep. (2) The outer layer of the anode itself.

anode polarization. See *polarization*.

anodic cleaning. Electrolytic cleaning in which the work is the anode. Also called reverse-current cleaning.

anodic coating. A film on a metal surface resulting from an electrolytic treatment at the anode.

anodic pickling. *Electrolytic pickling* in which the work is the anode.

anodic polarization. The change of the electrode potential in the noble (positive) direction due to current flow. See also *polarization*.

anodic protection. (1) A technique to reduce the corrosion rate of a metal by polarizing it into its passive region, where dissolution rates are low. (2) Imposing an external electrical potential to protect a metal from corrosive attack. (Applicable only to metals that show active-passive behavior.) Contrast with *cathodic protection*.

anodic reaction. Electrode reaction equivalent to a transfer of positive charge from the electronic to the ionic conductor. An anodic reaction is an oxidation process. An example common in corrosion is: $Me \rightarrow Me^{n+} + ne^-$.

anodizing. Forming a *conversion coating* on a metal surface by anodic oxidation; most frequently applied to aluminum.

anolyte. The electrolyte adjacent to the *anode* in an *electrolytic cell*.

antiferromagnetic material. A material wherein interatomic forced hold the elementary atomic magnets (electron spins) of a solid in alignment, a state similar to that of a *ferromagnetic material* but with the difference that equals numbers of elementary magnets (spins) face in opposite directions and are antiparallel, causing the solid to be weakly magnetic, that is, paramagnetic, instead of ferromagnetic.

antifriction material. A material that exhibits low-friction or self-lubricating properties.

antipitting agent. An *addition agent* for electroplating solutions to prevent the formation of pits or large pores in the electrodeposit.

anvil. A large, heavy metal block that supports the frame structure and holds the stationary die of a forging hammer. Also, the metal block on which blacksmith forgings are made.

anvil cap. Same as *sow block*.

apparent density. (1) The weight per unit volume of a powder, in contrast to the weight per unit volume of the individual particles. (2) The weight per unit volume of a porous solid, where the unit volume is determined from external dimensions of the mass. Apparent density is always less than the true density of the material itself.

$Ar_{cm}, Ar_1, Ar_3, Ar_4, Ar-, Ar.$ Defined under *transformation temperature*.

arbor. (1) In machine grinding, the spindle on which the wheel is mounted. (2) In machine cutting, a shaft or bar for holding and driving the cutter. (3) In founding, a metal shape embedded in green sand or dry sand cores to support the sand or the applied load during casting.

arbor press. A machine used for forcing arbors or mandrels into drilled or bored parts preparatory to turning or grinding. Also used for forcing bushings, shafts, or pins into or out of holes.

arbor-type cutter. A cutter having a hole for mounting on an arbor and usually having a keyway for a driving key.

arc. A luminous discharge of electrical current crossing the gap between two electrodes.

arc blow. The deflection of an electric arc from its normal path because of magnetic forces.

arc brazing. A brazing process in which the heat required is obtained from an electric arc.

arc cutting. A group of cutting processes which melts the metals to be cut with the heat of an arc between an electrode and the base metal. See *carbon arc cutting*, *metal arc cutting*, *gas metal arc cutting*, *gas tungsten arc cutting*, *plasma arc cutting*, and *air carbon arc cutting*. Compare with *oxygen arc cutting*.

arc furnace. A furnace in which metal is melted either directly by an electric arc between an electrode and the work or indirectly by an arc between two electrodes adjacent to the metal.

arc gouging. An arc cutting process variation used to form a bevel or groove.

arc melting. Melting metal in an electric arc furnace.

arc oxygen cutting. See preferred term *oxygen arc cutting*.

arc plasma. See *plasma-arc cutting*.

arc time. The time an arc is maintained in making an arc weld. Also known as *weld time*.

arc welding (AW). A group of welding processes which produces coalescence of metals by heating them with an arc, with or without the application of pressure, and with or without the use of filler metal.

argon oxygen decarburization (AOD). A secondary refining process for the controlled oxidation of carbon in a steel melt. In the AOD process, oxygen, argon, and nitrogen are injected into a molten metal bath through submerged, side-mounted tuyeres.

artifact. A feature of artificial character, such as a scratch or a piece of dust on a metallographic specimen, that can be erroneously interpreted as a real feature.

artificial aging. Aging above room temperature. See *aging (heat treatment)*. Compare with *natural aging*.

as-cast condition. Castings as removed from the mold without subsequent heat treatment.

athermal transformation. A reaction that proceeds without benefit of thermal fluctuations—that is, thermal activation is not required. Such reactions are diffusionless and can take place with great speed when the driving force is sufficiently high. For example, many martensitic transformations occur athermally on cooling, even at relatively low temperatures, because of

the progressively increasing drive force. In contrast, a reaction that occurs at constant temperature is an *isothermal transformation*; thermal activation is necessary in this case and the reaction proceeds as a function of time.

atmospheric corrosion. The gradual degradation or alteration of a material by contact with substances present in the atmosphere, such as oxygen, carbon dioxide, water vapor, and sulfur and chlorine compounds.

atmospheric riser. A riser that uses atmospheric pressure to aid feeding. Essentially, a *blind riser* into which a small core or rod protrudes; the function of the core or rod is to provide an open passage so that the molten interior of the riser will not be under a partial vacuum when metal is withdrawn to feed the casting but will always be under atmospheric pressure.

atomic hydrogen welding. An arc welding process that fuses metals together by heating them with an electric arc maintained between two metal electrodes enveloped in a stream of hydrogen. Shielding is provided by the hydrogen, which also carries heat by molecular dissociation and subsequent recombination. Pressure may or may not be used and filler metal may or may not be used. (This process is now of limited industrial significance.)

atomic percent. The number of atoms of an element in a total of 100 representative atoms of a substance.

atomization. The disintegration of a molten metal into particles by a rapidly moving gas or liquid stream or by other means.

attritious wear. Wear of abrasive grains in grinding such that the sharp edges gradually become rounded. A grinding wheel that has undergone such wear usually has a glazed appearance.

attritor. A high-intensity ball mill whose drum is stationary and whose balls are agitated by rotating baffles, paddles, or rods at right angle to the drum axis.

attritor grinding. The intensive grinding or alloying in an attritor. Examples: milling of carbides and binder metal powders and mechanical alloying of hard dispersoid particles with softer metal or alloy powders. See also *mechanical alloying*.

ausforming. Thermomechanical treatment of steel in the metastable austenitic condition below the recrystallization temperature followed by quenching to obtain martensite and/or bainite.

austempered ductile iron. A moderately alloyed *ductile iron* that is austempered for high strength with appreciable ductility. See also *austempering*.

austempering. A heat treatment for ferrous alloys in which a part is quenched from the austenitizing temperature at a rate fast enough to avoid formation of ferrite or pearlite and then held at a temperature just above M_s until transformation to bainite is complete. Although designated as bainite in both austempered steel and austempered ductile iron (ADI), austempered steel consists of two phase mixtures containing ferrite and carbide, while austempered ductile iron consists of two phase mixtures containing ferrite and austenite.

austenite. A solid solution of one or more elements in face-centered cubic iron (gamma iron). Unless otherwise designated (such as nickel austenite), the solute is generally assumed to be carbon.

austenitic grain size. The size attained by the grains in steel when heated to the austenitic region. This may be revealed by appropriate etching of cross sections after cooling to room temperature.

austenitic manganese steel. A cast, wear-resistant material containing about 1.2% C and 12% Mn. Used primarily in the fields of earthmoving, mining, quarrying, railroading, ore processing, lumbering, and in the manufacture of cement and clay products. Also known as Hadfield steel.

austenitic steel. An alloy steel whose structure is normally austenitic at room temperature.

austenitizing. Forming austenite by heating a ferrous alloy into the transformation range (partial austenitizing) or above the transformation range (complete austenitizing). When used without qualification, the term implies complete austenitizing.

autogenous weld. A fusion weld made without the addition of filler metal.

automatic press. A press in which the work is fed mechanically through the press in synchronism with the press action. An automation press is an automatic press that, in addition, is provided with built-in electrical and pneumatic control equipment.

autoradiography. An inspection technique in which radiation spontaneously emitted by a material is recorded photographically. The radiation is emitted by radioisotopes that are (a) produced in a metal by bombarding it with neutrons, (b) added to a metal such as by alloying, or (c) contained within a cavity in a metal part. The technique serves to locate the position of the radioactive element or compound.

auxiliary anode. In electroplating, a supplementary *anode* positioned so as to raise the current density on a certain area of the *cathode* and thus obtain better distribution of plating.

auxiliary electrode. An *electrode* commonly used in polarization studies to pass current to or from a test electrode. It is usually made from a noncorroding material.

axial rake. For angular (not helical) flutes, the angle between a plane containing the tooth face and the axial plane through the tooth point. See also *face mill* for definition of nomenclature.

axial relief. The relief or clearance behind the end cutting edge of a milling cutter. See also *face mill*.

axial rolls. In *ring rolling*, vertically displaceable, taped rolls mounted in a horizontally displaceable frame opposite to, but on the same centerline as, the main roll and rolling mandrel. The axial rolls control ring height during rolling.

axis (weld). A line through the length of a weld, perpendicular to and at the geometric center of its cross section.

B

Babbitt metal. A nonferrous bearing alloy originated by Isaac Babbitt in 1839. Currently, the term includes several tin-base alloys consisting mainly of various amounts of copper, antimony, tin, and lead. Lead-base Babbitt metals are also used.

back draft. A reverse taper on a casting pattern or a forging die that prevents the pattern or forged stock from being removed from the cavity.

backfire. The momentary recession of the flame into the welding tip or cutting tip followed by immediate reappearance or complete extinction of the flame. See also *flashback*.

backhand welding. A welding technique in which the welding torch or gun is directed opposite to the progress of welding. Sometimes referred to as the "pull gun technique" in gas metal arc welding and flux-cored arc welding. Compare with

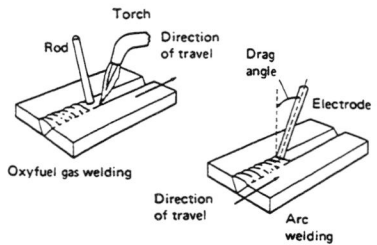

Backhand welding

forehand welding. See also *travel angle*, *work angle*, and *drag angle*.

backing. (1) In grinding, the material (paper, cloth, or fiber) that serves as the base for coated abrasives. (2) In welding, a material placed under or behind a joint to enhance the quality of the weld at the root. It may be a metal backing ring or strip; a pass of weld metal; or a nonmetal such as carbon, granular flux or a protective gas. (3) In plain bearings, that part of the bearing to which the bearing alloy is attached, normally by a metallurgical bond.

backoff. A rapid withdrawal of a grinding wheel or cutting tool from contact with a workpiece.

back rake. The angle on a single-point turning tool corresponding to axial rake in milling. It is the angle measured between the plane of the tool face and the reference plane and lies in a plane perpendicular to the axis of the work material and the base of the tool. See figure accompanying *single-point tool*.

backstep sequence. A longitudinal sequence in which the weld bead increments are deposited in the direction opposite to the progress of welding the joint. See also *block sequence* and *cascade sequence*.

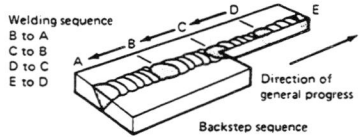

Backstep sequence

backward extrusion. Same as *indirect extrusion*. See *extrusion*.

back weld. A weld deposited at the back of a single groove weld.

bainite. A metastable aggregate of *ferrite* and *cementite* resulting from the transformation of *austenite* at temperatures below the *pearlite* range but above M_s, the martensite start temperature. Upper bainite is an aggregate that contains parallel lath-shape units of ferrite, produces the so-called "feathery" appearance in optical microscopy, and is formed above approximately 350 °C (660 °F). Lower bainite, which has an acicular appearance similar to tempered martensite, is formed below approximately 350 °C (660 °F).

bainitic hardening. Quench-hardening treatment resulting principally in the formation of *bainite*.

Bakelite. A proprietary name for a phenolic thermosetting resin used as a plastic mounting material for metallographic samples.

baking. (1) Heating to a low temperature in order to remove gases. (2) Curing or hardening surface coatings such as paints by exposure to heat. (3) Heating to drive off moisture, as in baking of sand cores after molding.

ball burnishing. (1) Same as *ball sizing.* (2) Removing burrs and polishing small stampings and small machined parts by *tumbling* in the presence of metal balls.

ball mill. A machine consisting of a rotating hollow cylinder partly filled with metal balls (usually hardened steel or white cast iron) or sometimes pebbles; used to pulverize crushed ores or other substances such as pigments or ceramics.

ball milling. A method of grinding and mixing material, with or without liquid, in a rotating cylinder or conical mill partially filled with grinding media such as balls or pebbles.

ball sizing. Sizing and finishing a hole by forcing a ball of suitable size, finish, and hardness through the hole or by using a burnishing bar or broach consisting of a series of spherical lands of gradually increasing size coaxially arranged. Also called *ball burnishing,* and sometimes ball broaching.

banded structure. A segregated structure consisting of alternating nearly parallel bands of different composition, typically aligned in the direction of primary hot working.

banding. Inhomogeneous distribution of alloying elements or phases aligned in filaments or plates parallel to the direction of working. See also *banded structure, ferrite-pearlite banding,* and *segregation banding.*

band mark. An indentation in carbon steel or strip caused by external pressure on the packaging band around cut lengths or coils; it may occur in handling, transit, or storage.

bands. Hot-rolled steel strip, usually produced for rerolling into thinner sheet or strip. Also known as hot bands or band steel.

bar. (1) A section hot rolled from a *billet* to a form, such as round, hexagonal, octagonal, square, or rectangular, with sharp or rounded corners or edges and a cross-sectional area of less than 105 cm^2 (16 in.^{2}). (2) A solid section that is long in relationship to its cross-sectional dimensions, having a completely symmetrical cross section and a width or greatest distance between parallel faces of 9.5 mm (3/8) in. or more. (3) An obsolete unit of pressure equal to 100 kPa.

bare electrode. A filler metal electrode consisting of a single metal or alloy that has been produced into a wire, strip, or bar form and that has had no coating or cover-

ing applied to it other than that which was incidental to its manufacture or preservation.

bar folder. A machine in which a folding bar or wing is used to bend a metal sheet whose edge is clamped between the upper folding leaf and the lower stationary jaw into a narrow, sharp, close, and accurate fold along the edge. It is also capable of making rounded folds such as those used in wiring. A universal folder is more versatile in that it is limited to width only by the dimensions of the sheet.

bark. The decarburized layer just beneath the scale that results from heating steel in an oxidizing atmosphere.

Barkhausen effect. The sequence of abrupt changes in magnetic induction occurring when the magnetizing force acting on a ferromagnetic specimen is varied.

barrel cleaning. Mechanical or electrolytic cleaning of metal in rotating equipment.

barrel finishing. Improving the surface finish of workpieces by processing them in rotating equipment along with abrasive particles that may be suspended in a liquid. The barrel is normally loaded about 60% full with a mixture of parts, media, compound, and water.

barreling. Convexity of the surfaces of cylindrical or conical bodies, often produced unintentionally during upsetting or as a natural consequence during compression testing. See also *compression test.*

barrel plating. Plating articles in a rotating container, usually a perforated cylinder that operates at least partially submerged in a solution.

barstock. Same as *bar.*

base. (1) A chemical substance that yields hydroxyl ions (OH) when dissolved in water. Compare with *acid.* (2) The surface on which a single-point tool rests when held in a tool post. Also known as heel. (3) In forging, see *anvil.*

base material. The material to be welded, brazed, soldered, or cut. See also *base metal* and *substrate.*

base metal. (1) The metal present in the largest proportion in an alloy; brass, for example, is a copper-base alloy. (2) The metal to be brazed, cut, soldered, or welded. (3) After welding, that part of the metal which was not melted. (4) A metal that readily oxidizes, or that dissolves to form ions. Contrast with *noble metal* (2).

basic bottom and lining. The inner bottom and lining of a melting furnace, consisting of materials such as crushed burned dolomite, magnesite, magnesite bricks, or basic slag that give a basic reaction at the operating temperature.

basic oxygen furnace. A large tiltable vessel lined with basic refractory material which is the principal type of furnace for modern steelmaking. After the furnace is charged with molten pig iron (which usually comprises 65 to 75% of the charge), scrap steel, and fluxes, a lance is brought down near the surface of the molten metal and a jet of high-velocity oxygen impinges on the metal. The oxygen reacts with carbon and other impurities in the steel to form liquid compounds that dissolve in the slag and gases that escape from the top of the vessel.

basic refractories. Refractories whose major constituent is lime, magnesia, or both, and which may react chemically with acid refractories, acid slags, or acid fluxes at high temperatures. Basic refractories are used for furnace linings. Compare with *acid refractory.*

basic steel. Steel melted in a furnace with a *basic bottom and lining* and under a slag containing an excess of a basic substance such as magnesia or lime.

basin. Same as *pouring basin.*

basis metal. The original metal to which one or more coatings are applied.

batch. A quantity of materials formed during the same process or in one continuous process and having identical characteristics throughout. See also *lot.*

batch furnace. A furnace used to heat treat a single load at a time. Batch-type furnaces are necessary for large parts such as heavy forgings and are preferred for complex alloy grades requiring long cycles.

Bauschinger effect. The phenomenon by which plastic deformation increases yield strength in the direction of plastic flow and decreases it in other directions.

bauxite. A whitish to reddish mineral composed largely of hydrates of alumina having a composition of $Al_2O_3 \cdot 2H_2O$. It is the most important ore (source) of aluminum, alumina abrasives, and alumina-based refractories.

Bayer process. A process for extracting alumina from bauxite ore before the electrolytic reduction. The bauxite is digested in a solution of sodium hydroxide, which converts the alumina to soluble aluminate. After the "red mud" residue has been filtered out, aluminum hydroxide is precipitated, filtered out, and calcined to alumina.

beach marks. Macroscopic progression marks on a fatigue fracture or stress-corrosion cracking surface that indicate successive positions of the advancing crack front. The classic appearance is of irregular elliptical or semielliptical rings, radiating outward from one or more origins.

Beach marks (also known as clamshell marks or arrest marks) are typically found on service fractures where the part is loaded randomly, intermittently, or with periodic variations in mean stress or alternating stress. See also *striation*.

bead. (1) Half-round cavity in a mold, or half-round projection or molding on a casting. (2) A single deposit of weld metal produced by fusion.

beaded flange. A flange reinforced by a low ridge, used mostly around a hole.

beading. Raising a ridge or projection on sheet metal.

bead weld. See preferred term *surfacing weld*.

bearing bronzes. Bronzes used for bearing applications. Two common types of bearing bronzes are copper-base alloys containing 5 to 20 wt% tin and a small amount of phosphorus (*phosphor bronzes*) and copper-base alloys containing up to 10 wt% tin and up to 30 wt% lead (*leaded bronzes*).

bearing steels. *Alloy steels* used to produce rolling-element bearings. Typically, bearings have been manufactured from both high-carbon (1.00%) and low-carbon (0.20%) steels. The high-carbon steels are used in either a through-hardened or a surface induction-hardened condition. Low-carbon bearing steels are carburized to provide the necessary surface hardness while maintaining desirable core properties.

bearing strength. The maximum bearing stress that can be sustained. Also, the bearing stress at that point on the stress-strain curve at which the tangent is equal to the bearing stress divided by n% of the bearing hole diameter.

bearing stress. The shear load on a mechanical joint (such as a pinned or riveted joint) divided by the effective bearing area. The effective bearing area of a riveted joint, for example, is the sum of the diameters of all rivets times the thickness of the loaded member.

bearing test. A method of determining the response to stress (load) of sheet products that are subjected to riveting, bolting, or a similar fastening procedure. The purpose of the test is to determine the *bearing strength* of the material and to measure the *bearing stress* versus the deformation of the hole created by a pin or rod of circular cross section that pierces the sheet perpendicular to the surface.

bed. (1) The stationary portion of a press structure that usually rests on the floor or foundation, forming the support for the remaining parts of the press and the pressing load. The *bolster* and sometimes the lower die are mounted on the top surface of the bed. (2) For machine tools, the portion of the main frame that supports the tool, the work, or both. (3) Stationary part of the shear frame that supports the material being sheared and the fixed blade.

Beilby layer. A layer of metal disturbed by mechanical working, wear, or mechanical polishing presumed to be without regular crystalline structure (amorphous); originally applied to grain boundaries.

belt furnace. A continuous-type furnace which uses a mesh-type or cast-link belt to carry parts through the furnace.

belt grinding. Grinding with an *abrasive belt*.

bench molding. Casting sand molds by hand tamping loose or production patterns at a bench without the assistance of air or hydraulic action.

bench press. Any small press that can be mounted on a bench or table.

bend allowance. The length of the arc of the neutral axis between the tangent points of a bend.

bend angle. The angle through which a bending operation is performed, that is, the supplementary angle to that formed by the two bend tangent lines or planes.

bending. The straining of material, usually flat sheet or strip metal, by moving it around a straight axis lying in the neutral plane. Metal flow takes place within the plastic range of the metal, so that the bent part retains a *permanent set* after removal of the applied stress. The cross section of the bend inward from the neutral plane is in compression; the rest of the bend is in tension.

bending brake. A form of open-frame single-action press that is comparatively wide between the housings, with a bed designed for holding long, narrow forming edges or dies. Used for bending and forming strip, plate, and sheet (into boxes, panels, roof decks, and so on). Also known as *press brake*.

bending dies. Dies used in presses for bending sheet metal or wire parts into various shapes. The work is done by the punch pushing the stock into cavities or depressions of similar shape in the die or by auxiliary attachments operated by the descending punch.

bending moment. The algebraic sum of the couples or the moments of the external forces, or both, to the left or right of any section on a member subjected to bending by couples or transverse forces, or both.

bending rolls. Various types of machinery equipped with two or more rolls to form curved sheet and sections.

bend or twist (defect). Distortion similar to warpage generally caused during forging or trimming operations. When the distortion is along the length of the part, it is termed bend; when across the width, it is termed twist. When bend or twist exceeds tolerance, it is considered a defect. Corrective action consists of hand straightening, machine straightening, or cold restriking.

bend radius. (1) The inside radius of a bend section. (2) The radius of a tool around which metal is bent during fabrication.

bend tangent. A tangent point at which a bending arc ceases or changes.

bend test. A test for determining relative ductility of metal that is to be formed (usually sheet, strip, plate, or wire) and for determining soundness and toughness of metal (after welding, for example). The specimen is usually bent over a specified diameter through a specified angle for a specified number of cycles.

beneficiation. Concentration or other preparation of ore for smelting.

bentonite. A colloidal claylike substance derived from the decomposition of volcanic ash composed chiefly of the minerals of the montmorillonite family. It is used for bonding molding sand.

beryllium bronze. See preferred term *beryllium-copper*.

beryllium-copper. Copper-base alloys containing not more than 3% Be. Available in both cast and wrought forms, these alloys rank high among copper alloys in attainable strength, while retaining useful levels of electrical and thermal conductivity.

beryllium-nickel. Age-hardenable nickel-base alloys containing up to 2.75% Be. Wrought beryllium nickel alloys are used primarily as mechanical and electrical/electronic components. Cast alloys are used in molds and cores for glass and polymer molding, diamond drill bit matrices, and cast turbine parts.

bessemer process. A process for making steel by blowing air through molten pig iron contained in a refractory lined vessel so as to remove by oxidation most of the carbon, silicon, and manganese. This process is essentially obsolete in the United States.

beta ray. A ray of electrons emitted during the spontaneous disintegration of certain atomic nuclei.

beta structure. A Hume-Rothery designation for structurally analogous body-centered cubic phases (similar to beta brass) or electron compounds that have ratios of three valence electrons to two atoms. Not to be confused with a beta phase on a phase diagram.

bevel. See preferred term, *corner angle*, and also the figure accompanying *face mill*.

bevel angle. The angle formed between the prepared edge of a member and a plane perpendicular to the surface of the member.

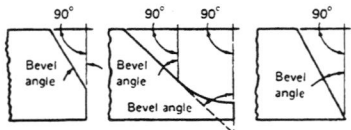

bevel flanging. Same as *flaring*.

biaxiality. In a *biaxial stress* state, the ratio of the smaller to the larger principal stress.

biaxial stress. A state of stress in which only one of the *principal stresses* is zero, the other two usually being in tension.

billet. (1) A semifinished section that is hot rolled from a metal *ingot*, with a rectangular cross section usually ranging from 105 to 230 cm^2 (16 to 36 in.^{2}), the width being less than twice the thickness. Where the cross section exceeds 230 cm^2 (36 in.^{2}), the term *bloom* is properly but not universally used. Sizes smaller than 105 cm^2 (16 in.^{2}) are usually termed bars. (2) A solid semifinished round or square product that has been hot worked by forging, rolling, or extrusion. See also *bar*.

billet mill. A primary rolling mill used for making steel billets.

binary alloy. An alloy containing only two component elements.

binary system. The complete series of compositions produced by mixing a pair of components in all proportions.

binder. (1) In founding, a material, other than water, added to foundry sand to bind the particles together, sometimes with the use of heat. (2) In powder technology, a cementing medium: either a material added to the powder to increase the green strength of the compact, which is expelled during sintering; or a material (usually of relatively low melting point) added to a powder mixture for the specific purpose of cementing together powder particles that alone would not sinter into a strong body.

binder metal. A metal used as a binder. An example would be cobalt in cemented carbides.

biological corrosion. Deterioration of metals as a result of the metabolic activity of microorganisms. Also known as biofouling.

bipolar electrode. An *electrode* in an *electrolytic cell* that is not mechanically connected to the power supply, but is so placed in the electrolyte, between the *anode* and *cathode*, that the part nearer the anode becomes cathodic and the part nearer the cathode becomes anodic. Also called intermediate electrode.

bipolar field. A longitudinal magnetic field that creates two magnetic poles within a piece of material. Compare with *circular field*.

biscuit. (1) An upset blank for drop forging. (2) A small cake of primary metal (such as uranium made from uranium tetrafluoride and magnesium by bomb reduction). Compare with *derby* and *dingot*.

black annealing. Box annealing or pot annealing ferrous alloy sheet, strip, or wire impart a black color to the oxidized surface. See also *box annealing*.

blackheart malleable. See *malleable iron*.

blacking. Carbonaceous materials, such as graphite or powdered carbon, usually mixed with a binder and frequently carried in suspension in water or other liquid used as a thin facing applied to surfaces of molds or cores to improve casting finish.

black oxide. A black finish on a metal produced by immersing it in hot oxidizing salts or salt solutions.

blank. (1) In forming, a piece of sheet metal, produced in cutting dies, that is usually subjected to further press operations. (2) A pressed, presintered, or fully sintered powder metallurgy compact, usually in the unfinished condition and requiring cutting, machining, or some other operation to produce the final shape. (3) A piece of stock from which a forging is made, often called a *slug* or *multiple*.

blank carburizing. Simulating the carburizing operation without introducing carbon. This is usually accomplished by using an inert material in place of the carburizing agent, or by applying a suitable protective coating to the ferrous alloy.

blankholder. (1) The part of a drawing or forming die that holds the workpiece against the draw ring to control metal flow. (2) The part of a drawing or forming die that restrains the movement of the workpiece to avoid wrinkling or tearing of the metal.

blanking. The operation of punching, cutting, or shearing a piece out of stock to a predetermined shape.

blank nitriding. Simulating the nitriding operation without introducing nitrogen. This is usually accomplished by using an inert material in place of the nitriding agent or by applying a suitable protective coating to the ferrous alloy.

blast furnace. A shaft furnace in which solid fuel is burned with an air blast to smelt ore in a continuous operation. Where the temperature must be high, as in the production of pig iron, the air is preheated. Where the temperature can be lower, as in smelting of copper, lead, and tin ores, a smaller furnace is economical, and preheating of the blast is not required.

blasting or blast cleaning. A process for cleaning or finishing metal objects with an air blast or centrifugal wheel that throws abrasive particles against the surface of the workpiece. Small, irregular particles of metal are used as the abrasive in gritblasting; sand, in sandblasting; and steel, in shotblasting.

blended sand. A mixture of sands of different grain size and clay content that provides suitable characteristics for foundry use.

blending. In powder metallurgy, the thorough intermingling of powders of the same nominal composition (not to be confused with *mixing*).

blind riser. A *riser* that does not extend through the top of the mold.

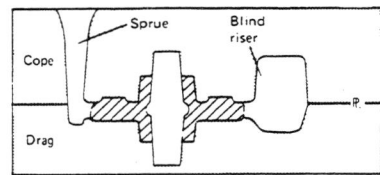

blister. (1) A casting defect, on or near the surface of the metal, resulting from the expansion of gas in a subsurface zone. It is characterized by a smooth bump on the surface of the casting and a hole inside the casting directly below the bump. (2) A raised area, often dome shaped, resulting from loss of adhesion between a coating or deposit and the basis metal.

blister copper. An impure intermediate product in the refining of copper, produced by blowing copper *matte* in a converter, the name being derived from the large blisters on the cast surface that result from the liberation of SO_2 and other gases.

block. A preliminary forging operation that roughly distributes metal preparatory for *finish*.

block and finish. The forging operation in which a part to be forged is blocked and finished in one heat through the use of tooling having both a block impression and a finish impression in the same die block.

blocker. The impression in the dies (often one of a series of impressions in a single die set) that imparts to the forging an intermediate shape, preparatory to forging of the final shape. Also called blocking impression.

blocker dies. Forging dies having generous contours, large radii, draft angles of 7° or more, and liberal finish allowances. See also *finish allowance*.

blocker-type forging. A forging that approximates the general shape of the final part with relatively generous *finish allowance* and radii. Such forgings are sometimes specified to reduce die costs where only a small number of forgings are described and the cost of machining each part to its final shape is not excessive.

blocking. In forging, a preliminary operation performed in closed dies, usually hot, to position metal properly so that in the finish operation the dies will be filled correctly. Blocking can ensure proper working of the material and can increase die life.

blocking impression. Same as *blocker*.

block sequence. A combined longitudinal and buildup sequence for a continuous multiple-pass weld in which separated lengths are completely or partially built up in cross section before intervening lengths are deposited. See also *backstep sequence* and *longitudinal sequence*.

Unwelded spaces filled after deposition of intermittent blocks

Block sequence

bloom. (1) A semifinished hot rolled product, rectangular in cross section, produced on a blooming mill. See also *billet*. For steel, the width of a bloom is not more than twice the thickness, and the cross-sectional area is usually not less than about 230 cm^2 (36 in.^{2}). Steel blooms are sometimes made by forging. (2) A visible exudation or efflorescence on the surface of an electroplating bath. (3) A bluish fluorescent cast to a painted surface caused by deposition of a thin film of smoke, dust, or oil. (4) A loose, flowerlike corrosion product that forms when certain metals are exposed to a moist environment.

bloomer. The mill or other equipment used in reducing steel ingots to blooms.

blooming mill. A primary rolling mill used to make blooms.

blowhole. A hole in a casting or a weld caused by gas entrapped during solidification. See also *porosity*.

blowpipe. See preferred terms *welding torch* and *cutting torch*.

blue annealing. Heating hot-rolled ferrous sheet in an open furnace to a temperature within the transformation range, then cooling in air to soften the metal. A bluish oxide surface layer forms.

blue brittleness. Brittleness exhibited by some steels after being heated to some temperature within the range of about 205 to 370 °C (400 to 700 °F), particularly if the steel is worked at the elevated temperature. Killed steels are virtually free of this kind of brittleness.

bluing. Subjecting the scale-free surface of a ferrous alloy to the action of air, steam, or other agents at a suitable temperature, thus forming a thin blue film of oxide and improving the appearance and resistance to corrosion. This term is ordinarily applied to sheet, strip, or finished parts. It is used also to denote the heating of springs after fabrication to improve their properties.

board hammer. A type of forging hammer in which the upper die and ram are attached to "boards" that are raised to the striking position by power-driven rollers and let fall by gravity. See also *gravity hammer*.

bolster. A plate to which dies may be fastened, the assembly being secured to the top surface of a press bed. In mechanical forging, such a plate is also attached to the ram.

bond. (1) In grinding wheels and other relatively rigid abrasive products, the material that holds the abrasive grains together. (2) In welding, brazing, or soldering, the junction of joined parts. Where filler metal is used, it is the junction of the fused metal and the heat-affected base metal. (3) In an adhesive bonded or diffusion bonded joint, the line along which the faying surfaces are joined together. (4) In thermal spraying, the junction between the material deposited and the substrate, or its strength.

book mold. A split permanent mold hinged like a book.

bore. A hole or cylindrical cavity produced by a single-point or multipoint tool other than a drill.

boriding. Thermochemical treatment involving the enrichment of the surface layer of an object with borides. This surface-hardening process is performed below the Ac$_1$ temperature. Also referred to as boronizing.

boring. Enlarging a hole by removing metal with a single- or occasionally a multiple-point cutting tool moving parallel to the axis of rotation of the work or tool.

bort. (1) Natural diamond of a quality not suitable for gem use. (2) Industrial diamond.

bosh. (1) The section of a blast furnace extending upward from the tuyeres to the plane of maximum diameter. (2) A lining of quartz that builds up during the smelting of copper ores and that decreases the diameter of the furnace at the tuyeres. (3) A tank, often with sloping sides, used for washing metal parts or for holding cleaned parts.

boss. A relatively short protrusion or projection from the surface of a forging or casting, often cylindrical in shape. Usually intended for drilling and tapping for attaching parts.

bottom board. In casting, a flat base for holding the *flask* in making sand molds.

bottom drill. A flat-ended twist drill used to convert a cone at the bottom of a drilled hole into a cylinder.

bottoming tap. A tap with a *chamfer* of 1 to 1-1/2 threads in length.

bottom pipe. An oxide-lined fold or cavity at the butt end of a slab, bloom, or billet; formed by folding the end of an ingot over on itself during primary rolling. Bottom pipe is not *pipe*, in that it is not a shrinkage cavity, and in that sense, the term is a misnomer. Bottom pipe is similar to *extrusion pipe*. It is normally discarded when the slab, bloom, or billet is cropped following primary reduction.

bowing. Deviation from flatness.

box annealing. Annealing a metal or alloy in a sealed container under conditions that minimize oxidation. In box annealing a ferrous alloy, the charge is usually heated slowly to a temperature below the transformation range, but sometimes above or within it, and is then cooled slowly; this process is also called close annealing or pot annealing. See also *black annealing*.

boxing. The continuation of a fillet weld around a corner of a member as an extension of the principal weld.

brake. A device for bending sheet metal to a desired angle.

brale indenter. A conical 120° diamond indenter with a conical tip (a 0.2 mm tip radius is typical) used in certain types of Rockwell and scratch hardness tests.

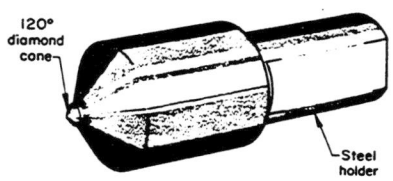

Brale indenter

10

brass. A copper-zinc alloy containing up to 40% Zn, to which smaller amounts of other elements may be added.

braze. A weld produced by heating an assembly to suitable temperatures and by using a filler metal having a liquidus above 450 °C (840 °F) and below the solidus of the base metal. The filler metal is distributed between the closely fitted faying surfaces of the joint by capillary action.

brazeability. The capacity of a metal to be brazed under the fabrication conditions imposed into a specific suitably designed structure and to perform satisfactorily in the intended service.

braze welding. A method of welding by using a filler metal having a liquidus above 450 °C (840 °F) and below the solidus of the base metals. Unlike *brazing*, in braze welding, the filler metal is not distributed in the joint by capillary attraction.

brazing. A group of welding processes that join solid materials together by heating them to a suitable temperature and using a filler metal having a liquidus above 450 °C (840 °F) and below the solidus of the base materials. The filler metal is distributed between the closely fitted surfaces of the joint by capillary attraction.

brazing alloy. See preferred term *brazing filler metal*.

brazing filler metal. (1) The metal which fills the capillary gap and has a liquidus above 450 °C (840 °F) but below the solidus of the base materials. (2) A nonferrous filler metal used in *brazing* and *braze welding*.

brazing sheet. Brazing filler metal in sheet form.

breakdown. (1) An initial rolling or drawing operation, or a series of such operations, for the purpose of reducing a casting or extruded shape prior to the finish reduction to desired size. (2) A preliminary press-forging operation.

breaking stress. Same as *fracture stress* (1).

breaks. Creases or ridges usually in "untempered" or in aged material where the yield point has been exceeded. Depending on the origin of the breaks, they may be termed *cross breaks, coil breaks, edge breaks,* or *sticker breaks.*

bridge die. A two-section extrusion die capable of producing tubing or intricate hollow shapes without the use of a separate mandrel. Metal separates into two streams as it is extruded past a bridge section, which is attached to the main die section and holds a stub mandrel in the die opening; the metal then is rewelded by extrusion pressure before it enters the die opening.

bridging. (1) Premature solidification of metal across a mold section before the metal below or beyond solidifies. (2) Solidification of slag within a cupola at or just above the tuyeres. (3) Welding or mechanical locking of the charge in a downfeed melting or smelting furnace. (4) In powder metallurgy, the formation of arched cavities in a powder mass. (5) In soldering, an unintended solder connection between two or more conductors, either securely or by mere contact. Also called a crossed joint or solder short.

bright annealing. Annealing in a protective medium to prevent discoloration of the bright surface.

bright dip. A solution that produces, through chemical action, a bright surface on an immersed metal.

brightener. An agent or combination of agents added to an electroplating bath to produce a lustrous deposit.

bright finish. A high-quality finish produced on ground and polished rolls. Suitable for electroplating.

bright nitriding. Nitriding in a protective medium to prevent discoloration of the bright surface. Compare with *blank nitriding*.

bright plate. An electrodeposit that is lustrous in the as-plated condition.

Brinell hardness number (HB). A number related to the applied load and to the surface area of the permanent impression made by a ball indenter computed from:

$$HB = \frac{2P}{\pi D(D - \sqrt{D^2 - d^2})}$$

where *P* is applied load, kgf; *D* is diameter of ball, mm; and *d* is mean diameter of the impression, mm.

Brinell hardness test. A test for determining the hardness of a material by forcing a hard steel or carbide ball of specified diameter (typically, 10 mm) into it under a specified load. The result is expressed as the *Brinell hardness number*.

Brinelling. (1) Indentation of the surface of a solid body by repeated local impact or impacts, or static overload. Brinelling may occur especially in a rolling-element bearing. (2) Damage to a solid bearing surface characterized by one or more plastically formed indentations brought about by overload. See also *false Brinelling*.

brine quenching. A quench in which brine (salt water-chlorides, carbonates, and cyanides) is the quenching medium. The salt addition improves the efficiency of water at the vapor phase or hot stage of the quenching process.

brittle crack propagation. A very sudden propagation of a crack with the absorption of no energy except that stored elastically in the body. Microscopic examination may reveal some deformation even though it is not noticeable to the unaided eye. Contrast with *ductile crack propagation.*

brittle fracture. Separation of a solid accompanied by little or no macroscopic plastic deformation. Typically, brittle fracture occurs by rapid crack propagation with less expenditure of energy than for *ductile fracture.* Brittle tensile fractures have a bright, granular appearance and exhibit little or no necking. A *chevron pattern* may be present on the fracture surface, pointing toward the origin of the crack, especially in brittle fractures in flat platelike components. Examples of brittle fracture include *transgranular cracking* (*cleavage* and *quasi-cleavage fracture*) and *intergranular cracking* (*decohesive rupture*).

brittleness. The tendency of a material to fracture without first undergoing significant *plastic deformation*. Contrast with *ductility*.

broaching. Cutting with a tool which consists of a bar having a single edge or a series of cutting edges (i.e., teeth) on its surface. The cutting edges of multiple-tooth, or successive single-tooth, broaches increase in size and/or change in shape. The broach cuts in a straight line or axial direction when relative motion is produced in relation to the workpiece, which may also be rotating. The entire cut is made in single or multiple passes over the workpiece to shape the required surface contour.

bronze. A copper-rich copper-tin alloy with or without small proportions of other elements such as zinc and phosphorus. By extension, certain copper-base alloys containing considerably less tin than other alloying elements, such as manganese bronze (copper-zinc plus manganese, tin, and iron) and leaded tin bronze (copper-lead plus tin and sometimes zinc). Also, certain other essentially binary copper-base alloys containing no tin, such as aluminum bronze (copper-aluminum), silicon bronze (copper-silicon), and beryllium bronze (copper-beryllium). Also, trade designations for certain specific copper-base alloys that are actually brasses, such as architectural bronzes (57 Cu, 40 Zn, 3 Pb) and commercial bronze (90 Cu, 10 Zn).

bronzing. (1) Applying a chemical finish to copper or copper-alloy surfaces to alter the color. (2) Plating a copper-tin alloy on various materials.

brush anodizing. An *anodizing* process similar to *brush plating*.

brush plating. Plating with a concentrated solution or gel held in or fed to an absorbing medium, pad, or brush carrying the anode (usually insoluble). The brush is moved back and forth over the area of the cathode to be plated.

buckle. (1) Bulging of a large, flat face of a casting; in investment casting, caused by *dip coat* peeling from the pattern. (2) An indentation in a casting, resulting from expansion of the sand, can be termed the start of an expansion defect. (3) A local waviness in metal bar or sheet, usually transverse to the direction of rolling.

buckling. (1) A mode of failure generally characterized by an unstable lateral material deflection due to compressive action on the structural element involved. (2) In metal forming, a bulge, bend, kink, or other wavy condition of the workpiece caused by compressive stresses. See also *compressive stress*.

buffer. (1) A substance which by its addition or presence tends to minimize the physical and chemical effects of one or more of the substances in a mixture. Properties often buffered include pH, oxidation potential, and flame or plasma temperatures. (2) A substance whose purpose is to maintain a constant hydrogen-ion concentration in water solutions, even where acids or alkalis are added. Each buffer has a characteristic limited range of pH over which it is effective.

buffing. Developing a lustrous surface by contacting the work with a rotating *buffing wheel*.

buffing wheel. Buff sections assembled to the required face width for use on a rotating shaft between flanges. Sometimes called a buff.

buildup. Excessive electrodeposition that occurs on high-current-density areas, such as corners or edges.

buildup sequence. The order in which the weld beads of a multiple-pass weld are deposited with respect to the cross section of the joint. See also *block sequence* and *longitudinal sequence*.

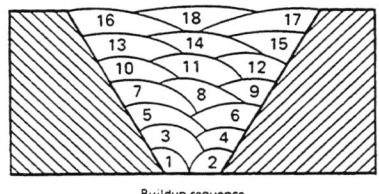

Buildup sequence

built-up edge. (1) Chip material adhering to the tool face adjacent to the cutting edge during cutting. (2) Material from the workpiece, especially in machining, which is stationary with respect to the tool.

bulging. (1) Expanding the walls of a cup, shell, or tube with an internally expanded segmented punch or a punch composed of air, liquids, or semiliquids such as waxes, rubber, and other elastomers. (2) The process of increasing the diameter of a cylindrical shell (usually to a spherical shape) or of expanding the outer walls of any shell or box shape whose walls were previously straight.

bulk forming. Forming processes, such as extrusion, forging, rolling, and drawing, in which the input material is in billet, rod, or slab form and a considerable increase in surface-to-volume ratio in the formed part occurs under the action of largely compressive loading. Compare with *sheet forming*.

bulk modulus of elasticity (K). The measure of resistance to change in volume; the ratio of hydrostatic stress to the corresponding unit change in volume. This elastic constant can be expressed by:

$$K = \frac{\sigma_m}{\Delta} = \frac{-p}{\Delta} = \frac{1}{\beta}$$

where K is the bulk modulus of elasticity, σ_m is hydrostatic or mean stress tensor, p is hydrostatic pressure, and is compressibility. Also known as bulk modulus, compression modulus, hydrostatic modulus, and volumetric modulus of elasticity.

bull block. A machine with a power-driven revolving drum for cold drawing wire through a drawing die as the wire winds around the drum.

bulldozer. Slow-acting horizontal *mechanical press* with a large bed used for bending and straightening. The work is done between dies and can be performed hot or cold. The machine is closely allied to a forging machine.

bullion. (1) A semirefined alloy containing sufficient precious metal to make recovery profitable. (2) Refined gold or silver, uncoined.

bull's-eye structure. The microstructure of malleable or ductile cast iron when graphite nodules are surrounded by a ferrite layer in a pearlitic matrix.

bumper. A machine used for packing molding sand in a flask by repeated jarring or jolting. See also *jolt ramming*.

bumping. (1) Forming a dish in metal by means of many repeated blows. (2) Forming a head. (3) Setting the seams on sheet metal parts. (4) Ramming sand in a flask by repeated jarring and jolting.

burned deposit. A dull, nodular electrodeposit resulting from excessive plating current density.

burned-in sand. A defect consisting of a mixture of sand and metal cohering to the surface of a casting.

burned-on sand. A mixture of sand and cast metal adhering to the surface of a casting. In some instances, may resemble *metal penetration*.

burned plating. See *burned deposit*.

burning. (1) Permanently damaging a metal or alloy by heating to cause either incipient melting or intergranular oxidation. See also *overheating*. (2) During subcritical annealing, particularly in continuous annealing, production of a severely decarburized and grain-coarsened surface layer that results from excessively prolonged heating to an excessively high temperature. (3) In grinding, getting the work hot enough to cause discoloration or to change the microstructure by tempering or hardening. (4) In sliding contacts, the oxidation of a surface due to local heating in an oxidizing environment.

burnishing. Finish sizing and smooth finishing of surfaces (previously machined or ground) by displacement, rather than removal, of minute surface irregularities with smooth-point or line-contact fixed or rotating tools.

burnoff. (1) Unintentional removal of an autocatalytic deposit from a nonconducting substrate, during subsequent electroplating operations, owing to the application of excessive current or a poor contact area. (2) Removal of volatile lubricants such as metallic stearates from metal powder compacts by heating immediately prior to sintering.

burr. (1) A thin ridge or roughness left on a workpiece (e.g., forgings or sheet metal blanks) resulting from cutting, punching, or grinding. (2) A rotary tool having teeth similar to those on hand files.

burring. Same as *deburring*.

bushing. A bearing or guide.

buster. A pair of shaped dies used to combine preliminary forging operations, such as edging and blocking, or to loosen scale.

butler finish. A semilustrous metal finish composed of fine, uniformly distributed parallel lines, usually produced with a soft abrasive buffing wheel; similar in appearance to the traditional hand-rubbed finish on silver.

buttering. A form of surfacing in which one or more layers of weld metal are deposited on the groove face of one member (for example, a high-alloy weld deposit on steel base metal that is to be welded to a dissimilar base metal). The buttering provides a suitable transition weld deposit for subsequent completion of the butt weld (joint).

butt joint. A joint between two abutting members lying approximately in the same plane. A welded butt joint may contain a variety of grooves. See also *groove weld*.

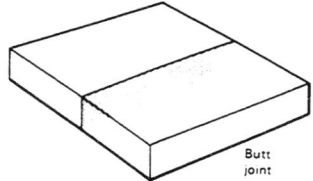
Butt joint

button. (1) A globule of metal remaining in an assaying crucible or cupel after fusion has been completed. (2) That part of a weld that tears out in destructive testing of a spot, seam, or projection welded specimen.

butt seam welding. See *seam welding*.

butt welding. Welding a *butt joint*.

C

cake. (1) A copper or copper alloy casting, rectangular in cross section, used for rolling into sheet or strip. (2) A coalesced mass of unpressed metal powder.

calcination. Heating ores, concentrates, precipitates, or residues to decompose carbonates, hydrates, or other compounds.

calcium silicon. An alloy of calcium, silicon, and iron containing 28 to 35% Ca, 60 to 65% Si, and 6% Fe (max), used as a deoxidizer and degasser for steel and cast iron; sometimes called calcium silicide.

calomel electrode. (1) An *electrode* widely used as a reference electrode of known potential in electrometric measurement of acidity and alkalinity, corrosion studies, voltammetry, and measurement of the potentials of other electrodes. (2) A secondary reference electrode of the composition: $Pt/Hg\text{-}Hg_2Cl_2/KCl$ solution. For 1.0 N KCl solution, its potential versus a hydrogen electrode at 25 °C (77 °F) and one atmosphere is +0.281 V.

calorizing. Imparting resistance to oxidation to an iron or steel surface by heating in aluminum powder at 800 to 1000 °C (1470 to 1830 °F).

camber. (1) Deviation from edge straightness, usually referring to the greatest deviation of side edge from a straight line. (2) The tendency of material being sheared from sheet to bend away from the sheet in the same plane. (3) Sometimes used to denote crown in rolls where the center diameter has been increased to compensate for deflection caused by the rolling pressure. (4) The planar deflection of a flat cable or flexible laminate from a straight line of specified length. A flat cable or flexible laminate with camber is similar to the curve of an unbanked race track.

cam press. A mechanical forming press in which one or more of the slides are operated by cams; usually a double-action press in which the blankholder slide is operated by cams through which the dwell is obtained.

can. A sheathing of soft metal that encloses a sintered metal billet for the purpose of hot working (hot isostatic pressing, hot extrusion) without undue oxidation.

canning. (1) A dished distortion in a flat or nearly flat sheet metal surface, sometimes referred to as oil canning. (2) Enclosing a highly reactive metal within a relatively inert material for the purpose of hot working without undue oxidation of the active metal.

capillary action. (1) The phenomenon of intrusion of a liquid into interconnected small voids, pores, and channels in a solid, resulting from surface tension. (2) The force by which liquid, in contact with a solid, is distributed between closely fitted faying surfaces of the joint to be brazed or soldered.

capillary attraction. (1) The combined force of adhesion and cohesion that causes liquids, including molten metals, to flow between very closely spaced and solid surfaces, even against gravity. (2) In powder metallurgy, the driving force for the infiltration of the pores of a sintered compact by a liquid.

capped steel. A type of steel similar to rimmed steel, usually cast in a bottle-top ingot mold, in which the application of a mechanical or a chemical cap renders the rimming action incomplete by causing the top metal to solidify. The surface condition of capped steel is much like that of rimmed steel, but certain other characteristics are intermediate between those of *rimmed steel* and those of *semikilled steel*.

capping (of powder metallurgy compacts). Partial or complete separation of a powder metallurgy compact into two or more portions by cracks that originate near the edges of the punch faces and that proceed diagonally into the compact.

carbide. A compound of carbon with one or more metallic elements.

carbide tools. Cutting or forming tools, usually made from tungsten, titanium, tantalum, or niobium carbides, or a combination of them, in a matrix of cobalt, nickel, or other metals. Carbide tools are characterized by high hardnesses and compressive strengths and may be coated to improve wear resistance. See also *cemented carbide*.

carbon arc cutting (CAC). An arc cutting process in which metals are severed by melting them with the heat of an arc between a carbon electrode and the base metal.

carbon arc welding (CAW). An arc welding process which produces coalescence of metals by heating them with an arc between a carbon electrode and the work. No shielding is used. Pressure and filler metal may or may not be used.

carbon dioxide welding. Gas metal arc welding using carbon dioxide as the shielding gas.

carbon edges. Carbonaceous deposits in a wavy pattern along the edges of a steel sheet or strip; also known as snaky edges.

carbon electrode. A nonfiller material electrode used in arc welding or cutting, consisting of a carbon or graphite rod, which may be coated with copper or other coatings.

carbon equivalent. (1) For cast iron, an empirical relationship of the total carbon, silicon, and phosphorus contents expressed by the formula:

$$CE = \%C + 0.3(\%Si) + 0.33(\%P) - 0.027(\%Mn) + 0.4(\%S)$$

(2) For rating of weldability:

$$CE = C + \frac{Mn}{6} + \frac{Ni}{15} + \frac{Cu}{15} + \frac{Cr}{5} + \frac{Mo}{5} + \frac{V}{5}$$

carbonitriding. A *case hardening* process in which a suitable ferrous material is heated above the lower transformation temperature in a gaseous atmosphere of such composition as to cause simultaneous absorption of carbon and nitrogen by the surface and, by diffusion, create a concentration gradient. The heat-treating process is completed by cooling at a rate that produces the desired properties in the workpiece.

carbonization. The conversion of an organic substance into elemental carbon in an inert atmosphere at temperatures ranging from 800 to 1600 °C (1470 to 2910 °F) and higher, but usually at about 1315 °C (2400 °F). Range is influenced by precursor, processing of the individual manufacturer, and properties desired. Should not be confused with *carburization*.

carbonizing flame. See preferred term *reducing flame*.

carbon potential. A measure of the ability of an environment containing active carbon to alter or maintain, under prescribed conditions, the carbon level of a steel. In any particular environment, the carbon level attained will depend on such factors as temperature, time, and steel composition.

carbon steel. Steel having no specified minimum quantity for any alloying element—other than the commonly accepted amounts of manganese (1.65%), silicon (0.60%), and copper (0.60%)—and containing only an incidental amount of any element other than carbon, silicon, manganese, copper, sulfur, and phosphorus. Low-carbon steels contain up to 0.30% C, medium-carbon steels contain from 0.30 to 0.60% C, and high-carbon steels contain from 0.60 to 1.00% C.

carbonyl powder. Metal powders prepared by the thermal decomposition of a metal carbonyl compound such as nickel tetra- carbonyl $Ni(CO)_4$ or iron pentacarbonyl $Fe(CO)_5$. See also *thermal decomposition.*

carburizing. Absorption and diffusion of carbon into solid ferrous alloys by heating, to a temperature usually above Ac_3, in contact with a suitable carbonaceous material. A form of *case hardening* that produces a carbon gradient extending inward from the surface, enabling the surface layer to be hardened either by quenching directly from the carburizing temperature or by cooling to room temperature, then reaustenitizing and quenching.

carburizing flame. A gas flame that will introduce carbon into some heated metals, as during a gas welding operation. A carburizing flame is a *reducing flame,* but a reducing flame is not necessarily a carburizing flame.

cascade sequence. A welding sequence in which a continuous multiple-pass weld is built up by depositing weld beads in overlapping layers, usually laid in a *backstep sequence.* Compare with *block sequence.*

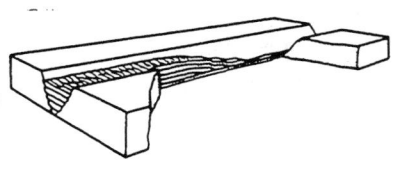

Cascade sequence

case. In heat treating, that portion of a ferrous alloy, extending inward from the surface, whose composition has been altered during *case hardening.* Typically considered to be the portion of an alloy (a) whose composition has been measurably altered from the original composition, (b) that appears light when etched, or (c) that

has a higher hardness value than the core. Contrast with *core.*

case crushing. A term used to denote longitudinal gouges arising from fracture in case-hardened gears.

case hardening. A generic term covering several processes applicable to steel that change the chemical composition of the surface layer by absorption of carbon, nitrogen, or a mixture of the two and, by diffusion, create a concentration gradient. The processes commonly used are *carburizing* and *quench hardening; cyaniding; nitriding;* and *carbonitriding.* The use of the applicable specific process name is preferred.

CASS test. Abbreviation for *copper-accelerated salt-spray test.*

castable. In casting, a combination of refractory grain and suitable bonding agent that, after the addition of a proper liquid, is generally poured into place to form a refractory shape or structure which becomes rigid because of chemical action.

castability. (1) A complex combination of liquid-metal properties and solidification characteristics that promotes accurate and sound final castings. (2) The relative ease with which a molten metal flows through a mold or casting die.

cast-alloy tool. A cutting tool made by casting a cobalt-base alloy and used at machining speeds between those for high-speed steels and cemented carbides.

casting. (1) Metal object cast to the required shape by pouring or injecting liquid metal into a mold, as distinct from one shaped by a mechanical process. (2) Pouring molten metal into a mold to produce an object of desired shape.

casting copper. Fire-refined tough pitch copper usually cast from melted secondary metal into ingot bars only, and used for making foundry castings but not wrought products.

casting defect. Any imperfection in a casting that does not satisfy one or more of the required design or quality specifications. This term is often used in a limited sense for those flaws formed by improper casting solidification.

casting shrinkage. The amount of dimensional change per unit length of the casting as it solidifies in the mold or die and cools to room temperature after removal from the mold or die. There are three distinct types of casting shrinkage. Liquid shrinkage refers to the reduction in volume of liquid metal as it cools to the liquidus. Solidification shrinkage is the reduction in volume of metal from the beginning to the end of solidification. Solid shrinkage involves the reduction in vol-

ume of metal from the solidus to room temperature.

casting strains. Strains in a casting caused by *casting stresses* that develop as the casting cools.

casting stresses. Residual stresses set up when the shape of a casting impedes contraction of the solidified casting during cooling.

cast iron. A generic term for a large family of cast ferrous alloys in which the carbon content exceeds the solubility of carbon in austenite at the eutectic temperature. Most cast irons contain at least 2% carbon, plus silicon and sulfur, and may or may not contain other alloying elements. See also *compacted graphite iron, ductile iron, gray iron, malleable iron,* and *white iron.*

cast steel. Steel in the form of a *casting.*

cast structure. The metallographic structure of a *casting* evidenced by shape and orientation of grains and by segregation of impurities.

catalyst. A substance capable of changing the rate of a reaction without itself undergoing any net change.

catastrophic failure. Sudden failure of a component or assembly that frequently results in extensive secondary damage to adjacent components or assemblies.

cathode. The negative *electrode* of an *electrolytic cell* at which reduction is the principal reaction. (Electrons flow toward the cathode in the external circuit.) Typical cathodic processes are cations taking up electrons and being discharged, oxygen being reduced, and the reduction of an element or group of elements from a higher to a lower valence state. Contrast with *anode.*

cathode copper. Copper deposited at the cathode in electrolytic refining.

cathode efficiency. Current efficiency at the *cathode.*

cathode film. The portion of solution in immediate contact with the *cathode* during *electrolysis.*

cathodic cleaning. *Electrolytic cleaning* in which the work is the *cathode.*

cathodic corrosion. Corrosion resulting from a cathodic condition of a structure usually caused by the reaction of an amphoteric metal with the alkaline products of *electrolysis.*

cathodic pickling. Electrolytic pickling in which the work is the *cathode.*

cathodic polarization. The change of the *electrode potential* in the active (negative) direction due to current flow. See also *polarization.*

cathodic protection. (1) Reduction of corrosion rate by shifting the *corrosion po-*

tential of the electrode toward a less oxidizing potential by applying an external *electromotive force*. (2) Partial or complete protection of a metal from corrosion by making it a *cathode*, using either a galvanic or an impressed current. Contrast with *anodic protection*.

cathodic reaction. Electrode reaction equivalent to a transfer of negative charge from the electronic to the ionic conductor. A cathodic reaction is a reduction process. An example common in corrosion is: $Ox + ne^- \rightarrow$ Red.

catholyte. The *electrolyte* adjacent to the cathode of an electrolytic cell.

cation. A positively charged ion that migrates through the electrolyte toward the *cathode* under the influence of a potential gradient. See also *anion* and *ion*.

caustic. (1) Burning or corrosive. (2) A hydroxide of a light metal, such as sodium hydroxide or potassium hydroxide.

caustic cracking. A form of *stress-corrosion cracking* most frequently encountered in carbon steels or iron-chromium-nickel alloys that are exposed to concentrated hydroxide solutions at temperatures of 200 to 250 °C (400 to 480 °F). Also known as caustic embrittlement.

caustic dip. A strongly alkaline solution into which metal is immersed for etching, for neutralizing acid, or for removing organic materials such as greases or paints.

caustic embrittlement. An obsolete historical term denoting a form of *stress-corrosion cracking* most frequently encountered in carbon steels or iron-chromium-nickel alloys that are exposed to concentrated hydroxide solutions at temperatures of 200 to 250 °C (400 to 480 °F).

caustic quenching. Quenching with aqueous solutions of 5 to 10% sodium hydroxide (NaOH).

cavitation. The formation and collapse, within a liquid, of cavities or bubbles that contain vapor or gas or both. In general, cavitation originates from a decrease in the static pressure in the liquid. It is distinguished in this way from boiling, which originates from an increase in the liquid temperature. There are certain situations where it may be difficult to make a clear distinction between cavitation and boiling, and the more general definition that is given here is therefore to be preferred. In order to erode a solid surface by cavitation, it is necessary for the cavitation bubbles to collapse on or close to that surface.

cavitation corrosion. A process involving conjoint *corrosion* and *cavitation*.

cavitation damage. The degradation of a solid body resulting from its exposure to *cavitation*. This may include loss of material, surface deformation, or changes in properties or appearance.

cavitation erosion. Progressive loss of original material from a solid surface due to continuing exposure to *cavitation*.

cavity. The mold or die impression that gives a casting its external shape.

CCT diagram. See *continuous cooling transformation diagram*.

cell (electrochemistry). Electrochemical system consisting of an *anode* and a *cathode* immersed in an *electrolyte*. The anode and cathode may be separate metals or dissimilar areas on the same metal. The cell includes the external circuit, which permits the flow of electrons from the anode toward the cathode. See also *electrochemical cell*.

cementation. The introduction of one or more elements into the outer portion of a metal object by means of diffusion at high temperature.

cement copper. Impure copper recovered by *chemical deposition* when iron (most often shredded steel scrap) is brought into prolonged contact with a dilute copper sulfate solution.

cemented carbide. A solid and coherent mass made by pressing and sintering a mixture of powders of one or more metallic carbides, such as tungsten carbide, and a much smaller amount of a metal, such as cobalt, to serve as a binder.

cementite. A hard (800 HV), brittle compound of iron and carbon, known chemically as iron carbide and having the approximate chemical formula Fe_3C. It is characterized by an orthorhombic crystal structure. When it occurs as a phase in steel, the chemical composition will be altered by the presence of manganese and other carbide-forming elements. The highest cementite contents are observed in white cast irons, which are used in applications where high wear resistance is required.

center drilling. Drilling a short, conical hole in the end of a workpiece—a hole to be used to center the workpiece for turning on a lathe.

centerless grinding. Grinding the outside or inside diameter of a cylindrical piece which is supported on a work support blade instead of being held between centers and which is rotated by a so-called regulating or feed wheel.

centrifugal casting. The process of filling molds by (1) pouring metal into a sand or permanent mold that is revolving about either its horizontal or its vertical

axis or (2) pouring metal into a mold that is subsequently revolved before solidification of the metal is complete. See also *centrifuge casting*.

centrifuge casting. A casting technique in which mold cavities are spaced symmetrically about a vertical axial common downgate. The entire assembly is rotated about that axis during pouring and solidification.

ceramic tools. Cutting tools made from sintered, hot-pressed, or hot isostatically pressed alumina-based or silicon nitride-based ceramic materials.

cermet. A powder metallurgy product consisting of ceramic particles bonded with a metal.

C-frame press. Same as *gap-frame press*.

CG iron. Same as *compacted graphite cast iron*.

chain-intermittent fillet welding. Depositing a line of intermittent fillet welds on each side of a member at a joint so that the increments on one side are essentially opposite those on the other. Contrast with *staggered-intermittent fillet welding*.

chamfer. (1) A beveled surface to eliminate an otherwise sharp corner. (2) A relieved angular cutting edge at a tooth corner.

chamfer angle. (1) The angle between a reference surface and the bevel. (2) On a milling contour, the angle between a beveled surface and the axis of the cutter.

chamfering. Making a sloping surface on the edge of a member. Also called beveling. See also *bevel angle*.

chaplet. Metal support that holds a core in place within a casting mold; molten metal solidifies around a chaplet and fuses it into the finished casting.

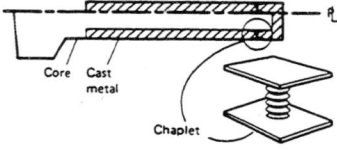

charge. (1) The materials fed into a furnace. (2) Weights of various liquid and solid materials put into a furnace during one feeding cycle.

charging. (1) For a lap, impregnating the surface with fine abrasive. (2) Placing materials into a furnace.

Charpy test. An impact test in which a V-notched, keyhole-notched, or U-notched specimen, supported at both ends, is struck behind the notch by a striker mounted at the lower end of a bar that can swing as a pendulum. The energy that is absorbed in fracture is calculated from the height to which the striker would have risen had there been no specimen and the height to which it actually rises after fracture of the specimen. Contrast with *Izod test*.

chase (machining). To make a series of cuts each, except for the first, following in the path of the cut preceding it, as in chasing a thread.

chatter. In machining or grinding, (1) a vibration of the tool, wheel, or workpiece producing a wavy surface on the work and (2) the finish produced by such vibration. (3) In tribology, elastic vibrations resulting from frictional or other instability.

chatter marks. Surface imperfections on the work being ground, usually caused by vibrations transferred from the wheel-work interface during grinding.

check. The intermediate section of a flask that is used between the *cope* and the *drag* when molding a shape that requires more than one parting plane.

checked edges. Sawtooth edges seen after hot rolling and/or cold rolling.

checkers. In a chamber associated with a metallurgical furnace, bricks stacked openly so that heat may be absorbed from the combustion products and later transferred to incoming air when the direction of flow is reversed.

checks. (1) Numerous, very fine cracks in a coating or at the surface of a metal part. Checks may appear during processing or during service and are most often associated with thermal treatment or thermal cycling. Also called check marks, or *heat checks*. (2) Minute cracks in the surface of a casting caused by unequal expansion or contraction during cooling. (3) Cracks in a die impression corner, generally due to forging strains or pressure, localized at some relatively sharp corner. Die blocks too hard for the depth of the die impression have a tendency to check or develop cracks in impression corners. (4) A series of small cracks resulting from thermal fatigue of hot forging dies.

chelating agent. (1) An organic compound in which atoms form more than one coordinate bond with metals in solution. (2) A substance used in metal finishing to control or eliminate certain metallic ions present in undesirable quantities.

chemical conversion coating. A protective or decorative nonmetallic coating produced *in situ* by chemical reaction of a metal with a chosen environment. It is often used to prepare the surface prior to the application of an organic coating.

chemical deposition. The precipitation or plating-out of a metal from solutions of its salts through the introduction of another metal or reagent to the solution.

chemical flux cutting. An oxygen cutting process in which metals are severed using a chemical flux to facilitate cutting.

chemically precipitated powder. A metal powder that is produced as a fine precipitate by chemical displacement.

chemical machining. Removing metal stock by controlled selective chemical dissolution.

chemical metallurgy. See *process metallurgy*.

chemical milling. The machining process in which metal is formed into intricate shapes by masking certain portions and then etching away the unwanted material.

chemical polishing. A process that produces a polished surface by the action of a chemical etching solution. The etching solution is compounded so that peaks in the topography of the surface are dissolved preferentially.

chemical vapor deposition (CVD). A coating process, similar to gas carburizing and carbonitriding, whereby a reactant atmosphere gas is fed into a processing chamber where it decomposes at the surface of the workpiece, liberating one material for either absorption by, or accumulation on, the workpiece. A second material is liberated in gas form and is removed from the processing chamber, along with excess atmosphere gas.

chemical wear. See *corrosive wear*.

chevron pattern. A fractographic pattern of radial marks (shear ledges) that look like nested letters "V"; sometimes called a herringbone pattern. Chevron patterns are typically found on brittle fracture surfaces in parts whose widths are considerably greater than their thicknesses. The points of the chevrons can be traced back to the fracture origin.

chill. (1) A metal or graphite insert embedded in the surface of a casting sand mold or core or placed in a mold cavity to increase the cooling rate at that point. (2) White iron occurring on a gray or ductile iron casting, such as the chill in the wedge test. See also *chilled iron*. Compare with *inverse chill*.

chilled iron. Cast iron that is poured into a metal mold or against a mold insert so as to cause the rapid solidification that often tends to produce a white iron structure in the casting.

Chinese-script eutectic. A configuration of eutectic constituents, found particularly in some cast alloys of aluminum containing iron and silicon and in magnesium al-

loys containing silicon, that resembles the characters in Chinese script.

chip breaker. (1) Notch or groove in the face of a tool parallel to the cutting edge, designed to break the continuity of the chip. (2) A step formed by an adjustable component clamped to the face of the cutting tool.

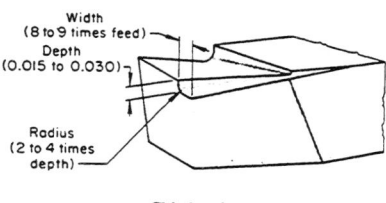

Chip breaker

chipping. (1) Removing seams and other surface imperfections in metals manually with a chisel or gouge, or by a continuous machine, before further processing. (2) Similarly, removing excessive metal.

chips. Pieces of material removed from a workpiece by cutting tools or by an abrasive medium.

chlorination. (1) Roasting ore in contact with chlorine or a chloride salt to produce chlorides. (2) Removing dissolved gases and entrapped oxides by passing chlorine gas through molten metal such as aluminum and magnesium.

chromadizing. Improving paint adhesion on aluminum or aluminum alloys, mainly aircraft skins, by treatment with a solution of chromic acid. Also called chromidizing or chromatizing. Not to be confused with *chromating* or *chromizing*.

chromate treatment. A treatment of metal in a solution of a hexavalent chromium compound to produce a *conversion coating* consisting of trivalent and hexavalent chromium compounds.

chromating. Performing a *chromate treatment*.

Chromel. (1) A 90Ni-10Cr alloy used in thermocouples. (2) A series of nickel-chromium alloys, some with iron, used for heat-resistant applications.

chromizing. A surface treatment at elevated temperature, generally carried out in pack, vapor, or salt baths, in which an alloy is formed by the inward diffusion of chromium into the base metal.

chuck. A device for holding work or tools on a machine so that the part can be held or rotated during machining or grinding.

CIP. The acronym for *cold isostatic pressing*.

circle grid. A regular pattern of circles, often 2.5 mm (0.1 in.) in diameter, marked on a sheet metal blank.

circle-grid analysis. The analysis of deformed circles to determine the severity with which a sheet metal blank has been deformed.

circle grinding. Either *cylindrical grinding* or *internal grinding*; the preferred terms.

circle shear. A shearing machine with two rotary disk cutters mounted on parallel shafts driven in unison and equipped with an attachment for cutting circles where the desired piece of material is inside the circle. It cannot be employed to cut circles where the desired material is outside the circle.

circular field. The magnetic field that (a) surrounds a nonmagnetic conductor of electricity, (b) is completely contained within a magnetic conductor of electricity, or (c) both exists within and surrounds a magnetic conductor. Generally applied to the magnetic field within any magnetic conductor resulting from a current being passed through the part or through a section of the part. Compare with *bipolar field*.

cladding. (1) A layer of material, usually metallic, that is mechanically or metallurgically bonded to a substrate. Cladding may be bonded to the substrate by any of several processes, such as roll-cladding and explosive forming. (2) A relatively thick layer (1 mm, or 0.04 in.) of material applied by surfacing for the purpose of improved corrosion resistance or other properties. See also *coating, surfacing*, and *hardfacing*.

clad metal. A composite metal containing two or more layers that have been bonded together. The bonding may have been accomplished by co-rolling, co-extrusion, welding, diffusion bonding, casting, heavy chemical deposition, or heavy electroplating.

clamshell marks. Same as *beach marks*.

classification. (1) The separation of ores into fractions according to size and specific gravity, generally in accordance with Stokes' law of sedimentation. (2) Separation of a metal powder into fractions according to particle size.

clearance. (1) The gap or space between two mating parts. (2) Space provided between the relief of a cutting tool and the surface that has been cut.

clearance angle. The angle between a plane containing the flank of the tool and a plane passing through the cutting edge in the direction of relative motion between the cutting edge and the work. See also the figures accompanying *face mill* and *single-point tool*.

cleavage. (1) Fracture of a crystal by crack propagation across a crystallographic plane of low index. (2) The tendency to cleave or split along definite crystallographic planes.

cleavage fracture. A fracture, usually of a polycrystalline metal, in which most of the grains have failed by cleavage, resulting in bright reflecting facets. It is one type of *crystalline fracture* and is associated with low-energy *brittle fracture*. Contrast with *shear fracture*.

cleavage plane. A characteristic crystallographic plane or set of planes in a crystal on which *cleavage fracture* occurs easily.

climb cutting. Analogous to *climb milling*.

climb milling. Milling in which the cutter moves in the direction of feed at the point of contact.

close annealing. Same as *box annealing*.

closed-die forging. The shaping of hot metal completely within the walls or cavities of two dies that come together to enclose the workpiece on all sides. The impression for the forging can be entirely either die or divided between the top and bottom dies. Impression-die forgings, often used interchangeably with the term closed-die forging, refers to a closed-die operation in which the dies contain a provision for controlling the flow of excess material, or *flash*, that is generated. By contrast, in flashless forging, the material is deformed in a cavity that allows little or no escape of excess material.

closed dies. Forging or forming impression dies designed to restrict the flow of metal to the cavity within the die set, as opposed to open dies, in which there is little or no restriction to lateral flow.

closed pass. A pass of metal through rolls where the bottom roll has a groove deeper than the bar being rolled and the top roll has a collar fitting into the groove, thus producing the desired shape free from *flash* or fin.

close-tolerance forging. A forging held to unusually close dimensional tolerances so that little or no machining is required after forging. See also *precision forging*.

cluster mill. A rolling mill in which each of the two working rolls of small diameter is supported by two or more backup rolls.

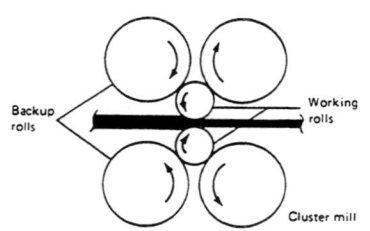

Cluster mill

coalesced copper. Massive oxygen-free copper made by briquetting ground, brittle cathode copper, then sintering the briquets in a pressurized reducing atmosphere, followed by hot working.

coalescence. (1) The union of particles of a dispersed phase into larger units, usually effected at temperatures below the fusion point. (2) Growth of grains at the expense of the remainder by absorption or the growth of a phase or particle at the expense of the remainder by absorption or reprecipitation.

coarsening. An increase in grain size, usually, but not necessarily, by *grain growth*.

coated abrasive. An abrasive product (sandpaper, for example) in which a layer of abrasive particles is firmly attached to a paper, cloth, or fiber backing by means of glue or synthetic-resin adhesive.

coated electrode. See preferred terms *covered electrode* and *lightly coated electrode*.

coating. A relatively thin layer (mm, or 0.04 in.) of material applied by surfacing for the purpose of corrosion prevention, resistance to high-temperature scaling, wear resistance, lubrication, or other purposes.

coaxing. Improvement of the fatigue strength of a specimen by the application of a gradually increasing stress amplitude, usually starting below the fatigue limit.

coefficient of friction. The dimensionless ratio of the friction force (F) between two bodies to the normal force (N) pressing these bodies together: (or $f) = (F/N)$.

coercive force. The magnetizing force that must be applied in the direction opposite to that of the previous magnetizing force in order to reduce magnetic flux density to zero; thus, a measure of the magnetic retentivity of magnetic materials.

cogging. The reducing operation in working an ingot into a billet with a forging hammer or a forging press.

cogging mill. A *blooming mill*.

coherent precipitate. A crystalline precipitate that forms from solid solution with an orientation that maintains continuity between the crystal lattice of the precipitate and the lattice of the matrix, usually accompanied by some strain in both lattices. Because the lattices fit at the interface between precipitate and matrix, there is no discernible phase boundary.

cohesion. (1) The state in which the particles of a single substance are held together by primary or secondary valence forces. As used in the adhesive field, the state in which the particles of the adhesive (or adherend) are held together. (2) Force of attraction between the molecules (or atoms) within a single phase. Contrast with *adhesion*.

cohesive strength. (1) The hypothetical stress causing tensile fracture without plastic deformation. (2) The stress corresponding to the forces between atoms.

coil. (1) An assembly consisting of one or more magnet wire windings. (2) Rolled metal sheet or strip.

coil breaks. Creases or ridges in sheet or strip that appear as parallel lines across the direction of rolling and that generally extend the full width of the sheet or strip.

coining. (1) A closed-die squeezing operation, usually performed cold, in which all surfaces of the work are confined or restrained, resulting in a well-defined imprint of the die upon the work. (2) A *restriking* operation used to sharpen or change an existing radius or profile. (3) The final pressing of a sintered powder metallurgy compact to obtain a definite surface configuration (not to be confused with *sizing*).

coin silver. An alloy containing 90% silver, with copper being the usual alloying element.

coke. A porous, gray, infusible product resulting from the dry distillation of bituminous coal, petroleum, or coal tar pitch that drives off most of the volatile matter. Used as a fuel in cupola melting.

cold box process. In foundry practice, a two-part organic resin binder system mixed in conventional mixers and blown into shell or solid core shapes at room temperature. A vapor mixed with air is blown into the core, permitting instant setting and immediate pouring of metal around it.

cold chamber machine. A die casting machine with an injection system that is charged with liquid metal from a separate furnace. Compare with *hot chamber machine*.

cold compacting. See preferred term *cold pressing*.

cold cracking. (1) Cracks in cold or nearly cold cast metal due to excessive internal stress caused by contraction. Often brought about when the mold is too hard or the casting is of unsuitable design. (2) A type of weld cracking that usually occurs below 205 °C (400 °F). Cracking may occur during or after cooling to room temperature, sometimes with a considerable time delay. Three factors combine to produce cold cracks; stress (for example, from thermal expansion and contraction), hydrogen (from hydrogen-containing welding consumables), and a susceptible microstructure (plate martensite is most susceptible to cracking, ferritic and baini-

tic structures least susceptible). See also *hot cracking*, *lamellar tearing*, and *stress-relief cracking*.

cold die quenching. A quench utilizing cold, flat, or shaped dies to extract heat from a part. Cold die quenching is slow, expensive, and is limited to smaller parts with large surface areas.

cold heading. Working metal at room temperature such that the cross-sectional area of a portion or all of the stock is increased. See also *heading* and *upsetting*.

cold inspection. A visual (usually final) inspection of forgings for visible imperfections, dimensions, weight, and surface condition at room temperature. The term may also be used to describe certain nondestructive tests such as magnetic-particle, dye-penetrant, and sonic inspection.

cold isostatic pressing. Forming technique in which high fluid pressure is applied to a powder (metal or ceramic) part at ambient temperature. Water or oil is used as the pressure medium.

cold lap. (1) Wrinkled markings on the surface of an ingot or casting from incipient freezing of the surface and too low a casting temperature. (2) A flaw that results when a workpiece fails to fill the die cavity during the first forging. A seam is formed as subsequent dies force metal over this gap to leave a seam on the workpiece surface. See also *cold shut*.

cold mill. A mill for cold rolling of sheet or strip.

cold pressing (powder metallurgy). Forming a powder metallurgy *compact* at a temperature low enough to avoid *sintering*, usually room temperature. Contrast with *hot pressing*.

cold rolled sheets. A metal mill product produced from a hot rolled pickled coil that has been given substantial cold reduction at room temperature. The resulting product usually requires further processing to make it suitable for most common applications. The usual end product is characterized by improved surface, greater uniformity in thickness, and improved mechanical properties compared with hot rolled sheet.

cold-setting process. In foundry practice, any of several systems for bonding mold or core aggregates by means of organic binders, relying on the use of catalysts rather than heat for polymerization (setting).

cold shortness. Brittleness that exists in some metals at temperatures below the recrystallization temperature.

cold shot. (1) A portion of the surface of an ingot or casting showing premature so-

lidification; caused by splashing of molten metal onto a cold mold wall during pouring. (2) Small globule of metal embedded in, but not entirely fused with, the casting.

cold shut. (1) A discontinuity that appears on the surface of cast metal as a result of two streams of liquid meeting and failing to unite. (2) A lap on the surface of a forging or billet that was closed without fusion during deformation. (3) Freezing of the top surface of an ingot before the mold is full.

Coldstream process. In powder metallurgy, a method of producing cleavage fractures in hard particles through particle impingements in a high-velocity cold gas stream. Also referred to as impact crushing.

cold treatment. Exposing steel to suitable subzero temperatures (85 °C, or 120 °F) for the purpose of obtaining desired conditions or properties such as dimensional or microstructural stability. When the treatment involves the transformation of retained austenite, it is usually followed by tempering.

cold trimming. The removal of flash or excess metal from a forging at room temperature in a trimming press.

cold welding. A solid-state welding process in which pressure is used at room temperature to produce coalescence of metals with substantial deformation at the weld. Compare with *hot pressure welding*, *diffusion welding*, and *forge welding*.

cold work. Permanent strain in a metal accompanied by strain hardening.

cold-worked structure. A microstructure resulting from plastic deformation of a metal or alloy below its recrystallization temperature.

cold working. Deforming metal plastically under conditions of temperature and strain rate that induce *strain hardening*. Usually, but not necessarily, conducted at room temperature. Contrast with *hot working*.

collapsibility. The tendency of a sand mixture to break down under the pressures and temperatures developed during casting.

collet. A split sleeve used to hold work or tools during machining or grinding.

color buffing. Producing a final high luster by buffing. Sometimes called *coloring*.

coloring. Producing desired colors on metal by a chemical or electrochemical reaction. See also *color buffing*.

columnar structure. A coarse structure of parallel elongated grains formed by unidirectional growth, most often observed in castings, but sometimes seen in structures resulting from diffusional growth accompanied by a solid-state transformation.

combination die. (1) A die-casting die having two or more different cavities for different castings. (2) For forming, see *compound die.*

combination mill. An arrangement of a continuous mill for roughing and a *guide mill* or *looping mill* for shaping.

combined carbon. Carbon in iron or steel that is combined chemically with other elements; not in the free state as graphite or temper carbon. The difference between the total carbon and the graphite carbon analyses. Contrast with *free carbon.*

combined cyanide. The cyanide of a metal-cyanide complex ion.

combined stresses. Any state of stress that cannot be represented by a single component of stress; that is, one that is more complicated than simple tension, compression, or shear.

comet tails. A group of comparatively deep unidirectional scratches that form adjacent to a microstructural discontinuity during mechanical polishing. They have the general shape of a comet tail. Comet tails form only when a unidirectional motion is maintained between the surface being polished and the polishing cloth.

comminution. (1) Breaking up or grinding an ore into small fragments. (2) Reducing metal to powder by mechanical means. (3) The act or process of reduction of powder particle size, usually but not necessarily by grinding or milling. See also *pulverization.*

compact (noun). The object produced by the compression of metal powder, generally while confined in a die.

compact (verb). The operation or process of producing a compact; sometimes called pressing.

compacted graphite iron. Cast iron having a graphite shape intermediate between the flake form typical of gray cast iron and the spherical form of fully spherulitic ductile cast iron. An acceptable compacted graphite iron structure is one that contains no flake graphite, % spheroidal graphite, and 80% compacted graphite (ASTM A 247, type IV). Also known as CG iron or vermicular iron, compacted graphite cast iron is produced in a manner similar to that for ductile cast iron, but using a technique that inhibits the formation of fully spherulitic graphite nodules. Typical nominal compositions of CG irons contain 3.1 to 4.0% C, 1.7 to 3.0% Si, and 0.1 to 0.6% Mn.

compacting pressure. In powder metallurgy, the specific compacting force related to the area of contact with the press punch expressed in megapascals, meganewtons per square meter, or tons per square inch.

compaction. (1) The act of forcing particulate or granular material together (consolidation) under pressure or impact to yield a relatively dense mass or formed object. (2) In powder metallurgy, the preparation of a compact or object produced by the compression of a powder, generally while confined in a die, with or without the inclusion of lubricants, binders, etc., and with or without the concurrent applications of heat.

compatibility. A measure of the extent to which materials are mutually soluble in the solid state.

complete fusion. Fusion which has occurred over the entire base material surfaces intended for welding and between all layers and weld beads.

complexing agent. A substance that is an electron donor and that will combine with a metal ion to form a soluble complex ion.

complex ion. An ion that may be formed by the addition reaction of two or more other ions.

component. (1) One of the elements or compounds used to define a chemical (or alloy) system, including all phases, in terms of the fewest substances possible. (2) One of the individual parts of a vector as referred to a system of coordinates. (3) An individual functional element in a physically independent body that cannot be further reduced or divided without destroying its stated function, for example, a resistor, capacitor, diode, or transistor.

composite coating. A coating on a metal or nonmetal that consists of two or more components, one of which is often particulate in form. Example: a cermet composite coating on a cemented carbide cutting tool. Also known as multilayer coating.

composite electrode. A welding electrode made from two or more distinct components, at least one of which is filler metal. A composite electrode may exist in any of various physical forms, such as stranded wires, filled tubes, or covered wire.

composite joint. A joint in which welding is used in conjunction with mechanical joining.

composite material. A combination of two or more materials (reinforcing elements, fillers, and composite matrix binder), differing in form or composition on a macroscale. The constituents retain their identities, that is, they do not dissolve or merge completely into one another although they act in concert. Normally, the components can be physically identified and exhibit an interface between one another. Examples are *cermets* and *metal-matrix composites.*

composite plate. An electrodeposit consisting of layers of at least two different compositions.

composite powder. A powder in which each particle consists of two or more different materials.

composite structure. A structural member (such as a panel, plate, pipe, or other shape) that is built up by bonding together two or more distinct components, each of which may be made of a metal, alloy, nonmetal, or *composite material.* Examples of composite structures include: honeycomb panels, clad plate, electrical contacts, sleeve bearings, carbide-tipped drills or lathe tools, and weldments constructed of two or more different alloys.

compound compact. A powder metallurgy *compact* consisting of mixed metals, the particles of which are joined by pressing or sintering, or both, with each metal particle retaining substantially its original composition.

compound die. Any die designed to perform more than one operation on a part with one stroke of the press, such as blanking and piercing, in which all functions are performed simultaneously within the confines of the blank size being worked.

compressibility. (1) The ability of a powder to be formed into a compact having well-defined contours and structural stability at a given temperature and pressure; a measure of the plasticity of powder particles. (2) A density ratio determined under definite testing conditions. Also referred to as compactibility.

compression ratio (powder metallurgy). The ratio of the volume of the loose powder to the volume of the compact made from it.

compression test. A method for assessing the ability of a material to withstand compressive loads. Analyses of structural behavior or metal forming require knowledge of compression stress-strain properties.

compressive strength. The maximum compressive stress that a material is capable of developing, based on original area of cross section. If a material fails in compression by a shattering fracture, the compressive strength has a very definite value. If a material does not fail in compression by a shattering fracture, the value obtained for compressive strength is an arbitrary value depending upon the degree of distortion that is regarded as indicating complete failure of the material.

compressive stress. A stress that causes an elastic body to deform (shorten) in the direction of the applied load. Contrast with *tensile stress*.

concave fillet weld. A fillet weld having a concave face.

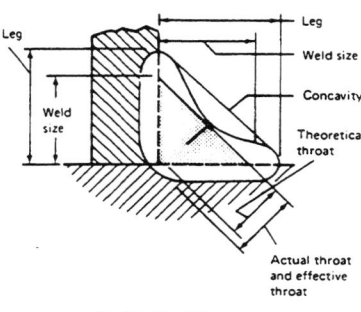

Concave fillet weld

concentration. (1) The mass of a substance contained in a unit volume of sample, for example, grams per liter. (2) A process for enrichment of an ore in valuable mineral content by separation and removal of waste material, or *gangue*.

concentration cell. An *electrolytic cell*, the *electromotive force* of which is caused by a difference in concentration of some component in the electrolyte. This difference leads to the formation of discrete *cathode* and *anode* regions.

concentration polarization. That portion of the *polarization* of a cell produced by concentration changes resulting from passage of current through the electrolyte.

concurrent heating. The application of supplemental heat to a structure during a welding or cutting operation.

conditioning heat treatment. A preliminary heat treatment used to prepare a material for a desired reaction to a subsequent heat treatment. For the term to be meaningful, the exact heat treatment must be specified.

cone angle. The angle that the cutter axis makes with the direction along which the blades are moved for adjustment, as in adjustable-blade reamers where the base of the blade slides on a conical surface.

conformal coating. A coating that covers and exactly fits the shape of the coated object.

congruent melting. An isothermal or isobaric melting in which both the solid and liquid phases have the same composition throughout the transformation.

congruent transformation. An isothermal or isobaric phase change in metals in which both of the phases concerned have the same composition throughout the process.

conjugate phases. In microstructural analysis, those states of matter of unique composition that coexist at equilibrium at a single point in temperature and pressure. For example, the two coexisting phases of a two-phase equilibrium.

constantan. A group of copper-nickel alloys containing 45 to 60% copper with minor amounts of iron and manganese and characterized by relatively constant electrical resistivity irrespective of temperature; used in resistors and thermocouples.

constant life fatigue diagram. In failure analysis, a plot (usually on rectangular coordinates) of a family of curves, each of which is for a single fatigue life (number of cycles), relating alternating stress, maximum stress, minimum stress, and mean stress. The constant life fatigue diagram is generally derived from a family of *S-N* curves, each of which represents a different stress ratio for a 50% probability of survival. See also *nominal stress, maximum stress, minimum stress, S-N curve, fatigue life,* and *stress ratio*.

constituent. (1) One of the ingredients that make up a chemical system. (2) A phase or a combination of phases that occurs in a characteristic configuration in an alloy microstructure.

constitution diagram. See *phase diagram*.

constraint. Any restriction that limits the transverse contraction normally associated with a longitudinal tension, and that hence causes a secondary tension in the transverse direction; usually used in connection with welding. Contrast with *restraint*.

consumable electrode. A general term for any arc welding electrode made chiefly of filler metal. Use of specific names such as *covered electrode*, bare electrode, flux-cored electrode, and *lightly coated electrode* is preferred.

consumable-electrode remelting. A process for refining metals in which an electric current passes between an electrode made of the metal to be refined and an ingot of the refined metal, which is contained in a water-cooled mold. As a result of the passage of electric current, droplets of molten metal form on the electrode and fall to the ingot. The refining action occurs from contact with the atmosphere, vacuum, or slag through which the drop falls. See also *electroslag* and vacuum *arc remelting*.

contact corrosion. A term primarily used in Europe to describe *galvanic corrosion* between dissimilar metals.

contact fatigue. Cracking and subsequent pitting of a surface subjected to alternating Hertzian stresses such as those produced under rolling contact or combined rolling and sliding. The phenomenon of contact fatigue is encountered most often in rolling-element bearings or in gears, where the surface stresses are high due to the concentrated loads and are repeated many times during normal operation.

contact plating. A metal plating process wherein the plating current is provided by galvanic action between the work metal and a second metal, without the use of an external source of current.

contact potential. In corrosion technology, the potential difference at the junction of two dissimilar substances.

container. The chamber into which an ingot or billet is inserted prior to extrusion. The container for backward extrusion of cups or cans is sometimes called a die.

contaminant. An impurity or foreign substance present in a material or environment that affects one or more properties of the material.

continuous casting. A casting technique in which a cast shape is continuously withdrawn through the bottom of the mold as it solidifies, so that its length is not determined by mold dimensions. Used chiefly to produce semifinished mill products such as billets, blooms, ingots, slabs, strip, and tubes. See also *strand casting*.

continuous cooling transformation (CCT) diagram. Set of curves drawn using logarithmic time and linear temperature as coordinates, which define, for each cooling curve of an alloy, the beginning and end of the transformation of the initial phase.

continuous mill. A rolling mill consisting of a number of strands of synchronized rolls (in tandem) in which metal undergoes successive reductions as it passes through the various strands.

continuous phase. In an alloy or portion of an alloy containing more than one phase, the phase that forms the matrix in which the other phase or phases are dispersed.

continuous precipitation. Precipitation from a supersaturated solid solution in which the precipitate particles grow by long-range diffusion without recrystallization of the matrix. Continuous precipitates grow from nuclei distributed more or less uniformly throughout the matrix. They usually are randomly oriented, but may form a *Widmanstätten structure*. Also called general precipitation. Compare with *discontinuous precipitation* and *localized precipitation*.

continuous-type furnace. A furnace used for heat treating materials that progress continuously through the furnace, entering one door and being discharged from another.

continuous weld. A weld extending continuously from one end of a joint to the other or, where the joint is essentially circular, completely around the joint. Contrast with *intermittent weld*.

contour forming. See *roll forming*, *stretch forming*, *tangent bending*, and *wiper forming*.

contour machining. Machining of irregular surfaces, such as those generated in tracer turning, tracer boring, and *tracer milling*.

contour milling. Milling of irregular surfaces. See also *tracer milling*.

contraction. The volume change that occurs in metals and alloys upon solidification and cooling to room temperature.

controlled atmosphere. (1) A specified inert gas or mixture of gases at a predetermined temperature in which selected processes take place. (2) As applied to sintering, to prevent oxidation and destruction of the powder compacts.

controlled cooling. Cooling a metal or alloy from an elevated temperature in a predetermined manner to avoid hardening, cracking, or internal damage, or to produce desired microstructure or mechanical properties.

controlled-pressure cycle. A forming cycle during which the hydraulic pressure in the forming cavity is controlled by an adjustable cam that is coordinated with the punch travel.

controlled rolling. A hot-rolling process in which the temperature of the steel is closely controlled, particularly during the final rolling passes, to produce a fine-grain microstructure.

conventional forging. A forging characteristic by design complexity and tolerances that fall within the broad range of general forging practice.

conventional milling. Milling in which the cutter moves in the direction opposite to the feed at the point of contact. Contrast with *climb milling*.

conventional strain. See *engineering strain* and *strain*.

conventional stress. See *engineering stress* and *stress*.

conversion coating. A coating consisting of a compound of the surface metal, produced by chemical or electrochemical treatments of the metal. Examples include chromate coatings on zinc, cadmium, magnesium, and aluminum, and oxide and phosphate coatings on steel. See also *chromate treatment* and *phosphating*.

converter. A furnace in which air is blown through a bath of molten metal or matte, oxidizing the impurities and maintaining the temperature through the heat produced by the oxidation reaction. A typical converter is the *argon oxygen decarburization* vessel.

convex fillet weld. A fillet weld having a convex face.

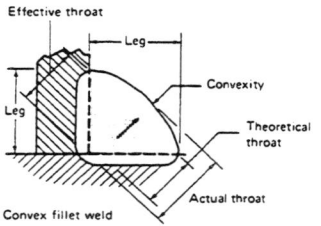

Convex fillet weld

coolant. The liquid used to cool the work during grinding and to prevent it from rusting. It also lubricates, washes away chips and grits, and aids in obtaining a finer finish. In metal cutting, the preferred term is cutting fluid.

cooling curve. A graph showing the relationship between time and temperature during the cooling of a material. It is used to find the temperatures at which phase changes occur. A property or function other than time may occasionally be used—for example, thermal expansion.

cooling rate. The average slope of the time-temperature curve taken over a specified time and temperature interval.

cooling stresses. Residual stresses in castings resulting from nonuniform distribution of temperature during cooling.

cope. In casting, the upper or topmost section of a *flask*, *mold*, or *pattern*.

copper-accelerated salt-spray (CASS) test. An *accelerated corrosion test* for some electrodeposits and for anodic coatings on aluminum.

copper brazing. A term improperly used to denote brazing with a copper filler metal. See preferred terms *furnace brazing* and *braze welding*.

copperhead. A reddish spot in a porcelain enamel coating caused by iron pickup during enameling, iron oxide left on poorly cleaned basis metal, or burrs on iron or steel basis metal that protrude through the coating and are oxidized during firing.

core. (1) A specially formed material inserted in a mold to shape the interior or other part of a casting that cannot be shaped as easily by the pattern. (2) In a ferrous alloy prepared for *case hardening*, that portion of the alloy that is not part of the *case*. Typically considered to be the portion that (a) appears dark (with certain etchants) on an etched cross section, (b) has an essentially unaltered chemical composition, or (c) has a hardness, after hardening, less than a specified value.

core assembly. In casting, a complex core consisting of a number of sections.

core binder. In casting, any material used to hold the grains of core sand together.

core blow. A gas pocket in a casting adjacent to a cored cavity and caused by entrapped gases from the core.

core blower. A machine for making foundry cores using compressed air to blow and pack the sand into the core box.

core box. In casting, a wood, metal, or plastic structure containing a shaped cavity into which sand is packed to make a core.

cored bars. In powder metallurgy, a compact of bar shape heated by its own electrical resistance to a temperature high enough to melt its interior.

core forging. (1) Displacing metal with a punch to fill a die cavity. (2) The product of such an operation.

core knockout machine. In casting, a mechanical device for removing cores from castings.

coreless induction furnace. An electric induction furnace for melting or holding molten metals that does not utilize a steel core to direct the magnetic field which stirs the melt.

core rod. In powder metallurgy, a member of a die assembly used in molding a hole in a compact.

core sand. In casting, sand for making cores to which a binding material has been added to obtain good cohesion and permeability after drying; usually low in clays.

coring. (1) A condition of variable composition between the center and surface of a unit of microstructure (such as a dendrite, grain, carbide particle); results from nonequilibrium solidification, which occurs over a range of temperature. (2) A central cavity at the butt end of a rod extrusion, sometimes called *extrusion pipe*.

corner angle. On face milling cutters, the angle between an angular cutting edge of a cutter tooth and the axis of the cutter, measured by rotation into an axial plane. See the figure accompanying *face mill*.

corner joint. A joint between two members located approximately at right angles to each other in the form of an "L."

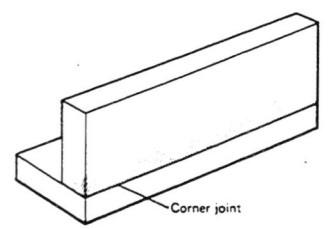

Corner joint

corona (resistance welding). The area sometimes surrounding the nugget of a spot weld at the faying surfaces which provides a degree of solid-state welding.

corrodkote test. An *accelerated corrosion test* for electrodeposits.

corrosion. The chemical or electrochemical reaction between a material, usually a metal, and its environment that produces a deterioration of the material and its properties.

corrosion embrittlement. The severe loss of ductility of a metal resulting from corrosive attack, usually *intergranular* and often not visually apparent.

corrosion-erosion. See *erosion-corrosion*.

corrosion fatigue. The process in which a metal fractures prematurely under conditions of simultaneous corrosion and repeated cyclic loading at lower stress levels or fewer cycles than would be required in the absence of the corrosive environment.

corrosion inhibitor. See *inhibitor*.

corrosion potential (E_{corr}). The *potential* of a corroding surface in an electrolyte, relative to a *reference electrode*. Also called rest potential, open-circuit potential, or freely corroding potential.

corrosion product. Substance formed as a result of *corrosion*.

corrosion protection. Modification of a *corrosion system* so that corrosion damage is mitigated.

corrosion rate. *Corrosion effect* on a metal per unit of time. The type of corrosion rate used depends on the technical system and on the type of corrosion effect. Thus, corrosion rate may be expressed as an increase in corrosion depth per unit of time (penetration rate, for example, mils/yr) or the mass of metal turned into corrosion products per unit area of surface per unit of time (weight loss, for example, $g/m^2/yr$). The corrosion effect may vary with time and may not be the same at all points of the corroding surface. Therefore, reports of corrosion rates should be accompanied by information on the type, time dependency, and location of the corrosion effect.

corrosion resistance. The ability of a material to withstand contact with ambient natural factors or those of a particular, artificially created atmosphere, without degradation or change in properties. For metals, this could be pitting or rusting; for organic materials, it could be crazing.

corrosion system. System consisting of one or more metals and all parts of the environment that influence *corrosion*.

corrosive wear. *Wear* in which chemical or electrochemical reaction with the environment is significant. See also *oxidative wear*.

corrugating. The forming of sheet metal into a series of straight, parallel alternate ridges and grooves with a rolling mill equipped with matched roller dies or a *press brake* equipped with a specially shaped punch and die.

corrugations. In metal forming, transverse ripples caused by a variation in strip shape during hot or cold reduction.

Cottrell process. Removal of solid particulates from gases with electrostatic precipitation.

coulometer. An electrolytic cell arranged to measure the quantity of electricity by the chemical action produced in accordance with Faraday's law.

counterblow hammer. A forging hammer in which both the *ram* and the *anvil* are driven simultaneously toward each other by air or steam pistons.

counterboring. Removal of material to enlarge a hole for part of its depth with a rotary, pilot guided, end cutting tool having two or more cutting lips and usually having straight or helical flutes for the passage of chips and the admission of a cutting fluid.

countersinking. Beveling or tapering the work material around the periphery of a hole creating a concentric surface at an angle less than 90° with the centerline of the hole for the purpose of chamfering holes or recessing screw and rivet heads.

covered electrode. A composite filler metal electrode consisting of a core of a bare electrode or metal cored electrode to which a covering sufficient to provide a slag layer on the weld metal has been applied. The covering may contain materials providing such functions as shielding from the atmosphere, deoxidation, and arc stabilization and can serve as a source of metallic additions to the weld.

covering power. (1) The ability of a solution to give satisfactory plating at very low current densities, a condition that exists in recesses and pits. This term suggests an ability to cover, but not necessarily to build up, a uniform coating, whereas *throwing power* suggests the ability to obtain a coating of uniform thickness on an irregularly shaped object. (2) The degree to which a porcelain enamel coating obscures the underlying surface.

CO_2 welding. See preferred term *gas metal arc welding*.

"C" process. See *Croning process*.

crack. (1) A fracture type discontinuity characterized by a sharp tip and high ratio of length and width to opening displacement. (2) A line of fracture without complete separation.

crack extension (*a*). An increase in crack size. See also *crack length*, *effective crack size*, *original crack size*, and *physical crack size*.

crack growth. Rate of propagation of a crack through a material due to a static or dynamic applied load.

crack length (depth) (*a*). In *fatigue* and *stress-corrosion cracking*, the *physical crack size* used to determine the crack growth rate and the *stress-intensity factor*. For a compact-type specimen, crack length is measured from the line connecting the bearing points of load application. For a center-crack tension specimen, crack length is measured from the perpendicular bisector of the central crack. See also *crack size*.

crack mouth opening displacement (CMOD). See *crack opening displacement*.

crack opening displacement (COD). On a K_{Ic} specimen, the opening displacement of the notch surfaces at the notch and in the direction perpendicular to the plane of the notch and the crack. The displacement at the tip is called the crack tip opening displacement (CTOD); at the mouth, it is called the crack mouth opening displacement (CMOD). See also *stress-intensity factor* for definition of K_{Ic}.

crack size (*a*). A lineal measure of a principal planar dimension of a crack. This measure is commonly used in the calculation of quantities descriptive of the stress and displacement fields. In practice, the value of crack size is obtained from procedures for measurement of *physical crack size*, *original crack size*, or *effective crack size*, as appropriate to the situation under consideration. See also *crack length (depth)*.

crack tip opening displacement (CTOD). See *crack opening displacement*.

crank. Forging shape generally in the form of a "U" with projections at more or less right angles to the upper terminals. Crank shapes are designated by the number of throws (for example, two-throw crank).

crank press. A mechanical press whose slides are actuated by a crankshaft.

crater. In arc welding, a depression at the termination of a weld bead or in the molten weld pool.

crater crack. A crack in the crater of a weld bead.

crater wear. The wear that occurs on the rake face of a cutting tool due to contact with the material in the chip that is sliding along that face.

craze cracking. Irregular surface cracking of a metal associated with thermal cycling. This term is used more in the United Kingdom than in the United States, where the term checking is used instead.

creep. Time-dependent strain occurring under stress. The creep strain occurring at a diminishing rate is called primary creep; that occurring at a minimum and almost constant rate, secondary creep; and that occurring at an accelerating rate, tertiary creep.

creep-feed grinding. A *grinding* process that produces deeper cuts at slow traverse rates.

creep limit. (1) The maximum stress that will cause less than a specified quantity of creep in a given time. (2) The maximum nominal stress under which the creep strain rate decreases continuously with time under constant load and at constant temperature. Sometimes used synonymously with *creep strength*.

creep rate. The slope of the creep-time curve at a given time. Deflection with time under a given static load.

creep recovery. The time-dependent decrease in strain in a solid, following the removal of force.

creep-rupture embrittlement. *Embrittlement* under creep conditions of, for example, aluminum alloys and steels that results in abnormally low rupture ductility. In aluminum alloys, iron in amounts above the solubility limit is known to cause such embrittlement; in steels, the phenomenon is related to the amount of impurities (for example, phosphorus, sulfur, copper, arsenic, antimony, and tin) present. In either case, failure occurs by *intergranular cracking* of the embrittled material.

creep-rupture strength. The stress that causes fracture in a creep test at a given time, in a specified constant environment. This is sometimes referred to as the stress-rupture strength. In glass technology, this is termed the static fatigue strength.

creep-rupture test. A test in which progressive specimen deformation and the time for rupture are both measured. In general, deformation is much greater than that developed during a creep test. Also known as stress-rupture test.

creep strain. The time-dependent total strain (extension plus initial gage length) produced by applied stress during a creep test.

creep strength. The stress that will cause a given *creep strain* in a creep test at a given time in a specified constant environment.

creep stress. The constant load divided by the original cross-sectional area of the specimen.

creep test. A method of determining the extension of metals under a given load at a given temperature. The determination usually involves the plotting of time-elongation curves under constant load; a single test may extend over many months. The results are often expressed as the elongation (in millimeters or inches) per hour on a given gage length (e.g., 25 mm, or 1 in.).

crevice corrosion. *Localized corrosion* of a metal surface at, or immediately adjacent to, an area that is shielded from full exposure to the environment because of close proximity between the metal and the surface of another material.

crimping. The forming of relatively small *corrugations* in order to set down and lock a seam, to create an arc in a strip of metal, or to reduce an existing arc or diameter. See also *corrugating*.

critical cooling rate. The minimum rate of continuous cooling for preventing undesirable transformations. For steel, unless otherwise specified, it is the slowest rate at which austenite can be cooled from above critical temperature to prevent its transformation above the martensite start temperature.

critical current density. In an electrolytic process, a current density at which an abrupt change occurs in an operating variable or in the nature of an electrodeposit or electrode film.

critical flaw size. The size of a flaw (defect) in a structure that will cause failure at a particular stress level.

critical point. (1) The temperature or pressure at which a change in crystal structure, phase, or physical properties occurs. Also termed *transformation temperature*. (2) In an equilibrium diagram, that combination of composition, temperature, and pressure at which the phases of an inhomogeneous system are in equilibrium.

critical shear stress. The shear stress required to cause slip in a designated slip direction on a given slip plane. It is called the critical resolved shear stress if the shear stress is induced by tensile or compressive forces acting on the crystal.

critical strain. (1) In mechanical testing, the strain at the *yield point*. (2) The strain just sufficient to cause *recrystallization*; because the strain is small, usually only a few percent, recrystallization takes place from only a few nuclei, which produces a recrystallized structure consisting of very large grains.

critical stress intensity factor. See *stress-intensity factor*.

critical temperature. That temperature above which the vapor phase cannot be condensed to liquid by an increase in pressure. Synonymous with *critical point* if pressure is constant.

critical temperature range. Synonymous with *transformation ranges*, which is the preferred term.

Croning process. In casting, a *shell molding process* that uses a phenolic resin binder. Sometimes referred to as C process or Chronizing.

crop. (1) An end portion of an ingot that is cut off as scrap. (2) To shear a bar or billet.

cross breaks. Same as *coil breaks*.

cross-country mill. A rolling mill in which the mill stands are so arranged that their tables are parallel with a transfer (or crossover) table connecting them. Such a mill is used for rolling structural shapes, rails, and any special form of bar stock not rolled in the ordinary bar mill.

cross direction. See *transverse direction*.

cross forging. Preliminary working of forging stock in flat dies to develop mechanical properties, particularly in the center portions of heavy sections.

cross rolling. Rolling of metal or sheet or plate so that the direction of rolling is about 90° from the direction of a previous rolling.

cross-wire weld. A weld made at the junction between crossed wires or bars.

crown. (1) The upper part (head) of a forming press frame. On hydraulic presses, the crown usually contains the cylinder; on mechanical presses, the crown contains the drive mechanism. See also *hydraulic press* and *mechanical press*. (2) A shape (crown) ground into a flat roll to ensure flatness of cold (and hot) rolled sheet and strip. (3) A contour on a sheet or roll where the thickness or diameter increases from edge to center.

crucible furnace. A melting or holding furnace in which the molten metal is contained in a pot-shaped (hemispherical) shell. Electric heaters or fuel-fired burners outside the shell generate the heat that passes through the shell (crucible) to the molten metal.

crush. (1) Buckling or breaking of a section of a casting mold due to incorrect register when the mold is closed. (2) An indentation in the surface of a casting due to displacement of sand when the mold was closed.

crush forming. Shaping a grinding wheel by forcing a rotating metal roll into its face so as to reproduce the desired contour.

crushing. A process of comminuting large pieces of metal or ore into rough size fractions prior to grinding into powder. A typical machine for this operation is a jaw crusher.

crushing test. (1) A radial compressive test applied to tubing, sintered-metal bearings, or other similar products for determining radial crushing strength (maximum load in compression). (2) An axial compressive test for determining quality of tubing, such as soundness of weld in welded tubing.

cryogenic treatment. See *cold treatment*.

crystal. (1) A solid composed of atoms, ions, or molecules arranged in a pattern that is repetitive in three dimensions. (2) That form, or particle, or piece of a substance in which its atoms are distributed in one specific orderly geometrical array, called "lattice," essentially throughout. Crystals exhibit characteristic optical and other properties and growth or cleavage surfaces, in characteristic directions.

crystalline. That form of a substance which is comprised predominantly of (one or more) crystals, as opposed to glassy or amorphous.

crystalline fracture. A pattern of brightly reflecting crystal facets on the fracture surface of a polycrystalline metal, resulting from *cleavage fracture* of many individual crystals. Contrast with *fibrous fracture*, and *silky fracture*; see also *granular fracture*.

crystallization. (1) The separation, usually from a liquid phase on cooling, of a solid crystalline phase. (2) The progressive process in which crystals are first nucleated (started) and then grown in size within a host medium which supplies their atoms. The host may be gas, liquid, or of another crystalline form.

crystal orientation. See *orientation*.

crystal system. One of seven groups into which all crystals may be divided; triclinic, monoclinic, orthorhombic, hexagonal, rhombohedral, tetragonal, and cubic.

cubic plane. A plane perpendicular to any one of the three crystallographic axes of the cubic (isometric) system; the *Miller indices* are {100}.

cup. (1) A sheet metal part; the product of the first drawing operation. (2) Any cylindrical part or shell closed at one end.

cupellation. Oxidation of molten lead containing gold and silver to produce lead oxide, thereby separating the precious metals from the base metal.

cup-and-cone fracture. A mixed-mode fracture, often seen in tensile-test specimens of a ductile material, where the central portion undergoes *plane-strain* fracture and the surrounding region undergoes *plane-stress* fracture. It is called a

cup fracture (or cup-and-cone fracture) because one of the mating fracture surfaces looks like a miniature cup—that is, it has a central depressed flat-face region surrounded by a shear lip; the other fracture surface looks like a miniature truncated cone.

cupping. (1) The first step in deep drawing. (2) Fracture of severely worked rods or wire where one end has the appearance of a cup and the other that of a cone.

cupping test. A mechanical test used to determine the ductility and stretching properties of sheet metal. It consists of measuring the maximum part depth that can be formed before fracture. The test is typically carried out by stretching the test piece clamped at its edges into a circular die using a punch with a hemispherical end. See also *Erichsen test, Olsen ductility test*, and *Swift cup test*.

cupola. A cylindrical vertical furnace for melting metal, especially cast iron, by having the charge come in contact with the hot fuel, usually metallurgical coke.

Curie temperature. The temperature marking the transition between ferromagnetism and paramagnetism, or between the ferroelectric phase and the paraelectric phase. Also known as Curie point. See also *ferromagnetism* and *paramagnetism*.

curling. Rounding the edge of sheet metal into a closed or partly closed loop.

current. The net transfer of electric charge per unit time. Also called electric current. See also *current density*.

current decay. In spot, seam, or projection welding, the controlled reduction of the welding current from its peak amplitude to a lower value to prevent excessively rapid cooling of the weld nugget.

current density. The current flowing to or from a unit area of an electrode surface.

current efficiency. (1) The ratio of the electrochemical equivalent current density for a specific reaction to the total applied current density. (2) The proportion of current used in a given process to accomplish a desired result; in electroplating, the proportion used in depositing or dissolving metal.

cut (foundry practice). (1) To recondition molding sand by mixing on the floor with a shovel or blade-type machine. (2) To form the sprue cavity in a mold. (3) Defect in a casting resulting from erosion of the sand by metal flowing over the mold or cored surface.

cut edge. A mechanically sheared edge obtained by slitting, shearing, or blanking.

cut-off (casting). Removing a casting from the sprue by refractory wheel or saw, arc-air torch, or gas torch.

cut-off (metal forming). A pair of blades positioned in dies or equipment (or a section of the die milled to produce the same effect as inserted blades) used to separate the forging from the bar after forging operations are completed. Used only when forgings are produced from relatively long bars instead of from individual, precut multiples or blanks. See also *blank* and *multiple*.

cutoff wheel. A thin abrasive wheel for severing or slotting any material or part.

cutting down. Removing roughness or irregularities of a metal surface by abrasive action.

cutting edge. The leading edge of a cutting tool (such as a lathe tool, drill or milling cutter) where a line of contact is made with the work during machining. See also the figure accompanying *single-point tool*.

cutting fluid. A fluid used in metal cutting to improve finish, tool life, or dimensional accuracy. On being flowed over the tool and work, the fluid reduces friction, the heat generated, and tool wear, and prevents galling. It conducts the heat away from the point of generation and also serves to wash the *chips* away.

cutting process. A process which brings about the severing or removal of metals. See also *arc cutting* and *oxygen cutting*.

cutting speed. The linear or peripheral speed of relative motion between the tool and workpiece in the principal direction of cutting.

cutting tip. That part of an oxygen cutting torch from which the gases issue.

cutting torch (oxyfuel gas). A device used for directing the preheating flame produced by the controlled combustion of fuel gases and to direct and control the cutting oxygen.

cutting torch (plasma arc). A device used for plasma arc cutting to control the position of the electrode, to transfer current to the arc, and to direct the flow of plasma and shielding gas.

cyanic copper. Copper electrodeposited from an alkali-cyanide solution containing a complex ion made up of univalent copper and the cyanide radical; also the solution itself.

cyanide slimes. Finely divided metallic precipitates that are formed when precious metals are extracted from their ores using cyanide solutions.

cyaniding. A case-hardening process in which a ferrous material is heated above the lower transformation temperature range in a molten salt containing cyanide to cause simultaneous absorption of carbon and nitrogen at the surface and, by diffusion, create a concentration gradient. *Quench hardening* completes the process.

cycle (N). In fatigue, one complete sequence of values of applied load that is repeated periodically. See also *S-N curve*.

cycle annealing. An annealing process employing a predetermined and closely controlled time-temperature cycle to produce specific properties or microstructures.

cyclic load. (1) Repetitive loading, as with regularly recurring stresses on a part, that sometimes leads to fatigue fracture. (2) Loads that change value by following a regular repeating sequence of change.

cylindrical grinding. Grinding the outer cylindrical surface of a rotating part.

cylindrical land. *Land* having zero relief.

D

damage tolerance. (1) A design measure of crack growth rate. Cracks in damage-tolerant designed structures are not permitted to grow to critical size during expected service life. (2) The ability of a part component, such as an aerospace engine, to resist failure due to the presence of flaws, cracks, or other damage for a specified period of usage. The damage tolerance approach is used extensively in the aerospace industry.

damping. The loss in energy, as dissipated heat, that results when a material or material system is subjected to an oscillatory load or displacement.

damping capacity. The ability of a material to absorb vibration (cyclical stresses) by internal friction, converting the mechanical energy into heat.

daylight. The distance, in the open position, between the moving and the fixed tables or the platens of a hydraulic press. In the case of a multiplaten press, daylight is the distance between adjacent platens. Daylight provides space for removal of the molded/formed part from the mold/die.

dc casting. Same as *direct chill casting*.

dead soft. A *temper* of nonferrous alloys and some ferrous alloys corresponding to the condition of minimum hardness and tensile strength produced by *full annealing*.

dealloying. The selective corrosion of one or more components of a solid solution alloy. Also called parting or *selective leaching*. See also *decarburization, decobaltification, denickelification, dezincification*, and *graphitic-corrosion*.

deburring. Removing burrs, sharp edges, or fins from metal parts by filing, grinding, or rolling the work in a barrel containing abrasives suspended in a suitable liquid medium. Sometimes called burring.

decalescence. A phenomenon, associated with the transformation of alpha iron to gamma iron on the heating (superheating) of iron or steel, revealed by the darkening of the metal surface owing to the sudden decrease in temperature caused by the fast absorption of the latent heat of transformation. Contrast with *recalescence*.

decarburization. Loss of carbon from the surface layer of a carbon-containing alloy due to reaction with one or more chemical substances in a medium that contacts the surface.

decobaltification. Corrosion in which cobalt is selectively leached from cobalt-base alloys, such as Stellite, or from cemented carbides. See also *dealloying* and *selective leaching*.

decohesive rupture. A *brittle fracture* that exhibits little or no bulk plastic deformation and does not occur by dimple rupture, cleavage, or fatigue. This type of fracture is generally the result of a reactive environment or a unique microstructure and is associated almost exclusively with rupture along grain boundaries.

decomposition. Separation of a compound into its chemical elements or components.

decomposition potential (or voltage). The *potential* of a metal surface necessary to decompose the electrolyte of a cell or a component thereof.

deep drawing. Forming deeply recessed parts by forcing sheet metal to undergo plastic flow between dies, usually without substantial thinning of the sheet.

deep etching. In metallography, *macroetching*, especially for steels, to determine the overall character of the material, that is, the presence of imperfections, such as seams, forging bursts, shrinkage-void remnants, cracks, and coring.

defect. (1) A discontinuity whose size, shape, orientation, or location makes it detrimental to the useful service of the part in which it occurs. (2) A discontinuity or discontinuities which by nature or accumulated effect (for example, total crack length) render a part or product unable to meet minimum applicable acceptance standards or specifications. This term designates rejectability. See also *discontinuity* and *flaw*.

defective. A quality control term, describing a unit of product or service containing at least one *defect*, or having several lesser imperfections that, in combination, cause the unit not to fulfill its anticipated function.

deflection. In metal forming and forging, the amount of deviation from a straight line or plane when a force is applied to a press member. Generally used to specify the allowable bending of the bed, slide, or frame at rated capacity with a load of predetermined distribution.

deformation. A change in the form of a body due to stress, thermal change, change in moisture, or other causes. Measured in units of length.

deformation bands. Parts of a crystal that have rotated differently during deformation to produce bands of varied orientation without individual grains.

deformation limit. In *drawing*, the limit of deformation is reached when the load required to deform the flange becomes greater than the load-carrying capacity of the cup wall. The deformation limit (limiting drawing ratio, LDR) is defined as the ratio of the maximum blank diameter that can be drawn into a cup without failure, to the diameter of the punch.

degasifier. A substance that can be added to molten metal to remove soluble gases that might otherwise be occluded or entrapped in the metal during solidification.

degassing. (1) A chemical reaction resulting from a compound added to molten metal to remove gases from the metal. Inert gases are often used in this operation. (2) A fluxing procedure used for aluminum alloys in which nitrogen, chlorine, chlorine and nitrogen, and chlorine and argon are bubbled up through the metal to remove dissolved hydrogen gases and oxides from the alloy. See also *flux*.

degradation. A deleterious change in the chemical structure, physical properties, or appearance of a material.

degreasing. Removing oil or grease from a surface. See also *vapor degreasing*.

delayed yield. A phenomenon involving a delay in time between the application of a stress and the occurrence of the corresponding yield-point strain.

delta ferrite. See *ferrite*.

Demarest process. A *fluid forming* process in which cylindrical and conical sheet metal parts are formed by a modified rubber bulging punch. The punch, equipped with a hydraulic cell, is placed inside the workpiece, which in turn is placed inside the die. Hydraulic pressure expands the punch.

dendrite. A crystal that has a treelike branching pattern, being most evident in cast metals slowly cooled through the solidification range.

dendritic powder. Particles usually of electrolytic origin typically having the appearance of a pine tree.

denickelification. Corrosion in which nickel is selectively leached from nickel-containing alloys. Most commonly observed in copper-nickel alloys after extended service in fresh water. See also *dealloying* and *selective leaching*.

density, absolute. The mass per unit volume of a solid material, expressed in g/cm^3, kg/m^3, or lb/ft^3.

density ratio. The ratio of the determined density of a powder compact to the absolute density of metal of the same composition, usually expressed as a percentage. Also referred to as percent theoretical density.

deoxidation. Removal of excess oxygen from the molten metal; usually accomplished by adding materials with a high affinity for oxygen.

deoxidation products. Those nonmetallic inclusions that form as a result of adding deoxidizing agents to molten metal.

deoxidized copper. Copper from which cuprous oxide has been removed by adding a *deoxidizer*, such as phosphorus, to the molten bath.

deoxidizer. A substance that can be added to molten metal to remove either free or combined oxygen.

deoxidizing. (1) The removal of oxygen from molten metals through the use of a suitable *deoxidizer*. (2) Sometimes refers to the removal of undesirable elements other than oxygen through the introduction of elements or compounds that readily react with them. (3) In metal finishing, the removal of oxide films from metal surfaces by chemical or electrochemical reaction.

dephosphorization. The elimination of phosphorus from molten steel.

depolarization. A decrease in the *polarization* of an electrode.

deposit corrosion. Corrosion occurring under or around a discontinuous deposit on a metallic surface. Also called poultice corrosion.

deposition efficiency (arc welding). The ratio of the weight of deposited metal to the net weight of filler metal consumed, exclusive of stubs.

deposition sequence. The order in which the increments of weld metal are deposited. See also *buildup sequence* and *longitudinal sequence*.

depth of cut. The thickness of material removed from a workpiece in a single machining part.

depth of fusion. The distance that fusion extends into the base metal or previous pass from the surface melted during welding.

depth of penetration (welding). See *joint penetration* and *root penetration*.

derby. A massive piece (intermediate in size, extending to more than 45 kg, or 100 lb, and usually cylindrical) of primary metal made by bomb reduction (such as uranium from uranium tetrafluoride reduced with magnesium). Compare with *biscuit* and *dingot*.

descaling. (1) Removing the thick layer of oxides formed on some metals at elevated temperatures. (2) A chemical or mechanical process for removing scale or investment material from castings.

deseaming. Analogous to *chipping*, the surface imperfections being removed by gas cutting.

desulfurizing. The removal of sulfur from molten metal by reaction with a suitable slag or by the addition of suitable compounds.

detonation flame spraying. A thermal spraying process variation in which the controlled explosion of a mixture of fuel gas, oxygen, and powdered coating material is utilized to melt and propel the material to the workpiece.

detritus. See *wear debris*.

developed blank. A sheet metal blank that yields a finished part without trimming or with the least amount of trimming.

dewaxing. In casting, the process of removing the expendable wax pattern from an investment mold or shell mold; usually accomplished by melting out the application of heat or dissolving the wax with an appropriate solvent.

dezincification. Corrosion in which zinc is selectively leached from zinc-containing alloys leaving a relatively weak layer of copper and copper oxide. Most commonly found in copper-zinc alloys containing less than 85% copper after extended service in water containing dissolved oxygen. See also *dealloying* and *selective leaching*.

diamagnetic material. A material whose specific permeability is less than unity and is therefore repelled weakly by a magnet. Compare with *ferromagnetic material* and *paramagnetic material*.

diamond pyramid hardness test. See *Vickers hardness test*.

diamond tool. (1) A diamond, shaped or formed to the contour of a single-point cutting tool, for use in precision machining of nonferrous or nonmetallic materials. (2) An insert made from polycrystalline diamond compacts.

diamond wheels. A grinding wheel in which crushed and sized industrial diamonds are held in a resinoid, metal, or vitrified bond.

diaphragm. (1) A porous or permeable membrane separating anode and cathode compartments of an electrolytic cell from each other or from an intermediate compartment. (2) Universal die member made of rubber or similar material used to contain hydraulic fluid within the forming cavity and to transmit pressure to the part being formed.

dichromate treatment. A chromate *conversion coating* produced on magnesium alloys in a boiling solution of sodium dichromate.

didymium. A natural mixture of the rare-earth elements praseodymium and neodymium, often given the quasichemical symbol Di.

die. A tool, usually containing a cavity, that imparts shape to solid, molten, or powdered metal primarily because of the shape of the tool itself. Used in many press operations (including blanking, drawing, forging, and forming), in die casting and in forming green powder metallurgy compacts. Die-casting and powder metallurgy dies are sometimes referred to as *molds*. See also *forging dies*.

die block. A block, often made of heat-treated steel, into which desired impressions are machined or sunk and from which closed-die forgings or sheet metal stampings are produced using hammers or presses. In forging, die blocks are usually used in pairs, with part of the impression in one of the blocks and the rest of the impression in the other. In sheet metal forming, the female die is used in conjunction with a male punch. See also *closed-die forging*.

die body. The stationary or fixed part of a powder pressing die.

die casting. (1) A casting made in a die. (2) A casting process in which molten metal is forced under high pressure into the cavity of a metal mold. See also *cold chamber machine* and *hot chamber machine*.

die cavity. The machined recess that gives a forging or stamping its shape.

die clearance. Clearance between a mated punch and die; commonly expressed as clearance per side. Also called clearance or punch-to-die clearance.

die cushion. A press accessory placed beneath or within a *bolster* plate or *die block* to provide an additional motion or pressure for stamping or forging operations; actuated by air, oil, rubber, springs, or a combination of these.

die forging. A forging that is formed to the required shape and size through working in machined impressions in specially prepared dies.

die forming. The shaping of solid or powdered metal by forcing it into or through the *die cavity*.

die holder. A plate or block, on which the die block is mounted, having holes or slots for fastening to the *bolster* plate or the *bed* of the press.

die impression. The portion of the die surface that shapes a forging or sheet metal part.

die insert. A relatively small die that contains part or all of the impression of a forging or sheet metal part and is fastened to the master *die block*.

die life. The productive life of a *die impression*, usually expressed as the number of units produced before the impression has worn beyond permitted tolerances.

die lubricant. (1) A lubricant applied to the working surfaces of dies and punches to facilitate drawing, pressing, stamping, and/or ejection. In powder metallurgy, the die lubricant is sometimes mixed into the powder before pressing into a compact. (2) A compound that is sprayed, swabbed, or otherwise applied on die surfaces or the workpiece during the forging or forming process to reduce friction. Lubricants also facilitate release of the part from the dies and provide thermal insulation. See also *lubricant*.

die match. The condition where dies, after having been set up in a press or other equipment, are in proper alignment relative to each other.

die opening. (1) In flash or upset welding, the distance between the electrodes, usually measured with the parts in contact before welding has commenced or immediately upon completion of the cycle but before upsetting. (2) In powder metallurgy, the entrance to the die cavity.

die proof. A casting of a *die impression* made to confirm the accuracy of the impression.

die radius. The radius on the exposed edge of a deep-drawing die, over which the sheet flows in forming drawn shells.

die set. A tool or tool holder consisting of a die base and punch plate for the attachment of a die and punch, respectively.

die shift. The condition that occurs after the dies have been set up in a forging unit in which a portion of the impression of one die is not in perfect alignment with the corresponding portion of the other die. This results in a mismatch in the forging, a condition that must be held within the specified tolerance.

die sinking. The machining of the die impressions to produce forgings of required shapes and dimensions.

die stamping. The general term for a sheet metal part that is formed, shaped, or cut by a die in a press in one or more operations.

die welding. See preferred terms *forge welding* and *cold welding*.

differential aeration cell. An *electrolytic cell*, the *electromagnetic force* of which is due to a difference in air (oxygen) concentration at one electrode as compared with that at another electrode of the same material. See also *concentration cell*.

differential coating. A coated product having a specified coating on one surface and a significantly lighter coating on the other surface (such as a hot dip galvanized product or electrolytic tin plate).

differential flotation. Separating a complex ore into two or more valuable minerals and *gangue* by *flotation*. Also called selective flotation.

differential heating. Heating that intentionally produces a temperature gradient within an object such that, after cooling, a desired stress distribution or variation in properties is present within the object.

diffusion. (1) Spreading of a constituent in a gas, liquid, or solid, tending to make the composition of all parts uniform. (2) The spontaneous movement of atoms or molecules to new sites within a material. (3) The movement of a material, such as a gas or liquid, in the body of a plastic. If the gas or liquid is absorbed on one side of a piece of plastic and given off on the other side, the phenomenon is called permeability. Diffusion and permeability are not due to holes or pores in the plastic but are caused and controlled by chemical mechanisms.

diffusion aid. A solid filler metal sometimes used in *diffusion welding*.

diffusion bonding. See preferred terms *diffusion welding* and *diffusion brazing*.

diffusion brazing. A brazing process which produces coalescence of metals by heating them to suitable temperatures and by using a filler metal or an *in situ* liquid phase. The filler metal may be distributed by capillary action or may be placed or formed at the faying surfaces. The filler metal is diffused with the base metal to the extent that the joint properties have been changed to approach those of the base metal. Pressure may or may not be applied.

diffusion coating. Any process whereby a base metal or alloy is either (1) coated with another metal or alloy and heated to a sufficient temperature in a suitable environment or (2) exposed to a gaseous or liquid medium containing the other metal or alloy, thus causing *diffusion* of the coating or of the other metal or alloy into the base metal with resultant changes in the composition and properties of its surface.

diffusion coefficient. A factor of proportionality representing the amount of substance diffusing across a unit area through a unit concentration gradient in unit time.

diffusion welding. A solid-state welding process that produces coalescence of the faying surfaces by the application of pressure at elevated temperature. The process does not involve macroscopic deformation, melting, or relative motion of parts. A solid filler metal (*diffusion aid*) may or may not be inserted between the faying surfaces. See also *forge welding*, *hot pressure welding*, and *cold welding*.

digging. A sudden erratic increase in cutting depth, or in the load on a cutting tool, caused by unstable conditions in the machine setup. Usually the machine is stalled, or either the tool or the workpiece is destroyed.

dilatometer. An instrument for measuring the linear expansion or contraction in a metal resulting from changes in such factors as temperature and allotropy.

dimple rupture. A fractographic term describing *ductile fracture* that occurs through the formation and coalescence of microvoids along the fracture path. The fracture surface of such a ductile fracture appears dimpled when observed at high magnification and usually is most clearly resolved when viewed in a scanning electron microscope.

dimpling. (1) The stretching of a relatively small, shallow indentation into sheet metal. (2) In aircraft, the stretching of metal into a conical flange for a countersunk head rivet.

dingot. An oversized *derby* (possibly a ton or more) of a metal produced in a bomb reaction (such as uranium from uranium tetrafluoride reduced with magnesium). For these metals, the term "ingot" is reserved for massive units produced in vacuum melting and casting. See also *biscuit* and *derby*.

dinking. Cutting of nonmetallic materials or light-gage soft metals by using a hollow punch with a knifelike edge acting against a wooden fiber or resiliently mounted metal plate.

dip brazing. A brazing process in which the heat required is furnished by a molten chemical or metal bath. When a molten chemical bath is used, the bath may act as a flux. When a molten metal bath is used, the bath provides the filler metal.

dip coat. (1) In the solid mold technique of investment casting, an extremely fine ceramic precoat applied as a slurry directly to the surface of the pattern to reproduce maximum surface smoothness. This coating is surrounded by coarser, less expensive, and more permeable investment to form the mold. (2) In the shell mold technique of investment casting, an extremely fine ceramic coating called the first coat, applied as a slurry directly to the surface of the pattern to reproduce maximum surface smoothness. The first coat is followed by other dip coats of different viscosity and usually containing different grading of ceramic particles. After each dip, coarser stucco material is applied to the still-wet coating. A buildup of several coats forms an investment shell mold. See also *investment casting.*

diphase cleaning. Removing soil by an emulsion that produces two phases in the cleaning tank: a solvent phase and an aqueous phase. Cleaning is effected by both solvent action and emulsification.

dip plating. Same as *immersion plating.*

dip soldering. A soldering process in which the heat required is furnished by a molten metal bath which provides the solder filler metal.

direct chill casting. A continuous method of making ingots for rolling or extrusion by pouring the metal into a short mold. The base of the mold is a platform that is gradually lowered while the metal solidifies, the frozen shell of metal acting as a retainer for the liquid metal below the wall of the mold. The ingot is usually cooled by the impingement of water directly on the mold or on the walls of the solid metal as it is lowered. The length of the ingot is limited by the depth to which the platform can be lowered; therefore, it is often called semicontinuous casting.

direct current arc furnace. An electric-arc furnace in which a single electrode positioned at the center of the furnace roof is the *cathode* of the system. Current passes from the electrode through the *charge* or bath to a cathode located at the bottom of the furnace. Current from the bottom of the furnace then passes through the furnace refractories to a copper base plate to outside cables. Used in the production of ferroalloys, carbon and alloy steels, and stainless steels. See also *arc furnace.*

direct current electrode negative (DCEN). The arrangement of direct current arc welding leads in which the work is the positive pole and the electrode is the negative pole of the welding arc. See also *straight polarity.*

direct current electrode positive (DCEP). The arrangement of direct current arc welding leads in which the work is the negative pole and the electrode is the

positive pole of the welding arc. See also *reverse polarity.*

direct current reverse polarity (DCRP). See *reverse polarity* and *direct current electrode positive.*

direct current straight polarity (DCSP). See *straight polarity* and *direct current electrode negative.*

directional property. Property whose magnitude varies depending on the relation of the test axis to a specific direction within the metal. The variation results from preferred orientation or from fibering of constituents or inclusions.

directional solidification. Controlled solidification of molten metal in a casting so as to provide feed metal to the solidifying front of the casting.

direct quenching. (1) Quenching carburized parts directly from the carburizing operation. (2) Also used for quenching pearlitic malleable parts directly from the malleabilizing operation.

discontinuity. (1) Any interruption in the normal physical structure or configuration of a part, such as cracks, laps, seams, inclusions, or porosity. A discontinuity may or may not affect the utility of the part. (2) An interruption of the typical structure of a weldment, such as a lack of homogeneity in the mechanical, metallurgical, or physical characteristics of the material or weldment. A discontinuity is not necessarily a defect. See also *defect* and *flaw.*

discontinuous precipitation. Precipitation from a supersaturated solid solution in which the precipitate particles grow by short-range diffusion, accompanied by recrystallization of the matrix in the region of precipitation. Discontinuous precipitates grow into the matrix from nuclei near grain boundaries, forming cells of alternate lamellae of precipitate and depleted (and recrystallized) matrix. Often referred to as cellular or nodular precipitation. Compare with *continuous precipitation* and *localized precipitation.*

discontinuous yielding. The nonuniform plastic flow of a metal exhibiting a yield point in which plastic deformation is inhomogeneously distributed along the gage length. Under some circumstances, it may occur in metals not exhibiting a distinct yield point, either at the onset of or during plastic flow.

dishing. Forming a shallow concave surface, the area being large compared to the depth.

disk grinding. Grinding with the flat side of an abrasive disk or segmented wheel. Also called vertical-spindle surface grinding.

dislocation. A linear imperfection in a crystalline array of atoms. Two basic types are recognized: (1) an edge dislocation corresponds to the row of mismatched atoms along the edge formed by an extra, partial plane of atoms within the body of a crystal; (2) a screw dislocation corresponds to the axis of a spiral structure in a crystal, characterized by a distortion that joins normally parallel planes together to form a continuous helical ramp (with a pitch of one interplanar distance) winding about the dislocation. Most prevalent is the so-called mixed dislocation, which is any combination of an edge dislocation and a screw dislocation.

disordered structure. The crystal structure of a solid solution in which the atoms of different elements are randomly distributed relative to the available lattice sites. Contrast with *ordered structure.*

disordering. Forming a lattice arrangement in which the solute and solvent atoms of a solid solution occupy lattice sites at random. See also *ordering* and *superlattice.*

dispersing agent. A substance that increases the stability of a suspension of particles in a liquid medium by deflocculation of the primary particles.

dispersion hardening. See *dispersion strengthening.*

dispersion-strengthened material. A metallic material that contains a fine dispersion of nonmetallic phase(s), such as Al_2O_3, MgO, SiO_2, CdO, ThO_2, Y_2O_3, or ZrO_2 singly or in combination, to increase the hot strength of the metallic matrix. Examples include dispersion-strengthened copper (Al_2O_3) used for welding electrodes, silver (CdO) used for electrical contacts, and nickel-chromium (Y_2O_3) superalloys used for gas turbine components. See also *mechanical alloying.*

dispersion strengthening. The strengthening of a metal or alloy by incorporating chemically stable submicron size particles of a nonmetallic phase that impede dislocation movement at elevated temperature.

dispersoid. Finely divided particles of relatively insoluble constituents visible in the microstructure of certain metallic alloys.

disruptive strength. The stress at which a metal fractures under hydrostatic tension.

distortion. Any deviation from an original size, shape, or contour that occurs because of the application of stress or the release of residual stress.

disturbed metal. The cold worked metal layer formed at a polished surface during the process of mechanical grinding and polishing.

divided cell. A cell containing a diaphragm or other means for physically separating the *anolyte* from the *catholyte*.

divorced eutectic. A metallographic appearance in which the two constituents of a eutectic structure appear as massive phases rather than the finely divided mixture characteristic of normal eutectics. Often, one of the constituents of the eutectics is continuous and indistinguishable from an accompanying proeutectic constituent.

domain, magnetic. A substructure in a ferromagnetic material within which all the elementary magnets (electron spins) are held aligned in one direction by interatomic forces; if isolated, a domain would be a saturated permanent magnet.

double-acting hammer. A forging hammer in which the ram is raised by admitting steam or air into a cylinder below the piston, and the blow intensified by admitting steam or air above the piston on the downward stroke.

double-action die. A die designed to perform more than one operation in a single stroke of the press.

double-action forming. Forming or drawing in which more than one action is achieved in a single stroke of the press.

double-action mechanical press. A press having two independent parallel movements by means of two slides, one moving within the other. The inner slide or plunger is usually operated by a crankshaft; the outer or blankholder slide, which dwells during the drawing operation, is usually operated by a toggle mechanism or by cams. See also *slide*.

double aging. Employment of two different aging treatments to control the type of precipitate formed from a supersaturated matrix in order to obtain the desired properties. The first aging treatment, sometimes referred to as intermediate or stabilizing, is usually carried out at higher temperature than the second.

double-bevel groove weld. A groove weld in which the joint edge of one member is beveled from both sides.

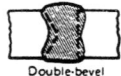
Double-bevel

double-J groove weld. A groove weld in which the joint edge of one member is in the form of two J's, one from either side.

Double-J

double tempering. A treatment in which a quench-hardened ferrous metal is sub-

jected to two complete tempering cycles, usually at substantially the same temperature, for the purpose of ensuring completion of the tempering reaction and promoting stability of the resulting microstructure.

double-U groove weld. A groove weld in which each joint edge is in the form of two J's or two half-U's, one from either side of the member.

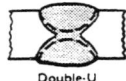
Double-U

double-V groove weld. A groove weld in which each joint edge is beveled from both sides.

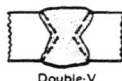
Double-V

double-welded joint. In arc and oxyfuel gas welding, any joint welded from both sides.

down cutting. See preferred term *climb cutting*.

downgate. Same as *sprue*.

downhand welding. See *flat-position welding*.

down milling. See preferred term *climb milling*.

downsprue. Same as *sprue*.

Dow process. A process for the production of magnesium by electrolysis of molten magnesium chloride.

draft. (1) An angle or taper on the surface of a pattern, core box, punch, or die (or of the parts made with them) that facilitates removal of the parts from a mold or die cavity, or a core from a casting. (2) The change in cross section that occurs during rolling or cold drawing.

drag. The bottom section of a *flask, mold,* or *pattern*.

drag (thermal cutting). The offset distance between the actual and the theoretical exit points of the cutting oxygen stream measured on the exit surface of the material.

drag angle. In welding, between the axis of the electrode or torch and a line normal to the plane of the weld when welding is being done with the torch positioned ahead of the weld puddle. See also the figure accompanying *backhand welding*.

drawability. A measure of the *formability* of a sheet metal subject to a drawing process. The term is usually used to indicate the ability of a metal to be deep drawn. See also *drawing* and *deep drawing*.

draw bead. An insert or riblike projection on the draw ring or hold-down surfaces

that aids in controlling the rate of metal flow during deep draw operations. Draw beads are especially useful in controlling the rate of metal flow in irregularly shaped stampings.

drawbench. The stand that holds the die and draw head used in drawing of wire, rod, and tubing.

draw forging. See *radial forging*.

draw forming. A method of curving bars, tubes, or rolled or extruded sections in which the stock is bent around a rotating *form block*. Stock is bent by clamping it to the form block, then rotating the form block while the stock is pressed between the form block and a pressure die held against the periphery of the form block.

draw head. Set of rolls or dies mounted on a draw-bench for forming a section from strip, tubing, or solid stock. See also *Turk's-head rolls*.

drawing. A term used for a variety of forming operations, such as *deep drawing* a sheet metal blank; *redrawing* a tubular part; and drawing rod, wire, and tube. The usual drawing process with regard to sheet metal working in a press is a method for producing a cuplike form from a sheet metal disk by holding it firmly between blankholding surfaces to prevent the formation of wrinkles while the punch travel produces the required shape.

drawing compound. (1) A substance applied to prevent *pickup* and *scoring* during drawing or pressing operations by preventing metal-to-metal contact of the work and die. Also known as *die lubricant*. (2) In metalworking, a lubricant having extreme-pressure properties. See also *extreme-pressure lubricant*.

drawing out. A stretching operation resulting from forging a series of upsets along the length of the workpiece.

draw marks. See *scoring, galling,* and *pickup*.

drawn shell. An article formed by drawing sheet metal into a hollow structure having a predetermined geometrical configuration.

draw plate. (1) In metal forming, a circular plate with a hole in the center contoured to fit a forming punch; used to support the *blank* during the forming cycle. (2) In casting, a plate attached to a pattern to facilitate drawing of a pattern from the mold.

draw radius. The radius at the edge of a die or punch over which sheet metal is drawn.

draw ring. A ring-shaped die part (either the die ring itself or a separate ring) over which the inner edge of sheet metal is drawn by the punch.

draw stock. The forging operation in which the length of a metal mass (stock) is increased at the expense of its cross section; no *upset* is involved. The operation includes converting ingot to pressed bar using "V," round, or flat dies.

dressing. (1) Cutting, breaking down or crushing the surface of a grinding wheel to improve its cutting ability and accuracy. (2) Removing dulled grains from the cutting face of a grinding wheel to restore cutting quality.

drift. (1) A flat piece of steel of tapering width used to remove taper shank drills and other tools from their holders. (2) A tapered rod used to force mismated holes into line for riveting or bolting. Sometimes called a drift pin.

drill. A rotary end-cutting tool used for making holes; it has one or more cutting lips and an equal number of helical or straight flutes for passage of chips and admission of cutting fluid.

drilling. Hole making with a rotary end-cutting tool having one or more cutting lips and one or more helical or straight flutes or tubes for the ejection of chips and the passage of a cutting fluid.

drop forging. The forging obtained by hammering metal in a pair of closed dies to produce the form in the finishing impression under a *drop hammer*; forging method requiring special dies for each shape.

drop hammer. A term generally applied to forging hammers in which energy for forging is provided by gravity, steam, or compressed air. See also *air-lift hammer*, *board hammer*, and *steam hammer*.

drop hammer forming. A process for producing shapes by the progressive deformation of sheet metal in matched dies under the repetitive blows of a gravity-drop or power-drop hammer. The process is restricted to relatively shallow parts and thin sheet from approximately 0.6 to 1.6 mm (0.024 to 0.064 in.).

droplet erosion. Erosive wear caused by the impingement of liquid droplets on a solid surface. See also *erosion*.

drop-through. An undesirable sagging or surface irregularity, usually encountered when brazing or welding near the solidus of the base metal, caused by overheating with rapid diffusion or alloying between the filler metal and the base metal.

dross. (1) The scum that forms on the surface of molten metal largely because of oxidation but sometimes because of the rising of impurities to the surface. (2) Oxide and other contaminants that form on the surface of molten solder.

dry corrosion. See *gaseous corrosion*.

dry cyaniding (obsolete). Same as *carbonitriding*.

dry sand mold. A casting mold made of sand and then dried at 100 °C (212 °F) or above before being used. Contrast with *green sand mold*.

dry strength (casting). The maximum strength of a molded sand specimen that has been thoroughly dried at 100 to 110 °C (220 to 230 °F) and cooled to room temperature. Also known as dry bond strength.

dual-phase steels. A new class of *high-strength low-alloy steels* characterized by a tensile strength value of approximately 550 MPa (80 ksi) and by a microstructure consisting of about 20% hard martensite particles dispersed in a soft ductile ferrite matrix. The term dual phase refers to the predominance in the microstructure of two phases, ferrite and martensite. However, small amounts of other phases, such as bainite, pearlite, or retained austenite, may also be present.

ductile crack propagation. Slow crack propagation that is accompanied by noticeable *plastic deformation* and requires energy to be supplied from outside the body. Contrast with *brittle crack propagation*.

ductile fracture. Fracture characterized by tearing of metal accompanied by appreciable gross plastic deformation and expenditure of considerable energy. Contrast with *brittle fracture*.

ductile iron. A *cast iron* that has been treated while molten with an element such as magnesium or cerium to induce the formation of free graphite as nodules or spherulites, which imparts a measurable degree of ductility to the cast metal. Ductile irons typically contain 3.0 to 4.0% C, 1.8 to 2.8% Si, 0.1 to 1.0% Mn, 0.01 to 0.1% P, and 0.01 to 0.03% S. Also known as nodular cast iron, spherulitic graphite cast iron, and spheroidal graphite (SG) iron.

ductility. The ability of a material to deform plastically without fracturing.

dummy block. In *extrusion*, a thick unattached disk placed between the ram and the billet to prevent overheating of the ram.

dummy cathode. (1) A *cathode*, usually corrugated to give variable current densities, that is plated at low current densities to preferentially remove impurities from a plating solution. (2) A substitute cathode that is used during adjustment of operating conditions.

dummying. Plating with *dummy cathodes*.

duplex grain size. The simultaneous presence of two grain sizes in substantial amounts, with one grain size appreciably larger than the others. Also termed mixed grain size.

duplexing. Any two-furnace melting or refining process. Also called duplex melting or duplex processing.

duplex microstructure. A two-phase structure.

duplex stainless steels. Stainless steels having a fine-grained mixed microstructure of ferrite and austenite with a composition centered around 26Cr-6.5Ni. The corrosion resistance of duplex stainless steels is like that of austenitic stainless steels. However, duplex stainless steels possess higher tensile and yield strengths and improved resistance to stress-corrosion cracking than their austenitic counterparts.

duralumin (obsolete). A term frequently applied to the class of age-hardenable aluminum-copper alloys containing manganese, magnesium, or silicon.

Durville process. A casting process that involves rigid attachment of the mold in an inverted position above the crucible. The melt is poured by tilting the entire assembly, causing the metal to flow along a connecting *launder* and down the side of the mold.

dusting. (1) A phenomenon, usually affecting carbon-base electrical motor brushes or other current-carrying contacts, wherein at low relative humidity or high applied current density, a powdery "dust" is produced during operation. (2) Applying a powder, such as sulfur to molten magnesium or graphite to a mold surface.

dynamic. Moving, or having high velocity. Frequently used with high strain rate (0.1 s^1) testing of metal specimens.

dynamic creep. Creep that occurs under conditions of fluctuating load or fluctuating temperature.

E

earing. The formation of ears or scalloped edges around the top of a drawn shell, resulting from directional differences in the plastic-working properties of rolled metal, with, across, or at angles to the direction of rolling. A main press-drive gear with an eccentric(s) as an integral part. The unit rotates about a common shaft, with the eccentric transmitting the rotary motion of the gear into the vertical motion of the slide through a connection.

eccentric press. A *mechanical press* in which an eccentric, instead of a crankshaft, is used to move the *slide*.

ECM. An abbreviation for *electrochemical machining*.

eddy-current testing. An electromagnetic nondestructive testing method in which eddy-current flow is induced in the test object. Changes in flow caused by variations in the object are reflected into a nearby coil or coils where they are detected and measured by suitable instrumentation.

edge dislocation. See *dislocation*.

edge joint. A joint between the edges of two or more parallel or nearly parallel members.

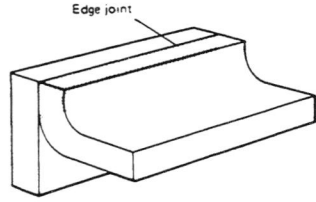

Edge joint

edger (edging impression). The portion of a die impression that distributes metal during forging into areas where it is most needed in order to facilitate filling the cavities of subsequent impressions to be used in the forging sequence. See also *fuller (fullering impression)*.

edge strain. Transverse strain lines or Lüders lines ranging from 25 to 300 mm (1 to 12 in.) in from the edges of cold rolled steel sheet or strip. See also *Lüders lines*.

edging. (1) In sheet metal forming, reducing the flange radius by retracting the forming punch a small amount after the stroke but before release of the pressure. (2) In rolling, the working of metal in which the axis of the roll is parallel to the thickness dimension. Also called edge rolling. (3) The forging operation of working a bar between contoured dies while turning it 90° between blows to produce a varying rectangular cross section. (4) In a forging, removing flash that is directed upward between dies, usually accomplished using a lathe.

EDM. Abbreviation for *electrical discharge machining*.

effective crack size (a_e). The *physical crack size* augmented for the effects of crack-tip plastic deformation. Sometimes the effective crack size is calculated from a measured value of a physical crack size plus a calculated value of a plastic-zone adjustment. A preferred method for calculation of effective crack size compares compliance from the secant of a load-deflection trace with the elastic compliance from a calibration for the type of specimen.

effective draw. The maximum limits of forming depth that can be achieved with a multiple-action press; sometimes called maximum draw or maximum depth of draw.

effective rake. The angle between a plane containing a tooth face and the axial plane through the tooth point as measured in the direction of chip flow through the tooth point. Thus, it is the rake resulting from both cutter configuration and direction of chip flow.

885 °F (475 °C) embrittlement. *Embrittlement* of stainless steels upon extended exposure to temperatures between 400 and 510 °C (750 and 950 °F). This type of embrittlement is caused by fine, chromium-rich precipitates that segregate at grain boundaries; time at temperature directly influences the amount of segregation. Grain-boundary segregation of the chromium-rich precipitates increases strength and hardness, decreases ductility and toughness, and changes corrosion resistance. This type of embrittlement can be reversed by heating above the precipitation range.

ejector. A device mounted in such a way that it removes or assists in removing a formed part from a die.

ejector half. The movable half of a die-casting die containing the ejector pins.

ejector rod. A rod used to push out a formed piece.

elastic constants. The factors of proportionality that relate elastic displacement of a material to applied forces. See also *bulk modulus of elasticity, modulus of elasticity, Poisson's ratio,* and *shear modulus*.

elastic deformation. A change in dimensions directly proportional to and in phase with an increase or decrease in applied force.

elastic hysteresis. A misnomer for an anelastic strain that lags a change in applied stress, thereby creating energy loss during cyclic loading. More properly termed *mechanical hysteresis*.

elasticity. The property of a material by virtue of which deformation caused by stress disappears upon removal of the stress. A perfectly elastic body completely recovers its original shape and dimensions after release of stress.

elastic limit. The maximum stress which a material is capable of sustaining without any permanent strain (deformation) remaining upon complete release of the stress. A material is said to have passed its elastic limit when the load is sufficient to initiate plastic, or nonrecoverable, deformation. See also *proportional limit*.

elastic modulus. Same as *modulus of elasticity*.

elastic ratio. *Yield point* divided by *tensile strength*.

elastic strain. See *elastic deformation*.

elastic strain energy. The energy expended by the action of external forces in deforming a body elastically. Essentially all the work performed during elastic deformation is stored as elastic energy, and this energy is recovered upon release of the applied force.

elastic waves. Mechanical vibrations in an elastic medium.

electrical discharge grinding. Grinding by spark discharges between a negative electrode grinding wheel and a positive workpiece separated by a small gap containing a dielectric fluid such as petroleum oil.

electrical discharge machining (EDM). Metal removed by a rapid spark discharge between different polarity electrodes, one on the workpiece and the other the tool separated by a gap distance of 0.013 to 0.9 mm (0.0005 to 0.035 in.). The gap is filled with dielectric fluid and metal particles which are melted, in part vaporized and expelled from the gap.

electrical discharge wire cutting. A special form of electrical discharge machining wherein the electrode is a continuous moving conductive wire. Also referred to as traveling wire electrical discharge machining.

electrical disintegration. Metal removal by an electrical spark acting in air. It is not subject to precise control, the most common application being the removal of broken tools such as taps and drills.

electrical pitting. The formation of surface cavities by removal of metal as a result of an electrical discharge across an interface.

electric arc furnace. See *arc furnace*.

electric arc spraying. A *thermal spraying* process using as a heat source an electric arc between two consumable electrodes of a coating material and a compressed gas which is used to atomize and propel the material to the substrate.

electric furnace. A metal melting or holding furnace that produces heat from electricity. It may operate on the resistance or induction principle. See also *induction furnace*.

electrochemical cell. An electrochemical system consisting of an *anode* and a *cathode* in metallic contact and immersed in an *electrolyte*. The anode and cathode may be different metals or dissimilar areas on the same metal surface. See also *cathodic protection*.

electrochemical corrosion. Corrosion that is accompanied by a flow of electrons between cathodic and anodic areas on metallic surfaces.

electrochemical discharge machining. Metal removal by a combination of the processes of *electrochemical machining* and *electrical discharge machining*. Most of the metal removal occurs via anodic dissolution (i.e., ECM action). Oxide films which form as a result of electrolytic action through an electrolytic fluid are removed by intermittent spark discharges (i.e., EDM action). Hence, the combination of the two actions.

electrochemical equivalent. The weight of an element or group of elements oxidized or reduced at 100% efficiency by the passage of a unit quantity of electricity. Usually expressed as grams per coulomb.

electrochemical grinding. A process where-by metal is removed by deplating. The workpiece is the anode; the cathode is a conductive aluminum oxide-copper or metal-bonded diamond grinding wheel with abrasive particles. Most of the metal is removed by deplating; 0.05 to 10% is removed by abrasive cutting.

electrochemical machining. Controlled metal removal by anodic dissolution. Direct current passes through flowing film of conductive solution which separates the workpiece from the electrode-tool. The workpiece is the *anode*, and the tool is the *cathode*.

electrochemical potential. The partial derivative of the total electrochemical free energy of a constituent with respect to the number of moles of this constituent where all factors are kept constant. It is analogous to the *chemical potential* of a constituent except that it includes the electric as well as chemical contributions to the free energy. The *potential* of an electrode in an electrolyte relative to a *reference electrode* measured under open circuit conditions.

electrochemical reaction. A reaction caused by passage of an electric current through a medium which contains mobile ions (as in electrolysis); or, a spontaneous reaction made to cause current to flow in a conductor external to this medium (as in a galvanic cell). In either event, electrical connection is made to the external portion of the circuit via a pair of electrodes. See also *electrolyte*.

electrochemical series. Same as *electromotive force series*.

electrode. Compressed graphite or carbon cylinder or rod used to conduct electric current in electric arc furnaces, arc lamps, and so forth.

electrode (electrochemistry). One of a pair of conductors introduced into an electrochemical cell, between which the ions in the intervening medium flow in opposite directions and on whose surfaces reactions occur (when appropriate external connection is made). In direct current operation, one electrode or "pole" is positively charged, the other negatively. See also *anode*, *cathode*, *electrochemical reaction*, and *electrolyte*.

electrode (welding). (1) In arc welding, a current-carrying rod that supports the arc between the rod and work, or between two rods as in twin carbon-arc welding. It may or may not furnish filler metal. See also *bare electrode*, *covered electrode*, and *lightly coated electrode*. (2) In resistance welding, a part of a resistance welding machine through which current and, in most instances, pressure are applied directly to the work. The electrode may be in the form of a rotating wheel, rotating roll, bar, cylinder, plate, clamp, chuck, or modification thereof. (3) In arc and plasma spraying, the current-carrying components which support the arc.

electrode cable. Same as *electrode lead*.

electrode deposition. The weight of weld-metal deposit obtained from a unit length of electrode.

electrode force. The force between electrodes in a spot, seam, and projection weld.

electrode lead. The electrical conductor between the source of arc welding current and the electrode holder.

electrode polarization. Change of *electrode potential* with respect to a reference value. The change may be caused, for example, by the application of an external electrical current or by the addition of an oxidant or reductant.

electrodeposition. (1) The deposition of a conductive material from a plating solution by the application of electrical current. (2) The deposition of a substance on an electrode by passing electric current through an electrolyte. *Electroplating, electroforming, electrorefining,* and *electrotwinning* result from electrodeposition.

electrode potential. The *potential* of an *electrode* in an *electrolysis* as measured against a *reference electrode*. The electrode potential does not include any resistance losses in potential in either the solution or external circuit. It represents the reversible work to move a unit charge from the electrode surface through the solution to the reference electrode.

electrode reaction. Interfacial reaction equivalent to a transfer of charge between

electronic and ionic conductors. See also *anodic reaction* and *cathodic reaction*.

electroforming. Making parts by electrodeposition on a removable form.

electrogalvanizing. The *electroplating* of zinc upon iron or steel.

electrogas welding (EGW). A *vertical position* arc welding process which produces coalescence of metals by heating them with an arc between a continuous filler metal (consumable) electrode and the work. Copper dams (molding shoes) are used to confine the molten weld metal. The electrodes may be either flux cored or solid wire. Shielding may or may not be obtained from an externally supplied gas or mixture. See also *gas metal arc welding* and *flux cored arc welding*.

electroless plating. (1) A process in which metal ions in a dilute aqueous solution are plated out on a substrate by means of autocatalytic chemical reduction. (2) The deposition of conductive material from an autocatalytic plating solution without the application of electrical current.

electrolysis. (1) Chemical change resulting from the passage of an electric current through an *electrolyte*. (2) The separation of chemical components by the passage of current through an electrolyte.

electrolyte. (1) A chemical substance or mixture, usually liquid, containing ions that migrate in an electric field. (2) A chemical compound or mixture of compounds which when molten or in solution will conduct an electric current.

electrolytic brightening. Same as *electropolishing*.

electrolytic cell. An assembly, consisting of a vessel, electrodes, and an electrolyte, in which *electrolysis* can be carried out.

electrolytic cleaning. A process of removing soil, scale, or corrosion products from a metal surface by subjecting it as an *electrode* to an electric current in an electrolytic bath.

electrolytic copper. Copper that has been refined by the electrolytic deposition, including cathodes that are the direct product of the refining operation, refinery shapes cast from melted cathodes, and, by extension, fabricators' products made therefrom. Usually when this term is used alone, it refers to electrolytic tough pitch copper without elements other than oxygen being present in significant amounts. See also *tough pitch copper*.

electrolytic deposition. Same as *electrodeposition*.

electrolytic grinding. A combination of grinding and machining wherein a metal-bonded abrasive wheel, usually diamond, is the *cathode* in physical contact with the anodic workpiece, the contact being made beneath the surface of a suitable electrolyte. The abrasive particles produce grinding act as nonconducting spacers permitting simultaneous machining through electrolysis.

electrolytic machining. Controlled removal of metal by use of an applied potential and a suitable electrolyte to produce the shapes and dimensions desired.

electrolytic pickling. *Pickling* in which electric current is used, the work being one of the electrodes.

electrolytic polishing. An electrochemical polishing process in which the metal to be polished is made the *anode* in an electrolytic cell where preferential dissolution at high points in the surface topography produces a specularly reflective surface. Also referred to as *electropolishing*.

electrolytic powder. Powder produced by electrolytic deposition or by pulverizing of an electrodeposit.

electrolytic protection. See preferred term *cathodic protection*.

electrolytic tough pitch. A term describing the method of raw copper preparation to ensure a good physical- and electrical-grade copper-finished product.

electromagnetic forming. A process for forming metal by the direct application of an intense, transient magnetic field. The workpiece is formed without mechanical contact by the passage of a pulse of electric current through a forming coil. Also known as magnetic pulse forming.

electromechanical polishing. An attack-polishing method in which the chemical action of the polishing fluid is enhanced or controlled by the application of an electric current between the specimen and the polishing wheel.

electrometallurgy. Industrial recovery or processing of metals and alloys by electric or electrolytic methods.

electromotive force. (1) The force that determines the flow of electricity; a difference of electric potential. (2) Electrical potential; voltage.

electromotive force series (emf series). A series of elements arranged according to their *standard electrode potentials*, with "noble" metals such as gold being positive and "active" metals such as zinc being negative. In corrosion studies, the analogous but more practical *galvanic series* of metals is generally used. The relative positions of a given metal are not necessarily the same in the two series.

electron bands. Energy states for the free electrons in a metal, as described by the use of the band theory (zone theory) of electron structure. Also called Brillouin zones.

electron beam cutting. A cutting process that uses the heat obtained from a concentrated beam composed primarily of high-velocity electrons, which impinge upon the workpieces to be cut; it may or may not use an externally supplied gas.

electron beam heat treating. A selective surface hardening process that rapidly heats a surface by direct bombardment with an accelerated stream of electrons.

electron beam machining. Removing material by melting and vaporizing the workpiece at the point of impingement of a focused high-velocity beam of electrons. The machining is done in high vacuum to eliminate scattering of the electrons due to interaction with gas molecules. The most important use of electron beam machining is for hole drilling.

electron beam welding. A welding process that produces coalescence of metals with the heat obtained from a concentrated beam composed primarily of high-velocity electrons impinging upon the surfaces to be joined. Welding can be carried out at atmospheric pressure (nonvacuum), medium vacuum (approximately 10^3 to 25 torr), or high vacuum (approximately 10^6 to 10^3 torr).

electrophoresis. Transport of charged colloidal or macromolecular materials in an electric field.

electroplate. The application of a metallic coating on a surface by means of electrolytic action.

electroplating. The electrodeposition of an adherent metallic coating on an object serving as a *cathode* for the purpose of securing a surface with properties or dimensions different from those of the basis metal.

electropolishing. A technique commonly used to prepare metallographic specimens, in which a high polish is produced making the specimen the *anode* in an *electrolytic cell*, where preferential dissolution at high points smooths the surface. Also referred to as *electrolytic polishing*.

electrorefining. Using electric or electrolytic methods to convert impure metal to purer metal, or to produce an alloy from impure or partly purified raw materials.

electroslag remelting (ESR). A *consumable-electrode remelting* process in which heat is generated by the passage of electric current through a conductive slag. The droplets of metal are refined by contact with the slag.

electroslag welding. A fusion welding process in which the welding heat is provided by passing an electric current through a layer of molten conductive slag (flux) contained in a pocket formed by water-cooled dams that bridge the gap between the members being welded. The resistance heated slag not only melts filler-metal electrodes as they are fed into the slag layer, but also provides shielding for the massive weld puddle characteristic of the process.

electrostrictive effect. The reversible interaction, exhibited by some crystalline materials, between an elastic strain and an electric field. The direction of the strain is independent of the polarity of the field. Compare with *piezoelectric effect*.

electrotinning. *Electroplating* tin on an object.

electrotyping. The production of printing plates by *electroforming*.

electrowinning. Recovery of a metal from an ore by means of electrochemical processes.

elongation. (1) A term used in mechanical testing to describe the amount of extension of a test piece when stressed. (2) In tensile testing, the increase in the gage length, measured after fracture of the specimen within the gage length, usually expressed as a percentage of the original gage length. See also *elongation, percent*.

elongation, percent. The extension of a uniform section of a specimen expressed as a percentage of the original gage length:

$$\text{Elongation, \%} = \frac{(L_x - L_o)}{L_o} \cdot 100$$

where L_o is the original gage length and L_x is the final gage length.

elutriation. A test for particle size in which the speed of a liquid or gas is used to suspend particles of a desired size, with larger sizes settling for removal and weighing, while smaller sizes are removed, collected, and weighed at certain time intervals.

embossing. (1) Technique used to create depressions of a specific pattern in plastic film and sheeting. Such embossing in the form of surface patterns can be achieved on molded parts by the treatment of the mold surface with photoengraving or another process. (2) Raising a design in relief against a surface.

embossing die. A die used for producing embossed designs.

embrittlement. The severe loss of *ductility* or *toughness* or both, of a material, usually a metal or alloy. Many forms of embrittlement can lead to *brittle fracture*. Many forms can occur during thermal treatment or elevated-temperature service (thermally induced embrittlement). Some of these forms of embrittlement, which affect steels, include *blue brittleness, 885 °F (475 °C) embrittlement, quench-age embrittlement, sigma-phase embrittlement, strain-age embrittlement, temper embrittlement, tempered martensite embrittlement,* and *thermal embrittlement.* In addition, steels and other metals and alloys can be embrittled by environmental conditions (environmentally assisted embrittlement). The forms of environmental embrittlement include *acid embrittlement, caustic embrittlement, corrosion embrittlement, creep-rupture embrittlement, hydrogen embrittlement, liquid metal embrittlement, neutron embrittlement, solder embrittlement, solid metal embrittlement,* and *stress-corrosion cracking.*

emf. An abbreviation for *electromotive force.*

emissivity. Ratio of the amount of energy or of energetic particles radiated from a unit area of a surface to the amount radiated from a unit area of an ideal emitter under the same conditions.

emulsion. (1) A two-phase liquid system in which small droplets of one liquid (the internal phase) are immiscible in, and are dispersed uniformly throughout, a second continuous liquid phase (the external phase). The internal phase is sometimes describe as the dispense phase. (2) A stable dispersion of one liquid in another, generally by means of an emulsifying agent that has affinity for both the continuous and discontinuous phases. The emulsifying agent, discontinuous phase, and continuous phase can together produce another phase that serves as an enveloping (encapsulating) protective phase around the discontinuous phase.

emulsion cleaner. A cleaner-consisting of organic solvents dispersed in an aqueous medium with the aid of an emulsifying agent.

enameling iron. A low-carbon, cold-rolled sheet steel, produced specifically for use as a base metal for porcelain enamel.

enantiotropy. The relation of crystal forms of the same substance in which one form is stable above a certain temperature and the other form is stable below that temperature. For example, ferrite and austenite are enantiotropic in ferrous alloys.

end clearance angle. See *clearance angle* and the figures accompanying *face mill* and *single-point tool.*

end cutting-edge angle. The angle of concavity between the face cutting edge and the face plane of the cutter. It serves as re-

lief to prevent the face cutting edges from rubbing in the cut. See also the figures accompanying *face mill* and *single-point tool.*

end mark. A roll mark caused by the end of a sheet marking the roll during hot or cold rolling.

end milling. A method of machining with a rotating cutting tool with cutting edges on both the face end and the periphery. See also *face milling* and *milling.*

endothermic atmosphere. A gas mixture produced by the partial combustion of a hydrocarbon gas with air in an endothermic reaction. Also known as endogas.

endothermic reaction. Designating or pertaining to a reaction that involves the absorption of heat. See also *exothermic reaction.*

end-quench hardenability test. A laboratory procedure for determining the hardenability of a steel or other ferrous alloy; widely referred to as the Jominy test. Hardenability is determined by heating a standard specimen above the upper critical temperature, placing the hot specimen in a fixture so that a stream of cold water impinges on one end, and, after cooling to room temperature is completed, measuring the hardness near the surface of the specimen at regularly spaced intervals along its length. The data are normally plotted as hardness versus distance from the quenched end.

end relief. Defined by the figure accompanying *single-point tool.*

endurance limit. The maximum stress that a material can withstand for an infinitely large number of fatigue cycles. See also *fatigue limit* and *fatigue strength.*

endurance ratio. The ratio of the *endurance limit* for completely reversed flexural stress to the tensile strength of a given material.

engineering strain (*e*). A term sometimes used for average linear strain or conventional strain in order to differentiate it from *true strain.* In tension testing it is calculated by dividing the change in the gage length by the original gage length.

engineering stress (*s*). A term sometimes used for conventional stress in order to differentiate it from *true stress.* In tension testing, it is calculated by dividing the breaking load applied to the specimen by the original cross-sectional area of the specimen.

environmental cracking. *Brittle fracture* of a normally ductile material in which the corrosive effect of the environment is a causative factor. Environmental cracking is a general term that includes *corrosion fatigue, high-temperature hydrogen*

attack, hydrogen blistering, hydrogen embrittlement, liquid metal embrittlement, solid metal embrittlement, stress-corrosion cracking, and *sulfide stress cracking.* The following terms have been used in the past in connection with environmental cracking, but are becoming obsolete: caustic embrittlement, delayed fracture, season cracking, static fatigue, stepwise cracking, sulfide corrosion cracking, and sulfide stress-corrosion cracking. See also *embrittlement.*

epitaxy. Growth of an electrodeposit or vapor deposit in which the orientation of the crystals in the deposit are directly related to crystal orientations in the underlying crystalline substrate.

epsilon (ε). Designation generally assigned to intermetallic, metal-metalloid, and metal-nonmetallic compounds found in ferrous alloy systems, for example, Fe_3Mo_2, $FeSi$, and Fe_3P.

epsilon carbide. Carbide with hexagonal close-packed lattice that precipitates during the first stage of tempering of primary martensite. Its composition corresponds to the empirical formula $Fe_{2.4}C$.

epsilon structure. Structurally analogous close-packed phases or electron compounds that have ratios of seven valence electrons to four atoms.

equiaxed grain structure. A structure in which the grains have approximately the same dimensions in all directions.

equilibrium. The dynamic condition of physical, chemical, mechanical, or atomic balance which appears to be a condition of rest rather than one of change.

equilibrium diagram. A graph of the temperature, pressure, and composition limits of phase fields in an alloy system as they exist under conditions of thermodynamical equilibrium. In metal systems, pressure is usually considered constant. Compare with *phase diagram.*

Erichsen test. A *cupping test* used to assess the ductility of sheet metal. The method consists of forcing a conical or hemispherical-ended plunger into the specimen and measuring the depth of the impression at fracture.

erosion. (1) Loss of material from a solid surface due to relative motion in contact with a fluid that contains solid particles. Erosion in which the relative motion of particles is nearly parallel to the solid surface is called *abrasive erosion.* Erosion in which the relative motion of the solid particles is nearly normal to the solid surface is called *impingement erosion* or impact erosion. (2) Progressive loss of original material from a solid surface due to mechanical interaction between that surface and a fluid, a multicomponent fluid, and impinging liquid, or solid particles. (3) Loss of material from the surface of an electrical contact due to an electrical discharge (arcing). See also *cavitation erosion*, *electrical pitting*, and *erosion-corrosion*.

erosion-corrosion. A conjoint action involving *corrosion* and *erosion* in the presence of a moving corrosive fluid, leading to the accelerated loss of material.

erosivity. The characteristic of a collection of particles, liquid stream, or a slurry that expresses its tendency to cause erosive wear when forced against a solid surface under relative motion.

etchant. A chemical solution used to etch a metal to reveal structural details. See also *etching*.

etch cleaning. Removing soil by dissolving away some of the underlying metal.

etch cracks. Shallow cracks in hardened steel containing high residual surface stresses, produced by etching in an embrittling acid.

etch figures. Characteristic markings produced on crystal surfaces by chemical attack, usually having facets parallel to low-index crystallographic planes.

etching. (1) Subjecting the surface of a metal to preferential chemical or electrolytic attack in order to reveal structural details for metallographic examination. (2) Chemically or electrochemically removing tenacious films from a metal surface to condition the surface for a subsequent treatment, such as painting or electroplating.

eutectic. (1) An isothermal reversible reaction in which a liquid solution is converted into two or more intimately mixed solids on cooling, the number of solids formed being the same as the number of components in the system. (2) An alloy having the composition indicated by the eutectic point on a phase diagram. (3) An alloy structure of intermixed solid constituents formed by a eutectic reaction often in the form of regular arrays of lamellae or rods.

eutectic carbide. Carbide formed during freezing as one of the mutually insoluble phases participating in the eutectic reaction of a hypereutectic tool steel. See also *hypereutectic alloy*.

eutectic melting. Melting of localized microscopic areas whose composition corresponds to that of the eutectic in the system.

eutectic point. The composition of a liquid phase in univariant equilibrium with two or more solid phases; the lowest melting alloy of a composition series.

eutectoid. (1) An isothermal reversible reaction in which a solid solution is converted into two or more intimately mixed solids on cooling, the number of solids formed being the same as the number of components in the system. (2) An alloy having the composition indicated by the eutectoid point on a phase diagram. (3) An alloy structure of intermixed solid constituents formed by a eutectoid reaction.

eutectoid point. The composition of a solid phase that undergoes univariant transformation into two or more other solid phases upon cooling.

exfoliation. Corrosion that proceeds laterally from the sites of initiation along planes parallel to the surface, generally at grain boundaries, forming corrosion products that force metal away from the body of the material, giving rise to a layered appearance. Most commonly associated with wrought aluminum alloys.

exogenous inclusion. An *inclusion* that is derived from external causes. Slag, dross, entrapped mold materials, and refractories are examples of inclusions that would be classified as exogenous. In most cases, these inclusions are macroscopic or visible to the naked eye. Compare with *indigenous inclusion*.

exothermic. Characterized by the liberation of heat.

exothermic atmosphere. A gas mixture produced by the partial combustion of a hydrocarbon gas with air in an exothermic reaction. Also known as exogas.

exothermic reaction. A reaction which liberates heat, such as the burning of fuel or when certain plastic resins are cured chemically.

expanding. A process used to increase the diameter of a cup, shell, or tube. See also *bulging*.

expendable pattern. A *pattern* that is destroyed in making a casting. It is usually made of wax (*investment casting*) or expanded polystyrene (*lost foam casting*).

explosion welding. A solid-state welding process effected by a controlled detonation, which causes the parts to move together at high velocity. The resulting bond zone has a characteristic wavy appearance.

explosive forming. The shaping of metal parts in which the forming pressure is generated by an explosive charge which takes the place of the punch in conventional forming. A single-element die is used with a blank held over it, and the explosive charge is suspended over the blank at a predetermined distance (standoff distance). The complete assembly is

often immersed in a tank of water. See also *high-energy-rate forming*.

extensometer. An instrument for measuring changes in length over a given gage length caused by application or removal of a force. Commonly used in tension testing.

extractive metallurgy. The branch of *process metallurgy* dealing with the winning of metals from their ores. Compare with *refining*.

extra hard. A *temper* of nonferrous alloys and some ferrous alloys characterized by values of tensile strength and hardness about one-third of the way from those of *full hard* to those of *extra spring* temper.

extra spring. A *temper* of nonferrous alloys and some ferrous alloys corresponding approximately to a cold worked state above *full hard* beyond which further cold work will not measurably increase strength or hardness.

extreme-pressure lubricant. A lubricant that imparts increased load-carrying capacity to rubbing surfaces under severe operating conditions. Extreme-pressure lubricants usually contain sulfur, halogens, or phosphorus.

extruded hole. A hole formed by a punch that first cleanly cuts a hole and then is pushed farther through to form a flange with an enlargement of the original hole.

extrusion. The conversion of an ingot or billet into lengths of uniform cross section by forcing metal to flow plastically through a die orifice. In forward (direct) extrusion, the die and ram are at opposite ends of the extrusion stock, and the product and ram travel in the same direction. Also, there is relative motion between the extrusion stock and the die. In backward (indirect) extrusion, the die is at the ram end of the stock and the product travels in the direction opposite that of the ram, either around the ram (as in the impact extrusion of cylinders such as cases for dry cell batteries) or up through the center of a hollow ram. See also *hydrostatic extrusion* and *impact extrusion*.

extrusion billet. A metal slug used as *extrusion stock*.

extrusion defect. See preferred term *extrusion pipe*.

extrusion forging. (1) Forcing metal into or through a die opening by restricting flow in other directions. (2) A part made by the operation.

extrusion ingot. A cast metal slug used as *extrusion stock*.

extrusion pipe. A central oxide-lined discontinuity that occasionally occurs in the last 10 to 20% of an extruded metal bar. It is caused by the oxidized outer surface of the billet flowing around the end of the billet and into the center of the bar during the final stages of extrusion. Also called *coring*.

extrusion stock. A rod, bar, or other section used to make extrusions.

eyeleting. The displacing of material about an opening in sheet or plate so that a lip protruding above the surface is formed.

F

face. In a lathe tool, the surface against which the chips bear as they are formed. See also the figure accompanying *single-point tool*.

face mill. See definition of nomenclature in accompanying figure.

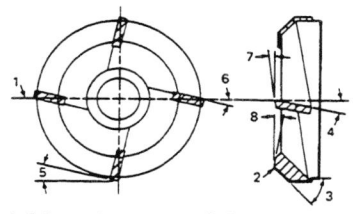

1. Reference plane
2. Tooth point
3. Corner angle (bevel)
4. Axial rake (positive)
5. Peripheral clearance angle
6. Radial rake (negative)
7. End clearance angle
8. End cutting edge angle

Face mill

face milling. Milling a surface that is perpendicular to the cutter axis. See also *milling*.

face of weld. The exposed surface of an arc or gas weld on the side from which the welding was done. See also the figure accompanying *fillet weld*.

face reinforcement. Reinforcement of a weld at the side of the joint from which welding was carried out.

face-type cutters. Cutters that can be mounted directly on and driven from the machine spindle nose.

facing. (1) In machining, generating a surface on a rotating workpiece by the traverse of a tool perpendicular to the axis of rotation. (2) In foundry practice, any material applied in a wet or dry condition to the face of a mold or core to improve the surface of the casting. See also *mold wash*. (3) For abrasion resistance, see preferred term *hardfacing*.

fagot. In forging work, a bundle of iron bars that will be heated and then hammered and welded to form a single bar.

failure. A general term used to imply that a part in service (a) has become completely inoperable, (b) is still operable but incapable of satisfactorily performing its intended function, or (c) has deteriorated seriously, to the point that it has become unreliable or unsafe for continued use.

failure mechanism. A structural or chemical process, such as corrosion or fatigue, that causes failure.

false bottom. An *insert* put in either member of a die set to increase the strength and improve the life of the die.

false Brinelling. (1) Damage to a solid bearing surface characterized by indentations not caused by plastic deformation resulting from overload, but thought to be due to other causes such as *fretting corrosion*. (2) Local spots appearing when the protective film on a metal is broken continually by repeated impacts, usually in the presence of corrosive agents. The appearance is generally similar to that produced by *Brinelling* but corrosion products are usually visible. It may result from fretting corrosion. This term should be avoided when a more precise description is possible. False Brinelling (race fretting) can be distinguished from true *Brinelling* because in false Brinelling, surface material is removed so that original finishing marks are removed. The borders of a false Brinell mark are sharply defined, whereas a dent caused by a rolling element does not have sharp edges and the finishing marks are visible in the bottom of the dent.

fatigue. The phenomenon leading to fracture under repeated or fluctuating stresses having a maximum value less than the ultimate tensile strength of the material. *Fatigue failure* generally occurs at loads which applied statically would produce little perceptible effect. Fatigue fractures are progressive, beginning as minute cracks that grow under the action of the fluctuating stress.

fatigue crack growth rate *(da/dN)*. The rate of crack extension caused by constant-amplitude fatigue loading, expressed in terms of crack extension per cycle of load application, and plotted logarithmically against the *stress intensity factor* range, K.

fatigue failure. Failure that occurs when a specimen undergoing *fatigue* completely fractures into two parts of has softened or been otherwise significantly reduced in stiffness by thermal heating or cracking.

fatigue life *(N)*. (1) The number of cycles of stress or strain of a specified character that a given specimen sustains before failure of a specified nature occurs. (2) The number of cycles of deformation required to bring about failure of a test specimen under a given set of oscillating conditions (stresses or strains). See also *S-N curve*.

fatigue limit. The maximum stress that presumably leads to fatigue fracture in a specified number of stress cycles. The

value of the *maximum stress* and the *stress ratio* also should be stated. See also *endurance limit*.

fatigue notch factor (K_f). The ratio of the *fatigue strength* of an unnotched specimen to the fatigue strength of a notched specimen of the same material and condition; both strengths are determined at the same number of *stress cycles*.

fatigue notch sensitivity (q). An estimate of the effect of a notch or hole of a given size and shape on the fatigue properties of a material, measured by $q = (K_f 1)/(K_t 1)$, where K_f is the *fatigue notch factor* and K_t is the *stress-concentration factor*. A material is said to be fully notch sensitive if q approaches a value of 1.0; it is not notch sensitive if the ratio approaches 0.

fatigue ratio. The ratio of fatigue strength to tensile strength. Mean stress and alternating stress must be stated.

fatigue strength. The maximum cyclical stress a material can withstand for a given number of cycles before failure occurs.

fatigue strength at N cycles (S_N). A hypothetical value of stress for failure at exactly N cycles as determined from an *S-N* curve. The value of S_N thus determined is subject to the same conditions as those which apply to the *S-N* curve. The value of S_N that is commonly found in the literature is the hypothetical value of maximum stress, S_{max}, minimum stress S_{min}, or stress amplitude, S_a, at which 50% of the specimens of a given sample could survive N stress cycles in which the mean stress $S_m = 0$. This is also known as the *median fatigue strength at N cycles*. See also *S-N curve*.

fatigue-strength reduction factor. The ratio of the fatigue strength of a member or specimen with no stress concentration to the fatigue strength with stress concentration. This factor has no meaning unless the stress range and the shape, size, and material of the member or specimen are stated.

fatigue striation. Parallel lines frequently observed in electron microscope fractographs or fatigue fracture surfaces. The lines are transverse to the direction of local crack propagation; the distance between successive lines represents the advance of the crack front during the one cycle of stress variation.

fatigue test. A method for determining the range of alternating (fluctuating) stresses a material can withstand without failing.

fatigue wear. (1) Removal of particles detached by fatigue arising from cyclic stress variations. (2) Wear of a solid surface caused by fracture arising from material fatigue. See also *spalling*.

faying surface. The surfaces of materials in contact with each other and joined or about to be joined.

feed. The rate at which a cutting tool or grinding wheel advances along or into the surface of a workpiece, the direction of advance depending on the type of operation involved.

feeder (feeder head, feedhead). In foundry practice, a *riser*.

feeding. (1) In casting, providing molten metal to a region undergoing solidification, usually at a rate sufficient to fill the mold cavity ahead of the solidification front and to compensate for any shrinkage accompanying solidification. (2) Conveying metal stock or workpieces to a location for use or processing, such as wire to a consumable electrode, strip to a die, or workpieces to an assembler.

feed lines. Linear marks on a machined or ground surface that are spaced at intervals equal to the *feed* per revolution or per stroke.

ferrimagnetic material. (1) A material that macroscopically has properties similar to those of a *ferromagnetic material* but that microscopically also resembles an antiferromagnetic material in that some of the elementary magnetic moments are aligned antiparallel. If the moments are of different magnitudes, the material may still have a large resultant magnetization. (2) A material in which unequal magnetic moments are lined up antiparallel to each other. Permeabilities are of the same order of magnitude as those of ferromagnetic materials, but are lower than they would be if all atomic moments were parallel and in the same direction. Under ordinary conditions the magnetic characteristics of ferrimagnetic materials are quite similar to those of ferromagnetic material.

ferrite. (1) A solid solution of one or more elements in body-centered cubic iron. Unless otherwise designated (for instance, as chromium ferrite), the solute is generally assumed to be carbon. On some equilibrium diagrams, there are two ferrite regions separated by an austenite area. The lower area is alpha ferrite; the upper, delta ferrite. If there is no designation, alpha ferrite is assumed. (2) An essentially carbon-free solid solution in which alpha iron is the solvent, and which is characterized by a body-centered cubic crystal structure. Fully ferritic steels are only obtained when the carbon content is quite low. The most obvious microstructural features in such metals are the ferrite grain boundaries.

ferrite banding. Parallel bands of free ferrite aligned in the direction of working. Sometimes referred to as ferrite streaks.

ferrite number. An arbitrary, standardized value designating the ferrite content of an austenitic stainless steel weld metal. This value directly replaces percent ferrite or volume percent ferrite and is determined by the magnetic test described in AWS A4.2

ferrite-pearlite banding. Inhomogeneous distribution of ferrite and pearlite aligned in filaments or plates parallel to the direction of working.

ferrite streaks. Same as *ferrite banding*.

ferritic grain size. The grain size of the ferritic matrix of a steel.

ferritic malleable. See *malleable iron*.

ferritizing anneal. A treatment given as-cast gray or ductile (nodular) iron to produce an essentially ferritic matrix. For the term to be meaningful, the final microstructure desired or the time-temperature cycle used must be specified.

ferroalloy. An alloy of iron that contains a sufficient amount of one or more other chemical elements to be useful as an agent for introducing these elements into molten metal, especially into steel or cast iron.

ferroelectric. A crystalline material that exhibits spontaneous electrical polarization, hysteresis, and piezoelectric properties.

ferroelectric effect. The phenomena whereby certain crystals may exhibit a spontaneous dipole moment (which is called ferroelectric by analogy with ferromagnetism exhibiting a permanent magnetic moment). Ferroelectric crystals often show several Curie points, domain structures, and hysteresis, much as do ferromagnetic crystals.

ferrograph. An instrument used to determine the size distribution of wear particles in lubricating oils of mechanical systems. The technique relies on the debris being capable of being attracted to a magnet.

ferromagnetic material. A material that in general exhibits the phenomena of hysteresis and saturation, and whose permeability is dependent on the magnetizing force. Microscopically, the elementary magnets are aligned parallel in volumes called *domains*. The unmagnetized condition of a ferromagnetic material results from the overall neutralization of the magnetization of the domains to produce zero external magnetization.

ferromagnetism. A property exhibited by certain metals, alloys, and compounds of the transition (iron group), rare-earth, and actinide elements in which, below a certain temperature termed the Curie temperature, the atomic magnetic moments tend to line up in a common direction. Ferromagnetism is characterized by the strong attraction of one magnetized body for another. See also *Curie temperature*. Compare with *paramagnetism*.

ferrous. Metallic materials in which the principal component is iron.

fiber. (1) The characteristic of wrought metal that indicates *directional properties* and is revealed by etching of a longitudinal section or is manifested by the fibrous or woody appearance of a fracture. It is caused chiefly by extension of the constituents of the metal, both metallic and nonmetallic, in the direction of working. (2) The pattern of preferred orientation of metal crystals after a given deformation process, usually wiredrawing. See also *fibering* and *preferred orientation*.

fibering. Elongation and alignment of internal boundaries, second phases, and inclusions in particular directions corresponding to the direction of metal flow during deformation processing.

fiber metallurgy. The technology of producing solid bodies from fibers or chopped filaments, with or without a metal matrix. The fibers may consist of such nonmetals as graphite or aluminum oxide, or of such metals as tungsten or boron. See also *metal-matrix composites*.

fiber stress. Local stress through a small area (a point or line) on a section where the stress is not uniform, as in a beam under a bending load.

fibrous fracture. A gray and amorphous *fracture* that results when a metal is sufficiently ductile for the crystals to elongate before fracture occurs. When a fibrous fracture is obtained in an impact test, it may be regarded as definite evidence of toughness of the metal. See also *crystalline fracture* and *silky fracture*.

fibrous structure. (1) In forgings, a structure revealed as laminations, not necessarily detrimental, on an etched section or as a ropy appearance on a fracture. It is not to be confused with silky or ductile fracture of a clean metal. (2) In wrought iron, a structure consisting of slag fibers embedded in ferrite. (3) In rolled steel plate stock, a uniform, fine-grained structure on a fractured surface, free of laminations or shaletype discontinuities.

filamentary shrinkage. A fine network of shrinkage cavities, occasionally found in steel castings, that produces a radiographic image resembling lace.

file hardness. Hardness as determined by the use of a steel file of standardized hardness on the assumption that a material that cannot be cut with the file is as hard as, or harder than, the file. Files covering a range of hardnesses may be employed;

the most common are files heat treated to approximately 67 to 70 HRC.

filiform corrosion. Corrosion that occurs under some coatings in the form of randomly distributed threadlike filaments.

filler metal. Metal added in making a brazed, soldered, or welded joint.

fillet. (1) Concave corner piece usually used at the intersection of casting sections. Also the radius of metal at such junctions as opposed to an abrupt angular junction. (2) A radius (curvature) imparted to inside meeting surfaces.

fillet weld. A weld, approximately triangular in cross section, joining two surfaces, essentially at right angles to each other in a lap, tee, or corner joint.

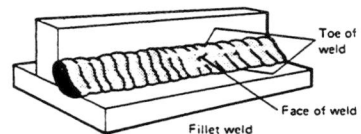
Fillet weld

final annealing. An imprecise term used to denote the last anneal given to a nonferrous alloy prior to shipment.

final polishing. A polishing process in which the primary objective is to produce a final surface suitable for microscopic examination.

fineness. A measure of the purity of gold or silver expressed in parts per thousand.

fines. (1) The product that passes through the finest screen in sorting crushed or ground material. (2) Sand grains that are substantially smaller than the predominating size in a batch or lot of foundry sand. (3) The portion of a powder composed of particles smaller than a specified size, usually 44 m (325 mesh).

fine silver. Silver with a fineness of 999; equivalent to a minimum content of 99.9% Ag with the remaining content unrestricted.

finish. (1) Surface condition, quality, or appearance of a metal. (2) Stock on a forging or casting to be removed in finish machining. (3) The forging operation in which the part is forged into its final shape in the finish die. If only one finish operations is scheduled to be performed in the finish die, this operation will be identified simply as finish; first, second, or third finish designations are so termed when one or more finish operations are to be performed in the same finish die.

finish allowance. (1) The amount of excess metal surrounding the intended final configuration of a formed part; sometimes called forging envelope, machining allowance, or cleanup allowance. (2)

Amount of stock left on the surface of a casting for machining.

finish annealing. A *subcritical annealing* treatment applied to cold-worked low- or medium-carbon steel. Finish annealing, which is a compromise treatment, lowers residual stresses, thereby minimizing the risk of distortion in machining while retaining most of the benefits to machinability contributed by cold working. Compare with *final annealing*.

finished steel. Steel that is ready for the market and has been processed beyond the stages of billets, blooms, sheet bars, slabs, and wire rods.

finisher (finishing impression). The *die impression* that imparts the final shape to a forged part.

finish grinding. The final grinding action on a workpiece, of which the objectives are surface finish and dimensional accuracy.

finishing die. The die set used in the last forging step.

finishing temperature. The temperature at which *hot working* is completed.

finish machining. A machining process analogous to *finish grinding*.

fire-refined copper. Copper that has been refined by the use of a furnace process only, including refinery shapes and, by extension, fabricators' products made therefrom. Usually, when this term is used alone it refers to fire-refined tough pitch copper without elements other than oxygen being present in significant amounts.

fir-tree crystal. A type of *dendrite*.

fisheye. An area on a steel fracture surface having a characteristic white crystalline appearance.

fisheye (weld defect). A discontinuity found on the fracture surface of a weld in steel that consists of a small pore or inclusion surrounded by an approximately round, bright area.

fishmouthing. See *alligatoring*.

fishscale. A scaly appearance in a porcelain enamel coating in which the evolution of hydrogen from the basis metal (iron or steel) causes loss of adhesion between the enamel and the basis metal. Individual scales are usually small, but have been observed in sizes up to 25 mm (1 in.) or more in diameter. The scales are somewhat like blisters that have cracked partway around the perimeter but still remain attached to the coating around the rest of the perimeter.

fishtail. (1) In roll forging, the excess trailing end of a forging. It is often used, before being trimmed off, as a tong hold for a subsequent forging operation. (2) In hot rolling or extrusion, the imperfectly

shaped trailing end of a bar or special section that must be cut off and discarded as mill scrap.

fissure. A small crack-like weld discontinuity with only slight separation (opening displacement) of the fracture surfaces. The prefixes macro or micro indicate relative size.

fixed-feed grinding. Grinding in which the wheel is fed into the work, or vice versa, by given increments or at a given rate.

fixed position welding. Welding in which the work is held in a stationary position.

fixture. A device designed to hold parts to be joined in proper relation to each other.

flake. A short, discontinuous internal crack in ferrous metals attributed to stresses produced by localized transformation and hydrogen-solubility effects during cooling after hot working. In fracture surfaces, flakes appear as bright, silvery areas with a coarse texture. In deep acid-etched transverse sections, they appear as discontinuities that are usually in the midway to center location of the section. Also termed hairline cracks and shatter cracks.

flake graphite. Graphitic carbon, in the form of platelets, occurring in the microstructure of *gray iron*.

flaking. (1) The removal of material from a surface in the form of flakes or scalelike particles. (2) A form of pitting resulting from fatigue. See also *spalling*.

flame annealing. Annealing in which the heat is applied directly by a flame.

flame cleaning. Cleaning metal surfaces of scale, rust, dirt, and moisture by use of a gas flame.

flame cutting. See preferred term *oxygen cutting*.

flame hardening. A process for hardening the surfaces of hardenable ferrous alloys in which an intense flame is used to heat the surface layers above the upper transformation temperature, whereupon the workpiece is immediately quenched.

flame spraying. *Thermal spraying* in which a coating material is fed into an oxyfuel gas flame, where it is melted. Compressed gas may or may not be used to atomize the coating material and propel it onto the substrate. The sprayed material is originally in the form of wire or powder. The term flame spraying is usually used when referring to a combustion-spraying process, as differentiated from *plasma spraying*.

flame straightening. Correcting distortion in metal structures by localized heating with a gas flame.

flank. The end surface of a tool that is adjacent to the cutting edge and below it when the tool is in a horizontal position, as for turning. See figure accompanying *single-point tool*.

flank wear. The loss of relief on the flank of the tool behind the cutting edge due to rubbing contact between the work and the tool during cutting; measured in terms of linear dimension behind the original cutting edge.

flare test. A test applied to tubing, involving tapered expansion over a cone. Similar to *pin expansion test*.

flaring. (1) Forming an outward acute-angle flange on a tubular part. (2) Forming a flange by using the head of a hydraulic press.

flash. (1) In forging, metal in excess of that required to fill the blocking or finishing forging impression of a set of dies completely. Flash extends out from the body of the forging as a thin plate at the line where the dies meet and is subsequently removed by trimming. Because it cools faster than the body of the component during forging, flash can serve to restrict metal flow at the line where dies meet, thus ensuring complete filling of the impression. See also *closed-die forging*. (2) In casting, a fin of metal that results from leakage between mating mold surfaces. (3) In welding, the material which is expelled or squeezed out of a weld joint and which forms around the weld.

flashback. A recession of the welding or cutting torch flame into or back of the mixing chamber of the torch.

flash butt welding. See preferred term *flash welding*.

flash extension. That portion of *flash* remaining on a forged part after trimming; usually included in the normal forging tolerances.

flashing. In *flash welding*, the heating portion of the cycle, consisting of a series of rapidly recurring localized short circuits followed by molten metal expulsions, during which time the surfaces to be welded are moved one toward the other at a predetermined speed.

flash land. Configuration in the blocking or finishing impression of forging dies designed to restrict or to encourage the growth of *flash* at the parting line, whichever may be required in a particular case to ensure complete filling of the impression.

flash line. The line left on a forging after the flash has been trimmed off.

flash plate. A very thin final electrodeposited film of metal.

flash welding. A resistance welding process that joins metals by first heating abutting surfaces by passage of an electric current across the joint, then forcing the surfaces together by the application of pressure. Flashing and upsetting are accompanied by expulsion of metal from the joint.

flask. A metal or wood frame used for making and holding a sand mold. The upper part is called the *cope*; the lower, the *drag*. See also *blind riser*.

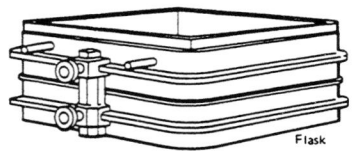
Flask

flat-die forging. Forging metal between flat or simple-contour dies by repeated strokes and manipulation of the workpiece. Also known as *open-die forging*, hand forging, or smith forging.

flat drill. A rotary end-cutting tool constructed from a flat piece of material, provided with suitable cutting lips at the cutting end.

flat edge trimmer. A machine for trimming notched edges on shells. The slide is cam driven so as to obtain a brief dwell at the bottom of the stroke, at which time the die, sometimes called a shimmy die, oscillates to trim the part.

flat-position welding. Welding from the upper side, the face of the weld being horizontal.

flattening. (1) A preliminary operation performed on forging stock to position the metal for a subsequent forging operation. (2) The removal of irregularities or distortion in sheets or plates by a method such as *roller leveling* or *stretcher leveling*.

flattening dies. Dies used to flatten sheet metal hems; that is, dies that can flatten a bend by closing it. These dies consist of a top and bottom die with a flat surface that can close one section (flange) to another (hem, seam).

flattening test. A quality test for tubing in which a specimen is flattened to a specified height between parallel plates.

flat wire. A roughly rectangular or square mill product, narrower than *strip*, in which all surfaces are rolled or drawn without any previous slitting, shearing, or sawing.

flaw. A nonspecific term often used to imply a crack-like discontinuity. See also preferred terms *discontinuity*, *and defect*.

flexible cam. An adjustable pressure-control cam of spring steel strips used to obtain varying pressure during a forming cycle.

flex roll. A movable jump roll designed to push up against a metal sheet as it passes through a roller leveler. The flex roll can be adjusted to deflect the sheet any amount up to the roll diameter.

flex rolling. Passing metal sheets through a *flex roll* unit to minimize yield-point elongation in order to reduce the tendency for *stretcher strains* to appear during forming.

flexural strength. A property of solid material that indicates its ability to withstand a flexural or transverse load.

floating die. (1) In metal forming, a die mounted in a die holder or punch mounted in its holder such that a slight amount of motion compensates for tolerance in the die parts, the work, or the press. (2) A die mounted on heavy springs to allow vertical motion in some trimming, shearing, and forming operations.

floating plug. In tube drawing, an unsupported mandrel that locates itself at the die inside the tube, causing a reduction in wall thickness while the die is reducing the outside diameter of the tube.

flop forging. A forging in which the top and bottom die impressions are identical, permitting the forging to be turned upside down during the forging operation.

flospinning. Forming cylindrical, conical and curvilinear shaped parts by power spinning over a rotating mandrel. See also *spinning*.

flotation. The concentration of valuable minerals from ores by agitation of the ground material with water, oil, and flotation chemicals. The valuable minerals are generally wetted by the oil, lifted to the surface by clinging air bubbles, and then floated off.

flow. Movement (slipping or sliding) of essentially parallel planes within an element of a material in parallel directions; occurs under the action of *shear stress*. Continuous action in this manner, at constant volume and without disintegration of the material, is termed *yield*, *creep*, or *plastic deformation*.

flowability. (1) In casting, a characteristic of a foundry sand mixture that enables it to move under pressure or vibration so that it makes intimate contact with all surfaces of the pattern or core box. (2) In welding, brazing, or soldering, the ability of molten filler metal to flow or spread over a metal surface.

flow brazing. Brazing by pouring hot molten nonferrous filler metal over a joint until the brazing temperature is attained. The filler metal is distributed in the joint by capillary action.

flow brightening. (1) Melting of an electrodeposit, followed by solidification, especially of tin plate. (2) Fusion (melting) of a chemically or mechanically deposited metallic coating on a substrate, particularly as it pertains to soldering.

flow lines. (1) Texture showing the direction of metal flow during hot or cold working. Flow lines can often be revealed by etching the surface or a section of a metal part. See accompanying macrograph. (2) In mechanical metallurgy, paths followed by minute volumes of metal during deformation.

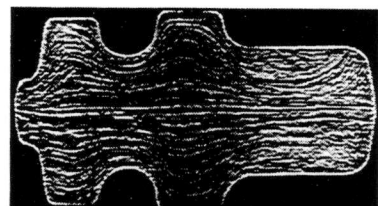

Flow lines

flow stress. The stress required to produce *plastic deformation* in a solid metal.

flow through. A forging defect caused by metal flow past the base of a rib with resulting rupture of the grain structure.

flow welding. A welding process which produces coalescence of metals by heating them with molten filler metal poured over the surfaces to be welded until the welding temperature is attained and until the required filler metal has been added. The filler metal is not distributed in the joint by capillary action.

fluid-cell process. A modification of the *Guerin process* for forming sheet metal, the fluid-cell process uses higher pressure and is primarily designed for forming slightly deeper parts, using a rubber pad as either the die or punch. A flexible hydraulic fluid cell forces an auxiliary rubber pad to follow the contour of the form block and exert a nearly uniform pressure at all points on the workpiece. See also *fluid forming* and *rubber-pad forming.*

fluid forming. A modification of the *Guerin process*, fluid forming differs from the fluid-cell process in that the die cavity, called a pressure dome, is not completely filled with rubber, but with hydraulic fluid retained by cup-shaped rubber diaphragm. See also *rubber-pad forming.*

fluidity. The ability of liquid metal to run into and fill a mold cavity.

fluidized bed. A contained mass of a finely divided solid that behaves like a fluid when brought into suspension in a moving gas or liquid.

fluorescent magnetic-particle inspection. Inspection with either dry magnetic particles or those in a liquid suspension, the particles being coated with a fluorescent substance to increase the visibility of the indications.

fluorescent penetrant inspection. Inspection using a fluorescent liquid that will penetrate any surface opening; after the surface has been wiped clean, the location of any surface flaws may be detected by the fluorescence, under ultraviolet light, of back-seepage of the fluid.

fluoroscopy. An inspection procedure in which the radiographic image of the subject is viewed on a fluorescent screen, normally limited to low-density materials or thin sections of metals because of the low light output of the fluorescent screen at safe levels of radiation.

flute. (1) As applied to drills, reamers, and taps, the channels or grooves formed in the body of the tool to provide cutting edges and to permit passage of cutting fluid and chips. (2) As applied to milling cutters and hobs, the chip space between the back of one tooth and the face of the following tooth.

flutes. Elongated grooves or voids that connect widely spaced cleavage planes.

fluting. (1) Forming longitudinal recesses in a cylindrical part, or radial recesses in a conical part. (2) A series of sharp parallel kinks or creases occurring in the arc when sheet metal is roll formed into a cylindrical shape. (3) Grinding the grooves of a twist drill or tap.

flux. (1) In metal refining, a material added to a melt to remove undesirable substances, like sand, ash, or dirt. Fluxing of the melt facilitates the agglomeration and separation of such undesirable constituents from the melt. It is also used as a protective covering for certain molten metal baths. Lime or limestone is generally used to remove sand, as in iron smelting; sand, to remove iron oxide in copper refining. (2) In brazing, cutting, soldering, or welding, material used to prevent the formation of, or to dissolve and facilitate removal of, oxides and other undesirable substances.

flux cored arc welding (FCAW). An arc welding process that joins metal by heating them with an arc between a continuous tubular filler-metal electrode and the work. Shielding is provided by a flux contained within the consumable tubular electrode. Additional shielding may or may not be obtained from an externally supplied gas or gas mixture. See also *electrogas welding.*

flux density. In magnetism, the number of *flux lines* per unit area passing through a cross section at right angles. It is given by

$B = H$, where and H are permeability and magnetic-field intensity, respectively.

flux lines. Imaginary lines used as a means of explaining the behavior of magnetic and other fields. Their concept is based on the pattern of lines produced when magnetic particles are sprinkled over a permanent magnet. Sometimes called magnetic lines of force.

fly cutting. Cutting with a single-tooth milling cutter.

flying shear. A machine for cutting continuous rolled products to length that does not require a halt in rolling, but rather moves along the runout table at the same speed as the product while performing the cutting, and then returns to the starting point in time to cut the next piece.

fog quenching. Quenching in a fine vapor or mist.

foil. Metal in sheet form less than 0.15 mm (0.006 in.) thick.

fold. (1) A defect in metal, usually on or near the surface, caused by continued fabrication of overlapping surfaces. (2) A forging defect caused by folding metal back onto its own surface during its flow in the die cavity. See also *lap.*

follow board. In foundry practice, a board contoured to a pattern to facilitate the making of a sand mold.

follow die. A *progressive die* consisting of two or more parts in a single holder; used with a separate lower die to perform more than one operation (such as piercing and blanking) on a part in two or more stations.

forced-air quench. A quench utilizing blasts of compressed air against relatively small parts such as a gear.

forehand welding. Welding in which the palm of the principal hand (torch or electrode hand) of the welder faces the direction of travel. It has special significance in oxyfuel gas welding in that the flame is directed ahead of the weld bead, which provides *preheating.* Contrast with *backhand welding.*

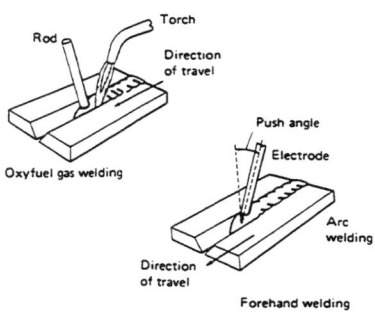

Forehand welding

forgeability. Term used to describe the relative ability of material to deform without fracture. Also describes the resistance to flow from deformation. See also *formability*.

forged roll Scleroscope hardness number (HFRSc or HFRSd). A number related to the height of rebound of a diamond-tipped hammer dropped on a forged steel roll. It is measured on a scale determined by dividing into 100 units the average rebound of a hammer from a forged steel roll of accepted maximum hardness. See also *Scleroscope hardness number* and *Scleroscope hardness test*.

forged structure. The macrostructure through a suitable section of a forging that reveals direction of working. See also the figure accompanying *flow lines*.

forge welding. Solid-state welding in which metals are heated in a forge (in air) and then welded together by applying pressure or blows sufficient to cause permanent deformation at the interface. The process is most commonly applied to the butt welding of steel.

forging. The process of working metal to a desired shape by impact or pressure in hammers, forging machines (upsetters), presses, rolls, and related forming equipment. Forging hammers, counterblow equipment, and high-energy-rate forging machines apply impact to the workpiece, while most other types of forging equipment apply squeeze pressure in shaping the stock. Some metals can be forged at room temperature, but most are made more plastic for forging by heating. Specific forging processes defined in this glossary include *closed-die forging*, *high-energy-rate forging*, *hot upset forging*, *isothermal forging*, *open-die forging*, *powder forging*, *precision forging*, *radial forging*, *ring rolling*, *roll forging*, *rotary forging*, and *rotary swaging*.

forging billet. A wrought metal slug used as *forging stock*.

forging dies. Forms for making forgings; they generally consist of a top and bottom die. The simplest will form a completed forging in a single impression; the most complex, consisting of several die inserts, may have a number of impressions for the progressive working of complicated shapes. Forging dies are usually in pairs, with part of the impression in one of the blocks and the rest of the impression in the other block.

forging envelope. See *finish allowance*.

forging ingot. A cast metal slug used as *forging stock*.

forging machine (upsetter or header). A type of forging equipment, related to the *mechanical press*, in which the principal forming energy is applied horizontally to the workpiece, which is gripped and held by prior action of the dies. See also *heading*, *hot upset forging*, and *upsetting*.

forging plane. In forging, the plane that includes the principal die face and that is perpendicular to the direction of ram travel. When parting surfaces of the dies are flat, the forging plane coincides with the *parting line*.

forging range. Temperature range in which a metal can be forged successfully.

forging rolls. Power-driven rolls used in preforming bar or billet stock that have shaped contours and notches for introduction of the work. See also *roll forging*.

forging stock. A wrought rod, bar, or other section suitable for subsequent change in cross section by forging.

formability. The ease with which a metal can be shaped through plastic deformation. Evaluation of the formability of a metal involves measurement of strength, ductility, and the amount of deformation required to cause fracture. The term workability is used interchangeably with formability; however, formability refers to the shaping of sheet metal, while workability refers to shaping materials by *bulk forming*. See also *forgeability*.

form block. Tooling, usually the male part, used for forming sheet metal contours; generally used in *rubber-pad forming*.

form cutter. Any cutter, profile sharpened or cam relieved, shaped to produce a specified form on the work.

form die. A die used to change the shape of a sheet metal blank with minimal plastic flow.

form grinding. Grinding with a wheel having a contour on its cutting face that is a mating fit to the desired form.

forming. (1) Making a change, with the exception of shearing or blanking, in the shape or contour of a metal part without intentionally altering its thickness. (2) The plastic deformation of a billet or a blanked sheet between tools (dies) to obtain the final configuration. Metalforming processes are typically classified as *bulk forming* and *sheet forming*. Also referred to as metalworking.

forming limit diagram (FLD). A diagram in which the major strains at the onset of necking in sheet metal are plotted vertically and the corresponding minor strains are plotted horizontally. The onset-of-failure line divides all possible strain combinations into two zones: the safe zone (in which failure during forming is not expected) and the failure zone (in which failure during forming is expected).

form-relieved cutter. A cutter so relieved that by grinding only the tooth face the original form is maintained throughout its life.

form rolling. Hot rolling to produce bars having contoured cross sections; not to be confused with *roll forming* of sheet metal or with *roll forging*.

form tool. A single-edge, nonrotating cutting tool, circular or flat, that produces its inverse or reverse form counterpart upon a workpiece.

forward extrusion. Same as direct extrusion. See *extrusion*.

fouling. An accumulation of deposits. This term includes accumulation and growth of marine organisms on a submerged metal surface and also includes the accumulation of deposits (usually inorganic) on heat exchanger tubing. See also *biological corrosion*.

foundry. A commercial establishment or building where metal castings are produced.

foundry returns. Metal in the form of gates, sprues, runners, risers, and scrapped castings of known composition returned to the furnace for remelting.

four-high mill. A type of rolling mill, commonly used for flat-rolled mill products, in which two large-diameter backup rolls are employed to reinforce two smaller work rolls, which are in contact with the product. Either the work rolls or the backup rolls may be driven. Compare with *two-high mill* and *cluster mill*.

four-point press. A press whose slide is actuated by four connections and four cranks, eccentrics, or cylinders, the chief merit being to equalize the pressure at the corners of the slides.

fractography. Descriptive treatment of fracture of materials, with specific reference to photographs of the fracture surface. Macrofractography involves photographs at low magnification.

fracture. The irregular surface produced when a piece of metal is broken. See also *brittle fracture*, *cleavage fracture*, *crystalline fracture*, *decohesive rupture*, *dimple rupture*, *ductile fracture*, *fibrous fracture*, *granular fracture*, *intergranular fracture*, *silky fracture*, and *transgranular fracture*.

fracture grain size. Grain size determined by comparing a fracture of a specimen with a set of standard fractures. For steel, a fully martensitic specimen is generally used, and the depth of hardening and the prior austenitic grain size are determined.

fracture mechanics. A quantitative analysis for evaluating structural behavior in terms of applied stress, crack length, and specimen or machine component geometry. See also *linear elastic fracture mechanics*.

fracture strength. The normal stress at the beginning of fracture. Calculated from the load at the beginning of fracture during a tension test and the original cross-sectional area of the specimen.

fracture stress. The true, normal stress on the minimum cross-sectional area at the beginning of fracture. The term usually applies to tension tests of unnotched specimens.

fracture surface markings. Fracture surface features that may be used to determine the fracture origin location and the nature of the stress that produced the fracture.

fracture test. Test in which a specimen is broken and its fracture surface is examined with the unaided eye or with a low-power microscope to determine such factors as composition, grain size, case depth, or discontinuities.

fracture toughness. A generic term for measures of resistance to extension of a crack. The term is sometimes restricted to results of *fracture mechanics* tests, which are directly applicable in fracture control. However, the term commonly includes results from simple tests of notched or precracked specimens not based on fracture mechanics analysis. Results from tests of the latter type are often useful for fracture control, based on either service experience or empirical correlations with fracture mechanics tests. See also *stress-intensity factor*.

fragmentation. The subdivision of a grain into small, discrete crystallite outlined by a heavily deformed network of intersecting slip bands as a result of cold working. These small crystals or fragments differ in orientation and tend to rotate to a stable orientation determined by the slip systems.

freckling. A type of segregation revealed as dark spots on a macroetched specimen of a consumable-electrode vacuum-arc-remelted alloy.

free bend. The bend obtained by applying forces to the ends of a specimen without the application of force at the point of maximum bending.

free carbon. The part of the total carbon in steel or cast iron that is present in elemental form as graphite or *temper carbon*. Contrast with *combined carbon*.

free ferrite. (1) Ferrite that is formed directly from the decomposition of hypoeutectoid austenite during cooling, without the simultaneous formation of cementite. (2) Ferrite formed into separate grains and not intimately associated with carbides as in pearlite. Also called *proeutectoid ferrite*.

free machining. Pertains to the machining characteristics of an alloy to which one or more ingredients have been introduced to produce small broken chips, lower power consumption, better surface finish, and longer tool life; among such additions are sulfur or lead to steel, lead to brass, lead and bismuth to aluminum, and sulfur or selenium to stainless steel.

freezing point. See preferred term *liquidus* and *solidus*. See also *melting point*.

freezing range. That temperature range between *liquidus* and *solidus* temperatures in which molten and solid constituents coexist.

fretting. A type of wear that occurs between tight-fitting surfaces subjected to cyclic relative motion of extremely small amplitude. Usually, fretting is accompanied by corrosion, especially of the very fine wear debris. Also referred to as *fretting corrosion* and *false Brinelling* (in rolling-element bearings).

fretting corrosion. (1) The accelerated deterioration at the interface between contacting surfaces as the result of corrosion and slight oscillatory movement between the two surfaces. (2) A form of *fretting* in which chemical reaction predominates. Fretting corrosion is often characterized by the removal of particles and subsequent formation of oxides, which are often abrasive and so increase the wear. Fretting corrosion can involve other chemical reaction products, which may not be abrasive.

fretting fatigue. (1) Fatigue fracture that initiates at a surface area where fretting has occurred. The progressive damage to a solid surface that arises from fretting. *Note*: If particles of wear debris are produced, then the term *fretting wear* may be applied.

fretting wear. Wear arising as a result of *fretting*.

friction. The resisting force tangential to the common boundary between two bodies when, under the action of an external force, one body moves or tends to move relative to the surface of the other.

friction coefficient. See *coefficient of friction*.

friction material. A sintered material exhibiting a high coefficient of friction designed for use where rubbing or frictional wear is encountered—for example, aircraft brake linings and clutch facings on tractors, heavy trucks, and earth-moving equipment. Friction materials consist of a dispersion of friction-producing ingredients in a metallic matrix.

friction welding. A solid-state process in which welds are made by holding a non-rotating workpiece in contact with a rotating workpiece under constant or gradually increasing pressure until the interface reaches the welding temperature and rotation can be stopped.

fuel gases. Gases usually used with oxygen for heating such as acetylene, natural gas, hydrogen, propane, methylacetylene propadiene stabilized, and other synthetic fuels and hydrocarbons.

full annealing. An imprecise term that denotes an annealing cycle to produce minimum strength and hardness. For the term to be meaningful, the composition and starting condition of the material and the time-temperature cycle used must be stated.

full-automatic plating. Electroplating in which the work is automatically conveyed through the complete cycle.

full center. Mild waviness down the center of a metal sheet or strip.

fuller (fullering impression). Portion of the die used in hammer forging primarily to reduce the cross section and to lengthen a portion of the forging stock. The fullering impression is often used in conjunction with an *edger (edging impression)*.

full hard. A *temper* of nonferrous alloys and some ferrous alloys corresponding approximately to a cold-worked state beyond which the material can no longer be formed by bending. In specifications, a full hard temper is commonly defined in terms of minimum hardness or minimum tensile strength (or, alternatively, a range of hardness or strength) corresponding to a specific percentage of cold reduction following a full anneal. For aluminum, a full hard temper is equivalent to a reduction of 75% from *dead soft*; for austenitic stainless steels, a reduction of about 50 to 55%.

full mold. A trade name for an expendable pattern casting process in which the polystyrene pattern is vaporized by the molten metal as the mold is poured. See also *lost foam casting*.

furnace brazing. A mass-production *brazing* process in which the filler metal is preplaced on the joint, then the entire assembly is heated to brazing temperature in a furnace. Usually, a protective furnace atmosphere is required, and wetting of the joint surfaces is accomplished without using a brazing flux.

fused spray deposit. A self-fluxing spray deposit which is deposited by conventional *thermal spraying* and subsequently fused using either a heating torch or a furnace. The coatings are usually made of nickel and cobalt alloys to which hard particles, such as tungsten carbide may be added for increased wear resistance.

fused zone. See preferred terms *fusion zone*, *nugget*, and *weld interface*.

fusible alloys. A group of binary, ternary, quaternary, and quinary alloys containing bismuth, lead, tin, cadmium, and indium. The term "fusible alloy" refers to any of more than 100 alloys that melt at relatively low temperatures, that is, below the melting point of tin-lead solder (183 °C, or 360 °F). The melting points of these alloys range as low as 47 °C (116 °F).

fusion welding. Any welding process in which the filler metal and base metal (substrate), or base metal only, are melted together to complete the weld.

fusion zone. The area of base metal melted as determined on the cross section of a weld.

G

gag. A metal spacer inserted so as to render a floating tool or punch inoperative.

gage. (1) The thickness of sheet or the diameter of wire. The various standards are arbitrary and differ with regard to ferrous and nonferrous products as well as sheet and wire. (2) An aid for visual inspection that enables an inspector to determine more reliably whether the size or contour of a formed part meets dimensional requirements. (3) An instrument used to measure thickness or length.

gage length. The original length of that portion of the specimen over which strain, change of length and other characteristics are measured.

gall. To damage the surface of a powder metallurgy compact or die part, caused by adhesion of powder to the die cavity wall or a punch surface.

galling. (1) A condition whereby excessive friction between high spots results in localized welding with subsequent *spalling* and a further roughening of the rubbing surfaces of one or both of two mating parts. (2) A severe form of scuffing associated with gross damage to the surfaces or failure. Galling has been used in many ways in tribology; therefore, each time it is encountered its meaning must be ascertained from the specific context of the usage. See also *scoring* and *scuffing*.

galvanic cell. (1) A cell in which chemical change is the source of electrical energy. It usually consists of two dissimilar conductors in contact with each other and with an electrolyte, or of two similar conductors in contact with each other and with dissimilar electrolytes. (2) A cell or system in which a spontaneous oxidation-reduction reaction occurs, the resulting flow of electrons being conducted in an external part of the circuit.

galvanic corrosion. Corrosion associated with the current of a galvanic cell consisting of two dissimilar conductors in an electrolyte or two similar conductors in dissimilar electrolytes. Where the two dissimilar metals are in contact, the resulting reaction is referred to as couple action.

galvanic couple. A pair of dissimilar conductors, commonly metals, in electrical contact. See also *galvanic corrosion*.

galvanic current. The electric current that flows between metals or conductive nonmetals in a *galvanic couple*.

galvanic series. A list of metals and alloys arranged according to their relative corrosion potentials in a given environment. Compare with *electromotive force series*.

galvanize. To coat a metal surface with zinc using any of various processes.

galvanneal. To produce a zinc-iron alloy coating on iron or steel by keeping the coating molten after hot-dip galvanizing until the zinc alloys completely with the basis metal.

gamma iron. The face-centered cubic form of pure iron, stable from 910 to 1400 °C (1670 to 2550 °F).

gamma structure. Structurally analogous phases or electron compounds having ratios of 21 valence electrons to 13 atoms. This is generally a large, complex cubic structure.

gang milling. Milling with several cutters mounted on the same arbor or with workpieces similarly positioned for cutting either simultaneously or consecutively during a single setup.

gang slitter. A machine with a number of pairs of rotary cutters spaced on two parallel shafts, used for *slitting* metal into strips or for trimming the edges of sheets.

gangue. The worthless portion of an ore that is separated from the desired part before smelting is commenced.

gap. The root opening in a weld joint.

gap-frame press. A general classification of press in which the uprights or housings are made in the form of a letter C, thus making three sides of the die space accessible.

gas atomization. An *atomization* process whereby molten metal is broken up into particles by a rapidly moving inert gas stream.

gas carbon arc welding. A carbon arc welding process variation which produces coalescence of metals by heating them with an electric arc between a single carbon electrode and the work. Shielding is obtained from a gas or gas mixture.

gas classification. The separation of a powder into its particle size fractions by means of a gas stream of controlled velocity flowing counterstream to the gravity-induced fall of the particles. The method is used to classify submesh-size particles.

gaseous corrosion. Corrosion with gas as the only corrosive agent and without any aqueous phase on the surface of the metal. Also called dry corrosion. See also *hot corrosion* and *sulfidation*.

gas holes. Holes in castings or welds that are formed by gas escaping from molten metal as it solidifies. Gas holes may occur individually, in clusters, or throughout the solidified metal.

gas metal arc cutting. An arc cutting process used to sever metals by melting them with the heat of an arc between a continuous metal (consumable) electrode and the work. Shielding is obtained entirely from an externally supplied gas or gas mixture.

gas metal arc welding (GMAW). An arc welding process which produces coalescence of metals by heating them with an arc between a continuous filler metal (consumable) electrode and the work. Shielding is obtained entirely from an externally supplied gas or gas mixture. Variations of the process include short-circuit arc GMAW, in which the consumable electrode is deposited during repeated short circuits, and pulsed arc GMAW, in which the current is pulsed. See also *globular transfer*.

gas pocket. A cavity caused by entrapped gas.

gas porosity. Fine holes or pores within a metal that are caused by entrapped gas or by the evolution of dissolved gas during solidification.

gas shielded arc welding. A general term used to describe gas metal arc welding, gas tungsten arc welding, and flux cored arc welding when gas shielding is employed. Typical gases employed include argon, helium, argon-hydrogen mixture, or carbon dioxide.

gassing. (1) Absorption of gas by a metal. (2) Evolution of gas from a metal during melting operations or upon solidification. (3) Evolution of gas from an electrode during electrolysis.

gas torch. See preferred terms welding torch and *cutting torch*.

gas tungsten arc cutting. An arc-cutting process in which metals are severed by melting them with an arc between a single tungsten (nonconsumable) electrode and the work. Shielding is obtained from a gas or gas mixture.

gas tungsten arc welding (GTAW). An arc welding process which produces coalescence of metals by heating them with an arc between a tungsten (nonconsumable) electrode and the work. Shielding is obtained from a gas or gas mixture. Pressure may or may not be used and filler metal may or may not be used.

gate. The portion of the runner in a mold through which molten metal enters the mold cavity. The generic term is sometimes applied to the entire network of connecting channels that conduct metal into the mold cavity. See also *gating system*.

gated pattern. In foundry practice, a *pattern* that includes not only the contours of the part to be cast but also the *gates*.

gathering. A forging operation that increases the cross section of part of the stock; usually a preliminary operation.

gathering stock. Any operation whereby the cross section of a portion of the forging stock is increased beyond its original size.

gating system. The complete assembly of sprues, runners, and gates in a mold through which metal flows to enter casting cavity. The term is also applied to equivalent portions of the *pattern*.

gear cutting. Producing tooth profiles of equal spacing on the periphery, internal surface, or face of a workpiece by means of an alternate shear gear-form cutter or a gear generator.

geared press. A press whose main crank or eccentric shaft is connected by gears to the driving source.

gear hobbing. Gear cutting by use of a tool resembling a worm gear in appearance, having helically spaced cutting teeth. In a single-thread hob, the rows of teeth advance exactly one pitch as the hob makes one revolution. With only one hob, it is possible to cut interchangeable gears of a given pitch of any number of teeth within the range of the hobbing machine.

gear milling. Gear cutting with a milling cutter that has been formed to the shape of the tooth space to be cut. The tooth spaces are machined one at a time.

gear shaping. Gear cutting with a reciprocating gear-shaped cutter rotating in mesh with the work blank.

general corrosion. (1) A form of deterioration that is distributed more or less uniformly over a surface. (2) *Corrosion* dominated by uniform thinning that proceeds without appreciable localized attack. See also *uniform corrosion*.

ghost lines. Lines running parallel to the rolling direction that appear in a sheet metal panel when it is stretched. These lines may not be evident unless the panel has been sanded or painted. Not to be confused with *leveler lines*.

gibs. Guides or shoes that ensure the proper parallelism, squareness, and sliding fit between metal forming press components such as the slide and the frame. They are usually adjustable to compensate for wear and to establish operating clearance.

glazing. Dulling the abrasive grains in the cutting face of a wheel during grinding.

glide. (1) Same as *slip*. (2) A noncrystallographic shearing movement, such as of one grain over another.

globular transfer. In consumable-electrode arc welding, a type of metal transfer in which molten filler metal passes across the arc as large droplets.

gold filled. Covered on one or more surfaces with a layer of gold alloy to form a clad or composite material. Gold-filled dental restorations are an example of such materials.

gooseneck. In die casting, a spout connecting a molten metal holding pot, or chamber, with a nozzle or sprue hole in the die and containing a passage through which molten metal is forced on its way to the die. It is the metal injection mechanism in a *hot chamber machine*.

gouging. In welding practice, the forming of a bevel or groove by material removal. See also *arc gouging* and *oxygen gouging*.

gouging abrasion. A form of high-stress abrasion in which easily observable grooves or gouges are created on the surface. See also *abrasion*.

G-P zone. A *Guinier-Preston zone*.

graded abrasive. An abrasive powder in which the sizes of the individual particles are confined to certain specified limits. See also *grit size*.

grain. An individual crystal in a polycrystalline material; it may or may not contain twinned regions and subgrains.

grain boundary. A narrow zone in a metal or ceramic corresponding to the transition from one crystallographic orientation to another, thus separating one *grain* from another; the atoms in each grain are arranged in an orderly pattern.

grain-boundary corrosion. Same as *intergranular corrosion*. See also *interdendritic corrosion*.

grain-boundary sulfide precipitation. An intermediate state of overheating of metals in which sulfide inclusions are redistributed to the austenitic grain boundaries by partial solution at the overheating temperature and reprecipitation during subsequent cooling.

grain coarsening. A heat treatment that produces excessively large austenitic grains in metals.

grain flow. Fiberlike lines on polished and etched sections of forgings caused by orientation of the constituents of the metal in the direction of working during forging. Grain flow produced by proper die design can improve required *mechanical properties* of forgings. See also *flow lines* and *forged structure*.

grain growth. (1) An increase in the average size of the grains in polycrystalline material, usually as a result of heating at elevated temperature. (2) In polycrystalline materials, a phenomenon occurring fairly close below the melting point in which the larger grains grow still larger while the smallest ones gradually diminish and disappear. See also *recrystallization*.

grain refinement. The manipulation of the solidification process to cause more (and therefore smaller) grains to be formed and/or to cause the grains to form in specific shapes. The term refinement is usually used to denote a chemical addition to the metal but can refer to control of the cooling rate.

grain refiner. A material added to a molten metal to induce a finer-than-normal grain size in the final structure.

grain size. (1) For metals, a measure of the areas or volumes of grains in a polycrystalline material, usually expressed as an average when the individual sizes are fairly uniform. In metals containing two or more phases, grain size refers to that of the matrix unless otherwise specified. Grain size is reported in terms of number of grains per unit area or volume, in terms of average diameter, or as a grain-size number derived from area measurements. (2) For grinding wheels, see preferred term *grit size*.

grain size distribution. Measures of the characteristic grain or crystallite dimensions (usually, diameters) in a polycrystalline solid; or of their populations by size increments from minimum to maximum. Usually determined by microscopy.

granular fracture. A type of irregular surface produced when metal is broken that is characterized by a rough, grainlike appearance, rather than a smooth or fibrous one. It can be subclassified as *transgranular fracture* or *intergranular fracture*. This type of fracture is frequently called *crystalline fracture*; however, the inference that the metal broke because it "crystallized" is not justified, because all metals are crystalline in the solid state. See also *fibrous fracture* and *silky fracture*.

granulated metal. Small pellets produced by pouring liquid metal through a screen or by dropping it onto a revolving disk, and, in both instances, chilling with water.

graphitic carbon. Free carbon in steel or cast iron.

graphitic corrosion. Corrosion of gray iron in which the iron matrix is selectively leached away, leaving a porous mass of graphite behind it occurs in relatively mild aqueous solutions and on buried pipe and fittings.

graphitic steel. Alloy steel made so that part of the carbon is present as graphite.

graphitization. The formation of graphite in iron or steel. Where graphite is formed during solidification, the phenomenon is termed primary graphitization; where formed later by heat treatment, secondary graphitization.

graphitizing. Annealing a ferrous alloy such that some or all the carbon precipitates as graphite.

gravity hammer. A class of forging hammer in which energy for forging is obtained by the mass and velocity of a freely falling ram and the attached upper die. Examples are the *board hammer* and *airlift hammer*.

gravity segregation. Variable composition of a casting or ingot caused by settling out of heavy constituents, or rising of light constituents, before or during solidification.

gray cast iron. See *gray iron.* ·

gray iron. A broad class of ferrous casting alloys (*cast irons*) normally characterized by a microstructure of flake graphite in a ferrous matrix. Gray irons usually contain 2.5 to 4% C, 1 to 3% Si, and additions of manganese, depending on the desired microstructure (as low as 0.1% Mn in ferritic gray irons and as high as 1.2% in pearlitics). Sulfur and phosphorus are also present in small amounts as residual impurities. See also *flake graphite*.

green compact. An unsintered powder metallurgy or ceramic compact.

green density. The density of a *green compact.*

green rot. A form of high temperature attack on stainless steels, nickel-chromium alloys and nickel-chromium-iron alloys subjected to simultaneous oxidation and carburization. Basically, attack occurs first by precipitation of chromium as chromium carbide, then by oxidation of the carbide particles.

green sand. A naturally bonded sand, or a compounded molding sand mixture, that

has been "tempered" with water and that is used while still moist.

green sand core. (1) A *core* made of *green sand* and used as-rammed. (2) A sand core that is used in the unbaked condition.

green sand mold. A casting mold composed of moist prepared molding sand. Contrast with *dry sand mold*.

green strength. The strength of a tempered foundry sand mixture at room temperature.

green strength. (1) The ability of a *green compact* to maintain its size and shape during handling and storage prior to *sintering*. (2) The tensile or compressive strength of a green compact.

grindability. Relative ease of grinding, analogous to *machinability*.

grindability index. A measure of the grindability of a material under specified grinding conditions, expressed in terms of volume of material removed per unit volume of wheel wear.

grinding. Removing material from a workpiece with a grinding wheel or abrasive belt.

grinding burn. See *burning* .

grinding cracks. Shallow cracks formed in the surfaces of relatively hard materials because of excessive grinding heat or the high sensitivity of the material. See also *grinding sensitivity*.

grinding fluid. An oil- or water-based fluid introduced into grinding operations to (1) reduce and transfer heat during grinding, (2) lubricate during chip formation, (3) wash loose chips or swarf from the grinding belt or wheel, and (4) chemically aid the grinding action or machine maintenance.

grinding oil. An oil-type grinding fluid; it may contain additives, but not water.

grinding relief. A groove or recess located at the boundary of a surface to permit the corner of the wheel to overhang during grinding.

grinding sensitivity. Susceptibility of a material to surface damage such as *grinding cracks*; it can be affected by such factors as hardness, microstructure, hydrogen content, and residual stress.

grinding stress. *Residual stress*, generated by grinding, in the surface layer of work. It may be tensile or compressive, or both.

grinding wheel. A cutting tool of circular shape made of abrasive grains bonded together. See also the figure accompanying the term *diamond wheel*.

grit. Crushed ferrous or synthetic abrasive material in various mesh sizes that is used in abrasive blasting equipment to clean castings. For materials used for grinding

belts or grinding wheels, the term *abrasive* is preferred.

grit blasting. *Abrasive blasting* with small irregular pieces of steel, malleable cast iron, or hard nonmetallic materials.

grit size. Nominal size of abrasive particles in a grinding wheel, corresponding to the number of openings per linear inch in a screen through which the particles can pass.

groove angle. The total included angle of the groove between parts to be joined. Thus, the sum of two bevel angles, either or both of which may be zero degrees.

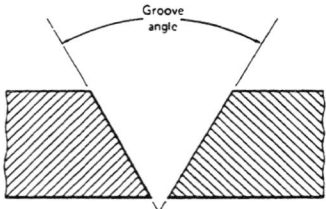

groove face. The portion of a surface or surfaces of a member included in a groove. See also the figure accompanying *root of joint*.

groove weld. A weld made in the groove between two members. The standard types are square, single-bevel, single flare-bevel, single flare-V, single-J, single-U, single-V, double-bevel, double flare-bevel, double flare-V, double-J, double-U and double-V.

Grossmann number (*H*). A ratio describing the ability of a quenching medium to extract heat from a hot steel work-piece in comparison to still water defined by the following equation:

$$H = \frac{h}{2k}$$

where *h* is the heat transfer coefficient and *k* is the conductivity of the metal.

gross porosity. In weld metal or in a casting, pores, gas holes or globular voids that are larger and in much greater numbers than those obtained in good practice.

groundbed. A buried item, such as junk steel or graphite rods, that serves as the *anode* for the *cathodic protection* of pipelines or other buried structures.

ground connection. In arc welding, a device used for attaching the work lead (ground cable) to the work.

growth (cast iron). A permanent increase in the dimensions of cast iron resulting from repeated or prolonged heating at temperatures above 480 °C (900 °F) due either to graphitizing of carbides or oxidation.

Guerin process. A *rubber-pad forming* process for forming sheet metal. The principal tools are the rubber pad and form block, or punch.

guided bend. The bend obtained by use of a plunger to force the specimen into a die in order to produce the desired contour of the outside and inside surfaces of the specimen.

guided bend test. A test in which the specimen is bent to a definite shape by means of a punch (mandrel) and a bottom block.

guide mill. A small hand mill with several stands in a train and with guides for the work at the entrance to the rolls.

Guinier-Preston (G-P) zone. A small precipitation domain in a supersaturated metallic solid solution. A G-P zone has no well-defined crystalline structure of its own and contains an abnormally high concentration of solute atoms. The formation of G-P zones constitutes the first stage of precipitation and is usually accompanied by a change in properties of the solid solution in which they occur.

gun drill. A drill, usually with one or more flutes and with coolant passages through the drill body, used for deep hole drilling.

gutter. A depression around the periphery of a forging *die impression* outside the *flash pan* that allows space for the excess metal; surrounds the finishing impression and provides room for the excess metal used to ensure a sound forging. A shallow impression outside the parting line.

H

habit plane. The plane or system of planes of a crystalline phase along which some phenomenon, such as twinning or transformation, occurs.

Hadfield steel. See *austenitic manganese steel*.

half cell. An *electrode* immersed in a suitable *electrolyte*, designed for measurements of *electrode potential*.

half hard. A *temper* of nonferrous alloys and some ferrous alloys characterized by tensile strength about midway between those of *dead soft* and *full hard* tempers.

Hall process. A commercial process for winning aluminum from alumina by electrolytic reduction of a fused bath of alumina dissolved in cryolite.

hammer. A machine that applies a sharp blow to the work area through the fall of a ram onto an anvil. The ram can be driven by gravity or power. See also *gravity hammer*.

hammer forging. Forging in which the work is deformed by repeated blows.

hammering. The working of metal sheet into a desired shape over a form or on a high-speed hammer and a similar anvil to produce the required dishing or thinning.

hammer welding. *Forge welding* by hammering.

hand brake. A small manual folding machine designed to bend sheet metal, similar in design and purpose to a *press brake*.

hand forge (smith forge). A forging operation in which forming is accomplished on dies that are generally flat. The piece is shaped roughly to the required contour with little or no lateral confinement; operations involving mandrels are included. The term hand forge refers to the operation performed, while hand forging applies to the part produced.

handling breaks. Irregular *breaks* caused by improper handling of metal sheets during processing. These breaks result from bending or sagging of the sheets during handling.

Hansgirg process. A process for producing magnesium by reduction of magnesium oxide with carbon.

hard chromium. Chromium electrodeposited for engineering purposes (such as to increase the wear resistance of sliding metal surfaces) rather than as a decorative coating. It is usually applied directly to basis metal and is customarily thicker (1.2 m or 0.05 mils) than a decorative deposit, but not necessarily harder.

hard drawn. An imprecise term applied to drawn products, such as wire and tubing, that indicates substantial cold reduction without subsequent annealing. Compare with *light drawn*.

hardenability. The relative ability of a ferrous alloy to form martensite when quenched from a temperature above the upper critical temperature. Hardenability is commonly measured as the distance below a quenched surface at which the metal exhibits a specific hardness (50 HRC, for example) or a specific percentage of martensite in the microstructure.

hardener. An alloy rich in one or more alloying elements that is added to a melt to permit closer control of composition than is possible by the addition of pure metals, or to introduce refractory elements not readily alloyed with the base metal. Sometimes called *master alloy* or rich alloy.

hardening. Increasing hardness of metals by suitable treatment, usually involving heating and cooling. When applicable, the following more specific terms should be used: *age hardening, case hardening, flame hardening, induction hardening, precipitation hardening* and *quench hardening*.

hardfacing. The application of a hard, wear-resistant material to the surface of a component by welding, spraying, or allied welding processes to reduce wear or loss of material by abrasion, impact, erosion, galling, and cavitation. See also *surfacing*.

hardfacing alloys. Wear-resistant materials available as bare welding rod, flux-coated rod, long-length solid wires, long-length tubular wires, or powders that are deposited by *hardfacing*.

hard metal. A collective term that designates a sintered material with high hardness, strength, and wear resistance, and is characterized by a tough metallic binder phase and particles of carbides, borides, or nitrides of the refractory metals. The term is in general use abroad, while for the carbides the term *cemented carbide* is preferred in the U.S., and the boride and nitride materials are usually categorized as *cermets*.

hardness. A measure of the resistance of a material to surface indentation or abrasion; may be thought of as a function of the stress required to produce some specified type of surface deformation. There is no absolute scale for hardness; therefore, to express hardness quantitatively, each type of test has its own scale of arbitrarily defined hardness. Indentation hardness can be measured by *Brinell, Rockwell, Vickers, Knoop,* and *Scleroscope hardness tests*.

hard solder. A term erroneously used to denote silver-base brazing filler metals.

hard surfacing. See preferred terms *surfacing* or *hardfacing*.

hard temper. Same as *full hard* temper.

Haring cell. A four-electrode cell for measurement of electrolyte resistance and electrode polarization during electrolysis.

H-band steel. Carbon, carbon-boron, or alloy steel produced to specified limits of hardenability; the chemical composition range may be slightly different from that of the corresponding grade of ordinary carbon or alloy steel.

heading. The *upsetting* of wire, rod, or bar stock in dies to form parts that usually contain portions that are greater in cross-sectional area than the original wire, rod, or bar.

hearth. The bottom portions of certain furnaces, such as blast furnaces, air furnaces, and other reverberatory furnaces, that support the charge and sometimes collect and hold molten metal.

heat. A stated tonnage of metal obtained from a period of continuous melting in a cupola or furnace, or the melting period required to handle this tonnage.

heat-affected zone (HAZ). That portion of the base metal that was not melted during brazing, cutting, or welding, but whose microstructure and mechanical properties were altered by the heat.

heat check. A pattern of parallel surface cracks that are formed by alternate rapid heating and cooling of the extreme surface metal, sometimes found on forging dies and piercing punches. There may be two sets of parallel cracks, one set perpendicular to the other.

heat-resistant alloy. An alloy developed for very-high-temperature service where relatively high stresses (tensile, thermal, vibratory, or shock) are encountered and where oxidation resistance is frequently required.

heat sink. A material that absorbs or transfers heat away from a critical element or part.

heat tinting. Coloration of a metal surface through oxidation by heating to reveal details of the microstructure.

heat treatable alloy. An alloy that can be hardened by heat treatment.

heat treating film. A thin coating or film, usually an oxide, formed on the surface of a metal during heat treatment.

heat treatment. Heating and cooling a solid metal or alloy in such a way as to obtain desired conditions or properties. Heating for the sole purpose of hot working is excluded from the meaning of this definition.

heavy metal. A sintered tungsten alloy with nickel, copper, and/or iron, the tungsten content being at least 90 wt% and the density being at least 16.8 g/cm^3.

heel. Synonymous with *base*.

heel block. A block or plate usually mounted on or attached to a lower die in a forming or forging press that serves to prevent or minimize the deflection of punches or cams.

hemming. A bend of 180° made in two steps. First, a sharp-angle bend is made; next the bend is closed using a flat punch and a die.

HERF. A common abbreviation for *high-energy-rate forging* or *high-energy-rate forming*.

herringbone pattern. Same as *chevron pattern*.

high-conductivity copper. Copper that, in the annealed condition, has a minimum electrical conductivity of 100% *IACS* as determined by ASTM test methods.

high-cycle fatigue. *Fatigue* that occurs at relatively large numbers of cycles. The arbitrary, but commonly accepted, dividing

line between high-cycle fatigue and *low-cycle fatigue* is considered to be about 10^4 to 10^5 cycles. In practice, this distinction is made by determining whether the dominant component of the *strain* imposed during cyclic loading is elastic (high cycle) or plastic (low cycle), which in turn depends on the properties of the metal and on the magnitude of the nominal *stress*.

high-energy-rate forging (HERF). A closed-die hot- or cold-forging process in which the stored energy of high-pressure gas is used to accelerate a ram to unusually high velocities in order to effect deformation of the workpiece. Ideally, the final configuration of the forging is developed in one blow or, at most, a few blows. In high-energy-rate forging, the velocity of the ram, rather than its mass, generates the major forging force. Also known as HERF processing, high-velocity forging, and high-speed forging.

high-energy-rate forming. A group of forming processes that applies a high rate of strain to the material being formed through the application of high rates of energy transfer. See also *explosive forming*, *high-energy-rate forging*, and *electromagnetic forming*.

high-frequency resistance welding. A resistance welding process that produces coalescence of metals with the heat generated from the resistance of the workpieces to a high-frequency alternating current in the 10 to 500 kHz range and the rapid application of an upsetting force after heating is substantially completed. The path of the current in the workpiece is controlled by use of the proximity effect (the feed current follows closely the return current conductor).

highlighting. Buffing or polishing selected areas of a complex shape to increase the luster or change the color of those areas.

high residual phosphorus copper. Deoxidized copper with residual phosphorus present in amounts (usually 0.013 to 0.04%) generally sufficient to decrease appreciably the conductivity of the copper.

high-speed machining. High-productivity machining processes which achieve cutting speeds in excess of 600 m/min (2000 sfm) and up to 18,000 m/min (60,000 sfm).

high-strength low-alloy (HSLA) steels. Steels designed to provide better mechanical properties and/or greater resistance to atmospheric corrosion than conventional carbon steels. They are not considered to be alloy steels in the normal sense because they are designed to meet specific mechanical properties rather than

a chemical composition (HSLA steels have yield strengths greater than 275 MPa, or 40 ksi). The chemical composition of a specific HSLA steel may vary for different product thicknesses to meet mechanical property requirements. The HSLA steels have low carbon contents (0.05 to 0.25% C) in order to produce adequate formability and weldability, and they have manganese contents up to 2.0%. Small quantities of chromium, nickel, molybdenum, copper, nitrogen, vanadium, niobium, titanium, and zirconium are used in various combinations.

high-temperature hydrogen attack. A loss of strength and ductility of steel by high-temperature reaction of absorbed hydrogen with carbides in the steel resulting in *decarburization* and internal fissuring.

hindered contraction. Contraction where the shape will not permit a metal casting to contract in certain regions in keeping with the coefficient of expansion.

HIP. See *hot isostatic pressing*.

hob. A rotary cutting tool with its teeth arranged along a helical thread, used for generating gear teeth or other evenly spaced forms on the periphery of a cylindrical workpiece. The hob and the workpiece are rotated in timed relationship to each other while the hob is fed axially or tangentially across or radially into the workpiece. Hobs should not be confused with multiple-thread milling cutters, rack cutters, and similar tools, where the teeth are not arranged along a helical thread.

hogging. Machining a part from bar stock, plate, or a simple forging in which much of the original stock is removed.

holddown plate (pressure pad). A pressurized plate designed to hold the workpiece down during a press operation. In practice, this plate often serves as a *stripper* and is also called a stripper plate.

holding. In heat treating of metals, that portion of the thermal cycle during which the temperature of the object is maintained constant.

holding furnace. A furnace into which molten metal can be transferred to be held at the proper temperature until it can be used to make castings.

holding temperature. In heat treating of metals, the constant temperature at which the object is maintained.

holding time. Time for which the temperature of the heat treated metal object is maintained constant.

hole expansion test. A simulative test in which a flat metal sheet specimen with a circular hole in its center is clamped between annular die plates and deformed by a punch, which expands and ultimately cracks the edge of the hole.

hole flanging. The forming of an integral collar around the periphery of a previously formed hole in a sheet metal part.

holidays. Discontinuities in a coating (such as porosity, cracks, gaps, and similar flaws) that allow areas of basis metal to be exposed to any corrosive environment that contacts the coated surface.

homogeneous carburizing. Use of a carburizing process to convert a low-carbon ferrous alloy to one of uniform and higher carbon content throughout the section.

homogenizing. A heat treating practice whereby a metal object is held at high temperature to eliminate or decrease chemical segregation by diffusion.

honing. A low-speed finishing process used chiefly to produce uniform high dimensional accuracy and fine finish, most often on inside cylindrical surfaces. In honing, very thin layers of stock are removed by simultaneously rotating and reciprocating a bonded abrasive stone or stick that is pressed against the surface being honed with lighter force than is typical of grinding.

Hooke's law. A generalization applicable to all solid material, which states that stress is directly proportional to strain and is expressed as:

$$\frac{Stress}{Strain} = \frac{\sigma}{\varepsilon} = constant = E$$

where E is the modulus of elasticity or Young's modulus. The constant relationship between stress and strain applies only below the proportional limit. See also *modulus of elasticity*.

Hoopes process. An electrolytic refining process for aluminum, using three liquid layers in the reduction cell.

horizontal-position welding. (1) Making a fillet weld on the upper side of the intersection of a vertical surface and a horizontal surface. (2) Making a horizontal groove weld on a vertical surface.

horizontal fixed position (pipe welding). The position of a pipe joint in which the axis of the pipe is essentially horizontal and the pipe is not rotated during welding. See also *vertical position* (pipe welding).

horizontal rolled position (pipe welding). The position of a pipe joint in which the axis of the pipe is essentially horizontal, and welding is carried out as the pipe is rotated about its axis.

horn. (1) In a resistance welding machine, a cylindrical arm or beam that transmits the electrode pressure and usually conducts the welding current. (2) A cone-shaped member that transmits ultrasonic energy from a transducer to a welding or machining tool. See also *ultrasonic impact grinding* and *ultrasonic welding*.

horn press. A mechanical metal forming press equipped with or arranged for a cantilever block or horn that acts as the die or support for the die, used in forming, piercing, setting down, or riveting hollow cylinders and odd-shaped work.

horn spacing. The distance between adjacent surfaces of the horns of a resistance welding machine.

hot box process. In foundry practice, resin-base (furan or phenolic) binder process for molding sands similar to shell coremaking; cores produced with it are solid unless mandrelled out.

hot chamber machine. A *die casting* machine in which the metal chamber under pressure is immersed in the molten metal in a furnace. The chamber is sometimes called a gooseneck, and the machine is sometimes called a gooseneck machine.

hot-cold working. (1) A high-temperature thermomechanical treatment consisting of deforming a metal above its transformation temperature and cooling fast enough to preserve some or all of the deformed structure. (2) A general term synonymous with *warm working*.

hot corrosion. An accelerated corrosion of metal surfaces that results from the combined effect of oxidation and reactions with sulfur compounds and other contaminants, such as chlorides, to form a molten salt on a metal surface that fluxes, destroys, or disrupts the normal protective oxide. See also *gaseous corrosion*.

hot cracking. (1) A crack formed in a weldment caused by the segregation at grain boundaries of low-melting constituents in the weld metal. This can result in grain-boundary tearing under thermal contraction stresses. Hot cracking can be minimized by the use of low-impurity welding materials and proper joint design. (2) A crack formed in a cast metal because of internal stress developed upon cooling following solidification. A hot crack is less open than a *hot tear* and usually exhibits less oxidation and decarburization along the fracture surface. See also *cold cracking, lamellar tearing*, and *stress-relief cracking*.

hot-die forging. A hot forging process in which both the dies and the forging stock are heated; typical die temperatures are 110 to 225 °C (200 to 400 °F) lower than the temperature of the stock. Compare with *isothermal forging*.

hot dip. Covering a surface by dipping the surface to be coated into a molten bath of the coating material. See also *hot dip coating*.

hot dip coating. A metallic coating obtained by dipping the basis metal into a molten metal.

hot extrusion. A process whereby a heated *billet* is forced to flow through a shaped die opening. The temperature at which extrusion is performed depends on the material being extruded. Hot extrusion is used to produce long, straight metal products of constant cross section, such as bars, solid and hollow sections, tubes, wires, and strips, from materials that cannot be formed by cold extrusion.

hot forging. (1) A forging process in which the die and/or forging stock are heated. See also *hot-die forging* and *isothermal forging*. (2) The plastic deformation of a pressed and/or sintered powder compact in at least two directions at temperatures above the recrystallization temperature.

hot forming. See *hot working*.

hot isostatic pressing. (1) A process for simultaneously heating and forming a compact in which the powder is contained in a sealed flexible sheet metal or glass enclosure and the so-contained powder is subjected to equal pressure from all directions at a temperature high enough to permit plastic deformation and sintering to take place. (2) A process that subjects a component (casting, powder forgings, etc.) to both elevated temperature and isostatic gas pressure in an autoclave. The most widely used pressurizing gas is argon. When castings are hot isostatically pressed, the simultaneous application of heat and pressure virtually eliminates internal voids and microporosity through a combination of plastic deformation, creep, and diffusion.

hot mill. A production line or facility for hot rolling of metals.

hot press forging. Plastically deforming metals between dies in presses at temperatures high enough to avoid strain hardening.

hot pressing. Simultaneous heating and forming of a powder compact. See also *pressure sintering*.

hot pressure welding. A solid-state welding process that produces coalescence of materials with heat and application of pressure sufficient to produce macrodeformation of the base material. Vacuum or other shielding media may be used. See also *forge welding* and *diffusion welding*. Compare with *cold welding*.

hot quenching. An imprecise term for various quenching procedures in which a quenching medium is maintained at a prescribed temperature above 70 °C (160 °F).

hot shortness. A tendency for some alloys to separate along grain boundaries when stressed or deformed at temperatures near the melting point. Hot shortness is caused by a low-melting constituent, often present only in minute amounts, that is segregated at grain boundaries.

hot tear. A fracture formed in a metal during solidification because of *hindered contraction*. Compare with *hot cracking*.

hot top. (1) A reservoir, thermally insulated or heated, that holds molten metal on top of a mold for feeding of the ingot or casting as it contracts on solidifying, thus preventing formation of *pipe* or *voids*. (2) A refractory-lined steel or iron casting that is inserted into the tip of the mold and is supported at various heights to feed the ingot as it solidifies.

hot trimming. The removal of *flash* or excess metal from a hot part (such as a forging) in a trimming press.

hot upset forging. A *bulk forming* process for enlarging and reshaping some of the cross-sectional area of a bar, tube, or other product form of uniform (usually round) section. It is accomplished by holding the heated forging stock between grooved dies and applying pressure to the end of the stock, in the direction of its axis, by the use of a heading tool, which spreads (upsets) the end by metal displacement. Also called hot heading or hot upsetting. See also *heading* and *upsetting*.

hot wire welding. A variation of arc welding processes in which a filler metal wire is resistance heated as it is fed into the molten weld pool.

hot-worked structure. The structure of a material worked at a temperature higher than the recrystallization temperature.

hot working. (1) The plastic deformation of metal at such a temperature and strain rate that recrystallization takes place simultaneously with the deformation, thus avoiding any *strain hardening*. Also referred to as hot forging and hot forming. (2) Controlled mechanical operations for shaping a product at temperatures above the recrystallization temperature. Contrast with *cold working*.

hub. A *boss* that is in the center of a forging and forms a part of the body of the forging.

hubbing. The production of forging die cavities by pressing a male master plug, known as a *hub*, into a block of metal.

Hull cell. A special electrodeposition cell giving a range of known current densities for test work.

hydraulic hammer. A gravity-drop forging hammer that uses hydraulic pressure to lift the hammer between strokes.

hydraulic-mechanical press brake. A mechanical *press brake* that uses hydraulic cylinders attached to mechanical linkages to power the ram through its working stroke.

hydraulic press. A press in which fluid pressure is used to actuate and control the ram. Hydraulic presses are used for both open- and closed-die forging.

hydrodynamic machining. Removal of material by the impingement of a high-velocity fluid against a workpiece. See also *waterjet/abrasive waterjet machining*.

hydrogen-assisted cracking (HAC). See *hydrogen embrittlement*.

hydrogen-assisted stress-corrosion cracking (HSCC). See *hydrogen embrittlement*.

hydrogen blistering. The formation of blisters on or below a metal surface from excessive internal hydrogen pressure. Hydrogen may be formed during cleaning, plating, or corrosion.

hydrogen brazing. A term sometimes used to denote brazing in a hydrogen-containing atmosphere, usually in a furnace; use of the appropriate process name is preferred.

hydrogen damage. A general term for the embrittlement, cracking, blistering, and hydride formation that can occur when hydrogen is present in some metals.

hydrogen embrittlement. A process resulting in a decrease of the *toughness* or *ductility* of a metal due to the presence of atomic hydrogen. Hydrogen embrittlement has been recognized classically as being of two types. The first, known as internal hydrogen embrittlement, occurs when the hydrogen enters molten metal which becomes supersaturated with hydrogen immediately after solidification. The second type, environmental hydrogen embrittlement, results from hydrogen being absorbed by solid metals. This can occur during elevated-temperature thermal treatments and in service during electroplating, contact with maintenance chemicals, corrosion reactions, cathodic protection, and operating in high-pressure hydrogen. In the absence of residual stress or external loading, environmental hydrogen embrittlement is manifested in various forms, such as blistering, internal cracking, hydride formation, and reduced ductility. With a tensile stress or stress-intensity factor exceeding a specific threshold, the atomic hydrogen interacts with the metal to induce subcritical crack growth leading to fracture. In the absence of a corrosion reaction (polarized cathodically), the usual term used is hydrogen-assisted cracking (HAC) or hydrogen stress cracking (HSC). In the presence of active corrosion, usually as pits or crevices (polarized anodically), the cracking is generally called *stress-corrosion cracking* (SCC), but should more properly be called hydrogen-assisted stress-corrosion cracking (HSCC). Thus, HSC and electrochemically anodic SCC can operate separately or in combination (HSCC). In some metals, such as high-strength steels, the mechanism is believed to be all, or nearly all, HSC. The participating mechanism of HSC is not always recognized and may be evaluated under the generic heading of SCC.

hydrogen-induced cracking (HIC). Same as *hydrogen embrittlement*.

hydrogen-induced delayed cracking. A term sometimes used to identify a form of *hydrogen embrittlement* in which a metal appears to fracture spontaneously under a steady stress less than the *yield stress*. There is usually a delay between the application of stress (or exposure of the stressed metal to hydrogen) and the onset of cracking. Also referred to as static fatigue.

hydrogen loss. The loss in weight of metal powder or a compact caused by heating a representative sample according to a specified procedure in a purified hydrogen atmosphere. Broadly, a measure of the oxygen content of the sample when applied to materials containing only such oxides as are reducible with hydrogen and no hydride-forming element.

hydrogen overvoltage. In electroplating, *overvoltage* associated with the liberation of hydrogen gas.

hydrogen stress cracking (HSC). See *hydrogen embrittlement*.

hydrometallurgy. Industrial *winning* or *refining* of metals using water or an aqueous solution.

hydrostatic extrusion. A method of extruding a *billet* through a die by pressurized fluid instead of the ram used in conventional *extrusion*.

hydrostatic pressing. A special case of isostatic pressing that uses a liquid such as water or oil as a pressure transducing medium and is therefore limited to near room temperature operation.

hydrostatic tension. Three equal and mutually perpendicular tensile stresses.

hypereutectic alloy. In an alloy system exhibiting a *eutectic*, any alloy whose composition has an excess of alloying element compared with the eutectic composition and whose equilibrium microstructure contains some eutectic structure.

hypereutectoid alloy. In an alloy system exhibiting a *eutectoid*, any alloy whose composition has an excess of alloying element compared with the eutectoid composition, and whose equilibrium microstructure contains some eutectoid structure.

hypoeutectic alloy. In an alloy system exhibiting a *eutectic*, any alloy whose composition has an excess of base metal compared with the eutectic composition, and whose equilibrium microstructure contains some eutectic structure.

hypoeutectoid alloy. In an alloy system exhibiting a *eutectoid*, any alloy whose composition has an excess of base metal compared with the eutectoid composition, and whose equilibrium microstructure contains some eutectoid structure.

hysteresis (magnetic). The lag of the magnetization of a substance behind any cyclic variation of the applied magnetizing field.

hysteresis (mechanical). The phenomenon of permanently absorbed or lost energy that occurs during any cycle of loading or unloading when a material is subjected to repeated loading.

I

IACS. International annealed copper standard; a standard reference used in reporting electrical conductivity. The conductivity of a material, in %IACS, is equal to 1724.1 divided by the electrical resistivity of the material in n2 m.

ideal critical diameter (D_I). Under an ideal quench condition, the bar diameter that has 50% martensite at the center of the bar when the surface is cooled at an infinitely rapid rate (that is, when $H = \infty$, where H is the quench severity factor or *Grossmann number*).

idiomorphic crystal. An individual crystal that has grown without restraint so that the habit planes are clearly developed. Compare with *allotriomorphic crystal*.

immersed-electrode furnace. A furnace used for liquid carburizing of parts by heating molten salt baths with the use of electrodes immersed in the liquid. See also *submerged-electrode furnace*.

immersion cleaning. Cleaning in which the work is immersed in a liquid solution.

immersion coating. A coating produced in a solution by chemical or electrochemical action without the use of external current.

immersion plating. Depositing a metallic coating on a metal immersed in a liquid solution, without the aid of an external electric current. Also called dip plating.

impact energy. The amount of energy, usually given in joules or foot-pound force, required to fracture a material, usually measured by means of an *Izod test* or *Charpy test*. The type of specimen and test conditions affect the values and therefore should be specified.

impact extrusion. The process (or resultant product) in which a punch strikes a slug (usually unheated) in a confining die. The metal flow may be either between punch and die or through another opening. The impact extrusion of unheated slugs is often called cold extrusion.

impact line. A blemish on a drawn sheet metal part caused by a slight change in metal thickness. The mark is called an impact line when it results from the impact of the punch on the blank; it is called a recoil line when it results from transfer of the blank from the die to the punch during forming, or from a reaction to the blank being pulled sharply through the *draw ring*.

impact load. An especially severe shock load such as that caused by instantaneous arrest of a falling mass, by shock meeting of two parts (in a mechanical hammer, for example), or by explosive impact, in which there can be an exceptionally rapid buildup of stress.

impact strength. A measure of the resiliency or toughness of a solid. The maximum force or energy of a blow (given by a fixed procedure) which can be withstood without fracture, as opposed to fracture strength under a steady applied force.

impact test. A test for determining the energy absorbed in fracturing a test piece at high velocity, as distinct from static test. The test may be carried out in tension, bending, or torsion, and the test bar may be notched or unnotched. See also *Charpy test, impact energy*, and *Izod test*.

impact wear. Wear of a solid surface resulting from repeated collisions between that surface and another solid body. The term *erosion* is preferred in the case of multiple impacts and when the impacting body or bodies are very small relative to the surface being impacted.

impingement. A process resulting in a continuing succession of impacts between liquid or solid particles and a solid surface.

impingement attack. *Corrosion* associated with turbulent flow of liquid. May be accelerated by entrained gas bubbles. See also *erosion-corrosion* and *impingement corrosion*.

impingement corrosion. A form of *erosion-corrosion* generally associated with the local impingement of a high-velocity, flowing fluid against a solid surface.

impingement erosion. Loss of material from a solid surface due to liquid impingement. See also *erosion*.

impregnation. (1) Treatment of porous castings with a sealing medium to stop pressure leaks. (2) The process of filling the pores of a sintered compact, usually with a liquid such as a lubricant. (3) The process of mixing particles of a nonmetallic substance in a cemented carbide matrix, as in diamond-impregnated tools.

impression-die forging. A forging that is formed to the required shape and size by machined impressions in specially prepared dies that exert three-dimensional control on the workpiece.

impurities. (1) Elements or compounds whose presence in a material is undesirable. (2) In a chemical or material, minor constituent(s) or component(s) not included deliberately; usually to some degree or above some level, undesirable.

inclinable press. A press that can be inclined to facilitate handling of the formed parts. See also *open-back inclinable press*.

inclusion. (1) A physical and mechanical discontinuity occurring within a material or part, usually consisting of solid, encapsulated foreign material. Inclusions are often capable of transmitting some structural stresses and energy fields, but to a noticeably different degree than from the parent material. (2) Particles of foreign material in a metallic matrix. The particles are usually compounds, such as oxides, sulfides, or silicates, but may be of any substance that is foreign to (and essentially insoluble in) the matrix. See also *exogenous inclusion, indigenous inclusion*, and *stringer*.

inclusion count. Determination of the number, kind, size, and distribution of nonmetallic inclusions in metals.

incomplete fusion. In welding, fusion which is less than complete.

indentation hardness. (1) The resistance of a material to indentation. This is the usual type of hardness test, in which a pointed or rounded indenter is pressed into a surface under a substantially static load. (2) Resistance of a solid surface to the penetration of a second, usually harder, body under prescribed conditions. Numerical values used to express indentation hardness are not absolute physical quantities, but depend on the hardness scale used to express hardness. See also *Brinell hardness test, Knoop hardness* *test*, *nanohardness test*, *Rockwell hardness test*, and *Vickers hardness test*.

indenter. In hardness testing, a solid body of prescribed geometry, usually chosen for its high hardness, that is used to determine the resistance of a solid surface to penetration.

indigenous inclusion. An *inclusion* that is native, innate, or inherent in the molten metal treatment. Indigenous inclusions include sulfides, nitrides, and oxides derived from the chemical reaction of the molten metal with the local environment. Such inclusions are small and require microscopic magnification for identification. Compare with *exogenous inclusion*.

indirect-arc furnace. An electric-arc furnace in which the metallic charge is not one of the poles of the arc.

indirect (backward) extrusion. See *extrusion*.

induction brazing. A brazing process in which the surfaces of components to be joined are selectively heated to brazing temperature by electrical energy transmitted to the workpiece by induction, rather than by a direct electrical connection, using an inductor or work coil.

induction furnace. An alternating current electric furnace in which the primary conductor is coiled and generates, by electromagnetic induction, a secondary current that develops heat within the metal charge. There are two classifications of induction furnaces: coreless and channel. In a coreless furnace, the refractory-lined crucible is completely surrounded by a water-cooled copper coil, while in the channel furnace the coil surrounds only a small appendage of the unit, called an inductor. The term "channel" refers to the channel that the molten metal forms as a loop within the inductor. It is this metal loop that forms the secondary of the electrical circuit, with the surrounding copper coil being the primary. In a coreless furnace, the entire metal content of the crucible is the secondary. See also *coreless induction furnace*.

induction hardening. A surface-hardening process in which only the surface layer of a suitable ferrous workpiece is heated by electromagnetic induction to above the upper critical temperature and immediately quenched.

induction heating. Heating by combined electrical resistance and hysteresis losses induced by subjecting a metal to the varying magnetic field surrounding a coil carrying alternating current.

induction melting. Melting in an *induction furnace*.

induction tempering. Tempering of steel using low-frequency electrical *induction heating*.

induction welding. Welding in which the required heat is generated by subjecting the workpiece to electromagnetic induction.

induction work coil. The *inductor* used when induction heating and melting as well as induction welding, brazing, and soldering.

inductor. A device consisting of one or more associated windings, with or without a magnetic core, for introducing inductance into an electric circuit.

industrial atmosphere. An atmosphere in an area of heavy industry with soot, fly ash, and sulfur compounds as the principal constituents.

inert anode. An *anode* that is insoluble in the *electrolyte* under the conditions prevailing in the *electrolysis*.

inert gas. (1) A gas, such as helium, argon, or nitrogen, which is stable, does not support combustion, and does not form reaction products with other materials. (2) In welding, a gas which does not normally combine chemically with the base metal or filler metal.

infiltration. The process of filling the pores of a sintered or unsintered compact with a metal or alloy of lower melting temperature.

ingate. Same as *gate*.

ingot. A casting of simple shape, suitable for hot working or remelting.

ingot iron. Commercially pure iron.

inhibitor. A substance that retards some specific chemical reaction. Pickling inhibitors retard the dissolution of metal without hindering the removal of scale from steel.

inoculant. Materials that, when added to molten metal, modify the structure and thus change the physical and mechanical properties to a degree not explained on the basis of the change in composition resulting from their use. Ferrosilicon-base alloys are commonly used to inoculate gray irons and ductile irons.

inoculation. The addition of a material to molten metal to form nuclei for crystallization. See also *inoculant*.

insert. (1) A part formed from a second material, usually a metal, which is placed in the molds and appears as an integral structural part of the final casting. (2) A removable portion of a die or mold.

insert die. A relatively small die which contains part or all of the impression of a forging, and which is fastened to a master die block.

inserted-blade cutters. Cutters having replaceable blades that are either solid or tipped and are usually adjustable.

instrumented impact test. An impact test in which the load on the specimen is continually recorded as a function of time and/or specimen deflection prior to fracture.

intense quenching. Quenching in which the quenching medium is cooling the part at a rate at least two and a half times faster than still water. See also *Grossmann number*.

intercept method. A quantitative metallographic technique in which the desired quantity, such as grain size or inclusion content, is expressed as the number of times per unit length a straight line on a metallographic image crosses particles of the feature being measured.

interconnected porosity. A network of connecting pores in a sintered object that permits a fluid or gas to pass through the object. Also referred to as interlocking or open porosity.

intercritical annealing. Any annealing treatment that involves heating to, and holding at, a temperature between the upper and lower critical temperatures to obtain partial austenitization, followed by either slow cooling or holding at a temperature below the lower critical temperature.

intercrystalline. Between the crystals, or grains, of a polycrystalline material.

intercrystalline corrosion. See *intergranular corrosion*.

intercrystalline cracking. See *intergranular cracking*.

interdendritic corrosion. Corrosive attack that progresses preferentially along interdendritic paths. This type of attack results from local differences in composition, such as coring commonly encountered in alloy castings.

interdendritic porosity. Voids occurring between the dendrites in cast metal.

interface. The boundary between any two phases. Among the three phases (gas, liquid, and solid), there are five types of interfaces: gas-liquid, gas-solid, liquid-liquid, liquid-solid, and solid-solid.

interfacial tension. The contractile force of an interface between two phases.

intergranular. Between crystals or grains. Also called intercrystalline. Contrast with *transgranular*.

intergranular corrosion. Corrosion occurring preferentially at grain boundaries, usually with slight or negligible attack on the adjacent grains. See also *interdendritic corrosion*.

intergranular cracking. Cracking or fracturing that occurs between the grains or crystals in a polycrystalline aggregate. Also called *intercrystalline cracking.* Contrast with *transgranular cracking.*

intergranular fracture. *Brittle fracture* of a polycrystalline material in which the fracture is between the grains, or crystals, that form the material. Also called *intercrystalline fracture.* Contrast with *transgranular fracture.*

intergranular penetration. In welding, the penetration of a filler metal along the grain boundaries of a base metal.

intergranular stress-corrosion cracking (IGSCC). *Stress-corrosion cracking* in which the cracking occurs along grain boundaries.

intermediate annealing. Annealing wrought metals at one or more stages during manufacture and before final treatment.

intermediate electrode. Same as *bipolar electrode.*

intermediate phase. In an alloy or a chemical system, a distinguishable homogeneous phase whose composition range does not extend to any of the pure components of the system.

intermetallic compound. An intermediate phase in an alloy system, having a narrow range of homogeneity and relatively simple stoichiometric proportions; the nature of the atomic binding can be of various types, ranging from metallic to ionic.

intermetallic phases. Compounds, or intermediate solid solutions, containing two or more metals, which usually have compositions, characteristic properties, and crystal structures different from those of the pure components of the system.

intermittent weld. A weld in which the continuity is broken by recurring unwelded spaces.

internal friction. The conversion of energy into heat by a material subjected to fluctuating stress.

internal grinding. Grinding an internal surface such as that inside a cylinder or hole.

internal oxidation. The formation of isolated particles of corrosion products beneath the metal surface. This occurs as the result of preferential oxidation of certain alloy constituents by inward diffusion of oxygen, nitrogen, sulfur, and so forth. Also called subscale formation.

internal shrinkage. A void or network of voids within a casting caused by inadequate feeding of that section during solidification.

internal stress. See preferred term *residual stress.*

interpass temperature. In a multiple-pass weld, the temperature (minimum or maximum as specified) of the deposited weld metal before the next pass is started.

interrupted aging. Aging at two or more temperatures, by steps, and cooling to room temperature after each step. See also *aging,* and compare with *progressive aging* and *step aging.*

interrupted-current plating. Plating in which the flow of current is discontinued for periodic short intervals to decrease anode polarization and elevate the *critical current density.* It is most commonly used in cyanide copper plating.

interrupted quenching. A quenching procedure in which the workpiece is removed from the first quench at a temperature substantially higher than that of the quenchant and is then subjected to a second quenching system having a different cooling rate than the first.

interstitial solid solution. A type of solid solution that sometimes forms in alloy systems having two elements of widely different atomic sizes. Elements of small atomic size, such as carbon, hydrogen, and nitrogen, often dissolve in solid metals to form this solid solution. The space lattice is similar to that of the pure metal, and the atoms of carbon, hydrogen, and nitrogen occupy the spaces or interstices between the metal atoms.

intracrystalline. Within or across the crystals or grains of a metal; same as transcrystalline and transgranular.

intracrystalline cracking. See *transgranular cracking.*

inverse chill. The condition in a casting section in which the interior is mottled or white, while the other sections are gray iron. Also known as reverse chill, internal chill, and inverted chill.

inverse segregation. A concentration of low-melting constituents in those regions of an alloy in which solidification first occurs.

investing. In *investment casting,* the process of pouring the investment slurry into a flask surrounding the pattern to form the mold.

investment. A flowable mixture, or slurry, of a graded refractory filler, a binder, and a liquid vehicle that, when poured around the patterns, conforms to their shape and subsequently sets hard to form the investment mold.

investment casting. (1) Casting metal into a mold produced by surrounding, or *investing,* an expendable pattern with a refractory slurry coating that sets at room temperature, after which the wax or plastic pattern is removed through the use of heat prior to filling the mold with liquid metal. Also called *precision casting* or *lost wax process.* (2) A part made by the investment casting process.

investment compound. A mixture of a graded refractory filler, a binder and a liquid vehicle, used to make molds for *investment casting.*

investment precoat. An extremely fine investment coating applied as a thin slurry directly to the surface of the pattern to reproduce maximum surface smoothness. The coating is surrounded by a coarser, cheaper, and more permeable investment to form the mold. See also *dip coat* and *investment casting.*

investment shell. Ceramic mold obtained by alternately dipping a pattern set up in dip coat slurry and stuccoing with coarse ceramic particles until the shell of desired thickness is obtained. See also *investment casting.*

ion. An atom, or group of atoms, which by loss or gain of one or more electrons has acquired an electric charge. If the ion is formed from an atom of hydrogen or an atom of a metal, it is usually positively charged; if the ion is formed from an atom of a nonmetal or from a group of atoms, it is usually negatively charged. The number of electronic charges carried by an ion is termed its electrovalence. The charges are denoted by superscripts that give their sign and number; for example, a sodium ion, which carries one positive charge, is denoted by Na^+; a sulfate ion, which carries two negative charges, by SO.

ion carburizing. A method of surface hardening in which carbon ions are diffused into a workpiece in a vacuum through the use of high-voltage electrical energy. Synonymous with plasma carburizing or glow-discharge carburizing.

ion exchange. The reversible interchange of ions between a liquid and solid, with no substantial structural changes in the solid.

ion implantation. The process of modifying the physical or chemical properties of the near surface of a solid (target) by embedding appropriate atoms into it from a beam of ionized particles.

ion nitriding. A method of surface hardening in which nitrogen ions are diffused into a workpiece in a vacuum through the use of high-voltage electrical energy. Synonymous with plasma nitriding or glow-discharge nitriding.

ion plating. A generic term applied to atomistic film deposition processes in which the substrate surface and/or the depositing film is subjected to a flux of high-energy particles (usually gas ions) sufficient to cause changes in the interfacial region or film properties.

iron casting. A part made of *cast iron*.

ironing. An operation used to increase the length of a tube or cup through reduction of wall thickness and outside diameter, the inner diameter remaining unchanged.

iron-powder electrode. A welding electrode with a covering containing up to about 50% iron powder, some of which becomes part of the deposit.

iron rot. Deterioration of wood in contact with iron-base alloys.

irradiation. The exposure of a material or object to x-rays, gamma rays, ultraviolet rays, or other ionizing radiation.

isocorrosion diagram. A graph or chart that shows constant corrosion behavior with changing solution (environment) composition and temperature.

isomorphous. Having the same crystal structure. This usually refers to intermediate phases that form a continuous series of solid solutions.

isostatic pressing. A process for forming a powder metallurgy compact by applying pressure equally from all directions to metal powder contained in a sealed flexible mold. See also *cold isostatic pressing* and *hot isostatic pressing*.

isothermal annealing. Austenitizing a ferrous alloy, then cooling to and holding at a temperature at which austenite transforms to a relatively soft ferrite-carbide aggregate. See also *austenitizing*.

isothermal forging. A hot-forging process in which a constant and uniform temperature is maintained in the workpiece during forging by heating the dies to the same temperature as the workpiece. The process permits the use of extremely slow strain rates, thus taking advantage of the strain rate sensitivity of flow stress for certain alloys (for example, titanium- and nickel-base alloys). The process is capable of producing net shape forgings that are ready to use without machining or near-net shape forgings that require minimal secondary machining.

isothermal transformation. A change in phase that takes place at a constant temperature. The time required for transformation to be completed, and in some instances the time delay before transformation begins, depends on the amount of supercooling below (or superheating above) the equilibrium temperature for the same transformation.

isothermal transformation (IT) diagram. A diagram that shows the isothermal time required for transformation of austenite to begin and to finish as a function of temperature. Same as time-temperature-transformation (TTT) diagram or S-curve.

isotropic. Having uniform properties in all directions. The measured properties of an isotropic material are independent of the axis of testing.

isotropy. The condition of having the same values of properties in all directions.

Izod test. A type of impact test in which a V-notched specimen, mounted vertically, is subjected to a sudden blow delivered by the weight at the end of a pendulum arm. The energy required to break off the free end is a measure of the impact strength or toughness of the material. Contrast with *Charpy test*.

J

jaw crusher. A machine for the primary disintegration of metal pieces, ores, or agglomerates into coarse powder. See also *crushing*.

jig. A mechanism for holding a part and guiding the tool during machining or assembly operation.

jig boring. Boring with a single-point tool where the work is positioned upon a table that can be located so as to bring any desired part of the work under the tool. Thus, holes can be accurately spaced. This type of boring can be done on milling machines or jig borers.

J-integral. A mathematical expression; a line or surface integral that encloses the crack front from one crack surface to the other, used to characterize the *fracture toughness* of a material having appreciable plasticity before fracture. The J-integral eliminates the need to describe the behavior of the material near the crack tip by considering the local stress-strain field around the crack front; J_{Ic} is the critical value of the J-integral required to initiate growth of a preexisting crack.

joint. The location where two or more members are to be or have been fastened together mechanically or by welding, brazing, soldering, or adhesive bonding.

joint clearance. The distance between the faying surfaces of a joint. In brazing, this distance is referred to as that which is present before brazing, at the brazing temperature, or after brazing is completed.

joint efficiency. The strength of a welded joint expressed as a percentage of the strength of the unwelded base metal.

joint penetration. The minimum depth to which a groove or flange weld extends from its face into the joint, exclusive of reinforcement. Joint penetration may include *root penetration*.

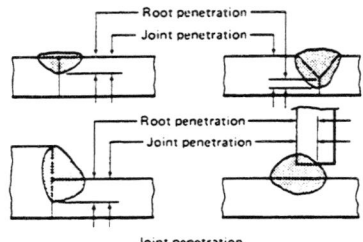

jolt ramming. Packing sand in a mold by raising and dropping the sand, pattern, and flask on a table. Jolt-type, jolt squeezers, jarring machines, and jolt rammers are machines using this principle. Also called jar ramming.

Jominy test. See *end-quench hardenability test*.

K

karat. A unit for designating the fineness of gold in an alloy. In this system, 24 karat (24 k) is 1000 fine or pure gold. The most popular jewelry golds are:

Karat designation	Gold content
24k	100% Au (99.5% min)
18k	18/24ths, or 75% Au
14k	14/24ths, or 58.33% Au
10k	10/24ths, or 41.67% Au

keel block. A standard test casting, for steel and other high-shrinkage alloys, consisting of a rectangular bar that resembles the keel of a boat, attached to the bottom of a large riser, or shrinkhead. Keel blocks that have only one bar are often called Y-blocks; keel blocks having two bars, double keel blocks. Test specimens are machined from the rectangular bar, and the shrinkhead is discarded.

kerf. The width of the cut produced during a cutting process.

keyhole. A technique of welding in which a concentrated heat source, such as a plasma arc, penetrates completely through a workpiece forming a hole at the leading edge of the molten weld metal. As the heat source progresses, the molten metal fills in behind the hold to form the weld bead.

keyhole specimen. A type of specimen containing a hole-and-slot notch, shaped like a keyhole, usually used in impact bend tests. See also *Charpy test* and *Izod test*.

killed steel. Steel treated with a strong deoxidizing agent such as silicon or aluminum in order to reduce the oxygen content to such a level that no reaction occurs between carbon and oxygen during solidification.

kiln. A large furnace used for baking, drying, or burning firebrick or refractories, or for calcining ores or other substances.

K_{ISCC}. Abbreviation for the critical value of the plane strain *stress-intensity factor* that will produce crack propagation by *stress-corrosion cracking* of a given material in a given environment.

kish. Free graphite that forms in molten hypereutectic cast iron as it cools. In castings, the kish may segregate toward the cope surface, where it lodges at or immediately beneath the casting surface.

knife-line attack. *Intergranular corrosion* of an alloy, usually stabilized stainless steel, along a line adjoining or in contact with a weld after heating into the sensitization temperature range.

knockout. (1) Removal of sand cores from a casting. (2) Jarring of an investment casting mold to remove the casting and investment from the flask. (3) A mechanism for freeing formed parts from a die used for stamping, blanking, drawing, forging or heading operations. (4) A partially pierced hole in a sheet metal part, where the slug remains in the hole and can be forced out by hand if a hole is needed.

Knoop hardness number (HK). A number related to the applied load and to the projected area of the permanent impression made by a rhombic-based pyramidal diamond indenter having included edge angles of 172° 30′ and 130° 0′ computed from the equation:

$$HK = \frac{P}{0.07028d^2}$$

where P is applied load, kgf; and d is the long diagonal of the impression, mm. In reporting Knoop hardness numbers, the test load is stated.

Knoop hardness test. An indentation hardness test using calibrated machines to force a rhombic-based pyramidal diamond indenter having specified edge angles, under specified conditions, into the surface of the material under test and to measure the long diagonal after removal of the load.

knuckle-lever press. A heavy short-stroke press in which the slide is directly actuated by a single toggle joint that is opened and closed by a connection and crack. It is used for embossing, coining, sizing, heading, swaging, and extruding.

knurling. Impressing a design into a metallic surface, usually by means of small, hard rollers that carry the corresponding design on their surfaces.

Kroll process. A process for the production of metallic titanium sponge by the reduction of titanium tetrachloride with a more active metal, such as magnesium or sodium. The sponge is further processed to granules or powder.

L

lack of fusion (LOF). A condition in a welded joint in which fusion is less than complete.

lack of penetration (LOP). A condition in a welded joint in which joint penetration is less than that specified.

ladle. Metal receptacle frequently lined with refractories used for transporting and pouring molten metal.

ladle metallurgy. Degassing processes for steel carried out in a *ladle*.

lamellar tearing. Occurs in the base metal adjacent to weldments due to high through-thickness strains introduced by weld metal shrinkage in highly restrained joints. Tearing occurs by decohesion and linking along the working direction of the base metal; cracks usually run roughly parallel to the fusion line and are steplike in appearance. See also *cold cracking*, *hot cracking*, and *stress-relief cracking*.

lamination. (1) A type of discontinuity with separation or weakness generally aligned parallel to the worked surface of a metal. May be the result of pipe, blisters, seams, inclusions, or segregation elongated and made directional by working. Laminations may also occur in powder metallurgy compacts. (2) In electrical products such as motors, a blanked piece of electrical sheet that is stacked up with several other identical pieces to make a stator or rotor.

lancing. (1) A press operation in which a single-line cut is made in strip stock without producing a detached slug. Chiefly used to free metal for forming, or to cut partial contours for blanked parts, particularly in progressive dies. (2) A misnomer for *oxyfuel gas cutting*.

land. (1) For profile-sharpened milling cutters, the relieved portion immediately behind the cutting edge. (2) For reamers, drills, and taps, the solid section between the flutes. (3) On punches, the portion adjacent to the nose that is parallel to the axis and of maximum diameter.

lap. A surface imperfection, with the appearance of a seam, caused by hot metal, fins, or sharp corners being folded over and then being rolled or forged into the surface but without being welded.

lap joint. A joint made between two overlapping members.

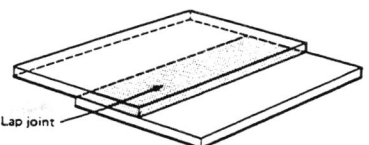

Lap joint

lapping. A finishing operation using fine abrasive grits loaded into a lapping material such as cast iron. Lapping provides major refinements in the workpiece including extreme accuracy of dimension, correction of minor imperfections of shape, refinement of surface finish, and close fit between mating surfaces.

laser. A device that emits a concentrated beam of electromagnetic radiation (light). Laser beams are used in metalworking to melt, cut, or weld metals; in less concentrated form they are sometimes used to inspect metal parts.

laser alloying. See *laser surface processing*.

laser beam cutting. A cutting process which severs materials with the heat obtained from the application of a concentrated coherent light beam impinging upon the workpiece to be cut. The process can be used with (gas-assisted laser beam cutting) or without an externally supplied gas.

laser beam machining. Use of a highly focused monofrequency collimated beam of light to melt or sublime material at the point of impingement on a workpiece.

laser beam welding. A welding process that joins metal parts using the heat obtained by directing a beam from a *laser* onto the weld joint.

laser hardening. A surface-hardening process which uses a laser to quickly heat a surface. Heat conduction into the interior of the part will quickly cool the surface, leaving a shallow martensitic layer.

laser surface processing. The use of lasers with continuous outputs of 0.5 to 10 kW to modify the metallurgical structure of a surface and to tailor the surface properties without adversely affecting the bulk properties. The surface modification can take the following three forms. The first is transformation hardening in which a surface is heated so that thermal diffusion and solid-state transformations can take place. The second is surface melting, which results in a refinement of the structure due to the rapid quenching from the melt. The third is surface (laser) alloying, in which alloying elements are added to the melt pool to change the composition of the surface. The novel structures produced by laser surface melting and alloying can exhibit improved electrochemical and tribological behavior.

latent heat. Thermal energy absorbed or released when a substance undergoes a phase change.

lateral extrusion. An operation in which the product is extruded sideways through an orifice in the container wall.

lath martensite. Martensite formed partly in steels containing less than approximately 1.0% C and solely in steels containing less than approximately 0.5% C as parallel arrays of packets of lath-shape units 0.1 to 0.3 m thick.

lattice constants. See *lattice parameter*.

lattice parameter. The length of any side of a unit cell of a given crystal structure. The term is also used for the fractional coordinates x, y, and z of lattice points when these are variable.

launder. (1) A channel for transporting molten metal. (2) A box conduit conveying particles suspended in water.

lay. Direction of predominant surface pattern remaining after cutting, grinding, lapping, or other processing.

lead. (1) The axial advance of a helix in one complete turn. (2) The slight bevel at the outer end of a face cutting edge of a face mill.

lead angle. In cutting tools, the helix angle of the flutes.

leak testing. A nondestructive test for determining the escape or entry of liquids or gases from pressurized or into evacuated components or systems intended to hold these liquids. Leak testing systems, which employ a variety of gas detectors, are used for locating (detecting and pinpointing) leaks, determining the rate of leakage from one leak or from a system, or monitoring for leakage.

ledeburite. The eutectic of the iron-carbon system, the constituents of which are *austenite* and *cementite*. The austenite decomposes into *ferrite* and cementite on cooling below Ar₁, the temperature at which transformation of austenite to ferrite or ferrite plus cementite is completed during cooling.

left-hand cutting tool. A cutter all of whose flutes twist away in a counterclockwise direction when viewed from either end.

leg of fillet weld. (1) Actual: The distance from the root of the joint to the toe of the

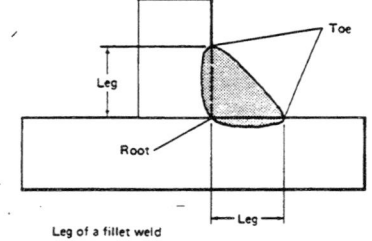

Leg of a fillet weld

fillet weld. (2) Nominal: The length of a side of the largest right triangle that can be inscribed in the cross section of the weld. See also the figures accompanying *concave fillet weld* and *convex fillet weld*.

Leidenfrost phenomenon. Slow cooling rates associated with a hot vapor blanket that surrounds a part being quenched in a liquid medium such as water. The gaseous vapor envelope acts as an insulator, thus slowing the cooling rate.

leveler lines. Lines on sheet or strip running transverse to the direction of *roller leveling*. These lines may be seen upon stoning or light sanding after leveling (but before drawing) and can usually be removed by moderate stretching.

leveling. Flattening of rolled sheet, strip, or plate by reducing or eliminating distortions. See also *stretcher leveling* and *roller leveling*.

levigation. (1) Separation of fine powder from coarser material by forming a suspension of the fine material in a liquid. (2) A means of classifying a material as to particle size by the rate of settling from a suspension.

levitation melting. An *induction melting* process in which the metal being melted is suspended by the electromagnetic field and is not in contact with a container.

light drawn. An imprecise term, applied to drawn products such as wire and tubing, that indicates a lesser amount of cold reduction than for *hard drawn* products.

lightly coated electrode. A filler-metal electrode used in arc welding, consisting of a metal wire with a light coating, usually of metal oxides and silicates, applied subsequent to the drawing operation primarily for stabilizing the arc. Contrast with *covered electrode*.

light metal. One of the low-density metals, such as aluminum, magnesium, titanium, beryllium, or their alloys.

limiting current density. The maximum current density that can be used to obtain a desired electrode reaction without undue interference such as from *polarization*.

limiting dome height (LDH) test. A mechanical test, usually performed unlubricated on sheet metal, that simulates the fracture conditions in a practical press-forming operation.

lineage structure. (1) Deviations from perfect alignment of parallel arms of a columnar dendrite as a result of interdendritic shrinkage during solidification from a liquid. This type of deviation may vary in orientation from a few minutes to as much as two degrees of arc.

(2) A type of substructure consisting of elongated subgrains.

linear elastic fracture mechanics. A method of fracture analysis that can determine the stress (or load) required to induce fracture instability in a structure containing a cracklike flaw of known size and shape. See also *fracture mechanics* and *stress-intensity factor*.

linear (tensile or compressive) strain. The change per unit length due to force in an original linear dimension. An increase in length is considered positive.

liner. (1) The slab of coating metal that is placed on the core alloy and is subsequently rolled down to clad sheet as a composite. (2) In extrusion, a removable alloy steel cylindrical chamber, having an outside longitudinal taper firmly positioned in the container or main body of the press, into which the billet is placed for extrusion.

line reaming. Simultaneous *reaming* of coaxial holes in various sections of a workpiece with a reamer having cutting faces or piloted surfaces with the desired alignment.

lip-pour ladle. Ladle in which the molten metal is poured over a lip, much as water is poured out of a bucket.

liquation. (1) The separation of a low melting constituent of an alloy from the remaining constituents, usually apparent in alloys having a wide melting range. (2) Partial melting of an alloy, usually as a result of *coring* or other compositional heterogeneities.

liquation temperature. The lowest temperature at which partial melting can occur in an alloy that exhibits the greatest possible degree of segregation.

liquid carburizing. Surface hardening of steel by immersion into a molten bath consisting of cyanides and other salts.

liquid honing. Producing a finely polished finish by directing an air-ejected chemical emulsion containing fine abrasives against the surface to be finished.

liquid metal embrittlement (LME). Catastrophic brittle failure of a normally ductile metal when in contact with a liquid metal and subsequently stressed in tension. See also *solid metal embrittlement*.

liquid nitriding. A method of surface hardening in which molten nitrogen-bearing, fused-salt baths containing both cyanides and cyanates are exposed to parts at subcritical temperatures.

liquid nitrocarburizing. A nitrocarburizing process (where both carbon and nitrogen are absorbed into the surface) utilizing molten liquid salt baths below the lower critical temperature.

liquid penetrant inspection. A type of nondestructive inspection that locates discontinuities that are open to the surface of a metal by first allowing a penetrating dye or fluorescent liquid to infiltrate the discontinuity, removing the excess penetrant, and then applying a developing agent that causes the penetrant to seep back out of the discontinuity and register as an indication. Liquid penetrant inspection is suitable for both ferrous and nonferrous materials, but is limited to the detection of open surface discontinuities in nonporous solids.

liquid phase sintering. Sintering of a compact or loose powder aggregate under conditions where a liquid phase is present during part of the sintering cycle.

liquid shrinkage. The reduction in volume of liquid metal as it cools to the liquidus.

liquidus. (1) The lowest temperature at which a metal or an alloy is completely liquid. (2) In a *phase diagram*, the locus of points representing the temperatures at which the various compositions in the system begin to freeze on cooling or finish melting on heating. See also *solidus*.

loading. (1) In cutting, building up of a cutting tool back of the cutting edge by undesired adherence of material removed from the work. (2) In grinding, filling the pores of a grinding wheel with material from the work, usually resulting in a decrease in production and quality of finish. (3) In powder metallurgy, filling of the die cavity with powder.

loam. A molding material consisting of sand, silt, and clay, used over brickwork or other structural backup material for making massive castings, usually of iron or steel.

local action. Corrosion due to the action of "local cells," that is, galvanic cells resulting from inhomogeneities between adjacent areas on a metal surface exposed to an *electrolyte*.

local cell. A *galvanic cell* resulting from inhomogeneities between areas on a metal surface in an *electrolyte*. The inhomogeneities may be of physical or chemical nature in either the metal or its environment.

local current density. Current density at a point or on a small area.

localized corrosion. Corrosion at discrete sites, for example, *crevice corrosion*, *pitting*, and *stress-corrosion cracking*.

localized precipitation. Precipitation from a supersaturated solid solution similar to *continuous precipitation*, except that the precipitate particles form at preferred locations, such as along slip planes, grain boundaries, or incoherent twin boundaries.

lock. In forging, a condition in which the flash line is not entirely in one plane. Where two or more plane changes occur, it is called compound lock. Where a lock is placed in the die to compensate for die shift caused by a steep lock, it is called a counterlock.

longitudinal direction. That direction parallel to the direction of maximum elongation in a worked material. See also *normal direction* and *transverse direction*.

longitudinal field. A magnetic field that extends within a magnetized part from one or more poles to one or more other poles and that is completed through a path external to the part.

longitudinal resistance seam welding. The making of a resistance seam weld in a direction essentially parallel to the throat depth of a resistance welding machine.

longitudinal sequence. The order in which the increments of a continuous weld are deposited with respect to its length. See also *backstep sequence* and *block sequence*.

looping mill. An arrangement of hot rolling stands such that a hot bar, while being discharged from one stand, is fed into a second stand in the opposite direction.

loose metal. Refers to an area in a formed panel that is not stiff enough to hold its shape, may be confused with *oil canning*.

lost foam casting. An *expendable pattern* process in which an expandable polystyrene pattern surrounded by the unbonded sand, is vaporized during pouring of the molten metal.

lost wax process. An *investment casting* process in which a wax pattern is used.

lot. (1) A specific amount of material produced at one time using one process and constant conditions of manufacture, and offered for sale as a unit quantity. (2) A quantity of material that is thought to be uniform in one or more stated properties such as isotopic, chemical, or physical characteristics. (3) A quantity of bulk material of similar composition whose properties are under study. Compare with *batch*.

low-alloy steels. A category of ferrous materials that exhibit mechanical properties superior to plain carbon steels as the result of additions of such alloying elements as nickel, chromium, and molybdenum. Total alloy content can range from 2.07% up to levels just below that of stainless steels, which contain a minimum of 10% Cr.

low-cycle fatigue. *Fatigue* that occurs at relatively small numbers of cycles (^{4} cycles). Low-cycle fatigue may be accompanied by some plastic, or permanent, deformation. Compare with *high-cycle fatigue*.

lower ram. The part of a pneumatic or hydraulic press that is moving in a lower cylinder and transmits pressure to the lower punch.

low-hydrogen electrode. A covered arc welding electrode that provides an atmosphere around the arc and molten weld metal that is low in hydrogen.

low-residual-phosphorus copper. Deoxidized copper with residual phosphorus present in amounts (usually 0.004 to 0.012%) generally too small to decrease appreciably the electrical conductivity of the copper.

low shaft furnace. A short shaft-type blast furnace used to produce pig iron and ferroalloys from low-grade ores, using low-grade fuel. The air blast is often enriched with oxygen. Also used for making a variety of other products such as alumina, cementmaking slags, and ammonia synthesis gas.

lubricant. (1) Any substance interposed between two surfaces in relative motion for the purpose of reducing the friction or wear between them. (2) A material applied to dies, molds, plungers, or workpieces that promotes the flow of metal, reduces friction and wear, and aids in the release of the finished part.

lubrication. (1) The reduction of frictional resistance and wear, or other forms of surface deterioration, between two load-bearing surfaces by the application of a *lubricant*. (2) Mixing or incorporating a lubricant with a powder to facilitate compacting and ejecting of the compact from the die cavity; also, applying a lubricant to die walls and/or punch surfaces.

Lüders lines. Elongated surface markings or depressions in sheet metal, often visible with the unaided eye, caused by discontinuous (inhomogeneous) yielding. Also known as Lüders bands, Hartmann lines, Piobert lines, or stretcher strains.

luster finish. A bright as-rolled finish, produced on ground metal rolls; it is suitable for decorative painting or plating, but usually must undergo additional surface preparation after forming.

M

machinability. The relative ease of machining a metal.

machinability index. A relative measure of the machinability of an engineering material under specified standard conditions. Also known as machinability rating.

machine forging. Forging performed in upsetters or horizontal forging machines.

machining. Removing material from a metal part, usually using a cutting tool, and usually using a power-driven machine.

machining allowance. See *finish allowance*.

machining damage. Irregularities or changes on the surface of a material due to machining or grinding operations that may deleteriously affect the performance of the material/part.

machining stress. *Residual stress* caused by machining.

macrograph. A graphic representation of the surface of a prepared specimen at a magnification not exceeding 25×. When photographed, the reproduction is known as a photomacrograph.

macrohardness test. A term applied to such hardness testing procedures as the Rockwell or Brinell hardness tests to distinguish them from microindentation hardness tests such as the Knoop or Vickers tests. See also *microindentation* and *microindentation hardness number*.

macroscopic stress. Residual stress in a material in a distance comparable to the gage length of strain measurement devices (as opposed to stresses within very small, specific regions, such as individual grains). Compare with *microscopic stress*.

macroshrinkage. Isolated, clustered, or interconnected voids in a casting that are detectable macroscopically. Such voids are usually associated with abrupt changes in section size and are caused by feeding that is insufficient to compensate for solidification shrinkage.

macrostructure. The structure of metals as revealed by macroscopic examination of the etched surface of a polished specimen.

magnetically hard alloy. See *permanent magnet material*.

magnetically soft alloy. See *soft magnetic material*.

magnetic-analysis inspection. A nondestructive method of inspection to determine the existence of variations in magnetic flux in ferromagnetic materials of constant cross section, such as might be caused by discontinuities and variations in hardness. The variations are usually indicated by a change in pattern on an oscilloscope screen.

magnetic-particle inspection. A nondestructive method of inspection for determining the existence and extent of surface cracks and similar imperfections in ferromagnetic materials. Finely divided magnetic particles, applied to the magnetized part, are attracted to and outline the pattern of any magnetic-leakage fields created by discontinuities.

magnetic pole. The area on a magnetized part at which the magnetic field leaves or enters the part. It is a point of maximum attraction in a magnet.

magnetic separator. A device used to separate magnetic from less magnetic or nonmagnetic materials. The crushed material is conveyed on a belt past a magnet.

magnetizing force. A force field, resulting from the flow of electric currents or from magnetized bodies, that produces magnetic induction.

magnetostriction. Changes in dimensions of a body resulting from application of a magnetic field.

malleability. The characteristic of metals that permits *plastic deformation* in compression without fracture. See also *ductility*.

malleable iron. A cast iron made by prolonged annealing of *white iron* in which decarburization, graphitization, or both take place to eliminate some or all of the cementite. The graphite is in the form of temper carbon. If decarburization is the predominant reaction, the product will exhibit a light fracture surface; hence whiteheart malleable. Otherwise, the fracture surface will be dark; hence blackheart malleable. Only the blackheart malleable is produced in the United States. Ferritic malleable has a predominantly ferritic matrix; pearlitic malleable may contain pearlite, spheroidite, or tempered martensite, depending on heat treatment and desired hardness.

malleabilizing. Annealing *white iron* in such a way that some or all of the combined carbon is transformed into graphite or, in some cases, so that part of the carbon is removed completely.

mandrel. (1) A blunt-ended tool or rod used to retain the cavity in a hollow metal product during working. (2) A metal bar around which other metal may be cast, bent, formed, or shaped. (3) A shaft or bar for holding work to be machined. (4) A form, such as a mold or matrix, used as a cathode in electroforming.

mandrel forging. The process of rolling or forging a hollow blank over a mandrel to produce a weldless, seamless ring or tube. See also *radial forging*.

Mannesmann process. A process for piercing tube billets in making seamless tubing. The billet is rotated between two heavy rolls mounted at an angle and is forced over a fixed mandrel.

maraging. A precipitation-hardening treatment applied to a special group of iron-base alloys to precipitate one or more intermetallic compounds in a matrix of essentially carbon-free martensite. See also *maraging steels*.

maraging steels. A special class of high-strength steels that differ from conventional steels in that they are hardened by a metallurgical reaction that does not involve carbon. Instead, these steels are strengthened by the precipitation of intermetallic compounds at temperatures of about 480 °C (900 °F). The term maraging is derived from martensite age hardening of a low-carbon, iron-nickel lath martensite matrix.

Marforming process. A *rubber-pad forming* process developed to form wrinkle-free shrink flanges and deep-drawn shells. It differs from the *Guerin process* in that the sheet metal blank is clamped between the rubber pad and the blankholder before forming begins.

marquenching. See *martempering*.

martempering. (1) A hardening procedure in which an austenitized ferrous material is quenched into an appropriate medium at a temperature just above the martensite start temperature of the material, held in the medium until the temperature is uniform throughout, although not long enough for bainite to form, then cooled in air. The treatment is frequently followed by tempering. (2) When the process is applied to carburized material, the controlling martensite start temperature is that of the case. This variation of the process is frequently called marquenching.

martensite. A generic term for microstructures formed by diffusionless phase transformation in which the parent and product phases have a specific crystallographic relationship. Martensite is characterized by an acicular pattern in the microstructure in both ferrous and nonferrous alloys. In alloys where the solute atoms occupy interstitial positions in the martensitic lattice (such as carbon in iron), the structure is hard and highly strained; but where the solute atoms occupy substitutional positions (such as nickel in iron), the martensite is soft and ductile. The amount of high-temperature phase that transforms to martensite on cooling depends to a large extent on the lowest temperature attained, there being a rather distinct beginning temperature (Ms) and a temperature at which the transformation is essentially complete (M_f). See also *lath martensite*, *plate martensite*, and *tempered martensite*.

martensite range. The interval between the martensite start (M_s) and the martensite finish (M_f) temperatures.

martensitic. A platelike constituent having an appearance and a mechanism of formation similar to that of martensite. See also *lath martensite* and *plate martensite*.

martensitic transformation. A reaction that takes place in some metals on cooling, with the formation of an acicular structure called *martensite*.

mash resistance seam welding. *Resistance seam welding* in which the weld is made in a lap joint, the thickness at the lap being reduced plastically to approximately the thickness of one of the lapped parts.

master alloy. An alloy, rich in one or more desired addition elements, that is added to a metal melt to raise the percentage of a desired constituent.

master alloy powder. A prealloyed metal powder of high concentration of alloy content, designed to be diluted when mixed with a base powder to produce the desired composition. See also *prealloyed powder*.

master pattern. In foundry practice, a pattern embodying a double contraction allowance in its construction, used for making castings to be employed as patterns in production work.

match. A condition in which a point in one metal forming or forging die half is aligned properly with the corresponding point in the opposite die half within specified tolerance.

matched edges. Two edges of the die face that are machined exactly at 90° to each other, and from which all dimensions are taken in laying out the die impression and aligning the dies in the forging equipment. Also referred to as match lines.

match plate. A plate of metal or other material on which patterns for metal casting are mounted (or formed as an integral part) to facilitate molding. The pattern is divided along its parting plane by the plate.

materials characterization. The use of various analytical methods (spectroscopy, microscopy, chromatography, etc.) to describe those features of composition (both bulk and surface) and structure (including defects) of a material that are significant for a particular preparation, study of properties, or use. Test methods that yield information primarily related to materials properties, such as thermal, electrical, and mechanical properties, are excluded from this definition.

matrix. The continuous or principal phase in which another constituent is dispersed.

matte. An intermediate product of *smelting*; an impure metallic sulfide mixture made by melting a roasted sulfide ore, such as an ore of copper, lead, or nickel.

matte finish. (1) A dull texture produced by rolling sheet or strip between rolls that have been roughened by blasting. (2) A dull finish characteristic of some electrodeposits, such as cadmium or tin.

maximum stress (S_{max}). The stress having the highest algebraic value in the stress cycle, tensile stress being considered positive and compressive stress negative. The *nominal stress* is used most commonly.

maximum stress intensity factor (K_{max}). The maximum value of the *stress-intensity factor* in a fatigue cycle.

McQuaid-Ehn grain size. The austenitic grain size developed in steels by carburizing at 927 °C (1700 °F) followed by slow cooling. Eight standard McQuaid-Ehn grain sizes rate the structure, from No. 8, the finest, to No. 1, the coarsest. The use of standardized ASTM methods for determining grain size is recommended.

mean stress (S_m). The algebraic average of the maximum and minimum stresses in one cycle, that is, $S_m = (S_{max} + S_{min})/2$. Also referred to as steady component of stress.

mechanical alloying (MA). An alternate cold welding and shearing of particles of two or more species of greatly differing hardness. The operation is carried out in high-intensity ball mills, such as attritors, and is the preferred method of producing oxide-dispersion-strengthened (ODS) materials. See also *attritor grinding* and *dispersion-strengthened material*.

mechanical hysteresis. Energy absorbed in a complete cycle of loading and unloading within the elastic limit and represented by the closed loop of the stress-strain curves for loading and unloading. Sometimes referred to as elastic, but more properly, mechanical.

mechanical metallurgy. The science and technology dealing with the behavior of metals when subjected to applied forces; often considered to be restricted to plastic working or shaping of metals.

mechanical plating. Plating wherein fine metal powders are peened onto the work by *tumbling* or other means. The process is used primarily to provide ferrous parts with coatings of zinc, cadmium, tin, and alloys of these metals in various combinations.

mechanical polishing. A process that yields a specularly reflecting surface entirely by the action of machining tools, which are usually the points of abrasive particles suspended in a liquid among the fibers of a polishing cloth.

mechanical press. A press whose slide is operated by a crank, eccentric, cam, toggle links, or other mechanical device.

mechanical properties. The properties of a material that reveal its elastic and inelastic behavior when force is applied, thereby indicating its suitability for mechanical applications; for example, modulus of elasticity, tensile strength, elongation, hardness, and fatigue limit. Compare with *physical properties*.

mechanical testing. The methods by which the *mechanical properties* of a metal are determined.

mechanical twin. A *twin* formed in a crystal by simple shear under external heating.

mechanical working. The subjecting of metals to pressure exerted by rolls, hammers, or presses in order to change the shape or physical properties of the metal.

median fatigue life. The middle value when all of the observed fatigue life values of the individual specimens in a group tested under identical conditions are arranged in order of magnitude. When an even number of specimens are tested, the average of the two middlemost values is used. Use of the sample median rather than the arithmetic mean (that is, the average) is usually preferred.

median fatigue strength at N cycles. An estimate of the stress level at which 50% of the population would survive N cycles. The estimate is derived from a particular point of the fatigue life distribution, since there is no test procedure by which a frequency distribution of fatigue strengths at N cycles can be directly observed. Also known as *fatigue strength at N cycles*.

melting point. The temperature at which a pure metal, compound, or eutectic changes from solid to liquid; the temperature at which the liquid and the solid are at equilibrium.

melting range. The range of temperatures over which an alloy other than a compound or eutectic changes from solid to liquid; the range of temperatures from *solidus* to *liquidus* at any given composition on a *phase diagram*.

melting rate. In electric arc welding, the weight or length of electrode melted in a unit of time. Sometimes called melt-off rate or burn-off rate.

melting temperature. See *melting point*.

melt-through. Complete joint penetration for a joint welded from one side.

merchant mill (obsolete). A mill, consisting of a group of stands of three rolls each arranged in a straight line and driven by one power unit, used to roll rounds, squares or flats of smaller dimensions than would be rolled on a bar mill.

mesh. (1) The number of screen openings per linear inch of screen; also called *mesh size*. (2) The screen number on the finest screen of a specified standard screen scale through which almost all of the particles of a powder sample will pass. See also *sieve analysis* and *sieve classification*.

mesh-belt conveyor furnace. A continuously operating furnace that uses a conveyor belt for the transport of the charge.

metal. (1) An opaque lustrous elemental chemical substance that is a good conductor of heat and electricity and, when polished, a good reflector of light. Most elemental metals are malleable and ductile and are, in general, denser than the other elemental substances. (2) As to structure, metals may be distinguished from nonmetals by their atomic binding and electron availability. Metallic atoms tend to lose electrons from the outer shells, the positive ions thus formed being held together by the electron gas produced by the separation. The ability of these "free electrons" to carry an electric current, and the fact that this ability decreases as temperature increases, establish the prime distinctions of a metallic solid. (3) From a chemical viewpoint, an elemental substance whose hydroxide is alkaline. (4) An *alloy*.

metal-arc cutting. Any of a group of arc cutting processes which severs metals by melting them with the heat of an arc between a metal electrode and the base metal. See also *shielded metal arc cutting* and *gas metal arc cutting*.

metal-arc welding. Any of a group of arc welding processes in which metals are fused together using the heat of an arc between a metal electrode and the work. Use of the specific process name is preferred.

metal cored electrode. A composite filler metal welding electrode consisting of a metal tube or other hollow configuration containing alloying ingredients. Minor amounts of ingredients facilitate arc stabilization and fluxing of oxides. External shielding gas may or may not be used.

metal dusting. Accelerated deterioration of metals in carbonaceous gases at elevated temperatures to form a dustlike corrosion product.

metal electrode. An electrode used in arc welding or cutting which consists of a metal wire or rod that is either bare or covered with a suitable covering or coating.

metal inert-gas welding. *Gas metal arc welding* using an inert gas such as argon as the shielding gas.

metallic glass. A noncrystalline metal or alloy, commonly produced by drastic supercooling of a molten alloy, by molecular deposition, which involves growth from the vapor phase (e.g., thermal evaporation and sputtering) or from a

liquid phase (e.g., electroless deposition and electrodeposition), or by external action techniques (e.g., ion implantation and ion beam mixing).

metallizing. Forming a metallic coating by atomized spraying with molten metal or by *vacuum deposition*. Also called spray metallizing.

metallograph. An optical instrument designed for visual observation and photomicrography of prepared surfaces of opaque materials at magnifications of 25 to approximately 2000×. The instrument consists of a high-intensity illuminating source, a microscope, and a camera bellows. On some instruments, provisions are made for examination of specimen surfaces using polarized light, phase contrast, oblique illumination, dark-field illumination, and bright-field illumination.

metallography. The study of the structure of metals and alloys by various methods, especially by optical and electron microscopy.

metallurgical coke. A coke, usually low in sulfur, having a very high compressive strength at elevated temperatures; used in metallurgical furnaces not only as fuel, but also to support the weight of the charge.

metallurgy. The science and technology of metals and alloys. Process metallurgy is concerned with the extraction of metals from their ores and with refining of metals; physical metallurgy, with the physical and mechanical properties of metals as affected by composition, processing, and environmental conditions; and mechanical metallurgy, with the response of metals to applied forces.

metal-matrix composite. A material that consists of a nonmetallic reinforcement, such as ceramic fibers or filaments, incorporated into a metallic matrix.

metal penetration. A surface condition in metal castings in which metal or metal oxides have filled voids between sand grains without displacing them.

metal powder. Elemental metals or alloy particles, usually in the size range of 0.1 to 1000 m.

metal powder cutting. A technique that supplements an oxyfuel torch with a stream of iron or blended iron-aluminum powder to facilitate flame cutting of difficult-to-cut materials. The powdered material propagates and accelerates the oxidation reaction, as well as the melting and spalling action of the materials to be cut.

metal spraying. Coating metal objects by spraying molten metal against their sur-

faces. See also *thermal spraying* and *flame spraying*.

metastable. (1) Of a material not truly stable with respect to some transition, conversion, or reaction but stabilized kinetically either by rapid cooling or by some molecular characteristics as, for example, by the extremely high viscosity of polymers. (2) Possessing a state of pseudoequilibrium that has a free energy higher than that of the true equilibrium state.

M_f temperature. For any alloy system, the temperature at which martensite formation on cooling is essentially finished. See also *transformation temperature* for the definition applicable to ferrous alloys.

microcrack. A crack of microscopic proportions. Also termed microfissure.

microfissure. A crack of microscopic proportions.

micrograph. A graphic reproduction of the surface of a specimen at a magnification greater than 25×. If produced by photographic means it is called a photomicrograph (not a microphotograph).

microhardness. The hardness of a material as determined by forcing an indenter such as a Vickers or Knoop indenter into the surface of a material under very light load; usually, the indentations are so small that they must be measured with a microscope. Capable of determining hardnesses of different microconstituents within a structure, or of measuring steep hardness gradients such as those encountered in *case hardening*. See also *microhardness test*.

microhardness number. A commonly used term for the more technically correct term *microindentation hardness number*.

microhardness test. A microindentation hardness test using a calibrated machine to force a diamond indenter of specific geometry, under a test load of 1 to 1000 gram-force, into the surface of the test material and to measure the diagonal or diagonals optically. See also *Knoop hardness test* and *Vickers hardness test*.

microindentation. (1) In hardness testing, the small residual impression left in a solid surface when an indenter, typically a pyramidal diamond stylus, is withdrawn after penetrating the surface. Typically, the dimensions of the microindentations are measured to determine microindentation hardness number. (2) The process of indenting a solid surface, using a hard stylus of prescribed geometry and under a slowly applied normal force, usually for the purpose of determining its microindentation hardness number. See also *Knoop hardness number*, *microindenta-*

tion hardness number, and *Vickers hardness number*.

microindentation hardness number. A numerical quantity, usually stated in units of pressure (kg/mm^2), that expresses the resistance to penetration of a solid surface by a hard indenter of prescribed geometry and under a specified, slowly applied normal force. The prefix "micro" indicates that the indentations produced are typically between 10.0 and 200.0 m across. See also *Knoop hardness number, nanohardness test*, and *Vickers hardness number*.

microscopic. Visible at magnifications above 25×.

microscopic stress. Residual stress in a material within a distance comparable to the grain size. See also *macroscopic stress*.

microsegregation. *Segregation* within a grain, crystal, or small particle. See also *coring*.

microshrinkage. A casting imperfection, not detectable microscopically, consisting of interdendritic voids. Microshrinkage results from contraction during solidification where the opportunity to supply filler material is inadequate to compensate for shrinkage. Alloys with wide ranges in solidification temperature are particularly susceptible.

microstrain. The strain over a gage length comparable to interatomic distances. These are the strains being averaged by the *macrostrain* measurement. Microstrain is not measurable by existing techniques. Variance of the microstrain distribution can, however, be measured by x-ray diffraction.

microstress. Same as *microscopic stress*.

microstructure. The structure of an object, organism, or material as revealed by a microscope at magnifications greater than 25×.

middling. A product intermediate between concentrate and tailing and containing enough of a valuable mineral to make retreatment profitable.

MIG welding. See preferred term *gas metal arc welding*.

mild steel. *Carbon steel* with a maximum of about 0.25% C and containing 0.4 to 0.7% Mn, 0.1 to 0.5% Si, and some residuals of sulfur, phosphorus, and/or other elements.

mill. (1) A factory in which metals are hot worked, cold worked, or melted and cast into standard shapes suitable for secondary fabrication into commercial products. (2) A production line, usually of four or more *stands*, for hot or cold rolling metal into standard shapes such as bar, rod, plate, sheet, or strip. (3) A single machine for hot rolling, cold rolling, or extruding metal; examples include *blooming mill, cluster mill, four-high mill*, and *Sendzimir mill*. (4) A shop term for a milling cutter. (5) A machine or group of machines for grinding or crushing ores and other minerals. (6) A machine for grinding or mixing material, for example, a ball mill and a paint mill. (7) Grinding or mixing a material, for example, milling a powder metallurgy material.

mill edge. The normal edge produced in hot rolling of sheet metal. This edge is customarily removed when hot rolled sheets are further processed into cold rolled sheets.

Miller indices. A system for identifying planes and directions in any crystal system by means of sets of integers. The indices of a plane are related to the intercepts of that plane with the axes of a unit cell; the indices of a direction, to the multiples of lattice parameter that represent the coordinates of a point on a line parallel to the direction and passing through the arbitrarily chosen origin of a unit cell.

mill finish. A nonstandard (and typically nonuniform) surface finish on mill products that are delivered without being subjected to a special surface treatment (other than a corrosion-preventive treatment) after the final working or heat-treating step.

milling (machining). Removing metal with a *milling cutter*.

milling (powder technology). The mechanical comminution of a material, usually in a ball mill, to alter the size or shape of the individual particles, to coat one component of a mixture with another, or to create uniform distributions of components.

milling cutter. A rotary cutting tool provided with one or more cutting elements, called teeth, which intermittently engage the workpiece and remove material by relative movement of the workpiece and cutter.

mill product. Any commercial product of a *mill*.

mill scale. The heavy oxide layer that forms during the hot fabrication or heat treatment of metals.

mineral dressing. Physical and chemical concentration of raw ore into a product from which a metal can be recovered at a profit.

minimized spangle. A hot dip galvanized coating of very small grain size, which makes the *spangle* less visible when the part is subsequently painted.

minimum bend radius. The minimum radius over which a metal product can be bent to a given angle without fracture.

minimum stress (S_{min}). In fatigue, the stress having the lowest algebraic value in the cycle, tensile stress being considered positive and compressive stress negative.

minimum stress-intensity factor (K_{min}). In fatigue, the minimum value of the *stress-intensity factor* in a cycle. This value corresponds to the *minimum load* when the *load ratio* 0 and is taken to be zero when the *load ratio* is 0.

minus sieve. The portion of a powder sample that passes through a standard sieve of a specified number. See also *plus sieve* and *sieve analysis*.

mischmetal. An natural mixture of rare-earth elements (atomic numbers 57 through 71) in metallic form. It contains about 50% cerium, the remainder being principally lanthanum and neodymium. Mischmetal is used as an alloying additive in ferrous alloys to scavenge sulfur, oxygen, and other impurities and in magnesium alloys to improve high-temperature strength.

mismatch. The misalignment or error in register of a pair of forging dies; also applied to the condition of the resulting forging.

misrun. Denotes an irregularity on a cast metal surface caused by incomplete filling of the mold due to low pouring temperatures, gas back pressure from inadequate venting of the mold, and inadequate gating.

mixed potential. The *potential* of a specimen (or specimens in a *galvanic couple*) when two or more electrochemical reactions are occurring. Also called galvanic couple potential.

mixing. In powder metallurgy, the thorough intermingling of powders of two or more different materials (not *blending*).

modification. Treatment of molten hypoeutectic (8 to 13% Si) or hypereutectic (13 to 19% Si) aluminum-silicon alloys to improve mechanical properties of the solid alloy by refinement of the size and distribution of the silicon phase. Involves additions of small percentages of sodium, strontium, or calcium (hypoeutectic alloys) or of phosphorus (hypereutectic alloys).

modulus of elasticity (E). (1) The measure of rigidity or stiffness of a material; the ratio of stress, below the proportional limit, to the corresponding strain. If a tensile stress of 13.8 MPa (2.0 ksi) results in an elongation of 1.0%, the modulus of elasticity is 13.8 MPa (2.0 ksi) divided by 0.01, or 1380 MPa (200 ksi). (2) In terms of the *stress-strain curve*, the modulus of elasticity is the slope of the stress-strain curve in the range of linear proportionality of stress to strain. Also known as *Young's modulus*. For materials that do not conform to Hooke's law throughout the elastic range, the slope of either the tangent to the stress-strain curve at the origin or at low stress, the secant drawn from the origin to any specified point on the stress-strain curve, or the chord connecting any two specific points on the stress-strain curve is usually taken to be the modulus of elasticity. In these cases, the modulus is referred to as the tangent modulus, secant modulus, or chord modulus, respectively.

modulus of resilience. The amount of energy stored in a material when loaded to its elastic limit. It is determined by measuring the area under the *stress-strain curve* up to the *elastic limit*. See also resilience, and *strain energy*.

modulus of rigidity. See *shear modulus*.

modulus of rupture. Nominal stress at fracture in a bend test or torsion test. In bending, modulus of rupture is the bending moment at fracture divided by the section modulus. In torsion, modulus of rupture is the torque at fracture divided by the polar section modulus.

Mohs hardness. The hardness of a body according to a scale proposed by Mohs, based on ten minerals, each of which would scratch the one below it. These minerals, in decreasing order of hardness, are:

Diamond	10
Corundum	9
Topaz	8
Quartz	7
Othoclase (feldspar)	6
Apatite	5
Fluorite	4
Calcite	3
Gypsum	2
Talc	1

mold. (1) The form, made of sand, metal, or refractory material, that contains the cavity into which molten metal is poured to produce a casting of desired shape. (2) A die.

mold cavity. The space in a mold that is filled with liquid metal to form the casting upon solidification. The channels through which liquid metal enters the mold cavity (sprue, runner, gates) and reservoirs for liquid metal (risers) are not considered part of the mold cavity proper.

molding machine. A machine for making sand molds by mechanically compacting sand around a pattern.

molding press. A press used to form powder metallurgy *compacts*.

molding sands. Foundry sands containing over 5% natural clay, usually between 8 and 20%.

mold jacket. Wood or metal form that is slipped over a sand mold for support during pouring of a casting.

mold wash. An aqueous or alcoholic emulsion or suspension of various materials used to coat the surface of a casting mold cavity.

molten metal flame spraying. A thermal spraying process variation in which the metallic material to be sprayed is in the molten condition. See also *flame spraying*.

molten weld pool. The liquid state of a weld prior to solidification as weld metal.

Mond process. A process for extracting and purifying nickel. The main features consist of forming nickel carbonyl by reaction of finely divided reduced metal with carbon monoxide, then decomposing the nickel carbonyl to deposit purified nickel on small nickel pellets.

monotectic. An isothermal reversible reaction in a binary system, in which a liquid on cooling decomposes into a second liquid of a different composition and a solid. It differs from a *eutectic* in that only one of the two products of the reaction is below its freezing range.

monotropism. The ability of a solid to exist in two or more forms (crystal structures), but in which one form is the stable modification at all temperatures and pressures. *Ferrite* and *martensite* are a monotropic pair below the temperature at which *austenite* begins to form, for example, in steels. Alternate spelling is monotrophism.

morphology. The characteristic shape, form, or surface texture or contours of the crystals, grains, or particles of (or in) a material, generally on a microscopic scale.

mosaic structure. In crystals, a substructure in which adjoining regions have only slightly different orientations.

mottled cast iron. Iron that consists of a mixture of variable proportions of gray cast iron and white cast iron; such a material has a mottled fracture appearance.

mounting. A means by which a specimen for metallographic examination may be held during preparation of a section surface. The specimen can be embedded in plastic or secured mechanically in clamps.

mounting resin. Thermosetting or thermoplastic resins used to mount metallographic specimens.

M_s temperature. For any alloy system, the temperature at which martensite starts to

form on cooling. See *transformation temperature* for the definition applicable to ferrous alloys.

mulling. The mixing and kneading of foundry molding sand with moisture and clay to develop suitable properties for molding.

multiaxial stresses. Any stress state in which two or three principal stresses are not zero.

multiple. A piece of stock for forging that is cut from bar or billet lengths to provide the exact amount of material for a single workpiece.

multiple-pass weld. A weld made by depositing filler metal with two or more successive passes.

multiple-slide press. A press with individual slides, built into the main slide or connected to individual eccentrics on the main shaft, that can be adjusted to vary the length of stroke and the timing. See also *slide*.

multiple spot welding. Spot welding in which several spots are made during one complete cycle of the welding machine.

m-value. See *strain-rate sensitivity*.

N

nanohardness test. An indentation hardness testing procedure, usually relying on indentation force versus tip displacement data, to make assessments of the resistance of surfaces to penetrations of the order of 10 to 1000 nm deep.

native metal. (1) Any deposit in the earth's crust consisting of uncombined metal. (2) The metal in such a deposit.

natural aging. Spontaneous aging of a supersaturated solid solution at room temperature. See also *aging*. Compare with *artificial aging*.

natural strain. See *true strain*.

NDE. See *nondestructive evaluation*.

NDI. See *nondestructive inspection*.

NDT. See *nondestructive testing*.

near-net shape. See *net shape*.

necking. (1) The reduction of the cross-sectional area of a material in a localized area by uniaxial tension or by stretching. (2) The reduction of the diameter of a portion of the length of a cylindrical shell or tube.

necking down. Localized reduction in area of a specimen during tensile deformation.

negative rake. Describes a tooth face in rotation whose cutting edge lags the surface of the tooth face. See also the figure accompanying *face mill*.

net shape. The shape of a powder metallurgy part, casting, or forging that conforms closely to specified dimensions. Such a part requires no secondary machining or finishing. A near-net shape part can be either one in which some but not all of the surfaces are net or one in which the surfaces require only minimal machining or finishing.

Neumann band. *Mechanical twin* in ferrite.

neutral flame. (1) A gas flame in which there is no excess of either fuel or oxygen in the inner flame. Oxygen from ambient air is used to complete the combustion of CO_2 and H_2 produced in the inner flame. (2) An oxyfuel gas flame in which the portion used is neither oxidizing nor reducing. See also *carburizing flame*, *oxidizing flame*, and *reducing flame*.

neutron embrittlement. *Embrittlement* resulting from bombardment with neutrons, usually encountered in metals that have been exposed to a neutron flux in the core of the reactor. In steels, neutron embrittlement is evidenced by a rise in the ductile-to-brittle *transition temperature*.

nibbling. Contour cutting of sheet metal by use of a rapidly reciprocating punch that makes numerous small cuts.

nip angle. See *angle of bite*.

nitriding. Introducing nitrogen into the surface layer of a solid ferrous alloy by holding at a suitable temperature (below Ac_1 for ferritic steels) in contact with a nitrogenous material, usually ammonia or molten cyanide of appropriate composition. Quenching is not required to produce a hard case. See also *bright nitriding* and *liquid nitriding*.

nitrocarburizing. Any of several processes in which both nitrogen and carbon are absorbed into the surface layers of a ferrous material at temperatures below the lower critical temperature and, by diffusion, create a concentration gradient. Nitrocarburizing is performed primarily to provide an antiscuffing surface layer and to improve fatigue resistance. Compare with *carbonitriding*.

noble. The positive direction of *electrode potential*, thus resembling noble metals such as gold and platinum.

noble metal. (1) A metal whose *potential* is highly positive relative to the hydrogen electrode. (2) A metal with marked resistance to chemical reaction, particularly to oxidation and to solution by inorganic acids. The term as often used is synonymous with *precious metal*.

noble potential. A *potential* more cathodic (positive) than the standard hydrogen potential.

no-draft (draftless) forging. A forging with extremely close tolerances and little or no *draft* that requires minimal machining to produce the final part. Mechanical properties can be enhanced by closer control of grain flow and by retention of surface material in the final component.

nodular graphite. Graphite in the nodular form as opposed to flake form (see *flake graphite*). Nodular graphite is characteristic of *malleable iron*. The graphite of nodular or *ductile iron* is spherulitic in form, but called nodular.

nodular iron. See preferred term *ductile iron*.

nodular pearlite. Pearlite that has grown as a colony with an approximately spherical morphology.

nominal stress. The stress at a point calculated on the net cross section without taking into consideration the effect on stress of geometric discontinuities, such as holes, grooves, fillets, and so forth. The calculation is made using simple elastic theory.

nondestructive evaluation (NDE). Broadly considered synonymous with nondestructive inspection (NDI). More specifically, the quantitative analysis of NDI findings to determine whether the material will be acceptable for its function, despite the presence of discontinuities. With NDE, a discontinuity can be classified by its size, shape, type, and location, allowing the investigator to determine whether or not the flaw(s) is acceptable. Damage tolerant design approaches are based on the philosophy of ensuring safe operation in the presence of flaws.

nondestructive inspection (NDI). A process or procedure, such as ultrasonic or radiographic inspection, for determining the quality or characteristics of a material, part, or assembly, without permanently altering the subject or its properties. Used to find internal anomalies in a structure without degrading its properties or impairing its serviceability.

nondestructive testing (NDT). Broadly considered synonymous with *nondestructive inspection (NDI)*.

nonmetallic inclusions. See *inclusions*.

normal direction. That direction perpendicular to the plane of working in a worked material. See also *longitudinal direction* and *transverse direction*.

normalizing. Heating a ferrous alloy to a suitable temperature above the transformation range and then cooling in air to a temperature substantially below the transformation range.

normal segregation. Concentration of alloying constituents that have low melting points in those portions of a casting that solidify last. Compare with *inverse segregation*.

normal solution. An aqueous solution containing one gram equivalent of the active reagent in 1 L of the solution.

normal stress. The stress component that is perpendicular to the plane on which the forces act. Normal stress may be either *tensile* or *compressive*.

nose radius. The radius of the rounded portion of the cutting edge of a tool. See the figure accompanying *single-point tool*.

notch acuity. Relates to the severity of the *stress concentration* produced by a given notch in a particular structure. If the depth of the notch is very small compared with the width (or diameter) of the narrowest cross section, acuity may be expressed as the ratio of the notch depth to the notch root radius. Otherwise, acuity is defined as the ratio of one-half the width (or diameter) of the narrowest cross section to the notch root radius.

notch brittleness. Susceptibility of a material to brittle fracture at points of stress concentration. For example, in a notch tensile test, the material is said to be notch brittle if the *notch strength* is less than the tensile strength of an unnotched specimen. Otherwise, it is said to be notch ductile.

notch depth. The distance from the surface of a test specimen to the bottom of the notch. In a cylindrical test specimen, the percentage of the original cross-sectional area removed by machining an annular groove.

notch ductility. The percentage reduction in area after complete separation of the metal in a tensile test of a *notched specimen*.

notched specimen. A test specimen that has been deliberately cut or notched, usually in a V-shape, to induce and locate point of failure.

notch factor. Ratio of the resilience determined on a plain specimen to the resilience determined on a notched specimen.

notching. Cutting out various shapes from the edge of a strip, blank, or part.

notching press. A mechanical press used for notching internal and external circumferences and also for notching along a straight line. These presses are equipped with automatic feeds because only one notch is made per stroke.

notch rupture strength. The ratio of applied load to original area of the minimum cross section in a *stress-rupture test* of a *notched specimen*.

notch sensitivity. The extent to which the sensitivity of a material to fracture is increased by the presence of a *stress concentration*, such as a notch, a sudden change in cross section, a crack, or a scratch. Low notch sensitivity is usually associated with ductile materials, and high notch sensitivity is usually associated with brittle materials.

notch strength. The maximum load on a notched tension-test specimen divided by the minimum cross-sectional area (the area at the root of the notch). Also called notch tensile strength.

nuclear grade. Material of a quality adequate for use in nuclear application.

nucleation. The initiation of a phase transformation at discrete sites, with the new phase growing on the nuclei. See also *nucleus (2)*.

nucleus. (1) The heavy central core of an atom, in which most of the mass and the total positive electric charge are concentrated. (2) The first structurally stable particle capable of initiating recrystallization of a phase or the growth of a new phase and possessing an interface with the parent metallic matrix. The term is also applied to a foreign particle that initiates such action.

nugget. (1) A small mass of metal, such as gold or silver, found free in nature. (2) The weld metal in a spot, seam, or projection weld.

n-value. See *strain-hardening exponent*.

O

offhand grinding. Grinding where the operator manually forces the wheel against the work, or vice versa. It often implies casual manipulation of either grinder or work to achieve the desired result. Dimensions and tolerances frequently are not specified, or are only loosely specified; the operator relies mainly on visual inspection to determine how much grinding should be done. Contrast with *precision grinding*.

offset. The distance along the strain coordinate between the initial portion of a stress-strain curve and a parallel line that intersects the stress-strain curve at a value of stress (commonly 0.2%) that is used as a measure of the *yield strength*. Used for materials that have no obvious *yield point*.

offset yield strength. The stress at which the strain exceeds by a specific amount (the *offset*) an extension of the initial, approximately linear, proportional portion of the stress-strain curve. It is expressed in force per unit area.

oil canning. See *canning*.

oil quenching. Hardening of carbon steel in an oil bath.

Olsen ductility test. A *cupping test* in which a piece of sheet metal, restrained except at the center, is deformed by a standard steel ball until fracture occurs. The height of the cup at the time of fracture is a measure of the ductility.

open-back inclinable press. A vertical crank press that can be inclined so that the bed will have an inclination generally varying from 0° to 30°. The formed parts slide off through an opening in the back. It is often called an OBI press.

open-die forging. The hot mechanical forming of metals between flat or shaped dies in which metal flow is not completely restricted. Also known as hand or smith forging. See also *hand forge (smith forge)*.

open dies. Dies with flat surfaces that are used for preforming stock or producing hand forgings.

open-gap upset welding. A form of *forge welding* in which the weld interfaces are heated with a fuel gas flame, then forced into intimate contact by the application of force. Not to be confused with *upset welding*, which is a resistance welding process.

open hearth furnace. A reverberatory melting furnace with a shallow hearth and a low roof. The flame passes over the charge on the hearth, causing the charge to be heated both by direct flame and by radiation from the roof and sidewalls of the furnace. See also *reverberatory furnace*.

open rod press. A *hydraulic press* in which the slide is guided by vertical, cylindrical rods (usually four) that also serve to hold the crown and bed in position.

orange peel. A surface roughening in the form of a pebble-grained pattern that occurs when a metal of unusually coarse grain size is stressed beyond its elastic limit. Also called pebbles and alligator skin.

orbital forging. See *rotary forging*.

ordered structure. The crystal structure of a *solid solution* in which the atoms of different elements seek preferred lattice positions. Contrast with *disordered structure*.

order hardening. A low-temperature *annealing* treatment for metals that permits short-range ordering of solute atoms within a matrix, which greatly impedes dislocation motion.

ore. A natural mineral that may be mined and treated for the extraction of any of its components, metallic or otherwise, at a profit.

ore dressing. Same as *mineral dressing*.

orientation. Arrangements in space of the axes of the lattice of a crystal with respect to a chosen reference or coordinate system. See also *preferred orientation*.

original crack size (a_o). The *physical crack size* at the start of testing.

oscillating die press. A small high-speed metal forming press in which the die and punch move horizontally with the strip during the working stroke. Through a reciprocating motion, the die and punch return to their original positions to begin the next stroke.

overaging. *Aging* under conditions of time and temperature greater than those required to obtain maximum change in a certain property, so that the property is altered in the direction of the initial value.

overbending. Bending metal through a greater arc than that required in the finished part to compensate for springback.

overdraft. A condition wherein a metal curves upward on leaving the rolls because of the higher speed of the lower roll.

overhead-drive press. A mechanical press with the driving mechanism mounted in or on the crown or upper parts of the uprights.

overhead-position welding. Welding that is performed from the underside of the joint.

overheating. Heating a metal or alloy to such a high temperature that its properties are impaired. When the original properties cannot be restored by further heat treating, by mechanical working, or by a combination of working and heat treating, the overheating is known as *burning*.

overlap. (1) Pultrusion of weld metal beyond the toe, face, or root of a weld. (2) In resistance seam welding, the area in a given weld remelted by the succeeding weld. See also *face of weld*, *root of weld*, and *toe of weld*.

oversize powder. Powder particles larger than the maximum permitted by a particle size specification.

overstressing. In fatigue testing, cycling at a stress level higher than that used at the end of the test.

oxidation. (1) A reaction in which there is an increase in valence resulting from a loss of electrons. Contrast with *reduction*. (2) A corrosion reaction in which the corroded metal forms an oxide; usually applied to reaction with a gas containing elemental oxygen, such as air. (3) A chemical reaction in which one substance is changed to another by oxygen combining with the substance. Much of the dross from holding and melting furnaces is the result of oxidation of the alloy held in the furnace.

oxidation losses. Reduction in the amount of metal or alloy through *oxidation*. Such losses are usually the largest factor in melting loss.

oxidative wear. (1) A *corrosive wear* process in which chemical reaction with oxygen or oxidizing environment predominates. (2) A type of *wear* resulting from the sliding action between two metallic components that generates oxide films on the metal surfaces. These oxide films prevent the formation of a metallic bond between the sliding surfaces, resulting in fine wear debris and low wear rates.

oxidized steel surface. Surface having a thin, tightly adhering oxidized skin (from straw to blue in color), extending in from the edge of a coil or sheet.

oxidizing agent. A compound that causes *oxidation*, thereby itself being reduced.

oxidizing atmosphere. A furnace atmosphere with an oversupply of oxygen that tends to oxidize materials placed in it.

oxidizing flame. A gas flame produced with excess oxygen in the inner flame that has an oxidizing effect.

oxyacetylene cutting. An *oxyfuel gas cutting* process in which the fuel gas is acetylene.

oxyacetylene welding. An *oxyfuel gas welding* process in which the fuel gas is acetylene.

oxyfuel gas cutting (OFC). Any of a group of processes used to sever metals by means of chemical reaction between hot base metal and a fine stream of oxygen. The necessary metal temperature is maintained by gas flames resulting from combustion of a specific fuel gas such as acetylene, hydrogen, natural gas, propane, propylene, or Mapp gas (stabilized methylacetylene-propadiene).

oxyfuel gas welding (OFW). Any of a group of processes used to fuse metals together by heating them with gas flames resulting from combustion of a specific fuel gas such as acetylene, hydrogen, natural gas, or propane. The process may be used with or without the application of pressure to the joint, and with or without adding any filler metal.

oxygas cutting. See preferred term *oxygen cutting*.

oxygen arc cutting. An oxygen cutting process used to sever metals by means of the chemical reaction of oxygen with the base metal at elevated temperatures. The necessary temperature is maintained by an arc between a consumable tubular electrode and the base metal.

oxygen cutting. Metal cutting by directing a fine stream of oxygen against a hot metal. The chemical reaction between oxygen and the base metal furnishes heat for localized melting, hence, cutting. In the case of oxidation-resistant metals, the reaction is facilitated by the use of a chemical flux or metal powder. See also *metal powder cutting*.

oxygen deficiency. A form of *crevice corrosion* in which galvanic corrosion proceeds because oxygen is prevented from diffusing into the crevice.

oxygen-free copper. Electrolytic copper free from cuprous oxide, produced without the use of residual metallic or metalloidal deoxidizers.

oxygen gouging. Oxygen cutting in which a bevel or groove is formed.

oxygen lance. A length of pipe used to convey oxygen either beneath or on top of the melt in a steelmaking furnace, or to the point of cutting in *oxygen lance cutting*.

oxygen lance cutting. An oxygen cutting process used to sever metals with oxygen supplied through a consumable lance; the preheat to start the cutting is obtained by other means.

oxygen probe. An atmosphere-monitoring device that electronically measures the difference between the partial pressure of oxygen in a furnace or furnace supply atmosphere and the external air.

oxyhydrogen cutting. An *oxyfuel gas cutting* process in which the fuel gas is hydrogen.

oxyhydrogen welding. An *oxyfuel gas welding* process in which the fuel gas is hydrogen.

oxynatural gas cutting. An *oxyfuel gas cutting* process in which the fuel gas is natural gas.

oxynatural gas welding. An *oxyfuel gas welding* process in which the fuel gas is natural gas.

oxypropane cutting. An *oxyfuel gas cutting* process in which the fuel gas is propane.

oxypropane welding. An *oxyfuel gas welding* process in which the fuel gas is propane.

P

pack carburizing. A method of surface hardening of steel in which parts are packed in a steel box with a carburizing compound and heated to elevated temperatures. This process has been largely supplanted by gas and liquid carburizing processes.

pack nitriding. A method of surface hardening of steel in which parts are packed in a steel box with a nitriding compound and heated to elevated temperatures.

pack rolling. Hot rolling a pack of two or more sheets of metal; scale prevents their being welded together.

pancake forging. A rough forged shape, usually flat, that can be obtained quickly with minimal tooling. Considerable machining is usually required to attain the finish size.

pancake grain structure. A metallic structure in which the lengths and widths of individual grains are large compared to their thicknesses.

paramagnetic material. (1) A material whose specific permeability is greater than unity and is practically independent of the magnetizing force. (2) Material with a small positive susceptibility due to the interaction and independent alignment of permanent atomic and electronic magnetic moments with the applied field. Compare with *ferromagnetic material*.

paramagnetism. A property exhibited by substances that, when placed in a magnetic field, are magnetized parallel to the field to an extent proportional to the field (except at very low temperatures or in extremely large magnetic fields). Compare with *ferromagnetism*.

Parkes process. A process used to recover precious metals from lead and based on the principle that if 1 to 2% Zn is stirred into the molten lead, a compound of zinc with gold and silver separates out and can be skimmed off.

partial annealing. An imprecise term used to denote a treatment given cold-worked metallic material to reduce its strength to a controlled level or to effect stress relief. To be meaningful, the type of material, the degree of cold work, and the time-temperature schedule must be stated.

particle shape. The appearance of a metal particle, such as spherical, rounded, angular, acicular, dendritic, irregular, porous, fragmented, blocky, rod, flake, nodular, or plate.

particle size. The controlling lineal dimension of an individual particle as determined by analysis with screens or other suitable instruments. See also *sieve analysis* and *sieve classification*.

particle size distribution. The percentage, by weight or by number, of each fraction into which a powder or sand sample has been classified with respect to sieve number or *particle size*.

particle sizing. Segregation of granular material into specified particle size ranges.

parting. (1) In the recovery of precious metals, the separation of silver from gold. (2) The zone of separation between *cope* and *drag* portions of the mold or flask in sand casting. (3) A composition sometimes used in sand molding to facilitate the removal of the pattern. (4) Cutting simultaneously along two parallel lines or along two lines that balance each other in side thrust. (5) A shearing operation used to produce two or more parts from a stamping.

parting compound. A material dusted or sprayed on foundry (casting) patterns to prevent adherence of sand and to promote easy separation of *cope* and *drag* parting surfaces when the cope is lifted from the drag.

parting line. (1) The intersection of the parting plane of a casting or plastic mold or the parting plane between forging dies with the mold or die cavity. (2) A raised line or projection on the surface of a casting, plastic part, or forging that corresponds to said intersection.

parting plane. (1) In forging, the dividing line between dies. (2) In casting, the dividing line between mold halves.

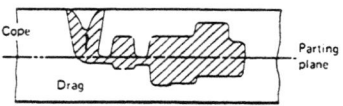

parting sand. In foundry practice, a fine sand for dusting on sand mold surfaces that are to be separated.

pass. (1) A single transfer of metal through a *stand* of rolls. (2) The open space between two grooved rolls through which metal is processed. (3) The weld metal deposited in one trip along the axis of a weld. See also *weld pass*.

passivation. (1) A reduction of the anodic reaction rate of an electrode involved in corrosion. (2) The process in metal corrosion by which metals become *passive*. (3) The changing of a chemically active surface of a metal to a much less reactive state. Contrast with *activation*.

passive. (1) A metal corroding under the control of a surface reaction product. (2) The state of the metal surface characterized by low corrosion rates in a potential region that is strongly oxidizing for the metal.

passive-active cell. A corrosion cell in which the *anode* is a metal in the *active* state and the *cathode* is the same metal in the *passive* state.

passivity. A condition in which a piece of metal, because of an impervious covering of oxide or other compound, has a *poten-*tial much more positive than that of the metal in the active state.

patenting. In wiremaking, a heat treatment applied to medium-carbon or high-carbon steel before drawing of wire or between drafts. This process consists of heating to a temperature above the transformation range and then cooling to a temperature below Ae_1 in air or in a bath of molten lead or salt.

patent leveling. Same as *stretcher leveling*.

patina. The coating, usually green, that forms on the surface of metals such as copper and copper alloys exposed to the atmosphere. Also used to describe the appearance of a weathered surface of any metal.

pattern. (1) A form of wood, metal, or other material around which molding material is placed to make a mold for casting metals. (2) A form of wax- or plastic-base material around which refractory material is placed to make a mold for casting metals. (3) A full-scale reproduction of a part used as a guide in cutting.

pearlite. A metastable lamellar aggregate of *ferrite* and *cementite* resulting from the transformation of *austenite* at temperatures above the *bainite* range.

pearlitic malleable. See *malleable iron*.

pearlitic structure. A microstructure resembling that of the pearlite constituent in steel. Therefore, it is a lamellar structure of varying degrees of coarseness.

peeling. The detaching of one layer of a coating from another, or from the basis metal, because of poor adherence.

peel test. A destructive method of inspection which mechanically separates a lap joint by peeling.

peening. Mechanical working of metal by hammer blows or shot impingement.

penetrant. A liquid with low surface tension used in *liquid penetrant inspection* to flow into surface openings of parts being inspected.

penetrant inspection. See preferred term *liquid penetrant inspection*.

penetration. (1) In founding, an imperfection on a casting surface caused by metal running into voids between sand grains; usually referred to as *metal penetration*. (2) In welding, the distance from the original surface of the base metal to that point at which fusion ceased. See also *joint penetration* and *root penetration*.

penetration hardness. Same as *indentation hardness*.

percussion welding. A resistance welding process which produces coalescence of abutting surfaces using heat from an arc produced by a rapid discharge of electrical energy. Pressure is applied percus-sively during or immediately following the electrical discharge.

perforating. The punching of many holes, usually identical and arranged in a regular pattern, in a sheet, workpiece blank, or previously formed part. The holes are usually round, but may be any shape. The operation is also called multiple punching. See also *piercing*.

peripheral milling. Milling a surface parallel to the axis of the cutter.

peritectic. An isothermal reversible reaction in metals in which a liquid phase reacts with a solid phase to produce a single (and different) solid phase on cooling.

peritectoid. An isothermal reversible reaction in which a solid phase reacts with a second solid phase to produce a single (and different) solid phase on cooling.

permanent magnet material. A ferromagnetic alloy capable of being magnetized permanently because of its ability to retain induced magnetization and magnetic poles after removal of externally applied fields; an alloy with high coercive force. The name is based on the fact that the quality of the early permanent magnets was related to their hardness.

permanent mold. A metal, graphite, or ceramic mold (other than an ingot mold) of two or more parts that is used repeatedly for the production of many *castings* of the same form. Liquid metal is usually poured in by gravity.

permanent set. The deformation remaining after a specimen has been stressed a prescribed amount in tension, compression, or shear for a specified time period and released for a specified time period. For creep tests, the residual unrecoverable deformation after the load causing the creep has been removed for a substantial and specified period of time. Also, the increase in length, expressed as a percentage of the original length, by which an elastic material fails to return to its original length after being stressed for a standard period of time.

permeability. (1) The passage or diffusion (or rate of passage) of a gas, vapor, liquid, or solid through a material (often porous) without physically or chemically affecting it; the measure of fluid flow (gas or liquid) through a material. (2) A general term used to express various relationships between magnetic induction and magnetizing force. These relationships are either "absolute permeability," which is a change in magnetic induction divided by the corresponding change in magnetizing force, or "specific (relative) permeability," the ratio of the absolute permeability to the permeability of free space. (3) In metal casting, the characteristics of molding materials that permit gases to pass through them. "Permeability number" is determined by a standard test.

pewter. A tin-base *white metal* containing antimony and copper. Originally, pewter was defined as an alloy of tin and lead, but to avoid toxicity and dullness of finish, lead is excluded from modern pewter. These modern compositions contain 1 to 8% Sb and 0.25 to 3% Cu.

pH. The negative logarithm of the hydrogen-ion activity; it denotes the degree of acidity or basicity of a solution. At 25 °C (77 °F), 7.0 is the neutral value. Decreasing values below 7.0 indicates increasing acidity; increasing values above 7.0, increasing basicity. The pH values range from 0 to 14.

phase. A physically homogeneous and distinct portion of a material system.

phase change. The transition from one physical state to another, such as gas to liquid, liquid to solid, gas to solid, or vice versa.

phase diagram. A graphical representation of the temperature and composition limits of phase fields in an alloy or ceramic system as they actually exist under the specific conditions of heating or cooling. A phase diagram may be an equilibrium diagram, an approximation to an equilibrium diagram, or a representation of metastable conditions or phases. Synonymous with constitution diagram. Compare with *equilibrium diagram.*

phase rule. The maximum number of phases (P) that may coexist at equilibrium is two, plus the number of components (C) in the mixture, minus the number of degrees of freedom (F): $P + F = C + 2$.

phosphating. Forming an adherent phosphate coating on a metal by immersion in a suitable aqueous phosphate solution. Also called phosphatizing. See also *conversion coating.*

phosphorized copper. General term applied to copper deoxidized with phosphorus. The most commonly used deoxidized copper.

photoelasticity. An optical method for evaluating the magnitude and distribution of stresses, using a transparent model of a part, or a thick film of photoelastic material bonded to a real part.

photomacrograph. A *macrograph* produced by photographic means.

photomicrograph. A *micrograph* produced by photographic means.

physical crack size (a_p). In fracture mechanics, the distance from a reference plane to the observed crack front. This distance may represent an average of several measurements along the crack front. The reference plane depends on the specimen form, and it is normally taken to be either the boundary or a plane containing either the load line or the centerline of a specimen or plate.

physical metallurgy. The science and technology dealing with the properties of metals and alloys, and of the effects of composition, processing, and environment on those properties.

physical properties. Properties of a material that are relatively insensitive to structure and can be measured without the application of force; for example, density, electrical conductivity, coefficient of thermal expansion, magnetic permeability, and lattice parameter. Does not include chemical reactivity. Compare with *mechanical properties.*

physical testing. Methods used to determine the entire range of a material's *physical properties.* In addition to density and thermal, electrical, and magnetic properties, physical testing methods may be used to assess simple fundamental physical properties such as color, crystalline form, and melting point.

physical vapor deposition (PVD). A coating process whereby the deposition species are transferred and deposited in the form of individual atoms or molecules. The most common PVD methods are sputtering and evaporation. Sputtering, which is the principal PVD process, involves the transport of a material from a source (target) to a substrate by means of the bombardment of the target by gas ions that have been accelerated by a high voltage. Evaporation, which was the first PVD process used, involves the transfer of material to form a coating by physical means alone, essentially vaporization. PVD coatings are used to improve the wear, friction, and hardness properties of cutting tools and as corrosion-resistant coatings.

pickle. The chemical removal of surface oxides (scale) and other contaminants such as dirt from iron and steel by immersion in an aqueous acid solution. The most common pickling solutions are sulfuric and hydrochloric acids.

pickle liquor. A spent acid-pickling bath.

pickle patch. A tightly adhering oxide or scale coating not properly removed during *pickling.*

pickle stain. Discoloration of metal due to chemical cleaning without adequate washing and drying.

pickling. Removing surface oxides from metals by chemical or electrochemical reaction.

pickoff. An automatic device for removing a finished part from the press die after it has been stripped.

pickup. (1) Transfer of metal from tools to part or from part to tools during a forming operation. (2) Small particles of oxidized metal adhering to the surface of a *mill product.*

Pidgeon process. A process for production of magnesium by reduction of magnesium oxide with ferrosilicon.

piercing. The general term for cutting (shearing or punching) openings, such as holes and slots, in sheet material, plate, or parts. This operation is similar to *blanking*; the difference is that the slug or pierce produced by piercing is scrap, while the blank produced by blanking is the useful part.

piezoelectric effect. The reversible interaction, exhibited by some crystalline materials, between an elastic strain and an electric field. The direction of the strain depends on the polarity of the field or vice versa. Compare with *electrostrictive effect.*

pig. A metal casting used in remelting.

pig iron. (1) High-carbon iron made by reduction of iron ore in the blast furnace. (2) Cast iron in the form of *pigs.*

Pilger tube-reducing process. See *tube reducing.*

pin (for bend testing). The plunger or tool used in making semiguided, guided, or wraparound bend tests to apply the bending force to the inside surface of the bend. In free bends or semiguided bends to an angle of 180°, a shim or block of the proper thickness may be placed between the legs of the specimen as bending is completed. This shim or block is also referred to as a pin or mandrel. See also *mandrel.*

pinchers. Surface disturbances on metal sheet or strip that result from rolling processes and that ordinarily appear as fernlike ripples running diagonally to the direction of rolling.

pinch pass. A pass of sheet metal through rolls to effect a very small reduction in thickness.

pinch trimming. The trimming of the edge of a tubular metal part or shell by pushing or pinching the flange or lip over the cutting edge of a stationary punch or over the cutting edge of a draw punch.

pin expansion test. A test for determining the ability of a tube to be expanded or for revealing the presence of cracks or other longitudinal weaknesses, made by forcing a tapered pin into the open end of the tube.

pinholes. (1) Very small holes that are sometimes found as a type of porosity in a casting because of the microshrinkage or gas evolution during solidification. In wrought products, due to removal of inclusions or microconstituents during macroetching of transverse sections. (2) Small cavities that penetrate the surface of a cured composite or plastic part. (3) In photography, a very small circular aperture.

Piobert lines. See *Lüders lines*.

pipe. (1) The central cavity formed by contraction in metal, especially ingots, during solidification. (2) An imperfection in wrought or cast products resulting from such a cavity. (3) A tubular metal product, cast or wrought. See also *extrusion pipe*.

pipe tap. A *tap* for making internal *pipe threads* within pipe fittings or holes.

pipe threads. Internal or external machine threads, usually tapered, of a design intended for making pressure-tight mechanical joints in piping systems.

pit. A small, regular or irregular crater in the surface of a material created by exposure to the environment, for example, corrosion, wear, or thermal cycling. See also *pitting*.

pitting. (1) Forming small sharp cavities in a surface by corrosion, wear, or other mechanically assisted degradation. (2) *Localized corrosion* of a metal surface, confined to a point or small area, that takes the form of cavities.

plane strain. The stress condition in *linear elastic fracture mechanics* in which there is zero strain in a direction normal to both the axis of applied tensile stress and the direction of crack growth (that is, parallel to the crack front); most nearly achieved in loading thick plates along a direction parallel to the plate surface. Under plane-strain conditions, the plane of fracture instability is normal to the axis of the principal tensile stress.

plane-strain fracture toughness (K_{Ic}). The crack extension resistance under conditions of *crack-tip plane strain*. See also *stress-intensity factor*.

plane stress. The stress condition in *linear elastic fracture mechanics* in which the stress in the thickness direction is zero; most nearly achieved in loading very thin sheet along a direction parallel to the surface of the sheet. Under plane-stress conditions, the plane of fracture instability is inclined 45° to the axis of the principal tensile stress.

plane-stress fracture toughness (K_c). In *linear elastic fracture mechanics*, the value of the crack-extension resistance at the instability condition determined from the tangency between the *R-curve* and the critical crack-extension force curve of the specimen. See also *stress-intensity factor*.

planimetric method. A method of measuring grain size in which the grains within a definite area are counted.

planing. Producing flat surfaces by linear reciprocal motion of work and the table to which it is attached, relative to a stationary single-point cutting tool.

planishing. Producing a smooth finish on metal by a rapid succession of blows delivered by highly polished dies or by a hammer designed for the purpose, or by rolling in a planishing mill.

plasma-arc cutting. An arc cutting process that severs metals by melting a localized area with heat from a constricted arc and removing the molten metal with a high-velocity jet of hot, ionized gas issuing from the plasma torch.

plasma arc welding (PAW). An arc welding process that produces coalescence of metals by heating them with a constricted arc between an electrode and the workpiece (transferred arc) or the electrode and the constricting nozzle (nontransferred arc). Shielding is obtained from hot, ionized gas issuing from an orifice surrounding the electrode and may be supplemented by an auxiliary source of shielding gas, which may be an inert gas or a mixture of gases. Pressure may or may not be used, and filler metal may or may not be supplied.

plasma carburizing. Same as *ion carburizing*.

plasma nitriding. Same as *ion nitriding*.

plasma spraying. A *thermal spraying* process in which the coating material is melted with heat from a plasma torch that generates a nontransferred arc; molten powder coating material is propelled against the base metal by the hot, ionized gas issuing from the torch.

plaster molding. Molding in which a gypsum-bonded aggregate flour in the form of a water slurry is poured over a pattern, permitted to harden, and, after removal of the pattern, thoroughly dried. This technique is used to make smooth nonferrous castings of accurate size.

plastic deformation. The permanent (inelastic) distortion of materials under applied stresses that strain the material beyond its *elastic limit*.

plastic flow. The phenomenon that takes place when metals are stretched or compressed permanently without rupture.

plasticity. The property of a material which allows it to be repeatedly deformed without rupture when acted upon by a force sufficient to cause deformation and which allows it to retain its shape after the applied force has been removed.

plastic-strain ratio (r-value). In formability testing of metals, the ratio of the true width strain to the true thickness strain in a sheet tensile test, $r = \varepsilon_w \varepsilon_t$. A formability parameter that relates to drawing, it is also known as the anisotropy factor. A high r-value indicates a material with good drawing properties.

plate. A flat-rolled metal product of some minimum thickness and width arbitrarily dependent on the type of metal. Plate thicknesses commonly range from 6 to 300 mm (0.25 to 12 in.); widths from 200 to 2000 mm (8 to 80 in.).

plate martensite. Martensite formed partly in steel containing more than approximately 0.5% C and solely in steel containing more than approximately 1.0% C that appears as lenticular-shape plates (crystals).

platen. (1) The sliding member, *slide*, or *ram* of a metal forming press. (2) The mounting plates of a plastic forming press, to which the entire mold assembly is bolted. (3) A part of a resistance welding, mechanical testing, or other machine with a flat surface to which dies, fixtures, backups, or electrode holders are attached and that transmits pressure or force.

plating. Forming an adherent layer of metal on an object; often used as a shop term for *electroplating*. See also *electrodeposition* and *electroless plating*.

plating rack. A fixture used to hold work and conduct current to it during *electroplating*.

platinum black. A finely divided form of platinum of a dull black color, usually, but not necessarily produced by reduction of salts in an aqueous solution.

plug. (1) A rod or mandrel over which a pierced tube is forced. (2) A rod or mandrel that fills a tube as it is drawn through a die. (3) A punch or mandrel over which a cup is drawn. (4) A protruding portion of a die impression for forming a corresponding recess in the forging. (5) A false bottom in a die.

plug tap. A *tap* with *chamfer* extending from three to five threads.

plug weld. A circular weld made through a hole in one member of a lap or tee joint. Neither a fillet-welded hole nor a spot weld is to be construed as a plug weld. The hole may be partially or completely filled with weld metal.

plumbage. A special quality of powdered graphite used to coat molds and, in a mixture of clay, to make crucibles.

plunge grinding. *Grinding* wherein the only relative motion of the wheel is radially toward the work.

plus mesh. The powder sample retained on a screen of stated size, identified by the retaining mesh number. See also *sieve analysis* and *sieve classification*.

plus sieve. The portion of a sample of a granular substance (such as metal powder) retained on a standard sieve of specified number. Contrast with *minus sieve*. See also *sieve analysis* and *sieve classification*.

plymetal. Sheet consisting of bonded layers of dissimilar metals.

P/M. The acronym for *powder metallurgy*.

pneumatic press. A press that uses air or a gas to deliver the pressure to the upper and lower rams.

point angle. In general, the angle at the point of a cutting tool. Most commonly, the included angle at the point of a twist drill, the general-purpose angle being 118°.

Poisson's ratio (). The absolute value of the ratio of transverse (lateral) strain to the corresponding axial strain resulting from uniformly distributed axial stress below the *proportional limit* of the material.

polarity (welding). See *direct current electrode negative*, *direct current electrode positive*, *straight polarity*, and *reverse polarity*.

polarization. (1) The change from the open-circuit electrode potential as the result of the passage of current. (2) A change in the *potential* of an electrode during electrolysis, such that the potential of an *anode* becomes more noble, and that of a *cathode* more active, than their respective reversible potentials. Often accomplished by formation of a film on the electrode surface.

polarization curve. A plot of *current density* versus *electrode potential* for a specific electrode-electrolyte combination.

pole. (1) A means of designating the orientation of a crystal plane by stereographically plotting its normal. For example, the north pole defines the equatorial plane. Either of the two regions of a permanent magnet or electromagnet where most of the lines of induction enter or leave.

pole figure. A stereoscopic projection of a polycrystalline aggregate showing the distribution of poles, or plane normals, of a specific crystalline plane, using specimen axes as reference axes. Pole figures are used to characterize preferred orientation in polycrystalline materials.

poling. A step in the fire refining of copper to reduce the oxygen content to tolerable limits by covering the bath with coal or coke and thrusting green wood poles below the surface. There is a vigorous evolution of reducing gases, which combine with the oxygen contained in the metal.

polishing. (1) A surface-finishing process for ceramics and metals utilizing successive grades of abrasive. (2) Smoothing metal surfaces, often to a high luster, by rubbing the surface with a fine abrasive, usually contained in a cloth or other soft lap. Results in microscopic flow of some surface metal together with actual removal of a small amount of surface metal. (3) Removal of material by the action of abrasive grains carried to the work by a flexible support, generally either a wheel or a coated abrasive belt. (4) A mechanical, chemical, or electrolytic process or combination thereof used to prepare a smooth, reflective surface suitable for microstructural examination that is free of artifacts or damage introduced during prior sectioning or grinding. See also *electrolytic polishing* and *electropolishing*.

polycrystalline. Pertaining to a solid comprised of many crystals or crystallites, intimately bonded together. May be homogeneous (one substance) or heterogeneous (two or more crystal types or compositions).

polymorphism. A general term for the ability of a solid to exist in more than one form. In metals, alloys, and similar substances, this usually means the ability to exist in two or more crystal structures, or in an amorphous state and at least one crystal structure. See also *allotropy*, *enantiotropy*, and *monotropism*.

pop-off. Loss of small portions of a porcelain enamel coating. The usual cause is outgassing of hydrogen or other gases from the basis metal during firing, but pop-off may also occur because of oxide particles or other debris on the surface of the basis metal. Usually, the pits are minute and cone shaped, but when pop-off is the result of severe *fishscale* the pits may be much larger and irregular.

porcelain enamel. A substantially vitreous or glassy, inorganic coating (borosilicate glass) bonded to metal by fusion at a temperature above 425 °C (800 °F). Porcelain enamels are applied primarily to components made of sheet iron or steel, cast iron, aluminum, or aluminum-coated steels.

pore. (1) A small opening, void, interstice, or channel within a consolidated solid mass or agglomerate, usually larger than atomic or molecular dimensions. (2) A minute cavity in a powder metallurgy compact, sometimes added intentionally. (3) A minute perforation in an electroplated coating.

porosity. (1) Fine holes or pores within a solid; the amount of these pores is expressed as a percentage of the total volume of the solid. (2) Cavity-type discontinuities in weldments formed by gas entrapment during solidification. (3) A characteristic of being porous, with voids or pores resulting from trapped air or shrinkage in a casting. See also *gas porosity* and *pinholes*.

positioned weld. A weld made in a joint that has been oriented to facilitate making the weld.

positive rake. Describes a tooth face in rotation whose cutting edge leads the surface of the tooth face. See also the figure accompanying *face mill*.

postheating. Heating weldments immediately after welding, for tempering, for stress relieving, or for providing a controlled rate of cooling to prevent formation of a hard or brittle structure. See also *postweld heat treatment*.

postweld heat treatment. Any heat treatment that follows the welding operation.

pot. (1) A vessel for holding molten metal. (2) The electrolytic reduction cell used to make such metals as aluminum from a fused electrolyte.

pot annealing. Same as *box annealing*.

pot die forming. Forming products from sheet or plate through the use of a hollow die and internal pressure which causes the preformed workpiece to assume the contour of the die.

potential. Any of various functions from which intensity or velocity at any point in a field may be calculated. The driving influence of an electrochemical reaction.

poultice corrosion. A term used in the automotive industry to describe the corrosion of vehicle body parts due to the collection of road salts and debris on ledges and in pockets that are kept moist by weather and washing. Also called deposit corrosion or attack.

pouring. The transfer of molten metal from furnace to ladle, ladle to ladle, or ladle into molds.

pouring basin. In metal casting, a basin on top of a mold that receives the molten metal before it enters the sprue or downgate.

powder. An aggregate of discrete particles that are usually in the size range of 1 to 1000 m.

powder cutting. See preferred term *chemical flux cutting* and *metal powder cutting*.

powder flame spraying. A thermal spraying process variation in which the material to be sprayed is in powder form. See also *flame spraying*.

powder forging. The plastic deformation of a powder metallurgy *compact* or *preform* into a fully dense finished shape by using compressive force; usually done hot and within closed dies.

powder lubricant. In powder metallurgy, an agent or component incorporated into a mixture to facilitate compacting and ejecting of the compact from its mold.

powder metallurgy (P/M). The technology and art of producing metal powders and utilizing metal powders for production of massive materials and shaped objects.

powder metallurgy forging. See *powder forging.*

powder metallurgy part. A shaped object that has been formed from metal powders and sintered by heating below the melting point of the major constituent. A structural or mechanical component made by the powder metallurgy process.

prealloyed powder. A metallic powder composed of two or more elements that are alloyed in the powder manufacturing process and in which the particles are of the same nominal composition throughout.

precharge. In metal forming, the pressure introduced into the cavity prior to forming of the part.

precious metals. Relatively scarce, highly corrosion resistant, valuable metals found in periods 5 and 6 (groups VIII and Ib) of the periodic table. They include ruthenium, rhodium, palladium, silver, asmium, iridium, platinum, and gold. See also *noble metal.*

precipitation. In metals, the separation of a new phase from solid or liquid solution, usually with changing conditions of temperature, pressure, or both.

precipitation hardening. Hardening in metals caused by the precipitation of a constituent from a supersaturated solid solution. See also *age hardening* and *aging.*

precipitation heat treatment. *Artificial aging* of metals in which a constituent precipitates from a supersaturated solid solution.

precision casting. A metal casting of reproducible, accurate dimensions, regardless of how it is made. Often used interchangeably with *investment casting.*

precision forging. A forging produced to closer tolerances than normally considered standard by the industry. With precision forging, a net shape, or at least a near-net shape, can be produced in the as-forged condition. See also *net shape.*

precision grinding. Machine grinding to specified dimensions and low *tolerances.*

precoat. (1) In investment casting, a special refractory slurry applied to a wax or plastic expendable pattern to form a thin coating that serves as a desirable base for application of the main slurry. See also *investment casting.* (2) To make the thin coating. (3) The thin coating itself.

precoated metal products. Mill products that have a metallic, organic, or conversion coating applied to their surfaces before they are fabricated into parts.

precracked specimen. A mechanical test specimen that is notched and subjected to alternating stresses until a crack has developed at the root of the notch.

preferred orientation. A condition of a polycrystalline aggregate in which the crystal orientations are not random, but rather exhibit a tendency for alignment with a specific direction in the bulk material, commonly related to the direction of working. See also *texture.*

preforming. (1) The initial pressing of a metal powder to form a compact that is to be subjected to a subsequent pressing operation other than coining or sizing. (2) Preliminary forming operations, especially for impression-die forging.

preheating. (1) Heating before some further thermal or mechanical treatment. For tool steel, heating to an intermediate temperature immediately before final austenitizing. For some nonferrous alloys, heating to a high temperature for a long time, in order to homogenize the structure before working. (2) In welding and related processes, heating to an intermediate temperature for a short time immediately before welding, brazing, soldering, cutting, or thermal spraying. (3) In powder metallurgy, an early stage in the sintering procedure when, in a continuous furnace, lubricant or binder burnoff occurs without atmosphere protection prior to actual sintering in the protective atmosphere of the high heat chamber.

presintering. Heating a powder metallurgy compact to a temperature below the final sintering temperature, usually to increase the ease of handling or shaping of a compact or to remove a lubricant or binder (*burnoff*) prior to sintering.

press. A machine tool having a stationary bed and a slide or ram that has reciprocating motion at right angles to the bed surface, the slide being guided in the frame of the machine. See also *hydraulic press, mechanical press, and slide.*

press brake. An open-frame single-action press used to bend, blank, corrugate, curl, notch, perforate, pierce, or punch sheet metal or plate.

press-brake forming. A metal forming process in which the workpiece is placed over an open die and pressed down into the die by a punch that is actuated by the ram portion of a *press brake.* The process is most widely used for the forming of relatively long, narrow parts that are not adaptable to *press forming* and for applications in which production quantities are too small to warrant the tooling cost for contour *roll forming.*

pressed density. The weight per unit volume of an unsintered compact. Same as green density.

press forming. Any sheet metal forming operation performed with tooling by means of a mechanical or hydraulic press.

pressing area. The clear distance (left to right) between housings, stops, gibs, gibways, or shoulders of strain rods, multiplied by the total distance from front to back on the bed of a metal forming *press.* Sometimes called working area.

pressing crack. A rupture in a green powder metallurgy compact that develops during ejection of the compact from the die. Sometimes referred to as a slip crack.

press quenching. A quench in which hot dies are pressed and aligned with a part before the quenching process begins. Then the part is placed in contact with a quenching medium in a controlled manner. This process avoids part distortion.

pressure casting. (1) Making castings with pressure on the molten or plastic metal, as in *die casting, centrifugal casting,* cold chamber pressure casting, and *squeeze casting.* (2) A casting made with pressure applied to the molten or plastic metal.

pressure gas welding. An oxyfuel gas welding process that produces coalescence simultaneously over the entire area of abutting surfaces by heating them with gas flames obtained from combustion of a fuel gas with oxygen and by application of pressure, without the use of filler metal.

pressure sintering. A hot pressing technique that usually employs low loads, high sintering temperatures, continuous or discontinuous sintering, and simple molds to contain the powder. Although the terms pressure sintering and *hot pressing* are used interchangeably, distinct differences exist between the two processes. In pressure sintering, the emphasis is on thermal processing; in hot pressing, applied pressure is the main process variable.

pressure welding. See preferred terms *cold weldng, diffusion welding, forge welding, hot pressure welding, pressure gas welding,* and *solid-state welding.*

primary creep. The first, or initial, stage of *creep*, or time-dependent deformation.

primary crystals. The first type of crystals that separate from a melt during solidification.

primary metal. Metal extracted from minerals and free of reclaimed metal scrap. Compare with *native metal*.

primary mill. A mill for rolling ingots or the rolled products of ingots to blooms, billets, or slabs. This type of mill is often called a *blooming mill* and sometimes called a cogging mill.

primes. Metal products, principally sheet and plate, of the highest quality and free from blemishes or other visible imperfections.

principal stress (normal). The maximum or minimum value of the *normal stress* at a point in a plane considered with respect to all possible orientations of the considered plane. On such principal planes the shear stress is zero. There are three principal stresses on three mutually perpendicular planes. The state of stress at a point may be (1) uniaxial, a state of stress in which two of the three principal stresses are zero, (2) biaxial, a state of stress in which only one of the three principal stresses is zero, and (3) triaxial, a state of stress in which none of the principal stresses is zero. Multiaxial stress refers to either biaxial or triaxial stress.

process annealing. A heat treatment used to soften metal for further cold working. In ferrous sheet and wire industries, heating to a temperature close to but below the lower limit of the transformation range and subsequently cooling for working. In the nonferrous industries, heating above the recrystallization temperatures at a time and temperature sufficient to permit the desired subsequent cold working.

process metallurgy. The science and technology of winning metals from their ores and purifying metals; sometimes referred to as chemical metallurgy. Its two chief branches are *extractive metallurgy* and *refining*.

proeutectoid phase. Particles of a phase in ferrous alloys that precipitate during cooling after austenitizing but before the eutectoid transformation takes place. See also *eutectoid*.

profiling. Any operation that produces an irregular contour on a workpiece, for which a tracer or template-controlled duplicating equipment usually is employed.

progressive aging. Aging by increasing the temperature in steps or continuously during the aging cycle. See also *aging* and compare with *interrupted aging* and *step aging*.

progressive die. A *die* with two or more stations arranged in line for performing two or more operations on a part; one operation is usually performed at each station.

progressive forming. Sequential forming at consecutive stations with a single die or separate dies.

projection welding. A resistance welding process which produces coalescence of metals with the heat obtained from resistance to electric current through the work parts held together under pressure by electrodes. The resulting welds are localized at predetermined points by projections, embossments, or intersections.

proof. (1) To test a component or system at its peak operating load or pressure. (2) Any reproduction of a *die impression* in any material; often a lead or plaster cast. See also *die proof*.

proof load. A predetermined load, generally some multiple of the service load, to which a specimen or structure is submitted before acceptance for use.

proof stress. (1) A specified stress to be applied to a member or structure to indicate its ability to withstand service loads. (2) The stress that will cause a specified small *permanent set* in a material.

proportional limit. The greatest stress a material is capable of developing without a deviation from straight-line proportionality between stress and strain. See also *elastic limit* and *Hooke's law*.

protective atmosphere. (1) A gas envelope surrounding the part to be brazed, welded, or thermal sprayed, with the gas composition controlled with respect to chemical composition, dew point, pressure, flow rate, and so forth. Examples are inert gases, combusted fuel gases, hydrogen, and vacuum. (2) The atmosphere in a heat treating or sintering furnace designed to protect the parts or compacts from oxidation, nitridation, or other contamination from the environment.

pseudobinary system. (1) A three-component or ternary alloy system in which an intermediate phase acts as a component. (2) A vertical section through a ternary diagram.

pseudocarburizing. See *blank carburizing*.

pseudonitriding. See *blank nitriding*.

puckering. Wrinkling or buckling in a drawn shell in an area originally inside the draw ring.

pull cracks. In a casting, cracks that are caused by residual stresses produced during cooling, and that result from the shape of the object.

pulverization. The process of reducing metal powder particle sizes by mechanical means; also called comminution or mechanical disintegration.

punch. (1) The male part of a die—as distinguished from the female part, which is called the die. The punch is usually the upper member of the complete die assembly and is mounted on the *slide* or in a *die set* for alignment (except in the inverted die). (2) In double-action draw dies, the punch is the inner portion of the upper die, which is mounted on the plunger (inner slide) and does the drawing. (3) The act of piercing or punching a hole. Also referred to as *punching*. (4) The movable tool that forces material into the die in powder molding and most metal forming operations. (5) The movable die in a trimming press or a forging machine. (6) The tool that forces the stock through the die in rod and tube extrusion and forms the internal surface in can or cup extrusion.

punching. (1) The die shearing of a closed contour in which the sheared out sheet metal part is scrap. (2) Producing a hole by die shearing, in which the shape of the hole is controlled by the shape of the punch and its mating die. Multiple punching of small holes is called *perforating*. See also *piercing*.

punch press. (1) In general, any mechanical press. (2) In particular, an endwheel gap-frame press with a fixed bed, used in piercing.

punch radius. The radius on the end of the punch that first contacts the work, sometimes called *nose radius*.

push bench. Equipment used for drawing moderately heavy-gage tubes by cupping sheet metal and forcing it through a die by pressure exerted against the inside bottom of the cup.

pusher furnace. A type of continuous furnace in which parts to be heated are periodically charged into the furnace in containers, which are pushed along the hearth against a line of previously charged containers thus advancing the containers toward the discharge end of the furnace, where they are removed.

push welding. Spot or projection welding in which the force is applied manually to one electrode, and the work or backing plate takes the place of the other electrode.

pyramidal plane. In noncubic crystals, any plane that intersects all three axes.

pyrometallurgy. High-temperature *winning* or *refining* of metals.

pyrometer. A device for measuring temperatures above the range of liquid thermometers.

Q

quality. (1) The totality of features and characteristics of a product or service that bear on its ability to satisfy a given need (fitness-for-use concept of quality). (2) Degree of excellence of a product or service (comparative concept). Often determined subjectively by comparison against an ideal standard or against similar products or services available from other sources. (3) A quantitative evaluation of the features and characteristics of a product or service (quantitative concept).

quantitative metallography. Determination of specific characteristics of a microstructure by quantitative measurements on micro-graphs or metallographic images. Quantities so measured include volume concentration of phases, grain size, particle size, mean free path between like particles or secondary phases, and surface area to volume ratio of microconstituents, particles, or grains.

quarter hard. A *temper* of nonferrous alloys and some ferrous alloys characterized by tensile strength about midway between that of *dead soft* and *half hard* tempers.

quasi-binary system. In a ternary or higher-order system, a linear composition series between two substances each of which exhibits congruent melting, wherein all equilibria, at all temperatures or pressures, involve only phases having compositions occurring in the linear series, so that the series may be represented as a binary on a *phase diagram*.

quasi-cleavage fracture. A fracture mode that combines the characteristics of *cleavage fracture* and *dimple fracture*. An intermediate type of fracture found in certain high-strength metals.

quench-age embrittlement. *Embrittlement* of low-carbon steels resulting from precipitation of solute carbon at existing dislocations and from precipitation hardening of the steel caused by differences in the solid solubility of carbon in ferrite at different temperatures. Quench-age embrittlement usually is caused by rapid cooling of the steel from temperatures slightly below Ac_1 (the temperature at which austenite begins to form), and can be minimized by quenching from lower temperatures.

quench aging. *Aging* induced by rapid cooling after *solution heat treatment*.

quench annealing. Annealing an austenitic ferrous alloy by *solution heat treatment* followed by rapid quenching.

quench cracking. Fracture of a metal during quenching from elevated temperature. Most frequently observed in hardened carbon steel, alloy steel, or tool steel parts of high hardness and low toughness. Cracks often emanate from fillets, holes, corners, or other stress raisers and result from high stresses due to the volume changes accompanying transformation to martensite.

quench hardening. (1) Hardening suitable alpha-beta alloys (most often certain copper to titanium alloys) by solution treating and quenching to develop a martensitic-like structure. (2) In ferrous alloys, hardening by austenitizing and then cooling at a rate such that a substantial amount of austenite transforms to martensite.

quenching. Rapid cooling of metals (often steels) from a suitable elevated temperature. This generally is accomplished by immersion in water, oil, polymer solution, or salt, although forced air is sometimes used. See also *brine quenching*, *caustic quenching*, *direct quenching*, *fog quenching*, *forced-air quenching*, *hot quenching*, *intense quenching*, *interrupted quenching*, *oil quenching*, *press quenching*, *selective quenching*, *spray quenching*, *time quenching*, and *water quenching*.

quenching crack. A crack formed in a metal as a result of thermal stresses produced by rapid cooling from a high temperature.

quenching oil. Oil used for quenching metals during a heat treating operation.

R

rabbit ear. Recess in the corner of a metal forming die to allow for wrinkling or folding of the blank.

racking. A term used to describe the placing of metal parts to be heat treated on a rack or tray. This is done to keep parts in a proper position to avoid heat-related distortions and to keep the parts separated.

radial draw forming. The forming of sheet metals by the simultaneous application of tangential stretch and radial compression forces. The operation is done gradually by tangential contact with the die member. This type of forming is characterized by very close dimensional control.

radial forging. A process using two or more moving anvils or dies for producing shafts with constant or varying diameters along their length or tubes with internal or external variations. Often incorrectly referred to as *rotary forging*.

radial marks. Lines on a fracture surface that radiate from the fracture origin and are visible to the unaided eye or at low magnification. Radial marks result from the intersection and connection of brittle fractures propagating at different levels. Also known as shear ledges. See also *chevron pattern*.

radial rake. The angle between the tooth face and a radial line passing through the cutting edge in a plane perpendicular to the cutter axis. See also the figure accompanying *face mill*.

radiation damage. A general term for the alteration of properties of a material arising from exposure to ionizing radiation (penetrating radiation), such as x-rays, gamma rays, neutrons, heavy-particle radiation, or fission fragments in nuclear fuel material. See also *neutron embrittlement*.

radioactive element. An element that has at least one isotope that undergoes spontaneous nuclear disintegration to emit positive alpha particles, negative beta particles, or gamma rays.

radioactivity. (1) The property of the nuclei of some isotopes to spontaneously decay (lose energy). Usual mechanisms are emission of α, β, or other particles and splitting (fissioning). Gamma rays are frequently, but not always, given off in the process. (2) A particular component from a radioactive source, such as β radioactivity.

radiograph. A photographic shadow image resulting from uneven absorption of penetrating radiation in a test object. See also *radiography*.

radiography. A method of nondestructive inspection in which a test object is exposed to a beam of x-rays or gamma rays and the resulting shadow image of the object is recorded on photographic film placed behind the object, or displayed on a viewing screen or television monitor (real-time radiography). Internal discontinuities are detected by observing and interpreting variations in the image caused by differences in thickness, density, or absorption within the test object. Variations of radiography include computed tomography, *fluoroscopy*, and neutron radiography. See also *real-time radiography*.

radius of bend. The radius of the cylindrical surface of the pin or mandrel that comes in contact with the inside surface of the bend during bending. In the case of free or semiguided bends to 180° in which a shim or block is used, the radius of bend is one-half the thickness of the shim or block.

rake. The angular relationship between the tooth face, or a tangent to the tooth face at a given point, and a given reference plane or line. See also the figures accompanying *face mill* and *single-point tool*.

ram. The moving or falling part of a drop hammer or press to which one of the dies is attached; sometimes applied to the upper flat die of a steam hammer. Also referred to as the *slide*.

ramming. (1) Packing foundry sand, refractory, or other material into a compact mass. (2) The compacting of molding (foundry) sand in forming a mold.

random sequence. A longitudinal welding sequence wherein the weld-bead increments are deposited at random to minimize distortion.

range of stress (S_r). The algebraic difference between the maximum and minimum stress in one cycle—that is

$$S_r = S_{max} - S_{min}$$

rapid solidification. The cooling or quenching of liquid (molten) metals at rates that range from 104 to 108 °C/s.

rare earth metal. A group of 17 chemically similar metals that includes the elements scandium and yttrium (atomic numbers 21 and 39, respectively) and the lanthanide elements (atomic numbers 57 through 71).

ratcheting. Progressive cyclic inelastic deformation (growth, for example) that occurs when a component or structure is subjected to a cyclic secondary stress superimposed on a sustained primary stress. The process is called thermal ratcheting when cyclic strain is induced by cyclic changes in temperature, and isothermal ratcheting when cyclic strain is mechanical in origin (even though accompanied by cyclic changes in temperature).

ratchet marks. Lines or markings on a fatigue fracture surface that results from the intersection and connection of fatigue fractures propagating from multiple origins. Ratchet marks are parallel to the overall direction of crack propagation and are visible to the unaided eye or at low magnification.

rate of strain hardening. Rate of change of *true stress* with respect to *true strain* in the plastic range.

rattail. A surface imperfection on a casting, occurring as one or more irregular lines, caused by expansion of sand in the mold. Compare with *buckle* (2).

reaction sintering. The sintering of a metal powder mixture consisting of at least two components that chemically react during the treatment.

reactive metal. A metal that readily combines with oxygen at elevated temperatures to form very stable oxides, for example, titanium, zirconium, and beryllium. Reactive metals may also become embrittled by the interstitial absorption of oxygen, hydrogen, and nitrogen.

real-time radiography. A method of nondestructive inspection in which a two-dimensional radiographic image can be immediately displayed on a viewing screen or television monitor. This technique does not involve the creation of a latent image; instead, the unabsorbed radiation is converted into an optical or electronic signal, which can be viewed immediately or can be processed in near real time with electronic and video equipment. See also *radiography*.

reaming. An operation in which a previously formed hole is sized and contoured accurately by using a rotary cutting tool (*reamer*) with one or more cutting elements (teeth). The principal support for the reamer during the cutting action is obtained from the workpiece.

recalescence. (1) The increase in temperature that occurs after undercooling, because the rate of liberation of heat during transformation of a material exceeds the rate of dissipation of heat. (2) A phenomenon, associated with the transformation of gamma iron to alpha iron on cooling (supercooling) of iron or steel, that is revealed by the brightening (reglowing) of the metal surface owing to the sudden increase in temperature caused by the fast liberation of the latent heat of transformation. Contrast with *decalescence*.

recarburize. (1) To increase the carbon content of molten cast iron or steel by adding carbonaceous material, high-carbon pig iron, or a high-carbon alloy. (2) To carburize a metal part to return surface carbon lost in processing; also known as carbon restoration.

recess. A groove or depression in a surface.

recovery. (1) The time-dependent portion of the decrease in strain following unloading of a specimen at the same constant temperature as the initial test. Recovery is equal to the total decrease in strain minus the instantaneous recovery. (2) Reduction or removal of work-hardening effects in metals without motion of large-angle grain boundaries. (3) The proportion of the desired component obtained by processing an ore, usually expressed as a percentage.

recrystallization. (1) The formation of a new, strain-free grain structure from that existing in cold-worked metal, usually accomplished by heating. (2) The change from one crystal structure to another, as occurs on heating or cooling through a critical temperature. (3) A process, usually physical, by which one crystal species is grown at the expense of another or at the expense of others of the same substance but smaller in size. See also *crystallization*.

recrystallization annealing. Annealing cold worked metal to produce a new grain structure without phase change.

recrystallization temperature. (1) The lowest temperature at which the distorted grain structure of a cold-worked metal is replaced by a new, strain-free grain structure during prolonged heating. Time, purity of the metal, and prior deformation are important factors. (2) The approximate minimum temperature at which complete recrystallization of a cold-worked metal occurs within a specified time.

recrystallized grain size. (1) The grain size developed by heating cold-worked metal. The time and temperature are selected so that, although recrystallization is complete, essentially no grain growth occurs. (2) In aluminum and magnesium alloys, the grain size after recrystallization, without regard to grain growth or the recrystallized conditions. See also *recrystallization*.

recuperator. (1) Equipment for transferring heat from gaseous products of combustion to incoming air or fuel. The incoming material passes through pipes surrounded by a chamber through which the outgoing gases pass. (2) A continuous heat exchanger in which heat is conducted from the products of combustion to incoming air through flue walls.

red mud. A residue, containing a high percentage of iron oxide, obtained in purifying bauxite in the production of alumina in the *Bayer process*.

redox potential. This *potential* of a reversible oxidation-reduction electrode measured with respect to a *reference electrode*, corrected to the hydrogen electrode, in a given *electrode*.

redrawing. The second and successive deep-drawing operations in which cup-like shells are deepened and reduced in cross-sectional dimensions. See also *deep drawing*.

reducing agent. (1) A compound that causes *reduction*, thereby itself becoming oxidized. (2) A chemical that, at high temperatures, lowers the state of oxidation of other batch chemicals.

reducing atmosphere. (1) A furnace atmosphere which tends to remove oxygen from substances or materials placed in the furnace. (2) A chemically active protective atmosphere which at elevated temperature will reduce metal oxides to their metallic state. Reducing atmosphere is a relative term and such an atmosphere may be reducing to one oxide but not to another oxide.

reducing flame. (1) A gas flame produced with excess fuel in the inner flame. (2) A gas flame resulting from combustion of a mixture containing too much fuel or too little air.

reduction. (1) In cupping and deep drawing, a measure of the percentage decrease from blank diameter to cup diameter, or of diameter reduction in redrawing. (2) In forging, rolling and drawing, either the ratio of the original to final cross-sectional area or the percentage decrease in cross-sectional area. (3) A reaction in which there is a decrease in valence resulting from a gain in electrons. Contrast with *oxidation.*

reduction cell. A pot or tank in which either a water solution of a salt or a fused salt is reduced electrolytically to form free metals or other substances.

reduction in area (RA). The difference between the original cross-sectional area of a tensile specimen and the smallest area at or after fracture as specified for the material undergoing testing. Also known as reduction of area.

reel. (1) A spool or hub for coiling or feeding wire or strip. (2) To straighten and planish a round bar by passing it between contoured rolls.

reel breaks. Transverse breaks or ridges on successive inner laps of a coil that results from crimping of the lead end of the coil into a gripping segmented mandrel. Also called reel kinks.

reference electrodes. A nonpolarizable *electrode* with a known and highly reproducible *potential* used for potentiometric and voltammetric analyses. See also *calomel electrode.*

reference material. In materials characterization, a material of definite composition that closely resembles in chemical and physical nature the material with which an analyst expects to deal; used for calibration or standardization. See also *standard reference material.*

refining. The branch of *process metallurgy* dealing with the purification of crude or impure metals. Compare with *extractive metallurgy.*

reflowing. Melting of an electrodeposit followed by solidification. The surface has the appearance and physical characteristics of a hot dipped surface (especially tin or tin alloy plates). Also called flow brightening.

refractory. (1) A material (usually an inorganic, nonmetallic, ceramic material) of very high melting point with properties that make it suitable for such uses as furnace linings and kiln construction. (2) The quality of resisting heat.

refractory alloy. (1) A heat-resistant alloy. (2) An alloy having an extremely high melting point. See also *refractory metal.* (3) An alloy difficult to work at elevated temperatures.

refractory metal. A metal having an extremely high melting point and low vapor pressure; for example, niobium (columbium), tantalum, molybdenum, tungsten, and rhenium.

regenerator. Same as *recuperator* except that the gaseous products or combustion heat brick checkerwork in a chamber connected to the exhaust side of the furnace while the incoming air and fuel are being heated by the brick checkerwork in a second chamber, connected to the entrance side. At intervals, the gas flow is reversed so that incoming air and fuel contact hot checkerwork while that in the second chamber is being reheated by exhaust gases.

regulator. A device for controlling the delivery of welding or cutting gas at some substantially constant pressure.

reliability. A quantitative measure of the ability of a product or service to fulfill its intended function for a specified period of time.

relief. The result of the removal of tool material behind or adjacent to the cutting edge to provide clearance and prevent rubbing (heel drag). See also *relief angle.*

relief angle. The angle formed between a relieved surface and a given plane tangent to a cutting edge or to a point on a cutting edge. Also known as clearance angle. See also the figure accompanying *single-point tool.*

relieving. Buffing or other abrasive treatment of the high points of an embossed metal surface to produce highlights that contrast with the finish in the recesses.

remanence. The magnetic induction remaining in a magnetic circuit after removal of the applied magnetizing force. Sometimes called remanent induction.

repressing. The application of pressure to a previously pressed and sintered powder metallurgy compact, usually for the purpose of improving some physical or mechanical property or for dimensional accuracy.

residual elements. Small quantities of elements unintentionally present in an alloy.

residual stress. (1) The stress existing in a body at rest, in equilibrium, at uniform temperature, and not subjected to external forces. Often caused by the forming or thermal processing curing process. (2) An internal stress not depending on external forces resulting from such factors as cold working, phase changes, or temperature gradients. (3) Stress present in a body that is free of external forces or thermal gradients. (4) Stress remaining in a structure or member as a result of thermal or mechanical treatment or both. Stress arises in fusion welding primarily because the weld metal contracts on cooling from the solidus to room temperature.

resilience. (1) The amount of energy per unit volume released on unloading. (2) The capacity of a material, by virtue of high yield strength and low elastic modulus, to exhibit considerable elastic recovery on release of load.

resinoid wheel. A grinding wheel bonded with a synthetic resin.

resist. (1) Coating material used to mask or protect selected areas of a substrate from the action of an etchant, solder, or plating. (2) A material applied to prevent flow of brazing filler metal into unwanted areas.

resistance brazing. A resistance joining process in which the workpieces are heated locally and filler metal that is preplaced between the workpieces is melted by the heat obtained from resistance to the flow of electric current through the electrodes and the work. In the usual application of resistance brazing, the heating current is passed through the joint itself.

resistance seam welding. A resistance welding process which produces coalescence at the faying surfaces by the heat obtained from resistance to electric current through workpieces that are held together under pressure by electrode wheels. The resulting weld is a series of overlapping resistance spot welds made progressively along a joint by rotating the electrodes.

resistance soldering. Soldering in which the joint is heated by electrical resistance. Filler metal is either face fed into the joint or preplaced in the joint.

resistance spot welding. A process in which faying surfaces are joined in one or more spots by the heat generated by resistance to the flow of electric current through workpieces that are held together under force by electrodes. The contacting surfaces in the region of current concentration are heated by a short-time pulse of low-voltage, high-amperage current to form a fused nugget of weld metal. When the flow of current ceases, the electrode force is maintained while the weld metal rapidly cools and solidifies. The electrodes are retracted after each weld, which usually is completed in a fraction of a second.

resistance welding. A group of welding processes which produces coalescence of metals with resistance heating and pressure. See also *flash welding, projection welding, resistance seam welding,* and *resistance spot welding.*

resistance welding die. The part of a resistance welding machine, usually shaped to the work contour, in which the parts being welded are held and which conducts the welding current.

restraint. Any external mechanical force that prevents a part from moving to accommodate changes in dimension due to thermal expansion or contraction. Often applied to weldments made while clamped in a fixture. Compare with *constraint.*

restriking. (1) The striking of a trimmed but slightly misaligned or otherwise faulty forging with one or more blows to improve alignment, improve surface condition, maintain close tolerances, increase hardness, or effect other improvements. (2) A *sizing* operation in which coining or stretching is used to correct or alter profiles and to counteract distortion. (3) A salvage operation following a primary forging operation in which the parts involved are rehit in the same forging die in which the pieces were last forged.

resultant rake. The angle between the tooth face and an axial plane through the tooth point measured in a plane perpendicular to the cutting edge. The resultant rake of a cutter is a function of three other angles, radial rake, axial rake, and corner angle.

retort. A vessel used for distillation of volatile materials, as in separation of some metals and in destructive distillation of coal.

reverberatory furnace. A furnace in which the flame used for melting the metal does not impinge on the metal surface itself, but is reflected off the walls of the root of the furnace. The metal is actually melted by the generation of heat from the walls and the roof of the furnace.

reverse-current cleaning. *Electrolytic cleaning* in which a current is passed between electrodes through a solution, and the part is set up as the anode. Also called *anodic cleaning.*

reverse drawing. *Redrawing* of a sheet metal part in a direction opposite to that of the original drawing.

reverse polarity. Direct-current arc welding circuit arrangement in which the electrode is connected to the positive terminal. A synonym for *direct current electrode positive (DCEP).* Contrast with *straight polarity.*

reverse redrawing. A second drawing operation in a direction opposite to that of the original drawing.

rheocasting. Casting of a continuously stirred semisolid metal slurry. The process involves vigorous agitation of the melt during the early stages of solidification to break up solid dendrites into small spherulites. See also *semisolid metal forming.*

rib. A long V-shaped or radiused indentation used to strengthen large sheet metal panels. (2) A long, usually thin protuberance used to provide flexural strength to a forging (as in a rib-web forging).

rigging. The engineering design, layout, and fabrication of pattern equipment for producing castings; including a study of the casting solidification program, feeding and gating, risering, skimmers, and fitting flasks.

rimmed steel. A low-carbon steel containing sufficient iron oxide to give a continuous evolution of carbon monoxide while the ingot is solidifying, resulting in a case or rim of metal virtually free of voids. Sheet and strip products made from rimmed steel ingots have very good surface quality.

ring and circle shear. A cutting or shearing machine with two rotary-disk cutters driven in unison and equipped with a circle attachment for cutting inside circles or rings from sheet metal, where it is impossible to start the cut at the edge of the sheet. One cutter shaft is inclined to the other to provide cutting clearance so that the outside section remains flat and usable. See also *circle shear* and *rotary shear.*

ring rolling. The process of shaping weldless rings from pierced disks or shaping thick-wall ring-shaped blanks between rolls that control wall thickness, ring diameter, height, and contour.

riser. (1) A reservoir of molten metal connected to a casting to provide additional metal to the casting, required as the result of shrinkage before and during solidification. (2) That section of pipeline extending from the ocean floor up the offshore oil-drilling platform. Also, the vertical tube in a steam generator convection bank that circulates water and steam upward.

riser blocks. (1) Plates or pieces inserted between the top of a metal forming press bed or bolster and the die to decrease the height of the die space. (2) Spacers placed between bed and housings to increase *shut height* on a four-piece tie-rod straight-side press.

river pattern. A term used in fractography to describe a characteristic pattern of cleavage steps running parallel to the local direction of crack propagation on the fracture surfaces of grains that have separated by *cleavage.*

riveting. Joining of two or more members of a structure by means of metal rivets, the unheaded end being upset after the rivet is in place.

roasting. Heating an ore to effect some chemical change that will facilitate *smelting.*

robber. An extra cathode or cathode extension that reduces the current density on what would otherwise be a high-current-density area on work being electroplated.

Rochelle copper. (1) A copper electrodeposit obtained from copper cyanide plating solution to which Rochelle salt (sodium potassium tartrate) has been added for grain refinement, better anode corrosion, and cathode efficiency. (2) The solution from which a Rochelle copper electrodeposit is obtained.

rock candy fracture. A fracture that exhibits separated-grain facets; most often used to describe an *intergranular fracture* in a large-grained metal.

rocking shear. A type of guillotine shear that utilizes a curved blade to shear sheet metal progressively from side to side by a rocker motion.

Rockwell hardness number. A number derived from the net increase in the depth of impression as the load on an indentor is increased from a fixed minor load to a major load and then returned to the minor load. Various scales of Rockwell hardness numbers have been developed based on the hardness of the materials to be evaluated. The scales are designated by alphabetic suffixes to the hardness designation. For example, 64 HRC represents the Rockwell hardness number of 64 on the Rockwell C scale. See also *Rockwell superficial hardness number.*

Rockwell hardness test. An indentation hardness test using a calibrated machine that utilizes the depth of indentation, under constant load, as a measure of hardness. Either a 120° diamond cone with a slightly rounded point or a 1.6 or 3.2 mm (1/16 or 1/8 in.) diam steel ball is used as the indenter. See also the figure accompanying *brale indenter.*

Rockwell superficial hardness number. Like the *Rockwell hardness number,* the superficial Rockwell number is expressed by the symbol HR followed by a scale designation. For example, 81 HR30N represents the Rockwell superficial hardness number of 81 on the Rockwell 30N scale.

Rockwell superficial hardness test. The same test as used to determine the Rockwell hardness number except that smaller minor and major loads are used. In Rockwell testing, the minor load is 10 kgf, and the major load is 60, 100, or 150 kgf. In superficial Rockwell testing, the minor load is 3 kgf, and major loads are 15, 30, or 45 kgf. In both tests, the indenter may be either a diamond cone or a steel ball, depending principally on the characteristics of the material being tested.

rod. A solid round metal section 9.5 mm (3/8 in.) or greater in diameter, whose length is great in relation to its diameter.

rod mill. (1) A *hot mill* for rolling rod. (2) A mill for fine grinding, somewhat similar to a *ball mill*, but employing long steel rods instead of balls to effect grinding.

roll bending. Curving sheets, bars, and sections by means of rolls. See also *bending rolls*.

roll compacting. Progressive compacting of metal powders by use of a rolling mill.

roller hearth furnace. A modification of the pusher-type continuous furnace that provides for rollers in the hearth or muffle of the furnace whereby friction is greatly reduced and lightweight trays can be used repeatedly without risk of unacceptable distortion and damage to the work. See also *pusher furnace*.

roller leveler breaks. Obvious transverse *breaks* usually about 3 to 6 mm (1/8 to 1/4 in.) apart caused by the sheet metal fluting during *roller leveling*. These will not be removed by stretching.

roller leveler lines. Same as *leveler lines*.

roller leveling. *Leveling* by passing flat sheet metal stock through a machine having a series of small-diameter staggered rolls that are adjusted to produce repeated reverse bending.

roller stamping die. An engraved roller used for impressing designs and markings on sheet metal.

roll flattening. The flattening of metal sheets that have been rolled in packs by passing them separately through a two-high cold mill with virtually no deformation. Not to be confused with *roller leveling*.

roll forging. A process of shaping stock between two driven rolls that rotate in opposite directions and have one or more matching sets of grooves in the rolls; used to produce finished parts or preforms for subsequent forging operations.

roll forming. Metal forming through the use of power-driven rolls whose contour determines the shape of the product; sometimes used to denote power *spinning*.

rolling. The reduction of the cross-sectional area of metal stock, or the general shaping

of metal products, through the use of rotating rolls. See also *rolling mills*.

rolling-contact fatigue. Repeated stressing of a solid surface due to rolling contact between it and another solid surface or surfaces. Continued rolling-contact fatigue of bearing or gear surfaces may result in rolling-contact damage in the form of subsurface fatigue cracks and/or material pitting and spallation.

rolling mills. Machines used to decrease the cross-sectional area of metal stock and to produce certain desired shapes as the metal passes between rotating rolls mounted in a framework comprising a basic unit called a *stand*. Cylindrical rolls produce flat shapes; grooved rolls produce rounds, squares, and structural shapes. See also *four-high mill*, *Sendzimir mill*, and *two-high mill*.

roll resistance spot welding. Process for making separated resistance spot welds with one or more rotating circular electrodes. The rotation of the electrodes may or may not be stopped during the making of a weld.

roll straightening. The straightening of metal stock of various shapes by passing it through a series of staggered rolls, the rolls usually being in horizontal and vertical planes, or by reeling in two-roll straightening machines.

roll threading. See preferred term *thread rolling*.

roll welding. Solid-state welding in which metals are heated, then welded together by applying pressure, with rolls, sufficient to cause deformation at the faying surfaces. See also *forge welding*.

root crack. A crack in either the weld or heat-affected zone at the root of a weld.

root face. The portion of a weld groove face adjacent to the root of the joint.

root of joint. The portion of a weld joint where the members are closest to each other before welding. In cross section, this may be a point, a line, or an area.

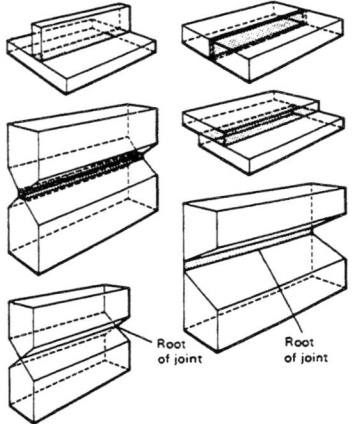

root of weld. The points, as shown in cross section, at which the weld bead intersects the base-metal surfaces either nearest to or coincident with the *root of joint*.

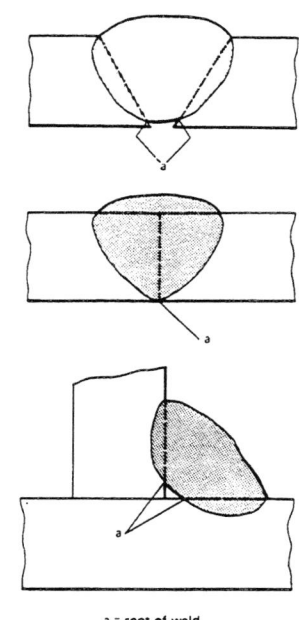

a = root of weld

root opening. In a weldment, the separation between the members at the *root of joint* prior to welding.

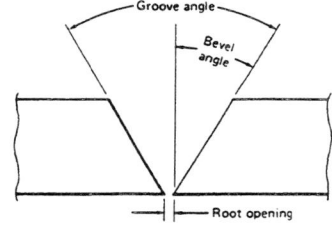

root pass. The first bead of a *multiple-pass weld*, laid in the *root of joint*.

root penetration. The depth that a weld extends into the *root of joint*, measured on the centerline of the root cross section. See also the figure accompanying *joint penetration*.

rosette. (1) Rounded configuration of microconstituents in metals arranged in whorls or radiating from a center. (2) Strain gages arranged to indicate at a single position strains in three different directions.

rotary forging. A process in which the workpiece is pressed between a flat anvil and a swiveling (rocking) die with a conical working face; the platens move toward each other during forging. Also called orbital forging. Compare with *radial forging*.

rotary furnace. A circular furnace constructed so that the hearth and workpieces rotate around the axis of the furnace during heating. Also called rotary hearth furnace.

rotary press. A machine for forming powder metallurgy parts that is fitted with a rotating table carrying multiple die assemblies in which powder is compacted.

rotary retort furnace. A continuous-type furnace in which the work advances by means of an internal spiral, which gives good control of the retention time within the heated chamber.

rotary shear. A sheet metal cutting machine with two rotating-disk cutters mounted on parallel shafts driven in unison.

rotary swager. A swaging machine consisting of a power-driven ring that revolves at high speed, causing rollers to engage cam surfaces and force the dies to deliver hammerlike blows on the work at high frequency. Both straight and tapered sections can be produced.

rotary swaging. A *bulk forming* process for reducing the cross-sectional area or otherwise changing the shape of bars, tubes, or wires by repeated radial blows with one or more pairs of opposed dies.

rouge finish. A highly reflective finish produced with rouge (finely divided, hydrated iron oxide) or other very fine abrasive, similar in appearance to the bright polish or mirror finish on sterling silver utensils.

rough blank. A *blank* for a metal forming or drawing operation, usually of irregular outline, with necessary stock allowance for process metal, which is trimmed after forming or drawing to the desired size.

rough grinding. Grinding without regard to finish, usually to be followed by a subsequent operation.

roughing stand. The first stand (or several stands) of rolls through which a reheated *billet* passes in front of the finishing stands. See also *rolling mills* and *stand*.

rough machining. Machining without regard to finish, usually to be followed by a subsequent operation.

roughness. (1) Relatively finely spaced surface irregularities, the heights, widths, and directions of which establish the predominant surface pattern. (2) The microscopic peak-to-valley distances of surface protuberances and depressions. See also *surface roughness* and the figure accompanying *waviness*.

rubber forming. Forming a sheet metal wherein rubber or another resilient material is used as a functional die part. Processes in which rubber is employed only to contain the hydraulic fluid are not classified as rubber forming.

rubber-pad forming. A sheet metal forming operation for shallow parts in which a confined, pliable rubber pad attached to the press slide (ram) is forced by hydraulic pressure to become a mating die for a punch or group of punches placed on the press bed or baseplate. Developed in the aircraft industry for the limited production of a large number of diversified parts, the process is limited to the forming of relatively shallow parts, normally not exceeding 40 mm (1.5 in.) deep. Also known as the Guerin process. Variations of the *Guerin process* include the *Marforming process*, the *fluid-cell process*, and *fluid forming*.

rubber wheel. A grinding wheel made with a rubber bond.

runner. (1) A channel through which molten metal flows from one receptacle to another. (2) The portion of the gate assembly of a casting that connects the sprue with the gate(s). (3) Parts of patterns and finished castings corresponding to the portion of the gate assembly described in (2).

runner box. A distribution box that divides molten metal into several streams before it enters the casting mold cavity.

runout. (1) The unintentional escape of molten metal from a mold, crucible, or furnace. (2) An imperfection in a casting caused by the escape of metal from the mold.

rupture stress. The stress at failure. Also known as *breaking stress* or *fracture stress*.

rust. A visible corrosion product consisting of hydrated oxides of iron. Applied only to ferrous alloys. See also *white rust*.

S

sacrificial protection. Reduction of corrosion of a metal in an *electrolyte* by galvanically coupling it to a more anodic metal; a form of *cathodic protection*.

saddling. Forming a seamless metal ring by forging a pierced disk over a mandrel (or saddle).

sag. An increase or decrease in the section thickness of a casting caused by insufficient strength of the mold sand of the cope or of the core.

salt bath heat treatment. Heat treatment for metals carried out in a bath of molten salt. See also *immersed-electrode furnaces* and *submerged-electrode furnaces*.

salt fog test. An *accelerated corrosion test* in which specimens are exposed to a fine mist of a solution usually containing sodium chloride, but sometimes modified with other chemicals. Also known as salt spray test.

salt spray test. See *salt fog test*.

sample. (1) One or more units of a product (or a relatively small quantity of a bulk material) withdrawn from a *lot* or process stream and then tested or inspected to provide information about the properties, dimensions, or other quality characteristics of the lot or process stream. (2) A portion of a material intended to be representative of the whole.

sand. A granular material naturally or artificially produced by the disintegration or crushing of rocks or mineral deposits. In casting, the term denotes an aggregate, with an individual particle (grain) size of 0.06 to 2 mm (0.002 to 0.08 in.) in diameter, that is largely free of finer constituents, such as silt and clay, which are often present in natural sand deposits. The most commonly used foundry sand is silica; however, zircon, olivine, aluminum silicates, and other crushed ceramics are used for special applications.

sandblasting. Abrasive blasting with sand. See also *blasting or blast cleaning* and compare with *shotblasting*.

sand casting. Metal castings produced in sand molds.

sand hole. A pit in the surface of a sand casting resulting from a deposit of loose sand on the surface of the mold.

sandwich rolling. Rolling two or more strips of metal in a pack, sometimes to form a roll-welded composite.

satin finish. A diffusely reflecting surface finish on metals, lustrous but not mirrorlike. One type is a *butler finish*.

saw gumming. In saw manufacture, grinding away of punch marks or milling marks in the gullets (spaces between the teeth) and, in some cases, simultaneous sharpening of the teeth; in reconditioning of worn saws, restoration of the original gullet size and shape.

sawing. Using a toothed blade or disc to sever parts or cut contours.

scab. A defect on the surface of a casting that appears as a rough, slightly raised surface blemish, crusted over by a thin porous layer of metal, under which is a honeycomb or cavity that usually contains a layer of sand; defect common to thin-wall portions of the casting or around hot areas of the mold.

scale. Surface oxidation, consisting of partially adherent layers of corrosion products, left on metals by heating or casting in air or in other oxidizing atmospheres.

scale pit. (1) A surface depression formed on a forging due to scale remaining in the dies during the forging operation. (2) A pit in the ground in which scale (such as that carried off by cooling water from rolling mills) is allowed to settle out as one step in the treatment of effluent waste water.

scaling. (1) Forming a thick layer of oxidation products on metals at high temperature. Scaling should be distinguished from rusting, which involves the formation of hydrated oxides. See also *rust.* (2) Depositing water-insoluble constituents on a metal surface, as in cooling tubes and water boilers.

scalping. Removing surface layers from an *ingot, billet,* or *slab.*

scarfing. Cutting surface areas of metal objects, ordinarily by using an oxyfuel gas torch. The operation permits surface imperfections to be cut from ingots, billets, or the edges of plate that are to be beveled for butt welding. See also *chipping.*

scarf joint. A butt joint in which the plane of the joint is inclined with respect to the main axis of the members.

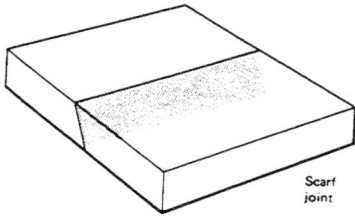

Scarf joint

Sceleroscope hardness number (HSc or HSd). A number related to the height of rebound of a diamond-tipped hammer dropped on the material being tested. It is measured on a scale determined by dividing into 100 units the average rebound of the hammer from a quenched (to maximum hardness) and untempered AISI W-5 tool steel test block.

Scleroscope hardness test. A dynamic indentation hardness test using a calibrated instrument that drops a diamond-tipped hammer from a fixed height onto the surface of the material being tested. The height of rebound of the hammer is a measure of the hardness of the material.

scorification. Oxidation, in the presence of fluxes, of molten lead containing precious metals, to partly remove the lead in order to concentrate the precious metals.

scoring. (1) The formation of severe scratches in the direction of sliding. (2) The act of producing a scratch or narrow groove in a surface by causing a sharp instrument to move along that surface. (3) The marring or scratching of any formed metal part by metal pickup on the punch or die. (4) The reduction in thickness of a material along a line to weaken it intentionally along that line.

scouring. (1) A wet or dry cleaning process involving mechanical scrubbing. (2) A wet or dry mechanical finishing operation, using fine abrasive and low pressure, carried out by hand or with a cloth or wire wheel to produce *satin* or *butler*-type finishes.

scrap. (1) Products that are discarded because they are defective or otherwise unsuitable for sale. (2) Discarded metallic material, from whatever source, that may be reclaimed through melting and refining.

scratch hardness. The hardness of a metal determined by the width of a scratch made by drawing a cutting point across the surface under a given pressure.

screen. (1) The woven wire or fabric cloth, having square openings, used in a sieve for retaining particles greater than the particular mesh size. U.S. standard, ISO, or Tyler screen sizes are commonly used. (2) One of a set of sieves, designated by the size of the openings, used to classify granular aggregates such as sand, ore, or coke by particle size. (3) A perforated sheet placed in the gating system of a mold to separate impurities from the molten metal.

screw dislocation. See *dislocation.*

screw press. A high-speed press in which the ram is activated by a large screw assembly powered by a drive mechanism.

scruff. A mixture of tin oxide and iron-tin alloy formed as dross on a tin-coating bath.

scuffing. (1) Localized damage caused by the occurrence of solid-phase welding between sliding surfaces, without local surface melting. (2) A mild degree of *galling* that results from the welding of asperities due to frictional heat. The welded asperities break, causing surface degradation.

sealing. (1) Closing pores in anodic coatings to render them less absorbent. (2) Plugging leaks in a casting by introducing thermosetting plastics into porous areas and subsequently setting the plastic with heat.

seal weld. Any weld used primarily to obtain lightness and prevent leakage.

seam. (1) On a metal surface, an unwelded fold or lap that appears as a crack, usually resulting from a discontinuity. (2) A surface defect on a casting related to but of lesser degree than a *cold shut.* (3) A ridge on the surface of a casting caused by a crack in the mold face.

seam weld. A continuous weld made between or upon overlapping members, in which coalescence may start and occur on the faying surfaces, or may have proceeded from the surface of one member.

The continuous weld may consist of a single weld bead or a series of overlapping spot welds. Common seam weld types include (1) lap seam welds joining flat sheets, (2) flange-joint lap seam welds with at least one flange overlapping the mating piece, and (3) mash seam welds with work metal compressed at the joint to reduce joint thickness.

seam welding. (1) Arc or resistance welding in which a series of overlapping spot welds is produced with rotating electrodes or rotating work, or both. (2) Making a longitudinal weld in sheet metal or tubing. See also *resistance seam welding.*

season cracking. An obsolete historical term usually applied to *stress-corrosion cracking* of brass.

secondary alloy. Any alloy whose major constituent is obtained from recycled scrap metal.

secondary creep. See *creep.*

secondary metal. Metal recovered from scrap by remelting and refining.

segment die. A die made of parts that can be separated for ready removal of the workpiece. Synonymous with *split die.*

segregation. (1) Nonuniform distribution of alloying elements, impurities, or microphases in metals and alloys. (2) A casting defect involving a concentration of alloying elements at specific regions, usually as a result of the primary crystallization of one phase with the subsequent concentration of other elements in the remaining liquid. Microsegregation refers to normal segregation on a microscopic scale in which material richer in an alloying element freezes in successive layers on the dendrites (*coring*) and in constituent network. Macrosegregation refers to gross differences in concentration (for example, from one area of a casting to another). See also *inverse segregation* and *normal segregation.*

segregation banding. Inhomogeneous distribution of alloying elements aligned in filaments or plates parallel to the direction of working.

seizing. The stopping of a moving part by a mating surface as a result of excessive friction.

seizure. The stopping of relative motion as the result of interfacial friction. Seizure may be accompanied by gross surface welding. The term is sometimes used to denote *scuffing.*

Sejournet process. See *Ugine-Sejournet process.*

selective heating. Intentionally heating only certain portions of a workpiece.

selective leaching. Corrosion in which one element is preferentially removed from an alloy, leaving a residue (often porous) of the elements that are more resistant to the particular environment. Also called *dealloying* or parting. See also *decarburization*, *decobaltification*, *denickelification*, *dezincification*, and *graphitic corrosion*.

selective quenching. Quenching only certain portions of an object.

self-diffusion. Thermally activated movement of an atom to a new site in a crystal of its own species, as, for example, a copper atom within a crystal of copper.

self-hardening steel. See preferred term *air-hardening steel*.

self-lubricating material. Any solid material that shows low friction without application of a lubricant.

semiautomatic plating. *Plating* in which prepared cathodes are mechanically conveyed through the plating baths, with intervening manual transfers.

semiconductor. A solid crystalline material whose electrical resistivity is intermediate between that of a metal and an insulator, ranging from about 10^3 2 cm to 10^8 2 cm, and is usually strongly temperature-dependent.

semifinisher. An impression in a series of forging dies that only approximates the finish dimensions of the forging. Semifinishers are often used to extend die life or the finishing impression, to ensure proper control of grain flow during forging, and to assist in obtaining desired tolerances.

semifinishing. Preliminary operations performed prior to finishing.

semiguided bend. The bend obtained by applying a force directly to the specimen in the portion that is to be bent. The specimen is either held at one end and forced around a pin or rounded edge, or is supported near the ends and bent by a force applied on the side of the specimen opposite the supports and midway between them. In some instances, the bend is started in this manner and finished in the manner of a *free bend*.

semikilled steel. Steel that is incompletely deoxidized and contains sufficient dissolved oxygen to react with the carbon to form carbon monoxide and thus offset solidification shrinkage.

semipermanent mold. A *permanent mold* in which sand cores or plaster are used.

semisolid metal forming. A two-step casting/forging process in which a billet is cast in a mold equipped with a mixer that continuously stirs the thixotropic melt, thereby breaking up the dendritic structure of the casting into a fine-grained spherical structure. After cooling, the bil-

let is stored for subsequent use. Later, a slug from the billet is cut, heated to the semisolid state, and forged in a die. Normally the cast billet is forged when 30 to 40% is in the liquid state. See also *rheocasting*.

sensitization. In austenitic stainless steels, the precipitation of chromium carbides, usually at grain boundaries, on exposure to temperatures of about 540 to 845 °C (about 1000 to 1550 °F), leaving the grain boundaries depleted of chromium and therefore susceptible to preferential attack by a corroding medium. Welding is the most common cause of sensitization. Weld decay (sensitization) caused by carbide precipitation in the weld heat-affected zone leads to *intergranular corrosion*.

sensitizing heat treatment. A heat treatment, whether accidental, intentional, or incidental (as during welding), that causes precipitation of constituents at grain boundaries, often causing the alloy to become susceptible to *intergranular corrosion* or *intergranular stress-corrosion cracking*. See also *sensitization*.

Sendzimir mill. A type of *cluster mill* with small-diameter work rolls and larger-diameter backup rolls, backed up by bearings on a shaft mounted eccentrically so that it can be rotated to increase the pressure between the bearing and the backup rolls. Used to roll precision and very thin sheet and strip.

series submerged arc welding. A submerged arc welding process variation in which electric current is established between two (consumable) electrodes which meet just above the surface of the work. The work is not in the electrical circuit. See also *submerged arc welding*.

series welding. Resistance welding in which two or more spot, seam, or projection welds are made simultaneously by a single welding transformer with three or more electrodes forming a series circuit.

set. The shape of the solidifying surface of a metal, especially copper, with respect to concavity or convexity. May also be called pitch.

set copper. An intermediate copper product containing about 3.5% cuprous oxide, obtained at the end of the oxidizing portion of the fire-refining cycle.

settling. (1) Separation of solids from suspension in a fluid of lower density, solely by gravitational effects. (2) A process for removing iron from liquid magnesium alloys by holding the melt at a low temperature after manganese has been added to it.

severity of quench. Ability of quenching medium to extract heat from a hot steel workpiece; expressed in terms of the *Grossmann number (H)*.

shadowing. Directional deposition of carbon or a metallic film on a plastic replica so as to highlight features to be analyzed by transmission electron microscopy. Most often used to provide maximum detail and resolution of the features of fracture surfaces.

shakeout. Removal of castings from a sand mold. See also *knockout*.

shaker-hearth furnace. A continuous type furnace that uses a reciprocating shaker motion to move the parts along the hearth.

shank. (1) The portion of a die or tool by which it is held in position in a forging unit or press. (2) The handle for carrying a small ladle or crucible. (3) The main body of a lathe tool. If the tool is an inserted type, the shank is the portion that supports the insert.

shank-type cutter. A cutter having a straight or tapered shank to fit into a machine-tool spindle or adapter.

shape memory alloys. A group of metallic materials that demonstrate the ability to some previously defined shape or size when subjected to the appropriate thermal procedure.

shaping. Producing flat surfaces using single-point tools. The work is held in a vise or fixture, or is clamped directly to the table. The ram supporting the tool is reciprocated in a linear motion past the work.

shatter crack. See *flake*.

shaving. (1) As a finishing operation, the accurate removal of a thin layer of a work surface by straightline motion between a cutter and the surface. (2) Trimming parts such as stampings, forgings, and tubes to remove uneven sheared edges or to improve accuracy.

shear. (1) The type of force that causes or tends to cause two contiguous parts of the same body to slide relative to each other in a direction parallel to their plane of contact. (2) A machine or tool for cutting metal and other material by the closing motion of two sharp, closely adjoining edges; for example, squaring shear and circular shear. (3) An inclination between two cutting edges, such as between two straight knife blades or between the punch cutting edge and the die cutting edge, so that a reduced area will be cut each time. This lessens the necessary force, but increases the required length of the working stroke. This method is referred to as angular shear. (4) The act of cutting by shearing dies or blades, as in shearing lines.

shear angle. The angle that the *shear plane*, in metal cutting, makes with the work surface.

shear bands. (1) Bands of very high shear strain that are observed during rolling of sheet metal. During rolling, these form at approximately 35° to the rolling plane, parallel to the transverse direction. They are independent of grain orientation and at high strain rates traverse the entire thickness of the rolled sheet. (2) Highly localized deformation zones in metals that are observed at very high strain rates, such as those produced by high velocity (100 to 3600 m/s, or 330 to 11,800 ft/s) projectile impacts or explosive rupture.

shear fracture. A mode of fracture in crystalline materials resulting from translation along slip planes that are preferentially oriented in the direction of the shearing stress.

shear ledges. See *radial marks*.

shear lip. A narrow, slanting ridge along the edge of a fracture surface. The term sometimes also denotes a narrow, often crescent-shaped, fibrous region at the edge of a fracture that is otherwise of the cleavage type, even though this fibrous region is in the same plane as the rest of the fracture surface.

shear modulus (*G*). The ratio of shear stress to the corresponding shear strain for shear stresses below the proportional limit of the material. Values of shear modulus are usually determined by torsion testing. Also known as modulus of rigidity.

shear plane. A confined zone along which shear takes place in metal cutting. It extends from the cutting edge to the work surface.

shear strain. The tangent of the angular change, caused by a force between two lines originally perpendicular to each other through a point in a body. Also called angular strain.

shear stress. (1) The stress component tangential to the plane on which the forces act. (2) A stress that exists when parallel planes in metal crystals slide across each other.

sheet. A flat-rolled metal product of some maximum thickness and minimum width arbitrarily dependent on the type of metal. It has a width-to-thickness ratio greater than about 50. Generally, such flat products under 6.5 mm (1/4 in.) thick are called sheets, and those 6.5 mm (1/4 in.) thick and over are called plates. Occasionally, the limiting thickness for steel to be designated as sheet steel is No. 10 Manufacturer's Standard Gage for sheet steel, which is 3.42 mm (0.1345 in.) thick.

sheet forming. The plastic deformation of a piece of sheet metal by tensile loads into a three-dimensional shape, often without significant changes in sheet thickness or surface characteristics. Compare with *bulk forming*.

shelf roughness. Roughness on upward-facing surfaces where undissolved solids have settled on parts during a plating operation.

shell. (1) A hollow structure or vessel. (2) An article formed by deep drawing. (3) The metal sleeve remaining when a billet is extruded with a dummy block of somewhat smaller diameter. (4) In shell molding, a hard layer of sand and thermosetting plastic or resin formed over a pattern and used as the mold wall. (5) A tubular casting used in making seamless drawn tube. (6) A pierced forging.

shell core. A shell-molded sand core.

shell hardening. A surface-hardening process in which a suitable steel workpiece, when heated through and quench hardened, develops a martensite layer or shell that closely follows the contour of the piece and surrounds a core of essentially pearlitic transformation product. This result is accomplished by a proper balance among section size, steel hardenability, and severity of quench.

shelling. (1) A term used in railway engineering to describe an advanced phase of *spalling*. (2) A mechanism of deterioration of coated abrasive products in which entire abrasive grains are removed from the cement coating that held the abrasive to the backing layer of the product.

shell molding. A foundry process in which a mold is formed from thermosetting resin-bonded sand mixtures brought in contact with preheated (150 to 260 °C, or 300 to 500 °F) metal patterns, resulting in a firm shell with a cavity corresponding to the outline of the pattern. Also called *Croning process*.

shielded metal arc welding (SMAW). A manual arc welding process in which the heat for welding is generated by an arc established between a flux-covered consumable electrode and a workpiece. The electrode tip, molten weld pool, arc, and adjacent areas of the workpiece are protected from atmospheric contamination by a gaseous shield obtained from the combustion and decomposition of the electrode covering. Additional shielding is provided for the molten metal in the weld pool by a covering of molten flux or slag. Filler metal is supplied by the core of the consumable electrode and from metal powder mixed with the electrode covering of certain electrodes. Shielded metal arc welding is often referred to as arc

welding with stick electrodes, manual metal arc welding, and stick welding.

shielding. (1) A material barrier that prevents radiation or a flowing fluid from impinging on an object or a portion of an object. (2) Placing an object in an electrolytic bath so as to alter the current distribution on the cathode. A nonconductor is called a shield; a conductor is called a *robber*, a thief, or a guard.

shielding gas. (1) Protective gas used to prevent atmospheric contamination during welding. (2) A stream of inert gas directed at the substrate during thermal spraying so as to envelop the plasma flame and substrate; intended to provide a barrier to the atmosphere in order to minimize oxidation.

shift. A casting imperfection caused by mismatch of cope and drag or of cores and molds.

shim. A thin piece of material used between two surfaces to obtain a proper fit, adjustment, or alignment.

shimmy die. See *flat edge trimmer*.

shock load. The sudden application of an external force that results in a very rapid build-up of stress—for example, piston loading in internal combustion engines.

shoe. (1) A metal block used in a variety of bending operations to form or support the part being processed. (2) An anvil cap or *sow block*.

Shore hardness. A measure of the resistance of material to indentation by a spring-loaded indenter during Sceleroscope hardness testing. The higher the number, the greater the resistance. Normally used for rubber materials. See also *Scleroscope hardness test*.

shortness. A form of brittleness in metal. It is designated as *cold shortness* or *hot shortness* to indicate the temperature range in which the brittleness occurs.

short transverse. See *transverse*.

shot. (1) Small, spherical particles of metal. (2) The injection of molten metal into a die casting die. The metal is injected so quickly that it can be compared to the shooting of a gun.

shotblasting. Blasting with metal *shot*; usually used to remove deposits or mill scale more rapidly or more effectively than can be done by *sandblasting*.

shot peening. A method of cold working metals in which compressive stresses are induced in the exposed surface layers of parts by the impingement of a stream of *shot*, directed at the metal surface at high velocity under controlled conditions.

shotting. The production of *shot* by pouring molten metal in finely divided streams.

Solidified spherical particles are formed during descent in a tank of water.

shrinkage. (1) The contraction of metal during cooling after hot forging. Die impressions are made oversize according to precise shrinkage scales to allow the forgings to shrink to design dimensions and tolerances. (2) See *casting shrinkage*.

shrinkage cavity. A void left in cast metal as a result of solidification shrinkage. Shrinkage cavities can appear as either isolated or interconnected irregularly shaped voids. See also *casting shrinkage*.

shrinkage cracks. Cracks that form in metal as a result of the pulling apart of grains by contraction before complete solidification. See also *hot tear*.

shrinkage rule. A measuring ruler with graduations expanded to compensate for the change in the dimensions of the solidified casting as it cools in the mold.

shroud. A protective, refractory-lined metal-delivery system to prevent reoxidation of molten steel when it is poured from ladle to tundish to mold during continuous casting.

shut height. For a metal forming press, the distance from the top of the bed to the bottom of the slide with the stroke down and adjustment up. In general, it is the maximum die height that can be accommodated for normal operation, taking the bolster plate into consideration. See also *bolster*.

side cutting-edge angle. Defined by the figure accompanying *single-point tool*.

side milling. Milling with cutters having peripheral and side teeth. They are usually profile sharpened but may be form relieved.

side rake. In a single-point turning tool, the angle between the tool face and a reference plane, corresponding to radial rake in milling. It lies in a plane perpendicular to the tool base and parallel to the rotational axis of the work. See also the figure accompanying *single-point tool*.

sieve. A standard wire mesh or screen used in graded sets to determine the mesh size or particle size distribution of particulate and granular solids. See also *sieve analysis*.

sieve analysis. A method of determining *particle size distribution*, usually expressed as the weight percentage retained upon each of a series of standard screens of decreasing mesh size.

sieve classification. The separation of powder into particle size ranges by the use of a series of graded sieves. Also called screen analysis.

sieve fraction. That portion of a powder sample that passes through a sieve of specified number and is retained by some finer mesh sieve of specified number. See also *sieve analysis*.

sigma phase. A hard, brittle, nonmagnetic intermediate phase with a tetragonal crystal structure, containing 30 atoms per unit cell, space group, *P4/mnm*, occurring in many binary and ternary alloys of the transition elements. The composition of this phase in the various systems is not the same, and the phase usually exhibits a wide range in homogeneity. Alloying with a third transition element usually enlarges the field homogeneity and extends it deep into the ternary section.

sigma-phase embrittlement. *Embrittlement* of iron-chromium alloys (most notably austenitic stainless steels) caused by precipitation at grain boundaries of the hard, brittle intermetallic *sigma phase* during long periods of exposure to temperatures between approximately 560 and 980 °C (1050 and 1800 °F). Sigma-phase embrittlement results in severe loss in *toughness* and *ductility*, and can make the embrittled material susceptible to *intergranular corrosion*. See also *sensitization*.

siliconizing. Diffusing silicon into solid metal, usually low-carbon steels, at an elevated temperature in order to improve corrosion or wear resistance.

silky fracture. A metal fracture in which the broken metal surface has a fine texture, usually dull in appearance. Characteristic of tough and strong metals. Contrast with *crystalline fracture* and *granular fracture*.

silver soldering. Nonpreferred term used to denote brazing with a silver-base filler metal. See preferred terms *furnace brazing*, *induction brazing*, and *torch brazing*.

single-action press. A metal forming press that provides pressure from one side.

single-bevel groove weld. A groove weld in which the joint edge of one member is beveled from one side.

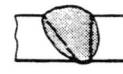

Single-bevel
groove weld

single-impulse welding. Spot, projection, or upset welding by a single impulse of current. Where alternating current is used, an impulse may be any fraction or number of cycles.

single-J groove weld. A groove weld in which the joint edge of one member is prepared in the form of a J, from one side. See

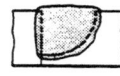

Single-J
groove weld

also the figure accompanying *single-bevel groove weld*.

single-point tool. See definition of nomenclature in the accompanying figure.

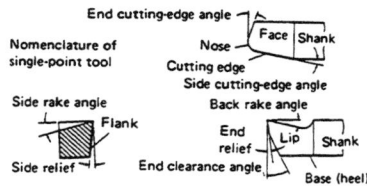

single-stand mill. A rolling mill designed such that the product contacts only two rolls at a given moment. Contrast with *tandem mill*.

single-U groove weld. A groove weld in which each joint edge is prepared in the form of a J or half-U from one side. See figure.

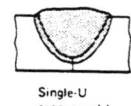

Single-U
groove weld

single-V groove weld. A groove weld in which each member is beveled from the same side. See figure.

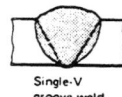

Single-V
groove weld

single welded joint. In arc and gas welding, any joint welded from one side only.

sinkhead. Same as *riser*.

sinking. (1) The operation of machining the impression of a desired forging into die blocks. (2) See *tube sinking*.

sintered density. The quotient of the mass (weight) over the volume of the sintered body expressed in grams per cubic centimeter.

sintering. The bonding of adjacent surfaces of particles in a mass of powder or a compact by heating. Sintering strengthens a powder mass and normally produces densification and, in powdered metals, recrystallization. See also *liquid phase sintering* and *solid-state sintering*.

size effect. Effect of the dimensions of a piece of metal on its mechanical and other properties and on manufacturing variables such as forging reduction and heat treatment. In general, the mechanical properties are lower for a larger size.

size of weld. (1) The joint penetration in a groove weld. (2) The lengths of the nominal legs of a fillet weld. (3) The weld metal thickness measured at the root of a flange weld. See accompanying figures.

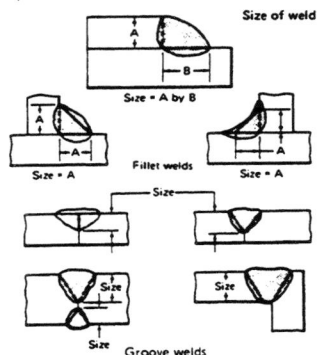

sizing. (1) Secondary forming or squeezing operations needed to square up, set down, flatten, or otherwise correct surfaces to produce specified dimensions and tolerances. See also *restriking*. (2) Some burnishing, broaching, drawing, and shaving operations are also called sizing. (3) A finishing operation for correcting ovality in tubing. (4) Final pressing of a sintered powder metallurgy part to obtain a desired dimension.

skelp. The starting stock for making welded pipe or tubing; most often it is strip stock of suitable width, thickness, and edge configuration.

skim gate. In foundry practice, a gating arrangement designed to prevent the passage of slag and other undesirable materials into a casting.

skimming. Removing or holding back dirt or slag from the surface of the molten metal before or during pouring.

skin. A thin outside metal layer, not formed by bonding as in cladding or electroplating, that differs in composition, structure, or other characteristics from the main mass of metal.

skin lamination. In flat-rolled metals, a surface rupture resulting from the exposure of a subsurface lamination by rolling.

skin pass. See *temper rolling*.

skiving. (1) Removal of a material in thin layers or chips with a high degree of shear or slippage, or both, of the cutting tool. (2) A machining operation in which the cut is made with a form tool with its face so angled that the cutting edge progresses from one end of the work to the other as the tool feeds tangentially past the rotating workpiece.

skull. (1) A layer of solidified metal or dross on the walls of a pouring vessel after the metal has been poured. (2) The unmelted residue from a liquated weld filler metal.

slab. A flat-shaped semifinished rolled metal ingot with a width not less than 250 mm (10 in.) and a cross-sectional area not less than 105 cm^2 (16 in.^{2}).

slabbing mill. A primary mill that produces slabs.

slab milling. See preferred term *peripheral milling*.

slack quenching. The incomplete hardening of steel due to quenching from the austenitizing temperature at a rate slower than the critical cooling rate for the particular steel, resulting in the formation of one or more transformation products in addition to martensite.

slag. A nonmetallic product resulting from the mutual dissolution of flux and nonmetallic impurities in smelting, refining, and certain welding operations (see, for example, *electroslag welding*). In steelmaking operations, the slag serves to protect the molten metal from the air and to extract certain impurities.

slag inclusion. (1) Slag or dross entrapped in a metal. (2) Nonmetallic solid material entrapped in weld metal or between weld metal and base metal.

slant fracture. A type of fracture in metals, typical of *plane-stress* fractures, in which the plane of separation is inclined at an angle (usually about 45°) to the axis of applied stress.

slide. The main reciprocating member of a metal forming press, guided in the press frame, to which the punch or upper die is fastened; sometimes called the *ram*. The inner slide of a double-action press is called the plunger or punch-holder slide; the outer slide is called the blankholder slide. The third slide of a triple-action press is called the lower slide, and the slide of a hydraulic press is often called the platen.

slime. (1) A material of extremely fine particle size encountered in ore treatment. (2) A mixture of metals and some insoluble compounds that forms on the anode in electrolysis.

slip. *Plastic deformation* by the irreversible shear displacement (translation) of one part of a crystal relative to another in a definite crystallographic direction and usually on specific crystallographic plane. Sometimes called glide.

slip band. A group of parallel slip lines so closely spaced as to appear as a single line when observed under an optical microscope. See also *slip line*.

slip direction. The crystallographic direction in which the translation of slip takes place.

slip flask. A tapered *flask* that depends on a movable strip of metal to hold foundry sand in position. After closing the mold, the strip is refracted and the flask can be removed and reused. Molds thus made are

usually supported by a *mold jacket* during pouring.

slip line. Visible traces of slip planes on metal surfaces; the traces are (usually) observable only if the surface has been polished before deformation. The usual observation on metal crystals (under a light microscope) is of a cluster of slip lines known as a *slip band*.

slip plane. The crystallographic plane in which *slip* occurs in a crystal.

slitting. Cutting or shearing along single lines to cut strips from a metal sheet or to cut along lines of a given length or contour in a sheet or workpiece.

sliver. An imperfection consisting of a very thin elongated piece of metal attached by only one end to the parent metal into whose surface it has been worked.

slot furnace. A common batch furnace for heat treating metals where stock is charged and removed through a slot or opening.

slotting. Cutting a narrow aperture or groove with a reciprocating tool in a vertical shaper or with a cutter, broach, or grinding wheel.

slot weld. A weld made in an elongated hole in one member of a lap or T-joint joining that member to that portion of the surface of the other member which is exposed through the hole. The hole may be open at one end and may be partially or completely filled with weld metal. A fillet welded slot should not be construed as conforming to this definition.

slow strain rate technique. An experimental technique for evaluating susceptibility to *stress-corrosion cracking*. It involves pulling the specimen to failure in uniaxial tension at a controlled slow strain rate while the specimen is in the test environment and examining the specimen for evidence of stress-corrosion cracking.

slug. (1) A short piece of metal to be placed in a die for forging or extrusion. (2) A small piece of material produced by piercing a hole in sheet material. See also *blank*.

slugging. The unsound practice of adding a separate piece of material in a joint before or during welding, resulting in a welded joint in which the weld zone is not entirely built up by adding molten filler metal or by melting and recasting base metal, and which therefore does not comply with design, drawing, or specification requirements.

slush casting. A hollow casting usually made of an alloy with a low but wide melting temperature range. After the desired thickness of metal has solidified in the mold, the remaining liquid is poured out. Considered an obsolete practice.

smelting. Thermal processing wherein chemical reactions take place to produce liquid metal from a beneficiated ore.

smith forging. See *hand forge (smith forge)*.

smut. A reaction product sometimes left on the surface of a metal after pickling, electroplating, or etching.

snagging. (1) Heavy stock removal of superfluous material from a workpiece by using a portable or swing grinder mounted with a coarse grain abrasive wheel. (2) *Offhand grinding* on castings and forgings to remove surplus metal such as gate and riser pads, fins, and parting lines.

snake. (1) The product formed by twisting and bending of hot metal rod prior to its next rolling process. (2) Any crooked surface imperfection in a plate, resembling a snake. (3) A flexible mandrel used in the inside of a shape to prevent flattening or collapse during a bending operation.

snap flask. A foundry flask hinged on one corner so that it can be opened and removed from the mold for reuse before the metal is poured.

snap temper. A precautionary interim stress-relieving treatment applied to high-hardenability steels immediately after quenching to prevent cracking because of delay in tempering them at the prescribed higher temperature.

S-N curve. A plot of stress (S) against the number of cycles to failure (N). The stress can be the maximum stress (S_{max}) or the alternating stress amplitude (S_a). The stress values are usually nominal stress; i.e., there is no adjustment for stress concentration. The diagram indicates the *S-N* relationship for a specified value of the mean stress (S_m) or the stress ratio (A or R) and a specified probability of survival. For N a log scale is almost always used. For S a linear scale is used most often, but a log scale is sometimes used. Also known as *S-N diagram*.

soak cleaning. *Immersion cleaning* without electrolysis.

soaking. In heat treating of metals, prolonged holding at a selected temperature to effect homogenization of structure or composition. See also *homogenizing*.

soft magnetic material. A ferromagnetic alloy that becomes magnetized readily upon application of a field and that returns to practically a nonmagnetic condition when the field is removed; an alloy with the properties of high magnetic permeability, low coercive force, and low magnetic hysteresis loss.

soft soldering. See preferred term *soldering*.

soft temper. Same as *dead soft* temper.

solder. A filler metal used in soldering which has a liquidus not exceeding 450 °C (840 °F). The most commonly used solders are tin-lead alloys. Other solder alloys include tin-antimony, tin-silver, tin-zinc, cadmium-silver, cadmium-zinc, zinc-aluminum, indium-base alloys, bismuth-base alloys (*fusible alloys*), and gold-base solders.

solderability. The relative ease and speed with which a surface is wetted by molten solder.

solder embrittlement. Reduction in mechanical properties of a metal as a result of local penetration of solder along grain boundaries.

soldering. A group of processes that join metals by heating them to a suitable temperature below the solidus of the base metals and applying a filler metal having a liquidus not exceeding 450 °C (840 °F). Molten filler metal is distributed between the closely fitted surfaces of the joint by capillary action. See also *solder*.

soldering flux. See *flux*.

solid cutters. Cutters made of a single piece of material rather than a composite of two or more materials.

solidification. The change in state from liquid to solid upon cooling through the melting temperature or melting range.

solidification range. The temperature between the liquidus and the solidus.

solidification shrinkage. The reduction in volume of metal from beginning to end of solidification. See also *casting shrinkage*.

solidification shrinkage crack. A crack that forms, usually at elevated temperature, because of the internal (shrinkage) stresses that develop during solidification of a metal casting. Also termed hot crack.

solid lubricant. Any solid used as a powder or thin film on a surface to provide protection from damage during relative movement and to reduce friction and wear. Examples include molybdenum disulfide, graphite, polytetrafluoroethylene (PTFE), and mica.

solid-metal embrittlement. The occurrence of *embrittlement* in a material below the melting point of the embrittling species. See also *liquid-metal embrittlement*.

solid solution. A single, solid, homogeneous crystalline phase containing two or more chemical species.

solid-state sintering. A sintering procedure for compacts or loose powder aggregates during which no component melts. Contrast with *liquid phase sintering*.

solid-state welding. A group of welding processes that join metals at temperatures essentially below the melting points of the base materials, without the addition of a brazing or soldering filler metal. Pressure may or may not be applied to the joint. Examples include *cold welding, diffusion welding, forge welding, hot pressure welding*, and *roll welding*.

solidus. (1) The highest temperature at which a metal or alloy is completely solid. (2) In a *phase diagram*, the locus of points representing the temperatures at which various compositions stop freezing upon cooling or begin to melt upon heating. See also *liquidus*.

solute. The component of either a liquid or solid solution that is present to a lesser or minor extent; the component that is dissolved in the *solvent*.

solution heat treatment. Heating an alloy to a suitable temperature, holding at that temperature long enough to cause one or more constituents to enter into *solid solution*, and then cooling rapidly enough to hold these constituents in solution.

solution potential. *Electrode potential* where half-cell reaction involves only the metal electrode and its ion.

solvent. The component of either a liquid or solid solution that is present to a greater or major extent; the component that dissolves the *solute*.

solvus. In a phase or equilibrium diagram, the locus of points representing the temperature at which solid phases with various compositions coexist with other solid phases, that is, the limits of solid solubility.

sorbite (obsolete). A fine mixture of ferrite and cementite produced either by regulating the rate of cooling of steel or by tempering steel after hardening. The first type is very fine pearlite that is difficult to resolve under the microscope; the second type is tempered martensite.

sour gas. A gaseous environment containing hydrogen sulfide and carbon dioxide in hydrocarbon reservoirs. Prolonged exposure to sour gas can lead to *hydrogen damage, sulfide-stress cracking*, and/or *stress-corrosion cracking* in ferrous alloys.

sow block. A block of heat-treated steel placed between the anvil of the hammer and the forging die to prevent undue wear to the anvil. Sow blocks are occasionally used to hold insert dies. Also called anvil cap.

space lattice. A regular, periodic array of points (lattice points) in space that represents the locations of atoms of the same kind in a perfect crystal. The concept may be extended, where appropriate, to crystalline compounds and other substances, in which case the lattice points often represent locations of groups of atoms of identical composition, arrangement, and orientation.

spacer strip. A metal strip or bar inserted in the root of a joint prepared for groove welding to serve as a backing and to maintain root opening throughout the course of the welding operation.

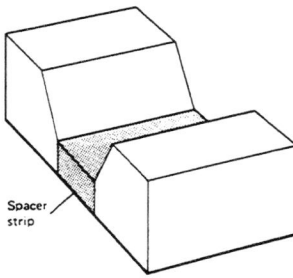

Spacer strip

spade drill. See preferred term *flat drill.*

spalling. (1) Separation of particles from a surface in the form of flakes. The term spalling is commonly associated with rolling-element bearings and with gear teeth. Spalling is usually a result of subsurface fatigue and is more extensive than pitting. (2) The spontaneous chipping, fragmentation, or separation of a surface or surface coating. (3) A chipping or flaking of a surface due to any kind of improper heat treatment or material dissociation.

spangle. The characteristic crystalline form in which a hot dipped zinc coating solidifies on steel strip.

spark testing. A method used for the classification of ferrous alloys according to their chemical compositions, by visual examination of the spark pattern or stream that is thrown off when the alloys are held against a grinding wheel rotating at high speed.

spatter. The metal particles expelled during arc or gas welding. They do not form part of the weld.

spatter loss. The metal lost due to *spatter.*

specific energy. In cutting or grinding, the energy expended or work done in removing a unit volume of material.

specimen. A test object, often of standard dimensions and/or configuration, that is used for destructive or nondestructive testing. One or more specimens may be cut from each unit of a *sample.*

speed of travel. In welding, the speed with which a weld is made along its longitudinal axis, usually measured in meters per second or inches per minute.

speiss. Metallic arsenides and antimonides that result from smelting metal ores such as those of cobalt or lead.

spelter. Crude zinc obtained in smelting zinc ores.

spelter solder. A brazing filler metal of approximately equal parts of copper and zinc.

spheroidal graphite. Graphite of spheroidal shape with a polycrystalline radial structure. This structure can be obtained, for example, by adding cerium or magnesium to the melt. See also *ductile iron* and *nodular graphite.*

spheroidite. An aggregate of iron or alloy carbides of essentially spherical shape dispersed throughout a matrix of *ferrite.*

spheroidized structure. A microstructure consisting of a matrix containing spheroidal particles of another constituent.

spheroidizing. Heating and cooling to produce a spheroidal or globular form of carbide in steel. Spheroidizing methods frequently used are:

1. Prolonged holding at a temperature just below Ae_1.
2. Heating and cooling alternatively between temperatures that are just above and just below Ae_1.
3. Heating to a temperature above Ae_1 or Ae_3 and then cooling very slowly in the furnace or holding at a temperature just below Ae_1.
4. Cooling at a suitable rate from the minimum temperature at which all carbide is dissolved to prevent the reformation of a carbide network, and then reheating in accordance with method 1 or 2 above. (Applicable to hypereutectoid steel containing a carbide network.)

spiegeleisen (spiegel). A pig iron containing 15 to 30% Mn and 4.5 to 6.5% C.

spindle. (1) Shaft of a machine tool on which a cutter or grinding wheel may be mounted. (2) Metal shaft to which a mounted wheel is cemented.

spinning. The forming of a seamless hollow metal part by forcing a rotating blank to conform to a shaped mandrel that rotates concentrically with the blank. In the typical application, a flat-rolled metal blank is forced against the mandrel by a blunt, rounded tool; however, other stock (notably, welded or seamless tubing) can be formed. A roller is sometimes used as the working end of the tool.

spinodal structure. A fine, homogeneous mixture of two phases that form by the growth of composition waves in a solid solution during suitable heat treatment. The phases of a spinodal structure differ in composition from each other and from the parent phase, but have the same crystal structure as the parent phase.

spline. Any of a series of longitudinal, straight projections on a shaft that fit into slots on a mating part to transfer rotation to or from the shaft.

split die. A die made of part that can be separated for ready removal of the workpiece. Also known as segment die.

split punch. A segmented punch or a set of punches in a powder metallurgy forming press that allow(s) a separate positioning for different powder fill heights and compact levels in dual-step and multistep parts. See also *stepped compact.*

sponge. A form of metal characterized by a porous condition that is the result of the decomposition or reduction of a compound without fusion. The term is applied to forms of iron, titanium, zirconium, uranium, plutonium, and the platinum-group metals.

sponge iron. A coherent, porous mass of substantially pure iron produced by solid state reduction of iron oxide (mill scale or iron ore).

spot drilling. Making an initial indentation in a work surface, with a drill, to serve as a centering guide in a subsequent machining process.

spotfacing. Using a rotary, hole-piloted end-facing tool to produce a flat surface normal to the axis of rotation of the tool on or slightly below the workpiece surface.

spot weld. A weld made between or upon overlapping members in which coalescence may start and occur on the faying surfaces or may proceed from the surface of one member. The weld cross section is approximately circular.

spot welding. Welding of lapped parts in which fusion is confined to a relatively small circular area. It is generally resistance welding, but may also be gas tungsten-arc, gas metal-arc, or submerged-arc welding.

spray quenching. A quenching process using spray nozzles to spray water or other liquids on a part. The quench rate is controlled by the velocity and volume of liquid per unit area per unit of time of impingement.

springback. (1) The elastic recovery of metal after stressing. (2) The extent to which metal tends to return to its original shape or contour after undergoing a forming operation. This is compensated for by overbending or by a secondary operation of *restriking.* (3) In flash, upset, or pressure welding, the deflection in the welding machine caused by the upset pressure.

spring temper. A *temper* of nonferrous alloys and some ferrous alloys characterized by tensile strength and hardness about two-thirds of the way from *full hard* to *extra spring* temper.

sprue. (1) The mold channel that connects the *pouring basin* with the runner or, in the absence of a pouring basin, directly into which molten metal is poured. Sometimes referred to as downsprue or downgate. (2) Sometimes used to mean all gates, risers, runners, and similar scrap that are removed from castings after shakeout.

sputtering. The bombardment of a solid surface with a flux of energetic particles (ions) that results in the ejection of atomic species. The ejected material may be used as a source for deposition. See also *physical vapor deposition*.

square drilling. Making square holes by means of a specially constructed drill made to rotate and also to oscillate so as to follow accurately the periphery of a square guide bushing or template.

square groove weld. A groove weld in which the abutting surfaces are square.

 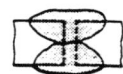

Square groove weld

squaring shear. A machining tool, used for cutting sheet metal or plate, consisting essentially of a fixed cutting knife (usually mounted on the rear of the bed) and another cutting knife mounted on the front of a reciprocally moving crosshead, which is guided vertically in side housings. Corner angles are usually 90°.

squeeze casting. A hybrid liquid-metal forging process in which liquid metal is forced into a permanent mold by a hydraulic press.

stabilizing treatment. (1) Before finishing to final dimensions, repeatedly heating a ferrous or nonferrous part to or slightly above its normal operating temperature and then cooling to room temperature to ensure dimensional stability in service. (2) Transforming retained austenite in quenched hardenable steels, usually by *cold treatment*. (3) Heating a solution-treated stabilized grade of austenitic stainless steel to 870 to 900 °C (1600 to 1650 °F) to precipitate all carbon as TiC, NbC, or TaC so that *sensitization* is avoided on subsequent exposure to elevated temperature.

stack cutting. Thermal cutting of stacked metal plates arranged so that all the plates are severed by a single cut.

stack molding. A foundry practice that makes use of both faces of a mold section,

one face acting as the drag and the other as the cope. Sections, when assembled to other similar sections, form several tiers of mold cavities, all castings being poured together through a common sprue.

stack welding. *Resistance spot welding* of stacked plates, all being joined simultaneously.

staggered-intermittent fillet welding. Making a line of intermittent fillet welds on each side of a joint so that the increments on one side are not opposite those on the other. Contrast with *chain-intermittent fillet welding*.

staggered-tooth cutters. Milling cutters with alternate flutes of oppositely directed helixes.

stainless steel. Any of several steels containing at least 10.5% chromium as the principal alloying element; they usually exhibit *passivity* in aqueous environments.

staking. Fastening two parts together permanently by recessing one part within the other and then causing plastic flow at the joint.

stamping. The general term used to denote all sheet metal pressworking. It includes blanking, shearing, hot or cold forming, drawing, bending, or coining.

stand. A piece of rolling mill equipment containing one set of work rolls. In the usual sense, any pass of a cold- or hot-rolling mill. See also *rolling mills*.

standard electrode potential. The reversible potential for an electrode process when all products and reactions are at unit activity on a scale in which the potential for the standard hydrogen half-cell is zero.

standard gold. A gold alloy containing 10% copper; at one time used for legal coinage in the United States.

standard reference material. A reference material, the composition or properties of which are certified by a recognized standardizing agency or group.

stardusting. An extremely fine form of roughness on the surface of a metal deposit.

starting sheet. A thin sheet of metal used as the cathode in electrolyte refining.

state of strain. A complete description of the deformation within a homogeneously deformed volume or at a point. The description requires, in general, the knowledge of the independent components of *strain*.

state of stress. A complete description of the stresses within a homogeneously stressed volume or at a point. The description requires, in general, the knowledge of the independent components of *stress*.

static fatigue. A term sometimes used to identify a form of hydrogen embrittlement in which a metal appears to fracture spontaneously under a steady stress less than the yield stress. There almost always is a delay between the application of stress (or exposure of the stressed metal to hydrogen) and the onset of cracking. More properly referred to as *hydrogen-induced delayed cracking*.

steadite. A hard structural constituent of cast iron that consists of a binary eutectic of ferrite, containing some phosphorus in solution, and iron phosphide (Fe_3P). The eutectic consists of 10.2% P and 89.8% Fe. The melting temperature is 1050 °C (1920 °F).

Stead's brittleness. A condition of brittleness that causes transcrystalline fracture in the coarse grain structure that results from prolonged annealing of thin sheets of low-carbon steel previously rolled at a temperature below about 705 °C (1300 °F). The fracture usually occurs at about 45° to the direction of rolling.

steam hammer. A type of *drop hammer* in which the ram is raised for each stroke by a double-action steam cylinder and the energy delivered to the workpiece is supplied by the velocity and weight of the ram and attached upper die driven downward by steam pressure. The energy delivered during each stroke can be varied.

steam treatment. The treatment of a sintered ferrous part in steam at temperatures between 510 and 595 °C (950 to 1100 °F) in order to produce a layer of black iron oxide (magnetite, or ferrous-ferric oxide, $FeO \cdot Fe_2O_3$) on the exposed surface for the purpose of increasing hardness and wear resistance.

Steckel mill. A cold reducing mill having two working rolls and two backup rolls, none of which is driven. The strip is drawn through the mill by a power reel in one direction as far as the strip will allow and then reversed by a second power reel, and so on until the desired thickness is attained.

steel. An iron-base alloy, malleable in some temperature ranges as initially cast, containing manganese, usually carbon, and often other alloying elements. In carbon steel and low-alloy steel, the maximum carbon is about 2.0%; in high-alloy steel, about 2.5%. The dividing line between low-alloy and high-alloy steels is generally regarded as being at about 5% metallic alloying elements.

Steel is said to be differentiated from two general classes of "irons": the cast irons, on the high-carbon side, and the relatively pure irons such as ingot iron, carbonyl iron, and electrolytic iron, on the low-carbon side. In some steels containing extremely low carbon, the manganese content is the principal differentiating factor, steel usually containing at least 0.25% and ingot iron considerably less.

step aging. Aging of metals at two or more temperatures, by steps, without cooling to room temperature after each step. See also *aging*, and compare with *interrupted aging* and *progressive aging*.

stepped compact. A powder metallurgy compact with one (dual step) or more (multistep) abrupt cross-sectional changes, usually obtained by pressing with split punches, each section of which uses a different pressure and a different rate of compaction. See also *split punch*.

stepped extrusion. See *extrusion*.

step fracture. Cleavage fractures that initiate on many parallel cleavage planes.

stereoscopic micrographs. A pair of micrographs (or fractographs) of the same area, but taken from different angles so that the two micrographs when properly mounted and viewed reveal the structures of the objects in their three-dimensional relationships.

sterling silver. A silver alloy containing at least 92.5% Ag, the remainder being unspecified but usually copper. Sterling silver is used for flat and hollow tableware and for various items of jewelry.

stick electrode. A shop term for *covered electrode*.

stick welding. See preferred term *shielded metal arc welding*.

sticker breaks. Arc-shaped *coil breaks*, usually located near the center of sheet or strip.

stiffness. (1) The rate of stress with respect to strain; the greater the stress required to produce a given strain, the stiffer the material is said to be. (2) The ability of a material or shape to resist elastic deflection. For identical shapes, the stiffness is proportional to the modulus of elasticity. For a given material, the stiffness increases with increasing moment of inertia, which is computed from cross-sectional dimensions.

stock. A general term used to refer to a supply of metal in any form or shape and also to an individual piece of metal that is formed, forged, or machined to make parts.

stopoff. A material used on the surfaces adjacent to the joint to limit the spread of soldering or brazing filler metal. See also *resist*.

stopper rod. A device in a bottom-pour ladle for controlling the flow of metal through the nozzle into a mold. The stop-per rod consists of a steel rod, protective refractory sleeves, and a graphite stopper head.

stopping off. (1) Applying a *resist*. (2) Depositing a metal (copper, for example) in localized areas to prevent carburization, decarburization, or nitriding in those areas. (3) Filling in a portion of a mold cavity to keep out molten metal.

straddle milling. Face milling a workpiece on both sides at once using two cutters spaced as required.

straightening. (1) Any bending, twisting, or stretching operation to correct any deviation from straightness in bars, tubes, or similar long parts or shapes. This deviation can be expressed as either camber (deviation from a straight line) or as total indicator reading (TIR) per unit of length. (2) A finishing operation for correcting misalignment in a forging or between various sections of a forging. See also *roll straightening*.

straight polarity. Direct-current arc welding circuit arrangement in which the electrode is connected to the negative terminal. A synonym for *direct current electrode negative (DCEN)*. Contrast with *reverse polarity*.

strain. The unit of change in the size or shape of a body due to force. Also known as nominal strain. The term is also used in a broader sense to denote a dimensionless number that characterizes the change in dimensions of an object during a deformation or flow process. See also *engineering strain*, and *true strain*.

strain-age embrittlement. A loss in *ductility* accompanied by an increase in hardness and strength that occurs when low-carbon steel (especially rimmed or capped steel) is aged following *plastic deformation*. The degree of *embrittlement* is a function of aging time and temperature, occurring in a matter of minutes at about 200 °C (400 °F), but requiring a few hours to a year at room temperature.

strain aging. (1) *Aging* following plastic deformation. (2) The changes in ductility, hardness, yield point, and tensile strength that occur when a metal or alloy that has been cold worked is stored for some time. In steel, strain aging is characterized by a loss of ductility and a corresponding increase in hardness, yield point, and tensile strength.

strain energy. The potential energy stored in a body by virtue of elastic deformation, equal to the work that must be done to produce this deformation.

strain hardening. An increase in hardness and strength of metals caused by plastic deformation at temperatures below the re-crystallization range. Also known as work hardening.

strain-hardening coefficient. See *strain-hardening exponent*.

strain-hardening exponent. The value of n in the relationship:

$$\sigma = K \varepsilon^n$$

where σ is the *true stress*, is the *true strain*, and K, which is called the strength coefficient, is equal to the true stress at a true strain of 1.0. The strain-hardening exponent, also called "n-value," is equal to the slope of the true stress/true strain curve up to maximum load, when plotted on log-log coordinates. The n-value relates to the ability of a sheet material to be stretched in metalworking operations. The higher the n-value, the better the formability (stretchability).

strain rate. The time rate of straining for the usual tensile test. Strain as measured directly on the specimen gage length is used for determining strain rate. Because strain is dimensionless, the units of strain rate are reciprocal time.

strain-rate sensitivity (m-value). The increase in stress (σ) needed to cause a certain increase in plastic strain rate (ε) at a given level of plastic strain (ε) and a given temperature (T).

$$\text{Strain-rate sensitivity} = m = \left(\frac{\Delta \log \sigma}{\Delta \log \dot{\varepsilon}} \right)_{\varepsilon T}$$

strain rods. (1) Rods sometimes used on gapframe metal forming presses to lessen the frame deflection. (2) Rods used to measure elastic strain, and thus stresses, in frames of metal forming presses.

strain state. See *state of strain*.

strand casting. A generic term describing *continuous casting* of one or more elongated shapes such as billets, blooms, or slabs; if two or more shapes are cast simultaneously, they are often of identical cross section.

stray current. (1) Current flowing through paths other than the intended circuit. (2) Current flowing in electrodeposition by way of an unplanned and undesired bipolar electrode that may be the tank itself or a poorly connected electrode.

stray-current corrosion. Corrosion resulting from direct current flow through paths other than the intended circuit. For example, by an extraneous current in the earth.

stress. The intensity of the internally distributed forces or components of forces that resist a change in the volume or shape of a material that is or has been subjected to external forces. Stress is expressed in force per unit area. Stress can be normal (tension or compression) or shear. See also *compressive stress, engineering stress, mean stress, nominal stress, normal stress, residual stress, shear stress, tensile stress,* and *true stress.*

stress amplitude. One-half the algebraic difference between the maximum and minimum stresses in one cycle of a repetitively varying stress.

stress concentration. On a macromechanical level, the magnification of the level of an applied stress in the region of a notch, void, hole, or inclusion.

stress concentration factor (K_t). A multiplying factor for applied stress that allows for the presence of a structural discontinuity such as a notch or hole; K_t equals the ratio of the greatest stress in the region of the discontinuity to the nominal stress for the entire section. Also called theoretical stress concentration factor.

stress corrosion. Preferential attack of areas under stress in a corrosive environment, where such an environment alone would not have caused corrosion.

stress-corrosion cracking (SCC). A cracking process that requires the simultaneous action of a corrodent and sustained tensile stress. This excludes corrosion-reduced sections that fail by fast fracture. It also excludes intercrystalline or transcrystalline corrosion, which can disintegrate an alloy without applied or residual stress. Stress-corrosion cracking may occur in combination with *hydrogen embrittlement.*

stress-intensity factor. A scaling factor, usually denoted by the symbol K, used in *linear-elastic fracture mechanics* to describe the intensification of applied stress at the tip of a crack of known size and shape. At the onset of rapid crack propagation in any structure containing a crack, the factor is called the critical stress-intensity factor, or the *fracture toughness.* Various subscripts are used to denote different loading conditions or fracture toughnesses:

K_c. Plane-stress fracture toughness. The value of stress intensity at which crack propagation becomes rapid in sections thinner than those in which plane-strain conditions prevail.

K_I. Stress-intensity factor for a loading condition that displaces the crack faces in a direction normal to the crack plane (also known as the opening mode of deformation).

K_{Ic}. Plane-strain fracture toughness. The minimum value of K_c for any given material and condition, which is attained when rapid crack propagation in the opening mode is governed by plane-strain conditions.

K_{Id}. Dynamic fracture toughness. The fracture toughness determined under dynamic loading conditions; it is used as an approximation of K_{Ic} for very tough materials.

K_{ISCC}. Threshold stress intensity factor for stress-corrosion cracking. The critical plane-strain stress intensity at the onset of stress-corrosion cracking under specified conditions.

K_Q. Provisional value for plane-strain fracture toughness.

K_{th}. Threshold stress intensity for stress-corrosion cracking. The critical stress intensity at the onset of stress-corrosion cracking under specified conditions.

K. The range of the stress-intensity factor during a fatigue cycle. See also *fatigue crack growth rate.*

stress-intensity factor range (K). In fatigue, the variation in the *stress-intensity factor* in a cycle, that is, K_{max} K_{min}. See also *fatigue crack growth rate.*

stress raisers. Design features (such as sharp corners) or mechanical defects (such as notches) that act to intensify the stress at these locations.

stress range. See *range of stress.*

stress ratio (A or R). The algebraic ratio of two specified stress values in a stress cycle. Two commonly used stress ratios are: (1) the ratio of the alternating stress amplitude to the mean stress, $A = S_a/S_m$; and (2) the ratio of the minimum stress to the maximum stress. $R = S_{min}/S_{max}$.

stress relaxation. The time-dependent decrease in stress in a solid under constant constraint at constant temperature.

stress-relaxation curve. A plot of the remaining or relaxed stress as a function of time. The relaxed stress equals the initial stress minus the remaining stress. Also known as stress-time curve.

stress-relief cracking. Cracking in the *heat-affected zone* or weld metal that occurs during the exposure of weldments to elevated temperatures during postweld heat treatment, in order to reduce residual stresses and improve toughness, or high temperature service. Stress-relief cracking occurs only in metals that can precipitation-harden during such elevated-temperature exposure; it usually occurs as *stress raisers,* is intergranular in nature, and is generally observed in the coarse-grained region of the weld heat-affected zone. Also called postweld heat treatment cracking or stress relief embrittlement.

stress-relief heat treatment. Uniform heating of a structure or a portion thereof to a sufficient temperature to relieve the major portion of the residual stresses, followed by uniform cooling.

stress relieving. Heating to a suitable temperature, holding long enough to reduce residual stresses, and then cooling slowly enough to minimize the development of new residual stresses.

stress-rupture strength. See *creep-rupture strength.*

stress-rupture test. See *creep-rupture test.*

stress state. See *state of stress.*

stress-strain curve. A graph in which corresponding values of stress and strain from a tension, compression, or torsion test are plotted against each other. Values of stress are usually plotted vertically (ordinates or *y*-axis) and values of strain horizontally (abscissas or *x*-axis). Also known as deformation curve and stress-strain diagram.

stretcher leveling. The leveling of a piece of sheet metal (that is, removing warp and distortion) by gripping it at both ends and subjecting it to a stress higher than its yield strength.

stretcher straightening. A process for straightening rod, tubing, and shapes by the application of tension at the ends of the stock. The products are elongated a definite amount to remove warpage.

stretcher strains. Elongated markings that appear on the surface of some sheet materials when deformed just past the yield point. These markings lie approximately parallel to the direction of maximum shear stress and are the result of localized yielding. See also *Lüders lines.*

stretch former. (1) A machine used to perform *stretch forming* operations. (2) A device adaptable to a conventional press for accomplishing stretch forming.

stretch forming. The shaping of a metal sheet or part, usually of uniform cross section, by first applying suitable tension or stretch and then wrapping it around a die of the desired shape.

stretching. The extension of the surface of a metal sheet in all directions. In stretching, the flange of the flat blank is securely clamped. Deformation is restricted to the area initially within the die. The stretching limit is the onset of metal failure.

striation. A fatigue fracture feature, often observed in electron micrographs, that indicates the position of the crack front after each succeeding cycle of stress. The distance between striations indicates the advance of the crack front across that crystal during one stress cycle, and a line normal to the striations indicates the direction of local crack propagation. See also *beach marks.*

strike. (1) A thin electrodeposited film of metal to be overlaid with other plated coatings. (2) A plating solution of high covering power and low efficiency designed to electroplate a thin, adherent film of metal.

striking. Electrodepositing, under special conditions, a very thin film of metal that will facilitate further plating with another metal or with the same metal under different conditions.

striking surface. Those areas on the faces of a set of metal forming dies that are designed to meet when the upper die and lower die are brought together. The striking surface helps protect impressions from impact shock and aids in maintaining longer die life.

stringer. In wrought materials, an elongated configuration of microconstituents or foreign material aligned in the direction of working. The term is commonly associated with elongated oxide or sulfide inclusions in steel.

stringer bead. A continuous weld bead made without appreciable transverse oscillation. Contrast with *weave bead*.

strip. (1) A flat-rolled metal product of some maximum thickness and width arbitrarily dependent on the type of metal; narrower than *sheet*. (2) A roll-compacted metal powder product. See also *roll compacting*. (3) Removal of a powder metallurgy compact from the die. An alternative to ejecting or knockout.

stripper. A plate designed to remove, or strip, sheet metal stock from the punching members during the withdrawal cycle. Strippers are also used to guide small precision punches in close-tolerance dies, to guide scrap away from dies, and to assist in the cutting action. Strippers are made in two types: fixed and movable.

stripper punch. A punch that serves as the top or bottom of a metal forming die cavity and later moves farther into the die to eject the part or compact. See also *ejector rod* and *knockout* (3).

stripping. (1) Removing a coating from a metal surface. (2) Removing a foundry pattern from the mold or the core box from the core.

structural shape. A piece of metal of any of several designs accepted as standard by the structural branch of the iron and steel industries.

structure. As applied to a crystal, the shape and size of the unit cell and the location of all atoms within the unit cell. As applied to microstructure, the size, shape, and arrangement of phases. See also *unit cell*.

stud welding. An arc welding process in which the contact surfaces of a stud, or similar fastener, and a workpiece are heated and melted by an arc drawn between them. The stud is then plunged rapidly onto the workpiece to form a weld. Partial shielding may be obtained by the use of a ceramic ferrule surrounding the stud. Shielding gas or flux may or may not be used. The two basic methods of stud welding are known as stud arc welding, which produces a large amount of weld metal around the stud base and a relatively deep penetration into the base metal, and capacitor discharge stud welding, which produces a very small amount of weld metal around the stud base and shallow penetration into the base metal.

styrofoam pattern. An expendable pattern of foamed plastic, especially expanded polystyrene, used in manufacturing castings by the lost foam process. See also *lost foam casting*.

sub-boundary structure (subgrain structure). A network of low-angle boundaries, usually with misorientations less than 1° within the main grains of a microstructure.

subcritical annealing. An annealing treatment in which a steel is heated to a temperature below the A_1 temperature, then cooled slowly to room temperature. See also *transformation temperature*.

subgrain. A portion of a crystal or *grain*, with an orientation slightly different from the orientation of neighboring portions of the same crystal.

submerged arc welding. Arc welding in which the arc, between a bare metal electrode and the work, is shielded by a blanket of granular, fusible material overlying the joint. Pressure is not applied to the joint, and filler metal is obtained from the consumable electrode (and sometimes from a supplementary welding rod).

submerged-electrode furnace. A furnace used for liquid carburizing of parts by heating molten salt baths with the use of electrodes submerged in the ceramic lining. See also *immersed-electrode furnace*.

subsieve fraction. Particles that will pass through a 44-m (325-mesh) screen.

subsieve size. See preferred term *subsieve fraction*.

substitutional element. An alloying element with an atomic size and other features similar to the solvent that can replace or substitute for the solvent atoms in the lattice and form a significant region of solid solution in the *phase diagram*.

substitutional solid solution. A *solid solution* in which the solvent and solute atoms are located randomly at the atom sites in the crystal structure of the solution. See also *interstitial solid solution*.

substrate. The material, workpiece, or substance on which the coating is deposited.

substructure. Same as *sub-boundary structure*.

subsurface corrosion. Formation of isolated particles of corrosion products beneath a metal surface. This results from the preferential reactions of certain alloy constituents to inward diffusion of oxygen, nitrogen, or sulfur.

sulfidation. The reaction of a metal or alloy with a sulfur-containing species to produce a sulfur compound that forms on or beneath the surface on the metal or alloy.

sulfide stress cracking (SSC). Brittle fracture by cracking under the combined action of *tensile stress* and *corrosion* in the presence of water and hydrogen sulfide. See also *environmental cracking*.

sulfur dome. An inverted container, holding a high concentration of sulfur dioxide gas, used in die casting to cover a pot of molten magnesium to prevent burning.

sulfur print. A macrographic method of examining for distribution of sulfide inclusions by placing a sheet of wet acidified photographic paper in contact with the polished sheet surface to be examined.

superabrasives. Synthetically produced diamond and cubic boron nitride (CBN) used in a wide variety of cutting and grinding applications.

superalloys. Heat-resistant alloys based on nickel, iron-nickel, or cobalt that exhibit high strength and resistance to surface degradation at elevated temperatures.

superconductivity. A property of many metals, alloys, compounds, oxides, and organic materials at temperatures near absolute zero by virtue of which their electrical resistivity vanishes and they become strongly diamagnetic.

supercooling. Cooling of a substance below the temperature at which a change of state would ordinarily take place without such a change of state occurring, for example, the cooling of a liquid below its freezing point without freezing taking place; this results in a *metastable* state.

superficial hardness test. See *Rockwell superficial hardness test*.

superfines. The portion of a metal powder that is composed of particles smaller than a specified size, usually 10 m.

superfinishing. An abrasive process utilizing either a curved bonded honing stick (stone) for a cylindrical workpiece or a cup wheel for flat and spherical work. A large contact area, 30% approximately, exists between workpiece and abrasive. The object of superfinishing is to remove surface fragmentation and to correct inequalities in geometry, such as grinding feed marks and chatter marks. Also known as microhoning. See also *honing*.

superheating. (1) Heating of a substance above the temperature at which a change of state would ordinarily take place without a change of state occurring, for example, the heating of a liquid above its boiling point without boiling taking place; this results in a metastable state. (2) Any increment of temperature above the melting point of a metal; sometimes construed to be any increment of temperature above normal casting temperatures introduced for the purpose of refining, alloying, or improving fluidity.

superlattice. See *ordered structure*.

superplastic forming (SPF). A strain rate sensitive sheet metal forming process that uses characteristics of materials exhibiting high tensile elongation. Superplastic forming methods include: blow molding, in which gas pressure is imposed on a superplastic diaphragm, causing the material to form into the die configuration; vacuum forming, a process similar to blow molding except that the forming pressure is limited to atmospheric pressure (100 kPa, or 15 psi) versus the maximum pressure of 700 to 3400 kPa (100 to 500 psi) for blow molding; thermoforming methods adopted from plastics technology, which involve a moving or adjustable die member in conjunction with gas pressure or vacuum; and superplastic forming/diffusion bonding (SPF/DB), which combines blow molding and solid-state bonding. See also *superplasticity*.

superplasticity. The ability of certain metals (most notably aluminum- and titanium-base alloys) to develop extremely high tensile elongations at elevated temperatures and under controlled rates of deformation.

supersaturated. A metastable solution in which the dissolved material exceeds the amount the solvent can hold in normal equilibrium at the temperature and other conditions that prevail.

support pins. Rods or pins of precise length used to support the overhang of irregularly shaped punches in metal forming presses.

support plate. A plate that supports a draw ring or draw plate in a sheet metal forming press. It also serves as a spacer. See also *draw plate* and *draw ring*.

surface alterations. Irregularities or changes on the surface of a material due to machining or grinding operations. The types of surface alterations associated with metal removal practices include mechanical (for example, plastic deformation, hardness variations, cracks, etc.), metallurgical (for example, phase transformations, twinning, recrystallization, and untempered or overtempered martensite), chemical (for example, intergranular attack, embrittlement, and pitting), thermal (heat-affected zone, recast, or redeposited metal, and resolidified material), and electrical surface alterations (conductivity change or resistive heating).

surface checking. Same as *checks*.

surface damage. In *tribology*, damage to a solid surface resulting from mechanical contact with another substance, surface, or surfaces moving relatively to it and involving the displacement or removal of material. In certain contexts, *wear* is a form of surface damage in which material is progressively removed. In another context, surface damage involves a deterioration of function of a solid surface even though there is no material loss from that surface. Surface damage may therefore precede wear.

surface finish. (1) The geometric irregularities in the surface of a solid material. Measurement of surface finish shall not include inherent structural irregularities unless these are the characteristics being measured. (2) Condition of a surface as a result of a final treatment. See also *roughness*.

surface grinding. Producing a plane surface by grinding.

surface hardening. A generic term covering several processes applicable to a suitable ferrous alloy that produces, by quench hardening only, a surface layer that is harder or more wear resistant than the core. There is no significant alteration of the chemical composition of the surface layer. The processes commonly used are *carbonitriding, carburizing, induction hardening, flame hardening, nitriding,* and *nitrocarburizing*. Use of the applicable specific process name is preferred.

surface modification. The alteration of surface composition or structure by the use of energy or particle beams. Two types of surface modification methods commonly employed are *ion implantation* and *laser surface processing*.

surface roughness. Fine irregularities in the *surface texture* of a material, usually including those resulting from the inherent action of the production process. Surface roughness is usually reported as the arithmetic roughness average, R_a, and is given in micrometers or microinches. See also the figure accompanying *waviness*.

surface texture. The roughness, waviness, lay, and flaws associated with a surface. See also *lay*.

surfacing. The deposition of filler metal (material) on a base metal (substrate) to obtain desired properties or dimensions. See also *buttering, cladding, coating,* and *hardfacing*.

surfacing weld. A type of weld composed of one or more stringer or weave beads deposited on an unbroken surface to obtain desired properties or dimensions.

swage. (1) The operation of reducing or changing the cross-section area of stock by the fast impact of revolving dies. (2) The tapering of bar, rod, wire, or tubing by forging, hammering, or squeezing; reducing a section by progressively tapering lengthwise until the entire section attains the smaller dimension of the taper.

swaging. Tapering bar, rod, wire, or tubing by forging, hammering, or squeezing; reducing a section by progressively tapering lengthwise until the entire section attains the smaller dimension of the taper. See also *rotary swaging*.

swarf. Intimate mixture of grinding chips and fine particles of abrasive and bond resulting from a grinding operation.

sweep. A type of foundry pattern that is a template cut to the profile of the desired mold shape that, when revolved around a stake or spindle, produces that shape in the mold.

sweeps. Floor and table sweepings containing precious metal particles.

Swift cup test. A simulative test for determining formability of sheet metal in which circular blanks of various diameters are clamped in a die ring and deep drawn into a cup by a flat-bottomed cylindrical punch. The ratio of the largest blank diameter that can be drawn successfully to the cup diameter is known as the *limiting drawing ratio (LDR)* or *deformation limit*.

swing forging machine. Equipment for continuously hot reducing ingots, blooms, or billets to square flats, rounds, or rectangles by the crank-driven oscillating action of paired dies.

swing frame grinder. A grinding machine suspended by a chain at the center point so that it may be turned and swung in any direction for grinding of billets, large castings, or other heavy work. Principal use is removing surface imperfections and roughness.

synthetic cold rolled sheet. A hot rolled pickled sheet given a sufficient final temper pass to impart a surface approximating that of cold rolled steel.

T

tacking. Making *tack welds.*

tack welds. (1) Small, scattered welds made to hold parts of a weldment in proper alignment while the final welds are being made. (2) Intermittent welds to secure weld backing bars. See also *backing* (2).

tailings. The discarded portion of a crushed ore, separated during concentration.

tandem mill. A rolling mill consisting of two or more stands arranged so that the metal being processed travels in a straight line from stand to stand. In continuous rolling, the various stands are synchronized so that the strip can be rolled in all stands simultaneously. Contrast with *single-stand mill.* See also *rolling mills.*

tandem welding. Arc welding in which two or more electrodes are in a plane parallel to the line of travel.

tangent bending. The forming of one or more identical bends having parallel axes by wiping sheet metal around one or more radius dies in a single operation. The sheet, which may have side flanges, is clamped against the radius die and then made to conform to the radius die by pressure from a rocker-plate die that moves along the periphery of the radius die. See also *wiper forming (wiping).*

tap. A cylindrical or conical thread-cutting tool with one or more cutting elements having threads of a desired form on the periphery. By a combination of rotary and axial motions, the leading end cuts an internal thread, the tool deriving its principal support from the thread being produced.

tap density. The apparent density of a powder, obtained when the volume receptacle is tapped or vibrated during loading under specified conditions.

tapping. (1) Producing internal threads with a cylindrical cutting tool having two or more peripheral cutting elements shaped to cut threads of the desired size and form. By a combination of rotary and axial motion, the leading end of the tap cuts the thread while the tap is supported mainly by the thread it produces. See also *tap.* (2) Opening the outlet of a melting furnace to remove molten metal. (3) Removing molten metal from a furnace.

tarnish. Surface discoloration of a metal caused by formation of a thin film of corrosion product.

Taylor process. A process for making extremely fine metal wire by inserting a piece of larger-diameter wire into a glass tube and stretching the two together at high temperature.

teapot ladle. A ladle in which, by means of an external spout, metal is removed from the bottom rather than the top of the ladle.

tee joint. A joint in which the members are oriented in the form of a T.

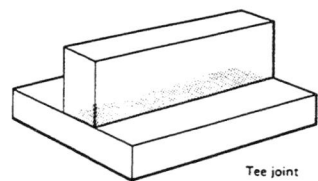
Tee joint

teeming. Pouring molten metal from a ladle into ingot molds. The term applies particularly to the specific operation of pouring either iron or steel into ingot molds.

temper. (1) In heat treatment, reheating hardened steel or hardened cast iron to some temperature below the eutectoid temperature for the purpose of decreasing hardness and increasing toughness. The process also is sometimes applied to normalized steel. (2) In tool steels, temper is sometimes used, but inadvisedly, to denote the carbon content. (3) In nonferrous alloys and in some ferrous alloys (steels that cannot be hardened by heat treatment), the hardness and strength produced by mechanical or thermal treatment, or both, and characterized by a certain structure, mechanical properties, or reduction in area during cold working. (4) To moisten *green sand* for casting molds with water.

temper brittleness. See *temper embrittlement.*

temper carbon. Clusters of finely divided graphite, such as that found in malleable iron, that are formed as a result of decomposition of cementite, for example, by heating white cast iron above the ferrite-austenite transformation temperature and holding at these temperatures for a considerable period of time. Also known as annealing carbon.

temper color. A thin, tightly adhering oxide skin (only a few molecules thick) that forms when steel is tempered at a low temperature, or for a short time, in air or a mildly oxidizing atmosphere. The color, which ranges from straw to blue depending on the thickness of the oxide skin, varies with both tempering time and temperature.

tempered layer. A surface or subsurface layer in a steel specimen that has been tempered by heating during some stage of the metallographic preparation sequence (usually grinding). When observed in a section after etching, the layer appears darker than the base material.

tempered martensite. The decomposition products that result from heating marten-

site below the ferrite-austenite transformation temperature. Under the optical microscope, darkening of the martensite needles is observed in the initial stages of tempering. Prolonged tempering at high temperatures produces spheroidized carbides in a matrix of ferrite. At the higher resolution of the electron microscope, the initial stage of tempering is observed to result in a structure containing a precipitate of fine iron carbide particles. At approximately 260 °C (500 °F), a transition occurs to a structure of larger and elongated cementite particles in a ferrite matrix. With further tempering at higher temperatures, the cementite particles become spheroidal, decreased in number, and increased in size.

tempered martensite embrittlement. *Embrittlement* of high-strength alloy steels caused by tempering in the temperature range of 205 to 370 °C (400 to 700 °F); also called 350 °C or 500 °F embrittlement. Tempered martensite embrittlement is thought to result from the combined effects of cementite precipitation on prior-austenite grain boundaries or interlath boundaries and the segregation of impurities at prior-austenite grain boundaries. It differs from *temper embrittlement* in the strength of the material and the temperature exposure range. In temper embrittlement, the steel is usually tempered at a relatively high temperature, producing lower strength and hardness, and embrittlement occurs upon slow cooling after tempering and during service at temperatures within the embrittlement range. In tempered martensite embrittlement, the steel is tempered within the embrittlement range, and service exposure is usually at room temperature.

temper embrittlement. *Embrittlement* of low-alloy steels caused by holding within or cooling slowly through a temperature range (generally 300 to 600 °C, or 570 to 1110 °F) just below the transformation range. Embrittlement is the result of the segregation at grain boundaries of impurities such as arsenic, antimony, phosphorus, and tin; it is usually manifested as an upward shift in ductile-to-brittle transition temperature. Temper embrittlement can be reversed by retempering above the critical temperature range, then cooling rapidly. Compare with *tempered martensite embrittlement.*

tempering. In heat treatment, reheating hardened steel to some temperature below the eutectoid temperature to decrease hardness and/or increase toughness.

temper rolling. Light cold rolling of sheet steel to improve flatness, to minimize the formation of *stretcher strains*, and to obtain a specified hardness or temper.

tensile strength. In tensile testing, the ratio of maximum load to original cross-sectional area. Also called *ultimate strength*. Compare with *yield strength*.

tensile stress. A stress that causes two parts of an elastic body, on either side of a typical stress plane, to pull apart. Contrast with *compressive stress*.

tensile testing. See *tension testing*.

tension. The force or load that produces elongation.

tension testing. A method of determining the behavior of materials subjected to uniaxial loading, which tends to stretch the material. A longitudinal specimen of known length and diameter is gripped at both ends and stretched at a slow, controlled rate until rupture occurs. Also known as tensile testing.

terminal phase. A solid solution having a restricted range of compositions, one end of the range being a pure component of an alloy system.

terminal solid solution. In a multicomponent system, any solid phase of limited composition range that includes the composition of one of the components of the system. See also *solid solution*.

ternary alloy. An alloy that contains three principal elements.

ternary system. The complete series of compositions produced by mixing three components in all proportions.

terne. An alloy of lead containing 3 to 15% Sn, used as a *hot dip coating* for steel sheet or plate. The term long terne is used to describe terne-coated sheet, whereas short terne is used for terne-coated plate. Terne coatings, which are smooth and dull in appearance (terne means dull or tarnished in French), give the steel better corrosion resistance and enhance its ability to be formed, soldered, or painted.

tertiary creep. See *creep*.

texture. In a polycrystalline aggregate, the state of distribution of crystal orientations. In the usual sense, it is synonymous with *preferred orientation*, in which the distribution is not random. Not to be confused with *surface texture*. See also *fiber*.

thermal aging. Exposure of a material or component to a given thermal condition or a programmed series of conditions for prescribed periods of time.

thermal analysis. A method for determining transformations in a metal by noting the temperatures at which thermal arrests occur. These arrests are manifested by changes in slope of the plotted or mechanically traced heating and cooling curves. When such data are secured under nearly equilibrium conditions of heating and cooling, the method is commonly used for determining certain critical temperatures required for the construction of phase diagrams.

thermal cutting. A group of cutting processes which melts the metal (material) to be cut. See also *air carbon arc cutting*, *arc cutting*, *carbon arc cutting*, *electron beam cutting*, *laser beam cutting*, *metal powder cutting*, *oxyfuel gas cutting*, *oxygen arc cutting*, *oxygen cutting*, and *plasma arc cutting*.

thermal decomposition. (1) The decomposition of a compound into its elemental species at elevated temperatures. (2) A process whereby fine solid particles can be produced from a gaseous compound. See also *carbonyl powder*.

thermal electromotive force. The *electromotive force* generated in a circuit containing two dissimilar metals when one junction is at a temperature different from that of the other. See also *thermocouple*.

thermal embrittlement. *Intergranular fracture* of maraging steels with decreased toughness resulting from improper processing after hot working. Thermal embrittlement occurs upon heating above 1095 °C (2000 °F) and then slow cooling through the temperature range of 980 to 815 °C (1800 to 1500 °F), and has been attributed to precipitation of titanium carbides and titanium carbonitrides at austenite grain boundaries during cooling through the critical temperature range. See also *maraging steels*.

thermal fatigue. Fracture resulting from the presence of temperature gradients that vary with time in such a manner as to produce cyclic stresses in a structure.

thermal inspection. A nondestructive test method in which heat-sensing devices are used to measure temperature variations in components, structures, systems, or physical processes. Thermal methods can be useful in the detection of subsurface flaws or voids, provided the depth of the flaw is not large compared to its diameter. Thermal inspection becomes less effective in the detection of subsurface flaws as the thickness of an object increases, because the possible depth of the defects increases.

thermally induced embrittlement. See *embrittlement*.

thermal-mechanical treatment. See *thermomechanical working*.

thermal shock. The development of a steep temperature gradient and accompanying high stresses within a material or structure.

thermal spraying. A group of coating or welding processes in which finely divided metallic or nonmetallic materials are deposited in a molten or semimolten condition to form a coating. The coating material may be in the form of powder, ceramic rod, wire, or molten materials. See also *electric arc spraying*, *flame spraying*, *plasma spraying*, and *powder flame spraying*.

thermal stresses. Stresses in a material resulting from nonuniform temperature distribution.

thermal wear. Removal of material due to softening, melting, or evaporation during sliding or rolling. Thermal shock and high-temperature erosion may be included in the general description of thermal wear. Wear by diffusion of separate atoms from one body to the other, at high temperatures, is also sometimes denoted as thermal wear.

thermit reactions. Strongly exothermic self-propagating reactions such as that where finely divided aluminum reacts with a metal oxide. A mixture of aluminum and iron oxide produces sufficient heat to weld steel, the filler metal being produced in the reaction. See also *thermit welding*.

thermit welding. A welding process which produces coalescence of metals by heating them with superheated liquid metal from a chemical reaction between a metal oxide and aluminum, with or without the application of pressure. Filler metal, when used, is obtained from the liquid metal. The process is used primarily for welding railroad track.

thermochemical machining. Removal of workpiece material—usually only burrs and fins—by exposure to hot fuel gases which are formed by igniting an explosive, combustible mixture of natural gas and oxygen. Also known as the thermal energy method.

thermochemical treatment. Heat treatment for steels carried out in a medium suitably chosen to produce a change in the chemical composition of the object by exchange with the medium.

thermocouple. A device for measuring temperatures, consisting of lengths of two dissimilar metals or alloys that are electrically joined at one end and connected to a voltage-measuring instrument at the other end. When one junction is hotter than the other, a *thermal electromotive force* is produced that is roughly proportional to the difference in temperature between the hot and cold junctions.

thermomechanical working. A general term covering a variety of metal forming processes combining controlled thermal and deformation treatments to obtain synergistic effects, such as improvement in strength without loss of toughness. Same as thermal-mechanical treatment.

thief. A racking device or nonfunctional pattern area used in the electroplating process to provide a more uniform current density on plated parts. Thieves absorb the unevenly distributed current on irregularly shaped parts, thereby ensuring that the parts will receive an electroplated coating of uniform thickness. See also *robber.*

thin-wall casting. A term used to define a casting that has the minimum wall thickness to satisfy its service function.

Thomas converter. A Bessemer converter having a basic bottom and lining, usually dolomite, and employing a basic slag.

threading. Producing external threads on a cylindrical surface.

thread rolling. The production of threads by rolling the piece between two grooved die plates, one of which is in motion, or between rotating grooved circular rolls. Also known as roll threading.

three-quarters hard. A *temper* of nonferrous alloys and some ferrous alloys characterized by tensile strength and hardness about midway between those of *half hard* and *full hard* tempers.

three-point bending. The bending of a piece of metal or a structural member in which the object is placed across two supports and force is applied between and in opposition to them. See also *V-bend die.*

threshold stress. Threshold stress for *stress-corrosion cracking.* The critical gross section stress at the onset of stress-corrosion cracking under specified conditions.

throat depth. The distance from the centerline of the electrodes or platens to the nearest point of interference for flat sheets in a resistance welding machine. In the case of a resistance seam welding machine with a universal head, the throat depth is measured with the machine arranged for transverse welding.

throat height. The unobstructed dimension between arms throughout the throat depth in a resistance welding machine.

throat of a fillet weld. A term that includes the theoretical throat, the actual throat, and the effective throat. (1) The theoretical throat is the distance from the beginning of the root of the joint perpendicular to the hypotenuse of the largest right triangle that can be inscribed within the fillet weld cross section. This dimension is based on the assumption that the root opening is equal to zero. (2) The actual throat is the shortest distance from the root of the weld to its face. (3) The effec-

tive throat is the minimum distance minus any reinforcement from the root of the weld to its face. See also the figures accompanying *concave fillet weld* and *convex fillet weld.*

throat of a groove weld. See preferred term *size of weld.*

through weld. A nonpreferred term sometimes used to indicate a weld of substantial length made by melting through one member of a lap or tee joint and into the other member.

throwing power. (1) The relationship between the *current density* at a point on a surface and its distance from the *counterelectrode.* The greater the ratio of the surface resistivity shown by the electrode reaction to the volume resistivity of the electrolyte, the better is the throwing power of the process. (2) The ability of a plating solution to produce a uniform metal distribution on an irregularly shaped *cathode.* Compare with *covering power.*

tiger stripes. Continuous bright lines on sheet or strip in the rolling direction.

TIG welding. Tungsten inert-gas welding; see preferred term *gas tungsten-arc welding.*

tilt boundary. A subgrain boundary consisting of an array of edge *dislocations.*

tilt mold. A casting mold, usually a book (permanent) mold, that rotates from a horizontal to a vertical position during pouring, which reduces agitation and thus the formation and entrapment of oxides.

tilt mold ingot. An ingot made in a *tilt mold.*

time quenching. A term used to describe a quench in which the cooling rate of the part being quenched must be changed abruptly at some time during the cooling cycle.

time-temperature curve. A curve produced by plotting time against temperature.

time-temperature-transformation (TTT) diagram. See *isothermal transformation (IT) diagram.*

tinning. Coating metal with a very thin layer of molten solder or brazing filler metal.

tin pest. A polymorphic modification of tin that causes it to crumble into a powder known as gray tin. It is generally accepted that the maximum rate of transformation occurs at about 40 °C (40 °F), but transformation can occur at as high as about 13 °C (55 °F).

tint etching. Immersing metallographic specimens in specially formulated chemical etchants in order to produce a stable film on the specimen surface. When viewed under an optical microscope,

these surface films produce colors which correspond to the various phases in the alloy. Also known as color etching.

tin tossing. Oxidizing impurities in molten tin by pouring it from one vessel to another in air, forming a dross that is mechanically separable.

TIR. Abbreviation for *total indicator reading.*

toe crack. A crack in the base metal occurring at the toe of a weld.

toe of weld. The junction between the face of a weld and the base metal. See also the figure accompanying *fillet weld.*

toggle press. A *mechanical press* in which the *slide* is actuated by one or more toggle links or mechanisms.

tolerance. The specified permissible deviation from a specified nominal dimension, or the permissible variation in size or other quality characteristic of a part.

tolerance limits. The extreme values (upper and lower) that define the range of permissible variation in size or other quality characteristic of a part.

tonghold. The portion of a forging billet, usually on one end, that is gripped by the operator's tongs. It is removed from the part at the end of the forging operation. Common to drop hammer and press-type forging.

tooling. A generic term applying to die assemblies and related items used for forming and forging metals.

tool steel. Any of a class of carbon and alloy steels commonly used to make tools. Tool steels are characterized by high hardness and resistance to abrasion, often accompanied by high toughness and resistance to softening at elevated temperature. These attributes are generally attained with high carbon and alloy contents.

tooth. (1) A projection on a multipoint tool (such as on a saw, milling cutter, or file) designed to produce cutting. (2) A projection on the periphery of a wheel or segment thereof—as on a gear, spline, or sprocket, for example—designed to engage another mechanism and thereby transmit force or motion, or both. A similar projection on a flat member such as a rack.

tooth point. On a *face mill,* the chamfered cutting edge of the blade, to which a flat is sometimes added to produce a shaving effect and to improve finish.

top-and-bottom process. A process for separating copper and nickel, in which their molten sulfides are separated into two liquid layers by the addition of sodium sulfide. The lower layer holds most of the nickel.

torch. See preferred terms *cutting torch* and *welding torch*.

torch brazing. A brazing process in which the heat required is furnished by a fuel gas flame.

torch soldering. A soldering process in which the heat required is furnished by a fuel gas flame.

torsion. (1) A twisting deformation of a solid or tubular body about an axis in which lines that were initially parallel to the axis become helices. (2) A twisting action resulting in shear stresses and strains.

torsional moment. In a body being twisted, the algebraic sum of the couples or the moments of the external forces about the axis of twist, or both.

total carbon. The sum of the *free carbon* and *combined carbon* (including carbon in solution) in a ferrous alloy.

total cyanide. Cyanide content of an electroplating bath (including both simple and complex ions).

total elongation. The total amount of permanent extension of a test piece broken in a tensile test usually expressed as a percentage over a fixed gage length. See also *elongation, percent*.

total indicator reading. See *total indicator variation*.

total indicator variation. The difference between the maximum and minimum indicator readings during a checking cycle.

toughness. Ability of a material to absorb energy and deform plastically before fracturing. Toughness is proportional to the area under the *stress-strain curve* from the origin to the breaking point. In metals, toughness is usually measured by the energy absorbed in a notch impact test. See also *impact test*.

tough pitch copper. Copper containing from 0.02 to 0.04% oxygen, obtained by refining copper in a reverberatory furnace.

tracer milling. Duplication of a three-dimensional form by means of a cutter controlled by a tracer that is directed by a master form.

tramp alloys. Residual alloying elements that are introduced into steel when unidentified alloy steel is present in the scrap charge to a steelmaking furnace.

tramp element. Contaminant in the components of a furnace charge, or in the molten metal or castings, whose presence is thought to be either unimportant or undesirable to the quality of the casting. Also called trace element.

transcrystalline. See *transgranular*.

transcrystalline cracking. Cracking or fracturing that occurs through or across a crystal. Also termed intracrystalline cracking.

transformation hardening. Heat treatment of steels comprising austenitization followed by cooling under conditions such that the austenite transforms more or less completely into martensite and possibly into bainite.

transformation-induced plasticity. A phenomenon, occurring chiefly in certain highly alloyed steels that have been heat treated to produce metastable austenite or metastable austenite plus martensite, whereby, on subsequent deformation, part of the austenite undergoes strain-induced transformation to martensite. Steels capable of transforming in this manner, commonly referred to as TRIP steels, are highly plastic after heat treatment, but exhibit a very high rate of strain hardening and thus have high tensile and yield strengths after plastic deformation at temperatures between about 20 and 500 °C (70 and 930 °F). Cooling to 195 °C (320 °F) may or may not be required to complete the transformation to martensite. Tempering usually is done following transformation.

transformation ranges. Those ranges of temperature within which austenite forms during heating and transforms during cooling. The two ranges are distinct, sometimes overlapping but never coinciding. The limiting temperatures of the ranges depend on the composition of the alloy and on the rate of change of temperature, particularly during cooling. See also *transformation temperature*.

transformation temperature. The temperature at which a change in phase occurs. This term is sometimes used to denote the limiting temperature of a transformation range. The following symbols are used for irons and steels:

Ac_{cm}. In hypereutectoid steel, the temperature at which a solution of cementite in austenite is completed during heating.

Ac_1. The temperature at which austenite begins to form during heating.

Ac_3. The temperature at which transformation of ferrite to austenite is completed during heating.

Ac_4. The temperature at which austenite transforms to delta ferrite during heating.

Ae_{cm}, Ae_1, Ae_3, Ae_4. The temperatures of phase changes at equilibrium.

Ar_{cm}. In hypereutectoid steel, the temperature at which precipitation of cementite starts during cooling.

Ar_1. The temperature at which transformation of austenite to ferrite or to ferrite plus cementite is completed during cooling.

Ar_3. The temperature at which austenite begins to transform to ferrite during cooling.

Ar_4. The temperature at which delta ferrite transforms to austenite during cooling.

Ar-. The temperature at which transformation of austenite to pearlite starts during cooling.

M_f. The temperature at which transformation of austenite to martensite is completed during cooling.

M_s (or Ar.). The temperature at which transformation of austenite to martensite starts during cooling.

Note: All these changes, except formation of martensite, occur at lower temperatures during cooling than during heating, and depend on the rate of change of temperature.

transgranular. Through or across crystals or grains. Also called intracrystalline or transcrystalline.

transgranular cracking. Cracking or fracturing that occurs through or across a crystal or grain. Also called transcrystalline cracking. Contrast with *intergranular cracking*.

transgranular fracture. Fracture through or across the crystals or grains of a material. Also called transcrystalline fracture or intracrystalline fracture. Contrast with *intergranular fracture*.

transition lattice. An unstable crystallographic configuration that forms as an intermediate step in a solid-state reaction such as precipitation from solid solution or eutectoid decomposition.

transition metal. A metal in which the available electron energy levels are occupied in such a way that the *d*-band contains less than its maximum number of ten electrons per atom, for example, iron, cobalt, nickel, and tungsten. The distinctive properties of the transition metals result from the incompletely filled *d*-levels.

transition phase. A nonequilibrium state that appears in a chemical system in the course of transformation between two equilibrium states.

transition point. At a stated pressure, the temperature (or at a stated temperature, the pressure) at which two solid phases exist in equilibrium—that is, an allotropic transformation temperature (or pressure).

transition structure. In precipitation from solid solution, a metastable precipitate that is coherent with the matrix.

transition temperature. (1) An arbitrarily defined temperature that lies within the temperature range in which metal fracture characteristics (as usually determined by tests of notched specimens) change rapidly, such as the ductile-to-brittle transition temperature (DBTT). The DBTT can be assessed in several ways, the most common being the temperature for 50% ductile and 50% brittle fracture (50% fracture appearance transition temperature, or FATT), or the lowest temperature at which the fracture is 100% ductile (100% fibrous criterion). The DBTT is commonly associated with *temper embrittlement* and *radiation damage* (neutron irradiation) of low-alloy steels. (2) Sometimes used to denote an arbitrarily defined temperature within a range in which the ductility changes rapidly with temperature.

transverse direction. Literally, "across," usually signifying a direction or plane perpendicular to the direction of working. In rolled plate or sheet, the direction across the width is often called long transverse; the direction through the thickness, short transverse.

transverse rolling machine. Equipment for producing complex preforms or finished forgings from round billets inserted transversely between two or three rolls that rotate in the same direction and drive the billet. The rolls, carrying replaceable die segments with appropriate impressions, make several revolutions for each rotation of the workpiece.

transverse rupture strength. The stress, as calculated from the flexure formula, required to break a sintered powder metallurgy specimen. The test for determining the transverse rupture strength involves applying the load at the center of a 31.8 by 12.7 by 6.4 mm (1.25 by 0.5 by 0.25 in.) beam which is supported near its ends.

travel angle. The angle that a welding electrode makes with a reference line perpendicular to the axis of the weld in the plane of the weld axis. This angle can be used to define the position of welding guns, welding torches, high-energy beams, welding rods, thermal cutting and thermal spraying torches, and thermal spraying guns. See also the figures accompanying *backhand welding* and *forehand welding*.

trees. Visible projections of electrodeposited metal formed at sites of high current density.

trepanning. A machining process for producing a circular hole or groove in solid stock, or for producing a disk, cylinder, or tube from solid stock, by the action of a tool containing one or more cutters (usually single-point) revolving around a center.

triaxiality. In a *triaxial stress* state, the ratio of the smallest to the largest principal stress, all stresses being tensile.

triaxial stress. A state of stress in which none of the three principal stresses is zero. See also *principal stress (normal)*.

tribology. (1) The science and technology of interacting surfaces in relative motion and of the practices related thereto. (2) The science concerned with the design, friction, lubrication, and wear of contacting surfaces that move relative to each other (as in bearings, cams, or gears, for example).

trimmer blade. The portion of the trimmers through which a forging is pushed to shear off the flash.

trimmer die. The punch press die used for trimming flash from a forging.

trimmer punch. The upper portion of the trimmer that contacts the forging and pushes it through the trimmer blades; the lower end of the trimmer punch is generally shaped to fit the surface of the forging against which it pushes.

trimmers. The combination of *trimmer punch, trimmer blades*, and perhaps *trimming shoe* used to remove the flash from the forging.

trimming. (1) In forging, removing any parting-line flash or excess material from the part with a trimmer in a trim press; can be done hot or cold. (2) In drawing, shearing the irregular edge of the drawn part. (3) In casting, the removal of gates, risers, and fins.

trimming press. A power press suitable for trimming flash from forgings.

trimming shoe. The holder used to support *trimmers*. Sometimes called trimming chair.

triple-action press. A mechanical or hydraulic press having three slides with three motions properly synchronized for triple-action drawing, redrawing, and forming. Usually, two slides—the blankholder slide and the plunger—are located above and a lower slide is located within the bed of the press. See also *hydraulic press*, *mechanical press*, and *slide*.

triple point. (1) A point on a phase diagram where three phases of a substance coexist in equilibrium. (2) The intersection of the boundaries of three adjoining grains, as observed in a metallographic section.

TRIP steel. A commercial steel product exhibiting *transformation-induced plasticity*.

troostite (obsolete). A previously unresolvable, rapidly etching, fine aggregate of carbide and ferrite produced either by tempering martensite at low temperature or by quenching a steel at a rate slower than the critical cooling rate. Preferred terminology for the first product is tempered martensite; for the latter, fine pearlite.

Troy ounce. A unit of weight for *precious metals* that is equal to 31.1034768 g (1.0971699 oz avoirdupois).

true current density. See preferred term *local current density*.

true rake. See preferred term *effective rake*.

true strain. (1) The ratio of the change in dimension, resulting from a given load increment, to the magnitude of the dimension immediately prior to applying the load increment. (2) In a body subjected to axial force, the natural logarithm of the ratio of the gage length at the moment of observation to the original gage length. Also known as natural strain.

true stress. The value obtained by dividing the load applied to a member at a given instant by the cross-sectional area over which it acts.

truing. The removal of the outside layer of abrasive grains on a grinding wheel for the purpose of restoring its face.

tuberculation. The formation of *localized corrosion* products scattered over the surface in the form of knoblike mounds called tubercles. The formation of tubercles is usually associated with *biological corrosion*.

tube reducing. Reducing both the diameter and wall thickness of tubing with a mandrel and a pair of rolls. See also *spinning*.

tube sinking. Drawing tubing through a die or passing it through rolls without the use of an interior tool (such as a mandrel or plug) to control inside diameter; sinking generally produces a tube of increased wall thickness and length.

tube stock. A semifinished tube suitable for subsequent reduction and finishing.

tumbling. Rotating workpieces, usually castings or forgings, in a barrel partly filled with metal slugs or abrasives, to remove sand, scale, or fins. It may be done dry, or with an aqueous solution added to the contents of the barrel. See also *barrel finishing*.

tungsten inert-gas welding. See preferred term *gas tungsten arc welding*.

Turk's-head rolls. Four undriven working rolls, arranged in a square or rectangular pattern, through which metal strip, wire, or tubing is drawn to form square or rectangular sections.

turning. Removing material by forcing a single-point cutting tool against the surface of a rotating workpiece. The tool may or may not be moved toward or along the axis of rotation while it cuts away material.

tuyere. An opening in a cupola, blast furnace, or converter for the introduction of air or inert gas.

twin. Two portions of a crystal with a definite orientation relationship; one may be regarded as the parent, the other as the twin. The orientation of the twin is a mirror image of the orientation of the parent across a twinning plane or an orientation that can be derived by rotating the twin portion about a twinning axis. See also *annealing twin* and *mechanical twin*.

twin bands. Bands across a crystal grain, observed on a polished and etched section, where crystallographic orientations have a mirror-image relationship to the orientation of the matrix grain across a composition plane that is usually parallel to the sides of the band.

twist boundary. A subgrain boundary consisting of an array of screw *dislocations*.

two-high mill. A type of rolling mill in which only two rolls, the working rolls, are contained in a single housing. Compare with *four-high mill* and *cluster mill*.

type metal. Any of a series of alloys containing lead (58.5 to 95%), antimony (2.5 to 25%), and tin (2.5 to 20%) used to make printing type. Small amounts of copper (1.5 to 2.0%) are added to increase hardness in some applications.

U

U-bend die. A die, commonly used in press-brake forming, that is machined horizontally with a square or rectangular cross-sectional opening that provides two edges over which metal is drawn into a channel shape.

Ugine-Sejournet process. A direct extrusion process for metals that uses molten glass to insulate the hot billet and to act as a lubricant.

ultimate elongation. The elongation at rupture.

ultimate strength. The maximum stress (tensile, compressive, or shear) a material can sustain without fracture; determined by dividing maximum load by the original cross-sectional area of the specimen. Also known as nominal strength or maximum strength.

ultimate tensile strength. The ultimate or final (highest) stress sustained by a specimen in a tension test.

ultrahard tool materials. Very hard, wear-resistant materials—specifically, polycrystalline diamond and polycrystalline cubic boron nitride—that are fabricated into solid or layered cutting tool blanks for machining applications.

ultrahigh-strength steels. Structural steels with minimum yield strengths of 1380 MPa (200 ksi). Such steels include medium-carbon low-alloy steels, medium-alloy air-hardening steels, and high

fracture toughness (K_{Ic} of 100 MPa, or 91 ksi) steels.

ultrasonic beam. A beam of acoustical radiation with a frequency higher than the frequency range for audible sound —that is, above about 20 kHz.

ultrasonic cleaning. Immersion cleaning aided by ultrasonic waves that cause microagitation.

ultrasonic frequency. A frequency, associated with elastic waves, that is greater than the highest audible frequency, generally regarded as being higher than 20 kHz.

ultrasonic impact grinding. Material removal by means of an abrasive slurry and the ultrasonic vibration of a nonrotating tool. The abrasive slurry flows through a gap between the workpiece and the vibrating tool. Material removal occurs when the abrasive particles, suspended in the slurry, are struck on the downstroke of the vibrating tool. The velocity imparted to the abrasive particles causes microchipping and erosion as the particles impinge on the workpiece. See also *ultrasonic machining*.

ultrasonic inspection. A nondestructive method in which beams of high-frequency sound waves are introduced into materials for the detection of surface and subsurface flaws in the material. The sound waves travel through the material with some attendant loss of energy (attenuation) and are reflected at interfaces. The reflected beam is displayed and then analyzed to define the presence and location of flaws or discontinuities. Most ultrasonic inspection is done at frequencies between 0.1 and 25 MHz—well above the range of human hearing, which is about 20 Hz to 20 kHz.

ultrasonic machining. A process for machining of hard, brittle, nonmetallic materials that involves the ultrasonic vibration of a rotating diamond core drill or milling tool. Rotary ultrasonic machining is similar to the conventional drilling of glass and ceramic with diamond core drills, except that the rotating core drill is vibrated at an ultrasonic frequency of 20 kHz. Rotary ultrasonic machining does not involve the flow of an abrasive slurry through a gap between the workpiece and the tool. Instead, the tool contacts and cuts the workpiece, and a liquid coolant, usually water, is forced through the bore of the tube to cool and flush away the removed material. See also *ultrasonic impact grinding*.

ultrasonic testing. See *ultrasonic inspection*.

ultrasonic welding. A solid-state process in which materials are welded by locally ap-

plying high-frequency vibratory energy to a joint held together under pressure. Ultrasonic energy is produced through a transducer, which converts high-frequency electrical vibrations to mechanical vibrations at the same frequency, usually above 15 kHz (above the audible range). Mechanical vibrations are transmitted through a coupling system to the welding tip and into the workpieces. The tip vibrates laterally, essentially parallel to the weld interface, while static force is applied perpendicular to the interface.

underbead crack. A crack in the heat-affected zone of a weld generally not extending to the surface of the base metal.

undercooling. Same as *supercooling*.

undercut. (1) In weldments, a groove melted into the base metal adjacent to the toe or root of a weld and left unfilled by weld metal. (2) For castings or forgings, same as *back draft*.

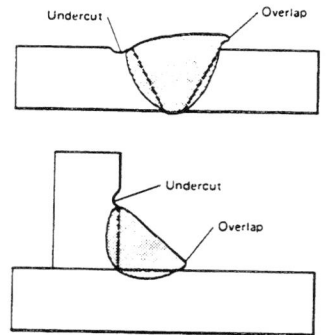

underdraft. A condition wherein a metal curves downward on leaving a set of rolls because of higher speed in the upper roll.

underfill. (1) In weldments, a depression on the face of the weld or root surface extending below the surface of the adjacent base metal. (2) A portion of a forging that has insufficient metal to give it the true shape of the impression.

underfilm corrosion. Corrosion that occurs under organic films in the form of randomly distributed threadlike filaments or spots. In many cases this is identical to *filiform corrosion*.

understressing. Applying a cyclic stress lower than the *endurance limit*. This may improve fatigue life if the member is later cyclically stressed at levels above the endurance limit.

uniaxial stress. A state of stress in which two of the three principal stresses are zero. See also *principal stress (normal)*.

uniform corrosion. (1) A type of corrosion attack (deterioration) uniformly distributed over a metal surface. (2) Corrosion that proceeds at approximately the same rate over a metal surface. Also called general corrosion.

uniform elongation. The elongation at maximum load and immediately preceding the onset of necking in a tensile test.

uniform strain. The strain occurring prior to the beginning of localization of strain (necking); the strain to maximum load in the tension test.

unit cell. A parallelepiped element of crystal structure, containing a certain number of atoms, the repetition of which through space will build up the complete crystal.

unit power. The net amount of power required during machining or grinding to remove a unit volume of material in unit time.

universal forging mill. A combination of four hydraulic presses arranged in one plane equipped with billet manipulators and automatic controls, used for radial or draw forging.

universal mill. A rolling mill in which rolls with a vertical axis roll the edges of the metal stock between some of the passes through the horizontal rolls.

upset. (1) The localized increase in cross-sectional area of a workpiece or weldment resulting from the application of pressure during mechanical fabrication or welding. (2) That portion of a welding cycle during which the cross-sectional area is increased by the application of pressure. (3) Bulk deformation resulting from the application of pressure in welding. The upset may be measured as a percent increase in interfacial area, a reduction in length, or a percent reduction in thickness (for lap joints).

upset forging. A forging obtained by *upset* of a suitable length of bar, billet, or bloom.

upsetting. The working of metal so that the cross-sectional area of a portion or all of the stock is increased. See also *heading.*

upset welding. A resistance welding process in which the weld is produced, simultaneously over the entire area of abutting surfaces or progressively along a joint, by applying mechanical force (pressure) to the joint, then causing electrical current to flow across the joint to heat the abutting surfaces. Pressure is maintained throughout the heating period. See also *open-gap upset welding.*

V

vacancy. A structural imperfection in which an individual atom site is temporarily unoccupied.

vacuum arc remelting (VAR). A consumable-electrode remelting process in which heat is generated by an electric arc between the electrode and the ingot. The process is performed inside a vacuum chamber. Exposure of the droplets of molten metal to the reduced pressure reduces the amount of dissolved gas in the metal. See also *consumable-electrode remelting.*

vacuum carburizing. A high-temperature gas carburizing process using furnace pressures between 13 and 67 kPa (0.1 to 0.5 torr) during the carburizing portion of the cycle. Steels undergoing this treatment are austenitized in a rough vacuum, carburized in a partial pressure of hydrocarbon gas, diffused in a rough vacuum, and then quenched in either oil or gas.

vacuum casting. A casting process in which metal is melted and poured under very low atmospheric pressure; a form of permanent mold casting in which the mold is inserted into liquid metal, vacuum is applied, and metal is drawn up into the cavity.

vacuum degassing. The use of vacuum techniques to remove dissolved gases from molten alloys.

vacuum deposition. Deposition of a metal film onto a substrate in a vacuum by metal evaporation techniques.

vacuum furnace. A furnace using low atmospheric pressures instead of a protective gas atmosphere like most heat-treating furnaces.

vacuum hot pressing. A method of processing materials (especially metal and ceramic powders) at elevated temperatures, consolidation pressures, and low atmospheric pressures.

vacuum induction melting (VIM). A process for remelting and refining metals in which the metal is melted inside a vacuum chamber by induction heating. The metal can be melted in a crucible and then poured into a mold.

vacuum melting. Melting in a vacuum to prevent contamination from air and to remove gases already dissolved in the metal; the solidification can also be carried out in a vacuum or at low pressure.

vacuum nitrocarburizing. A subatmospheric *nitrocarburizing* process using a basic atmosphere of 50% ammonia/50% methane, containing controlled oxygen additions of up to 2%.

vacuum refining. Melting in a vacuum to remove gaseous contaminants from the metal.

vacuum sintering. Sintering of ceramics or metals at subatmospheric pressure.

vapor degreasing. Degreasing of work in the vapor over a boiling liquid solvent, the vapor being considerably heavier than air. At least one constituent of the soil must be soluble in the solvent. Modifications of this cleaning process include vapor-spray-vapor, warm liquid-vapor, boiling liquid-warm liquid-vapor, and ultrasonic degreasing.

vapor deposition. See *chemical vapor deposition, physical vapor deposition,* and *sputtering.*

vapor plating. Deposition of a metal or compound on a heated surface by reduction or decomposition of a volatile compound at a temperature below the melting points of the deposit and the base material. The reduction is usually accomplished by a gaseous reducing agent such as hydrogen. The decomposition process may involve thermal dissociation or reaction with the base material. See also *vacuum deposition.*

V-bend die. A die commonly used in press-brake forming, usually machined with a triangular cross-sectional opening to provide two edges as fulcrums for accomplishing *three-point bending.*

vent. A small opening in a foundry mold for the escape of gases.

vertical position welding. The position of welding in which the axis of the weld is approximately vertical.

vibratory finishing. A process for deburring and surface finishing in which the product and an abrasive mixture are placed in a container and vibrated.

Vickers hardness number (HV). A number related to the applied load and the surface area of the permanent impression made by a square-based pyramidal diamond indenter having included face angles of 136°, computed from:

$$HV = 2P \sin \frac{\alpha/2}{d^2} = \frac{1.8544P}{d^2}$$

where P is applied load (kgf), d is mean diagonal of the impression (mm), and is the face angle of the indenter (136°).

Vickers hardness test. A microindentation hardness test employing a 136° diamond pyramid indenter (Vickers) and variable loads, enabling the use of one hardness scale for all ranges of hardness—from very soft lead to tungsten carbide. Also known as diamond pyramid hardness test. See also *microindentation* and *microindentation hardness number.*

virgin metal. Same as *primary metal.*

void. (1) A *shrinkage cavity* produced in castings or weldments during solidification. (2) A term generally applied to paints to describe *holidays*, holes, and skips in a film.

V process. A molding (casting) process in which the sand is held in place in the mold by vacuum. The mold halves are covered with a thin sheet of plastic to retain the vacuum.

W

walking-beam furnace. A continuous-type heat treating or sintering furnace consisting of two sets of rails, one stationary and the other movable, that lift and advance parts inside the hearth. With this system, the moving rails lift the work from the stationary rails, move it forward, and then lower it back onto the stationary rails. The moving rails then return to the starting position and repeat the process to advance the parts again.

Wallner lines. A distinct pattern of intersecting sets of parallel lines, sometimes producing a set of V-shaped lines, sometimes observed when viewing brittle fracture surfaces at high magnification in an electron microscope. Wallner lines are attributed to interaction between a shock wave and a brittle crack front propagating at high velocity. Sometimes Wallner lines are misinterpreted as *fatigue striations*.

warm working. Deformation of metals at elevated temperatures below the recrystallization temperature. The flow stress and rate of strain hardening are reduced with increasing temperature; therefore, lower forces are required than in cold working. See also *cold working* and *hot working*.

warpage. (1) Deformation other than contraction that develops in a casting between solidification and room temperature. (2) The distortion that occurs during annealing, stress relieving, and high-temperature service.

wash. (1) A coating applied to the face of a mold prior to casting. (2) An imperfection at a cast surface similar to a *cut (3)*.

wash metal. Molten metal used to wash out a furnace, ladle, or other container.

waterjet/abrasive waterjet machining. A hydrodynamic machining process that uses a high-velocity stream of water as a cutting tool. This process is limited to the cutting of nonmetallic materials when the jet stream consists solely of water. However, when fine abrasive particles are injected into the water stream, the process can be used to cut harder and denser materials. Abrasive waterjet machining has expanded the range of fluid jet machining to include the cutting of metals, glass, ceramics, and composite materials. Water pressures up to 410 MPa (60 ksi) are used. The coherent jet of water is propelled at speeds up to approximately 850 m/s (2800 ft/s).

water quenching. A quench in which water is the quenching medium. The major disadvantage of water quenching is its poor efficiency at the beginning or hot stage of the quenching process. See also *quenching*.

waviness. A wavelike variation from a perfect surface, generally much larger and wider than the roughness caused by tool or grinding marks.

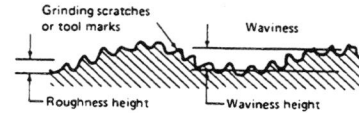

wax pattern. A precise duplicate, allowing for shrinkage, of the casting and required gates, usually formed by pouring or injecting molten wax into a die or mold. See also *investment casting*.

wear. Damage to a solid surface, generally involving progressive loss of material, due to a relative motion between that surface and a contacting surface or substance. Compare with *surface damage*.

wear debris. Particles that become detached in a wear process.

wear pad. In forming, an expendable pad of rubber or rubberlike material of nominal thickness that is placed against the diaphragm to lessen the wear on it. See also *diaphragm (2)*.

weathering. Exposure of materials to the outdoor environment.

weathering steels. Copper-bearing *high-strength low-alloy steels* that exhibit high resistance to atmospheric corrosion in the unpainted condition.

web. (1) A relatively flat, thin portion of a forging that effects an interconnection between ribs and bosses; a panel or wall that is generally parallel to the forging plane. See also *rib*. (2) For twist drills and reamers, the central portion of the tool body that joins the lands. (3) A plate or thin portion between stiffening ribs or flanges, as in an I-beam, H-beam or other similar section.

weight percent. Percentage composition by weight. Contrast with *atomic percent*.

weld. A localized coalescence of metals or nonmetals produced either by heating the materials to suitable temperatures, with or without the application of pressure, or by the application of pressure alone and with or without the use of filler material.

weldability. A specific or relative measure of the ability of a material to be welded under a given set of conditions. Implicit in this definition is the ability of the completed weldment to fulfill all functions for which the part was designed.

weld bead. A deposit of filler metal from a single welding *pass (3)*.

weld crack. A crack in weld metal. See also *crater crack, root crack, toe crack,* and *underbead crack*.

weld decay. *Intergranular corrosion,* usually of stainless steels or certain nickel-base alloys, that occurs as the result of sensitization in the *heat-affected zone* during the welding operation. See also *sensitization*.

welding. (1) Joining two or more pieces of material by applying heat or pressure, or both, with or without filler material, to produce a localized union through fusion or recrystallization across the interface. The thickness of the filler material is much greater than the capillary dimensions encountered in *brazing*. (2) May also be extended to include brazing and soldering. (3) In *tribology*, adhesion between solid surfaces in direct contact at any temperature.

welding current. The current in the welding circuit during the making of a weld.

welding cycle. The complete series of events involved in the making of a weld.

welding electrode. See preferred term *electrode* (welding).

welding ground. Same as *work lead*.

welding leads. The electrical cables that serve as either *work lead* or *electrode lead* of an arc welding circuit.

welding machine. Equipment used to perform the welding operation. For example, spot welding machine, arc welding machine, seam welding machine, etc.

welding process. A materials joining process which produces coalescence of materials by heating them to suitable temperatures, with or without the application of pressure or by the application of pressure alone, and with or without the use of filler metal.

welding rod. A form of filler metal used for welding or brazing which does not conduct the electrical current, and which may be either fed into the weld pool or preplaced in the joint.

welding sequence. The order in which the various component parts of a weldment or structure are welded.

welding stress. *Residual stress* caused by localized heating and cooling during welding.

welding tip. A welding torch tip designed for welding.

welding torch (arc). A device used in the gas tungsten and plasma arc welding processes to control the position of the electrode, to transfer current to the arc, and to direct the flow of shielding and plasma gas. See also *gas tungsten arc welding* and *plasma arc welding*.

welding torch (oxyfuel gas). A device used in oxyfuel gas welding, torch brazing, and torch soldering for directing the heating flame produced by the controlled combustion of fuel gases. See also *oxyfuel gas welding*.

welding wire. See preferred terms *electrode* (welding) and *welding rod*.

weld interface. The interface between weld metal and base metal in a fusion weld, between base metals in a solid-state weld without filler metal, or between filler metal and base metal in a solid-state weld with a filler metal and in a braze.

weld line. See preferred term *weld interface*.

weldment. An assembly whose component parts are joined by welding.

weld metal. That portion of a weld which has been melted during welding.

weld nugget. The weld metal in spot, seam or projection welding. See also *nugget* and *resistance spot welding*.

weld pass. A single progression of a welding or surfacing operation along a joint, weld deposit, or substrate. The result of a pass is a weld bead, layer, or spray deposit.

weld penetration. See preferred terms *joint penetration* and *root penetration*.

Wenstrom mill. A rolling mill similar to a universal mill but where the edges and sides of a rolled section are acted on simultaneously.

wet blasting. A process for cleaning or finishing by means of a slurry of abrasive in water directed at high velocity against the workpieces.

wetting. (1) The spreading, and sometimes absorption, of a fluid on or into a surface. (2) A condition in which the interface tension between a liquid and a solid is such that the contact angle if 0° to 90°. (3) The phenomenon whereby a liquid filler metal or flux spreads and adheres in a thin continuous layer on a solid base metal. (4) The formation of a relatively uniform, smooth, unbroken, and adherent film of solder to a basis metal.

wetting agent. (1) A substance that reduces the surface tension of a liquid, thereby causing it to spread more readily on a solid surface. (2) A surface-active agent that produces *wetting* by decreasing the *cohesion* within the liquid.

whisker. (1) A short single crystal fiber or filament used as a reinforcement in a matrix. Whisker diameters range from 1 to 25 m, with aspect ratios (length to diameter ratio) generally between 50 and 150. (2) Metallic filamentary growths, often microscopic, sometimes formed during electrodeposition and sometimes spontaneously during storage or service, after finishing.

white-etching layer. A surface layer in a steel that, as viewed in a section after etching, appears whiter than the base metal. The presence of the layer may be due to a number of causes, including plastic deformation induced by machining or surface rubbing, heating during a metallographic preparation stage to such an extent that the layer is austenitized and then hardened during cooling, and diffusion of extraneous elements into the surface.

whiteheart malleable. See *malleable iron*.

white iron. A *cast iron* that is essentially free of graphite, and most of the carbon content is present as separate grains of hard Fe_3C. White iron exhibits a white, crystalline fracture surface because fracture occurs along the iron carbide platelets. White cast irons have alloy contents well above 4%.

white layer. (1) Compound layer that forms in steels as a result of the *nitriding* process. (2) In tribology, a *white-etching layer*, typically associated with ferrous alloys, that is visible in metallographic cross sections of bearing surfaces. See also *Beilby layer*.

white metal. (1) A general term covering a group of white-colored metals of relatively low melting points based on tin or lead. These materials are used for bearings and jewelry. (2) A copper matte of about 77% Cu obtained from smelting of sulfide copper ores.

white rust. Zinc oxide; the powder product of corrosion of zinc or zinc-coated surfaces.

Widmanstätten structure. A structure characterized by a geometrical pattern resulting from the formation of a new phase along certain crystallographic planes of the parent solid solution. The orientation of the lattice in the new phase is related crystallographically to the orientation of the lattice in the parent phase. The structure was originally observed in meteorites, but is readily produced in many alloys, such as titanium, by appropriate heat treatment.

wildness. A condition that exists when molten metal, during cooling, evolves so much gas that it becomes violently agitated, forcibly ejecting metal from the mold or other container.

winning. Recovering a metal from an ore or chemical compound using any suitable hydrometallurgical, pyrometallurgical, or electrometallurgical method.

wiped coat. A hot dipped galvanized coating from which virtually all free zinc is removed by wiping prior to solidification, leaving only a thin zinc-iron alloy layer.

wiped joint. A joint made with solder having a wide melting range and with the heat supplied by the molten solder poured onto the joint. The solder is manipulated with a hand-held cloth or paddle so as to obtain the required size and contour.

wiper forming, wiping. Method of curving sheet metal sections or tubing over a form block or die in which this form block is moved relative to a wiper block or slide block.

wiping effect. Activation of a metal surface by mechanical rubbing or wiping to enhance the formation of conversion coatings, such as phosphate coatings.

wire. (1) A thin, flexible, continuous length of metal, usually of circular cross section, and usually produced by drawing through a die. The size limits for round wire sections range from approximately 0.13 mm (0.005 in.) to 25 mm (1 in.). Larger rounds are commonly referred to as *bars*. See also *flat wire*. (2) A length of single metallic electrical conductor, it may be of solid, stranded or tinsel construction, and may be either bare or insulated.

wire bar. A cast shape, particularly of tough pitch copper, that has a cross section approximately square with tapered ends, designed for hot rolling to rod for subsequent drawing into wire.

wire drawing. Reducing the cross section of wire by pulling it through a die.

wire flame spraying. A *thermal spraying* process variation in which the material to be sprayed is in wire or rod form. See also *flame spraying*.

wire rod. Hot-rolled coiled stock that is to be cold drawn into wire.

wiring. Formation of a curl along the edge of a shell, tube, or sheet and insertion of a rod or wire within the curl for stiffening the edge. See also *curling*.

woody structure. A macrostructure, found particularly in wrought iron and in extruded rods of aluminum alloys, that shows elongated surfaces of separation when fractured.

work angle. The angle that the electrode makes with the referenced plane or surface of the base metal in a plane perpendicular to the axis of the weld.

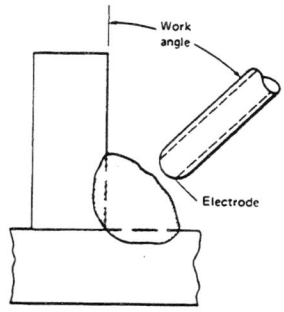

Work angle

Electrode

work hardening. Same as *strain hardening*.

working electrode. The test or specimen electrode in an *electrochemical cell*.

work lead. The electrical conductor connecting the source of arc welding current to the work. Also called work connection, welding ground, or ground lead.

worm. An exudation (sweat) of molten metal forced through the top crust of solidifying metal by gas evolution. See also *zinc worms*.

wrap forming. See *stretch forming*.

wrinkling. A wavy condition obtained in deep drawing of sheet metal, in the area of the metal between the edge of the flange and the draw radius. Wrinkling may also occur in other forming operations when unbalanced compressive forces are set up.

wrought iron. A commercial iron consisting of slag (iron silicate) fibers entrained in a ferrite matrix.

X

x-ray. A penetrating electromagnetic radiation, usually generated by accelerating electrons to high velocity and suddenly stopping them by collision with a solid body. Wavelengths of x-rays range from about 10^1 to 10^2 Å, the average wavelength used in research being about 1 Å. Also known as roentgen ray or x-radiation.

Y

Y-block. A single *keel block*.

yellow brass. A name sometimes used in reference to the 65Cu-35Zn type of *brass*.

yield. (1) Evidence of plastic deformation in structural materials. Also known as plastic flow or creep. See also *flow*. (2) The ratio of the number of acceptable items produced in a production run to the total number that were attempted to be produced. (3) Comparison of casting weight to the total weight of metal poured into the mold.

yield point. The first stress in a material, usually less than the maximum attainable stress, at which an increase in strain occurs without an increase in stress. Only certain materials—those which exhibit a localized, heterogeneous type of transition from elastic to plastic deformation—produce a yield point. If there is a decrease in stress after yielding, a distinction may be made between upper and lower yield points. The load at which a sudden drop in the flow curve occurs is called the upper yield point. The constant load shown on the flow curve is the lower yield point.

yield point elongation. In materials that exhibit a yield point, the difference between the elongation at the completion and at the start of discontinous yield.

yield strength. The stress at which a material exhibits a specified deviation from proportionality of stress and strain. An offset of 0.2% is used for many materials, particularly metals. Compare with *tensile strength*.

yield stress. The stress level of highly ductile materials at which large strains take place without further increase in stress.

Young's modulus. A term used synonymously with modulus of elasticity. The ratio of tensile or compressive stresses to the resulting strain. See also *modulus of elasticity*.

Z

zinc worms. Surface imperfections, characteristic of high-zinc brass castings, that occur when zinc vapor condenses at the mold/metal interface, where it is oxidized and then becomes entrapped in the solidifying metals.

zincrometal. A steel coil-coated product consisting of a mixed-oxide underlayer containing zinc particles and a zinc-rich organic (epoxy) topcoat. It is weldable, formable, paintable, and compatible with commonly used adhesives. Zincrometal is used to protect outer body door panels in automobiles from corrosion.

zone melting. Highly localized melting, usually by induction heating, of a small volume of an otherwise solid metal piece, usually a metal rod. By moving the induction coil along the rod, the melted zone can be transferred from one end to the other. In a binary mixture where there is a large difference in composition on the liquidus and solidus lines, high purity can be attained by concentrating one of the constituents in the liquid as it moves along the rod.

Selection of Structural Materials

Selection of Structural Materials

The process of selecting materials for design applications in commercial and military aerospace vehicles can best be shown by a flow chart. Two examples are presented in Fig. 1 and 2. Figure 1 indicates the interrelationships among aerospace material properties, design geometry, and manufacturing processes. Figure 2 presents a more-detailed flowchart for use in selecting aerospace materials, in which the design engineer first determines the application and configuration (shape) of the structure or assembly. The environmental stresses that the structure will encounter during shipment, storage, and service are then identified. Finally, a comprehensive material property database is developed, which includes mechanical and corrosion-resistance properties required for optimum performance under the expected vehicle operating conditions. These properties include:

- Ultimate tensile strength
- 0.2% offset yield strength
- Impact strength
- Hardness
- Fracture toughness
- Corrosion and wear resistance
- Weight (this factor is very important for air, land, and sea vehicles)

Once these initial parameters are defined and data are collected, the design engineer constructs a list of the most suitable candidate materials from which the best material will be chosen for the specific structural application. This list should include the manufacturing and quality assurance practices implemented during the fabrication of each candidate material, because manufacturing and quality assurance have a significant influence on the material properties and cost of the final product. For example, in the case of steel structures, the production process may include various types of thermomechanical treatments in order to meet structural requirements for substantially increased strength (Ref. 2). Two applications that require very high strength are aircraft main landing gears and helicopter main rotor blades.

The design engineer determines the best material for the intended application, based on the cost of each candidate

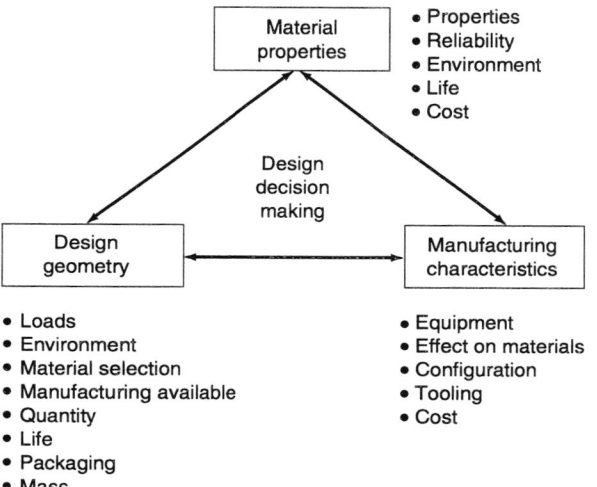

Fig. 1 Relationship among material properties, design geometry, and manufacturing characteristics. The overall cost, mass, and performance of commercial and military equipment components are determined by three design factors. Reprinted with permission from Ref. 1

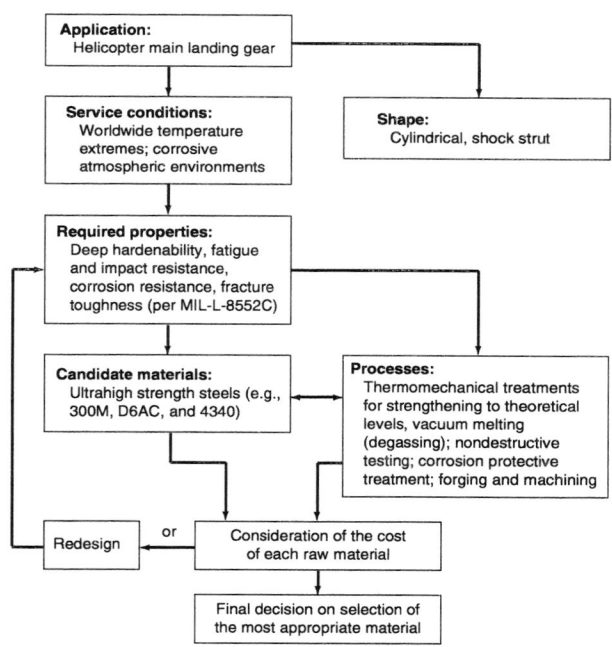

Fig. 2 Flowchart of the materials selection process

material. Farag (Ref. 3) has addressed this aspect of materials selection by comparing the costs of various engineering materials—metallics, nonmetallics, and composites. Figures 3 and 4 illustrate the comparative cost data for materials of construction, using the cost of hot-rolled carbon steel as a basis. The cost per unit volume per unit ultimate tensile strength is given for several engineering materials in Fig. 5, which uses the cost of structural steel as a basis. Figure 5 is useful in selecting materials for construction of structures that will be highly stressed in service and that will therefore require high ultimate tensile strengths.

Figures 6 and 7 can be used especially in the selection of advanced technology materials for aerospace, automotive, and marine vehicles—which often demand that the final product use materials possessing a high strength-to-weight ratio and the lowest possible weight. Figure 6 compares some engineering materials based on their cost per unit volume per unit specific tensile strength, using structural steel as the basis. Figure 7 compares the strength versus the density for some engineering materials. For metals and polymers, this strength is the initial yield strength, because these materials have been mechanically worked. For brittle ceramics, the strength is the crushing strength in compression. For elastomers, the strength is the tear

strength. For composite materials, the strength is the tensile failure strength (Ref. 5).

The design engineer has two economic alternatives in the materials selection process: one may determine the material that is of least expense for the given task; or one may select a material whose cost is greater than the least expensive material, but is easier (and, therefore, cheaper) to manufacture and process. For example, the use of prehardened steel sheets is one instance in which a more expensive material provides the user with an end product that is lower in cost. This cost savings results from a reduction of material processing requirements—specifically, the elimination of heat treatment.

The ten material value analysis questions listed below may be used by the designer to determine the most cost effective material for a specific application. In this analysis, each question applies to specific materials and processes. The value of a candidate material is considered to be greatest if the response to all ten questions is no.

- Can we do without it?
- Does it do more than is required?
- Does it cost more than it is worth?

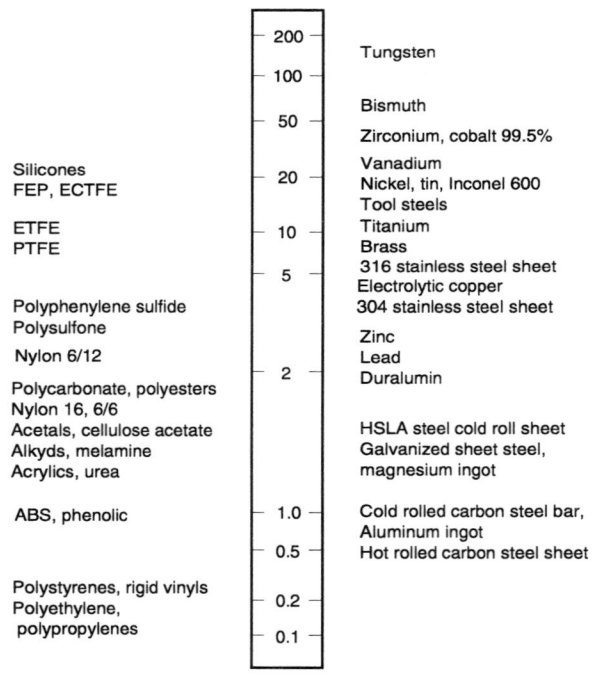

Fig. 3 Comparison of some engineering materials on the basis of cost per unit volume relative to the cost of hot-rolled carbon steel. This comparison is based on prices at the end of 1976. Reprinted with permission from Ref. 3

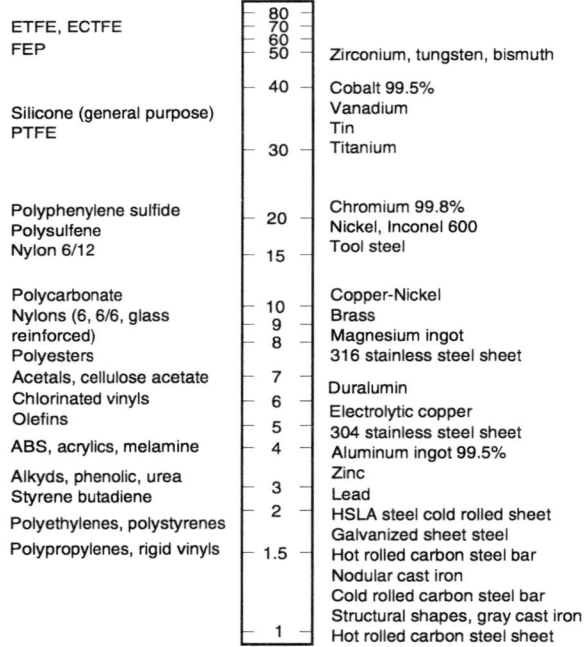

Fig. 4 Comparison of some engineering materials on the basis of cost per unit weight relative to the cost of hot-rolled carbon steel. This comparison is based on prices at the end of 1976. Reprinted with permission from Ref. 3

- Is there something that does the job better?
- Can it be made by a less costly method?
- Can a standard item be used?
- Considering the quantities used, could a less costly tooling method be used?
- Does it cost more than the total of reasonable labor, overhead, material, and profit?
- Can someone else provide it at less cost without affecting dependability?
- If it were your money, would you refuse to buy the item because it costs too much?

Final material selection decisions must be based on all of the data gathered in the course of the decision making process. For both military and commercial design applications, design engineers should be particularly aware of the need for adequate strength, fracture toughness, and corrosion and wear resistance. The primary material selection factors listed below should be used in every case in which a decision must be made for or against the selection of a structural material.

- Functional requirements and constraints
- Mechanical properties
- Design configuration
- Available and alternative materials
- Fabricability
- Corrosion and degradation resistance
- Stability
- Properties of unique interest
- Cost

Specifications and Standards

Throughout the design process, the engineer who is responsible for making the final material selection decision must be aware of, and make reference to, pertinent performance and product specifications and standards. Especially in cases that involve military construction, conformance to standards and specifications indicates that the structural end product has been properly designed, so that it will endure the environmental and mechanical stresses that the part is expected to encounter in service. The design engineer must ensure that the standard applies to the product being fabricated and must be certain the standard guarantees that the end product will perform satisfactorily.

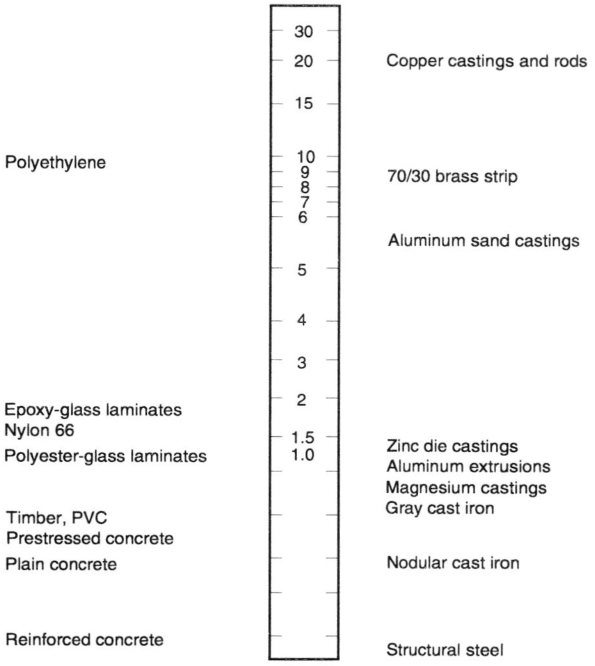

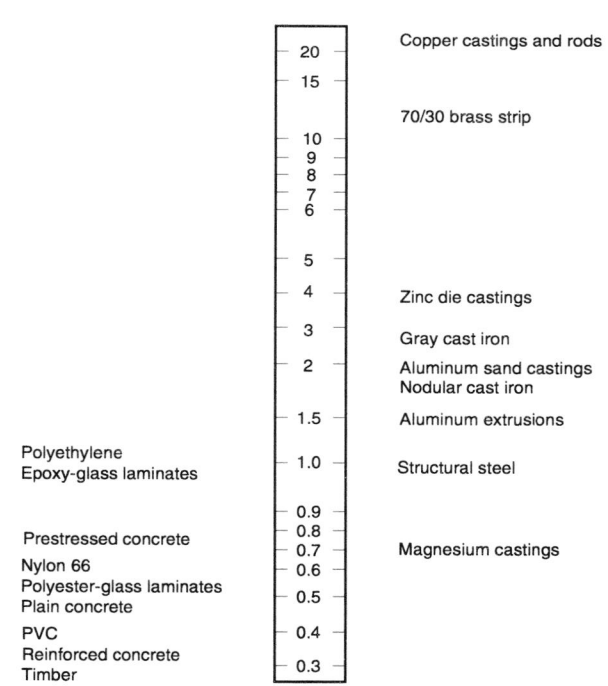

Fig. 5 Comparison of some engineering materials on the basis of cost per unit volume per unit ultimate tensile strength relative to the cost of structural steel. The comparison is based on prices at the end of 1976. Reprinted with permission from Ref. 3

Fig. 6 Comparison of some engineering materials on the basis of cost per unit volume per unit specific strength relative to the cost of structural steel. The comparison is based on prices at the end of 1976. Reprinted with permission from Ref. 3

For a comprehensive description of functional requirements for a properly manufactured finished product, the design engineer should refer to product (material) specifications. Product specification requirements pertain to configurations, materials, tolerances, methods of fabrication, and other material and process requirements. The design engineer must relate these product performance requirements directly to the mechanical, physical, and chemical properties to which the final product must conform, so that the product can be considered further for structural application.

Structural Life Cycle and Failure Modes

Design engineers can improve military structures by understanding the overall performance of these components, based on actual service experience. Various types of material deterioration and failure have occurred in components used in military and commercial equipment. These failure modes include corrosion, fracture, and fatigue.

Corrosion is defined as the degradation of material by a reaction with its environment. The material is usually a metal, and the reaction, in most cases, is electrochemical in nature (Ref. 6-8). Among the most important forms of corrosion that have been observed in military equipment are:

Crevice and pitting corrosion. In stainless steels, pitting attack occurs in localized areas, especially at crevices. Pitting and crevice corrosion usually are observed in passive metals, such as aluminum and aluminum alloys, the stainless steels, and nickel-base alloys (Ref. 9).

Stress corrosion. In this process, the simultaneous effects of a corrosive agent and a sustained tensile stress cause the material to develop cracks (Ref. 7). Materials that fail by stress corrosion do not deform prior to failure. This brittle failure is usually perpendicular to the applied stress (Ref.

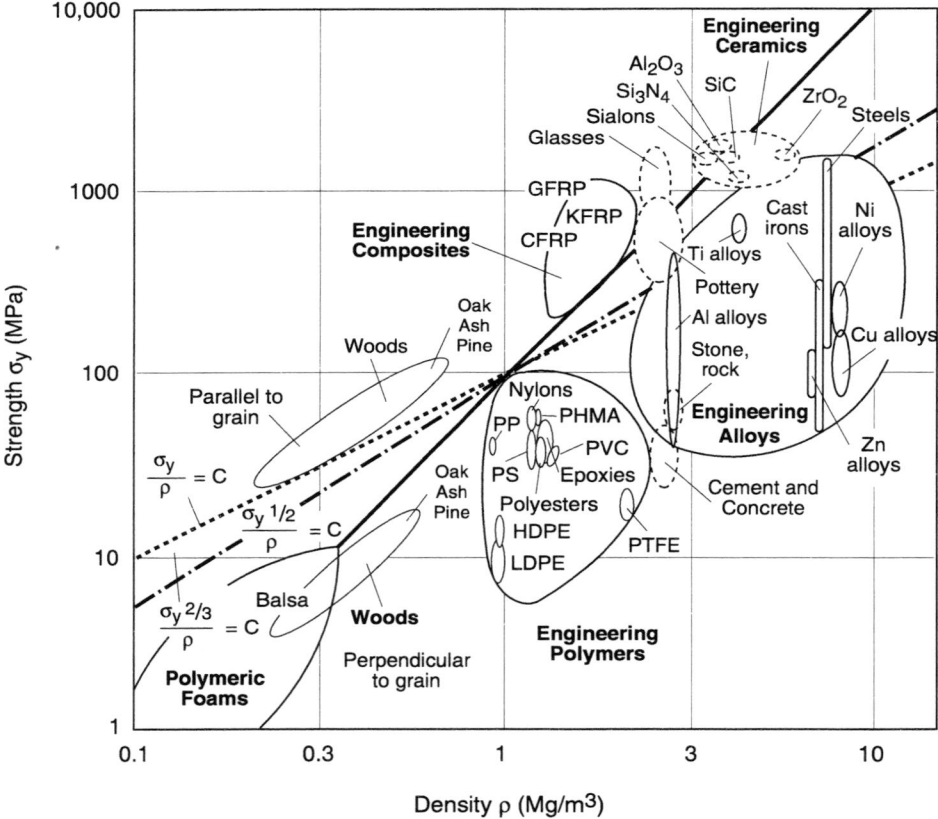

Fig. 7 Strength, σ_y, versus density, ρ. Yield strength is plotted for metals and polymers, compressive strength for ceramics, tear strength for elastomers, and tensile strength for composites. The guidelines of constant σ_y/ρ, $\sigma_y^{2/3}/\rho$, and $\sigma_y^{1/2}/\rho$ are used in minimum-weight, yield-limited design. Reprinted with permission from Ref. 4

10). In most aircraft parts that fail by stress corrosion, the path of the crack is intergranular (Ref. 11).

Erosion-corrosion. In this process, failure is caused by a combination of corrosion and erosion. Erosion is the gradual loss of material from a solid surface because of a mechanical interaction between the surface and liquids, gases, or solid particles, or a combination of these substances (Ref. 12). In Army materiel, erosion-corrosion can occur in cannon tubes, gas turbine engines, and rocket nozzles (Ref. 13, 14).

Corrosion of metals under organic coatings. For military equipment, the most important types of this degradation are blistering and anodic undermining (Ref. 15). Blistering involves the development of local areas in which the coating has separated from the metal. Water accumulates in these localized regions, and corrosion can take place. The principal causative agent of blistering and coating delamination is water (Ref. 16). In anodic undermining, the coating is removed by anodic corrosion reactions occurring underneath the coating. Filiform corrosion, which appears as thin filaments, is a specific form of anodic undermining.

Two important types of coating failure are adhesive and cohesive (Ref. 17). In adhesive failure, which tends to develop with strong coatings, the film breaks the adhesive bond and peels from the surface. Cohesive failure is a phenomenon in which the coating only partly adheres to the substrate. Cohesive failure occurs when the coating material has lower internal (cohesive) strength compared with its adhesive strength.

Microbiological corrosion. This type of degradation is induced by microorganisms, particularly bacteria and fungi. Bacteria that have been recently isolated (using synthetic culture media) were observed to be active corrodents of aluminum (Ref. 18). Fungi may cause corrosion of organic coatings, especially alkyd systems (Ref. 19). It is important to recognize that the most severe corrosion problems de-

Table I   Corrosion-resistant materials

Material	Characteristics
Coated steel	Ultimate tensile strength, 80 to 125 ksi. Low-carbon, medium-carbon, and low-alloy steels can be made resistant to atmospheric corrosion by coating them. **Examples:** A325, A490, SAE J429 materials
Stainless steel:	
Austenitic	Ultimate tensile strength, 75 to 120 ksi. Most common of the stainless steels, and more corrosion resistant than the three listed below. Nonmagnetic. Cannot be heat-treated, but can be cold-worked. Good high- and low-temperature properties. 321 can be used up to 816 °C (1500 °F), for example. **Examples:** A193 B8 series, A320 B8 series, any of the 300 or 18-8 series materials, such as 303, 304, 316, 321, 347
Ferritic	Ultimate tensile strength, 70 ksi. Cannot be heat-treated or cold-worked. Magnetic. **Examples** include 430 and 430F
Martensitic	Ultimate tensile strength ,70 to 180 ksi. Heat treatable, magnetic. Can experience stress corrosion if not properly treated. **Examples:** 410, 416, 431
Precipitation hardening	Ultimate tensile strength, 135 ksi. Heat treatable. More ductile than martensitic stainless steels. **Examples:** 630, 17-4 PH, Custom 455, PH 13-8 Mo, ASTM A453-B17B, AISI 660
Nickel-base alloys:	
Nickel-copper	Ultimate tensile strength, 70 to 80 ksi. Can be cold-worked, but not heat-treated. **Example:** Monel
Nickel-copper-aluminum	Ultimate tensile strength, 130 ksi. Can be both cold-worked and heat-treated. Good low-temperature material. **Example:** K-monel
Titanium	Ultimate tensile strength, 135 to 200 ksi. Good corrosion resistance. Low coefficient of expansion. Has a tendency to gall more readily than some other corrosion-resistant materials. Expensive. **Example:** Ti-6Al-4V.
Superalloys	Ultimate tensile strength, 145 to 286 ksi. High-strength materials with excellent properties at high and/or low temperatures. Primarily used in aerospace applications. Expensive. Some, such as MP35N, are virtually immune to marine environments and to stress-corrosion cracking. **Examples:** H-11, Inconel, MP35N, A286, Nimonic 80A. MP35N, Inconel 718 and A286 are especially recommended for cryogenic applications.
Nonferrous materials	Many nonferrous fastener materials can provide outstanding corrosion resistance in applications which would rapidly destroy more common bolt materials. The main drawback to these materials is a general lack of strength, but lack of strength can sometimes be made up by using fasteners of larger diameter, or by using more fasteners. **Examples:** silicon bronze, ultimate tensile strength, 70 to 80 ksi; aluminum, ultimate tensile strength, 13 to 55 ksi; nylon, ultimate tensile strength, 11 ksi

Source: Ref. 21

velop from both environmental and mechanical factors (Ref. 20).

Table I lists comparative corrosion-resistance properties for materials used in bolted joints.

Fracture is a type of catastrophic failure that can be either brittle or ductile. Brittle fracture occurs without plastic deformation (Ref. 6, 22). Brittle fracture is most likely to occur in large, thick structures composed of high strength materials under high stresses (Ref. 7). Failures by stress-corrosion cracking may be characterized as brittle fracture when strain does not occur before failure. The features of ductile fracture, also known as ductile overload failure, are that fracture occurs at 45° to the applied stress, and the failing metal deforms prior to fracture (Ref. 8).

Fatigue contributes to a major portion of the total number of failures in engineering structures (Ref. 9). Fatigue failures occur because of repeated loading and unloading of structural members (Ref. 10). Design fatigue diagrams, called S-N curves, are used by engineers to predict the life of components subjected to alternating stresses. The nominal stress (S) is plotted against the number of cycles to failure (N) in these S-N curves (Ref. 17).

Materials Properties and Design

The importance of material properties in determining the life of a specific component during the design process is indicated in Fig. 8, which summarizes some of the most common failure modes observed in structural materials. Figure 8 also correlates individual failure modes with those material properties that have the greatest significance in determining whether the specific failure mode will occur during the life cycle of the structural material.

Design configuration is an important factor in materials selection because it contributes substantially to the service demands that are placed on aerospace structures. The probability of structural failure due to wear or corrosion, for example, is very dependent on the shape (configuration) of the part. This is the reason design engineers specify part con-

Failure Mode	Ultimate tensile strength	Yield strength	Compressive yield strength	Shear yield strength	Fatigue properties	Ductility	Impact energy	Transition temperature	Modulus of elasticity	Creep rate	Fracture toughness K_{IC}	Stress corrosion fracture toughness K_{ISCC}	Electrochemical potential	Hardness	Coefficient of thermal expansion
Gross yielding		▨		▨											
Buckling			▨						▨						
Creep										▨					
Brittle fracture											▨				
Fatigue, low-cycle					▨	▨									
Fatigue, high-cycle	▨				▨										
Contact fatigue															
Fretting			▨											▨	
Corrosion													▨		
Stress corrosion cracking	▨											▨			
Galvanic corrosion													▨		
Hydrogen embrittlement	▨														
Wear														▨	
Thermal fatigue															▨
Corrosion fatigue					▨										▨

Fig. 8 Relationships between failure modes and material properties. Shaded block at intersection of material property and failure mode indicates that a particular material property is influential in controlling a particular failure mode.

figurations that include contoured (radiused) edges, instead of sharp edges which are difficult to protect against wear and corrosion.

Material availability. The availability of a material for construction of military equipment is another significant factor in materials selection. The material selected for a component must be readily available (or producible) in the required configuration and quantity.

One of the elements that complicates the supply of essential materials for military construction requirements is the occurrence of material shortages during wartime. This situation, in which supply pressures are placed on essential (strategic) alloying materials, occasionally develops in peacetime and requires the use of substitute materials. The

development and marketing of EX steels, a new grade of wrought steels, for the purpose of replacing certain Society of Automotive Engineers (SAE-grade) steels during material shortages, typifies how material supply problems have been solved in the United States. EX steels not only reduce costs (because of the absence of expensive alloying elements such as nickel and molybdenum), but also maintain hardness, strength, ductility, and the preferred metallurgical response to heat treatment (Ref. 23). Figure 9 shows that several low-nickel EX steels have the same hardenability characteristics as SAE grades 8620, 8640, and 52100. EX steels have a higher manganese content than corresponding SAE-grade steels in order to compensate for lower percentages of nickel, chromium, and molybdenum. Table II lists the compositions and corresponding SAE-grade steels for various EX steels.

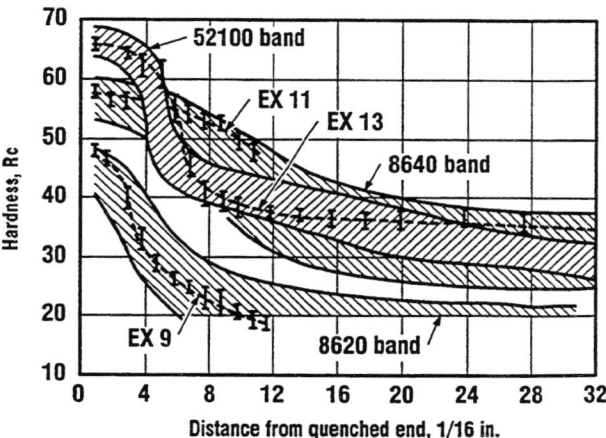

Fig. 9 Jominy curves for EX 9, 11, and 13 (low-silicon versions of EX 10, 12, and 14). Hardenabilities of representative heats fall within hardenability bands of SAE 8620, 8640, and 52100. Vertical bars indicate hardness ranges at respective points along the curves. Samples were taken from top and bottom of the first, middle, and last ingots of each heat. Reprinted with permission from Ref. 23

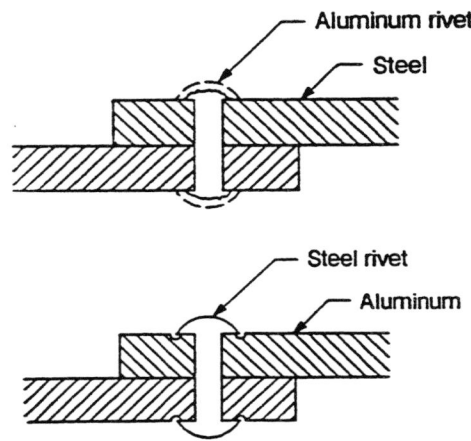

Fig. 10 Example of galvanic corrosion. Source: Ref. 25

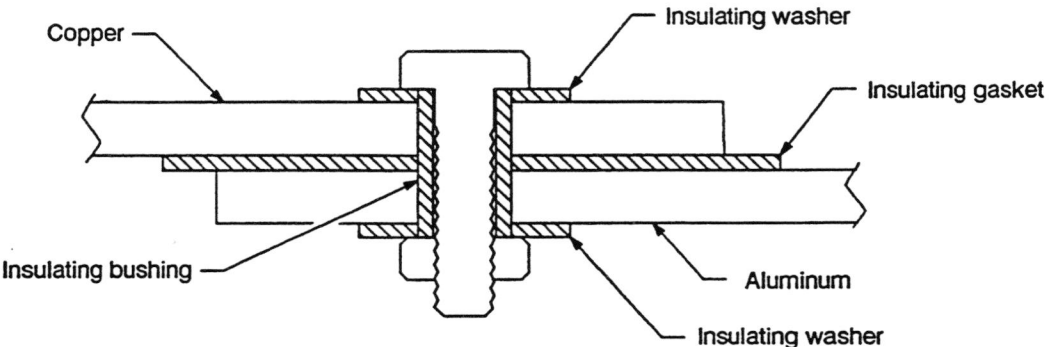

Fig. 11 Method of avoiding galvanic corrosion between dissimilar metals. Source: Ref. 25

The fabricability of a material must be considered by the design engineer in order to determine materials that offer the greatest number of manufacturing process alternatives. Materials that can be fabricated using the desired processes (without special precautions, such as toxicity of materials or processing byproducts) are generally desirable for military or civilian construction requirements.

Fabricability and mechanical properties are usually inversely related. One example is the relationship between the alloying element content, strength, and weldability of steel. As the quantity of carbon and other alloying elements increases, the corresponding weldability and machinability decrease.

Susceptibility to corrosion or other types of degradative processes is a significant factor for design engineers to consider while determining a suitable material for construction of the final product. Design engineers should be aware of appropriate engineering techniques that can be applied to prevent or minimize corrosion in service (Ref. 25). This corrosion-preventive approach is illustrated in Fig. 10 to 13. An example of galvanic (electrochemical) corrosion, which occurs when there is contact between dissimilar metals, is illustrated in Fig. 10. Figure 11 shows the logical preventive technique of applying insulating materials (gaskets, washers, and bushings) to prevent dissimilar metal contact and consequent galvanic corrosion.

Crevice corrosion, which has been observed many times in military and commercial aerospace equipment, is illustrated in Fig. 12. Corrosion develops inside the crevices because of oxygen concentration cells. Metal ion concentration cells can cause corrosion immediately outside of the crevice. Pitting corrosion—the formation of deep pits in specific areas of a structure—can develop around crevices, particularly in corrosion-resistant (stainless) steels. Therefore, it is important to design with the objective of avoiding pitting corrosion. Figure 13 illustrates the application of welding to prevent crevice (oxygen concentration cell) corrosion. Sealing the small opening between the two bars in Fig. 13 prevents this type of corrosion, since the weldment effectively prevents the development of any oxygen concentration differences in the region of the crevices.

Proper design measures must also be taken to prevent stress-corrosion cracking (SCC) and hydrogen embrittlement. In corrosion-resistant steels, SCC results from a susceptible alloy-environment combination, a sustained tensile stress, and a high-temperature environment. Table III

Table II   The EX steels and equivalent standard grades

| EX No. | Composition, % | | | | | Equivalent SAE Grade |
	C	Mn	Cr	Mo	Other	
10	0.19 to 0.24	0.95 to 1. 25	0.25 to 0.40	0.05 to 0.10	0.20 to 0.40 Ni	8620
15	0.18 to 0.34	0.90 to 1.20	0.40 to 0.60	0.13 to 0.20	...	8620
16	0.20 to 0.25	0.90 to 1.20	0.40 to 0.60	0.13 to 0.20	...	8622
17	0.23 to 0.28	0.90 to 1.20	0.40 to 0.60	0.13 to 0.20	...	8625
18	0.25 to 0.30	0.90 to 1.20	0.40 to 0.60	0.13 to 0.20	...	8627
19	0.18 to 0.23	0.90 to 1.20	0.40 to 0.60	0.08 to 0.15	0.0005 B min	94B17
20	0.13 to 0.18	0.90 to 1.20	0.40 to 0.60	0.13 to 0.20	...	8615
21	0.15 to 0.20	0.90 to 1.20	0.40 to 0.60	0.13 to 0.20	...	8617
24	0.18 to 0.23	0.75 to 1.00	0.45 to 0.65	0.20 to 0.30	...	8620
30	0.13 to 0.18	0.70 to 0.90	0.45 to 0.65	0.45 to 0.60	0.70 to 1.00 Ni	4815
31	0.15 to 0.20	0.70 to 0.90	0.45 to 0.65	0.45 to 0.60	0.70 to 1.00 Ni	4817
32	0.18 to 0.23	0.70 to 0.90	0.45 to 0.65	0.45 to 0.60	0.20 to 0.40 Ni	4820
33	0.17 to 0.24	0.85 to 1.25	0.45 to 0.65	0.05 min	...	4027
34	0.28 to 0.33	0.90 to 1.20	0.20 min	0.13 to 0.20	...	8630
36	0.38 to 0.43	0.90 to 1.20	0.40 to 0.60	0.13 to 0.20	...	8640
38	0.43 to 0.48	0.90 to 1.20	0.45 to 0.65	0.13 to 0.20	...	8645
39	0.48 to 0.53	0.90 to 1.20	0.45 to 0.65	0.13 to 0.20	...	8650
40	0.51 to 0.59	0.95 to 1.20	0.45 to 0.65	0.13 to 0.20	...	8655
54	0.19 to 0.25	0.70 to 1.05	0.40 to 0.70	0.05 min	...	4118
55	0.15 to 0.20	0.70 to 1.00	0.45 to 0.65	0.65 to 0.80	1.65 to 2.00 Ni	4817
56	0.08 to 0.13	0.70 to 1.00	0.45 to 0.65	0.65 to 0.80	1.65 to 2.00 Ni	9310
57	0.08 max	1.25 max	17 to 19	1.75 to 2.25	...	30303
58	0.16 to 0.21	1.00 to 1.30	0.45 to 0.65	...	...	4118
59	0.18 to 0.23	1.00 to 1.30	0.70 to 0.90	...	...	8620
60	0.20 to 0.25	1.00 to 1.30	0.70 to 0.90	...	...	8622
61	0.23 to 0.38	1.00 to 1.30	0.70 to 0.90	...	...	8625
62	0.25 to 0.30	1.00 to 1.30	0.70 to 0.90	...	...	8627
63	0.31 to 0.38	0.75 to 1.10	0.45 to 0.65	...	0.0005 to 0.003 B	...
64	0.16 to 0.21	1.00 to 1.30	0.70 to 0.90	...	...	...
65	0.21 to 0.26	1.00 to 1.30	0.70 to 0.90	...	...	...

All steels contain 0.035% P max except EX57 (0.040% P max); 0.040% S max except EX57 (0.15-0.35% S); and 0.20-0.35% Si except EX54 (0.33% Si max), EX57 (1.00% Si max), EX58, EX59, EX60, EX61, and EX62 (0.15-0.30% Si), and EX63, EX64, and EX65 (0.15-0.35% Si). Source: Ref. 24

lists various types of environments that may cause SCC in some materials.

Three common methods for preventing SCC are: thermal stress relief (for the 300-series stainless steels) to eliminate fabrication stresses; shot peening, which produces a protective residual compressive stress in a thin section of the surface of the material (Ref. 28); and heat treatment, especially tempering, to reduce internal stresses caused by the quenching process (Ref. 29). Figure 14 indicates the effectiveness of glass bead peening in helping to prevent failure by SCC. Almen A Intensity refers to the force of the peening process, as measured by an Almen gage.

The problem of hydrogen embrittlement is illustrated in Fig. 15. It has been stated that the microstructure of steel materials and lattice defects are directly related to the tendency of the material to fail by hydrogen embrittlement (Ref. 16, 30). Sulfur content in these steels also has a direct influence on the occurrence of hydrogen attack (Ref. 16).

Hardness. The selection of ultrahard materials for components that can be fabricated from lower hardenability steels should be avoided. When specifying ultrahigh strength or low-alloy steels for aerospace construction, the

design engineer should ensure that the heat treatment cycle promotes the greatest degree of hydrogen egress from these materials to avoid embrittlement.

Stability pertains primarily to the effects of temperature, including temperature variations, on a material exposed for specific lengths of time. Material stability also can

Table III Environments that may cause stress-corrosion cracking

Material	Environment
Aluminum alloys	$NaCl/H_2O_2$ solutions NaCl solutions Seawater Air Water vapor
Carbon steels	NaOH solutions Nitrate solutions H_2SO_4/HNO_3 HCN solutions Molten Na/Pb alloys
Copper alloys	Ammonia vapors and solutions Amines Water, water vapor
Gold alloys	$FeCl_3$ solutions Acetic acid/salt solutions
Alloy 600	NaOH solutions
Lead	Lead acetate solutions
Alloy 400	Fused NaOH Hydrofluoric acid Hydrofluosilicic acid Boiling NaOH solutions
Nickel	Mercury Sulfur
Types 304, 316 stainless steels	Chloride solutions H_2S NaOH solutions
Titanium alloys	Red fuming nitric acid Seawater N_2O_4; Methanol/HCl

Source: Ref. 27

Fig. 12 Example of crevice corrosion. Source: Ref. 26

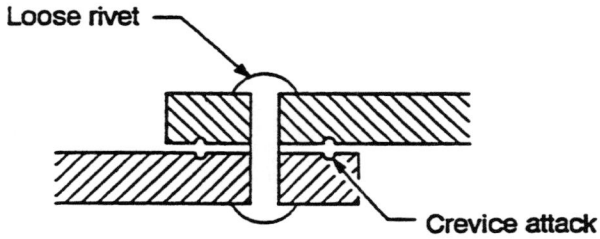

Fig. 13 Welding as a method of preventing crevice corrosion. Source: Ref. 25

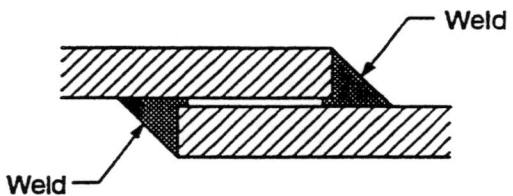

be affected by radiation, radioactive fluids, chemicals, microorganisms, or other environmental factors.

Figure 16 shows the relationship between ultimate tensile strength (UTS)/density ratio and temperature. Filament-reinforced plastics have an excellent UTS/density and modulus/density ratios at room temperature, and up to about 370 °C (700 °F). However, the whisker-reinforced metals are generally useful in the 540 to 1100 °C (1000 to 2000 °F) range (Ref. 31). Figures 17 and 18 provide additional design information about the relationships between specific strength and modulus to temperature for various materials. For the indicated temperature range of –18 to 820 °C (0 to 1500 °F), alloys of nickel and titanium have excellent temperature stability with respect to specific modulus and specific UTS (Ref. 31, 32).

Economics as a basis for material selection. The prices of specific grades of structural materials can be obtained either by referring to purchasing manuals or by contacting the manufacturer directly for price quotations. Costs of some major automotive materials are presented in Table IV.

Cost savings can usually be realized in materials selection by changing either the fabrication procedures or the configuration of the final product. A simple method for reducing material costs is to change to an alternative material, without substantially changing the form or processing procedure. If greater strength is desired in the material to be used, the cost of a substitute material will be expected to be higher. The magnitude of the cost increase will depend on the new material, as shown in Table V for high-strength low-alloy (HSLA) steels. As another example, Fig. 19 presents a cost comparison between metallic (magnesium) and short-fiber composite parts used in a radar antenna system. The use of long-fiber composites for some of the components has resulted in greater cost savings.

Figure 20 shows the weight savings required to change from SAE 1010 HSLA steel to another HSLA steel. The per-

Table IV Average material costs for major automotive materials

Material	Cost per pound (1982 US $)
Mild steel	0.25
High-strength steel(a)	0.30
Cast iron	0.15
Plastics	0.75
Wrought aluminum	1.15
Cast aluminum	0.90
Others	0.50

(a) Not galvanized or treated with anticorrosion coatings. Source: Reprinted with permission from Ref. 33

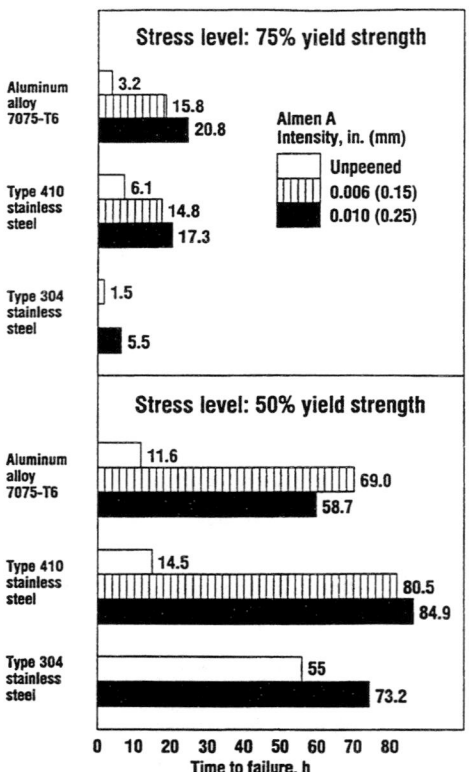

Fig. 14 Effect of shot peening intensity on failure rates of different materials exposed to stress-corrosive environments. Source: Ref. 28

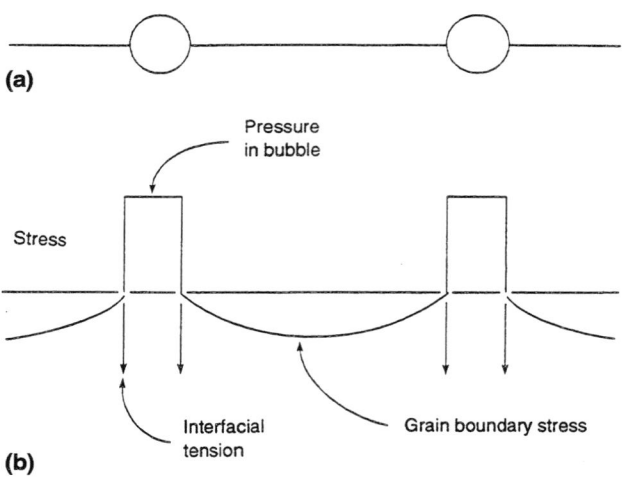

Fig. 15 Hydrogen embrittlement due to the formation of methane bubbles at internal boundaries. (a) The carbon and hydrogen diffuse the surface of the bubble, where they react to form methane. (b) The methane pressure is balanced by the surface tension of the bubble plus the tensile stress induced in the matrix. The rate of growth of the bubbles is controlled by the diffusion of the iron atoms along the boundary due to the stress gradient. Reprinted with permission from Ref. 30

cent savings in weight must be achieved in order to break even with the cost of the original SAE 1010 material. For example, if a part is manufactured with SAE 1010 and SAE 950X is selected as the substitute material, the weight savings percentage required to break even on material costs will be 19%.

Ensuring that low-cost and reliable aerospace materials are selected for military and commercial systems will always be a significant and challenging task for design engineers. However, a viable balance must exist between the expenses and properties of structural materials. For example, resistance to corrosion, abrasion, fracture, and fatigue must be specified for materials that will be exposed to severe environments and mechanical loading. The cost of these desirable structural materials should also be reasonable, compared to other materials with similar properties.

Mechanical properties of structural materials. Some of the mechanical properties usually considered in material selection are strength (ultimate tensile and yield), hardness, and ductility. The design engineer also should make the fracture mechanics approach an integral part of the process of considering mechanical properties. This will ensure the design of a safe and fracture-resistant structure by properly proportioning the final product, so that it does not fail either by tensile overload or by compressive instability, and designing to prevent brittle fracture due to unstable crack propagation (Ref. 35).

An essential component of the fracture mechanics design approach is the use of plane-strain fracture toughness (K_{Ic}) data. Consideration of the plane-strain fracture toughness can assist the design engineer in obtaining a reasonable estimate of the total fatigue life of a particular structure. Knowledge of the specific stress or stress intensity range required for crack growth, as well as the determination of the proper crack growth rate expression, is essential for the effective application of K_{Ic} data. K_{Ic} is directly related to the amount of energy required to initiate crack propagation. Another significant characteristic of K_{Ic} is that it does not depend on specimen configuration; it is a material property. Other measures of toughness (such as the notch toughness) are affected by sample geometry. The most important characteristic of K_{Ic} (for most engineering materials) is the inverse relationship of K_{Ic} to yield strength.

Materials selection for unique mechanical and physical properties. Selection from a series of candidate materials also may be made more efficient by careful consideration of unique mechanical and physical properties. In military and commercial aerospace applications, desirable properties for construction materials include lightness (low density), wear resistance, high yield or ultimate tensile strength, high elastic modulus, and special thermal expansion characteristics, especially in relation to metal-to-glass seals.

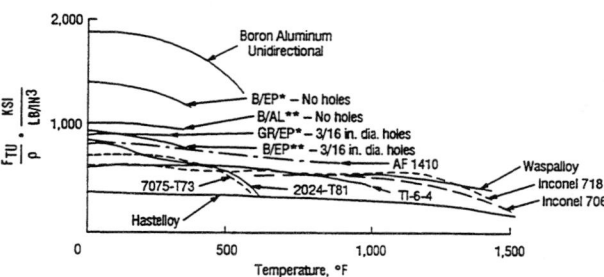

Fig. 16 Relationship between structural efficiency (UTS/density ratio) and temperature for various materials. B/EP is boron epoxy; B/AL is boron aluminum; GR/EP is graphite epoxy. Reprinted with permission from Ref. 31

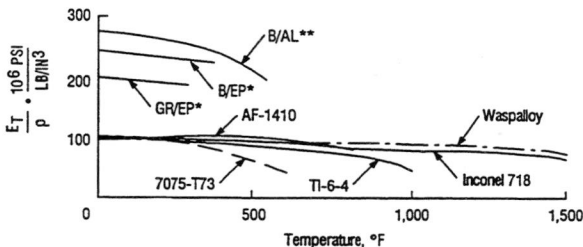

Fig. 17 Specific modulus vs. temperature for various military materials of construction. B/EP is boron epoxy; B/AL is boron aluminum; GR/EP is graphite epoxy. Reprinted with permission from Ref. 32

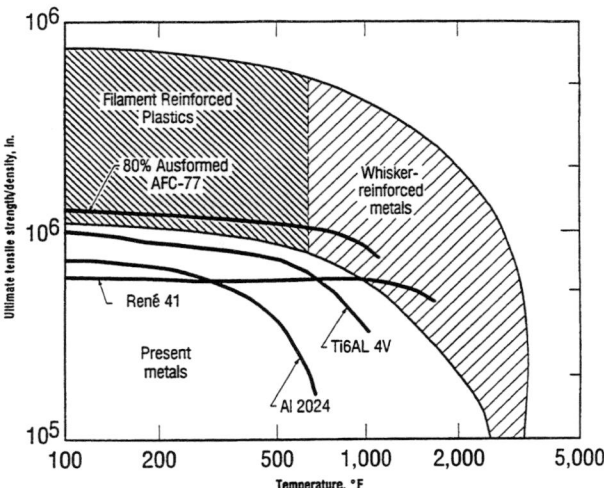

Fig. 18 Variation of specific strength with temperature for selected materials used in military construction. Reprinted with permission from Ref. 32

Therefore, military and commercial aerospace design engineers are usually interested in selecting lightweight materials that have high strength-to-density or modulus-to-density ratios. A comparison of different materials, based on modulus-to-density and yield-strength-to-density ratios, is given in Fig. 21.

Table VI provides a material property comparison between AF 1410 and Ti-6Al-4V. AF 1410 is a high-strength, high-fracture toughness steel that was developed by the U.S. Air Force. Annealed Ti-6Al-4V is a titanium-base material commonly used in military and commercial aerospace forgings. Table VI shows that AF 1410 is superior to TI-6Al-4V in all properties (except bearing-related properties) on a strength-to-density basis (Ref. 36).

One of the primary objectives in advanced composite material development for aerospace structures has been the optimization of weight savings. The actual application of advanced composites in aircraft structures has resulted in

substantial weight savings, mostly in military air vehicles. For example, the substitution of boron-epoxy composite material for aluminum on the F-14 horizontal stabilizer resulted in weight savings of over 26% (Ref. 32).

Properties and Applications of Structural Materials

Ferrous Materials

For steel and other ferrous materials selected for military structures, alloying elements are significant in conferring desirable physical, mechanical, and corrosion resistance properties. For example, additions of niobium, vanadium, or aluminum improve the strength and notch toughness of carbon steels by reducing the ferritic grain size. Niobium and vanadium additions (0.10% max) may be preferred over aluminum because the former elements also increase the hardenability and high-temperature strength of these steels. However, niobium and vanadium additions are made only for special types of steel, because these elements may cause embrittlement during stress-relieving heat treatments at approximately 590 °C (1100 °F). Chemical modifications also assist in improving the strength of heat-treated steels (Ref. 37, 38).

Comparison of metal to composite part cost (see Fig. 19)

Part	Cost ratio (metal: composite)
Base	1:2
Elevation arm	1:4
Azimuth gimbal	1:4
Elevation stop	2:1
Elevation cover	5:1
Guides	1.5:1

Includes material, manufacturing, and tooling costs.

Anatomy of a radar antenna

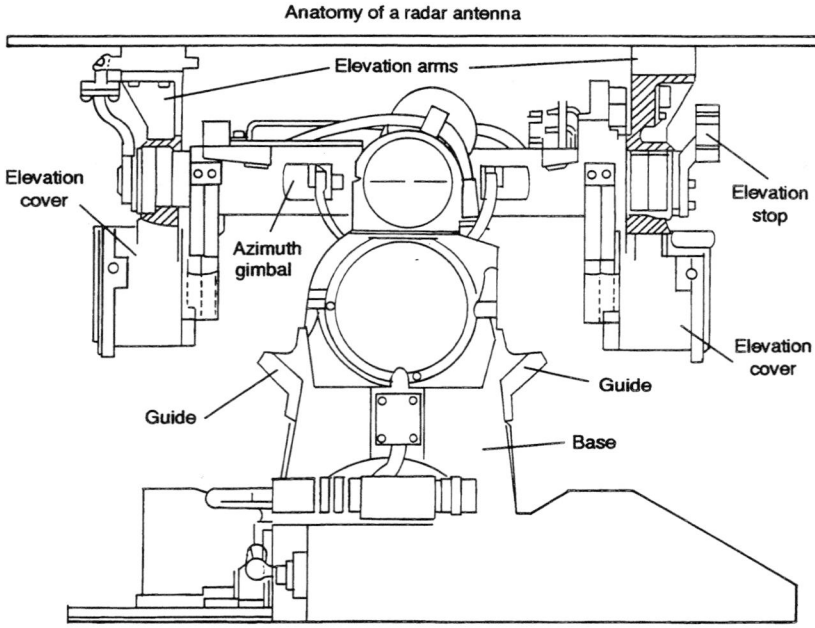

Fig. 19 Cost comparison of metal (magnesium) to short-fiber composite parts used in a radar antenna system. Refer to corresponding table above. Reprinted from Ref. 34

Table V   Strength and cost relationships for various HSLA steels

Specification	Type	Minimum yield point		Strength increase over A36, %	Cost increase over A36, %(a)	Other features
		MPa	ksi			
ASTM A36	Plain carbon	248	36	...	...	Lowest cost per pound
ASTM A572, Grade 50	Nb/V microalloy	345	50	38.9	9.4	Good weldability
ASTM A440	Mn-Cu	345	50	38.9	13.4	Not for welded structures. Corrosion resistance twice that of A36
ASTM A441	Mn-V-Cu	345	50	38.9	13.4	For welded structures. Atmospheric corrosion resistance twice that of A36
ASTM A242	Multiple alloy, weathering steel	345	50	38.9	38.3	Weathering steel. Atmospheric corrosion resistance 4 times that of A36 (often 5 to 8 times that for Type 1). Type 2 has excellent toughness.

(a) Based on 1970 U.S. dollars. Source: Reprinted with permission from Ref. 22

Table VI   AF1410 properties compared to those of Ti-6Al-4V

Property	AF1410		Ti-6Al-4V annealed	
	Value	Value/ density ratio	Value	Value/ density ratio
Ultimate tensile strength, ksi	235	839	130	812
Tensile yield strength, ksi	220	786	120	750
Compressive yield strength, ksi	231	825	126	787
Ultimate shear strength, ksi	139	496	79	494
Ultimate bearing strength, ksi (e/D = 2.0)	438	1,564	256	1,600
Bearing yield strength, ksi (e/D = 2.0)	324	1,157	208	1,300
Modulus of elasticity, 10^6 psi	28	100	15	100
K_{Ic}, ksi$\sqrt{in}$.	130	464	70	437

Source: Reprinted with permission from Ref. 36

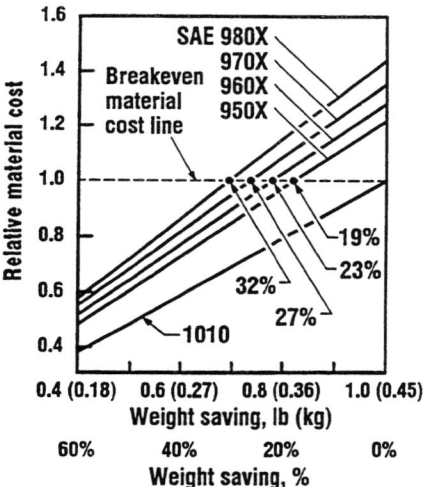

Fig. 20 Percentages of weight saving needed to break even on material costs when replacing SAE 1010 steel. Based on 1974 prices. Reprinted with permission from Ref. 22

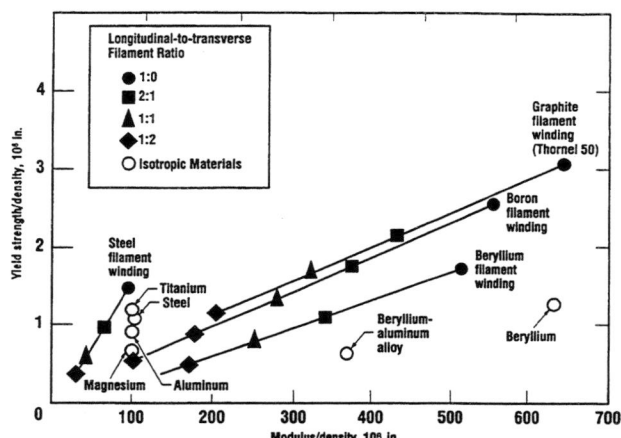

Fig. 21 Relationship between yield-strength-to-density and modulus-to-density ratios for various materials. Reprinted with permission from Ref. 31

Fracture properties of iron and steel structural materials also are important aspects to consider during the material selection process. Fracture mechanics principles have been used since the early 1960s for the development of rocket motor chambers (Ref. 39). Concepts of fracture mechanics also have been incorporated into military standards for determining the fracture toughness of metals and for designing airframes and engine components. For example, in military specification MIL-A-83444, the U.S. Air Force has established damage tolerance requirements that can be applied in the design of military aircraft (Ref. 40). The fracture toughness of carbon and alloy steels is reduced by nonmetallic inclusions, either singly or in clusters. In addition, thermomechanical treatments can be used to enhance strength and toughness of ultrahigh-strength steels during manufacture.

A primary application of carbon and alloy steels is gears. Gear steels must be easy to machine, heat treatable, and wear resistant. The normal hardness of carburized gears should be between HRC 62 and HRC 63 (Ref. 37). Carburizing is necessary to form a sufficiently thick case on the material to prevent case crushing (also referred to as spalling, core yielding, and core deformation).

High-strength low-alloy steels are characterized primarily by their mechanical properties in the as-rolled condition. HSLA steels were developed to meet design criteria demanding improved weight savings, formability, fatigue strength, weldability, and notch toughness. Their good resistance to abrasion and atmospheric corrosion are additional attractive features for design considerations (Ref. 22). The most significant feature of HSLA steels is their high strength-to-weight ratio. This permits the designer to reduce weight without sacrificing strength by specifying the use of thinner sections.

Automotive engineers have selected HSLA steels to attain desirable weight reduction while maintaining required strength and fatigue properties. Typical applications include bumper reinforcements, lock and hinge components, steering and suspension components, bumper backup, and seatback head-restraint supports (Ref. 22).

Ductile (nodular) iron is a versatile engineering material which can fulfill fabrication requirements in military equipment. These materials can be used where high strength, toughness, and wear resistance are desired. The graphite present in ductile iron confers good resistance to mechanical wear, as well as improved machinability. Ductile irons also are noteworthy for their high ductility and hardenability.

For high-temperature service, austenitic ductile iron should be considered (Ref. 41). This class of ductile irons contains a nickel-stabilized matrix that is austenitic at all service temperatures. Approximately half of all ductile iron applications are military automotive castings. Some of the most common applications for this material are crankshafts, differential carriers, and bearing caps (Ref. 2).

Nodular iron is not without its problems. Patches or areas in castings have been observed to contain either deteriorated graphite or flake graphite (Ref. 42). These types of defects have been referred to as flotation and chunky graphite, and they have a detrimental effect on mechanical properties.

Hardfacing alloys. Iron- and steel-base materials are used in an antiwear process known as hardfacing. Hardfacing alloys are welded or fused to an inexpensive substrate metal to provide protection against abrasion, heat, corrosion, and impact (Ref. 43). Tungsten carbide (WC) is among the hardest materials known. When used as a reinforcement in steel, it enhances resistance to abrasive wear. The principal alloying elements for hardfacing materials include chromium (Cr), molybdenum (Mo), nickel (Ni), and cobalt (Co). Nickel- and cobalt-base materials also are used for hardfacing applications.

Stainless steels can be used to advantage in commercial and military design because of their high strength and stiffness, as well as their excellent wear and corrosion resistance. Chromium imparts corrosion resistance to stainless steels by combining with oxygen to form a protective (passive) surface film (Ref. 44). Formation of this chemically-inactive surface layer is essential for maintaining the excellent corrosion resistance properties of stainless steels. The passive film will regenerate, during exposure to oxygen in the environment, on a thoroughly-cleaned stainless steel surface where the film has been disrupted or completely removed (Ref. 44, 45).

Additional favorable properties of stainless steels include their ability to be used over a wide temperature range and their relatively low cost (Ref. 46). These properties should be considered during the initial step of any material selection process in the design of military equipment.

Stainless steels are classified by their metallurgical characteristics. The types of stainless steels are described as follows:

Ferritic. Included in the 400 series, these grades contain low carbon and high chromium. The amount of chromium ranges from 14 to 18% in Type 430 to a high of 23 to 27% in Type 446. Although higher amounts of chromium increase resistance to corrosion and scaling, there is some sacrifice in other properties, such as notch impact strength (Ref. 47). Molybdenum may be added to ferritic stainless steels to enhance corrosion resistance. These steels are similar to carbon and alloy steels in their ferromagnetic properties. Ferritic stainless steels have a balance in the amount of carbon and chromium, which tends to suppress the development of austenite (gamma-iron, or γ-Fe) at high temperatures. Because of the small amount of γ-Fe present, ferritic steels do not transform to martensite upon cooling, but remain ferritic throughout their operating temperature range. Ferritic stainless steels are nonhardenable by the usual methods of heating and quenching.

Martensitic. These steels also are in the 400 series. They can be heat treated to a wide range of useful hardness and strength levels. The major alloying element in martensitic stainless steels is chromium. The quantities of chromium and carbon are balanced to ensure that the soft phase of γ-Fe, which develops at high temperatures (greater than 723 °C, or 1333 °F), transforms into the hard martensitic phase. The amount of carbon in martensitic steels is usually under 0.15%, while the chromium is between 11.5 and 14%, such as in Types 403, 410, 414, and 416. Up to 1.2% C is used with 16 to 18% chromium in some types, including the 440 series. Martensitic steels provide hardness up to approximately 62 HRC, with tensile strengths to about 1.965 MPa (285 ksi). They are magnetic and are suitable for applications requiring wear and abrasion resistance, such as turbine parts and bearings. These steels are generally not as corrosion-resistant as the ferritic and austenitic types. Martensitic steels usually require annealing to prevent cracking, followed by hardening to develop high strength and good corrosion resistance.

Austenitic. This group of 200 and 300 series stainless steels contains nickel as a second major alloying element in addition to chromium. Austenitic stainless steels retain an austenitic structure (γ-Fe) during cooling from elevated temperatures. Type 302 contains 18% Cr and 8% Ni. The nickel content in austenitic stainless steels may vary between 7 and 20%. The 200 series austenitic steels contain less nickel than 300 series steels because of the addition of manganese and nitrogen. Nickel additions improve the stability of the austenite over a wide temperature range. Corrosion resistance is also enhanced by nickel, which is the primary reason the austenitic types provide the best corrosion resistance of all stainless steels, particularly when these steels have been annealed to dissolve chromium carbides, then rapidly quenched to retain carbon in solution (Ref. 47). Austenitic stainless steels also have high ductility, low yield strength, and high ultimate strength, which contribute to their excellent behavior in deep drawing and forming. In the as-annealed condition, the austenitic types are nonmagnetic.

Depending on composition, these steels may become slightly magnetic when cold worked because they deform to martensite. Austenitics also are readily welded. They may be work hardened to high levels, although not as high as the level attained by heat treating the hardenable (martensitic) types of the 400 series stainless steels.

Precipitation-hardened stainless steels. These materials include the semi-austenitic and martensitic groups. Strengthening of these alloys occurs by precipitation hardening (aging) in the 480 to 650 °C (896 to 1202 °F) range (Ref. 48). This heat treatment, in the semi-austenitic group, strengthens the material by causing aluminum to combine with nickel to form NiAl and Ni₃Al precipitates.

PH 13-8 Mo stainless steel can be used when there is a short supply of high strength, cobalt-bearing maraging steels. The PH 13-8 Mo stainless alloy can be precipitation-hardened to a tensile strength of 1620 kPa (235 ksi) (Ref. 49). PH 13-8 Mo stainless steel may be used when greater strength than 17-4 PH stainless is required, but the specific application does not need the higher strength of maraging steels. However, 17-4 PH and other stainless steels are included in the high-strength steel category. This class, which is also called the ultrahigh strength steels, typically includes steels having tensile and yield strengths exceeding 1034 kPa (150 ksi).

Pitting and crevice corrosion are two major modes of degradation that material selectors consider when designing with stainless steels. Pitting occurs when the passive film is destroyed in small, isolated areas on the surface of these materials. Local differences in oxygen concentration lead to an increase in the number of metal cations around the structure. These attract chloride ions, which decrease the pH and promote crevice corrosion. This type of deterioration occurs in areas providing narrow openings (crevices) where small, active (anodic) sites develop because the passive film becomes unstable due to oxygen starvation in the crevice. Chloride ions as well as high concentrations of hydrogen ions will adversely affect the stability of the passive layer on stainless steel surfaces (Ref. 22). Despite the higher propensity of failure by SCC with yield strengths surpassing 1379 kPa (200 ksi) (Ref. 50), these high strength materials have very good fracture toughness and fatigue resistance. For high-strength steels, significant growth of fatigue cracks has been observed at stress intensity (K) levels which are higher than K levels of high-strength aluminum and titanium alloys (Ref. 51).

Steel wire is the strongest form of steel, and it is used in military equipment systems (Ref. 52). The maximum strength of this material is about 4140 kPa (600 ksi). Applications include nails, bolts, rivets, screws, and welds. Aircraft cord wire and music spring wire are also among the primary uses for steel wire.

Steel wire is classified into four major subgroups based on carbon content, as follows:

- *Low carbon* contains a maximum of 0.15% C, and it is designated AISI 1005 to AISI 1015. (In these designations, the last two integers indicate the approximate C level, and the first two integers mean that the steel is nonresulfurized with less than 1% Mn.) Characteristics of these low grades include good formability, weldability, and moderate strength.

- *Medium low carbon* contains from 0.16 to 0.23% C. Designations for this type are AISI 1016 to AISI 1023. These grades have higher strength values compared to low-carbon steel wire.

- *Medium high carbon:* Designated as AISI 1024 to AISI 1044, this type contains 0.24 to 0.44% C. Strength and hardness values for this type are greater than the above types. The ductility of these grades often is improved by thermal treatment.

- *High carbon:* Designated as AISI 1045 to AISI 1095, these grades contain over 0.44% C. They are used in applications under demanding conditions that require high toughness and extra strength. Typical applications include wire rope, bridge rope, and music wire.

Besides these four major steel wire classes, there are 110 special grades of steel wire grouped as commodity wire. Each grade of commodity wire is produced to perform one specific application.

Corrosion protection for steel wire is provided either by electrolytic or hot dip galvanizing (Ref. 52, 53). Both of these processes produce tenacious zinc coatings on the steel wire. Electrodeposited coatings are usually lighter and smoother than the hot dip galvanized coatings.

Aluminum, Magnesium, and Zinc

Aluminum (Al) is one of the most economical and structurally effective materials used for commercial and military equipment applications. Some of its attractive properties include:

- Good appearance
- Ease of fabrication
- Good corrosion resistance (especially in the newer 7000 series aluminum alloys, which perform very well, but are not completely immune, in environments that promote hydrogen embrittlement and intergranular corrosion) (Ref. 54)

- Low density and high strength-to-weight ratio (Ref. 23, 55)
- Higher strength by alloying, heat treatment, or cold working, or a combination of all three (Ref. 55)
- High fracture toughness (Ref. 56)

Methods for manufacturing aluminum alloys vary considerably and have a significant effect on material properties. The principal strengthening treatment for aluminum alloys is cold working. Another strengthening treatment is solution heat treatment and precipitation hardening. For example, additions of 4 to 8% Zn and 1 to 3% Mg enhance the development of excellent precipitation hardening characteristics in the 7000 series.

For aluminum, the most commonly used temper is T6, which provides good mechanical properties as well as enhanced machinability and corrosion resistance. When resistance to stress-corrosion cracking is required, the T7 temper is employed. Artificial aging times and temperatures that are higher than those used for maximum hardness produce the T7 tempers (Ref. 57).

Fracture mechanics can be effective in selecting aluminum-base materials for military and commercial construction applications. A widely consulted parameter for materials selection is the plane strain fracture toughness (K_{Ic}), a conservative lower limit for defining the toughness of a structural material (Ref. 58). It is important to understand that K_{Ic} values are valid only at sufficient material thicknesses for plane strain conditions (Ref. 59).

Another significant property to consider during the selection of aluminum alloys is the critical stress-intensity factor K_c. This is the magnitude of stress intensity at the crack tip, at which unstable fracture occurs (Ref. 7). The K_c of 7000 series aluminum alloys may be improved by overaging, a double heat treatment which reduces aluminum alloy susceptibility to SCC and exfoliation (Ref. 60).

Aluminum alloy 7075-T73 has been used for large structural forgings on the McDonnell Douglas DC-10 because of its resistance to SCC and fatigue crack propagation, and because of its high fracture toughness. Another attractive feature of 7075-T73 is its excellent resistance to exfoliation corrosion. Alloys 7075-T76 and 7178-T76 have also demonstrated superior resistance to exfoliation in long-term (11-year) seacoast exposure tests (Ref. 61, 62).

Alloy 6061 has been used in vertical takeoff and landing aircraft for low pressure hydraulic tubing and rotor blades, because of its good corrosion resistance. 6061 is also useful in master cylinder pistons, valve parts, and aircraft wheels (Ref. 2, 63).

Many variables affect the mechanical properties of aluminum castings. The most significant are the specific alloy and the variation of chemical composition, metal soundness, metallurgical characteristics such as macrograin size, solidification rate, and heat treatment (Ref. 64).

Some of the major objectives for improvement of aluminum alloys include enhancement of strength, toughness, and corrosion resistance (Ref. 65). Wrought aluminum powder metallurgy (P/M) alloys have been developed to attain these goals. For example, the fracture toughness of P/M aluminum alloys has been observed to be superior to that of many 7000 series alloys produced by ingot metallurgy (I/M) (Ref. 66, 67). P/M forgings can be used in parts now produced by conventional aluminum forgings, including connecting rods, pistons, gears, and other automotive structures (Ref. 2).

To enhance cost and strength-to-weight characteristics, AISI 434 stainless steel has been clad to alloy 5052 for use in aircraft firewalls (Ref. 68). Cladding is essentially a combination of thin layers of different metals. The production of clad metal strip, by continuous roll bonding of two or more metallic strips, is based on solid state welding.

Various aluminum alloys can be used for fastening airframe skins. Alloys 1100, 5052, 2117, and 5056 may be used for rivet bodies. Mandrels, which are used in blind fastening, can be manufactured of 5056, 2017, 2024, 7178, 7075, or other materials (Ref. 69).

Magnesium (Mg) is the lightest of the commercially available metals, and it has a high strength-to-weight ratio when alloyed with other elements. Typical materials alloyed with magnesium are aluminum and zinc (Ref. 7, 70). Because magnesium is about two-thirds the weight of aluminum, magnesium alloys have higher strength-to-weight ratios than many aluminum alloys, making it useful in aerospace structures.

Important properties of magnesium and magnesium alloys include:

- Resistance to alkalis, oils, solvents, some acids, and most organic chemicals under ordinary atmospheric conditions
- Ease of machinability and fabrication by most metalworking processes, such as welding
- Good static fatigue and ductility characteristics
- Good toughness, impact resistance, and thermal diffusivity

Certain magnesium alloys, such as AZ31B, can be aged or heat treated, but strengthening is not attained to the same degree as in aluminum alloys.

In general, magnesium alloys are the most electrochemically reactive materials among the major structural airframe metals. Therefore, magnesium alloy structures require damage-tolerant anodic and organic (primer and topcoat) coating systems for optimum corrosion resistance.

Because the driving force for the galvanic corrosion reaction is the difference in electrical potential between two metals, severe corrosion is expected when magnesium is joined to materials that are much higher on the galvanic series. Designers must ensure that magnesium structures in contact with dissimilar metals are adequately protected. Two important precautions to take in minimizing galvanic corrosion of magnesium are preventing the accumulation of water at dissimilar metal contacts and choosing metals that are electrochemically compatible with magnesium (Ref. 71). Some typical cases of magnesium corrosion problems that have been observed in aircraft are pitting and intergranular corrosion of magnesium alloy extrusions on the trailing edges of wings and wing control surfaces, and pitting and intergranular corrosion of magnesium alloy landing gear components (Ref. 72). Solutions to these problems include replacement of the magnesium alloy extrusions by parts fabricated from aluminum alloys or composite materials.

Magnesium and magnesium alloys are prohibited in components that come into contact with fuel (Ref. 73, 74). Water, which is an electrolyte, can accumulate in fuel systems, and this will generate an aggressive corrosion reaction if magnesium or magnesium alloys are present.

Some of the more significant magnesium alloys to consider in commercial and military equipment design and construction are:

- *AZ31B:* This alloy is used primarily for applications where temperatures do not exceed 149 °C (300 °F).
- *AZ61A:* This wrought magnesium-base material contains more aluminum than AZ31B. The greater aluminum content increases the strength but slightly decreases the ductility of this alloy. AZ61A must be stress relieved after welding to prevent SCC.
- *HK31A:* This magnesium alloy is available in wrought and cast forms. It has relatively high strength between 149 °C (300 °F) and 371 °C (700 °F) and is used primarily for components requiring good strength-to-weight ratios in this temperature range. HK31A has been applied to the manufacture of missile and aircraft skins.
- *HM21A:* Compared to other magnesium alloys in the 260 °C (500 °F) to 427 °C (800 °F) range, HM21A possesses superior strength.

Zinc (Zn) is used primarily in galvanizing, a coating process that is performed either by immersing steel parts or structures into molten zinc or by electroplating (Ref. 2, 75). The zinc coating serves as a sacrificial, corrosion-preventive coating for the substrate material. Because of its sacrificial corrosion behavior, zinc is an important constituent of zinc-rich coatings, which are based on a binder of ethyl silicate (approximately 40% silicon dioxide, SiO_2) (Ref. 76). These zinc-rich coatings have been widely utilized in preventing the degradation of military equipment.

In structural applications, zinc alloys are used mostly for pressure die casting. These die casting alloys use a special, high-purity grade of zinc to ensure stability of the dimensions and properties of each component. Zinc die castings are used mostly in the automotive industry. Some of the most important applications are parts for carburetors, fuel pumps, grilles, and hydraulic brakes. Zinc die castings are also used for parts that perform both structural and decorative purposes in automotive equipment, such as brackets, instrument panels, and body molding (Ref. 7).

The addition of higher quantities of aluminum (8 to 27%, compared with approximately 4% for conventional zinc die castings) has led to the development of zinc casting alloys with significantly better mechanical properties. These alloys are ZA-8, ZA-12, and ZA-27. Some of the advantages of these alloys include:

- Excellent castability
- Excellent wear characteristics, significant in bearing and bushing applications
- Good mechanical properties, especially strength and toughness
- Nonincendivity (applied to ZA-8 and ZA-12); light metals, such as Al, Mg, and titanium (Ti), can spark and ignite hazardous fuel-air mixtures, vapors, or combustible particulates (Ref. 77)
- Finished cost (including machining and tooling) is usually lower than for cast iron

Some applications of high-performance zinc alloys are lower-speed bearings and bushings (Ref. 77, 78). This is an important application for zinc alloys ZA-12 and ZA-27, which possess the appropriate hardnesses and coefficients of friction. These materials also have potential use in automotive housings, piston-to-crankshaft links, and gear boxes (Ref. 78).

Titanium

Titanium alloys have proven to be very effective structural materials, especially in aerospace applications. Titanium is used primarily in components that require a combination of high strength, low density, and corrosion resistance (Ref. 79). Titanium has two allotropic crystal forms: αTi [up to 882 °C (1620 °F)], having a hexagonal close-packed (hcp) lattice; and βTi, which is body-centered cubic (bcc).

Due to this allotropic nature of titanium, alloys with significantly different yield strengths can be formed. Considerable increases in the strength of titanium alloys are attained by alloying with several elements, primarily aluminum. Some of the properties that make titanium alloys advantageous for structural applications are: good-strength-to-weight ratio, low density, low coefficient of thermal expansion, good corrosion resistance, good performance in oxidizing atmospheres up to about 550 °C (1020 °F), good notch toughness, and low heat-treating temperature during hardening.

In applications that are considered fracture-critical—which includes most aerospace equipment components—the following material and process factors that influence the fracture toughness of titanium and titanium alloys should be carefully monitored (Ref. 80):

- Chemistry variations
- Heat treatment
- Microstructure
- Product thickness
- Yield strength

Regarding chemical variations in titanium products, interstitial elements (which increase the strength of titanium) are detrimental to its toughness (Ref. 50). Therefore, commercial extra-low interstitial (ELI) titanium alloys are produced for specific applications that demand high toughness.

Variations in titanium alloy chemistry, microstructure, and crystallography texture can change the magnitude of fracture toughness by a factor of two or three (Ref. 79). The plane strain fracture toughness (K_{Ic}) of titanium alloys is very useful in determining the loads that can be sustained by a structural titanium member in the presence of a crack. The fatigue crack propagation behavior of titanium alloys tends to parallel K_{Ic}; for a specific titanium alloy, conditions that generate a high K_{Ic} will give the lowest fatigue crack growth rates.

Another factor in the selection of titanium alloys is corrosion resistance. The low susceptibility of titanium alloys to corrosive deterioration is due to their formation of a protective passive surface film (Ref. 81, 82). Generally, titanium is resistant to metallic chlorides, organic acids, and oxidizing inorganic acids over a broad range of temperatures and concentrations (Ref. 83). An example of the corrosion resistance of titanium is given in several studies of corrosion by galvanic coupling. Ti-6Al-4V (6-4) and Ti-6Al-2Sn-4Zr-2Mo (6-2-4-2) can be used in design applications where these alloys must be in direct contact with graphite epoxy composite material (GECM) (Ref. 17). The resistance of Ti-6-4 to electrochemical corrosion when coupled to GECM has been confirmed in other investigations (Ref. 84, 85). In these tests, Ti-6-4 was used for fasteners, which were connected to GECM and exposed to humidity and salt spray environments. A rivet that has been developed for composite material fastening consists of a Ti-6-4 shank and a Ti-Nb tail section. A Ti-Nb rivet has been used on the control surfaces of the F-18 and on the AV8B vertical short takeoff and landing aircraft (Ref. 15).

When titanium alloys are exposed to chemical environments containing chlorides or methyl alcohol, these materials may fail by SCC. The SCC failure will produce cracking at stress levels that are significantly lower than the specific titanium alloy tensile strength (Ref. 86).

Certain titanium alloys have been observed to crack under stress at about 370 °C (700 °F). Several types of chloride salts can induce failure, and this type of degradation has been referred to as hot salt stress-corrosion cracking (HSSCC). Alpha-phase titanium alloys that are most susceptible to HSSCC include Ti-5Al-2.5Sn, Ti-7Al-12Zr, and Ti-5Al-5Sn-5Zr. The most HSSCC-resistant titanium alloys include Ti-4Al-3Mo-lV, Ti-2.5Al-11Sn-5Zr-1Mo-0.2Si, and the experimental alloy Ti-2Al-4Mo-4Zr (Ref. 87).

Compared with other hcp metals, pure titanium is highly ductile. At room temperature, it can be cold rolled without significant cracking to greater than 90% reduction in thickness (Ref. 80). Aluminum is the most important substitutional additive in binary α-stabilized titanium alloys, because it adds to the ductility and light weight of titanium.

Advanced titanium materials have been fabricated using superplastic forming/diffusion bonding (SPF/DB) and hot isostatic pressing (HIP) (Ref. 88). Superplastic forming involves the extensive deformation of metals within a specific temperature range. The superplastic temperature range of Ti-6-4, for example, is 900 to 980 °C (1650 to 1700 °F) (Ref. 89). When Ti-6-4 and other medium-strength titanium alloys undergo SPF/DB, they elongate up to 1000%. They also diffusion-bond in the same temperature range, which is below the α–β transformation temperature (Ref. 90). SPF/DB has been utilized in the manufacture of parts for the B-1 bomber, including the large lower engine access door (Ref. 91). SPF/DB technology also has been applied in the fabrication of a cylindrical missile section for the U.S. Navy. Ti-6-4 is the material that has been used in the manufacture of these missile components. With the SPF/DB method, the cylindrical missile parts have been bonded and formed in a single set of operations. Generally, the use of SPF/DB in the manufacture of military hardware has enhanced the desirability of using titanium alloys, because the end products are lower in cost (Ref. 88).

HIP can be used for titanium, as well as for other metals and ceramics (Ref. 92). In this process, materials are pressurized with an inert gas, usually argon, at high temperatures (up to 2000 °C, or 3630 °F). Pressures of up to 204 MPa (30 ksi) are applied isostatically to the parts. The HIP process produces castings with dense, homogeneous microstructures, which generally increase fatigue life.

For high performance and structurally efficient aerospace and automotive components, the most widely used titanium alloys belong to the two-phase (α-β) microstructural classification. These titanium alloys are attractive because of their high strength and light weight. A major application of titanium is in military aircraft gas turbine engines (Ref. 93). Future use of titanium alloys in engines may increase through the implementation of rapid solidification technology (RST) (Ref. 94).

Ti-6-6-2 is a forgeable alloy with greater hardenability than Ti-6-4. For example, uniform hardening can be accomplished in 100-mm (4 in.) thick sections with Ti-6-6-2, compared with only 25 mm (1 in.) thick sections when Ti-6-4 is used.

Ti-8Mo-8V-2Fe-3Al (Ti-8-8-2-3) is another candidate material for use in heavy sections and forgings. Based on comparison of some of the mechanical and physical characteristics of these materials, Ti-8-8-2-3 is similar to Ti-6-6-2. Ti-8-8-2-3, for example, can be deep-hardened, and it is able to retain its high strength to 340 to 370 °C (650 to 700 °F) (Ref. 31).

Ti-6Al-2Sn-4Zr-2Mo (Ti-6-2-4-2) may be used in applications requiring high strength and toughness at temperatures up to 500 to 540 °C (940 to 1000 °F). This alloy may be used for forgings and flat-rolled products in gas turbine engine and airframe applications.

Ti-6Al-2Sn-4Zr-6Mo (Ti-6-2-4-6) is another high-strength titanium alloy that can be used in gas turbine engines, particularly in the disk and fan blade components of compressors. Ti-6-2-4-6 should be considered for long-duration load-carrying applications at temperatures up to 400 °C (750 °F), and for short-duration load-carrying structures at temperatures to 540 °C (1000 °F).

Refractory Metals

Refractory metals are elements that have melting points greater that 2204 °C (4000 °F). For example, tungsten melts at 3410 °C (6170 °F). In addition to tungsten, the following are also refractory metals: niobium (Nb), molybdenum (Mo), hafnium (Hf), rhenium (Re), and tantalum (Ta). All of these metals have substantial potential for use in military equipment, especially aerospace and electronics components.

In addition to their very high melting points, the desirable characteristics of refractory metals include:

- Resistance to electrochemical corrosion (applies especially to tantalum, whose corrosion resistance is comparable to that of glass) (Ref. 95)
- Excellent high-temperature strength
- High resistance to thermal shock (applies to tungsten)
- Excellent resistance to wear and abrasion
- Good electrical and heat-conducting properties

The most easily fabricated refractory metals are niobium, tantalum, and their alloys. Fabricability is often used as a deciding factor in the selection of a specific refractory metal (Ref. 96).

Refractory metals are often used as alloying elements (Ref. 97). For example, the currently used wear-resistant alloys may include either tungsten or molybdenum combined with cobalt and chromium. Also, large amounts of tungsten have been used in cobalt-base alloys requiring high-temperature strength and resistance to high radial stresses (Ref. 98). Molybdenum has been used as an alloying element in a stress-corrosion-resistant material, multiphase MP35N, which has been used for aircraft fasteners.

Development of the tantalum-base refractory alloys T-111 and T-222 was the result of extensive government-sponsored work to improve the creep resistance of tantalum alloys. T-222 is the standard material to which other tantalum alloys are compared. T-222 has attractive mechanical properties at pyrogenic and cryogenic temperatures. This alloy is also easily welded, has a ductile-to-brittle transition temperature well below –18 °C (0 °F), possesses good fabricability, and is easy to produce (Ref. 99). T-222 also has substantially better creep rupture properties in comparison to other refractory materials.

Refractory metal alloys satisfactorily retain their mechanical strength at elevated temperatures. Minor additions of zirconium and hafnium to Ta-W-Mo alloys have produced significant strengthening up to 1930 °C (3500 °F)

(Ref. 99, 100). This strengthening effect due to alloying of refractory metals has been reported at 1480 °C (2700 °F).

Some of the unsatisfactory characteristics of refractory metals include their high cost and poor fabricability (except for niobium and tantalum, which can be formed, machined, and joined by conventional methods) (Ref. 96). Refractory metals are susceptible to accelerated atmospheric oxidation, which occurs at about 200 °C (400 °F) for tungsten, 400 °C (750 °F) for molybdenum, and 430 °C (800 °F) for tantalum and niobium. Coatings which consist of more than one element, referred to as composite coating systems, have been tested for high-temperature oxidation protection of refractory metals. A composite coating consisting of Hf-20Ta was observed to be resistant to erosion, thermal cycling, and oxidizing environments from about 1760 to 2200 °C (3200 to 4000 °F). However, oxidation resistance from 1090 to 1650 °C (2000 to 3000 °F) needs further improvement. Additional corrosion protection is required to prevent degradation of the substrate metal. Examples of possible mechanisms in substrate metal degradation include an increase in the ductile-to-brittle transition temperature and poor oxidation in certain localized areas, particularly edges and corners (Ref. 96).

For the application of niobium alloys in high-temperature conditions, the use of various types of silicide or aluminide coatings is necessary (Ref. 96, 100). Elevated-temperature applications for niobium alloys B-66 and Cb-752 include gas turbine blades and vanes, shrouds, and rocket engine nozzles. B-66 and silicide-coated Cb-752 also are used in the lifting and guidance structures of glide re-entry vehicles (Ref. 31).

Refractory metals have been used in the alloying of heat-resistant nickel-base superalloys. Niobium, molybdenum, tungsten, and tantalum are often added to nickel-base superalloys for solid solution strengthening of the superalloy matrix (Ref. 101, 102). In nickel-cobalt-chromium (Ni-Co-Cr) base alloys that are used for single crystal gas turbine blades, 12% tantalum has been added. Tantalum is also widely used in the manufacture of electrolytic capacitors (Ref. 103).

For military electronic equipment applications, tungsten has substantial potential. An example is the use of tungsten in very large scale integrated circuits. Tungsten serves as a diffusion barrier (interface) between aluminum interconnecting lines and silicon chips (Ref. 104). This tungsten interface prevents the shorting of metal-oxide semiconductor circuits.

Superalloys

Superalloys are metallic materials that contain chromium and other elements to enhance resistance to oxidation and hot corrosion, and to maintain strength at high temperatures. Oxidation is defined as the reaction of an alloy with oxygen in the presence of combustion products from an uncontaminated fuel. Hot corrosion is aggressive corrosive degradation produced by oxidation and by reactions with sulfur and other contaminants (Ref. 105).

Superalloys are classified as follows:

- *Iron-base:* Iron-base superalloys are hardened by carbides, intermetallic precipitates, and/or solid solution strengthening elements such as molybdenum and tungsten. Typical alloys in this class include A-286, MULTIMET alloys' and Haynes 556.

- *Cobalt-base:* These materials contain cobalt as the major constituent with some solid solution strengthening secondary phase carbides or intermetallic compounds. Some of the cobalt-base superalloys that can be used in the 650 to 1150 °C (1200 to 2100 °F) temperature range are Haynes 25, Haynes 188, UMCo-50, and S-816.

- *Nickel-base:* These heat-resistant alloys contain approximately 30 to 75% Ni, with up to 30% Cr. Small amounts of aluminum, titanium, niobium, molybdenum, and tungsten are added to nickel-base superalloys to increase strength, corrosion resistance, or both. In order to conserve cobalt, Ni-Cr alloys have been developed. These Ni-Cr materials have the same abrasion resistance as Co-Cr alloys, but the galling and cavitation erosion resistance of Ni-Cr alloys is less than that of the Co-Cr alloys (Ref. 106).

For superalloys used mainly in high-temperature gas turbine applications, the most significant physical properties are density, expansion coefficient, and thermal conductivity.

The development of iron-base and iron-nickel-base superalloys has been much slower than that of cobalt-base and nickel-base superalloys. Nickel-base superalloys are the largest class of these materials, but cobalt-base superalloys have received considerable attention for high-temperature applications because the melting point of cobalt is 40 °C (100 °F) higher than that of nickel (Ref. 31). Most of the cobalt-base alloys that are produced, therefore, are used in high-temperature gas turbine parts (Ref. 97).

Advanced cobalt-base and nickel-base superalloys, including the mechanically alloyed (MA), oxide dispersion-strengthened (ODS) MA 6000, may be used up to 1090 °C (2000 °F) (Ref. 31, 107). The nickel-base ODS alloys, such as MA 6000, derive their strength from uniformly dispersed and stable oxides. In mechanical alloying ODS alloys are produced by a constant fracturing and rewelding of a mixture of metal powders and oxide particles.

Nickel-base superalloys also have been strengthened by precipitation of a gamma-prime (γ') phase or a gamma double prime (γ'') phase (Ref. 102, 108). Two examples of the γ phase are Ni_3Al and Ni_3Ti. The addition of cobalt to nickel-base superalloys may produce $(Ni,Co)_3(Al, Ti)$, in which cobalt substitutes for some of the nickel (Ref. 50).

The γ' phase hardens the nickel-base superalloy matrix as temperatures increase to approximately 800 °C (1470 °F). Inconel X-750 is a precipitation-hardened nickel-base superalloy. At 540 °C (1000 °F), its yield strength is about three times that of Inconel 600, which is not strengthened by intermetallic precipitation.

The creep-rupture strength of superalloys is one of the most significant properties to consider during the material selection process. The high-temperature (760 to 1200 °C, or 1400 to 2200 °F) creep-rupture characteristics of nickel-base superalloys are superior to those of either the cobalt-base or nickel-iron-base superalloys. Cast nickel-base superalloys, such as MAR-M246, maintain the highest stress-rupture strengths at elevated temperatures. Multistage heat treatments increase both the strength and ductility of cast nickel-base superalloys used in aircraft engines (Ref. 48).

Both nickel- and cobalt-base materials can be used for hardfacing applications, where resistance to abrasion and wear is necessary (Ref. 97, 106, 109, 110). These hardfacing alloys are used in equipment in which nonlubricated components move relative to each other. Nickel-base hardfacing alloys generally are used in applications where both corrosion and abrasion resistance are required. Cobalt-base hardfacing alloys, which are used in diesel engine components, are used for parts that must endure high stress metal-to-metal wear. These cobalt-base alloys are used particularly in nonlubricated parts.

The fatigue crack propagation (FCP) behavior of nickel-base superalloys is another important characteristic to consider in the selection process for these materials. The stress intensity factor (K) can be correlated to the FCP behavior of nickel-base superalloys (Ref. 111). For FCP rates exceeding about 2.5×10^{-5} mm/cycle (10^{-6} in./cycle), the data are very well represented by the Paris equation:

$$\frac{da}{dN} = C(K)^n$$

where da/dN is the measured crack growth rate, and C and n are material constants at certain temperatures, frequencies, and load ratios. For Waspaloy, the FCP rates (with respect to K) were best with material that had coarse grains and fine γ' particles (Ref. 111).

Table VII Results of 1008 h cyclic oxidation test in flowing air at temperatures from 980 °C to 1205 °C (1800 °F to 2200 °F)
Specimens were cycled to room temperature once a week.

	Oxidation rate at temperature															
	980 °C (1800 °F)				1095 °C (2000 °F)				1150 °C (2100 °F)				1205 °C (2200 °F)			
	Metal loss		Average metal affected(a)		Metal loss		Average metal affected(a)		Metal loss		Average metal affected(a)		Metal loss		Average metal affected(a)	
Alloy	mm	mils	mm	mils	mm	mils	mm	mils	mm	mils	mm	mils	mm	mils	mm	mils
Haynes alloy 214	0.0025	0.1	0.005	0.2	0.0025	0.1	0.0025	0.1	0.005	0.2	0.0075	0.3	0.005	0.2	0.018	0.7
Haynes alloy 230	0.0075	0.3	0.018	0.7	0.013	0.5	0.033	1.3	0.058	2.3	0.086	3.4	0.11	4.5	0.2	7.9
Hastelloy alloy S	0.005	0.2	0.013	0.5	0.01	0.4	0.033	1.3	0.025	1.0	0.043	1.7	>0.81	>31.7(b)	>0.81	>31.7
Haynes alloy 188	0.005	0.2	0.015	0.6	0.01	0.4	0.033	1.3	0.18	7.2	0.2	8.0	>0.55	>21.7	>0.55	>21.7
Inconel alloy 600	0.0075	0.3	0.023	0.9	0.028	1.1	0.041	1.6	0.043	1.7	0.074	2.9	0.13	5.1	0.21	8.4
Inconel alloy 617	0.0075	0.3	0.033	1.3	0.015	0.6	0.046	1.8	0.028	1.1	0.086	3.4	0.27	10.6	0.32	12.5
AISI type 310	0.01	0.4	0.028	1.1	0.025	1.0	0.058	2.3	0.075	3.0	0.11	4.4	0.2	8.0	0.26	10.3
RA333	0.0075	0.3	0.025	1.0	0.025	1.0	0.058	2.3	0.05	2.0	0.1	4.0	0.18	7.1	0.45	17.7
Haynes alloy 556	0.01	0.4	0.028	1.1	0.025	1.0	0.067	2.6	0.24	9.3	0.29	11.6	>3.8	>150.0	>3.8	>150.0
Inconel alloy 601	0.013	0.5	0.033	1.3	0.03	1.2	0.067	2.6	0.061	2.4	0.135	5.3	0.11	4.4	0.19	7.5
Hastelloy alloy X	0.0075	0.3	0.023	0.9	0.038	1.5	0.069	2.7	0.11	4.5	0.147	5.8	>0.9	>35.4	>0.9	>35.4
Inconel alloy 625	0.0075	0.3	0.018	0.7	0.084	3.3	0.12	4.8	0.41	16.0	0.46	18.2	>1.21	>47.6	>1.21	>47.6
RA330	0.01	0.4	0.11	4.3	0.02	0.8	0.17	6.7	0.041	1.6	0.22	8.7	0.096	3.8	0.21	8.3
Incoloy alloy 800H	0.023	0.9	0.046	1.8	0.14	5.4	0.19	7.4	0.19	7.5	0.23	8.9	0.29	11.3	0.35	13.6
Haynes alloy 25	0.01	0.4	0.018	0.7	0.23	9.2	0.26	10.2	0.43	16.8	0.49	19.2	>0.96	>37.9	>0.96	>37.9
Multimet	0.01	0.4	0.033	1.3	0.226	8.9	0.29	11.6	>1.2	>47.2	>1.2	>47.2	>3.72	>146.4	>3.72	>146.4
AISI type 446	0.033	1.3	0.058	2.3	0.33	13.1	0.37	14.5	>0.55	>21.7	>0.55	>21.7	>0.59	>23.3	>0.59	>23.3
AISI type 304	0.14	5.5	0.21	8.1	>0.69	>27.1	>0.69	>27.1	>0.6	>23.6	>0.6	>23.6	>1.7	>68.0	>1.73	>68.0
AISI type 316	0.315	12.4	0.36	14.3	>1.7	>68.4	>1.7	>68.4	>2.7	>105.0	>2.7	>105.0	>3.57	>140.4	>3.57	>140.4

Note: All figures shown as greater than stated value represent extrapolation of tests in which samples were consumed in less than 1008 h. (a) Average metal affected = metal loss + internal penetration.

Superalloy resistance to oxidation and high-temperature corrosion is important to consider during material selection. In Fig. 22, the relative oxidation rates of several conventional superalloys and dispersion-strengthened Ni-Cr alloys are illustrated. Dispersion-strengthened alloys contain uniformly distributed inert particles, which confer greater high-temperature strength and stability to the material (Ref. 99). The superalloys shown in this figure are both nickel- and cobalt-base.

Table VII shows oxidation data generated in flowing air covering a wide variety of commercial alloys which include stainless steels, Fe-Ni-Cr alloys (e.g., alloy 800H, RA 330), Ni-Cr-Fe alloys (e.g., alloys 600 and 601), and iron-, nickel-, and cobalt-base superalloys (Ref. 112). Many superalloys exhibited little oxidation attack at temperatures up to 1095 °C (2000 °F). At 1150 and 1205 °C (2100 and 2200 °F), however, most alloys suffered unacceptable oxidation rates. One nickel-base superalloy, Haynes 214 (Ni-16Cr-3Fe-4.5Al-Y), which relies on an aluminum oxide scale for protection, exhibited negligible oxidation (less than 1.0 mil of total depth of attack) after 1008 hours at all temperatures (980 to 1205 °C, or 1800 to 2200 °F). Alloy 214 (an alumina former) is significantly better than all the other alloys relying on chromium oxide scales for protection.

The burner rig is a widely accepted test system in the gas turbine industry for determining the oxidation resistance of alloys under dynamic conditions (*i.e.,* high combustion gas velocity). Table VIII summarizes such data for superalloys, Fe-Ni-Cr alloys, and stainless steels (Ref. 113). The specimens were exposed at the test temperature to a combustion gas stream with a velocity of about 85 m/s (280 ft/sec) generated by combustion of fuel oil, with an air to fuel ratio of about 50:1. During the test, the specimens were automatically cycled out of the hot gas steam every 30 minutes and fan cooled to less than 260 °C (500 °F) for 2 minutes before re-insertion into the hot zone. Under these dynamic conditions, superalloys in general performed better than Ni-Cr-Fe alloys (e.g., 600, 601), Fe-Ni-Cr alloys (e.g., 800H), and stainless steels. Again, alloy 214 is significantly better than all the chromia-forming alloys. Among the chromia formers, Haynes 230 (Ni-22Cr-14W-2Mo-La) performed significantly better than other superalloys.

Nickel-base superalloys with over 42% Ni are known for their resistance to chloride-induced SCC (Ref. 114). This grouping includes alloy C-276, which also is serviceable in environments which contain oxidizing agents (Ref. 115). Specific nickel-base alloys have different degrees of resistance to SCC because of variations in nickel content. Compared with austenitic stainless steels, the SCC resistance of nickel-base alloys generally is superior (Ref. 116).

Generally, superalloys have been used for construction of aircraft gas turbine parts, such as combustors and rotors. These materials also have been selected for rocket engine turbopumps. Nickel-base Inconel alloy MA754 is being used for turbine vanes in the General Electric F404 engine (Ref. 117). Inconel alloy 625 has been used in aircraft for thrust reverser plugs and ducting systems. Alloy 625 also has been applied in fuel and hydraulic line tubing, turbine shroud rings, and trust-chamber tubing for rocket motors. Inconel alloy 601 is useful for turbine engine afterburner components, exhaust liners, and bleed-air ducting systems. Because of the extremely high strength of Inconel alloy 718 from 430 to 650 °C (800 to 1200 °F), this material is applied in compressor disks and blades, compressor and turbine

Table VIII   Burner rig dynamic oxidation data for superalloys, Fe-Ni-Cr alloys' and stainless steels

| | Oxidation attack, mils | | | | | |
| | 1800 °F/1000 h | | | 2000 °F/500 h | | |
Material	Metal loss	Average metal affected	Maximum metal affected	Metal loss	Average metal affected	Maximum metal affected
Alloy 214	0.4	1.0	1.2	0.5	1.2	1.8
Alloy 230	0.8	2.8	3.5	2.2	5.2	5.7
Alloy 188	1.1	3.5	4.2	7.5	9.8	10.7
Alloy 556	1.7	4.9	6.2	8.7	10.8	11.7
Alloy X	2.7	5.6	6.4	9.0	12.9	13.5
Alloy S	3.1	5.9	6.6	11.8	13.7	15.2
Alloy RA333	2.5	6.0	7.0	4.0	8.0	8.7
Alloy 625	4.9	7.1	7.6	>31.0	>31.0	>31.0
Alloy 617	2.7	9.8	10.7	12.4	>24.0(a)	>24.0(a)
Alloy RA330	7.8	10.6	11.8	10.9	12.9	13.6
MULTIMET alloy	11.8	14.4	14.8	49.1(b)	53.8(b)	55.8(b)
Alloy 800H	12.3	14.5	15.3	30.5(c)	33.4(c)	34.0(c)
Type 310 stainless	13.7	16.2	16.5	21.2	23.7	24.1
Alloy 600	12.3(d)	14.4(d)	17.8(d)	17.2	19.5	20.7
Alloy 601	3.0	18.8	20.0	10.7	>24.0(a)	>24.0(a)
Type 304 stainless	...	...	>>23.0(e)	...	...	...
Type 316 stainless	...	...	>>23.0(e)	...	...	...

(a) Internal penetration through thickness of the sample. (b) Extrapolated from 225 h. (c) Extrapolated from 400 h. (d) Extrapolated from 917 h. (e) Consumed in 65 h.

shafts, and turbine disks. In the wrought copper combustion chamber of the space shuttle's main engine, electroformed nickel is used as the coolant channel sheath (closure) material (Ref. 118). The iron-nickel-cobalt (Fe-Ni-Co) alloy Inconel 903, which is noted for its low coefficient of expansion, has been used for turbine shrouds and turbine cases. Alloy 903 also has been specified for different parts of the space shuttle's reusable engines (Ref. 119). Inconel alloy 617 has a major application in afterburners. Nickel 200, which contains 99.5% Ni, is used for rocket motor cases (Ref. 120).

For hot section combustors in gas turbines, Hastelloy X is the most widely used alloy, particularly in commercial aircraft engines. For gas turbines in modern military aircraft, such as the F-15 and the F-16, cobalt-base Haynes 188, which offers a significant performance advantage over Hastelloy X, has been used in combustor and afterburner applications. The temperature advantage of Haynes 188 over Hastelloy X in creep strength is about 83 °C (150 °F) from 760 to 980 °C (1400 to 1800 °F). Haynes 230, a nickel-base combustor alloy, exhibits strength capabilities approaching those of alloy 188, and its oxidation and thermal stability characteristics are better than those of alloy 188 (Ref. 121).

Some applications of cobalt-base superalloys include: hot sections of engines, burner liners, high-temperature valves and springs, gas turbine rotors and buckets, turbine blades, combustion chambers, aircraft afterburner parts, rocket chambers, manifolds, post-entry lips for jet engines, nozzle rings, rocket nozzles, thrust reversers, shafts, and nozzle guide vanes (Ref. 98).

Beryllium and Copper

Beryllium is a lightweight metal used in aerospace structural applications. The strength-to-density ratio of beryllium compares favorably with other structural materials. For example, fine-grain forged beryllium has a greater strength-to-density ratio than titanium or high-strength steels. Beryllium is also selected by designers because of its high modulus-to-density ratio, thermal expansion compatibility, high heat of combustion, and good thermal conductivity. Beryllium also has several undesirable characteristics, including a very low resistance to corrosion (however, beryllium-copper alloys have very good corrosion resistance), low fracture toughness, brittle behavior, inability to produce high-strength and ductile welds, high cost, and high toxicity due to very small (about 10 µm) particles that can be ingested by the lungs (Ref. 122-124).

Pure beryllium is highly anisotropic. For example, structural-grade block beryllium has ultimate tensile strengths of 365 kPa (53 ksi) longitudinal and 386 kPa (56 ksi) transverse. Percent elongation is 2.3 longitudinal and 3.6 transverse (Ref. 125). Beryllium retains its useful mechanical properties up to moderately high temperatures. For instance, the room temperature tensile properties of beryllium SR-200 sheet begin a rapid decline at about 650 °C (1200 °F) (Ref. 31).

Beryllium is used in commercial, precipitation-hardenable copper-beryllium alloys. Content in these alloys ranges from 0.6 to 2% Be. Copper-beryllium alloys can be heat treated to produce tensile strengths up to 1462 MPa (212 ksi) (Ref. 48). Some characteristics of these casting alloys are better strength and hardness than other copper alloys, good electrical conductivity, excellent resistance to wear and corrosion, resistance to sparking, and nonmagnetism (Ref. 126).

Copper-beryllium is useful for spring material in aerospace equipment because of its high strength, high elastic limit, high fatigue strength, and good electrical conductivity (Ref. 80). The high resistance of copper-beryllium to stress relaxation at operating stresses up to 448 kPa (65 ksi) helps make this material very attractive for spring applications (Ref. 127). Copper-beryllium has been used in the manufacture of electronic interconnection devices for 127 µm (5 mil) fiber optic waveguides (Ref. 128). Although copper-beryllium alloys are higher in cost than other copper-base alloys, design engineers have selected copper-beryllium to significantly lengthen the expected service life of these electrical parts.

Pure beryllium is very attractive in military equipment, especially aerospace applications, because of its very low specific weight (Ref. 129). It has been selected for a spacecraft gimballed telescope assembly and for the primary optics of space vehicles (Ref. 130). This material also has been selected for spacecraft navigational components such as gyroscopes and accelerometers (Ref. 123).

Ceramics

Ceramic materials are solid compounds of both metallic and nonmetallic elements. Inorganic nonmetallic materials are the major constituents of ceramics, which include abrasives, porcelain, porcelain enamels, cements, glass (silicates), and carbon (graphite). The most significant characteristics of ceramic materials are high hardness, brittleness, greater resistance to high temperatures and other severe environments than metals or polymers, high compressive strength, high dielectric constant (which contributes to their good performance as electrical and thermal insulators), and low fracture toughness as compared to structural metallic materials.

Semiconductive ceramics have been developed for use in electronic components. The good electrical insulation properties of two semiconductive ceramics, silicon carbide (SiC) and zinc oxide (ZnO), make these materials useful varistors (Ref. 131). Varistors are variable (nonlinear) resistors that can be used as circuit protectors, shunts across contacts to prevent sparking, and control devices for current or voltage regulation (Ref. 132). The SiC-type semiconductors are high-temperature tolerant, as shown by the capability of these devices to operate reliably at 600 °C (1110 °F). These SiC semiconductors are useful in instrumentation and control electronics which are placed inside and on aircraft engines (Ref. 133).

High-alumina ceramics contain 80% (or more) Al_2O_3 and are used as bases for semiconductor transistors, integrated circuits, and other electronic components (Ref. 134, 135). Some of the specific types and applications of high Al_2O_3 ceramics are (Ref. 133):

- *85% Al_2O_3:* used in electronic equipment, shaft seal rings, valve trim, chokes, and nozzles
- *90% Al_2O_3, opaque:* used as a packaging material for light-sensitive electronics
- *94% Al_2O_3:* applied in transistor substrates, bases and headers; also, this ceramic is used for hybrid and integrated circuit packages (housings)
- *99.5% Al_2O_3, extra smooth:* used for electronic circuit substrates
- *99.9% Al_2O_3, translucent:* used for critical components in electronic parts

Alumina (Al_2O_3) is the oldest and most familiar of the engineering ceramics. This compound has a high degree of resistance to wear and corrosion. Along with SiC, it has been used as fiber in aluminum matrices to produce composite materials having high strength-to-weight ratios (Ref. 136). The use of ceramic whiskers or particles in matrices is increasing because of significant improvements in structural efficiency. For example, SiC-aluminum composites have an ultimate tensile strength over 690 kPa (100 ksi), a tensile modulus of 117 to 138 GPa (17 to 20×10^6 psi), and a density which is not much higher than aluminum (Ref. 137). Graphite and glass also have been used as fiber reinforcements. These reinforced materials are ideal for the reduced weight requirements of aircraft components.

Glass is an amorphous, or noncrystalline, inorganic material composed of silica (SiO_2), lime ($CaCO_3$), and sodium carbonate (Na_2CO_3). In addition to the technical ceramics

such as Al_2O_3, the nonsilicate glass systems include phosphate glasses, which are resistant to hydrofluoric acid; borate glasses, formed by the addition of alkali or alkaline earth oxides to diboron trioxide (B_2O_3) (Ref. 138); heat absorbing glasses, which are made with iron monoxide (FeO); and glasses based on oxides of aluminum, vanadium (V), germanium (Ge), and other metals (Ref. 136).

High-silica glasses increase in strength as temperature rises. Increased SiO_2 content also correlates with higher Young's modulus values for certain glass materials (fused silica and 96% silica).

Two glass materials that are used as fiber reinforcements for aerospace composites are S-glass and E-glass. The more frequently used of these materials is E-glass, which is based on lime ($CaCO_3$), Al_2O_3, and borosilicate. S-glass was developed for improved tensile strength, and it consists of silica (SiO_2), Al_2O_3, and magnesia (MgO) (Ref. 139). E-glass has a room-temperature tensile strength of 2413 MPa (350 ksi), equivalent to that of high-strength steel (Ref. 140). The room-temperature tensile strength of S-glass (3450 MPa, or 500 ksi) is significantly greater than that of high-strength steel.

Another type of glass fiber is C-glass, which was developed for acid resistance. C-glass is not used for reinforcement of aerospace composite materials.

Ceramic (vitreous) coatings also are referred to as ceramic or porcelain enamels. These coatings are applied for good protection against high-temperature oxidation and corrosion. Silicon-base ceramic coatings can be used from 540 to 760 °C (1000 to 1400 °F) (Ref. 141). This reflects much greater thermal stability and oxidation resistance compared to organic coatings, which can be used up to approximately 150 °C (300 °F).

Ceramics have been tested for many years for application in gas turbine engines as thermal barrier coatings. The most frequently tested ceramic material has been yttria-stabilized zirconia (Ref. 142-144). Duplex ceramic-metallic coatings offer the best protection against oxidation and thermal diffusion for gas turbine parts. These coatings consist of a ceramic layer (ZrO_2-Y_2O_3) over the metal substrate and a nickel-chromium-aluminum-yttrium coating over the ceramic. The layer between the substrate and outer coating may, alternatively, consist of a cermet (NiCrAlY-Y_2O_3), which has been observed to be very effective in hot corrosion protection (Ref. 145, 146).

High-temperature applications. Ceramic materials under consideration for use in heat engines are Si_3N_4, SiC, ZrO_2, and lithium-aluminum-silicate (Li-Al-SiO_2). Si_3N_4

and SiC are used in components which are subjected to severe thermal and stress-corrosion environments. Both of these ceramics are candidates for rotors, stators, and other engine components (Ref. 147). ZrO_2 is an excellent insulator with good thermal expansion characteristics, suitable for coatings on cylinder liners, piston caps, and intake/exhaust ports. $Li-Al-SiO_2$ has good thermal properties, but its strength and fatigue life are poor. Applications for $Li-Al-SiO_2$ are thus limited to nonstructural (non-load-bearing) parts in heat engines.

Plastics and Elastomers

Plastics are highly polymeric materials that will flow when heat and pressure are applied. These versatile materials are composed of repeating organic chemical units called monomers. The structural chain length of a polymer is important in determining many of the properties of a plastic. Increasing chain length produces higher toughness, greater creep resistance, better resistance to SCC, increased melt temperature and melt viscosity, and greater difficulty in processing (Ref. 148). Crystallinity in polymers, such as polyethylene and polypropylene, can significantly increase strength and rigidity (Ref. 136, 148).

Polyethylene is a thermoplastic. A thermoplastic material will soften when heat is applied to it, flow under the influence of stress, and return to its original texture (solid or rubbery) when cooled. However, some thermoplastics, including polytetrafluoroethylene and ultrahigh molecular weight polyethylene, do not flow when heat is applied.

Thermosets are cured, or hardened, into a permanent shape via an irreversible chemical reaction known as crosslinking. Heat is usually applied to carry this reaction to completion (Ref. 140). Therefore, thermosetting plastics cannot be resoftened and remolded, as thermoplastics can.

Elastomers are all of the elastic, or rubberlike, polymers. At room temperature, elastomers can be stretched (under low stress) to at least two times their original length. When the stress is removed, the elastomer returns to its normal configuration. These materials also are referred to as rubbers and thermoplastic elastomers.

The emphasis in current automotive and aerospace design is on weight and cost reduction by the replacement of metallic parts with plastics and with reinforced plastic composite materials. Some of the manufacturing savings that can be attained when plastics are substituted for metallic die castings are:

- Greater design freedom versus die castings, because thinner walls and better detail can be incorporated
- Fewer steps in the assembly process
- Elimination of anticorrosion treatments
- Elimination of expansive automatic machining lines
- Up to 50% weight reduction may be achieved in comparison with the use of a metallic material

Certain plastics can be used in structures exposed to high temperatures. Polyimides and polyamideimides are two classes of thermoplastics that can be used for parts that require high heat resistance (Ref. 149). An example of this type of application is on the wing panels of supersonic reconnaissance planes, which fly at Mach 3 (three times the speed of sound). These wing panels must withstand temperatures up to about 250 °C (480 °F).

Engineering thermoplastics are also candidates for service where resistance to flame propagation is a requirement. An excellent choice for this application is rigid vinyl. The combustion resistance of this material is very high in comparison with other organic construction materials. Once a flame source is removed, rigid vinyl does not continue to burn. Rigid vinyl also does not form flaming droplets that promote fire propagation (Ref. 150).

Acetals. The characteristics of these plastics include high stiffness, resistance to creep, and impact strengths that do not change significantly with temperature. The stiffness of acetal resins is very high among the thermoplastics. Because of their resistance to surface friction and abrasion, acetals are used in moving parts, such as gears and bearings (Ref. 151, 152).

Acrylics. These plastics are dominated by methylmethacrylate, which is of significance because it is transparent. The transparency of colorless acrylic is about the same as the best quality optical glass. Acrylics withstand weathering and are resistant to acids and alkalies. However, acrylics are not resistant to organic solvents. These thermoplastics are also subject to creep, and they have low scratch and abrasion resistance. Acrylics have been combined with PVC to form polymers that are tough, resistant to chemicals and impact, and self-extinguishing. Acrylics are used in the formulation of paints for military equipment. Acrylic paints include the low infrared reflective topcoat conforming to MIL-C-46159 (Ref. 153). An acrylic lacquer meeting the requirements of MIL-L-81352A has also been used for surface protection of military weapon systems (Ref. 154). Another typical application for acrylics is in aircraft canopies (Ref. 151).

Acrylonitrile-butadiene-styrene (ABS) resins. These materials are known for their toughness and chemical resistance. ABS plastics generally have good impact and abrasive resistance. The low temperature limit for the impact resistance of ABS plastics is –40 °C (–40 °F). One problem with ABS materials is weathering. Prolonged exposure to sunlight reduces surface gloss, impact strength, and flex resistance. A potential application for ABS resins is in electrical connectors (Ref. 151).

Alkyds. These thermosetting materials are formed by a condensation reaction between a dibasic acid (or anhydride) and a polyhydric alcohol. Following this initial reaction, a fatty acid (usually long chain) is introduced into the resin molecule (Ref. 155). Alkyds have excellent weatherability and toughness, excellent fungus resistance, and excellent dimensional stability. Alkyd resins also have excellent dielectric strength (good insulating properties). In the radio frequency and ultrahigh frequency ranges, their electrical properties are stable up to 120 to 150 °C (250 to 300 °F). Disadvantages of these materials include their low impact strength and their susceptibility to degradation by strong acids and bases. Alkyds have been used in surface coatings, circuit breaker insulation, and electrical insulation.

Chlorinated polyvinyl chloride (CPVC). This thermoplastic possesses electrical resistance superior to that of rigid PVC. It is an excellent choice for parts subjected to corrosive high-temperature service up to about 105 °C (220 °F). The thermal insulation afforded by CPVC compared with copper and steel is excellent.

Diallyl phthalates (allyls). These high-performance thermosets are important for effective electrical insulation. The high dielectric strength of allyls is maintained up to 200 °C (400 °F) and in high humidity environments (Ref. 155). Other engineering properties of allyls include greater moisture resistance than any of the other thermosets, excellent resistance to corrosion (including microbiological degradation) and chemicals, good resistance to creep at room and elevated temperatures, and ability to be self-extinguishing with addition of appropriate flame retardants. Potential applications for allyls include electrical connectors, insulators, and resistors; nose cones; radomes; and aircraft leading edges.

Epoxies. Most of these thermosetting resins are derived from bisphenol A, which is produced by a reaction between phenol and acetone (Ref. 140). Epoxy-terminated molecules, with two epoxide groups, are produced by reacting bisphenol A with epichlorhydrin. Epoxies have superior thermal resistance and dimensional stability. They are also highly immune to solvents and other chemicals. Epoxy resins are widely used in the encapsulation (casting) of electronic components. As the plastic substrates for reinforcement with glass fibers or graphite fibers, epoxies are being used for reinforced plastic composite materials.

Additional epoxy applications include surface coatings, such as the epoxy based primer that meets MIL-P-23377 (Ref. 156); structural adhesives; and foams. Widely used in electronics applications, epoxy resins have superior thermal resistance and dimensional stability. Epoxies also are highly immune to solvents and other chemicals.

Fluorocarbons and fluorosilicones. Tetrafluoroethylene (TFE), or polytetrafluoroethylene (PTFE), is the most widely known material in this classification. TFE is noted for its excellent electrical and thermal properties. The serviceable temperature range for PTFE is –100 to 260 °C (–150 to 500 °F) (Ref. 157). PTFE has good electrical properties over a wide frequency range, and these favorable electrical characteristics are similar to those of polyethylene. Fluorocarbon plastics and elastomers are highly resistant to chemicals, except gaseous fluorine at high temperatures and pressures, and molten alkali (Ref. 155, 157). These materials also have good impact strength, good weather and abrasion resistance, and relatively high thermal and hydrolytic stability.

Typical applications for TFE include nonlubricated bearings, high-temperature electronic parts, packings, gaskets, seals, and rings (Ref. 158, 159). Chlorotrifluoroethylene (CTFE) is used in fuel sight lenses, electrical insulators, and inserts. Fluorinated ethylene-propylene (FEP) finds use in wire insulation and jacketing, coils, gaskets, electrical terminals, tube sockets, and terminal insulators. Vinylidene fluoride is used in electrical insulation, seals, and gaskets. Applications for ethylene trifluoroethylene (ETFE) include electrical connectors and electrical insulation (Ref. 160). Ethylene-chlorotrifluoroethylene (E-CTFE) is used in wire and cable coatings (for applications requiring high-performance wire) and film for laminates used in aircraft interiors (Ref. 151). Aerospace applications for the fluorosilicone elastomers are O-rings for fuel systems, electrical connectors, fuel seals, and channel sealants and faying surface sealants (Ref. 158).

Furan polymers. These thermosets are products of a condensation reaction initiated by either furfural or furfuryl alcohol. The engineering characteristics of furan polymers include excellent resistance to acids, alkalies, and solvents; good wetting properties; and low viscosity, which permits these resins to penetrate semiporous products and thus add strength, weatherability, and heat resistance. Furan polymers are relatively inexpensive, and they can be used for adhesives (Ref. 155).

Methylmethacrylate has a high strength-to-weight ratio, good dimensional stability, and low water absorption. It also has good weatherability and good electrical insulation characteristics. Methylmethacrylate is easily shaped and has been widely used for aircraft canopies and windows (Ref. 157).

Nylon. This group of polymeric materials is also referred to as the polyamides. Their noteworthy characteristics include high tensile strengths (55 to 76 MPa, or 8 to 11 ksi, in some grades); high melting point; good resistance to abrasion; good toughness and impact strength; excellent high temperature performance; poor resistance to water absorption, making nylon unsuitable in damp environments when good dimensional stability or good electrical properties is required; good resistance to alkalies and organic chemicals; and poor resistance to acids and polar solvents, such as alcohols and glycols (Ref. 151, 155).

Nylon loses its physical properties and oxidizes with prolonged exposure in air above 100 °C (212 °F). Nylon is typically used in gears, cams, and other sliding contact parts; wire insulation, gaskets, high pressure flexible tubing, and rope used by cargo helicopters for lifting and transporting military equipment (Ref. 151, 159).

Phenolic resins. Important characteristics of these plastics include superior heat resistance, good flame resistance, excellent dimensional stability, good resistance to chemical agents, long shelf life, good impact strength, and high flexural modulus (Ref. 139, 155, 160).

Commercial phenolic resins, which are primarily manufactured from phenol and formaldehyde, are referred to as phenolformaldehyde resins. Phenolics have been used in the electrical components of aircraft and guided missiles (Ref. 139, 161, 162).

Polycarbonates are produced by the condensation of bisphenol A with phosgene. Polycarbonates are tough and transparent, resistant to heat and flames, and dimensionally stable. Humidity and short-term exposure to boiling water produce negligible effects on the dimensions and properties of molded polycarbonates. Polycarbonates are also noted for their creep resistance and electrical insulating properties. In general, polycarbonates are not affected by greases, oils, or acids. However, continuous long-term exposure to boiling water produces embrittlement of this resin. Crazing polycarbonates are induced by exposure to aromatic solvents, esters, and ketones.

Polycarbonate resins have been used for ducting and trim in aircraft. Additional polycarbonate applications include protective covers for electrical relays, electrical switch components, and electrical connects (Ref. 155, 162, 163).

Polyesters can be either thermosetting or thermoplastic. The thermosetting polyesters can be either very hard (brittle) and tough (resilient) or soft and flexible. One or more of the following compounds can be added to thermosetting polyesters to confer fire retardance: chlorendic anhydride, tetrabromophthalic anhydride, tetrachlorophthalic anhydride, dibromoneopentyl glycol, and chlorostryrene (Ref. 152). Thermosetting polyesters can be made resistant to chemicals by the addition of the following chemicals: neopentyl glycol, isophthalic acid, hydrogenated bisphenol A, and trimethyl pentanediol. Neopentyl glycol and methylmethacrylate can be added to thermosetting polyesters for increased weathering resistance.

Unreinforced thermoplastic polyesters have good hardness, strength, and toughness; high abrasion resistance; good chemical resistance and low moisture absorption; good resistance to fatigue and SCC; and good electrical characteristics. Typical applications for high-temperature aromatic polyesters include self-lubricating bearings, high-temperature circuit boards, and encapsulation for electronic parts, such as diodes and transistors (Ref. 151, 163). Thermoplastic polyesters can be used in gears and bearings.

Polyimides are thermoplastic resins suitable for relatively high-temperature applications. For continuous service at 260 °C (500 °F), polyimides are capable of maintaining 50 to 60% of their room-temperature strength. These polymers can be used for fabrication of high-temperature foams. Additional applications for polyimides are in gears, bushings, turbofan engine backing rings, piston rings, valve seats, and high-temperature bearings of jet engines (Ref. 164).

Polysulfone. Significant properties of polysulfone resins include a heat-deflection temperature of 175 °C (345 °F) at 1.8 MPa (260 psi), and a long-term-use temperature of 150 to 175 °C (300 to 345 °F). These resins show environmental stress-cracking resistance which is greatly improved by the addition of glass fibers (5 to 10% concentration). In addition, they have high tensile strength (70 MPa, or 10 ksi) at yield; a flexural modulus of elasticity of close to 2800 MPa (400 ksi) at room temperature; total strain (creep) well below 2% at 100 °C (210 °F) and 20 MPa (3 ksi); stable electrical properties up to 175 °C (350 °F); and susceptibility to attack by polar organic solvents such as ketones, chlorinated hydrocarbons, and aromatic hydrocarbons.

Potential applications of polysulfone resins include injection-molded printed circuit boards; housings for meters, switches, and electronic components, aircraft cabin interior parts (because of the self-extinguishing and low smoke density characteristics of polysulfone); and high-temperature curing adhesives (Ref. 151, 155, 164, 165).

Polyurethanes may be classified as either plastics or elastomers. They are extremely tough and resistant to tearing, impact, puncturing, and abrasion. The tensile strength of polyurethanes is relatively high. Polyurethanes are also noted for resistance to oil and chemical solvents. Polyurethane coating systems are used extensively in military and commercial equipment. These coatings generally consist of a mixture of pigmented polyesters and aliphatic isocyanates. Polyurethanes can also be used for the manufacture of very tough adhesives, which can be used from about –270 to 120 °C (–450 to 250 °F) (Ref. 140, 166, 167).

Polyvinyl chloride (PVC) has a high fatigue strength relative to other thermoplastics. See also the section on chlorinated polyvinyl chloride (CPVC).

Silicones have a characteristic polymeric structure containing alternating silicon and oxygen atoms. The Si-O matrix, which constitutes the backbone of silicone polymers, is responsible for the unique characteristics of these materials. The Si-O bond is significantly stronger than a C-C or C-O bond.

Significant design properties of silicones include thermal stability and a mild dependence of physical properties on temperature. For example, general purpose silicone rubber performs between 100 and 260 °C (–150 and 500 °F) without physical property deterioration. In addition, these materials show good resistance to weathering, good dielectric properties and resistance to glow discharges, resistance to ozone and ultraviolet light attack, and resistance to most chemical agents at normal ambient temperatures (Ref. 139, 168, 169). The heat resistance of silicones is superior to that of other plastics elastomers. Up to 200 °C (400 °F), it has been observed that silicone rubber maintains its tensile strength, ductility, and hardness better than other organic rubber materials (Ref. 168).

Reinforced silicones can be used for electronic component encapsulation of such items as transistors, diodes, resistors, and capacitors. Laminated silicones are useful for aircraft radomes (Ref. 160).

Synthetic rubbers. This group of elastomers includes a broad variety of compounds. Some of the most significant of these materials for military applications are described below.

Polychloroprene (neoprene) is noted for its outstanding oil and oxidation resistance. Depending on the particular formulation of neoprene, the mechanical and physical properties of this material will vary. Tensile strengths can be up to 26 MPa (4 ksi) and elongation values up to 900%. Neoprene has about the same degree of resilience and abrasion resistance as natural rubber. Applications for solid neoprene include wire and cable insulation, tube and hose covers, and power transmission belts (Ref. 157, 170, 171).

Nitrile butadiene rubber (NBR) has good resistance to oils and fuels, good tensile and elongation characteristics, heat resistance, and low compression set. The weathering properties of NBR are improved substantially by the addition of anti-ozonants. Ozone oxidizes the carbon atoms of various types of elastomers, producing deep fissures (narrow cracks in the material). This degradation is induced only when the elastomer is under a tensile stress (Ref. 172). Nitrile rubbers have very good resistance to petroleum-based materials, including aircraft fuels. An application for NBR is fuel and hydraulic components, such as hoses (Ref. 171).

Styrene butadiene rubber (SBR), also known as Buna-S, has high elongation and elastic recovery, high tensile strength, and good flexibility down to –80 °C (–120 °F), but SBR materials do soften with increasing temperature. SBRs are impervious to water, alcohol, dilute acids, or dilute alkalies, but they are soluble in ketones, esters, and many hydrocarbons. These materials show excellent stability at temperatures from 120 to 200 °C (250 to 400 °F). Applications for SBR include adhesives, sealants, tire cord, and tubing (Ref. 155, 171).

Acrylic rubbers (AR) are copolymers of acrylic esters and olefins. These rubbers do not include acrylonitrile. AR can tolerate temperatures from –40 to 200 °C (–40 to 400 °F). They are used in seals, O-rings, packings, and adhesives (Ref. 171).

Epichlorohydrin rubber (CO and ECO) has a polyether backbone to which chloromethyl groups are attached. These elastomers have excellent resistance to the diffusive action of gases. CO and ECO can be used where resistance to solvents, fuel, oil, and ozone is required (Ref. 171).

Ethylene-propylene-diene monomer (EPDM) is a fairly new class of elastomers. Although they have poor resistance to oil and hydrocarbons, EPDM elastomers have outstanding resistance to ozone and weathering. EPDM is useful for coverings on electrical wires (Ref. 171, 172).

Fluorocarbon rubbers (FCRs) have especially favorable properties as a result of their high fluorine content. They are resistant to strong acids, oils, chemical solvents, and ozone. FCRs are generally not resistant to strong alkalies. FCRs are also stable up to 220 to 230 °C (425 to 450 °F), and they have good toughness and electrical properties. Applications for FCRs include seals, O-rings, and foam (Ref. 171).

Composite Materials

Composite materials are in use for an increasing number of commercial and military equipment applications. Based on a conservative extrapolation of this growing trend in composites usage, between 35 and 45% of the next generation of military aircraft structures will be made of composite materials.

One major purpose for the selection of composite materials is to obtain desirable structural engineering properties. Among these favorable characteristics are reduced weight (especially for aircraft), higher strength:weight ratios, improved stiffness (modulus):density ratios, and better corrosion resistance. Attainment of these properties will ensure that the equipment system performs with greater mechanical efficiency and durability. Various types of structural composite materials have been developed, primarily for use in aerospace equipment.

Fiber-reinforced plastic (FRP) is any polymeric (resinous) material reinforced with fibers. Reinforcement of plastic materials will improve or modify the properties of these materials. One type of FRP composite material is a fiber-reinforced thermosetting plastic known as sheet molding compound (SMC). SMC does not require processing (such as drying of volatiles or advancement of cure) after manufacture to prepare it for use at the molding press, and it can be molded without producing reaction byproducts. The amount of pressure required to mold SMC is just enough to make the material flow and compact. (High pressure is unnecessary.) SMC has been produced as a polyester-glass fiber (Ref. 173, 174).

Another type of FRP is premix, which also has been called dough molding compound, flow mix, or bulk molding compound. Premix is similar to SMC because both compounds can be molded without excessive pressure or reaction byproducts.

Composite laminates are manufactured by bonding two or more superimposed sheets of reinforcing fibers. Laminates are usually formed by applying heat and pressure to the layered structure (Ref. 155, 175).

Structural sandwich is formed by bonding two thin facings (layers) to a thick core. Nearly all the bending rigidity of the structural sandwich is provided by the facings. The core material separates the facings, transmits shear between the facings, and provides most of the shear rigidity of the sandwich construction. Structural sandwich composites having high stiffness:weight ratios can be manufactured through proper selection of facing and core materials. Lightweight sandwich cores are usually made of low-density material (Ref. 176).

Adhesives are used in sandwich composite construction to bond facings to inserts, such as reinforcing plates and edge strips. Most of these adhesives are formulations of organic resins, which provide high-strength bonds over a wide range of mechanical loading conditions. Inorganic (ceramic) adhesives have also been used in the bonding of composite sandwiches.

Structural sandwich facing material may consist of any thin sheet that is capable of carrying the major applied structural loads. Some of the materials used for facings are metals, including aluminum alloys, steel, titanium alloys, magnesium alloys, nickel-base alloys, cobalt-base alloys, niobium alloys, molybdenum alloys, and beryllium; reinforced plastic materials; and advanced composite materials.

Metal-matrix composite (MMC) materials are advanced composites that consist of a metallic matrix reinforced with high-strength and high-stiffness fibers. MMCs are called advanced composites because they have potential for use in high-performance military equipment.

Current MMCs generally consist of either a reinforced aluminum or magnesium matrix. Aluminum-matrix composites are usually reinforced with either boron (B), silicon carbide (SiC), borsic (Si-coated B), or graphite filaments. Magnesium-matrix composites are usually reinforced with alumina (Al_2O_3). The use of MMCs is minor in comparison to other advanced reinforced plastic composites, particularly graphite and aramid-Kevlar composites (Ref. 177-180).

Fiber-reinforced ceramic composites are currently in the experimental phase of development. The most successful efforts in the manufacture of ceramic composites have been with glass or glass-ceramic matrices and ceramic fiber reinforcements. Recent tests on graphite-fiber, glass-matrix composites have shown that the high strength and fracture toughness of these materials can be retained up to 593 °C (1110 °F). Reinforcing fibers, composed of alumina (Al_2O_3) and SiC, have been effective in improving the strength and toughness of glass matrices. However, ceramic matrices such as Al_2O_3 and magnesium oxide reinforced with carbon fibers have suffered from reduced strength despite their high toughness. Low strengths also have been observed in SiC-reinforced silicon nitride (Si_3N_4) ceramic composites (Ref. 181).

Carbon-carbon (CC) composite materials have been developed for very high-temperature applications up to 2800 °C (5000 °F). Because the mechanical properties of CC composites can be tailored to particular applications, these materials can be manufactured in various forms. CC composites are well suited for high-temperature applications such as the thermal protection system of the space shuttle because these materials do not lose their strength as the temperature increases to about 2200 °C (4000 °F) (Ref. 182-184).

References

1. L.M. Thompson, A. Bauer, and D. Brodowsky, "Rear Axle Designed in Oriented FRP," *Automotive Engineering*, Vol. 90, No. 8, 1982, p 74

2. R.B. Gunia, Ed., *Source Book on Materials Selection,* Vol. 1, American Society for Metals, 1977

3. M.M. Farag, *Materials and Process Selection in Engineering*, Applied Science Publishers, Ltd., 1979

4. M.F. Ashby, "Materials Selection in Conceptual Design," *Materials and Engineering Design: The Next Decade*, B.F. Dyson and D.R. Hayhurst, Ed., The Institute of Metals, 1989

5. M.F. Ashby, "On the Engineering Properties of Materials," *Acta Metallurgica*, Vol. 37, No. 5, 1989, pp 1273-1293

6. C.J. McMahon, Jr., "The Microstructural Aspects of Tensile Fracture," *Fundamental Phenomena in the Materials Sciences, Vol. 4, Fracture of Metals, Polymers, and Glasses*, L.J. Bonis, J.J. Duga, and J.J. Gilman, Ed., Plenum Press, 1967, pp 247-284

7. E.T. Wessel, W.G. Clark, and W.K. Wilson, "Engineering Methods for the Design and Selection of Materials Against Fracture," US Army Tank-Automotive Center, June 1966

8. R.H. Williams, "Failure Analysis and Corrosion Control," *National SAMPE Technical Conference, Vol. I, Aircraft Structures and Materials Application*, September 9-11, 1969, pp 123-139

9. W.T. Kirkby, "Design Against Fatigue as a Contribution to Reliability," Technical Report 69063, Royal Aircraft Establishment, March 1969

10. I. Granet, *Modern Materials Science*, Reston Publishing Company, Inc., p 36

11. B.J. Pendley, S.P. Henslee, and S.D. Manning, "Durability Methods Development, Volume 3, Structural Durability Survey: State-of-the-Art Assessment," Technical Report AFFDL-TR-79-3118, Air Force Flight Dynamics Laboratory, Wright-Patterson Air Force Base, 1979

12. D.T. Curry, *Machine Design*, Vol. 54, No. 12, 1982, pp 79-82

13. L.C. Rowe, *Automotive Engineering*, Vol. 82, No. 2, 1974, pp 40-45

14. M. Emrich, *Assembly Engineering*, Vol. 25, No. 10, 1982, pp 36-40

15. B. Cole and E.J. Bateh, "Special Fastener Development for Composite Structure," Report AFWAL-TR-82-3049, Air Force Wright Aeronautical Laboratories, Wright-Patterson Air Force Base, 1982

16. J.A. Kargol, N.F. Fiore, and W.T. Ebihara, "Hydrogen Behavior in Coated and Uncoated Low Alloy Steels," AFWAL-TR-81-4019, *Proceedings of the 1980 Tri-Service Corrosion Conference*, F.H. Meyer, Jr., Ed., 1981

17. B.A. Miller, "The Galvanic Corrosion of Graphite Epoxy Composite Materials Coupled with Alloys," Report GAE/MC/75D-8, AFIT Thesis, Air Force Institute of Technology, Wright-Patterson Air Force Base, 1975

18. M. Doruk, "Some Observations on the Corrosion of Aircraft at the Air Force Base in Bandirma, Turkey," Paper 4, *AGARD Conference Proceedings No. 315, Aircraft Corrosion*, 1981

19. H.J. Versteegen and M.J.M. Versteeg, "Design and Maintenance Against Corrosion of Aircraft Structures," *AGARD Conference Proceedings No. 315, Aircraft Corrosion*, 52nd Meeting of the AGARD Structures and Materials Panel, Cesme, Turkey, April 5-10, 1981

20. J.F. Jenkins, "Evaluation of Objects Exposed to Deep Ocean Environments," *Proceedings of the Tri-Service Corrosion of Military Equipment Conference*, AFML-TR-75-42, Vol. 2, Air Force Materials Laboratory, Wright-Patterson Air Force Base, 1975, pp 75-97

21. J.H. Bickford, *An Introduction to the Design and Behavior of Bolted Joints*, 2nd ed., pp 86-87

22. E.E. Fletcher, *High-Strength, Low-Alloy Steels: Status, Selection, and Physical Metallurgy*, Battelle Press, Battelle Columbus Laboratories, Inc., 1979

23. C.R. Weymueller, "Choosing the Right EX Steels," *Metal Progress*, Vol. 98, No. 4, 1970, pp 130-134

24. W.T. Groves and L.J. Vande Walle, in *Metal Progress*, June 1981

25. S.K. Coburn, "Designing to Prevent Corrosion," *Materials Protection*, Vol. 6, No. 2, 1967, pp 33-39

26. *Chemical Engineering*, Vol. 89, No. 13, 1982

27. *Chemical Engineering*, Vol. 87, No. 22, 1984, pp 82-131

28. M. Woeful and R. Mulhall, "Glass Bead Impact Blasting," *Metal Progress*, Vol. 122, No. 4, 1982, pp 57-59

29. D.R. McIntyre, "How to Prevent Stress-Corrosion Cracking in Stainless Steels—II," *Chemical Engineering*, Vol. 87, No. 9, 1980, pp 131-136

30. R.O. Williams, "The Effect of Microstructure on the Susceptibility of Low-Alloy Steels to Hydrogen Attack," ORNL-5781, Oak Ridge National Laboratory, August 1981

31. J.K. Stanley and H. Smallen, "Advanced Metallic Materials for Aerospace Applications," Report SAMSO-TR-69-355, The Aerospace Corp., September 1969

32. C.A. Paez, "Present and Future Developments in Aerospace Materials and Structures," *Proceedings of the 10th National SAMPE Technical Conference*, Kiamesha Lake, NY, Vol. 10, October 1978, pp 51-62

33. C.N. Cochran and R.H.G. McClure, "Material Decisions Based on Economics and Energy," *Automotive Engineering*, Vol. 90, No. 7, 1982, pp 29-34

34. J. Reardon, "New Image for Polymer Composites," *Machine Design*, 1990, Vol. 62, No. 1, pp 48-56

35. S.T. Rolfe and J.M. Barsom, *Fracture and Fatigue Control in Structures: Applications of Fracture Mechanics*, Prentice-Hall, 1977

36. G.V. Bennett, "Lower-Cost Structure by Substituting AF1410 Steel for Titanium," *Proceedings of the 10th National SAMPE Technical Conference*, Kiamesha Lake, NY, Vol. 10, October 1978, pp 91-99

37. R.B. Ross, *Metallic Materials Specifications Handbook*, 3rd ed., F.N. Span, London, 1980

38. C.H. Samans, *Chemical Engineering*, Vol. 74, No. 4, 1968, pp 150-160

39. W.W. Gerberich, R.H. Van Stone, and A.W. Gunderson, "Fracture Properties of Carbon and Alloy Steels," *Application of Fracture Mechanics for Selection of Metallic Structural Materials*, J.E. Campbell, W.W. Gerberich, and J.H. Underwood, Ed., American Society for Metals, 1982

40. Military Specification MIL-A-83444, "Airplane Damage Tolerance Requirements," U.S. Air Force, Aeronautical Systems Division, ASD/ENFS, Wright-Patterson Air Force Base, 1974

41. D.L. Torkington, *Metal Progress*, Vol. 119, No. 6, 1981, pp 38-43

42. W.H. Moore, *Casting Engineering and Foundry World*, Vol. 14, No. 3, 1982, pp 42-50

43. R.L. Baker, *Casting Engineering and Foundry World*, Vol. 14, No. 3, 1982, pp 15-18

44. "Stainless Steels for Bulk Materials Handling," Committee of Stainless Steel Producers, American Iron and Steel Institute, 1980

45. "Standard Recommended Practice for Cleaning and Descaling Stainless Steel Parts, Equipment and Systems," ANSI/ASTM A 380-78, ASTM, 1978

46. C.A. Smith, *Steel Times,* Vol. 208, No. 4, 1980, pp 274-282

47. "Selection of Stainless Steels," Universal-Cyclops Specialty Steel Division, Cyclops Corp., 1966

48. W.F. Smith, *Structure and Properties of Engineering Alloys,* McGraw-Hill, 1981

49. E.L. AuBuchon and R.V. London, *Metal Progress,* Vol. 119, No. 6, 1981, pp 35-37

50. D.B. Dawson, M. Levy, and D.W. Seitz, Jr., "Stress Corrosion Cracking of High Hardness Steel Armor," *Proceedings of the Tri-Service Corrosion of Military Equipment Conference,* Report AFM-TR-75-42, Vol. 2, 1974

51. T. Crooker, "Crack Propagation in Aluminum Alloys Under High Amplitude Cyclic Load," NRL Report 7286, Naval Research Laboratory, 1971

52. "Steel Wire: Review of a Versatile Design Material," *Design News,* Vol. 39, No. 5, 1983, pp 94-98

53. "Galvanizing Steel: New Lease on Life for Mature Process," *Modern Metals,* Vol. 40, No. 1, 1983, pp 64-74

54. S.J. Ketcham and J.J. DeLuccia, "Recent Developments in Materials and Processes for Aircraft Corrosion Control," *North Atlantic Treaty Organization Advisory Group for Aerospace Research and Development (AGARD) Conference Proceedings No. 315,* Paper 11, 1981

55. "The Selection and Application of Aluminum and Aluminum Alloys," ASM Committee on Applications of Aluminum, *ASM Handbook,* 10th ed., Vol. 1: *Properties and Selection,* ASM International, 1990

56. M.A. Steinberg, "Materials for the New Generation of Aircraft," *The Science of Materials Used in Advanced Technology,* E.R. Parker and U. Colombo, Ed., John Wiley and Sons, 1973

57. *Aluminum Standards and Data 1979,* 6th ed., The Aluminum Association, Inc., March 1979

58. "Designing with Aluminum Alloy Forgings," *Precision Metal,* Vol. 41, No. 3, 1983, pp 15-18

59. R.W. Hertzberg, *Deformation and Fracture Mechanics of Engineering Materials,* 3rd ed., John Wiley and Sons, 1990

60. R.F. Simenz and M.K. Guess, "Structural Aluminum Materials for the 1980's," *Journal of Aircraft,* Vol. 17, No. 17, 1980, pp 415-520

61. W. Minter, "Corrosion/Stress Corrosion Suspect Materials List," Sikorsky Aircraft Div., United Technologies Corp., 1983

62. D.O. Sprowls, "High Strength Aluminum Alloys with Improved Resistance to Corrosion and Stress-Corrosion Cracking," *Proceedings of the 1976 Tri-Service Conference on Corrosion,* S.J. Ketcham, Ed., Metals and Ceramics Information Center, MCIC-77-33, Battelle Columbus Laboratories, December 1977, pp 89-120

63. Reynolds Aluminum Mill Products, Aluminum Distributors, Inc., 1977

64. "Mechanical Properties of Sand Cast Aluminum Alloy Test Castings," *Casting Engineering and Foundry World,* Vol. 14, No. 3, 1982, pp 67-71

65. M.B. Hyatt, W.E. Quist, and J.T. Quinlivan, "Improved Aluminum Alloys for Airframe Applications," *Metal Progress,* Vol. 111, No. 3, 1977, pp 56-59

66. "Wrought Aluminum P/M Alloys," *Metal Progress,* Vol. 122, No. 6, 1982, pp 47-49

67. "Wrought P/M Aluminum Alloys," Aluminum Company of America, 1980

68. J.T. Skelly, *Metal Progress,* Vol. 124, No. 3, 1983, pp 35-38

69. "The Universal Fastener," *Design News,* Vol. 39, No. 18, 1983, pp 48-56

70. D.P. Hanley, *Introduction to the Selection of Engineering Materials,* Van Nostrand Reinhold Company, 1980

71. "Magnesium—Designing Around Corrosion," The Dow Chemical Company, Inorganic Chemicals Department, 1982

72. I.M. Scolaris, "Corrosion Preventing Methods Developed from Direct Experience with Aerospace Structures," *AGARD Conference Proceedings No. 315,* Aircraft Corrosion, Paper 9, April 1981

73. Military Specification MIL-F-8615D, "Fuel System Components, General Specification for," US Dept. of the Air Force, March 19, 1976

74. Military Specification MIL-F-38363B, "Fuel System, Aircraft General Specification for," US Dept. of the Air Force, October 26, 1981

75. "NASA Shuttles with Zinc," *Zinc!,* No. 2, Zinc Institute, 1982

76. J.R. Steinmetz, "The Use of Ethyl Silicate in Zinc-Rich Paints," *Modern Paint and Coatings,* June 1983, pp 48-49

77. "Zinc Aluminum Alloys ZA-8, ZA-12, ZA-27," Certified Alloys Company, 1982

78. K.J. Altorfer, *Design News,* Vol. 38, No. 15, 1982, pp 43-50

79. H.W. Rosenberg, J.C. Chesnutt, and H. Margolin, "Fracture Properties of Titanium Alloys," *Application of Fracture Mechanics for Selection of Metallic Structural Materials,* J.E. Campbell, W.W. Gerberich, and J.H. Underwood, Ed., American Society for Metals, 1982, pp 213, 252

80. Military Standardization Handbook, MIL-HDBK-5D, *Metallic Materials and Elements for Aerospace Vehicle Structures,* Vol. 2, Department of Defense, June 1983

81. "Titanium: The Solution to the Corrosion Problem in FGD Scrubber Systems," Titanium Industries, 1982

82. A.E. Leykin, E.S. Porotsky, and B.I. Rodin, *Science of Aviation Material,* Foreign Technology Division, Wright-Patterson Air Force Base, 1967

83. N.F. Harper, *The Metallurgist and Materials Technologist,* Vol. 13, No. 5, 1981, pp 259-263

84. D.G. Treadway, "Corrosion Control at Graphite/Epoxy-Aluminum and Titanium Interfaces," Technical Report AFML-TR-74-150, Air Force Materials Laboratory, Wright-Patterson Air Force Base, 1974

85. B.A. Miller and S.G. Lee, "The Effect of Graphite-Epoxy Composites on the Galvanic Corrosion of Aerospace Alloys," Technical Report AFML-TR-76-121, Wright-Patterson Air Force Base, 1976

86. J.L. Gossett, *Chemical Engineering,* Vol. 89, No. 23, 1982, pp 143-146

87. R.A. Wood and R.J. Favor, *Titanium Alloys Handbook,* MCIC-HB-02, Metals and Ceramics Information Center, Battelle Columbus Laboratories, 1972

88. C.A. Paez and R. Gordon, *Journal of Aircraft,* Vol. 18, No. 9, 1981, pp 712-717

89. M. Bond, *Engineering,* Vol. 219, No. 10, 1979, pp I-VI

90. J.F. Collins and W.T. Highberger, *Metal Progress,* Vol. 119, No. 4, 1979, pp 79-83

91. H.E. Chandler, *Metal Progress,* Vol. 117, No. 4, 1977, pp 41-49

92. H.D. Hanes and J. McFadden, *Metal Progress,* Vol. 123, No. 5, 1983, pp 23-31

93. R.A. Wood, "Titanium Utilization and Availability, Part I," *Current Awareness Bulletin,* Issue 102, Metals and Ceramics Information Center, Battelle Columbus Laboratories, August 31, 1981, pp 1-4

94. J.L. McCall, "Highlights of the Briefing/Workshop on Rapid Solidification Technology," *Current Awareness Bulletin,* Issue 103, Metals and Ceramics Information Center, Battelle Columbus Laboratories, September 25, 1981, p 304

95. F.S. Shuker, *Chemical Engineering,* Vol. 90, No. 9, 1983, pp 81-84

96. "Refractory Metals and Alloys," *ASM Handbook,* 10th ed., Vol. 2, ASM International, 1990, pp 557-582

97. J.A. Ford, *Industrial Research and Development,* Vol. 24, No. 12, 1982, pp 80-83

98. F.R. Morral, "Cobalt-Base Alloys in Aerospace," *Aircraft Structures and Materials Application,* National SAMPE Technical

Conference, Vol. 1, Society of Aerospace Material and Process Engineers, 1969

99. K.I. Collier, C.L. Ramsey, C.L. Barnett, and J.C. Ingram, Jr., "Aerospace Structural Potential of Beryllium, Dispersion-Strengthened Metals, and Tantalum," Air Force Flight Dynamics Laboratory (AFFDL), Report AFFDL-TR-68-51, Wright-Patterson Air Force Base, 1968

100. A. Olevitch, Ed., "Emerging Aerospace Materials and Fabrication Techniques," Technical Report AFML-TR-67-1, Air Force Materials Laboratory (AFML), Wright-Patterson Air Force Base, 1967

101. *High-Temperature, High-Strength Nickel-Base Alloys,* 3rd ed., International Nickel Company, Inc., 1977

102. R.F. Decker, *Strengthening Mechanisms in Nickel-Base Superalloys,* International Nickel Company, Inc., 1970

103. B.W. Gonser, "Tantalum Utilization and Availability," *Current Awareness Bulletin,* No. 109, Metals and Ceramics Information Center, Battelle Columbus Laboratories, March 26, 1982, pp 1-3

104. P.A. Gargini, *Industrial Research and Development,* Vol. 25, No. 3, 1983, pp 141-147

105. M.J. Wahll, D.J. Maykuth, and H.J. Hucek, *Handbook of Superalloys,* Metals and Ceramics Information Center, Battelle Press, 1979

106. P. Crook and A. Asphahani, *Chemical Engineering,* Vol. 90, No. 1, 1983, pp 127-132

107. Y.G. Kim and H.F. Merrick, "Characterization of an Oxide-Dispersion-Strengthened Superalloy, MA6000E, for Turbine Blade Applications," International Nickel Company, 1979

108. N.S. Stoloff, Ed., "Wrought and P/M Superalloys," *ASM Handbook,* 10th ed., Vol. 1, ASM International, 1990, pp 951-980

109. L.H. Price, *Metal Progress,* Vol. 124, No. 3, 1983, pp 21-27

110. K.J. Bhansali and W.L. Silence, *Metal Progress,* Vol. 115, No. 11, 1977, pp 39-43

111. S.D. Antolovich and J.E. Campbell, "Fracture Properties of Superalloys," *Application of Fracture Mechanics for Selection of Metallic Structural Materials,* T.E. Campbell, W.W. Gerberich and J.H. Underwood, Ed., American Society for Metals, 1982, pp 253-310

112. *Metals Handbook,* 9th ed., Vol. 13, *Corrosion,* ASM International, 1987, p 1312

113. G.Y. Lai, Haynes International, Inc., unpublished data, 1987

114. R.W. Kirchner, *Chemical Engineering,* Vol. 90, No. 19, 1983, pp 81-86

115. F.G. Hodge, *Industrial Research and Development,* Vol. 25, No. 7, 1983, pp 82-85

116. J. Kolts, J.B.C. Wu, and A.I Asphahani, *Metal Progress,* Vol. 124, No. 4, 1983, pp 25-36

117. H.E. Chandler, "Superalloy Update," *Metal Progress,* Vol. 123, No. 7, 1983, pp 21-28

118. "Case History Study, North American Rockwell," Carboline Company, Inc.

119. "Huntington Alloys Creating Change in Aerospace Applications," Huntington Alloys, Inc.

120. "Quick Reference Guide to High-Nickel Alloys," Huntington Alloys, Inc.

121. M.F. Rothman, *World Aerospace Profile 1988,* Arthur Reed and Roy Allen, Ed., Sterling Publications, Ltd., 1988, p 83

122. L.L. Soffa and G.J. Basl, "High-Strength Forged Beryllium for Structural Applications," Report REK DH-005, Rocketdyne Corp., 1965

123. M.L. Bauccio, "Properties and Applications of Beryllium, Beryllium-Copper and Beryllium Oxide Aerospace Materials," CORROSION/84, Preprint 109, National Association of Corrosion Engineers, 1984

124. M. Hunt, "Surprising Beryllium," *Materials Engineering,* Vol. 105, No. 11, 1988, pp 46-49

125. N.P. Pinto, "Properties," *Beryllium Science and Technology,* Vol. 2, D.R. Floyd and J.N. Lowe, Ed., Plenum Press, 1979, pp 319-347

126. C. Lorenz, "Beryllium-Copper Usage Growing in Auto Industry," *Design News,* Vol. 38, No. 15, 1982, pp 52-56

127. "Berylco® Alloys, A Guide to Their Selection and Use," Bulletin 106 PD1, Cabot Berylco, October 1982, p 3

128. D. Goldstein and J. Tydings, "A Connector-Like Device for Joining Optical Fibers," *Twelfth Annual Connector Symposium Proceedings,* Cherry Hill, NJ, October 1979, pp 214-218

129. F.E. Stone, "Design Considerations," *Beryllium Science and Technology,* Vol. 2, D.R. Floyd and J.N. Lowe, Ed., Plenum Press, 1979, pp 379-415

130. M. Garin and L.A. Grant, "Design and Fabrication of Brazed Beryllium Assemblies," 83-0868-CP, 24th Annual AIAA ADM Conference, Lake Tahoe, NV, May 1983

131. S. Hayakawa, *Industrial Research and Development,* Vol. 25, No. 2, 1983, pp 142-147

132. M. Sapoff, "Varistor," *The Encyclopedia of Electronics,* C. Susskind, Ed., Reinhold Publishing Corp., 1962, p 911

133. W.C. Nieberding, *Industrial Research and Development,* Vol. 25, No. 9, 1983, pp 148-150

134. D.J. Godfrey, M.W. Lindley, E.R.W. May, and R.L. Brown, *Engineering Ceramics,* Admiralty Materials Laboratory, U.K., 1971

135. L.E. Ferreira, D.D. Briggs, and R.G. Barnhart, "High-Alumina Ceramics," *Source Book on Materials Selection, Vol. II,* R.B. Gunia, Ed., American Society for Metals, 1977, pp 404-408

136. "Materials Reference Issue," *Machine Design,* Vol. 49, No. 6, 1977

137. H.E. Chandler and D.F. Baxter, *Metal Progress,* Vol. 123, No. 1, 1983

138. W.D. Kingery, H.K. Bowen, and D.R. Uhlmann, *Introduction to Ceramics,* 2nd ed., John Wiley and Sons, Inc., 1976

139. Military Handbook MIL-HDBK-17A, *Plastics for Aerospace Vehicles: Part 1: Reinforced Plastics,* Department of Defense, 1971

140. W.E. Driver, *Plastics Chemistry and Technology,* Van Nostrand Reinhold Company, 1977

141. P.E. France, *Chemical Engineering,* Vol. 90, No. 13, 1983, pp 61-63

142. S.K. Anderson, S.K. Lau, R.J. Bratton, S.Y. Lee, K.L. Rieke, J. Allen, and K.E. Munson, "Advanced Ceramic Coating Development for Industrial/Utility Gas Turbine Applications," NASA CR-165619, National Aeronautics and Space Administration, 1982

143. R.J. Keller, "Research and Development for Improved Thermal Barrier Coatings," ER-8216-1, Air Force Systems Command, Aeronautical Systems Division, Wright-Patterson Air Force Base, 1982

144. J.W. Vogan and A.R. Stetson, "Advanced Ceramic Coating Development for Industrial/Utility Gas Turbines," NASA CR 169852, National Aeronautics and Space Administration, 1982

145. D.J. Bak, *Design News,* Vol. 39, No. 1, 1983

146. M.A. Gedwill, T.K. Glasgow, and S.R. Levine, *Thin Solid Films,* Vol. 95, 1982, pp 66-72

147. M.R. Pasucci, "The Role of Ceramics in Engines—An Assessment," *Current Awareness Bulletin,* Metals and Ceramics Information Center, Battelle Columbus Laboratories, Issue 126, August 1983, pp 1-4

148. S.L. Rosen, *Fundamental Principles of Polymeric Materials,* John Wiley and Sons, Inc., NY, 1982, p 42

149. M. Bakker, *Plastics Design Forum,* Vol. 8, No. 6, 1983, pp 83-85

150. "Designing with Thermoplastics," Chemical Group, The BF Goodrich Company, 1980

151. *Plastics Engineering: Handbook of the Society of the Plastics Industry, Inc.,* 4th ed., J. Frados, Ed., Van Nostrand Reinhold Company, 1976, p 29

152. *Handbook of Plastics and Elastomers,* C.A. Harper, Ed., McGraw-Hill, Inc., 1975

153. Military Specification MIL-L-46159A, "Lacquer, Acrylic, Low Reflective, Olive Drab," U.S. Army Materials and Mechanics Research Center, 1977

154. Military Specification MIL-L-81352, "Lacquer, Acrylic (for Naval Weapons Systems)," U.S. Naval Air Systems Command, 1973

155. Military Standardization Handbook MIL-HDBK-700A, *Plastics,* Department of Defense, March 1975

156. Military Specification MIL-P-23377, "Primer Coating, Epoxy Polyamide, Chemical and Solvent Resistant," U.S. Army Materials and Mechanics Research Center, March 1978

157. H.J. Sharp, *Engineering Materials, Selection and Value Analysis,* American Elsevier Publishing Company, Inc., 1966

158. D.A. Strivers, "Fluoroelastomers," *The Vanderbilt Rubber Handbook,* R.O. Babbit, Ed., R.T. Vanderbilt Company, Inc., 1978, pp 244-258

159. R.C. Beercheck, *Machine Design,* Vol. 54, No. 5, 1982, pp 157-161

160. "Mechanical and Physical Properties of Engineering Plastics," *Source Book on Materials Selection, Vol. II,* R.B. Gunia, Ed., 1977, pp 400-403

161. *Plastics Design Forum,* Vol. 8, No. 6, 1983, pp 75-76

162. D.V. Rosato and G. Lubin, "Plastics in Aircraft and Aerospace," *Handbook of Fiberglass and Advanced Plastics Composites,* G. Lubin, Ed., Van Nostrand Reinhold Company, 1969, pp 801, 832

163. H.L. Thomsa, *Industrial Research and Development,* Vol. 25, No. 6, 1983, pp 96-99

164. J.N. Anderson, "Designing with High Performance Plastics in Off the Road and Automotive Equipment Applications," Society of Automotive Engineering, Technical Paper 810-968, 1981

165. L.T. Manzione, *Plastics World,* Vol. 41, No. 11, 1983, pp 57-59

166. Military Handbook MIL-HDBK-17, *Plastics for Aerospace Vehicles, Part II: Transparent Glazing Materials,* Air Force Materials Laboratory, Wright-Patterson Air Force Base, June 1977

167. Military Specification MIL-C-46168B, "Coating, Aliphatic Polyurethane, Chemical Agent Resistant," U.S. Army Mobility Equipment Research and Development Command, DRDME-DS, 1982

168. Military Specification MIL-C-83286B, "Coating, Urethane, Aliphatic Isocyanate, for Aerospace Applications," US Air Force Materials Laboratory, Wright-Patterson Air Force Base, 1980

169. B.B. Hardman and R.W. Shade, *Materials Technology,* General Electric Company, 1980

170. A.R. Mersberg and J.W. Lee, *Materials Performance,* Vol. 19, No. 12, 1980, pp 13-17

171. S.W. Schmitt, "The Neoprenes," *The Vanderbilt Rubber Handbook,* R.O. Babbit, Ed., R.T. Vanderbilt Company, 1978, pp 137-146

172. Military Handbook MIL-HDBK-149A, *Rubber and Rubber-Like Materials,* U.S. Army Materials and Mechanics Research Center, June 1965, pp 63-64

173. D.A. Riegner and J.C. Hau, *Automotive Engineering,* Vol. 90, No. 5, 1982, pp 51-57

174. P.R. Young, "Reinforced Molding Compounds," *Handbook of Fiberglass and Advanced Plastics Composites,* G. Lubin, Ed., Van Nostrand Reinhold Company, 1969, pp 369-420

175. C.A. Harper, Ed., *Handbook of Plastics and Elastomers,* McGraw-Hill Book Co., 1975

176. Military Handbook MIL-HDBK-23A, *Structural Sandwich Composites,* Department of Defense, 1968

177. E.L. Foster, "Technological Development of Metal Matrix Composites for DOD Application Requirements; Part 2: Findings and Recommendations, Including the Proceedings of the Second and Third MMC Workshops," Report P-1177, Defense Advanced Research Projects Agency, February 1977

178. "Advanced Composite Repair Guide," Report NOR 82-60, Air Force Wright Aeronautical Laboratories, Air Force Systems Command, Wright-Patterson Air Force Base, March 1982

179. *Advanced Composites Design Guide, Vol. 1: Design,* Structures Division, Air Force Flight Dynamics Laboratory, Air Force Systems Command, Wright-Patterson Air Force Base, September 1976

180. E.M. Lenoe, "Comments on the Status of Composite Structures Technology," *Fibrous Composites in Structural Design,* E.M. Lenoe, D.W. Oplinger, and J.J. Burke, Ed., Plenum Press, 1980

181. D.K. Shelty, "Ceramic Matrix Composites," *Current Awareness Bulletin,* No. 118, Metals and Ceramics Information Center, Battelle Columbus Laboratories, 1982

182. D.J. Holt, *Aerospace Engineering,* Vol. 3, No. 4, 1983, pp 14-18

183. D.R. Rummler, *Machine Design,* Vol. 55, No. 23, 1983, pp 127-128

184. D.R. Rummler, "Recent Advances in Carbon-Carbon Materials Systems," *Advanced Materials Technology,* NASA Conference Publication 2251, National Aeronautics and Space Administration, Langley Research Center, 1982, pp 293-312

Physical Data on the Elements and Alloys

Periodic table of the elements

		Orbit
	2 0 He 4.00260 2	K
	10 0 Ne 10.179 2-8	K–L
	18 0 Ar 39.948 2-8-8	K–L–M
	36 0 Kr 83.80 8 18 8	–L–M–N
	54 0 Xe 131.30 -18-18-8	–M–N–O
	86 0 Rn (222) -32-18-8	N–O–P

Nonmetals — Metals

Key to chart

50	+2 +4
Sn	
118.69	
-18-18-4	

Atomic Number → 50
Symbol → Sn
Atomic Weight → 118.69
Oxidation States
Electron Configuration

Transition Elements

Group Iᵃ

Iᵃ	IIᵃ	IIIᵇ	IVᵇ	Vᵇ	VIᵇ	VIIᵇ	VIII			Iᵇ	IIᵇ	IIIᵃ	IVᵃ	Vᵃ	VIᵃ	VIIᵃ	0
1 H +1 −1 1.0079 1																	2 He
3 Li +1 6.939 2-1	4 Be +2 9.0122 2 2											5 B +3 10.81 2-3	6 C +2 +4 −4 12.011 2-4	7 N +1 +2 +3 +4 +5 −1 −2 −3 14.0067 2-5	8 O −2 15.9994 2-6	9 F −1 18.998403 2-7	10 Ne
11 Na +1 22.9898 2-8-1	12 Mg +2 24.312 2-8-2											13 Al +3 26.98154 2-8-3	14 Si +2 +4 −4 28.08 2-8-4	15 P +3 +5 −3 30.97376 2-8-5	16 S +4 +6 −2 32.06 2-8-6	17 Cl +1 +5 +7 −1 35.453 2-8-7	18 Ar
19 K +1 39.09 -8-8-1	20 Ca +2 40.08 -8-8-2	21 Sc +3 44.9559 -8-9-2	22 Ti +2 +3 +4 47.9 -8-10-2	23 V +2 +3 +4 +5 50.941 -8-11-2	24 Cr +2 +3 +6 51.996 -8-13-1	25 Mn +2 +3 +4 +7 54.9380 -18-13-2	26 Fe +2 +3 55.847 -18-14-2	27 Co +2 +3 58.9332 -18-15-2	28 Ni +2 +3 58.71 -18-16-2	29 Cu +1 +2 63.54 -18-18-1	30 Zn +2 65.38 -18-18-2	31 Ga +3 69.72 -18-18-3	32 Ge +2 +4 72.59 -18-18-4	33 As +3 +5 −3 74.9216 -18-18-5	34 Se +4 +6 −2 78.96 -18-18-6	35 Br +1 +5 −1 79.904 -18-18-7	36 Kr
37 Rb +1 85.467 -18-8-1	38 Sr +2 87.62 -18-8-2	39 Y +3 88.9059 -18 9 2	40 Zr +4 91.22 -18-10-2	41 Nb +3 +5 92.9064 -18-12-1	42 Mo +6 95.94 -18-13-1	43 Tc +4 +6 +7 98.9062 -18-13-2	44 Ru +3 101.07 -18-15-1	45 Rh +3 102.905 -18-16-1	46 Pd +2 +4 106.4 -18-18-0	47 Ag +1 107.868 -18-18-1	48 Cd +2 112.40 -18-18-2	49 In +3 114.82 -18-18-3	50 Sn +2 +4 118.69 -18-18-4	51 Sb +3 +5 −3 121.75 -18-18-5	52 Te +4 +6 −2 127.60 -18-18-6	53 I +1 +5 +7 −3 126.9045 -18-18-7	54 Xe
55 Cs +1 132.9054 -18-8-1	56 Ba +2 137.3 -18-8-2	57* La +3 138.9055 -18-9-2	72 Hf +4 178.49 -32-10-2	73 Ta +5 180.948 -32-11-2	74 W +6 183.85 -32-12-2	75 Re +4 +6 +7 186.207 -32-13-2	76 Os +3 +4 190.2 -32-14-2	77 Ir +3 +4 192.9 -32-15-2	78 Pt +2 +4 195.09 -32-16-2	79 Au +1 +3 196.9665 -32-18-1	80 Hg +1 +2 200.59 -32-18-2	81 Tl +1 +3 204.37 -32-18-3	82 Pb +2 +4 207.19 -32-18-4	83 Bi +3 +5 208.980 -32-18-5	84 Po +2 +4 (209) -32-18-6	85 At −1 (210) -32-18-7	86 Rn
87 Fr +1 (223) -18-8-1	88 Ra +2 226.0254 -18-8-2	89** Ac +3 (227) -18-9-2	104 Rf (261) -32-10-2	105 Ha (262) -32-11-2	106 (263) -32-12-2												

***Lanthanides**

| 58 Ce +3 +4 140.12 -20-8-2 | 59 Pr +3 140.9077 -21-8-2 | 60 Nd +3 144.24 -22-8-2 | 61 Pm +3 147 -23-8-2 | 62 Sm +2 +3 150.4 -24-8-2 | 63 Eu +2 +3 151.96 -25-8-2 | 64 Gd +3 157.25 -25-9-2 | 65 Tb +3 158.925 -27-8-2 | 66 Dy +3 162.50 -28-8-2 | 67 Ho +3 164.9304 -29-8-2 | 68 Er +3 167.26 -30-8-2 | 69 Tm +3 168.9342 -31-8-2 | 70 Yb +2 +3 173.04 -32-8-2 | 71 Lu +3 174.967 -32-9-2 |

****Actinides**

| 90 Th +4 232.038 -18-10-2 | 91 Pa +5 +4 231.0359 -20-9-2 | 92 U +3 +4 +5 +6 238.029 -21-9-2 | 93 Np +3 +4 +5 +6 237.0482 -22-9-2 | 94 Pu +3 +4 +5 +6 239.052 -24-8-2 | 95 Am +3 +4 +5 +6 (243) -25-8-2 | 96 Cm +3 (247) -25-9-2 | 97 Bk +3 +4 (247) -27-8-2 | 98 Cf +3 (251) -28-8-2 | 99 Es (254) -29-8-2 | 100 Fm (257) -30-8-2 | 101 Md +2 +3 (258) -31-8-2 | 102 No +2 +3 (259) -32-8-2 | 103 Lr +3 (260) -32-9-2 |

Numbers in parentheses are mass numbers of most stable isotope of that element

Periodic system for ferrous metallurgists

The basic principles of alloying were applied to develop the periodic system shown here, which illustrates the fundamental alloying nature of iron. The solid solubility of each element in iron can be resolved accurately with few exceptions by considering only atomic size. Alloying valence, crystal structure and electronegativity are useful supplementary factors of varying significance. The tendency to form compounds, intermetallic or ionic, is related to the difference in electronegativity which, in general, increases in a sweep from lower left to upper right of the periodic system.

Adapted Primarily for Ferrous Metallurgists

Atomic size factors (in parentheses) are % smaller (−) or larger (+) than gamma (FCC) iron at 75 F. Lattice environment (Coordination No.) is taken into account; CN is 12 except 6 for interstitials H, B, C, N & O. Groups VI, VIb, VII and VIIb form ionic compounds with the metals.

Nonmetal block

IV	V	VI	VII
C-6 (−34) ▲ ⊗ XX	N-7 (−36) ▲ ⊗ XX	O-8 (−33) ▲ ⊗ XX	F-9 XX
Si-14 (+7) ● XX	P-15 (+2) ⊗ XX	S-16 (+1) ⊗ XX	Cl-17 XX

H-1 (−58) ▲ ⊗ XX

Light elements block

0	I	II	III
He-2 FCC (Others)			B-5 (−29) ▲ ⊗ XX
Ne-10 FCC	Na-11 (+50) ⊗ BCC* HCP†	Mg-12 (+27) ◐ HCP	Al-13 (+14) ◐ FCC

Main table

0	Ia	IIa	IIIa	IVa	Va	VIa	VIIa	VIII	VIII	VIII	Ib	IIb	IIIb	IVb	Vb	VIb	VIIb
Ar-18 FCC	K-19 (+86) ⊗ BCC	Ca-20 (+56) ⊗ FCC* BCC	Sc-21 (+29) ⊗ HCP* BCC	Ti-22 (+16) ◐ HCP* BCC	V-23 (+6) ● BCC	Cr-24 (+1) ● BCC	Mn-25 (+1) ◐ XX* FCC ‡	Fe-26 (0) ● BCC* FCC	Co-27 (−1) ● HCP* FCC	Ni-28 (−1) ● FCC	Cu-29 (+1) ● FCC	Zn-30 (+6) ● HCP	Ga-31 (+12) ● XX	Ge-32 (+9) ● XX	As-33 (+11) ⊗ XX	Se-34 (+11) ⊗ XX	Br-35 XX
Kr-36 FCC	Rb-37 (+97) ⊗ BCC	Sr-38 (+71) ⊗ FCC* HCP ‡	Y-39 (+42) ⊗ HCP* BCC	Zr-40 (+27) ⊗ HCP* BCC	Cb-41 (+15) ◐ BCC	Mo-42 (+10) ● BCC	Tc-43 (+8) ● HCP	Ru-44 (+6) ● HCP	Rh-45 (+6) ● FCC	Pd-46 (+9) ● FCC	Ag-47 (+14) ◐ FCC	Cd-48 (+20) ⊗ HCP	In-49 (+25) ⊗ XX	Sn-50 (+23) ⊗ XX	Sn-51 (+27) ⊗ XX	Te-52 (+27) ⊗ XX	I-53 XX
Xe-54 FCC	Cs-55 (+112) ⊗ BCC	Ba-56 (+76) ⊗ BCC	La-57 (+48) ⊗ HCP* FCC ‡	Hf-72 (+26) ⊗ HCP* BCC	Ta-73 (+16) ◐ BCC	W-74 (+11) ● BCC	Re-75 (+9) ● HCP	Os-76 (+7) ● HCP	Ir-77 (+8) ● FCC	Pt-78 (+10) ● FCC	Au-79 (+14) ◐ FCC	Hg-80 (+25) ⊗ XX	Tl-81 (+36) ⊗ HCP* BCC	Pb-82 (+39) ⊗ FCC	Bi-83 (+35) ⊗ XX	Po-84 (+40) ⊗ XX	At-85
Rn-86 FCC	Fr-87	Ra-88	Ac-89 (+49) ⊗ FCC														
Alloying Valence	1	2	3	4	5	6	6	6	6	6	5.56	4.56	3.56	2.56 Note 2	1.56 Note 2	(2) Note 3	(1) Note 3

Note 1: The rare-earth (lanthanide, 58-71) and actinide (90-103) series are omitted.
Note 2: Valence is 4 for C; 3 for N and P.
Note 3: (1) and (2) are not alloying valences.

Legend

Substitutional solid solutions

● FAVORABLE SIZE FACTOR: 0 TO ± 13%
◐ BORDERLINE SIZE FACTOR: ± 14 TO ± 16%
⊗ UNFAVORABLE SIZE FACTOR: > ±16%

Interstitial solid solutions

▲ FAVORABLE SIZE FACTOR: > (−40%)
▲ BORDERLINE SIZE FACTOR: (−30) TO (−40%)
▲ UNFAVORABLE SIZE FACTOR: < (−30%)

Structure

BCC – BODY CENTERED CUBIC
FCC – FACE CENTERED CUBIC
HCP – HEXAGONAL CLOSE PACKED
XX – NOT BCC, FCC OR HCP
* – USUALLY MORE COMPLEX
* – STRUCTURE AT 75 F
† – ALSO FCC ‡ ALSO BCC

Type of gamma iron (FCC) field if alloyed with iron

GAMMA LOOP, LIKE Cr
LIMITED GAMMA LOOP, LIKE B
OPEN GAMMA REGION, LIKE Ni
LIMITED GAMMA REGION, LIKE C

Standard atomic weights of the elements

Symbol	Atomic number	Atomic weight	Symbol	Atomic number	Atomic weight	Symbol	Atomic number	Atomic weight
Ac	89	227.0278	He	2	4.00260	Ra	88	226.0254
Al	13	26.98154	Ho	67	164.9304	Rn	86	(222)
Am	95	(243)	H	1	1.0079	Re	75	186.207
Sb	51	121.75*	In	49	114.82	Rh	45	102.9055
Ar	18	39.948	I	53	126.9045	Rb	37	85.4678*
As	33	74.9216	Ir	77	192.22*	Ru	44	101.07*
At	85	(210)	Fe	26	55.847*	Sm	62	150.36*
Ba	56	137.33	Kr	36	83.80	Sc	21	44.9559
Bk	97	(247)	La	57	138.9055*	Se	34	78.96*
Be	4	9.01218	Lr	103	(260)	Si	14	28.0855*
Bi	83	208.9804	Pb	82	207.2	Ag	47	107.868
B	5	10.81	Li	3	6.941*			
Br	35	79.904	Lu	71	174.967*	Na	11	22.98977
Cd	48	112.41	Mg	12	24.305	Sr	38	87.62
Cs	55	132.9054	Mn	25	54.9380	S	16	32.06
Ca	20	40.08	Md	101	(258)	Ta	73	180.9479
Cf	98	(251)	Hg	80	200.59*	Tc	43	(98)
C	6	12.011	Mo	42	95.94	Te	52	127.60*
Ce	58	140.12	Nd	60	144.24*	Tb	65	158.9254
Cl	17	35.453	Ne	10	20.179	Tl	81	204.383
Cr	24	51.996	Np	93	237.0482	Th	90	232.0381
Co	27	58.9332	Ni	28	58.69	Tm	69	168.9342
Cu	29	63.546*	Nb	41	92.9064	Sn	50	118.69*
Cm	96	(247)	N	7	14.0067	Ti	22	47.88*
Dy	66	162.50*	No	102	(259)			
Es	99	(252)	Os	76	190.2	W	74	183.85*
Er	68	167.26*	O	8	15.9994*	(Unh)	106	(263)
Eu	63	151.96	Pd	46	106.42	(Unp)	105	(262)
Fm	100	(257)	P	15	30.97376	(Unq)	104	(261)
F	9	18.998403	Pt	78	195.08*	U	92	238.0289
Fr	87	(223)	Pu	94	(244)	V	23	50.9415
Gd	64	157.25*	Po	84	(209)	Xe	54	131.29*
Ga	31	69.72	K	19	39.0983	Yb	70	173.04*
Ge	32	72.59*	Pr	59	140.9077	Y	39	88.9059
Au	79	196.9665	Pm	61	(145)	Zn	30	65.38
Hf	72	178.49*	Pa	91	231.0359	Zr	40	91.22

Note: The atomic weights are in atomic mass units (amu) relative to $^{12}C = 12$. The atomic weights of many elements are not invariant but depend on the origin and treatment of the material. The values given apply to elements as they exist naturally on earth and to certain artificial elements. They are considered reliable to ±1 in the last digit or ±3 when followed by an asterisk (*). Values in parentheses are used for radioactive elements whose atomic weights cannot be quoted precisely without knowledge of the origin of the elements; the value given is the atomic mass number of the isotope of that element of longest known half-life

Melting points of the elements

Symbol	Atomic number	Melting point K	Melting point °C	Estimated error; footnotes	Symbol	Atomic number	Melting point K	Melting point °C	Estimated error; footnotes
Ac	89	1324	1051	±50	Mn	25	1519	1246	±5
Ag	47	1235.08	961.93	a	Mo	42	2896	2623	b
Al	13	933.602	660.452	b	N	7	63.1458	−210.0042	±0.0002; d, g
Am	95	1449	1176	c	Na	11	371.0	97.8	±0.1
Ar	18	83.798	−189.352	a, d	Nb	41	2742	2469	b
As	33	876	603	e	Nd	60	1294	1021	c
At	85	(575)	(302)	est.; f	Ne	10	24.563	−248.587	±0.002; b, d
Au	79	1337.58	1064.43	a	Ni	28	1728	1455	b
B	5	2365	2092	f	No	102	(1100)	(827)	est.; f
Ba	56	1002	729	±2	Np	93	910	637	±2
Be	4	1562	1289	±5	O	8	54.361	−218.789	a, d
Bi	83	544.592	271.442	b	Os	76	3306	3033	±20
Bk	97	1256	983	c	P	15	317.29	44.14	±0.1; j
Br	35	265.90	−7.25	d	Pa	91	1848	1575	f
C	6	4100	3826	e	Pb	82	600.652	327.502	b
Ca	20	1113	840	±2	Pd	46	1828	1555	±0.4; g
Cd	48	594.258	321.108	b	Pm	61	1315	1042	c
Ce	58	1071	798	±3; c	Po	84	527	254	f
Cf	98	1213	940	c	Pr	59	1204	931	c
Cl	17	172.18	−100.97	d	Pt	78	2042.1	1769.0	b
Cm	96	1613	1340	f	Pu	94	913	640	±1
Co	27	1768	1495	b	Ra	88	973	700	f
Cr	24	2133	1860	±20	Rb	37	312.63	39.48	±0.5
Cs	55	301.54	28.39	±0.05	Re	75	3459	3186	±20
Cu	29	1358.02	1084.87	±0.04; g	Rh	45	2236	1963	b
Dy	66	1685	1412	c	Rn	86	202	−71	…
Er	68	1802	1529	c	Ru	44	2607	2334	±10; g
Es	99	1093	820	c	S	16	388.37	115.22	…
Eu	63	1095	822	c	Sb	51	903.905	630.755	b
F	9	53.48	−219.67	d	Sc	21	1814	1541	c
Fe	26	1808	1535	b	Se	34	494	221	…
Fm	100	(1800)	(1527)	est.; f	Si	14	1687	1414	±2
Fr	87	(300)	(27)	est.; f	Sm	62	1347	1074	c
Ga	31	302.9241	29.7741	±0.001; d, g	Sn	50	505.1181	231.9681	a
Gd	64	1586	1313	c	Sr	38	1042	769	…
Ge	32	1211.5	938.3	…	Ta	73	3293	3020	…
H	1	13.81	−259.34	a, d	Tb	65	1629	1356	c
He	2	4.215	−268.935	h	Tc	43	2477	2204	±50
Hf	72	2504	2231	±20	Te	52	722.72	449.57	±0.3
Hg	80	234.314	−38.836	b	Th	90	2031	1758	±10
Ho	67	1747	1474	c	Ti	22	1943	1670	±6; g
I	53	386.7	113.6	d	Tl	81	577	304	±2
In	49	429.784	156.634	b	Tm	69	1818	1545	c
Ir	77	2720	2447	b	U	92	1407	1134	…
K	19	336.34	63.19	±0.5	V	23	2202	1929	±6
Kr	36	115.765	−157.385	±0.001; g	W	74	3695	3422	b
La	57	1191	918	c	Xe	54	161.3918	−111.7582	±0.0002; d, g
Li	3	453.7	180.6	±0.5	Y	39	1795	1522	c
Lr	103	(1900)	(1627)	est.; f	Yb	70	1092	819	c
Lu	71	1936	1663	c	Zn	30	692.73	419.58	a
Md	101	(1100)	(827)	est.; f	Zr	40	2128	1855	±5
Mg	12	922	649	±0.5					

Note: The melting points, except those footnoted to indicate otherwise, are derived from R. Hultgren, P.D. Desai, D.T. Hawkins, M. Gleiser, K.K. Kelley and D.D. Wagman, *Selected Values of the Thermodynamic Properties of the Elements*, American Society for Metals, 1973, which are based on the 1948 International Practical Temperature Scale. Values have been corrected to the 1968 scale (IPTS-68). Except for triple points, values are for a pressure of one atmosphere. (Note that melting and freezing points should be identical for pure elements.) (a) Defined fixed point on 1968 International Practical Temperature Scale (IPTS-68): Amended Edition of 1975, *Metrologia, 12, p 7-17* (1976). (b) Secondary reference point in Extended List of Secondary Reference Points on 1968 International Practical Temperature Scale (IPTS-68), *Metrologia, 13*, p 197-206 (1977). (c) From *Metals Handbook*, 9th Ed., Vol. 2, American Society for Metals (1979). (d) Triple point. (e) Sublimation point at atmospheric pressure. (f) From R.H. Lamoreaux, Melting Point, Gram-Atomic Volumes and Enthalpies of Atomization for Liquid Elements, LBL Report 4995 (1976). (g) Secondary reference point 1980 supplement to 1977 Extended List of Secondary Reference Points on 1968 International Practical Temperature Scale (IPTS-68): Amended Edition of 1975, Report 5 of Working Group II (April 1980). (h) Boiling point at 1 atm; there are various triple points; see Hultgren *et al.* (1973). (j) Melting point for white α-P; red P sublimes without melting at atmospheric pressure and has a triple point of 862.8 K (589.7 °C)

Atomic size parameters for the elements

Element	Atomic number	Pearson symbol	Atomic volume (Ω), nm^3	Interatomic distance (S_0), nm	Equivalent atomic radius (r_0), nm	Notes	Element	Atomic number	Pearson symbol	Atomic volume (Ω), nm^3	Interatomic distance (S_0), nm	Equivalent atomic radius (r_0), nm	Notes
Ac	89	cF4	0.037451	0.3755	0.2076	...	αN	7	cP8	0.022653	(0.1098)	(0.1755)	a, b
Ag	47	cF4	0.017056	0.2889	0.1597	...	βNa	11	cI2	0.039493	0.3716	0.2113	...
Al	13	cF4	0.016603	0.2864	0.1583	...	Nb	41	cI2	0.017980	0.2859	0.1625	...
αAm	95	hP4	0.029271	0.3451	0.1911	a	αNd	60	hP4	0.034179	0.3322	0.2013	...
Ar	18	cF4	0.037473	0.3756	0.2076	a	Ne	10	cF4	0.022212	0.3155	0.1744	a
αAs	33	hR2	0.021518	0.2517	0.1726	...	Ni	28	cF4	0.010942	0.2492	0.1377	...
Au	79	cF4	0.016959	0.2884	0.1594	...	αNp	93	oP8	0.019224	0.2599	0.1662	...
γB	5	tP50	0.007786	(0.1624)	(0.1230)	c	αO	8	mC4	0.017360	(0.115)	(0.1606)	a, b
Ba	56	cI2	0.063367	0.4350	0.2473	...	Os	76	hP2	0.013988	0.2735	0.1495	...
αBe	4	hP2	0.008108	0.2225	0.1246	...	P (black)	15	oC8	0.018993	(0.2224)	(0.1655)	c
αBi	83	hR2	0.035384	0.3071	0.2037	...	αPa	91	tI2	0.025212	0.3214	0.1819	...
αBk	97	hP4	0.027965	0.3398	0.1883	...	Pb	82	cF4	0.030326	0.3500	0.1935	...
Br	35	oC8	0.03277	(0.227)	(0.199)	a, b	Pd	46	cF4	0.014717	0.2751	0.1520	...
C (graph)	6	hP4	0.008800	(0.1421)	(0.1281)	c	Pm	61	hP4	0.03360	0.330	0.200	...
αCa	20	cF4	0.043631	0.3952	0.2184	...	αPo	84	cP1	0.038137	0.3366	0.2088	...
Cd	48	hP2	0.021581	0.2979	0.1727	...	αPr	59	hP4	0.034545	0.3338	0.2020	...
αCe	58	cF4	0.034367	0.3650	0.2017	...	Pt	78	cF4	0.015097	0.2774	0.1533	...
Cl	17	oC8	0.02886	(0.198)	(0.190)	a, b	αPu	94	mP16	0.019998	0.257	0.1684	...
αCm	96	hP4	0.029984	0.3479	0.1927	...	Ra	88	cI2	0.068216	0.4459	0.2535	...
αCo	27	hP2	0.011076	0.2497	0.1383	...	Rb	37	cI2	0.092743	0.4939	0.2808	...
Cr	24	cI2	0.012003	0.2498	0.1420	...	Re	75	hP2	0.014713	0.2740	0.1520	...
Cs	55	cI2	0.115794	0.5318	0.3024	...	Rh	45	cF4	0.013753	0.2689	0.1486	...
Cu	29	cF4	0.011809	0.2238	0.1413	...	Ru	44	hP2	0.013568	0.2650	0.1480	...
αDy	66	hP2	0.031558	0.3504	0.1960	...	αS	16	oF128	0.025754	(0.2037)	(0.1832)	c
αEr	68	hP2	0.030636	0.3467	0.1941	...	αSb	51	hR2	0.030201	0.2908	0.1932	...
Eu	63	cI2	0.048121	0.3969	0.2256	...	αSc	21	hP2	0.024974	0.3254	0.1813	...
αF	9	mC8	0.01605	(0.149)	(0.197)	a, b	γSe	34	hP3	0.027274	0.2374	0.1867	...
αFe	26	cI2	0.011777	0.2483	0.1411	...	Si	14	cF8	0.020020	0.2352	0.1684	...
αGa	31	oC8	0.019580	0.2484	0.1672	...	αSm	62	hR3	0.033202	0.3587	0.1994	...
αGd	64	hP2	0.033050	0.3572	0.1991	...	βSn	50	tI4	0.027049	0.3022	0.1862	...
αGe	32	cF8	0.022634	0.2450	0.1755	...	αSr	38	cF4	0.056299	0.4302	0.2378	...
αH	1	hP2	0.037882	(0.3768)	(0.2083)	a, b	Ta	73	cI2	0.018019	0.2861	0.1626	...
αHe	2	hP2	0.032367	0.3577	0.1977	a	αTb	65	hP2	0.032066	0.3528	0.1971	...
αHf	72	hP2	0.022321	0.3127	0.1747	...	Tc	43	hP2	0.014264	0.2707	0.1505	...
αHg	80	hR1	0.023354	0.2993	0.1773	a	αTe	52	hP3	0.033969	0.2834	0.2009	...
αHo	67	hP2	0.031139	0.3487	0.1952	...	αTh	90	cF4	0.032873	0.3596	0.1987	...
I	53	oC8	0.042696	(0.269)	(0.2168)	b	αTi	22	hP2	0.017653	0.2986	0.1615	...
In	49	tI2	0.026158	0.3252	0.1842	...	αTl	81	hP2	0.028586	0.3408	0.1897	...
Ir	77	cF4	0.014146	0.2715	0.1500	...	αTm	69	hP2	0.030006	0.3447	0.1928	...
K	19	cI2	0.075327	0.4608	0.2620	...	αU	92	oC4	0.020747	0.2753	0.1705	...
Kr	36	cF4	0.044992	0.3992	0.2206	a	V	23	cI2	0.013824	0.2619	0.1489	...
αLa	57	hP4	0.037532	0.3456	0.2077	...	W	74	cI2	0.015844	0.2741	0.1558	...
βLi	3	cI2	0.021609	0.3309	0.1728	...	Xe	54	cF4	0.057463	0.4336	0.2396	a
αLu	71	hP2	0.029524	0.3434	0.1917	...	αY	39	hP2	0.033033	0.3557	0.1991	...
Mg	12	hP2	0.023239	0.3197	0.1770	...	αYb	70	cF4	0.041250	0.3878	0.2143	...
αMn	25	cI58	0.012245	0.2258	0.1430	...	Zn	30	hP2	0.015214	0.2664	0.1537	...
Mo	42	cI2	0.015583	0.2745	0.1550	...	αZr	40	hP2	0.023279	0.3179	0.1771	...

Note: The atomic size parameters were derived from the crystal structure data for the Elements given in *Bull. Alloy Phase Diagrams*, 2(3), p 401-402 (1981). The volume per atom of the structure, Ω, was derived from the room temperature lattice parameter data by calculating the volume of the unit cell and dividing by the number of atoms contained within the unit cell, which is given by the numerals in the Pearson symbol. The closest distance of approach, S_0, was derived from the minimum interatomic distances in the unit cell, except for the diatomic gases and the nonmetallic elements that exist in molecular form. The atomic radius, r_0, was derived from the volume per atom data, using the relationship $4/3\,\pi r_0^3$ = volume per atom. (a) These elements are gaseous, or liquid, at room temperature. The structural data for Ar, H, Kr, Ne, and Xe refer to 4.2 K, whereas that for Br, Cl, He, Hg, N, and O refer to 123 K, 113 K, 1.5 K, 227 K, 20 K, and 23 K, respectively. (b) These elements form diatomic gases, and the basis of the crystal structure is, therefore, the molecular unit R_2. The volume per atom data are thus more meaningful if considered in terms of two times the volume per molecule, whereas the values of the parameter r_0 (listed in parentheses) should not be equated with atomic radii in this context. The values of S_0 (also listed in parentheses) refer to interatomic distances in the covalently bonded molecule R_2, rather than distance between equipositioned neighbors in a crystal. (c) The chemistry of these elements permits them to form a number of allotropes at room temperature, with crystal structures based on different molecular bases. The comments above on volume per atom and r_0 are, thus, also applicable to these elements, because they form crystal structures based on chains or network layers

Heats of transition of the elements

Element	Atomic number	Transformation	Enthalpy (ΔH), J/mol	Temperature, °C	Element	Atomic number	Transformation	Enthalpy (ΔH), J/mol	Temperature, °C
Ag	47	L↔S	11 300	961.93(b)	Nd	60	β↔α	3030	855
Al	13	L↔S	10 700	660.457(d)	Ne	10	L↔S	331.7	24.561 K(c)
Am	95	L↔γ	14 395	1176	Ni	28	L↔S	17 470	1455(d)
		γ↔β	5 860	1077	Np	93	L↔γ	5 190	639
		β↔α	775	650			γ↔β	5 270	576
Ar	18	L↔S	1190	83.798 K(a)			β↔α	5 605	280
Au	79	L↔S	13 000	1064.43(b)	O	8	L↔γ	223	54.361 K(a)
B	5	L↔S	50 200	2077			γ↔β	371.3	43.801 K
Ba	56	L↔S	7 120	727			β↔α	48.4	23.867 K
Be	4	L↔β	(12 600)	1287	Os	76	L↔S	(31 800)	3025
		β↔α	(2 100)	1277	P(white α)	15	L↔α	629	44
Bi	83	L↔S	11 300	271.442(c)	Pa	91	L↔β	12 340	1572
Br	35	L↔S	5 286	265.9 K			β↔α	6 640	1170
Ca	20	L↔β	8 540	842	Pb	82	L↔S	4 800	327.502(d)
		β↔α	842	443	Pd	46	L↔S	(17 560)	1554(d)
Cd	48	L↔S	6 200	321.108(d)	Pm	61	L↔S	(7 550)	...
Ce	58	L↔δ	5 460	800			B↔A	(2 900)	...
		δ↔γ	2 990	725	Pr	59	L↔β	6 890	930
		γ↔β	190	...			β↔α	3 170	795
		β↔α	1 950	...	Pt	78	L↔S	(19 650)	1769(d)
Cl	17	L↔S	3 203	172.16 K	Pu	94	L↔ε	2 825	640
Cm	96	L↔β	14 645	1345			ε↔δ′	1 840	479
		β↔γ	3 245	1277			δ′↔δ	80	457
Co	27	L↔β	16 200	1495(d)			δ↔γ	585	315
		β↔α	450	427			γ↔β	565	207
Cr	24	L↔S	(20 500)	1857			β↔α	3 375	122
Cs	55	L↔S	2 090	28.44	Rb	37	L↔S	2 190	39.32
Cu	29	L↔S	13 050	1084.88(d)	Re	75	L↔S	(33 230)	3180
Dy	66	L↔β	11 060	1409	Rh	45	L↔S	(21 490)	1963(d)
		β↔α	4 160	1385	Rn	86	L↔S	(2 890)	−71
Er	68	L↔S	19 900	1522	Ru	44	L↔S	(24 280)	2250
Eu	63	L↔S	9 210	817	S	16	L↔β	1 718	115
F	9	L↔β	255	53.48 K			β↔α	402	95
		β↔α	364	45.55 K	Sb	51	L↔S	19 900	630.775(d)
Fe	26	L↔δ	13 800	1535(d)	Sc	21	L↔β	14 100	1539
		δ↔γ	840	1392			β↔α	4 010	1335
		γ↔α	900	911	Se	34	L↔S	6 700	220
Ga	31	L↔S	5 565	29.771(d)	Si	14	L↔S	50 210	1417
Gd	64	L↔β	10 050	1312	Sm	62	L↔β	8 620	1072
		β↔α	3 910	1260			β↔α	3 110	917
Ge	32	L↔S	37 030	937	Sn	50	L↔β	7 195	231.9681(b)
H	1	L↔S	58.68	13.81 K(a)	Sr	38	L↔γ	7 431	777
Hf	72	L↔S	(29 300)	2227			γ↔α	837	547
		β↔α	(5 910)	1781	Ta	73	L↔S	36 570	2985
Hg	80	L↔α	2 295	−38.836(d)	Tb	65	L↔β	10 800	1355
Ho	67	L↔β	(16 900)	1470			β↔α	5 020	1285
I	53	L↔S	7 820	113.5	Te	52	L↔S	17 490	449.5
In	49	L↔S	3 280	156.634(d)	Th	90	L↔β	13 807	1750
Ir	77	L↔S	(26 140)	2447(d)			β↔α	3 599	1360
K	19	L↔S	2 320	63.71	Ti	22	L↔β	14 150	1663
Kr	36	L↔S	1 638	115.770 K(c)			β↔α	4 170	893
La	57	L↔γ	6 200	920	Tl	81	L↔β	4200	303
		γ↔β	3 120	860			β↔α	360	234
		β↔α	360	275	Tm	69	L↔S	16 840	1545
Li	3	L↔β	3 000	180.54	U	92	L↔γ	9 142	1135
Lu	71	L↔S	(18 650)	1663			β→γ	4 757	776
Mg	12	L↔S	8 477	650			α→β	2 791	669
Mn	25	L↔δ	(12 060)	1245	V	23	L↔S	22 845	1917
		δ↔γ	1 880	1135	W	74	L↔S	46 000	3422(d)
		γ↔β	2 120	1085	Xe	54	L↔S	2300	161.388 K(c)
		β↔α	2 230	700	Y	39	L↔β	11 400	1525
Mo	42	L↔S	35 980	2623(d)			β↔α	4 990	1480
N	7	L↔β	360.4	63.146 K(c)	Yb	70	L↔β	7 660	824
		β↔α	116	35.61 K			β↔α	1 750	760
Na	11	L↔β	2 600	97.86	Zn	30	L↔S	7 320	419.58(b)
Nb	41	L↔S	(26 900)	2473(d)	Zr	40	L↔S	20 920	1855(d)
Nd	60	L↔β	7 140	1015			β↔α	4 015	862

(a) Triple point values, which are defined fixed points of IPTS-68. (b) Melting points or freezing points, which are defined fixed points of IPTS-68. (c) Triple point values, which are secondary reference points of IPTS-68. (d) Melting points or freezing points, which are secondary reference points of IPTS-68

Physical properties of the elements

Element	Density(a), g/cm³ (lb/in.³)	Boiling point, °C (°F)	Specific heat(b), cal/g · °C (J/kg · K)	Heat of fusion, cal/g (Btu/lb)
Actinium (Ac)	... (...)	... (...)	... (...)	... (...)
Aluminum (Al)	2.70 (0.0974)	2450 (4442)	0.215 (900)	94.5 (170)
Americium (Am)	11.87 (0.4285)	... (...)	... (...)	... (...)
Antimony (Sb)	6.65 (0.240)	1380 (2516)	0.049 (205)	38.3 (68.9)
Argon (A)	1.784(g) (0.06440)(g)	−185.8 (−302.4)	0.125 (523)	6.7 (12)
Arsenic (As)	5.72 (0.206)	613(j) (1135)(j)	0.082 (343)	88.5 (159.3)
Astatine (At)	... (...)	... (...)	... (...)	... (...)
Barium (Ba)	3.6 (0.13)	1640 (2980)	0.068 (285)	... (...)
Berkelium (Bk)	... (...)	... (...)	... (...)	... (...)
Beryllium (Be)	1.85 (0.0668)	2770 (5020)	0.45 (190)	260 (470)
Bismuth (Bi)	9.80 (0.354)	1560 (2840)	0.0294 (123)	12.5 (22.5)
Boron (B)	2.45 (0.0884)	... (...)	0.309 (1290)	... (...)
Bromine (Br)	3.12 (0.113)	58 (136)	0.070 (290)	16.2 (29.2)
Cadmium (Cd)	8.65 (0.312)	765 (1409)	0.055 (230)	13.2 (23.8)
Calcium (Ca)	1.55 (0.0560)	1440 (2625)	0.149(s) (624)(s)	52 (93.6)
Californium (Cf)	... (...)	... (...)	... (...)	... (...)
Carbon, graphite (C)	2.25 (0.0812)	4830 (8730)	0.165 (691)	... (...)
Cerium (Ce)	6.77 (0.244)	3470 (6280)	0.045 (190)	8.5 (15.9)
Cesium (Cs)	1.87 (0.0675)	690 (1273)	0.04817 (201.7)	3.8 (6.8)
Chlorine (Cl)	3.214(g) (0.1160)(g)	−34.7 (−30.5)	0.116 (486)	21.6 (38.9)
Chromium (Cr)	7.19 (0.260)	2665 (4829)	0.11 (460)	96 (173)
Cobalt (Co)	8.85 (0.319)	2900 (5250)	0.099 (410)	58.4 (105)
Copper (Cu)	9.86 (0.323)	2595 (4703)	0.092 (380)	50.6 (91.1)
Curium (Cm)	7 (0.3)	... (...)	... (...)	... (...)
Dysprosium (Dy)	8.55 (0.309)	2330 (4230)	0.041 (170)	25.2 (45.4)
Einsteinium (E)	... (...)	... (...)	... (...)	... (...)
Erbium (Er)	9.15 (0.330)	2630 (4770)	0.040 (170)	24.5 (44.1)
Europium (Eu)	5.24 (0.189)	1490 (2710)	0.039 (160)	16.5 (29.6)
Fermium (Fm)	... (...)	... (...)	... (...)	... (...)
Fluorine (F)	1.696(g) (0.06123)(g)	−188.2 (−306.8)	0.18 (750)	10.1 (18.2)
Francium (Fr)	... (...)	... (...)	... (...)	... (...)
Gadolinium (Gd)	7.86 (0.284)	2730 (4950)	0.071 (300)	23.5 (42.4)
Gallium (Ga)	5.91 (0.213)	2237 (4059)	0.079 (330)	19.16 (34.49)
Germanium (Ge)	5.32 (0.192)	2830 (5125)	0.073 (310)	... (...)
Gold (Au)	19.3 (0.0697)	2970 (5380)	0.0312(ee) (131)(ee)	16.1 (29.0)

(continued)

Symbol	Coefficient of linear thermal expansion(c), μin./in. °C (μin./in. °F)	Thermal conductivity(c), cal/cm²/cm/s/°C	Electrical resistivity, μΩ · cm	Modulus of elasticity in tension, 10⁶ psi
Ac	... (...)	...	...	...
Al	23.6(d) (13.1)(d)	0.53	2.6548(b)	9
Am	... (...)	...	...	...
Sb	8.5 to 10.8(e) (4.7 to 6)(e)	0.045	39.0(f)	11.3
A	... (...)	0.406×10^{-4}	...	...
As	4.7 (2.6)	...	33.3(b)	...
At	... (...)	...	...	...
Ba	... (...)	...	...	...
Bk	... (...)	...	...	...
Be	11.6(m) (6.4)(m)	0.35	4(b)(n)	40 to 44
Bi	13.3 (7.4)	0.020	106.8(f)	4.6
B	8.3(q) 4.6(q)	...	1.8×10^{12} (f)	...
Br	... (...)	...	...	...
Cd	29.8 (16.55)	0.22	6.83(f)	8(r)
Ca	22.3(t) (12.4)(t)	0.3	3.91(f)	3.2 to 3.8(u)
Cf	... (...)	...	...	...
C	0.6 to 4.3(d) (0.3 to 2.4)(d)	0.057	1375(f)	0.7
Ce	8 (4.44)	0.026(v)	75(w)	6(z)
Cs	97(y) (54)(y)	...	20(b)	...
Cl	... (...)	0.172×10^{-4}	...	...
Cr	6.2 (3.4)	0.16	12.9(f)	36
Co	13.8 (7.66)	0.165	6.24(b)	30
Cu	16.5 (9.2)	0.941 ± 0.005	1.6730(b)	16
Cm	... (...)	...	...	...
Dy	9 (5)	0.024(v)	57(w)	10 to 14(x)
E	... (...)	...	...	...
Er	9 (5)	0.023(v)	107(w)	16(x)
Eu	26 (14.44)	...	90(w)	...
Fm	... (...)	...	...	...
F	... (...)	...	...	...
Fr	... (...)	...	...	...
Gd	4(z) (2.22)(z)	0.021(v)	140.5(w)	8 to 14(x)
Ga	18(aa) (10)(aa)	0.07 to 0.09(bb)	17.4(cc)	...
Ge	5.75 (3.19)	0.14	46(dd)	...
Au	14.2 (7.9)	0.71	2.35(b)	11.6

(continued)

Physical properties of the elements (continued)

Element	Density(a), g/cm³ (lb/in.³)	Boiling point, °C (°F)	Specific heat(b), cal/g · °C (J/kg · K)	Heat of fusion, cal/g (Btu/lb)
Hafnium (Hf)	13.1 (0.473)	5400 (9750)	0.0351 (147)	… (…)
Helium (He)	0.1785(g) (0.006444)(g)	−268.9 (−452.0)	1.25 (5230)	… (…)
Holmium (Ho)	6.79 (0.245)	2330 (4230)	0.039 (160)	24.9 (44.7)
Hydrogen (H)	0.0899(g) (0.00325)(g)	−252.7 (−422.9)	3.45 (14 400)	15.0 (27.0)
Indium (In)	7.31 (0.264)	2000 (3632)	0.057 (240)	6.8 (12.2)
Iodine (I)	4.94 (0.178)	183 (361)	0.052 (220)	14.2 (25.6)
Iridium (Ir)	22.65 (0.8177)	5300 (9570)	0.0307 (129)	… (…)
Iron (Fe)	7.87 (0.284)	3000 ± 150 (5430 ± 270)	0.11 (460)	65.5 (117.9)
Krypton (Kr)	3.743(g) (0.1351)(g)	−152 (−242)	… (…)	… (…)
Lanthanum (La)	6.15 (0.222)	3470 (6280)	0.048 (200)	17.3 (31.1)
Lawrencium (Lw)	… (…)	… (…)	… (…)	… (…)
Lead (Pb)	11.34 (0.4094)	1725 (3137)	0.0309(f) (129)(f)	6.26 (11.27)
Lithium (Li)	0.534 (0.193)	1330 (2426)	0.79 (3300)	104.2 (187.6)
Lutetium (Lu)	9.85 (0.356)	1930 (3510)	0.037 (150)	26.29 (47.32)
Magnesium (Mg)	1.74 (0.0628)	1107 ± 10 (2025 ± 20)	0.245 (1030)	88 ± 2 (158 ± 4)
Manganese (Mn)	7.43 (0.268)	2150 (3900)	0.115(qq) (481)(qq)	63.7 (114.7)
Mendelevium (Mv)	… (…)	… (…)	… (…)	… (…)
Mercury (Hg)	13.55 (0.4892)	357 (675)	0.033 (140)	2.8 (5.0)
Molybdenum (Mo)	10.2 (0.368)	5560 (10 040)	0.066 (280)	69.8(k) (125.6)(k)
Neodymium (Nd)	7.00 (0.253)	3180 (5756)	0.045 (190)	11.78 (21.20)
Neon (Ne)	0.8999(g) (0.03249)(g)	−246.0 (−410.8)	… (…)	… (…)
Neptunium (Np)	20.5 (0.740)	… (…)	… (…)	… (…)
Nickel (Ni)	8.9 (0.32)	2730 (4950)	0.105 (440)	73.8 (132.8)
Niobium (Nb)	8.57 (0.309)	4927 (8901)	0.065(f) (270)(f)	69 (124.2)
Nitrogen (N)	1.250(g) (0.04513)(g)	−195.8 (−320.4)	0.247 (1030)	6.2 (11.2)
Nobelium (No)	… (…)	… (…)	… (…)	… (…)
Osmium (Os)	22.61 (0.8162)	5500 (9950)	0.031 (130)	… (…)
Oxygen (O)	1.429(g) (0.05159)(g)	−183.0 (−297.4)	0.218 (913)	3.3 (5.9)
Palladium (Pd)	12.02 (0.4339)	3980 (7200)	0.0584(f) (245)(f)	34.2 (61.6)
Phosphorus, white (P)	1.83 (0.0661)	280 (536)	0.177 (741)	5.0 (9.0)
Platinum (Pt)	21.45 (0.7743)	4530 (8185)	0.0314(f) (131)(f)	26.9 (48.4)
Plutonium (Pu)	19.4 (0.700)	3235 (6000)	0.033(aaa) (140)(aaa)	… (…)
Polonium (Po)	9.40 (0.339)	… (…)	… (…)	… (…)
Potassium (K)	0.86 (0.031)	760 (1400)	0.177 (741)	14.6 (26.3)
Praseodymium (Pr)	6.77 (0.244)	3020 (5468)	0.045 (188)	11.71 (21.08)
Promethium (Pm)	… (…)	… (…)	… (…)	… (…)

(continued)

Symbol	Coefficient of linear thermal expansion(c), μin./in. °C (μin./in. °F)	Thermal conductivity(c), cal/cm²/cm/s/ °C	Electrical resistivity, μΩ · cm	Modulus of elasticity in tension, 10⁶ psi
Hf	519(ff) (288)(ff)	0.223(gg)	35.1(w)	…
He	… (…)	3.32×10^{-4}	…	…
Ho	… (…)	…	87(w)	11(x)
H	… (…)	4.06×10^{-4}	…	…
In	33 (18)	0.057	8.37(b)	1.57
I	93 (52)	10.4×10^{-4}	1.3×10^{15}(b)	…
Ir	6.8 (3.8)	0.14	5.3(b)	76
Fe	11.76(jj) (6.53)(jj)	0.18(kk)	9.71(b)	28.5 ± 0.5
Kr	… (…)	0.21×10^{-4}	…	…
La	5 (2.77)	0.033(u)	57(y)	10 to 11(x)
Lw	… (…)	…	…	…
Pb	29.3(mm) (16.3)(mm)	0.083(f)	20.648(b)	2
Li	56 (31)	0.17	8.55(f)	…
Lu	… (…)	…	79(w)	…
Mg	27.1(nn) (15.05)(nn)	0.367	4.45(b)	6.35(pp)
Mn	22(rr) (12.22)(rr)	…	185(ss)	23
Mv	… (…)	…	…	…
Hg	… (…)	0.0196(f)	98.4(tt)	…
Mo	4.9(d) (2.7)(d)	0.34	5.3(f)	47
Nd	6 (3.33)	0.031(uu)	64(w)	…
Ne	… (…)	0.00011	…	…
Np	… (…)	…	…	…
Ni	13.3(s) (7.39)(s)	0.22(w)	6.84(b)	30(vv)
Nb	7.31 (4.06)	0.125(f)	12.5(f)	…
N	… (…)	0.000060	…	…
No	… (…)	…	…	…
Os	4.6(ww) (2.6)(ww)	…	9.5(b)	81
O	… (…)	0.000059	…	…
Pd	11.76 (6.53)	1.68(ee)	10.8(b)	16.3
P	125 (70)	…	10^{17}(xx)	…
Pt	8.9 (4.9)	0.165(yy)	10.6(b)	21.3(zz)
Pu	55(bbb) (30.55)(bbb)	0.020(w)	141.4(ccc)	14(ddd)
Po	… (…)	…	…	…
K	83 (46)	0.24	6.15(f)	…
Pr	4 (2.22)	0.28(uu)	68(w)	7 to 14(x)
Pm	… (…)	…	…	…

(continued)

Physical properties of the elements (continued)

Element	Density(a), g/cm³ (lb/in.³)	Boiling point, °C (°F)	Specific heat(b), cal/g · °C (J/kg · K)	Heat of fusion, cal/g (Btu/lb)
Protactinium (Pa)	15.4 (0.556)	... (...)	... (...)	... (...)
Radium (Ra)	5.0 (0.18)	... (...)	... (...)	... (...)
Radon (Rn)	9.960(g) (0.3956)(g)	−61.8 (−79.2)	... (...)	... (...)
Rhenium (Re)	21.0 (0.76)	5900 (10 650)	0.033 (140)	... (...)
Rhodium (Rh)	12.41 (0.4480)	4500 (8130)	0.059(f) (250)(f)	... (...)
Rubidium (Rb)	1.53 (0.0552)	688 (1270)	0.080 (330)	6.5 (11.79)
Ruthenium (Ru)	12.45 (0.4494)	4900 (8850)	0.057(f) 240(f)	... (...)
Samarium (Sm)	7.49 (0.270)	1630 (2966)	0.042(hhh) (180)(hhh)	17.29 (31.12)
Scandium (Sc)	2.9 (0.10)	2730 (4946)	0.134 (561)	84.52 (152.14)
Selenium (Se)	4.8 (0.17)	685 ± 1 (1265 ± 2)	0.084(u) (350)(u)	16.4 (29.5)
Silicon (Si)	2.33 (0.0841)	2680 (4860)	0.162(f) (678)(f)	432 (778)
Silver (Ag)	10.49 (0.3787)	2210 (4010)	0.0559(f) (234)(f)	25 (45)
Sodium (Na)	0.9712 (0.03506)	892 (1638)	0.295 (1240)	27.5 (49.5)
Strontium (Sr)	2.60 (0.0939)	1380 (2520)	0.176 (737)	25 (45)
Sulfur, yellow (S)	2.07 (0.0747)	444.6 (832.3)	0.175 (733)	9.3 (16.7)
Tantalum (Ta)	16.6 (0.599)	5425 ± 100 (9800 ± 200)	0.034(w) (140)	38 (68)
Technetium (Tc)	11.5 (0.415)	... (...)	... (...)	... (...)
Tellurium (Te)	6.24 (0.225)	989.8 ± 3.8 (1813.6 ± 6.8)	0.047 (200)	32 (58)
Terbium (Tb)	8.25 (0.298)	2530 (4586)	0.044 (180)	24.54 (44.17)
Thallium (Tl)	11.85 (0.4278)	1457 (2655)	0.031 (130)	5.04 (9.07)
Thorium (Th)	11.5 (0.415)	3850 ± 350 (7000 ± 600)	0.034 (140)	<19.82 (<35.68)
Thulium (Tm)	9.31 (0.336)	1720(ggg) (3130)(ggg)	0.038 (160)	26.04 (46.87)
Tin (Sn)	7.30 (0.264)	2270 (4120)	0.054 (230)	14.5 (26.1)
Titanium (Ti)	4.51 (0.163)	3260 (5900)	0.124 (519)	104(k) (188)(k)
Tungsten (W)	19.3 (0.697)	5930 (10 706)	0.033 (140)	44 (70)
Uranium (U)	19.07 (0.6884)	3818 (6904)	0.02709(vvv) (113.4)(vvv)	... (...)
Vanadium (V)	6.11 (0.221)	3400 (6150)	0.119(t) (498)(t)	... (...)
Xenon (Xe)	5.896(g) (0.2128)(g)	−108.0 (−162.4)	... (...)	... (...)
Ytterbium (Yb)	6.96 (0.251)	1530 (2786)	0.035 (150)	12.71 (22.88)
Yttrium (Y)	4.47 (0.161)	3030 (5490)	0.071 (300)	46 (83)
Zinc (Zn)	7.13 (0.257)	906 (1663)	0.0915 (383)	24.09 (43.36)
Zirconium (Zr)	6.49 (0.234)	3580 (6470)	0.067 ± 0.001 (280 ± 4)	60(k) (110)(k)

(a) Density may depend considerably on previous treatment. (b) At 20 °C (68 °F). (c) Near 20 °C (68 °F). (d) From 20 to 100 °C (68 to 212 °F). (e) From 20 to 60 °C (68 to 140 °F). (f) At 0 °C (32 °F). (g) Gas, grams per litre at 20 °C (68 °F) and 760 mm (30 in.). (h) 28 atm. (j) Sublimes. (k) Estimated. (m) From 25 to 100 °C (77 to 212 °F). (n) Annealed, commercial purity. (p) Approximate. (q) From 20 to 750 °C (68 to 1380 °F). (r) Sand cast. (s) From 0 to 100 °C (32 to 212 °F). (t) For α at 0 to 400 °C (32 to 750 °F). (u) Annealed. (v) At 28 °C (82 °F). (w) At 25 °C (77 °F). (x) Measured from stress-strain relationship on as-cast metal. (y) From 0 to 26 °C (32 to 70 °F). (z) Near 40 °C (105 °F); the coefficient of expansion of gadolinium changes rapidly between −100 and +100 °C (−150 and +212 °F). (aa) From 0 to 30 °C (32 to 86 °F). (bb) At melting point. (cc) For a-axis; 8.1 for b-axis and 54.3 for c-axis. (dd) Ohm · cm of intrinsic germanium at 300 K. (ee) At 18 °C (64 °F). (ff) From 20 to 200 °C (68 to 390 °F). (gg) W/cm/°C at 50 °C (120 °F). (jj) At 25 °C (77 °F) for high-purity k iron. (kk) For ingot iron at 0 °C (32 °F). (mm) From 17 to 100 °C (63 to 212 °F). (nn) Along a-axis; 24.3 along c-axis. (pp) Dynamic; static, 5.77; both for 99.98% magnesium. (qq) α; γ 0.120; both at 25.2 °C (77.3 °F). (rr) α; γ, 14; both from 0 to 100 °C (32 to 212 °F).

(continued)

Symbol	Coefficient of linear thermal expansion(c), μin./in. °C (μin./in/ °F)	Thermal conductivity(c), cal/cm²/cm/s/°C	Electrical resistivity, μ Ω · cm	Modulus of elasticity in tension, 10⁶ psi
Pa	... (...)	...	...	...
Ra	... (...)	...	...	...
Rn	... (...)	...	...	...
Re	6.7(eee) (3.7)(eee)	0.17	19.3(b)	66.7(b)
Rh	8.3 (4.6)	0.21(yy)	4.51(b)	42.5(fff)
Rb	90 (50)	...	12.5(b)	...
Ru	9.1 (5.1)	...	7.6(f)	60(p)
Sm	... (...)	...	88(w)	8(x)
Sc	... (...)	...	61(jjj)	...
Se	37 (21)	7 to 18.3 × 10⁻⁴	12(f)	8.4
Si	2.8 to 7.3 (1.6 to 4.1)	0.20	10(f)	16.35(kkk)
Ag	19.68(s) (10.9)(s)	1.0(f)	1.59(b)	11
Na	71 (39)	0.32	4.2(f)	...
Sr	... (...)	...	23(b)	...
S	64 (36)	6.31 × 10⁻⁴	2 × 10²³(b)	...
Ta	6.5 (3.6)	0.130	12.45(w)	27(b)
Tc	... (...)	...	...	...
Te	16.75 (9.3)	0.014	436 000(mmm)	6
Tb	7 (3.88)	...	...	...
Tl	28 (16)	0.093	18(f)	...
Th	12.5(nnn) (6.9)(nnn)	0.090(ppp)	13(f)	...
Tm	... (...)	...	79(w)	...
Sn	23(qqq) (13)(qqq)	1.50(e)	11(rrr)	6 to 6.5(sss)
Ti	8.41 (4.67)	6.6(ttt)	42(b)	16.8
W	4.6 (2.55)	0.397(e)	5.65(uuu)	50
U	6.8 to 14.1(www) (3.8 to 7.8)(www)	0.07(xxx)	30(yyy)	24
V	8.3(zzz) (4.6)(zzz)	0.074(ppp)	24.8 to 26.0(b)	18 to 20
Xe	... (...)	1.24 × 10⁻⁴	...	...
Yb	25 (13.9)	...	29(w)	...
Y	... (...)	0.035(uu)	57(aaaa)	17(x)
Zn	39.7(aaaa) (22.0)(aaaa)	0.27(w)	5.916(b)	(bbb)
Zr	5.85(cccc) (3.2)(cccc)	0.211(dddd)	40	13.7

(ss) α at 20 °C (68 °F). (tt) At 50 °C (122 °F). (uu) At –2.22 °C (28 °F). (vv) At 0 °C (32 °F), unmagnetized. (ww) At 50 °C (122 °F), parallel to a-axis, mean value; parallel to c-axis at 50 °C (122 °F), 5.8. (xx) At 11 °C (51.8 °F). (yy) At 17 °C (63 °F). (zz) For small cyclic strains. (aaa) For α at 25 °C (77 °F). (bbb) From 21 to 104 °C (70 to 219 °F). (ccc) At 107 °C (224.6 °F). (ddd) At 25 °C (77 °F), for cast metal. (eee) From 20 to 500 °C (68 to 930 °F). (fff) For hard wire. (ggg) At –173 °C (–279 °F). (hhh) Calculated. (jjj) Average value at 22 °C (72 °F), zone-refined bar. (kkk) Chill cast specimen 90.2 by 24.6 by 24.6 mm (3.55 by 0.97 by 0.97 in.). (mmm) At 23 °C (73 °F). (nnn) From 25 to 1000 °C (77 to 1830 °F), for iodide thorium. (ppp) At 100 °C (212 °F). (qqq) From 0 to 100 °C (32 to 212 °F), for polycrystalline metal. (rrr) At 0 °C (32 °F), for white tin. (sss) Cast tin. (ttt) Btu · ft/h · ft² · °F at –400 °F. (uuu) At 27 °C (80.6 °F). (vvv) At 27 °C (80 °F). (www) Rolled rods. (xxx) At 70 °C (158 °F). (yyy) Crystallographic average. (zzz) From 23 to 100 °C (73 to 212 °F). (aaaa) From 20 to 250 °C (68 to 480 °F), for polycrystalline metal. (bbbb) Pure zinc has no clearly defined modulus of elasticity. (cccc) α, polycrystalline. (dddd) W/cm/°C at 27 °C (80.6 °F)

Temperature-dependent allotropic structures of the elements(a)

Allotrope	Pearson symbol	Lattice parameters, nm a	b	c	c/a, or α or β	Stability range(b)
αAm	hP4	0.3468	...	1.1241	2×1.621	RT
βAm	cF4	0.4894	...	...	...	>605 °C
αAr	cF4	0.5312	...	...	...	<83.8 K
βAr	hP2	0.3760	...	0.6141	1.633	<83.8 K
αAs	hR2	0.41320	....	...	α = 54.12°	RT
εAs	oC8	0.362	1.085	0.448	...	>448 °C
αBe	hP2	0.22857	...	0.35839	1.5680	RT
βBe	cI2	0.25515	...	...	...	>1250 °C
αBk	hP4	0.3416	...	1.1069	2×1.620	RT
βBk	cF4	0.4997	...	...	...	~RT
αCa	cF4	0.55884	...	...	...	RT
γCa	cI2	0.4480	...	...	...	>737 °C
αCe	cF4	0.51610	...	...	...	RT
βCe	hP4	0.3673	...	1.1802	2×1.607	<263 K
γCe	cF4	0.485	...	...	...	<95 K
αCm	hP4	0.3496	...	1.1331	2×1.621	RT
βCm	cF4	0.4382	...	...	...	~RT
αCo	hP2	0.25071	...	0.40694	1.6232	RT
βCo	cF4	0.35445	...	...	...	>388 °C
αDy	hP2	0.35915	...	0.56501	1.5732	RT
βDy	cI2	?	...	...	...	>970 °C
γDy	oF4	0.3595	0.6184	0.5678	...	<86 K
αEr	hP2	0.35592	...	0.55850	1.5692	RT
βEr	cI2	?	...	...	...	HT
αF	mC8	0.550	0.338	0.728	β = 102.17°	4.2 K
βF	cP16	0.667	...	...	...	>45.6 K
αFe	cI2	0.28665	...	...	...	RT
γFe	cF4	0.36467	...	...	...	>910 °C
δFe	cI2	0.29135	...	...	...	>1390 °C
αGd	hP2	0.36336	...	0.57810	1.5910	RT
βGd	cI2	0.406	...	...	...	>1262 °C
αH	hP2	0.3771	...	0.6152	1.631	4.2 K
βH	cF4	0.5334	...	...	...	<1.3 K
αHf	hP2	0.31946	...	0.50511	1.5811	RT
βHf	cI2	0.3610	...	...	...	>1995 °C
αHo	hP2	0.35778	...	0.56178	1.5702	RT
βHo	cI2	?	...	...	...	HT
αLa	hP4	0.37740	...	1.2171	2×1.6125	RT
βLa	cF4	0.53045	...	...	...	>340 °C
γLa	cI2	0.4265	...	...	...	>868 °C
αLi	hP2	0.3111	...	0.5093	1.637	<72 K
βLi	cI2	0.35093	...	...	...	RT
αLu	hP2	0.35052	...	0.55494	1.5832	RT
βLu	cI2	?	...	...	...	HT
αMn	cI58	0.89219	...	...	...	RT
βMn	cP20	0.63152	...	...	...	>727 °C
γMn	cF4	0.38624	...	...	...	>1095 °C
δMn	cI2	0.30806	...	...	...	>1135 °C
αN	cP8	0.5659	...	...	...	4.2 K
βN	hP4	0.4046	...	0.6629	1.638	>35.6 K
αNa	hP2	0.3767	...	0.6154	1.634	<36 K
βNa	cI2	0.42096	...	...	...	RT
αNd	hP4	0.36582	...	1.17966	2×1.6124	RT
βNd	cI2	0.413	...	...	...	>862 °C
αNp	oP8	0.6683	0.4723	0.4887	...	RT
βNp	tP4	0.4896	...	0.3387	0.692	>280 °C
γNp	cI2	0.352	...	...	...	>577 °C
αO	mC4	0.5403	0.3429	0.5086	β = 132.53°	4.2 K
βO	hR2	0.4210	...	...	α = 46.27°	>23.9 K
γO	cP16	0.683	...	...	...	>43.6 K
αPa	tI2	0.3945	...	0.3242	0.822	RT
βPa	cI2	0.381	...	...	...	>1170 °C
αPo	cP1	0.3366	...	...	...	RT
βPo	hR1	0.3373	...	...	α = 98.08°	>54 °C
αPr	hP4	0.36721	...	1.18326	2×1.6111	RT
βPr	cI2	0.413	...	...	...	>821 °C
αPu	mP16	0.6183	0.4822	1.0968	α = 101.78°	RT
βPu	mI34	0.9284	1.0463	0.7859	α = 92.13°	>122 °C
γPu	oF8	0.31587	0.57682	1.0162	...	>235 °C
δPu	cF4	0.46371	...	0.3279	...	>319 °C
δ'Pu	tI2	0.33261	...	0.44630	1.3418	>450 °C
εPu	cI2	0.5703	...	...	...	>471 °C
αSc	hP2	0.33088	...	0.52680	1.5921	RT
βSc	cI2	?	...	...	...	>1334 °C
αSe	mP32	0.9054	0.9083	0.2336	β = 90.82°	RT
βSe	mP64	1.5018	1.4713	0.8879	β = 93.6°	RT
γSe	hP3	0.43655	...	0.49576	1.1356	RT
αSm	hR3	0.36290	...	2.6207	4.5×1.6084	RT
βSm	cI2	?	...	...	...	>917 °C
αSn	cF8	0.64892	...	...	...	<18 °C
βSn	tI4	0.58316	...	0.31815	0.5456	RT
αSr	cF4	0.6084	...	...	...	RT
βSr	hP2	0.428	...	0.705	1.647	>213 °C
γSr	cI2	0.487	...	...	...	>605 °C
αTb	hP2	0.36055	...	0.56966	1.5800	RT
βTb	cI2	?	...	...	...	>1316 °C
αTh	cF4	0.50851	...	...	...	RT
βTh	cI2	0.411	...	...	...	>1400 °C
αTi	hP2	0.29503	...	0.46836	1.5875	RT
βTi	cI2	0.33065	...	...	...	>900 °C
αTl	hP2	0.34563	...	0.55263	1.5989	RT
βTl	cI2	0.3879	...	...	...	>230 °C
αTm	hP2	0.3575	...	0.55540	1.5700	RT
βTm	cI2	?	...	...	...	HT
αU	oC4	0.28538	0.58680	0.49557	...	RT
βU	tP30	1.0759	...	0.5654	0.526	>662 °C
γU	cI2	0.3524	...	...	...	>772 °C
αYb	cF4	0.54848	...	...	...	RT
βYb	cI2	0.444	...	...	...	>732 °C
γYb	hP2	0.38799	...	0.63859	1.6459	<270 K
αZr	hP2	0.32317	...	0.51476	1.5928	RT
βZr	cI2	0.3609	...	...	...	>865 °C

(a) The accuracy of the data in this table is considered to be reliable to ±2 in the last reported digit. (b) RT, room temperature; HT, high temperature

Pressure-dependent allotropic structures of the elements(a)

Allotrope	Pearson symbol	Lattice parameters, nm			c/a, or α or β	Stability range(b)
		a	b	c		
Al-I	cF4	0.40496	...	...	...	RTP
Al-II	hP2	0.2693	...	0.4398	1.633	>205 kB
αAm	hP4	0.3468	...	1.1241	2×1.621	RTP
γAm	oC4	0.3063	0.5968	0.5169	...	>150 kB
αBa	cI2	0.5023	...	...	...	RTP
βBa	hP2	0.3901	...	0.6154	1.578	>53.3 kB
γBa	?	...	...	...	...	>230 kB
αBi	hR2	0.4760	...	...	α = 57.23°	RTP
βBi	?	...	...	...	...	>28 kB
γBi	mP3	0.605	0.420	0.465	β = 85.33°	>30 kB
δBi	?	...	...	...	...	>43 kB
εBi	?	...	...	...	...	>65 kB
ζBi	cI2	3.800	...	...	...	>90 kB
C (graph)	hP4	0.24612	...	0.67090	2.7259	RTP
C (dia)	cF8	0.35669	...	...	...	600 kB
αCe	cF4	0.51610	...	...	...	RTP
α'Ce	cF4	0.482	...	...	...	>15 kB
Ce-III	mI2	0.4762	0.3170	0.3169	β = 91.7°	>51 kB
αCr	cI2	0.28847	...	...	...	RTP
α'Cr	tI2	0.2882	...	0.2887	1.002	HP
Cs-I	cI2	0.6141	...	...	...	RTP
Cs-II	cF4	0.5984	...	...	...	>23.7 kB
Cs-III	cF4	0.5800	...	...	...	>42.2 kB
αDy	hP2	0.35915	...	0.56501	1.5731	RTP
γDy	hR3	0.3436	...	2.4830	4.5×1.606	>75 kB
αFe	cI2	0.28665	...	...	...	RTP
εFe	hP2	0.2485	...	0.3990	1.606	>130 kB
αGa	oC8	0.45192	0.76586	0.45258	...	RTP
βGa	tI2	0.2808	...	0.4458	1.587	>12 kB
γGa	oC40	1.0593	1.3523	0.5203	...	>30 kB; 220 K
αGd	hP2	0.36336	...	0.57810	1.5910	RTP
γGd	hR3	0.361	...	2.603	4.5×1.60	>30 kB
αGe	cF8	0.56574	...	...	...	RTP
βGe	tI4	0.4884	...	0.2692	0.551	>120 kB
γGe	tP12	0.593	...	0.698	0.18	Decompressed βGe
δGe	cI16	0.692	...	...	...	>120 kB
αHe	hP2	0.3577	...	0.5842	1.633	4.2 K
βHe	cF4	4.240	...	...	...	1.25 kB; 1.6 K
γHe	cI2	4.110	...	...	...	0.3 kB; 1.73 K
αHg	hR1	0.3005	...	...	α = 70.53°	227 K
βHg	tI2	0.3995	...	0.2825	0.707	HP; 77 K
αHo	hP2	0.35778	...	0.56178	1.5702	RTP
γHo	hR3	0.334	...	2.45	4.5×1.63	>40 kB
K-I	cI2	0.5321	...	...	...	RTP
K-II	?	...	...	...	...	280 kB; 77 K
K-III	?	...	...	...	...	360 kB; 77 K
αLa	hP4	0.37740	...	1.2171	2×1.6125	RTP
β'La	cF4	0.517	...	...	...	>20 kB
αLi	cI2	0.35093	...	...	...	RTP
γLi	cF4	0.4388	...	...	...	CW at <72 K
αN	cP8	0.5659	...	...	...	4.2 K
N-II	tP4	0.3957	...	5.101	1.289	>33 kB; 20 K
αNd	hP4	0.36582	...	1.17966	2×1.6124	RTP
γNd	cF4	0.480	...	...	...	>50 kB
Pb-I	cF4	0.49502	...	...	...	RTP
Pb-II	hP2	0.3265	...	0.5387	1.653	>103 kB
αPr	hP4	0.36721	...	1.18326	2×1.6111	RTP
γPr	cF4	0.488	...	...	...	>40 kB
αRb	cI2	0.5703	...	...	...	RTP
βRb	?	...	...	...	...	>10.8 kB
γRb	?	...	...	...	...	>20.5 kB
Sb-I	hR2	0.45065	...	...	α = 57.11°	RTP
Sb-II	cP1	0.2992	...	...	...	>50 kB
Sb-III	hP2	0.3376	...	0.5341	1.582	>75 kB
Sb-IV	mP4	0.556	0.404	0.422	β = 86.0°	>140 kB
αSi	cF8	0.54306	...	...	...	RTP
βSi	tI4	0.4686	...	0.2585	0.551	>95 kB
γSi	cI16	0.636	...	...	...	>160 kB
δSi	hP4	0.380	...	0.628	1.635	Decompressed βSi
αSm	hR3	0.36290	...	2.607	4.5×1.6048	RTP
γSm	hP4	0.3618	...	1.166	2×1.611	>40 kB
βSn	tI4	0.58316	...	0.31815	0.5456	RTP
γSn	tI2	0.370	...	0.337	0.911	>90 kB
αSr	cF4	0.6084	...	...	...	RTP
Sr-II	cI2	0.4437	...	...	...	>35 kB
αTb	cP2	0.36055	...	0.56966	1.5800	RTP
Tb-II	hR3	0.341	...	2.45	4.5×1.60	>60 kB
αTe	hP3	0.44561	...	0.59271	1.3301	RTP
βTe	hR2	0.469	...	...	α = 53.30°	>30 kB
γTe	hR1	0.3002	...	...	α = 103.3°	>70 kB
αTi	hP2	0.34563	...	0.55263	1.5989	RTP
ωTi	hP2	0.4625	...	0.2813	0.608	Decompressed
αTl	hP2	0.34563	...	0.55540	1.5700	RTP
γTl	cF4	...	...	...	...	HP
αZr	hP2	0.32217	...	0.51476	1.5928	RTP
ωZr	hP2	0.506	...	0.3109	0.617	Decompressed

(a) The accuracy of the data in this table is considered to be reliable to ± 2 in the last reported digit. (b) RTP, room temperature and pressure; HP, high pressure. High-pressure data refers to pressures within ± 1 of the last reported digit in kilobars.

Magnetic phase transition temperatures of the elements

Chemical symbol	Atomic number	Allotrope	Phase transition temperature (T_c), K	Type of magnetic ordering(a)	Phase transition temperature (T_{c2}), K	Type of magnetic ordering(a)	Phase transition temperature (T_{c3}), K	Type of magnetic ordering(a)	Saturation magnetic moment, μ_B
Ce(b)	58	β-dcph	13.7	AC?	12.5	AC?	...	...	2.61
		γ-fcc	14.4	AC?	...	...	...	...	~2.5
Cm	96	...	52	AC	...	...	...	...	...
Co	27	fcc	1388	FM	...	...	...	...	1.715
Cr	24	...	312.7	AI	...	...	...	...	0.45
Dy	66	α-cph	179.0	AI	89.0	FM	...	...	10.33
Er	68	...	85.0	AI	53	AC	20.0	CF	9.1
Eu	63	...	90.4	AC	...	...	...	...	5.9
Fe(c)	26	α-bcc	1044	FM	...	...	...	...	2.216
		γ-fcc	67	AC	...	...	...	...	0.75
Gd	64	α-cph	293.4	FM	...	...	...	...	7.63
Ho	67	...	132.0	AI	20.0	CF	...	...	10.34
Mn	25	α-bcc	100	AC	...	...	...	...	(d)
Nd	60	α-dcph	19.9	AI	7.5	AC	...	...	1.84
Ni	28	...	627.4	FM	...	...	...	...	0.616
Pm	61	α-dcph	98	FM?	...	...	...	...	0.24
Pr	59	α-dcph	0.06	AC	...	...	...	...	0.36
Sm	62	α-rhomb.	106	h, A(e)	13.8	c, A(e)	...	...	0.1
Tb	65	α-cph	230.0	AI	219.5	FM	...	...	9.34
Tm	69	...	58.0	AI	40 to 32	FI	...	...	7.14

(a) FM, transition from paramagnetic to ferromagnetic state; AC, transition to periodic (antiferromagnetic) state that is commensurate with the lattice periodicity (e.g., spins on three atom layers directed up followed by three layers down, etc.); AI, transition to periodic (antiferromagnetic) state that is generally not commensurate with the lattice periodicity (e.g., helical spin ordering); CF, transition to conical ferromagnetic state (combination of planar helical antiferromagnet plus ferromagnetic component); FI, transition to ferromagnetic periodic structure (unequal number of up and down spin layers). (b) Ce exists in five crystal structures, two of which are magnetic (γ, fcc; and β, dcph). γCe is estimated to be antiferromagnetic below 14.4 K by extrapolation from fcc Ce-La alloys. (αCe does not exist in pure form below ~100 K.) βCe is thought to exhibit antiferromagnetism on the hexagonal lattice sites below 13.7 K and on the cubic sites below 12.5 K. (c) Magnetic measurements quoted in table below for γFe are for fcc Fe precipitated in copper. (d) The magnetic moment assignments of Mn are complex; see *Magnetic Materials*, R.S. Tebble and D.J. Craik, London: Wiley Interscience, 1969, p 60-62. (e) h, A; c, A indicate that sites of hexagonal and cubic point symmetry order antiferromagnetically, but at different temperatures

Properties of the superconductive elements

Chemical symbol	Critical temperature (T_c)(a), K	Pressure(b), kbar	Critical field (H_o), oersted	Debye temperature (θ_D)(c), K	Electronic specific heat(γ) (c), mJ/mol·K	Notes
Al	1.75	...	104.9	420	1.35	...
	1.98-0.075	0-62	...	...	...	d
	1.15-~5.7	...	...	...	...	e, f
Am						
(α, ?)	0.6	...	...	...	...	...
(β, ?)	1.0	...	...	...	...	...
As	0.31-0.5	220-140	...	...	...	d
	0.2-0.25	~140-100	...	...	...	d
BaII	~1-1.8	~55-85	...	...	...	d
III	1.8-5	~85-144	...	...	...	d
IV	4.5-5.4	144-190	...	...	...	d
Be	0.026	...	...	...	0.21	...
	5-9.75	...	...	...	...	e, f
Bi	6.17-2.6	...	...	...	...	e
II	3.9	25-27	...	...	...	d
III	6.55-7.25	28-38	...	...	...	d
IV	7.0, 8.7-6.0	43, 43-62	...	...	...	d
V	6.7, 8.3	48-80	...	...	...	d
VI	8.55	90, 92-101	...	...	...	d
VII(?)	8.2	30	...	...	...	d
Cd	0.517	...	28	209	0.69	...
(g)	0.79-0.91	...	...	...	...	e
(h)	0.53-0.59	...	...	...	...	e
αCe	0.020-0.045	20-35	...	...	...	d
α'Ce	1.9-1.3	45-125	...	...	...	d
Cs V	~1.5	>125	...	...	...	d
Ga	1.083	...	58.3	325	0.60	...
	2.5-8.5	...	...	...	...	e, f
βGa	5.9, 6.2	...	560	...	...	...
γGa	7	...	950	...	...	f
ΔGa	7.85	...	815	...	...	f
Ga II	6.38	≥35	...	...	...	d
II'	7.5	≥35	...	...	...	d, j
Ge	5.35	115	...	...	...	d
Hf	0.128	...	12.7	...	2.21	...
αHg	4.154	...	411	87, 71.9	1.81	...
βHg	3.949	...	339	93	1.37	...
Hg	3.87-4.5	...	...	...	...	e
In	3.408	...	281.5	109	1.672	...
	2.2-5.6	...	...	...	...	e, f
Ir	0.1125	...	16	425	3.19	...
αLa	4.88	...	800	151	9.8	...
βLa	6.00	...	1096, 1600	139	11.3	...
La	~5.5-12.9	0-210	...	...	...	d
	3.55-6.74	...	...	...	...	e
Lu	0.1	...	350	...	...	...
	0.022-1.0	45-190	...	...	...	d
Mo	0.915	...	96	460	1.83	...
Nb	3.3-8.0	...	...	...	...	e
	9.25	...	2060	276	7.80	f
	2.0-10.1	...	...	...	...	e
Os	0.66	...	70	500	2.35	...
P	5.8	170	...	...	...	d
Pa	1.4	...	...	...	...	...
Pb	7.196	...	803	96	3.1	...
	1.8-7.5	...	...	...	...	e
II	3.55	160	...	...	...	d
Re	1.697	...	200	4.5	2.35	...
	1.7-~7	...	...	...	...	e
II	2.3 max	...	...	...	...	d, k
Ru	0.49	...	69	580	2.8	...
Sb	2.6-2.7	...	...	...	...	d, m
III	3.55-3.40	85-~150	...	...	...	d
Se II	6.75, 6.95	~130	...	...	...	d
Si	6.7-7.1	120-130	...	...	...	d
Sn	3.722	...	305	195	1.78	...
	3.5-~6	...	...	...	...	e
II	5.2-4.85	125-160	...	...	...	d
III	5.30	113	...	...	...	d
Ta	4.47	...	829	258	6.15	...
	<1.7-4.51	...	...	...	...	e, f
Tc	7.8	...	1410	411	6.28	f
	4.6-7.7	...	...	...	...	e
Te II	2.4-5.1	38-55	...	...	...	d
III	4.1-4.2	~53-62	...	...	...	d
IV	4.72-4	63-80	...	...	...	d
(?)	3.3-2.8	100-260	...	...	...	d
Th	1.38	...	160	165	4.32	...
Ti	0.40	...	56	415	3.3	...
	1.3 max	...	...	...	...	e
Tl	2.38	...	178	78.5	1.47	...
	1.45	35	...	...	...	d, n
	1.95	35	...	...	...	d, p
	2.33-2.96	...	...	...	...	e
U	2.4-0.4	10-85	...	...	...	d
V	5.40	...	1408	383	9.82	...
	1.8-6.02	...	...	...	...	e
W	0.0154	...	1.15	383	0.90	...
	<1.0-4.1	...	...	...	...	e
Y	2.3-1.7-2.5	110-125-160	...	...	...	d
Zn	0.85	...	54	310	0.66	...
	0.77-1.70, ~1.9	...	...	...	...	e
Zr	0.61	...	47	290	2.77	...
ωZr	0.65, 0.95	...	...	...	...	...
	1-1.7	60-~130	...	...	...	d, q

(a) Range is given where the element exhibits superconductivity when prepared by the application of high pressure. (b) 1 kbar = 10^8 newton/meter^2 = 0.987 katm. (c) For a complete data set, see Phillips, N.E., "Low Temperature Heat Capacity of Metals," *Critical Reviews in Solid State Sciences, 2,* 467-554 (1972). Also Mendelssohn, K., in *Cryophysics,* Interscience, New York, p 178 (1960); Gschneidner, K.A., Jr., in *Solid State Physics, 16,* 275-426 (1964); Parkinson, D.H., *Rep. Progr. Phys., 21,* 226 (1958); and Heiniger, F., Bucher, E., and Muller, J., "Low Temperature Specific Heat of Transition Metals and Alloys," *Phys. Kondens. Materie, 5,* 243-284 (1966). (d) Elements exhibiting superconductivity under or after application of high pressure. (e) Elements in thin films condensed usually at low temperatures. (f) For high magnetic field superconductive properties, see sources listed in footnote (c). (g) Disordered. (h) Ordered. (j) 35 kbar, then P removed. (k) "Plastic" compression. (m) Prepared at 120 kbar, held below 77 K. (n) Cubic form. (p) Hexagonal form. (q) Omega form, metastable

Metals and Alloys

Density of metals and alloys

Metal or alloy	Density g/cm³	lb/in.³
Aluminum and aluminum alloys		
Aluminum (99.996%)	2.6989	0.0975
Wrought alloys		
EC, 1060 alloys	2.70	0.098
1100	2.71	0.098
2011	2.82	0.102
2014	2.80	0.101
2024	2.77	0.100
2218	2.81	0.101
3003	2.73	0.099
4032	2.69	0.097
5005	2.70	0.098
5050	2.69	0.097
5052	2.68	0.097
5056	2.64	0.095
5083	2.66	0.096
5086	2.65	0.096
5154	2.66	0.096
5357	2.70	0.098
5456	2.66	0.096
6061, 6063	2.70	0.098
6101, 6151	2.70	0.098
7075	2.80	0.101
7079	2.74	0.099
7178	2.82	0.102
Casting alloys		
A13	2.66	0.096
43	2.69	0.097
108, A108	2.79	0.101
A132	2.72	0.098
D132	2.76	0.100
F132	2.74	0.099
138	2.95	0.107
142	2.81	0.101
195, B195	2.81	0.101
214	2.65	0.096
220	2.57	0.093
319	2.79	0.101
355	2.71	0.098
356	2.68	0.097
360	2.64	0.095
380	2.71	0.098
750	2.88	0.104
40E	2.81	0.101
Copper and copper alloys		
Wrought coppers		
Pure copper	8.96	0.324
Electrolytic tough pitch copper (ETP)	8.89	0.321
Deoxidized copper, high residual phosphorus (DHP)	8.94	0.323
Free-machining copper		
0.5% Te	8.94	0.323
1.0% Pb	8.94	0.323
Wrought alloys		
Gilding, 95%	8.86	0.320
Commercial bronze, 90%	8.80	0.318
Jewelry bronze, 87.5%	8.78	0.317
Red brass, 85%	8.75	0.316
Low brass, 80%	8.67	0.313
Cartridge brass, 70%	8.53	0.308
Yellow brass	8.47	0.306
Muntz metal	8.39	0.303
Leaded commercial bronze	8.83	0.319

Metal or alloy	Density g/cm³	lb/in.³
Low-leaded brass (tube)	8.50	0.307
Medium-leaded brass	8.47	0.306
High-leaded brass (tube)	8.53	0.308
High-leaded brass	8.50	0.307
Extra-high-leaded brass	8.50	0.307
Free-cutting brass	8.50	0.307
Leaded Muntz metal	8.41	0.304
Forging brass	8.44	0.305
Architectural bronze	8.47	0.306
Inhibited admiralty	8.53	0.308
Naval brass	8.41	0.304
Leaded naval brass	8.44	0.305
Manganese bronze (A)	8.36	0.302
Phosphor bronze, 5% (A)	8.86	0.320
Phosphor bronze, 8% (C)	8.80	0.318
Phosphor bronze, 10% (D)	8.78	0.317
Phosphor bronze, 1.25%	8.89	0.321
Free-cutting phosphor bronze	8.89	0.321
Cupro-nickel, 30%	8.94	0.323
Cupro-nickel, 10%	8.94	0.323
Nickel silver, 65-18	8.73	0.315
Nickel silver, 55-18	8.70	0.314
High-silicon bronze (A)	8.53	0.308
Low-silicon bronze (B)	8.75	0.316
Aluminum bronze, 5% Al	8.17	0.294
Aluminum bronze, (3)	7.78	0.281
Aluminum-silicon bronze	7.69	0.278
Aluminum bronze, (1)	7.58	0.274
Aluminum bronze, (2)	7.58	0.274
Beryllium copper	8.23	0.297
Casting alloys		
Chromium copper (1% Cr)	8.7	0.31
88Cu-10Sn-2Zn	8.7	0.31
88Cu-8Sn-4Zn	8.8	0.32
89Cu-11Sn	8.78	0.317
88Cu-6Sn-1.5Pb-4.5Zn	8.7	0.31
87Cu-8Sn-1Pb-4Zn	8.8	0.32
87Cu-10Sn-1Pb-2Zn	8.8	0.32
80Cu-10Sn-10Pb	8.95	0.323
83Cu-7Sn-7Pb-3Zn	8.93	0.322
85Cu-5Sn-9Pb-1Zn	8.87	0.320
78Cu-7Sn-15Pb	9.25	0.334
70Cu-5Sn-25Pb	9.30	0.336
85Cu-5Sn-5Pb-5Zn	8.80	0.318
83Cu-4Sn-6Pb-7Zn	8.6	0.31
81Cu-3Sn-7Pb-9Zn	8.7	0.31
76Cu-2.5Sn-6.5Pb-15Zn	8.77	0.317
72Cu-1Sn-3Pb-24Zn	8.50	0.307
67Cu-1Sn-3Pb-29Zn	8.45	0.305
61Cu-1Sn-1Pb-37Zn	8.40	0.304
Manganese bronze		
60 ksi	8.2	0.30
65 ksi	8.3	0.30
90 ksi	7.9	0.29
110 ksi	7.7	0.28
Aluminum bronze Alloy 9A	7.8	0.28

Metal or alloy	Density g/cm³	lb/in.³
Alloy 9B	7.55	0.272
Alloy 9C	7.5	0.27
Alloy 9D	7.7	0.28
Nickel silver		
12% Ni	8.95	0.323
16% Ni	8.95	0.323
20% Ni	8.85	0.319
25% Ni	8.8	0.32
Silicon bronze	8.30	0.300
Silicon brass	8.30	0.300
Iron and iron alloys		
Pure iron	7.874	0.2845
Ingot iron	7.866	0.2842
Wrought iron	7.7	0.28
Gray cast iron	7.15(a)	0.258(a)
Malleable iron	7.27(b)	0.262(b)
0.06% C steel	7.871	0.2844
0.23% C steel	7.859	0.2839
0.435% C steel	7.844	0.2834
1.22% C steel	7.830	0.2829
Low-carbon chromium-molybdenum steels		
0.5% Mo steel	7.86	0.283
1Cr-0.5Mo steel	7.86	0.283
1.25Cr-0.5Mo steel	7.86	0.283
2.25Cr-1.0Mo steel	7.86	0.283
5Cr-0.5Mo steel	7.78	0.278
7Cr-0.5Mo steel	7.78	0.278
9Cr-1Mo steel	7.67	0.276
Medium-carbon alloy steels		
1Cr-0.35Mo-0.25V steel	7.86	0.283
H11 die steel (5Cr-1.5Mo-0.4V)	7.79	0.281
Other iron-base alloys		
A-286	7.94	0.286
16-25-6 alloy	8.08	0.292
RA-330	8.03	0.290
Incoloy	8.02	0.290
Incoloy T	7.98	0.288
Incoloy 901	8.23	0.297
T1 tool steel	8.67	0.313
M2 tool steel	8.16	0.295
H41 tool steel	7.88	0.285
20W-4Cr-2V-12Co steel	8.89	0.321
Invar (36% Ni)	8.00	0.289
Hipernik (50% Ni)	8.25	0.298
4% Si	7.6	0.27
10.27% Si	6.97	0.252
Stainless steels and heat-resistant alloys		
Corrosion-resistant steel castings		
CA-15	7.612	0.2750
CA-40	7.612	0.2750
CB-30	7.53	0.272
CC-50	7.53	0.272
CE-30	7.67	0.277
CF-8	7.75	0.280
CF-20	7.75	0.280
CF-8M, CF-12M	7.75	0.280
CF-8C	7.75	0.280
CF-16F	7.75	0.280
CH-20	7.72	0.279
CK-20	7.75	0.280
CN-7M	8.00	0.289

(a) 6.95 to 7.35 g/cm³ (0.251 to 0.265 lb/in.³). (b) 7.20 to 7.34 g/cm³ (0.260 to 0.265 lb/in.³). (c) Annealed. (d) As cast. (e) Face-centered cubic. (f) Hexagonal. (g) Body-centered cubic. (h) Close-packed hexagonal. (j) Rhombohedral

(continued)

Density of metals and alloys (continued)

Metal or alloy	Density g/cm³	lb/in.³
Heat-resistant alloy castings		
HA	7.72	0.279
HC	7.53	0.272
HD	7.58	0.274
HE	7.67	0.277
HF	7.75	0.280
HH	7.72	0.279
HI	7.72	0.279
HK	7.75	0.280
HL	7.72	0.279
HN	7.83	0.283
HT	7.92	0.286
HU	8.04	0.290
HW	8.14	0.294
HX	8.14	0.294
Wrought stainless and heat-resisting steels		
Type 301	7.9	0.29
Type 302	7.9	0.29
Type 302B	8.0	0.29
Type 303	7.9	0.29
Type 304	7.9	0.29
Type 305	8.0	0.29
Type 308	8.0	0.29
Type 309	7.9	0.29
Type 310	7.9	0.29
Type 314	7.72	0.279
Type 316	8.0	0.29
Type 317	8.0	0.29
Type 321	7.9	0.29
Type 347	8.0	0.29
Type 403	7.7	0.28
Type 405	7.7	0.28
Type 410	7.7	0.28
Type 416	7.7	0.28
Type 420	7.7	0.28
Type 430	7.7	0.28
Type 430F	7.7	0.28
Type 431	7.7	0.28
Types 440A, 440B, 440C	7.7	0.28
Type 446	7.6	0.27
Type 501	7.7	0.28
Type 502	7.8	0.28
19-9DL	7.97	0.29
Precipitation-hardening stainless steels		
PH15-7 Mo	7.804	0.2819
17-4 PH	7.8	0.28
17-7 PH	7.81	0.282
Nickel-base alloys		
D-979	8.27	0.299
Nimonic 80A	8.25	0.298
Nimonic 90	8.27	0.299
M-252	8.27	0.298
Inconel	8.51	0.307
Inconel "X" 550	8.30	0.300
Inconel 700	8.17	0.295
Inconel "713C"	7.913	0.2859
Waspaloy	8.23	0.296
René 41	8.27	0.298
Hastelloy alloy B	9.24	0.334
Hastelloy alloy C	8.94	0.323
Hastelloy alloy X	8.23	0.297
Udimet 500	8.07	0.291
GMR-235	8.03	0.290

Metal or alloy	Density g/cm³	lb/in.³
Cobalt-chromium-nickel-base alloys		
N-155 (HS-95)	8.23	0.296
S-590	8.36	0.301
Cobalt-base alloys		
S-816	8.68	0.314
V-36	8.60	0.311
HS-25	9.13	0.330
HS-36	9.04	0.327
HS-31	8.61	0.311
HS-21	8.30	0.300
Molybdenum-base alloy		
Mo-0.5Ti	10.2	0.368
Lead and lead alloys		
Chemical lead (99.90+% Pb)	11.34	0.4097
Corroding lead (99.73+% Pb)	11.36	0.4104
Arsenical lead	11.34	0.4097
Calcium lead	11.34	0.4097
5-95 solder	11.0	0.397
20-80 solder	10.2	0.368
50-50 solder	8.89	0.321
Antimonial lead alloys		
1% antimonial lead	11.27	0.407
Hard lead (96Pb-4Sb)	11.04	0.399
Hard lead (94Pb-6Sb)	10.88	0.393
8% antimonial lead	10.74	0.388
9% antimonial lead	10.66	0.385
Lead-base babbitt alloys		
Lead-base babbitt		
SAE 13	10.24	0.370
SAE 14	9.73	0.352
Alloy 8	10.04	0.363
Arsenical lead		
Babbitt (SAE 15)	10.1	0.365
"G" Babbitt	10.1	0.365
Magnesium and magnesium alloys		
Magnesium (99.8%)	1.738	0.06279
Casting alloys		
AM100A	1.81	0.065
AZ63A	1.84	0.066
AZ81A	1.80	0.065
AZ91A, B, C	1.81	0.065
AZ92A	1.82	0.066
HK31A	1.79	0.065
HZ32A	1.83	0.066
ZH42, ZH62A	1.86	0.067
ZK51A	1.81	0.065
ZE41A	1.82	0.066
EZ33A	1.83	0.066
EK30A	1.79	0.065
EK41A	1.81	0.065
Wrought alloys		
M1A	1.76	0.064
A3A	1.77	0.064
AZ31B	1.77	0.064
PE	1.76	0.064
AZ61A	1.80	0.065
AZ80A	1.80	0.065
ZK60A, B	1.83	0.066
ZE10A	1.76	0.064
HM21A	1.78	0.064
HM31A	1.81	0.065

Metal or alloy	Density g/cm³	lb/in.³
Nickel and nickel alloys		
Nickel (99.95% Ni + Co)	8.902	0.322
"A" Nickel	8.885	0.321
"D" Nickel	8.78	0.317
Duranickel	8.26	0.298
Cast nickel	8.34	0.301
Monel	8.84	0.319
"K" Monel	8.47	0.306
Monel (cast)	8.63	0.312
"H" Monel (cast)	8.5	0.31
"S" Monel (cast)	8.36	0.302
Inconel	8.51	0.307
Inconel (cast)	8.3	0.30
Ni-o-nel	7.86	0.294
Nickel-molybdenum-chromium-iron alloys		
Hastelloy B	9.24	0.334
Hastelloy C	8.94	0.323
Hastelloy D	7.8	0.282
Hastelloy F	8.17	0.295
Hastelloy N	8.79	0.317
Hastelloy W	9.03	0.326
Hastelloy X	8.23	0.297
Nickel-chromium-molybdenum-copper alloys		
Illium G	8.58	0.310
Illium R	8.58	0.310
Electrical resistance alloys		
80Ni-20Cr	8.4	0.30
60Ni-24Fe-16Cr	8.247	0.298
35Ni-45Fe-20Cr	7.95	0.287
Constantan	8.9	0.32
Tin and tin alloys		
Pure tin	7.3	0.264
Soft solder (30% Pb)	8.32	0.301
Soft solder (37% Pb)	8.42	0.304
Tin babbitt		
Alloy 1	7.34	0.265
Alloy 2	7.39	0.267
Alloy 3	7.46	0.269
Alloy 4	7.53	0.272
Alloy 5	7.75	0.280
White metal	7.28	0.263
Pewter	7.28	0.263
Titanium and titanium alloys		
99.9% Ti	4.507	0.1628
99.2% Ti	4.507	0.1628
99.0% Ti	4.52	0.163
Ti-6Al-4V	4.43	0.160
Ti-5Al-2.5Sn	4.46	0.161
Ti-2Fe-2Cr-2Mo	4.65	0.168
Ti-8Mn	4.71	0.171
Ti-7Al-4Mo	4.48	0.162
Ti-4Al-4Mn	4.52	0.163
Ti-4Al-3Mo-1V	4.507	0.1628
Ti-2.5Al-16V	4.65	0.168
Zinc and zinc alloys		
Pure zinc	7.133	0.2577
AG40A alloy	6.6	0.24
AC41A alloy	6.7	0.24

(a) 6.95 to 7.35 g/cm³ (0.251 to 0.265 lb/in.³). (b) 7.20 to 7.34 g/cm³ (0.260 to 0.265 lb/in.³). (c) Annealed. (d) As cast. (e) Face-centered cubic. (f) Hexagonal. (g) Body-centered cubic. (h) Close-packed hexagonal. (j) Rhombohedral

(continued)

Density of metals and alloys (continued)

Metal or alloy	Density g/cm³	lb/in.³	Metal or alloy	Density g/cm³	lb/in.³	Metal or alloy	Density g/cm³	lb/in.³
Commercial rolled zinc			**Permanent magnet materials**			Rhodium	12.44	0.447
0.08% Pb	7.14	0.258	Cunico	8.30	0.300	Ruthenium	12.2	0.441
0.06 Pb, 0.06 Cd	7.14	0.258	Cunife	8.61	0.311	Selenium	4.79	0.174
0.3 Pb, 0.3 Cd	7.14	0.258	Comol	8.16	0.295	Silicon	2.33	0.084
Copper-hardened,			Alnico I	6.89	0.249	Silver	10.49	0.379
rolled zinc (1% Cu)	7.18	0.259	Alnico II	7.09	0.256	Sodium	0.97	0.035
Rolled zinc alloy			Alnico III	6.89	0.249	Tantalum	16.6	0.600
(1 Cu, 0.010 Mg)	7.18	0.259	Alnico IV	7.00	0.253	Thalium	11.85	0.428
Zn-Cu-Ti alloy			Alnico V	7.31	0.264	Thorium	11.72	0.423
(0.8 Cu, 0.15 Ti)	7.18	0.259	Alnico VI	7.42	0.268	Tungsten	19.3	0.697
Precious metals			Barium ferrite	4.7	0.17	Uranium	19.07	0.689
Silver	10.49	0.379	Vectolite	3.13	0.113	Vanadium	6.1	0.22
Gold	19.32	0.698	**Pure metals**			Zirconium	6.5	0.23
70Au-30Pt	19.92	...	Antimony	6.62	0.239	**Rare earth metals**		
Platinum	21.45	0.775	Beryllium	1.848	0.067	Cerium	8.23(c)	...
Pt-3.5Rh	20.9	...	Bismuth	9.80	0.354	Cerium	6.66(d)	...
Pt-5Rh	20.65	...	Cadmium	8.65	0.313	Cerium	6.77(e)	...
Pt-10Rh	19.97	...	Calcium	1.55	0.056	Dysprosium	8.55(f)	...
Pt-20Rh	18.74	...	Cesium	1.903	0.069	Erbium	9.15(f)	...
Pt-30Rh	17.62	...	Chromium	7.19	0.260	Europium	5.245(e)	...
Pt-40Rh	16.63	...	Cobalt	8.85	0.322	Gadolinium	7.86(f)	...
Pt-5Ir	21.49	...	Gallium	5.907	0.213	Holmium	6.79(f)	...
Pt-10Ir	21.53	...	Germanium	5.323	0.192	Lanthanum	6.19(d)	...
Pt-15Ir	21.57	...	Hafnium	13.1	0.473	Lanthanum	6.18(c)	...
Pt-20Ir	21.61	...	Indium	7.31	0.264	Lanthanum	5.97(e)	...
Pt-25Ir	21.66	...	Iridium	22.5	0.813	Lutetium	9.85(f)	...
Pt-30Ir	21.70	...	Lithium	0.534	0.019	Neodymium	7.00(d)	...
Pt-35Ir	21.79	...	Manganese	7.43	0.270	Neodymium	6.80(e)	...
Pt-5Ru	20.67	...	Mercury	13.546	0.489	Praseodymium	6.77(d)	...
Pt-10Ru	19.94	...	Molybdenum	10.22	0.369	Praseodymium	6.64(e)	...
Palladium	12.02	0.4343	Niobium	8.57	0.310	Samarium	7.49(g)	...
60Pd-40Cu	10.6	0.383	Osmium	22.583	0.816	Scandium	2.99(f)	...
95.5Pd-4.5Ru	12.07(a)	...	Plutonium	19.84	0.717	Terbium	8.25(f)	...
95.5Pd-4.5Ru	11.62(b)	...	Potassium	0.86	0.031	Thulium	9.31(f)	...
			Rhenium	21.04	0.756	Ytterbium	6.96(c)	...
						Yttrium	4.47(f)	...

(a) 6.95 to 7.35 g/cm³ (0.251 to 0.265 lb/in.³). (b) 7.20 to 7.34 g/cm³ (0.260 to 0.265 lb/in.³). (c) Face-centered cubic. (d) Hexagonal. (e) Body-centered cubic. (f) Close-packed hexagonal. (g) Rhombohedral

Linear thermal expansion of metals and alloys

Metal or alloy	Temp., °C	Coefficient of expansion, µin./in. · °C	Metal or alloy	Temp., °C	Coefficient of expansion, µin./in. · °C	Metal or alloy	Temp., °C	Coefficient of expansion, µin./in. · °C
Aluminum and aluminum alloys			7075	20 to 100	23.2	40E	21 to 93	24.7
Aluminum			7079, 7178	20 to 100	23.4	**Copper and copper alloys**		
(99.996%)	20 to 100	23.6	**Casting alloys**			**Wrought coppers**		
Wrought alloys			A13	20 to 100	20.4	Pure copper	20	16.5
EC, 1060, 1100	20 to 100	23.6	43 and 108	20 to 100	22.0	Electrolytic tough		
2011, 2014	20 to 100	23.0	A108	20 to 100	21.5	pitch copper (ETP)	20 to 100	16.8
2024	20 to 100	22.8	A132	20 to 100	19.0	Deoxidized copper,		
2218	20 to 100	22.3	D132	20 to 100	20.05	high residual		
3003	20 to 100	23.2	F132	20 to 100	20.7	phosphorus (DHP)	20 to 300	17.7
4032	20 to 100	19.4	138	20 to 100	21.4	Oxygen-free copper	20 to 300	17.7
5005, 5050, 5052	20 to 100	23.8	142	20 to 100	22.5	Free machining		
5056	20 to 100	24.1	195	20 to 100	23.0	copper, 0.5% Te		
5083	20 to 100	23.4	B195	20 to 100	22.0	or 1% Pb	20 to 300	17.7
5086	60 to 300	23.9	214	20 to 100	24.0	**Wrought alloys**		
5154	20 to 100	23.9	220	20 to 100	25.0	Gilding, 95%	20 to 300	18.1
5357	20 to 100	23.7	319	20 to 100	21.5	Commercial bronze,		
5456	20 to 100	23.9	355	20 to 100	22.0	90%	20 to 300	18.4
6061, 6063	20 to 100	23.4	356	20 to 100	21.5	Jewelry bronze,		
6101, 6151	20 to 100	23.0	360	20 to 100	21.0	87.5%	20 to 300	18.6
			750	20 to 100	23.1			

(a) Longitudinal; 23.4 transverse. (b) Longitudinal; 21.1 transverse. (c) Longitudinal; 19.4 transverse

(continued)

Linear thermal expansion of metals and alloys (continued)

Metal or alloy	Temp., °C	Coefficient of expansion, μin./in. · °C
Red brass, 85%	20 to 300	18.7
Low brass, 80%	20 to 300	19.1
Cartridge brass, 70%	20 to 300	19.9
Yellow brass	20 to 300	20.3
Muntz metal	20 to 300	20.8
Leaded commercial bronze	20 to 300	18.4
Low-leaded brass	20 to 300	20.2
Medium-leaded brass	20 to 300	20.3
High-leaded brass	20 to 300	20.3
Extra-high-leaded brass	20 to 300	20.5
Free-cutting brass	20 to 300	20.5
Leaded Muntz metal	20 to 300	20.8
Forging brass	20 to 300	20.7
Architectural bronze	20 to 300	20.9
Inhibited admiralty	20 to 300	20.2
Naval brass	20 to 300	21.2
Leaded naval brass	20 to 300	21.2
Manganese bronze (A)	20 to 300	21.2
Phosphor bronze, 5% (A)	20 to 300	17.8
Phosphor bronze, 8% (C)	20 to 300	18.2
Phosphor bronze, 10% (D)	20 to 300	18.4
Phosphor bronze, 1.25%	20 to 300	17.8
Free-cutting phosphor bronze	20 to 300	17.3
Cupro-nickel, 30%	20 to 300	16.2
Cupro-nickel, 10%	20 to 300	17.1
Nickel silver, 65-18	20 to 300	16.2
Nickel silver, 55-18	20 to 300	16.7
Nickel silver, 65-12	20 to 300	16.2
High-silicon bronze (a)	20 to 300	18.0
Low-silicon bronze (b)	20 to 300	17.9
Aluminum bronze (3)	20 to 300	16.4
Aluminum-silicon bronze	20 to 300	18.0
Aluminum bronze (1)	20 to 300	16.8
Beryllium copper	20 to 300	17.8
Casting alloys		
88Cu-8Sn-4Zn	21 to 177	18.0
89Cu-11Sn	20 to 300	18.4
88Cu-6Sn-1.5Pb-4.5Zn	21 to 260	18.5
87Cu-8Sn-1Pb-4Zn	21 to 177	18.0
87Cu-10Sn-1Pb-2Zn	21 to 177	18.0
80Cu-10Sn-10Pb	21 to 204	18.5
78Cu-7Sn-15Pb	21 to 204	18.5
85Cu-5Sn-5Pb-5Zn	21 to 204	18.1
72Cu-1Sn-3Pb-24Zn	21 to 93	20.7
67Cu-1Sn-3Pb-29Zn	21 to 93	20.2
61Cu-1Sn-1Pb-37Zn	21 to 260	21.6
Manganese bronze		
60 ksi	21 to 204	20.5
65 ksi	21 to 93	21.6
110 ksi	21 to 260	19.8

Metal or alloy	Temp., °C	Coefficient of expansion, μin./in. · °C
Aluminum bronze		
Alloy 9A	...	17
Alloy 9B	20 to 250	17
Alloys 9C, 9D	...	16.2
Iron and iron alloys		
Pure iron	20	11.7
Fe-C alloys		
0.06% C	20 to 100	11.7
0.22% C	20 to 100	11.7
0.40% C	20 to 100	11.3
0.56% C	20 to 100	11.0
1.08% C	20 to 100	10.8
1.45% C	20 to 100	10.1
Invar (36% Ni)	20	0-2
13Mn-1.2C	20	18.0
13Cr-0.35C	20 to 100	10.0
12.3Cr-0.4Ni-0.09C	20 to 100	9.8
17.7Cr-9.6Ni-0.06C	20 to 100	16.5
18W-4Cr-1V	0 to 100	11.2
Gray cast iron	0 to 100	10.5
Malleable iron (pearlitic)	20 to 400	12
Lead and lead alloys		
Corroding lead (99.73 + % Pb)	17 to 100	29.3
5-95 solder	15 to 110	28.7
20-80 solder	15 to 110	26.5
50-50 solder	15 to 110	23.4
1% antimonial lead	20 to 100	28.8
Hard lead (96Pb-4Sb)	20 to 100	27.8
Hard lead (94Pb-6Sb)	20 to 100	27.2
8% antimonial lead	20 to 100	26.7
9% antimonial lead	20 to 100	26.4
Lead-base babbitt		
SAE 14	20 to 100	19.6
Alloy 8	20 to 100	24.0
Magnesium and magnesium alloys		
Magnesium (99.8%)	20	25.2
Casting alloys		
AM100A	18 to 100	25.2
AZ63A	20 to 100	26.1
AZ91A, B, C	20 to 100	26
AZ92A	18 to 100	25.2
HZ32A	20 to 200	26.7
ZH42	20 to 200	27
ZH62A	20 to 200	27.1
ZK51A	20	26.1
EZ33A	20 to 100	26.1
EK30A, EK41A	20 to 100	26.1
Wrought alloys		
M1A, A3A	20 to 100	26
AZ31B, PE	20 to 100	26
AZ61A, AZ80A	20 to 100	26
ZK60A, B	20 to 100	26
HM31A	20 to 93	26.1
Nickel and nickel alloys		
Nickel (99.95% Ni + Co)	0 to 100	13.3
Duranickel	0 to 100	13.0
Monel	0 to 100	14.0

Metal or alloy	Temp., °C	Coefficient of expansion, μin./in. · °C
Monel (cast)	25 to 100	12.9
Inconel	20 to 100	11.5
Ni-o-nel	27 to 93	12.9
Hastelloy B	0 to 100	10.0
Hastelloy C	0 to 100	11.3
Hastelloy D	0 to 100	11.0
Hastelloy F	20 to 100	14.2
Hastelloy N	21 to 204	10.4
Hastelloy W	23 to 100	11.3
Hastelloy X	26 to 100	13.8
Illium G	0 to 100	12.19
Illium R	0 to 100	12.02
80Ni-20Cr	20 to 1000	17.3
60Ni-24Fe-16Cr	20 to 1000	17.0
35Ni-45Fe-20Cr	20 to 500	15.8
Constantan	20 to 1000	18.8
Tin and tin alloys		
Pure tin	0 to 100	23
Solder (70Sn-30Pb)	15 to 110	21.6
Solder (63Sn-37Pb)	15 to 110	24.7
Titanium and titanium alloys		
99.9% Ti	20	8.41
99.0% Ti	93	8.55
Ti-5Al-2.5Sn	93	9.36
Ti-8Mn	93	8.64
Zinc and zinc alloys		
Pure zinc	20 to 250	39.7
AG40A alloy	20 to 100	27.4
AC41A alloy	20 to 100	27.4
Commercial rolled zinc		
0.08 Pb	20 to 40	32.5
0.3 Pb, 0.3 Cd	20 to 98	33.9(a)
Rolled zinc alloy (1 Cu, 0.010 Mg)	20 to 100	34.8(b)
Zn-Cu-Ti alloy (0.8 Cu, 0.15 Ti)	20 to 100	24.9(c)
Pure metals		
Beryllium	25 to 100	11.6
Cadmium	20	29.8
Calcium	0 to 400	22.3
Chromium	20	6.2
Cobalt	20	13.8
Gold	20	14.2
Iridium	20	6.8
Lithium	20	56
Manganese	0 to 100	22
Palladium	20	11.76
Platinum	20	8.9
Rhenium	20 to 500	6.7
Rhodium	20 to 100	8.3
Ruthenium	20	9.1
Silicon	0 to 1400	5
Silver	0 to 100	19.68
Tungsten	27	4.6
Vanadium	23 to 100	8.3
Zirconium	...	5.85

(a) Longitudinal; 23.4 transverse. (b) Longitudinal; 21.1 transverse. (c) Longitudinal; 19.4 transverse

Thermal conductivity of metals and alloys

Metal or alloy	Thermal conductivity near room temperature, cal/cm$^2 \cdot$ cm $\cdot$ s $\cdot$ °C
Aluminum and aluminum alloys	
Wrought alloys	
EC (O)	0.57
1060 (O)	0.56
1100	0.53
2011 (T3)	0.34
2014 (O)	0.46
2024 (O)	0.45
2218 (T72)	0.37
3003 (O)	0.46
4032 (O)	0.37
5005	0.48
5050 (O)	0.46
5052 (O)	0.33
5056 (O)	0.28
5083	0.28
5086	0.30
5154	0.30
5357	0.40
5456	0.28
6061 (O)	0.41
6063 (O)	0.52
6101 (T6)	0.52
6151 (O)	0.49
7075 (T6)	0.29
7079 (T6)	0.29
7178	0.29
Casting alloys	
A13	0.29
43 (F)	0.34
108 (F)	0.29
A108	0.34
A132 (T551)	0.28
D132 (T5)	0.25
F132	0.25
138	0.24
142 (T21, sand)	0.40
195 (T4, T62)	0.33
B195 (T4, T6)	0.31
214	0.33
200 (T4)	0.21
319	0.26
355 (T51, sand)	0.40
356 (T51, sand)	0.40
360	0.35
380	0.23
750	0.44
40E	0.33
Copper and copper alloys	
Wrought coppers	
Pure copper	0.941
Electrolytic tough pitch copper (ETP)	0.934
Deoxidized copper, high residual phosphorus (DHP)	0.81
Free-machining copper (0.5% Te)	0.88
Free-machining copper (1% Pb)	0.92
Wrought alloys	
Gilding, 95%	0.56
Commercial bronze, 90%	0.45
Jewelry bronze, 87.5%	0.41
Red brass, 85%	0.38
Low brass, 80%	0.33
Cartridge brass, 70%	0.29
Yellow brass	0.28
Muntz metal	0.29
Leaded-commercial bronze	0.43
Low-leaded brass (tube)	0.28
Medium-leaded brass	0.28
High-leaded brass (tube)	0.28
High-leaded brass	0.28
Extra-high-leaded brass	0.28
Leaded Muntz metal	0.29
Forging brass	0.28
Architectural bronze	0.29
Inhibited admiralty	0.26
Naval brass	0.28

Metal or alloy	Thermal conductivity near room temperature, cal/cm$^2 \cdot$ cm $\cdot$ s $\cdot$ °C
Leaded naval brass	0.28
Manganese bronze (A)	0.26
Phosphor bronze, 5% (A)	0.17
Phosphor bronze, 8% (C)	0.15
Phosphor bronze, 10% (D)	0.12
Phosphor bronze, 1.25%	0.49
Free-cutting phosphor bronze	0.18
Cupro-nickel, 30%	0.07
Cupro-nickel, 10%	0.095
Nickel silver, 65-18	0.08
Nickel silver, 55-18	0.07
Nickel silver, 65-12	0.10
High-silicon bronze (A)	0.09
Low-silicon bronze (B)	0.14
Aluminum bronze, 5% Al	0.198
Aluminum bronze, (3)	0.18
Aluminum-silicon bronze	0.108
Aluminum bronze, (1)	0.144
Aluminum bronze, (2)	0.091
Beryllium copper	0.20(a)
Casting alloys	
Chromium copper (1% Cr)	0.4(a)
89cu-11Sn	0.121
88Cu-6Sn-1.5Pb-4.5Zn	(b)
87Cu-8Sn-1Pb-4Zn	(c)
87Cu-10Sn-1Pb-2Zn	(c)
80Cu-10Sn-10Pb	(c)
Manganese bronze, 110 ksi	(d)
Aluminum bronze	
Alloy 9A	(e)
Alloy 9B	(f)
Alloy 9C	(b)
Alloy 9D	(c)
Propeller bronze	(g)
Nickel silver	
12% Ni	(h)
16% Ni	(h)
20% Ni	(j)
25% Ni	(k)
Silicon bronze	(h)
Iron and iron alloys	
Pure iron	0.178
Cast iron (3.16 C, 1.54 Si, 0.57 Mn)	0.112
Carbon steel (0.23 C, 0.64 Mn)	0.124
Carbon steel (1.22 C, 0.35 Mn)	0.108
Alloy steel (0.34 C, 0.55 Mn, 0.78 Cr, 3.53 Ni, 0.39 Mo, 0.05 Cu)	0.079
Type 410	0.057
Type 304	0.036
T1 tool steel	0.058
Lead and lead alloys	
Corroding lead (99.73 + % Pb)	0.083
5-95 solder	0.085
20-80 solder	0.089
50-50 solder	0.111
1% antimonial lead	0.080
Hard lead (96Pb-4Sb)	0.073
Hard lead (94Pb-6Sb)	0.069
8% antimonial lead	0.065
9% antimonial lead	0.064
Lead-base babbitt (SAE 14)	0.057
Lead-base babbitt (alloy 8)	0.058
Magnesium and magnesium alloys	
Magnesium (99.8%)	0.367
Casting alloys	
AM100A	0.17
AZ63A	0.18
AZ81A (T4)	0.12
AZ91A, B, C	0.17
AZ92A	0.17
HK31A (T6, sand cast)	0.22

Metal or alloy	Thermal conductivity near room temperature, cal/cm$^2 \cdot$ cm $\cdot$ s $\cdot$ °C
HZ32A	0.26
ZH42	0.27
ZH62A	0.26
ZK51A	0.26
ZE41A (T5)	0.27
EZ33A	0.24
EK30A	0.26
EK41A (T5)	0.24
Wrought alloys	
M1A	0.33
AZ31B	0.23
AZ61A	0.19
AZ80A	0.18
ZK60A, B (F)	0.28
ZE10A (O)	0.33
HM21A (O)	0.33
HM31A	0.25
Nickel and nickel alloys	
Nickel (99.95% Ni + Co)	0.22
"A" nickel	0.145
"D" nickel	0.115
Monel	0.062
"K" Monel	0.045
Inconel	0.036
Hastelloy B	0.027
Hastelloy C	0.03
Hastelloy D	0.05
Illium G	0.029
Illium R	0.031
60Ni-24Fe-16Cr	0.032
35Ni-45Fe-20Cr	0.031
Constantan	0.051
Tin and tin alloys	
Pure tin	0.15
Soft solder (63Sn-37Pb)	0.12
Tin foil (92Sn-8Zn)	0.14
Titanium and titanium alloys	
Titanium (99.0%)	0.043
Ti-5Al-2.5Sn	0.019
Ti-2Fe-2Cr-2Mo	0.028
Ti-8Mn	0.026
Zinc and zinc alloys	
Pure zinc	0.27
AG40A alloy	0.27
AC41A alloy	0.26
Commercial rolled zinc	
0.08 Pb	0.257
0.06 Pb, 0.06 Cd	0.257
Rolled zinc alloy (1 Cu, 0.010 Mg)	0.25
Zn-Cu-Ti alloy (0.8 Cu, 0.15 Ti)	0.25
Pure metals	
Beryllium	0.35
Cadmium	0.22
Chromium	0.16
Cobalt	0.165
Germanium	0.14
Gold	0.71
Indium	0.057
Iridium	0.14
Lithium	0.17
Molybdenum	0.34
Niobium	0.13
Palladium	0.168
Platinum	0.165
Plutonium	0.020
Rhenium	0.17
Rhodium	0.21
Silicon	0.20
Silver	1.0
Sodium	0.32
Tantalum	0.130
Thallium	0.093
Thorium	0.090
Tungsten	0.397
Uranium	0.071
Vanadium	0.074
Yttrium	0.035

(a) Depends on processing. (b) 18% of Cu. (c) 12% of Cu. (d) 9.05% of Cu. (e) 15% of Cu. (f) 16% of Cu. (g) 11% of Cu. (h) 7% of Cu. (j) 6% of Cu. (k) 6.5% of Cu

Electrical conductivity and resistivity of metals and alloys

Metal or alloy	Conductivity, % IACS	Resistivity, μΩ · cm	Metal or alloy	Conductivity, % IACS	Resistivity, μΩ · cm	Metal or alloy	Conductivity, % IACS	Resistivity, μΩ · cm
Aluminum and aluminum alloys			65Pt-35Ir	5	36	85Cu-10Mn-4Ni		
Aluminum (99.996%)	64.95	2.65	95Pt-5Ru	5.5	31.5	(shunt manganin)	4.5	38.23
EC (O, H19)	62	2.8	90Pt-10Ru	4	43	70Cu-20Ni-10Mn	3.6	48.88
5052 (O, H38)	35	4.93	89Pt-11Ru	4	43	67Cu-5Ni-27Mn	1.8	99.74
5056 (H38)	27	6.4	86Pt-14Ru	3.5	46	**Ni-base alloys**		
6101 (T6)	56	3.1	96Pt-4W	5	36	99.8 Ni	23	7.98
Copper and copper alloys			**Palladium and palladium alloys**			71Ni-29Fe	9	19.95
Wrought copper			Palladium	16	10.8	80Ni-20Cr	1.5	112.2
Pure copper	103.06	1.67	95.5Pd-4.5Ru	7	24.2	75Ni-20Cr-3Al+Cu		
Electrolytic (ETP)	101	1.71	90Pd-10Ru	6.5	27	or Fe	1.3	132.98
Oxygen-free copper			70Pd-30Ag	4.3	40	76Ni-17Cr-4Si-3Mn	1.3	132.98
(OF)	101	1.71	60Pd-40Ag	4.0	43	60Ni-16Cr-24Fe	1.5	112.2
Free-machining copper			50Pd-50Ag	5.5	31.5	35Ni-20Cr-45Fe	1.7	101.4
0.5% Te	95	1.82	72Pd-26Ag-2Ni	4	43	**Fe-Cr-Al alloy**		
1.0% Pb	98	1.76	60Pd-40Cu	5	35(c)	72Fe-23Cr-5Al-0.5Co	1.3	135.48
Wrought alloys			45Pd-30Ag-20Au-5Pt	4.5	39	**Pure metals**		
Cartridge brass,			35Pd-30Ag-14Cu-			Iron (99.99%)	17.75	9.71
70%	28	6.2	10Pt-10Au-1Zn	5	35	**Thermostat metals**		
Yellow brass	27	6.4	**Gold and gold alloys**			75Fe-22Ni-3Cr	3	78.13
Leaded commercial			Gold	75	2.35	72Mn-18Cu-10Ni	1.5	112.2
bronze	42	4.1	90Au-10Cu	16	10.8	67Ni-30Cu-1.4Fe-		
Phosphor bronze,			75Au-25Ag	16	10.8	1Mn	3.5	56.52
1.25%	48	3.6	72.5Au-14Cu-8.5Pt-			75Fe-22Ni-3Cr	12	15.79
Nickel silver, 55-18	5.5	31	4Ag-1Zn	10	17	66.5Fe-22Ni-8.5Cr	3.3	58.18
Low-silicon bronze			69Au-25Ag-6Pt	11	15	**Permanent magnet materials**		
(B)	12	14.3	41.7Au-32.5Cu-			Carbon steel		
Beryllium copper	22 to 30(a)	5.7 to 7.8(a)	18.8Ni-7Zn	4.5	39	(0.65% C)	9.5	18
Casting alloys			**Electrical heating alloys**			Carbon steel (1% C)	8	20
Chromium copper			**Ni-Cr and Ni-Cr-Fe alloys**			Chromium steel		
(1% Cr)	80 to 90(a)	2.10	78.5Ni-20Cr-1.5Si			(3.5% Cr)	6.1	29
88Cu-8Sn-4Zn	11	15	(80-20)	1.6	108.05	Tungsten steel		
87Cu-10Sn-1Pb-2Zn	11	15	73.5Ni-20Cr-5Al-			(6% W)	6	30
Electrical contact materials			1.5Si	1.2	137.97	Cobalt steel (17% Co)	6.3	28
Copper alloys			68Ni-20Cr-8.5Fe-2Si	1.5	116.36	Cobalt steel (36% Co)	6.5	27
0.04 oxide	100	1.72	60Ni-16Cr-22.5Fe-			**Intermediate alloys**		
1.25 Sn + P	48	3.6	1.5Si	1.5	112.20	Cunico	7.5	24
5 Sn + P	18	11	35Ni-20Cr-43.5Fe-			Cunife	9.5	18
8 Sn + P	13	13	1.5Si	1.7	101.4	Comol	3.6	45
15 Zn	37	4.7	**Fe-Cr-Al alloys**			**Alnico alloys**		
20 Zn	32	5.4	72Fe-23Cr-5Al	1.3	138.8	Alnico I	3.3	75
35 Zn	27	6.4	55Fe-37.5Cr-7.5Al	1.2	166.23	Alnico II	3.3	65
2 Be + Ni or Co(b)	17 to 21	9.6 to 11.5	**Pure metals**			Alnico III	3.3	60
Silver and silver alloys			Molybdenum	34	5.2	Alnico IV	3.3	75
Fine silver	106	1.59	Platinum	16	10.64	Alnico V	3.5	47
92.5Ag-7.5Cu	85	2	Tantalum	13.9	12.45	Alnico VI	3.5	50
90Ag-10Cu	85	2	Tungsten	30	5.65	**Magnetically soft materials**		
72Ag-28Cu	87	2	**Nonmetallic heating element materials**			**Electrical steel sheet**		
72Ag-26Cu-2Ni	60	2.9	Silicon carbide, SiC	1 to 1.7	100 to 200	M-50	9.5	18
85Ag-15Cd	35	4.93	Molybdenum			M-43	6 to 9	20 to 28
97Ag-3Pt	50	3.5	disilicide, MoSi2	4.5	37.24	M-36	5.5 to 7.5	24 to 33
97Ag-3Pd	60	2.9	Graphite	…	910.1	M-27	3.5 to 5.5	32 to 47
90Ag-10Pd	30	5.3	**Instrument and control alloys**			M-22	3.5 to 5	41 to 52
90Ag-10Au	40	4.2	**Cu-Ni alloys**			M-19	3.5 to 5	41 to 56
60Ag-40Pd	8	23	98Cu-2Ni	35	4.99	M-17	3 to 3.5	45 to 58
70Ag-30Pd	12	14.3	94Cu-6Ni	17	9.93	M-15	3 to 3.5	45 to 69
Platinum and platinum alloys			89Cu-11Ni	11	14.96	M-14	3 to 3.5	58 to 69
Platinum	16	10.6	78Cu-22Ni	5.7	29.92	M-7	3 to 3.5	45 to 52
95Pt-5Ir	9	19	55Cu-45Ni			M-6	3 to 3.5	45 to 52
90Pt-10Ir	7	25	(constantan)	3.5	49.87	M-5	3 to 3.5	45 to 52
85Pt-15Ir	6	28.5	**Cu-Mn-Ni alloys**			**Moderately high-permeability materials(d)**		
80Pt-20Ir	5.6	31	87Cu-13Mn			Thermenol	0.5	162
75Pt-25Ir	5.5	33	(manganin)	3.5	48.21	16 Alfenol	0.7	153
70Pt-30Ir	5	35	83Cu-13Mn-4Ni			Sinimax	2	90
			(manganin)	3.5	48.21			

(a) Precipitation hardened; depends on processing. (b) A heat-treatable alloy. (c) Annealed and quenched. (d) At low field strength and high electrical resistance. (e) At higher field strength; annealed for optimum magnetic properties

(continued)

Electrical conductivity and resistivity of metals and alloys (continued)

Metal or alloy	Conductivity, % IACS	Resistivity, μΩ · cm
Monimax	2.5	80
Supermalloy	3	65
4-79 Moly Permalloy, Hymu 80	3	58
Mumetal	3	60
1040 alloy	3	56
High Permalloy 49, A-L 4750, Armco 48	3.6	48
45 Permalloy	3.6	45
High-permeability materials(e)		
Supermendur	4.5	40
2V Permendur	4.5	40
35% Co, 1% Cr	9	20
Ingot iron	17.5	10
0.5% Si steel	6	28
1.75% Si steel	4.6	37
3.0% Si steel	3.6	47
Grain-oriented 3.0% Si steel	3.5	50

Metal or alloy	Conductivity, % IACS	Resistivity, μΩ · cm
Grain-oriented 50% Ni iron	3.6	45
50% Ni iron	3.5	50
Relay steels and alloys after annealing		
Low-carbon iron and steel		
Low-carbon iron	17.5	10
1010 steel	14.5	12
Silicon steels		
1% Si	7.5	23
2.5% Si	4	41
3% Si	3.5	48
3% Si, grain-oriented	3.5	48
4% Si	3	59
Stainless steels		
Type 410	3	57
Type 416	3	57
Type 430	3	60
Type 443	3	68
Type 446	3	61

Metal or alloy	Conductivity, % IACS	Resistivity, μΩ · cm
Nickel irons		
50% Ni	3.5	48
78% Ni	11	16
77% Ni (Cu, Cr)	3	60
79% Ni (Mo)	3	58
Stainless and heat-resisting alloys		
Type 302	3	72
Type 309	2.5	78
Type 316	2.5	74
Type 317	2.5	74
Type 347	2.5	73
Type 403	3	57
Type 405	3	60
Type 501	4.5	40
HH	2.5	80
HK	2	90
HT	1.7	100

(a) Precipitation hardened; depends on processing. (b) A heat-treatable alloy. (c) Annealed and quenched. (d) At low field strength and high electrical resistance. (e) At higher field strength; annealed for optimum magnetic properties

Vapor pressures of the metallic elements

Element	Pressure, atm											
	0.0001		0.001		0.01		0.1		0.50		1.0	
	°C	°F	°C	°F	°C	°F	°C	°F	°C	°F	°C	°F
Aluminum	1110	2030	1263	2305	1461	2662	1713	3115	1940	3524	2056	3733
Antimony	759	1398	872	1602	1013	1855	1196	2185	1359	2478	1440	2624
Arsenic	308	586	363	685	428	802	499	930	578	1072	610	1130
Bismuth	914	1677	1008	1846	1121	2050	1254	2289	1367	2493	1420	2588
Cadmium	307(a)	585(a)	384(b)	723(b)	471	880	594	1101	708	1306	765	1409
Calcium	688	1270	802(c)	1476(c)	958(b)	1756(b)	1175	2147	1380	2516	1487	2709
Carbon	3257	5895	3547	6417	3897	7047	4317	7803	4667	8433	4827	8721
Chromium	1420(a)	2588(a)	1594(b)	2901(b)	1813	3295	2097	3807	2351	4264	2482	4500
Copper	1412	2574	1602	2916	1844	3351	2162	3924	2450	4442	2595	4703
Gallium	1178	2152	1329	2424	1515	2759	1751	3184	1965	3569	2071	3760
Gold	1623	2953	1839	3342	2115	3839	2469	4476	2796	5065	2966	5371
Iron	1564	2847	1760	3200	2004	3639	2316	4201	2595	4703	2735	4955
Lead	815	1499	953	1747	1135	2075	1384	2523	1622	2952	1744	3171
Lithium	592	1098	707	1305	858	1576	1064	1947	1266	2311	1372	2502
Magnesium	516	961	608(a)	1126(a)	725(b)	1337(b)	886	1627	1030	1886	1107	2025
Manganese	1115(d)	2039(d)	1269(b)	2316(b)	1476	2889	1750	3182	2019	3666	2151	3904
Mercury	77.9(b)	172.2(b)	120.8	249.4	176.1	349.0	251.3	484.3	321.5	610.7	357	675
Molybdenum	2727	4941	3057	5535	3477	6291	4027	7281	4537	8199	4804	8679
Nickel	1586	2887	1782	3240	2025	3677	2321	4210	2593	4699	2732	4950
Platinum	2367	4293	2687	4869	3087	5589	3637	6579	4147	7497	4407	7965
Potassium	261	502	332	630	429	804	565	1051	704	1299	774	1425
Rubidium	223	433	288	550	377	711	497	927	617	1143	679	1254
Selenium	282	540	347	657	430	806	540	1004	634	1173	680	1256
Silicon	1572	2862	1707	3105	1867	3393	2057	3735	2217	4023	2287	4149
Silver	1169	2136	1334	2433	1543	2809	1825	3317	2081	3778	2212	4014
Sodium	349	660	429	804	534	993	679	1254	819	1506	892	1638
Strontium	…	…	(a)	(a)	877(b)	1629(b)	1081	1978	1279	2334	1384	2523
Tellurium	(a)	(a)	509(b)	948(b)	632	1170	810	1490	991	1816	1087	1989
Thallium	692	1277	809	1488	962	1764	1166	2131	1359	2478	1457	2655
Tin	…	…	…	…	…	…	1932(b)	3510(b)	2163	3925	2270	4118
Tungsten	3547	6417	3937	7119	4437	8019	5077	9171	5647	10197	5927	10701
Zinc	399(a)	750(a)	477(b)	891(b)	579	1074	717	1323	842	1548	907	1665

(a) In the solid state. (b) In the liquid state. (c) β. (d) γ

Standard reduction potentials of metals

Add the correction given at 18 °C (64 °F) if the activity (effective concentration of metallic ions) is $C \times$ standard; if activity is less than standard, $\log C$ will be negative and the potential will be reduced

Metal	Ion considered	Standard reduction potential, V	Correction(a)
Gold	Au^{3+}	+1.50	0.019 $\log C$
Platinum	Pt^{2+}	+1.2	0.029 $\log C$
Silver	Ag^+	+0.799	0.058 $\log C$
Mercury	$(Hg)_2^{2+}$	+0.789	0.029 $\log C$
Copper	Cu^{2+}	+0.337	0.020 $\log C$
Hydrogen (1 atm)	H^+	+0.000	0.058 $\log C$
Lead	Pb^{2+}	−0.126	0.029 $\log C$
Tin	Sn^{2+}	−0.136	0.029 $\log C$
Nickel	Ni^{2+}	−0.250	0.029 $\log C$
Cobalt	Co^{2+}	−0.28	0.029 $\log C$
Cadmium	Cd^{2+}	−0.403	0.029 $\log C$
Iron	Fe^{2+}	−0.440	0.029 $\log C$
Chromium	Cr^{3+}	−0.74	0.019 $\log C$
Zinc	Zn^{2+}	−0.763	(a)
Manganese	Mn^{2+}	−1.18	(a)
Titanium(b)	Ti^{2+}	−1.63	(a)
Aluminum(b)	Al^{3+}	−1.66	(a)
Magnesium(b)	Mg^{2+}	−2.37	(a)
Sodium(b)	Na^{2+}	−2.71	(a)
Calcium(b)	Ca^{2+}	−2.87	(a)
Potassium(b)	K^+	−2.92	(a)
Lithium(b)	Li^+	−3.02	(a)

(a) Potential almost independent of original concentration of metal ions. (b) Calculated values, of theoretical interest only. Aluminum, unless amalgamated, gives more positive values owing to the presence of an oxide film. The other metals evolve hydrogen freely, and potential measurements made directly would not represent equilibrium values.

The 45 most abundant elements in the earth's crust

Relative abundance	Element	Abundance, ppm wt
1	Oxygen	466 000
2	Silicon	277 000
3	Aluminum	81 300
4	Iron	50 000
5	Calcium	36 300
6	Sodium	28 300
7	Potassium	25 900
8	Magnesium	20 900
9	Titanium	4 400
10	Hydrogen	1 400
11	Phosphorus	1 050
12	Manganese	950
13	Fluorine	625
14	Barium	425
15	Strontium	375
16	Sulfur	260
17	Carbon	200
18	Zirconium	165
19	Vanadium	135
20	Chlorine	130
21	Chromium	100
22	Rubidium	90
23	Nickel	75
24	Zinc	70
25	Cerium	60
26	Copper	55
27	Yttrium	33
28	Lanthanum	30
29	Neodymium	28
30	Cobalt	25
31	Scandium	22
32	Lithium	20
33	Niobium	20
34	Nitrogen	20
35	Gallium	15
36	Lead	13
37	Radium	13
38	Boron	10
39	Krypton	9.8
40	Praseodymium	8.2
41	Protoactinium	8.0
42	Thorium	7.2
43	Neon	7.0
44	Samarium	6.0
45	Gadolinium	5.4

The electrochemical series

In this table, the elements are electropositive to the ones which follow them and will displace them from solutions of their salts

1	Cesium	34	Silver
2	Rubidium	35	Mercury
3	Potassium	36	Palladium
4	Sodium	37	Ruthenium
5	Lithium	38	Rhodium
6	Barium	39	Platinum
7	Strontium	40	Iridium
8	Calcium	41	Osmium
9	Magnesium	42	Gold
10	Beryllium	43	Hydrogen
11	Ytterbium	44	Tin
12	Erbium	45	Silicon
13	Scandium	46	Titanium
14	Aluminum	47	Niobium
15	Zirconium	48	Tantalum
16	Thorium	49	Tellurium
17	Cerium	50	Antimony
18	Didymium	51	Carbon
19	Lanthanum	52	Boron
20	Manganese	53	Tungsten
21	Zinc	54	Molybdenum
22	Iron	55	Vanadium
23	Nickel	56	Chromium
24	Cobalt	57	Arsenic
25	Thallium	58	Phosphorus
26	Cadmium	59	Selenium
27	Lead	60	Iodine
28	Germanium	61	Bromine
29	Indium	62	Chlorine
30	Gallium	63	Fluorine
31	Bismuth	64	Nitrogen
32	Uranium	65	Sulfur
33	Copper	66	Oxygen

Metal melting range and color scale

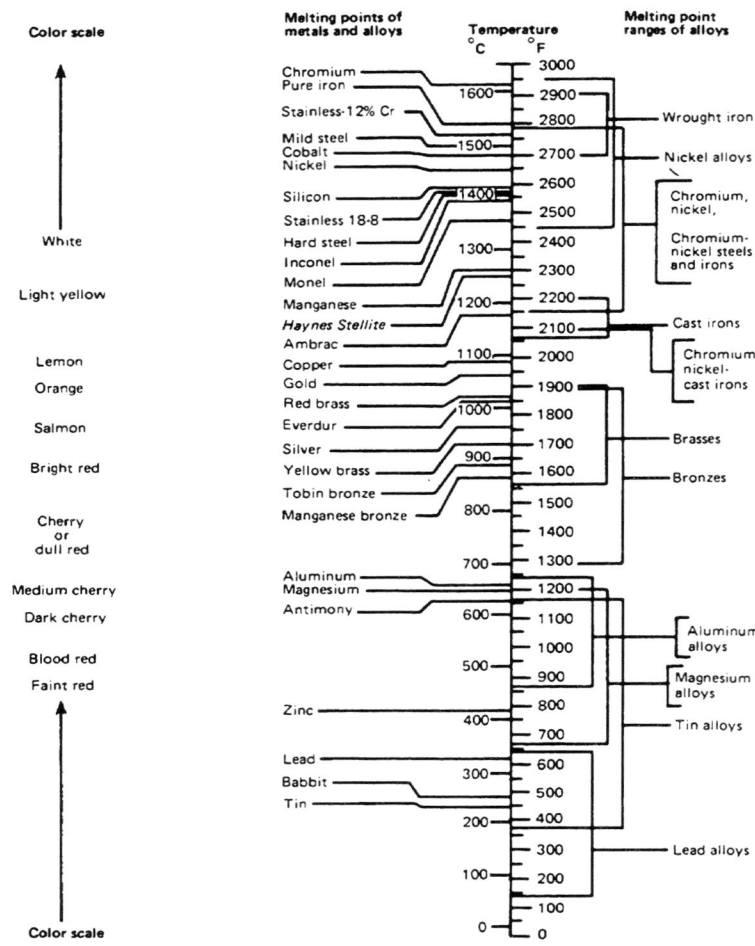

Predominant flame colors of metallic elements

Element	Color
Lithium	Deep red
Strontium	Crimson
Calcium	Yellow-red
Sodium	Bright yellow
Barium	Yellow-green
Molybdenum	Green-yellow
Zinc	Light green
Boron	Green
Tellurium	Deep green
Thallium	Greenish blue
Antimony	Blue-green
Copper	Green-blue
Arsenic	Light blue
Lead	Light blue
Selenium	Blue
Indium	Deep blue
Potassium	Purple-red
Rubidium	Violet
Cesium	Bluish purple

Average percentage of metals in igneous rocks

Metal	Percentage
Silicon	27.72
Aluminum	8.13
Iron	5.01
Calcium	3.63
Sodium	2.85
Potassium	2.60
Titanium	0.63
Manganese	0.10
Barium	0.05
Chromium	0.037
Zirconium	0.026
Nickel	0.020
Vanadium	0.017
Rare earths	0.015
Copper	0.010
Tungsten	0.005
Lithium	0.004
Zinc	0.004
Niobium, tantalum	0.003
Hafnium	0.003
Thorium	0.002
Lead	0.002−
Cobalt	0.001
Beryllium	0.001
Strontium	0.001−
Uranium	0.001−

Energy requirements for production of metal from ore

Metal	Energy			
	Theoretical		Actual	
	10^6 Btu/ton	TJ/Mg	10^6 Btu/ton	TJ/Mg
Aluminum	15.7	18.3	203	236
Magnesium	4.1	4.8	309	359
Iron	3.2	3.7	16.7	19.4
Steel	6.0	7	47.5	55
Copper	1.4	1.6	46.1	54
Titanium	9.9	12	431	501

Crystal Structures

Crystal structures of the elements at 25 °C

Element	Pearson symbol	Lattice parameters, n/m			c/a, or α or β	Notes
		a	b	c		
Ac	cF4	0.5311	...	...	...	...
Ag	cF4	0.40861	...	...	...	...
Al	cF4	0.40496	...	...	...	...
αAm	hP4	0.3468	...	1.1241	2 × 1.621	...
Ar	cF4	0.5312	...	...	...	a
αAs	hR2	0.41320	...	...	α = 54.12°	c
Au	cF4	0.40784	...	...	...	...
γB	tP50	0.8756	...	0.5078	0.580	c
Ba	cI2	0.5023	...	...	...	...
αBe	hP2	0.22857	...	0.35839	1.5680	...
αBi	hR2	0.47460	...	...	α = 57.23°	...
αBk	hP4	0.3416	...	1.1069	2 × 1.620	...
Br	oC8	0.668	0.449	0.874	...	a, b
C (graph.)	hP4	0.24612	...	0.67090	2.7259	c
αCa	cF4	0.55884	...	...	...	...
Cd	hP2	0.29788	...	0.56167	1.8856	...
αCe	cF4	0.51610	...	...	...	...
Cl	oC8	0.624	0.448	0.826	...	a, b
αCm	hP4	0.3496	...	1.1331	2 × 1.621	...
αCo	hP2	0.25071	...	0.40694	1.6232	...
Cr	cI2	0.28847	...	...	...	...
Cs	cI2	0.6141	...	...	...	...
Cu	cF4	0.36149	...	...	...	...
αDy	hP2	0.35915	...	0.56501	1.5732	...
αEr	hP2	0.35592	...	0.55850	1.5692	...
Eu	cI2	0.45827	...	...	...	...
αF	mC8	0.550	0.328	0.728	β = 102.17°	...
αFe	cI2	0.28665	...	...	...	...
αGa	oC8	0.45192	0.76586	0.45258	...	...
αGd	hP2	0.36336	...	0.57810	1.5910	...
αGe	cF8	0.56574	...	...	...	...
αH	hP2	0.3771	...	0.6152	1.631	a,b
αHe	hP2	0.3577	...	0.5842	1.633	a
αHf	hP2	0.31946	...	0.50511	1.5811	...
αHg	hR1	0.3005	...	...	α = 70.53°	a
αHo	hP2	0.35778	...	0.56178	1.5702	...
I	oC8	0.7268	0.4797	0.9797	...	b
In	tI2	0.45990	...	0.49470	1.0757	...
Ir	cF4	0.38391	...	...	...	...
K	cI2	0.5321	...	...	...	...
Kr	cF4	0.56459	...	...	...	a
αLa	hP4	0.37740	...	1.2171	2 × 1.6125	...
βLi	cI2	0.35093	...	...	...	...
αLu	hP2	0.35052	...	0.55494	1.5832	...
Mg	hP2	0.32093	...	0.52107	1.6236	...
αMn	cI58	0.89219	...	...	...	...
Mo	cI2	0.31470	...	...	...	...
αN	cP8	0.5659	...	...	...	a, b
βNa	cI2	0.42096	...	...	...	...
Nb	cI2	0.33007	...	...	...	...
αNd	hP4	0.36582	...	1.17966	2 × 1.6124	...
Ne	cF4	0.44622	...	...	...	a
Ni	cF4	0.35241	...	...	...	...
αNp	oP8	0.6663	0.4723	0.4887	...	...
αO	mC4	0.5403	0.3429	0.5086	β = 132.53°	a, b
Os	hP2	0.27348	...	0.43913	1.5316	...
P (black)	oC8	0.33136	1.0478	0.43763	...	c
αPa	tI2	0.3945	...	0.3242	0.822	...
Pb	cF4	0.49502	...	...	...	...
Pd	cF4	0.38901	...	...	...	...
Pm	hP4	0.365	...	1.165	2 × 1.596	...
αPo	cP1	0.3366	...	...	...	...
αPr	hP4	0.36721	...	1.18326	2 × 1.6111	...
Pt	cF4	0.29233	...	...	...	...
αPu	mP16	0.6183	0.4822	1.0968	β = 101.78°	...
Ra	cI2	0.5148	...	...	...	...

Note: The accuracy of the data presented in this table is considered to be reliable to ±2 in the last reported digit. (a) These elements are gaseous, or liquid, at room temperature. The structural data for Ar, H, Kr, Ne and Xe refer to 4.2 K, whereas that for Br, Cl, He, Hg, N and O refer to 123 K, 113 K, 1.5 K, 225 K, 20 K and 23 K, respectively. (b) These elements form diatomic gases and the basis of the crystal structure is thus the molecular unit R_2 rather than a single atom of the species. (c) The chemistry of these elements permits them to form a number of allotropes at room temperature with crystal structures based on different molecular bases. The allotrope quoted in the table is the most commonly observed form

(continued)

Crystal structures of the elements at 25 °C (continued)

Element	Pearson symbol	Lattice parameters, n/m			c/a, or α or β	Notes
		a	b	c		
Rb	cI2	0.5703	...	...	...	...
Re	hP2	0.27608	...	0.44580	1.6147	...
Rh	cF4	0.28032	...	...	...	...
Ru	hP2	0.27053	...	0.42814	1.5826	...
αS	oF128	1.0464	1.28660	2.44860	...	c
αSb	hR2	0.45065	...	...	α = 57.11°	...
αSc	hP2	0.33088	...	0.52680	1.5921	...
γSe	hP3	0.43655	...	0.49576	1.1356	...
Si	cF8	0.54306	...	...	...	...
αSm	hR3	0.36290	...	2.6207	4.5 × 1.6048	...
βSn	tI4	0.58316	...	0.31815	0.5456	...
αSr	cF4	0.6084	...	...	...	...
Ta	cI2	0.33031	...	...	...	...
αTb	hP2	0.36055	...	0.56966	1.5800	...
Tc	hP2	0.2738	...	0.4394	1.605	...
αTe	hP3	0.44561	...	0.59271	1.3301	...
αTh	cF4	0.50851	...	...	...	...
αTi	hP2	0.29503	...	0.46836	1.5875	...
αTl	hP2	0.34563	...	0.55263	1.5989	...
αTm	hP2	0.35375	...	0.55540	1.5700	...
αU	oC4	0.28538	0.58680	0.49557	...	...
V	cI2	0.30238	...	...	...	...
W	cI2	0.31651	...	...	...	...
Xe	cF4	0.6132	...	...	...	a
αY	hP2	0.36482	...	0.57318	1.5711	...
αYb	cF4	0.54848	...	...	...	...
Zn	hP2	0.26644	...	0.49494	1.8576	...
αZr	hP2	0.32317	...	0.51476	1.5928	...

Note: The accuracy of the data presented in this table is considered to be reliable to ±2 in the last reported digit. (a) These elements are gaseous, or liquid, at room temperature. The structural data for Ar, H, Kr, Ne and Xe refer to 4.2 K, whereas that for Br, Cl, He, Hg, N and O refer to 123 K, 113 K, 1.5 K, 225 K, 20 K and 23 K, respectively. (b) These elements form diatomic gases and the basis of the crystal structure is thus the molecular unit R_2 rather than a single atom of the species. (c) The chemistry of these elements permits them to form a number of allotropes at room temperature with crystal structures based on different molecular bases. The allotrope quoted in the table is the most commonly observed form

Crystal structure of metals

A1 Copper type. Face-centered cubic: *Fm3m*; *cF*4. Four atoms per cell, at 0, 0, 0; ½, O, ½; 0, ½, ½ and ½, ½, O. For Cu, *a* = 3.61 Å. **Examples:** Ag, Al, Au, αCa, αCe, βCo, Cu, γFe, Ir, Ni, Pb, Pd, Pt, Rh, αSr, αTh

A2 Tungsten type. Body-centered cubic: *Im3m*; *cI*2. Two atoms per cell, at 0, 0, 0 and ½, ½, ½. For W, *a* = 3.16 Å. **Examples:** Ba, Cb, Cr, Cs, β Cu-Zn (HT), αFe, δFe, K, βLi, Mo, βNa, Rb, Ta, V, W

A3 Magnesium type. Hexagonal close-packed: *P*6_3*/mmc*; *hP*2. Two atoms per cell, at 0, 0, 0 and ⅓, ⅔, ½. (The atoms are at the positions of the zinc atoms of the *B*4 structure shown.) For Mg, *a* = 3.21 Å and *c* = 5.20 Å. **Examples:** αBe, Cd, αCo, Mg, αTi, Zn, αZr

A4 Carbon (diamond) type. Face-centered cubic: *Fd3m*; *cF*8. Eight atoms per cell, at 0, 0, 0; 0, ½, ½; ½, 0, ½; ½, ½, 0; ¼, ¼, ¼; ¼, ¾, ¾; ¾, ¼, ¾ and ¾, ¾, ¼. For C (diamond), *a* = 3.57 Å. **Examples:** C (diamond), Ge, Si, αSn

A5 Tin type (βSn, white). Body-centered tetragonal; *I*4_1*/amd*; *tI*4. Four atoms per cell, at 0, 0, 0; ½, ½, ½; 0, ½, ¼ and ½, 0, ¾. For βSn, *a* = 5.83 Å and *c* = 3.18 Å. **Examples:** AlSb II (HP), InSb II (HP), βSn (white)

A6 Indium type. Body-centered tetragonal: *I4/mmm*; *tI*2. Two atoms per cell, at 0, 0, 0 and ½, ½, ½. It is conventional, however, to use the cell that has four atoms, at 0, 0, 0; 0, ½, ½; ½, 0, ½ and ½, ½, 0. For In, *a* = *b* = 4.60 Å and *c* = 4.95 Å. At room temperature, the unit cell resembles *A*1. **Examples:** δ$GaNi_2$, In, $InPd_3$

A7 Arsenic type (αAs). Rhombohedral: *R*$\bar{3}$*m*; *hR*2. In the cell based on hexagonal axes, there are six atoms, at 0, 0, *z*; ⅓, ⅔, ⅔ + *z*; ⅔, ⅓, ⅓ + *z*; 0, 0, $\bar{z}$; ⅓, ⅔, ⅔ − *z* and ⅔, ⅓, ⅓ − *z*; where *z* = 0.226, *a* = 3.76 Å and *c* = 10.55 Å for αAs. A cell based on rhombohedral axes contains two atoms, at *x*, *x*, *x* and $\bar{x}$, $\bar{x}$, $\bar{x}$; where *x* = 0.276, *a* = 4.13 Å and α = 54 ° 8′ for αAs. **Examples:** αAs, Bi, Sb

A8 Selenium type (γSe). Hexagonal: *P*$3_1$21 or *P*$3_2$21; *hP*3. Three atoms per cell, at *x*, 0, ⅓; 0, *x*, ⅔ and $\bar{x}$, $\bar{x}$, 0 (or at *x*, 0, ⅔; 0, *x*, ⅓ and $\bar{x}$, $\bar{x}$, 0). For Se, *a* = 4.36 Å, *c* = 4.96 Å and *x* = 0.217. **Example:** γSe

A9 Carbon (graphite). Hexagonal: *P*6_3*/mmc*; *hP*4. Four atoms per cell, at 0, 0, *x*; 0, 0, *x* + ½; ⅓, ⅔, *y* and ⅔, ⅓, *y* + ½. For C (graphite), *x* = *y* = 0, *a* = 2.46 Å and *c* = 6.70 Å. There is a less common rhombohedral structure in which *a* = 2.46 Å and *c* = 6.70 Å

A10 Mercury type. Rhombohedral: *R*$\bar{3}$*m*; *hR*1. One atom per cell, at 0, 0, 0. For Hg, *a* = 3.005 Å and *a* = 70 ° 32 ′. A hexagonal cell, where *a* = 3.47 Å and *c* = 6.74 Å, has three atoms per cell, at 0, 0, 0; ⅓, ⅔, ⅔ and ⅔, ⅓, ⅓. **Example:** Hg

A11 Gallium type. Orthorhombic: *Cmca*; *oC*8. Eight atoms per cell, at 0, *y*, *z*; 0, $\bar{y}$, $\bar{z}$; ½, *y*, ½ − *z*; ½, $\bar{y}$, ½ + *z*; ½, ½ + *y*, *z*; ½, ½ − *y*, $\bar{z}$; 0, ½ + *y*, ½ − *z* and 0, ½ − *y*, ½ + *z*. For Ga, *a* = 2.90 Å, *b* = 8.13 Å, *c* = 3.17 Å, *y* = 0.1549 and *z* = 0.081

A12 Alpha-manganese type. Cubic: *I*$\bar{4}$3*m*; *cI*58. Fifty-eight atoms per cell. Alpha-manganese appears to be an ordered array of either two or three physically distinguishable types of manganese atoms located on four crystallographically different sets of positions. These sets of positions have an ordered array of atoms in the chi-phase structure ($Fe_{36}Cr_{12}Mo_{12}$ and $Al_{12}Mg_{17}$). Other closely related structures are the mu, *P*, *R* and delta phases. **Examples:** γ$Al_{12}Mg_{17}$, χ$Co_5Cr_3Si_2$, $CrMn_2$, $Fe_{36}Cr_{12}Mo_{12}$, $Fe_5Si_2V_3$, αMn

A13 Beta-manganese type. Cubic: *P*$4_1$32; *cP*20. Twenty atoms per cell. **Examples:** βAg_3Al, *T* C-Cr-Fe-W, γCu_5Si, βMn

A15 W_3O or Cr_3Si type. Cubic: *Pm3n*; *cP*8. Atom I, two at 0, 0, 0 and ½, ½, ½; Atom II, six at ¼, 0, ½; ½, ¼, 0; 0, ½, ¼; ¾, 0, ½; ½, ¾, 0 and 0, ½, ¾. For W_3O, *a* = 5.04 Å. The prototype structure originally was attributed to βW. This has since been shown to be the oxide , W_3O, having random distribution of atoms. **Examples:** AlV_3, $AuTi_3$, CoV_3, Cr_3O, Cr_3Si, Mo_3O, V_3Si, W_3O, W_3Si

A20 Alpha-uranium type. Orthorhombic: *Cmcm*; *oC*4. Four atoms per cell, at 0, *y*, ¼; 0, $\bar{y}$, ¾; ½, ½ + *y*, ¼ and ½, ½ − *y*, ¾. For αU, *a* = 2.85 Å, *b* = 5.87 Å, *c* = 4.95 Å and *y* = 0.1024

Af $HgSn_{10}$ type. Hexagonal: *P6/mmm*. One Hg or Sn atom per cell, at 0, 0, 0

Ag Boron (alpha-boron) type. Tetragonal: *P*4_2*/nnm*; *tP*50. Fifty atoms per cell

Ah Alpha-polonium type. Primitive cubic: *Pm3m*; *cP*1. One atom per cell, at 0, 0, 0. For αPo, *a* = 3.34 Å. **Examples:** Ag-Te (metastable), Au-Te (metastable), αPo, Sb II (HP)

Ai Beta-polonium type. Rhombohedral: *R*$\bar{3}$*m*; *hR*1. One atom per cell, at 0, 0, 0. For βPo, *a* = 3.36 Å and α = 98 ° 13 ′. A cell based on a hexagonal axis is like the *A*10 (Hg-type) structure

Ak Alpha-selenium type (a metastable form). Monoclinic: *P*2_1*/c*; *mP*32. Each cell contains 32 atoms

... Samarium type (αSm). Rhombohedral: *R*$\bar{3}$*m*; *hR*3. In a cell based on a hexagonal axis, there are nine atoms per cell, at 0, 0, 0; ⅓, ⅔, ⅔; ⅔, ⅓, ⅓; 0, 0, *z*; 0, 0, $\bar{z}$; ⅓, ⅔, ⅔ + *z*; ⅓, ⅔, ⅔ − *z*; ⅔, ⅓, + *z* and ⅔, ⅓, ⅓ − *z*. For αSm, *a* = 3.621 Å, *c* = 26.25 Å and *z* = 2/9. A cell based on rhombohedral axes contains three atoms per cell, at 0, 0, 0; *x*, *x*, *x* and $\bar{x}$, $\bar{x}$, $\bar{x}$; with *a* = 9.00 Å, α = 23 ° 19′ and *x* = 2/9. **Examples:** αSm, Ce-Y, δNd-Tm, δPr-Y

... Lanthanum (αLa) type. Hexagonal: *P*6_3*/mmc*; *hP*4. Four atoms per cell, at 0, 0, 0; 0, 0, ½ and ±(⅓, ⅔, ¼) or ±(⅓, ⅔, ¾). **Examples:** Am, βCe (LT), αLa, αNd, αPr

... Beta-uranium type. Tetragonal: *P*4_2*/mnm*; *tP*30. Thirty atoms per cell

... Alpha-plutonium type. Monoclinic: *P*2_1*/m*; *mP*16. Sixteen atoms per cell. For αPu, *a* = 6.183 Å, *b* = 4.822 Å, *c* = 10.963 Å and β = 101.79 °

B1 NaCl type. Face-centered cubic: *Fm3m*; *cF*8. Four sodium atoms at 0, 0, 0; 0, ½, ½; ½, 0, ½ and ½, ½, 0; four chlorine atoms at ½, ½, ½; ½, 0, 0; 0, ½, 0 and 0, 0, ½. For NaCl, *a* 5.64 Å. **Examples:** BaS, CdO, CdS, CrN, HfC, HfN, NaCl, NiO (HT), PbS, PbSe, TiO, UC, UO, UP, US, VO, ZrO

B2 CsCl or β′Cu-Zn type. Cubic: *Pm3m*; *cP*2. One cesium atom at 0, 0, 0; and one chlorine atom at ½, ½, ½. For CsCl, *a* = 4.11 Å. **Examples:** AgCd, CoTi, CsCl, FeAl, FeCo, FeTi, FeV, βNiAl, βNiGa, δNiIn, NiTi, β′Cu-Zn

(continued)

Crystal structure of metals (continued)

B3 ZnS (sphalerite, or zinc blende) type. Face-centered cubic: $F\bar{4}3m$; $cF8$. Four zinc atoms at 0, 0, 0; 0, $\frac{1}{2}$, $\frac{1}{2}$; $\frac{1}{2}$, 0, $\frac{1}{2}$ and $\frac{1}{2}$, $\frac{1}{2}$, 0; four sulfur atoms at $\frac{1}{4}$, $\frac{1}{4}$, $\frac{1}{4}$; $\frac{1}{4}$, $\frac{3}{4}$, $\frac{3}{4}$; $\frac{3}{4}$, $\frac{1}{4}$, $\frac{3}{4}$ and $\frac{3}{4}$, $\frac{3}{4}$, $\frac{1}{4}$. For ZnS (sphalerite), $a = 5.42$ Å. **Examples:** CdS, CdSe, CdTe, CuFeS$_2$(HT), GaP, GaSb, InAs, InP, InSb, βMnS, βSiC, ZnO, ZnS (sphalerite), ZnSe

B4 ZnS (wurtzite) type. Hexagonal: $P6_3mc$; $hP4$. Two zinc atoms at $\frac{1}{3}$, $\frac{2}{3}$, z, and $\frac{2}{3}$, $\frac{1}{3}$, $\frac{1}{2}$, $+z$ (with $z = 0$); two sulfur atoms at $\frac{1}{3}$, $\frac{2}{3}$, z and $\frac{2}{3}$, $\frac{1}{3}$, $\frac{1}{2} + z$ (with $z = 0.371$). For ZnS (wurtzite), $a = 3.82$ Å and $c = 6.26$ Å. Equivalent positions for Zn, 0, 0, 0 and $\frac{1}{3}$, $\frac{2}{3}$, $\frac{1}{2}$; for S, 0, 0, z and $\frac{1}{3}$, $\frac{2}{3}$, $\frac{1}{2} + z$ (with $z = 0.371$). **Examples:** AlN, BeO, CdS, CdSe, CuH, InN, InSb, γMnS, ZnO, ZnS (wurtzite), ZnSe

B8₁ NiAs type. Hexagonal: $P6_3/mmc$; $hP4$. Two nickel atoms at 0, 0, 0 and 0, 0, $\frac{1}{2}$; two arsenic atoms at $\frac{1}{3}$, $\frac{2}{3}$, $\frac{1}{4}$ and $\frac{2}{3}$, $\frac{1}{3}$, $\frac{3}{4}$. **Examples:** CoSb, CoSe, CoTe, CrH, CrSe, α″FeS, MnSb, NiAs, NiSb, NiTe, TiS, VS, VSb

B8₂ Ni₂In type. Hexagonal: $P6_3/mmc$; $hP6$. Two nickel atoms at 0, 0, 0 and 0, 0, $\frac{1}{2}$; two nickel atoms at $\frac{1}{3}$, $\frac{2}{3}$, $\frac{3}{4}$ and $\frac{2}{3}$, $\frac{1}{3}$, $\frac{1}{4}$; two indium atoms at $\frac{1}{3}$, $\frac{2}{3}$, $\frac{1}{4}$ and $\frac{2}{3}$, $\frac{1}{3}$, $\frac{3}{4}$/ For Ni₂In, $a = 4.18$ Å and $c = 5.13$ Å. **Examples:** AlZr₂, CoNiSn, Cu₂In, In₂Bi, Ni₂In, Ni$_{1.4}$Sn, Ti₂Sn

B10 PbO type. Tetragonal: $P4/nmm$; $tP4$. Two oxygen atoms at 0, 0, 0 and $\frac{1}{2}$, $\frac{1}{2}$, 0; two lead atoms at 0, $\frac{1}{2}$, z and $\frac{1}{2}$, 0, $\bar{z}$ (with $z = 0.237$). **Examples:** FeS, βFeTe$_{0.9}$, PbO, SnO

B11 Gamma CuTi type. Tetragonal: $P4/nmm$; $tP4$. Two copper atoms at 0, $\frac{1}{2}$, z and $\frac{1}{2}$, 0, $\bar{z}$ (with $z = 0.10$); two titanium atoms at 0, $\frac{1}{2}$, z and $\frac{1}{2}$, 0, $\bar{z}$ (with $z = 0.65$). For γCuTi, $a = 3.12$ Å and $c = 5.92$ Å. **Examples:** AgZr, AuTi (LT), γCuTi

B19 β′AuCd type. Orthorhombic: $Pmma$; $oP4$. Two gold atoms at $\frac{1}{4}$, $\frac{1}{2}$, z and $\frac{3}{4}$, $\frac{1}{2}$, $\bar{z}$ (with $z = 0.812$); two cadmium atoms at $\frac{1}{4}$, 0, z and $\frac{3}{4}$, 0, $\bar{z}$ (with $z = 0.313$). For β′AuCd, $a = 4.76$ Å, $b = 3.15$ Å and $c = 4.86$ Å. **Examples:** β″AgCd, β′AuCd, CdMg, IrMo, IrW

B20 FeSi type. Cubic: $P2_13$; $cP8$. Four iron atoms at x, x, x; $\frac{1}{2}+x$, $\frac{1}{2}-x$, $\bar{x}$; $\bar{x}$, $\frac{1}{2}+x$, $\frac{1}{2}-x$ and $\frac{1}{2}-x$, $\bar{x}$, $\frac{1}{2}+x$ (with $x = 0.137$); four silicon atoms at x, x, x; $\frac{1}{2}+x$, $\frac{1}{2}-x$, $\bar{x}$; $\bar{x}$, $\frac{1}{2}+x$, $\frac{1}{2} - x$ and $\frac{1}{2}-x$, $\bar{x}$, $\frac{1}{2}+x$ (with $x = 0.842$). **Examples:** CoSi, FeSi, MnSi

B31 MnP type. Orthorhombic: $Pnma$; $oP8$. Four manganese atoms at x, $\frac{1}{4}$, z; $\bar{z}$, $\frac{3}{4}$, $\bar{z}$, $\frac{1}{2}-x$, $\frac{3}{4}$, $\frac{1}{2}+z$ and $\frac{1}{2}+x$, $\frac{1}{4}$, $\frac{1}{2}-z$ (with $x = 0.20$ and $z = 0.005$); four phosphorus atoms at x, $\frac{1}{4}$, z; $\bar{x}$, $\frac{3}{4}$, $\bar{z}$; $\frac{1}{2}-x$, $\frac{3}{4}$, $\frac{1}{2}+z$ and $\frac{1}{2}+x$, $\frac{1}{4}$, $\frac{1}{2}-z$ (with $x = 0.57$ and $z = 0.19$). **Examples:** CoP, CrP, FeP, MnP, WP

Bh WC type. Hexagonal: $P\bar{6}m2$; $hP2$. One tungsten atom at 0, 0, 0; one carbon atom at $\frac{1}{3}$, $\frac{2}{3}$, $\frac{1}{2}$ or at $\frac{2}{3}$, $\frac{1}{3}$, $\frac{1}{2}$. For WC, $a = 2.91$ Å and $c = 2.84$ Å. **Examples:** γMoC, TiS, WC, WN, Zr₃S₂

C1 CaF₂ (fluorite) type. Face-centered cubic; $Fm3m$; $cF12$. Four calcium atoms at 0, 0, 0; 0, $\frac{1}{2}$, $\frac{1}{2}$; $\frac{1}{2}$, 0, $\frac{1}{2}$ and $\frac{1}{2}$, $\frac{1}{2}$, 0; eight fluorine atoms at $\frac{1}{4}$, $\frac{1}{4}$, $\frac{1}{4}$; $\frac{1}{4}$, $\frac{3}{4}$, $\frac{3}{4}$; $\frac{3}{4}$, $\frac{1}{4}$, $\frac{3}{4}$; $\frac{3}{4}$, $\frac{3}{4}$, $\frac{1}{4}$; $\frac{3}{4}$, $\frac{3}{4}$, $\frac{3}{4}$; $\frac{3}{4}$, $\frac{1}{4}$, $\frac{1}{4}$; $\frac{1}{4}$, $\frac{3}{4}$, $\frac{1}{4}$ and $\frac{1}{4}$, $\frac{1}{4}$, $\frac{3}{4}$. For CaF₂, $a = 5.46$ Å. **Examples:** Be₂B, Be₂C, CaF₂, CoSi₂, rare-earth hydrides, K₂O, K₂S, Mg₂Pb, Mg₂Si, ζNiSi₂, UN₂, UO₂

C2 FeS₂ (pyrite) type. Cubic: $Pa3$; $cP12$. Four iron atoms at 0, 0, 0; 0, $\frac{1}{2}$, $\frac{1}{2}$; $\frac{1}{2}$, 0, $\frac{1}{2}$ and $\frac{1}{2}$, $\frac{1}{2}$, 0; eight sulfur atoms at x, x, x; $\frac{1}{2}+x$, $\frac{1}{2}-x$, $\bar{x}$; $\bar{x}$, $\frac{1}{2}+x$, $\frac{1}{2}-x$; $\frac{1}{2}-x$, $\bar{x}$, $\frac{1}{2}+x$; $\bar{x}$, $\bar{x}$, $\bar{x}$; $\frac{1}{2}-x$, $\frac{1}{2}+x$, x; x, $\frac{1}{2}-x$, $\frac{1}{2}+x$ and $\frac{1}{2}+x$, x, $\frac{1}{2}-x$. For FeS₂ (pyrite), $x = 0.386$ and $a = 5.42$ Å. **Examples:** CoPS, CoS₂, CoSe₂, FeS₂ (pyrite), MnS₂, MnTe₂, NiS$_{2+x}$, NiSe₂

C3 Cu₂O type. Cubic: $Pn3m$; $cP6$. Four copper atoms at $\frac{1}{4}$, $\frac{1}{4}$, $\frac{1}{4}$; $\frac{1}{4}$, $\frac{3}{4}$, $\frac{3}{4}$; $\frac{3}{4}$, $\frac{1}{4}$, $\frac{3}{4}$ and $\frac{3}{4}$, $\frac{3}{4}$, $\frac{1}{4}$; two oxygen atoms at 0, 0, 0 and $\frac{1}{2}$, $\frac{1}{2}$, $\frac{1}{2}$. For Cu₂O, $a = 4.26$ Å. **Examples:** Ag₂O, Cu₂O

C4 TiO₂ (rutile) type. Tetragonal: $P4_2/mnm$; $tP6$. Two titanium atoms at 0, 0, 0 and $\frac{1}{2}$, $\frac{1}{2}$, $\frac{1}{2}$; four oxygen atoms at x, x, 0; $\bar{x}$, $\bar{x}$ 0; $\frac{1}{2}+x$, $\frac{1}{2}-x$, $\frac{1}{2}$ and $\frac{1}{2}-x$, $\frac{1}{2}+x$, $\frac{1}{2}$. For TiO₂, $x = 0.3056$, $a = 4.59$ Å and $c = 2.96$ Å. **Examples:** CrO₂, βMnO₂, PbO₂, SnO₂, TaO₂, TeO₂, TiO₂ (rutile), VO₂ (HT), WO₂

C11b MoSi₂ type. Tetragonal: $I4/mmm$; $tI6$. Two molybdenum atoms at 0, 0, 0 and $\frac{1}{2}$, $\frac{1}{2}$, $\frac{1}{2}$; four silicon atoms at 0, 0, z; 0, 0, $\bar{z}$;

$\frac{1}{2}$, $\frac{1}{2}$, $\frac{1}{2}+z$ and $\frac{1}{2}$, $\frac{1}{2}$, $\frac{1}{2}-z$ (with $z = 0.333$). For MoSi₂, $a = 3.20$ Å and $c = 7.86$ Å. **Examples:** AgZr₂, AlCr₂, Au₂Be, Au₂Mn, CuTi₂, Hg₂Mg, MoSi₂, Ni₂Ta, Si₂W

C14 MgZn₂ type. Hexagonal: $P6_3/mmc$; $hP12$. Four magnesium atoms at $\frac{1}{3}$, $\frac{2}{3}$, z; $\frac{2}{3}$, $\frac{1}{3}$, $\bar{z}$; $\frac{2}{3}$, $\frac{1}{3}$, $\frac{1}{2}+z$ and $\frac{1}{3}$, $\frac{2}{3}$, $\frac{1}{2}-z$ (with $z = 0.062$); two zinc atoms at 0, 0, 0 and 0, 0, $\frac{1}{2}$; six zinc atoms at x, $2x$, $\frac{1}{4}$; $2\bar{x}$, $\bar{x}$, $\frac{1}{4}$; x, $\bar{x}$, $\frac{1}{4}$; $\bar{x}$, $2\bar{x}$, $\frac{3}{4}$; $2x$, x, $\frac{3}{4}$ and $\bar{x}$, x, $\frac{3}{4}$ (with $x = 0.83$). For MgZn₂, $a = 5.18$ Å and $c = 8.52$ Å. **Examples:** Al₂Zr, Be₂Mo, CaCd₂, CaMg₂, CdCu₂, Fe₂Mo, FeSiW, Fe₂Ta, Fe₂Ti, Fe₂W, MgZn₂, TiZn₂

C15 Cu₂Mg type. Face-centered cubic: $Fd3m$; $cF24$. Eight magnesium atoms at 0, 0, 0; 0, $\frac{1}{2}$, $\frac{1}{2}$; $\frac{1}{2}$, 0, $\frac{1}{2}$; $\frac{1}{2}$, $\frac{1}{2}$, 0; $\frac{1}{4}$, $\frac{1}{4}$, $\frac{1}{4}$; $\frac{1}{4}$, $\frac{3}{4}$, $\frac{3}{4}$; $\frac{3}{4}$, $\frac{1}{4}$, $\frac{3}{4}$ and $\frac{3}{4}$, $\frac{3}{4}$, $\frac{1}{4}$; 16 copper atoms at $\frac{5}{8}$, $\frac{5}{8}$, $\frac{5}{8}$; $\frac{5}{8}$, $\frac{1}{8}$, $\frac{1}{8}$; $\frac{1}{8}$, $\frac{5}{8}$, $\frac{1}{8}$; $\frac{1}{8}$, $\frac{1}{8}$, $\frac{5}{8}$; $\frac{7}{8}$, $\frac{7}{8}$, $\frac{7}{8}$; $\frac{5}{8}$, $\frac{3}{8}$, $\frac{3}{8}$; $\frac{1}{8}$, $\frac{7}{8}$, $\frac{3}{8}$; $\frac{1}{8}$, $\frac{3}{8}$, $\frac{7}{8}$; $\frac{7}{8}$, $\frac{5}{8}$, $\frac{7}{8}$; $\frac{7}{8}$, $\frac{1}{8}$, $\frac{3}{8}$; $\frac{3}{8}$, $\frac{5}{8}$, $\frac{3}{8}$; $\frac{3}{8}$, $\frac{1}{8}$, $\frac{7}{8}$; $\frac{1}{8}$, $\frac{7}{8}$, $\frac{5}{8}$; $\frac{7}{8}$, $\frac{3}{8}$, $\frac{1}{8}$; $\frac{3}{8}$, $\frac{7}{8}$, $\frac{1}{8}$ and $\frac{3}{8}$, $\frac{3}{8}$, $\frac{5}{8}$. For Cu₂Mg, $a = 7.05$ Å. **Examples:** Al₂Ca, Al₂U, CdCuZn, Co₂U, Co₂Zr, Cr₂Ti, Cu₂Mg, FeNiTa, Fe₂U, Fe₂Zr, MgNiZn, αTiCo₂, ZrW₂

C16 CuAl₂ type. Body-centered tetragonal: $I4/mcm$; $tI12$. Four copper atoms at 0, 0, $\frac{1}{4}$; $\frac{1}{2}$, $\frac{1}{2}$, $\frac{3}{4}$; 0, 0, $\frac{3}{4}$ and $\frac{1}{2}$, $\frac{1}{2}$, $\frac{1}{4}$; eight aluminum atoms at x, $\frac{1}{2}+x$, 0; $\bar{x}$, $\frac{1}{2}-x$, 0; $\frac{1}{2}+x$, $\bar{x}$, 0; $\frac{1}{2}+x$, x, 0; $\frac{1}{2}+x$, x, $\frac{1}{2}$; $\frac{1}{2}-x$, x, $\frac{1}{2}$; x, $\frac{1}{2}-x$, $\frac{1}{2}$ and x, $\frac{1}{2}+x$, $\frac{1}{2}$ (with $x = 0.158$). For CuAl₂, $a = 6.07$ Å and $c = 4.87$ Å. **Examples:** Co₂B, Cr₂B, θCuAl₂, Fe₂B, FeSn₂, Mo₂B, Ni₂B, W₂B

C18 FeS₂ (marcasite) type. Orthorhombic: $Pnnm$; $oP6$. Two iron atoms at 0, 0, 0 and $\frac{1}{2}$, $\frac{1}{2}$, $\frac{1}{2}$; four sulfur atoms at x, y, 0; $\bar{x}$, $\bar{y}$, 0; $\frac{1}{2}+x$, $\frac{1}{2}-y$, $\frac{1}{2}$ and $\frac{1}{2}-x$, $\frac{1}{2}+y$, $\frac{1}{2}$ (with $x = 0.200$ and $y = 0.378$). For FeS₂ (marcasite), $a = 4.44$ Å, $b = 5.42$ Å and $c = 3.39$ Å. **Examples:** γCrSb₂, FeP₂, FeS₂ (marcasite), FeSe₂, FeTe₂, NiSb₂

C22 Fe₂P type. Hexagonal: $P\bar{6}2m$: $hP9$. Three iron atoms at x, 0, 0; 0, x, 0 and $\bar{x}$, $\bar{x}$, 0 (with $x = 0.256$); three iron atoms at x, 0, $\frac{1}{2}$; 0, x $\frac{1}{2}$ and $\bar{x}$, $\bar{x}$, $\frac{1}{2}$ (with $x = 0.594$); three phosphorus atoms at 0, 0, $\frac{1}{2}$; $\frac{1}{3}$, $\frac{2}{3}$, 0 and $\frac{2}{3}$, $\frac{1}{3}$, 0. For Fe₂P, $a = 5.93$ Å and $c = 3.45$ Å. **Examples:** Fe₂P, Mn₂P, Ni₂P, Pt₂Si (HT)

C38 Cu₂Sb type. Tetragonal: $P4/nmm$; $tP6$. Four copper atoms at 0, 0, 0; $\frac{1}{2}$, $\frac{1}{2}$, 0; 0, $\frac{1}{2}$, z and $\frac{1}{2}$, 0, z (with $z = 0.27$; two antimony at-

(continued)

Crystal structure of metals (continued)

oms at 0, ½, z and ½, 0, $\bar{z}$ (with $z = 0.70$). For Cu_2Sb, $a = 3.99$ Å and $c = 6.09$ Å. **Examples:** Al-NaSi$_4$, AsCr$_2$, AsMn$_2$, Bi$_2$U, Cu$_2$Sb, Pu$_2$U, Sb$_2$U

D0$_3$ BiF or BiLi$_3$ type. Face-centered cubic superlattice: $Fm3m$; $cF16$. Four bismuth atoms at 0, 0, 0; 0, ½, ½; ½, 0, ½ and ½, ½, 0; 12 fluorine (or lithium) atoms at ½, ½, ½; ½, 0, 0; 0, ½, 0; 0, 0, ½; ¼, ¼, ¼; ¼, ¾, ¾; ¾, ¼, ¾; ¾, ¾, ¼; ¾, ¾, ¾; ¾, ¼, ¼; ¼, ¾, ¼ and ¼, ¼, ¾. For BiLi$_3$, $a = 6.71$ Å. **Examples:** BiF$_3$, BiLi$_3$, Fe$_3$Al, γCu$_3$Zn (HT), αFe$_3$Si, Mn$_3$Si, Ni$_3$Sn (HT)

D0$_{11}$ Fe$_3$C (cementite) type. Orthorhombic: $Pnma$; $oP16$. Sixteen atoms per cell. **Examples:** Co$_3$B, Co$_3$C, Fe$_3$C, Mn$_3$C, Ni$_3$C, Pd$_3$P

D0$_{19}$ Ni$_3$Sn type. Hexagonal: $P6_3/mmc$; $hP8$. Two tin atoms at ⅓, ⅔, ¼ and ⅔, ⅓, ¾; six nickel atoms at x, $2x$, ¼; $2\bar{x}$, $\bar{x}$, ¼; x, $\bar{x}$, ¼; $\bar{x}$, $2\bar{x}$, ¾; $2x$, x, ¾ and $\bar{x}$, x, ¾ (with $x = 0.833$). For Ni$_3$Sn, $a = 5.29$ Å and $c = 4.24$ Å. **Examples:** AlTi$_{2-3}$, Cd$_3$Mg, CdMg$_3$, Co$_3$Mo, Co$_3$W, β''Fe$_3$Sn, γNi$_3$In, Ni$_3$Sn, Ti$_4$Pb

D0$_{24}$ Ni$_3$Ti type. Hexagonal: $P6_3/mmc$; $hP16$. Four titanium atoms at 0, 0, 0; 0, 0, ½; ⅓, ⅔, ¼ and ⅔, ⅓, ¾; 12 nickel atoms at ½, 0, 0; 0, ½, 0; ½, ½, 0; ½, 0, ½; 0, ½, ½; ½, ½, ½; x, $2x$, ¼; $2\bar{x}$, $\bar{x}$, ¼; x, $\bar{x}$, ¼; $\bar{x}$, $2\bar{x}$, ¾; $2x$, x, ¾ and x, x, ¾ (with $x = 0.833$). For Ni$_3$Ti, $a = 2.55$ Å and $c = 8.31$ Å. **Examples:** Co$_3$Ti, Ni$_3$Ti, Pd$_3$Zr

D5$_1$ Alpha alumina (αAl$_2$O$_3$) type. Rhombohedral-hexagonal: $R\bar{3}c$; $hR10$. Ten atoms per unit rhombohedral cell or 30 atoms per unit hexagonal cell. (There are also other structures of alumina.) **Examples:** αAl$_2$O$_3$, αFe$_2$O$_3$, Rh$_2$O$_3$, Ti$_2$O$_3$, V$_2$O$_3$ (HT)

D8$_2$ Gamma brass (γCu$_5$Zn$_8$) type. Body-centered cubic: $I\bar{4}3m$; $cI52$. Fifty-two atoms per cell. **Examples:** γAg$_5$Cd$_8$, γAg$_5$Zn$_8$, Al$_8$V$_5$, δCd$_8$Cu$_5$, γCu$_5$Zn$_8$

D8$_4$ Cr$_{23}$C$_6$ type. Face-centered cubic: $Fm3m$; $cF116$. One hundred sixteen atoms per cell. **Examples:** Cr$_{23}$C$_6$, Fe$_{21}$Mo$_2$C$_6$, Fe$_{21}$W$_2$C$_6$, Mn$_{23}$C$_6$, Ni$_{19.5}$Zr$_{3.5}$B$_6$

D8$_5$ Fe$_7$W$_6$ (μ-phase) type. Rhombohedral: $R\bar{3}m$; $hR13$. Thirteen atoms per cell. **Examples:** μCo$_7$Mo$_6$, Co$_7$W$_6$, Fe$_7$Mo$_6$, Fe$_7$W$_6$, NiTa

D8$_b$ Sigma FeCr (σ-phase) type. Tetragonal: $P4_2/mnm$; $tP30$. Thirty atoms per cell. **Examples:** σCoCr, Co$_2$Mo$_3$, CrMn$_3$, σFeCr, σFeMo, σFeV, σMn-Mo, TaV

E2$_1$ CaTiO$_3$ (perovskite) type. Cubic: $Pm3m$; $cP5$. One calcium atom at 0, 0, 0; one titanium atom at ½, ½, ½; three oxygen atoms at 0, ½, ½; ½, 0, ½ and ½, ½, 0. **Examples:** AlCFe$_3$, AlCMn$_3$, AlCTi$_3$, Ca-TiO$_3$, Fe$_3$C$_x$In, Fe$_3$NNi, Fe$_3$NPd, Fe$_3$NSn

E9$_3$ Fe$_3$W$_3$C (η-carbide) type. Face-centered cubic: $Fd3m$; $cF112$. One hundred twelve atoms per cell. **Examples:** CoCb$_2$(C, N, O)$_x$, Co$_2$Mo$_4$C, Cr$_3$Cb$_3$C, ηFe$_2$Cb$_3$, Fe$_3$Mo$_3$C, Fe$_3$Mo$_3$N, Fe$_3$W$_3$C, Mn$_3$Mo$_3$C, Mo$_3$Ni$_3$C, NiTi$_2$, Ni$_3$W$_3$C

H1$_1$ Spinel (Al$_2$MgO$_4$ or Fe$_3$O$_4$) type. Face-centered cubic: $Fd3m$; $cF56$. Fifty-six atoms per cell. **Examples:** Al$_2$CrS$_4$, Al$_2$MgO$_4$, Co$_2$NiS$_4$, Co$_3$O$_4$, Co$_3$S$_4$, CuS$_4$Ti$_2$, FeNi$_2$S$_4$, Fe$_3$O$_4$, Fe$_3$S$_4$ (greigite), Ni$_3$S$_4$ (LT)

... R-phase (in Co-Cr-Mo) type. Rhombohedral-hexagonal: $R\bar{3}$; $hR53$. Fifty-three atoms per cell. See A12 type, above. **Examples:** R-(Co-Cr-Mo), Co$_3$Cr$_3$Si$_2$, R-(Co-Mn-Mo), Fe$_2$SiV$_2$, Mn$_{78}$Mo$_3$Si$_{19}$, Mn$_6$Si, Ni$_3$SiV$_6$

... P-phase (in Mo-Cr-Ni) type. Orthorhombic: $Pbnm$; $oP56$. Fifty-six atoms per cell. See A12 type, above. **Examples:** P-Cr$_{18}$Mo$_{42}$Ni$_{40}$, P-(Mo-Fe-Ni), P-(Mo-Mn-Co)

L1$_0$ AuCu (I) type. Tetragonal superlattice: $P4/mmm$; $tP4$. Two gold atoms at 0, 0, 0 and ½, ½, 0; two copper atoms at 0, ½, ½ and ½, 0, ½. **Examples:** AgTi, AlTi, AuCu I, θCdPt, FePd, γ''FePt, θMnNi, NiPt

L1$_1$ CuPt type. Rhombohedral superlattice: $R\bar{3}m$; $hR32$. **Example:** CuPt

L1$_2$ AuCu$_3$ (I) type. Cubic superlattice: $Pm3m$; $cP4$. One gold atom at 0, 0, 0; three copper atoms at 0, ½, ½; ½, 0, ½ and ½, ½, 0. **Examples:** α'AlNi$_3$, AlZr$_3$, Au$_3$Cu, AuCu$_3$ (I), CoPt$_3$, Cr$_3$Pt, Fe$_3$Ga, FePd$_3$, Ni$_3$Fe, Ni$_3$Mn, Sn$_3$U

L'2 Martensite (Fe-C) type. Tetragonal: $I4/mmm$. In the unit cell there are iron atoms at 0, 0, 0 and ½, ½, ½; the carbon atoms are random, at ½, ½, 0 and or 0, 0, ½, to provide two iron atoms and up to 0.12 carbon atoms per cell. **Examples:** Fe-C martensite, α'Fe-N martensite

L'3 Fe$_2$N, or W$_2$C, type. Hexagonal: $P6_3/mmc$; $hP3$. Two iron atoms at ⅓, ⅔, ¼ and ⅔, ⅓, ¾; one nitrogen atom at either 0, 0, 0 or 0, 0, ½. **Examples:** Fe$_2$N, ζMn$_2$N, βTa$_2$C, Ta$_{-2}$N, V$_2$C, W$_2$C (LT)

Crystal structure of metals

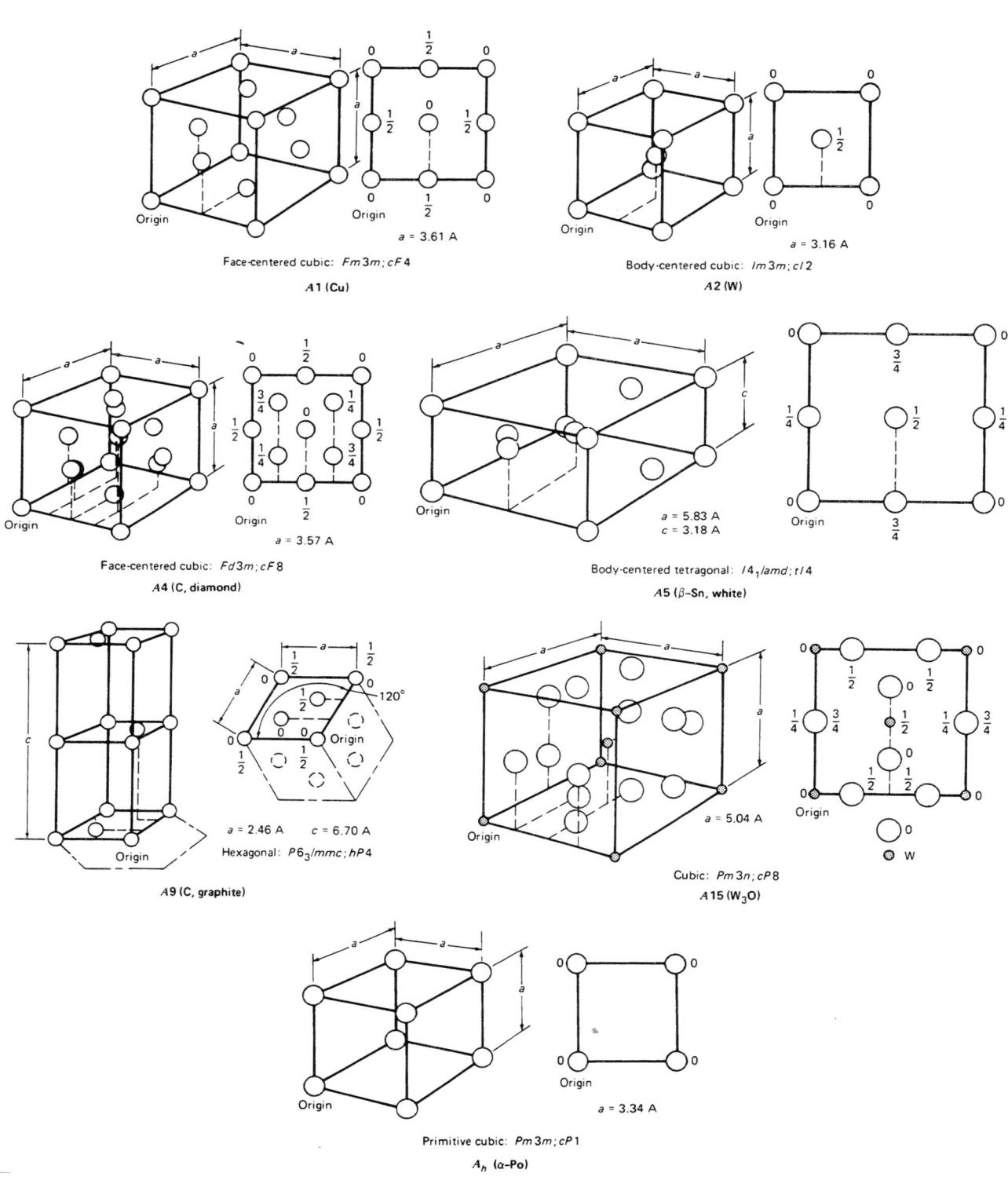

Face-centered cubic: $Fm3m$; $cF4$

$a = 3.61$ A

$A1$ (Cu)

Body-centered cubic: $Im3m$; $cI2$

$a = 3.16$ A

$A2$ (W)

Face-centered cubic: $Fd3m$; $cF8$

$a = 3.57$ A

$A4$ (C, diamond)

Body-centered tetragonal: $I4_1/amd$; $tI4$

$a = 5.83$ A
$c = 3.18$ A

$A5$ (β-Sn, white)

Hexagonal: $P6_3/mmc$; $hP4$

$a = 2.46$ A $c = 6.70$ A

$A9$ (C, graphite)

Cubic: $Pm3n$; $cP8$

$a = 5.04$ A

○ 0   ⊙ W

$A15$ (W$_3$O)

Primitive cubic: $Pm3m$; $cP1$

$a = 3.34$ A

A_h (α-Po)

(continued)

Crystal structure of metals (continued)

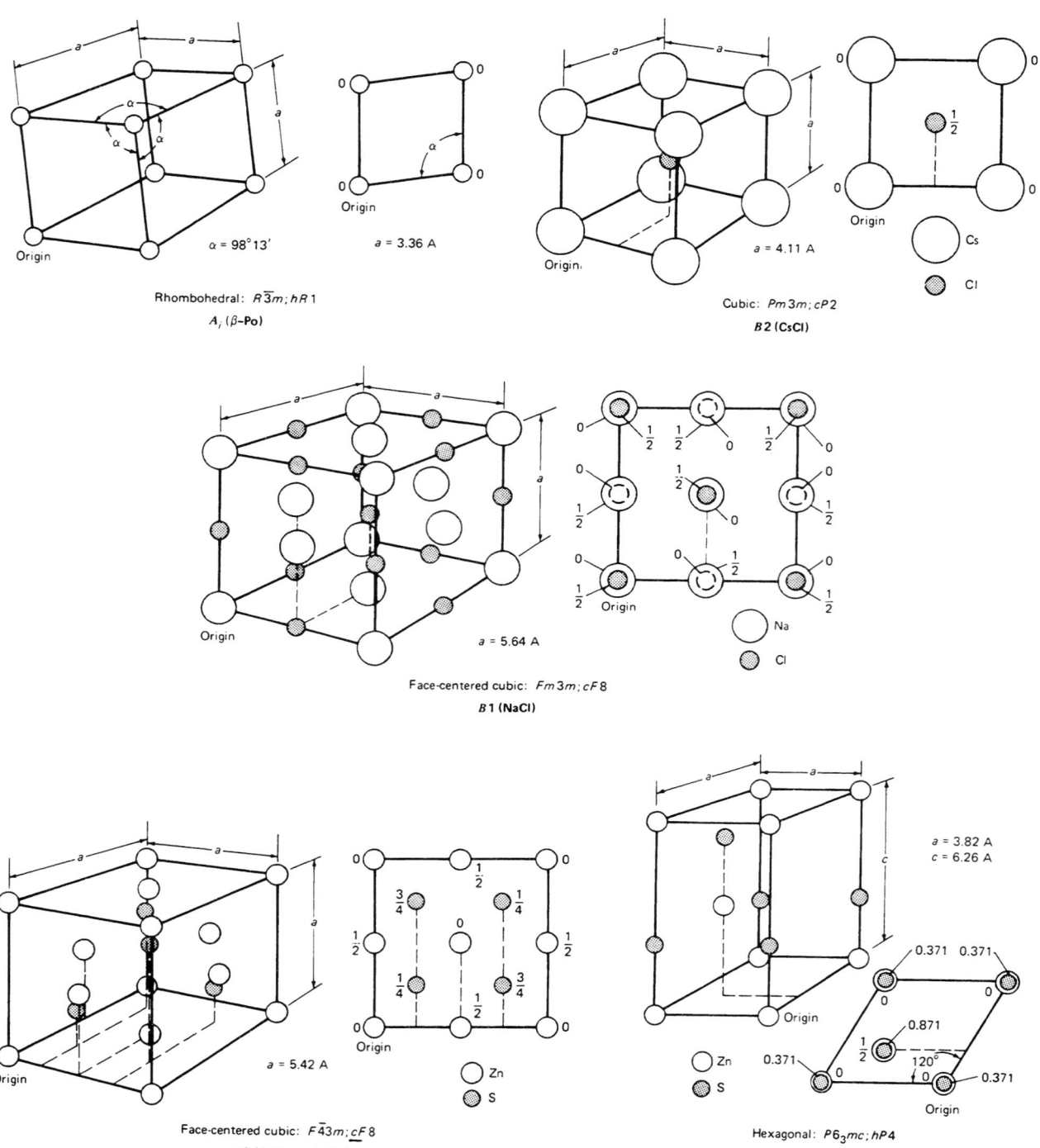

Rhombohedral: $R\overline{3}m; hR1$
A_i (β–Po)

$\alpha = 98°13'$

$a = 3.36$ A

Cubic: $Pm3m; cP2$
$B2$ (CsCl)

$a = 4.11$ A

○ Cs

◉ Cl

Face-centered cubic: $Fm3m; cF8$
$B1$ (NaCl)

$a = 5.64$ A

○ Na

◉ Cl

Face-centered cubic: $F\overline{4}3m; \underline{cF8}$
$B3$ (ZnS, sphalerite)

$a = 5.42$ A

○ Zn

◉ S

Hexagonal: $P6_3mc; hP4$
$B4$ (ZnS, wurtzite)

$a = 3.82$ A
$c = 6.26$ A

○ Zn

◉ S

(continued)

Crystal structure of metals (continued)

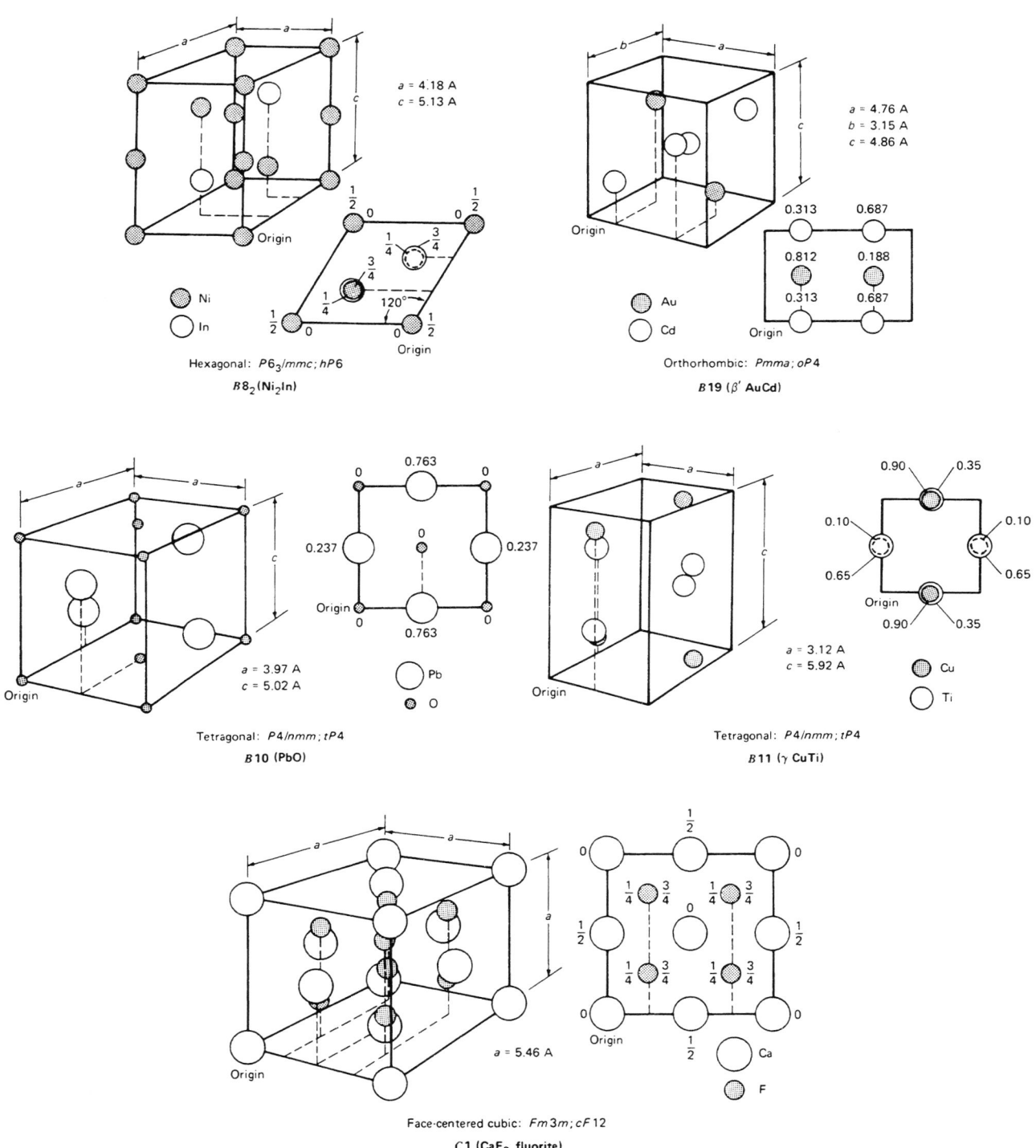

Hexagonal: $P6_3/mmc$; $hP6$

$B8_2 (Ni_2In)$

$a = 4.18$ A
$c = 5.13$ A

Ni

In

Orthorhombic: $Pmma$; $oP4$

$B19$ (β' **AuCd**)

$a = 4.76$ A
$b = 3.15$ A
$c = 4.86$ A

Au

Cd

Tetragonal: $P4/nmm$; $tP4$

$B10$ **(PbO)**

$a = 3.97$ A
$c = 5.02$ A

Pb

O

Tetragonal: $P4/nmm$; $tP4$

$B11$ (γ **CuTi**)

$a = 3.12$ A
$c = 5.92$ A

Cu

Ti

Face-centered cubic: $Fm3m$; $cF12$

$C1$ **(CaF$_2$, fluorite)**

$a = 5.46$ A

Ca

F

(continued)

Crystal structure of metals (continued)

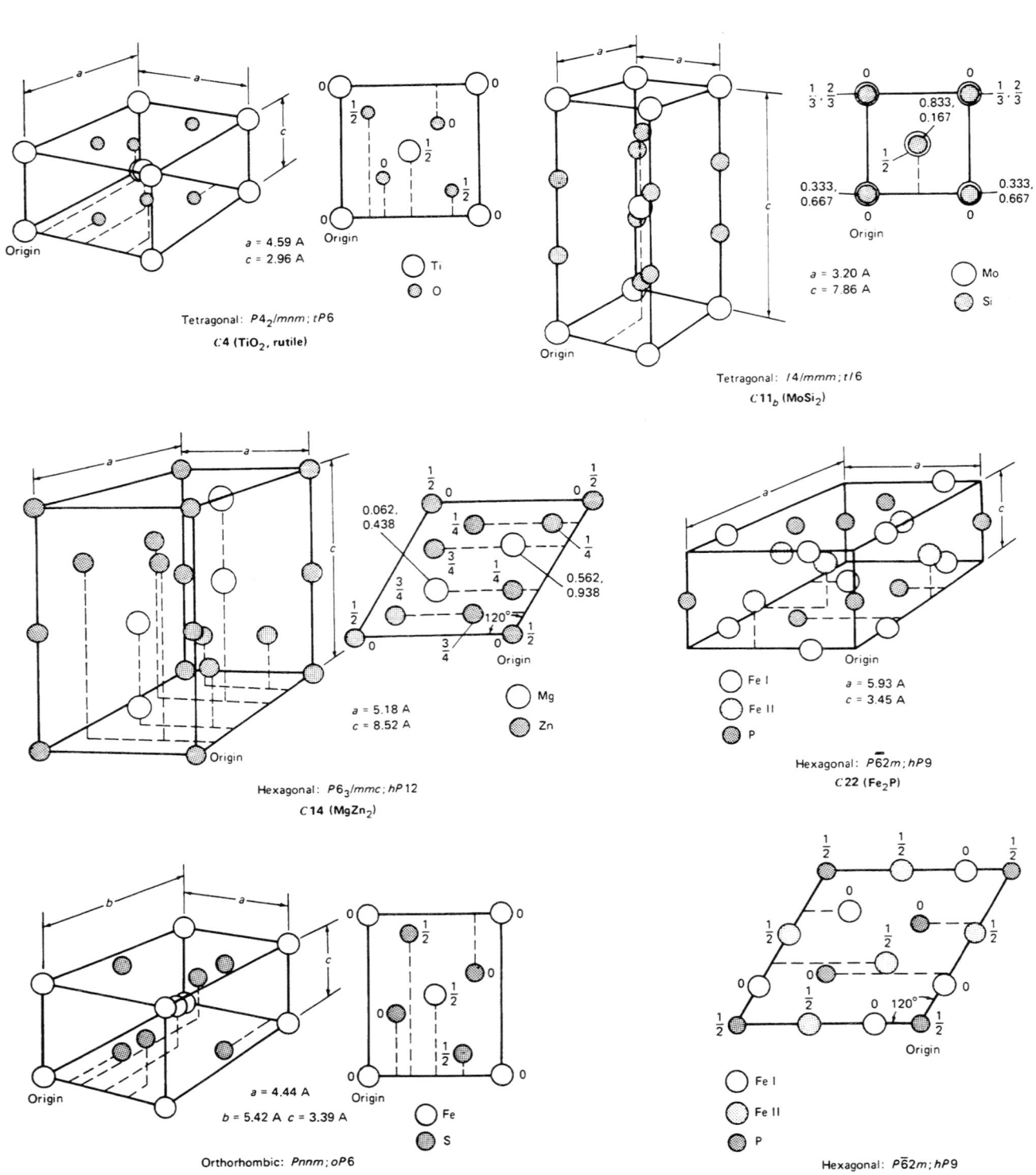

a = 4.59 A
c = 2.96 A

○ Ti
◍ O

Tetragonal: $P4_2/mnm$; $tP6$

$C4$ (TiO_2, rutile)

a = 3.20 A
c = 7.86 A

○ Mo
◍ Si

Tetragonal: $I4/mmm$; $tI6$

$C11_b$ ($MoSi_2$)

a = 5.18 A
c = 8.52 A

○ Mg
◍ Zn

Hexagonal: $P6_3/mmc$; $hP12$

$C14$ ($MgZn_2$)

○ Fe I
◔ Fe II
◍ P

a = 5.93 A
c = 3.45 A

Hexagonal: $P\bar{6}2m$; $hP9$

$C22$ (Fe_2P)

a = 4.44 A
b = 5.42 A c = 3.39 A

○ Fe
◍ S

Orthorhombic: $Pnnm$; $oP6$

$C18$ (FeS_2, marcasite)

○ Fe I
◔ Fe II
◍ P

Hexagonal: $P\bar{6}2m$; $hP9$

$C22$ (Fe_2P)

(continued)

Crystal structure of metals (continued)

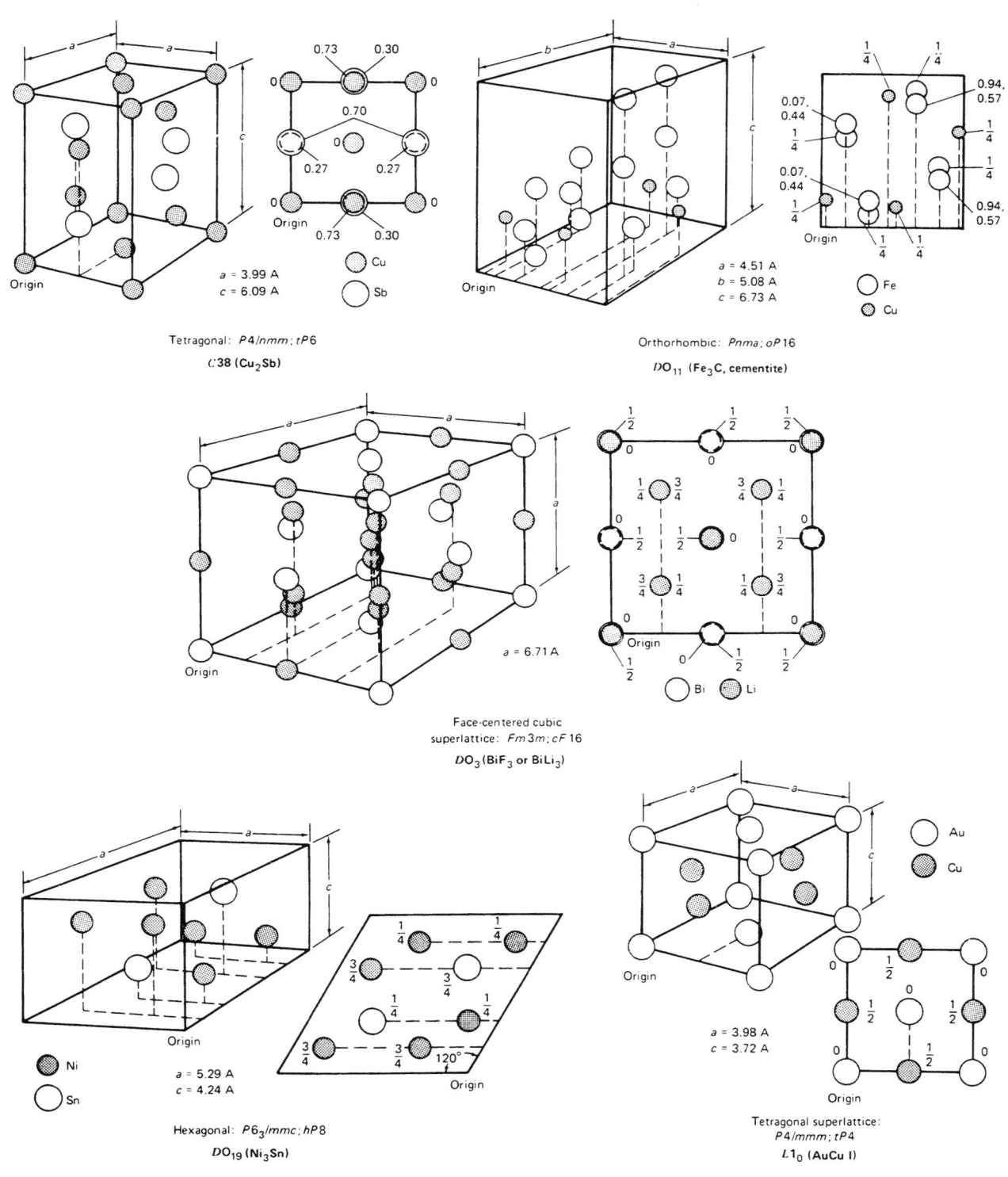

$a = 3.99$ A
$c = 6.09$ A

Cu
Sb

Tetragonal: $P4/nmm$; $tP6$

$C38$ (Cu_2Sb)

$a = 4.51$ A
$b = 5.08$ A
$c = 6.73$ A

Fe
Cu

Orthorhombic: $Pnma$; $oP16$

DO_{11} (Fe_3C, cementite)

$a = 6.71$ A

Bi   Li

Face-centered cubic
superlattice: $Fm3m$; $cF16$

DO_3 (BiF_3 or $BiLi_3$)

Ni
Sn

$a = 5.29$ A
$c = 4.24$ A

Hexagonal: $P6_3/mmc$; $hP8$

DO_{19} (Ni_3Sn)

Au
Cu

$a = 3.98$ A
$c = 3.72$ A

Tetragonal superlattice:
$P4/mmm$; $tP4$

$L1_0$ (AuCu I)

(continued)

Crystal structure of metals (continued)

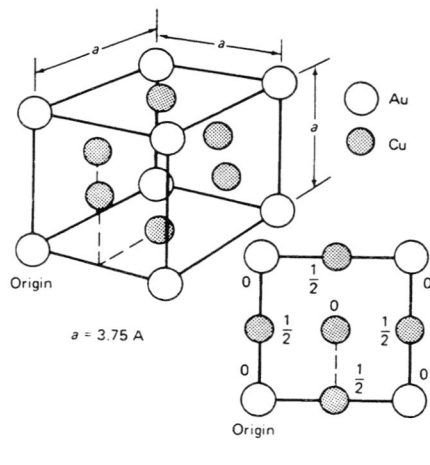

Au

Cu

$a = 3.75$ A

Origin

$\frac{1}{2}$ 0 0

0 $\frac{1}{2}$ $\frac{1}{2}$ 0

0 $\frac{1}{2}$ 0

Origin

Cubic superlattice:
$Pm3m$; $cP4$

$l.1_2$ (AuCu₃I)

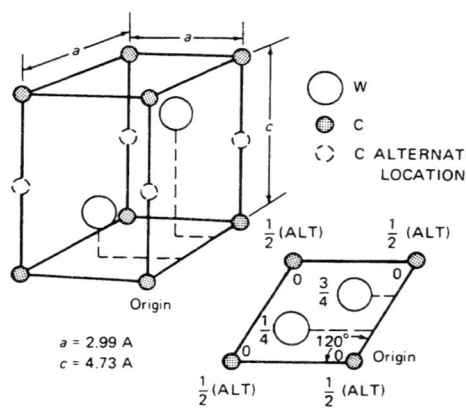

W

C

⟨⟩ C ALTERNATE
LOCATION

$a = 2.99$ A
$c = 4.73$ A

Origin

$\frac{1}{2}$ (ALT) $\frac{1}{2}$ (ALT)

0 $\frac{3}{4}$ 0

$\frac{1}{4}$ 120°

0 0 Origin

$\frac{1}{2}$ (ALT) $\frac{1}{2}$ (ALT)

Hexagonal: $P6_3/mmc$; $hP3$

$l.'3$ (Fe₂N or W₂C)

Testing and Inspection

Materials Characterization

Nondestructive methods for evaluating materials

Method	Measures or detects	Applications	Advantages	Limitations
Acoustic emission	Crack initiation and growth rate Internal cracking in welds during cooling Boiling or cavitation Friction or wear Plastic deformation Phase transformations	Pressure vessels Stressed structures Turbine or gear boxes Fracture mechanics research Weldments Sonic signature analysis	Remote and continuous surveillance Permanent record Dynamic (rather than static) detection of cracks Portable Triangulation techniques to locate flaws	Transducers must be placed on part surface Highly ductile materials yield low amplitude emissions Part must be stressed or operating Test system noise needs to be filtered out
Acoustic impact (tapping)	Debonded areas or delaminations in metal or laminates Cracks in turbine wheels or turbine blades Loose rivets or fasteners Crushed core	Brazed or adhesive-bonded structures Bolted or riveted assemblies Turbine blades Turbine wheels Composite structures Honeycomb assemblies	Portable Easy to operate May be automated Permanent record or positive meter readout No couplant required	Part geometry and mass influences test results Impactor and probe must be repositioned to fit geometry of part Reference standards required Pulser impact rate is critical for repeatability
Barkhausen noise analysis	Residual stress in ferromagnetic steels Anoding burns	Jet engine components such as compressor blades, discs, diffuser cases Ground chromium plated steel	Nondestructive stress analysis Permanent record Fully automatic Portable	Expensive Requires reference standard Need trained operator
Eddy current (100 Hz to 10 kHz) Eddy current (10 kHz to 6 MHz)	Subsurface cracks around fastener holes in aircraft structure Surface and subsurface cracks and seams Alloy content Heat treatment variations Wall thickness, coating thickness Crack depth Conductivity Permeability	Aluminum and titanium structure Tubing Wire Ball bearings "Spot checks" on all types of surfaces Proximity gage Metal detector Metal sorting Measure conductivity in % IACS	Detect subsurface cracks not detectable by radiography No special operator skills required High speed, low cost Automation possible for symmetrical parts Permanent record capability for symmetrical parts No couplant or probe contact required	Part geometry Will not detect short cracks Conductive materials Shallow depth of penetration (thin walls only) Masked or false indications caused by sensitivity to variations, such as part geometry, lift-off Reference standards required Permeability variations
Eddy-sonic	Debonded areas in metal-core or metal-faced honeycomb structures Delaminations in metal laminates or composites Crushed core	Metal-core honeycomb Metal-faced honeycomb Conductive laminates such as boron or graphite fiber composites Bonded metal panels	Portable Simple to operate No couplant required May be automated	Specimen or part must contain conductive materials to establish eddy-current field Reference standards required Part geometry
Electric current (direct current conduction method)	Cracks Crack depth Resistivity Wall thickness Corrosion-induced wall-thinning	Metallic materials Electrically conductive materials Train rails Nuclear fuel elements Bars, plates, other shapes	Access to only one surface required Battery or dc source Portable	Edge effect Surface contamination Good surface contact required Difficult to automate Electrode spacing Reference standards required
Electrified particle	Surface defects in non-conducting material Through-to-metal pinholes on metal-backed material Tension, compression, cyclic cracks Brittle-coating stress cracks	Glass Porcelain enamel Nonhomogeneous materials such as plastic or asphalt coatings Glass-to-metal seals	Portable Useful on materials not practical for penetrant inspection	Poor resolution on thin coatings False indications from moisture streaks or lint Atmospheric conditions High voltage discharge
Exo-electron emission	Fatigue in metals	Metals	Access to only one surface required Permanent record Quantitative	No surface films or contamination Geometry limitations Skilled technician required

(continued)

Nondestructive methods for evaluating materials (continued)

Method	Measures or detects	Applications	Advantages	Limitations
Filtered particle	Cracks Porosity Differential absorption	Porous materials such as clay, carbon, powdered metals, concrete Grinding wheels High-tension insulators Sanitary ware	Colored or fluorescent particles Leaves no residue after baking part over 205 °C (400 °F) Quickly and easily applied Portable	Size and shape of particles must be selected before use Penetrating power of suspension medium is critical Particle concentration must be controlled Skin irritation
Fluoroscopy (cine-fluorography, kine-fluorography)	Level of fill in containers Foreign objects Internal components Density variations Voids, thickness Spacing or position	Particles in liquid flow Presence of cavitation Operation of valves and switches Burning in small solid-propellant rocket motors	High-brightness images Real-time viewing Image magnification Permanent record Moving subject can be observed	Costly equipment Geometric unsharpness Thick specimens Speed of event to be studied Viewing area
Holography (acoustical-liquid surface levitation)	Lack of bond Delaminations Voids Porosity Resin-rich or resin-starved areas Inclusions Density variations	Metals Plastics Composites Laminates Honeycomb structures Ceramics Biological specimens	No hologram film development required Real-time imaging provided Liquid-surface responds rapidly to ultrasonic energy	Through-transmission techniques only Object and reference beams must superimpose on special liquid surface Immersion test only Laser required
Holography (interferometry (see shearography)	Strain Plastic deformation Cracks Debonded areas Voids and inclusions Vibration	Bonded and composite structures Automotive or aircraft tires Three-dimensional imaging	Surface of test object can be uneven No special surface preparations or coatings required No physical contact with test specimen	Vibration-free environment is required Heavy base to dampen vibrations Difficult to identify type of flaw detected
Infrared (radiometers)	Lack of bond Hot spots Heat transfer Isotherms Temperature ranges	Brazed joints Adhesive-bonded joints Metallic plating or castings; debonded areas or thickness Electrical assemblies Temperature monitoring	Sensitive to 0.85 °C (1.5 °F) temperature variation Permanent record or thermal picture Quantitative Remote sensing; need not contact part Portable	Emissivity Liquid-nitrogen-cooled detector Critical time-temperature relationship Poor resolution for thick specimens Reference standards required
Leak testing	Leaks: helium, ammonia, smoke, water, air bubbles, radioactive gas, and halogens	Joints: welded, brazed, and adhesive-bonded Sealed assemblies Pressure or vacuum chambers Fuel or gas tanks	High sensitivity to extremely small, tight separations not detectable by NDT methods Sensitivity related to method selected	Accessibility to both surfaces or part required Smeared metal or contaminants may prevent detection Cost related to sensitivity
Magnetic field	Cracks Wall thickness Hardness Coercive force Magnetic anistropy Magnetic field Nonmagnetic coating thickness on steel	Ferromagnetic materials Ship degaussing Liquid level control Treasure hunting Wall thickness of nonmetallic materials Material sorting	Measurement of magnetic material properties May be automated Easily detects magnetic objects in nonmagnetic material Portable	Permeability Reference standards required Edge-effect Probe lift-off
Magnetic particle	Surface and slightly subsurface defect; cracks, seams, porosity, inclusions Permeability variations Extremely sensitive for locating small tight cracks	Ferromagnetic materials, bar, forgings, weldments, extrusions, etc.	Advantage over penetrant in that it indicates defects open to the surface Relatively fast and low cost May be portable	Alignment of magnetic field is critical Demagnetization of parts required after tests Parts must be cleaned before and after inspection Masking by surface coatings
Magnetic perturbation	Cracks Crack depth Broken strands in steel cables Permeability effects Nonmetallic inclusions Grinding burns and cracks under chromium plating	Ferromagnetic metals Broken steel cables in reinforced concrete	May be automated Easily detects magnetic objects in nonmagnetic materials Detect subsurface defects	Requires reference standard Need trained operator Part geometry Expensive equipment

(continued)

Nondestructive methods for evaluating materials (continued)

Method	Measures or detects	Applications	Advantages	Limitations
Microwave (300MHz–300GHz)	Cracks, holes, debonded areas, etc., in non-metallic parts; Changes in composition, degree of cure, moisture content; Thickness measurement; Dielectric constant; Loss tangent	Reinforced plastics; Chemical products; Ceramics; Resins; Rubber; Liquids; Polyurethane foam; Radomes	Between radio waves and infrared in the electromagnetic spectrum; Portable; Contact with part surface not normally required; Can be automated	Will not penetrate metals; Reference standards required; Horn to part spacing critical; Part geometry; Wave interference; Vibration
Mössbauer effect	Nuclear magnetic resonance in materials, most common being iron-57; Polarization of magnetic domains in steel	Detect and identify iron in specimen or sample; Detect iron films on stainless steel; Measure retained austenite (2 to 35%) in steels; Determine nitrided surfaces on steel; Interaction of domains with dislocation in ferromagnetic materials	Provide unique information about the surroundings of the iron-57 nuclei	Radiation hazard; Nonportable; Precision equipment for vibrating source and spectrum analysis
Neutron activation analysis (reactor, accelerator, or radio-isotope)	Radiation emission resulting from neutron activation; Oxygen in steel; Nitrogen in food products; Silicon in metals and ores	Metallurgical; Prospecting; Well logging; Oceanography; On-line process control of liquid or solid materials	Automatic systems; Accurate (ppm range); Fast; No contact with sample; Sample preparation minimal	Radiation hazard; Fast decay time; Radioactive reference standard required; Sensitivity varies with irradiation time
Penetrants (dye or fluorescent)	Defects open to surface of parts; cracks, porosity, seams, laps, etc.; Through-wall leaks	All parts with non-absorbing surfaces; forgings, weldments, castings, etc. (Note: bleed-out from porous surfaces can mask indications of defects)	Low cost; Portable; Indications may be further examined visually; Results easily interpreted	Surface films, such as coatings, scale, and smeared metal may prevent detection of defects; Parts must be cleaned before and after inspection; Defect must be open to surface
Radiography (thermal neutrons from reactor, accelerator, or Californium-252)	Hydrogen contamination of titanium or zirconium alloys; Defective or improperly loaded pyrotechnic devices; Improper assembly of metal, nonmetal parts; Corrosion products	Pyrotechnic devices; Metallic, nonmetallic assemblies; Biological specimens; Nuclear reactor fuel elements and control rods; Adhesive bonded structures	High neutron absorption by hydrogen, boron, lithium, cadmium, uranium, plutonium; Low neutron absorption by most metals; Complement to X-ray or gamma-ray radiography	Very costly equipment; Nuclear reactor or accelerator required; Trained physicists required; Radiation hazard; Nonportable; Indium or gadolinium screens required
Radiography (gamma rays), Cobalt-60, Indium-192	Internal defects and variations; porosity, inclusions, cracks, lack of fusion, geometry variations, corrosion thinning; Density variations; Thickness, gap and position	Usually where X-ray machines are not suitable because source cannot be placed in part with small openings and/or power source not available; Panoramic imaging	Low initial cost; Permanent records: film; Small sources can be placed in parts with small openings; Portable; Low contrast	One energy level per source; Source decay; Radiation hazard; Trained operators needed; Lower image resolution; Cost related to source strength
Radiography (X-rays, film)	Internal defects and variations; porosity; inclusions; cracks; lack of fusion; geometry variations; corrosion thinning; Density variations; Thickness, gap and position; Misassembly; Misalignment	Castings; Electrical assemblies; Weldments; Small, thin, complex wrought products; Nonmetallics; Solid propellant rocket motors; Composites	Permanent records; film; Adjustable energy levels (5 kv to 25 meV); High sensitivity to density changes; No couplant required; Geometry variations do not affect direction of X-ray beam	High initial costs; Orientation of linear defects in part may not be favorable; Radiation hazard; Depth of defect not indicated; Sensitivity decreases with increase in scattered radiation
Radiometry (X-gamma-, or beta-ray; transmission or backscatter)	Wall thickness; Plating thickness; Variations in density or composition; Fill level in cans or containers; Inclusions or voids	Sheet, plate, foil, strip, tubing; Nuclear reactor fuel rods; Cans or containers; Plated parts; Composites	Fully automatic; Fast; Extremely accurate; In-line process control; Portable	Radiation hazard; Beta-ray useful for ultrathin coatings only; Source decay; Reference standards required

(continued)

Nondestructive methods for evaluating materials (continued)

Method	Measures or detects	Applications	Advantages	Limitations
Shearography -electronic (see holographic interferometry)	Debonded areas or delaminations in metal or nonmetal composites or laminates and honeycomb Crushed core Surface strain	Bonded assemblies, composite and honeycomb structures	One-sided access Inspect structures not inspectable by ultrasonics Production or in-service inspection Real-time Insensitive to environmental vibration	Structure requires stressing (heat or pressure) to induce strain Costly equipment Trained operator
Sonic (less than 0.1 MHz)	Debonded areas or delaminations in metal or nonmetal composites or laminates Cohesive bond strength under controlled conditions Crushed or fracture core Bond integrity of metal insert fasteners	Metal or nonmetal composite or laminates brazed or adhesive-bonded Plywood Rocket motor nozzles Honeycomb	Portable Easy to operate Locates far-side debonded areas May be automated Access to only one surface required	Surface geometry influences test results Reference standards required Adhesive or core thickness variations influence results
Thermal (thermochromic paint, liquid crystals	Lack of bond Hot spots Heat transfer Isotherms Temperature ranges Blockage in coolant passages	Brazed joints Adhesive-bonded joints Metallic platings or coatings Electrical assemblies Temperature monitoring	Very low initial cost Can be readily applied to surfaces which may be difficult to inspect by other methods No special operator skills	Thin-walled surfaces only Critical time-temperature relationship Image retentivity affected by humidity Reference standards required
Thermoelectric probe	Thermoelectric potential Coating thickness Physical properties Thompson effect P-N junctions in semiconductors	Metal sorting Ceramic coating thickness on metals Semiconductors	Portable Simple to operate Access to only one surface required	Hot probe Difficult to automate Reference standards required Surface contaminants Conductive coatings
Tomography X-ray (Cat Scan)	Boundaries Surface reconstruction Crack size, location, and orientation	Metals research Medicine Rocket motors Rocket nozzles	Pinpoint defect location Image display is computer controlled High contrast	Very expensive Need highly trained operator
Ultrasonic (0.1 to 25 MHz)	Internal defects and variations; cracks, lack of fusion porosity, inclusions, delaminations, lack of bond, texturing Thickness or velocity Poisson's ratio, elastic modulus	Wrought metals Welds Brazed joints Adhesive-bonded joints Nonmetallics In-service parts	Most sensitive to cracks Test results known immediately Automating and permanent record capability Portable High penetration capability	Couplant required Small, thin, complex parts may be difficult to check Reference standards required Trained operators for manual inspection Special probes
Ultrasonic angle reflectivity	Elastic properties, acoustic attenuation in solids Near-surface metallic property gradients, e.g. carburization in steel Metallic grain structure and size	Metals Nonmetals	Access to only one surface required Permanent record Quantitative No physical contact of sample required Sample preparation minimal	Test parts must be immersed Geometry limitations: test part must have a flat, smooth area Goniometer device required Skilled technician required

Tables and Flow Charts

The tables and flow charts in this section have been developed as tools to provide information about the most widely used methods of analysis for different classes of materials. These tables and charts are *not* intended to be all-inclusive but to identify the most commonly used techniques for the types of materials to be characterized and the types of information needed. As a result, many techniques that require special modifications or conditions to perform the desired analysis are omitted. The previous section of this article describes how to use these tools. After examining the tables or charts, the reader is encouraged to refer to the appropriate articles in the Handbook for additional information prior to consultation with an analytical specialist.

Abbreviations used in the headings of the tables and charts are:

Elem	Elemental analysis
Alloy ver	Alloy verification
Iso/Mass	Isotopic or mass analysis
Qual	Qualitative analysis (identification of constituents)
Semiquant	Semiquantitative analysis (order of magnitude)
Quant	Quantitative analysis (precision of ±20% relative standard deviation)
Macro/Bulk	Macroanalysis or bulk analysis
Micro	Microanalysis (≤10 μm)
Surface	Surface analysis
Major	Major component (>10 wt%)
Minor	Minor component (0.1 to 10 wt%)
Trace	Trace component (1 to 1000 ppm or 0.0001 to 0.1 wt%)
Ultratrace	Ultratrace component (<1 ppm or <0.0001 wt%)

The acronyms listed below are used in the tables and charts:

AAS	Atomic absorption spectrometry		MFS	Molecular fluorescence spectroscopy
AES	Auger electron spectroscopy		NAA	Neutron activation analysis
COMB	High-temperature combustion		NMR	Nuclear magnetic resonance
EFG	Elemental and functional group analysis		OES	Optical emission spectroscopy
EPMA	Electron probe x-ray microanalysis		OM	Optical metallography
ESR	Electron spin resonance		RBS	Rutherford backscattering spectrometry
FT-IR	Fourier transform infrared spectroscopy		RS	Raman spectroscopy
GC/MS	Gas chromatography/mass spectrometry		SAXS	Small-angle x-ray scattering
GMS	Gas mass spectrometry		SEM	Scanning electron microscopy
IA	Image analysis		SIMS	Secondary ion mass spectroscopy
IC	Ion chromatography		SSMS	Spark source mass spectrometry
ICP-AES	Inductively coupled plasma atomic emission spectroscopy		TEM	Transmission electron microscopy
IGF	Inert gas fusion		UV/VIS	Ultraviolet/visible absorption spectroscopy
IR	Infrared spectroscopy		XPS	X-ray photoelectron spectroscopy
ISE	Ion selective electrode		XRD	X-ray diffraction
LC	Liquid chromatography		XRS	X-ray spectrometry
LEISS	Low-energy ion-scattering spectroscopy			

INORGANIC SOLIDS: Metals, alloys, semiconductors

Wet analytical chemistry, electrochemistry, ultraviolet/visible absorption spectroscopy, and molecular fluorescence spectroscopy can generally be adapted to perform many of the bulk analyses listed. • = generally usable; N or † = limited number of elements or groups; G = carbon, nitrogen, hydrogen, sulfur, or oxygen: see boxed summary in article for details; S or * = under special conditions; D = after dissolution; Z or ** = semiconductors only

Method	Elem	Alloy ver	Iso/Mass	Qual	Semiquant	Quant	Macro/Bulk	Micro	Surface	Major	Minor	Trace	Phase ID	Structure	Morphology
AAS	D					D	D			D	D	D			
AES	•			•	•			•	•	•	•	S			S
COMB	G	G				G	G				G	G			
EPMA	•	S		•	•	•		•		•	•	N	S	N	•
ESR	N			N	N	N	N				N	N		N	
IA							•	•							•
IC	D, N			D, N	D, N	D, N	D, N			D, N	D, N	D, N			
ICP-AES	D	D		D	D	D	D			D	D	D			
IGF	G	G					G	G			G	G			
IR/FT-IR	Z			Z	Z	Z	Z				Z	Z			
LEISS	•			•	•			S	•	•	•	•			
NAA	•		N	•	•	•	•				•	•			
OES	•	•		•	•	•	•				•	•			
OM							•	•							•
RBS	•			•	•	•			•	•	S	S			
RS	Z			Z	Z	Z	Z			Z	Z	Z			
SEM	•			•	•	S		•		•	•		S		•
SIMS	•		•	•					•	•	•	S			
SSMS	•	•	•		•	•	•			•	•	•			
TEM	•			•	•	S		•		•	•		•	•	•
XPS	•			•	•				•	•	•				
XRD				•	•	S	•			•	•		•	•	
XRS	•	•		•	•	•	•			•	•	N			

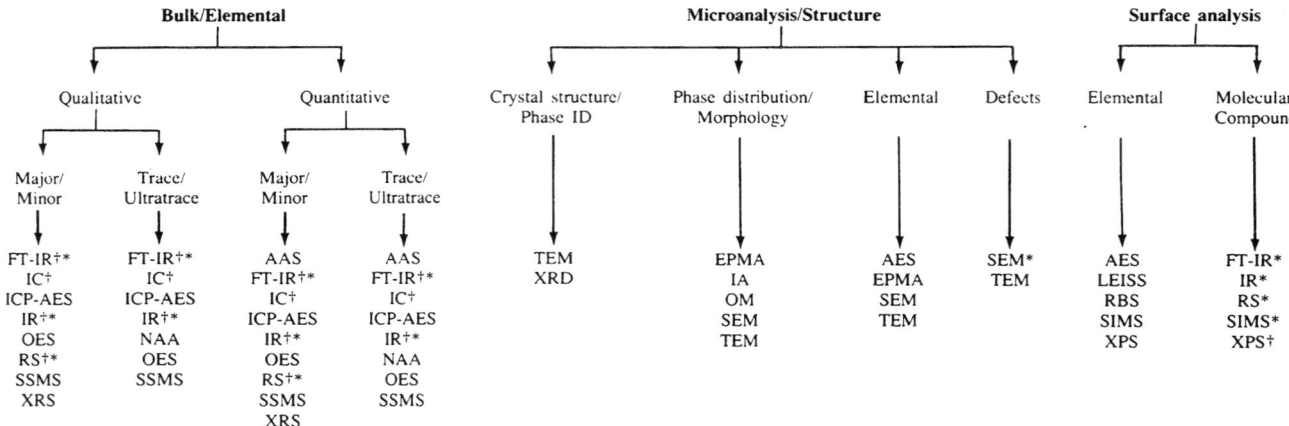

INORGANIC SOLIDS: Glasses, ceramics

Wet analytical chemistry, ultraviolet/visible absorption spectroscopy, and molecular fluorescence spectroscopy can generally be adapted to perform many of the bulk analyses listed. • = generally usable; N or † = limited number of elements or groups; S or * = under special conditions; D = after dissolution

Method	Elem	Speci-ation	Iso/Mass	Qual	Semiquant	Quant	Macro/Bulk	Micro	Surface	Major	Minor	Trace	Phase ID	Struc-ture	Morph-ology
AAS	D					D	D			D	D	D			
AES	•			•	•			•	•	•	•	S			S
EPMA	•			•	•	•		•		•	•	S	S		S
IA							•	•							•
IC	D, N			D, N	D, N	D, N	D, N			D, N	D, N	D, N			
ICP-AES	D			D	D	D	D			D	D	D			
IR/FT-IR	S	S		S	S	S	S			S	S	S	S	S	
LEISS	•			•	•			S	•	•	•	•			
NAA	•		N	•	•	•	•			S	•	•			
OES	•			•	•	•	•			•	•	•			
OM							•	•							•
RBS	•			•	•	•			•	•	S	S			
RS	S	S		S	S	S	S	S		S	S	S	S		
SEM	•			•	•			•		•	•		S		•
SIMS	•		•	•	•				•	•	•	S			•
SSMS	•		•	•	•		•			•	•	•			
TEM	•			•	•	S		•		•	•				•
XPS	•	N		•	•				•	•	•		•	•	•
XRD				•	•	S	•			•	•		•	•	
XRS	•			•	•	•	•			•	•	N			

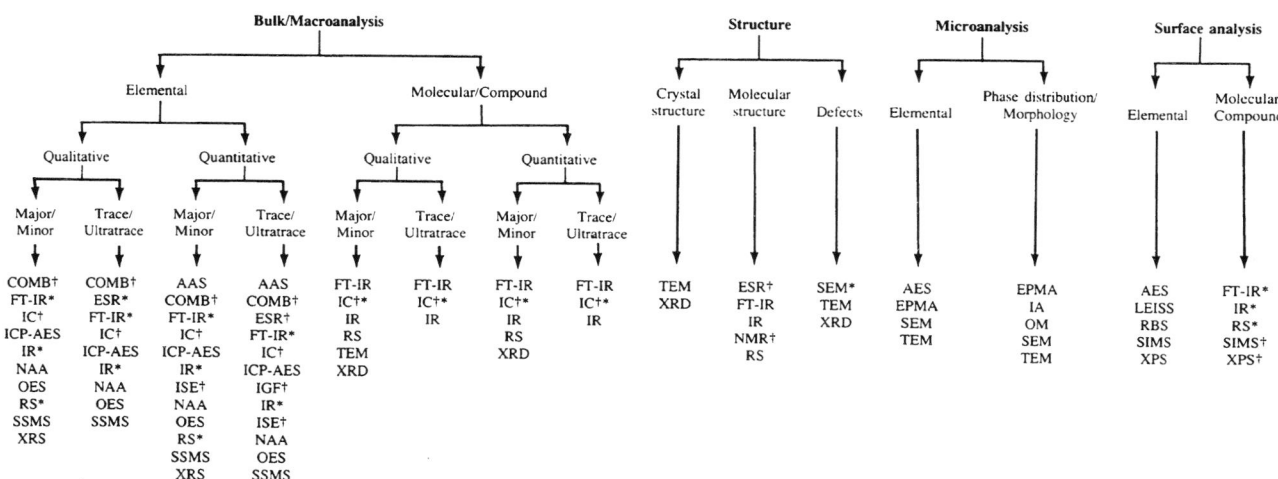

INORGANIC SOLIDS: Minerals, ores, slags, pigments, inorganic compounds, effluents, chemical reagents, composites, catalysts

Wet analytical chemistry, electrochemistry, ultraviolet/visible absorption spectroscopy, and molecular fluorescence spectroscopy can generally be adapted to perform many of the bulk analyses listed. • = generally usable; G = carbon, nitrogen, hydrogen, sulfur, or oxygen: see boxed summary in article for details; N or † = limited number of elements or groups; S or * = under special conditions; D = after dissolution

Method	Elem	Speciation	Iso/Mass	Qual	Semiquant	Quant	Macro/Bulk	Micro	Surface	Major	Minor	Trace	Compound/Phase	Structure	Morphology
AAS	D					D	D			D	D	D			
AES	•		•		•			S	•	•	•	S			S
COMB	G			G		G	G			G	G	G			
EPMA	•			•	•	•		•		•	•	S	S		•
ESR	N	N		N	N	N	N				N	N		N	
IA							•	•							
IC	D	S		D	D	D	D			D	D	D			
ICP-AES	D			D	D	D	D			D	D	D			
IGF	G					G	G				G	G			
IR/FT-IR	S, D			S, D	S, D	S, D	S, D		S	S, D	S, D	S, D	S, D	S, D	
ISE	D, N				D, N	D, N	D, N			D, N	D, N	D, N			
LEISS	•			•	•				•	•	•	•			
NAA	•		N	•	•	•	•			•	•	•			
OES	•			•	•	•	•			•	•	•			
OM							•	•							•
RBS	•				•	•			S	•	•	•			
RS	S, D			S, D	S, D	S, D	S, D		S	S, D	S, D	S, D	S, D	S, D	
SEM	•			•	•			•		•	•				•
SIMS	•		•	•					•	•	•	S	S		
SSMS	•		•	•	•		•			•	•				
TEM	•			•		S		•		•	•		•	•	•
XPS	•	•		•					•	•	•		S		
XRD				•	•	S	•			•	•		•	•	
XRS	•			•	•	•	•			•	•	N			

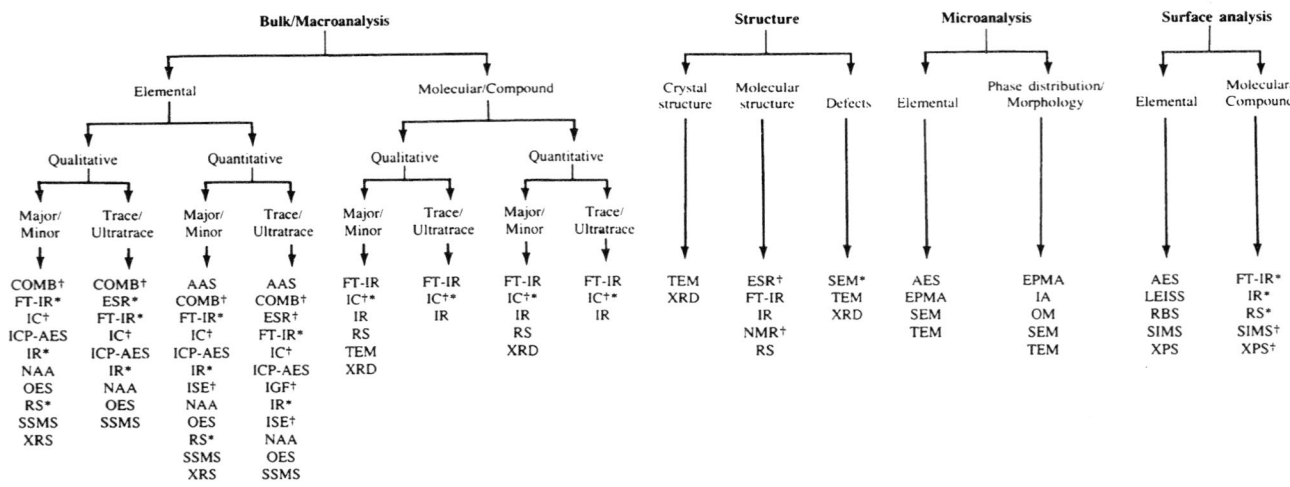

INORGANIC LIQUIDS AND SOLUTIONS: Water, effluents, leachates, acids, bases, chemical reagents

Wet analytical chemistry, electrochemistry, ultraviolet/visible absorption spectroscopy, and molecular fluorescence spectroscopy can generally be adapted to perform the bulk analyses listed. Most of the techniques listed for inorganic solids can be used on the residue after the solution is evaporated to dryness. • = generally usable; N or † = limited number of elements or groups; S = under special conditions; V or * = volatile liquids or components

Method	Elem	Speci-ation	Com-pound	Iso/Mass	Qual	Semi-quant	Quant	Macro/Bulk	Major	Minor	Trace	Structure
AAS	•						•	•	•	•	•	
EFG	N	N	N			N	N	N	N	N	N	
ESR	N	N			N	N	N	N		N	N	N
GC/MS	V, N		V	V	V	V	V	V		V	V	
GMS	V, N		V	V	V	V	V	V	V	V	V	
IC	•	•	S		•	•	•	•	•	•	•	
ICP-AES	•				•	•	•	•	•	•	•	
IR/FT-IR	•	•	•		•	•	•	•	•	•	•	
ISE	•	S			•	•	•	•	•	•	•	
NAA	•			N	•	•	•	•	•	•	•	
NMR	N		N		N	N		N		N		N
RS	•	•			•	•	•	•	•	•	S	
XRS	•				•	•	•	•	•	•	N	

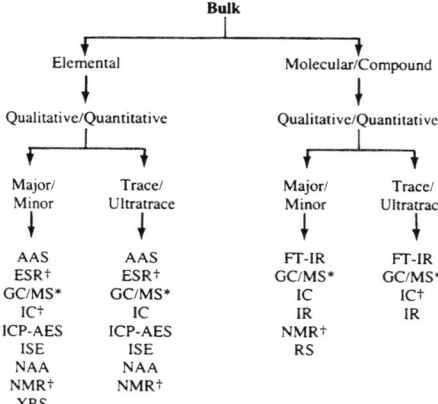

INORGANIC GASES: Air, effluents, process gases

Most of the techniques listed for inorganic solids and inorganic liquids can be used if the gas is sorbed onto a solid or into a liquid. • = generally usable

Method	Elem	Speci-ation	Com-pound	Iso/Mass	Qual	Semi-quant	Quant	Macro/Bulk	Major	Minor	Trace
GC/MS	•		•	•	•	•	•	•	•	•	•
GMS	•		•	•	•	•	•	•	•	•	•
IR/FT-IR	•	•	•		•	•	•	•	•	•	•
RS	•	•	•		•	•	•	•	•	•	

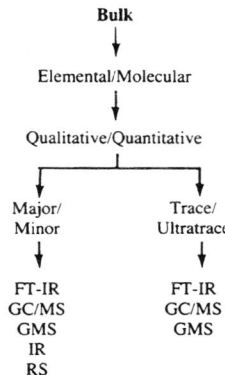

ORGANIC SOLIDS: Polymers, plastics, epoxies, long-chain hydrocarbons, esters, foams, resins, detergents, dyes, organic composites, coal and coal derivatives, wood products, chemical reagents, organometallics

Most of the techniques for inorganic solids and inorganic liquids can be used on any residue after ashing. • = generally usable; N or † = limited number of elements or groups; S or * = under special conditions; D = after dissolution/extraction; V = volatile solids or components (can also be analyzed by GC/MS), pyrolyzed solids; C = crystalline solids

Method	Elem	Speci-ation	Com-pound	Iso/Mass	Qual	Semi-Quant	Quant	Macro/Bulk	Micro	Sur-face	Major	Minor	Trace	Struc-ture	Morph-ology
AES	•				•	•			•	•	•	•			
COMB	N						N	N			N	N	N		
EFG	•		•		•	•	•	•			•	•			
EPMA	N				N	N	N		N		N	N	N		N
ESR	N	N			N	N	N	N				N	N	N	
GC/MS	V		V	V	V	V	V	V			V	V	V		
IA								•	•						•
IC	D, N		D, N		D, N	D, N	D, N	D, N			D, N	D, N	D, N		
IR/FT-IR	D, N	D, •	D, •		D, •	D, •	D, •	D, •		D, S	D, •	D, •	D, •	D, S	
LC			D		D	D	D	D			D	D	D		
LEISS	•				•	•				•	•	•	S		
MFS	D, N	D, N	D, N		D, N	D, N	D, N	D, N			D, N	D, N	D, N		
NAA	N			N	N	N	N	N			N	N	N		
NMR	N		N		N	N		N			N	N		N	
OM								•	•						•
RS	D, N	D, •	D, •		D, •	D, •	D, •	D, •		S	D, •	D, •		D, S	
SAXS	•		•		•			•			•	•		•	
SEM	N				N	N			•		N	N			•
SIMS	•		S	•	•				•	•	•	•	•		
TEM	S		C		N	N			•		N	N		C	•
UV/VIS	D, •	D, •	D, •		D, •	D, •	D, •	D, •			D, •	D, •	D, •		
XPS	•	N	S		•	•				•	•	•			
XRD			C		C	C	C, S	C			C	C		C	
XRS	N				N	N	N	N			N	N	N		

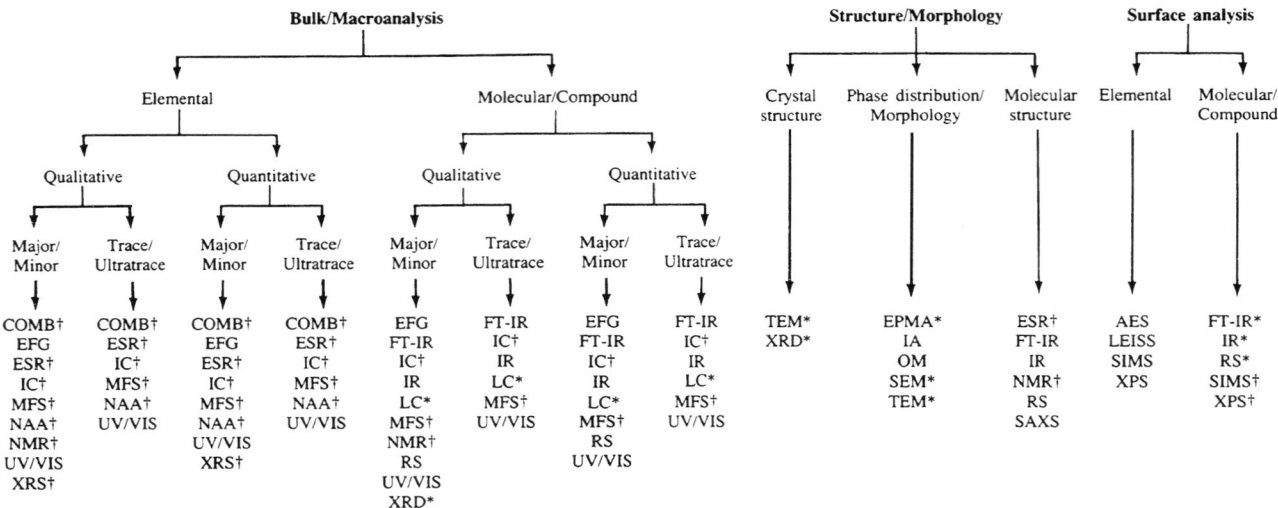

ORGANIC LIQUIDS AND SOLUTIONS: Hydrocarbons, petroleum and petroleum derivatives, solvents, reagents

Most of the techniques listed for inorganic solids and inorganic liquids can be used on any residue after ashing. Many wet chemical techniques can be adapted to perform the analyses listed. • = generally usable; N or † = limited number of elements or groups; S = under special conditions; V or * = volatile liquids

Method	Elem	Speci-ation	Com-pound	Iso/Mass	Qual	Semi-quant	Quant	Macro/Bulk	Major	Minor	Trace	Structure
EFG	•		•		•	•	•	•		•	•	
ESR	N	N			N	N	N	N	•	N	N	N
GC/MS			V	V	V	V	V	V	V	V	V	
GMS			V	V	V	V	V	V	V	V	V	
IR/FT-IR	S	S	•		•	•	•	•	•	•	•	•
LC			•		•	•	•	•	•	•	•	
MFS	N		N		N	N	N	N	N	N	N	
NAA	N			N	N	N	N	N	N	N	N	
NMR	N		N		N	N	N	N	N	N		N
RS	S	S	•		•	•	•	•	•	•		•
UV/VIS			•		•	•	•	•	•	•	•	
XRS	N				N	N	N	N	N	N	N	

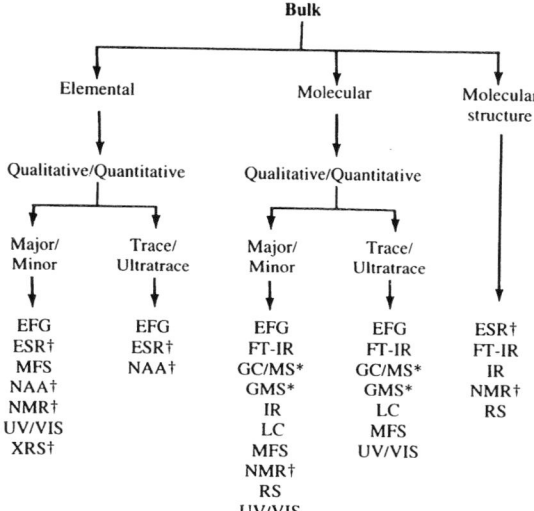

ORGANIC GASES: Natural gas, effluents, pyrolysis products, process gas

Most of the techniques listed for organic solids and organic liquids can be used if the gas is sorbed onto a solid or into a liquid. • = generally usable; S = under special conditions; L = after sorption onto a solid or into a liquid

Method	Elem	Speci-ation	Com-pound	Iso/Mass	Qual	Semi-quant	Quant	Macro/Bulk	Major	Minor	Trace	Structure
GC/MS	•		•	•	•	•	•	•	•	•	•	
GMS	•		•	•	•	•	•	•	•	•	•	
IR/FT-IR	•	S	•		•	•	•	•	•	•	•	•
LC			L		L	L	L	L	L	L	L	
RS	•	S	•		•	•	•	•	•	•		•

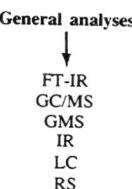

Optical and X-Ray Spectroscopy

Optical emission spectroscopy

General Uses

- Quantitative determination of major and trace elemental constituents in various sample types
- Qualitative elemental analysis

Examples of Applications

- Rapid determination of concentrations of alloying elements in steels and other alloys
- Elemental analysis of geological materials
- Determination of trace impurity concentrations in semiconductor materials
- Wear metals analysis in oils
- Determination of alkali and alkaline earth concentrations in aqueous samples
- Determination of calcium in cement

Samples

- *Form:* Conducting solids (arcs, sparks, glow discharges), powders (arcs), and solutions (flames)
- *Size:* Depends on specific technique; from approximately 10^{-6} g to several grams
- *Preparation:* Machining or grinding (metals), dissolution (for flames), and digestion of ashing (organic samples)

Limitations

- Some elements are difficult or impossible to determine, such as nitrogen, oxygen, hydrogen, halogens, and noble gases
- Sample form must be compatible with specific technique
- All methods provide matrix-dependent responses

Estimated Analysis Time

- 30 s to several hours, depending on sample preparation requirements

Capabilities of Related Techniques

- *X-ray fluorescence:* Bulk and minor constituent elemental analysis; requires sophisticated data reduction for quantitative analysis; not useful for light elements (atomic number ≤ 9)
- *Inductively coupled plasma emission spectroscopy:* Rapid quantitative elemental analysis with parts per billion detection limits; samples must be in solution; not useful for hydrogen, nitrogen, oxygen, halides, and noble gases
- *Direct-current plasma emission spectroscopy:* Similar in performance to inductively coupled plasma emission spectroscopy
- *Atomic absorption spectroscopy:* Favorable sensitivity and precision for most elements; single-channel technique; inefficient for multielement analysis

Inductively coupled plasma atomic emission spectroscopy

General Use

- Simultaneous multielement analysis
- Quantitative and qualitative analysis for over 70 elements with detection limits in the parts per billion (ng/mL) to parts per million (µg/mL) range
- Determination of major, minor, and trace elemental components

Examples of Applications

- Composition of metal alloys
- Trace impurities in alloys, metals, reagents, and solvents
- Analysis of geological, environmental, and biological materials
- Water analysis
- Process control

Samples

- *Form:* Liquids, gases, and solids; liquids are most common
- *Size:* 5 to 50 mL of solution, 10 to 500 mg of solids
- *Preparation:* Most samples are analyzed as solutions; solutions can be analyzed as received, diluted, or preconcentrated as required; solids must usually be dissolved to form solutions; gases may be analyzed directly

Limitations

- Detection limits parts per billion to parts per million
- Cannot analyze for noble gases
- Halogens and some nonmetals require vacuum spectrometer and optics
- Sensitivity poor for alkali elements, especially rubidium; cannot determine cesium

Estimated Analysis Time

- Dissolution of solids in sample preparation may require up to 16 h
- Analysis may require minutes to several hours

Capabilities of Related Techniques

- *Direct-current arc emission spectrography:* Samples may be analyzed directly as solids; sensitivity and quantitative precision poorer; longer analysis time required

- *Atomic absorption spectroscopy:* Single-element analysis; better sensitivity for most elements, especially by using electrothermal atomization, but not as good for refractory elements; more limited dynamic range

Atomic absorption spectrometry

General Use

- Quantitative analyses of approximately 70 elements

Examples of Applications

- Trace impurities in alloys and process reagents
- Water analysis
- Direct air sampling/analysis
- Direct solids analysis of ores and finished metals

Samples

- *Form:* Solids, solutions, and gaseous (mercury)
- *Size:* Depends on technique used—from a milligram (solids by graphite furnace atomic absorption spectrometry) to 10 mL of solution for conventional flame work
- *Preparation:* Depends on the type of atomizer used; usually a solution must be prepared

Limitations

- Detection limits range from subparts per billion to parts per million
- Cannot analyze directly for noble gases, halogens, sulfur, carbon, or nitrogen
- Poorer sensitivity for refractory oxide or carbide-forming elements than plasma atomic emission spectrometry
- Basically a single-element technique

Estimated Analysis Time

- Highly variable, depending on the type of atomizer and technique used
- Sample dissolution may take 4 to 8 h or as little as 5 min
- Typical analysis times range from approximately 1 min (flames) to several minutes (furnaces)

Capabilities of Related Techniques

- *Inductively coupled plasma atomic emission spectrometry and direct current plasma atomic emission spectrometry* are simultaneous multielement techniques with a wider dynamic analytical range and sensitivities complementing those of atomic absorption spectrometry. They cost considerably more to set up and require more expert attention to potential matrix interference (spectral) problems

Ultraviolet/visible absorption spectrometry

General Uses

- Quantitative analysis
- Qualitative analysis, especially of organic compounds
- Fundamental studies of the electronic structure of atomic and molecular species

Examples of Applications

- Quantitative determination of the principal and trace constituents in metals and alloys
- Quantitative determination of trace constituents in environmental (air and water) samples. These determinations are often conducted on site
- Measurement of the rates of chemical reactions
- Identification of the functional groups in organic molecules
- Detection of species in the effluent of liquid chromatographs
- Online monitoring of species in process streams
- Quantitative analysis of electroplating and chemical treatment baths
- Analysis of wastewater streams before and after treatment

Samples

- *Form:* Gas, liquid, or solid. Analyses are most commonly performed on liquid solutions
- *Size:* For solutions, typical sample volumes range from approximately 0.1 to 30 mL
- *Preparation:* Often quite complex. Complexity increases with the difficulty of placing the analyte in solution and the number of interferences

Advantages

- For quantitative analysis of inorganic ions, spectroscopic samples may contain as little as 0.01 mg/L in the case of species that form highly absorbing complexes
- For qualitative analysis of organic compounds, concentrations of spectroscopic samples may be as small as 100 nanomolar

- The initial capital outlay for ultraviolet/visible spectrophotometric techniques is usually far less than that for related techniques

Limitations

- The analyte must absorb radiation from 200 to 800 nm, or be capable of being converted into a species that can absorb radiation in this region
- Additional steps are often necessary to eliminate or account for interferences by species (other than the analyte) that also absorb radiation near the analytical wavelength

Estimated Analysis Time

- The actual time required to analyze each sample is a matter of minutes. However, it may take several hours to prepare the sample, make the standards, and create a calibration curve

Capabilities of Related Techniques

- *Molecular fluorescence spectroscopy:* For molecules that fluoresce strongly, this technique offers significantly greater sensitivities and freedom from interferences
- *Atomic absorption spectroscopy, optical emission spectroscopy:* For quantitative analysis of metals and some nonmetals, these techniques usually offer better sensitivity and almost complete freedom from interferences

Molecular fluorescence spectrometry

General Uses

- Fundamental studies of electronic transitions of organic and inorganic fluorescent molecules as well as of some atomic species
- Qualitative and quantitative chemical analysis

Examples of Applications

- Correlation of structural features of crystalline materials with their fluorescence spectral properties to study the nature of the interactions between lattice constituents
- Spectral fingerprinting for oil spill identification
- Determination of carcinogenic polynuclear aromatic compounds in environmental and biological samples
- Use of fluorescent labels for immunoassays
- Determination of metals that exhibit native fluorescence or that can induce fluorescence in appropriate crystallophor matrices
- Determination of electronic excited-state lifetimes

- Studies of protein dynamics, protein-ligand interactions and transport, and membrane structure and function
- Detection for high-performance liquid chromatography

Samples

- *Form:* Gas, liquid, or solid. Analyses are most commonly performed on liquid solutions
- *Size:* For solutions, sample size ranges from several milliliters to subnanoliter volumes using laser excitation
- *Preparation:* Requirements are often minimal and may include extraction, addition of reagents to induce or modify fluorescence, and so on. Special techniques, such as matrix isolation or Shpol'skii spectroscopy, require more complex sample preparation

Limitations

- The molecular (or atomic) species of interest must exhibit native fluorescence or must be coupled to or modified by chemical or physical interactions with other chemical species
- Detection limits for intensely fluorescent chemical species can extend from nanomolar to subpicomolar levels, depending on the type of instrumentation used

Estimated Analysis Time

- Usually ranges from minutes to hours. Actual measurement time per sample is generally on the order of seconds or less per measurement, and the analysis time depends on the number of measurements conducted per wavelength and the number of wavelengths used as well as other experimental variables and data-analysis requirements

Capabilities of Related Techniques

- *Ultraviolet/visible absorption spectroscopy:* Directly applicable to a wider range of molecules because fluorescent and nonfluorescent absorbers can be studied, but absorption is therefore less selective. Absorption spectra are more featureless than fluorescence spectra, and absorption detection limits are at least three orders of magnitude poorer and linear dynamic ranges much shorter than those of fluorescence
- *Atomic absorption and emission techniques (flame, furnace, inductively coupled plasma):* For atomic species, these techniques often afford improved detection limits and applicability to a wider range of elements

X-ray spectrometry

General Use

- Qualitative and quantitative elemental determination in solids and liquids
- Applications to materials and thin films

Examples of Applications

- Qualitative identification of elemental composition of materials for elements of atomic number greater than 11; identification at concentrations greater than a few ppm requires only a few minutes
- Support of phase identification using x-ray powder diffraction patterns
- Selection of alternate methods of quantitative analysis
- Quantitative determination of elements without regard to form or oxidation state in various solid and liquid materials and compositions
- Determination of thickness of thin films of metal on various substrates

Samples

- *Form:* Samples may be bulk solids, powders, pressed pellets, glasses, fused disks, or liquids
- *Size:* Typical samples are 32 mm (1 $\frac{1}{4}$ in.) in diameter or placed in special cups, holders, and mounts
- *Sampling depth* may range from a few micrometers to a millimeter or more, depending on x-ray energy used and matrix composition of the sample
- *Sample preparation* may involve none, polishing to obtain a flat surface, grinding and pelletizing, or fusion in a flux

Advantages

- Applicable to various samples, including solids
- Relatively rapid and easy to learn
- Semiquantitative results can be obtained from many samples without use of standards; most standards may be kept for long periods of time, because most applications are for solids
- Instrumentation is relatively inexpensive

Limitations

- Detection limits for bulk determinations are normally a few ppm to a few tens of ppm, depending on the x-ray energy used and the sample matrix composition
- For thin-film samples, detection limits are approximately 100 ng/cm^2
- Not suitable for elements of atomic number less than 11 unless special equipment is available, in which case elements down to atomic number 6 may be determined

Capabilities of Related Techniques

- *Inductively coupled plasma optical emission spectroscopy and atomic absorption spectrometry* have better detection limits for most elements than x-ray spectrometry and are often better choices for liquid samples; elements of low atomic number can be determined using these techniques

Particle-induced x-ray emission

General Uses

- Nondestructive multielemental analysis of thin samples, sodium through uranium, to approximately 1 ppm or 10^{-9} g/cm^2
- Nondestructive multielemental analysis of thick samples for medium and heavy elements
- Semiquantitative analysis of elements versus depth
- Elemental analyses of large and/or fragile objects through external beam proton milliprobe
- Elemental analyses using proton microprobes, spatial resolution to a few microns, and mass detection limits below 10^{-16} g

Examples of Applications

- Analysis of air filters for a wide range of elements
- Analysis of atmospheric aerosols by particle size for source transport, removal, and effect studies
- Analysis of powdered plant materials and geological powders for broad elemental content
- Analysis of elemental content of waters, solute, and particulate phases, including suspended particles
- Medical analysis for elemental content, including toxicology and epidemiology
- Analysis of materials for the semiconductor industry and for coating technology
- Archaeological and historical studies of books and artifacts, often using external beams
- Forensic studies

Samples

- *Form:* Thin samples (generally no more than a 10-mg/cm^2 thick solid) are analyzed in vacuum, as are stabilized powders and evaporated fluids. Thick samples can be any solid and thickness, but proton beam penetration is typically 30 mg/cm^2 or approximately 0.15 mm (0.006 in.) in a geological sample
- *Size:* The sample area analyzed is on the order of millimeters to centimeters, except in microprobes, in which beam spot sizes approaching 1 μm are available

- *Preparation:* None for air filters and many materials. Powders and liquids must be stabilized, dried, and generally placed on a substrate, such as plastic. Thick samples can be pelletized

Limitations

- Access to an ion accelerator of a few mega electron volts is necessary
- Generally, no elements below sodium are quantified
- Elements must be present above approximately 1 ppm
- Sample damage is more likely than with some alternate methods
- No chemical information is generated
- Computer codes are necessary for large numbers of analyses

Estimated Analysis Time

- 30 s to 5 min in most cases; thousands of samples can be handled in a few days

Capabilities of Related Techniques

- *X-ray fluorescence:* With repeated analyses at different excitation energies, essentially equivalent or somewhat superior results can be obtained when sample size and mass are sufficient
- *Neutron activation analysis:* Variable elemental sensitivity to neutron trace levels for some elements, essentially none for other elements. Neutron activation analysis is generally best for detecting the least common elements, but performs the poorest on the most common elements, complementing x-ray techniques
- *Electron microprobe:* Excellent spatial resolution (approximately 1 μm), but elemental mass sensitivity only approximately one part per thousand
- *Optical methods:* Atomic absorption or emission spectroscopy, for example, are generally applicable to elements capable of being dissolved or dispersed for introduction into a plasma

Raman spectroscopy

General Uses

- Molecular analysis of bulk samples and surface or near-surface species as identified by their characteristic vibrational frequencies
- Low-frequency vibrational information on solids for metal-ligand vibrations and lattice vibrations
- Determination of phase composition of solids

Examples of Applications

- Identification of effects of preparation on glass structure

- Structural analysis of polymers
- Determination of structural disorder in graphites
- Determination of surface structure of metal oxide catalysts
- Identification of corrosion products on metals
- Identification of surface adsorbates on metal electrodes

Samples

- *Form:* Solid, liquid, or gas
- *Size:* Single crystal of material to virtually any size the Raman spectrometer can accommodate

Limitations

- *Sensitivity:* Poor to fair without enhancement
- Raman spectroscopy requires concentrations greater than approximately 1 to 5%
- Analysis of surface or near-surface species difficult but possible
- Sample fluorescence or impurity fluorescence may prohibit Raman characterization

Estimated Analysis Time

- 30 min to 8 h per sample

Capabilities of Related Techniques

- *Infrared spectroscopy and Fourier-transform infrared spectroscopy:* Molecular vibrational identification of materials; lacks sensitivity to surface species; difficult on aqueous systems
- *High-resolution electron energy loss spectroscopy:* Vibrational analysis of surface species in ultrahigh-vacuum environment; extremely sensitive; requires ultrahigh-vacuum setup; low resolution compared to Raman spectroscopy; cannot be used for *in situ* studies

Infrared spectroscopy

General Uses

- Identification and structure determination of organic and inorganic materials
- Quantitative determination of molecular components in mixtures
- Identification of molecular species adsorbed on surfaces
- Identification of chromatographic effluents
- Determination of molecular orientation
- Determination of molecular conformation and stereochemistry

Examples of Applications

- Identification of chemical reaction species; reaction kinetics

- Quantitative determination of nontrace components in complex matrices

- Determination of molecular orientation in stretched polymer films

- Identification of flavor and aroma components

- Determination of molecular structure and orientation of thin films deposited on metal substrates (oxidation and corrosion products, soils, adsorbed surfactants, and so on)

- Depth profiling of solid samples (granules, powders, fibers, and so on)

- Characterization and identification of different phases in solids or liquids

Samples

- *Form:* Almost any solid, liquid, or gas sample

- *Size (minimum):* Solids—10 ng if it can be ground in a transparent matrix, such as potassium bromide; 10-μm diameter for a single particle; 1 to 10 ng if soluble in a volatile solvent (methanol, methylene chloride, chloroform, and so on). Flat metal surfaces—1 by 1 cm (0.4 by 0.4 in.) or larger. Liquids—10 μL if neat, considerably less if soluble in a transparent solvent. Gases—1 to 10 ng

- *Preparation:* Minimal or none; may have to grind in a potassium bromide matrix or dissolve in a volatile or infrared-transparent solvent

Limitations

- Little elemental information

- Molecule must exhibit a change in dipole moment in one of its vibrational modes upon exposure to infrared radiation

- Background solvent or matrix must be relatively transparent in the spectral region of interest

Estimated Analysis Time

- 1 to 10 min per sample

Capabilities of Related Techniques

- *Raman spectroscopy:* Complementary molecular vibrational information

- *X-ray fluorescence:* Elemental information on bulk samples

- *X-ray photoelectron spectroscopy:* Elemental information on adsorbed species

- *High-resolution electron energy loss spectroscopy:* Molecular vibrational surface information

- *Mass spectrometry:* Molecular weight information

- *Nuclear magnetic resonance:* Additional molecular structure information

Mass Spectroscopy

Spark source mass spectrometry

General Uses

- Qualitative and quantitative analysis of inorganic elements

- Measurements of trace impurities in materials

Examples of Applications

- Analysis of impurities in high-purity silicon for semiconductors

- Determination of precious metals in geological ores

- Measurement of toxic elements in natural water samples

- Verification of alloy compositions

Samples

- *Form:* Solid, solid residues from evaporation of liquids

- *Size:* Milligrams to micrograms, depending on impurity levels

- *Preparation:* If conductive, sawing or machining into electrodes. If nonconductive, grinding or mixing with high-purity conducting matrix, such as graphite or silver powder

Limitations

- Not generally used to measure gaseous elements

- Detection limits for most elements at parts per billion levels

- Chemical preparation can introduce significant contamination

Estimated Analysis Time

- Sample preparation requires 1 to 6 h

- Analysis requires 30 min to 1 h

- Data reduction requires 30 min to 1 h

Capabilities of Related Techniques

- *Laser ionization mass spectrometry:* Quicker; less sample preparation, but not as quantitative

- *Inductively coupled plasma atomic emission spectrometry:* Less expensive; can be easily automated for large number of samples per day. Requires dissolution of samples; does not measure all elements equally well

Gas analysis by mass spectrometry

General Uses

- Qualitative and quantitative analysis of inorganic and organic compounds and mixtures

Examples of Applications

- Analysis of internal atmospheres of sealed components
- Analysis of gas inclusions in ceramic or glass-to-metal seals
- Quantification of specific compounds in volatile liquids or gaseous mixtures
- Analysis of gases in inorganic or geologic materials
- Analysis of high-purity gases for contaminants
- Gas isotope ratios

Samples

- *Form:* Primarily gases
- *Size:* 1×10^{-4} mL (STP) or larger
- *Preparation:* Sample must be collected in a clean glass or steel bottle; air must not be allowed into the same bottle

Limitations

- Low part per million detection limits for inorganic and organic gases
- Difficulty in identifying components in complex organic mixtures
- Components of interest must be volatile

Estimated Analysis Time

- 30 min to 1 h per analysis, not including sample preparation and calibration

Capabilities of Related Techniques

- *Gas chromatography/mass spectrometry:* Greater capability for compound identification in complex organic mixtures; not quantitative
- *Gas chromatography:* Quantitative, but requires larger sample size
- *Infrared/Raman spectroscopy:* Requires larger sample size; infrared spectroscopy does not detect some inorganic gases

Classical, Electrochemical and Radiochemical Analysis

Classical wet analytical chemistry

General Uses

- Quantitative elemental composition analysis
- Qualitative identification of material type
- Qualitative detection of component moieties
- "Umpire" check on quantitative instrumental methods
- Isolation and characterization of inclusions and phases
- Characteristics of coatings and surfaces
- Determination of oxidation state

Examples of Applications

- Quantitative determination of alloy matrix elements for which instrumental methods are unavailable or unreliable
- Complete characterization of a homogeneous sample for use as an instrument calibration standard
- Quantitative determination of composition when sample is too small or of unsuitable shape for instrumental approaches
- Gross average composition determination of inhomogeneous samples
- Isolation of compounds and stoichiometric phases from metal alloy matrices for compositional analysis or examination by instrumental techniques
- Coating weight determination of plated metals, lubricant films, and other surface layers
- Partitioning of element oxidation states
- Sorting of mixed materials based on qualitative detection of one or more key matrix components

Samples

- *Form:* Crystalline or amorphous solids (metals, ceramics, glasses, ores, and so on) and liquids (pickling and plating baths, lubricants, and so on)
- *Size:* Depends on extent of required analyses—generally 1 to 2 g per element for solids and 20 mL total for process liquids, although requirements vary widely with technique
- *Preparation:* Solids are milled, drilled, crushed, or similarly dissociated into particles typically 2 mm (0.08 in.) in diameter or smaller. Machined alloys are solvent degreased. Materials lacking high order homogeneity require special care to ensure a representative laboratory sample

Limitations

- Slow compared to alternate instrumental techniques
- Except in rare cases, relatively large sample weights are required.

Estimated Analysis Time

- 2 to 80 h per element (8 h per element is typical)

Capabilities of Related Techniques

- *X-ray spectrometry:* Rapid for major component elements with $Z > 9$
- *Optical emission spectroscopy:* Can be rapid for minor and trace components
- *Atomic absorption spectrophotometry:* Somewhat more rapid; better ultimate detection limits for trace components
- *Electrochemical analysis:* Better ultimate detection limits for trace components
- *Electron and ion microprobes: In situ* characterization of inclusions
- *Secondary ion mass spectroscopy, x-ray photoelectron spectroscopy, and Auger spectroscopy:* Much more sensitive studies of surface layers
- *Mössbauer spectroscopy:* Sensitive to certain oxidation states stable only in condensed solids

Potentiometric membrane electrodes

General Uses

- Quantification of cationic and anionic substances
- Quantification of gaseous species in aqueous solutions
- Detector for analytical titrations

Examples of Applications

- Activity or concentration determination of selected cationic, anionic, or gaseous species in a variety of materials, including supply waters, waste waters, plating baths, mineral ores, biological fluids, soils, food products, and sewage
- Elemental analysis of organic compounds, especially for nitrogen and halide content
- Titrimetric determination of major components in metal alloys
- Detector for chromatographic processes
- Detector to follow chemical reaction kinetics

Samples

- *Form:* The measurement must ultimately be made in a solution; aqueous solutions are generally used
- *Size:* Several milliliters of sample are typically needed, but samples of less than a microliter can be analyzed using a microelectrode. Samples as small as 0.5 mL can be measured using commercial gas sensors
- *Preparation:* Depending on the system, sample preparation can be extensive to remove interferents and to release the ion of interest from binding agents in the sample, but is sometimes unnecessary

Advantages

- Insensitive to sample turbidity
- Short analysis times
- Small sample volume requirement
- Simple to operate
- Inexpensive
- Portable
- Easy to automate
- Measures activity

Limitations

- Potential drift
- Interferences
- Often requires aqueous solution

Estimated Analysis Time

- Several minutes per sample after dissolution or other sample preparation

Capabilities of Related Techniques

- *Amperometric gas sensors:* Quantification of oxygen and hydrogen peroxide
- *Ultraviolet/visible (UV/VIS) spectroscopy:* Direct or indirect determination of cation, anion, gaseous, and molecular species
- *Ion chromatography:* Determination of cation and anion species concentrations
- *Atomic spectroscopies:* Quantification of sample elemental components
- *Voltammetry:* Quantification of cations, anions, and certain organics

Voltammetry

General Uses

- Qualitative and quantitative analysis of metals and nonmetals in solutions of concentration 10^{-2} to 10^{-9} mol/L
- Multicomponent, effectively nondestructive repeatable analysis

- Elucidation of solute-solute and solute-solvent equilibria
- Kinetic investigations
- Structure determination in solution

Examples of Applications

- Analysis and characterization of metals in commercial chemicals, pharmaceuticals, high-purity metals, and alloys
- Monitoring of pollutant metals and nonmetals in foodstuffs, water, effluents, herbage, biological/medical systems, and petroleum
- Detection of herbicide and pesticide residues in plant and animal tissue
- Continuous monitoring of major and minor metallic and nonmetallic compounds in commercial electroplating baths

Samples

- *Form:* Solution, mostly aqueous
- *Size:* Cell capacities from 10 to 100 mL are common, but cells less than 1 mL have been devised for special purposes
- *Preparation:* Bulk samples (solids or liquids) must be pretreated to obtain required species in an acceptable and manageable concentration range together with an excess concentration of electroinactive background electrolyte

Limitations

- Sample preparation may sometimes be time consuming relative to the usually short probe time
- Interference from electrochemical signals of species other than those whose analysis is required
- Restricted anodic or cathodic ranges of some electrode materials
- Perfect renewal of electrode surface between analyses is not always feasible
- Complexities in electrochemical or chemical behavior of required species in solution may prevent straightforward detection

Estimated Analysis Time

- 15 min to 3 h per sample, depending on sample preparation time. In batch analysis, the probing of individual solutions for several components may require only minutes

Capabilities of Related Techniques

- *Electrogravimetry:* Extremely accurate but more time consuming; more skill required to eliminate interferences from codeposition; deposition reaction allowed to continue to completion
- *Coulometry:* Small amounts of reagents may be electrogenerated without the need for standardization and storage of dilute solutions. Analysis possible in same lower regions as voltammetry but without the same versatility of multielement analysis
- *Potentiometry:* The exceptional sensitivity of many ion-sensitive systems makes them widely and often simply used as direct probes and in titration methods. Some redox couples may be slow in establishing equilibrium at indicator electrodes
- *Amperometry:* Offers more flexibility in selection of convenient solid electrode materials, because electrode history is less significant in titration techniques. May be used at low concentrations at which other titration methods are inaccurate ($\sim10^{-4}$ mol/L), but cannot reach the low levels attained by voltammetry for direct analysis
- *Conductometry:* May not be used in the presence of high concentrations of electrolyte species other than that required
- *Classical wet chemistry:* Generally more accurate, but electrochemical methods offer better detection limits for trace analyses

Electrogravimetry

General Uses

- Removal of easily reduced ions before analysis by another method
- Quantitative determination of ions after removal of interfering ions
- Quantitative determination of metal ions in the presence of other metal ions
- Quantitative determination of metal ions using electrogravimetry in conjunction with other techniques

Examples of Applications

- Quantitative analysis of metals in alloys
- Precision analysis of metallurgical products and samples
- Quantitative measurement of microgram amounts of metals
- Quantitative determination of metals in the presence of other ions, such as chloride

Samples

- *Form:* Solution
- *Size:* Down to decigram amounts of solid
- *Preparation:* Solutions of analyte

Limitations

- Limited to the analysis of ionizable species; complete deposition (99.5%) is essential for precise results

Estimated Analysis Time

- After sample preparation, electrolysis requires 15 min to well over an hour, depending on conditions

Capabilities of Related Techniques

- *Controlled-potential coulometry:* Quantitative determination of metal ions
- *Coulometric titrations:* Applicable for all volumetric reactions
- *Atomic absorption spectrophotometry:* Quantitative and qualitative determination of metal ions in the presence of other ions
- *X-ray fluorescence:* Qualitative and quantitative determination of elements

Electrometric titration

General Uses

- Widely applicable as a branch of volumetric analysis
- Automated determinations
- High-precision determinations
- Potentiometric continuous monitoring and process control

Examples of Applications

- Determinations in colored, turbid, or very dilute solutions that preclude use of chemical indication
- By electrogeneration, use of titrants that are unstable if stored or have no real lifetime
- Assay of primary standards
- Analyses that must be controlled remotely, for example, of radioactive samples
- Maintenance of constant conditions, for example, of pH or component concentrations, during fermentation, sludge treatment, and so on

Samples

- *Form:* Any
- *Size:* Small, unless lacking homogeneity or of very low analyte concentration
- *Preparation:* If a solution, frequently none. Solids must be made into a suitable solution without loss of analyte; interfering substances must be masked or removed

Advantages

- Submicrogram amounts of analyte can be determined, because a single drop of solution can be titrated
- Electrical quantities need not be measured absolutely, except in coulometric titration. Because results are determined from changes in these quantities, instrumental requirements and conditions are usually less rigid than indirect-measurement determinations

Limitations

- Precision and accuracy are usually better when the titration determines a single analyte, although the successive titration of several analytes may be possible
- Concentration and reaction speed must be high enough to permit rapid equilibration and to yield reproducible end points. Approximately 10^{-5} M may be taken as the lowest practical limit, although titration at lower concentrations is sometimes possible

Estimated Analysis Time

- At least several minutes after sample preparation

Capabilities of Related Techniques

- *Controlled-potential coulometry:* Determination of major constituents, metals, and certain organics; study of electrochemical reaction. Suitable for determinations requiring greater selectivity than obtainable by constant-current techniques. Small-scale preparations
- *Catalytic techniques:* Determinations by effect of a substance upon the speed of a normally slow reaction. The effect can be monitored electrochemically, or electrochemical titration can be used to maintain constancy of the reacting system
- *Voltammetry and polarography:* Provides direct electrochemical measurement of metals and nonmetals and is the basis of amperometric titration
- *Stripping analysis:* Quantitative determination of analytes, usually metals in very dilute solutions
- *Electrographic analysis:* Qualitative or at best semiquantitative analysis of metallic samples

Controlled-potential coulometry

General Uses

- Quantitative chemical analysis; major constituent assay of solutions, alloys, nonmetallic materials, and compounds
- Accuracy and precision are typically 0.1%
- Primarily applicable to the transition and heavier elements
- Studies of electrochemical reaction pathways and mechanisms

Examples of Applications

- High-accuracy assays of nuclear fuels for uranium and plutonium

- Assays of electroplating solutions and alloys for gold, silver, palladium, or iridium

- Measurement of Fe^{3+}/Fe^{2+} ratios in ceramics

- Verification of standards for other analytical techniques, for example, titanium in titanium dioxide

- Determination of molybdenum in molybdenum-tungsten alloys

- Assays of organic compounds containing nitro groups

Samples

- *Form:* Samples must be dissolved in a solvent suitable for electrolysis—usually aqueous solutions

- *Size:* Quantity sufficient for 1- to 10-mg analyte per determination

Limitations

- Not recommended for trace or minor impurity analysis

- Requires dissolution of samples—destructive analysis technique

- Not useful for determination of alkali and alkaline earth metals

- Qualitative knowledge of sample composition is required

- For inorganic constituent analysis, organic constituents must be completely destroyed

- Requires good electrolysis cell design and potential control

- May require elimination of oxygen from sample solutions

Estimated Analysis Time

- 5 to 30 min per measurement, after solution preparation

Capabilities of Related Techniques

- *Controlled-potential electrolysis:* Can be used as a separation technique before other measurement techniques

- *Polarography and solid-electrode voltammetry:* These techniques measure electrolysis current-potential characteristics of solution samples, provide more global qualitative information on sample composition, and are more suitable for minor constituent and trace analysis but not as accurate or precise

Elemental and functional group analysis

General Uses

- Identification of organic compounds

- Determination of the empirical formula of organic compounds

- Determination of the composition of a mixture

- Determination of purity

Examples of Applications

- Identification of the product of a reaction or process

- Identification (or determination of the composition) of plastics, fibers, fuels, or other organic materials

- Determination of the approximate composition of a mixture of organic substances or of a mixture of organic and inorganic substances

- Determination of water in a sample

- Determination of unsaturation in polymers

Samples

- *Form:* Solid or liquid. For solutions, some form of chromatography (gas chromatography or high performance liquid chromatography) is usually preferable

- *Size:* From a few milligrams for carbon, hydrogen, and nitrogen to a few tenths of a gram for most functional groups

- *Preparation:* Careful purification and drying of the sample is essential if the purpose of the analysis is identification or determination of an empirical formula

Limitations

- See the individual methods discussed in this article

Estimated Analysis Time

- If the proper apparatus or equipment and a trained operator are available, the time required can be less than 1 h for carbon and hydrogen or several hours for some functional group determinations. Results can be obtained from a commercial laboratory in a few days. Purification of the sample may add hours to the times given above

Capabilities of Related Techniques

- *Mass spectroscopy:* High-resolution methods may provide elemental composition

- *X-ray spectroscopy:* May provide complete structure of a crystalline compound

- *Infrared spectroscopy, nuclear magnetic resonance:* May provide supplemental information about the sample

- *Gas chromatography:* Mixtures of acids can be separated and quantitatively determined using this method, although the acids are usually converted to their esters before the chromatographic separation

High-temperature combustion

General Use

- Determination of carbon and sulfur in metals and organics

Samples

- *Form:* Solids, chips, or powders
- *Size:* 1 g or less, depending on type of material
- *Preparation:* Bulk samples must be cut to prescribed size required for determination. Specimen should not be contaminated with carbon or sulfur before analysis

Limitations

- Specimen must be homogeneous
- Graphite-bearing specimens require special handling
- Method is destructive

Estimated Analysis Time

- *Sample preparation:* 2 to 3 min
- *Analysis time:* 40 s to 2 min

Capabilities of Related Techniques

- *Optical emission:* Determination of total carbon and sulfur in metals
- *X-ray fluorescence:* Determination of sulfur in most metals

Inert gas fusion

General Use

- Quantitative determination of oxygen, nitrogen, and hydrogen in ferrous and nonferrous materials

Samples

- *Form:* Solids, chips, or powders
- *Size:* Usually 2 g or less, depending on material type and the expected amount of gases present
- *Preparation:* Bulk materials must be cut to prescribed size. Care must be taken not to contaminate specimens with nitrides, oxides, or hydrides. Materials for oxygen and hydrogen must be kept cool during sectioning to prevent hydrogen diffusion and surface oxidation

Limitations

- Metals with low boiling points require special precautions
- Materials with stable nitrides or oxides require addition of fluxes
- Method is destructive to the material

Estimated Analysis Time

- 1 to 10 min after sample preparation

Capabilities of Related Techniques

- *Vacuum fusion:* Total nitrogen, oxygen, and hydrogen in some metals without fusing the specimen
- *Hot extraction:* Total nitrogen, oxygen, and hydrogen in some metals without fusing the specimen
- *Optical and mass spectroscopy:* Nitrogen, oxygen, and hydrogen by x-ray bombardment on some metals

Neutron activation analysis

General Uses

- Nondestructive trace-element assay of essentially any material
- Ultrasensitive (as low as 10^{-12} g/g) destructive quantitative analysis
- Measurement of isotope ratios in favorable cases

Examples of Applications

- Quality control for purity or composition of materials
- Element-abundance measurements in geochemical and cosmochemical research
- *Resource evaluations:* Assay of surface materials, drill cores, ore samples, and so on
- *Pollution studies:* Assay of air, water, fossil fuels, and chemical wastes for toxic elements
- *Biological studies:* Determination of the retention of toxic elements in humans and laboratory animals, in some cases by *in vivo* measurements
- *Forensic investigations:* Trace-element assay of automobile paint, human hair, and so on

Samples

- *Form:* Virtually any solid or liquid
- *Size:* Typically 0.1 to a few grams, but the full range extends from a microgram to at least 100 kg (220 lb)
- *Preparation:* None required in many cases

Limitations

- Several elements are unobservable except through sample irradiation with high-energy neutrons or use of radiochemical (destructive) techniques

- Access to a nuclear reactor or some other high-intensity neutron source is required

- Turnaround time can be long (up to 3 weeks)

- The samples usually become somewhat radioactive—objectionable in some cases

- Intense radioactivity induced in one or more of the major elements in a sample may mask the presence of some or all of the trace elements

Estimated Analysis Time

- Counting times can range from under 1 min (for short half-life radioisotopes) to many hours (for low-intensity long-lived radioisotopes)

Capabilities of Related Techniques

- *X-ray fluorescence:* Useful for determining major and minor elements (concentrations > 0.1%), but several elements with $Z \geq 11$ can be observed at 5 to 20 ppm

- *Atomic absorption:* Can determine most elements individually. Requires dissolution of the sample. Some elements observed at levels of 10^{-9} g/mL, which usually corresponds to $\geq ^{-7}$ g/g of the original solid sample

- *Inductively coupled plasma emission spectroscopy:* Requires dissolution of the sample. Typical detection limits are a few parts per million

- *Inductively coupled plasma mass spectroscopy:* Can determine trace elements in the presence of high levels of major and minor elements in the sample matrix, such as in alloy and metal analysis. Requires dissolution of the sample. Detection limits for most elements are from 0.05 to 0.1 ng/mL of solution

- *Isotope dilution mass spectrometry:* Can quantitatively measure as few as 10^6 atoms of most elements individually. Isotope ratios are used. Analysis time is at least one day per element

- *Spark source mass spectrometry:* Simultaneous detection of most elements at levels down to approximately 1 ppm. Data are recorded on a photoplate, limiting accuracy. Sample preparation may take several hours

- *Particle-induced x-ray emission spectroscopy:* Measures element concentrations near the surface of a sample. Several elements can be measured at a few parts per million. Access to a proton accelerator is required.

Radioanalysis

General Uses

- Quantitative determination of radioactive isotopes

- Measurement of efficiency in separations for chemical analysis

- Tracer for diffusion measurements or for chemical reactions

- Dating of prehistoric materials

- Measurement of natural radioactive elements in geological materials

- Quantitative measurement of chemical elements by activation analysis

Examples of Applications

- Measurement of the diffusion of nickel into iron

- Measurement of chemical reaction rates under various conditions of temperature and concentration

- Measurement of trace impurity concentrations in nonradioactive materials

- Identification and measurement of radioactive pollutants in environmental materials

- Measurement of the efficiency of chemical separation procedures, such as precipitations, extractions, and distillations

- Determination of radioactive elements in chemical recovery processes for nuclear materials

- Measurement of segregation of uranium isotopes and their daughters in geologic processes

Samples

- *Form:* Solid, liquid, or gas

- *Size:* Limited by minimum and maximum count rates that can be handled by the radioactivity detector, which ordinarily has a range of 100 to 100,000 cpm (counts per minute)

- *Preparation:* None when the radioactive species is the only one in the sample and has penetrating radiation. In other cases, chemical separations are required

Limitations

- Half-life of the radioactive element should be greater than several hours and less than approximately 1,000,000 years

- Accuracy and precision are usually not less than 5%, although in some cases 1 to 2% can be achieved

- Radioactive elements that are pure α-particle or β-particle emitters (no γ-ray emission) generally require chemical separations before measurement

Estimated Analysis Time

- 1 h to several days, depending on whether chemical separations are needed

- Chemical separations may require additional procedures to ensure chemical equivalence of the radioactive isotope tracer with its stable element

Capabilities of Related Techniques

- *Isotope dilution—mass spectrometry:* Accuracy and precision are generally better—less than 1%. Sample preparation is required, and suitable stable or long-lived radioactive isotopes must be available

- *X-ray spectrometry:* Useful for trace analysis of stable elements or of radioactive elements with half-lives greater than 10,000,000 years. May be an alternative for neutron activation analysis

Resonance Methods

Electron spin resonance

General Uses

- Identification of elements of the various transition series in solids and solutions

- Identification of the valence states of transition-element ions

- Detection of color centers and defects in crystalline solids

- Characterization of local crystal environments around transition ions in solids

- Identification of magnetic states of materials, such as ferromagnetic, antiferromagnetic, ferrimagnetic, and spin glass

- Detection of defect centers and radiation damage

Examples of Applications

- Determination of the main and trace-level transition-ion content and of the crystallinity of minerals

- Study of catalyst surfaces and their free-radical reactions

- Characterization of the paramagnetic properties and the kinetic reactions of inorganic and organic free radicals

- Determination of the transition-ion and free-radical content and of the crystallinity and viscosity of fossil fuels

Samples

- *Form:* Crystalline, semicrystalline, or amorphous solids in the form of crystals, powders, or films; also liquid crystals, liquids, and occasionally gases

- *Size:* Crystals—typically $2 \times 1 \times 1$ mm ($0.08 \times 0.04 \times 0.04$ in.) to 0.1% of this volume. Powders—typically 1 mg to 1 g. Solutions—1 mL, aqueous samples with high dielectric losses require use of a special flat sample cell

- *Preparation:* Generally none; large crystals must be cut to size; some solutions require deoxygenation; some samples require cryogenic temperatures

Limitations

- Sample must be paramagnetic; that is, it must contain transition ions, free radicals, defect centers, and so on

- Paramagnetic centers must be sufficiently high in concentration; typical amounts are 0.1 nmole

Estimated Analysis Time

- 10 min to several hours per sample, depending on circumstances

Capabilities of Related Techniques

- *Nuclear magnetic resonance:* Identifies molecular structure, mainly of organic molecules in solution; less sensitive than electron spin resonance by two or three orders of magnitude

- *Mössbauer resonance:* Similar information gained, but less versatile than electron spin resonance because in most applications iron or perhaps tin must be present

- *Quadrupole resonance:* Related to nuclear magnetic resonance, but much less versatile; rarely used routinely

- *Microwave spectroscopy:* Identification of relatively small molecules in the gas phase

Ferromagnetic resonance

General Uses

- Identification of magnetic state

- Quantitative determination of static magnetic parameters

- Determination of microwave losses

Examples of Applications

- Measurement of magnetization

- Study of magnetocrystalline anisotropy

- Investigation of exchange stiffness

Samples

- *Form:* Crystalline or amorphous solids—metals and alloys
- *Size and shape:* Thin films, needles, ribbons, disks with thickness small in comparison to lateral dimensions

Limitations

- Data from other techniques should be available for unequivocal conclusions

Estimated Analysis Time

- A few hours per specimen

Capabilities of Related Techniques

- *Mössbauer spectroscopy:* Magnetic structure analysis, phase analysis, and surface analysis; limited to relatively few isotopes
- *Electron spin resonance:* Identification of magnetic states of materials; identification of valence states of transition element ions. Samples must be paramagnetic

Nuclear magnetic resonance

General Uses

- Phase analysis
- Electronic structure of metals
- Near-neighbor environment of atoms in solids
- Measures rate of kinetic processes, for example molecular reorientation or diffusion
- Magnetic structural studies
- Defect and annealing studies
- Molecular structure of organic compounds
- Quantitative analysis of specific component and functional groups

Examples of Applications

- Detection of phase changes
- Study of hydrogen diffusion in metals
- Studies of long-range order in intermetallic compounds
- Spin wave studies in ferromagnetic materials
- Effect of pressure on electronic structure
- Isomer identification and quantification
- Determination of copolymer ratios

Samples

- *Form:* Inorganic powders, thin wires, or thin foils, with one dimension small compared with the radio frequency skin depth, generally 10 μm or less. Special shapes and single crystals are used in some cases. Organic solids are usually dissolved in an appropriate solvent; organic liquids can be run directly or diluted. For conventional nuclear magnetic resonance, samples must generally be nonmagnetic. For ferromagnetic nuclear resonance, samples must generally be strongly magnetic
- *Size:* Several grams (inorganic) to 0.1 g (organic)

Estimated Analysis Time

- 30 min to 48 h

Capabilities of Related Techniques

- *Optical metallography:* Shows morphology and number of phases present
- *X-ray diffraction:* Gives related crystal structure information
- *Mössbauer effect:* Provides a detectable effect in the presence of many defects
- *Infrared/Fourier transform infrared spectroscopy:* More sensitive for compound identification; applicable to gases; easier data interpretation
- *Gas chromatography/mass spectrometry:* Useful for identification of complex mixtures; more sensitive

Mössbauer spectroscopy

General Uses

- Phase analysis
- Study of atomic arrangements
- Study of critical-point phenomena
- Magnetic-structure analysis
- Diffusion studies
- Surface and corrosion analysis

Examples of Applications

- Measurements of retained austenite in steel
- Analysis of corrosion products on steel
- Effects of grinding of the surface of carbon steel
- Measurement of δ-ferrite in stainless steel weld metal
- Curie and Neel point measurements

Samples

- *Form:* Solids (metal, ferrites, geological materials, and so on)
- *Size:* For transmission—powders or foils. Size varies, but of the order of 50 μm thick and 50 mg of material. For scattering—films, or bulk metals with an area of the order of 1 cm^2 (0.15 $in.^2$) or greater. As source—of the order of 1 to 100 mCi of radioactive material incorporated within approximately 50 μm of the sample surface

Limitations

- Limited to relatively few isotopes, notably ^{57}Fe, ^{119}Sn, ^{121}Sb, ^{186}W
- Maximum temperature of analysis is usually only a fraction of the melting temperature
- Phase identification is sometimes ambiguous

Estimated Analysis Time

- 30 min to 48 h

Capabilities of Related Techniques

- *Optical metallography:* Shows morphology of the phases present
- *X-ray diffraction:* Faster; provides crystal structure information
- *Nuclear magnetic resonance:* Applicable to a wider range of isotopes, faster, more sensitive for nonmagnetic materials, applicable to liquids

Metallographic Techniques

Optical metallography

General Uses

- Imaging of topographic or microstructural features on polished and etched surfaces at magnifications of 1 to 1500×
- Characterization of grain and phase structures and dimensions

Examples of Applications

- Determination of fabrication and heat-treatment history
- Determination of braze- and weld-joint integrity
- Failure analysis
- Characterization of the effects of processing on microstructure and properties

Samples

- *Form:* Metals, ceramic, composites, and geologic materials
- *Size:* Dimensions ranging from 10^{-5} to 10^{-1} m
- *Preparation:* Specimens are usually sectioned and mounted, ground, and polished to produce a flat, scratch-free surface, then etched to reveal microstructural features of interest

Limitations

- *Resolution limit:* Approximately 1 μm
- Limited depth of field (cannot focus on rough surfaces)
- Does not give direct chemical or crystallographic information about microstructural features

Estimated Analysis Time

- 30 min to several hours per specimen, including preparation

Capabilities of Related Techniques

- *Scanning electron microscopy:* Provides better resolution (higher magnifications); greater depth of field (can image rough surfaces); qualitative elemental microanalysis
- *Electron probe x-ray microanalysis:* Provides quantitative elemental microanalysis
- *Transmission electron microscopy:* Provides much better resolution (much higher magnifications) on specially prepared specimens; semiquantitative elemental microanalysis; crystallographic information on microstructural features

Image analysis

General Uses

- Quantification of the morphological aspects of images obtained by optical metallography, scanning electron microscopy, and transmission electron microscopy

Examples of Applications

- Quantitative determination of grain size, grain shape, grain boundary area per unit volume, and so on, in single-phase metals and ceramics
- Quantitative determination of second-phase volume fractions, sizes, interfacial areas per unit volume, spacings, and so on, in multiphase metals and ceramics
- Quantitative determination of particle size distributions in powders

Samples

- Instrument can be interfaced with an optical microscope or a scanning electron microscope to permit direct analysis (without taking photographs)
- Instrument also can be interfaced with macroviewer (epidiascope) for quantification of features on photographs (from optical microscope, scanning electron microscope, transmission electron microscope, and so on). Photographs must have sufficient contrast

Limitations

- Does not provide direct chemical information on microstructural features or particles and cannot generally discriminate between microstructural features or particles of different compositions

- Measures two-dimensional geometric quantities; third-dimensional parameters must be inferred

Estimated Analysis Time

- 5 min to several hours per sample

Capabilities of Related Techniques

- *Electron probe x-ray microanalysis:* Provides quantitative elemental compositions but no quantitative geometric characterization

- *Scanning electron microscopy:* Image contrast arises from different phenomena than for optical microscopy; discrimination of phases of differing compositions is possible

Diffraction Methods

X-ray powder diffraction

General Uses

- Identification of crystalline phases contained in unknown samples

- Quantitative determination of the weight fraction of crystalline phases in multiphase materials

- Characterization of solid-state phase transformations

- Lattice-parameter and lattice-type determinations

- Orientation of single crystals

- Stereographic projections

- Alignment for cutting crystallographic planes

Examples of Applications

- Qualitative and quantitative analysis of crystalline phases in coal ash, ceramic powders, corrosion products, and so on

- Determination of phase diagrams

- Determination of pressure- and/or temperature-induced phase transformations

- Quantitative analysis of solid solutions from lattice-parameter measurements

- Determination of anisotropic thermal expansion coefficients

Samples

- *Form:* Crystalline solids (metals, ceramics, geological materials, and so on)

- *Size:* For powder samples 1 mg is usually adequate

- *Preparation:* Sometimes none; sample may require crushing to fit into the sample holder

Limitations

- Must be crystalline for phase identification

- Identification requires existence of standard patterns: JCPDS powder diffraction file of inorganic and organic phases, NBS Crystal Data (contains lattice constants for inorganic and organic phases), and Cambridge File of Organic Single Crystal Structural Data

Estimated Analysis Time

- Qualitative analysis requires less than 1 h for major phases, up to 16 h for trace phase confirmation

- Quantitative analysis, after a procedure is set up, requires several minutes to several hours

Capabilities of Related Techniques

- *X-ray spectrometry, inductively coupled plasma atomic emission spectroscopy, atomic absorption spectrometry, classical wet chemical analysis:* Quantitative and qualitative elemental information

- *Auger electron spectroscopy:* Elemental and structural data on small portions of the samples

- *Single-crystal x-ray diffraction:* Crystal structure using small single crystals

- *Infrared and Raman spectroscopy:* Molecular structure and sometimes crystal structure

- *Neutron diffraction:* Similar information, but can be applied in some cases in which x-ray powder diffraction fails

Single-crystal x-ray diffraction

General Uses

- Unit cell identification

- Formula weight determination from crystal class, space group, and density information

- Location of atoms within the unit cell, that is, the determination of the crystal structure: (1) atom coordination numbers and geometries, (2) interatomic bond distances and angles, (3) interactions between molecules and/or ions, and (4) absolute configurations

Examples of Applications

- Identification of previously studied crystalline phases by comparison of unit cell information

- Characterization of new crystalline compounds or phases
- Understanding physical properties of crystalline phases in terms of interatomic interactions

Samples

- *Form:* Single crystal, preferably with well-formed faces and, if transparent, without visible flaws when viewed under a microscope
- *Size:* Approximately the size of a salt grain, that is, 0.1 to 0.4 mm on a side
- *Preparation:* Generally, suitable crystals can be grown using various slow crystallization techniques. Air-stable crystals are usually mounted on glass fibers; air-sensitive crystals are sealed inside glass or quartz capillaries

Limitations

- Sample must be a single crystal. Crystals that appear to be single may consist of several domains with different orientations. In addition, the crystal selected for investigation may not be representative of the bulk sample
- For unit cell identification, literature or data base must be available
- Success in the determination of the atomic structure of a crystal can be impeded by subtle twinning effects, ambiguities in the assignment of the crystal class and/or the space group, the presence of a superlattice, disorder in the atoms at a particular site, partial occupancy of sites, and a host of other problems that may be related to data collection and reduction. Although a structure may appear well determined, the presumed precision is often much less than that suggested, for example, by the estimated standard deviations of the atomic coordinates

Estimated Analysis Time

- May vary from a few hours for a simple determination of the unit cell geometry to a few days for an average crystal structure determination. Problem structures can quickly turn into research projects

Capabilities of Related Techniques

- *Neutron single-crystal diffraction* can be used to determine the atomic structures of crystals. Advantages include higher accuracy for the positions of atoms with low atomic number and information about magnetic structure. Disadvantages include the requirement for larger single crystal (0.5 to 1.0 mm), limited number of neutron facilities, and longer data-collection times
- *Rietveld refinement of x-ray* or *neutron powder diffraction data* can also be used to determine crystal structures. Advantages include use of a crystalline powder

rather than a single crystal. Disadvantages include limited number of atoms to be located and refined and less favorable precision

Crystallographic texture measurement and analysis

General Uses

- Quantitative determination of crystallographic preferred orientations in polycrystalline samples

Examples of Applications

- Quantitative description of deformation and recrystallization textures in metals and ceramics, including geological materials

Samples

- *Form:* Samples must be flat sections taken from the polycrystal. For transmission, samples must be thinned to a uniform thickness. The orientation of the surface normal must be precisely specified relative to the polycrystalline processing axes; another reference direction, such as the rolling direction, usually must also be known
- *Size:* The grain size of the polycrystal must be small compared to the x-ray beam diameter so that the beam irradiates a minimum of approximately 5000 grains; the sample may be oscillated to increase the area covered
- *Preparation:* The surface must be polished carefully to remove any possible disturbance from milling or cutting
- *Other requirements:* A random powder sample of the same shape as the textured sample usually must be prepared

Limitations

- Obtaining valid measurements of intensity is generally not possible when the x-ray source beam lies $\geq 70°$ from the direction normal to the sample surface because of geometric defocusing aberrations
- The crystallographic pole figure does not describe completely the orientation distribution for the sample; a complete description may be computed from a set of pole figures (the orientation distribution function)

Estimated Analysis Time

- Reliable pole figures can typically be measured in 10 to 100 min, depending on the diffraction equipment available. Because at least three pole figures are generally required to compute the orientation distribution function, this may require 30 to 180 min

- Computational time is usually insignificant compared to measurement time

Capabilities of Related Techniques

- *Neutron diffraction:* Increasingly prevalent as an alternative to x-ray measurement of pole figures, this technique can penetrate depths of up to four to five orders of magnitude larger than x-rays. Thus, true volume-averaged textures are measured. Sample preparation is simple in neutron experiments, and defocusing problems can usually be eliminated. However, suitable neutron sources are not readily accessible, and beam time may be expensive

- Some pole figures and orientation distribution functions can be determined from many single-crystal orientations, for example, single grain orientation by selected area channeling analysis using a scanning electron microscope. These methods are tedious, because measuring statistically reliable numbers of single orientations may involve over 3000 grains

X-ray topography

General Uses

- Imaging of individual lattice defects, such as dislocations, twins, and stacking faults, in near-perfect crystals

- Nondestructive characterization of surface relief, texture, lattice distortion, and strain fields due to defects and defect accumulations in imperfect single crystals and polycrystalline aggregates

- Measurement of crystal defect densities as well as crystallite/subgrain size and shapes

- Evaluation of tilt angles across subgrain boundaries, interfacial defects and strains, domain structures, and other substructural entities

Examples of Applications

- Study of crystal growth, recrystallization, and phase transformations, focusing on crystal perfection and attendant defects

- Characterization of deformation processes and fracture behavior

- Correlation between crystal defects and electronic properties in solid-state device materials

- Synchrotron radiation extends the use of topography to permit the study of dynamic processes, such as magnetic domain motion, *in situ* transformations (solidification, polymerization, recrystallization), radiation damage, and yielding

Samples

- *Form:* For defect imaging (transmission or reflection case), flat, relatively perfect ($<10^6$ dislocations/cm^2) single crystals with uniform thickness or wedge shape. Evaluation of lattice distortions, texture, substructure, and surface relief in monocrystals, polycrystalline aggregates, ceramic or metal alloys, or composites

- *Size:* 1×1 cm (0.4×0.4 in.), 1 μm to several millimeters thick, up to 5-cm (2-in.) diam or larger wafers; thin films 100 nm and thicker

- *Preparation:* Usually desirable to remove surface damage due to cutting, abrading, and so on, from virgin material by chemical or electrolytic polishing

Limitations

- Sample must be crystalline

- Relatively defect-free crystals required for defect imaging techniques

- Thickness of single-crystal or polycrystalline samples that can be studied in transmission arrangement is limited by intensity and wavelength of incident radiation used as well as absorption by the sample

- Direct images are actual size. Further magnification must be obtained optically; that is, grain size of photographic plate emulsion must be small enough to allow substantial enlargement

Estimated Analysis Time

- Several minutes to hours exposure time for conventional photographic (plate or film) methods, in addition to developing/enlarging time

- Milliseconds to several seconds using synchrotron radiation and/or electronic or electro-optical imaging systems

Capabilities of Related Techniques

- *Optical metallography:* Characterization of grain size and shape, subgrains, phase morphology, and slip traces using suitable etchants; estimation of low dislocation densities and determination of slip systems by etch pit techniques

- *Scanning electron microscopy:* Observation of irregular surfaces, surface relief, and various features induced by deformation, such as slip bands and rumpling; examination of fracture surfaces to evaluate crack initiation and propagation

- *Electron channeling:* Qualitative evaluation of crystal perfection over shallow surface layer of crystals with high symmetry orientations

- *Transmission electron microscopy:* Imaging of line and planar defects and estimation of defect densities; substructural and morphological characterization of thin foils prepared from bulk sample or replicas taken from the surface

- *Neutron diffraction and topography:* Study of very thick or heavy metals in transmission arrangement and of magnetic domain structure

X-ray diffraction residual stress techniques

General Uses

Macrostress measurement

- Nondestructive surface residual stress measurement for quality control
- Determination of subsurface residual stress distributions
- Measurement of residual stresses associated with failures caused by fatigue or stress corrosion

Microstress measurement

- Determination of the percent cold work at and below the surface
- Measurement of hardness in steels in thin layers

Examples of Applications

- Determination of the depth and magnitude of the compressive layer and hardness produced by carburizing steels
- Investigation of the uniformity of the surface compressive residual stresses produced by shot peening in complex geometries
- Measurement of surface residual stresses and hardness on the raceway of ball and roller bearings as functions of hours of service
- Study of the alteration of residual stress and percent cold work distributions caused by stress-relieving heat treatment or forming
- Measurement of surface and subsurface residual stresses parallel and perpendicular to a weld fusion line as a function of distance from the weld
- Determination of the direction of maximum residual stress and percent cold work gradient caused by machining

Samples

- *Form:* Polycrystalline solids, metallic or ceramic, moderate to fine grained
- *Size:* Various, with limitations dictated by the type of apparatus, the stress field to be examined, and x-ray optics
- *Preparation:* Generally, none. Large samples and inaccessible areas may require sectioning with prior strain gaging to record the resulting stress relaxation. Careful handling or protective coatings may be required to preserve surface stresses

Limitations

- Expensive, delicate apparatus generally limited to a laboratory or shop
- Only a shallow (<0.025-mm, or 0.001-in.) surface layer is measured, requiring electrolytic polishing to remove layers for subsurface measurement
- Samples must be polycrystalline, of reasonably fine grain size, and not severely textured

Estimated Analysis Time

- 1 min to 1 h per measurement, depending on the diffracted x-ray intensity and technique used. Typically, 1 h per measurement for subsurface work, including material removal and sample repositioning

Capabilities of Related Techniques

- *General dissection techniques:* Generally good for determination of gross residual stress distributions extending over large distances or depths. Restricted to simple geometries
- *Hole drilling:* Applicable to a variety of samples with stress fields uniform over dimensions larger than the strain-gage rosette and depth of the drilled hole and with magnitudes less than nominally 60% of yield strength. Serious errors are possible due to local yielding for higher stresses, variation in the stress field beneath the rosettes, eccentricity of the hole, or as a result of residual stresses induced in drilling the holes
- *Ultrasonic methods:* Require relatively long gage lengths and stress-free references standards. Of limited practical application due to errors caused by transducer coupling, preferred orientation, cold work, temperature, and grain size. Sensitivity varies greatly with material
- *Magnetic (Barkhausen or magnetostrictive) methods:* Limited to ferromagnetic materials and subject to many of the limitations and error sources of ultrasonic methods. Highly nonlinear response with low sensitivity to tensile stresses

Radial distribution function analysis

General Uses

- Determination of interatomic distance distributions and coordination numbers of amorphous materials, polycrystalline materials, liquids, and gases
- Determination of long-range order in amorphous materials

Examples of Applications

- Structural ordering in amorphous carbon
- Coordination numbers for liquids
- Bonding topologies in silicate glasses
- Variation of long-range order of glasses as a function of preparation
- Three-dimensional structure of gas molecules

Samples

- *Form:* Solid, liquid, or gas

- *Size:* 1 mm^3 (0.00006 in.^{3}) for x-ray diffraction; 1 cm^3 (0.06 in.^{3}) for neutron diffraction; 100 Å thick for electron diffraction

- *Preparation:* Flat for reflection geometry; cylindrical for transmission geometry; weakly scattering thin-wall container for polycrystalline or liquid samples. For example, Mylar is used for flat samples

Limitations

- The radial distribution function is a one-dimensional display of distances in a three-dimensional sample. Therefore, the peaks in the radial distribution function beyond approximately 4 Å usually consist of several components or different types of distances. Consequently, the radial distribution function does not generally contain enough information to determine uniquely the arrangement of atoms in a sample. However, the radial distribution function provides a stringent test for the accuracy of any proposed model

Estimated Analysis Time

- X-ray or neutron diffraction data can be collected in several days with any powder or single-crystal diffractometer

- High-intensity sources, such as synchrotrons, permit faster data collection or use of smaller samples

- Electron diffraction data can be collected within seconds

- Obtaining an accurate radial distribution function usually requires a day of interactive computation on a mainframe computer

- Interpretation of the radial distribution function may take several days to several weeks, depending on the number of model radial distribution functions considered

Capabilities of Related Techniques

- *Single-crystal diffraction study:* This should almost always be the method of choice if a single crystal can be obtained; it takes less time and yields much more structural information

- *Extended x-ray absorption fine structure:* Can be useful for studying the stereochemistry of a minor constituent; such a study requires use of a radiation source, such as a synchrotron, that can be tuned to an absorption edge of the element of interest and the collection of data for model compounds

Small-angle x-ray and neutron scattering

General Uses

- Detection of heterogeneities in solids and liquids

- Monitoring of phase separations

Examples of Applications

- Phase separation in multicomponent metallic and polymer alloys, ceramics, and glasses

- Microphase separation in block copolymers

- Nucleation in metals

- Structure of crazes in glassy polymers

- Defects in metals

Samples

- *Form:* Solid or liquid

- *Size:* 1 to 20 mm (0.04 to 0.8 in.) on a side; thickness depends on mass-absorption coefficient

Limitations

- Sensitivity depends on wavelength, mass-absorption coefficient, and electron density contrast between scattering entities and surroundings

- The cause of the scattering and the shape of the scattering species cannot be determined uniquely in most materials without extensive modifications to the sample. Alternative methods must be used

Estimated Analysis Time

- 1 to 8 h

Capabilities of Related Techniques

- *Scanning electron microscopy:* Imaging and qualitative identification of microstructural features in solids as small as 3×10^{-8} m

- *Transmission electron microscopy:* Imaging and qualitative analysis of features smaller than those resolved using scanning electron microscopy; crystallographic data can be obtained

- *X-ray diffraction:* Identification and quantification of phases or compounds in unknown samples; measurement of crystal defect densities, residual stresses, and so on

Extended x-ray absorption fine structure

General Uses

- Determination of local structure (short-range order) about a given atomic center in all states of matter
- For crystals and oriented surfaces, structural information on next-nearest and further-out neighbors can also be determined
- Identification of phases and compounds containing a specific element
- Orientation of adsorbed molecules on single-crystal surfaces
- Combined with near-edge structure, bonding and site symmetry of a constituent element in a material can be determined

Examples of Applications

- Bond distance, coordination, and type of nearest neighbors about a given constituent atomic species in disordered systems, such as glasses, liquids, solutions, and random alloys, or complex systems, such as catalysts, biomolecules, and minerals
- In ordered systems, such as crystals and oriented surfaces, structural information on the next-nearest neighbors can also be obtained
- Structural evolution in amorphous to crystalline transformation can be followed
- Geometry of chemisorbed atoms or molecules on single-crystal surfaces
- *In situ* structure determination of active sites in catalysts
- *In vivo* structure determination of active sites in metalloproteins
- Combined with x-ray absorption near-edge structure, bonding and local structure of trace impurities in natural materials (for example, coal) and synthetic materials (for example, diamond) can be determined

Samples

- *Form:* Solids, liquids, and gases. Solids should ideally be in thin foil (metals and alloys) or uniform films of fine powders (~400 mesh or finer)
- *Size:* Area—25 × 5 mm (1 × 0.2 in.) minimum. Thickness—bulk samples: 1 to 2 absorption lengths at the absorption edge of the element of interest; dilute samples: up to a few millimeters thick
- *Preparation:* Mainly required for solid samples— rolled metal foil, sputtered or evaporated thin films, uniformly dispersed powder films. Aerobic or anaerobic environments must be used for biological materials and checks for radiation damage (loss of biological activity)

Limitations

- Nonunique results if the element of interest exists in multiple nonequivalent sites or valence states
- Structural results can depend strongly on model systems used in quantitative analysis
- Low concentration limit is near 100 ppm or a few millimolar in favorable cases
- For dilute and surface systems, synchrotron radiation is necessary

Estimated Experimental Scan Time

- 30 min to 1 h for bulk samples (single scan)
- 5 to 10 h or more for dilute samples and surfaces (multiple scans)

Estimated Data Analysis Time

- 2 to 20 h or more per spectrum

Capabilities of Related Techniques

- *X-ray or neutron diffraction:* Identification of bulk crystalline phases; radial distribution functions of bulk amorphous phases
- *X-ray anomalous scattering:* Characterization of bulk amorphous materials
- *Electron energy loss spectroscopy:* Extended fine structure in solids; samples must withstand vacuum
- *Magic-angle spinning nuclear magnetic resonance:* For solid-state studies

Neutron diffraction

General Uses

- Determination of atomic arrangements (for example, crystal structure, short- and long-range order), especially for structures containing light (low atomic number) atoms in the presence of heavier atoms or neighboring elements
- Determination of magnetic structures
- Determination of structural changes as a function of temperature, pressure, magnetic field, etc.
- Quantitative analysis of multiphase materials
- Determination of residual stress and texture in polycrystalline engineering materials

Examples of Applications

- Location of hydrogen atoms in organometallic compounds or hydrogen-containing intermetallic compounds
- Observation of site preference for metals in intermetallic compounds
- Observation of ferri-, ferro-, antiferro-, or complex magnetic ordering

- Observation of phase transitions
- Refinement of structural parameters and phase fractions in mixed catalyst systems
- Determination of three-dimensional residual and applied stress tensors as a function of depth in engineering materials
- Study of grain interaction stresses in composites and hexagonal metals
- Quantitative measurement of texture and texture gradients
- Determination of radial distribution functions in liquid and amorphous materials

Samples

- *Form:* Polycrystalline solids, powders, single crystals, amorphous solids, liquids
- *Size:* 1 to 10 cm^3 for powder structure determination, single crystals of 0.5 to several hundred mm^3. Low absorption allows large samples of powders, liquids, and amorphous solids (for radial distribution function studies), typically 1 to 50 g
- *Preparation:* Minimal preparation necessary; insensitive to sample geometry; often tall, narrow specimens desirable (5×0.2 cm)

Limitations

- Relatively low intensity, which may require large sample volumes and/or long counting times
- Strong absorbers (Gd, Cd, Sm, Li, B) must be avoided or kept at low concentrations
- X-ray studies usually must precede neutron diffraction analysis for structure modeling
- Substitution of deuterium for hydrogen may be necessary to reduce incoherent scattering contributions to backgrounds
- In the U.S., only seven centers provide neutron beams of sufficient strength for scattering research; worldwide, about 30 centers are active in neutron scattering

Estimated Analysis Time

- *Acquisition:* Typically 12 h for powder diffraction with optimized instrumentation; days to weeks for single-crystal and texture studies
- *Analysis:* Time is highly variable. Rietveld and single-crystal studies require hours to days to weeks

Capabilities of Related Techniques

- *X-ray diffraction:* Highly complementary but most sensitive to heavy elements, insensitive to magnetism. X-ray diffraction primarily a near-surface probe compared with bulk penetration by neutrons

Electron Optical Methods

Analytical transmission electron microscopy

General Uses

- Imaging of microstructural features at 1000 to 450,000×. Microstructural detail resolution of <1 nm
- Qualitative and quantitative elemental analysis of microstructural features as small as 30 nm
- Crystal structure and orientation determination of microstructural features as small as 30 nm
- Lattice imaging of crystals with interplanar spacings >0.12 nm

Examples of Applications

- Very high magnification characterization of the microstructure of metals, ceramics, geologic materials, polymers, and biological materials
- Identification (composition and crystal structure) of inorganic phases, precipitates, and contaminants

Samples

- *Form:* Solids (metals, ceramics, minerals, polymers, biological, and so on)
- *Size:* An approximately 5-μm-thick, 3-mm-diam disk
- *Preparation:* Bulk specimens must be sectioned and electrothinned or ion milled to produce regions that permit transmission of the electron beam. Electron-transparent regions typically less than 100 nm thick. Powdered samples are often dispersed on a thin carbon substrate

Limitations

- *Sample preparation* is tedious, and development of a suitable procedure may take weeks
- *Imaging resolution* is approximately 0.12 nm
- *Elemental microanalysis:* Typical minimum size of region analyzed is 30 nm in diameter. Threshold sensitivity is approximately 0.5 to 1 wt%. Accuracy of quantification routinely is 5 to 15% (relative). Quantification is possible only for elements with atomic number ≥11. Some instruments can detect (qualitatively) only elements with atomic number ≥11
- *Electron diffraction:* Minimum size of region analyzed is approximately 30 nm in diameter. Crystal structure identification is limited to phases or compounds tabulated in powder diffraction file (approximately 40,000 phases or compounds). Orientation relationship between two coexisting phases can be determined only if phase crystal structures are known or can be determined. Determination of full space and point groups is possible only using specialized microdiffraction techniques

Estimated Analysis Time

- 3 to 30 h per specimen (does not include sample preparation)

Capabilities of Related Techniques

- *X-ray diffraction:* Gives bulk crystallographic information

- *Optical metallography:* Faster; lower magnification (up to approximately 1000×) overview of sample microstructure. No chemical information

- *Scanning electron microscopy:* Faster; lower magnification than transmission electron microscopy (up to approximately 20,000× in bulk samples), with image resolution routinely as good as approximately 50 nm in bulk samples. Qualitative chemical information only. Limited crystallographic information through electron channeling

- *Electron probe x-ray microanalysis:* Faster; gives more accurate (better than 1%, relative) quantitative elemental analysis for elements as light as boron. Poorer spatial resolution (approximately 1 μm). No crystallographic information

Scanning electron microscopy

General Uses

- Imaging of surface features at 10 to 100,000×. Resolution of features down to 3 to 100 nm, depending on sample

- When equipped with a backscattered detector, microscope allows (1) observation of grain boundaries on unetched samples, (2) domain observation in ferromagnetic materials, (3) evaluation of the crystallographic orientations of grains with diameters down to 2 to 10 μm, and (4) imaging of a second phase on unetched surfaces when the second phase has a different average atomic number

- When suitably modified, the microscope can be used for defect and quality control of semiconductor devices

Examples of Applications

- Examinations of metallographically prepared samples at magnifications well above the useful magnification of the optical microscope

- Examination of fracture surfaces and deeply etched surfaces requiring depth of field well beyond that possible with the optical microscope

- Evaluation of crystallographic orientation of features on a metallographically prepared surface, for example, individual grains, precipitate phases, and dendrites

- Identification of the chemistry of features down to micron sizes on the surface of bulk samples, for example, inclusions, precipitate phases, and wear debris

- Evaluation of chemical composition gradients on the surface of bulk samples over distances approaching 1 μm

- Examination of semiconductor devices for failure analysis, function control, and design verification

Samples

- *Form:* Any solid or liquid having a low vapor pressure ($\geq 10^{-3}$ torr, or 0.13 Pa)

- *Size:* Limited by the scanning electron microscope available. Generally, samples as large as 15 to 20 cm can be placed in the microscope, but regions on such samples that can be examined without repositioning are limited to approximately 4 to 8 cm

- *Preparation:* Standard metallographic polishing and etching techniques are adequate for electrically conducting materials. Nonconducting materials are generally coated with a thin layer of carbon, gold, or gold alloy. Samples must be electrically grounded to the holder, and fine samples, such as powders, can be dispersed on an electrically conducting film, such as a silver paint that has been thoroughly dried. Samples must be free from high vapor pressure liquids, such as water, organic cleaning solutions, and remnant oil-base films

Limitations

- Image quality on relatively flat samples, such as metallographically polished and etched samples, is generally inferior to the optical microscope below 300 to 400×

- Feature resolution, although much better than the optical microscope, is inferior to the transmission electron microscope and the scanning transmission electron microscope

Capabilities of Related Techniques

- *X-ray diffraction:* Provides bulk crystallographic information

- *Optical microscopy:* Faster, less expensive, and provides superior image quality on relatively flat samples at less than 300 to 400×

- *Scanning transmission electron microscopy, Auger electron microscopy:* See page 221 for comparison

- *Transmission electron microscopy:* Provides information from within the volume of material, such as dislocation images, small angle boundary distribution, and vacancy clusters. Superior resolution, but requires thin samples

Electron probe x-ray microanalysis

General Uses

- Qualitative and quantitative elemental analysis of solids for elements with atomic number of 11 (sodium) or greater, detection limits of the order of 100 ppm, and at a lateral spatial resolution of the order of 1 μm

- Qualitative elemental analysis for light elements with atomic numbers from 5 (boron) to 10 (neon)
- Elemental compositional mapping of areas with dimensions as large as millimeters with spatial resolution to 1 μm

Examples of Applications

- Compositional analysis of individual phases at the microstructural level in multiphase samples, for example, analysis of individual inclusions in steels and other alloys
- Analysis of compositional gradients at boundaries
- Determination of compositional homogeneity or heterogeneity at the micrometer scale in single-phase materials
- Compositional mapping of heterogeneous specimens to produce maps of elemental location and concentration

Samples

- *Form:* A bulk solid sample metallographically polished to a mirror finish is ideal for optimum analysis. Other forms include rough surfaces, individual particles, and films on substrates
- *Size:* Typically 25 mm in diameter by 10 mm thick, but can be larger depending on the configuration of the instrument stage

Limitations

- Detects elements with atomic number ≥5 (boron); quantitative analysis is successful for atomic number ≥11 (sodium)
- Sensitivity is 100 ppm with x-ray measurement by wavelength-dispersive spectrometry; 1000 ppm (0.1 wt%) with x-ray measurement by energy-dispersive spectrometry. Poorer sensitivity for light elements in a heavy matrix
- Lateral and depth spatial resolution approximately 1 μm, limited by electron scattering in specimen and not the focused electron beam diameter
- Quantitative analysis limited to flat, polished specimens. Unusual geometries such as fracture surfaces, individual particles, and films on substrates can be analyzed, but with greater uncertainty

Estimated Analysis Time

- 100-s spectrum accumulation time per analysis point; 10 s for on-line computerized data reduction to yield quantitative analysis results

Capabilities of Related Techniques

- *Analytical electron microscopy:* Extends spatial resolution of electron probe analysis down to approximately 10 nm and provides high-resolution imaging and electron diffraction for crystallographic information
- *Secondary ion mass spectrometry, laser microprobe mass analysis:* Provide coverage of the entire periodic table at part per million sensitivity for most elements, but destructively and with larger uncertainties in quantitative analysis
- *Auger microprobe analysis:* Provides elemental analysis of the surface (1 to 5 nm deep) of the sample at the same lateral spatial resolution, and has particular sensitivity to the light elements
- *Raman microprobe analysis:* Provides molecular analysis at the micrometer spatial level
-

Low-energy electron diffraction

General Uses

- Surface crystallography and microstructure
- Surface phase identification (adsorption, segregation, reconstruction)
- Analysis of surface dynamic processes (growth kinetics, thermal vibration)
- Determination of surface atom positions to 0.1 Å

Examples of Applications

- Reconstruction of semiconductor, metal, and alloy surfaces
- Analysis of chemical reactions at surfaces (chemisorbed layers)
- Influence of surface structure on catalytic processes
- Evolution of crystal structure in epitaxial growth
- Grain size determination in thin oriented films

Samples

- *Form:* Solids (metals and semiconductors; insulators in special cases). Single crystals or oriented films; polycrystalline samples with large grain size can be analyzed under special circumstances
- *Size:* 1 mm^2 to 25 cm^2 (0.0015 to 3.9 in.^{2})
- *Preparation:* Samples must be polished carefully to expose the appropriate surface orientation. Surface contaminants must be removed by annealing in vacuum, annealing in a low-pressure (≤10^{-6} torr) oxidizing or reducing atmosphere to clean the surface chemically, or ion beam etching and annealing *in situ*. Some samples can be cleaved on appropriate crystallographic planes *in situ*; in this case, no further preparation is required

Limitations

- Samples must be at least slightly electrically conductive. Electrical charging of nonconductors is a problem

- Ultrahigh vacuum is required

- Determination of sizes of ordered regions (grains, islands, terraces, and so on) is limited by instrumental parameters to sizes less than approximately 500 nm

- Surface preparation is extensive and can be difficult

Estimated Analysis Time

- 10 min to 3 months, depending on information desired and initial condition of sample

Capabilities of Related Techniques

- *Reflection high-energy electron diffraction:* High-energy (~10 keV) analog of low-energy electron diffraction. Provides basically the same information as low-energy electron diffraction. However, some measurements—for example, atom positions—are more difficult than with low-energy electron diffraction; others—for example, three-dimensional crystal growth on surfaces—are facilitated

- *Glancing-angle x-ray diffraction:* More limited than low-energy electron diffraction, unless synchrotron radiation is used. Interpretation of atom positions is simpler. The structure of internal interfaces can be determined in special cases. Flat surfaces and large sample areas are required

- *Transmission electron microscopy, scanning transmission electron microscopy:* Many transmission electron or scanning transmission electron microscopes allow reflection diffraction. This is analogous to reflection of high-energy electron diffraction, but the use of higher-energy electron beams (of the order of 100 keV) requires very shallow angles of incidence for surface sensitivity. Standard transmission electron and scanning transmission electron microscopes do not operate under ultrahigh vacuum conditions

Electron or X-Ray Spectroscopic Methods

Auger electron spectroscopy

General Uses

- Compositional analysis of the 0- to 3-nm region near the surface for all elements except H and He

- Depth-compositional profiling and thin film analysis

- High lateral resolution surface chemical analysis and inhomogeneity studies to determine compositional variations in areas ≥100 nm

- Grain-boundary and other interface analyses facilitated by fracture

- Identification of phases in cross sections

Examples of Applications

- Analysis of surface contamination of materials to investigate its role in such properties as corrosion, wear, secondary electron emission, and catalysis

- Identification of chemical-reaction products, for example, in oxidation and corrosion

- In-depth composition evaluation of surface films, coatings, and thin films used for various metallurgical surface modifications and microelectronic applications

- Analysis of grain-boundary chemistry to evaluate the role of boundary precipitation and solute segregation on mechanical properties, corrosion, and stress corrosion cracking phenomena

Samples

- *Form:* Solids (metals, ceramics, and organic materials) with relatively low vapor pressures ($<10^{-8}$ torr at room temperature). Higher vapor pressure materials can be handled by sample cooling. Similarly, many liquid samples can be handled by sample cooling or by applying a thin film onto a conductive substrate

- *Size:* Individual powder particles as small as 1 μm in diameter can be analyzed. The maximum sample size depends on the specific instrument; 1.5 cm (0.6 in.) in diameter by 0.5 cm (0.2 in.) high is not uncommon

- *Surface topography:* Flat surfaces are preferable, but rough surfaces can be analyzed in selected small areas (~1 μm) or averaged over large areas (0.5 mm in diameter)

- *Preparation:* Frequently none. Samples must be free of fingerprints, oils, and other high vapor pressure materials

Limitations

- Insensitivity to hydrogen and helium

- The accuracy of quantitative analysis is limited to ±30% of the element present when calculated using published elemental sensitivity factors (Ref. 1). Better quantification (±10%) is possible by using standards that closely resemble the sample

- Electron beam damage can severely limit useful analysis of organic and biological materials and occasionally ceramic materials

- Electron beam charging may limit analysis when examining highly insulating materials

- Quantitative detection sensitivity for most elements is from 0.1 to 1.0 at.%

Estimated Analysis Time

- Usually under 5 min for a complete survey spectrum from 0 to 2000 eV. Selected peak analyses for studying chemical effects, Auger elemental imaging, and depth profiling generally take much longer

Capabilities of Related Techniques

- *X-ray photoelectron spectroscopy:* Provides compositional and chemical binding state information, relatively nondestructive

- *Ion scattering spectroscopy:* Provides superb top atomic layer information, specificity of surface atomic bonding in selected cases, and surface composition and depth profiling information

- *Secondary ion mass spectroscopy:* High elemental detection sensitivity from part per million to part per billion levels; surface compositional information; depth profiling capability; sensitivity for all elements, including hydrogen and helium

- *Electron probe:* Analysis to 1-μm depth in conventional operation, quantitative and nondestructive

- *Analytical electron microscopy:* Chemical analysis in conjunction with high-resolution microscopy

X-ray photoelectron spectroscopy

General Uses

- Elemental analysis of surfaces of all elements except hydrogen

- Chemical state identification of surface species

- In-depth composition profiles of elemental distribution in thin films

- Composition analysis of samples when destructive effects of electron beam techniques must be avoided

Examples of Applications

- Determination of oxidation states of metal atoms in metal oxide surface films

- Identification of surface carbon as graphitic or carbide

Samples

- *Form:* Solids (metals, glasses, semiconductors, low vapor pressure ceramics)

- *Size:* ≤ 6.25 cm^3 (≤ 0.4 in.^{3})

- *Preparation:* Must be free of fingerprints, oils, or other surface contamination

Limitations

- Data collection is slow compared with other surface analysis techniques, but analysis time can be decreased substantially when high resolution or chemical state identification is not needed

- Poor lateral resolution

- Surface sensitivity comparable to other surface analysis techniques

- Charging effects may be a problem with insulating samples. Some instruments are equipped with charge-compensation devices

- The accuracy of quantitative analysis is limited

Estimated Analysis Time

- Requires an overnight vacuum pumpdown before analysis

- Qualitative analysis can be performed in 5 to 10 min

- Quantitative analysis requires 1 h to several hours, depending on information desired

Capabilities of Related Techniques

- *Auger electron spectoscropy:* Compositional analysis of surfaces. Faster, with better lateral resolution than XPS. Has depth-profiling capabilities. Electron beam can be very damaging; bonding and other chemical state information are not easily interpreted

- *Low-energy ion-scattering spectroscopy:* Sensitive to the top atom layer of the surface and has profiling capabilities. Quantitative analysis requires use of standards; no chemical state information; poor mass resolution for high-Z elements

- *Secondary ion mass spectroscopy:* The most sensitive of all surface analysis techniques. Can detect hydrogen, and depth profiling is possible. Has pronounced matrix effects that can cause orders of magnitude variations in elemental sensitivity and make quantitative analysis difficult

Methods Based on Sputtering or Scattering Phenomena

Field ion microscopy and atom probe microanalysis

General Uses

- Observation of the microstructure of materials in atomic detail

- Chemical microanalysis of materials at the atomic level, with equal sensitivity to all chemical elements

Examples of Applications

- Study of point defects in radiation damage, dislocations, stacking faults, grain boundaries, and interphase interfaces in metals and alloys

- Study of the nucleation, growth, and coarsening of precipitates in age-hardening materials, including aluminum, iron, and nickel-base alloys

- Study of alloying element partitioning, phase stability, and phase transformation phenomena, for example, in steels and turbine alloys

- Study of order-disorder reactions and of spinodal decomposition in magnetic materials

- Study of segregation of alloy elements and impurities to dislocations and interfaces

- Study of metal surfaces, including surface diffusion and reconstruction, surface segregation, surface reactions, adsorption, heterogeneous catalysis, oxidation, nucleation and growth of thin films, and depth profiling through surface layers

- Study of semiconductor materials, including oxidation and metallization, interdiffusion, and investigation of local composition variations in thin films

Samples

- *Form:* Solids (metals or semiconductors)
- *Size:* Needlelike, end radius below 100 nm, overall length to 10 mm (0.4 in.)
- *Preparation:* Chemical or electropolishing, or ion milling, from round or square cross-section sample blanks

Limitations

- Samples must possess some degree of electrical conductivity
- Samples must have appreciable mechanical strength
- Field of view is restricted to areas of approximately 200 nm in diameter
- Needlelike geometry restricts sample tests that can be studied
- Method of analysis is destructive; material is removed from surface

Estimated Analysis Time

- 3 to 30 h per sample

Capabilities of Related Techniques

- *Transmission electron microscopy, scanning transmission electron microscopy, x-ray analytical electron microscopy, electron energy loss spectroscopy:* Microscopy and microanalysis of solids with larger field of view. Faster analysis, but with lower spatial resolution, and sensitivity to different chemical elements varies

- *Low-energy electron diffraction, Auger electron spectroscopy, x-ray photoelectron spectroscopy:* Determination of structure and chemistry of surfaces, providing more crystallographic data and information on chemical state

- *Secondary ion mass spectroscopy, Rutherford backscattering spectroscopy:* Depth profiling through surface layers, with lower spatial and depth resolution. Analytical sensitivity to different chemical elements varies widely in secondary ion mass spectroscopy

- *Laser microprobe mass analysis:* Time-of-flight mass spectrometric analysis of solids. Relatively fast, suitable for a wide range of materials, and sensitive to all chemical elements, but much lower spatial resolution and less quantitative than the atom probe

Low-energy ion-scattering spectroscopy

General Uses

- Identification of elements present on solid surfaces
- Semiquantitative determination of the atomic concentration of the elements present on the surface

Examples of Applications

- Identification of surface stains and corrosion products
- Determination of composition depth profiles and film thicknesses when combined with inert gas ion sputtering
- Study of the segregation of alloy and compound constituents to the surface
- Study of oxidation using ^{18}O
- Determination of the extent of coverage of ultrathin films
- Study of desorption of adsorbed layers
- Identification of the faces of polar crystals

Samples

- *Form:* Solids (metals, ceramics, ores, corrosion products, thin films, and so on) as powders or flat solid surfaces
- *Size:* Flat surface or distributed powders—$2 \times 1 \times 0.5$ cm ($0.8 \times 0.4 \times 0.2$ in.) maximum. Minimum size is determined by the probing beam size (0.05 cm, or 0.02 in. typical)
- *Preparation:* None; samples must be handled with clean instruments to avoid contamination

Limitations

- For high atomic number materials, elements must be present at a level of >0.1 at.% of a monolayer. For low atomic number materials, elements must be present at a level of >10 at.% of a monolayer
- Samples must be vacuum worthy
- For high atomic number materials, adjacent elements cannot be separated due to insufficient mass resolution
- Minimum beam size is 150 μm

Estimated Analysis Time

- 10 min for a single surface scan
- Several hours for depth profiles up to several thousand angstroms

Capabilities of Related Techniques

- *Auger electron spectroscopy:* Identification of elements present in the first 1 to 10 nm (10 to 100 Å)
- *X-ray photoelectron spectroscopy:* Identification of elements present in the first 1 to 10 nm (10 to 100 Å) over a 3- to 10-mm (0.12- to 0.40-in.) area
- *Secondary ion mass spectroscopy:* Mass/charge identification of elements present in the first 0.2 to 2 nm (2 to 20 Å)

Secondary ion mass spectroscopy

General Uses

- Surface compositional analysis with approximately 5- to 10-nm depth resolution
- Elemental in-depth concentration profiling
- Trace element analysis at the parts per billion to parts per million range
- Isotope abundances
- Hydrogen analysis
- Spatial distribution of elemental species

Examples of Applications

- Identification of inorganic or organic surface layers on metals, glasses, ceramics, thin films, or powders
- In-depth composition profiles of oxide surface layers, corrosion films, leached layers, and diffusion profiles
- In-depth concentration profiles of low-level dopants (≤1000 ppm) diffused or ion implanted in semiconductor materials
- Hydrogen concentration and in-depth profiles in embrittled metal alloys, vapor-deposited thin films, hydrated glasses, and minerals
- Quantitative analysis of trace elements in solids

- Isotopic abundances in geological and lunar samples
- Tracer studies (for example, diffusion and oxidation) using isotope-enriched source materials
- Phase distribution in geologic minerals, multiphase ceramics, and metals
- Second-phase distribution due to grain-boundary segregation, internal oxidation, or precipitation

Samples

- *Form:* Crystalline or noncrystalline solids, solids with modified surfaces, or substrates with deposited thin films or coatings; flat, smooth surfaces are desired; powders must be pressed into a soft metal foil (for example, indium) or compacted into a pellet
- *Size:* Variable, but typically 1 cm × 1 cm × 1 mm
- *Preparation:* None for surface or in-depth analysis; polishing for microstructural or trace element analysis

Limitations

- Analysis is destructive
- Qualitative and quantitative analyses are complicated by wide variation in detection sensitivity from element to element and from sample matrix to sample matrix
- The quality of the analysis (precision, accuracy, sensitivity, and so on) is a strong function of the instrument design and the operating parameters for each analysis

Estimated Analysis Time

- One to a few hours per sample

Capabilities of Related Techniques

- *Auger electron spectroscopy:* Qualitative and quantitative elemental surface and in-depth analysis is straightforward, but the detection sensitivity is limited to >1000 ppm; microchemical analysis with spatial resolution to <100 nm
- *Rutherford backscattering spectroscopy:* Nondestructive elemental in-depth profiling; quantitative determination of film thickness and stoichiometry
- *Electron microprobe analysis:* Quantitative elemental analysis and imaging with depth resolution ≥1 μm

Rutherford backscattering spectroscopy

General Uses

- Quantitative compositional analysis of thin films, layered structures, or bulks
- Quantitative measurements of surface impurities of heavy elements on substrates of lighter elements

- Defect distribution depth profile in single-crystal sample
- Surface atom relaxation in single crystal
- Interfacial studies on heteroepitaxy layers
- Lattice location of impurities in single crystal

Examples of Applications

- Analysis of silicide or alloy formation; identification of reaction products; obtaining reaction kinetics, activation energy, and moving species
- Composition analysis of bulk garnets
- Depth distribution of heavy ion implantation and/or diffusion in a light substrate
- Surface damage and contamination on reactive ion etched samples
- Providing calibration samples for other instrumentation, such as secondary ion mass spectroscopy and Auger electron microscopy
- Defect depth distribution due to ion implantation damage or residue damage from improper annealing
- Lattice location of impurities in single crystal
- Surface atom relaxation of single crystal
- Lattice strain measurements of heteroepitaxy layers or superlattices

Samples

- *Form:* Solid samples with smooth surfaces, thin films on smooth substrates, self-supporting thin foils, and so on
- *Size:* Typically 1 cm × 1 cm × 1 mm; can accept sample as small as 2 × 2 mm
- *Preparation:* No special preparation required other than the surface must be smooth

Limitations

- Composition information may be obtained, but not chemical bonding information
- Poor lateral resolution. Typical beam spot is 1 × 1 mm. With attachment, beam spot may be reduced to 1 × 1 μm
- Poor mass resolution for heavy (high-Z) elements, cannot distinguish surface impurities of gold from platinum, tantalum, tungsten, and so on; mass resolution is better for low- and mid-Z elements, for example, can distinguish ^{37}Cl from ^{35}Cl
- Poor sensitivity for low-Z elements on substrates with elements heavier than the impurity
- Depth resolution is generally approximately 20 nm. Glancing angle Rutherford backscattering spectrometry provides 1 to 2 nm

Estimated Analysis Time

- For routine Rutherford backscattering spectrometry, up to four samples per hour; for channeling Rutherford backscattering spectrometry, about 1 h per sample

Capabilities of Related Techniques

- *Particle-induced x-ray emission:* Mass distinction
- *Nuclear reaction:* Low-Z element sensitivity
- *Low-energy ion-scattering spectroscopy:* Low-energy Rutherford backscattering spectrometry using keV ions rather than typical Rutherford backscattering spectrometry, which uses MeV ions
- *Cross-section transmission electron microscopy:* Depth and types of defect

Chromatography

Gas chromatography/mass spectrometry

General Uses

- Analysis of complex mixtures of volatile compounds
- In mass spectrometry/mass spectrometry, analysis of nonvolatile compounds
- In pyrolysis gas chromatography/mass spectrometry, analysis and quality control of polymers
- In liquid chromatography/mass spectrometry, analysis of heat-sensitive and degradable compounds, such as biological materials

Examples of Applications

- *Gas chromatography/mass spectrometry:* Mixtures of volatile compounds in petroleum oil, coal gasification and liquefaction products, oil shale, and tar sands; pollutants in air, waste water, and solid waste; drugs and metabolites; pesticides; and additives, such as antioxidants and plasticizers in plastics
- *Liquid chromatography/mass spectrometry:* Mixtures of nonvolatile and heat-sensitive compounds
- *Mass spectrometry/mass spectrometry:* Mixtures of nonvolatile and high molecular weight solids
- *Pyrolysis gas chromatography/mass spectrometry:* Analysis of polymers and their additives

Samples

- *Form:* Solids, liquids, and gases; all organics and some inorganics

- *Size:* For gas chromatography/mass spectrometry, a 1- to 5-μL injection in which each compound of interest is in the 20- to 200-ng range or, when split in the injection port, is in that range. For selected ion monitoring gas chromatography/mass spectrometry, 1.5 μL in which each compound of interest is in the 100- to 500-pg range; in some cases, down to 0.5 pg. For liquid chromatography/mass spectrometry and mass spectrometry/mass spectrometry, sample sizes are 20 to 500 ng per compound of interest and 10 to 500 ng per compound of interest, respectively
- *Preparation:* Samples should be prepared to conform to the sample size restrictions given above

Limitations

- Compound(s) must be ionizable
- Detection limit is from 5 to 20 ng, depending on the compound. In selected ion monitoring gas chromatography/mass spectrometry, the detection limit can be as low as 0.5 pg

Estimated Analysis Time

- When analyzing one compound, direct introduction of sample takes 10 to 20 min per analysis. For gas chromatography/mass spectrometry, analysis of 1 to 2 compounds takes approximately 15 min, while analysis of 20 or more compounds takes 180 min or more
- *Analysis and interpretation of data:* Variable (15 min to days depending on the number of compounds analyzed)

Capabilities of Related Techniques

- *Gas chromatography/Fourier-transform infrared spectroscopy:* Functional group analysis, but at least an order of magnitude less sensitive
- *Nuclear magnetic resonance:* Only single compounds and at least two orders of magnitude less sensitive
- *Secondary ion mass spectroscopy:* A mass spectrometry method for looking only at surfaces of materials

Liquid chromatography

General Uses

- Separation and quantitative analysis of components in organic, inorganic, pharmaceutical, and biochemical mixtures
- Analysis of organic and inorganic compounds for impurities
- Isolation of pure compounds from mixtures

Examples of Applications

- Analysis of solvents for low-level organic contaminants

- Monitoring the stability of polymers during aging tests
- Analysis of foods and natural products for high molecular weight sugars
- Analysis of thermally unstable pesticides
- Isolation of microgram amounts of material for identification purposes
- Isolation of large quantities (1 to 10 g) of purified compounds for synthetic purposes

Samples

- *Form:* Solids (dissolved in a suitable solvent) or liquids
- *Size:* 0.1 to 1 g generally required for quantitative work; however, as little as 10^{-5} g may be analyzed. Larger quantities may be required for preparative work
- *Preparation:* Sample must be dissolved in a suitable solvent at concentrations of 0.1 to 100 mg/mL (although neat liquids may be analyzed). Filtration or extraction of the solution may be necessary. Injection volume ≤0.1 mL is typical

Limitations

- Solids must be freely soluble in carrier solvent, and liquids must be miscible with it
- Difficult to make unambiguous identification of a particular component; subsequent analysis by infrared spectroscopy or mass spectrometry may be necessary

Estimated Analysis Time

- Requires 15 to 60 min for an analysis; several replicate analyses are usually performed
- May require one-half day to change column, solvents, or detector for a particular analysis
- Lengthy development may be required if method is not available

Capabilities of Related Techniques

- *Gas chromatography:* Restricted to volatile or pyrolyzable samples
- *Mass spectrometry, infrared spectroscopy, nuclear magnetic resonance:* Used to identify components isolated by liquid chromatography
- *Ion chromatography:* A type of liquid chromatography restricted to the analysis of ionic species

Ion chromatography

General Uses

- Qualitative and quantitative analyses of a wide range of inorganic and organic anions and certain cations in aqueous solutions

Examples of Applications

- Aqueous solutions, such as leachates, brines, well waters, and condensates
- Organically bound halides and sulfur following Schöniger flask combustion and adsorption techniques
- Determination of anions on contaminated surfaces
- Plating bath solution analysis

Samples

- *Form:* Solids or aqueous solutions
- *Size:* Minimum of 1 to 5 mg for solids; minimum of 1 mL for solutions; 0.5 μg can be detected on surfaces
- *Materials:* Inorganic and organic materials, geological samples, glasses, ceramics, leachates, explosives, alloys, and pyrotechnics
- *Preparation:* Aqueous solutions can be analyzed as received or after dilution; analysis of solids must follow a sample preparation and dissolution procedure

Limitations

- Detection limits below the part per million level for many ions; part per billion under ideal conditions

- *Cations:* Limited to alkali and alkaline earths, ammonia, and low molecular weight amines if suppressed conductivity detection is used
- Must be ionic in solution
- Must be water soluble
- Limited work has been done in organic solvents

Estimated Analysis Time

- Requires 15 min to 1 h per sample if already in aqueous solutions
- Requires 1 h per sample for organics
- Times for other sample matrices are not well established

Capabilities of Related Techniques

- *Wet analytical chemistry:* Much slower and less sensitive when mixtures of ions are to be analyzed
- *Ion selective electrode:* Less senstiive and limited to one ion at a time
- *Atomic absorption spectroscopy:* Not capable of analyzing anions; more versatile for analysis of cations

Microscopes for use in materials characterization

	Acronym	Magnification range, times	Resolution, nm	Analysis time(a), s	Nondestructive?	Operating voltage, eV	Environment
Light microscope	LM	5 to 2,000	200	10	Yes	0	Ambient
Scanning electron microscope	SEM	10,000 to 250,000	5	60	Yes(b)	500 to 40,000	Vacuum/ambient(c)
Scanning electron microscope/field emitter	SEM/FE	10,000 to 400,000	1.0	90	Yes(b)	100 to 300,000	Vacuum
Transmission electron microscope	TEM	1,000 to 1,000,000	0.15	300	No	30,000 to 1,000,000	Vacuum
Scanning transmission electron microscope	STEM	100,000 to 1,000,000	0.5	300	No	1,000 to 50,000	Vacuum
Scanning TEM with selective area diffraction	STEM/SAD	1,000 to 1,000,000	0.15	30	No	30,000 to 1,000,000	Vacuum
Scanning laser acoustic microscope	SLAM	10 to 500	1,000	60	Yes	0	Ambient
Confocal scanning microscope	CSM	100 to 10,000	0.1	20	Yes	0	Ambient
Scanning tunneling microscope	STM	5,000 to 10,000,000	0.0001	300	Yes(d)	0 to 1 V	Vacuum/ambient
Atomic force microscope	AFM	5,000 to 2,000,000	0.0005	300	Yes	0	Ambient

(a) Time listed is to acquire an image with no analysis required. (b) Sample preparation may be destructive. (c) Most SEMs are kept under vacuum, all measurements are made in vacuum. (d) STMs are generally nondestructive, but its probe beam may do damage

Tension Testing

Percentage reduction of area for tension test specimens

Diam(a), in.	Area, in.²	Reduction of area, % for initial diam of:			Diam(a), in.	Area, in.²	Reduction of area, % for initial diam of:		
		0.505 in.	0.506 in.	0.504 in.			0.505 in.	0.506 in.	0.504 in.
0.251	0.0494	75.3	75.4	75.2	0.311	0.0759	62.1	62.2	62.0
0.252	0.0498	75.1	75.2	75.0	0.312	0.0764	61.8	62.0	61.7
0.253	0.0502	74.9	75.0	74.8	0.313	0.0769	61.6	61.7	61.5
0.254	0.0506	74.7	74.8	74.6	0.314	0.0774	61.3	61.5	61.2
0.255	0.0510	74.5	74.6	74.4	0.315	0.0779	61.1	61.2	61.0
0.256	0.0514	74.3	74.4	74.2	0.316	0.0784	60.8	61.0	60.7
0.257	0.0518	74.1	74.2	74.0	0.317	0.0789	60.6	60.7	60.5
0.258	0.0522	73.9	74.0	73.8	0.318	0.0794	60.3	60.5	60.2
0.259	0.0526	73.7	73.8	73.6	0.319	0.0799	60.1	60.2	60.0
0.260	0.0530	73.5	73.6	73.4	0.320	0.0804	59.8	60.0	59.7
0.261	0.0535	73.3	73.4	73.2	0.321	0.0809	59.6	59.8	59.4
0.262	0.0539	73.0	73.2	73.0	0.322	0.0814	59.4	59.5	59.2
0.263	0.0543	72.9	73.0	72.8	0.323	0.0819	59.1	59.3	58.9
0.264	0.0547	72.7	72.8	72.6	0.324	0.0824	58.8	59.0	58.7
0.265	0.0551	72.5	72.6	72.4	0.325	0.0829	58.6	58.8	58.4
0.266	0.0555	72.3	72.4	72.2	0.326	0.0834	58.3	58.5	58.2
0.267	0.0559	71.9	72.2	72.0	0.327	0.0839	58.1	58.3	57.9
0.268	0.0564	71.8	71.9	71.7	0.328	0.0844	57.8	58.0	57.7
0.269	0.0568	71.6	71.7	71.5	0.329	0.0850	57.5	57.7	57.4
0.270	0.0572	71.4	71.5	71.3	0.330	0.0855	57.3	57.5	57.1
0.271	0.0576	71.2	71.3	71.1	0.331	0.0860	57.0	57.2	56.9
0.272	0.0581	71.0	71.1	70.9	0.332	0.0865	56.8	57.0	56.6
0.273	0.0585	70.8	70.9	70.7	0.333	0.0870	56.5	56.7	56.4
0.274	0.0589	70.6	70.7	70.5	0.334	0.0876	56.2	56.4	56.1
0.275	0.0593	70.4	70.5	70.3	0.335	0.0881	56.0	56.2	55.8
0.276	0.0598	70.1	70.2	70.0	0.336	0.0886	55.7	55.9	55.6
0.277	0.0602	69.9	70.0	69.8	0.337	0.0891	55.5	55.7	55.3
0.278	0.0606	69.7	69.9	69.6	0.338	0.0897	55.2	55.4	55.0
0.279	0.0611	69.5	69.6	69.4	0.339	0.0902	54.9	55.1	54.8
0.280	0.0615	69.3	69.4	69.2	0.340	0.0907	54.7	54.9	54.5
0.281	0.0620	69.0	69.2	69.0	0.341	0.0913	54.4	54.6	54.2
0.282	0.0624	68.8	69.0	68.7	0.342	0.0918	54.1	54.3	54.0
0.283	0.0629	68.6	68.7	68.5	0.343	0.0924	53.8	54.0	53.7
0.284	0.0633	68.4	68.5	68.3	0.344	0.0929	53.6	53.8	53.4
0.285	0.0637	68.2	68.3	68.1	0.345	0.0934	53.3	53.5	53.2
0.286	0.0642	67.9	68.1	67.8	0.346	0.0940	53.0	53.2	52.9
0.287	0.0646	67.7	67.9	67.6	0.347	0.0945	52.8	53.0	52.6
0.288	0.0651	67.5	67.6	67.4	0.348	0.0951	52.5	52.7	52.3
0.289	0.0655	67.3	67.4	67.2	0.349	0.0956	52.2	52.4	52.1
0.290	0.0660	67.0	67.2	66.9	0.350	0.0962	51.9	52.1	51.8
0.291	0.0665	66.8	66.9	66.7	0.351	0.0967	51.7	51.9	51.5
0.292	0.0670	66.5	66.7	66.4	0.352	0.0973	51.4	51.6	51.2
0.293	0.0674	66.3	66.5	66.2	0.353	0.0978	51.1	51.3	51.0
0.294	0.0679	66.1	66.2	66.0	0.354	0.0984	50.8	51.0	50.7
0.295	0.0683	65.9	66.0	65.8	0.355	0.0989	50.6	50.8	50.4
0.296	0.0688	65.6	65.8	65.5	0.356	0.0995	50.3	50.5	50.1
0.297	0.0692	65.4	65.6	65.3	0.357	0.1000	50.0	50.2	49.9
0.298	0.0697	65.2	65.3	65.1	0.358	0.1006	49.8	50.0	49.6
0.299	0.0702	64.9	65.1	64.8	0.359	0.1012	49.5	49.7	49.3
0.300	0.0707	64.7	64.8	64.6	0.360	0.1017	49.2	49.4	49.0
0.301	0.0712	64.4	64.6	64.3	0.361	0.1023	48.9	49.1	48.7
0.302	0.0716	64.2	64.4	64.1	0.362	0.1029	48.6	48.8	48.4
0.303	0.0721	64.0	64.1	63.9	0.363	0.1034	48.4	48.6	48.2
0.304	0.0725	63.8	63.9	63.7	0.364	0.1040	48.1	48.3	47.9
0.305	0.0730	63.5	63.7	63.4	0.365	0.1046	47.8	48.0	47.6
0.306	0.0735	63.3	63.4	63.2	0.366	0.1052	47.5	47.7	47.3
0.307	0.0740	63.0	63.2	62.9	0.367	0.1057	47.2	47.4	47.0
0.308	0.0745	62.8	62.9	62.7	0.368	0.1063	46.9	47.1	46.7
0.309	0.0749	62.6	62.7	62.5	0.369	0.1069	46.6	46.8	46.4
0.310	0.0754	62.3	62.5	62.2	0.370	0.1075	46.3	46.5	46.1

This table was compiled by Arthur W.F. Green, and was first published in the 1939 *Metals Handbook*. It is designed to save time in computing the percentage reduction of area of standard 0.505-in. diam tension test specimens after testing. The 0.505-in. diam is a nominal dimension; specimens are usually machined to a ±0.001-in. tolerance, hence the utility of a table that gives values for 0.506 and 0.504, as well as the nominal 0.505 in. Percentage reduction of area is obtained from measurement of the necked-down diameter of the specimen (first column of the table) and reading the percentage of reduction of area in the applicable column for initial diameter of test specimen. For example, if the initial diameter of the test specimen were 0.506 in. and the diameter at the fracture after testing were 0.441 in., the reduction of area would be 24.0%, as shown in the table. (a) At the fracture after testing

(continued)

Percentage reduction of area for tension test specimens (continued)

Diam(a), in.	Area, in.²	Reduction of area, % for initial diam of:			Diam(a), in.	Area, in.²	Reduction of area, % for initial diam of:		
		0.505 in.	0.506 in.	0.504 in.			0.505 in.	0.506 in.	0.504 in.
0.371	0.1081	46.0	46.2	45.8	0.436	0.1493	25.4	25.7	25.2
0.372	0.1086	45.8	46.0	45.6	0.437	0.1499	25.1	25.4	24.9
0.373	0.1092	45.5	45.7	45.3	0.438	0.1506	24.8	25.1	24.5
0.374	0.1098	45.2	45.4	45.0	0.439	0.1513	24.4	24.7	24.2
0.375	0.1104	44.9	45.1	44.5	0.440	0.1520	24.1	24.4	23.9
0.376	0.1110	44.6	44.8	44.4	0.441	0.1527	23.7	24.0	23.5
0.377	0.1116	44.3	44.5	44.1	0.442	0.1534	23.4	23.7	23.2
0.378	0.1122	44.0	44.2	43.8	0.443	0.1541	23.0	23.3	22.8
0.379	0.1128	43.7	43.9	43.5	0.444	0.1548	22.7	23.0	22.5
0.380	0.1134	43.4	43.6	43.2	0.445	0.1555	22.3	22.6	22.1
0.381	0.1140	43.1	43.3	42.9	0.446	0.1562	22.0	22.3	21.7
0.382	0.1146	42.8	43.0	42.6	0.447	0.1569	21.6	21.9	21.4
0.383	0.1152	42.5	42.7	42.3	0.448	0.1576	21.3	21.6	21.0
0.384	0.1158	42.2	42.4	42.0	0.449	0.1583	20.9	21.2	20.7
0.385	0.1164	41.9	42.1	41.7	0.450	0.1590	20.6	20.9	20.3
0.386	0.1170	41.6	41.8	41.4	0.451	0.1597	20.2	20.5	20.0
0.387	0.1176	41.3	41.5	41.1	0.452	0.1604	19.9	20.2	19.6
0.388	0.1182	41.0	41.2	40.8	0.453	0.1611	19.5	19.9	19.2
0.389	0.1188	40.7	40.9	40.5	0.454	0.1618	19.2	19.5	18.9
0.390	0.1194	40.4	40.6	40.2	0.455	0.1625	18.8	19.2	18.5
0.391	0.1200	40.1	40.3	39.9	0.456	0.1633	18.4	18.8	18.1
0.392	0.1206	39.8	40.0	39.6	0.457	0.1640	18.1	18.4	17.8
0.393	0.1213	39.4	39.7	39.2	0.458	0.1647	17.7	18.1	17.4
0.394	0.1219	39.1	39.4	38.9	0.459	0.1654	17.4	17.7	17.1
0.395	0.1225	38.8	39.1	38.6	0.460	0.1661	17.0	17.4	16.7
0.396	0.1231	38.5	38.8	38.3	0.461	0.1669	16.6	17.0	16.3
0.397	0.1237	38.2	38.5	38.0	0.462	0.1676	16.3	16.6	16.0
0.398	0.1244	37.9	38.1	37.6	0.463	0.1683	15.9	16.3	15.6
0.399	0.1250	37.6	37.8	37.3	0.464	0.1690	15.6	15.9	15.3
0.400	0.1256	37.3	37.5	37.0	0.465	0.1698	15.2	15.5	14.9
0.401	0.1262	37.0	37.2	36.7	0.466	0.1705	14.8	15.2	14.5
0.402	0.1269	36.6	36.9	36.4	0.467	0.1712	14.5	14.8	14.2
0.403	0.1275	36.3	36.6	36.1	0.468	0.1720	14.1	14.4	13.8
0.404	0.1281	36.0	36.3	35.8	0.469	0.1727	13.7	14.1	13.4
0.405	0.1288	35.7	35.9	35.4	0.470	0.1734	13.4	13.7	13.1
0.406	0.1294	35.4	35.6	35.1	0.471	0.1742	13.0	13.3	12.7
0.407	0.1301	35.0	35.3	34.8	0.472	0.1749	12.6	13.0	12.3
0.408	0.1307	34.7	35.0	34.5	0.473	0.1757	12.2	12.6	11.9
0.409	0.1313	34.4	34.7	34.2	0.474	0.1764	11.9	12.2	11.6
0.410	0.1320	34.1	34.3	33.8	0.475	0.1772	11.5	11.8	11.2
0.411	0.1326	33.8	34.0	33.5	0.476	0.1779	11.1	11.5	10.8
0.412	0.1333	33.4	33.7	33.2	0.477	0.1787	10.7	11.1	10.4
0.413	0.1339	33.1	33.4	32.9	0.478	0.1794	10.4	10.8	10.1
0.414	0.1346	32.8	33.0	32.5	0.479	0.1802	10.0	10.3	9.7
0.415	0.1352	32.5	32.8	32.2	0.480	0.1809	9.6	10.0	9.3
0.416	0.1359	32.1	32.4	31.9	0.481	0.1817	9.2	9.6	8.9
0.417	0.1365	31.8	32.1	31.6	0.482	0.1824	8.9	9.3	8.6
0.418	0.1372	31.5	31.7	31.2	0.483	0.1832	8.5	8.9	8.2
0.419	0.1378	31.2	31.4	30.9	0.484	0.1839	8.1	8.5	7.8
0.420	0.1385	30.8	31.1	30.6	0.485	0.1847	7.7	8.1	7.4
0.421	0.1392	30.5	30.7	30.2	0.486	0.1855	7.3	7.7	7.0
0.422	0.1398	30.2	30.4	29.9	0.487	0.1862	7.0	7.4	6.7
0.423	0.1405	29.8	30.1	29.6	0.488	0.1870	6.6	7.0	6.3
0.424	0.1411	29.5	29.8	29.3	0.489	0.1878	6.2	6.6	5.9
0.425	0.1418	29.2	29.5	28.9	0.490	0.1885	5.8	6.2	5.5
0.426	0.1425	28.8	29.1	28.6	0.492	0.1901	5.0	5.4	4.7
0.427	0.1432	28.5	28.8	28.2	0.494	0.1916	4.3	4.7	4.0
0.428	0.1438	28.2	28.5	27.9	0.496	0.1932	3.5	3.9	3.2
0.429	0.1445	27.8	28.1	27.6	0.498	0.1947	2.7	3.1	2.4
0.430	0.1452	27.5	27.8	27.4	0.500	0.1963	1.9	2.3	1.6
0.431	0.1458	27.2	27.5	26.9	0.502	0.1979	1.1	1.5	0.8
0.432	0.1465	26.8	27.1	26.6	0.504	0.1995	0.4	0.7	0.0
0.433	0.1472	26.5	26.8	26.2	0.505	0.2002	0.0	0.4	...
0.434	0.1479	26.1	26.4	25.9	0.506	0.2010	...	0.0	...
0.435	0.1486	25.8	26.1	25.5					

This table was compiled by Arthur W.F. Green, and was first published in the 1939 *Metals Handbook*. It is designed to save time in computing the percentage reduction of area of standard 0.505-in. diam tension test specimens after testing. The 0.505-in. diam is a nominal dimension; specimens are usually machined to a ±0.001-in. tolerance, hence the utility of a table that gives values for 0.506 and 0.504, as well as the nominal 0.505 in. Percentage reduction of area is obtained from measurement of the necked-down diameter of the specimen (first column of the table) and reading the percentage of reduction of area in the applicable column for initial diameter of test specimen. For example, if the initial diameter of the test specimen were 0.506 in. and the diameter at the fracture after testing were 0.441 in., the reduction of area would be 24.0%, as shown in the table. (a) At the fracture after testing

Hardness Testing

Equivalent Hardness Numbers and Tensile Strengths for Steel

Rockwell B hardness numbers(a)

Rockwell B-scale hardness	Vickers hardness	Brinell hardness, 10-mm-diam ball, 500-kg load	Brinell hardness, 10-mm-diam ball, 3000-kg load	Rockwell hardness, A scale, 60-kg load, Brale indenter	Rockwell hardness, C scale, 150-kg load, Brale indenter	Rockwell hardness, F scale, 60-kg load, 1/16-in.-diam ball	Rockwell superficial hardness, 1/16-in.-diam ball, 15T scale, 15-kg load	30T scale, 30-kg load	45T scale, 45-kg load	Knoop hardness, 500-g load and greater	Scleroscope hardness	Tensile strength (approx), ksi	Rockwell B-scale hardness
98	228	189	228	60.2	(19.9)	...	92.5	81.8	70.9	241	34	107	98
97	222	184	222	59.5	(18.6)	...	92.1	81.1	69.9	236	33	104	97
96	216	179	216	58.9	(17.2)	...	91.8	80.4	68.9	231	32	102	96
95	210	175	210	58.3	(15.7)	...	91.5	79.8	67.9	226	...	99	95
94	205	171	205	57.6	(14.3)	...	91.2	79.1	66.9	221	31	97	94
93	200	167	200	57.0	(13.0)	...	90.8	78.4	65.9	216	30	94	93
92	195	163	195	56.4	(11.7)	...	90.5	77.8	64.8	211	...	92	92
91	190	160	190	55.8	(10.4)	...	90.2	77.1	63.8	206	29	90	91
90	185	157	185	55.2	(9.2)	...	89.9	76.4	62.8	201	28	88	90
89	180	154	180	54.6	(8.0)	...	89.5	75.8	61.8	196	27	86	89
88	176	151	176	54.0	(6.9)	...	89.2	75.1	60.8	192	...	84	88
87	172	148	172	53.4	(5.8)	...	88.9	74.4	59.8	188	26	82	87
86	169	145	169	52.8	(4.7)	...	88.6	73.8	58.8	184	26	81	86
85	165	142	165	52.3	(3.6)	...	88.2	73.1	57.8	180	25	79	85
84	162	140	162	51.7	(2.5)	...	87.9	72.4	56.8	176	...	78	84
83	159	137	159	51.1	(1.4)	...	87.6	71.8	55.8	173	24	76	83
82	156	135	156	50.6	(0.3)	...	87.3	71.1	54.8	170	24	75	82
81	153	133	153	50.0	...	...	86.9	70.4	53.8	167	...	73	81
80	150	130	150	49.5	...	...	86.6	69.7	52.8	164	23	72	80
79	147	128	147	48.9	...	...	86.3	69.1	51.8	161	...	70	79
78	144	126	144	48.4	...	...	86.0	68.4	50.8	158	22	69	78
77	141	124	141	47.9	...	...	85.6	67.7	49.8	155	22	68	77
76	139	122	139	47.3	...	...	85.3	67.1	48.8	152	...	67	76
75	137	120	137	46.8	...	99.6	85.0	66.4	47.8	150	21	66	75
74	135	118	135	46.3	...	99.1	84.7	65.7	46.8	148	21	65	74
73	132	116	132	45.8	...	98.5	84.3	65.1	45.8	145	...	64	73
72	130	114	130	45.3	...	98.0	84.0	64.4	44.8	143	20	63	72
71	127	112	127	44.8	...	97.4	83.7	63.7	43.8	141	20	62	71
70	125	110	125	44.3	...	96.8	83.4	63.1	42.8	139	...	61	70
69	123	109	123	43.8	...	96.2	83.0	62.4	41.8	137	20	60	69
68	121	107	121	43.3	...	95.6	82.7	61.7	40.8	135	19	59	68
67	119	106	119	42.8	...	95.1	82.4	61.0	39.8	133	19	58	67
66	117	104	117	42.3	...	94.5	82.1	60.4	38.7	131	19	57	66
65	116	102	116	41.8	...	93.9	81.8	59.7	37.7	129	...	56	65
64	114	101	114	41.4	...	93.4	81.4	59.0	36.7	127	18	...	64
63	112	99	112	40.9	...	92.8	81.1	58.4	35.7	125	18	...	63
62	110	98	110	40.4	...	92.2	80.8	57.7	34.7	124	18	...	62
61	108	96	108	40.0	...	91.7	80.5	57.0	33.7	122	...	...	61
60	107	95	107	39.5	...	91.1	80.1	56.4	32.7	120	17	...	60
59	106	94	106	39.0	...	90.5	79.8	55.7	31.7	118	...	...	59
58	104	92	104	38.6	...	90.0	79.5	55.0	30.7	117	...	...	58
57	103	91	103	38.1	...	89.4	79.2	54.4	29.7	115	...	...	57
56	101	90	101	37.7	...	88.8	78.8	53.7	28.7	114	...	...	56
55	100	89	100	37.2	...	88.2	78.5	53.0	27.7	112	...	...	55

(a) For carbon and alloy steels in the annealed, normalized, and quenched-and-tempered conditions; less accurate for cold worked condition and for austenitic steels. The values in parentheses are beyond normal range and are given for information only

Rockwell C hardness numbers(a)

Rockwell C-scale hardness	Vickers hardness	Brinell hardness, 3000-kg load, 10-mm ball		Rockwell hardness			Rockwell superficial hardness, superficial Brale indenter			Knoop hardness, 500-g load and greater	Scleroscope hardness	Tensile strength (approx), ksi	Rockwell C-scale hardness
		Standard ball	Tungsten carbide ball	A scale, 60-kg load, Brale indenter	B scale, 100-kg load, 1/16-in.-diam ball	D scale, 100-kg load, Brale indenter	15N scale, 15-kg load	30N scale, 30-kg load	45N scale, 45-kg load				
68	940	…	…	85.6	…	76.9	93.2	84.4	75.4	920	97	…	68
67	900	…	…	85.0	…	76.1	92.9	83.6	74.2	895	95	…	67
66	865	…	…	84.5	…	75.4	92.5	82.8	73.3	870	92	…	66
65	832	…	(739)	83.9	…	74.5	92.2	81.9	72.0	846	91	…	65
64	800	…	(722)	83.4	…	73.8	91.8	81.1	71.0	822	88	…	64
63	772	…	(705)	82.8	…	73.0	91.4	80.1	69.9	799	87	…	63
62	746	…	(688)	82.3	…	72.2	91.1	79.3	68.8	776	85	…	62
61	720	…	(670)	81.8	…	71.5	90.7	78.4	67.7	754	83	…	61
60	697	…	(654)	81.2	…	70.7	90.2	77.5	66.6	732	81	…	60
59	674	…	(634)	80.7	…	69.9	89.8	76.6	65.5	710	80	351	59
58	653	…	615	80.1	…	69.2	89.3	75.7	64.3	690	78	338	58
57	633	…	595	79.6	…	68.5	88.9	74.8	63.2	670	76	325	57
56	613	…	577	79.0	…	67.7	88.3	73.9	62.0	650	75	313	56
55	595	…	560	78.5	…	66.9	87.9	73.0	60.9	630	74	301	55
54	577	…	543	78.0	…	66.1	87.4	72.0	59.8	612	72	292	54
53	560	…	525	77.4	…	65.4	86.9	71.2	58.6	594	71	283	53
52	544	(500)	512	76.8	…	64.6	86.4	70.2	57.4	576	69	273	52
51	528	(487)	496	76.3	…	63.8	85.9	69.4	56.1	558	68	264	51
50	513	(475)	481	75.9	…	63.1	85.5	68.5	55.0	542	67	255	50
49	498	(464)	469	75.2	…	62.1	85.0	67.6	53.8	526	66	246	49
48	484	(451)	455	74.7	…	61.4	84.5	66.7	52.5	510	64	238	48
47	471	442	443	74.1	…	60.8	83.9	65.8	51.4	495	63	229	47
46	458	432	432	73.6	…	60.0	83.5	64.8	50.3	480	62	221	46
45	446	421	421	73.1	…	59.2	83.0	64.0	49.0	466	60	215	45
44	434	409	409	72.5	…	58.5	82.5	63.1	47.8	452	58	208	44
43	423	400	400	72.0	…	57.7	82.0	62.2	46.7	438	57	201	43
42	412	390	390	71.5	…	56.9	81.5	61.3	45.5	426	56	194	42
41	402	381	381	70.9	…	56.2	80.9	60.4	44.3	414	55	188	41
40	392	371	371	70.4	…	55.4	80.4	59.5	43.1	402	54	182	40
39	382	362	362	69.9	…	54.6	79.9	58.6	41.9	391	52	177	39
38	372	353	353	69.4	…	53.8	79.4	57.7	40.8	380	51	171	38
37	363	344	344	68.9	…	53.1	78.8	56.8	39.6	370	50	166	37
36	354	336	336	68.4	(109.0)	52.3	78.3	55.9	38.4	360	49	161	36
35	345	327	327	67.9	(108.5)	51.5	77.7	55.0	37.2	351	48	157	35
34	336	319	319	67.4	(108.0)	50.8	77.2	54.2	36.1	342	47	153	34
33	327	311	311	66.8	(107.5)	50.0	76.6	53.3	34.9	334	46	149	33
32	318	301	301	66.3	(107.0)	49.2	76.1	52.1	33.7	326	44	145	32
31	310	294	294	65.8	(106.0)	48.4	75.6	51.3	32.5	318	43	141	31
30	302	286	286	65.3	(105.5)	47.7	75.0	50.4	31.3	311	42	138	30
29	294	279	279	64.7	(104.5)	47.0	74.5	49.5	30.1	304	41	135	29
28	286	271	271	64.3	(104.0)	46.1	73.9	48.6	28.9	297	40	131	28
27	279	264	264	63.8	(103.0)	45.2	73.3	47.7	27.8	290	39	128	27
26	272	258	258	63.3	(102.5)	44.6	72.8	46.8	26.7	284	38	125	26
25	266	253	253	62.8	(101.5)	43.8	72.2	45.9	25.5	278	38	122	25
24	260	247	247	62.4	(101.0)	43.1	71.6	45.0	24.3	272	37	119	24
23	254	243	243	62.0	(100.0)	42.1	71.0	44.0	23.1	266	36	117	23
22	248	237	237	61.5	99.0	41.6	70.5	43.2	22.0	261	35	114	22
21	243	231	231	61.0	98.5	40.9	69.9	42.3	20.7	256	35	112	21

(a) For carbon and alloy steels in the annealed, normalized, and quenched-and-tempered conditions; less accurate for cold worked condition and for austenitic steels. The values in bold-faced type correspond to the values in the joint SAE-ASM-ASTM hardness conversions as printed in ASTM E140. The values in parentheses are beyond normal range and are given for information only

Vickers hardness numbers(a)

Vickers hardness	Brinell hardness, 3000-kg load, 10-mm ball Standard ball	Brinell hardness, 3000-kg load, 10-mm ball Tungsten carbide ball	Rockwell hardness A scale, 60-kg load, Brale indenter	Rockwell hardness B scale, 100-kg load, 1/16-in. diam ball	Rockwell hardness C scale, 150-kg load, Brale indenter	Rockwell hardness D scale, 100-kg load, Brale indenter	Rockwell superficial 15N scale, 15-kg load	Rockwell superficial 30N scale, 30-kg load	Rockwell superficial 45N scale, 45-kg load	Knoop hardness, 500-g load and greater	Scleroscope hardness	Tensile strength (approx), ksi	Vickers hardness
940	…	…	85.6	…	68.0	76.9	93.2	84.4	75.4	920	97	…	940
920	…	…	85.3	…	67.5	76.5	93.0	84.0	74.8	908	96	…	920
900	…	…	85.0	…	67.0	76.1	92.9	83.6	74.2	895	95	…	900
880	…	(767)	84.7	…	66.4	75.7	92.7	83.1	73.6	882	93	…	880
860	…	(757)	84.4	…	65.9	75.3	92.5	82.7	73.1	867	92	…	860
840	…	(745)	84.1	…	65.3	74.8	92.3	82.2	72.2	852	91	…	840
820	…	(733)	83.8	…	64.7	74.3	92.1	81.7	71.8	837	90	…	820
800	…	(722)	83.4	…	64.0	73.8	91.8	81.1	71.0	822	88	…	800
780	…	(710)	83.0	…	63.3	73.3	91.5	80.4	70.2	806	87	…	780
760	…	(698)	82.6	…	62.5	72.6	91.2	79.7	69.4	788	86	…	760
740	…	(684)	82.2	…	61.8	72.1	91.0	79.1	68.6	772	84	…	740
720	…	(670)	81.8	…	61.0	71.5	90.7	78.4	67.7	754	83	…	720
700	…	(656)	81.3	…	60.1	70.8	90.3	77.6	66.7	735	81	…	700
690	…	(647)	81.1	…	59.7	70.5	90.1	77.2	66.2	725	…	…	690
680	…	(638)	80.8	…	59.2	70.1	89.8	76.8	65.7	716	80	355	680
670	…	(630)	80.6	…	58.8	69.8	89.7	76.4	65.3	706	…	348	670
660	…	620	80.3	…	58.3	69.4	89.5	75.9	64.7	697	79	342	660
650	…	611	80.0	…	57.8	69.0	89.2	75.5	64.1	687	78	336	650
640	…	601	79.8	…	57.3	68.7	89.0	75.1	63.5	677	77	328	640
630	…	591	79.5	…	56.8	68.3	88.8	74.6	63.0	667	76	323	630
620	…	582	79.2	…	56.3	67.9	88.5	74.2	62.4	657	75	317	620
610	…	573	78.9	…	55.7	67.5	88.2	73.6	61.7	646	…	310	610
600	…	564	78.6	…	55.2	67.0	88.0	73.2	61.2	636	74	303	600
590	…	554	78.4	…	54.7	66.7	87.8	72.7	60.5	625	73	298	590
580	…	545	78.0	…	54.1	66.2	87.5	72.1	59.9	615	72	293	580
570	…	535	77.8	…	53.6	65.8	87.2	71.7	59.3	604	…	288	570
560	…	525	77.4	…	53.0	65.4	86.9	71.2	58.6	594	71	283	560
550	(505)	517	77.0	…	52.3	64.8	86.6	70.5	57.8	583	70	276	550
540	(496)	507	76.7	…	51.7	64.4	86.3	70.0	57.0	572	69	270	540
530	(488)	497	76.4	…	51.1	63.9	86.0	69.5	56.2	561	68	265	530
520	(480)	488	76.1	…	50.5	63.5	85.7	69.0	55.6	550	67	260	520
510	(473)	479	75.7	…	49.8	62.9	85.4	68.3	54.7	539	…	254	510
500	(465)	471	75.3	…	49.1	62.2	85.0	67.7	53.9	528	66	247	500
490	(456)	460	74.9	…	48.4	61.6	84.7	67.1	53.1	517	65	241	490
480	(448)	452	74.5	…	47.7	61.3	84.3	66.4	52.2	505	64	235	480
470	441	442	74.1	…	46.9	60.7	83.9	65.7	51.3	494	…	228	470
460	433	433	73.6	…	46.1	60.1	83.6	64.9	50.4	482	62	223	460
450	425	425	73.3	…	45.3	59.4	83.2	64.3	49.4	471	…	217	450
440	415	415	72.8	…	44.5	58.8	82.8	63.5	48.4	459	59	212	440
430	405	405	72.3	…	43.6	58.2	82.3	62.7	47.4	447	58	205	430
420	397	397	71.8	…	42.7	57.5	81.8	61.9	46.4	435	57	199	420
410	388	388	71.4	…	41.8	56.8	81.4	61.1	45.3	423	56	193	410
400	379	379	70.8	…	40.8	56.0	80.8	60.2	44.1	412	55	187	400
390	369	369	70.3	…	39.8	55.2	80.3	59.3	42.9	400	…	181	390

(a) For carbon and alloy steels in the annealed, normalized, and quenched-and-tempered conditions; less accurate for cold worked condition and for austenitic steels. The values in bold-faced type correspond to the values in the joint SAE-ASM-ASTM hardness conversions as printed in ASTM E140. The values in parentheses are beyond normal range and are given for information only

Vickers hardness numbers(a) (continued)

Vickers hardness	Brinell hardness, 3000-kg load, 10-mm ball — Standard ball	Brinell hardness, 3000-kg load, 10-mm ball — Tungsten carbide ball	Rockwell hardness — A scale, 60-kg load, Brale indenter	Rockwell hardness — B scale, 100-kg load, 1/16-in. diam ball	Rockwell hardness — C scale, 150-kg load, Brale indenter	Rockwell hardness — D scale, 100-kg load, Brale indenter	Rockwell superficial hardness, superficial Brale indenter — 15N scale, 15-kg load	Rockwell superficial hardness, superficial Brale indenter — 30N scale, 30-kg load	Rockwell superficial hardness, superficial Brale indenter — 45N scale, 45-kg load	Knoop hardness, 500-g load and greater	Scleroscope hardness	Tensile strength (approx), ksi	Vickers hardness
380	360	360	69.8	(110.0)	38.8	54.4	79.8	58.4	41.7	389	52	175	380
370	350	350	69.2	...	37.7	53.6	79.2	57.4	40.4	378	51	170	370
360	341	341	68.7	(109.0)	36.6	52.8	78.6	56.4	39.1	367	50	164	360
350	331	331	68.1	...	35.5	51.9	78.0	55.4	37.8	356	48	159	350
340	322	322	67.6	(108.0)	34.4	51.1	77.4	54.4	36.5	346	47	155	340
330	313	313	67.0	...	33.3	50.2	76.8	53.6	35.2	337	46	150	330
320	303	303	66.4	(107.0)	32.2	49.4	76.2	52.3	33.9	328	45	146	320
310	294	294	65.8	...	31.0	48.4	75.6	51.3	32.5	318	...	142	310
300	284	284	65.2	(105.5)	29.8	47.5	74.9	50.2	31.1	309	42	138	300
295	280	280	64.8	...	29.2	47.1	74.6	49.7	30.4	305	...	136	295
290	275	275	64.5	(104.5)	28.5	46.5	74.2	49.0	29.5	300	41	133	290
285	270	270	64.2	...	27.8	46.0	73.8	48.4	28.7	296	...	131	285
280	265	265	63.8	(103.5)	27.1	45.3	73.4	47.8	27.9	291	40	129	280
275	261	261	63.5	...	26.4	44.9	73.0	47.2	27.1	286	39	127	275
270	256	256	63.1	(102.0)	25.6	44.3	72.6	46.4	26.2	282	38	124	270
265	252	252	62.7	...	24.8	43.7	72.1	45.7	25.2	277	...	122	265
260	247	247	62.4	(101.0)	24.0	43.1	71.6	45.0	24.3	272	37	120	260
255	243	243	62.0	...	23.1	42.2	71.1	44.2	23.2	267	...	117	255
250	238	238	61.6	99.5	22.2	41.7	70.6	43.4	22.2	262	36	115	250
245	233	233	61.2	...	21.3	41.1	70.1	42.5	21.1	258	35	113	245
240	228	228	60.7	98.1	20.3	40.3	69.6	41.7	19.9	253	34	111	240
230	219	219	...	96.7	(18.0)	...	...	...	...	243	33	106	230
220	209	209	...	95.0	(15.7)	...	...	...	...	234	32	101	220
210	200	200	...	93.4	(13.4)	...	...	...	...	226	30	97	210
200	190	190	...	91.5	(11.0)	...	...	...	...	216	29	92	200
190	181	181	...	89.5	(8.5)	...	...	...	...	206	28	88	190
180	171	171	...	87.1	(6.0)	...	...	...	...	196	26	84	180
170	162	162	...	85.0	(3.0)	...	...	...	...	185	25	79	170
160	152	152	...	81.7	(0.0)	...	...	...	...	175	23	75	160
150	143	143	...	78.7	...	...	...	...	...	164	22	71	150
140	133	133	...	75.0	...	...	...	...	...	154	21	66	140
130	124	124	...	71.2	...	...	...	...	...	143	20	62	130
120	114	114	...	66.7	...	...	...	...	...	133	18	57	120
110	105	105	...	62.3	...	...	...	...	...	123	...	...	110
100	95	95	...	56.2	...	...	...	...	...	112	...	...	100
95	90	90	...	52.0	...	...	...	...	...	107	...	...	95
90	86	86	...	48.0	...	...	...	...	...	102	...	...	90
85	81	81	...	41.0	...	...	...	...	...	97	...	...	85

(a) For carbon and alloy steels in the annealed, normalized, and quenched-and-tempered conditions: less accurate for cold worked condition and for austenitic steels. The values in bold-faced type correspond to the values in the joint SAE-ASM-ASTM hardness conversions as printed in ASTM E140. The values in parentheses are beyond normal range and are given for information only

Brinell hardness numbers(a)

Brinell indentation diam, mm	Brinell 3000-kg, Standard ball	Brinell 3000-kg, Tungsten carbide ball	Vickers hardness	Rockwell A scale, 60-kg, Brale indenter	Rockwell B scale, 100-kg, 1/16-in. diam ball	Rockwell C scale, 150-kg, Brale indenter	Rockwell D scale, 100-kg, Brale indenter	Rockwell superficial 15N scale, 15-kg load	Rockwell superficial 30N scale, 30-kg load	Rockwell superficial 45N scale, 45-kg load	Knoop hardness, 500-g load and greater	Scleroscope hardness	Tensile strength (approx), ksi	Brinell indentation diam, mm
2.25	...	(745)	840	84.1	...	65.3	74.8	92.3	82.2	72.2	852	91	...	2.25
2.30	...	(712)	783	83.1	...	63.4	73.4	91.6	80.5	70.4	808	...	...	2.30
2.35	...	(682)	737	82.2	...	61.7	72.0	91.0	79.0	68.5	768	84	...	2.35
2.40	...	(653)	697	81.2	...	60.0	70.7	90.2	77.5	66.5	732	81	...	2.40
2.45	...	627	667	80.5	...	58.7	69.7	89.6	76.3	65.1	703	79	347	2.45
2.50	...	601	640	79.8	...	57.3	68.7	89.0	75.1	63.5	677	77	328	2.50
2.55	...	578	615	79.1	...	56.0	67.7	88.4	73.9	62.1	652	75	313	2.55
2.60	...	555	591	78.4	...	54.7	66.7	87.8	72.7	60.6	626	73	298	2.60
2.65	...	534	569	77.8	...	53.5	65.8	87.2	71.6	59.2	604	71	288	2.65
2.70	...	514	547	76.9	...	52.1	64.7	86.5	70.3	57.6	579	70	273	2.70
2.75	(495)	495	539	76.7	...	51.6	64.3	86.3	69.9	56.9	571	...	269	2.75
			528	76.3	...	51.0	63.8	85.9	69.4	56.1	558	68	263	
2.80	(477)	477	516	75.9	...	50.3	63.2	85.6	68.7	55.2	545	...	257	2.80
			508	75.6	...	49.6	62.7	85.3	68.2	54.5	537	66	252	
2.85	(461)	461	495	75.1	...	48.8	61.9	84.9	67.4	53.5	523	...	244	2.85
			491	74.9	...	48.5	61.7	84.7	67.2	53.2	518	65	242	
2.90	444	444	474	74.3	...	47.2	61.0	84.1	66.0	51.7	499	...	231	2.90
			472	74.2	...	47.1	60.8	84.0	65.8	51.5	496	63	229	
2.95	429	429	455	73.4	...	45.7	59.7	83.4	64.6	49.9	476	61	220	2.95
3.00	415	415	440	72.8	...	44.5	58.8	82.8	63.5	48.4	459	59	212	3.00
3.05	401	401	425	72.0	...	43.1	57.8	82.0	62.3	46.9	441	58	202	3.05
3.10	388	388	410	71.4	...	41.8	56.8	81.4	61.1	45.3	423	56	193	3.10
3.15	375	375	396	70.6	...	40.4	55.7	80.6	59.9	43.6	407	54	184	3.15
3.20	363	363	383	70.0	(110.0)	39.1	54.6	80.0	58.7	42.0	392	52	177	3.20
3.25	352	352	372	69.3	(109.0)	37.9	53.8	79.3	57.6	40.5	379	51	172	3.25
3.30	341	341	360	68.7	(108.5)	36.6	52.8	78.6	56.4	39.1	367	50	164	3.30
3.35	331	331	350	68.1	(108.0)	35.5	51.9	78.0	55.4	37.8	356	48	159	3.35
3.40	321	321	339	67.5	(108.0)	34.3	51.0	77.3	54.3	36.4	345	47	154	3.40
3.45	311	311	328	66.9	(107.5)	33.1	50.0	76.7	53.3	34.4	336	46	149	3.45
3.50	302	302	319	66.3	(107.0)	32.1	49.3	76.1	52.2	33.8	327	45	146	3.50
3.55	293	293	309	65.7	(106.0)	30.9	48.3	75.5	51.2	32.4	318	43	142	3.55
3.60	285	285	301	65.3	(105.5)	29.9	47.6	75.0	50.3	31.2	310	42	138	3.60
3.65	277	277	292	64.6	(105.0)	28.8	46.7	74.4	49.3	29.9	302	41	134	3.65
3.70	269	269	284	64.1	(104.5)	27.6	45.9	73.7	48.3	28.5	294	40	131	3.70
3.75	262	262	276	63.6	(104.0)	26.6	45.0	73.1	47.3	27.3	286	39	127	3.75

(a) For carbon and alloy steels in the annealed, normalized, and quenched-and-tempered conditions; less accurate for cold worked condition and for austenitic steels. Values in bold-faced type correspond to the values in the joint SAE-ASM-ASTM hardness conversions as printed in ASTM E140. Values in parentheses are beyond normal range and are given for information only. (b) Brinell numbers are based on the diameter of impressed indentation. If the ball distorts (flattens) during test, the relationship between Brinell and Vickers or Rockwell scales will vary in accordance with the degree of such distortion when related to hardnesses determined with a Vickers diamond pyramid, Rockwell Brale, or other indenter that does not sensibly distort. At high hardnesses, therefore, Brinell numbers will vary in accordance with the type of ball used. Standard steel balls tend to flatten slightly more than tungsten carbide balls, resulting in a larger indentation and a lower Brinell number than shown by a tungsten carbide ball. Thus, on a specimen of about 539 to 547 HV, a standard ball will leave a 2.75-mm indentation (495 HB), and a tungsten carbide ball a 2.70-mm indentation (514 HB). Conversely, identical indentation diameters for both types of ball will correspond to different Vickers and Rockwell values. Thus, if indentations in two different specimens both are 2.75 mm in diameter (495 HB), the specimen tested with a tungsten carbide ball has a Vickers hardness of 539, whereas the specimen tested with a standard ball has a Vickers hardness of 528

Brinell hardness numbers(a) (continued)

Brinell indentation diam, mm	Brinell hardness(b), 3000-kg load, 10-mm ball Standard ball	Tungsten carbide ball	Vickers hardness	A scale, 60-kg load, Brale indenter	B scale, 100-kg load, 1/16-in.-diam ball	C scale, 150-kg load, Brale indenter	D scale, 100-kg load, Brale indenter	Rockwell superficial hardness, superficial Brale indenter 15N scale, 15-kg load	30N scale, 30-Kg load	45N scale, 45-Kg load	Knoop hardness, 500-g load and greater	Scleroscope hardness	Tensile strength (approx), ksi	Brinell indentation diam, mm
3.80	255	255	269	63.0	(102.0)	25.4	44.2	72.5	46.2	26.0	279	38	123	3.80
3.85	248	248	261	62.5	(101.1)	24.2	43.2	71.7	45.1	24.5	272	37	120	3.85
3.90	241	241	253	61.8	100.0	22.8	42.0	70.9	43.9	22.8	265	36	116	3.90
3.95	235	235	247	61.4	99.0	21.7	41.4	70.3	42.9	21.5	259	35	114	3.95
4.00	229	229	241	60.8	98.2	20.5	40.5	69.7	41.9	20.1	253	34	111	4.00
4.05	223	223	234	…	97.3	(19.0)	…	…	…	…	247	…	107	4.05
4.10	217	217	228	…	96.4	(17.7)	…	…	…	…	242	33	105	4.10
4.15	212	212	222	…	95.5	(16.4)	…	…	…	…	237	32	102	4.15
4.20	207	207	218	…	94.6	(15.2)	…	…	…	…	232	31	100	4.20
4.25	201	201	212	…	93.7	(13.8)	…	…	…	…	227	…	98	4.25
4.30	197	197	207	…	92.8	(12.7)	…	…	…	…	222	30	95	4.30
4.35	192	192	202	…	91.9	(11.5)	…	…	…	…	217	29	93	4.35
4.40	187	187	196	…	90.9	(10.2)	…	…	…	…	212	…	90	4.40
4.45	183	183	192	…	90.0	(9.0)	…	…	…	…	207	28	89	4.45
4.50	179	179	188	…	89.0	(8.0)	…	…	…	…	202	27	87	4.50
4.55	174	174	182	…	88.0	(6.7)	…	…	…	…	198	…	85	4.55
4.60	170	170	178	…	87.0	(5.4)	…	…	…	…	194	26	83	4.60
4.65	167	167	175	…	86.0	(4.4)	…	…	…	…	190	…	81	4.65
4.70	163	163	171	…	85.0	(3.3)	…	…	…	…	186	25	79	4.70
4.75	159	159	167	…	83.9	(2.0)	…	…	…	…	182	…	78	4.75
4.80	156	156	163	…	82.9	(0.9)	…	…	…	…	178	24	76	4.80
4.85	152	152	159	…	81.9	…	…	…	…	…	174	…	75	4.85
4.90	149	149	156	…	80.8	…	…	…	…	…	170	23	73	4.90
4.95	146	146	153	…	79.7	…	…	…	…	…	166	…	72	4.95
5.00	143	143	150	…	78.6	…	…	…	…	…	163	22	71	5.00
5.10	137	137	143	…	76.4	…	…	…	…	…	157	21	67	5.10
5.20	131	131	137	…	74.2	…	…	…	…	…	151	…	65	5.20
5.30	126	126	132	…	72.0	…	…	…	…	…	145	20	63	5.30
5.40	121	121	127	…	69.8	…	…	…	…	…	140	19	60	5.40
5.50	116	116	122	…	67.6	…	…	…	…	…	135	18	58	5.50
5.60	111	111	117	…	65.4	…	…	…	…	…	131	17	56	5.60

(a) For carbon and alloy steels in the annealed, normalized, and quenched-and-tempered conditions; less accurate for cold worked condition and for austenitic steels. Values in bold-faced type correspond to the values in the joint SAE-ASM-ASTM hardness conversions as printed in ASTM E140. Values in parantheses are beyond normal range and are given for information only. (b) Brinell numbers are based on the diameter of impressed indentation. If the ball distorts (flattens) during test, Brinell numbers will vary in accordance with the degree of such distortion when related to hardnesses determined with a Vickers diamond pyramid, Rockwell Brale, or other indenter that does not sensibly distort. At high hardnesses, therefore, the relationship between Brinell and Vickers or Rockwell scales is affected by the type of ball used. Standard steel balls tend to flatten slightly more than tungsten carbide balls, resulting in a larger indentation and a lower Brinell number than shown by a tungsten carbide ball. Thus, on a specimen of about 539 to 547 HV, a standard ball will leave a 2.75-mm indentation (495 HB), and a tungsten carbide ball a 2.70-mm indentation (514 HB). Conversely, identical indentation diameters for both types of ball will correspond to different Vickers and Rockwell values. Thus, if indentations in two different specimens both are 2.75 mm in diameter (495 HB), the specimen tested with a standard ball has a Vickers hardness of 539, whereas the specimen tested with a tungsten carbide ball has a Vickers hardness of 528

Diamond Pyramid Hardness Numbers

Diamond pyramid hardness numbers are obtained when the 136° diamond pyramid indenter is used on microhardness testers at any test load. This indenter is cut in the shape of a square-based pyramid having an apex angle of 136°. Both diagonals of the indentation are measured and the average is used to compute the diamond pyramid number. This number is defined as the load per unit area of surface contact in kilograms per square millimeter and is calculated from the formula:

$$DPN = \frac{2L \sin\frac{a}{2}}{d^2}$$

where DPN is the diamond pyramid number; L is the load in kilograms applied to the indenter; a is the 136° apex angle; and d is the length of average diagonal in millimeters.

The diamond pyramid number corresponding to a measured average length of diagonal, d, for an applied load of 1 g may be obtained directly from the following table. To obtain the DPN for any other applied load, multiply the number obtained for length, d, for 1 g from the table by the actual applied load in grams used to make the indentation. As an alternative method and in order to avoid the necessity of making interpolations, it is possible to multiply the average length of the diagonals by 10 and then look up the number of the table corresponding to this length. Multiply this number by 100 times the applied load in grams used to make the indentation to find the proper diamond pyramid number.

Example: Direct Procedure. A specimen tested with an applied load of 1000 g is measured and shows an average length for the two diagonals of 50.5 μm. By interpolation a number is obtained from the table of 0.7272. Since an applied load of 1000 g was used while the table was computed using a load of 1 g, multiply this number by 1000 g giving a DPN of 727.2 for the material.

Example: Alternative Procedure. Multiply the length of the average diagonal, 50.5 μm, by 10, giving a length of 505 μm. This length gives a number of 0.00727 on the table and multiplying this number by 100×1000 g obtains a DPN of 727.

Diamond pyramid hardness numbers
Computed using a 1-g load

Average diagonal length, μm	Applied load, g									
	0	1	2	3	4	5	6	7	8	9
00	...	1854.	463.6	206.0	115.9	74.16	51.51	37.84	28.97	22.89
10	18.54	15.33	12.88	10.97	9.461	8.242	7.244	6.416	5.723	5.137
20	4.636	4.205	3.831	3.505	3.219	2.967	2.743	2.544	2.365	2.205
30	2.060	1.930	1.811	1.703	1.604	1.514	1.431	1.355	1.284	1.219
40	1.159	1.103	1.051	1.003	0.9578	0.9157	0.8764	0.8395	0.8048	0.7723
50	0.7416	0.7128	0.6857	0.6600	0.6358	0.6129	0.5912	0.5706	0.5511	0.5326
60	0.5150	0.4983	0.4823	0.4671	0.4526	0.4388	0.4256	0.4130	0.4010	0.3894
70	0.3784	0.3678	0.3576	0.3479	0.3386	0.3296	0.3210	0.3127	0.3047	0.2971
80	0.2897	0.2826	0.2757	0.2691	0.2628	0.2566	0.2507	0.2449	0.2394	0.2341
90	0.2289	0.2239	0.2190	0.2144	0.2098	0.2054	0.2012	0.1970	0.1930	0.1892
100	0.1854	0.1817	0.1782	0.1748	0.1714	0.1682	0.1650	0.1619	0.1590	0.1560
110	0.1533	0.1505	0.1478	0.1452	0.1427	0.1402	0.1378	0.1354	0.1332	0.1310
120	0.1288	0.1267	0.1246	0.1226	0.1206	0.1187	0.1168	0.1150	0.1132	0.1115
130	0.1097	0.1081	0.1064	0.1048	0.1033	0.1018	0.1003	0.0988	0.974	0.0960
140	0.0649	0.0933	0.0920	0.0907	0.0894	0.0882	0.0870	0.0858	0.0847	0.0835
150	0.0824	0.0813	0.0803	0.0792	0.0782	0.0772	0.0762	0.0752	0.0743	0.0734
160	0.0724	0.0715	0.0707	0.0698	0.0690	0.0681	0.0673	0.0665	0.0657	0.0647
170	0.0642	0.0634	0.0627	0.0620	0.0613	0.0606	0.0599	0.0592	0.0585	0.0579
180	0.0572	0.0566	0.0560	0.0554	0.0548	0.0542	0.0536	0.0530	0.0525	0.0519
190	0.0514	0.0508	0.0503	0.0498	0.0493	0.0488	0.0483	0.0478	0.0473	0.0468
200	0.0464	0.0459	0.0455	0.0450	0.0446	0.0442	0.0437	0.0433	0.0429	0.0425
210	0.0421	0.0417	0.0413	0.0409	0.0405	0.0401	0.0397	0.0394	0.0390	0.0387
220	0.0383	0.0380	0.0376	0.0373	0.0370	0.0366	0.0363	0.0360	0.0357	0.0354
230	0.0351	0.0348	0.0345	0.0342	0.0339	0.0336	0.0333	0.0330	0.0327	0.0325
240	0.0322	0.0319	0.0317	0.0314	0.0312	0.0309	0.0306	0.0304	0.0302	0.0299
250	0.0297	0.0294	0.0292	0.0289	0.0287	0.0285	0.0283	0.0281	0.0279	0.0276
260	0.0274	0.0272	0.0270	0.0268	0.0266	0.0264	0.0262	0.0260	0.0258	0.0256
270	0.0254	0.0253	0.0251	0.0249	0.0247	0.0245	0.0243	0.0242	0.0240	0.0238
280	0.0236	0.0235	0.0233	0.0232	0.0230	0.0228	0.0227	0.0225	0.0224	0.0222
290	0.0221	0.0219	0.0218	0.0216	0.0215	0.0213	0.0212	0.0210	0.0209	0.0207
300	0.0206	0.0205	0.0203	0.0202	0.0201	0.0199	0.0198	0.0197	0.0196	0.0194
310	0.0193	0.0192	0.0191	0.0189	0.0188	0.0187	0.0186	0.0185	0.0183	0.0182
320	0.0181	0.0180	0.0179	0.0178	0.0177	0.0176	0.0175	0.0173	0.0172	0.0171

(continued)

Diamond pyramid hardness numbers (continued)
Computed using a 1-g load

Average diagonal length, μm	Applied load, g									
	0	1	2	3	4	5	6	7	8	9
330	0.0170	0.0169	0.0168	0.0167	0.0166	0.0165	0.0164	0.0163	0.0162	0.0161
340	0.0160	0.0160	0.0159	0.0158	0.0157	0.0156	0.0155	0.0154	0.0153	0.0152
350	0.01514	0.01505	0.01497	0.01488	0.01480	0.01471	0.01463	0.01455	0.01447	0.01439
360	0.01431	0.01423	0.01415	0.01407	0.01400	0.01392	0.01384	0.01377	0.01369	0.01362
370	0.01355	0.01347	0.01340	0.01333	0.01326	0.01319	0.01312	0.01305	0.01298	0.01291
380	0.01284	0.01277	0.01271	0.01264	0.01258	0.01251	0.01245	0.01238	0.01232	0.01226
390	0.01219	0.01213	0.01207	0.01201	0.01195	0.01189	0.01183	0.01177	0.01171	0.01165
400	0.01159	0.01153	0.01148	0.01142	0.01136	0.01131	0.01125	0.01119	0.01114	0.01109
410	0.01103	0.01098	0.01093	0.01087	0.01082	0.01077	0.01072	0.01066	0.01061	0.01056
420	0.01051	0.01046	0.01041	0.01036	0.01031	0.01027	0.01022	0.01017	0.01012	0.01008
430	0.01003	0.00998	0.00994	0.00989	0.00985	0.00980	0.00976	0.00971	0.00967	0.00962
440	0.00958	0.00953	0.00949	0.00945	0.00941	0.00936	0.00932	0.00928	0.00924	0.00920
450	0.00916	0.00912	0.00908	0.00904	0.00900	0.00896	0.00892	0.00888	0.00884	0.00880
460	0.00876	0.00873	0.00869	0.00865	0.00861	0.00858	0.00854	0.00850	0.00847	0.00843
470	0.00840	0.00836	0.00832	0.00829	0.00825	0.00822	0.00818	0.00815	0.00812	0.00808
480	0.00805	0.00802	0.00798	0.00795	0.00792	0.00788	0.00785	0.00782	0.00779	0.00776
490	0.00772	0.00769	0.00766	0.00763	0.00760	0.00757	0.00754	0.00751	0.00748	0.00745
500	0.00742	0.00739	0.00736	0.00733	0.00730	0.00727	0.00724	0.00721	0.00719	0.00716
510	0.00713	0.00710	0.00707	0.00705	0.00702	0.00699	0.00696	0.00694	0.00691	0.00688
520	0.00686	0.00683	0.00681	0.00678	0.00675	0.00673	0.00670	0.00668	0.00665	0.00663
530	0.00660	0.00658	0.00655	0.00653	0.00650	0.00648	0.00645	0.00643	0.00641	0.00638
540	0.00636	0.00634	0.00631	0.00629	0.00627	0.00624	0.00622	0.00620	0.00617	0.00615
550	0.00613	0.00611	0.00609	0.00606	0.00604	0.00602	0.00600	0.00598	0.00596	0.00593
560	0.00591	0.00589	0.00587	0.00585	0.00583	0.00581	0.00579	0.00577	0.00575	0.00573
570	0.00571	0.00569	0.00567	0.00565	0.00563	0.00561	0.00559	0.00557	0.00555	0.00553
580	0.00551	0.00549	0.00547	0.00546	0.00544	0.00542	0.00540	0.00538	0.00536	0.00534
590	0.00533	0.00531	0.00529	0.00527	0.00526	0.00524	0.00522	0.00520	0.00519	0.00517
600	0.00515	0.00513	0.00512	0.00510	0.00508	0.00507	0.00505	0.00503	0.00502	0.00500
610	0.00498	0.00497	0.00495	0.00494	0.00492	0.00490	0.00489	0.00487	0.00486	0.00484
620	0.00482	0.00481	0.00479	0.00478	0.00476	0.00475	0.00473	0.00472	0.00470	0.00469
630	0.00467	0.00466	0.00464	0.00463	0.00461	0.00460	0.00458	0.00457	0.00456	0.00454
640	0.00453	0.00451	0.00450	0.00448	0.00447	0.00446	0.00444	0.00443	0.00442	0.00440
650	0.00439	0.00438	0.00436	0.00435	0.00434	0.00432	0.00431	0.00430	0.00428	0.00427
660	0.00426	0.00424	0.00423	0.00422	0.00421	0.00419	0.00418	0.00417	0.00416	0.00414
670	0.00413	0.00412	0.00411	0.00409	0.00408	0.00407	0.00406	0.00405	0.00403	0.00402
680	0.00401	0.00400	0.00399	0.00398	0.00396	0.00395	0.00394	0.00393	0.00392	0.00391
690	0.00390	0.00388	0.00387	0.00386	0.00385	0.00384	0.00383	0.00382	0.00381	0.00380
700	0.00378	0.00377	0.00376	0.00375	0.00374	0.00373	0.00372	0.00371	0.00370	0.00369
710	0.00368	0.00367	0.00366	0.00365	0.00364	0.00363	0.00362	0.00361	0.00360	0.00359
720	0.00358	0.00357	0.00356	0.00355	0.00354	0.00353	0.00352	0.00351	0.00350	0.00349
730	0.00348	0.00347	0.00346	0.00345	0.00344	0.00343	0.00342	0.00341	0.00340	0.00340
740	0.00339	0.00338	0.00337	0.00336	0.00335	0.00334	0.00333	0.00332	0.00331	0.00331
750	0.00330	0.00329	0.00328	0.00327	0.00326	0.00325	0.00324	0.00324	0.00323	0.00322
760	0.00321	0.00320	0.00319	0.00318	0.00318	0.00317	0.00316	0.00315	0.00314	0.00314
770	0.00313	0.00312	0.00311	0.00310	0.00309	0.00309	0.00308	0.00307	0.00307	0.00306
780	0.00305	0.00304	0.00303	0.00303	0.00302	0.00301	0.00300	0.00299	0.00299	0.00298
790	0.00297	0.00296	0.00296	0.00295	0.00294	0.00293	0.00293	0.00292	0.00291	0.00291
800	0.00290	0.00289	0.00288	0.00288	0.00287	0.00287	0.00286	0.00285	0.00284	0.00283
810	0.00283	0.00282	0.00281	0.00280	0.00280	0.00279	0.00278	0.00278	0.00277	0.00277
820	0.00276	0.00275	0.00274	0.00274	0.00273	0.00273	0.00272	0.00271	0.00270	0.00270
830	0.00269	0.00268	0.00268	0.00267	0.00267	0.00266	0.00265	0.00265	0.00264	0.00263
840	0.00263	0.00262	0.00262	0.00261	0.00260	0.00260	0.00259	0.00258	0.00258	0.00257
850	0.00257	0.00256	0.00256	0.00255	0.00254	0.00254	0.00253	0.00253	0.00252	0.00251
860	0.00251	0.00250	0.00250	0.00249	0.00248	0.00248	0.00247	0.00247	0.00246	0.00246
870	0.00245	0.00244	0.00244	0.00243	0.00243	0.00242	0.00242	0.00241	0.00241	0.00240
880	0.00240	0.00239	0.00238	0.00238	0.00237	0.00237	0.00236	0.00236	0.00235	0.00235
890	0.00234	0.00234	0.00233	0.00233	0.00232	0.00232	0.00231	0.00230	0.00230	0.00229
900	0.00229	0.00228	0.00228	0.00227	0.00227	0.00226	0.00226	0.00225	0.00225	0.00224
910	0.00224	0.00223	0.00223	0.00223	0.00222	0.00222	0.00221	0.00221	0.00220	0.00220
920	0.00219	0.00219	0.00218	0.00218	0.00217	0.00217	0.00216	0.00216	0.00215	0.00215
930	0.00214	0.00214	0.00214	0.00213	0.00213	0.00212	0.00212	0.00211	0.00211	0.00210
940	0.00210	0.00209	0.00209	0.00208	0.00208	0.00208	0.00207	0.00207	0.00206	0.00206
950	0.00205	0.00205	0.00205	0.00204	0.00204	0.00203	0.00203	0.00202	0.00202	0.00202
960	0.00201	0.00201	0.00200	0.00200	0.00200	0.00199	0.00199	0.00198	0.00198	0.00198
970	0.00197	0.00197	0.00196	0.00196	0.00196	0.00195	0.00195	0.00194	0.00194	0.00194
980	0.00193	0.00193	0.00192	0.00192	0.00192	0.00191	0.00191	0.00190	0.00190	0.00190
990	0.00189	0.00189	0.00188	0.00188	0.00188	0.00187	0.00187	0.00187	0.00186	0.00186
1000	0.00185									

Source: Torsion Balance Co., Clifton, N.J.

Knoop Hardness Numbers

Knoop hardness numbers are obtained when the Knoop diamond indenter is used with microhardness testers at any test load. These numbers will vary for the same material according to the test load used. Therefore, in all cases the test load must be specified. The Kentron tester will apply loads of from 1 to 10,000 g, but a range of 1 to 1000 g is ordinarily used with the Knoop indenter.

The Knoop indenter is cut in the shape of a diamond-based pyramid giving a diamond-shaped impression in which the long diagonal is very close to seven times the length of the short diagonal. The included longitudinal angle measured from edge to edge is 172° 30′ and the transverse angle is 130° 00′. Because of the difference in the lengths of the two diagonals, almost all of the elastic recovery of the indentations made with the Knoop indenter takes place in the transverse direction. Hence, the measurement of the long diagonal together with the computed indenter constant gives a very close approximation of the unrecovered projected area of the indentation in square millimeters. The relationship between the applied load in kilograms and the approximate unrecovered projected area in square millimeters is called the Knoop hardness number for the specimen for that applied load.

The Knoop hardness number may be expressed by the formula:

$$KN = \frac{L}{A_p} - \frac{L}{l^2 C_p}$$

where KN is the Knoop hardness number; L is the load in kilograms applied to the indenter; A_p is the unrecovered projected area in square millimeters; l is the measured length of the long diagonal of the indentation in millimeters; C_p is the constant relating l to the unrecovered projected area of the indentation. For an indenter with a longitudinal angle of 172° 30′ and a transverse angle of 130° 00′. C_p is 7.028×10^{-2}.

The Knoop hardness number corresponding to a measured length l for an applied load of 1 g may be obtained directly from the table. To obtain the Knoop hardness number for any other applied load, multiply the number obtained from the table for 1 g by the actual applied load in grams used to make the indentation.

Example. A specimen tested on a microhardness tester with an applied load of 100 g is measured under the microscope and shows a length of 42 μm for the long diagonal. Reference to the table shows that this length of diagonal would give a Knoop hardness number of 8.066. However, because an applied load of 100 g was used while the table was computed using a load of 1 g, multiply the number obtained from the table by 100 giving a Knoop hardness number of 806.6 for this material for an applied test load of 100 g.

Knoop hardness numbers
Computed using a 1-g load, for a theoretical indenter having a longitudinal angle of 172°30′ and a transverse angle of 130°00′ giving a constant for projected area (C_p) of 7.028×10^{-2}; a constant correction is provided to correct the Knoop number obtained to simulate a test made with an indenter with perfect angles.

Length of diagonal, μm	Applied load, g									
	0.0	0.1	0.2	0.3	0.4	0.5	0.6	0.7	0.8	0.9
1.0	14 229	11.759	9 881	8 419	7 260	6 324	5 558	4 923	4 392	3 942
2.0	3 557	3 226	2 940	2 690	2 470	2 227	2 105	1 952	1 815	1 692
3.0	1 581	1 481	1 390	1 307	1 231	1 162	1 098	1 039	985.4	935.5
4.0	889.3	846.5	806.6	769.5	735.0	702.7	672.4	644.1	617.6	592.6
5.0	569.2	547.1	526.2	506.5	488.0	470.4	453.7	437.9	423.0	408.8
6.0	395.2	382.4	370.2	358.5	347.4	336.8	326.7	317.0	307.7	298.9
7.0	290.4	282.3	274.5	267.0	259.8	253.0	246.4	240.0	233.9	228.0
8.0	222.3	216.9	211.6	206.5	201.7	196.9	192.4	188.0	183.7	179.6
9.0	175.7	171.8	168.1	164.5	161.0	157.7	154.4	151.2	148.2	145.2
10.0	142.3	139.5	136.8	134.1	131.6	129.1	126.6	124.3	122.0	119.8
11.0	117.6	115.5	113.4	111.4	109.5	107.6	105.7	103.9	102.2	100.5
12.0	98.81	97.18	95.60	94.05	92.54	91.06	89.62	88.22	86.85	85.50
13.0	84.19	82.91	81.66	80.44	79.24	78.07	76.93	75.81	74.72	73.64
14.0	72.60	71.57	70.57	69.58	68.62	67.68	66.75	65.85	64.96	64.09
15.0	63.24	62.40	61.58	60.78	59.99	59.22	58.47	57.33	57.00	56.28
16.0	55.58	54.89	54.22	53.55	52.90	52.26	51.64	51.02	50.41	49.82
17.0	49.23	48.66	48.10	47.54	47.00	46.64	45.93	45.42	44.91	44.41
18.0	43.92	43.44	42.96	42.49	42.02	41.57	41.13	40.69	40.26	39.83
19.0	39.42	39.00	38.60	38.20	37.81	37.42	37.04	36.66	36.29	35.93
20.0	35.57	35.22	34.87	34.53	34.19	33.86	33.53	33.21	32.89	32.57
21.0	32.27	31.96	31.66	31.36	31.07	30.78	30.50	30.22	29.94	29.67
22.0	29.40	29.13	28.87	28.61	28.36	28.11	27.86	27.61	27.37	27.13
23.0	26.90	26.67	26.44	26.21	25.99	25.77	25.55	25.33	25.12	24.91
24.0	24.70	24.50	24.30	24.10	23.90	23.71	23.51	23.32	23.14	22.95

(continued)

Knoop hardness numbers (continued)

Computed using a 1-g load, for a theoretical indenter having a longitudinal angle of 172°30′ and a transverse angle of 130°00′ giving a constant for projected area (C_p) of 7.028×10^{-2}; a constant correction is provided to correct the Knoop number obtained to simulate a test made with an indenter with perfect angles.

Length of diagonal, μm	Applied load, g									
	0.0	0.1	0.2	0.3	0.4	0.5	0.6	0.7	0.8	0.9
25.0	22.77	22.59	22.41	22.23	22.05	21.88	21.71	21.54	21.38	21.21
26.0	21.05	20.89	20.73	20.57	20.42	20.26	20.11	19.96	19.81	19.66
27.0	19.52	19.37	19.23	19.09	18.95	18.82	18.68	18.54	18.41	18.28
28.0	18.15	18.02	17.89	17.77	17.64	17.52	17.40	17.27	17.15	17.04
29.0	16.92	16.80	16.69	16.57	16.46	16.35	16.24	16.13	16.02	15.92
30.0	15.81	15.71	15.60	15.50	15.40	15.30	15.20	15.10	15.00	14.90
31.0	14.81	14.71	14.62	14.52	14.43	14.34	14.25	14.16	14.07	13.98
32.0	13.90	13.81	13.72	13.64	13.55	13.47	13.39	13.31	13.23	13.15
33.0	13.07	12.99	12.91	12.83	12.75	12.68	12.60	12.53	12.45	12.38
34.0	12.31	12.24	12.17	12.09	12.02	11.95	11.89	11.82	11.75	11.68
35.0	11.62	11.55	11.48	11.42	11.35	11.29	11.23	11.16	11.10	11.04
36.0	10.93	10.92	10.86	10.80	10.74	10.68	10.62	10.56	10.51	10.45
37.0	10.39	10.34	10.28	10.23	10.17	10.12	10.06	10.01	9.958	9.906
38.0	9.854	9.802	9.751	9.700	9.650	9.600	9.550	9.501	9.452	9.403
39.0	9.355	9.307	9.260	9.213	9.166	9.120	9.074	9.028	8.983	8.938
40.0	8.893	...	...	...	...	...	...	...	...	...

Length of diagonal, μm	Applied load, g									
	0	1	2	3	4	5	6	7	8	9
10	142.3	117.6	98.81	84.19	72.60	63.24	55.58	49.23	43.92	39.42
20	35.57	32.36	29.40	26.90	24.70	22.77	21.05	19.52	18.15	16.92
30	15.81	14.81	13.90	13.07	12.31	11.62	10.98	10.39	9.854	9.355
40	8.893	8.465	8.066	7.695	7.350	7.027	6.724	6.441	6.176	5.926
50	5.692	5.471	5.262	5.065	4.880	4.704	4.537	4.379	4.230	4.088
60	3.952	3.824	3.702	3.585	3.474	3.368	3.267	3.170	3.077	2.989
70	2.904	2.823	2.745	2.670	2.598	2.530	2.463	2.400	2.339	2.280
80	2.223	2.169	2.116	2.065	2.017	1.969	1.924	1.880	1.837	1.796
90	1.757	1.718	1.681	1.645	1.610	1.577	1.544	1.512	1.482	1.452
100	1.423	1.395	1.368	1.341	1.316	1.291	1.266	1.243	1.220	1.198
110	1.176	1.155	1.134	1.114	1.095	1.076	1.057	1.039	1.022	1.005
120	0.9881	0.9718	0.9560	0.9405	0.9254	0.9107	0.8962	0.8822	0.8685	0.8550
130	0.8419	0.8291	0.8166	0.8044	0.7924	0.7807	0.7693	0.7581	0.7472	0.7364
140	0.7260	0.7157	0.7057	0.6958	0.6862	0.6768	0.6675	0.6585	0.6496	0.6409
150	0.6324	0.6240	0.6158	0.6078	0.5999	0.5922	0.5847	0.5773	0.5700	0.5628
160	0.5558	0.5489	0.5422	0.5355	0.5290	0.5226	0.5164	0.5102	0.5051	0.4982
170	0.4923	0.4866	0.4810	0.4754	0.4700	0.4664	0.4593	0.4542	0.4491	0.4441
180	0.4392	0.4344	0.4296	0.4249	0.4202	0.4157	0.4113	0.4069	0.4026	0.3983
190	0.3942	0.3900	0.3860	0.3820	0.3781	0.3742	0.3704	0.3666	0.3629	0.3593
200	0.3557	0.3522	0.3487	0.3453	0.3419	0.3386	0.3353	0.3321	0.3289	0.3257
210	0.3227	0.3196	0.3166	0.3136	0.3107	0.3078	0.3050	0.3022	0.2994	0.2967
220	0.2940	0.2913	0.2887	0.2861	0.2836	0.2811	0.2786	0.2761	0.2737	0.2713
230	0.2690	0.2667	0.2644	0.2621	0.2599	0.2577	0.2555	0.2533	0.2512	0.2491
240	0.2470	0.2450	0.2430	0.2410	0.2390	0.2371	0.2351	0.2332	0.2314	0.2295
250	0.2277	0.2259	0.2241	0.2223	0.2205	0.2188	0.2171	0.2154	0.2138	0.2121
260	0.2105	0.2089	0.2073	0.2057	0.2042	0.2026	0.2011	0.1996	0.1981	0.1966
270	0.1952	0.1937	0.1923	0.1909	0.1895	0.1882	0.1868	0.1854	0.1841	0.1828
280	0.1815	0.1802	0.1789	0.1777	0.1764	0.1752	0.1740	0.1727	0.1715	0.1704
290	0.1692	0.1680	0.1669	0.1657	0.1646	0.1635	0.1624	0.1613	0.1602	0.1592
300	0.1581	0.1571	0.1560	0.1550	0.1540	0.1530	0.1520	0.1510	0.1500	0.1490
310	0.1481	0.1471	0.1462	0.1452	0.1443	0.1434	0.1425	0.1416	0.1407	0.1398
320	0.1390	0.1381	0.1372	0.1364	0.1355	0.1347	0.1339	0.1331	0.1323	0.1315
330	0.1307	0.1299	0.1291	0.1283	0.1275	0.1268	0.1260	0.1253	0.1245	0.1238
340	0.1231	0.1224	0.1217	0.1209	0.1202	0.1195	0.1189	0.1182	0.1175	0.1168
350	0.1162	0.1155	0.1148	0.1142	0.1135	0.1129	0.1123	0.1116	0.1110	0.1104
360	0.1098	0.1092	0.1086	0.1080	0.1074	0.1068	0.1062	0.1056	0.1051	0.1045
370	0.1039	0.1034	0.1028	0.1023	0.1017	0.1012	0.1006	0.1001	0.09958	0.09906
380	0.09854	0.09802	0.09751	0.09700	0.09650	0.09600	0.09550	0.09501	0.09452	0.09403
390	0.09355	0.09307	0.09260	0.09213	0.09166	0.09120	0.09074	0.09028	0.08983	0.08938
400	0.08893	0.08849	0.08805	0.08761	0.08718	0.08675	0.08632	0.08590	0.08548	0.08506
410	0.08465	0.08423	0.08383	0.08342	0.08302	0.08262	0.08222	0.08183	0.08144	0.08105
420	0.08066	0.08028	0.07990	0.07952	0.07915	0.07878	0.07841	0.07804	0.07768	0.07731
430	0.07695	0.07660	0.07624	0.07589	0.07554	0.07520	0.07485	0.07451	0.07417	0.07383

(continued)

Knoop hardness numbers (continued)

Computed using a 1-g load, for a theoretical indenter having a longitudinal angle of 172°30′ and a transverse angle of 130°00′ giving a constant for projected area (C_p) of 7.028×10^{-2}; a constant correction is provided to correct the Knoop number obtained to simulate a test made with an indenter with perfect angles.

Length of diagonal, μm	Applied load, g									
	0.0	0.1	0.2	0.3	0.4	0.5	0.6	0.7	0.8	0.9
440	0.07350	0.07316	0.07283	0.07250	0.07218	0.07185	0.07153	0.07121	0.07090	0.07058
450	0.07027	0.06996	0.06965	0.06934	0.06903	0.06873	0.06843	0.06813	0.06783	0.06754
460	0.06724	0.06695	0.06666	0.06638	0.06609	0.06581	0.06552	0.06524	0.06497	0.06469
470	0.06441	0.06414	0.06387	0.06360	0.06333	0.06306	0.06280	0.06254	0.06228	0.06202
480	0.06176	0.06150	0.06125	0.06099	0.06074	0.06049	0.06024	0.06000	0.05975	0.05951
490	0.05926	0.05902	0.05878	0.05854	0.05831	0.05807	0.05784	0.05761	0.05737	0.05714
500	0.05692	0.05669	0.05646	0.05624	0.05602	0.05579	0.05557	0.05536	0.05514	0.05492
510	0.05471	0.05449	0.05428	0.05407	0.05386	0.05365	0.05344	0.05323	0.05303	0.05282
520	0.05262	0.05242	0.05222	0.05202	0.05182	0.05162	0.05143	0.05123	0.05104	0.05085
530	0.05065	0.05046	0.05027	0.05009	0.04990	0.04971	0.04953	0.04934	0.04916	0.04898
540	0.04880	0.04862	0.04844	0.04826	0.04808	0.04790	0.04773	0.04756	0.04738	0.04721
550	0.04704	0.04687	0.04670	0.04653	0.04626	0.04619	0.04603	0.04586	0.04570	0.04554
560	0.04537	0.04521	0.04505	0.04489	0.04473	0.04457	0.04442	0.04426	0.04410	0.04395
570	0.04379	0.04364	0.04349	0.04334	0.04319	0.04304	0.04289	0.04274	0.04259	0.04244
580	0.04230	0.04215	0.04201	0.04186	0.04172	0.04158	0.04144	0.04129	0.04115	0.04101
590	0.04088	0.04074	0.04060	0.04046	0.04033	0.04019	0.04006	0.03992	0.03979	0.03966
600	0.03952	0.03939	0.03926	0.03913	0.03900	0.03887	0.03875	0.03862	0.03849	0.03837
610	0.03824	0.03811	0.03799	0.03787	0.03774	0.03762	0.03750	0.03738	0.03726	0.03714
620	0.03702	0.03690	0.03678	0.03666	0.03654	0.03643	0.03631	0.03619	0.03608	0.03596
630	0.03585	0.03574	0.03562	0.03551	0.03540	0.03529	0.03518	0.03507	0.03496	0.03485
640	0.03474	0.03463	0.03452	0.03442	0.03431	0.03420	0.03410	0.03399	0.03389	0.03378
650	0.03368	0.03357	0.03347	0.03337	0.03327	0.03317	0.03306	0.03296	0.03286	0.03276
660	0.03267	0.03257	0.03247	0.03237	0.03227	0.03218	0.03208	0.03198	0.03189	0.03179
670	0.03170	0.03160	0.03151	0.03142	0.03132	0.03123	0.03114	0.03105	0.03095	0.03086
680	0.03077	0.03068	0.03059	0.03050	0.03041	0.03032	0.03024	0.03015	0.03006	0.02997
690	0.02989	0.02980	0.02971	0.02963	0.02954	0.02946	0.02937	0.02929	0.02921	0.02912
700	0.02904	0.02896	0.02887	0.02879	0.02871	0.02863	0.02855	0.02847	0.02839	0.02831
710	0.02823	0.02815	0.02807	0.02799	0.02791	0.02783	0.02776	0.02768	0.02760	0.02752
720	0.02745	0.02737	0.02730	0.02722	0.02715	0.02707	0.02700	0.02692	0.02685	0.02677
730	0.02670	0.02663	0.02656	0.02648	0.02641	0.02634	0.02627	0.02620	0.02613	0.02605
740	0.02598	0.02591	0.02584	0.02577	0.02571	0.02564	0.02557	0.02550	0.02543	0.02536
750	0.02530	0.02523	0.02516	0.02509	0.02503	0.02496	0.02490	0.02483	0.02476	0.02470
760	0.02463	0.02457	0.02451	0.02444	0.02438	0.02431	0.02425	0.02419	0.02412	0.02406
770	0.02400	0.02394	0.02387	0.02381	0.02375	0.02369	0.02363	0.02357	0.02351	0.02345
780	0.02339	0.02333	0.02327	0.02321	0.02315	0.02309	0.02303	0.02297	0.02292	0.02286
790	0.02280	0.02274	0.02268	0.02263	0.02257	0.02251	0.02246	0.02240	0.02234	0.02229
800	0.02223	0.02218	0.02212	0.02207	0.02201	0.02196	0.02190	0.02185	0.02179	0.02174
810	0.02169	0.02164	0.02158	0.02153	0.02147	0.02142	0.02137	0.02132	0.02127	0.02121
820	0.02116	0.02111	0.02106	0.02101	0.02096	0.02091	0.02086	0.02080	0.02075	0.02070
830	0.02065	0.02060	0.02056	0.02051	0.02046	0.02041	0.02036	0.02031	0.02026	0.02021
840	0.02017	0.02012	0.02007	0.02002	0.01998	0.01993	0.01988	0.01983	0.01979	0.01974
850	0.01969	0.01965	0.01960	0.01956	0.01951	0.01946	0.01942	0.01937	0.01933	0.01928
860	0.01924	0.01919	0.01915	0.01911	0.01906	0.01902	0.01897	0.01893	0.01889	0.01884
870	0.01880	0.01876	0.01871	0.01867	0.01863	0.01858	0.01854	0.01850	0.01846	0.01842
880	0.01837	0.01833	0.01829	0.01825	0.01821	0.01817	0.01813	0.01809	0.01804	0.01800
890	0.01796	0.01792	0.01788	0.01784	0.01780	0.01776	0.01772	0.01768	0.01764	0.01761
900	0.01757	0.01753	0.01749	0.01745	0.01741	0.01737	0.01733	0.01730	0.01726	0.01722
910	0.01718	0.01714	0.01711	0.01707	0.01703	0.01700	0.01696	0.01692	0.01688	0.01685
920	0.01681	0.01677	0.01674	0.01670	0.01667	0.01663	0.01659	0.01656	0.01652	0.01649
930	0.01645	0.01642	0.01638	0.01635	0.01631	0.01628	0.01624	0.01621	0.01617	0.01614
940	0.01610	0.01607	0.01604	0.01600	0.01597	0.01593	0.01590	0.01587	0.01583	0.01580
950	0.01577	0.01573	0.01570	0.01567	0.01563	0.01560	0.01557	0.01554	0.01550	0.01547
960	0.01544	0.01541	0.01538	0.01534	0.01531	0.01528	0.01525	0.01522	0.01519	0.01515
970	0.01512	0.01509	0.01506	0.01503	0.01500	0.01497	0.01494	0.01491	0.01488	0.01485
980	0.01482	0.01479	0.01476	0.01473	0.01470	0.01467	0.01464	0.01461	0.01458	0.01455
990	0.01452	0.01449	0.01446	0.01443	0.01440	0.01437	0.01434	0.01431	0.01429	0.01426
1000	0.01423	...	...	...	...	...	...	...	...	...

Metallography

General Data

Characteristics of minerals used in coated abrasives

Commercial name	Chemical composition	Mineral name	Origin	Specific gravity	Mohs	Knoop	Grain shape
Flint	SiO_2	Quartz	Natural	2.6	6.8 to 7.0	820	Light wedges
Emery	Al_2O_3, FeO	Impure corundum	Natural	3.7 to 4.3	8.5 to 9.0		Blocky
Garnet	SiO_2, FeO, Al_2O_3 complex	Almandite	Natural	3.4 to 4.3	7.5 to 8.5		Light wedges
Crocus	FeO	Iron oxide hematite	Synthetic and natural	4.0 to 5.3	6.0		Fine milled
Aluminum oxide(a)	Al_2O_3 fused	Corundum (alpha)	Synthetic	3.96	9.4	2050	Heavy wedges
Silicon carbide(a)	SiC	Moissanite (alpha)	Synthetic	3.2	9.6	2480	Sharp wedges, silvery

(a) Used most often in metallographic grinding

Lubrication selection chart

Type of material sanded	Grease stick	Straight mineral oil	Sulfurized and chlorinated oil	Soluble oil with water(a)	Straight water(a)(b)	10% lard oil
Ferrous metals	X	X	X	X		X
Nonferrous metals(c)	X	X		X		X
Plastics					X	
Glass-stone-marble					X	
Rubber					X	
Nickel and nickel chrome alloys of the heat-resisting type	X		X			

(a) Use waterproof coated abrasives only. (b) Use only nonrecirculating system. (c) Such as brass, bronze, and aluminum

Grit comparison

Grit	Aluminum oxide silicon carbide	Garnet	Flint	Emery
Very fine	600(a)	...	...	...
	500	...	...	...
	400(a)	400-10/0	...	...
	360	...	...	...
	320(a)	320-9/0	...	...
	280	280-8/0	...	...
	240(a)	240-7/0	...	...
	220	220-6/0	Extra fine	...
Fine	180(a)	180-5/0	...	...
	150	150-4/0	...	...
	120	120-3/0	Fine	Fine
	100	100-2/0	...	...
	80(a)	80-1/0	Medium	...
	60	60-1/2	...	Medium
Coarse	50	50-1	...	...
	40	40-1½	...	...
Very coarse	36	36-2	Extra coarse	...
	30	30-2½	...	Extra coarse
	24	24-3	...	...
	20	20-3½	...	...
	16	16-4	...	...
	12	12-4½	...	...

(a) Recommended for most metallographic grinding

Electrolytic Polishing

Electrolytes. Eight groups of electrolytes, together with conditions for their use in electropolishing of various metals and alloys, and the applicability of these electrolytes to electropolishing of specific metals are summarized in the following tables. Preferred (or sometimes required) characteristics of an electrolyte are:

- A somewhat viscous consistency
- Acts as a good solvent for the anode metal (the specimen) during electrolysis conditions
- Does not attack the anode metal when no current is flowing
- Contains one or more ions of large radii, such as $(PO_4)^{-3}$, $(ClO_4)^{-1}$, or $(SO_4)^{-2}$, and sometimes large organic molecules
- Simple to mix, stable, and safe to handle (many effective electrolytes are deficient in these respects)
- Effectively functions at room temperature and not sensitive to temperature changes

Advantages and Limitations of Electropolishing. When properly applied, electropolishing can be a useful tool for the metallographer. The principal advantages of electropolishing are:

- For some metals, electropolishing can produce a high-quality surface finish equivalent to the best obtained by mechanical methods
- Once a procedure has been established, good results can be obtained with less operator skill than that required for mechanical polishing
- There can be a marked saving of time if many specimens of the same material are to be polished sequentially
- Electropolishing is especially suited to the softer metals, which may be difficult to polish by mechanical methods
- No scratches are produced in electrolytic polishing—a definite advantage in viewing high-quality electropolished surfaces of optically active materials under polarized light
- Artifacts resulting from mechanical deformation, such as disturbed metal or mechanical twins, which are produced on the surface even by careful grinding and mechanical polishing, do not occur in electropolishing
- Surfaces resulting from electropolishing are completely unworked by the polishing procedure, an important feature in low-load hardness testing or X-ray studies
- In some applications, etching can be accomplished by simply reducing the voltage to approximately one-tenth the potential required for polishing, then continuing electrolysis for a few seconds
- Electropolishing is frequently useful in electron metallography (where high resolution is often important) because it can produce clean, undistorted metal surfaces

Metallographic preparation by electropolishing is subject to several limitations; these should be recognized to prevent misapplication of the method and disappointment in the results. The principal disadvantages include:

- Because the chemicals and combinations of chemicals used in electropolishing are poisonous and many are highly flammable or potentially explosive, only well-trained personnel who are thoroughly familiar with chemical laboratory procedures should be permitted to handle or mix the chemicals, or to operate the polishing baths
- The conditions and electrolytes required to obtain a satisfactorily polished surface differ for different alloys; hence, when appropriate procedures do not exist, considerable time may be required to develop a procedure for a new alloy, if it can be developed at all
- In multiphase alloys, the rates of polishing of different phases often are not the same. Polishing results depend heavily on whether the second or third phases are strongly cathodic or anodic with respect to the matrix. The matrix is dissolved preferentially if the other phases are to stand in relief. Preferential attack may also occur at the interface between two phases. These effects are most pronounced when phases other than the matrix are virtually unattacked by the polishing bath and are reversed when the matrix phase is relatively cathodic
- A large number of electrolytes may be needed to polish the variety of metals encountered by a given laboratory
- Plastic or metal mounting materials may react with the electrolyte
- Electropolished surfaces exhibit an undulating rather than a plane surface, and in some cases may not be suited for examination at all magnifications. Under some conditions, furrowing and pitting may be produced
- Edge effects limit applications involving small specimens, surface phenomena, coatings, interfaces, and cracks
- Attack around nonmetallic particles and adjacent metal, voids, and various inhomogeneities may not be the same as that of the matrix, thus exaggerating the size of the voids and inclusions
- Electropolished surfaces of certain materials may be passive and difficult to etch

Electrolytes for electropolishing of various metals and alloys (based on ASTM E3)

Class	Formula	Use	Cell voltage	Time	Notes
Group I—Electrolytes composed of perchloric acid and alcohol with or without organic additions					
I-1	800 ml ethanol (absolute)(a) 140 ml distilled water (optional), 60 ml perchloric acid (60%)	Al and Al alloys with less than 2% Si	30 to 80	15 to 60 s	...
		Carbon, alloy and stainless steels	35 to 65	15 to 60 s	...
		Pb, Pb-Sn, Pb-Sn-Cd, Pb-Sn-Sb	12 to 35	15 to 60 s	...
		Zn, Zn-Sn-Fe, Zn-Al-Cu	20 to 60	...	...
		Mg and high-Mg alloys	...	...	(b)
I-2	800 ml ethanol (absolute)(a), 200 ml perchloric acid (60%)	Stainless steel; aluminum	35 to 80	15 to 60 s	...
I-3	940 ml ethanol (absolute)(a), 6 ml distilled water, 54 ml perchloric acid (70%)	Stainless steel	30 to 45	15 to 60 s	...
		Thorium	30 to 40	15 to 45 s	...
I-4	700 ml ethanol (absolute)(a), 120 ml distilled water, 100 ml 2-butoxyethanol, 80 ml perchloric acid (60%)	Steel, cast iron, Al, Al alloys, Ni, Sn, Ag, Be, Ti, Zr, U, heat-resisting alloys	30 to 65	15 to 60 s	(c)
I-5	700 ml ethanol (absolute)(a), 120 ml distilled water, 100 ml glycerol, 80 ml perchloric acid (60%)	Stainless, alloys and high speed steels; Al, Fe, Fe-Si alloys, Pb, Zr	15 to 50	15 to 60 s	(d)
I-6	760 ml ethanol (absolute)(a), 30 ml distilled water, 190 ml ether, 20 ml perchloric acid (60%)	Aluminum, aluminum-silicon alloys, iron-silicon alloys	35 to 60	15 to 60 s	(e)
I-7	600 ml methanol (absolute), 370 ml 2-butoxyethanol, 30 ml perchloric acid (60%)	Molybdenum, titanium, zinc, zirconium, uranium-zirconium alloy	60 to 150	5 to 30 s	...
I-8	840 ml methanol (absolute), 4 ml distilled water, 125 ml glycerol, 31 ml perchloric acid (70%)	Aluminum, aluminum-silicon alloys, iron-silicon alloys	50 to 100	5 to 60 s	...
I-9	590 ml methanol (absolute), 6 ml distilled water, 350 ml 2-butoxyethanol, 54 ml perchloric acid (70%)	Germanium	25 to 35	30 to 60 s	...
		Titanium	58 to 66	45 s	(f)
		Vanadium	30	3 s	(g)
		Zirconium	70 to 75	15 s	(h)
I-10	950 ml methanol (absolute), 15 ml nitric acid, 50 ml perchloric acid (60%)	Aluminum	30 to 60	15 to 60 s	...
Group II—Electrolytes composed of perchloric acid (60%) and glacial acetic acid					
II-1	940 ml acetic acid, 60 ml perchloric acid	Cr, Ti, U, Zr, Fe, cast iron; carbon, alloy and stainless steels	20 to 60	1 to 5 min.	(j)
II-2	900 ml acetic acid, 100 ml perchloric acid	Zr, Ti, U, steels, superalloys	12 to 70	½ to 2 min.	...
II-3	800 ml acetic acid, 200 ml perchloric acid	U, Zr, Ti, Al, steels, superalloys	40 to 100	1 to 15 min.	...
II-4	700 ml acetic acid, 300 ml perchloric acid	Nickel, lead, lead-antimony alloys	40 to 100	1 to 5 min.	...
II-5	650 ml acetic acid, 350 ml of perchloric acid	3% silicon iron	...	5 min.	(k)
Group III—Electrolytes composed of phosphoric acid (85%) in water or organic solvent					
III-1	1000 ml phosphoric acid	Cobalt	1.2	3 to 5 min.	...
III-2	175 ml distilled water, 825 ml phosphoric acid	Pure copper	1.0 to 1.6	10 to 40 min.	(m)
III-3	300 ml water, 700 ml phosphoric acid	Stainless steel, brass, copper and copper alloys except tin-bronze	1.5 to 1.8	5 to 15 min.	(m)
III-4	600 ml water, 400 ml phosphoric acid	α or α + β brass, Cu-Fe, Cu-Co, Co, Cd	1 to 2	1 to 15 min.	(n)
III-5	1000 ml water, 580 g pyrophosphoric acid	Copper, copper-zinc	1 to 2	10 min.	(m)
III-6	500 ml diethylene glycol monoethyl ether, 500 ml phosphoric acid	Steel	5 to 20	5 to 15 min.	(p)
III-7	200 ml water, 380 ml ethanol (95%), 400 ml phosphoric acid	Aluminum, magnesium, silver	25 to 30	4 to 6 min.	(q)
III-8	300 ml ethanol (absolute), 300 ml glycerol (cp), 300 ml phosphoric acid	Uranium	...	...	...
III-9	500 ml ethanol (95%), 250 ml glycerol, 250 ml phosphoric acid	Manganese, manganese-copper alloys	18	...	...
III-10	500 ml distilled water, 250 ml ethanol (95%), 250 ml phosphoric acid	Copper and copper-base alloys	...	1 to 5 min.	...
III-11	Ethanol (absolute) to make 1000 ml of solution; 400 g pyrophosphoric acid	Stainless steel; all austenitic heat-resisting alloys	...	10 min.	(r)
III-12	625 ml ethanol (95%), 375 ml phosphoric acid	Magnesium-zinc	1.5 to 2.5	3 to 30 min.	...
III-13	445 ml ethanol (95%), 275 ml ethylene glycol, 275 ml phosphoric acid	Uranium	18 to 20	5 to 15 min.	(s)

Note: Chemical components of electrolytes are listed in the order of mixing. Except where noted otherwise, the electrolytes are intended for use at ambient temperatures, in the approximate range of 18 to 38 °C (65 to 100 °F), and with stainless steel cathodes. (a) In etchants I-1 through I-6, absolute 5D-3A or SD-30 ethanol can be substituted for absolute ethanol. (b) Nickel cathode. (c) One of the best electrolytes for universal use. (d) Universal electrolyte comparable to I-4. (e) Particularly good with aluminum-silicon alloys. (f) Polish only. (g) 3-s cycles repeated at least seven times to prevent heating. (h) Polish and etch simultaneously. (j) Good general-purpose electrolyte. (k) 0.06 A/cm². (m) Copper cathode. (n) Copper or stainless steel cathode. (p) 49 °C (120 °F). (q) Aluminum cathode; 38 to 43 °C (100 to 110 °F). (r) 38 °C (100 °F) plus. (s) 0.03 A/cm². (t) Particularly good for sintered molybdenum; 0 to 27 °C (32 to 80 °F). (u) 0 to 0.27 °C (32 to 80 °F). (v) 0.3 A/cm². (w) 0.1 to 0.2 A/cm². (x) 0.05 A/cm². (y) 0.1 A/cm². (z) 1 to 5 A/cm²; 38 °C (100 °F) plus. (aa) 1 A/cm²; 27 to 49 °C (80 to 120 °F). (bb) 0.6 A/cm²; 27 to 49 °C (80 to 120 °F). (cc) 0.5 A/cm²; 27 to 49 °C (80 to 120 °F). (dd) 0.5 A/cm²; 38 to 54 °C (100 to 130 °F). (ee) Graphite cathode; 0.1 A/cm²; 32 to 38 °C (90 to 100 °F); (ff) 0.5 A/cm²; 21 to 49 °C (70 to 120 °F). (gg) 0.002 A/cm²; 21 to 38 °C (70 to 100 °F). (hh) 0.5 A/cm². *Caution:* dangerous. (jj) 0.08 to 0.3 A/cm². (kk) 0.5 (approx) A/cm². *Caution:* this mixture will decompose vigorously after a short time; do not try to keep. (mm) Bath should be stirred. Cool below 2 °C (35 °F) with cracked ice. (nn) Mix slowly. Heat is developed. Avoid contamination with water. Use below 2 °C (35 °F). (pp) The chromic acid is dissolved in the water, and this solution is then added to the acetic acid. Electrolyte is used below 2 °C (35 °F). (qq) Electrolyte is used below 16 °C (60 °F). (rr) *Caution:* electrolyte will decompose on standing, and is dangerous if kept too long. (ss) Polish 30 s, but allow to remain in electrolyte until brown film is dissolved. (tt) Graphite cathode. (uu) Graphite cathode; 0.003 to 0.009 A/cm². (vv) Graphite cathode; 0.09 A/cm²; 38 to 49 °C (100 to 120 °F). (ww) Graphite cathode; 0.03 to 0.06 A/cm². (xx) Copper cathode; 0.01 to 0.2 A/cm². (yy) An extremely useful electrolyte for certain applications, but dangerous

(continued)

Electrolytes for electropolishing of various metals and alloys (based on ASTM E3) (continued)

Class	Formula	Use	Cell voltage	Time	Notes
Group IV—Electrolytes composed of sulfuric acid in water or organic solvent					
IV-1	250 ml water, 750 ml sulfuric acid	Stainless steel	1.5 to 6	1 to 2 min.	...
IV-2	400 ml water, 600 ml sulfuric acid	Stainless steel, iron, nickel	1.5 to 6	2 to 6 min.	...
IV-3	750 ml water, 250 ml sulfuric acid	Stainless steel, iron, nickel	1.5 to 6	2 to 10 min.	...
		Molybdenum	1.5 to 6	$\frac{1}{3}$ to 1 min.	(t)
IV-4	900 ml water, 100 ml sulfuric acid	Molybdenum	1.5 to 6	$\frac{1}{3}$ to 2 min.	(t)
IV-5	70 ml water, 20 ml glycerol, 720 ml sulfuric acid	Stainless steel	1.5 to 6	$\frac{1}{2}$ to 5 min.	...
IV-6	220 ml water, 200 ml glycerol, 580 ml sulfuric acid	Stainless steel, aluminum	1.5 to 12	1 to 20 min.	...
IV-7	875 ml methanol (absolute), 125 ml sulfuric acid	Molybdenum	6 to 18	$\frac{1}{2}$ to $1\frac{1}{2}$ min.	(u)
Group V—Electrolytes composed of chromic acid in water					
V-1	830 ml water, 620 g chromic acid	Stainless steel	1.5 to 9	2 to 10 min.	...
V-2	830 ml water, 170 g chromic acid	Zinc, brass	1.5 to 12	10 to 60 s	...
Group VI—Electrolytes composed of mixed acids or salts in water or organic solution					
VI-1	600 ml phosphoric acid (85%), 400 ml sulfuric acid	Stainless steel	...	...	...
VI-2	150 ml water, 300 ml phosphoric acid (85%), 550 ml sulfuric acid	Stainless steel	...	2 min.	(v)
VI-3	240 ml water, 420 ml phosphoric acid (85%), 340 ml sulfuric acid	Stainless and alloy steels	...	2 to 10 min.	(w)
VI-4	330 ml water, 550 ml phosphoric acid (85%), 120 ml sulfuric acid	Stainless steel	...	1 min.	(x)
VI-5	450 ml water, 390 ml phosphoric acid (85%), 160 ml sulfuric acid	Bronze (to 9% tin)	...	1 to 5 min.	(y)
VI-6	330 ml water, 580 ml phosphoric acid (85%), 90 ml sulfuric acid	Bronze (to 6% tin)	...	1 to 5 min.	(y)
VI-7	140 ml water, 100 ml glycerol, 430 ml phosphoric acid (85%), 330 ml sulfuric acid	Steel	...	1 to 5 min.	(z)
VI-8	200 ml water, 590 ml glycerol, 100 ml phosphoric acid (85%), 110 ml sulfuric acid	Stainless steel	...	5 min.	(aa)
VI-9	260 ml water, 175 g chromic acid, 175 ml phosphoric acid (85%), 580 ml sulfuric acid	Stainless steel	...	30 min.	(bb)
VI-10	175 ml water, 105 g chromic acid, 460 ml phosphoric acid (85%), 390 ml sulfuric acid	Stainless steel	...	60 min.	(cc)
VI-11	240 ml water, 80 g chromic acid, 650 ml phosphoric acid (85%), 130 ml sulfuric acid	Stainless and alloy steels	...	5 to 60 min.	(dd)
VI-12	100 ml hydrofluoric acid, 900 ml sulfuric acid	Tantalum	...	9 min.	(ee)
VI-13	210 ml water, 180 ml hydrofluoric acid, 610 ml sulfuric acid	Stainless steel	...	5 min.	(ff)
VI-14	800 ml water, 100 g chromic acid, 46 ml sulfuric acid, 310 g sodium dichromate, 96 ml acetic acid (glacial)	Zinc	...	...	(gg)
VI-15	260 ml hydrogen peroxide (30%), 240 ml hydrofluoric acid, 500 ml sulfuric acid	Stainless steel	...	5 min.	(hh)
VI-16	520 ml water, 80 ml hydrofluoric acid, 400 ml sulfuric acid	Stainless steel	...	$\frac{1}{2}$ to 4 min.	(jj)
VI-17	600 ml water, 180 g chromic acid, 60 ml nitric acid, 3 ml hydrochloric acid, 240 ml sulfuric acid	Stainless steel	...	...	...

Note: Chemical components of electrolytes are listed in the order of mixing. Except where noted otherwise, the electrolytes are intended for use at ambient temperatures, in the approximate range of 18 to 38 °C (65 to 100 °F), and with stainless steel cathodes. (a) In etchants I-1 through I-6, absolute 5D-3A or SD-30 ethanol can be substituted for absolute ethanol. (b) Nickel cathode. (c) One of the best electrolytes for universal use. (d) Universal electrolyte comparable to I-4. (e) Particularly good with aluminum-silicon alloys. (f) Polish only. (g) 3-s cycles repeated at least seven times to prevent heating. (h) Polish and etch simultaneously. (j) Good general-purpose electrolyte. (k) 0.06 A/cm^2. (m) Copper cathode. (n) Copper or stainless steel cathode. (p) 49 °C (120 °F). (q) Aluminum cathode; 38 to 43 °C (100 to 110 °F). (r) 38 °C (100 °F) plus. (s) 0.03 A/cm^2. (t) Particularly good for sintered molybdenum; 0 to 27 °C (32 to 80 °F). (u) 0 to 0.27 °C (32 to 80 °F). (v) 0.3 A/cm^2. (w) 0.1 to 0.2 A/cm^2. (x) 0.05 A/cm^2. (y) 0.1 A/cm^2. (z) 1 to 5 A/cm^2; 38 °C (100 °F) plus. (aa) 1 A/cm^2; 27 to 49 °C (80 to 120 °F). (bb) 0.6 A/cm^2; 27 to 49 °C (80 to 120 °F). (cc) 0.5 A/cm^2; 27 to 49 °C (80 to 120 °F). (dd) 0.5 A/cm^2; 38 to 54 °C (100 to 130 °F). (ee) Graphite cathode; 0.1 A/cm^2; 32 to 38 °C (90 to 100 °F); (ff) 0.5 A/cm^2; 21 to 49 °C (70 to 120 °F). (gg) 0.002 A/cm^2; 21 to 38 °C (70 to 100 °F). (hh) 0.5 A/cm^2. *Caution:* dangerous. (jj) 0.08 to 0.3 A/cm^2. (kk) 0.5 (approx) A/cm^2. *Caution:* this mixture will decompose vigorously after a short time; do not try to keep. (mm) Bath should be stirred. Cool below 2 °C (35 °F) with cracked ice. (nn) Mix slowly. Heat is developed. Avoid contamination with water. Use below 2 °C (35 °F). (pp) The chromic acid is dissolved in the water, and this solution is then added to the acetic acid. Electrolyte is used below 2 °C (35 °F). (qq) Electrolyte is used below 16 °C (60 °F). (rr) *Caution:* electrolyte will decompose on standing, and is dangerous if kept too long. (ss) Polish 30 s, but allow to remain in electrolyte until brown film is dissolved. (tt) Graphite cathode. (uu) Graphite cathode; 0.003 to 0.009 A/cm^2. (vv) Graphite cathode; 0.09 A/cm^2; 38 to 49 °C (100 to 120 °F). (ww) Graphite cathode; 0.03 to 0.06 A/cm^2. (xx) Copper cathode; 0.01 to 0.2 A/cm^2. (yy) An extremely useful electrolyte for certain applications, but dangerous

(continued)

Electrolytes for electropolishing of various metals and alloys (based on ASTM E3) (continued)

Class	Formula	Use	Cell voltage	Time	Notes
VI-18	750 ml glycerol, 125 ml acetic acid (glacial), 125 ml nitric acid	Bismuth	12	1 to 5 min.	(kk)
VI-19	900 ml ethylene glycol monoethyl ether, 100 ml hydrochloric acid	Magnesium	50 to 60	10 to 30 s	(mm)
VI-20	685 ml methanol (absolute), 225 ml hydrochloric acid, 90 ml sulfuric acid	Molybdenum, sintered and cast	19 to 35	20 to 35 s	(nn)
VI-21	855 ml ethanol (absolute), 100 ml n-butyl alcohol, 109 g $AlCl_3 \cdot 6H_2O$, 250 g zinc chloride (anhydrous)	Titanium	30 to 60	1 to 6 min.	...
VI-22	750 ml acetic acid (glacial), 210 ml distilled water, 180 g chromic acid	Uranium	80	5 to 30 min.	(pp)
VI-23	720 ml ethanol (95%), 90 g $AlCl_3 \cdot 6H_2O$, 225 g zinc chloride (anhydrous), 120 ml distilled water, 80 ml n-butyl alcohol	Pure zinc	25 to 40	$\frac{1}{2}$ to 3 min.	(qq)
VI-24	870 ml glycerol, 43 ml hydrofluoric acid, 87 ml nitric acid	Zirconium(h)	9 to 12	1 to 10 min.	(rr)
VI-25	980 ml saturated solution of potassium iodide in distilled water, 20 ml hydrochloric acid	Bismuth	7	30 s	(ss)
Group VII—Alkaline electrolytes					
VII-1	Water to make 1000 ml, 80 g potassium cyanide, 40 g potassium carbonate, 50 g gold chloride	Gold, silver	7.5	2 to 4 min.	(tt)
VII-2	Water to make 1000 ml, 100 g sodium cyanide, 100 g potassium ferrocyanide	Silver	2.5	To 1 min.	(tt)
VII-3	Water to make 1000 ml, 400 g potassium cyanide, 280 g silver cyanide, 280 g potassium dichromate	Silver	...	To 9 min.	(uu)
VII-4	Water to make 1000 ml, 160 g trisodium phosphate	Tungsten	...	10 min.	(vv)
VII-5	Water to make 1000 ml, 100 g sodium hydroxide	Tungsten, lead	...	8 to 10 min.	(ww)
VII-6	Water to make 1000 ml, 200 g potassium hydroxide	Zinc, tin	2 to 6	15 min.	(xx)
Group VIII—Electrolyte composed of methanol and nitric acid					
VIII-1	600 ml methanol (absolute), 330 ml nitric acid	Nickel, copper, zinc, Monel, brass, nickel-chrome, stainless steel	40 to 70	10 to 60 s	(yy)

Note: Chemical components of electrolytes are listed in the order of mixing. Except where noted otherwise, the electrolytes are intended for use at ambient temperatures, in the approximate range of 18 to 38 °C (65 to 100 °F), and with stainless steel cathodes. (a) In etchants I-1 through I-6, absolute 5D-3A or SD-30 ethanol can be substituted for absolute ethanol. (b) Nickel cathode. (c) One of the best electrolytes for universal use. (d) Universal electrolyte comparable to I-4. (e) Particularly good with aluminum-silicon alloys. (f) Polish only. (g) 3-s cycles repeated at least seven times to prevent heating. (h) Polish and etch simultaneously. (j) Good general-purpose electrolyte. (k) 0.06 A/cm^2. (m) Copper cathode. (n) Copper or stainless steel cathode. (p) 49 °C (120 °F). (q) Aluminum cathode; 38 to 43 °C (100 to 110 °F). (r) 38 °C (100 °F) plus. (s) 0.03 A/cm^2. (t) Particularly good for sintered molybdenum; 0 to 27 °C (32 to 80 °F). (u) 0 to 0.27 °C (32 to 80 °F). (v) 0.3 A/cm^2. (w) 0.1 to 0.2 A/cm^2. (x) 0.05 A/cm^2. (y) 0.1 A/cm^2. (z) 1 to 5 A/cm^2; 38 °C (100 °F) plus. (aa) 1 A/cm^2; 27 to 49 °C (80 to 120 °F). (bb) 0.6 A/cm^2; 27 to 49 °C (80 to 120 °F). (cc) 0.5 A/cm^2; 27 to 49 °C (80 to 120 °F). (dd) 0.5 A/cm^2; 38 to 54 °C (100 to 130 °F). (ee) Graphite cathode; 0.1 A/cm^2; 32 to 38 °C (90 to 100 °F); (ff) 0.5 A/cm^2; 21 to 49 °C (70 to 120 °F). (gg) 0.002 A/cm^2; 21 to 38 °C (70 to 100 °F). (hh) 0.5 A/cm^2. *Caution:* dangerous. (jj) 0.08 to 0.3 A/cm^2. (kk) 0.5 (approx) A/cm^2. *Caution:* this mixture will decompose vigorously after a short time; do not try to keep. (mm) Bath should be stirred. Cool below 2 °C (35 °F) with cracked ice. (nn) Mix slowly. Heat is developed. Avoid contamination with water. Use below 2 °C (35 °F). (pp) The chromic acid is dissolved in the water, and this solution is then added to the acetic acid. Electrolyte is used below 2 °C (35 °F). (qq) Electrolyte is used below 16 °C (60 °F). (rr) *Caution:* electrolyte will decompose on standing, and is dangerous if kept too long. (ss) Polish 30 s, but allow to remain in electrolyte until brown film is dissolved. (tt) Graphite cathode. (uu) Graphite cathode; 0.003 to 0.009 A/cm^2. (vv) Graphite cathode; 0.09 A/cm^2; 38 to 49 °C (100 to 120 °F). (ww) Graphite cathode; 0.03 to 0.06 A/cm^2. (xx) Copper cathode; 0.01 to 0.2 A/cm^2. (yy) An extremely useful electrolyte for certain applications, but dangerous

Applicability of electrolytes to electropolishing of various metals and alloys (based on ASTM E3)

Metal	Electrolyte	Metal	Electrolyte
Aluminum	I-1, I-2, I-4, I-5, I-6, I-8, I-10, II-3, III-7, IV-6	Magnesium	I-1, III-7, III-12, VI-19
Aluminum-silicon alloys	I-6, I-8	Manganese	III-9
Antimony	II-4	Molybdenum	I-7, IV-3, IV-4, IV-7, VI-20
Beryllium	I-4	Nickel	I-4, II-4, IV-2, VIII-1
Bismuth	VI-18, VI-25	Nickel-chromium alloys	II-4, VIII-1
Cadmium	III-4	Silver	I-4, III-7, VII-1, VII-2, VII-3
Cast iron	I-4, II-1	Steel: austenitic, stainless, and superalloys	I-1, I-2, I-3, I-4, I-5, II-1, II-2, II-3, III-3, III-6, III-11, IV-1, IV-2, IV-3, IV-5, IV-6, V-1, VI-1, VI-2, VI-3, VI-4, VI-7, VI-8, VI-9, VI-10, VI-11, VI-13, VI-15, VI-16, VI-17, VIII-1
Chromium	II-1, VIII-1		
Cobalt	I-5, III-1, III-4		
Copper	III-2, III-3, III-4, III-5, III-10, VIII-1	Steel: carbon and alloy	I-1, I-4, I-5, II-1, II-2, II-3, III-6, VI-3, VI-7, VI-11
Copper-nickel alloys	III-3, III-10, VIII-1	Tantalum	VI-12
Copper-tin alloys	III-10, VI-5, VI-6, VIII-1	Thorium	I-3
Copper-zinc alloys	III-3, III-4, III-5, III-10, V-2, VIII-1	Tin	I-4, VI-5, VI-6, VII-6
Germanium	I-9	Titanium	I-4, I-7, I-9, II-1, II-2, II-3, VI-21
Gold	VII-1	Tungsten	VII-4, VII-5
Iron, pure	I-5, II-1, IV-2, IV-3	Uranium	I-4, I-7, II-1, II-2, II-3, III-8, III-13, VI-22
Iron-copper alloys	III-3, III-4	Vanadium	I-9
Iron-nickel alloys	I-5, II-1, II-2, II-4, IV-3, VIII-1	Zinc	I-1, I-5, III-12, V-2, VI-14, VI-23, VII-6, VIII-1
Iron-silicon alloys	I-5, I-6, I-8, II-5	Zirconium	I-4, I-5, I-7, I-9, II-1, II-2, II-3, VI-24
Lead	I-1, I-5, II-4, VII-5		

Electrolytes and voltages for tampon-type local electropolishing of various metals

Electrolyte composition	Metal	Voltage
9 ml perchloric acid (60%), 91 ml butyl cellosolve	Steel, iron and iron-base alloys	35 to 40
	Aluminum and aluminum alloys	30 to 35
	Beryllium and beryllium alloys	43 to 46
10 ml perchloric acid (60%), 45 ml acetic acid (glacial), 45 ml butyl cellosolve	Steel	30 to 35
	Chromium-base alloys	32 to 37
	Nickel and nickel-base alloys	30 to 40
	Cobalt-base alloys	30 to 60
54 ml phosphoric acid (85%), 22 ml ethanol (absolute), 3 ml distilled water, 21 ml butyl cellosolve	Copper and copper alloys	4 to 6
11 ml perchloric acid (60%), 65 ml methanol (absolute), 24 ml butyl cellosolve	Titanium alloys	26 to 28

Characteristics of pure methanol and ethanol

Name	Active constituent	Nominal composition, vol%(a)
Methanol (methyl alcohol)	CH_3OH	99.5(b)
Methanol (methyl alcohol), 95%	CH_3OH	95(c)
Ethanol (ethyl alcohol), anhydrous	C_2H_5OH	99.5(d)(e)
Ethanol (ethyl alcohol), 95%	C_2H_5OH	95(e)

(a) Nominal percentage of the active constituent; remainder is water, unless otherwise specified. (b) Synthetic methanol; the commercial grade is of high purity and is satisfactory for use in all ordinary metallographic etchants where methanol is specified (wood alcohol has not been manufactured commercially in the United States since 1969). Methanol is available only as an anhydrous (also called absolute) grade containing less than 0.1 or 0.2% water as packaged, and usually not more than about 0.5% water at time of use, depending on storage and handling. (c) Where methanol, 95%, is called for, the ordinary anhydrous grade must be diluted by the user with 5% water by volume. (d) The anhydrous (also called absolute) grade of ethanol is ordinarily used only where no significant amount of water can be tolerated. It contains less than 0.1 or 0.2% water as packaged, and usually not more than about 0.5% water at time of use, depending on storage and handling. (e) Available only with special government permit

Characteristics of aqueous liquid chemicals used in many metallographic etchants
Except for sulfuric acid, all data apply to both laboratory and technical or commercial grades of chemicals

Name	Active constituent	Nominal composition, wt%(a)	Specific gravity	Degrees Baumé(b)
Aqueous acids				
Acetic acid, glacial	$HC_2H_3O_2$	99.5	1.05	7.0
Fluoboric acid	HBF_4	48	1.32	35
Hydrochloric acid(c)	HCl	37	1.18	22
Hydrofluoric acid	HF	48	1.15	19
Lactic acid	$HC_3H_5O_3$	85	1.20	24
Nitric acid	HNO_3	70	1.42	43
Perchloric acid	$NClO_4$	70	1.67	58
		60	1.53	50
Phosphoric acid (ortho)	H_3PO_4	85	1.70	60
Sulfuric acid	H_2SO_4	96(d)	1.84(e)	66(e)
Miscellaneous aqueous chemicals				
Ammonium hydroxide	NH_4OH	28(f)	0.90	26
Hydrogen peroxide	H_2O_2	3(g)	1.01	1.4
		30(h)	1.11	15
		50(j)	1.20	24

(a) Nominal percentage of the active constituent; remainder is water. Reagents made by different manufacturers may differ slightly in nominal concentration and allowable range of concentration. (b) Specific gravity as indicated on the Baumé scale; sometimes used for technical grades and in laboratory measurements. (c) Technical grade is also called muriatic acid. (d) Laboratory grade. Technical grade has concentration of 93%. (e) Specific gravity and degrees Baumé are nearly constant for 93 to 100% sulfuric acid. (f) Percent NH_3. (g) Sometimes called "10 volume". (h) Sometimes called '100 volume". (j) Sometimes called "170 volume"

Nominal compositions of various grades of denatured alcohol (ethanol) used in some metallographic etchants

Component	Parts by volume in specially denatured alcohol(a)					
	Formula SD-1(b)		Formula SD-3A		Formula SD-30	
	Anhydrous	95%(c)	Anhydrous	95%(c)	Anhydrous	95%(c)
Ethanol, anhydrous	100	95	100	95	100	95
Water	...	5	...	5	...	5
Methanol	4	4	5	5	10	10
Methyl isobutyl ketone	1	1	...	...	...	...

Component	Parts by volume in proprietary solvent(d)		Parts by volume in "reagent" alcohol(d)	
	Anhydrous	95%(c)	Anhydrous	95%(c)
SD-1, anhydrous(b)	100	...	...	...
SD-1, 95%(b)(c)	...	100	...	...
SD-3A, anhydrous	...	...	95	...
SE-3A, 95%(c)	...	...	...	95
Methyl isobutyl ketone	1	1	...	...
Hydrocarbon solvent or gasolene	1	1	...	...
Ethyl acetate	1	1	...	...
Isopropyl alcohol	...	...	5	5

(a) Specially denatured alcohol is available only with special government permit. (b) The formula shown here has replaced the old SD-1 formula in which wood alcohol was specified; wood alcohol has not been manufactured commercially in the United States since 1969. (c) The designation of type of denatured alcohol as 95% means that the denatured product contains 5 parts of water for every 95 parts of anhydrous (absolute) ethanol, plus denaturants as specified. (d) Available without government permit from suppliers of laboratory chemicals, for scientific and general laboratory purposes

Description of miscellaneous chemicals used in metallographic etchants

aluminum chloride, anhydrous. Solid; $AlCl_3$; reacts violently with water, evolving HCl gas; use of hydrated form, $AlCl_3 \cdot 6H_2O$, is preferred.

ammonium molybdate. Crystals; also called ammonium paramolybdate or heptamolybdate; $(NH_4)_6Mo_7O_{24} \cdot 4H_2O$; can be used interchangeably with "molybdic acid, 85%".

benzalkonium chloride. Crystals; essentially alkyl-dimethyl-benzyl-ammonium chloride. May not be readily available in this form; see *zephiran chloride*.

1-butanol. See *n-butyl alcohol*.

2-butoxyethanol. See *butyl cellosolve*.

n-butyl alcohol. Liquid; normal butyl alcohol; also called butyl alcohol and 1-butanol.

butyl carbitol. Liquid; diethylene glycol monobutyl ether.

butyl cellosolve. Liquid; ethylene glycol monobutyl ether; also called 2-butoxyethanol.

carbitol. Liquid; diethylene glycol mono-ethyl ether.

cellosolve. Liquid; ethylene glycol mono-ethyl ether.

chromic acid. Dark-red crystals or flakes; CrO_3; also called chromic anhydride, chromic acid anhydride, and chromium trioxide. See *chromic oxide*, Cr_2O_3.

chromic anhydride. See *chromic acid*.

chromic oxide. Fine green powder; Cr_2O_3; a polishing abrasive. Do not confuse with chromic acid (CrO_3), which is a strong acid and a component of many etchants.

cupric ammonium chloride. Crystals; a double salt, $CuCl_2 \cdot 2NH_4Cl \cdot 2H_2O$. If not available, substitute 0.6 g $CuCl_2 \cdot 2H_2O$ plus 0.4 g NH_4Cl for each gram of the double salt.

diethylene glycol. Syrupy liquid; also called 2,2'-oxydiethanol and dihydroxydiethyl ether; $(HOCH_2CH_2)_2O$. More viscous than ethylene glycol; otherwise similar in behavior.

diethylene glycol monobutyl ether. See *butyl carbitol*.

diethylene glycol monoethyl ether. See *carbitol*.

diethyl ether. See *ether*.

ether. Liquid; also called ethyl ether and diethyl ether; very low flash point, highly explosive; boiling point is 34.4 °C (94 °F).

ethylene glycol. Syrupy liquid, also called 1.2-ethanediol and dihydroxyethane; $(CH_2)_2(OH)_2$. Less viscous than diethylene glycol; otherwise similar in behavior.

ethylene glycol monobutyl ether. Liquid; also called 2-butoxyethanol or butyl cellosolve.

ethylene glycol monoethyl ether. See *cellosolve*.

ethyl ether. See *ether*.

ferric nitrate. Crystals; $Fe(NO_3)_3 \cdot 9H_2O$. There is no anhydrous form of this salt.

fluoboric acid, 48%. Liquid; HBF_4; is not readily available in small quantities, substitute 10.3 ml HF (48%) plus 4.4 g H_3BO_3, for each 10 ml of 48% fluoboric acid specified.

glycerol. Syrupy liquid; also called glycerin or glycereine; $C_3H_5(OH)_3$; contains up to 5% (by weight) water.

molybdic acid, 85%. Crystals or powder containing the equivalent of 85% MoO_3. This misnamed chemical consists mostly of ammonium molybdate (or paramolybdate), which is $(NH_4)_6 Mo_7O_{24} \cdot 4H_2O$. The two chemicals can be used interchangeably. See *ammonium molybdate*.

muriatic acid. Liquid; technocal grade HCl.

picric acid. Crystals; 2,4,6-trinitrophenol; crystals of laboratory chemical contain 10 to 15% water; explosive; its crystalline metallic salts are even more explosive. Do not use grades that do not have the 10 to 15% water content.

pyrophosphoric acid. Crystals or viscous liquid; $H_4P_2O_7$, anhydrous; hydrolyzes to phosphoric acid (H_3PO_4) slowly in cold water and rapidly in hot water.

zephiran chloride. Aqueous solution; a proprietary material produced in grades containing about 12% and 17% (by weight) benzalkonium chloride (alkyl-dimethyl-benzyl-ammonium chloride) as the active constituent, plus some ammonium acetate; also called sephiran chloride. Available from pharmacies or pharmaceutical distributors. See *benzalkonium chloride*.

Irons and Steels

Etchants and recommendations for macroetching of carbon and alloy steels

Etchant composition(a)	Etching time	Surface required(b)	Purpose, or characteristic revealed
Etchants for use at 71 to 82 °C (160 to 180 °F)			
1 part HCl, 1 part water	15 to 60 min.	A or B	Segregation, porosity, hardness penetration, cracks, inclusions, dendrites, flow lines, soft spots, structure, weld examination
Concentrated HCl	15 to 60 min.	A or B	Segregation, porosity, hardness penetration, cracks, inclusions, dendrites, flow lines, soft spots, structure, weld examination
2 parts H_2SO_4, 1 part HCl, 3 parts water	30 to 60 min.	A	Segregation, porosity, hardness penetration, cracks, inclusions, dendrites, flow lines, soft spots, structure, weld examination
50 parts HCl, 7 parts H_2SO_4, 18 parts water	30 to 60 min.	A	Segregation, porosity, hardness penetration, cracks, inclusions, dendrites, flow lines, soft spots, structure, weld examination
10 to 40 parts HNO_3, 4 to 10 parts HF (48%), 50 to 87 parts water(c)	Until desired etch is obtained	B or C	Segregation, porosity, hardness penetration, cracks, inclusions, dendrites, flow lines, soft spots, structure, weld examination
38 parts HCl, 12 parts H_2SO_4, 50 parts water	30 to 60 min.	B or C	Segregation, porosity, hardness penetration, cracks, inclusions, dendrites, flow lines, soft spots, structure, weld examination
10 parts H_2SO_4, 90 parts water	15 to 60 min.	A	Sulfide and oxide inclusions
Etchants for use at room temperature			
2 to 25% HNO_3 in water or ethanol	5 to 30 min.	B or C	Carburization and decarburization, hardness penetration, cracks, segregation, weld examination
2.5 g $CuCl_2 \cdot 2H_2O$, 20 g $MgCl_2 \cdot 6H_2O$, 10 ml HCl, 500 ml ethanol	Until coppery sheen appears	B or C	Phosphorus-rich areas, banding
50 g $(NH_4)_2S_2O_8$, 500 ml water	Swab until desired etch is obtained	C	Grain size, weld examination
40 g $FeCl_3$, 3 g $CuCl_2$, 40 ml HCl, 500 ml water	15 to 30 s	B or C	Dendritic structure of cast steel(d)
30 g $FeCl_3$, 1 g $CuCl_2$, 0.5 g $SnCl_2$, 50 ml HCl, 500 ml ethanol, 500 ml water	30 s to 2 min.	C	Dendritic structure of cast steel(e)
4 g picric acid in 100 ml methanol	3 to 5 h	C	Carbon segregation

(a) Parts are listed by volume. All acids are of concentrated strength; commercial grades ordinarily can be used instead of laboratory or reagent grades. Water or alcohol should never be poured into an acid; rather, the acid should always be poured and gradually stirred into the other liquid. (b) A indicates a saw-cut or machined surface; B, an average ground surface; C, a polished surface. (c) Ratio of HNO_3 to HF can vary as indicated. (d) Precede use of this etchant with etch in 10% nital for 10 to 20 s. (e) Overetching deposits excessive copper, which may obscure details of structure

Etchants for microscopic examination of carbon and alloy steels of medium carbon content

Etchant	Purpose, or characteristic revealed
Nital: 1 to 5 ml HNO_3 in 100 ml ethanol (95%) or methanol (95%)	General structure (most-used etchant for routine work)
Picral: saturated solution of picric acid in ethanol (95%) or methanol (95%)	General structure; provides better resolution of certain carbide structures than is obtained with nital
Vilella's reagent: 5 ml HCl, 1 g picric acid, 100 ml ethanol (95%) or methanol (95%)	Reveals outlines of prior austenite grain boundaries in quenched-and-tempered steels
Super picral: picral with a few drops of HCl or zephiran chloride per 25 ml of solution	General structure; for good resolution of carbide structures; often preferred for heat treated structures
Potassium metabisulfite solution: 10 g potassium metabisulfite in 100 ml water	For resolution of hardened structures; use should be preceded by an etch in nital or picral
Howarth's reagent: 10 ml H_2SO_4, 10 ml HNO_3, 80 ml water	Detection of overheating and burning
10 g tartaric acid, 100 ml water	Examination of inclusions
30 g potassium dichromate in 225 ml hot distilled water; add 30 ml acetic acid (glacial)	Reveals lead inclusions, causing them to appear yellow or gold under polarized light
Alkaline chromate solution: 16 g CrO_3 in 145 ml distilled water; add 80 g NaOH (slowly, with constant stirring)	Reveals intergranular oxidation; used for medium-carbon alloy steels that contain nickel

Checklist of principal macroetch observations to be recorded for semifinished steel products

Surface or subsurface	Center or central area	General
(A) Cracks	(a) Pipe	(α) Flakes or cooling cracks
(B) Seams or laps	(b) Porosity	(β) Dendritic pattern
(C) Decarburization	(c) Bursts	(γ) Ingot pattern
(D) Pinholes	(d) Segregations	(δ) Grain size
(E) Segregations		

Etchants for macroscopic examination of cast irons

Etchant	Composition	Etching technique	Application
Stead's reagent	10 g cupric chloride, 40 g magnesium chloride, 20 ml hydrochloric acid, 1 000 ml ethanol(a)	Immersion for up to 3 h	Used to reveal the eutectic cell number in gray cast irons
Rapid cell-etching reagent	10 g cupric chloride, 50 ml water, 100 ml hydrochloric acid	Dip etch for about 60 s	As above, but results are less distinct
Ammonium persulfate	10 g ammonium persulfate, 100 ml water, few drops concentrated H_2SO_4(b)	Immersion and swabbing	Reveals carbide and phosphide distribution
Nital	5 or 10% nitric acid, 95 or 90% ethanol	Dip etch for up to 3 min.	Used to reveal macrostructure in white irons
4% picral	4% picric acid, 96% ethanol	Dip etch for up to 3 min.	Used to reveal macrostructure in white irons

(a) Dissolve cupric chloride in a minimum quantity of hot water (10 to 15 ml); add magnesium chloride and dissolve; add ethanol, then hydrochloric acid. (b) Add H_2SO_4 just before use

Etchants for microscopic examination of cast irons

Etchant	Composition	Etching technique	Applications
Picral	4% picric acid, 96% ethanol	Dip etch for 2 to 10 s	General-purpose etching of all pearlitic gray, malleable and ductile cast irons; best etchant for pearlite; etches some austenitic cast irons, Ni-Hard and acicular irons
Nital, 5%	5% nitric acid, 95% ethanol	(1) Dip etch for 2 to 10 s	(1) General-purpose etching of all ferritic gray, malleable and ductile cast irons; etches grain boundaries; etches some austenitic irons and irons containing martensite
		(2) Electrolytic etch(a)	(2) High-chromium irons
Nital, 2%	2% nitric acid, 98% ethanol	Dip etch for 2 to 10 s	Observation of ferritic grain boundaries at high magnification
Ferric chloride	10 g ferric chloride, 100 ml water	Dip etch for 3 to 20 s	Austenitic cast irons
Mixed acid in glycerol	10 ml HNO_3, 20 ml HF, 40 ml glycerol	Dip etch for 10 to 40 s	High-silicon irons (14 to 16%) Si
Vilella's reagent	1 vol HNO_3, 2 vol HCl, 3 vol glycerol	Dip etch for up to 20 s	High-chromium irons
Potassium ferricyanide	10% alkaline aqueous solution of potassium ferricyanide	Dip etch for 5 to 30 s in etchant at 50 °C (122 °F)	High-chromium irons
Murakami's reagent	10 g KOH, 10 g $K_3Fe(CN)_6$	(1) Dip etch for 2 to 3 min.	(1) 30% chromium irons
		(2) Dip etch for 10 to 30 s	(2) High-phosphorus irons, to distinguish between iron phosphide and iron carbide
Alkaline sodium picrate	2 g picric acid, 25 g NaOh, 100 ml water; warm to dissolve	(1) Dip etch for 10 s to 2 min. at boiling point	(1) Blackens cementite
		(2) Electrolytic etch(b)	(2) Blackens cementite

(a) Specimen is anode; platinum cathode. Current density, 0.13 to 0.31 A/cm^2 (0.5 to 2.0 A/in.^{2}) for up to 2 min. (b) Specimen is anode; stainless steel cathode. Current density, 0.13 to 0.31 A/cm^2 (0.5 to 2.0 A/in.^{2}) for up to 2 min in cold solution

Etchants for microscopic examination of carbon and alloy steels

Etchant	Purpose, or characteristic revealed
Nital: 1 to 5 ml HNO_3 in 100 ml ethanol (95%) or methanol (95%)	Develops ferrite grain boundaries in low-carbon steels; produces maximum contrast between pearlite and a cementite or ferrite network; develops grain boundaries in 4% silicon steel; develops ferrite boundaries in structures consisting of martensite and ferrite; etches chromium-bearing low-alloy steels resistant to action of picral
Picral: saturated solution of picric acid in ethanol (95%) or methanol (95%)	Reveals maximum detail in pearlite, untempered and tempered martensite, and bainite; reveals undissolved carbide particles in martensite; differentiates ferrite, martensite, and massive carbide by coloration; differentiates bainite and fine pearlite; reveals carbide particles in grain boundaries of low-carbon steel and wrought iron
50 ml 1 to 2% nital, 50 ml 4% picral	Etches some alloy steels, such as 4340
Vilella's reagent: 5 ml HCl, 1 g picric acid, 100 ml ethanol (95%) or methanol (95%)	For contrast etching(a); reveals outlines of prior austenite grains in untempered and tempered martensite, and in austempered steels; reveals pearlite colonies
1 to 1.5 ml HCl (conc), 2 to 3 g picric acid, 100 ml ethanol (95%)	Reveals pearlite colonies(b)
10 g tartaric acid, 100 ml water	For grading inclusions(c)
30 g $K_2Cr_2O_7$ in 225 ml hot distilled water; add 30 ml acetic acid (glacial)	Reveals lead inclusions, causing them to appear yellow or gold when specimen is examined under polarized light(d)
16 g CrO_3 in 145 ml distilled water; add 80 g NaOH(e)	Reveals intergranular oxidation; used for medium-carbon alloy steels that contain nickel(f)
Super picral: picral with a few drops of HCl or zephiran chloride per 25 ml of solution	General structure; for good resolution of carbide structures; often preferred for heat treated structures
10 g potassium metabisulfite, 100 ml water	For resolution of hardened structures; use should be preceded by an etch in nital or picral
Howarth's reagent: 10 ml H_2SO_4, 10 ml HNO_3, 80 ml water	For detection of overheating and burning, and for examination of steel forgings
8 g sodium metabisulfite in 100 ml water	Produces good contrast in as-quenched martensitic structures
1 g KCN in 100 ml water, mixed with 0.25 g diphenylthiocarbazone in 10 ml chloroform	Reveals lead inclusions by coloring them red; coloration is most visible when specimens are viewed under polarized light

(a) Specimen should be tempered 20 to 30 min. at 316 °C (600 °F). (b) Immerse specimen 5 to 10 s in solution at room temperature. (c) Immerse specimen for 5 min., rinse in hot water, polish lightly with alumina to remove film, rinse in hot water and dry. (d) Etch 10 to 20 s in solution at room temperature, rinse in hot water and dry. (e) Sodium hydroxide (NaOH) must be added slowly, with constant stirring. (f) Immerse specimen in boiling solution for 10 to 30 min., rinse in hot water, dry in air blast

Compositions and applications of etchants for stainless steel casting alloys

Etchant no.	Etchant name	Composition
1	Oxalic acid (electrolytic, 6 V)	10 g oxalic acid, 100 ml water
2	Vilella's reagent	5 ml HCl, 1 g picric acid, 100 ml ethanol (95%) or methanol (95%)
3	Kalling's reagent 2	100 ml HCl, 5 g $CuCl_2$, 100 ml ethanol (95%)
4	Murakami's reagent (unheated)	1 to 4 g $K_3Fe(CN)_6$, 10 g KOH (or 7 g NaOH), 100 ml water
5	Murakami's reagent (boiling)	Same composition as etchant 4, above, but heated to boiling temperature for use
6	Chromic acid (electrolytic, 6 V)	10 g CrO_3, 100 ml water
7	10N potassium hydroxide (electrolytic, 6 V)	560 g KOH diluted with distilled water to a volume of 1000 ml
8	HCl, HNO_3, acetic acid	15 ml HCl, 10 ml HNO_3, 10 ml acetic acid
9	Acid ferric chloride	Saturated solution of $FeCl_3 \cdot 6H_2O$ in concentrated HCl; add a few drops HNO_3
10	Glyceregia	10 ml HNO_3, 20 to 50 ml HCl, 30 ml glycerol
11	Sodium cyanide (electrolytic, 6 V)	10 g NaCN, 90 ml water

Application of etchants to examination of specific stainless steel casting alloys

Etchant numbers correspond to those assigned in the table on compositions and applications of etchants for stainless steel casting alloys

Alloy	Normal heat treatment	General microstructure	Etchants for revealing: Ferrite	Carbide	Sigma phase
CA-6NM	Hardened and tempered(a)	2	3	4	...
CA-15	Hardened and tempered(b)	2 or 9	3	4	...
CD-4MCu	Annealed(b)	1, 2, or 6	...	...	2 then 7
CE-30	As cast	2	3	4	2 then 7; or 11
CF-3	Annealed(c)	7	3	4	2 then 7
CF-3M	Annealed(c)	8	3	4	2 then 7; or 11
CF-8	Annealed(d)	7 or 10	3	4	2 then 7
CF-8C	Annealed(d)	1 or 6	3	4	2 then 7; or 11
CF-8M	Annealed(d)	9	3	4	2 then 7
CF-20	Annealed(d)	1	3	4	...
CG-8M	Annealed(d)	1	3	4	5; or 7 then 11
CN-7M	Annealed(e)	1 or 6	...	...	...

(a) Heat to 955 °C (1750 °F) min, air cool and temper at 593 °C (1100 °F) min. (b) Heat to 1120 °C (2050 °F) min, furnace cool to 1040 °C (1900 °F), quench in water or oil. (c) Heat to 1040 °C (1900 °F) min, rapid cool. (d) Heat to 1040 °C (1900 °F) min, water quench. (e) Heat to 1120 °C (2050 °F) min, water quench

Compositions and procedures for use of etchants for microscopic examination of wrought stainless steels

Etchant	Composition	Procedure
1	**Vilella's reagent:** 5 ml HCl, 1 g picric acid, 100 ml ethanol (95%) or methanol (95%)	Immerse or swab specimen for a few seconds to 15 min.; reaction may be accelerated by adding a few drops of 3% H_2O_2
2(a)	**Glyceregia:** 10 ml HNO_3, 20 to 50 ml HCl, 30 ml glycerol	Mix HCl and glycerol thoroughly before adding HNO_3; discard before solution attains a dark orange color; immerse or swab specimen for a few seconds to a few minutes; higher percentage of HCl minimizes pitting
3(a)	10 ml HNO_3, 20 ml HCl, 30 ml water	Immerse specimen for a few seconds to a minute; produces much stronger reaction than etchant 2; discard before solution attains a dark orange color
4(a)	10 ml HNO_3, 10 ml acetic acid, 15 ml HCl, 2 to 5 drops glycerol	Immerse or swab specimen for a few seconds to a few minutes
5	10 ml HNO_3, 20 ml HF, 20 to 40 ml glycerol	Immerse specimen for 2 to 10 s
6	**Fry's reagent:** 40 ml HCl, 5 g $CuCl_2$, 30 ml water, 25 ml ethanol (95%) or methanol (95%)	Swab specimen for a few seconds to a minute
7	5 g $FeCl_3$, 50 ml HCl, 100 ml water	Immerse or swab for a few seconds to a few minutes; small additions of HNO_3 activate solution and minimize pitting Alternative procedure: Immerse or swab specimen for a few seconds at a time; repeat as necessary
8	5 g $FeCl_3$, 15 ml HCl, 60 ml ethanol (95%) or methanol (95%)	Immerse or swab specimen for a few seconds to a few minutes
9	10 ml HCl, 100 ml ethanol (95%) or methanol (95%)	Immerse specimen for 5 to 30 min., or electrolytic at 6 V for 3 to 5 s
10(a)	Concentrated HNO_3	Electrolytic at 0.2 A/cm^2 for a few seconds
11a	**Kalling's reagent 1:** 2 g $CuCl_2$, 40 ml HCl, 40 to 80 ml ethanol (95%) or methanol (95%), 40 ml water	Immerse or swab specimen for a few seconds to a few minutes
11b	**Kalling's reagent 2:** 2 g $CuCl_2$, 40 ml HCl, 40 to 80 ml ethanol (95%) or methanol (95%)	Submerged swabbing for a few seconds to several minutes; attacks ferrite more readily than austenite
12	85 g NaOH, 50 ml water	Electrolytic at 6 V for 5 to 10 s
13	45 g KOH, 60 ml water	Electrolytic at 2.5 V for a few seconds; usually stains sigma and chi yellow to red-brown, ferrite gray blue-gray, carbides barely touched, austenite not touched
14(a)(b)	**Murakami's reagent:** 10 g $K_3Fe(CN)_6$, 10 g KOH or NaOH, 100 ml water; use fresh solution	Immerse or swab specimen for 15 to 60 s; stains carbides and sigma(c) Immerse in fresh, hot solution 2 to 20 min.; stains carbides dark, ferrite yellow, sigma blue; austenite turns brown on overetching Swab 5 to 60 s; immersion will produce a stain etch Follow with water rinse, alcohol rinse, and dry
15	10 g ammonium persulfate, 100 ml water	Electrolytic at 6 V for a few seconds to a minute
16	25 ml HCl, 3 g ammonium bifluoride, 125 ml water, few grains potassium metabisulfite	Mix fresh; for stock solution, mix first three items; add potassium metabisulfite just before use Immerse specimen for a few seconds to a few minutes
17	10 g $FeCl_3$, 90 ml water	Immerse specimen for a few seconds
18	10 g oxalic acid, 100 ml water	Electrolytic at 6 V for a few to 60 s
19	10 g CrO_3, 100 ml water	Electrolytic at 6 V for 5 to 60 s; attacks carbides
20	2 g CrO_3, 20 ml HCl, 80 ml water	Immerse 5 to 60 s; CrO_3 may be increased up to 20 g for difficult alloys; staining and pitting increase as CrO_3 is increased
21(a)(b)	10 g NaCN, 100 ml water	Electrolytic at 6 V: 5 s for etching sigma, 30 s for ferrite and general structure, and up to 5 min. for carbides.
22	20 ml HNO_3, 4 ml HCl, 20 ml methanol	Immerse specimen for 10 to 60 s
23	5 ml HNO_3, 45 ml HCl, 50 ml water	Immerse specimen for 10 min. or longer
24	Concentrated NH_4OH	Electrolytic at 6 V for 30 to 60 s; attacks carbides only
25	**Marble's reagent:** 10 g $CuSO_4$, 50 ml HCl, 50 ml water	Immerse or swab specimen for 5 to 60 s

(a) Use exhaust hood; etchant can give off extremely poisonous or noxious fumes. (b) Poisonous by ingestion and by contact. To discard, neutralize or turn basic with ammonia and flush down an acid-disposal drain with a large amount of water. (c) To differentiate, etchant 15, electrolytic at 4 V, will attack sigma but not carbides. If pitting occurs, reduce voltage

Etchants suggested for microscopic examination of wrought stainless steels

Steel	Etchant no.(a)
200 and 300 series:	
General structure	18, 4, 2, 3, 19, 1, 11a, 11b, 20, 25
Sigma	13, 12, 14, 21
Carbides	18, 24
Carbides and sigma	22, 23
400 series:	
General structure	1, 2, 3, 4, 5, 6, 7, 8, 9, 10, 11a, 11b
Sigma	12, 13, 14
Carbides	15
600 series:	
General structure	1, 15, 4, 16, 17, 9
Carbides	15

(a) Numbers correspond to etchants for which compositions and procedures are presented in accompanying table. Where two or more etchants are given, they are listed in order of descending preference

Heat-Resisting Alloys

Microetching procedures for wrought iron-nickel-chromium heat-resisting alloys
Etchant numbers correspond to those assigned in the table on etchants for microscopic examination of wrought heat-resisting alloys

Etchant no.	Etching method	Etching time, s	Cell voltage	Purpose, or characteristics revealed
Alloy A-286 (AISI 660)				
1	Swab	3 to 20	...	General structure
2	Swab	5 to 60	...	General structure; may stain or pit
3	Immerse	10 to 60	...	General structure
Incoloy 800				
4	Electrolytic	15 to 30	5 to 10	General structure; grain boundaries
5	Electrolytic	10 to 20	5 to 10	Grain boundaries; carbide particles
6	Electrolytic	10 to 15	20	Carbide particles
7	Swab	15 to 30	...	Grain boundaries(a); carbide; no staining
8	Electrolytic	10 to 30	10	Preferential attack at grain boundaries
Incoloy 825				
4	Electrolytic	15 to 30	5 to 10	General structure; grain boundaries
5	Electrolytic	10 to 20	5 to 10	Grain boundaries; carbide particles
7	Swab	15 to 30	...	Grain boundaries; carbide; no staining
9	Swab or immerse	(b)	...	General structure
RA 330				
8	Electrolytic	5 to 10	5	For etch pitting
10	Electrolytic	2 to 10	3	General structure; precipitates

(a) Grain boundaries are faint if free of carbide particles. (b) Etching time varies from a few seconds to 12 min.

Microetching procedures for wrought cobalt-base heat-resisting alloys
Etchant numbers correspond to those assigned in the table on etchants for microscopic examination of wrought heat-resisting alloys

Etchant no.	Etching method	Etching time, s	Cell voltage	Purpose, or characteristics revealed
Haynes 25 (AISI 670), Haynes 188				
26	Electrolytic	2 to 5	6	General structure
Stellite 6B				
26	Electrolytic	2 to 5	6	General structure
27:				
Stage 1	Electrolytic	2 to 5	6	General structure
Stage 2	Immerse	5 to 10	...	Carbide particles

Microetching procedures for wrought nickel-base heat-resisting alloys
Etchant numbers correspond to those assigned in the table on etchants for microscopic examination of wrought heat-resisting alloys

Etchant no.	Etching method	Etching time, s	Cell voltage	Purpose, or characteristics revealed
Hastelloy C				
11	Electrolytic	2 to 10	3	General structure
Hastelloy W				
12	Immerse	...	...	General structure
Hastelloy X (AISI 680)				
13	Electrolytic	2 to 10	6	General structure; remove stains with HNO_3
Inconel 600 and 601				
4	Electrolytic	15 to 20	5 to 10	General structure. No pitting
5	Electrolytic	15 to 20	5 to 10	General structure; grain boundaries; carbide
6	Electrolytic	15 to 20	5 to 10	General structure; excellent for revealing carbide particles
7	Swab or immerse	(a)	...	Grian-boundary contrast fair; carbide
14	Immerse	(b)(c)	...	General structure; carbide particles
Inconel 625				
8	Electrolytic	1 to 2	50	Grain-boundary films; results vary with thermal history of specimen
15	Electrolytic	10 to 20	5 to 10	Grain boundaries; no staining; results vary with thermal history of specimen
16	Electrolytic	15 to 20	5 to 10	General structure; grain boundaries
17	Electrolytic	15 to 20	5 to 10	General structure
18	Electrolytic	8 to 20	2 to 10	Outlines phases; may cause pitting; poor results on cold worked metal

(a) $\frac{1}{2}$ to 5 min. (b) 1 to 5 min. (c) Heat specimen to reduce etching time. (d) 5 to 10 min. (e) Use well-prepared specimen

(continued)

Microetching procedures for wrought nickel-base heat-resisting alloys (continued)

Etchant numbers correspond to those assigned in the table on etchants for microscopic examination of wrought heat-resisting alloys

Etchant no.	Etching method	Etching time, s	Cell voltage	Purpose, or characteristics revealed
Inconel 706 and Alloy 718				
8	Electrolytic	1 to 2	50	Grain-boundary films; shows grain boundaries in relief
15	Electrolytic	10 to 20	5 to 10	Good for general structure and phase outline; grain boundaries
16	Electrolytic	15 to 20	5 to 10	Good for general structure, phase outline, and matrix segregation for most heat treated conditions; grain boundaries
17	Electrolytic	15 to 20	5 to 10	General structure; precipitate phases in fully heat treated material
19	Swab or immerse	(a)	...	General structure; microsegregation
20	Immerse	(d)	...	Carbide particles; chromium carbide particles darken faster than nitrides and Laves phase
Inconel X-750 (AISI 688)				
4	Electrolytic	15 to 20	5 to 10	General structure; no pitting
5	Electrolytic	15 to 20	5 to 10	Grain boundaries; carbide; no pitting
15	Electrolytic	10 to 20	5 to 10	Good for revealing grain boundaries and carbide particles
21	Swab	2 to 10	...	Excellent for showing details of overaged gamma prime
22	Swab	5 to 60	...	General structure; microsegregation
U-700 (AISA 687)				
23	Swab or immerse	10 to 20	...	Good for contrast(e)
24	Swab or immerse	(b)	...	General structure; grain boundaries; no staining
28	Electrolytic	5 to 20	5 to 10	General structure; grain boundaries
Waspaloy (AISI 685)				
1	Swab	3 to 20	...	General structure
24	Swab or immerse	(b)	...	General structure; no staining
25	Swab or immerse	5 to 30	...	General structure

(a) $\frac{1}{2}$ to 5 min. (b) 1 to 5 min. (c) Heat specimen to reduce etching time. (d) 5 to 10 min. (e) Use well-prepared specimen

Etchants for microscopic examination of wrought heat-resisting alloys

Etchant no.	Etchant name	Composition(a)	Remarks on preparation and use
1	HCl, HNO₃, acetic acid	15 ml HCl, 10 ml HNO$_3$, ml acetic acid	...
2	Chrome regia	2 g CrO$_3$, 20 ml HCl, 80 ml water	CrO$_3$ may be increased, but staining may result
3	Ferric chloride—hydrochloric	5 g FeCl$_3$, 15 ml HCl, 100 ml methanol	...
4	Nital	5 ml HNO$_3$, 95 ml methanol	Use colorless acid and absolute methanol
5	Oxalic acid	10 g oxalic acid, 100 ml water	Can be stored
6	Phosphoric acid	80 ml H$_3$PO$_4$, 20 ml water	Change to a 1-to-1 solution for specific results
7	Glyceregia	10 ml HNO$_3$, 20 ml HCl, 40 ml glycerol	Must be freshly prepared
8	Hydrochloric-methanol	10 ml HCl, 90 ml methanol	Water can be substituted for methanol to show segregation
9	Vilella's reagent	5 ml HCl, 1 g picric acid, 100 ml methanol	A few drops of 3% H$_2$O$_2$ will speed etching reaction
10	HCl-H₂O	5 ml HCl, 95 ml water	...
11	Chromic acid	2 to 10 g CrO$_3$, 100 ml water	...
12	Hydrochloric-chromic	80 ml HCl, 20 ml 50% chromic acid	Use fresh solution
13	Oxalic acid	10 g oxalic acid, 90 ml water	...
14	Nitric-hydrofluoric	20 ml HNO$_3$, 3 ml HF	Use colorless acids; remove thoroughly by water rinse
15	Chromic-acetic	25 g CrO$_3$, 7 ml water, 130 ml acetic acid	Can be stored for up to one month
16	Chromic acid	5 g CrO$_3$, 100 ml water	...
17	47-41-12	47 ml H$_2$SO$_4$, 41 ml HNO$_3$, 12 ml H$_3$PO$_4$	Add H$_2$SO$_4$ last, and slowly; produces noxious fumes and is highly corrosive
18	Hydrochloric-acetic	10 ml acetic acid, 3 drops HCl, 90 ml water	...
19	Inverted glyceregia	50 ml HCl, 10 ml glycerol, 10 ml HNO$_3$	...
20	Murakami's reagent	10 g KOH or NaOH, 10 g K$_3$Fe(CN)$_6$, 100 ml water	Dissolve KOH (or NaOH) and K$_3$Fe(CN)$_6$ in boiling water; etch specimen in boiling solution; prepare fresh for use
21	Nitric-hydrofluoric	50 ml HNO$_3$, 50 drops HF	Use colorless acids
22	Hydrochloric-hydrofluoric-nitric	80 ml HCl, 13 ml HF, 7 ml HNO$_3$	...
23	Marble's reagent	4 g CuSO$_4$ · 5H$_2$O, 20 ml HCl, 20 ml water	Dissolve CuSO$_4$ in water and add HCl
24	Kalling's reagent	2 g CuCl$_2$, 40 ml HCl, 80 ml methanol	Can be stored
25	92-5-3	92 ml HCl, 5 ml H$_2$SO$_4$, 3 ml HNO$_3$	Must be freshly prepared
26	Hydrochloric—hydrogen peroxide	97 ml HCl, 3 ml 3% H$_2$O$_2$	Must be freshly prepared
27	Grosbeck's reagent (two-stage)	Stage 1: 2 to 10% CrO$_3$ in water; stage 2: equal parts 20% KMnO$_3$, 8% NaOH	Mix second stage immediately before use
28	HCl-ethanol-H₂O₂	35 ml HCl, 65 ml ethanol (95%), 7 drops H$_2$O$_2$ (30%)	Must be freshly prepared

(a) Use concentrated acids, unless indicated otherwise. Use distilled water to avoid staining

Etchants for microscopic examination of iron-chromium-nickel heat-resistant casting alloys

Common name	Composition	Remarks on use
Etchants for delineating general structure		
Aqua regia	20 ml HNO_3, 60 ml HCl	Immerse specimen
Glyceregia	10 ml HNO_3, 20-50 ml HCl, 20 ml glycerol	Immerse specimen; use a hood
Hydrochloric acid (50%)	50 ml HCl, 50 ml water	Outlines ferrite; immerse specimen
Marble's reagent	10 g $CuSO_4$, 50 ml HCl, 50 ml water	Immerse specimen
Vilella's reagent	1 g picric acid, 5 ml HCl, 100 ml ethanol	Immerse specimen
Etchants for staining or film-forming		
Alkaline hydrogen peroxide	25 ml NH_4OH, 50 ml H_2O_2 (3%), 25 ml water	Ordinarily used after a delineating etchant; immerse specimen
Alkaline potassium ferricyanide	10 g $K_3Fe(CH)_6$, 10 g NaOH, 100 ml water	Same as above
Alkaline potassium permanganate	4 g NaOH, 10 g $KMnO_4$, 85 ml water	Same as above
Alkaline sodium picrate	2 g picric acid, 25 g NaOH, 100 ml water	Same as above
Emmanuel's reagent	30 g $K_3Fe(CN)_6$, 30 g KOH, 60 ml water	Attacks sigma phase with little or no effect on carbide particles; immerse specimen
Murakami's reagent	10 g $K_3Fe(CN)_6$, 10 g KOH, 100 ml water	Stains carbide particles without staining sigma phase(a); immerse specimen
Solutions for electrolytic etching		
Ammonium hydroxide	Concentrated NH_4OH	Final electrolytic etch after etching in Vilella's reagent and in 10N KOH (electrolytic)
Cadmium acetate	10 g cadmium acetate, 100 ml water	Attacks (Cr, Fe)$_{23}$C$_6$ carbide particles
Chromic acid	2 to 10 g Cr_2O_3, 100 ml water	Outlines carbide particles; extracts sigma phase
Lead acetate (2N)	38 g $Pb(C_2H_3O_2)_2 \cdot 3H_2O$, distilled water to make 100 ml	Stains austenite, then sigma phase, then carbide particles; 1.5 V for 30 s
Oxalic acid	10 g oxalic acid, 100 ml water	Outlines carbide and sigma; 6 V, 1 to 5 s
Potassium hydroxide (1N)	5.6 g KOH, 100 ml water	Blackens sigma phase without outlining other phases; 1.5 V for 1 s
Potassium hydroxide (10N)	56 g KOH, 100 ml water	Intermediate etch between Vilella's and ammonium hydroxide (electrolytic)
Sodium cyanide	10 g NaCN, 100 ml water	Used after glyceregia; outlines carbide particles, stains sigma phase; use at 1 A/in.^{2} for 1 to 5 s, under hood

(a) Sometimes sigma phase is stained. Behavior must be established on a given composition

Silicon Steels

Pitting etchants for determination of grain orientation in silicon steels by optical microscopy

Etchant	Composition (parts are by volume)	Conditions for use	Purpose
1	1 part HF, 1 part HNO_3, 4 parts water	Immerse for 10 s	Exposes {100} crystallographic faces in (110) [001] (cube-on-edge) oriented 3.25% Si steel
2	2 parts HF, 1 part HNO_3, 3 parts methanol, 4 parts glycerol	Swab for 1 min.	Same as for etchant 1
3	A: 6 ml H_2O_2 (30%), 0.1 ml HCl, 100 ml water B: 40 ml $FeCl_3 \cdot 6H_2O$, 40 ml ethanol, 20 ml water	Immerse in A for 10 s, rinse and dry; then immerse in B for 3 s, rinse and dry	Develops etch pits in (110)[001] (cube-on-edge) oriented 3.25% silicon steel
4	100 g ferric sulfate, 100 ml H_2SO_4, 1000 ml water	Immerse for 15 s in solution heated to 80 to 90 °C (175 to 195 °F)	Develops etch pits in (100)[001] (cube-on-face) oriented 3.25% silicon steel
5	A: 5 ml HF, 95 ml methanol B: 100 ml H_2O_2 (3%), 100 ml water, 2 drops HCl C: 5 ml HCl, 95 ml methanol	Polish, etch heavily in nital; repolish; etch in nital to reveal grain boundaries; immerse 10 s in A, rinse, dry; immerse 2 s in B, rinse, dry; immerse 30 s in C, rinse, dry	Exposes {100} crystallographic faces in primary recrystallized 3.25% silicon steel and nonoriented silicon steels
6	600 g $FeCl_3$, 10 g ammonium bisulfate, 600 ml HCl, 150 ml HNO_3, 1650 ml water	Immerse for 1 min in solution heated to 49 to 60 °C (120 to 140 °F)	Exposes {111} crystallographic faces in secondary recrystallized 50Ni-50Fe

Refractory Metals

Etchants for metallographic specimens of refractory metals

Etchant name or ASTM number (E407)	Composition
Etchants for tungsten and molybdenum and their alloys	
Murakami's reagent (etchant 98c)	10 g $K_3Fe(CN)_6$, 10 g KOH or NaOH, 100 ml water
Murakami's reagent (mod)	15 g $K_3Fe(CN)_6$, 2 g NaOH, 100 ml water
Etchant 131 (electrolytic)	5 ml H_2SO_4, 1 ml HF, 100 ml methanol (95%)
Etchant 132(a)	5 ml HF, 10 ml HNO_3, 30 ml lactic acid
Etchant 209	15 ml HNO_3, 3 ml HF, 80 ml water
Additional etchants for molybdenum and molybdenum alloys	
Etchant 129	10 ml HF, 30 ml HNO_3, 60 ml lactic acid
Etchant 130 (electrolytic)	25 ml HCl, 10 ml H_2SO_4, 75 ml methanol
Etchants for niobium and tantalum and their alloys	
Etchant 66	30 ml HF, 15 ml HNO_3, 30 ml HCl
Etchant 158	10 ml HF, 10 ml HNO_3, 20 ml glycerol
Etchant 159	5 ml HF, 20 ml HNO_3, 5 ml HF, 50 ml water
Etchant 161	25 ml HNO_3, 5 ml HF, 50 ml water
Etchant 163	30 ml H_2SO_4, 30 ml HF, 3 to 5 drops H_2O_2 (30%), 30 ml water
Etchant 164	50 ml HNO_3, 30 g ammonium bifluoride, 20 ml water
Additional etchants for niobium and niobium alloys	
Etchant 160	20 ml HF, 15 ml H_2SO_4, 5 ml HNO_3, 50 ml water
Etchant 162B	30 ml lactic acid, 10 ml HNO_3, 10 ml HF
HNO_3-HF-water	20 ml HNO_3, 10 ml HF, 70 ml water
Hcl-H_2SO_4-HNO_3-water	15 ml HCl, 15 ml H_2SO_4, 8 ml HNO_3, 62 ml water
Additional etchants for tantalum and tantalum alloys	
Etchant 177	10 g NaOH, 100 ml water
Etchant 178	20 ml HF, 20 ml HNO_3, 60 ml lactic acid
Etchant 179B (electrolytic)	10 ml HF, 90 ml H_2SO_4

(a) Procedure: Swab with heavy pressure for 5 to 10 s, water rinse, alcohol rinse, dry, etch with Murakami's reagent (etchant 98c)

Magnetic Alloys

Electrolytes and conditions for electropolishing of iron-nickel and iron-cobalt magnetic alloys

Alloy	Electrolyte	Conditions for use
Fe-Ni only	135 ml acetic acid (glacial), 25 g CrO_3, 7 ml water	80 V, 0.8 to 1.6 A/cm^2, 5 to 30 s at 7 °C (45 °F) max
Fe-Ni or Fe-Co	100 ml acetic acid (glacial), 10 ml perchloric acid	45 V, 0.2 A/cm^2, 3 to 4 min. at 25 °C (75 °F)

Etchants for microscopic examination of iron-nickel and iron-cobalt magnetic alloys

Etchant	Composition	Conditions for use(a)
Chemical etching		
1	100 ml HCl, 2 g $CuCl_2$, 7 g $FeCl_3$, 5 ml HNO_3, 200 ml methanol, 100 ml water	Immerse or swab for 10 to 15 s
2(b)	15 ml HCl, 5 g $FeCl_3$ (anhydrous), 60 ml ethanol	Immerse for 5 to 10 s
3	3 ml HCl, 1 ml HNO_3, saturated with $CuCl_3$	Swab for 2 to 3 s
4	15 ml HCl, 5 ml HNO_3, 10 ml glycerol	Swab for 10 to 15 s
5	Ammonium persulfate (saturated aqueous solution)	Immerse for 20 to 30 s
6	2 to 10% nital (HNO_3 in ethanol or methanol)	Immerse for 5 to 10 s
7	50 ml HCl, 10 g $CuSO_4$, 50 ml water (Marble's reagent)	Immerse or swab for 5 s
Electrolytic etching		
8	5 to 10 ml HCl, 100 ml water	2 to 5 s at 250 to 500 mA/cm^2
9	2 g CrO_3, 100 ml water	2 to 5 s at 100 to 200 mA/cm^2
10	3% sulfuric acid	5 to 10 s

(a) All etchants are used at room temperature. (b) Recommended for electron metallography

Electrolytes for electrolytic etching of permanent magnet materials other than Alnico alloys and hard ferrites

Magnetic material	Electrolyte	Characteristic revealed
Vicalloy	10 ml HNO_3, 90 ml water	General structure
Cunife	10 ml $CuSO_4$, 50 ml HCl, 50 ml water	Grain size and structure of the solid-solution alloy
Rare earth ($Co_3Cu_{1.6}Fe_{0.5}Ce$)	Nital (various strengths)	Identification of copper-rich phase
82Co-6Au-12Fe	3 ml HNO_3, 3 ml H_2SO_4, 94 ml water	Grain-boundary precipitates; use after electropolishing

Etchants recommended for microscopic examination of iron-nickel and iron-cobalt magnetic alloys

Etchant numbers correspond to those assigned in the table on etchants for microscopic examination of iron-nickel and iron-cobalt magnetic alloys

Etchant	Characteristics revealed
Iron-nickel alloys	
1, 2, 4, 5(a), 6	Grain size, structure
3(b), 7	Grain size
Iron-cobalt alloys	
1, 5, 6, 7, 8, 9, 10	Grain size, structure

(a) For etching 50Fe-50Ni. (b) For etching high-nickel alloys such as Moly Permalloy

Electrical Contact Materials

Etchants and etching procedures for electrical contact materials

Etchant No.	Composition	Procedure for use
1	20 ml NH_4OH, 10 to 20 ml H_2O_2 (30%), 10 to 20 ml water	Swab at room temperature, 3 to 10 s; use fresh; more water, less H_2O_2 for copper alloys; vice versa for silver alloys
2	2 g $K_2Cr_2O_7$, 1.5 g NaCl, 8 ml H_2SO_4 (conc), 100 ml water	Swab at room temperature, 5 to 10 s; good for etching hard-to-etch copper alloys
3	50 ml HN_4OH, 10 to 30 ml H_2O_2 (30%)	Swab at room temperature for 3 to 10 s; use fresh
4	10 g $FeCl_3$, 90 ml water	Swab or immerse
5	A: 100 ml saturated aqueous solution of $K_2Cr_2O_7$, 2 ml saturated aqueous solution of NaCl, 10 ml H_2SO_4 B: 1 part solution A, 10 parts water C: 98 ml water, 3 g CrO_3, 2 ml H_2SO_4	Use solution A, then solution B, then solution C; swab at room teperature for 15 to 20 s with each solution; rinse in water between solutions
6	20 g CrO_3, 4.5 g NH_4Cl, 18 ml HNO_3 (conc), 15 ml H_2SO_4 (conc), water to make ½ L (Waterbury reagent)	Dilute 2 to 1 with water at time of use; swab at room temperature for 3 to 10 s
7	A: 25 ml HNO_3, 1 g $K_2Cr_2O_7$, 100 ml water B: 40 g CrO_3, 3 g Na_2SO_4, 200 ml water	Mix equal parts of A and B; swab at room temperature for 5 to 10 s
8	20 ml HNO_3 (conc), 20 ml acetic acid (glacial), 20 ml glycerol	Swab at 38 to 42 °C (100 to 110 °F) for 3 to 10 s
9	0.2% CrO_3 and 0.2% H_2SO_4, in water	Swab for 1 min.
10	A: 200 ml HNO_3 (50%), 2 g $K_2Cr_2O_7$ B: 20 g CrO_3, 1.5 g Na_2SO_4, 100 ml water	Mix 1 part A with 20 parts B at time of use; swab at room teperature, 3 to 15 s
11	10 ml $K_3Fe(CN)_6$ (30%), 10 ml NaOH (10%) (Murakami's reagent)	Swab at room temperature for 5 to 15 s; use at half strength for more control
12	20 ml HNO_3 (conc), 20 ml acetic acid (glacial)	Immerse at room temperature, 10 to 20 s
13	20 ml KCN (10%), 20 ml $(NH_4)_2S_2O_8$ (10%)	Use in a hood; immerse at room temperature for 10 to 30 s
14	A: 5% nital B: 5% $FeCl_3$ in methanol	Immerse specimen alternately in A and B
15	10 ml HNO_3, 20 ml HCl, 10 ml glycerol	Swab at room temperature for 3 to 10 s
16	30 ml HCl, 10 ml water	Electrolytic; up to 5 V dc; 1.5 A/cm^2; room temperature, 1 to 3 min.

Suggested etchants for specific electrical contact materials

Etchant numbers correspond to those assigned in the table on etchants and etching procedures for electrical contact materials

Material	Etchant no.	Material	Etchant no.	Material	Etchant no.
Copper-graphite	4	Silver-graphite	6, 3, 8(a)	Gold-silver-platinum	13
Chromium-copper	1	Silver-nickel	9, 10	Gold-plated nickel-iron	11
Cadmium-copper	1	Silver-magnesium-nickel	9	Gold-silver clad palladium	12
Copper-cobalt-beryllium	1, 2	Silver-molybdenum	11	Platinum-ruthenium	16
Copper-cobalt-silicon	1, 2	Silver-tungsten	3, 11	Platinum-iridium	16
Copper-tungsten	1, 11	Silver-tunsten carbide	11	Palladium welded to nickel silver	12
Silver (99.9% Ag)	3, 4, 5	Tungsten	11	Palladium-rithenium	13
Silver-copper	6, 7	Tungsten-copper	11	Palladium-copper	13
Silver-copper-cadmium	7	Tungsten-nickel	11	Palladium-silver	13
Silver-cadmium	6, 3, 8(a)	Molybdenum	11	Palladium-platinum-gold-silver-copper-zinc	16
Silver-cadmium brazed to brass	6, 3, 8(a)	Molybdenum-silver	11		

(a) In the order given

Sleeve-Bearing Materials

Etchants for microscopic examination of sleeve-bearing materials

Etchant	Some applications
NH₄OH, H₂O₂(a)	Commercial bronze liner
	Copper-lead alloy liner
	Copper-lead tin alloy liner
	High-leaded tin bronze liner
	Leaded tin bronze liner
	Lead-tin-copper overlay on copper-lead alloy liner
	Nickel bronze infiltrated with lead-base babbitt
	Nickel-tin bronze infiltrated with lead-base babbitt
	Silver electroplate on steel
	Silver-lead alloy electroplate on steel
	Tin-base babbitt overlay on copper-lead-tin alloy liner
	Tin bronze infiltrated with lead-base babbitt
	Tin bronze infiltrated with Teflon
	Trimetal bearing: lead-tin-copper electroplated overlay, brass electroplated barrier, copper-lead alloy
0.5% HF	Aluminum alloy clad to steel
	Aluminum-silicon alloy clad to steel
	High-tin aluminum alloy clad with unalloyed aluminum
	Lead-tin-copper overlay on aluminum alloy liner
	Low-tin aluminum alloy clad to steel
	Trimetal bearing: lead-tin-copper electroplated overlay, copper electroplated barrier, aluminum-silicon-cadmium alloy
5% nital	High-tin aluminum alloy clad to nickel-plated steel
	Lead-base babbitt liner
	Tin-base babbitt liner
	Steel backing of any bearing only
Keller's reagent	Lead-tin-copper overlay on aluminum-cadmium alloy
Ferric chloride(b)	Cadmium alloy liner

(a) Equal parts of concentrated NH₄OH and water with 2 to 4 drops of H₂O₂ (30%) per 10 ml of solution. (b) 10 g FeCl₃, 90 ml ethanol

Aluminum

Etchants for use in macroscopic examination of aluminum alloys

Etchant	Composition(a)	Procedure for use
1 (caustic etch)	10 t NaOH to each 90 ml water	Immerse specimen 5 to 15 min. in solution heated to 60 to 70 °C (140 to 160 °F)(b), rinse in water, dip in 50% HNO₃ solution to desmut, rinse in water, dry
2 (Tucker's reagent)	45 parts HCl(c), 15 parts HNO₃(c), 15 parts HF (48%), 25 parts water	Mix fresh before using; immerse or swab specimen for 10 to 15 s, rinse in warm water, dry, and examine for desired effect; repeat as necessary until desired effect is obtained
3	1 part HF (48%), 9 parts water	Requires fairly smooth surface; immerse until desired effect is obtainec, hot-water rinse, dry
4 (Poulton's reagent)	12 parts HCl(c), 6 parts HNO₃(c), 1 part HF (48%), 1 part water	May be premixed and stored(d) for long periods; etch by brief immersion or by swabbing; rinse in cool water, and do not allow either the etchant or the specimen to heat up during etching
5	50 parts HCl(c), 15 parts HNO₃(c), 3 parts HF (48%), 5 parts FeCl₃ solution(c)	Mix fresh before use; cool solution to 10 to 15 °C (50 to 59 °F) with jacket of cold water; immerse a few seconds, rinse in cold water; repeat immersion and rinsing until desired effect is obtained

(a) Parts are by volume. (b) This etchant may be used without being heated, but the etching action will be slower. (c) Concentrated. (d) Solution should be stored in a vented container, preferably under a fume hood, to prevent buildup of gas pressure. The container should be made of polyethylene or be lined with wax

Etchants for use in microscopic examination of aluminum alloys

Etchant	Composition	Procedure for use
1 (hydrofluoric acid etch)	1 ml HR (48%), 200 ml water	Swab for 15 s or immerse for 30 to 45 s
2	1 g NaOH, 100 ml water	Swab for 5 to 10 s
3A (Keller's reagent)	2 ml HF (48%), 3 ml HCl(a), 5 ml HNO$_3$(a), 190 ml water	Immerse for 8 to 15 s, wash in stream of warm water, blow dry; do not remove etching products from surface
3B (dilute Keller's reagent)	20 ml etchant 3A, 80 ml water	Mix fresh before using; immerse specimen for 5 to 10 s
4 (modified Keller's reagent)	2 ml HF (48%), 3 ml HCl(a), 20 ml HNO$_3$(a), 175 ml water	Immerse for 10 to 60 s, wash in stream of warm water, blow dry; do not remove etching products from surface
5 (Barker's reagent)	4 to 5 ml HBF$_4$ (48%), 200 ml water	Electrolytic: use Al, Pb or stainless steel for cathode, specimen is anode; anodize 40 to 80 s at about 0.2 A/cm^2 (about 20 V dc); check results on microscope with crossed polarizers
6	25 ml HNO$_3$(a), 75 ml water	Immerse in solution at 70 °C (160 °F) for 40 to 60 s
7	20 ml H$_2$SO$_4$(a), 80 ml water	Immerse at 70 °C (160 °F) for 30 s; rinse in cold water
8	10 ml H$_3$PO$_4$ (85%), 90 ml water	Immerse at 50 °C (120 °F) 1 min., or 3 to 5 min.
9	5 ml HF (48%), 10 ml H$_2$SO$_4$, 85 ml water	Immerse for 30 s
10	4 g KMnO$_4$, 2 g Na$_2$CO$_3$, 94 ml water, few drops wetting agent	Specimen surface must be well polished, and be precleaned in 20% H$_3$PO$_4$ at 95 °C (205 °F) for uniform wettability(b); after precleaning, rinse in cold water and immediately immerse in etchant for 30 s
11	2 g NaOH, 5 g NaF, 93 ml water	Immerse for 2 to 3 min.
12	50 ml Poulton's reagent(b), 25 ml HNO$_3$(a), 40 ml solution of 3 g chromic acid per 10 ml of water	Put a few drops on as-rolled or as-extruded surface for 1 to 4 min., rinse, and swab to desmut; examine on microscope with crossed polarizers to show grains; repeat etching, if necessary; for some 5xxx alloys, increase amount of HNO$_3$ in solution to 50 ml
13	8 ml HNO$_3$(a), 2 ml HCl(a), 45 ml water, 45 ml methanol	Immerse for 10 s
14	5 ml acetic acid (glacial), 1 ml HNO$_3$(a), 94 ml water	Immerse for 20 to 30 min.

(a) Concentrated. (b) Etchant 4 in the table on etchants for use in macroscopic examination of aluminum alloys

Phases that may be present in various aluminum alloy systems

Alloy system	Examples of alloy	Alloy form	Phases
Al-Fe-Se	1100, EC	Ingot	FeAl$_3$, FeAl$_6$, Fe$_3$SiAl$_{12}$, Fe$_2$Si$_2$Al$_9$, Si
		Wrought	FeAl$_3$, Fe$_3$SiAl$_{12}$
Al-Fe-Mn-Si	3003	Ingot	(Fe,Mn)Al$_6$, α(Al-Fe,Mn-Si), Si
		Wrought	(Fe,Mn)Al$_6$, α(Al-Fe,Mn-Si)
Al-Fe-Mg-Si (Mg:Si $\cong$ 1.7:1)	6063	Ingot	FeAl$_3$, FeAl$_6$, Fe$_3$SiAl$_{12}$, Mg$_2$Si
		Wrought	FeAl$_3$, Fe$_3$SiAl$_{12}$, Mg$_2$Si
Al-Fe-Mg-Si (high Si)	356	Cast	Fe$_2$Si$_2$Al$_9$, Mg$_2$Si, Si
Al-Fe-Mg-Si (high Mg)	520	Cast	FeAl$_3$, Fe$_3$SiAl$_{12}$, Mg$_2$Si, Mg$_2$Al$_3$
Al-Cu-Fe-Si	295	Cast	FeAl$_3$, Fe$_3$SiAl$_{12}$, CuAl$_2$, Cu$_2$FeAl$_7$
Al-Fe-Mg-Si-Cr	6061	Ingot	(Fe,Cr)$_3$SiAl$_{12}$, Fe$_2$Si$_2$Al$_9$, FeMg$_3$Si$_6$Al$_8$, Mg$_2$Si, Si
		Wrought	(Fe,Cr)$_3$SiAl$_{12}$, Mg$_2$Si
Al-Cu-Fe-Si-Mg-Mn	2014	Ingot	(Fe,Mn)$_3$SiAl$_{12}$, CuAl$_2$, Cu$_2$Mg$_8$Si$_6$Al$_5$, Si
		Wrought	(Fe,Mn)$_3$SiAl$_{12}$, CuAl$_2$, Cu$_2$Mg$_8$Si$_6$Al$_5$
	2024	Ingot	(Fe,Mn)Al$_6$, (Fe,Mn)Al$_3$, (Fe,Mn)$_3$SiAl$_{12}$, Mg$_2$Si, CuAl$_2$, CuMgAl$_2$, Cu$_2$FeAl$_7$
		Wrought	(Fe,Mn)$_3$SiAl$_{12}$, Mg$_2$Si, CuMgAl$_2$, CU$_2$FeAl$_7$, Cu$_2$Mn$_3$Al$_{20}$(a)
Al-Cu-Mg-Ni-Fe-Si	2218, 2618	Ingot and wrought	In addition to others, Ni may cause NiAl$_3$, Ni$_2$Al$_3$, Cu$_3$NiAl$_6$ or FeNiAl$_9$ to appear
Al-Fe-Mg-Si-Mn-Cr	5083, 5086, 5456	Ingot	(Fe,Mn,Cr)Al$_6$, (Fe,Mn,Cr)$_3$SiAl$_{12}$, Mg$_2$Al$_3$, (Cr,Mn,Fe)Al$_7$(b)
		Wrought	(Fe,Mn,Cr)$_3$SiAl$_{12}$, Mg$_2$Si, Mg$_2$Al$_3$, Cr$_2$Mg$_3$Al$_{18}$(a)
Al-Cu-Mg-Zn-Fe-Si-Cr	7075	Ingot	(Fe,Cr)Al$_3$, (Fe,Cr)$_3$SiAl$_{12}$, Mg$_2$Si, Mg(Zn$_2$,AlCu), CrAl$_7$(b)
		Wrought	(Fe,Cr)$_3$SiAl$_{12}$, Cu$_2$FeAl$_7$, Mg$_2$Si, CuMgAl$_2$, Mg(Zn$_2$,AlCu), Cr$_2$Mg$_3$Al$_{18}$(a)

(a) May be identity of fine precipitate which comes out at elevated temperatures; not positively identified. (b) Only when chromium content is near high side of range

Applicability of etchants to macroscopic examination of aluminum alloys

Etchant numbers correspond to those assigned in the table on etchants for use in macroscopic examination of aluminum alloys

Alloy	Etchant
High-purity aluminum	4 or 5
Commercial-purity aluminum: 1xxx series	4, 2, or 1
All high-copper alloys: 2xxx series and casting alloys	1
Al-Mn alloys: 3xxx series	4, 2, or 1
Al-Si alloys: 4xxx series and casting alloys(a)	4, 2, or 3
Al-Mg alloys: 5xxx series and casting alloys	4, 2, or 1
Al-Mg-Si alloys: 6xxx series and casting alloys	4, 2, or 1
Al-Cu-Mg-Zn alloys: 7xxx series and casting alloys	1

(a) Also, welds and brazed joints made with the use of these alloys as filler metals

Applicability of etchants to microscopic examination of aluminum alloys

Etchant numbers correspond to those assigned in the table on etchants for use in microscopic examination of aluminum alloys

Alloy	Etchant	Evidence revealed
Examination for grain size and shape		
1xxx, 3xxx, 5xxx, 6xxx series; most casting alloys	5 or 12	Grain contrast when using crossed polarizers, with or without sensitive tint
2xxx, 7xxx series; Al-Cu or Al-Zn casting alloys	3A or 11	Grain contrast or grain-boundary lines
5xxx series alloys with more than 3% Mg	8 (3 to 5 min.)	Precipitation in grain boundaries
Examination for cold working		
1xxx, 3xxx, 5xxx, 6xxx series alloys	5 or 12	Deformation bands or markings that cause streaked effect when using crossed polarizers
2xxx, 7xxx series alloys	3A or 11	Deformation bands or markings that accompany relatively great amounts of cold working
5xxx series alloys with more than 3% Mg	8 (3 to 5 min.)	Precipitation in bands of slip
Examination for incomplete recrystallization		
1xxx, 3xxx, 5xxx, 6xxx series alloys	5 or 12	Even-toned, well-outlined grains that are recrystallized, otherwise streaked, or banded
2xxx series alloys, hot worked and heat treated	3A or 11	Unrecrystallized grains made up of multiple, very fine subgrains
6xxx series alloys, hot worked and heat treated	9	Unrecrystallized grains made up of multiple, very fine subgrains
7xxx series alloys, hot worked and heat treated	8 (3 to 5 min.) or 14	Unrecrystallized grains made up of multiple, very fine subgrains
Examination for preferred orientation		
1xxx, 3xxx, 5xxx, 6xxx series alloys	5 or 12	Predominance of certain gray tones when crossed polarizers are used, lack of randomness
2xxx series alloys in T4 temper	3A or 11	Lack of randomness in grain contrast
Examination for identification of constituents		
1xxx series alloys	1 or 7	(a)
2xxx, 3xxx series; Al-Cu and Al-Mn casting alloys	8 (1 min.)	(a)
7xxx series; Al-Zn casting alloys	3B	(a)
Examination for overheating (partial melting)		
2xxx series alloys	8 (1 min.)	Rosettes and grain-boundary eutectic
6xxx series alloys	2	Grain-boundary eutectic formations
7xxx series alloys	3B	Rosettes and grain-boundary eutectic formations
Examination for general constituent size and distribution		
All wrought alloys and casting alloys	1, 8 (1 min.)(b)	Coarse insoluble particles and fine precipitate particles. Longer etching time exaggerates size of fine particles
Examination for distinction between solution heat treated (T4) and artificially aged (T6) tempers		
2xxx series alloys	3A or 11	Loss of grain contrast, general darkening, in T6 compared with T4
6061	9	Clear outlining of grain boundaries in T6; faint outlining in T4
7075, recrystallized	4	More grain contrast, sharper grain-boundary outlining, in T4
Examination for overaging or poor quench of solution heat treated alloy		
2017 and 2024, in T4 temper	6	Faint dark precipitate at grain boundaries
Examination for cladding thickness		
Alclad 2014, 2024, 7075	3A or 11	Boundary between high grain contrast or outlining of alloy core and lighter-etching cladding
Brazing sheet	1 (swab) or 13	Boundary of high-silicon cladding alloy
Other clad alloys	1 (immerse), 2, 3A, 5 or 11	Any differences in structure that demarcate one layer from another
Examination for solid-solution coring or segregation, and diffusion effects		
3xxx, 5xxx series; Al-Mg casting alloys	10	Interference colors due to differences in thickness of tarnish films laid down on the surface
2xxx series alloys and others with more than 1% Cu	3A or 11	Brownish-colored films due to redeposition of copper

(a) See table on metallographic identification of phases in aluminum alloys. (b) Or any etchant that does not pit solid-solution matrix

Metallographic identification of phases in aluminum alloys

Basic and alternative phase designations(a)	Elements that enter in solution	External shape(b)	Appearance before etching(c)	Birefringence(d)	Etchants that aid identification(e)
Si	...	Cubic habit; primary particles form isometric polygons; eutectic may form script, blades, or very fine lamellae	Light bluish-gray	None	Generally best identified without etching; etchant 1 (swab) outlines particles and appears to lighten the color
Mg_2Si	...	Cubic habit; eutectic forms script that easily coalesces on heating	Natural color is darker bluish-gray than silicon, but usually tarnishes to bright blue, black or vari-colored	None (when not roughened or tarnished)	Easily identified without etching; caustic etchant 2 will not attack and may enhance blue color; acid etchants will attack and dissolve readily
$MgZn_2$ or $\eta(Mg-Zn)$	Isomorphous series with CuMgAl	Usually well-rounded or irregular, except in lamellar eutectic or precipitated from solid solution	White, watery; does not polish in relief	Slight change from light to dark gray	Etchant 3B gives a smooth, dark-gray to black color
$CrAl_7$	Fe as $(Cr,Fe)Al_7$ Mn as $(Cr,Mn)Al_7$	Primary crystals form elongated polygons	Light metallic gray	Weak, but will reveal twinning in large crystals	Resists attack by all common etchants
$CuAl_2$ or $\theta(Al-Cu)$	...	Usually well-rounded or irregular, except when precipitated from solid solution	Pale pinkish color	Strong, orange to greenish-blue; some orientations show little change	Remains light and clear in etchants 1 (swab), 3A, and 8 (1 min.); etchant 6 will darken and is good for detecting barely visible grain-boundary precipitate
$FeAl_3$	Cr as $(Fe,Cr)Al_3$ Mn as $(Fe,Mn)Al_3$ Possibly Cu	Elongated blades or star-shaped clusters when eutectic; resists coalescence	Light metallic gray; slightly darker than Fe_3SiAl_{12}	Weak and not easily detectable	Etchant 7 will dissolve and blacken; in high-copper alloys, etchant 8 (1 min.) will color it dark-brown to bluish-black; in Al-Cu-Mg-Zn alloys, etchant 3B will color it medium-brown or gray; rough and outlined
$FeAl_6$	A metastable phase in absence of Mn or Cu (see $MnAl_6$)	Isomorphous with $MnAl_6$, but usually found only under conditions of high solidification rate; forms fine lamellar eutectic	Not easily defined, because of fine particle size	Same as $MnAl_6$	Not attacked by etchant 7 but darkened by etchant 1 (swab)
Mg_2Al_3 or Mg_5Al_8, $\beta(Al-Mg)$	...	Usually well-rounded or irregular	White; lighter than aluminum but may tarnish to yellow or tan; not in relief	None (when not tarnished)	Caustic etchant such as 2 will not attack or color; acid etchants generally pit and dissolve it with varying rapidity
$MnAl_6$	Fe as $(Fe,Mn)Al_6$; isomorphous with $(Fe,Cu)(Al,Cu)_6$, or $(Fe,Cu)Al_6$	Primary or coarse eutectic forms solid or hollow parallelograms; fine eutectic may form script	Light metallic gray	Strong; light to dark gray; does not twin	Etchant 8 (1 min.) will neither attack nor darken this phase; however, it will attack companion phases such as $(Fe,Mn)Al_3$ or $(Fe,Mn)_3SiAl_{12}$

Note: There are some phases other than those listed in this table that are less common or that appear in such small amount or as such fine particulate that identification can be made only indirectly. These include $TiAl_3$, AlB_2 and TiB_2, Pb and Bi, $NiAl_3$, Ni_2Al_3, $FeNiAl_9$, Cu_3NiAl_6, and $Cu_2Mn_3Al_{20}$. Other phases that do not normally come into equilibrium with aluminum may occasionally be encountered as a result of incomplete melting or some other abnormality in practice. (a) There is no widely accepted manner of naming or designating phases as they are encountered in equilibrium phase diagrams or in description of alloy constitution. Even composition formulas are inexact, because many phases have broad homogeneity ranges or their actual composition may not coincide exactly with the ideal atomic arrangement upon which crystal structure is based. Phragmén (Phragmén, G. On the Phases Occurring in Alloys of Aluminum with Copper, Magnesium, Manganese, Iron, and Silicon, *J. Inst. Metals*, Vol 77, 1950, p 489-552) advocated the use of a lower-case-letter prefix indicating the basic crystal structure (c = cubic, h = hexagonal, and so on). Otherwise, Greek letters and upper-case English letters have been arbitrarily used, although "T" usually denotes a ternary phase and "Q" a quaternary phase. (b) Applies mainly to cast forms or to wrought alloys that have not been extensively worked. However, some iron-rich phases that are resistant to coalescence or spheroidization will retain dimensional ratios that are indicative of crystalline symmetry. (c) Applies to appearance after mechanical polishing. Electrolytic polishing is rarely suitable for making phase identification. (d) An exceptionally good flat polish with no tarnishing is required, because any element of the surface that is not parallel to the plane of the surface (that is, normal to the optical axis) will cause an apparent birefringence that is not due to crystal structure. The sensitivity of this technique will also vary with the quality of the optical system. A rotating stage is necessary. (e) Etchant numbers correspond to those assigned in the table on etchants for use in microscopic examination of aluminum alloys. (f) There are two crystal forms of α(Al-Fe-Si)—namely, Fe_3SiAl_{12} (cubic, also called α_1 and Fe_2SiAl_8) and $Fe_3Si_2Al_{12}$ (hexagonal, also called α_2). It was believed at one time that cubic Fe_3SiAl_{12} was isomorphous with analogous ternary phases $Cr_4Si_4Al_{13}$ and Mn_3SiAl_{12}, but the latter at least has since been found to be hexagonal. Nevertheless, the presence of even very small amounts of Mn, Cr_3 and Cu in α(Al-Fe-Si) seems to favor the cubic form normally encountered in commercial alloys. Metallographic distinction between the cubic and the hexagonal forms is very difficult to detect. When etched in etchant 3B (see the table on etchants for use in microscopic examination of aluminum alloys), complex alloys containing chromium and manganese (such as 5083 and 7075) may show etching contrasts within the scriptlike phase normally taken to be cubic Fe_3SiAl_{12}, but no separate identify has yet been established. (g) This phase and its variants can give a variety of etching responses for a given etch, depending on its composition and that of the matrix

(continued)

Metallographic identification of phases in aluminum alloys (continued)

Basic and alternative phase designations(a)	Elements that enter in solution	External shape(b)	Appearance before etching(c)	Birefringence(d)	Etchants that aid identification(e)
$Cr_2Mg_3Al_{18}$ or T(Al-Cr-Mg), E(Al-Cr-Mg)	...	Usually forms by precipitation or by peritectic reaction from $CrAl_7$	Very light metallic gray; not much in relief	None	Strongly attacked by etchants 6 and 7
$(Fe,Cu)(Al,Cu)_6$ or $(Fe,Cu)Al_6$, $\alpha(Al$-Cu-$Fe)$			(See $MnAl_6$)		
Cu_2FeAl_7 or $\beta(Al$-Cu-$Fe)$, N(Al-Cu-Fe)	...	Elongated blades when formed eutectically; also forms peritectically from $(Fe,Mn)_3SiAl_{12}$ and other iron-rich phases	Very light metallic gray; only slightly darker than $CuAl_2$	Moderate; light to dark gray	Outlined, but not colored, by etchants 3B and 8 (1 min.); hence, can be distinguished from other iron-rich phases with which it is associated
$CuMgAl_2$ or $Cu_2Mg_2Al_5$, S(Al-Cu-Mg)	...	Very much resembles $CuAl_2$	Slightly grayer than $CuAl_2$; tarnishes to brown or black very readily during polishing	Very strong; yellowish to purple or greenish-blue	Roughened and darkened to varying degrees by etchants 3B and 8 (1 min.), depending on polish; etchant 3A darkens this phase while leaving $CuAl_2$ uncolored; etchant 6 reveals barely visible grain-boundary precipitate
CuMgAl			(See $MgZn_2$)		
$Cr_4Si_4Al_{13}$ or $\alpha(Al$-Cr-$Si)$			(See Fe_3SiAl_{12})(f)		
$CuMg_4Al_5$ or T(Al-Cu-Mg), c(Al-Cu-Mg)	Isomorphous series with $Mg_3Zn_3Al_2$	Irregular rounded	Very light or slightly yellow	None	Behaves like other Mg-rich phases, attacked rapidly by acidic etchants, not attacked by caustic etchants
Fe_3SiAl_{12} or $Fe_3Si_2Al_{12}$, $\alpha(Al$-Fe-$Si)$, c(Al-Fe-Si); also $(Fe,Cu)_3SiAl_{12}$ or $\alpha(Al$-Fe,Cr-$Si)$; $(Fe,Mn)_3SiAl_{12}$ or $\alpha(Al$-Fe,Mn-$Si)$	(f); besides the apparent interchangeability of Fe, Cr and Mn, this phase can probably also contain Cu	Usually well-defined script when formed eutectically, especially when silicon is not low; may also form polyhedrons or irregular shapes, or precipitate as Widmanstätten type	Light metallic gray, slightly lighter than either $FeAl_3$ or $Fe_2Si_2Al_9$; often polishes in relief	None	Rarely attacked strongly, but can darken to shades of brown when copper is present, using etchant 8 (1 min.); in the absence of copper, etchant 8 (1 min.) will roughen and outline it, distinguishing it from $MnAl_6$; chromium makes it more resistant to etching(g)
$Fe_2Si_2Al_9$ or $FeSiAl_5$, $\beta(Al$-Fe-$Si)$	...	Bladelike when formed eutectically; retains flat shape in wrought alloys	Light metallic gray, intermediate between Fe_3SiAl_{12} and silicon	Moderate; light to dark gray	Etchant 1 (immerse) will attack and darken to varying degrees, depending on Fe-Si ratio; etchant 7 will attack and dissolve it out; in both cases, Fe_3SiAl_{12} is outlined but not appreciably darkened

Note: There are some phases other than those listed in this table that are less common or that appear in such small amount or as such fine particulate that identification can be made only indirectly. These include $TiAl_3$, AlB_2 and TiB_2, Pb and Bi, $NiAl_3$, Ni_2Al_3, $FeNiAl_9$, Cu_3NiAl_6, and $Cu_2Mn_3Al_{20}$. Other phases that do not normally come into equilibrium with aluminum may occasionally be encountered as a result of incomplete melting or some other abnormality in practice. (a) There is no widely accepted manner of naming or designating phases as they are encountered in equilibrium phase diagrams or in description of alloy constitution. Even composition formulas are inexact, because many phases have broad homogeneity ranges or their actual composition may not coincide exactly with the ideal atomic arrangement upon which crystal structure is based. Phragmén (Phragmén, G. On the Phases Occurring in Alloys of Aluminum with Copper, Magnesium, Manganese, Iron, and Silicon, *J. Inst. Metals*, Vol 77, 1950, p 489-552) advocated the use of a lower-case-letter prefix indicating the basic crystal structure (c = cubic, h = hexagonal, and so on). Otherwise, Greek letters and upper-case English letters have been arbitrarily used, although "T" usually denotes a ternary phase and "Q" a quaternary phase. (b) Applies mainly to cast forms or to wrought alloys that have not been extensively worked. However, some iron-rich phases that are resistant to coalescence or spheroidization will retain dimensional ratios that are indicative of crystalline symmetry. (c) Applies to appearance after mechanical polishing. Electrolytic polishing is rarely suitable for making phase identification. (d) An exceptionally good flat polish with no tarnishing is required, because any element of the surface that is not parallel to the plane of the surface (that is, normal to the optical axis) will cause an apparant birefringence that is not due to crystal structure. The sensitivity of this technique will also vary with the quality of the optical system. A rotating stage is necessary. (e) Etchant numbers correspond to those assigned in the table on etchants for use in microscopic examination of aluminum alloys. (f) There are two crystal forms of $\alpha(Al$-Fe-$Si)$—namely, Fe_3SiAl_{12} (cubic, also called α_1 and Fe_2SiAl_8) and $Fe_3Si_2Al_{12}$ (hexagonal, also called α_2). It was believed at one time that cubic Fe_3SiAl_{12} was isomorphous with analogous ternary phases $Cr_4Si_4Al_{13}$ and Mn_3SiAl_{12}, but the latter at least has since been found to be hexagonal. Nevertheless, the presence of even very small amounts of Mn, Cr_3 and Cu in $\alpha(Al$-Fe-$Si)$ seems to favor the cubic form normally encountered in commercial alloys. Metallographic distinction between the cubic and the hexagonal forms is very difficult to detect. When etched in etchant 3B (see the table on etchants for use in microscopic examination of aluminum alloys), complex alloys containing chromium and manganese (such as 5083 and 7075) may show etching contrasts within the scriptlike phase normally taken to be cubic Fe_3SiAl_{12}, but no separate identify has yet been established. (g) This phase and its variants can give a variety of etching responses for a given etch, depending on its composition and that of the matrix

(continued)

Metallographic identification of phases in aluminum alloys (continued)

Basic and alternative phase designations(a)	Elements that enter in solution	External shape(b)	Appearance before etching(c)	Birefringence(d)	Etchants that aid identification(e)
$Mg_3Zn_3Al_2$ or T(Al-Mg-Zn)			(See $CuMg_4Al_5$)		
Mn_3SiAl_{12} or α(Al-Mn-Si)			(See Fe_3SiAl_{12})(f)		
$Cu_2Mg_8Si_6Al_5$ or Q(Al-Cu-Mg-Si), λ(Al-Cu-Mg-Si), h(Al-Cu-Mg-Si)	...	This is true quaternary phase; forms irregular shapes in eutectics	Light metallic gray; darker than $CuAl_2$	Strong; changes from orange to blue	Etchant 8 (1 min.) does not attack it, but the color distinction between it and $CuAl_2$ remains the same as when not etched
$FeMg_3Si_6Al_8$ or Q(Al-Fe-Mg-Si), π(Al-Fe-Mg-Si), h(Al-Fe-Mg-Si)	...	This is a true quaternary phase; forms irregular shapes in eutectics; sometimes shows hexagonal symmetry	Very light metallic gray; not much in relief	Strong; changes from yellow to blue	Not attacked by etchant 1 (immerse), hence distinguished from $Fe_2Si_2Al_9$, with which it is usually associated

Note: There are some phases other than those listed in this table that are less common or that appear in such small amount or as such fine particulate that identification can be made only indirectly. These include $TiAl_3$, AlB_2 and TiB_2, Pb and Bi, $NiAl_3$, Ni_2Al_3, $FeNiAl_9$, Cu_3NiAl_6, and $Cu_2Mn_3Al_{20}$. Other phases that do not normally come into equilibrium with aluminum may occasionally be encountered as a result of incomplete melting or some other abnormality in practice. (a) There is no widely accepted manner of naming or designating phases as they are encountered in equilibrium phase diagrams or in description of alloy constitution. Even composition formulas are inexact, because many phases have broad homogeneity ranges or their actual composition may not coincide exactly with the ideal atomic arrangement upon which crystal structure is based. Phragmén (Phragmén, G. On the Phases Occurring in Alloys of Aluminum with Copper, Magnesium, Manganese, Iron, and Silicon, *J. Inst. Metals*, Vol 77, 1950, p 489-552) advocated the use of a lower-case-letter prefix indicating the basic crystal structure (c = cubic, h = hexagonal, and so on). Otherwise, Greek letters and upper-case English letters have been arbitrarily used, although "T" usually denotes a ternary phase and "Q" a quaternary phase. (b) Applies mainly to cast forms or to wrought alloys that have not been extensively worked. However, some iron-rich phases that are resistant to coalescence or spheroidization will retain dimensional ratios that are indicative of crystalline symmetry. (c) Applies to appearance after mechanical polishing. Electrolytic polishing is rarely suitable for making phase identification. (d) An exceptionally good flat polish with no tarnishing is required, because any element of the surface that is not parallel to the plane of the surface (that is, normal to the optical axis) will cause an apparant birefringence that is not due to crystal structure. The sensitivity of this technique will also vary with the quality of the optical system. A rotating stage is necessary. (e) Etchant numbers correspond to those assigned in the table on etchants for use in microscopic examination of aluminum alloys. (f) There are two crystal forms of α(Al-Fe-Si)—namely, Fe_3SiAl_{12} (cubic, also called α_1 and Fe_2SiAl_8) and $Fe_3Si_2Al_{12}$ (hexagonal, also called α_2). It was believed at one time that cubic Fe_3SiAl_{12} was isomorphous with analogous ternary phases $Cr_4Si_4Al_{13}$ and Mn_3SiAl_{12}, but the latter at least has since been found to be hexagonal. Nevertheless, the presence of even very small amounts of Mn, Cr, and Cu in α(Al-Fe-Si) seems to favor the cubic form normally encountered in commercial alloys. Metallographic distinction between the cubic and the hexagonal forms is very difficult to detect. When etched in etchant 3B (see the table on etchants for use in microscopic examination of aluminum alloys), complex alloys containing chromium and manganese (such as 5083 and 7075) may show etching contrasts within the scriptlike phase normally taken to be cubic Fe_3SiAl_{12}, but no separate identify has yet been established. (g) This phase and its variants can give a variety of etching responses for a given etch, depending on its composition and that of the matrix

Copper

Etching for macroscopic examination of coppers and copper alloys
Procedure for use: immerse at room temperature, rinse in warm water, dry

Composition of etchant	Copper or copper alloy	Purpose, or characteristic revealed
50 ml HNO_3, 0.5 g $AgNO_3$, 50 ml water	All coppers and copper alloys	Produces a brilliant, deep etch
10 ml HNO_3, 90 ml water	Coppers and all brasses	Grains; cracks and other defects
50 ml HNO_3, 50 ml water(a)	Coppers, all brasses, aluminum bronze(b)	Grains; cracks and other defects; reveals grain contrast
30 ml HCl, 10 ml $FeCl_3$, 120 ml water or methanol	Coppers and all brasses	Grains; cracks and other defects; reveals grain contrast(c)
20 ml acetic acid, 10 ml 5% CrO_3, 5 ml 10% $FeCl_3$, 100 ml water(d)	All brasses	Produces a brilliant, deep etch
2 g $K_2Cr_2O_7$, 4 ml saturated solution of NaCl, 8 ml H_2SO_4, 100 ml water(e)	Coppers, high-Cu alloys, phosphor bronze	Grain boundaries, oxide inclusions
40 g CrO_3, 7.5 g NH_4Cl, 50 ml HNO_3, 8 ml H_2SO_4, 100 ml water	Silicon brass, silicon bronze	General macrostructure

(a) Solution should be agitated during etching, to prevent pitting of some alloys. (b) Aluminum bronzes may form smut, which can be removed by brief immersion in concentrated HNO_3. (c) Excellent for grain contrast. (d) Amount of water can be varied as desired. (e) Immerse specimen for 15 to 30 min., then swab with fresh solution

Electrolytes and conditions for electrolytic polishing of coppers and copper alloys

Electrolyte composition	Voltage	Current density, A/dm^2	Cathode	Time	Copper or copper alloy
825 ml H_3PO_4, 175 ml water	1.0 to 1.6	2 to 10	Copper	10 to 40 min.	Unalloyed copper
250 ml H_3PO_4, 250 ml ethanol, 50 ml propanol, 500 ml distilled water, 3 g urea	3 to 6	40 to 80	Stainless	50 s	Coppers and copper alloys
700 ml H_3PO_4, 350 ml water	1.2 to 2.0	6 to 10	Copper	15 to 30 min.	Coppers; alpha, beta, alpha-beta brasses; Al, Si, Sn, and P bronzes; copper with less than 3% Be, Fe, Pb, or Cr
580 g $H_4P_2O_7$, 1000 ml water	1.2 to 1.9	8 to 12	Copper	10 to 15 min.	Coppers, brasses
300 ml HNO_3, 600 ml methanol	20 to 70 / 30 to 50	65 to 310 / 250 to 310	Stainless / Stainless	10 to 60 s / 5 to 10 s	Coppers, brasses / Silicon bronze, phosphor bronze
170 g CrO_3, 830 ml water	1.5 to 12	95 to 220	Stainless	10 to 60 s	Brasses
400 ml H_3PO_4, 600 ml water	1.0 to 2.0	6 to 15	Copper or stainless	1 to 15 min.	Alpha, alpha-beta brasses; Cu-Fe, Cu-Cr
30 ml HNO_3, 900 ml methanol, 300 g copper nitrate	45 to 50	105 to 125	Stainless	15 s	Bronzes (have tendency to etch)
670 ml H_3PO_4, 100 ml H_2SO_4, 300 ml distilled water	2 to 3	10	Copper	15 min.	Copper; Cu-Sn containing up to 6% Sn
470 ml H_3PO_4, 200 ml H_2SO_4, 400 ml distilled water	2 to 2.3	10	Copper	15 min.	Cu-Sn up to 9% Sn
350 ml H_3PO_4, 650 ml ethanol	2 to 5	2 to 7	Copper	10 to 15 min.	Copper alloys with high lead (up to 30%)
540 ml H_3PO_4, 460 ml water	2 / 2 to 2.2	0.65 to 0.75 / 10 to 15	Copper / Copper	5 to 15 min. / 15 min.	Copper / Nickel silver

Etchants and procedures for microetching of coppers and copper alloys

Etchant composition(a)	Procedure	Copper or copper alloy
20 ml HN$_4$OH, 0 to 20 ml water, 8 to 20 ml H$_2$O$_2$ (3%)	Immersion or swabbing for 1 min.: H$_2$O$_2$ content varies with copper content of alloy to be etched, use fresh H$_2$O$_2$ for best results(b)	Coppers and copper alloys; film on etched aluminum bronze can be removed with weak Grard's solution
1 g Fe(NO$_3$)$_3$, 100 ml water	Immersion	Etching and attack polishing of coppers and copper alloys
25 ml NH$_4$OH, 25 ml water, 50 ml (NH$_4$)$_2$S$_2$O$_8$ (2.5%)	Immersion	Attack polishing of coppers and some copper alloys
2 g K$_2$Cr$_2$O$_7$, 8 ml H$_2$SO$_4$, 4 ml NaCl (saturated solution), 100 ml water	Immersion (NaCl replaceable by 1 drop HCl per 25 ml soln; add just beore using); follow with FeCl$_3$ or other contrast etch	Coppers; copper alloys of beryllium, manganese and silicon; nickel silver; bronzes; chromium copper
CrO$_3$ (saturated aqueous solution)	Immersion or swabbing	Coppers, brasses, bronzes, nickel silver
50 ml CrO$_3$ (10 to 15%), 1 to 2 drops HCl	Immersion (add HCl at time of use)	Same as above. Color by electrolytic etching or with FeCl$_3$ etchants
10 g (NH$_4$)$_2$S$_2$O$_8$, 90 ml water	Immersion (use either cold or boiling)	Coppers, brasses, bronzes, nickel silver, aluminum bronze
10% aqueous solution of copper ammonium chloride plus ammonium hydroxide to neutrality or alkalinity	Immersion; wash specimen thoroughly	Coppers, brasses, nickel silver; darkening large areas of beta in alpha-beta brass

FeCl$_3$, g	HCl, ml	Water, ml		
5	50	100	Immersion or swabbing; etch lightly or by successive light etches to required results	Coppers, brasses, bronzes, aluminum bronze; darkens beta in brass; gives contrast following dichromate and other etches etches
20	5	100(c)(d)		
25	25	100		
1	20	100		
8	25	100		
5	10	100(e)(f)		

Etchant composition(a)	Procedure	Copper or copper alloy
5 g FeCl$_3$, 100 ml ethanol, 5 to 30 ml HCl	Immersion or swabbing for 1 s to several minutes	Coppers and copper alloys
Nitric acid (various concentrations)	Immersion or swabbing; AgNO$_3$ (0.15 to 0.3%) added to 1:1 solution gives a brilliant, deep etch	Coppers and copper alloys
Ammonium hydroxide (dilute solutions)	Immersion	Attack polishing of brasses and bronzes
50 ml HNO$_3$, 20 g CrO$_3$, 30 ml water	Immersion	Aluminum bronze, free-cutting brass; film from polishing can be removed with 10% HF
5 ml HNO$_3$, 20 g CrO$_3$, 75 ml water	Immersion	Same as above

(a) The use of concentrated etchants is intended unless otherwise specified. (b) This etchant may be alternated with FeCl$_3$. (c) Grard's No. 1 etchant. (d) Plus 1 g CrO$_3$. (e) Grard's No. 2 etchant. (f) Plus 1 g CuCl$_2$ and 0.05 g SnCl$_2$

Electrolytes and operating conditions for electrolytic etching of copper and copper alloys

Electrolyte composition	Operating conditions	Copper or copper alloy
5 to 14% H$_3$PO$_4$, 8% (sp gr 1.042) is preferred	1 to 4 V; etching time, 10 s	Coppers
Struer's D-2: 250 ml H$_3$PO$_4$ (85%), 250 ml ethanol (95%), 500 ml water, 2 ml wetting agent	1 to 3 V; current density, 10 to 15 A/dm^2; etching time, 30 to 60 s	Coppers
10 ml (NH$_4$)C$_2$H$_3$O$_2$, 30 ml Na$_2$S$_2$O$_3$, 30 ml NH$_4$OH, 30 ml water	30 A/dm^2 time varies with composition and previous treatment of specimen	Cold worked brasses
30 g FeSO$_4$, 4 g NaOH, 100 ml H$_2$SO$_4$, 1900 ml water	0.1 A at 8 to 10 V; for not over 15 s; do not swab surface after etching	Darkens beta in brasses and gives contrast after H$_2$O$_2$-NH$_4$OH etch; also for nickel silver and bronzes
1% CrO$_3$, 99% water	6 V; aluminum cathode; etching time, 3 to 6 s	Beryllium copper and aluminum bronze
5 ml acetic acid (glacial), 10 ml HNO$_3$, 30 ml water	½ to 1 V; current density, 20 to 50 A/dm^2; etching time, 5 to 15 s	Copper-nickel alloys; avoiding contrast associated with coring

Lead

Recommended etchants and procedures for macroscopic and microscopic examination of lead and lead alloys

Etchant no.	Composition(a)	Procedure	Use
1	1 part acetic acid (glacial), 1 part nitric acid(b), 4 parts glycerol	Use freshly prepared solution at 80 °C (175 °F); discard after use; for macroetching: etch several minutes, rinse in water; for microetching: etch several seconds; for best results, alternate etching with polishing	Macroetching of lead; development of microstructures and grain boundaries in lead, and in lead-calcium, lead-antimony, and lead-tin (low-tin) alloys
2	100 parts acetic acid (glacial), 10 parts hydrogen peroxide (30%)	Etch for 10 to 30 min., depending on the depth of the disturbed layer; dry and clean with concentrated nitric acid if required	Microetching of lead-antimony alloys containing up to 2% antimony
3	3 parts acetic acid (glacial), 1 part hydrogen peroxide (30%)	Etch by immersion specimen in solution for 6 to 15 s; dry with alcohol	Microetching of lead, Pb-Ca alloys, and Pb-Sb alloys containing more than 2% Sb. Also removes disturbed metal
4	Solution A: 15 g ammonium molybdate, 100 ml distilled water Solution B: 6 parts nitric acid(b), 4 parts distilled water	Mix equal quantities of solutions A and B; etch by alternately swabbing specimen and washing in running water	Macroetching of lead. A very rapid etchant; well suited for removing thick layers of disturbed metal from specimens
5	3 parts acetic acid (glacial), 4 parts nitric acid(b), 16 parts distilled water	Use freshly prepared solution at 40 to 42 °C (105 to 110 °F); immerse specimen for 4 to 30 min. until disturbed layer is removed; clean with coton in running water	Microetching of unalloyed lead, and lead-tin alloys containing up to 3% tin
6	2 parts acetic acid (glacial), 2 parts nitric acid(b), 2 parts hydrogen peroxide (30%), 5 parts distilled water	Etch for 2 to 10 s by swabbing; rinse specimen in running water and dry with alcohol	Macroetching of unalloyed lead, and of lead-bismuth, lead-tellurium and lead-nickel alloys
7	1 part nitric acid(b), 1 part distilled water	Etch for 5 to 10 min. by immersion; if thick layer of disturbed metal is to be removed, solution can be heated to boiling; rinse in running water, rinse in alcohol and dry	Developing macrostructure of welds and laminations in lead products
8	Solution A: 10% aqueous solution of ammonium persulfate Solution B: 30% aqueous solution of tartaric acid	Mix 5 ml of solution A with 2 ml of solution B; swab specimen for 5 to 10 s; rinse in running water	Microetching to distinguish cuboidal SbSn phase from Sb-rich phases in Pb-Sb-Sn alloys such as bearing alloys or type metals. Solution A blackens SbSn phase; solution B etches Sb-rich phases
9	6 parts perchloric acid (70%), 4 parts water	Immerse specimen (cathode) in electrolyte; anode is platinum spiral; etch 45 to 90 s at 6 V, 4 A, from a rectifier	Electrolytic etching of lead-antimony alloys containing more than 2% antimony
10	1 part hydrochloric acid(b), 9 parts water	Same as for etchant 9	Same as for etchant 9

(a) Parts are by volume. (b) Concentrated

Nickel

Electrolytes and current densities for electropolishing of nickel and nickel-copper alloys

Composition of electrolyte	Applicable alloys	Current density, A/in.^{2}
37 ml H$_3$PO$_4$(a), 56 ml glycerol, 7 ml water	Nickel 200	9 to 10
	Nickel 270	10 to 12
	Duranickel 301	8 to 10
	Monel 400	6 to 7
33 ml HNO$_3$(a), 66 ml methanol	Monel 400, R-405, K-500	10 to 15

(a) Concentrated

Etchants for microscopic examination of nickel and nickel-copper alloys for grain boundaries and general structure

Composition of etchant	Conditions for use
Etchants for Nickel 200 and 270; Duranickel 301; and Monel 400, R-405 and K-500	
1 part 10% aqueous solution of sodium cyanide, 1 part 10% aqueous solution of ammonium persulfate; mix solutions when ready to use	Immerse or swab specimen for 5 to 90 s (*Caution:* use fume hood; solutions release toxic fumes when mixed)
1 part concentrated nitric acid, 1 part acetic acid (glacial); use fresh solution	For revealing grain boundaries; immerse or swab specimen for 5 to 20 s
Alternative etchant for Monel K-500	
Glyceregia: 10 ml concentrated nitric acid, 20 ml concentrated hydrochloric acid, 30 to 40 ml glycerol	Etch by immersing or swabbing the specimen for 30 s to 5 min.

Magnesium

Selected etchants for macroscopic and microscopic examination of magnesium alloys

Etchant No.	Composition	Etching procedure	Characteristics and use
1	**Nital:** 1 to 5 ml HNO_3 (conc), 100 ml ethanol (95%) or methanol (95%)	Swab or immerse specimen for a few seconds to 1 min.; wash in water then alcohol and dry	Shows general structure
2	**Glycol:** 1 ml HNO_3 (conc), 24 ml water, 75 ml ethylene glycol	Immerse specimen face up and swab with cotton for 3 to 5 s for as-cast or aged metal, and up to 1 min. for heat treated metal; wash in water, then alcohol and dry	Shows general structure; reveals constituents in Mg-rare earth and Mg-Th alloys
3	**Acetic glycol:** 20 ml acetic acid, 1 ml HNO_3 (conc), 60 ml ethylene glycol, 20 ml water	Immerse specimen face up with gentle agitation for 1 to 3 s for as-cast or aged metal, and for 10 s for heat treated metal; wash in water, then alcohol and dry	Shows general structure and grain boundaries in heat treated castings; shows grain boundaries in Mg-rare earth and Mg-Th alloys
4	10 ml HF (48%), 90 ml water	Immerse specimen face up for 1 to 2 s; wash in water, then alcohol and dry	Darkens $Mg_{17}Al_{12}$ phase and leaves $Mg_{32}(Al,Zn)_{49}$ phase unetched and white
5	**Phospho-picral:** 0.7 ml H_3PO_4, 4 to 6 g picric acid, 100 ml ethanol (95%)	Immerse specimen face up for about 10 to 20 s or until polished surface is darkened; wash in alcohol and dry	For estimating the amount of massive phase; stains matrix and leaves phase white; staining improves as magnesium-ion content increases with use
6	**Acetic-picral:** 5 ml acetic acid, 6 g picric acid, 10 ml water, 100 ml ethanol (95%)	Immerse specimen face up with gentle agitation until face turns brown; wash in a stream of alcohol and dry with a blast of air	A universal etchant; defines grain boundaries in most alloys and tempers by etch rate and color of stain; reveals cold work and twinning readily
7	**Acetic-picral:** 20 ml acetic acid, 3 g picric acid, 20 ml water, 50 ml ethanol (95%)	Same as for etchant 6, above, but etch for at least 15 s to develop a heavy film	Orientation of crackled film is parallel to trace of basal plane; film crackles in high-alloy areas; distinguishes between fusion voids surrounded by normal level of alloy and microshrinkage with low alloy content
8	**Acetic-picral:** 10 ml acetic acid, 4.2 g picric acid, 10 ml water, 70 ml ethanol (95%)	Immerse specimen face up with gentle agitation until face turns brown; wash in a stream of alcohol and dry with a blast of air	Reveals grain boundaries more readily than etchant 6, above, especially in dilute alloys
9	0.6 g picric acid, 10 ml ethanol (95%), 90 ml water	Immerse specimen face up for 15 to 30 s; wash in alcohol and dry	Used after HF etchant to darken matrix to give better contrast between matrix and white ternary phase
10	2 ml HF (48%), 2 ml HNO_3 (conc), 96 ml water	Immerse specimen face up with gentle agitation; do not swab	Grain structure and coring in Mg-Zn-Zr alloys

Tin

Etchants for use in microscopic examination of tin and tin alloys

Etchant composition	Uses
5 ml HCl, 2 g $FeCl_3$, 30 ml water, 60 ml absolute alcohol	General use for tin and tin alloys
2 ml HCl, 98 ml methanol (95%) or ethanol (95%)	Grain-boundary etch for pure tin
10 ml HNO_3, 10 ml acetic acid, 80 ml glycerol	Darkens the lead in the eutectic of tin-rich tin-lead alloys
5% silver nitrate in water	Darkens primary and eutectic lead in lead-rich tin-lead alloys
2% nital	Recommended for etching tin-antimony alloys; darkens tin-rich matrix, leaving intermetallic compounds unattacked; often used for etching specimens of babbitted bearings
Picral	For etching tin-coated steel and tin-coated cast iron
1 drop concentrated HNO_3, 2 drops HF, 25 ml glycerol; then picral	For etching tin-coated steel
Dilute ammonium hydroxide with a few drops of 30% hydrogen peroxide	For etching tin-coated copper and copper alloys

Titanium

Etchants for microscopic examination of titanium and titanium alloys

Specimen meal	Etchant composition	Purpose
Unalloyed titanium	1 to 3 ml HF, 10 ml HNO_3, 30 ml lactic acid	Reveals hydrides
	1 ml HF, 30 ml HNO_3, 30 ml lactic acid	Reveals hydrides
Most titanium alloys	Kroll's reagent: 1 to 3 ml HF, 2 to 6 ml HNO_3, water to 1000 ml	General-purpose etch
	10 ml HF, 5 ml HNO_3, 85 ml water	General-purpose etch
	1 ml HF, 2 ml HNO_3, 50 ml H_2O_2, 47 ml water	Removes stain
	10 ml HF, 10 ml HNO_3, 30 ml lactic acid	Chemical polish and etch
	2 ml HF, 98 ml water	Reveals alpha case
	1 to 2 ml HF, 4 to 5 ml H_2O_2, water to 1000 ml	Nonstaining etch
Near-alpha titanium alloys	2 ml HF, 98 ml water; then 1 ml HF, 2 ml HNO_3, 97 ml water	General-purpose etch(a)
Alpha-beta titanium alloys	10 ml KOH (40%), 5 ml H_2O_2, 20 ml water	Stains alpha, transformed beta
Ti-Al-Zr and Ti-Si alloys	18.5 g benzalkonium chloride, 33 ml ethanol, 40 ml glycerol, 25 ml HF	General-purpose etch
Ti-3Al-8V-6Cr-4Mo-4Zr	30 ml H_2O_2, 3 drops HF	General-purpose etch
Ti-8Mn; aged Ti-13V-11Cr-3Al	2 ml HF, 4 ml HNO_3, 94 ml water	General-purpose etch
Ti-Si alloys	2 drops HF, 1 drop HNO_3, 3 ml HCl, 25 ml glycerol	General-purpose etch

(a) First etchant stains alpha phase; second etchant removes stain

Zinc

Etchants for zinc and zinc alloys

Etchant	Composition
1(a)	200 g CrO_3, 15 g Na_2SO_4, 1000 ml water
2(b)	50 g CrO_3, 4 g Na_2SO_4, 1000 ml water
3	200 g CrO_3, 1000 ml water

(a) For rolled zinc-copper alloys, the Na_2SO_4 content can be reduced to 7.5 g. If desired, a smoothly etched surface can be obtained by increasing the Na_2SO_4 to 30 g. (b) This etchant can be made by mixing one part (by volume) of etchant 1 and three parts of water

Etchants and etching times for zinc and zinc die-casting alloys

Specimen metal	Etchant(a)	Time, s, for examination at:	
		250×	1000×
Cast or rolled zinc	1	5	1
Alloy AC41A or AG40A	2	1	1

(a) Etchant numbers correspond to those assigned in the table on etchants for zinc and zinc alloys

Cast Irons

Gray Cast Irons

Typical base compositions and mechanical properties of SAE J431 automotive gray cast irons

UNS	SAE grade	Typical composition, %				
		TC(a)	Mn	Si	P	S
F10004	G1800(b)	3.40 to 3.70	0.50 to 0.80	2.80 to 2.30	0.15	0.15
F10005	G2500(b)	3.20 to 3.50	0.60 to 0.90	2.40 to 2.00	0.12	0.15
F10006	G3000(c)	3.10 to 3.40	0.60 to 0.90	2.30 to 1.90	0.10	0.15
F10007	G3500(c)	3.00 to 3.30	0.60 to 0.90	2.20 to 1.80	0.08	0.15
F10008	G4000(c)	3.00 to 3.30	0.70 to 1.00	2.10 to 1.80	0.07	0.15

SAE grade	Hardness, HB	Minimum transverse load		Minimum deflection		Minimum tensile strength	
		kg	lb	mm	in.	MPa	ksi
G1800	187 max	780	1720	3.6	0.14	118	18
G2500	170 to 229	910	2000	4.3	0.17	173	25
G3000	187 to 241	1000	2200	5.1	0.20	207	30
G3500	207 to 255	1110	2450	6.1	0.24	241	35
G4000	217 to 269	1180	2600	6.9	0.27	276	40

(a) Total carbon. If either carbon or silicon is on the high side of the range, the other should be on the low side. Properties determined from an as-cast test bar (1.2 in., or 30.5-mm, diam). (b) Ferritic-pearlitic microstructure. (c) Pearlitic microstructure

Typical base compositions and mechanical properties of SAE J431 automotive gray cast irons for heavy duty service

UNS	SAE grade	Typical composition, %				
		TC(a)	Mn	Si	P	S
F10009	G2500a(b)	3.40 min	0.60 to 0.90	1.60 to 2.10	0.12	0.12
F10010	G3500b(c)	3.40 min	0.60 to 0.90	1.30 to 1.80	0.08	0.12
F10011	G3500c(c)	3.50 min	0.60 to 0.90	1.30 to 1.80	0.08	0.12
F10012	G4000d(d)	3.10 to 3.60	0.60 to 0.90	1.95 to 2.40	0.07	0.12

SAE grade	Hardness, HB	Minimum transverse load		Minimum deflection		Minimum tensile stength	
		kg	lb	mm	in.	MPa	ksi
G2500a	170 to 229	910	2000	4.3	0.17	173	25
G3500b	207 to 255	1090	2400	6.1	0.24	241	35
G3500c	207 to 255	1090	2400	6.1	0.24	241	35
G4000d	241 to 321(e)	1180	2600	6.9	0.27	276	40

(a) Total carbon. If either carbon or silicon is on the high side of the range, the other should be on the low side. Alloying elements not listed in this table may be required. Properties determined from an as-cast test bar (30.5 mm, or 1.2-in. diam). (b) Microstructure: size 2 to 4 type A graphite in a matrix of lamellar pearlite containing not more than 15% free ferrite. (c) Microstructure: size 3 to 5 type A graphite in a matrix of lamellar pearlite containing not more than 5% free ferrite or free carbide. (d) Alloy gray iron containing 0.85 to 1.25 Cr, 0.40 to 0.60 Mo and 0.20 to 0.45 Ni or as agreed. Microstructure: primary carbides and size 4 to 7 type A or E graphite in a matrix of fine pearlite, as determined in a zone at least 3.2 mm (1/8 in.) deep at a specified location on a cam surface. (e) Determined on a specified bearing surface

Automotive applications of gray cast iron

Grade	Typical uses	Grade	Typical uses
G1800	Miscellaneous soft iron castings (as cast or annealed) in which strength is not a primary consideration	G3500	Diesel engine blocks, truck and tractor cylinder blocks and heads, heavy flywheels, tractor transmission cases, heavy gear boxes
G2500	Small cylinder blocks, cylinder heads, air-cooled cylinders, pistons, clutch plates, oil pump bodies, transmission cases, gear boxes, clutch housings, and light duty brake drums	G3500b	Brake drums and clutch plates for heavy duty service where both resistance to heat checking and higher strength are definite requirements
G2500a	Brake drums and clutch plates for moderate service requirements, where high carbon iron is desired to minimize heat checking	G3500c	Brake drums for extra heavy duty service
		G4000	Diesel engine castings, liners, cylinders, and pistons
G3000	Automobile and diesel cylinder blocks, cylinder heads, flywheels, differential carrier castings, pistons, medium duty brake drums and clutch plates	G4000d	Camshafts

Typical mechanical properties of standard gray iron test bars, as cast

ASTM class	Tensile strength		Torsional shear strength		Compressive strength		Reversed bending fatigue limit		Transverse load on test bar		Hardness, HB
	MPa	ksi	MPa	ksi	MPa	ksi	MPa	ksi	kg	lb	
20	152	22	179	26	572	83	69	10	839	1850	156
25	179	26	220	32	669	97	79	11.5	987	2175	174
30	214	31	276	40	752	109	97	14	1145	2525	210
35	252	36.5	334	48.5	855	124	110	16	1293	2850	212
40	293	42.5	393	57	965	140	128	18.5	1440	3175	235
50	362	52.5	503	73	1130	164	148	21.5	1638	3600	262
60	431	62.5	610	88.5	1293	187.5	169	24.5	1678	3700	302

Typical moduli of elasticity of standard gray iron test bars, as cast

ASTM class	Tensile modulus		Torsional modulus	
	GPa	10^6 psi	GPa	10^6 psi
20	66 to 97	9.6 to 14.0	27 to 39	3.9 to 5.6
25	79 to 102	11.5 to 14.8	32 to 41	4.6 to 6.0
30	90 to 113	13.0 to 16.4	36 to 45	5.2 to 6.6
35	100 to 119	14.5 to 17.2	40 to 48	5.8 to 6.9
40	110 to 138	16.0 to 20.0	44 to 54	6.4 to 7.8
50	130 to 157	18.8 to 22.8	50 to 55	7.2 to 8.0
60	141 to 162	20.4 to 23.5	54 to 59	7.8 to 8.5

Ductile Irons

Compositions and general uses for standard grades of ductile irons(a)

Specification no.	Grade or class	UNS	Typical composition, %					Description	General uses
			TC(b)	Si	Mn	P	S		
ASTM A395; ASME SA395	60-40-18	F32800	3.00 min	2.50 max(c)	...	0.08 max	...	Ferritic; annealed	Pressure-containing parts for use at elevated temperatures
ASTM A476; SAE AMS5316	80-60-03	F34100	3.00 min(d)	3.0 max	...	0.08 max	0.05 max	As cast	Paper mill dryer rolls, at temperatures up to 230 °C (450 °F)
ASTM A536; MIL-I-11466B(MR)	60-40-18(e)	F32800						Ferritic; may be annealed	Shock-resistant parts; low-temperature service
	65-45-12(e)	F33100						Mostly ferritic; as cast or annealed	General service
	80-55-06(e)	F33800						Ferritic/pearlitic; as cast	General service
	100-70-03(e)	F34800						Mostly pearlitic; may be normalized	Best combination of strength, wear resistance and response to surface hardening
	120-90-02(e)	F36200						Martensitic; oil quenched and tempered	Highest strength and wear resistance
SAE J434c	D4018(f)	F32800	3.20-4.10	1.80-3.00	0.10-1.00	0.015-0.10	0.005-0.035	Ferritic	Moderately stressed parts requiring good ductility and machinability
	D4512(f)	F33100						Ferritic/pearlitic	Moderately stressed parts requiring moderate machinability
	D5506(f)	F33800						Ferritic/pearlitic	Highly stressed parts requiring good toughness
	D7003(f)	F34800						Pearlitic	Highly stressed parts requiring very good wear resistance and good response to selective hardening
	DQ & T(f)	F30000						Martensitic	Highly stressed parts requiring uniformity of microstructure and close control of properties
MIL-I-24137 (Ships)	Class A	F33101	3.0 min	2.50 max(g)	...	0.08 max	...	Ferritic; annealed	General shipboard service
	Class B	F43020	2.40-3.00	1.80-3.20	0.80-1.50(h)	0.20 max	...	Austenitic(k)	Shipboard service requiring resistance to corrosion, heat or shock; or requiring nonmagnetic properties
	Class C	F43021	2.70-3.10	2.00-3.00	1.90-2.50(j)	0.15 max	...	Austenitic(k)	Same as Class B

(a) For mechanical properties and typical applications, see the following table. (b) Total carbon. (c) The silicon limit may be increased by 0.08% for each 0.01% reduction in phosphorus content, up to 2.75 Si. (d) Carbon equivalent (CE), 3.8 to 4.5; CE = TC + 0.3 (Si + P). (e) Composition subordinate to mechanical properties; composition range for any element may be specified by agreement between supplier and purchaser. (f) General composition given under grade D4018 for reference only. Typically, foundries will produce to narrower ranges than those shown and will establish different median compositions for different grades. (g) For castings with sections 13 mm (1/2 in.) and smaller, may have 2.75 Si max and 0.08 P max, or 3.00 Si max with 0.05 P max; for castings with sections 50 mm (2 in.) and greater, CE must not exceed 4.3. (h) Plus 18 to 22 Ni and 1.7 to 2.4 Cr. (j) Plus 20 to 23 Ni and 0.5 Cr max. (k) Stress relieved at 650 °C (1200 °F); or solution treated at 950 °C (1750 °F), if necessary to dissolve carbides

Mechanical properties and typical applications for standard grades of ductile irons (a)

Specification no.	Grade or class	Hardness, HB(b)	Tensile strength, min(c) MPa	ksi	Yield strength, min(c) MPa	ksi	Elongation in 50 mm, or 2 in., min, %(c)	Typical applications
ASTM A395-76; ASME SA395	60-40-18	143 to 187	414	60	276	40	18	Valves and fittings for steam and chemical-plant equipment
ASTM A476-70(d); SAE AMS5316	80-60-03	201 min	552	80	414	60	3	Paper-mill dryer rolls
ASTM A536-72-MIL-I-11466B(MR)	6-40-18	...	414	60	276	40	18	Pressure-containing parts such as valve and pump bodies
	65-45-12	...	448	65	310	45	12	Machine components subject to shock and fatigue loads
	80-55-06	...	552	80	379	55	6	Crankshafts, gears and rollers
	100-70-03	...	689	100	483	70	3	High-strength gears and machine components
	120-90-02	...	827	120	621	90	2	Pinions, gears, rollers and slides
SAE J434c	D4018	170 max	414	60	276	40	18	Steering knuckles
	D4512	156 to 217	448	65	310	45	12	Disc-brake calipers
	D5506	187 to 255	552	80	379	55	6	Crankshafts
	D7003	241 to 302	689	100	483	70	3	Gears
	DQ & T	(e)	(f)	(f)	(f)	(f)	(f)	Rocker arms
MIL-I-24137 (Ships)	Class A	190 max	414	60	310	45	15	Electric equipment, engine blocks, pumps, housings, gears, valve bodies, clamps and cylinders
	Class B	190 max	379	55	207	30	7	Pressure parts, machine components and propellers
	Class C	175 max	345	50	172	25	20	Pressure parts, machine components and propellers

(a) For compositions, descriptions and uses, see the preceding table. (b) Measured at a predetermined location on the casting. (c) Determined using a standard specimen taken from a separately cast test block, as set forth in the applicable specification. (d) Reapproved in 1976. (e) Range specified by mutual agreement between producer and purchaser. (f) Value must be compatible with minimum hardness specified for production castings

Average mechanical properties of ductile irons heat treated to various strength levels(a)

Nearest standard grade	Hardness, HB	Ultimate tensile strength MPa	ksi	Yield strength MPa	ksi	Elongation, %(b)	Elastic modulus GPa	10^6 psi	Poisson's ratio
Tension									
60-40-18	167	461	66.9	329(c)	47.7(c)	15.0	169	24.5	0.29
65-45-12	167	464	67.3	332(c)	48.2(c)	15.0	168	24.4	0.29
80-55-06	192	559	81.1	362(c)	52.5(c)	11.2	168	24.4	0.31
120-90-02	331	974	141.3	864(c)	125.3(c)	1.5	164	23.8	0.28
Compression									
60-40-18	167	...	...	359(c)	52.0(c)	...	164	23.8	0.26
65-45-12	167	...	...	362(c)	52.5(c)	...	163	23.6	0.31
80-55-06	192	...	...	386(c)	56.0(c)	...	165	23.9	0.31
120-90-02	331	...	...	920(c)	133.5(c)	...	164	23.8	0.27
Torsion									
60-40-18	167	472	68.5	195(d)	28.3(d)	...	63	9.1	...
65-45-12	167	475	68.9	297(d)	30.0(d)	...	65.5(e)	9.5(e)	...
80-55-06	192	504	73.1	193(d)	28.0(d)	...	64	9.3	...
120-90-02	331	875	126.9	492(d)	71.3(d)	...	65(e)	9.4(e)	...
							62	9.0	
							64(e)	9.3(e)	
							63.4	9.2	
							64(e)	9.3(e)	

(a) Determined for a single heat of ductile iron, heat treated to approximate various standard grades. Properties were obtained using test bars machined from 25-mm (1-in.) keel blocks. (b) In 50 mm, or 2 in. (c) 0.2% offset. (d) 0.0375% offset. (e) Calculated from tensile modulus and Poisson's ratio in tension

Short-time tensile properties of some high-silicon ductile (nodular) irons(a)

Nominal alloy content	Strength, MPa (ksi) at room temperature		Elongation, %	Strength, MPa (ksi) at 540 °C (1000 °F)		Elongation, %
	Yield(b)	Tensile		Yield(b)	Tensile	
4Si	445 (64.3)	565 (81.9)	19.5	225 (32.5)	245 (35.6)	45
4Si-0.5Mo	470 (68.2)	595 (86.4)	17	250 (36.3)	275 (40.1)	35.5
4Si-1.0Mo	475 (68.8)	610 (88.4)	14.5	270 (39.6)	300 (43.8)	24.5
4Si-1.5Mo	485 (70.3)	620 (90.2)	12	260 (38.0)	300 (43.3)	23
4Si-2.0Mo	480 (69.3)	635 (92.1)	10	280 (40.7)	320 (46.4)	21.5
4Si-2.5 Mo	490 (71.0)	650 (94.8)	10.5	270 (39.6)	320 (46.6)	20.5
2.8Si-1.0Mo	350 (51.0)	535 (77.8)	15	230 (33)	270 (39.3)	20
4Si-0.9V(c)	505 (73.3)	580 (84.3)	4.5	…	…	…

Nominal alloy content	Strength, MPa (ksi) at 650 °C (1200 °F)		Elongation, %	Strength, MPa (ksi) at 700 °C (1300 °F)		Elongation, %
	Yield(b)	Tensile		Yield(b)	Tensile	
4Si	65 (9.7)	80 (12.1)	59	50 (7.3)	60 (8.8)	87.5
4Si-0.5Mo	110 (16.3)	130 (18.8)	71.5	65 (9.8)	75 (11.1)	75.5
4Si-1.0Mo	110 (16.1)	130 (18.8)	53.0	75 (11.0)	90 (12.9)	59.0
4Si-1.5Mo	120 (17.3)	135 (20.0)	48.5	80 (11.5)	90 (13.0)	55.5
4Si-2.0Mo	120 (17.4)	145 (21.0)	42.5	85 (12.3)	95 (14.1)	51.5
4Si-2.5Mo	120 (17.8)	150 (21.5)	31.0	90 (12.8)	100 (14.4)	40.5
2.8Si-1.0Mo	120 (17.5)	140 (20.6)	45.0	70 (10.2)	90 (13.1)	51.5
4Si-0.9V(c)	95 (14.0)	105 (15.0)	34.0	45 (6.3)	45 (6.8)	37.5(d)

(a) All irons were subcritically annealed for 4 h at 790 °C (1450 °F), air cooled. (b) 0.2% offset. (c) Soak 3 h at 900 °C (1650 °F); cool to 700 °C (1300 °F), soak 5 h; furnace cool to below 425 °C (800 °F). (d) At 760 °C (1400 °F)

Compositions and properties of ASTM A439 austenitic ductile (nodular) irons at room temperature

Type	Composition, %(a)						Hardness, HB	Mechanical properties(a)		Elongation, %
	C	Si	Mn	Ni	Cr	P		Strength, MPa (ksi)		
								Tensile	Yield	
D2	3.00	1.50-3.00	0.70-1.25	18.00-22.0	1.75-2.75	0.08	139-202	400 (58)	210 (30)	8
D2B	3.00	1.50-3.00	0.70-1.25	18.00-22.00	2.75-4.00	0.08	148-211	400 (58)	210 (30)	7
D2C	2.90	1.00-3.00	1.80-2.40	21.00-24.00	0.50	0.08	121-171	400 (58)	190 (28)	20
D3	2.60	1.00-2.80	1.00	28.00-32.00	2.50-3.50	0.08	139-202	380 (55)	210 (30)	6
D3A	2.60	1.00-2.80	1.00	28.00-32.00	1.00-1.50	0.08	131-193	380 (55)	210 (30)	10
D4	2.60	5.00-6.00	1.00	28.00-32.00	4.50-5.50	0.08	202-273	410 (60)	…	…
D5	2.60	1.00-2.80	1.00	34.00-36.00	0.10	0.08	131-185	380 (55)	210 (30)	10
D5B	2.40	1.00-2.80	1.00	34.00-36.00	2.00-3.00	0.08	139-193	380 (55)	210 (30)	6
D5S(b)	2.0 nom	5.5 nom	0.5 nom	36.00 nom	2.50 nom	0.08	160	430 (63)	250 (36)	20

(a) Maximum unless expressed as a range or otherwise noted. (b) Inclusion in ASTM A439 is pending

Malleable Irons

Typical composition ranges for ferritic and pearlitic malleable irons(a)

Type	Composition, %				
	TC	Mn	Si	P, max	S
Ferritic					
Grade 32510	2.30 to 2.70	0.25 to 0.55	1.00 to 1.75	0.05	0.03 to 0.18
Grade 35018	2.00 to 2.45	0.25 to 0.55	1.00 to 1.35	0.05	0.03 to 0.18
Pearlitic	2.00 to 2.70	0.25 to 1.25	1.00 to 1.75	0.05	0.03 to 0.18

(a) Analyzed in the white iron condition

Applications of malleable iron castings(a)

Specification no.	Class or grade	Microstructure	Typical applications
Ferritic			
ASTM A47; ANSI G48.1; FED QQ-I-666C	32510 35018	Temper carbon and ferrite	General engineering service at normal and elevated temperatures for good machinability and excellent shock resistance
ASTM A338	32510 35018	Temper carbon and ferrite	Flanges, pipe fittings, and valve parts for railroad, marine, and other heavy duty service up to 345 °C (650 °F)
ASTM A197; ANSI G49.1	...	Free of primary graphite	Pipe fittings and valve parts for pressure service
Pearlitic and Martensitic			
ASTM A220; ANSI G48.2; MIL-I-11444B	40010 45008 45006 50005 60004 70003 80002 90001	Temper carbon in necessary matrix without primary cementite or graphite	General engineering service at normal and elevated temperatures Dimensional tolerance range for castings is stipulated
Automotive			
ASTM A602; SAE J158	M3210	Ferritic	For low-stress parts requiring good machinability: steering gear housings, carriers, and mounting brackets
	M4504	Ferrite and tempered pearlite(b)	Compressor crankshafts and hubs
	M5003	Ferrite and tempered pearlite(b)	For selective hardening: planet carriers, transmission gears, differential cases
	M5503	Tempered martensite	For machinability and improved response to induction hardening
	M7002	Tempered martensite	For high-strength parts: connecting rods and universal joint yokes
	M8501	Tempered martensite	For high strength plus good wear resistance: certain gears

(a) For mechanical properties, see the following table. (b) May be all tempered martensite for some applications

Properties of malleable iron castings(a)

Specification no.	Class or grade	Tensile strength MPa	ksi	Yield strength MPa	ksi	Hardness, HB	Elongation(b), %
Ferritic							
ASTM A47, A338;							
ANSI G48.1;							
FED QQ-I-666c	32510	345	50	224	32	156 max	10
	35018	365	53	241	35	156 max	18
ASTM A197	...	276	40	207	30	156 max	5
Pearlitic and Martensitic							
ASTM A220;							
ANSI G48.2;							
MIL-I-11444B	40010	414	60	276	40	149 to 197	10
	45008	448	65	310	45	156 to 197	8
	45006	448	65	310	45	156 to 207	6
	50005	483	70	345	50	179 to 229	5
	60004	552	80	414	60	197 to 241	4
	70003	586	85	483	70	217 to 269	3
	80002	655	95	552	80	241 to 285	2
	90001	724	105	621	90	269 to 321	1
Automotive							
ASTM A602; SAE J158	M3210(c)	345	50	224	32	156 max	10
	M4504(d)	448	65	310	45	163 to 217	4
	M5003(d)	517	75	345	50	187 to 241	3
	M5503(e)	517	75	379	55	187 to 241	3
	M7002(e)	621	90	483	70	229 to 269	2
	M8501(e)	724	105	586	85	269 to 302	1

(a) For microstructures and typical applications, see the preceding table. (b) Minimum in 50 mm (2 in.). (c) Annealed. (d) Air quenched and tempered. (e) Liquid quenched and tempered

Alloy Cast Irons: General

Ranges of alloy content for various types of alloy cast irons

Description	TC(b)	Mn	P	S	Si	Ni	Cr	Mo	Cu	Matrix structure, as-cast(c)
Abrasion-resistant white irons										
Low-C white iron(d)	2.2 to 2.8	0.2 to 0.6	0.15	0.15	1.0 to 1.6	1.5	1.0	0.5	(e)	CP
High-C, low-Si white iron	2.8 to 3.6	0.3 to 2.0	0.30	0.15	0.3 to 1.0	2.5	3.0	1.0	(e)	CP
Malleable white iron	2.2 to 2.5	0.3 to 0.5	0.15	0.15	1.0 to 1.6	...	...	...	...	CP
Martensitic nickel-chromium iron	2.5 to 3.7	1.3	0.30	0.15	0.8	2.7 to 5.0	1.1 to 4.0	1.0	...	M, A
Martensitic nickel, high-chromium iron	2.5 to 3.6	1.3	0.10	0.15	1.0 to 2.2	5 to 7	7 to 11	1.0	...	M, A
Martensitic chromium-molybdenum iron	2.0 to 3.6	0.5 to 1.5	0.10	0.06	1.0	1.5	11 to 23	0.5 to 3.5	1.2	M, A
High-chromium iron	2.3 to 3.0	0.5 to 1.5	0.10	0.06	1.0	1.5	23 to 28	1.5	1.2	M
Corrosion-resistant irons										
High-silicon iron(f)	0.4 to 1.1	1.5	0.15	0.15	14 to 17	...	5.0	1.0	0.5	F
High-chromium iron	1.2 to 4.0	0.3 to 1.5	0.15	0.15	0.5 to 3.0	5.0	12 to 35	4.0	3.0	M, A
Nickel-chromium gray iron(g)	3.0	0.5 to 1.5	0.08	0.12	1.0 to 2.8	13.5 to 36	1.5 to 6.0	1.0	7.0	A
Nickel-chromium ductile iron(h)	3.0	0.7 to 4.5	0.08	0.12	1.0 to 3.0	18 to 36	1.0 to 5.5	1.0	...	A
Heat-resistant gray irons										
Medium-silicon iron(j)	1.6 to 2.5	0.4 to 0.8	0.30	0.10	4.0 to 7.0	...	...	...	...	F
High-chromium iron	1.8 to 3.0	0.3 to 1.5	0.15	0.15	0.5 to 2.5	5.0	15 to 35	...	...	F, CP
Nickle-chromium iron(g)	1.8 to 3.0	0.4 to 1.5	0.15	0.15	1.0 to 2.75	13.5 to 36	1.8 to 6.0	1.0	7.0	A
Nickel-chromium-silicon iron(k)	1.8 to 2.6	0.4 to 1.0	0.10	0.10	5.0 to 6.0	13 to 43	1.8 to 5.5	1.0	10.0	A
High-aluminum iron	1.3 to 2.0	0.4 to 1.0	0.15	0.15	1.3 to 6.0	...	20 to 25 Al	...	...	F
Heat-resistant ductile irons										
Medium-silicon ductile iron	2.8 to 3.8	0.2 to 0.6	0.08	0.12	2.5 to 6.0	1.5	...	...	...	F
Nickel-chromium ductile iron(h)	3.0	0.7 to 2.4	0.08	0.12	1.75 to 5.5	18 to 36	1.75 to 3.5	1.0	...	A

(a) Where a single value is given rather than a range, that value is a maximum limit. (b) Total carbon. (c) CP, coarse pearlite; M, martensite; A, austenite; F, ferrite. (d) May be produced from a malleable-iron base composition. (e) Cu may replace all or part of the Ni. (f) Such as Duriron, Durichlor 51, Superchlor. (g) Such as Ni-Resist austenitic iron (ASTM A436). (h) Such as Ni-Resist austenitic ductile iron (ASTM A439. (j) Such as Silal. (k) Such as Nicrosilal

Physical properties of selected alloy cast irons

Description(a)	Density Mg/m³	Density lb/in.³	Coefficient of thermal expansion(b) 10⁻⁶ m/m · °C	Coefficient of thermal expansion(b) 10⁻⁶ in./in. · °F	Electrical resistivity, μΩ · m	Thermal conductivity W/m · K	Thermal conductivity Btu/ft · h · °F
Abrasion-resistant white irons							
Low-C white iron	7.6 to 7.8	0.275 to 0.282	12(c)	6.7(c)	0.53	22(d)	13(d)
Martensitic nickel-chromium iron	7.6 to 7.8	0.275 to 0.282	8 to 9(c)	4.4 to 5(c)	0.80	30(d)	17(d)
Corrosion-resistant irons							
High-silicon iron	7.0 to 7.05	0.252 to 0.254	12.4 to 13.1	6.9 to 7.3	0.50	...	...
High-chromium iron	7.3 to 7.5	0.264 to 0.271	9.4 to 9.9	5.2 to 5.5	...	...	...
High-nickel gray iron	7.4 to 7.6	0.267 to 0.275	8.1 to 19.3	4.5 to 10.7	1.0(d)	38 to 40	22 to 23
High-nickel ductile iron	7.4	0.267	12.6 to 18.7	7.0 to 10.4	1.0(d)	13.4	7.75
Heat-resistant gray irons							
Medium-silicon iron	6.8 to 7.1	0.246 to 0.256	10.8	6.0	...	37	21
High-chromium iron	7.3 to 7.5	0.264 to 0.271	9.3 to 9.9	5.2 to 5.5	...	20	12
High-nickel iron	7.3 to 7.5	0.264 to 0.271	8.1 to 19.3	4.5 to 10.7	1.4 to 1.7	37 to 40	21 to 23
Nickel-chromium-silicon iron	7.33 to 7.45	0.265 to 0.269	12.6 to 16.2	7.0 to 9.0	1.5 to 1.7	30	17
High-alluminum iron	5.5 to 6.4	0.20 to 0.23	15.3	8.5	2.4	...	...
Heat-resistant ductile irons							
Medium-silicon ductile iron	7.1	0.257	10.8 to 13.5	6.0 to 7.5	0.58 to 0.87	...	...
High-nickel ductile (20 Ni)	7.4	0.268	18.7	10.4	1.02	13	7.7
High-nickel ductile (23 Ni)	7.4	0.268	18.4	10.2	1.0(d)	...	...

(a) For compositions, see the preceding table. (b) At 21 °C (70 °F). (c) 10 to 260 °C (50 to 500 °F). (d) Estimated

Alloy Cast Irons: Abrasion-Resistant

Chemical composition of standard martensitic white cast irons(a)

Class	Type	Designation	TC(c)	Mn	P	S	Si	Cr	Ni	Mo	Cu
								Composition, wt %(b)			
I	A	Ni-Cr-HC	3.0 to 3.6	1.3	0.30	0.15	0.8	1.4 to 4.0	3.3 to 5.0	1.0	...
I	B	Ni-Cr-LC	2.5 to 3.0	1.3	0.30	0.15	0.8	1.4 to 4.0	3.3 to 5.0	1.0	...
I	C	Ni-Cr-GB	2.9 to 3.7	1.3	0.30	0.15	0.8	1.1 to 1.5	2.7 to 4.0	1.0	...
I	D	Ni-Hi-Cr	2.5 to 3.6	1.3	0.10	0.15	1.0 to 2.2	7 to 11	5 to 7	1.0	...
II	A	12% Cr	2.4 to 2.8	0.5 to 1.5	0.10	0.06	1.0	11 to 14	0.5	0.5 to 1.0	1.2
II	B	15% Cr-Mo-LC	2.4 to 2.8	0.5 to 1.5	0.10	0.06	1.0	14 to 18	0.5	1.0 to 3.0	1.2
II	C	15% Cr-Mo-HC	2.8 to 3.6	0.5 to 1.5	0.10	0.06	1.0	14 to 18	0.5	2.3 to 3.5	1.2
II	D	20% Cr-Mo-LC	2.0 to 2.6	0.5 to 1.5	0.10	0.06	1.0	18 to 23	1.5	1.5	1.2
II	E	20% Cr-Mo-HC	2.6 to 3.2	0.5 to 1.5	0.10	0.06	1.0	18 to 23	1.5	1.0 to 2.0	1.2
III	A	25% Cr	2.3 to 3.0	0.5 to 1.5	0.10	0.06	1.0	23 to 28	1.5	1.5	1.2

(a) From ASTM A532-75a. Certain specific compositions of alloys II-B, II-C, II-D and II-E are covered by U.S. Patent No. 3, 410, 682. (b) Where a single value is given rather than a range, that value is a maximum limit. (c) Total carbon

Mechanical properties of standard martensitic white cast irons(a)

Class	Type	Designation	Min value		Hardness, HB		Typical maximum section thickness	
			Sand cast	Chill cast	Max value		mm	in.
					Hardened	Annealed		
I	A	Ni-Cr-HC	550	600	...	...	200	8
I	B	Ni-Cr-LC	550	600	...	...	200	8
I	C	Ni-Cr-GB	550	600	...	...	75(b)	3(b)
I	D	Ni-Hi-Cr	550	500	600	...	300	12
II	A	12% Cr	550	...	600	400	25(b)	1(b)
II	B	15% Cr-Mo-LC	450	...	600	400	100	4
II	C	15% Cr-Mo-HC	550	...	600	400	75	3
II	D	20% Cr-Mo-LC	450	...	600	400	200	8
II	E	20% Cr-Mo-HC	450	...	600	400	300	12
III	A	25% Cr	450	...	600	400	200	8

(a) From ASTM A532-75a; for compositions, see the preceding table. (b) Ball diameter

Hardness conversions for white cast irons (from averaged data)

HB	HV	HRC	Scleroscope	HB	HV	HRC	Scleroscope
High-chromium irons				540	600	53	...
815	1000	68.5	...	520	575	51.5	...
800	975	68	...	490	550	50	...
790	950	67.5	...	475	525	48.5	...
775	925	67	...	440	500	47	...
760	900	66	...	420	475	45.5	...
745	875	65.0	...	395	450	43.5	...
730	850	64.5	...	370	425	41.7	...
720	825	63.5	...	...	400	40	...
700	800	62.5	...				
680	775	61.5	...				
660	750	61.0	...	**Chromium-nickel irons**			
640	725	59.5	...	750	830-860	...	90-93
625	700	58	...	700	740-770	...	84-87
610	675	57	...	650	690-720	...	79-82
585	650	56	...	600	630-660	...	75-78
560	625	54.5	...	550	570-610	...	70-73
				500	510-540	...	67-70

Transverse strengths and relative toughness of various pearlitic and martensitic white irons(a)

Type of iron	Basic composition	Transverse strength		Deflection		Toughness(b)	
		kg	lb	mm	in.	kg · m	lb · in.
Sand-cast pearlitic	3.2 to 3.5 C, 1 to 2 Cr	635 to 815	1400 to 1800	2.0 to 2.3	0.080 to 0.092	1.27 to 1.87	112 to 162
Sand-cast martensitic	2.8 to 3.6 C, 1.4 to 4 Cr, 3.3 to 5 Ni	1810 to 2490	4000 to 5500	2.0 to 3.0	0.08 to 0.12	3.62 to 7.47	320 to 660
	2.5 to 3.6 C, 7 to 11 Cr, 4.5 to 7 Ni	2270 to 2720	5000 to 6000	2.0 to 2.8	0.08 to 0.11	4.54 to 7.62	400 to 660
	2.8 to 3.4 C, 12 to 16 Cr, 2 to 4 Mo	1015 to 1370	2235 to 3015	3.2 to 3.6	0.125 to 0.14	3.25 to 4.93	279 to 422
	3.5 to 4.1 C, 12 to 16 Cr, 2.5 to 3Mo	800 to 1000	1760 to 2200	2.0 to 2.8	0.08 to 0.110	1.60 to 2.80	140 to 240
Chill-cast martensitic	2.8 to 3.6 C, 1.4 to 4 Cr, 3.3 to 5 Ni	2040 to 3180	4500 to 7000	2.0 to 3.0	0.08 to 0.12	4.08 to 9.54	360 to 840
	2.5 to 3.6 C, 7 to 11 Cr, 4.5 to 7 Ni	2500 to 3180	5500 to 7000	2.5 to 3.8	0.10 to 0.15	6.25 to 12.1	550 to 1050
	3.2 to 3.4 C, 12 to 16 Cr, 1.5 to 3 Mo	1980 to 2300	4360 to 5060	5.1 to 6.5	0.202 to 0.26	10.1 to 15.0	870 to 1320
	3.5 to 4.1 C, 12 to 16 Cr, 2.5 to 3 Mo	1270 to 1570	2800 to 3470	3.6 to 3.8	0.140 to 0.15	4.57 to 5.97	392 to 520

(a) Data from as-cast 30.5 mm (1.2-in.) diam test bars broken over a 457-mm (18-in.) span. (b) Relative toughness evaluated as product of transverse strength times deflection

Alloy Cast Irons: Corrosion- and Heat-Resistant

Typical mechanical properties of corrosion-resistant cast irons

Type of iron(a)	Hardness, HB	Tensile strength		Compressive strength		Impact energy		Transverse breaking load(b)		Transverse deflection(b)	
		MPa	ksi	MPa	ksi	J	ft · lb	kg	lb	mm	in.
High-silicon iron	480 to 520	90 to 180	13 to 26	690	100	2.7 to 5.4(c)	2 to 4(c)	545 to 1000	1200 to 2200	0.65	0.026
High-chromium iron	250 to 740	205 to 830	30 to 120	690	100	0.1 to 3(d) 27 to 47(c)	0.1 to 2(d) 20 to 35(c)	910 to 1590	2000 to 3500	1.5 to 3.8	0.06 to 0.15
High-nickel gray iron	120 to 250	170 to 310	25 to 45	690 to 1100	100 to 160	80 to 200(c)	60 to 150(c)	820 to 1590	1800 to 3500	5 to 25	0.20 to 1.00
High-nickel ductile iron	130 to 240	380 to 480	55 to 70	1240 to 1380	180 to 200	14 to 40(d)	10 to 30(d)	...	...	...	...

(a) For composition ranges, see the table in the section Alloy Cast Irons: General. (b) For as-cast 30.5-mm (1.2 in.) diam bar broken over a 457-mm (18-in.) span. (c) Unnotched 30.5-mm diam test bar broken over a 152-mm (6-in.) span in a Charpy testing machine. (d) Standard Charpy

Typical mechanical properties of heat-resistant alloy cast irons

Type of iron(a)	Hardness, HB	Tensile strength		Compressive strength		Impact energy		Transverse breaking load(b)		Transverse deflection(b)	
		MPa	ksi	MPa	ksi	J	ft · lb	kg	lb	mm	in.
Medium-silicon gray iron	170 to 250	170 to 310	25 to 45	620 to 1040	90 to 150	20 to 31(c)	15 to 23(c)	455 to 1090	1000 to 2400	4.6 to 8.9	0.18 to 0.35
High-chromium gray iron	250 to 500	210 to 620	30 to 90	690	100	27 to 47(c)	20 to 35(c)	910 to 1590	2000 to 3500	1.5 to 3.8	0.06 to 0.15
High-nickel gray iron	130 to 250	170 to 310	25 to 45	690 to 1100	100 to 160	80 to 200(c)	60 to 150(c)	820 to 1360	1800 to 3000	5 to 25	0.2 to 1.0
Ni-Cr-Si gray iron	110 to 210	140 to 310	20 to 45	480 to 690	70 to 100	110 to 200(c)	80 to 150(c)	820 to 1130	1800 to 2500	7 to 35	0.3 to 1.4
High-aluminum gray iron	180 to 350	235 to 620	34 to 90	...	...	...	...	...	...	...	...
Medium-silicon ductile iron	140 to 300	415 to 690	60 to 100(c)	...	...	7 to 155(d)	5 to 115(d)	...	...	...	...
High-nickel ductile iron (20 Ni)	140 to 200	380 to 415	55 to 60(e)	1240 to 1380	180 to 200	16(f)	12(f)	...	...	...	...
High-nickel ductile iron (23Ni)	130 to 170	400 to 450	58 to 65(g)	...	...	38(f)	28(f)	...	...	...	...

(a) For composition ranges, see the table in the section Alloy Cast Irons: General. (b) Unnotched 30.5-mm (1.2-in.) diam test bar broken on 152-mm (6-in.) supports in a Charpy testing machine. (c) Yield strength, 310 to 520 MPa (45 to 75 ksi); elongation, 0.2%. (d) Standard Charpy test on 10-mm unnotched specimen. (e) Yield strength, 210 to 240 MPa (30 to 35 ksi); elongation, 8 to 20%. (f) Standard Charpy test on 10-mm notched specimen. (g) Yield strength, 195 to 240 MPa (28 to 35 ksi); elongation, 20 to 40%

Oxidation of plain and alloy cast irons

Iron	Composition, %(a)				Oxide penetration				Growth at 815 °C (1500 °F)(c)	
	TC	Si	Cr	Ni	At 760 °C (1400 °F)(b)		At 815 °C (1500 °F)(c)			
					mm/yr	mil/yr	mm/yr	mil/yr	mm/yr	mil/yr
Austenitic(d)	2.69	1.96	2.05	13.96	4.7	184	9.5	374	0.4	15
Austenitic(e)	2.97	2.38	4.87	(14.0)	2.4	96	5.9	232	0.2	8
Austenitic	2.40	1.57	2.98	30.28	2.1	83	6.3	249	0.2	8
Austenitic	(1.8)	(6.0)	(5.0)	(30.0)	0.05	2	1.3	53	0.2	8
Austenitic	(2.8)	(1.7)	(2.0)	(20.0)	4.2	166	7.9	312	0.2	8
Austenitic	(2.7)	(2.5)	(5.0)	(20.0)	1.9	74	3.6	143	0.4	15
Plain ferritic	(3.2)	(2.2)	...	...	>20(f)	>800(f)	>85(f)	>3300(f)	2.0	78
Low-alloy ferritic	(3.3)	(1.5)	(0.6)	(1.5)	>20(f)	>800(f)	>90(f)	>3500(f)	1.4	54
Low-alloy ferritic	(3.3)	(2.2)	(1.0)	(1.0)	5.8	228	25.9	1020	1.2	47
Low-alloy ferritic	(3.1)	(2.2)	(0.9)	(1.5)	7.2	284	29.0	1140	1.6	62

(a) Figures enclosed in parentheses are estimated values. Phosphorus and sulfur contents in all iron samples were about 0.10% . TC, total carbon. (b) Exposure of 2000 h in electric furnace at 760 °C (1400 °F) with air atmosphere containing 17 to 19% O. (c) Exposed for 492 h in gas-fired heat treating furnace at 815 °C (1500 °F). (d) 6.05% copper. (e) 6.0% copper. (f) Specimen completely burned

Oxidation of ferritic and austenitic cast irons

Iron	Composition, %(a)				Growth		Oxide penetration	
	TC	Si	Cr	Ni	mm/yr	mil/yr	mm/yr	mil/yr
After 3723 h at 750 to 760 °C (1375 to 1400 °F) in electric furnace, air atmosphere								
Ferritic	3.05	2.67	0.90	1.55	2.0	78	(b)	(b)
Austenitic	2.97	1.63	1.89	20.02	0.8	31	6.9	270
Austenitic	2.52	2.67	5.16	20.03	nil	nil	0.2	6
Austenitic	2.32	1.86	2.86	30.93	nil	nil	2.0	78
Austenitic	1.86	5.84	5.00	29.63	nil	nil	< 0.1	< 3
After 1677 h at 815 to 925 °C (1500 to 1700 °F) in gas-fired furnace, slightly reducing atmosphere								
Ferritic	(3.2)	(2.2)	...	...	3.2	125	(b)	(b)
Austenitic(c)	(3.0)	(2.4)	(5.0)	(14.0)	0.4	15	8.4	330
Austenitic	(2.7)	(2.5)	(5.0)	(20.0)	0.4	15	5.6	220
Austenitic	(2.4)	(1.6)	(3.0)	(30.0)	0.4	15	6.9	270
Austenitic	(1.8)	(6.0)	(5.0)	(30.0)	0.4	15	0.1	5

(a) Figures enclosed in parentheses are estimated values. Phosphorus and sulfur contents in all iron samples were about 0.10%. TC, total carbon. (b) Sample was completely burned. (c) 6.0% copper (estimated)

Machining

Machinability of gray iron

Microstructure	ASTM class	Tensile strength		Hardness, HB	Cutting speed(a)	
		MPa	ksi		m/min.	ft/min.
Acicular iron	50	407	59	263	46	150
Fine pearlite, alloy	40	310	45	225	95	310
Ferrite (annealed)	...	108	15.7	100	293	960
Coarse pearlite, no alloy	35	241	35	195	99	325

(a) Cutting speed at which removal of 200 in.^{3} (3280 cm^3) produced 0.030-in. (0.75-mm) wear land on single-point carbide tools

Speeds for machining ductile iron

Ductile iron	High-speed steel tools, ft/min.(a)								Cemented carbide tools, ft/min.(a)		
	Turning(b)	Drilling(c)	Reaming(d)	Tapping	Thread chasing	Milling(e)	Shaping(f) and planing	Broaching	Turning(b)	Reaming(d)	Milling(e)
60-45-10 with full ferritic matrix	50-150	80-130	50-100	20-30	30-70	50-125	40-100	20-35	175-400	75-150	200-400
Semipearlitic matrix	40-90	50-100	40-70	15-20	20-50	35-65	30-75	15-25	100-300	50-90	175-350
80-60-03 with full pearlitic matrix	40-90	50-70	40-70	15-20	20-50	35-65	30-75	15-25	100-300	50-90	175-350

(a) Multiply listed values by 0.3 to find recommended speeds in m/min. In all instances, the longest tool life between regrinds will result when tools are operated at minimum to medium speeds for the range given. (b) Feeds of 0.25 to 0.50 mm/rev (0.010 to 0.020 ipr). Maximum speed is for cuts not more than 1.5 mm (1/16 in.) deep. (c) Feeds should be commensurate with drill diameter—light feeds for small-diameter drills and heavier feeds for larger drills. It is good practice to reduce speed as drill diameter increases; speeds at or near the maximum values given in this table are for drills 12 mm, or 1/2 in., in diameter or less. (d) Use feeds three to four times those used for drills of similar size. An allowance of 0.3 to 0.4 mm (0.012 to 0.015 in.) is sufficient for reaming. (e) Speeds cited are principally for face milling; however, they may be used for plain milling. (f) Depth of cut and feed vary with sturdiness of the setup. The operating speed for roughing should approach the minimum value cited

Recommended cutting conditions for turning malleable iron(a)

					Cutting speed, ft/min.(d)					
					Dry, for tool life of:			In soluble oil, for tool life of:		
Grade	Hardness, HB	Type of cut(b)	Feed, ipr/rev(c)	Depth of cut, in.(c)	20 min.	30 min.	40 min.	20 min.	30 min.	40 min.
32510	109	Rough skin	0.015	0.100	440	390	340	500	410	340
			0.030	0.100	350	300	250	410	360	600
		Coarse underskin	0.015	0.060	600	510	430	840	720	630
			0.030	0.060	440	380	350	600	530	460
		Finish	0.003	0.010	1380	1240	1150	1660	1520	1400
			0.007	0.010	1210	950	700	1380	1220	1080
48004	179	Rough skin	0.015	0.100	230	180	130	270	220	130
			0.030	0.100	160	120	70	200	150	110
		Coarse underskin	0.015	0.060	300	260	230	355	330	280
			0.030	0.060	230	185	150	260	230	210
		Finish	0.003	0.010	590	515	460	700	650	615
			0.007	0.010	510	470	450	670	610	550
60003	230	Rough skin	0.015	0.100	175	140	115	200	165	140
			0.030	0.100	130	115	100	150	120	100
		Coarse underskin	0.015	0.060	245	220	200	285	240	225
			0.030	0.060	170	145	125	210	165	135
		Finish	0.003	0.010	525	495	465	540	510	480
			0.007	0.010	470	430	400	480	440	415
80002	250	Rough skin	0.015	0.100	…	…	…	195	165	135
			0.030	0.100	…	…	…	145	115	90
		Coarse underskin	0.015	0.060	185	160	145	190	170	155
			0.030	0.060	160	125	100	160	125	100
		Finish	0.003	0.010	…	…	…	470	420	385
			0.007	0.010	…	…	…	365	330	305

(a) All tests were made under the direction of the Machining Subcommittee of the Malleable Founders Society (now the Iron Castings Society). Grade C2 cutting tools were used dry or with soluble-oil cutting fluid. Tests were discontinued at a uniform wear land of 0.015 in. Discontinued grades 48004 and 60003 are approximately equivalent to grades 50005 and 60004, respectively. (b) Tool configuration for rough skin and coarse underskin cuts: –5° BR; 15° SCEA; –5° SR; 15° ECEA; 5° relief. Tool configuration for finish cuts: 0° BR; 15° SCEA; 5° SR; 15° ECEA; 5° relief. (c) Multiply tabulated values by 25 to obtain an equivalent value in mm. (d) Multiply tabulated values by 0.3 to obtain an equivalent value in m/min

Carbon and Alloy Steels

Quality Descriptors

The need for communication among producers and between producers and users has resulted in the development of a group of terms known as "fundamental quality descriptors". These are names applied to various steel products to imply that the particular products possess certain characteristics that make them especially well suited for specific applications or fabrication processes. Some of the fundamental quality descriptors in common use are listed below.

Carbon steels

Semifinished for forging
 Forging quality
  Special hardenability
  Special internal soundness
  Nonmetallic inclusion requirement
  Special surface
Carbon steel structural sections
 Structural quality
Carbon steel plates
 Regular quality
 Structural quality
 Cold drawing quality
 Cold pressing quality
 Cold flanging quality
 Forging quality
 Pressure vessel quality
 Marine quality
Hot rolled carbon steel bars
 Merchant quality
 Special quality
  Special hardenability
  Special internal soundness
  Nonmetallic inclusion requirement
  Special surface
 Scrapless nut quality
 Axle shaft quality
 Cold extrusion quality
 Cold heading and cold forging quality
Cold finished carbon steel bars
 Standard quality
  Special hardenability
  Special internal soundness
  Nonmetallic inclusion requirement
  Special surface
 Cold heading and cold forging quality
 Cold extrusion quality
Hot rolled sheets
 Commercial quality
 Drawing quality
 Drawing quality special killed
 Structural quality
Cold rolled sheets
 Commercial quality
 Drawing quality
 Drawing quality special killed
 Structural quality
Porcelain enameling sheets
 Commercial quality

 Drawing quality
Long terne sheets
 Commercial quality
 Drawing quality
 Drawing quality special killed
 Structural quality
Galvanized sheets
 Commercial quality
 Drawing quality
 Drawing quality special killed
 Structural quality
 Lock forming quality
Electrolytic zinc-coated sheets
 Commercial quality
 Drawing quality
 Drawing quality special killed
 Structural quality
Hot rolled strip
 Commercial quality
 Drawing quality
 Drawing quality special killed
 Structural quality
Cold rolled strip
 Specific quality descriptors are not provided in cold rolled strip, since this product is largely produced for specific end use
Tin mill products
 Specific quality descriptors are not applicable to tin mill products
Carbon steel wire
 Industrial quality wire
 Cold extrusion wires
 Heading, forging and roll threading wires
 Mechanical spring wires
 Upholstery spring construction wires
 Welding wire
Carbon steel flat wire
 Stitching wire
 Stapling wire
Carbon steel pipe
Structural tubing
Line pipe
Oil country tubular goods
Steel specialty tubular products
 Pressure tubing
 Mechanical tubing
 Aircraft tubing
Hot rolled carbon steel wire rods
 Industrial quality
 Rods for manufacture of wire for

  electric welded chain
 Rods for heading, forging, and
  roll threading wire
 Rods for lock washer wire
 Rods for scrapless nut wire
 Rods for upholstery spring wire
 Rods for welding wire

Alloy steels

Alloy steel plates
 Regular quality or structural quality
 Drawing quality
 Pressure vessel quality
 Structural quality
 Aircraft quality
 Aircraft physical quality
Hot rolled alloy steel bars
 Regular quality
 Aircraft structural or steel subject to
  magnetic particle inspection
 Axle shaft quality
 Bearing quality
 Cold heading quality
 Special cold heading quality
 Rifle barrel quality, gun quality, shell
  or AP shot quality
Alloy steel wire
 Aircraft quality
 Bearing quality
 Special surface quality
Cold finished alloy steel bars
 Regular quality
 Aircraft quality or steel subject to mag-
  netic particle inspection
 Axle shaft quality
 Bearing shaft quality
 Cold heading quality
 Special cold heading quality
 Rifle barrel quality, gun quality, shell
  or AP shot quality
Line pipe
Oil country tubular goods
Steel specialty tubular goods
 Pressure tubing
 Mechanical tubing
 Stainless and heat resisting pipe, pressure
  tubing and mechanical tubing
 Aircraft tubing
 Pipe

Note: Detailed descriptions of many of the categories listed in this table appear in an appropriate section of the AISI Steel Products Manual

Composition Ranges and Tolerances

Alloy steel heat composition ranges and limits—bars, blooms, billets, and slabs

Element	Limit or max of specified range, %	Range, % Open hearth or basic oxygen steel	Range, % Electric furnace steel	Element	Limit or max of specified range, %	Range, % Open hearth or basic oxygen steel	Range, % Electric furnace steel
Carbon	To 0.55 incl	0.05	0.05				
	Over 0.55 to 0.70 incl	0.08	0.07	Molybdenum	To 0.10 incl	0.05	0.05
	Over 0.70 to 0.80 incl	0.10	0.009		0.10 to 0.20 incl	0.07	0.07
	Over 0.80 to 0.95 incl	0.12	0.11		Over 0.20 to 0.50 incl	0.10	0.10
	Over 0.95 to 1.35 incl	0.13	0.12		Over 0.50 to 0.80 incl	0.15	0.15
					Over 0.80 to 1.15 incl	0.20	0.20
Manganese	To 0.60 incl	0.20	0.15				
	Over 0.60 to 0.90 incl	0.20	0.20	Tungsten	To 0.50 incl	0.20	0.20
	Over 0.90 to 1.05 incl	0.25	0.25		Over 0.50 to 1.00 incl	0.30	0.30
	Over 1.05 to 1.90 incl	0.30	0.30		Over 1.00 to 2.00 incl	0.50	0.50
	Over 1.90 to 2.10 incl	0.40	0.35		Over 2.00 to 4.00 incl	0.60	0.60
Sulfur(a)	To 0.050 incl	0.015	0.015	Copper	To 0.60 incl	0.20	0.20
	Over 0.050 to 0.07 incl	0.02	0.02		Over 0.60 to 1.50 incl	0.30	0.30
	Over 0.07 to 0.10 incl	0.04	0.04		Over 1.50 to 2.00 incl	0.35	0.35
	Over 0.10 to 0.14 incl	0.05	0.05				
				Vanadium	To 0.25 incl	0.05	0.05
Silicon	To 0.15 incl	0.08	0.08		Over 0.25 to 0.50 incl	0.10	0.10
	Over 0.15 to 0.20 incl	0.10	0.10				
	Over 0.20 to 0.40 incl	0.15	0.15	Aluminum	Up to 0.10 incl	0.05	0.05
	Over 0.40 to 0.60 incl	0.20	0.20		Over 0.10 to 0.20 incl	0.10	0.10
	Over 0.60 to 1.00 incl	0.30	0.30		Over 0.20 to 0.30 incl	0.15	0.15
	Over 1.00 to 2.20 incl	0.40	0.35		Over 0.30 to 0.80 incl	0.25	0.25
					Over 0.80 to 1.30 incl	0.35	0.35
Chromium	To 0.40 incl	0.15	0.15		Over 1.30 to 1.80 incl	0.45	0.45
	Over 0.40 to 0.90 incl	0.20	0.20				
	Over 0.90 to 1.05 incl	0.25	0.25	**Steelmaking process**		**Lowest max, %(c)**	
	Over 1.05 to 1.60 incl	0.30	0.30	Phosphorus	Basic open hearth, basic oxygen or basic electric furnace steels	0.035(d)	
	Over 1.60 to 1.75 incl	(b)	0.35		Basic electric furnace "E" steels	0.025	
	Over 1.75 to 2.10 incl	(b)	0.40		Acid open hearth or electric furnace steel	0.050	
	Over 2.10 to 3.99 incl	(b)	0.50				
Nickel	To 0.50 incl	0.20	0.20	Sulfur	Basic open hearth, basic oxygen or basic electric furnace steels	0.040(d)	
	Over 0.50 to 1.50 incl	0.30	0.30		Basic electric furnace "E" steels	0.025	
	Over 1.50 to 2.00 incl	0.35	0.35		Acid open hearth or electric furnace steel	0.050	
	Over 2.00 to 3.00 incl	0.40	0.40				
	Over 3.00 to 5.30 incl	0.50	0.50				
	Over 5.30 to 10.00 incl	1.00	1.00				

(a) A range of sulfur content normally indicates a resulfurized steel. (b) Not normally produced by open hearth process. (c) Not applicable to rephosphorized or resulfurized steels. (d) Lower maximum limits on phosphorus and sulfur are required by certain quality descriptors

Carbon steel heat composition ranges and limits—semifinished products for forging, hot rolled and cold finished bars, wire rod, and seamless tubing

Element	Limit or max of specified range, %	Range,%	Element	Limit or max of specified range, %	Range,%
Carbon				Over 0.08 to 0.13 incl	0.05
(a)	To 0.25 incl	0.05	Sulfur		
	Over 0.25 to 0.40 incl	0.06	(b)	0.050 (c) to 0.09 incl	0.03
	Over 0.40 to 0.55 incl	0.07		Over 0.09 to 0.15	0.05
	Over 0.55 to 0.80 incl	0.10		Over 0.15 to 0.23 incl	0.07
	Over 0.80	0.13		Over 0.23 to 0.35 incl	0.09
Manganese			Silicon		
	To 0.40 incl	0.15	(d)	To 0.15 incl	0.08
	Over 0.40 to 0.50 incl	0.20		Over 0.15 to 0.20 incl	0.10
	Over 0.50 to 1.65 incl	0.30		Over 0.20 to 0.30 incl	0.15
				Over 0.30 to 0.60 incl	0.20
Phosphorus					
(b)	0.040 (c) to 0.08 incl	0.03			

(a) Add 0.01 to specified carbon ranges for steels with manganese contents exceeding 1.10%. (b) Lower maximum limits on phosphorus and sulfur are required by certain quality descriptors. (c) Lowest permissible maximum for this element. (d) Silicon content not normally specified for acid bessemer steels

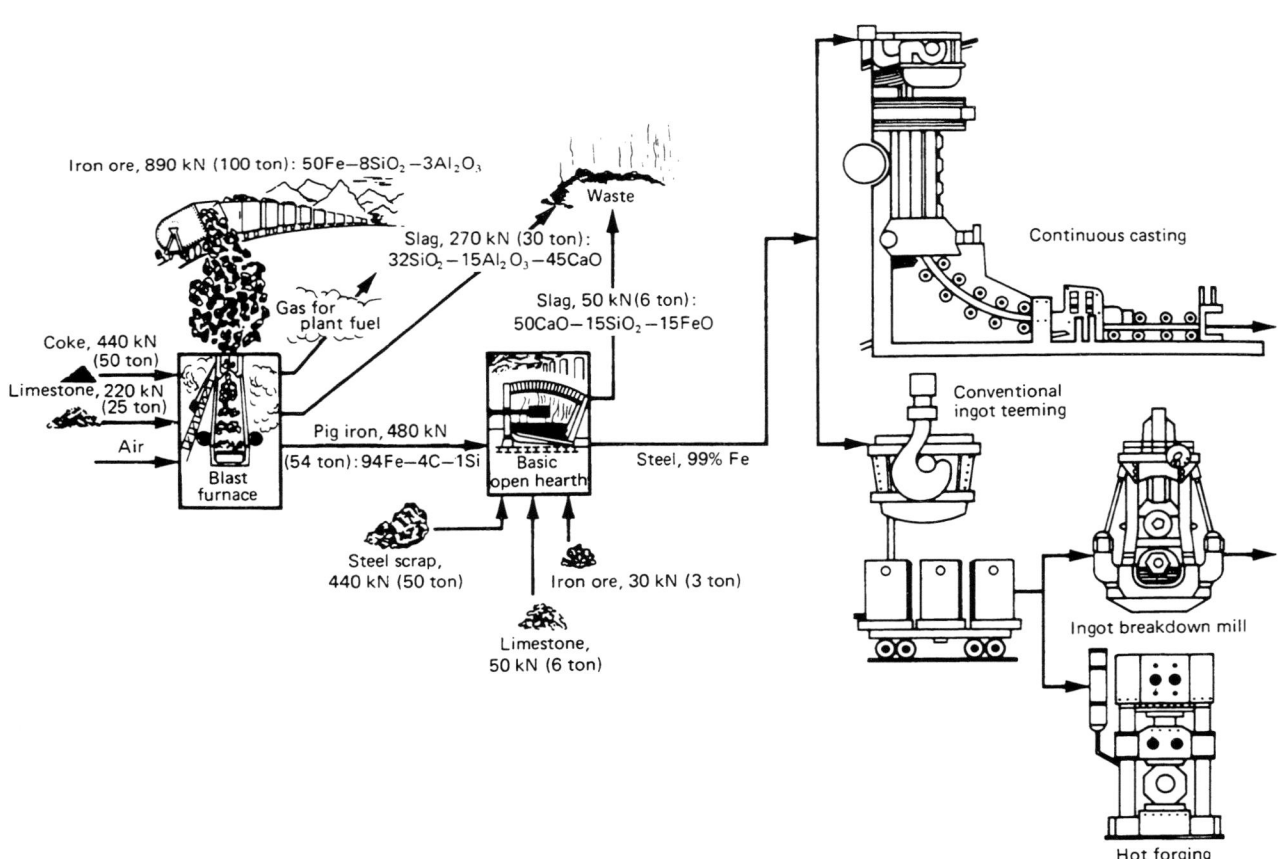

282

Carbon steel heat composition ranges and limits—structural shapes, plate, strip, sheet, and welded tubing

Element	Limit or max of specified range, %	Range, %	Element	Limit or max of specified range, %	Range, %
Carbon			Phosphorus		
(a)	0.08 (b)(c) to 0.15 incl	0.05	(d)	0.04 (b) to 0.08 incl	0.03
	Over 0.15 to 0.30 incl	0.06		Over 0.08 to 0.15 incl	0.05
	Over 0.30 to 0.40 incl	0.07			
	Over 0.40 to 0.60 incl	0.08	Sulfur		
	Over 0.60 to 0.80 incl	0.11	(d)	0.05 to 0.08 incl	0.03
	Over 0.80 to 1.35 incl	0.14		Over 0.08 to 0.15 incl	0.05
				Over 0.15 to 0.23 incl	0.07
Manganese				Over 0.23 to 0.33 incl	0.10
	0.40 (b) to 0.50 incl	0.20	Silicon	0.10 to 0.15 incl	0.08
	Over 0.50 to 1.15 incl	0.30		Over 0.15 to 0.30 incl	0.15
	Over 1.15 to 1.65 incl	0.35		Over 0.30 to 0.60 incl	0.30

(a) Add 0.01 to specified carbon range for steels with manganese contents exceeding 1.00%. (b) Lowest permissible maximum limit for this element. (c) 0.12% for structural shapes and plate. (d) Lower maximum limits on phosphorus and sulfur are required by certain quality descriptors

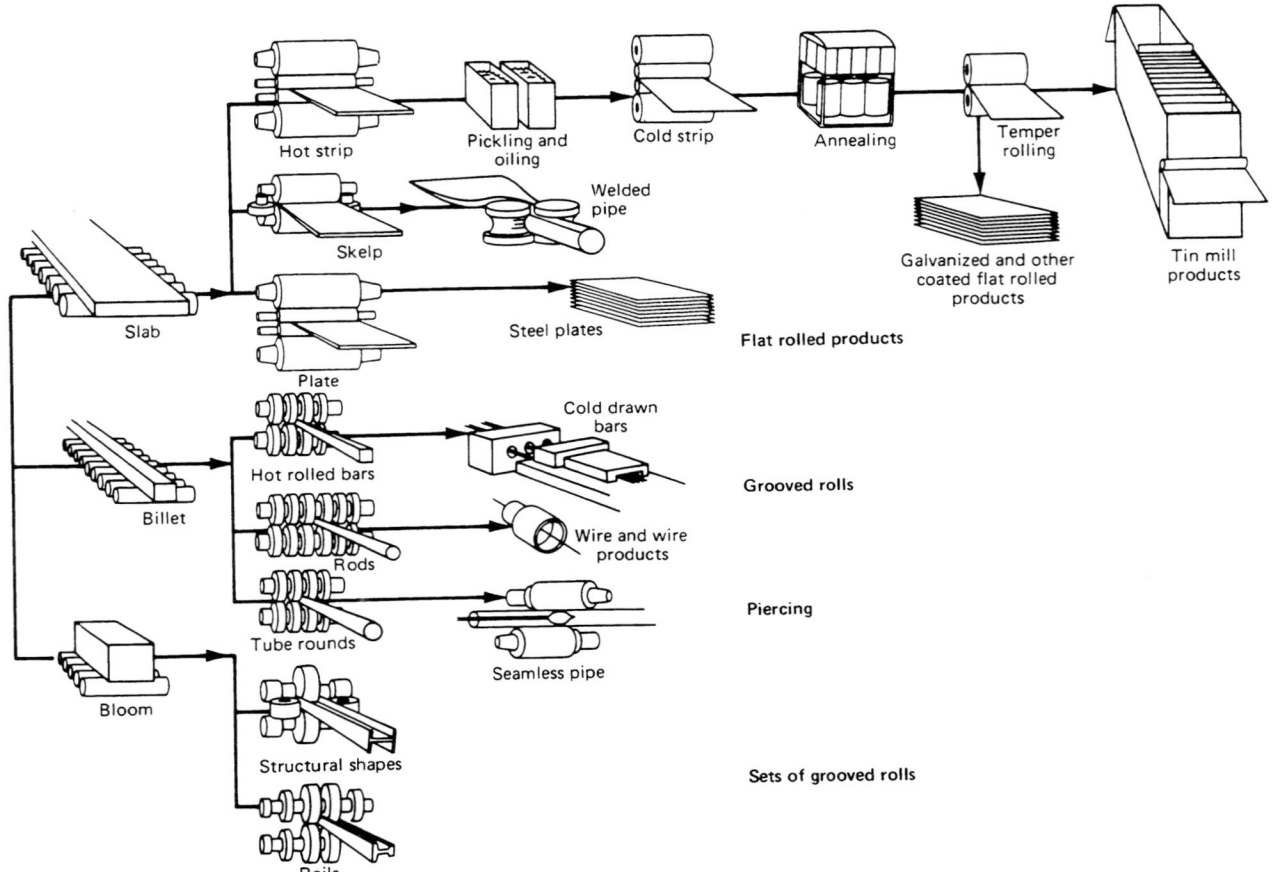

Steel processing flow lines

Flat rolled products are usually rolled from slabs by mills using sets of cylindrical rolls. Grooved rolls squeeze billets into different cross-sections (round, angles, etc.) in a sequence of operations. Piercing is the process used to make seamless pipe and tubing from a semifinished product called tube rounds. Sets of grooved rolls are used to roll blooms into heavy beams for construction or for rails. A small but significant percentage of heated ingot steel is squeezed in forging presses to make large shafts for power plants, nuclear plant components, and other products.
Source: Steel Processing Flow Lines, American Iron and Steel Institute

Alloy steel heat composition ranges and limits—plate

Element	Limit or max of specified range, %	Range, % Open hearth or basic oxygen steels	Electric furnace steels	Element	Limit or max of specified range, %	Range, % Open hearth or basic oxygen steels	Electric furnace steels
Carbon	To 0.25 incl	0.06	0.05		Over 5.30 to 10.00 incl	1.00	1.00
	Over 0.25 to 0.40 incl	0.07	0.06	Molybdenum	To 0.10 incl	0.05	0.05
	Over 0.40 to 0.55 incl	0.08	0.07		Over 0.10 to 0.20 incl	0.07	0.07
	Over 0.55 to 0.70 incl	0.11	0.10		Over 0.20 to 0.50 incl	0.10	0.10
	Over 0.70	0.14	0.13		Over 0.50 to 0.80 incl	0.15	0.15
					Over 0.80 to 1.15 incl	0.20	0.20
Manganese	To 0.45 incl	0.20	0.15				
	Over 0.45 to 0.80 incl	0.25	0.20	Tungsten	To 0.50 incl	0.20	0.20
	Over 0.80 to 1.15 incl	0.30	0.25		Over 0.50 to 1.00 incl	0.30	0.30
	Over 1.15 to 1.70 incl	0.35	0.30		Over 1.00 to 2.00 incl	0.50	0.50
	Over 1.70 to 2.10 incl	0.40	0.35		Over 2.00 to 4.00 incl	0.60	0.60
Sulfur(a)	To 0.060 incl	0.02	0.02	Copper	To 0.60 incl	0.20	0.20
	Over 0.060 to 0.100 incl	0.04	0.04		Over 0.60 to 1.50 incl	0.30	0.30
	Over 0.100 to 0.140 incl	0.05	0.05		Over 1.50 to 2.00 incl	0.35	0.35
Silicon	To 0.15 incl	0.08	0.08	Vanadium	To 0.25 incl	0.05	0.05
	Over 0.15 to 0.20 incl	0.10	0.10		Over 0.25 to 0.50 incl	0.10	0.10
	Over 0.20 to 0.40 incl	0.15	0.15				
	Over 0.40 to 0.60 incl	0.20	0.20	Aluminum	Up to 0.10 incl	0.05	0.05
	Over 0.60 to 1.00 incl	0.30	0.30		Over 0.10 to 0.20 incl	0.10	0.10
	Over 1.00 to 2.20 incl	0.40	0.35		Over 0.20 to 0.30 incl	0.15	0.15
					Over 0.30 to 0.80 incl	0.25	0.25
Chromium	To 0.40 incl	0.20	0.15		Over 0.80 to 1.30 incl	0.35	0.35
	Over 0.40 to 0.80 incl	0.25	0.20		Over 1.30 to 1.80 incl	0.45	0.45
	Over 0.80 to 1.05 incl	0.30	0.25				
	Over 1.05 to 1.25 incl	0.35	0.30	**Steelmaking process**		**Lowest max, %(b)**	
	Over 1.25 to 1.75 incl	0.50	0.40	Phosphorus	Basic open hearth or basic oxygen	0.035(c)	
	Over 1.75 to 3.99 incl	0.60	0.50		Basic electric furnace	0.025	
Nickel	To 0.50 incl	0.20	0.20	Sulfur	Basic open hearth or basic oxygen	0.040(c)	
	Over 0.50 to 1.50 incl	0.30	0.30		Basic electric furnace	0.025	
	Over 1.50 to 2.00 incl	0.35	0.35				
	Over 2.00 to 3.00 incl	0.40	0.40				
	Over 3.00 to 5.30 incl	0.50	0.50				

(a) A range of sulfur content normally indicates a resulfurized steel. (b) Not applicable to resulfurized or rephosphorized steels. (c) Lower maximum limits on phosphorus and sulfur are required by certain quality descriptors.

Carbon steel product composition tolerances

Element	Limit or max of specified range, %	Tolerance over max or under min limits, % Cross-sectional area of product: to 100 in.²	100 to 200 in.²	200 to 400 in.²	400 to 800 in.²
Carbon	To 0.25 incl	0.02	0.03	0.04	0.05
	Over 0.25 to 0.55 incl	0.03	0.04	0.05	0.06
	Over 0.55	0.04	0.05	0.06	0.07
Manganese	To 0.90 incl	0.03	0.04	0.06	0.07
	Over 0.90 to 1.65 incl	0.06	0.06	0.07	0.08
Phosphorus	Over maximum only, to 0.040 incl	0.008	0.008	0.010	0.015
Sulfur	Over maximum only, to 0.050 incl	0.008	0.010	0.010	0.015
Silicon	To 0.35 incl	0.02	0.02	0.03	0.04
	Over 0.35 to 0.60 incl	0.05	...	...	...
Copper	Under minimum only	0.02	0.03	...	...
Lead	0.15 to 0.35 incl	0.03	0.03	...	...

Note: Product composition requirements are not applicable to rimmed or capped steels, boron content of boron steels, or phosphorus and sulfur contents of rephosphorized and resulfurized steels

Alloy steel product composition tolerances—bars, billets, blooms, and slabs

Element	Limit or max of specified range, %	Tolerance over max or under min limits, %(a) Cross-sectional area of product:			
		to 100 in.^{2}	100 to 200 in.^{2}	200 to 400 in.^{2}	400 to 800 in.^{2}
Carbon	To 0.30 incl	0.01	0.02	0.03	0.04
	Over 0.30 to 0.75 incl	0.02	0.03	0.04	0.05
	Over 0.75	0.03	0.04	0.05	0.06
Manganese	To 0.90 incl	0.03	0.04	0.05	0.06
	Over 0.90 to 2.10 incl	0.04	0.05	0.06	0.07
Phosphorus	Over max only	0.005	0.010	0.010	0.010
Sulfur	Over max only	0.005	0.010	0.010	0.010
Silicon	To 0.40 incl	0.02	0.02	0.03	0.04
	Over 0.40 to 2.20 incl	0.05	0.06	0.06	0.07
Chromium	To 0.90 incl	0.03	0.04	0.04	0.05
	Over 0.90 to 2.10 incl	0.05	0.06	0.06	0.07
	Over 2.10 to 3.99 incl	0.10	0.10	0.12	0.14
Nickel	To 1.00 incl	0.03	0.03	0.03	0.03
	Over 1.00 to 2.00 incl	0.05	0.05	0.05	0.05
	Over 2.00 to 5.30 incl	0.07	0.07	0.07	0.07
	Over 5.30 to 10.00 incl	0.10	0.10	0.10	0.10
Molybdenum	To 0.20 incl	0.01	0.01	0.02	0.03
	Over 0.20 to 0.40 incl	0.02	0.03	0.03	0.04
	Over 0.40 to 1.15 incl	0.03	0.04	0.05	0.06
Tungsten	To 1.00 incl	0.04	0.05	0.05	0.06
	Over 1.00 to 4.00 incl	0.08	0.09	0.10	0.12
Copper(b)	To 1.00 incl	0.03	…	…	…
	Over 1.00 to 2.00 incl	0.05	…	…	…
Vanadium	To 0.10 incl	0.01	0.01	0.01	0.01
	Over 0.10 to 0.25 incl	0.02	0.02	0.02	0.02
	Over 0.25 to 0.50 incl	0.03	0.03	0.03	0.03
	Min value specified, check under min limit(b)	0.01	0.01	0.01	0.01
Niobium(b)	To 0.10 incl	0.01(c)	…	…	…
Titanium(b)	To 0.10 incl	0.01(c)	…	…	…
Zirconium (b)	To 0.15 incl	0.03	…	…	…
Aluminum(b)	Up to 0.10 incl	0.03	…	…	…
	Over 0.10 to 0.20 incl	0.04	…	…	…
	Over 0.20 to 0.30 incl	0.05	…	…	…
	Over 0.30 to 0.80 incl	0.07	…	…	…
	Over 0.80 to 1.80 incl	0.10	…	…	…
Lead(b)	0.15 to 0.35 incl	0.03	…	…	…
Nitrogen(b)	To 0.030 incl	0.005	…	…	…

(a) Product composition requirements are not applicable to boron content of boron steels or sulfur content of resulfurized steels. (b) Tolerances shown apply only to cross-sectional areas of 100 in.^{2} or less. (c) If the minimum of the range is 0.01%, the lower tolerance is 0.005%

Alloy steel product composition tolerances—plate

Element	Limit or max of specified range, %	Tolerance over max or under min limits, %	Element	Limit or max of specified range, %	Tolerance over max or under min limits, %
Carbon	To 0.30 incl	0.02	Molybdenum	To 0.20 incl	0.01
	Over 0.30 to 0.75 incl	0.03		Over 0.20 to 0.40 incl	0.03
	Over 0.75	0.04		Over 0.40 to 1.15 incl	0.04
Manganese	To 0.90 incl	0.04	Tungsten	To 1.00 incl	0.05
	Over 0.90 to 2.10 incl	0.05		Over 1.00 to 4.00 incl	0.09
Phosphorus (a)	Over max only	0.01	Copper	To 1.00 incl	0.03
				Over 1.00 to 2.00 incl	0.05
Sulfur (a)(b)	…	0.01	Vanadium	To 0.10 incl	0.01
				Over 0.10 to 0.25 incl	0.02
Silicon	To 0.40 incl	0.02		Over 0.25 to 0.50 incl	0.03
	Over 0.40 to 2.20 incl	0.06		Min value specified, check under min limit	0.01
Chromium	To 0.90 incl	0.04	Aluminum	Up to 0.10 incl	0.03
	Over 0.90 to 2.10 incl	0.06		Over 0.10 to 0.20 incl	0.04
	Over 2.10 to 3.99 incl	0.10		Over 0.20 to 0.30 incl	0.05
Nickel	To 1.00 incl	0.03		Over 0.30 to 0.80 incl	0.07
	Over 1.00 to 2.00 incl	0.05		Over 0.80 to 1.80 incl	0.10
	Over 2.00 to 5.30 incl	0.07			
	Over 5.30	0.10			

(a) For pressure-vessel quality plate, the specified composition includes product composition tolerances for phosphorus and sulfur. (b) Product composition requirements not applicable to sulfur content of resulfurized steel

AISI-SAE Designations

AISI-SAE system of designations

Numerals and digits	Type of steel and nominal alloy content	Numerals and digits	Type of steel and nominal alloy content
Carbon steels		**Nickel-chromium-molybdenum steels (continued)**	
10xx(a)	Plain carbon (Mn 1.00% max)	93xx	Ni 3.25; Cr 1.20; Mo 0.12
11xx	Resulfurized	94xx	Ni 0.45; Cr 0.40; Mo 0.12
12xx	Resulfurized and rephosphorized	97xx	Ni 0.55; Cr 0.20; Mo 0.20
15xx	Plain carbon (max Mn range—1.00 to 1.65%)	98xx	Ni 1.00; Cr 0.80; Mo 0.25
		Nickel-molybdenum steels	
Manganese steels		46xx	Ni 0.85 and 1.82; Mo 0.20 and 0.25
13xx	Mn 1.75	48xx	Ni 3.50; Mo 0.25
Nickel steels		**Chromium steels**	
23xx	Ni 3.50	50xx	Cr 0.27, 0.40, 0.50 and 0.65
25xx	Ni 5.00	51xx	Cr 0.80, 0.87, 0.92, 0.95, 1.00 and 1.05
Nickel-chromium steels		50xxx	Cr 0.50 ⎤
31xx	Ni 1.25; Cr 0.65 and 0.80	51xxx	Cr 1.02 ⎬ C 1.00 min
32xx	Ni 1.75; Cr 1.07	52xxx	Cr 1.45 ⎦
33xx	Ni 3.50; Cr 1.50 and 1.57		
34xx	Ni 3.00; Cr 0.77	**Chromium-vanadium steels**	
		61xx	Cr 0.60, 0.80 and 0.95; V 0.10 and 0.15 min
Molybdenum steels			
40xx	Mo 0.20 and 0.25	**Tungsten-chromium steels**	
44xx	Mo 0.40 and 0.52	72xx	W 1.75; Cr 0.75
Chromium-molybdenum steels		**Silicon-manganese steels**	
41xx	Cr 0.50, 0.80 and 0.95; Mo 0.12, 0.20, 0.25 and 0.30	92xx	Si 1.40 and 2.00; Mn 0.65, 0.82 and 0.85; Cr 0.00 and 0.65
Nickel-chromium-molybdenum steels		**High-strength low-alloy steels**	
43xx	Ni 1.82; Cr 0.50 and 0.80; Mo 0.25	9xx	Various SAE grades
43BVxx	Ni 1.82; Cr 0.50; Mo 0.12 and 0.25; V 0.03 min		
47xx	Ni 1.05; Cr 0.45; Mo 0.20 and 0.35	**Boron steels**	
81xx	Ni 0.30; Cr 0.40; Mo 0.12	xxBxx	B denotes boron steel
86xx	Ni 0.55; Cr 0.50; Mo 0.20		
87xx	Ni 0.55; Cr 0.50; Mo 0.25		
88xx	Ni 0.55; Cr 0.50; Mo 0.35		

(a) xx in the last two digits of these designations indicates that the carbon content (in hundredths of a percent) is to be inserted

Compositions ranges and limits for AISI-SAE standard carbon steels—structural shapes, plate, strip, sheet, and welded tubing

AISI-SAE designation	UNS designation	Heat composition ranges and limits, %(a)		AISI-SAE designation	UNS designation	Heat composition ranges and limits, %(a)	
		C	Mn			C	Mn
1006	G10060	0.08 max	0.25-0.45	1043	G10430	0.39-0.47	0.70-1.00
1008	G10080	0.10 max	0.25-0.50	1045	G10450	0.42-0.50	0.60-0.90
1009	G10090	0.15 max	0.60 max	1046	G10460	0.42-0.50	0.70-1.00
1010	G10100	0.8-0.13	0.30-0.60	1049	G10490	0.45-0.53	0.60-0.90
1012	G10120	0.10-0.15	0.30-0.60	1050	G10500	0.47-0.55	0.60-0.90
1015	G10150	0.12-0.18	0.30-0.60	1055	G10550	0.52-0.60	0.69-0.90
1016	G10160	0.12-0.18	0.60-0.90	1060	G10600	0.55-0.66	0.60-0.90
1017	G10170	0.14-0.20	0.30-0.60	1064	G10640	0.59-0.70	0.50-0.80
1018	G10180	0.14-0.20	0.60-0.90	1065	G10650	0.59-0.70	0.60-0.90
1019	G10190	0.14-0.20	0.70-1.00	1070	G10700	0.65-0.76	0.60-0.90
1020	G10200	0.17-0.23	0.30-0.60	1074	G10740	0.69-0.80	0.50-0.80
1021	G10210	0.17-0.23	0.60-0.90	1078	G10780	0.72-0.86	0.30-0.60
1022	G10220	0.17-0.23	0.70-1.00	1080	G10800	0.74-0.88	0.60-0.90
1023	G10230	0.19-0.25	0.30-0.60	1084	G10840	0.80-0.94	0.60-0.90
1025	G10250	0.22-0.28	0.30-0.60	1085	G10850	0.80-0.94	0.70-1.00
1026	G10260	0.22-0.28	0.60-0.90	1086	G10860	0.80-0.94	0.30-0.50
1030	G10300	0.27-0.34	0.60-0.90	1090	G10900	0.84-0.98	0.60-0.90
1033	G10330	0.29-0.36	0.70-1.00	1095	G10950	0.90-1.04	0.30-0.50
1035	G10350	0.31-0.38	0.60-0.90	1524(b)	G15240	0.18-0.25	1.30-1.65
1037	G10370	0.31-0.38	0.70-1.00	1527(b)	G15270	0.22-0.29	1.20-1.55
1038	G10380	0.34-0.42	0.60-0.90	1536(b)	G15360	0.30-0.38	1.20-1.55
1039	G10390	0.36-0.44	0.70-1.00	1541(b)	G15410	0.36-0.45	1.30-1.65
1040	G10400	0.36-0.44	0.60-0.90	1548(b)	G15480	0.43-0.52	1.05-1.40
1042	G10420	0.39-0.47	0.60-0.90	1552(b)	G15520	0.46-0.55	1.20-1.55

(a) Typical limits on phosphorus and sulfur contents are 0.040% maximum phosphorus and 0.050% maximum sulfur. Steels listed in this table can be produced with additions of lead or boron. Leaded steels typically contain 0.15 to 0.35% lead and are identified by inserting the letter "L" in the designation—11L17; boron steels can be expected to contain 0.0005 to 0.003% boron and are identified by inserting the letter "B" in the designation—15B41. (b) Formerly designated 10xx grade

Composition ranges and limits for AISI-SAE standard carbon steels containing less than 1.00% manganese—semifinished products for forging, hot rolled and cold finished bars, wire rod, and seamless tubing

AISI-SAE designation	UNS designation	Heat composition ranges and limits, %(a)		AISI-SAE designation	UNS designation	Heat composition ranges and limits, %(a)	
		C	Mn			C	Mn
1005	G10050	0.06 max	0.35 max	1042	G10420	0.40-0.47	0.60-0.90
1006	G10060	0.08 max	0.25-0.40	1043	G10430	0.40-0.47	0.70-1.00
1008	G10080	0.10 max	0.30-0.50	1044	G10440	0.43-0.50	0.30-0.60
1010	G10100	0.08-0.13	0.30-0.60	1045	G10450	0.43-0.50	0.60-0.90
1011(b)	G10110	0.08-0.13	0.60-0.90	1046	G10460	0.43-0.50	0.70-1.00
1012	G10120	0.10-0.15	0.30-0.60	1049	G10490	0.46-0.53	0.60-0.90
1013(b)	G10130	0.11-0.16	0.50-0.80	1050	G10500	0.48-0.55	0.60-0.90
1015	G10150	0.13-0.18	0.30-0.60	1053	G10530	0.48-0.55	0.70-1.00
1016	G10160	0.13-0.18	0.60-0.90	1055	G10550	0.50-0.60	0.60-0.90
1017	G10170	0.15-0.20	0.30-0.60	1059(c)	G10590	0.55-0.65	0.50-0.80
1018	G10180	0.15-0.20	0.60-0.90	1060	G10600	0.55-0.65	0.60-0.90
1019	G10190	0.15-0.20	0.70-1.00	1064	G10640	0.60-0.70	0.50-0.80
1020	G10200	0.18-0.23	0.30-0.60	1065	G10650	0.60-0.70	0.60-0.90
1021	G10210	0.18-0.23	0.60-0.90	1069(b)	G10690(b)	0.65-0.75	0.40-0.70
1022	G10220	0.18-0.23	0.70-1.00	1070	G10700	0.65-0.75	0.60-0.90
1023	G10230	0.20-0.25	0.30-0.60	1074(b)	G10740	0.70-0.80	0.50-0.80
1025	G10250	0.22-0.28	0.30-0.60	1075(b)	G10750	0.70-0.80	0.40-0.70
1026	G10260	0.22-0.28	0.60-0.90	1078	G10780	0.72-0.85	0.30-0.60
1029	G10290	0.25-0.31	0.60-0.90	1080	G10800	0.75-0.88	0.60-0.90
1030	G10300	0.28-0.34	0.60-0.90	1084	G10840	0.80-0.93	0.60-0.90
1035	G10350	0.32-0.38	0.60-0.90	1085(b)	G10850	0.80-0.93	0.70-1.00
1037	G10370	0.32-0.38	0.70-1.00	1086	G10860	0.80-0.93	0.30-0.50
1038	G10380	0.35-0.42	0.60-0.90	1090	G10900	0.85-0.98	0.60-0.90
1039	G10390	0.37-0.44	0.70-1.00	1095	G10950	0.90-1.03	0.30-0.50
1040	G10400	0.37-0.44	0.60-0.90				

(a) Typical limits on phosphorus and sulfur contents are 0.040% maximum phosphorus and 0.050% maximum sulfur. Steels listed in this table can be produced with additions of lead or boron. Leaded steels typically contain 0.15 to 0.35% lead and are identified by inserting the letter "L" in the designation—11L17; boron steels can be expected to contain 0.0005 to 0.003% boron and are identified by inserting the letter "B" in the designation—15B41. (b) SAE standard grade only. (c) AISI standard grade only

Composition ranges and limits for AISI-SAE standard resulfurized carbon steels

AISI-SAE designation	UNS designation	Heat composition ranges and limits, %(a)		
		C	Mn	S
1110	G11100	0.08-0.13	0.30-0.60	0.08-0.13
1117	G11170	0.14-0.20	1.00-1.30	0.08-0.13
1118	G11180	0.14-0.20	1.30-1.60	0.08-0.13
1137	G11370	0.32-0.39	1.35-1.65	0.08-0.13
1139	G11390	0.35-0.43	1.35-1.65	0.13-0.20
1140	G11400	0.37-0.44	0.70-1.00	0.08-0.13
1141	G11410	0.37-0.45	1.35-1.65	0.08-0.13
1144	G11440	0.40-0.48	1.35-1.65	0.24-0.33
1146	G11460	0.42-0.49	0.70-1.00	0.08-0.13
1151	G11510	0.48-0.55	0.70-1.00	0.08-0.13

(a) Typical limit on phosphorus content is 0.040% maximum. Because of the adverse effect of silicon on machinability, steels listed in this table are generally not deoxidized with silicon. Steel listed in this table can be produced as leaded steels, typically containing 0.15 to 0.35% lead and identified by inserting the letter "L" in the designation—11L17

Composition ranges and limits for AISI-SAE merchant-quality steels

AISI-SAE designation	Heat composition ranges and limits, % (a)	
	C	Mn
M1008	0.10 max	0.25-0.60
M1010	0.07-0.14	0.25-0.60
M1012	0.09-0.16	0.25-0.60
M1015	0.12-0.19	0.25-0.60
M1017	0.14-0.21	0.25-0.60
M1020	0.17-0.24	0.25-0.60
M1023	0.19-0.27	0.25-0.60
M1025	0.20-0.30	0.25-0.60
M1031	0.26-0.36	0.25-0.60
M1044	0.40-0.50	0.25-0.60

(a) Typical limits on phosphorus and sulfur contents are 0.040% maximum phosphorus and 0.050% maximum sulfur

Composition ranges and limits for AISI-SAE standard carbon H-steels

AISI-SAE designation	UNS designation	Heat composition ranges and limits, %(a)		
		C	Mn	Si
1038H	H10380	0.34-0.43	0.50-1.00	0.15-0.30
1045H	H10450	0.42-0.51	0.50-1.00	0.15-0.30
1522H	H15220	0.17-0.25	1.00-1.50	0.15-0.30
1524H	H15240	0.18-0.26	1.25-1.75	0.15-0.30
1526H	H15260	0.21-0.30	1.00-1.50	0.15-0.30
1541H	H15410	0.35-0.45	1.25-1.75	0.15-0.30
15B21H(b)	H15211	0.17-0.24	0.70-1.20	0.15-0.30
15B35H(b)	H15351	0.31-0.39	0.70-1.20	0.15-0.30
15B37H(b)	H15371	0.30-0.39	1.00-1.50	0.15-0.30
15B41H(b, c)	H15411	0.35-0.45	1.25-1.75	0.15-0.30
15B48H(b, c)	H15481	0.43-0.53	1.00-1.50	0.15-0.30
15B62H(b)	H15621	0.54-0.67	1.00-1.50	0.40-0.60

(a) Typical limits on phosphorus and sulfur contents are 0.040% maximum phosphorus and 0.050% maximum sulfur. (b) Can be expected to contain 0.0005 to 0.003% boron. (c) AISI grade only

Composition ranges and limits for AISI-SAE standard resulfurized and rephosphorized carbon steels

AISI-SAE designation	UNS designation	Heat composition ranges and limits, %(a)			
		C max	Mn	P	S
1211	G12110	0.13	0.60-0.90	0.07-0.12	0.10-0.15
1212	G12120	0.13	0.70-1.00	0.07-0.12	0.16-0.23
1213	G12130	0.13	0.70-1.00	0.07-0.12	0.24-0.33
12L14(b)	G12144	0.15	0.85-1.15	0.04-0.09	0.26-0.35
1215	G12150	0.09	0.75-1.05	0.04-0.09	0.26-0.35

(a) Because of the adverse effect of silicon on machinability, steels listed in this table are generally not deoxidized with silicon. (b) Contains 0.15 to 0.35% lead; other steels listed in this table can be produced with the same lead content

Composition ranges and limits for AISI-SAE standard carbon steels with a maximum manganese content exceeding 1.10%—semifinished products for forging, hot rolled and cold finished bars, wire rod, and seamless tubing

AISI-SAE designation	UNS designation	Heat composition ranges and limits, %(a)		Former AISI-SAE designation
		C	Mn	
1513	G15130	0.10-0.16	1.10-1.40	...
1518(b)	G15180	0.15-0.21	1.10-1.40	...
1522	G15220	0.18-0.24	1.10-1.40	...
1524	G15240	0.19-0.25	1.35-1.65	1024
1525(b)	G15250	0.23-0.29	0.80-1.10	...
1526	G15256	0.22-0.29	1.10-1.40	...
1527	G15270	0.22-0.29	1.20-1.50	1027
1536(b)	G15360	0.30-0.37	1.20-1.50	1036
1541	G15410	0.36-0.44	1.35-1.65	1041
1547(b)	G15470	0.43-0.51	1.35-1.65	1047
1548	G15480	0.44-0.52	1.10-1.40	1048
1551	G15510	0.45-0.56	0.85-1.15	1051
1552	G15520	0.47-0.55	1.20-1.50	1052
1561	G15610	0.55-0.65	0.75-1.05	1061
1566	G15660	0.60-0.71	0.85-1.15	1066
1572(b)	G15720	0.65-0.76	1.00-1.30	1072

(a) Typical limits on phosphorus and sulfur contents are 0.040% maximum phosphorus and 0.050% maximum sulfur. Killed steels commonly contain 0.15 to 0.30% silicon; other ranges are negotiable. Steels listed in this table can be produced with additions of lead or boron. Leaded steels typically contain 0.15 to 0.35% lead and are identified by inserting the letter "L" in the designation—11L17; boron steels can be expected to contain 0.0005 to 0.003% boron and are identified by inserting the letter "B" in the designation—15B41. (b) SAE standard grade only

Composition ranges and limits for AISI-SAE standard alloy steels—bars, billets, blooms, and slabs

AISI-SAE designation	UNS designation	Heat composition ranges and limits, %							
		C	Mn	P max(s)	S max(a)	Si	Cr	Ni	Mo
1330	G13300	0.28-0.33	1.60-1.90	0.035	0.040	0.15-0.30	...	...	...
1335	G13350	0.33-0.38	1.60-1.90	0.035	0.040	0.15-0.30	...	...	...
1340	G13400	0.38-0.43	1.60-1.90	0.035	0.040	0.15-0.30	...	...	...
1345	G13450	0.43-0.48	1.60-1.90	0.035	0.040	0.15-0.30	...	...	...
4012	G40120	0.09-0.14	0.75-1.00	0.035	0.040	0.15-0.30	...	...	0.15-0.25
4023	G40230	0.20-0.25	0.70-0.90	0.035	0.040	0.15-0.30	...	...	0.20-0.30
4024	G40240	0.20-0.25	0.70-0.90	0.035	0.035-0.050(b)	0.15-0.30	...	...	0.20-0.30
4027	G40270	0.25-0.30	0.70-0.90	0.035	0.040	0.15-0.30	...	...	0.20-0.30
4028	G40280	0.25-0.30	0.70-0.90	0.035	0.035-0.050(b)	0.15-0.30	...	...	0.20-0.30
4032	G40320	0.30-0.35	0.70-0.90	0.035	0.040	0.15-0.30	...	...	0.20-0.30
4037	G40370	0.35-0.40	0.70-0.90	0.035	0.040	0.15-0.30	...	...	0.20-0.30
4042(c)	G40420	0.40-0.45	0.70-0.90	0.035	0.040	0.15-0.30	...	...	0.20-0.30
4047	G40470	0.45-0.50	0.70-0.90	0.035	0.040	0.15-0.30	...	...	0.20-0.30
4118	G41180	0.18-0.23	0.70-0.90	0.035	0.040	0.15-0.30	0.40-0.60	...	0.08-0.15
4130	G41300	0.28-0.33	0.40-0.60	0.035	0.040	0.15-0.30	0.80-1.10	...	0.15-0.25
4135(c)	G41350	0.33-0.38	0.70-0.90	0.035	0.040	0.15-0.30	0.30-1.10	...	0.15-0.25
4137	G41370	0.35-0.40	0.70-0.90	0.035	0.040	0.15-0.30	0.80-1.10	...	0.15-0.25
4140	G41400	0.38-0.43	0.75-1.00	0.035	0.040	0.15-0.30	0.80-1.10	...	0.15-0.25
4142	G41420	0.40-0.45	0.75-1.00	0.035	0.040	0.15-0.30	0.80-1.10	...	0.15-0.25
4145	G41450	0.43-0.48	0.75-1.00	0.035	0.040	0.15-0.30	0.80-1.10	...	0.15-0.25
4147	G41470	0.45-0.50	0.75-1.00	0.035	0.040	0.15-0.30	0.80-1.10	...	0.15-0.25
4150	G41500	0.48-0.53	0.75-1.00	0.035	0.040	0.15-0.30	0.80-1.10	...	0.15-0.25
4161	G41610	0.56-0.64	0.75-1.00	0.035	0.040	0.15-0.30	0.70-0.90	...	0.25-0.35
4320	G43200	0.17-0.22	10.45-0.65	0.035	0.040	0.15-0.30	0.40-0.60	1.65-2.00	0.20-0.30
4340	G43400	0.38-0.43	0.60-0.80	0.035	0.040	0.15-0.30	0.70-0.90	1.65-2.00	0.20-0.30
E4340(d)	G43406	0.38-0.43	0.65-0.85	0.025	0.025	0.15-0.30	0.70-0.90	1.65-2.00	0.20-0.30
4419(c)	G44190	0.18-0.23	0.45-0.65	0.035	0.040	0.15-0.30	...	...	0.45-0.60
4422(c)	G44220	0.20-0.25	0.70-0.90	0.035	0.040	0.15-0.30	...	...	0.35-0.45
4427(c)	G44270	0.24-0.29	0.70-0.90	0.035	0.040	0.15-0.30	...	...	0.35-0.45
4615	G46150	0.13-0.18	0.45-0.65	0.035	0.040	0.15-0.30	...	1.65-2.00	0.20-0.30
4617(c)	G46170	0.15-0.20	0.45-0.65	0.035	0.040	0.15-0.30	...	1.65-2.00	0.20-0.30
4620	G46200	0.17-0.22	0.45-0.65	0.035	0.040	0.15-0.30	...	1.65-2.00	0.20-0.30
4621(c)	G46210	0.18-0.23	0.70-0.90	0.035	0.040	0.15-0.30	...	1.65-2.00	0.20-0.30
4626	G46260	0.24-0.29	0.45-0.65	0.035	0.040	0.15-0.30	...	0.70-1.00	0.15-0.25

(a) Limits for phosphorus and sulfur are for steel made by open hearth or basic oxygen processes. (b) A range of sulfur content normally indicates a resulfurized steel. (c) SAE standard grade only. (d) Prefix "E" indicates that the steel is made by electric furnace process. (e) Can be expected to contain 0.0005 to 0.003% boron. (f) AISI standard grade only. (g) Contains 0.10 to 0.15% vanadium. (h) Contains 0.15% min vanadium

(continued)

Composition ranges and limits for AISI-SAE standard alloy steels—bars, billets, blooms, and slabs (continued)

AISI-SAE designation	UNS designation	C	Mn	P max(s)	S max(a)	Si	Cr	Ni	Mo
						Heat composition ranges and limits, %			
4718(c)	G47180	0.16-0.21	0.70-0.90	…	…	…	0.35-0.55	0.90-1.20	0.30-0.40
4720	G47200	0.17-0.22	0.50-0.70	0.035	0.040	0.15-0.30	0.35-0.55	0.90-1.20	0.15-0.25
4815	G48150	0.13-0.18	0.40-0.60	0.035	0.040	0.15-0.30	…	3.25-3.75	0.20-0.30
4817	G48170	0.15-0.20	0.40-0.60	0.035	0.040	0.15-0.30	…	3.25-3.75	0.20-0.30
4820	G48200	0.18-0.23	0.50-0.70	0.035	0.040	0.15-0.30	…	3.25-3.75	0.20-0.30
5015(e)	G50150	0.12-0.17	0.30-0.50	0.035	0.040	0.15-0.30	0.30-0.50	…	…
50B40(c, e)	G50401	0.38-0.43	0.75-1.00	0.035	0.040	0.15-0.30	0.40-0.60	…	…
50B44(e)	G50441	0.43-0.48	0.75-1.00	0.035	0.040	0.15-0.30	0.40-0.60	…	…
5046(c)	G50460	0.43-0.48	0.75-1.00	0.035	0.040	0.15-0.30	0.20-0.35	…	…
50B46(e)	G50461	0.44-0.49	0.75-1.00	0.035	0.040	0.15-0.30	0.20-0.35	…	…
50B50(e)	G50501	0.48-0.53	0.75-1.00	0.035	0.040	0.15-0.30	0.40-0.60	…	…
5060(c)	G50600	0.56-0.64	0.75-1.00	0.035	0.040	0.15-0.30	0.40-0.60	…	…
50B60(e)	…	0.56-0.64	0.75-1.00	0.035	0.040	0.15-0.30	0.40-0.60	…	…
5115(c)	G51150	0.13-0.18	0.70-0.90	0.035	0.040	0.15-0.30	0.70-0.90	…	…
5117(f)	G51170	0.15-0.20	0.70-0.90	0.035	0.040	0.15-0.30	0.70-0.90	…	…
5120	G51200	0.17-0.22	0.70-0.90	0.035	0.040	0.15-0.30	0.70-0.90	…	…
5130	G51300	0.28-0.33	0.70-0.90	0.035	0.040	0.15-0.30	0.80-1.10	…	…
5132	G51320	0.30-0.35	0.60-0.80	0.035	0.040	0.15-0.30	0.75-1.00	…	…
5135	G51350	0.33-0.38	0.60-0.80	0.035	0.040	0.15-0.30	0.80-1.05	…	…
5140	G51400	0.38-0.43	0.70-0.90	0.035	0.040	0.15-0.30	0.70-0.90	…	…
5145(c)	G51450	0.43-0.48	0.70-0.90	0.035	0.040	0.15-0.30	0.70-0.90	…	…
5147(c)	G51470	0.46-0.51	0.70-0.95	0.035	0.040	0.15-0.30	0.85-1.15	…	…
5150	G51500	0.48-0.53	0.70-0.90	0.035	0.040	0.15-0.30	0.70-0.90	…	…
5155	G51550	0.51-0.59	0.70-0.90	0.035	0.040	0.15-0.30	0.70-0.90	…	…
5160	G51600	0.56-0.64	0.75-1.00	0.035	0.040	0.15-0.30	0.70-0.90	…	…
51B60(e)	G51601	0.56-0.64	0.75-1.00	0.035	0.040	0.15-0.30	0.70-0.90	…	…
50100	G50986	0.98-1.10	0.25-0.45	0.025	0.025	0.15-0.30	0.40-0.60	…	…
51100	G51986	0.98-1.10	0.25-0.45	0.025	0.025	0.15-0.30	0.90-1.15	…	…
52100	G52986	0.98-1.10	0.25-0.45	0.025	0.025	0.15-0.30	1.30-1.60	…	…
6118(g)	G61180	0.16-0.21	0.50-0.70	0.035	0.040	0.15-0.30	0.50-0.70	…	…
6150(h)	G61500	0.48-0.53	0.70-0.90	0.035	0.040	0.15-0.30	0.80-1.10	…	…
8115(c)	G81150	0.13-0.18	0.70-0.90	0.035	0.040	0.15-0.30	0.30-0.50	0.20-0.40	0.08-0.15
81B45(e)	G81451	0.43-0.48	0.75-1.00	0.035	0.040	0.15-0.30	0.35-0.55	0.20-0.40	0.08-0.15
8615	G86150	0.13-0.18	0.70-0.90	0.035	0.040	0.15-0.30	0.40-0.60	0.40-0.70	0.15-0.25
8617	G86170	0.15-0.20	0.70-0.90	0.035	0.040	0.15-0.30	0.40-0.60	0.40-0.70	0.15-0.25
8620	G86200	0.18-0.23	0.70-0.90	0.035	0.040	0.15-0.30	0.40-0.60	0.40-0.70	0.15-0.25
8622	G86220	0.20-0.25	0.70-0.90	0.035	0.040	0.15-0.30	0.40-0.60	0.40-0.70	0.15-0.25
8625	G86250	0.23-0.28	0.70-0.90	0.035	0.040	0.15-0.30	0.40-0.60	0.40-0.70	0.15-0.25
8627	G86270	0.25-0.30	0.70-0.90	0.035	0.040	0.15-0.30	0.40-0.60	0.40-0.70	0.15-0.25
8630	G86300	0.28-0.33	0.70-0.90	0.035	0.040	0.15-0.30	0.40-0.60	0.40-0.70	0.15-0.25
8637	G86370	0.35-0.40	0.75-1.00	0.035	0.040	0.15-0.30	0.40-0.60	0.40-0.70	0.15-0.25
8640	G86400	0.38-0.43	0.75-1.00	0.035	0.040	0.15-0.30	0.40-0.60	0.40-0.70	0.15-0.25
8642	G86420	0.40-0.45	0.75-1.00	0.035	0.040	0.15-0.30	0.40-0.60	0.40-0.70	0.15-0.25
8645	G86450	0.43-0.48	0.75-1.00	0.035	0.040	0.15-0.30	0.40-0.60	0.40-0.70	0.15-0.25
86B45(c, e)	G86451	0.43-0.48	0.75-1.00	0.035	0.040	0.15-0.30	0.40-0.60	0.40-0.70	0.15-0.25
8650(c)	G86500	0.48-0.53	0.75-1.00	0.035	0.040	0.15-0.30	0.40-0.60	0.40-0.70	0.15-0.25
8655	G86550	0.51-0.59	0.75-1.00	0.035	0.040	0.15-0.30	0.40-0.60	0.40-0.70	0.15-0.25
8660(c)	G86600	0.56-0.64	0.75-1.00	0.035	0.040	0.15-0.30	0.40-0.60	0.40-0.70	0.15-0.25
8720	G87200	0.18-0.23	0.70-0.90	0.035	0.040	0.15-0.30	0.40-0.60	0.40-0.70	0.20-0.30
8740	G87400	0.38-0.43	0.75-1.00	0.035	0.040	0.15-0.30	0.40-0.60	0.40-0.70	0.20-0.30
8822	G88220	0.20-0.25	0.75-1.00	0.035	0.040	0.15-0.30	0.40-0.60	0.40-0.70	0.30-0.40
9254(c)	G92540	0.51-0.59	0.60-0.80	0.035	0.040	1.20-1.60	0.60-0.80	…	…
9255(c)	G92550	0.51-0.59	0.70-0.95	0.035	0.040	1.80-2.20	…	…	…
9260	G92600	0.56-0.64	0.75-1.00	0.035	0.040	1.80-2.20	…	…	…
9310(c)	G93106	0.08-0.13	0.45-0.65	0.025	0.025	0.15-0.30	1.00-1.40	3.00-3.50	0.08-0.15
94B15(c, e)	G94151	0.13-0.18	0.75-1.00	0.035	0.040	0.15-0.30	0.30-0.50	0.30-0.60	0.08-0.15
94B17(c)	G94171	0.15-0.20	0.75-1.00	0.035	0.040	0.15-0.30	0.30-0.50	0.30-0.60	0.08-0.15
94B30(e)	G94301	0.28-0.33	0.75-1.00	0.035	0.040	0.15-0.30	0.30-0.50	0.30-0.60	0.08-0.15

(a) Limits for phosphorus and sulfur are for steel made by open hearth or basic oxygen processes. (b) A range of sulfur content normally indicates a resulfurized steel. (c) SAE standard grade only. (d) Prefix "E" indicates that the steel is made by electric furnace process. (e) Can be expected to contain 0.0005 to 0.003% boron. (f) AISI standard grade only. (g) Contains 0.10 to 0.15% vanadium. (h) Contains 0.15% min vanadium

Composition ranges and limits for AISI-SAE standard alloy steels—plate

AISI-SAE designation	UNS designation	Heat composition ranges and limits, %(a)				
		C	Mn	Cr	Ni	Mo
1330	G13300	0.27-0.34	1.50-1.90	...	...	...
1335	G13350	0.32-0.39	1.50-1.90	...	...	...
1340	G13400	0.36-0.44	1.50-1.90	...	...	...
1345	G13450	0.41-0.49	1.50-1.90	...	...	...
4118	G41180	0.17-0.23	0.60-0.90	0.40-0.65	...	0.08-0.15
4130	G41300	0.27-0.34	0.35-0.60	0.80-1.15	...	0.15-0.25
4135	G41350	0.32-0.39	0.65-0.95	0.08-1.15	...	0.15-0.25
4137	G41370	0.33-0.40	0.65-0.95	0.80-1.15	...	0.15-0.25
4140	G41400	0.36-0.44	0.70-1.00	0.08-1.15	...	0.15-0.25
4142	G41420	0.38-0.46	0.70-1.00	0.80-1.15	...	0.15-0.25
4145	G41450	0.41-0.49	0.70-1.00	0.80-1.15	...	0.15-0.25
4340	G43400	0.36-0.44	0.55-0.80	0.60-0.90	1.65-2.00	0.20-0.30
E4340(b)	G43406	0.37-0.44	0.60-0.85	0.65-0.90	1.65-2.00	0.20-0.30
4615	G46150	0.12-0.18	0.40-0.65	...	1.65-2.00	0.20-0.30
4617	G46170	0.15-0.21	0.40-0.65	...	1.65-2.00	0.20-0.30
4620	G46200	0.16-0.22	0.40-0.65	...	1.65-2.00	0.20-0.30
5160	G51600	0.54-0.65	0.70-1.00	0.60-0.90	...	...
6150(c)	G61500	0.46-0.54	0.60-0.90	0.80-1.15	...	...
8615	G86150	0.12-0.18	0.60-0.90	0.35-0.60	0.40-0.70	0.15-0.25
8617	G86170	0.15-0.21	0.60-0.90	0.35-0.60	0.40-0.70	0.15-0.25
8620	G86200	0.17-0.23	0.60-0.90	0.35-0.60	0.40-0.70	0.15-0.25
8622	G86220	0.19-0.25	0.60-0.90	0.35-0.60	0.40-0.70	0.15-0.25
8625	G86250	0.22-0.29	0.60-0.90	0.35-0.60	0.40-0.70	0.15-0.25
8627	G86270	0.24-0.31	0.60-0.90	0.35-0.60	0.40-0.70	0.15-0.25
8630	G86300	0.27-0.34	0.60-0.90	0.35-0.60	0.40-0.70	0.15-0.25
8637	G86370	0.33-0.40	0.70-1.00	0.35-0.60	0.40-0.70	0.15-0.25
8640	G86400	0.36-0.44	0.70-1.00	0.35-0.60	0.40-0.70	0.15-0.25
8655	G86550	0.49-0.60	0.70-1.00	0.35-0.60	0.40-0.70	0.15-0.25
8742	G87420	0.38-0.46	0.70-1.00	0.35-0.60	0.40-0.70	0.20-0.30

(a) Indicated ranges and limits apply to steels made by open hearth or basic oxygen processes; maximum content for phosphorus is 0.035% and for sulfur 0.040%. For steels made by electric furnace process, the ranges and limits are reduced as follows: C—0.01%; Mn—0.05%; Cr—0.05% (under 1.25%), 0.10% (over 1.25%); maximum content for either phosphorus or sulfur is 0.025%. Silicon content is 0.15 to 0.30%. Other silicon ranges may be negotiated. (b) Prefix "E" indicates that the steel is made by electric furnace process. (c) Contains 0.15% minimum vanadium

Composition ranges and limits for AISI-SAE standard alloy H-steels

AISI-SAE designation	UNS designation	Heat composition ranges and limits, %(a)				
		C	Mn	Cr	Ni	Mo
1330H	H13300	0.27-0.33	1.45-2.05	...	...	...
1335H	H13350	0.32-0.38	1.45-2.05	...	...	...
1340H	H13400	0.37-0.44	1.45-2.05	...	...	...
1345H	H13450	0.42-0.49	1.45-2.05	...	...	...
4027H	H40270	0.24-0.30	0.60-1.00	...	...	0.20-0.30
4028H(b)	H40280	0.24-0.30	0.60-1.00	...	...	0.20-0.30
4032H	H40320	0.29-0.35	0.60-1.00	...	...	0.20-0.30
4037H	H40370	0.34-0.41	0.60-1.00	...	...	0.20-0.30
4042H	H40420	0.39-0.46	0.60-1.00	...	...	0.20-0.30
4047H	H40470	0.44-0.51	0.60-1.00	...	...	0.20-0.30
4118H	H41180	0.17-0.23	0.60-1.00	0.30-0.70	...	0.08-0.15
4130H	H41300	0.27-0.33	0.30-0.70	0.75-1.20	...	0.15-0.25
4135H	H41350	0.32-0.38	0.60-1.00	0.75-1.20	...	0.15-0.25
4137H	H41370	0.34-0.41	0.60-1.00	0.75-1.20	...	0.15-0.25
4140H	H41400	0.37-0.44	0.65-1.10	0.75-1.20	...	0.15-0.25
4142H	H41420	0.39-0.46	0.65-1.10	0.75-1.20	...	0.15-0.25
4145H	H41450	0.42-0.49	0.65-1.10	0.75-1.20	...	0.15-0.25
4147H	H41470	0.44-0.51	0.65-1.10	0.75-1.20	...	0.15-0.25
4150H	H41500	0.47-0.54	0.65-1.10	0.75-1.20	...	0.15-0.25
4161H	H41610	0.55-0.65	0.65.1.10	0.65-0.95	...	0.25-0.35
4320H	H43200	0.17-0.23	0.40-0.70	0.35-0.65	1.55-2.00	0.20-0.30
4340H	H43400	0.37-0.44	0.55-0.90	0.65-0.95	1.55-2.00	0.20-0.30
E4340H(b)	H43406	0.37-0.44	0.60-0.95	0.65-0.95	1.55-2.00	0.20-0.30
4419H(c)	H44190	0.17-0.23	0.35-0.75	...	...	0.45-0.60

(a) Typical limits on phosphorus and sulfur contents are 0.035% maximum phosphorus and 0.040% maximum sulfur. Typical limit on silicon content is 0.15 to 0.30% silicon. (b) Electric furnace steel. (c) SAE standard grade only. (d) AISI standard grade only. (e) Can be expected to contain 0.0005 to 0.003% boron. (f) Contains 0.10 to 0.15% vanadium. (g) Contains 0.15% minimum vanadium. (h) 1.70 to 2.20% silicon

(continued)

Composition ranges and limits for AISI-SAE standard alloy H-steels (continued)

AISI-SAE designation	UNS designation	Heat composition ranges and limits, %(a)				
		C	Mn	Cr	Ni	Mo
4620H	H46200	0.17-0.23	0.35-0.75	...	1.55-2.00	0.20-0.30
4621H(c)	H46210	0.17-0.23	0.60-1.00	...	1.55-2.00	0.20-0.30
4626H(d)	H46260	0.23-0.29	0.40-0.70	...	0.65-1.05	0.15-0.25
4718H(c)	H47180	0.15-0.21	0.60-0.95	0.30-0.60	0.85-1.25	0.30-0.40
4720H	H47200	0.17-0.23	0.45-0.75	0.30-0.60	0.85-1.25	0.15-0.25
4815H	H48150	0.12-0.18	0.30-0.70	...	3.20-3.80	0.20-0.30
4817H	H48170	0.14-0.20	0.30-0.70	...	3.20-3.80	0.20-0.30
4820H	H48200	0.17-0.23	0.40-0.80	...	3.20-3.80	0.20-0.30
50B40H(e)	H50401	0.37-0.44	0.65-1.10	0.30-0.70	...	...
50B44H(e)	H50441	0.42-0.49	0.65-1.10	0.30-0.70	...	...
5046H	H50460	0.43-0.50	0.65-1.10	0.13-0.43	...	...
50B46H(e)	H50461	0.43-0.50	0.65-1.10	0.13-0.43	...	...
50B50H(e)	H50501	0.47-0.54	0.65-1.10	0.30-0.70	...	...
50B60H(e)	H50601	0.55-0.65	0.65-1.10	0.30-0.70	...	...
5120H	H51200	0.17-0.23	0.60-1.00	0.60-1.00	...	...
5130H	H51300	0.27-0.33	0.60-1.10	0.75-1.20	...	...
5132H	H51320	0.29-0.35	0.50-0.90	0.65-1.10	...	...
5135H	H51350	0.32-0.38	0.50-0.90	0.70-1.15	...	...
5140H	H51400	0.37-0.44	0.60-1.00	0.60-1.00	...	...
5145H(c)	H51450	0.42-0.49	0.60-1.00	0.60-1.00	...	...
5147H(c)	H51470	0.45-0.52	0.60-1.05	0.80-1.25	...	...
5150H	H51500	0.47-0.54	0.60-1.00	0.60-1.00	...	...
5155H	H51550	0.50-0.60	0.60-1.00	0.60-1.00	...	...
5160H	H51600	0.55-0.65	0.65-1.10	0.60-1.00	...	...
51B60H(e)	H51601	0.55-0.65	0.65-1.10	0.60-1.00	...	...
6118H(f)	H61180	0.15-0.21	0.40-0.80	0.40-0.80	...	...
6150H(g)	H61500	0.47-0.54	0.60-1.00	0.75-1.20	...	...
81B45H(e)	H81451	0.42-0.49	0.70-1.05	0.30-0.60	0.15-0.45	0.08-0.15
8617H	H86170	0.14-0.20	0.60-0.95	0.35-0.65	0.35-0.75	0.15-0.25
8620H	H86200	0.17-0.23	0.60-0.95	0.35-0.65	0.35-0.75	0.15-0.25
8622H	H86220	0.19-0.25	0.60-0.95	0.35-0.65	0.35-0.75	0.15-0.25
8625H	H86250	0.22-0.28	0.60-0.95	0.35-0.65	0.35-0.75	0.15-0.25
8627H	H86270	0.24-0.30	0.60-0.95	0.35-0.65	0.35-0.75	0.15-0.25
8630H	H86300	0.27-0.33	0.60-0.95	0.35-0.65	0.35-0.75	0.15-0.25
86B30H(e)	H86301	0.27-0.33	0.60-0.95	0.35-0.65	0.35-0.75	0.15-0.25
8637H	H86370	0.34-0.41	0.70-1.05	0.35-0.65	0.35-0.75	0.15-0.25
8640H	H86400	0.37-0.44	0.70-1.05	0.35-0.65	0.35-0.75	0.15-0.25
8642H	H86420	0.39-0.46	0.70-1.05	0.35-0.65	0.35-0.75	0.15-0.25
8645H	H86450	0.42-0.49	0.70-1.05	0.35-0.65	0.35-0.75	0.15-0.25
86B45H(e)	H86451	0.42-0.49	0.70-1.05	0.35-0.65	0.35-0.75	0.15-0.25
8650H	H86500	0.47-0.54	0.70-1.05	0.35-0.65	0.35-0.70	0.15-0.25
8655H	H86550	0.50-0.60	0.70-1.05	0.35-0.65	0.35-0.75	0.15-0.25
8660H	H86600	0.55-0.65	0.70-1.05	0.35-0.65	0.35-0.75	0.15-0.25
8720H	H87200	0.17-0.23	0.60-0.95	0.35-0.65	0.35-0.75	0.20-0.30
8740H	H87400	0.37-0.44	0.70-1.05	0.35-0.65	0.35-0.75	0.20-0.30
8822H	H88220	0.19-0.25	0.70-1.05	0.35-0.65	0.35-0.75	0.30-0.40
9260H(h)	H92600	0.55-0.65	0.65-1.10	...	...	...
9310H(b)	H93100	0.07-0.13	0.40-0.70	1.00-1.45	2.95-3.55	0.08-0.15
94B15H(e)	H94151	0.12-0.18	0.70-1.05	0.25-0.55	0.25-0.65	0.08-0.15
94B17H(e)	H94171	0.14-0.20	0.70-1.05	0.25-0.55	0.25-0.65	0.08-0.15
94B30H(e)	H94301	0.27-0.33	0.70-1.05	0.25-0.55	0.25-0.65	0.08-0.15

(a) Typical limits on phosphorus and sulfur contents are 0.035% maximum phosphorus and 0.040% maximum sulfur. Typical limit on silicon content is 0.15 to 0.30% silicon. (b) Electric furnace steel. (c) SAE standard grade only. (d) AISI standard grade only. (e) Can be expected to contain 0.0005 to 0.003% boron. (f) Contains 0.10 to 0.15% vanadium. (g) Contains 0.15% minimum vanadium. (h) 1.70 to 2.20% silicon

Composition ranges and limits for SAE HSLA steels

SAE designation(a)	Heat composition limits, %(b)			SAE designation(a)	Heat composition limits, %(b)		
	C max	Mn max	P max		C max	Mn max	P max
942X	0.21	1.35	0.04	950D	0.15	1.00	0.15
945A	0.15	1.00	0.04	950X	0.23	1.35	0.04
945C	0.23	1.40	0.04	955X	0.25	1.35	0.04
945X	0.22	1.35	0.04	960X	0.26	1.45	0.04
950A	0.15	1.30	0.04	965X	0.26	1.45	0.04
950B	0.22	1.30	0.04	970X	0.26	1.65	0.04
950C	0.25	1.60	0.04	980X	0.26	1.65	0.04

(a) Second and third digits of designation indicate minimum yield strength in ksi. Suffix "X" indicates that the steel contains niobium, vanadium, nitrogen or other alloying elements. A second suffix "K" indicates that the steel is produced fully killed using fine grain practice; otherwise, the steel is produced semikilled. (b) Maximum contents of sulfur and silicon for all grades: 0.050% S, 0.90% Si

ASTM Specifications

ASTM specifications that incorporate AISI-SAE designations

A29	Carbon and alloy steel bars, hot rolled and cold finished, generic	A547	Cold-heading-quality alloy steel wire for hexagon head bolts
A108	Standard-quality cold finished carbon steel bars	A548	Cold-heading-quality carbon steel wire for tapping or sheet metal screws
A295	High carbon-chromium ball and roller bearing steel	A549	Cold-heading-quality carbon steel wire for wood screws
A304	Alloy steel bars having hardenability requirements		
A322	Hot-rolled alloy steel bars	A575	Merchant-quality hot-rolled carbon steel bars
A331	Cold-finished alloy steel bars	A576	Special-quality hot-rolled carbon steel bars
A434	Hot-rolled or cold-finished quenched and tempered alloy steel bars	A634	Aircraft-quality hot-rolled and cold-rolled alloy steel sheet and strip
A505	Hot-rolled and cold-rolled alloy steel sheet and strip, generic	A646	Premium-quality alloy steel blooms and billets for aircraft and aerospace forgings
A506	Regular-quality hot-rolled and cold-rolled alloy steel sheet and strip		
A507	Drawing-quality hot-rolled and cold-rolled alloy steel sheet and strip	A659	Commercial-quality hot-rolled carbon steel sheet and strip
A510	Carbon steel wire rods and coarse round wire, generic	A680	Untempered spring-quality cold-rolled hard carbon steel strip
A534	Carburizing steels for antifriction bearings	A682	Cold-rolled spring-quality carbon steel strip, generic
A535	Special-quality ball and roller bearing steel	A684	Untempered spring-quality cold-rolled soft carbon steel strip
A544	Scrapless-nut-quality carbon steel wire	A689	Carbon and alloy steel bars for springs
A545	Cold-heading-quality carbon steel wire for machining screws	A711	Carbon and alloy steel blooms, billets and slabs for forging
A546	Cold-heading-quality medium-high-carbon steel wire for hexagon-head bolts	A713	High-carbon spring steel wire for heat treated components

Generic ASTM specifications

A6	Rolled steel structural plate, shapes, sheet piling and bars, generic	A568	Carbon and HSLA, hot rolled and cold rolled steel sheet and hot rolled strip, generic
A20	Steel plate for pressure vessels, generic		
A29	Carbon and alloy steel bars, hot rolled and cold finished, generic	A646	Premium-quality alloy steel blooms and billets for aircraft and aerospace forgings
A505	Alloy steel sheet and strip, hot rolled and cold rolled, generic		
A510	Carbon steel wire rod and coarse round wire, generic	A711	Carbon and alloy steel blooms, billets and slabs for forging

Composition ranges and limits for sheet and strip, plain carbon and HSLA grades (ASTM specifications)

ASTM specification	Description(a)	Composition(b), % C	Mn	P	S	Other
A611	CRSQ					
	Grades A, B, C	0.20	0.60	0.04	0.04	(c)
	Grade E	0.20	0.90	0.04	0.04	(c)
A366	CRCQ	0.15	0.60	0.035	0.04	(c)
A109	CR strip					
	Tempers 1, 2, 3	0.25	0.60	0.035	0.04	(c)
	Tempers 4, 5	0.15	0.60	0.035	0.04	(c)
A619	CRDQ	0.10	0.50	0.025	0.035	(c)
A620	CR DQSK	0.10	0.50	0.025	0.035	(d)
A570	HR SQ					
	Grades A, B, C	0.25	0.25-0.60	0.04	0.04	(c)
	Grades D, E	0.25	0.60-0.90	0.04	0.04	(c)
A569	HR CQ	0.15	0.60	0.035	0.04	(c)
A621	HR DQ	0.10	0.50	0.025	0.035	...
A622	HR DQSK	0.10	0.50	0.025	0.035	(d)
A414	Pressure vessel					
	Grade A	0.15	0.90	0.035	0.04	(c)
	Grade B	0.22	0.90	0.035	0.04	(c)
	Grade C	0.25	0.90	0.035	0.04	(c)
	Grade D	0.25	1.20	0.035	0.04	(c)
	Grade E	0.27	1.20	0.035	0.04	(c)
	Grade F	0.31	1.20	0.035	0.04	(c)
	Grade G	0.31	1.35	0.035	0.04	(c)
A606	HSLA	0.22	1.25	...	0.05	(e)
A607	Grade 45	0.22	1.35	0.04	0.05	(f)
	Grade 50	0.23	1.35	0.04	0.05	(f)
	Grade 55	0.25	1.35	0.04	0.05	(f)
	Grade 60	0.26	1.50	0.04	0.05	(f)
	Grade 65	0.26	1.50	0.04	0.05	(f)
	Grade 70	0.26	1.65	0.04	0.05	0.012 max N(f)
A715	Basic composition	0.15	1.65	0.025	0.035	0.012 max N

Type 1: 0.05 min Ti, 0.10 max Si(g)
Type 2: 0.02 min V, 0.60 max Si(h), 0.005 min N(g)(h)
Type 3: 0.005 min Nb, 0.08 max V(g), 0.60 max Si(h), 0.020 max N(g)(h)
Type 4: 0.05 min Zr, 0.90 max Si, 0.80 max Cr(h), 0.10 max Ti(h), 0.0025 max B(h), 0.005-0.06 Nb(g)(j)
Type 5: 0.03 min Nb(k), 0.20 min Mo(k), 0.30 max Si(g)
Type 6: 0.005-0.10 Nb, 0.90 max Si(g)
Type 7: 0.005 min Nb or V, or both, 0.60 max Si, 0.020 max N(g)

(a) CR, cold rolled; SQ, structural quality; DQ, drawing quality; DQSK, drawing quality special killed. (b) Maximum. (c) Cu when specified as Cu-bearing steel; 0.20% min. (d) Aluminum as deoxidizer usually exceeds 0.010% in the product. (e) Other elements may be added if necessary to meet mechanical and corrosion requirements. (f) 0.005 min Nb or 0.01 min V for all grades. (g) These elements are added to basic composition. (h) Not added to grades 50 and 60. (j) Might not be added to grade 50. (k) Available as grade 80 only

Composition ranges and limits for carbon steel structural shapes and plate (ASTM specifications)

ASTM specification	Form, type or grade	UNS designation	C max	Mn	Si	Cu(b)
A36	Plate	...	0.29	0.80-1.20	...	0.20
	Shapes	K02600	0.26	(c)	(d)	0.20
	Bars	...	0.29	0.60-0.90	...	0.20
A283	Plate	...	...	...	...	0.20
A284	Grade A	K01804	0.24	0.90 max	0.10-0.30	...
	Grade B	K02001	0.24	0.90 max	0.15-0.30	...
	Grade C	K02401	0.36	0.90 max	0.15-0.30	...
	Grade D	K02702	0.35	0.90 max	0.15-0.30	...
A529	Plate, bars and shapes	K02703	0.27	1.20 max	...	0.20
A573	Grade 58	K02301	0.23	0.60-0.90	0.10-0.35	...
	Grade 65	K02404	0.26	0.85-1.20	0.15-0.30	...
	Grade 70	K02701	0.28	0.85-1.20	0.15-0.30	...
A678	Grade A	K01600	0.16	0.90-1.50	0.15-0.50	0.20
	Grade B	K02002	0.20	0.70-1.60	0.15-0.50	0.20
	Grade C	K02204	0.22	1.00-1.60	0.20-0.50	0.20

(a) Typical limits on phosphorus and sulfur contents are 0.040% maximum phosphorus and 0.050% maximum sulfur. (b) Minimum copper content applicable only if copper-bearing steel is specified. (c) 0.85-1.35% manganese required for shapes heavier than 634 kg/m (426 lb/ft). (d) 0.15-0.30% silicon required for shapes heavier than 634 kg/m (426 lb/ft)

Composition ranges and limits for HSLA and alloy steel plate (ASTM specifications)

ASTM specification	Type or grade	UNS designation	C	Mn	P max	S max	Si	Cr	Ni	Mo	V	Other
A242	Type 1	K11510	0.15 max	1.00 max	0.45	0.05	...	...	...	...	...	0.20 min Cu
	Type 2	K12010	0.20 max	1.35 max	0.04	0.05	...	...	...	...	...	0.20 min Cu if both 0.5 Si and 0.5 Cr not present
A440	...	K12810	0.28 max	1.10-1.60	0.04	0.05	0.30 max	...	...	...	...	0.20 min Cu
A441	...	K12211	0.22 max	0.85-1.25	0.04	0.05	0.30 max	...	...	...	...	0.20 min Cu; 0.02 min V
A514	Type A	K11856	0.15-0.21	0.80-1.10	0.035	0.04	0.40-0.80	0.50-0.80	...	0.18-0.28	...	0.05-0.15 Zr; 0.0025 max B
	Type B	K11630	0.12-0.21	0.70-1.00	0.035	0.04	0.20-0.35	0.40-0.65	...	0.15-0.25	0.03-0.08	0.01-0.03 Ti; 0.0005-0.005 B
	Type C	K11511	0.10-0.20	1.10-1.50	0.035	0.04	0.15-0.30	...	...	0.20-0.30	...	0.001-0.005 B
	Type D	K11662	0.13-0.20	0.40-0.70	0.035	0.04	0.20-0.35	0.85-1.20	...	0.15-0.25	...	0.04-0.10 Ti; 0.20-0.40 Cu; 0.0015-0.005 B
	Type E	K21604	0.12-0.20	0.40-0.70	0.035	0.04	0.20-0.35	1.40-2.00	...	0.40-0.60	...	0.04-0.10 Ti; 0.20-0.40 Cu; 0.0015-0.005 B
	Type F	K11576	0.10-0.20	0.60-1.00	0.035	0.04	0.15-0.35	0.40-0.65	0.70-1.00	0.40-0.60	0.03-0.08	0.15-0.50 Cu; 0.0005-0.006 B
	Type G	K11872	0.15-0.21	0.80-1.10	0.035	0.04	0.50-0.90	0.50-0.90	...	0.40-0.60	...	0.05-0.15 Zr; 0.0025 max B
	Type H	K11646	0.12-0.21	0.95-1.30	0.035	0.04	0.20-0.35	0.40-0.65	0.30-0.70	0.20-0.30	0.03-0.08	0.0005-0.005 B
	Type J	K11625	0.12-0.21	0.45-0.70	0.035	0.04	0.20-0.35	...	...	0.50-0.65	...	0.001-0.005 B
	Type K	K11523	0.10-0.20	1.10-1.50	0.035	0.04	0.15-0.30	...	...	0.45-0.55	...	0.001-0.005 B
	Type L	K11682	0.13-0.20	0.40-0.70	0.035	0.04	0.20-0.35	1.15-1.65	...	0.25-0.40	...	0.04-0.10 Ti; 0.20-0.40 Cu; 0.0015-0.005 B
	Type M	K11683	0.12-0.21	0.45-0.70	0.035	0.04	0.20-0.35	...	1.20-1.50	0.45-0.60	...	0.001-0.005 B
	Type N	K11847	0.15-0.21	0.80-1.10	0.035	0.04	0.40-0.90	0.50-0.80	...	0.25 max	...	0.05-0.15 Zr; 0.0005-0.0025 B
	Type P	K21650	0.12-0.21	0.45-0.70	0.035	0.04	0.20-0.35	1.20-1.50	...	...	...	...
A572	Grade 42	...	0.21 max	1.35 max	0.04	0.05	0.30 max	...	...	...	...	0.20 min Cu(a)
	Grade 45	...	0.22 max	1.35 max	0.04	0.05	0.30 max	...	...	...	...	0.20 min Cu(a)
	Grade 50	...	0.23 max	1.35 max	0.04	0.05	0.30 max	...	...	...	...	0.20 min Cu(a)
	Grade 55	...	0.25 max	1.35 max	0.04	0.05	0.30 max	...	...	...	...	0.20 min Cu(a)
	Grade 60	...	0.26 max	1.35 max	0.04	0.05	0.30 max	...	...	...	...	0.20 min Cu(a)
	Grade 65	...	0.26 max	1.65 max	0.04	0.05	0.30 max	...	...	...	...	0.20 min Cu(a)
A588	Grade A	K11430	0.10-0.19	0.90-1.25	0.04	0.05	0.15-0.30	0.40-0.65	...	...	0.02-0.10	0.25-0.40 Cu
	Grade B	K12043	0.20 max	0.75-1.25	0.04	0.05	0.15-0.30	0.40-0.70	0.50 max	...	0.001-0.10	0.20-0.40 Cu
	Grade C	K11538	0.15 max	0.80-1.35	0.04	0.05	0.15-0.30	0.30-0.50	0.25-0.50	...	0.001-0.10	0.20-0.50 Cu
	Grade D	K11552	0.10-0.20	0.75-1.25	0.04	0.05	0.50-0.90	0.50-0.90	...	...	...	0.30 max Cu; 0.05-0.15 Zr; 0.04 max Nb
	Grade E	K11567	0.15 max	1.20 max	0.04	0.05	0.15-0.30	...	0.75-1.25	0.10-0.25	0.05 max	0.50-0.80 Cu
	Grade F	K11541	0.10-0.20	0.50-1.00	0.04	0.05	0.30 max	0.30 max	0.40-1.10	0.10-0.20	0.01-0.10	0.30-1.00 Cu
	Grade G	K12040	0.20 max	1.20 max	0.04	0.05	0.25-0.70	0.50-1.00	0.80 max	0.10 max	...	0.30-0.50 Cu; 0.07 max Ti
	Grade H	K12032	0.20 max	1.25 max	0.035	0.040	0.25-0.75	0.10-0.25	0.30-0.60	0.15 max	0.02-0.10	0.20-0.35 Cu; 0.005-0.030 Ti
	Grade J	K12044	0.20 max	0.60-1.00	0.04	0.05	0.30-0.50	...	0.50-0.70	...	...	0.30 min Cu; 0.03-0.05 Ti
A633	Grade A	K01802	0.18 max	1.00-1.35	0.04	0.05	0.15-0.50	...	...	...	...	0.05 max Nb
	Grade B	K01803	0.18 max	1.00-1.35	0.04	0.05	0.15-0.50	...	...	...	...	...
	Grade C	K12000	0.20 max	1.15-1.50	0.04	0.05	0.15-0.50	...	...	0.10 max	...	0.01-0.05 Nb
	Grade D	K02003	0.20 max	0.70-1.60	0.04	0.05	0.15-0.50	0.25 max	0.25 max	0.08 max	...	0.35 max Cu
	Grade E	K12202	0.22 max	1.15-1.50	0.04	0.05	0.15-0.50	...	...	...	0.04-0.11	0.01-0.03 N
A656	Grade 1	K11804	0.18 max	1.60 max	0.040	0.050	0.60 max	...	...	...	0.05-0.15	0.020 min Al; 0.005-0.030 N
	Grade 2	K11503	0.15 max	0.90 max	0.040	0.050	0.10 max	...	...	...	...	0.01 min Al; 0.05-0.50 Ti
A699	...	K10614	0.06 max	1.20-2.20	0.04	0.025	0.35 max	...	...	0.25-0.35	...	0.03-0.09 Nb; 0.20-0.35 Cu optional
A710	Grade A	K20747	0.07 max	0.40-0.70	0.025	0.025	0.35 max	0.60-0.90	0.70-1.00	0.15-0.25	...	1.00-1.30 Cu; 0.02 min Nb
	Grade B	K20622	0.06 max	0.40-0.65	0.025	0.025	0.20-0.35	...	1.20-1.50	...	...	1.00-1.30 Cu; 0.02 min Nb

(a) These grades may contain niobium, vanadium or nitrogen

Composition ranges and limits for alloy steel pressure-vessel plate (ASTM specifications)

ASTM specification	Type or grade	UNS designation	Heat composition ranges and limits, %(a)									
			C	Mn	P	S	Si	Cr	Ni	Mo	V	Other
A202	Grade A	K11742	0.17	1.05-1.40	0.035	0.04	0.60-0.90	0.35-0.60	...	...	...	...
	Grade B	K12542	0.25	1.05-1.40	0.035	0.04	0.60-0.90	0.35-0.60	...	...	...	...
A203	Grade A	K21703	0.23	0.80	0.035	0.04	0.15-0.30	...	2.10-2.50	...	...	...
	Grade B	K22103	0.25	0.80	0.035	0.04	0.15-0.30	...	2.10-2.50	...	...	...
	Grade D	K31718	0.20	0.80	0.035	0.04	0.15-0.30	...	3.25-3.75	...	...	...
	Grade E	K32018	0.23	0.80	0.035	0.04	0.15-0.30	...	3.25-3.75	...	...	...
A204	Grade A	K11820	0.25	0.90	0.035	0.04	0.15-0.30	...	0.45-0.60	0.45-0.60	...	...
	Grade B	K12020	0.27	0.90	0.035	0.04	0.15-0.30	...	0.45-0.60	0.45-0.60	...	...
	Grade C	K12320	0.28	0.90	0.035	0.04	0.15-0.30	...	...	0.45-0.60	...	...
A225	Grade A	K11803	0.18	1.45	0.035	0.04	0.15-0.30	...	...	...	0.09-0.14	...
	Grade B	K12003	0.20	1.45	0.035	0.04	0.15-0.30	...	...	...	0.09-0.14	...
	Grade C	K12524	0.25	1.60	0.035	0.04	0.13-0.32	0.37-0.73	...	...	0.11-0.20	...
A302	Grade A	K12021	0.25	0.95-1.30	0.035	0.040	0.15-0.30	...	...	0.45-0.60	...	...
	Grade B	K12022	0.25	1.15-1.50	0.035	0.040	0.15-0.30	...	...	0.45-0.60	...	...
	Grade C	K12039	0.25	1.15-1.50	0.035	0.040	0.15-0.30	...	0.40-0.70	0.45-0.60	...	...
	Grade D	K12054	0.25	1.15-1.50	0.035	0.040	0.15-0.30	...	0.70-1.00	0.45-0.60	...	...
A353	...	K81340	0.13	0.90	0.035	0.040	0.15-0.30	...	8.50-9.50	...	...	...
A387	Grade 2	K12143	0.21	0.55-0.80	0.035	0.040	0.15-0.30	0.50-0.80	...	0.45-0.60	...	...
	Grade 12	K11757	0.17	0.40-0.65	0.035	0.040	0.15-0.30	0.80-1.15	...	0.45-0.60	...	...
	Grade 11	K11789	0.17	0.40-0.65	0.035	0.040	0.50-0.80	1.00-1.50	...	0.45-0.65	...	...
	Grade 22	K21590	0.15	0.30-0.60	0.035	0.035	0.50	2.00-2.50	...	0.90-1.10	...	...
	Grade 21	K31545	0.15	0.30-0.60	0.035	0.035	0.50	2.75-3.25	...	0.90-1.10	...	...
	Grade 5	K41545	0.15	0.30-0.60	0.040	0.030	0.50	4.00-6.00	...	0.45-0.65	...	...
	Grade 7	S50300	0.15	0.30-0.60	0.030	0.030	1.00	6.00-8.00	...	0.45-0.65	...	...
	Grade 9	S50400	0.15	0.30-0.60	0.030	0.030	1.00	8.00-10.00	...	0.90-1.10	...	...
A517	Grade A	K11856	0.15-0.21	0.80-1.10	0.035	0.04	0.40-0.80	0.50-0.80	...	0.18-0.28	...	0.05-0.15 Zr; 0.0025 B
	Grade B	K11630	0.12-0.21	0.70-1.00	0.035	0.04	0.20-0.65	0.40-0.65	...	0.15-0.25	0.03-0.08	0.01-0.03 Ti; 0.0005-0.005 B
	Grade C	K11511	0.10-0.20	1.10-1.50	0.035	0.04	0.15-0.30	...	...	0.20-0.30	...	0.001-0.005 B
	Grade D	K11662	0.13-0.20	0.40-0.70	0.035	0.04	0.20-0.35	0.85-1.20	...	0.15-0.25	...	0.04-0.10 Ti; 0.20-0.40 Cu; 0.0015-0.005 B
	Grade E	K21604	0.12-0.20	0.40-0.70	0.035	0.04	0.20-0.35	1.40-2.00	...	0.40-0.60	...	0.04-0.10 Ti; 0.20-0.40 Cu; 0.0015-0.005 B
	Grade F	K11576	0.10-0.20	0.60-1.00	0.035	0.04	0.15-0.35	0.40-0.65	0.70-1.00	0.40-0.60	0.03-0.08	0.15-0.50 Cu; 0.002-0.006 B
	Grade G	K11872	0.15-0.21	0.80-1.10	0.035	0.04	0.50-0.90	0.50-0.90	...	0.40-0.60	...	0.050-0.15 Zr; 0.0025 B
	Grade H	K11646	0.12-0.21	0.95-1.30	0.035	0.04	0.20-0.35	0.40-0.650	0.30-0.70	0.20-0.30	0.03-0.08	0.0005-0.005 B
	Grade J	K11625	0.12-0.21	0.45-0.70	0.035	0.04	0.20-0.35	...	...	0.50-0.65	...	0.001-0.005 B
	Grade K	K11523	0.10-0.20	1.10-1.50	0.035	0.04	0.15-0.30	...	...	0.45-0.55	...	0.001-0.005 B
	Grade L	K11682	0.13-0.20	0.40-0.70	0.035	0.04	0.20-0.35	1.15-1.65	...	0.25-0.40	...	0.04-0.10 Ti; 0.20-0.40 Cu; 0.0015-0.005 B
	Grade M	K11683	0.12-0.21	0.45-0.70	0.035	0.04	0.20-0.35	...	1.20-1.50	0.45-0.60	...	0.001-0.005 B
	Grade P	K21650	0.12-0.21	0.45-0.70	0.035	0.04	0.20-0.35	0.85-1.20	1.20-1.50	0.45-0.60	...	0.001-0.005 B
A533	Type A	K12521	0.25	1.15-1.50	0.035	0.040	0.15-0.30	...	...	0.45-0.60	...	...
	Type B	K12539	0.25	1.15-1.50	0.035	0.040	0.15-0.30	...	0.40-0.70	0.45-0.60	...	...
	Type C	K12554	0.25	1.15-1.50	0.035	0.040	0.15-0.30	...	0.70-1.00	0.45-0.60	...	...
	Type D	K12529	0.25	1.15-1.50	0.035	0.040	0.15-0.30	...	0.20-0.40	0.45-0.60	...	...
A538	Grade A	K92810	0.03	0.10	0.010	0.010	0.10	...	17.0-19.0	4.0-4.5	...	0.10-0.25 Ti; 7.0-8.5 Co; 0.05-0.15 Al; 0.003 B; 0.02 Zr; 0.05 Ca
	Grade B	K92890	0.03	0.10	0.010	0.010	0.10	...	17.0-19.0	4.6-5.1	...	0.30-0.50 Ti; 7.0-8.5 Co; 0.05-0.15 Al
	Grade C	K93120	0.03	0.10x	0.010	0.010	0.10	...	18.0-19.0	4.6-5.2	...	0.55-0.80 Ti; 8.0-9.5 Co; 0.05-0.15 Al
A542	...	K21590	0.15	0.30-0.60	0.035	0.035	0.15-0.30	2.00-2.50	...	0.90-1.10	...	...
A543	Type A	K42338	0.23	0.40	0.035	0.040	0.20-0.35	1.50-2.00	2.60-4.00	0.45-0.60	0.03	...
	Type B	K42339	0.23	0.40	0.020	0.020	0.20-0.35	1.50-2.00	2.60-4.00	0.45-0.60	0.03	...
A553	Type I	K81340	0.13	0.90	0.035	0.040	0.15-0.30	...	8.50-9.50	...	...	...
	Type II	K71340	0.13	0.90	0.035	0.040	0.15-0.30	...	7.50-8.50	...	...	...
A562	...	K11224	0.12	1.20	0.04	0.05	0.15-0.50	...	...	...	...	(4 × %C) Ti; 0.15 Cu
A590	...	K91890	0.03	0.10	0.010	0.010	0.10	4.50-5.50	11.5-12.5	2.75-3.25	...	0.20-0.35 Ti; 0.40 Al
A605	...	K91401	0.13	0.20-0.40	0.010	0.010	0.10	0.65-0.85	8.5-9.5	0.90-1.10	0.06-0.12	4.25-5.00 Co
A645	...	K41583	0.30-0.60	0.30-0.60	0.025	0.025	0.20-0.35	...	4.75-5.25	0.20-0.35	...	0.020 N; 0.02-0.12 Al
A734	Type A	...	0.17	0.45-0.75	0.035	0.015	0.35	0.90-1.20	0.90-1.20	0.25-0.40	...	0.06 Al
	Type B	...	0.17	1.60	0.035	0.015	0.35	0.25	...	...	0.11	0.35 Cu; 0.030 N; 0.050 Nb; 0.06 Al
A735	...	...	0.06	1.20-2.20	0.04	0.025	0.35	...	...	0.23-0.47	...	0.20-0.35 Cu; 0.03-0.09 Nb
A736	...	...	0.07	0.40-0.70	0.025	0.025	0.35	0.60-0.90	0.70-1.00	0.15-0.25	...	1.00-1.30 Cu; 0.02 min Nb
A737	Grade A	...	0.20	1.00-1.35	0.035	0.030	0.15-0.50	...	...	...	0.10	...
	Grade B	...	0.20	1.15-1.50	0.035	0.030	0.15-0.50	...	...	...	...	0.05 Nb
	Grade C	...	0.20	1.15-1.50	0.035	0.030	0.15-0.50	...	...	...	0.04-0.11	0.03 N

(a) Maximum unless a range is specified

Composition ranges and limits for carbon steel pressure-vessel plate (ASTM specifications)

Specification	Type or grade	UNS designation	Heat composition ranges and limits, %(a)				
			C	Mn	P	S	Si
A285	Grade A	K01700	0.17	...	0.035	0.045	...
	Grade B	K02200	0.22	0.90	0.035	0.045	...
	Grade C	K02801	0.28	0.90	0.035	0.045	...
A288	...	K02803	0.30	0.90-1.50	0.035	0.040	0.15-0.30
A442	Grade 55	K02202	0.24	0.60-1.10	0.04	0.05	0.15-0.30
	Grade 60	K02402	0.27	0.60-1.10	0.04	0.05	0.15-0.30
A455	Type I	K03300	0.33	0.85-1.20	0.040	0.050	0.10
	Type II	K02802	0.28	0.85-1.20	0.040	0.050	0.15-0.30
A515	Grade 55	K02001	0.28	0.90	0.035	0.040	0.15-0.30
	Grade 60	K02401	0.31	0.90	0.035	0.040	0.15-0.30
A515	Grade 65	K02800	0.33	0.90	0.035	0.040	0.15-0.30
	Grade 70	K03101	0.35	0.90	0.035	0.040	0.15-0.30
A516	Grade 55	K01800	0.26	0.60-1.20	0.035	0.04	0.15-0.30
	Grade 60	K02100	0.27	0.60-1.20	0.035	0.04	0.15-0.30
	Grade 65	K02403	0.29	0.85-1.20	0.035	0.04	0.15-0.30
	Grade 70	K02700	0.31	0.85-1.20	0.035	0.04	0.15-0.30
A537	...	K02400	0.24	0.70-1.60	0.035	0.040	0.15-0.30
A612(b)	...	K02900	0.27	1.00-1.50	0.035	0.040	0.15-0.30
A662	Grade A	K01701	0.17	0.90-1.35	0.035	0.040	0.15-0.30
	Grade B	K02203	0.19	0.85-1.50	0.035	0.040	0.15-0.30
A724(b)	Grade A	...	0.18	1.00-1.60	0.035	0.040	0.55

(a) Maximum unless a range is specified. (b) Residual alloying elements restricted as follows: 0.35 max Cu; 0.25 max Ni; 0.25 max Cr; 0.08 max Mo; 0.08 max V

AMS Designations

Product descriptions and carbon contents for wrought carbon steels

AMS designation	Product form	Carbon content	Nearest AISI-SAE grade	AMS designation	Product form	Carbon content	Nearest AISI-SAE grade
5010E	Bars—screw machine stock	...	1112	5069A	Bars, forgings, tubing	0.15-0.20	1018
5020	Bars, forging, tubing	0.32-0.39	11L37	5070C	Bars, forgings (55 ksi tensile strength)	0.18-0.23	1022
5022G	Bars, forging, tubing	0.14-0.20(a)	1117	5075B	Tubing, seamless (55 ksi tensile strength)	0.22-0.28	1025
5024D	Bars, forging, tubing	0.32-0.39	1137				
5032B	Wire (annealed)	0.18-0.23	1020	5077B	Tubing, welded	0.22-0.28	1025
5040F	Sheet, strip (deep forming grade)	0.15 max	1010	5080D	Bars, forgings, tubing	0.31-0.38	1035
5041	Sheet, strip (cold rolled, extra deep drawing)	0.08 max	1006	5082A	Tubing, seamless (90 ksi tensile strength)	0.31-0.38	1035
5042F	Sheet, strip (forming grade)	0.15 max	1010	5085A	Plate, sheet, strip (annealed)	0.47-0.55	1050
5044D	Sheet, strip (half hard temper)	Low	1010	5110B	Music wire—commercial	...	1080
5045C	Sheet, strip (hard temper)	Low	1020	5112E	Music spring wire—best quality	...	1090
5047A	Sheet, strip (aluminum killed)	Low	1010	5115C	Wire—spring	0.60-0.75	1070
5050F	Tubing, seamless (annealed)	0.15 max	1010	5120F	Strip	0.69-0.80	1074
5053C	Tubing, welded (annealed)	0.13 max	1010	5121C	Strip, spring	0.89-1.04	1095
5060C	Bars, forgings, tubing	0.13-0.18	1015	5122C	Strip (hard temper)	0.89-1.04	1095
5061B	Bars, wire	0.08-0.20	...	5132D	Bars	0.90-1.30	1095
5062B	Bars, forgings, tubing, plate, sheet, strip	0.25 max	...				

(a) Contains 1.2% manganese

Product descriptions and nominal compositions for wrought alloy steels

AMS designation	Product form(a)	Nominal composition, %					Nearest AISI-SAE grade
		C	Cr	Ni	Mo	Other	
6242C	Bars, forgings	0.15-0.20	...	5	...	...	2517
6250F	Bars, forgings, tubing	0.07-0.13	1.5	3.5	...	...	3310
6260G	Bars, forgings, tubing	0.07-0.13	1.2	3.25	0.12	...	9310
6263D	Bars, forgings, tubing	0.11-0.17	1.2	3.25	0.12	...	9315
6264D	Bars, forgings, tubing	0.14-0.20	1.2	3.25	0.12	...	9317
6265C	Bars, forgings, tubing (P, VM)	0.07-0.13	1.2	3.25	0.12	...	9310
6266C	Bars, forgings, tubing	0.08-0.13	0.50	1.85	0.25	0.003 B	43BV12
6267A	Bars, forgings, tubing (P)	0.07-0.13	1.2	3.25	0.12	...	9310
6270G	Bars, forgings, tubing	0.11-0.17	0.50	0.55	0.20	...	8615
6272E	Bars, forgings, tubing	0.15-0.20	0.50	0.55	0.20	...	8617
6274G	Bars, forgings, tubing	0.18-0.23	0.50	0.55	0.20	...	8620
6275B	Bars, forgings, tubing	0.15-0.20	0.40	0.45	0.12	0.003 B	94B17
6276C	Bars, forgings, tubing (P,VM)	0.18-0.23	0.50	0.55	0.2	...	8620
6277A	Bars, forgings, tubing (P)	0.18-0.23	0.50	0.55	0.20	...	8620
6280E	Bars, forgings	0.28-0.33	0.50	0.55	0.20	...	8630
6281C	Tubing	0.28-0.33	0.5	0.55	0.2	...	8630
6282D	Tubing	0.33-0.38	0.50	0.55	0.25	...	8735
6290C	Bars, forgings	0.11-0.17	...	1.8	0.25	...	4615
6292C	Bars, forgings	0.15-0.20	...	1.8	0.25	...	4617
6294C	Bars, forgings	0.17-0.22	...	1.8	0.25	...	4620
6299A	Bars, forgings, tubing	0.17-0.23	0.50	1.8	0.25	...	4320H
6300	Bars, forgings	0.35-0.40	...	...	0.25	...	4037
6302B	Bars, forgings	0.28-0.33	1.25	...	0.50	0.65 Si, 0.25 V	...
6303A	Bars, forgings	0.25-0.30	1.25	...	0.5	0.65 Si, 0.85 V	...
6304C	Bars, forgings, tubing	0.40-0.50	0.95	...	0.55	0.30 V	...
6312A	Bars, forgings	0.38-0.43	...	1.8	0.25	...	4640
6317B	Bars, forgings (heat treated; 125 ksi tensile strength)	0.38-0.43	...	1.8	0.25	...	4640
6320F	Bars, forgings	0.33-0.38	0.50	0.55	0.25	...	8735
6321A	Bars, forgings, tubing	0.38-0.43	0.43	0.30	0.12	...	81B40
6322F	Bars, forgings	0.38-0.43	0.50	0.55	0.25	0.003 B	8740
6323D	Tubing	0.38-0.43	0.50	0.55	0.25	...	8740
6324C	Bars, forgings	0.38-0.43	0.65	0.70	0.25	...	8740 mod

(a) P, premium quality; VM, vacuum melted; CM, consumable electrode remelted

(continued)

Product descriptions and nominal compositions for wrought alloy steels (continued)

AMS designation	Product form(a)	C	Cr	Ni	Mo	Other	Nearest AISI-SAE grade
6325D	Bars, forgings (heat treated; 105 ksi tensile strength)	0.38-0.43	0.50	0.55	0.25	...	8740
6327D	Bars, forgings (heat treated; 125 ksi tensile strength)	0.38-0.43	0.50	0.55	0.25	...	8740
6328E	Bars, forgings, tubing	0.48-0.53	0.50	0.55	0.25	...	8750
6330A	Bars, forgings	0.33-0.38	0.6	1.25	...	...	3135
6342D	Bars, forgings, tubing	0.38-0.32	0.80	1.0	0.25	...	9840
6350D	Plate, sheet, strip (annealed)	0.28-0.33	0.95	...	0.20	...	4130
6351A	Plate, sheet, strip (spheroidized)	0.28-0.33	0.95	...	0.20	...	4130
6352B	Plate, sheet, strip (annealed)	0.32-0.39	0.95	...	0.2	...	4135H
6354	Plate, sheet, strip	0.10-0.17	0.6	...	0.2	0.75 Si 0.1 Zr	NAX 9115-AC
6355G	Plate, sheet, strip (annealed)	0.28-0.33	0.50	0.55	0.20	...	8630
6356A	Plate, sheet, strip	0.30-0.35	0.95	...	0.20	...	...
6357D	Plate, sheet, strip	0.33-0.38	0.50	0.55	0.25	...	8735
6358B	Plate, sheet, strip	0.38-0.43	0.50	0.55	0.25	...	8740
6359B	Plate, sheet, strip	0.38-0.43	0.80	1.8	0.25	...	4340
6360F	Tubing, seamless	0.28-0.33	0.95	...	0.20	...	4130
6361	Tubing, seamless (125 ksi tensile strength)	0.27-0.33	0.95	...	0.2	...	4130
6362	Tubing, seamless (150 ksi tensile strength)	0.27-0.33	0.95	...	0.2	...	4130
6365E	Tubing, seamless	0.33-0.38	0.95	...	0.20	...	4135
6370F	Bars, forgings, rings	0.28-0.33	0.95	...	0.2	...	4130
6371D	Tubing	0.28-0.33	0.95	...	0.20	...	4130
6372D	Tubing	0.33-0.38	0.95	...	0.20	...	4135
6373A	Tubing, welded	0.28-0.33	0.95	...	0.20	...	4130
6378	Bars (die drawn and tempered; 130 ksi yield strength)	0.38-0.45	0.95	...	0.2	...	4140
6379	Bars (die drawn and tempered; 165 ksi yield strength)	0.40-0.53	0.95	...	0.2	...	4140
6381B	Tubing	0.38-0.43	0.95	...	0.20	...	4140
6382G	Bars, forgings	0.38-0.43	0.95	...	0.20	...	4140
6385B	Plate, sheet, strip	0.27-0.33	1.25	...	0.50	0.65 Si 0.25 V	...
6386A	Plate, sheet (heat treated; 90 and 100 ksi yield strength)	...	...	...	...	...	...
6390A	Tubing (special quality)	0.38-0.43	0.95	...	0.20	...	4140
6395	Plate, sheet, strip	0.38-0.43	0.95	...	0.20	...	4140
6406A	Plate, sheet, strip	0.41-0.46	2.1	...	0.58	1.6 Si 0.05 V	...
6407B	Bars, forgings, tubing	0.27-0.33	1.2	2.05	0.45	...	...
6411	Bars, forgings, tubing (P, CM)	0.28-0.33	0.85	1.8	0.40	...	...
6412F	Bars, forgings	0.35-0.40	0.80	1.8	0.25	...	4337
6413D	Tubing	0.35-0.40	0.8	1.8	0.25	...	4337
6414A	Bars, forgings, tubing (P)	0.38-0.43	0.80	1.8	0.25	...	4340
6415G	Bars, forgings, tubing	0.38-0.43	0.80	1.8	0.25	...	4340
6416	Bars, forgings, tubing	0.41-0.46	0.8	1.8	0.4	1.6 Si 0.07 V	...
6417	Bars, forgings, tubing (P, CM)	0.38-0.43	0.82	1.8	0.40	1.6 Si 0.07 V	4340 mod
6418C	Bars, forgings, tubing	0.23-0.28	0.30	1.8	0.40	1.3 Mn 1.5 Si	Hy-Tuf
6419	Bars, forgings, tubing (P, CM)	0.41-0.46	0.82	1.8	0.40	1.6 Si 0.07 V	300 M
6421A	Bars, forgings, tubing	0.35-0.40	0.80	0.85	0.20	0.003 B	Mod 98B37
6422C	Bars, forgings, tubing	0.38-0.43	0.80	0.85	0.20	0.003 B	Mod 98B40
6423A	Bars, forgings, tubing	0.40-0.46	0.92	0.75	0.52	0.003 B	...
6426A	Bars, forgings, tubing (VM)	0.80-0.90	1.0	...	0.58	0.75 Si 0.07 V	...
6427D	Bars, forgings, tubing	0.28-0.33	0.85	1.8	0.40	0.07 V	...
6428A	Bars, forgings, tubing	0.32-0.38	0.80	1.8	0.35	0.20 V	4335 mod
6429A	Bars, forgings, tubing (CM or VM)	0.33-0.38	0.80	1.8	0.35	0.20 V	4335 mod
6430A	Bars, forgings, tubing (special grade)	0.33-0.38	0.80	1.8	0.35	0.20 V	4335 mod
6431A	Bars, forgings, tubing (P, VM)	0.45-0.50	1.05	0.55	1.0	0.11 V	D6AC
6432	Bars, forgings, tubing	0.43-0.49	1.05	0.55	1.0	0.11 V	D6A
6433A	Plate, sheet, strip (special grade)	0.33-0.38	0.80	1.8	0.35	0.20 V	...
6434A	Plate, sheet, strip	0.31-0.38	0.80	1.8	0.35	0.20 V	4335 mod
6435A	Plate, sheet, strip (P, CM) (annealed)	0.33-0.38	0.80	1.8	0.35	0.20 V	4335 mod

(a) P, premium quality; VM, vacuum melted; CM, consumable electrode remelted

(continued)

Product descriptions and nominal compositions for wrought alloy steels (continued)

AMS designation	Product form(a)	Nominal composition, %					Nearest AISI-SAE grade
		C	Cr	Ni	Mo	Other	
6436A	Plate, sheet, strip	0.20-0.25	1.25	...	0.5	0.65 Si 0.85 V	...
6437A	Plate, sheet, strip	0.38-0.43	5.0	...	1.3	0.5 V	H11
6438A	Plate, sheet, strip (P, CM)	0.45-0.50	1.05	0.55	1.0	0.11	D6
6440E	Bars, wire, forgings	0.98-1.10	1.45	...	...	...	52100
6441D	Tubing (bearing quality)	0.98-1.10	1.45	...	...	...	52100
6442C	Bars, wire, forgings	0.98-1.10	0.50	...	...	...	50100
6443B	Bars, wire, forgings (P, VM)	0.95-1.1	1.05	...	...	...	51100
6444B	Bars, wire, forgings, tubing (P, VM)	0.95-1.10	1.45	...	...	...	52100
6445A	Bars, wire, forgings, tubing (P, VM)	0.92-1.02	1.05	...	...	1.1 Mn	51100 mod
6446	Bars, forgings (P)	0.95-1.10	1.05	...	...	...	51100
6447	Bars, forgings, tubing (P)	0.95-1.10	1.45	...	...	...	52100
6448C	Bars, forgings	0.48-0.53	0.95	...	...	...	6150
6450C	Wire (spring-annealed)	0.48-0.53	0.95	...	...	0.22 V	6150
6455C	Plate, sheet, strip (spring)	0.48-0.53	0.95	...	...	0.22 V	6150
6470F	Bars, forgings, tubing for nitriding	0.38-0.43	1.6	...	0.35	0.22 V 1.15 Al	...
6471	Bars, forgings, tubing for nitriding (P)	0.38-0.43	1.6	...	0.35	1.15 Al	...
6472	Bars, forgings for nitriding (heat treated; 112 ksi tensile strength)	0.38-0.43	1.6	...	0.35	1.13 Al	...
6475C	Bars, forgings, tubing for nitriding	0.21-0.26	1.1	3.5	0.25	1.25 Al	...
6485B	Bars, forgings	0.38-0.43	5.0	...	1.3	0.50 V	H11
6487C	Bars, forgings (P, VM)	0.38-0.43	5.0	...	1.3	0.50 V	H11
6488	Bars, forgings (P)	0.38-0.43	5.0	...	1.3	0.50 V	H11
6490B	Bars, forgings, tubing (P, VM)	0.77-0.85	4.0	...	4.25	1.0 V	M50
6512	Bars, forgings, tubing, rings (CM) (annealed)	...	...	18	4.9	7.8 Co 0.40 Ti 0.10 Al	...
6514	Bars, forgings, tubing, rings (annealed) (CM)	...	...	18.5	4.9	9.0 Co 0.65 Ti 0.10 Al	...
6520	Plate, sheet, strip (solution heat treated) (CM)	...	...	18	4.9	7.8 Co 0.40 Ti 0.10 Al	...
6521	Plate, sheet, strip (solution heat treated) (CM)	...	...	18.5	4.9	9.0 Co 0.65 Ti 0.10 Al	...
6526	Bars, forgings, tubing (annealed) (P, CM)	0.29-0.34	1.0	7.5	1.0	4.5 Co 0.09 V	...
6530E	Tubing, seamless	0.28-0.33	0.50	0.55	0.20	...	8630
6535D	Tubing, seamless	0.33-0.38	0.50	0.55	0.25	...	8735
6540A	Bars, forgings, rings, tubing (annealed)	0.24-0.30	0.48	8.0	0.48	4.0 Co 0.09 V	...
6541A	Bars, forgings, tubing, rings (annealed) (P, CM)	0.24-0.30	0.48	8.0	0.48	4.0 Co 0.09 V	...
6542A	Bars, forgings, tubing (annealed) (P, CM)	0.42-0.48	0.27	7.75	0.27	4.0 Co 0.09 V	...
6545A	Plate, sheet, strip (annealed)	0.24-0.30	0.48	8.0	0.48	4.0 Co 0.09 V	...
6546A	Plate, sheet, strip (annealed) (P, CM)	0.24-0.30	0.48	8.0	0.48	4.0 Co 0.09 V	...
6550E	Tubing, welded	0.28-0.33	0.50	0.55	0.20	...	8630

(a) P, premium quality; VM, vacuum melted; CM, consumable electrode remelted

Physical Properties

Densities of steels

Nearest AISI-SAE grade	Chemical composition, %						Treatment or condition	Density	
	C	Mn	Si	Cr	Ni	Other		Mg/m³	lb/in.³
1008	0.06	0.38	0.01	...	...	...	Annealed	7.871	0.2844
1024	0.23	0.64	0.11	...	...	...	Annealed	7.859	0.2839
1042	0.44	0.69	0.20	...	...	...	Annealed	7.844	0.2834
(a)	1.22	0.35	0.16	...	...	...	Annealed	7.830	0.2829
5130	0.31	0.74	...	1.00	...	...	Hardened and tempered	7.84	0.283
52100	0.98	0.28	...	1.68	...	...	Annealed	7.81	0.282
(a)	0.51	0.22	...	1.72	3.52	...	Quenched in brine (BQ)	7.79	0.281
							BQ, tempered 190 °C (375 °F)	7.80	0.282
							BQ, tempered 365 °C (690 °F)	7.82	0.283
							BQ, tempered 600 °C (1110 °F)	7.835	0.2831
							Annealed	7.835	0.2831
18Ni250(b)	0.026	0.1	0.11	...	18.5	4.7 Mo 7.0 Co 0.22 Ti 0.003 B	...	8.0	0.289

(a) No AISI-SAE grade of similar composition. (b) Nominal composition

Specific damping capacity(a)

AISI-SAE grade	Treatment or condition	Specific damping capacity, %	AISI-SAE grade	Treatment or condition	Specific damping capacity, %	AISI-SAE grade	Treatment or condition	Specific damping capacity, %
1018	Normalized	1.5	1095	WQ, temper at 100 °C	0.2	1Cr-3Ni-0.3C	Not known	0.8
1095	Spheroidized	0.8				Gray cast		
1095	Water quench from 800 °C	0.5	4140	Not known	0.15	iron	Not known	7 to 20
			1Ni-0.4C	Not known	0.3			

(a) Surface shear stress, 34.5 MPa (5 ksi); data apply to room temperature

Specific heat of steels

| Nearest AISI-SAE grade | Chemical composition, % | | | | | | Treatment or condition | Mean apparent specific heat, J/Kg · K, at temperature range, °C | | | | | | | | | | | |
	C	Mn	Cr	Ni	Mo	Other		50 to 100	150 to 200	200 to 250	250 to 300	300 to 350	350 to 400	450 to 500	550 to 600	650 to 700	700 to 750	750 to 800	850 to 900
1008	0.06	0.38	...	...	...	...	Annealed	481	519	536	553	574	595	662	754	867	1105	875	846
1008	0.08	0.31	...	...	...	...	Annealed	481	523	544	557	569	595	662	741	858	1139	960	...
1010	0.10	0.42	...	...	...	0.008 P; 0.028 S	Not known	450	500	520	535	565	590	650	730	825	...	...	...
1025	0.23	0.64	...	...	...	...	Annealed	486	519	532	557	574	599	662	749	846	1432	950	...
1042	0.42	0.64	...	...	...	...	Annealed	486	515	528	548	569	586	649	708	770	1583	624	548
1078	0.80	0.32	...	...	...	...	Annealed	490	532	548	565	586	607	670	712	770	2081	615	...
(a)	1.22	0.35	...	...	...	...	Annealed	486	540	544	557	578	599	636	699	816	2089	649	...
1524	0.23	1.51	...	...	...	0.11 Cu	Annealed	477	511	528	544	565	590	649	741	837	1449	821	536
4130(b)	0.3	0.5	0.95	...	...	...	Hardened and tempered	477	515	...	544	...	595	657	737	825	...	833	...
4140	0.41	0.67	1.01	...	0.23	...	Hardened and tempered	...	473(c)	...	...	...	519(c)	...	561(c)	...	...	...	...
5132	0.32	0.69	1.09	0.073	...	...	Annealed	494	523	536	553	574	595	657	741	837	1499	934	574
5140	0.39	0.79	1.03	...	...	...	Hardened and tempered	452(c)	473(c)	...	...	...	519(c)	...	561(c)	...	...	...	...
(a)	0.35	0.59	0.88	0.26	.0.20		Annealed	477	515	528	544	569	595	657	737	825	1616	883	
(a)	0.33	0.55	0.17	3.47	...	...	Not known	481	523	536	548	569	590	662	749	1637	955	603	640
(a)	0.34	0.55	0.78	3.53	.0.39		Hardened and tempered	486	523	540	557	582	607	670	770	1051	1662	636	636
(a)	0.49	0.90	...	...	...	1.98 Si; 0.64 Cu	Not known	498	523	540	557	578	603	666	749	829	904	1365	...

(a) No equivalent grade. (b) Nominal composition. (c) Value presented is mean value for range of temperatures between room temperature and the higher of the cited temperatures

Coefficients of linear thermal expansion

Nearest AISI-SAE grade	Chemical composition, %									Treatment or condition
	C	Mn	P	S	Si	Cr	Ni	Mo	Other	
1008	0.06	0.38	0.017	0.035	0.01	0.02	0.55	0.03	0.08 Cu; 0.001 Al; 0.039 As	Annealed
1008	0.07	0.08	0.01	0.02	0.01	...	...	...	0.02 Cu	Annealed
1010	0.08	0.31	0.029	0.05	0.08	0.04	0.07	0.02	0.002 Al; 0.032 As	Annealed
1010	0.10	0.42	0.008	0.028	...	...	...	...	...	Not known
1015	0.17	0.42	0.012	0.035	...	...	...	...	Rolled	...
1020	0.22	0.12	0.01	0.03	0.01	...	...	...	...	Annealed
1022	0.22	0.90	0.01	0.017	0.25	...	...	...	...	Not known
1022	0.23	0.64	0.034	0.034	0.11	Trace	0.07	...	0.13 Cu; 0.010 Al; 0.036 As	...
1035	0.36	0.32	...	...	0.20	...	...	...	...	Annealed
1035	0.33	0.12	0.01	0.03	0.03	...	...	...	0.03 Cu	Annealed
1040	0.40	0.11	0.01	0.03	0.07	...	...	...	0.03 Cu	Annealed
1040	0.42	0.64	0.031	0.029	0.11	Trace	0.06	...	0.12 Cu; 0.006 Al; 0.033 As	Annealed
1045	0.44	0.69	0.037	0.038	0.20	0.03	0.04	...	0.06 Cu; 0.006 Al; 0.024 As	Annealed
1045	0.44	0.57	0.013	0.033	0.16	...	...	...	0.14 V	Annealed
1052	0.49	1.21	0.05	0.05	0.12	...	...	...	...	Annealed
1055	0.56	0.19	Trace	0.023	0.04	...	...	...	0.02 Cu	Annealed
1060	0.59	0.92	0.024	0.033	0.25	...	...	...	...	Annealed
1070	0.66	0.72	0.028	0.916	0.18	...	...	...	...	Rolled
1078	0.80	0.32	...	...	...	...	...	...	...	Annealed
1080	0.81	0.10	Trace	0.025	0.025	...	...	...	0.02 Cu	Annealed
(e)	0.82	1.65	...	...	0.20	0.03	...	...	...	Not known
1085	0.80	0.32	0.008	0.009	0.13	0.11	0.13	...	0.07 Cu; 0.004 Al; 0.021 As	Annealed
1095	1.1	0.3	0.025	0.025	0.2	...	...	...	...	Annealed Hardened
(e)	1.22	0.35	0.009	0.015	0.16	0.11	0.13	0.01	0.077 Cu; 0.006 Al; 0.025 As	Annealed
1145	0.42	0.64	0.031	0.029	0.11	Trace	0.06	...	0.12 Cu; 0.006 Al; 0.003 As	Annealed
1145	0.44	0.69	0.037	0.038	0.20	0.03	0.04	...	0.06 Cu; 0.006 Al; 0.024 As	Annealed
1524	0.23	1.51	...	...	...	...	...	...	...	Not known
(e)	0.82	1.65	...	...	0.20	0.03	...	...	...	Not known
(e)	0.40	0.67	...	...	...	...	0.80	...	...	Oil hardened, tempered 600 °C (1110 °F)
2330	0.33	0.78	0.014	0.035	0.09	...	3.59	...	...	Annealed
3140	0.40-0.50	0.50-0.80	...	...	...	0.45-0.75	1.00-1.50	...	...	Hardened and tempered
4137	0.39	0.51	0.015	0.029	0.19	0.87	...	0.21	...	Rolled
4140	0.41	0.67	...	...	...	1.01	...	0.23	...	Oil hardened, tempered 600 °C (1110 °F)
4340	0.41	...	...	...	...	1.07	1.43	0.26	...	Oil hardened, tempered 630 °C (1170 °F)
(e)	0.32	...	...	...	...	0.67	2.60	0.51	...	Oil hardened, tempered 650 °C (1200 °F)
4615	...	...	...	...	...	...	1.65	0.30	...	Not known
4617	0.18	...	...	...	...	...	1.76	0.20	...	Carburized and hardened
5140	0.35	0.31	...	...	0.19	0.75	...	...	...	Annealed
(e)	0.37	0.33	...	...	0.21	1.57	...	...	...	Annealed
52100	0.94	0.34	...	...	0.27	0.95	...	...	...	Annealed Hardened
6150	0.22	0.75	0.019	0.033	0.27	0.96	...	...	0.17 V	Annealed Hardened, tempered 205 °C (400 °F)
6150	0.35	1.42	0.013	0.057	0.20	1.00	...	...	0.11 V	Annealed
6150	0.53	0.80	0.015	0.020	0.15	1.02	...	...	0.17 V	Annealed Hardened, tempered 425 °C (800 °F) Hardened, tempered 650 °C (1200 °F)
18Ni250(e)	0.026	0.1	...	...	0.11	...	18.5	4.7	7.0 Co; 0.22 Ti; 0.003 B	Not known

(a) To obtain coefficients in µin./in. · °F multiply values in table by 0.556. (b) Stated value represents average coefficient between 0 °C (32 °F) and indicated temperature. (c) 10.3 µm/m · K from –100 °C to 20 °C, 9.8 µm/m · K from –150 to 20 °C. (d) Stated value represents average coefficient between 25 °C (75 °F) and indicated temperature. (e) Nominal composition. (f) 11.2 µm/m · K from –100 to 20 °C; 10.4 µm/m · K from –150 to 20 °C. (g) Stated value represents average coefficient between 24 and 284 °C (74 and 540 °F)

(continued)

Nearest AISI-SAE grade	Average coefficients of expansion, µm/m · K(a)								
	100 / 212	200 / 392	300 / 572	400 / 752	500 / 932	600 / 1112	700 / 1292	800 / 1472	1000 / 1832
1008	12.6(b)	13.1(b)	13.5(b)	13.8(b)	14.2(b)	14.6(b)	15.0(b)	14.7	13.8
1008	11.6	12.5	13.0	13.6	14.2	14.6	15.0	...	...
1010	12.2(b)	13.0(b)	13.5(b)	13.9(b)	14.3(b)	14.7(b)	15.0(b)	...	...
1010	11.9(c)	12.6	13.3	13.8	14.3	14.7	14.9	14.0	...
1015	11.9(b)	12.5(b)	13.0(b)	13.6(b)	14.2(b)	...	...	...	...
1020	11.7	12.1	12.8	13.4	13.9	14.4	14.8	...	...
1022	12.5	12.7	...	...	...	...	...	...	...
1022	12.2(b)	12.7(b)	13.1(b)	13.5(b)	13.9(b)	14.4(b)	14.9(b)	12.6	13.4
1035	...	12.6	13.3	13.8	14.3	14.8	15.2	...	...
1035	11.1	11.9	12.7	13.4	14.0	14.4	14.8	...	...
1040	11.3	12.0	12.5	13.3	13.9	14.4	14.8	...	...
1045	11.2(b)	12.1(b)	13.0(b)	13.6(b)	14.0(b)	14.6(b)	14.8(b)	11.8	13.6
	11.6(b)	12.3(b)	13.1(b)	13.7(b)	14.2(b)	14.7(b)	15.1(b)	...	...
	11.2(d)	11.9(d)	12.7(d)	13.5(d)	14.1(d)	14.5(d)	14.8(d)	...	...
1052	11.3(d)	11.8(d)	12.7(d)	13.7(d)	14.5(d)	14.7(d)	15.0(d)	...	...
1055	11.0	11.8	12.6	13.4	14.0	14.5	14.8	...	...
1060	11.1(d)	11.9(d)	12.9(d)	13.5(d)	14.1(d)	14.6(d)	14.9(d)	...	...
1070	11.8(b)	12.6(b)	13.3(b)	14.0(b)	...	...	...	...	...
1078	11.1	11.7	...	13.2	...	14.2	...	13.8	15.7
1080	11.0	11.6	12.4	13.2	13.8	14.2	14.7	...	...
(e)	8.8(b)	9.8(b)	11.3(b)	12.3(b)	13.1(b)	13.6(b)	14.2(b)	...	...
1085	11.1(b)	11.7(b)	12.5(b)	13.2(b)	13.6(b)	14.2(b)	14.7(b)	...	...
1095	11.4(b)	...	...	...	...	...	...	...	...
	13.0(b)	...	...	...	...	...	...	...	...
(e)	10.6(b)	11.2(b)	12.1(b)	12.9(b)	13.5(b)	14.2(b)	14.7(b)	14.3	16.8
1145	11.2(b)	12.1(b)	13.0(b)	13.6(b)	14.0(b)	14.6(b)	14.8(b)	...	...
	11.6(b)	12.3(b)	13.1(b)	13.7(b)	14.2(b)	14.7(b)	15.1(b)	...	...
1524	11.9	12.7	...	13.9	...	14.7	...	12.1	13.8
(e)	8.8(b)	9.8(b)	11.3(b)	12.3(b)	13.1(b)	13.6(b)	14.2(b)	...	...
	11.9	12.6	...	13.8	...	14.5	...	...	...
2330	10.9(d)	11.2(d)	12.1(d)	12.9(d)	13.4(d)	13.8(d)	...	...	...
3140	11.8(b)	12.3(b)	12.9(b)	13.4(b)	14.0(b)	...	...	...	...
4137	11.2(b)	11.8(b)	12.4(b)	13.0(b)	13.6(b)	...	...	...	...
4140	12.3	12.7	...	13.7	...	14.5	...	...	...
4340	(f)	12.4	...	13.6	...	14.3	...	...	...
(e)	...	11.6	...	13.1	...	13.9	...	...	...
4615	11.5	12.1	12.7	13.2	13.7	14.1	...	...	...
4617	12.5	13.1	...	...	...	...	...	...	...
5140	...	12.6	13.4	13.9	14.3	14.6	15.0	...	...
(e)	...	12.8	13.4	13.8	14.2	14.4	14.6	...	...
52100	11.9(b)	...	...	...	...	...	...	...	...
	12.6(b)	...	...	...	...	...	...	...	...
6150	12.2	12.7	13.3	13.7	14.1	14.4	...	...	...
	12.0	12.5	12.9	13.0	13.3	13.7	...	...	...
6150	12.4(d)	12.6(d)	13.3(d)	13.8(d)	14.2(d)	14.5(d)	14.7(d)	...	...
6150	12.4	12.8	13.4	13.9	14.2	14.5	...	...	...
	11.8	12.4	13.1	13.6	13.9	14.1	...	...	...
	12.3	12.7	13.4	13.9	14.3	14.7	...	...	...
18Ni250(e)	...	...	10.1(g)	...	10.1	...	...	...	...

(a) To obtain coefficients in µin./in. · °F multiply values in table by 0.556. (b) Stated value represents average coefficient between 0 °C (32 °F) and indicated temperature. (c) 10.3 µm/m · K from –100 °C to 20 °C, 9.8 µm/m · K from –150 to 20 °C. (d) Stated value represents average coefficient between 25 °C (75 °F) and indicated temperature. (e) Nominal composition. (f) 11.2 µm/m · K from –100 to 20 °C; 10.4 µm/m · K from –150 to 20 °C. (g) Stated value represents average coefficient between 24 and 284 °C (74 and 540 °F).

Electrical resistivities of steels

Nearest AISI-SAE grade	Resistivity, $\mu\Omega \cdot$ m, at indicated temperature											
	°C 20 / °F 68	100 / 212	200 / 392	400 / 752	600 / 1112	700 / 1292	800 / 1472	900 / 1652	1000 / 1832	1100 / 2012	1200 / 2192	1300 °C / 2372 °F
1008	0.130	0.178	0.252	0.448	0.725	0.898	1.073	1.124	1.160	1.189	1.216	1.241
1008	0.142	0.190	0.263	0.458	0.734	0.905	1.081	1.130	1.165	1.193	1.220	1.244
1025	0.169	0.219	0.292	0.487	0.758	0.925	1.094	1.136	1.167	1.194	1.219	1.239
1042	0.171	0.221	0.296	0.493	0.766	0.932	1.111	1.149	1.179	1.207	1.230	...
1078	0.180	0.232	0.308	0.505	0.772	0.935	1.129	1.164	1.191	1.214	1.231	1.246
(a)	0.196	0.252	0.333	0.540	0.802	0.964	1.152	1.196	1.226	1.249	1.271	1.287
1524	0.208	0.259	0.333	0.523	0.786	0.946	1.103	1.143	1.174	1.202	1.227	1.250
4130(b)	0.223	0.271	0.342	0.529	0.786	...	1.103	...	1.171	...	1.222	...
4140	0.222	0.263	0.326	0.475	0.646	...	...	...	...	...	...	...
4340	0.248	0.298	0.367	0.552	0.797	...	...	...	...	...	...	...
5132	0.210	0.259	0.330	0.517	0.778	0.934	1.106	1.145	1.177	1.205	1.230	1.251
5140	0.228	0.281	0.352	0.530	0.785	...	...	...	...	...	...	...
(a)	0.223	0.271	0.342	0.529	0.786	0.944	1.103	1.138	1.171	1.200	1.222	1.242
(a)	0.271	0.320	0.390	0.567	0.814	0.992	1.122	1.149	1.180	1.204	1.228	1.248
(a)	0.289	0.337	0.406	0.582	0.825	0.994	1.114	1.146	1.176	1.199	1.222	1.242
(a)	0.429	0.470	0.529	0.685	0.911	1.057	1.173	1.197	1.223	1.249	1.271	1.289
18Ni250(b)	0.6 to 0.7	...	...	...	...	...	...	...	...	...	...	...
	0.36 to 0.6	...	...	...	...	...	...	...	...	...	...	...

Nearest AISI-SAE grade	Chemical composition, %							Treatment or condition
	C	Mn	Si	Cr	Ni	Mo	Other	
1008	0.06	0.38	...	...	...	...	...	Annealed
1008	0.08	0.31	...	...	...	...	...	Annealed
1025	0.23	0.64	...	...	...	...	...	Annealed
1042	0.42	0.64	...	...	...	...	...	Annealed
1078	0.80	0.32	...	...	...	...	...	Annealed
(a)	1.22	0.35	...	...	...	...	...	Annealed
1524	0.23	1.51	...	...	...	...	0.11 Cu	Not known
4130(b)	0.3	0.5	0.3	0.95	...	0.2	...	Hardened and tempered
4140	0.41	0.67	...	1.01	...	0.23	...	Hardened and tempered
4340	0.41	...	...	1.07	1.43	0.26	...	Hardened and tempered
5132	0.32	0.69	...	1.09	0.073	...	...	Annealed
5140	0.39	0.79	...	1.03	...	...	...	Hardened and tempered
(a)	0.35	0.59	...	0.88	0.26	0.20	...	Annealed
(a)	0.33	0.55	...	0.17	3.47	...	...	Not known
(a)	0.34	0.55	...	0.78	3.53	0.39	...	Hardened and tempered
(a)	0.49	0.90	1.98	...	...	...	0.64 Cu	Not known
18Ni250(b)	0.026	0.1	0.11	...	18.5	4.7	7.0 Co; 0.22 Ti; 0.003 B	Annealed Aged

(a) No AISI-SAE standard grade of similar composition. (b) Nominal composition

Thermal conductivities of steels

Nearest AISI-SAE grade	Conductivity, W/m · K(a), at indicated temperature											
	°C 0 °F 32	100 212	200 392	300 572	400 752	500 932	600 1112	700 1292	800 1472	1000 1832	1200 2192	°C °F
1008	59.5	57.8	53.2	49.4	45.6	41.0	36.8	33.1	28.5	27.6	29.7	
1008	65.3(b)	60.3	54.9	...	45.2	...	36.4	...	28.5	27.6	...	
1010	65.2	60.2	55.5	50.7	46.0	41.5	36.9	32.9	28.9	...	...	(c)
1025	51.9	51.1	49.0	46.1	42.7	39.4	35.6	31.8	26.0	27.2	29.7	
1042	51.9	50.7	48.2	45.6	41.9	38.1	33.9	30.1	24.7	26.8	29.7	
1078	47.8	48.2	45.2	41.4	38.1	35.2	32.7	30.1	24.3	26.8	30.1	
(d)	45.2	44.8	43.5	41.0	38.5	36.0	33.5	31.0	23.9	26.0	28.5	
1524	46.0	45.8	45.0	42.6	40.1	37.4	34.4	30	26.6	27.2	...	
4037	...	48.2	45.6	...	39.4	...	33.9	...	...	...	...	
4130(e)	...	42.7	...	40.6	...	37.3	...	31.0	...	28.1	30.1	
4140	...	42.7	42.3	...	37.7	...	33.1	...	...	...	...	
5132	48.6	46.5	44.4	42.3	38.5	35.6	31.8	28.9	26.0	28.1	30.1	
5140	...	44.8	43.5	...	37.7	...	31.4	...	...	...	...	
(d)	42.7	42.7	41.9	40.6	38.9	36.4	33.9	31.0	26.4	28.1	30.1	
(d)	36.4	37.7	38.9	39.4	36.8	35.2	32.7	26.4	25.1	27.6	30.1	
(d)	33.1	33.9	35.2	35.6	35.6	33.5	30.6	28.1	26.8	28.5	30.1	
(d)	13.0	13.8	16.3	18.0	19.3	20.5	21.8	22.6	23.4	25.5	28.1	
18Ni250(e)	19.7(b)	20.9	...	...	...	...	...	...	...	...	...	

Nearest AISI-SAE grade	Chemical composition, %									Treatment or condition
	C	Mn	P	S	Si	Cr	Ni	Mo	Other	
1008	0.08	0.31	...	...	...	0.045	0.07	0.02	...	Not known
1008	0.06	0.4	...	...	...	...	...	...	...	Annealed
1010	0.10	0.42	0.008	0.028	...	...	...	...	...	Not known
1025	0.23	0.64	...	...	...	Trace	0.074	...	0.13 Cu	Annealed
1042	0.42	0.64	...	...	...	Trace	0.063	...	0.12 Cu	Annealed
1078	0.80	0.32	...	...	...	0.11	0.13	0.01	0.07 Cu	Annealed
(d)	1.22	0.35	...	...	...	0.11	0.13	0.01	0.08 Cu	Annealed
1524	0.23	1.51	0.037	0.038	0.12	0.06	0.04	0.025	0.105 Cu; 0.033 Co; 0.015 Al	Annealed
4037	0.37	1.56	...	...	...	...	...	0.26	...	Hardened and tempered
4130(e)	0.3	0.5	...	...	0.3	0.95	...	0.5	...	Hardened and tempered
4140	0.41	0.67	...	...	...	1.01	...	0.23	...	Hardened and tempered
5132	0.32	0.69	...	...	...	1.09	0.073	0.012	0.07 Cu	Annealed
5140	0.39	0.79	...	...	...	1.03	...	...	...	Hardened and tempered
(d)	0.35	0.59	...	...	...	0.88	0.26	0.20	0.12 Cu	Not known
(d)	0.33	0.55	...	...	...	0.17	3.47	0.04	0.09 Cu	Not known
(d)	0.34	0.55	...	...	...	0.78	3.53	0.39	0.05 Cu	Hardened and tempered
(d)	1.22	13.0	...	...	0.22	0.03	0.07	...	0.07 Cu	Not known
18Ni250(e)	0.026	0.1	...	...	0.11	...	18.5	4.5	7.0 Co; 0.22 Ti; 0.003 B	Not known

(a) To obtain conductivities in BTU/(ft · h · °F), multiply values in table by 0.5778; to obtain conductivities in cal/(cm · s · °C), multiply by 0.002388. (b) Thermal conductivity at 21 °C (70 °F). (c) 70.4 W/m · K at –100 °C (–148 °F). (d) No equivalent grade. (e) Nominal composition

Mechanical Properties

This guide shows what to expect from a given grade of steel in the indicated condition. Data were obtained from specimens 0.505 in. in diameter which were machined from 1-in. rounds; gage lengths were 2 in. Average properties of hot rolled, normalized, and annealed material are listed, while properties of quenched and tempered grades are for single heats. Sources of the data are Bethlehem Steel Corp. and Republic Steel Corp.

Because of the many variables that affect a steel's properties, however, these listed properties should not be considered either as average or typical. Both strengths and ductilities may range up and down from the values given, depending on the compositions of individual heats of the same grade, section sizes, and internal structures. Properties of carbon steels and many alloy steels are also affected by residual elements (particularly nickel, chromium and molybdenum), even though their amounts are limited to maximums by AISI and SAE specifications.

Fine-grained steels normally have better impact strength than coarse-grained types, a factor which should be considered when reviewing the results of Izod tests. Hardness values are not always related to corresponding tensile strengths. In particular, this effect occurs with carbon steels because they are shallow-hardening. Hardness tests were made on surfaces, and these hardnesses will not reflect the tensile strengths obtained with specimens representing bar centers. (Center hardnesses are usually lower than surface hardnesses.)

Hot rolled properties for alloy steels are not given because these grades are customarily heat treated. Because the samples were small enough to ensure full quenching, values indicate strengths and ductilities which may be obtained with hardened, fine-grained steels of a similar section size at room temperatures.

Mechanical properties of selected carbon and alloy steels in the hot rolled, normalized, and annealed condition

AISI no.(a)	Treatment	Austenitizing temperature		Tensile strength		Yield strength		Elongation, %	Reduction in area, %	Hardness, HB	Izod impact strength	
		°C	°F	MPa	ksi	MPa	ksi				J	ft · lb
1015	As-rolled	...	...	420.6	61.0	313.7	45.5	39.0	61.0	126	110.5	81.5
	Normalized	925	1700	424.0	61.5	324.1	47.0	37.0	69.6	121	115.5	85.2
	Annealed	870	1600	386.1	56.0	284.4	41.3	37.0	69.7	111	115.0	84.8
1020	As-rolled	...	...	448.2	65.0	330.9	48.0	36.0	59.0	143	86.8	64.0
	Normalized	870	1600	441.3	64.0	346.5	50.3	35.8	67.9	131	117.7	86.8
	Annealed	870	1600	394.7	57.3	294.8	42.8	36.5	66.0	111	123.4	91.0
1022	As-rolled	...	...	503.3	73.0	358.5	52.0	35.0	67.0	149	81.3	60.0
	Normalized	925	1700	482.6	70.0	358.5	52.0	34.0	67.5	143	117.3	86.5
	Annealed	870	1600	429.2	62.3	317.2	46.0	35.0	63.6	137	120.7	89.0
1030	As-rolled	...	...	551.6	80.0	344.7	50.0	32.0	57.0	179	74.6	55.0
	Normalized	925	1700	520.6	75.5	344.7	50.0	32.0	60.8	149	93.6	69.0
	Annealed	845	1550	463.7	67.3	341.3	49.5	31.2	57.9	126	69.4	51.2
1040	As-rolled	...	...	620.5	90.0	413.7	60.0	25.0	50.0	201	48.8	36.0
	Normalized	900	1650	589.5	85.5	374.0	54.3	28.0	54.9	170	65.1	48.0
	Annealed	790	1450	518.8	75.3	353.4	51.3	30.2	57.2	149	44.3	32.7
1050	As-rolled	...	...	723.9	105.0	413.7	60.0	20.0	40.0	229	31.2	23.0
	Normalized	900	1650	748.1	108.5	427.5	62.0	20.0	39.4	217	27.1	20.0
	Annealed	790	1450	636.0	92.3	365.4	53.0	23.7	39.9	187	16.9	12.5
1060	As-rolled	...	...	813.6	118.0	482.6	70.0	17.0	34.0	241	17.6	13.0
	Normalized	900	1650	775.7	112.5	420.6	61.0	18.0	37.2	229	13.2	9.7
	Annealed	790	1450	625.7	90.8	372.3	54.0	22.5	38.2	179	11.3	8.3
1080	As-rolled	...	...	965.3	140.0	586.1	85.0	12.0	17.0	293	6.8	5.0
	Normalized	900	1650	1010.1	146.5	524.0	76.0	11.0	20.6	293	6.8	5.0
	Annealed	790	1450	615.4	89.3	375.8	54.5	24.7	45.0	174	6.1	4.5
1095	As-rolled	...	...	965.3	140.0	572.3	83.0	9.0	18.0	293	4.1	3.0
	Normalized	900	1650	1013.5	147.0	499.9	72.5	9.5	13.5	293	5.4	4.0
	Annealed	790	1450	656.7	95.3	379.2	55.0	13.0	20.6	192	2.7	2.0
1117	As-rolled	...	...	486.8	70.6	305.4	44.3	33.0	63.0	143	81.3	60.0
	Normalized	900	1650	467.1	67.8	303.4	44.0	33.5	63.8	137	85.1	62.8
	Annealed	855	1575	429.5	62.3	279.2	40.5	32.8	58.0	121	93.6	69.0

(a) All grades are fine grained except for those in the 1100 series which are coarse-grained. Heat treated specimens were oil quenched unless otherwise indicated

(continued)

Mechanical properties of selected carbon and alloy steels in the hot rolled, normalized, and annealed condition (continued)

AISI no.(a)	Treatment	Austenitizing temperature		Tensile strength		Yield strength		Elongation, %	Reduction in area, %	Hardness, HB	Izod impact strength	
		°C	°F	MPa	ksi	MPa	ksi				J	ft·lb
1118	As-rolled	...	...	521.2	75.6	316.5	45.9	32.0	70.0	149	108.5	80.0
	Normalized	925	1700	477.8	69.3	319.2	46.3	33.5	65.9	143	103.4	76.3
	Annealed	790	1450	450.2	65.3	284.8	41.3	34.5	66.8	131	106.4	78.5
1137	As-rolled	...	...	627.4	91.0	379.2	55.0	28.0	61.0	192	82.7	61.0
	Normalized	900	1650	668.8	97.0	396.4	57.5	22.5	48.5	197	63.7	47.0
	Annealed	790	1450	584.7	84.8	344.7	50.0	26.8	53.9	174	49.9	36.8
1141	As-rolled	...	...	675.7	98.0	358.5	52.0	22.0	38.0	192	11.1	8.2
	Normalized	900	1650	706.7	102.5	405.4	58.8	22.7	55.5	201	52.6	38.8
	Annealed	815	1500	598.5	86.8	353.0	51.2	25.5	49.3	163	34.3	25.3
1144	As-rolled	...	...	703.3	102.0	420.6	61.0	21.0	41.0	212	52.9	39.0
	Normalized	900	1650	667.4	96.8	399.9	58.0	21.0	40.4	197	43.4	32.0
	Annealed	790	1450	584.7	84.8	346.8	50.3	24.8	41.3	167	65.1	48.0
1340	Normalized	870	1600	836.3	121.3	558.5	81.0	22.0	62.9	248	92.5	68.2
	Annealed	800	1475	703.3	102.0	436.4	63.3	25.5	57.3	207	70.5	52.0
3140	Normalized	870	1600	891.5	129.3	599.8	87.0	19.7	57.3	262	53.6	39.5
	Annealed	815	1500	689.5	100.0	422.6	61.3	24.5	50.8	197	46.4	34.2
4130	Normalized	870	1600	668.8	97.0	436.4	63.3	25.5	59.5	197	86.4	63.7
	Annealed	865	1585	560.5	81.3	360.6	52.3	28.2	55.6	156	61.7	45.5
4140	Normalized	870	1600	1020.4	148.0	655.0	95.0	17.7	46.8	302	22.6	16.7
	Annealed	815	1500	655.0	95.0	417.1	60.5	25.7	56.9	197	54.5	40.2
4150	Normalized	870	1600	1154.9	167.5	734.3	106.5	11.7	30.8	321	11.5	8.5
	Annealed	815	1500	729.5	105.8	379.2	55.0	20.2	40.2	197	24.7	18.2
4320	Normalized	895	1640	792.9	115.0	464.0	67.3	20.8	50.7	235	72.9	53.8
	Annealed	850	1560	579.2	84.0	424.7	61.6	29.0	58.4	163	109.8	81.0
4340	Normalized	870	1600	1279.0	185.5	861.8	125.0	12.2	36.3	363	15.9	11.7
	Annealed	810	1490	744.6	108.0	472.3	68.5	22.0	49.9	217	51.1	37.7
4620	Normalized	900	1650	574.3	83.3	366.1	53.1	29.0	66.7	174	132.9	98.0
	Annealed	855	1575	512.3	74.3	372.3	54.0	31.3	60.3	149	93.6	69.0
4820	Normalized	860	1580	755.0	109.5	484.7	70.3	24.0	59.2	229	109.8	81.0
	Annealed	815	1500	681.2	98.8	464.0	67.3	22.3	58.8	197	92.9	68.5
5140	Normalized	870	1600	792.9	115.0	472.3	68.5	22.7	59.2	229	38.0	28.0
	Annealed	830	1525	572.3	83.0	293.0	42.5	28.6	57.3	167	40.7	30.0
5150	Normalized	870	1600	870.8	126.3	529.5	76.8	20.7	58.7	255	31.5	23.2
	Annealed	825	1520	675.7	98.0	357.1	51.8	22.0	43.7	197	25.1	18.5
5160	Normalized	855	1575	957.0	138.8	530.9	77.0	17.5	44.8	269	10.8	8.0
	Annealed	815	1495	722.6	104.8	275.8	40.0	17.2	30.6	197	10.0	7.4
6150	Normalized	870	1600	939.8	136.3	615.7	89.3	21.8	61.0	269	35.5	26.2
	Annealed	815	1500	667.4	96.8	412.3	59.8	23.0	48.4	197	27.4	20.2
8620	Normalized	915	1675	632.9	91.8	357.1	51.8	26.3	59.7	183	99.7	73.5
	Annealed	870	1600	536.4	77.8	385.4	55.9	31.3	62.1	149	112.2	82.8
8630	Normalized	870	1600	650.2	94.3	429.5	62.3	23.5	53.5	187	94.6	69.8
	Annealed	845	1550	564.0	81.8	372.3	54.0	29.0	58.9	156	95.2	70.2
8650	Normalized	870	1600	1023.9	148.5	688.1	99.8	14.0	40.4	302	13.6	10.0
	Annealed	795	1465	715.7	103.8	386.1	56.0	22.5	46.4	212	29.4	21.7
8740	Normalized	870	1600	929.4	134.8	606.7	88.0	16.0	47.9	269	17.6	13.0
	Annealed	815	1500	695.0	100.8	415.8	60.3	22.2	46.4	201	40.0	29.5
9255	Normalized	900	1650	932.9	135.3	579.2	84.0	19.7	43.4	269	13.6	10.0
	Annealed	845	1550	774.3	112.3	486.1	70.5	21.7	41.1	229	8.8	6.5
9310	Normalized	890	1630	906.7	131.5	570.9	82.8	18.8	58.1	269	119.3	88.0
	Annealed	845	1550	820.5	119.0	439.9	63.8	17.3	42.1	241	78.6	58.0

(a) All grades are fine grained except for those in the 1100 series which are coarse-grained. Heat treated specimens were oil quenched unless otherwise indicated

Mechanical properties of selected carbon and alloy steels in the quenched and tempered condition

AISI no. (a)	Tempering temperature °C	°F	Tensile strength MPa	ksi	Yield strength MPa	ksi	Elongation, %	Reduction in area, %	Hardness, HB
1030(b)	205	400	848	123	648	94	17	47	495
	315	600	800	116	621	90	19	53	401
	425	800	731	106	579	84	23	60	302
	540	1000	669	97	517	75	28	65	255
	650	1200	586	85	441	64	32	70	207
1040(b)	205	400	896	130	662	96	16	45	514
	315	600	889	129	648	94	18	52	444
	425	800	841	122	634	92	21	57	352
	540	1000	779	113	593	86	23	61	269
	650	1200	669	97	496	72	28	68	201
1040	205	400	779	113	593	86	19	48	262
	315	600	779	113	593	86	20	53	255
	425	800	758	110	552	80	21	54	241
	540	1000	717	104	490	71	26	57	212
	650	1200	634	92	434	63	29	65	192
1050(b)	205	400	1124	163	807	117	9	27	514
	315	600	1089	158	793	115	13	36	444
	425	800	1000	145	758	110	19	48	375
	540	1000	862	125	655	95	23	58	293
	650	1200	717	104	538	78	28	65	235
1050	315	600	979	142	724	105	14	47	321
	425	800	938	136	655	95	20	50	277
	540	1000	876	127	579	84	23	53	262
	650	1200	738	107	469	68	29	60	223
1060	205	400	1103	160	779	113	13	40	321
	315	600	1103	160	779	113	13	40	321
	425	800	1076	156	765	111	14	41	311
	540	1000	965	140	669	97	17	45	277
	650	1200	800	116	524	76	23	54	229
1080	205	400	1310	190	979	142	12	35	388
	315	600	1303	189	979	142	12	35	388
	425	800	1289	187	951	138	13	36	375
	540	1000	1131	164	807	117	16	40	321
	650	1200	889	129	600	87	21	50	255
1095(b)	205	400	1489	216	1048	152	10	31	601
	315	600	1462	212	1034	150	11	33	534
	425	800	1372	199	958	139	13	35	388
	540	100	1138	165	758	110	15	40	293
	650	1200	841	122	586	85	20	47	235
1095	205	400	1289	187	827	120	10	30	401
	315	600	1262	183	813	118	10	30	375
	425	800	1213	176	772	112	12	32	363
	540	1000	1089	158	676	98	15	37	321
	650	1200	896	130	552	80	21	47	269
1137	205	400	1082	157	938	136	5	22	352
	315	600	986	143	841	122	10	33	285
	425	800	876	127	731	106	15	48	262
	540	1000	758	110	607	88	24	62	229
	650	1200	655	95	483	70	28	69	197
1137(b)	205	400	1496	217	1165	169	5	17	415
	315	600	1372	199	1124	163	9	25	375
	425	800	1103	160	986	143	14	40	311
	540	1000	827	120	724	105	19	60	262
	650	1200	648	94	531	77	25	69	187
1141	205	400	1634	237	1213	176	6	17	461
	315	600	1462	212	1282	186	9	32	415
	425	800	1165	169	1034	150	12	47	331
	540	1000	896	130	765	111	18	57	262
	650	1200	710	103	593	86	23	62	217
1144	205	400	876	127	627	91	17	36	277
	315	600	869	126	621	90	17	40	262
	425	800	848	123	607	88	18	42	248
	540	1000	807	117	572	83	20	46	235
	650	1200	724	105	503	73	23	55	217
1330(b)	205	400	1600	232	1455	211	9	39	459
	315	600	1427	207	1282	186	9	44	402

(a) All grades are fine grained except for those in the 1100 series which are coarse-grained. Heat treated specimens were oil quenched unless otherwise indicated. (b) Water quenched

(continued)

Mechanical properties of selected carbon and alloy steels in the quenched and tempered condition (continued)

AISI no. (a)	Tempering temperature °C	°F	Tensile strength MPa	ksi	Yield strength MPa	ksi	Elongation, %	Reduction in area, %	Hardness, HB
	425	800	1158	168	1034	150	15	53	335
	540	1000	876	127	772	112	18	60	263
	650	1200	731	106	572	83	23	63	216
1340	205	400	1806	262	1593	231	11	35	505
	315	600	1586	230	1420	206	12	43	453
	425	800	1262	183	1151	167	14	51	375
	540	1000	965	140	827	120	17	58	295
	650	1200	800	116	621	90	22	66	252
4037	205	400	1027	149	758	110	6	38	310
	315	600	951	138	765	111	14	53	295
	425	800	876	127	731	106	20	60	270
	540	1000	793	115	655	95	23	63	247
	650	1200	696	101	421	61	29	60	220
4042	205	400	1800	261	1662	241	12	37	516
	315	600	1613	234	1455	211	13	42	455
	425	800	1289	187	1172	170	15	51	380
	540	1000	986	143	883	128	20	59	300
	650	1200	793	115	689	100	28	66	238
4130(b)	205	400	1627	236	1462	212	10	41	467
	315	600	1496	217	1379	200	11	43	435
	425	800	1282	186	1193	173	13	49	380
	540	1000	1034	150	910	132	17	57	315
	650	1200	814	118	703	102	22	64	245
4140	205	400	1772	257	1641	238	8	38	510
	315	600	1551	225	1434	208	9	43	445
	425	800	1248	181	1138	165	13	49	370
	540	1000	951	138	834	121	18	58	285
	650	1200	758	110	655	95	22	63	230
4150	205	400	1931	280	1724	250	10	39	530
	315	600	1765	256	1593	231	10	40	495
	425	800	1517	220	1379	200	12	45	440
	540	1000	1207	175	1103	160	15	52	370
	650	1200	958	139	841	122	19	60	290
4340	205	400	1875	272	1675	243	10	38	520
	315	600	1724	250	1586	230	10	40	486
	425	800	1469	213	1365	198	10	44	430
	540	1000	1172	170	1076	156	13	51	360
	650	1200	965	140	855	124	19	60	280
5046	205	400	1744	253	1407	204	9	25	482
	315	600	1413	205	1158	168	10	37	401
	425	800	1138	165	931	135	13	50	336
	540	1000	938	136	765	111	18	61	282
	650	1200	786	114	655	95	24	66	235
50B46	205	400	…	…	…	…	…	…	560
	315	600	1779	258	1620	235	10	37	505
	425	800	1393	202	1248	181	13	47	405
	540	1000	1082	157	979	142	17	51	322
	650	1200	883	128	793	115	22	60	273
50B60	205	400	…	…	…	…	…	…	600
	315	600	1882	273	1772	257	8	32	525
	425	800	1510	219	1385	201	11	34	435
	540	1000	1124	163	1000	145	15	38	350
	650	1200	896	130	779	113	19	50	290
5130	205	400	1613	234	1517	220	10	40	475
	315	600	1496	217	1407	204	10	46	440
	425	800	1275	185	1207	175	12	51	379
	540	1000	1034	150	938	136	15	56	305
	650	1200	793	115	689	100	20	63	245
5140	205	400	1793	260	1641	238	9	38	490
	315	600	1579	229	1448	210	10	43	450
	425	800	1310	190	1172	170	13	50	365
	540	1000	1000	145	862	125	17	58	280
	650	1200	758	110	662	96	25	66	235
5150	205	400	1944	282	1731	251	5	37	525
	315	600	1737	252	1586	230	6	40	475

(a) All grades are fine grained except for those in the 1100 series which are coarse-grained. Heat treated specimens were oil quenched unless otherwise indicated. (b) Water quenched

(continued)

Mechanical properties of selected carbon and alloy steels in the quenched and tempered condition (continued)

AISI no. (a)	Tempering temperature		Tensile strength		Yield strength		Elongation, %	Reduction in area, %	Hardness, HB
	°C	°F	MPa	ksi	MPa	ksi			
	425	800	1448	210	1310	190	9	47	410
	540	1000	1124	163	1034	150	15	54	340
	650	1200	807	117	814	118	20	60	270
5160	205	400	2220	322	1793	260	4	10	627
	315	600	1999	290	1772	257	9	30	555
	425	800	1606	233	1462	212	10	37	461
	540	1000	1165	169	1041	151	12	47	341
	650	1200	896	130	800	116	20	56	269
51B60	205	400	...	...	...	...	...	...	600
	315	600	...	...	...	...	...	...	540
	425	800	1634	237	1489	216	11	36	460
	540	1000	1207	175	1103	160	15	44	355
	650	1200	965	140	869	126	20	47	290
6150	205	400	1931	280	1689	245	8	38	538
	315	600	1724	250	1572	228	8	39	483
	425	800	1434	208	1331	193	10	43	420
	540	1000	1158	168	1069	155	13	50	345
	650	1200	945	137	841	122	17	58	282
81B45	205	400	2034	295	1724	250	10	33	550
	315	600	1765	256	1572	228	8	42	475
	425	800	1407	204	1310	190	11	48	405
	540	1000	1103	160	1027	149	16	53	338
	650	1200	896	130	793	115	20	55	280
8630	205	400	1641	238	1503	218	9	38	465
	315	600	1482	215	1392	202	10	42	430
	425	800	1276	185	1172	170	13	47	375
	540	1000	1034	150	896	130	17	54	310
	650	1200	772	112	689	100	23	63	240
8640	205	400	1862	270	1669	242	10	40	505
	315	600	1655	240	1517	220	10	41	460
	425	800	1379	200	1296	188	12	45	400
	540	1000	1103	160	1034	150	16	54	340
	650	1200	896	130	800	116	20	62	280
86B45	205	400	1979	287	1641	238	9	31	525
	315	600	1696	246	1551	225	9	40	475
	425	800	1379	200	1317	191	11	41	395
	540	1000	1103	160	1034	150	15	49	335
	650	1200	903	131	876	127	19	58	280
8650	205	400	1937	281	1675	243	10	38	525
	315	600	1724	250	1551	225	10	40	490
	425	800	1448	210	1324	192	12	45	420
	540	1000	1172	170	1055	153	15	51	340
	650	1200	965	140	827	120	20	58	280
8660	205	400	...	...	...	...	...	...	580
	315	600	...	...	...	...	...	...	535
	425	800	1634	237	1551	225	13	37	460
	540	1000	1310	190	1213	176	17	46	370
	650	1200	1068	155	951	138	20	53	315
8740	205	400	1999	290	1655	240	10	41	578
	315	600	1717	249	1551	225	11	46	495
	425	800	1434	208	1358	197	13	50	415
	540	1000	1207	175	1138	165	15	55	363
	650	1200	986	143	903	131	20	60	302
9255	205	400	2103	305	2048	297	1	3	601
	315	600	1937	281	1793	260	4	10	578
	425	800	1606	233	1489	216	8	22	477
	540	1000	1255	182	1103	160	15	32	352
	650	1200	993	144	814	118	20	42	285
9260	205	400	...	...	...	...	...	...	600
	315	600	...	...	...	...	...	...	540
	425	800	1758	255	1503	218	8	24	470
	540	1000	1324	192	1131	164	12	30	390
	650	1200	979	142	814	118	20	43	295
94B30	205	400	1724	250	1551	225	12	46	475
	315	600	1600	232	1420	206	12	49	445
	425	800	1344	195	1207	175	13	57	382
	540	1000	1000	145	931	135	16	65	307
	650	1200	827	120	724	105	21	69	250

(a) All grades are fine grained except for those in the 1100 series which are coarse-grained. Heat treated specimens were oil quenched unless otherwise indicated. (b) Water quenched

Monotonic and cyclic stress-strain properties of selected steels

Grade(a)	Orientation(b)	Description(c)	Hardness, HB	Tensile strength MPa	ksi	Reduction in area, %	True strain at fracture	Modulus of elasticity GPa	10⁴ psi	Fatigue strength coefficient, σ′(d) MPa	ksi	Fatigue strength exponent, b(d)	Fatigue ductility coefficient, ε′f(d)	Fatigue ductility exponent, c(d)
A538A(c)	L	STA	405	1515	220	67	1.10	185	27	1655	240	−0.065	0.30	−0.62
A538B(e)	L	STA	460	1860	270	56	0.82	185	27	2135	310	−0.071	0.80	−0.71
A538C(e)	L	STA	480	2000	290	55	0.81	180	26	2240	325	−0.07	0.60	−0.75
AM-350(f)	L	HR, A		1315	191	52	0.74	195	28	2800	406	−0.14	0.33	−0.84
AM-350(f)	L	CD	496	1905	276	20	0.23	180	26	2690	390	−0.102	0.10	−0.42
Gainex(f)	LT	HR sheet		530	77	58	0.86	200	29.2	805	117	−0.07	0.86	−0.65
Gainex(f)	L	HR sheet		510	74	64	1.02	200	29.2	805	117	−0.071	0.86	−0.65
H-11	L	Ausformed	660	2585	375	33	0.40	205	30	3170	460	−0.077	0.08	−0.74
RQC-100(f)	LT	HR plate	290	940	136	43	0.56	205	30	1240	180	−0.07	0.66	−0.69
RQC-100(f)	L	HR plate	290	930	135	67	1.02	205	30	1240	180	−0.07	0.66	−0.69
10B62	L	Q&T	430	1640	238	38	0.89	195	28	1780	258	−0.067	0.32	−0.56
1005-1009	LT	HR sheet	90	360	52	73	1.3	205	30	580	84	−0.09	0.15	−0.43
1005-1009	LT	CD sheet	125	470	68	66	1.09	205	30	515	75	−0.059	0.30	−0.51
1005-1009	L	CD sheet	125	415	60	64	1.02	200	29	540	78	−0.073	0.11	−0.41
1005-1009	L	HR sheet	90	345	50	80	1.6	200	29	640	93	−0.109	0.10	−0.39
1015	L	Normalized	80	415	60	68	1.14	205	30	825	120	−0.11	0.95	−0.64
1020	L	HR plate	108	440	64	62	0.96	205	30	895	130	−0.12	0.41	−0.51
1040	L	As forged	225	620	90	60	0.93	200	29	1540	223	−0.14	0.61	−0.57
1045	L	Q&T	225	725	105	65	1.04	200	29	1225	178	−0.095	1.00	−0.66
1045	L	Q&T	410	1450	210	51	0.72	200	29	1860	270	−0.073	0.60	−0.70
1045	L	Q&T	390	1345	195	59	0.89	205	30	1585	230	−0.074	0.45	−0.68
1045	L	Q&T	450	1585	230	55	0.81	205	30	1795	260	−0.07	0.35	−0.69
1045	L	Q&T	500	1825	265	51	0.71	205	30	2275	330	−0.08	0.25	−0.68
1045	L	Q&T	595	2240	325	41	0.52	205	30	2725	395	−0.081	0.07	−0.60
1144	L	CDSR	265	930	135	33	0.51	195	28.5	1000	145	−0.08	0.32	−0.58
1144	L	DAT	305	1035	150	25	0.29	200	28.8	1585	230	−0.09	0.27	−0.53
1541F	L	Q&T forging	290	950	138	49	0.68	205	29.9	1275	185	−0.076	0.68	−0.65
1541F	L	Q&T forging	260	890	129	60	0.93	205	29.9	1275	185	−0.071	0.93	−0.65
4130	L	Q&T	258	895	130	67	1.12	220	32	1275	185	−0.083	0.92	−0.63
4130	L	Q&T	365	1425	207	55	0.79	200	29	1695	246	−0.081	0.89	−0.69
4140	L	Q&T, DAT	310	1075	156	60	0.69	200	29.2	1825	265	−0.08	1.2	−0.59
4142	L	DAT	310	1060	154	29	0.35	205	30	1450	210	−0.10	0.22	−0.51
4142	L	DAT	335	1250	181	28	0.34	200	28.9	1250	181	−0.08	0.06	−0.62
4142	L	Q&T	380	1415	205	48	0.66	205	29	1825	265	−0.08	0.45	−0.75
4142	L	Q&T and deformed	400	1550	225	47	0.63	200	29	1895	275	−0.09	0.50	−0.75
4142	L	Q&T	450	1760	255	42	0.54	200	29	2000	290	−0.08	0.40	−0.73
4142	L	Q&T and deformed	475	2035	295	20	0.22	200	29	2070	300	−0.082	0.20	−0.77
4142	L	Q&T and deformed	450	1930	280	37	0.46	200	29	2105	305	−0.09	0.60	−0.76
4142	L	Q&T	475	1930	280	35	0.43	205	30	2170	315	−0.081	0.09	−0.61
4142	L	Q&T	560	2240	325	27	0.31	205	30	2655	385	−0.089	0.07	−0.76
4340	L	HR, A	243	825	120	43	0.57	195	28	1200	174	−0.095	0.45	−0.54
4340	L	Q&T	409	1470	213	38	0.48	200	29	2000	290	−0.091	0.48	−0.60
4340	L	Q&T	350	1240	180	57	0.84	195	28	1655	240	−0.076	0.73	−0.62
5160	L	Q&T	430	1670	242	42	0.87	195	28	1930	280	−0.071	0.40	−0.57
52100	L	SH, Q&T	518	2015	292	11	0.12	205	30	2585	375	−0.09	0.18	−0.56
9262	L	A	260	925	134	14	0.16	205	30	1040	151	−0.071	0.16	−0.47
9262	L	Q&T	280	1000	145	33	0.41	195	28	1220	177	−0.073	0.41	−0.60
9262	L	Q&T	410	1565	227	32	0.38	200	29	1855	269	−0.057	0.38	−0.65
950C(g)	LT	HR plate	159	565	82	64	1.03	205	29.6	1170	170	−0.12	0.95	−0.61
950C(g)	L	HR bar	150	565	82	69	1.19	205	30	970	141	−0.11	0.85	−0.59
950X(g)	L	Plate channel	150	440	64	65	1.06	205	30	625	91	−0.075	0.35	−0.54
950X(g)	L	HR plate	156	530	77	72	1.24	205	29.5	1005	146	−0.10	0.85	−0.61
980X(g)	L	Plate channel	225	695	101	68	1.15	195	28.2	1055	153	−0.08	0.21	−0.53

(a) AISI/SAE grade, unless otherwise indicated. (b) Orientation of axis of specimen, relative to rolling direction; L is longitudinal (parallel to rolling direction); LT is long transverse (perpendicular to rolling direction). (c) STA, solution treated and aged; HR, hot rolled; CD, cold drawn; Q&T, quenched and tempered; CDSR, cold drawn strain relieved; DAT, drawn at temperature; A, annealed. (d) For use in the Manson-Coffin-Basquin strain life relationship: $\Delta\varepsilon = (\sigma'_F)\,(N_f)^b + \varepsilon'_f\,(N_f)^c$ where $\Delta\varepsilon$ is strain range, E is modulus of elasticity, and N_f is number of cycles to failure. (e) ASTM designation. (f) Proprietary designation. (g) SAE HSLA grade

Low-temperature impact properties of quenched and tempered alloy steels(a)

AISI no.	Composition, %					Heat treatment			
						Quenching temperature(b)		Tempering temperature	
	C	Mn	Ni	Cr	Mo	°C	°F	°C	°F
4320	0.21	0.74	1.53	1.09	0.19	900	1650	205	400
								540	1000
								650	1200
4330	0.30	0.84	1.69	1.10	0.20	855	1575	205	400
								425	800
								540	1000
								650	1200
4340	0.38	0.77	1.65	0.93	0.21	845	1550	205	400
								425	800
								540	1000
								650	1200
4360(c)	0.57	0.87	1.62	1.08	0.22	800	1475	425	800
								650	1200
4380(c)	0.76	0.91	1.67	1.11	0.21	790	1450	425	800
								540	1000
								650	1200
8620	0.20	0.89	0.60	0.68	0.20	900	1650	150	300
								425	800
								650	1200
8630	0.34	0.77	0.66	0.62	0.22	855	1575	425	800
								540	1000
								650	1200
8640	0.45	0.78	0.65	0.61	0.20	845	1550	425	800
								540	1000
								650	1200
8660(c)	0.56	0.81	0.70	0.56	0.25	800	1475	425	800
								540	1000
								650	1200
8680(c)	0.76	0.81	0.67	0.60	0.22	790	1450	425	800
								540	1000
								650	1200

Hardness, HRC	Impact energy(d)										Transition temperature (50% brittle)	
	−185 °C (−300 °F)		−130 °C (−200 °F)		−75 °C (−100 °F)		−20 °C (0 °F)		40 °C (100 °F)			
	J	ft · lb	J	ft · lb	J	ft · lb	J	ft · lb	J	ft · lb	°C	°F
43	12	9	23	17	34	25	37	27	38	28	...	...
33	11	8	22	16	38	28	54	40	61	45	−55	−65
21	26	19	47	35	108	80	122	90	122	90	−115	−175
50	7	5	19	14	22	16	27	20	27	20	...	...
43	8	6	14	10	19	14	26	19	28	21	20	70
35	23	17	23	17	31	23	41	30	46	34	−80	−110
27	23	17	47	35	61	45	72	53	76	56	−120	−185
52	15	11	20	15	27	20	28	21	28	21	...	...
44	12	9	18	13	22	16	28	21	34	25	...	...
38	20	15	24	18	38	28	47	35	49	36	−90	−130
30	20	15	38	28	75	55	75	55	75	55	−120	−185
48	7	5	7	5	14	10	15	11	19	14	...	...
30	16	12	20	15	34	25	57	42	58	43	−80	−110
49	5	4	7	5	11	8	12	9	14	10	...	...
42	11	8	11	8	14	10	16	12	20	15	15	60
31	7	5	15	11	26	19	45	33	52	38	−45	−50
43	15	11	22	16	31	23	47	35	47	35	...	...
36	11	8	18	13	27	20	47	35	61	45	−30	−20
21	14	10	115	85	145	107	156	115	159	117	−125	−195
41	9	7	16	12	23	17	34	25	42	31	−20	0
34	15	11	27	20	58	43	72	53	73	54	−105	−155
27	24	18	38	28	100	74	108	80	111	82	−110	−165
46	7	5	14	10	19	14	27	20	31	23	...	...
38	15	11	20	15	33	24	54	40	54	40	−80	−110
30	24	18	30	22	66	49	85	63	89	66	−95	−140
47	5	4	8	6	14	10	18	13	22	16	...	...
41	14	10	16	12	20	15	27	20	41	30	−25	−10
30	22	16	24	18	34	25	73	54	81	60	−70	−90
50	4	3	5	4	7	5	12	9	14	10	...	...
42	5	4	7	5	14	10	19	14	23	17	...	...
32	4	3	8	6	15	11	34	25	54	40	−35	−30

(a) Induction furnace laboratory heats normalized before hardening and tempering. (b) Specimens 1.14-cm- (0.45-in.-) square bars were quenched in oil. (c) Higher than standard carbon content. (d) Charpy V-notch values scaled from curves

Typical mechanical properties of normalized alloy steel sheet

Grade	Thickness		Tensile strength		Yield strength(a)		Elonga-tion(b), %	Hardness, HRC
	mm	in.	MPa	ksi	MPa	ksi		
4130	4.9	0.193	835	121	585	85	14	25
4335(c)	4.6	0.180	1725	250	1240	180	8	48
4340(c)	2.0	0.080	1860	270	1345	195	7	50

(a) At 0.2% offset. (b) In 50 mm or 2 in. (c) Modified: 0.40% Mo, 0.20% V

Approximate upper limit of strength at which selected steels are currently used

Type	Yield strength		Tensile strength		Applications
	MPa	ksi	MPa	ksi	
Low-alloy steel	1720	250	2070	300	Landing gear components
5Cr-Mo-V	1650	240	2000	90	Airframe structures
Stainless steel Martensitic	1100	160	1240	180	Bolts and rivets
Age hardenable Martensitic	1170	170	1310	190	Clamps and brackets
Extra-low-carbon, age hardenable martensitic	1690	245	1770	257	...
Cold-rolled austenitic	1240	180	1380	200	Missile tankage for liquid propellant
Semi-austenitic	1520	220	1620	235	Honeycomb structures
High-alloy steels 18% Ni maraging	1790	259	2070	300	Fasteners and shafts
9Ni-4Co	1650	240	1930	280	...
Matrix	2000	290	2480	360	Axles and shafting

Heat Treating

Effect of mass on hardness of normalized carbon and alloy steels

Grade	Normalizing temperature		Hardness, HB, for bar with diameter, mm (in.), of:			
	°C	°F	13(1/2)	25(1)	50(2)	100(4)
Carbon steels, carburizing grades						
1015	925	1700	126	121	116	116
1020	925	1700	131	131	126	121
1022	925	1700	143	143	137	131
1117	900	1650	143	137	137	126
1118	925	1700	156	143	137	131
Carbon steels, direct-hardening grades						
1030	925	1700	156	149	137	137
1040	900	1650	183	170	167	167
1050	900	1650	223	217	212	201
1060	900	1650	229	229	223	223
1080	900	1650	293	293	285	269
1095	900	1650	302	293	269	255
1137	900	1650	201	197	197	192
1141	900	1650	207	201	201	201
1144	900	1650	201	197	192	192
Alloy steels, carburizing grades						
3310	890	1630	269	262	262	248
4118	910	1670	170	156	143	137
4320	895	1640	248	235	212	201
4419	955	1750	149	143	143	143
4620	900	1650	192	174	167	163
4820	860	1580	235	229	223	212
8620	915	1675	197	183	179	163
9310	890	1630	285	269	262	255
Alloy steels, direct-hardening grades						
1340	870	1600	269	248	235	235
3140	870	1600	302	262	248	241
4027	905	1660	179	179	163	156
4063	870	1600	285	285	285	277
4130	870	1600	217	197	167	163
4140	870	1600	302	302	285	241
4150	870	1600	375	321	311	293
4340	870	1600	388	363	341	321
5140	870	1600	235	229	223	217
5150	870	1600	262	255	248	241
5160	860	1575	285	269	262	255
6150	870	1600	285	269	262	255
8630	870	1600	201	187	187	187
8650	870	1600	363	302	293	285
8740	870	1600	269	269	262	255
9255	900	1650	277	269	269	269

Typical normalizing temperatures for standard carbon and alloy steels

Based on production experience, normalizing temperature may vary from as much as 27 °C (50 °F) below to as much as 55 °C (100 °F) above indicated temperature. The steel should be cooled in still air from indicated temperature

Grade	Temperature °C	Temperature °F	Grade	Temperature °C	Temperature °F
Plain carbon steels			**Standard alloy steels (continued)**		
1015	915	1675	4817	925	1700
1020	915	1675	4820	925	1700
1022	915	1675	5046	870	1600
1025	900	1650	5120	925	1700
1030	900	1650	5130	900	1650
1035	885	1625	5132	900	1650
1040	860	1575	5135	870	1600
1045	860	1575	5140	870	1600
1050	860	1575	5145	870	1600
1060	830	1525	5147	870	1600
1080	830	1525	5150	870	1600
1090	830	1525	5155	870	1600
1095	845	1550	5160	870	1600
1117	900	1650	6118	925	1700
1137	885	1625	6120	925	1700
1141	860	1575	6150	900	1650
1144	860	1575	8617	925	1700
Standard alloy steels			8620	925	1700
1330	900	1650	8622	925	1700
1335	870	1600	8625	900	1650
1340	870	1600	8627	900	1650
3135	870	1600	8630	900	1650
3140	870	1600	8637	870	1600
3310	925	1700	8640	870	1600
4027	900	1650	8642	870	1600
4028	900	1650	8645	870	1600
4032	900	1650	8650	870	1600
4037	870	1600	8655	870	1600
4042	870	1600	8660	870	1600
4047	870	1600	8720	925	1700
4063	870	1600	8740	925	1700
4118	925	1700	8742	870	1600
4130	900	1650	8822	925	1700
4135	870	1600	9255	900	1650
4137	870	1600	9260	900	1650
4140	870	1600	9262	900	1650
4142	870	1600	9310	925	1700
4145	870	1600	9840	870	1600
4147	870	1600	9850	870	1600
4150	870	1600	50B40	870	1600
4320	925	1700	50B44	870	1600
4337	870	1600	50B46	870	1600
4340	870	1600	50B50	870	1600
4520	925	1700	60B60	870	1600
4620	925	1700	81B45	870	1600
4621	925	1700	86B45	870	1600
4718	925	1700	94B15	925	1700
4720	925	1700	94B17	925	1700
4815	925	1700	94B30	900	1650
			94B40	900	1650

Approximate critical temperatures for selected carbon and low-alloy steels

Steel	Critical temperatures on heating at 28 °C/h (50 °F/h)				Critical temperatures on cooling at 28 °C/h (50 °F/h)			
	Ac$_1$		Ac$_3$		Ar$_3$		Ar$_1$	
	°C	°F	°C	°F	°C	°F	°C	°F
1010	724	1335	877	1610	849	1560	682	1260
1020	724	1335	846	1555	816	1500	682	1260
1030	727	1340	813	1495	788	1450	677	1250
1040	727	1340	793	1460	757	1395	671	1240
1050	727	1340	768	1415	741	1365	682	1260
1060	727	1340	746	1375	727	1340	685	1265
1070	727	1340	732	1350	710	1310	691	1275
1080	729	1345	735	1355	699	1290	693	1280
1340	716	1320	777	1430	721	1330	621	1150
3140	735	1355	766	1410	721	1330	660	1220
4027	727	1340	807	1485	760	1400	671	1240
4042	727	1340	793	1460	732	1350	654	1210
4130	757	1395	810	1490	754	1390	693	1280
4140	732	1350	804	1480	743	1370	679	1255
4150	743	1370	766	1410	729	1345	671	1240
4340	724	1335	774	1425	710	1310	654	1210
4615	727	1340	810	1490	760	1400	649	1200
5046	716	1320	771	1420	732	1350	682	1260
5120	766	1410	838	1540	799	1470	699	1290
5140	738	1360	788	1450	727	1340	693	1280
5160	710	1310	766	1410	716	1320	677	1250
52100	727	1340	768	1415	716	1320	688	1270
6150	749	1380	788	1450	743	1370	693	1280
8115	721	1330	838	1540	788	1450	671	1240
8620	732	1350	829	1525	768	1415	660	1220
8640	732	1350	779	1435	727	1340	666	1230
9260	743	1370	816	1500	749	1380	713	1315

Recommended temperatures and cooling cycles for full annealing of small carbon steel forgings

Data are for forgings up to 75 mm (3 in.) in section thickness. Time at temperature usually is a minimum of 1 h for sections up to 25 mm (1 in.) thick; 1/2 h is added for each additional 25 mm (1 in.) of thickness

Steel	Annealing temperature		Cooling cycle(a)				Hardness range, HB
	°C	°F	°C		°F		
			From	To	From	To	
1018	855-900	1575-1650	855	705	1575	1300	111-149
1020	855-900	1575-1650	855	700	1575	1290	111-149
1022	855-900	1575-1650	855	700	1575	1290	111-149
1025	855-900	1575-1650	855	700	1575	1290	111-187
1030	845-885	1550-1625	845	650	1550	1200	126-197
1035	845-885	1550-1625	845	650	1550	1200	137-207
1040	790-870	1450-1600	790	650	1450	1200	137-207
1045	790-870	1450-1600	790	650	1450	1200	156-217
1050	790-870	1450-1600	790	650	1450	1200	156-217
1060	790-845	1450-1550	790	650	1450	1200	156-217
1070	790-845	1450-1550	790	650	1450	1200	167-229
1080	790-845	1450-1550	790	650	1450	1200	167-229
1090	790-830	1450-1525	790	650	1450	1200	167-229
1095	790-830	1450-1525	790	655	1450	1215	167-229

(a) Furnace cooling at 28 °C/h (50 °F/h)

Recommended annealing temperatures for alloy steels (furnace cooling)

Steel	Annealing temperature °C	°F	Hardness (max), HB	Steel	Annealing temperature °C	°F	Hardness (max), HB
1330	845-900	1550-1650	179	5140	815-870	1500-1600	187
1335	845-900	1550-1650	187	5145	815-870	1500-1600	197
1340	845-900	1550-1650	192	5147	815-870	1500-1600	197
1345	845-900	1550-1650	…	5150	815-870	1500-1600	201
3140	815-870	1500-1600	187	5155	815-870	1500-1600	217
4037	815-855	1500-1575	183	5160	815-870	1500-1600	223
4042	815-855	1500-1575	192	51B60	815-870	1500-1600	223
4047	790-845	1450-1550	201	50100	730-790	1350-1450	197
4063	790-845	1450-1550	223	51100	730-790	1350-1450	197
4130	790-845	1450-1550	174	52100	730-790	1350-1450	207
4135	790-845	1450-1550	…	6150	845-900	1550-1650	201
4137	790-845	1450-1550	192	81B45	845-900	1550-1650	192
4140	790-845	1450-1550	197	8627	815-870	1500-1600	174
4145	790-845	1450-1550	207	8630	790-845	1450-1550	179
4147	790-845	1450-1550	…	8637	815-870	1500-1600	192
4150	790-845	1450-1550	212	8640	815-870	1500-1600	197
4161	790-845	1450-1550	…	8642	815-870	1500-1600	201
4337	790-845	1450-1550	…	8645	815-870	1500-1600	207
4340	790-845	1450-1550	223	86B45	815-870	1500-1600	207
50B40	815-870	1500-1600	187	8650	815-870	1500-1600	212
50B44	815-870	1500-1600	197	8655	815-870	1500-1600	223
5046	815-870	1500-600	192	8660	815-870	1500-1600	229
50B46	815-870	1500-1600	192	8740	815-870	1500-1600	202
50B50	815-870	1500-1600	201	8742	815-870	1500-1600	…
50B60	815-870	1500-1600	217	9260	815-870	1500-1600	229
5130	790-845	1450-1550	170	94B30	790-845	1450-1550	174
5132	790-845	1450-1550	170	94B40	790-845	1450-1550	192
5135	815-870	1500-1600	174	9840	790-845	1450-1550	207

Recommended temperatures and time cycles for annealing of alloy steels

Steel	Austenitizing temperature °C	°F	Conventional cooling(a) Temperature °C From	To	°F From	To	Cooling rate °C/h	°F/h	Time, h	Isothermal method(b) Cool to °C	°F	Hold, h	Hardness (approx), HB
To obtain a predominantly pearlitic structure(c)													
1340	830	1525	735	610	1350	1130	11	20	11	620	1150	4.5	183
2340	800	1475	655	555	1210	1030	8.3	15	12	595	1100	6	201
2345	800	1475	655	550	1210	1020	9.1	15	12.7	595	1100	6	201
3120(d)	885	1625	…	…	…	…	…	…	…	650	1200	4	179
3140	830	1525	735	650	1350	1200	11	20	7.5	660	1225	6	187
3150	830	1525	705	645	1300	1190	11	20	5.5	660	1225	6	201
3310(e)	870	1600	…	…	…	…	…	…	…	595	1100	14	187
4042	830	1525	745	640	1370	1180	11	20	9.5	660	1225	4.5	197
4047	830	1525	735	630	1350	1170	11	20	9	660	1225	5	207
4062	830	1525	695	630	1280	1170	8.3	15	7.3	660	1225	6	223
4130	855	1575	765	665	1410	1230	20	35	5	675	1250	4	174
4140	845	1550	755	665	1390	1230	14	25	6.4	675	1250	5	197
4150	830	1525	745	670	1370	1240	8.4	15	8.6	675	1250	6	212
4320(d)	885	1625	…	…	…	…	…	…	…	660	1225	6	197
4340	830	1525	705	565	1300	1050	8.3	15	16.5	650	1200	8	223
4620(d)	885	1625	…	…	…	…	…	…	…	650	1200	6	187
4640	830	1525	715	600	1320	1110	7.6	15	15	620	1150	8	197
4820(d)	…	…	…	…	…	…	…	…	…	605	1125	4	192
5045	830	1525	755	665	1390	1230	11	20	8	660	1225	4.5	192
5120(d)	885	1625	…	…	…	…	…	…	…	690	1275	4	179
5132	845	1550	755	670	1390	1240	11	20	7.5	675	1250	6	183
5140	830	1525	740	670	1360	1240	11	20	6	675	1250	6	187
5150	830	1525	705	650	1300	1200	11	20	5	675	1250	6	201
52100(f)	…	…	…	…	…	…	…	…	…	…	…	…	…
6150	830	1525	760	675	1400	1250	8.4	15	10	675	1250	6	201
8620(d)	885	1625	…	…	…	…	…	…	…	660	1225	4	187
8630	845	1550	735	640	1350	1180	11	20	8.5	660	1225	6	192
8640	830	1525	725	640	1340	1180	11	20	8	660	1225	6	197
8650	830	1525	710	650	1310	1200	8.4	15	7.2	650	1200	8	212
8660	830	1525	700	655	1290	1210	8.4	15	8	650	1200	8	229
8720(d)	885	1625	…	…	…	…	…	…	…	660	1225	4	187
8740	830	1525	725	645	1340	1190	11	20	7.5	660	1225	7	201
8750	830	1525	720	630	1330	1170	8.4	15	10.7	660	1225	7	217
9260	860	1575	760	705	1400	1300	8.4	15	6.7	660	1225	6	229
9310(e)	870	1600	…	…	…	…	…	…	…	595	1100	14	187
9840	830	1525	695	640	1280	1180	8.4	15	6.6	650	1200	6	207
9850	830	1525	700	645	1290	1190	8.4	15	6.7	650	1200	8	223
To obtain a predominantly ferritic and spheroidized carbide structure													
1320(d)	805	1480	…	…	…	…	…	…	…	650	1200	8	170
1340	750	1380	735	610	1350	1130	5.5	10	22	640	1180	8	174
2340	715	1320	655	555	1210	1030	5.5	10	18	605	1125	10	192
2345	715	1320	655	550	1210	1020	5.5	10	19	605	1125	10	192
3120(d)	790	1450	…	…	…	…	…	…	…	650	1200	8	163
3140	745	1370	735	650	1350	1200	5.5	10	15	660	1225	10	174
3150	750	1380	705	645	1300	1190	5.5	10	11	660	1225	10	187
9840	745	1370	695	640	1280	1180	5.5	10	11	650	1200	10	192
9850	745	1370	700	645	1290	1190	5.5	10	11	650	1200	12	207

(a) The steel is cooled in the furnace at the indicated rate through the temperature range shown. (b) The steel is cooled rapidly to the temperature indicated and is held at that temperature for the time specified. (c) In isothermal annealing to obtain pearlitic structure, steels may be austenitized at temperatures up to 70 °C (125 °F) higher than temperatures listed. (d) Seldom annealed. Structures of better machinability are developed by normalizing or by transforming isothermally after rolling or forging. (e) Annealing is impractical by the conventional process of continuous slow cooling. The lower transformation temperature is markedly depressed, and excessively long cooling cycles are required to obtain transformation to pearlite. (f) Predominantly pearlitic structures are seldom desired in this steel

Austenitizing temperatures for direct-hardening carbon and alloy steels (SAE)

Steel	Temperature		Steel	Temperature	
	°C	°F		°C	°F
Carbon steels			**Alloy steels** (continued)		
1025	855-900	1575-1650	4130	815-870	1500-1600
1030	845-870	1550-1600	4135	845-870	1550-1600
1035	830-855	1525-1575	4137	845-870	1550-1600
1037	830-855	1525-1575	4140	845-870	1550-1600
1038(a)	830-855	1525-1575	4142	845-870	1550-1600
1039(a)	830-855	1525-1575	4145	815-845	1500-1550
1040(a)	830-855	1525-1575	4147	815-845	1500-1550
1042	800-845	1475-1550	4150	815-845	1500-1550
1043(a)	800-845	1475-1550	4161	815-845	1500-1550
1045(a)	800-845	1475-1550	4337	815-845	1500-1550
1046(a)	800-845	1475-1550	4340	815-845	1500-1550
1050(a)	800-845	1475-1550	50B40	815-845	1500-1550
1055	800-845	1475-1550	50B44	815-845	1500-1550
1060	800-845	1475-1550	5046	815-845	1500-1550
1065	800-845	1475-1550	50B46	815-845	1500-1550
1070	800-845	1475-1550	50B50	800-845	1475-1550
1074	800-845	1475-1550	50B60	800-845	1475-1550
1078	790-815	1450-1500	5130	830-855	1525-1575
1080	790-815	1450-1500	5132	830-855	1525-1575
1084	790-815	1450-1500	5135	815-845	1500-1550
1085	790-815	1450-1500	5140	815-845	1500-1550
1086	790-815	1450-1500	5145	815-845	1500-1550
1090	790-815	1450-1500	5147	800-845	1475-1550
1095	790-815(b)	1450-1500(b)	5150	800-845	1475-1550
Free-cutting carbon steels			5155	800-845	1475-1550
1137	830-855	1525-1575	5160	800-845	1475-1550
1138	815-845	1500-1550	51B60	800-845	1475-1550
1140	815-845	1500-1550	50100	775-800(c)	1425-1475(c)
1141	800-845	1475-1550	51100	775-800(c)	1425-1475(c)
1144	800-845	1475-1550	52100	775-800(c)	1425-1475(c)
1145	800-845	1475-1550	6150	845-885	1550-1625
1146	800-845	1475-1550	81B45	815-855	1500-1575
1151	800-845	1475-1550	8630	830-870	1525-1600
1536	815-845	1500-1550	8637	830-855	1525-1575
1541	815-845	1500-1550	8640	830-855	1525-1575
1548	815-845	1500-1550	8642	815-855	1500-1575
1552	815-845	1500-1550	8645	815-855	1500-1575
1566	855-885	1575-1625	86B45	815-855	1500-1575
Alloy steels			8650	815-855	1500-1575
1330	830-855	1525-1575	8655	800-845	1475-1550
1335	815-845	1500-1550	8660	800-845	1475-1550
1340	815-845	1500-1550	8740	830-855	1525-1575
1345	815-845	1500-1550	8742	830-855	1525-1575
3140	815-845	1500-1550	9254	815-900	1500-1650
4037	830-855	1525-1575	9255	815-900	1500-1650
4042	830-855	1525-1575	9260	815-900	1500-1650
4047	815-855	1500-1575	94B30	845-885	1550-1625
4063	800-845	1475-1550	94B40	845-885	1559-1625
			9840	830-855	1525-1575

(a) Commonly used on parts where induction hardening is employed. All steels from SAE 1030 up may have induction hardening applications. (b) This temperature range may be employed for 1095 steel that is to be quenched in water, brine or oil. For oil quenching, 1095 steel may alternatively be austenitized in the range 815 to 870 °C (1500 to 1600 °F). (c) This range is recommended for steel that is to be water quenched. For oil quenching, steel should be austenitized in the range 815 to 870 °C (1500 to 1600 °F)

Reheating (austenitizing) temperatures for hardening of carburized carbon and alloy steels (SAE)(a)

Steel	Temperature		Steel	Temperature	
	°C	°F		°C	°F
Carbon steels			**Alloy steels**		
1010	760-790	1400-1450	3310	790-830	1450-1525
1012	760-790	1400-1450	4320	830-845	1525-1550
1015	760-790	1400-1450	4615	815-845	1500-1550
1016	760-790	1400-1450	4617	815-845	1500-1550
1017	760-790	1400-1450	4620	815-845	1500-1550
1018	760-790	1400-1450	4621	815-845	1500-1550
1019	760-790	1400-1450	4626	815-845	1500-1550
1020	760-790	1400-1450	4718	815-845	1500-1550
1022	760-790	1400-1450	4720	815-845	1500-1550
1513	760-790	1600-1650	4815	800-830	1475-1525
1518	760-790	1600-1650	4817	800-830	1475-1525
1522	760-790	1600-1650	4820	800-830	1475-1525
1524	760-790	1600-1650	8115	845-870	1550-1600
1525	760-790	1600-1650	8615	845-870	1550-1600
1526	760-790	1600-1650	8617	845-870	1550-1600
1527	760-790	1600-1650	8620	845-870	1550-1600
Free-cutting carbon steels			8622	845-870	1550-1600
1109	760-790	1400-1450	8625	845-870	1550-1600
1115	760-790	1400-1450	8627	845-870	1550-1600
1117	760-790	1400-1450	8720	845-870	1550-1600
1118	760-790	1400-1450	8822	845-870	1550-1600
			9310	790-830	1450-1525

(a) Carburizing is commonly carried out at 900 to 925 °C (1650 to 1700 °F); slow cooled and reheated to given austenitizing temperature

Typical hardnesses of various carbon and alloy steels after tempering
Data were obtained on 25-mm (1-in.) bars adequately quenched to develop full hardness

Grade	Carbon content, %	Hardness, HRC, after tempering for 2 h at:									Heat treatment
		°C 205 / °F 400	260 / 500	315 / 600	370 / 700	425 / 800	480 / 900	540 / 1000	595 / 1100	650 °C / 1200 °F	
Carbon steels, water hardening											
1030	0.30	50	45	43	39	31	28	25	22	95(a)	Normalized at 900 °C (1650 °F); water quenched
1040	0.40	51	48	46	42	37	30	27	22	94(a)	from 830 to 845 °C (1525 to 1550 °F); average dew
1050	0.50	52	50	46	44	40	37	31	29	22	point, 16 °C (60 °F)
1060	0.60	56	55	50	42	38	37	35	33	26	Normalized at 885 °C (1625 °F); water quenched
1080	0.80	57	55	50	43	41	40	39	38	32	from 800 to 815 °C (1475 to 1500 °F); average dew
1095	0.95	58	57	52	47	43	42	41	40	33	point, 7 °C (45 °F)
1137	0.40	44	42	40	37	33	30	27	21	91(a)	Normalized at 900 °C (1650 °F); water quenched
1141	0.40	49	46	43	41	38	34	28	23	94(a)	from 830 to 855 °C (1525 to 1575 °F); average dew
1144	0.40	55	50	47	45	39	32	29	25	97(a)	point, 13 °C (55 °F)
Alloy steels, water hardening											
1330	0.30	47	44	42	38	35	32	26	22	16	Normalized at 900 °C (1650 °F); water quenched
2330	0.30	47	44	42	38	35	32	26	22	16	from 800 to 815 °C (1475 to 1500 °F); average dew
3130	0.30	47	44	42	38	35	32	26	22	16	point, 16 °C (60 °F)
4130	0.30	47	45	43	42	38	34	32	26	22	Normalized at 885 °C (1625 °F); water quenched
5130	0.30	47	45	43	42	38	34	32	26	22	from 800 to 855 °C (1475 to 1575 °F); average dew
8630	0.30	47	45	43	42	38	34	32	26	22	point, 16 °C (60 °F)
Alloy steels, oil hardening											
1340	0.40	57	53	50	46	44	41	38	35	31	Normalized at 870 °C (1600 °F); oil quenched
3140	0.40	55	52	49	47	41	37	33	30	26	from 830 to 845 °C (1525 to 1550 °F); average dew
4140	0.40	57	53	50	47	45	41	36	33	29	point, 16 °C (60 °F)
4340	0.40	55	52	50	48	45	42	39	34	31	Normalized at 870 °C (1600 °F); oil quenched
4640	0.40	52	51	50	47	42	40	37	31	27	from 830 to 845 °C (1525 to 1575 °F); average dew
8740	0.40	57	53	50	47	44	41	38	35	22	point, 13 °C (55 °F)
4150	0.50	56	55	53	51	47	46	43	39	35	Normalized at 870 °C (1600 °F); oil quenched
5150	0.50	57	55	52	49	45	39	34	31	28	from 830 to 870 °C (1525 to 1600 °F); average dew
6150	0.50	58	57	53	50	46	42	40	36	31	point, 13 °C (55 °F)
8650	0.50	55	54	52	49	45	41	37	32	28	Normalized at 870 °C (1600 °F); oil quenched
8750	0.50	56	55	52	51	46	44	39	34	32	from 815 to 845 °C (1500 to 1550 °F); average dew
9850	0.50	54	53	51	48	45	41	36	33	30	point, 13 °C (55 °F)

(a) Hardness, HRB

Machining

Machinability ratings of plain carbon steels, percent of cutting speed for B1112/1212

Grade(a)	Machinability rating, %	HB	Grade(a)	Machinability rating, %	HB	Grade(a)	Machinability rating, %	HB
1005	45	…	1038	65	163	1069	48	…
1006	50	95	1039	60	179	1070	55(d)	192
1008	55	95	1040	60	170	1074	55(d)	192
1010	55	105	1042	60	179	1075	48	…
1011	53	…		70(b)	179	1078	55(d)	192
1012	55	105	1043	60	179	1080	45(d)	192
1013	53	…		70(b)	179	1084	45(d)	192
1015	60	111	1045	55	179	1085	45(d)	192
1016	70	121		65(c)	170	1086	45(d)	192
1017	65	116	1046	55	187	1090	45(d)	197
1018	70	126		65(c)	179	1095	45(d)	197
1019	70	131	1049	45	197	1524	60	163
1020	65	121		55(c)	187	1527	65	163
1021	70	131	1050	45	197	1536	55	187
1022	70	137		55(c)	189	1541	45	207
1023	65	121	1053	55	…		60(c)	184
1025	65	126	1055	55(c)	197	1547	40	207
1026	75	143	1059	52	…		45(c)	187
1029	68	…	1060	60(d)	183	1548	45	217
1030	70	149	1064	60(d)	183		50(c)	192
1035	65	163	1065	60(d)	187	1552	50(c)	193
1037	65	167	1066	48	…			

(a) Unless otherwise indicated, all values are for cold drawn steels. (b) Normalized and cold drawn. (c) Annealed and cold drawn. (d) Spheroidized and cold drawn

Machinability ratings and recommended feeds and speeds for cold-drawn carbon steel bars(a)

Steel grade	Machinability rating(b), %	Form turning			Single-point turning			Drilling		
		Width of cut, in.	Speed, ft/min.	Feed, ipr	Depth of cut, in.	Speed, ft/min.	Feed, ipr	Size of hole, in.	Speed, ft/min.	Feed, ipr
12L14	158	0.500	260	0.0033	0.125	260	0.0093	0.250	145	0.0060
		1.000	240	0.0028	0.250	240	0.0088	0.500	145	0.0066
		1.500	240	0.0027	0.375	235	0.0071	0.750	160	0.0077
		2.000	235	0.0020	0.500	230	0.0060	1.000	160	0.0088
		2.500	230	0.0016	…	…	…	1.250	165	0.0099
1213	136	0.500	225	0.0030	0.125	225	0.0085	0.250	125	0.0054
		1.000	210	0.0025	0.250	210	0.0080	0.500	125	0.0060
		1.500	210	0.0025	0.375	205	0.0065	0.750	140	0.0070
		2.000	205	0.0018	0.500	200	0.0055	1.000	140	0.0080
		2.500	200	0.0015	…	…	…	1.250	145	0.0090
1119, 1212	100	0.500	165	0.0025	0.125	165	0.0070	0.250	105	0.0045
		1.000	160	0.0020	0.250	160	0.0065	0.500	105	0.0050
		1.500	160	0.0018	0.375	155	0.0055	0.750	115	0.0060
		2.000	155	0.0015	0.500	150	0.0045	1.000	115	0.0070
		2.500	150	0.0012	…	…	…	1.250	120	0.0080
1211	94	0.500	155	0.0023	0.125	155	0.0066	0.250	99	0.0042
		1.000	150	0.0019	0.250	150	0.0061	0.500	99	0.0047
		1.500	150	0.0017	0.375	146	0.0052	0.750	108	0.0056
		2.000	146	0.0014	0.500	141	0.0042	1.000	108	0.0066
		2.500	141	0.0011	…	…	…	1.250	113	0.0076

(a) All cutting speeds and feeds based on cutting with high-speed steel tools. (b) Based on a machinability rating of 100% for 1212 steel

(continued)

Machinability ratings and recommended feeds and speeds for cold-drawn carbon steel bars(a) (continued)

Steel grade	Machinability rating(b), %	Form turning			Single-point turning			Drilling		
		Width of cut, in.	Speed, ft/min.	Feed, ipr	Depth of cut, in.	Speed, ft/min.	Feed, ipr	Size of hole, in.	Speed, ft/min.	Feed, ipr
1117, 1118	91	0.500	150	0.0022	0.125	150	0.0064	0.250	95	0.0041
		1.000	145	0.0018	0.250	145	0.0059	0.500	95	0.0045
		1.500	145	0.0016	0.375	141	0.0050	0.750	105	0.0055
		2.000	141	0.0014	0.500	136	0.0041	1.000	105	0.0064
		2.500	136	0.0011	...	...	...	1.250	119	0.0073
1144, annealed	85	0.500	140	0.0021	0.125	140	0.0059	0.250	89	0.0040
		1.000	136	0.0017	0.250	136	0.0055	0.500	89	0.0045
		1.500	136	0.0015	0.375	132	0.0047	0.750	98	0.0055
		2.000	132	0.0013	0.500	127	0.0040	1.000	98	0.0064
		2.500	127	0.0010	...	...	...	1.250	102	0.0070
1141, annealed	81	0.500	135	0.0020	0.125	135	0.0057	0.250	86	0.0040
		1.000	130	0.0017	0.250	130	0.0053	0.500	86	0.0045
		1.500	130	0.0015	0.375	127	0.0045	0.750	94	0.0054
		2.000	127	0.0012	0.500	122	0.0037	1.000	94	0.0063
		2.500	122	0.0010	...	...	...	1.250	98	0.0072
1016, 1018, 1022	78	0.500	130	0.0019	0.125	130	0.0055	0.250	82	0.0038
		1.000	125	0.0016	0.250	125	0.0051	0.500	82	0.0043
		1.500	125	0.0014	0.375	121	0.0043	0.750	90	0.0052
		2.000	121	0.0012	0.500	117	0.0035	1.000	90	0.0060
		2.500	117	0.0009	...	...	...	1.250	94	0.0068
1144	76	0.500	125	0.0019	0.125	125	0.0052	0.250	79	0.0037
		1.000	121	0.0015	0.250	121	0.0049	0.500	79	0.0042
		1.500	121	0.0014	0.375	117	0.0041	0.750	87	0.0050
		2.000	117	0.0011	0.500	113	0.0034	1.000	87	0.0058
		2.500	113	0.0009	...	...	...	1.250	91	0.0066
1035, 1141, 1050, annealed	70	12.7	585	0.043	3.18	585	0.12	6.35	370	0.086
		25.4	570	0.036	6.35	570	0.11	12.7	370	0.097
		38.1	570	0.033	9.53	550	0.097	19.05	405	0.11
		50.8	550	0.028	12.7	535	0.079	25.4	405	0.13
		63.5	535	0.020	...	...	...	31.75	425	0.16
1040	64	12.7	535	0.038	3.18	535	0.11	6.35	340	0.081
		25.4	515	0.030	6.35	515	0.10	12.7	340	0.089
		38.1	515	0.028	9.53	500	0.086	19.05	370	0.11
		50.8	500	0.023	12.7	485	0.071	25.4	370	0.12
		63.5	485	0.018	...	...	...	31.75	385	0.14
1045	57	12.7	485	0.036	3.18	485	0.10	6.35	305	0.071
		25.4	460	0.030	6.35	460	0.094	12.7	305	0.079
		38.1	460	0.025	9.53	445	0.079	19.05	330	0.094
		50.8	445	0.023	12.7	430	0.066	25.4	330	0.11
		63.5	430	0.018	...	...	...	31.75	345	0.13
1050	54	12.7	455	0.036	3.18	455	0.097	6.35	290	0.071
		25.4	440	0.028	6.35	440	0.089	12.7	290	0.079
		38.1	440	0.025	9.53	425	0.076	19.05	315	0.094
		50.8	425	0.020	12.7	410	0.061	25.4	315	0.11
		63.5	410	0.018	...	...	...	31.75	330	0.13

(a) All cutting speeds and feeds based on cutting with high-speed steel tools. (b) Based on a machinability rating of 100% for 1212 steel

Machinability ratings for alloy steels

Grade	Machinability rating(a)	Typical hardness range, HB	Grade	Machinability rating(a)	Typical hardness range, HB	Grade	Machinability rating(a)	Typical hardness range, HB
1330	55 (b)	179-235	4617	65 (c)	174-223	52100	40 (d)	183-241
1335	55 (b)	179-235	4620	65 (c)	183-229	6118	60 (c)	179-217
1340	50 (b)	183-241	4621	60 (c)	183-229	6150	55 (d)	183-241
1345	45 (c)	183-241	4626	70 (c)	170-212	8115	65 (c)	163-202
4012	70 (c)	149-196	4718	60 (c)	187-229	81B45	65 (b)	179-223
4023	70 (c)	156-207	4720	65 (c)	187-229	8615	70 (c)	179-235
4024	75 (c)	156-207	4815	50 (e)	187-229	8617	70 (c)	179-235
4027	70 (c)	167-212	4817	50 (e)	187-229	8620	65 (c)	179-235
4028	75 (b)	167-212	4820	50 (e)	187-229	86L20	85 (f)	...
4032	70 (b)	174-217	5015	65 (c)	156-196	8622	65 (c)	179-235
4037	70 (b)	174-217	50B40	65 (b)	174-223	8625	60 (b)	179-223
4042	65 (b)	179-229	50B44	65 (b)	174-223	8627	60 (b)	170-223
4047	65 (b)	179-229	5046	60 (b)	174-223	8630	70 (b)	179-229
4118	60 (c)	170-207	50B46	60 (b)	174-223	8637	65 (b)	179-229
4130	70 (b)	187-229	50B50	55 (b)	183-235	8640	65 (b)	184-229
4135	70 (b)	187-229	5060	55 (d)	170-212	8642	65 (b)	184-229
4137	70 (b)	187-229	50B60	55 (d)	170-212	8645	65 (b)	184-235
4140	65 (b)	187-229	5115	65 (c)	163-201	86B45	65 (b)	184-235
41L40	85 (b)	...	5120	70 (c)	163-201	8650	60 (b)	187-248
4142	65 (b)	187-229	5130	70 (b)	174-212	8655	55 (b)	187-248
4145	60 (b)	187-229	5132	70 (b)	174-212	8660	55 (d)	179-217
4147	60 (b)	187-235	5135	70 (b)	179-217	8720	65 (c)	179-235
4150	55 (b)	187-241	5140	65 (b)	179-217	8740	65 (b)	184-235
4161	50 (d)	187-241	5145	65 (b)	179-229	8822	55 (d)	179-223
4320	60 (e)	187-229	5147	65 (b)	179-229	9254	45 (d)	187-241
4340	50 (d)	187-241	5150	60 (b)	183-235	9255	40 (d)	179-229
E4340	50 (d)	187-241	5155	55 (b)	183-235	9260	40 (d)	184-235
4419	65 (b)	170-212	5160	55 (d)	179-217	9310	50 (e)	184-229
4422	65 (c)	170-212	51B60	55 (d)	179-217	94B15	70 (c)	163-202
4427	65 (b)	170-212	50100	40 (d)	183-241	94B17	70 (c)	163-202
4615	65 (c)	174-223	51100	40 (d)	183-241	94B30	70 (b)	170-223

(a) All ratings apply to cold-finished bars. (b) Microstructure composed primarily of lamellar pearlite and ferrite. (c) Microstructure composed primarily of blocky or acicular pearlite and bainite, as found in hot rolled steels. (d) Microstructure compound primarily of spheroidite. (e) Microstructure resulting from subcritical anneal. (f) Microstructure not specified

Machinability ratings for cold-drawn alloy steel bars(a)

Steel grade	Condition before drawing	Machinability rating(b), %	Steel grade	Condition before drawing	Machinability rating(b), %	Steel grade	Condition before drawing	Machinability rating(b), %
1330	Annealed	55	4620	Hot rolled	65	6118	Hot rolled	60
1335	Annealed	55	4626	Hot rolled	70	6150	Annealed	55
1340	Annealed	50	4720	Hot rolled	65	81B45	Annealed	65
1345	Annealed	45	4815	Annealed	50	8615	Hot rolled	70
4023	Hot rolled	70	4817	Annealed	50	8617	Hot rolled	70
4024	Hot rolled	75	4820	Annealed	50	8620	Hot rolled	65
4027	Annealed	70	5015	Hot rolled	65	8622	Hot rolled	65
4028	Annealed	75	50B44	Annealed	65	8625	Annealed	60
4037	Annealed	70	50B46	Annealed	60	8627	Annealed	60
4047	Annealed	65	50B50	Annealed	55	8630	Annealed	70
4118	Hot rolled	60	50B60	Spheroidized	55	8637	Annealed	65
4130	Annealed	70	5120	Hot rolled	70	8640	Annealed	65
4137	Annealed	70	5130	Annealed	70	8642	Annealed	65
4140	Annealed	65	5132	Annealed	70	8645	Annealed	65
4142	Annealed	65	5135	Annealed	70	8655	Annealed	55
4145	Annealed	60	5140	Annealed	65	8720	Hot rolled	65
4147	Annealed	60	5150	Annealed	60	8740	Annealed	65
4150	Annealed	55	5155	Annealed	55	8822	Hot rolled	55
4161	Spheroidized	50	5160	Spheroidized	55	9260	Spheroidized	40
4320	Annealed	60	51B60	Spheroidized	55	94B17	Hot rolled	70
4340	Annealed	50	51100	Spheroidized	40	94B30	Annealed	70
4615	Hot rolled	65	52100	Spheroidized	40			

(a) Source: AISI Committee of Hot Rolled and Cold Finished Bar Producers. (b) Based on cutting with high-speed tool steels and a machinability rating of 100% for 1212 steel

Conditions recommended for various machining operations on wrought steels of different hardness levels

Grade(a)	Material condition(b)	Hardness, HB	Type of tool(c)	Broach	Tap	Deep drill	Bore	Form cut	Drill	Plane	Turn	End mill
12L13	HR, N, A, CD	150-200	HSS	40(d)	75	155(e)	165(f)	165(g)	130(h)	95(i)	195(j)	185(f)
			Carbide			275(d)	615(h)	525(f)		300(k)	700(l)	490(h)
41L40	HR, N, A, CD	200-260	HSS	25(d)	65	110(e)	100(f)	90(d)	90(f)	65(i)	120(j)	140(g)
			Carbide			250(e)	380(h)	310(g)		290(k)	475(l)	400(f)
1030	HR, N, A, CD	175-225	HSS	20(d)	65	90(h)	95(f)	85(d)	75(h)	60(i)	95(j)	120(g)
			Carbide			200(d)	350(h)	275(g)		250(k)	400(l)	375(f)
4140R	HR, N, A, CD	200-250	HSS	20(d)	55	95(h)	85(f)	75(d)	80(f)	55(i)	90(j)	125(g)
			Carbide			200(h)	345(h)	270(g)		300(k)	400(l)	375(f)
8620	HR, A, CD	175-225	HSS	20(d)	55	80(h)	85(d)	75(d)	65(h)	55(i)	110(j)	90(g)
			Carbide			200(d)	320(h)	250(g)		250(k)	440(l)	365(f)
1060	HR, N, A, CD	175-225	HSS	20(d)	55	85(h)	85(f)	80(d)	70(h)	60(m)	90(j)	115(g)
			Carbide			200(d)	325(h)	265(g)		250(k)	395(h)	375(f)
4140	HR, A, CD	175-225	HSS	20(d)	45	80(h)	75(f)	65(d)	65(h)	50(i)	90(j)	90(g)
			Carbide			200(d)	295(h)	225(g)		240(k)	400(l)	350(f)
4340	HR, A, CD	175-225	HSS	20(d)	45	80(h)	75(f)	65(d)	65(h)	50(i)	90(j)	90(g)
			Carbide			200(d)	295(h)	225(g)		240(k)	400(l)	350(f)
H11	A	225-300	HSS	20(d)	30	65(h)	55(g)	50(n)	50(f)	45(i)	65(h)	115(g)
			Carbide			200(d)	260(h)	175(d)		200(k)	350(l)	300(f)
4340	N, Q & T	275-325	HSS	15(d)	25	55(f)	55(g)	50(n)	45(g)	35(m)	60(j)	65(g)
			Carbide			150(d)	240(f)	175(d)		180(i)	330(l)	240(g)
4340	Q & T	375-425	HSS	8(d)	12	30(d)	30(d)	30(n)	25(d)	...	35(h)	45(d)
			Carbide	...	...	100(h)	150(g)	110(d)	...	...	225(j)	140(d)

(a) L, lead-bearing grade; R, resulfurized steel. (b) HR, hot rolled; N, normalized; A, annealed; CD, cold drawn; Q & T, quenched and tempered. (c) Dimensions of cut as follows: tap, deep drill and drill: 0.5-in. diam hole; bore and end mill: 0.05-in. deep cut; form cut: 0.5-in. wide cut; plane: 0.10-in. deep cut; turn: 0.15-in. deep cut. (d) Chip load in broaching or feed in milling: 0.003 to 0.004 in./tooth; feed in other operations: 0.003 to 0.004 in./rev. (e) Feed: 0.012 in./rev. (f) Feed: 0.007 to 0.008 in./rev. (g) Feed: 0.005 to 0.006 in./rev. (h) Feed: 0.009 to 0.010 in./rev. (i) Feed: 0.05 in./stroke. (j) Feed: 0.015 in./rev. (k) Feed: 0.08 in./stroke. (l) Feed: 0.020 in./rev. (m) Feed: 0.03 in./stroke. (n) Feed: 0.001 to 0.002 in./rev

Forming

Typical mechanical properties of steel sheet

Type or quality	Special feature	Yield strength MPa	ksi	Tensile strength MPa	ksi	Elongation in 50 mm or 2 in., %	Hardness, HRB	Strain-hardening exponent, n	Plastic strain ratio, r_m
Hot rolled									
Commercial	Standard properties	262	38	359	52	30	55	0.15	0.9
Drawing (rimmed)	Improved properties	241	35	345	50	35	50	0.18	1.0
Drawing (special killed)	Non-aging	241	35	345	50	40	50	0.20	1.0
Medium strength	Inclusion shape control	345	50	414	60	25	70	0.15	0.9
High-strength	Inclusion shape control	552	80	620	90	15	90	...	...
Cold rolled									
Commercial	Improved finish	234	34	317	46	35	45	0.20	1.0
Drawing (rimmed)	Stretchability	207	30	310	45	45	40	0.24	1.2
Drawing (special killed)	Deep drawing	172	25	296	43	40	40	0.22	1.6
Interstitial free	Extra deep drawing	152	22	317	46	45	45	0.24	2.0
Medium strength	Formable	414	60	483	70	25	85	0.20	1.2
High strength	Moderately formable	689	100	724	105	10	25(a)	...	...

(a) Hardness, HRC

Minimum bend radii for selected plain carbon and low-alloy steel sheet materials

Product	Production description(a) Quality temper or strength level	Thickness	Minimum bend radius Parallel to rolling direction	Across rolling direction
Cold rolled				
1008/1010	CQ	...	0.25 mm (0.01 in.)	...
1008/1010	DQ	...	0.25 mm (0.01 in.)	...
1008/1010	No. 3(b)	...	$1t$	$0.5t$
1008/1010	No. 2(c)	...	NR	$1t$
1008/1010	No. 1(d)	...	NR	NR
Hot rolled				
1008/1010	CQ	<2.25 mm (<0.09 in.)	$0.75t$	$0.5t$
		>2.25 mm (>0.09 in.)	$1.5t$	$1t$
1008/1010	DQ	<2.25 mm (<0.09 in.)	$0.5t$	$0.25t$
		>2.25 mm (>0.09 in.)	$0.75t$	$0.5t$
Annealed				
1020/1025	...	...	$1t$ to $2t$	...
4130, 8630	...	...	$1.5t$ to $2t$	...
1070, 1095	...	...	$2t$ to $3t$	...
ASTM A607 (HSLA)	345 MPa (50 ksi)	...	$1.5t$	$1t$
	415 MPa (60 ksi)	...	$3t$	$2t$
	480 MPa (70 ksi)	...	$4t$	$3t$

(a) CQ, commercial quality; DQ, drawing quality; t, sheet thickness; NR, not recommended. (b) Quarter hard. (c) Half hard. (d) Full hard

Suggested minimum inside bend radii for SAE J410c steels of various strengths and thicknesses

Grade	Min bend radius for thickness t of		
	<4.6 mm (<0.180 in.)	4.6 to 6.4 mm (0.180 to 0.250 in.)	6.4 to 12.7 mm (0.250 to 0.500 in.)
942X	...	t	$2t$
945A, 945C	t	2t	$2.5t$
945X	t	t	$2t$
950A, 950B, 950C, 950D	t	$2t$	$3t$
950X	$1.5t$	$2.5t$	$2.5t$
955X	$2t$	$3t$	$3t$
960X	$2.5t$	$3.5t$	$3.5t$
965X	$3t$	$4t$	$4t$
970X	$3.5t$	$4.5t$	$4.5t$
980X(a)	$3.5t$	$4.5t$	$4.5t$

(a) Available only in thicknesses to 9.5 mm (0.375 in.)

Specified bend test radii for inclusion shape-controlled ASTM A715 steel sheet

Grade	Bend test radius for	
	Transverse bends(a)	Longitudinal bends(a)
50	$0.5t$	0
60	$0.5t$	0
70	$0.75t$	$0.5t$
80	$0.75t$	$0.5t$

(a) For sheet thickness t up to 5.84 mm (0.2299 in.)

Tool Materials

Composition Limits

Composition limits of principal types of tool steels

AISI	SAE	UNS	Composition(a), %								
			C	Mn	Si	Cr	Ni	Mo	W	V	Co
Molybdenum high speed steels											
M1	M1	T11301	0.78-0.88	0.15-0.40	0.20-0.50	3.50-4.00	0.30	8.20-9.20	1.40-2.10	1.00-1.35	...
M2	M2	T11302	0.78-0.88; 0.95-1.05	0.15-0.40	0.20-0.45	3.75-4.50	0.30	4.50-5.50	5.50-6.75	1.75-2.20	...
M3, class 1	M3	T11313	1.00-1.10	0.15-0.40	0.20-0.45	3.75-4.50	0.30	4.75-6.50	5.00-6.75	2.25-2.75	...
M3, class 2	M3	T11323	1.15-1.25	0.15-0.40	0.20-0.45	3.75-4.50	0.30	4.75-6.50	5.00-6.75	2.75-3.75	...
M4	M4	T11304	1.25-1.40	0.15-0.40	0.20-0.45	3.75-4.75	0.30	4.25-5.50	5.25-6.50	3.75-4.50	...
M6	...	T11306	0.75-0.85	0.15-0.40	0.20-0.45	3.75-4.50	0.30	4.50-5.50	3.75-4.75	1.30-1.70	11.00-13.00
M7	...	T11307	0.97-1.05	0.15-0.40	0.20-0.55	3.50-4.00	0.30	8.20-9.20	1.40-2.10	1.75-2.25	...
M10	...	T11310	0.84-0.94; 0.95-1.05	0.10-0.40	0.20-0.45	3.75-4.50	0.30	7.75-8.50	...	1.80-2.20	...
M30	...	T11330	0.75-0.85	0.15-0.40	0.20-0.45	3.50-4.25	0.30	7.75-9.00	1.30-2.30	1.00-1.40	4.50-5.50
M33	...	T11333	0.85-0.92	0.15-0.40	0.15-0.50	3.50-4.00	0.30	9.00-10.00	1.30-2.10	1.00-1.35	7.75-8.75
M34	...	T11334	0.85-0.92	0.15-0.40	0.20-0.45	3.50-4.00	0.30	7.75-9.20	1.40-2.10	1.90-2.30	7.75-8.75
M36	...	T11336	0.80-0.90	0.15-0.40	0.20-0.45	3.75-4.50	0.30	4.50-5.50	5.50-6.50	1.75-2.25	7.75-8.75
M41	...	T11341	1.05-1.15	0.20-0.60	0.15-0.50	3.75-4.50	0.30	3.25-4.25	6.25-7.00	1.75-2.25	4.75-5.75
M42	...	T11342	1.05-1.15	0.15-0.40	0.15-0.65	3.50-4.25	0.30	9.00-10.00	1.15-1.85	0.95-1.35	7.75-8.75
M43	...	T11343	1.15-1.25	0.20-0.40	0.15-0.65	3.50-4.25	0.30	7.50-8.50	2.25-3.00	1.50-1.75	7.75-8.75
M44	...	T11344	1.10-1.20	0.20-0.40	0.30-0.55	4.00-4.75	0.30	6.00-7.00	5.00-5.75	1.85-2.20	11.00-12.25
M46	...	T11346	1.22-1.30	0.20-0.40	0.40-0.65	3.70-4.20	0.30	8.00-8.50	1.90-2.20	3.00-3.30	7.80-8.80
M47	...	T11347	1.05-1.15	0.15-0.40	0.20-0.45	3.50-4.00	0.30	9.25-10.00	1.30-1.80	1.15-1.35	4.75-5.25
Tungsten high speed steels											
T1	T1	T12001	0.65-0.80	0.10-0.40	0.20-0.40	3.75-4.00	0.30	1.00 max	17.25-18.75	0.90-1.30	...
T2	T2	T12002	0.80-0.90	0.20-0.40	0.20-0.40	3.75-4.50	0.30	...	17.50-19.00	1.80-2.40	...
T4	T4	T12004	0.70-0.80	0.10-0.40	0.20-0.40	3.75-4.50	0.30	0.40-1.00	17.50-19.00	0.80-1.20	4.25-5.75
T5	T5	T12005	0.75-0.85	0.20-0.40	0.20-0.40	3.75-5.00	0.30	0.50-1.25	17.50-19.00	1.80-2.40	7.00-9.50
T6	...	T12006	0.75-0.85	0.20-0.40	0.20-0.40	4.00-4.75	0.30	0.40-1.00	18.50-21.00	1.50-2.10	11.00-13.00
T8	T8	T12008	0.75-0.85	0.20-0.40	0.20-0.40	3.75-4.50	0.30	0.40-1.00	13.25-14.75	1.80-2.40	4.25-5.75
T15	...	T12015	1.50-1.60	0.15-0.40	0.15-0.40	3.75-5.00	0.30	1.00 max	11.75-13.00	4.50-5.25	4.75-5.25
Chromium hot work steels											
H10	...	T20810	0.35-0.45	0.25-0.70	0.80-1.20	3.00-3.75	0.30	2.00-3.00	...	0.25-0.75	...
H11	H11	T20811	0.33-0.43	0.20-0.50	0.80-1.20	4.75-5.50	0.30	1.10-1.60	...	0.30-0.60	...
H12	H12	T20812	0.30-0.40	0.20-0.50	0.80-1.20	4.75-5.50	0.30	1.25-1.75	1.00-1.70	0.50	...
H13	H13	T20813	0.32-0.45	0.20-0.50	0.80-1.20	4.75-5.50	0.30	1.10-1.75	...	0.80-1.20	...
H14	...	T20814	0.35-0.45	0.20-0.50	0.80-1.20	4.75-5.50	0.30	...	4.00-5.25	...	...
H19	...	T20819	0.32-0.45	0.20-0.50	0.20-0.50	4.00-4.75	0.30	0.30-0.55	3.75-4.50	1.75-2.20	4.00-4.50
Tungsten hot work steels											
H21	H21	T20821	0.26-0.36	0.15-0.40	0.15-0.50	3.00-3.75	0.30	...	8.50-10.00	0.30-0.60	...
H22	...	T20822	0.30-0.40	0.15-0.40	0.15-0.40	1.75-3.75	0.30	...	10.00-11.75	0.25-0.50	...
H23	...	T20823	0.25-0.35	0.15-0.40	0.15-0.60	11.00-12.75	0.30	...	11.00-12.75	0.75-1.25	...
H24	...	T20824	0.42-0.53	0.15-0.40	0.15-0.40	2.50-3.50	0.30	...	14.00-16.00	0.40-0.60	...
H25	...	T20825	0.22-0.32	0.15-0.40	0.15-0.40	3.75-4.50	0.30	...	14.00-16.00	0.40-0.60	...
H26	...	T20826	0.45-0.55(b)	0.15-0.40	0.15-0.40	3.75-4.50	0.30	...	17.25-19.00	0.75-1.25	...
Molybdenum hot work steels											
H42	...	T20842	0.55-0.70(b)	0.15-0.40	...	3.75-4.50	0.30	4.50-5.50	5.50-6.75	1.75-2.20	...

(a) Maximum unless a range is specified. All steels except group W contain 0.25 max Cu. All steels except group W steels contain 0.20 max Cu, 0.03 max P and 0.03 max S; group W steels contain 0.20 max Cu, 0.025 max P and 0.025 max S. Where specified, sulfur may be increased to 0.06 to 0.15% to improve machinability of group H, M and T steels. (b) Available in several carbon ranges. (c) Contains free graphite in the microstructure. (d) Optional. (e) Specified carbon ranges are designated by suffix numbers

(continued)

Composition limits of principal types of tool steels (continued)

AISI	SAE	UNS	C	Mn	Si	Cr	Ni	Mo	W	V	Co
	Designations		Composition(a), %								
Air-hardening medium-alloy cold work steels											
A2	A2	T30102	0.95-1.05	1.00	0.50	4.75-5.50	0.30	0.90-1.40	...	0.15-0.50	...
A3	...	T30103	1.20-1.30	0.40-0.60	0.50	4.75-5.50	0.30	0.90-1.40	...	0.80-1.40	...
A4	...	T30104	0.95-1.05	1.80-2.20	0.50	0.90-2.20	0.30	0.90-1.40	...	...	...
A6	...	T30106	0.65-0.75	1.80-2.50	0.50	0.90-1.20	0.30	0.90-1.40	...	...	...
A7	...	T30107	2.00-2.85	0.80	0.50	5.00-5.75	0.30	0.90-1.40	0.50-1.50	3.90-5.15	...
A8	...	T30108	0.50-0.60	0.50	0.75-1.10	4.75-5.50	0.30	1.15-1.65	1.00-1.50	...	...
A9	...	T30109	0.45-0.55	0.50	0.95-1.15	4.75-5.50	1.25-1.75	1.30-1.80	...	0.80-1.40	...
A10	...	T30110	1.25-1.50(c)	1.60-2.10	1.00-1.50	...	1.55-2.05	1.25-1.75	...	...	...
High-carbon, high-chromium cold work steels											
D2	D2	T30402	1.40-1.60	0.60	0.60	11.00-13.00	0.30	0.70-1.20	...	1.10	1.00
D3	D3	T30403	2.00-2.35	0.60	0.60	11.00-13.50	0.30	...	1.00	1.00	...
D4	...	T30404	2.05-2.40	0.60	0.60	11.00-13.00	0.30	0.70-1.20	...	1.00	...
D5	D5	T30405	1.40-1.60	0.60	0.60	11.00-13.00	0.30	0.70-1.20	...	1.00	2.50-3.50
D7	...	T30407	2.15-2.50	0.60	0.60	11.50-13.50	0.30	0.70-1.20	...	3.80-4.40	...
Oil-hardening cold work steels											
O1	O1	T31501	0.85-1.00	1.00-1.40	0.50	0.40-0.60	0.30	...	0.40-0.60	0.30	...
O2	O2	T31502	0.85-0.95	1.40-1.80	0.50	0.35	0.30	0.30	...	0.30	...
O6	O6	T31506	1.25-1.55(c)	0.30-1.10	0.55-1.50	0.30	0.30	0.20-0.30	...	...	...
O7	...	T31507	1.10-1.30	1.00	0.60	0.35-0.85	0.30	0.30	1.00-2.00	0.40	...
Shock-resisting steels											
S1	S1	T41901	0.40-0.55	0.10-0.40	0.15-1.20	1.00-1.80	0.30	0.50	1.50-3.00	0.15-0.30	...
S2	S2	T41902	0.40-0.55	0.30-0.50	0.90-1.20	...	0.30	0.30-0.60	...	0.50	...
S5	S5	T41905	0.50-0.65	0.60-1.00	1.75-2.25	0.35	...	0.20-1.35	...	0.35	...
S6	...	T41906	0.40-0.50	1.20-1.50	2.00-2.50	1.20-1.50	...	0.30-0.50	...	0.20-0.40	...
S7	...	T41907	0.45-0.55	0.20-0.80	0.20-1.00	3.00-3.50	...	1.30-1.80	...	0.20-0.30(d)	...
Low-alloy special-purpose tool steels											
L2	...	T61202	0.45-1.00(b)	0.10-0.90	0.50	0.70-1.20	...	0.25	...	0.10-0.30	...
L6	L6	T61206	0.65-0.75	0.25-0.80	0.50	0.60-1.20	1.25-2.00	0.50	...	0.20-0.30(d)	...
Low-carbon mold steels											
P2	...	T51602	0.10	0.10-0.40	0.10-0.40	0.75-1.25	0.10-0.50	0.15-0.40	...	...	...
P3	...	T51603	0.10	0.20-0.60	0.40	0.40-0.75	1.00-1.50	...	...	...	...
P4	...	T51604	0.12	0.20-0.60	0.10-0.40	4.00-5.25	...	0.40-1.00	...	...	...
P5	...	T51605	0.10	0.20-0.60	0.40	2.00-2.50	0.35	...	...	...	...
P6	...	T51606	0.05-0.15	0.35-0.70	0.10-0.40	1.25-1.75	3.25-3.75	...	...	...	...
P20	...	T51620	0.28-0.40	0.60-1.00	0.20-0.80	1.40-2.00	...	0.30-0.55	...	...	...
P21	...	T51621	0.18-0.22	0.20-0.40	0.20-0.40	0.20-0.30	3.90-4.25	...	...	0.15-0.25	1.05-1.25Al
Water-hardening tool steels											
W1	W108, W109, W110, W112	T72301	0.70-1.50(e)	0.10-0.40	0.10-0.40	0.15	0.20	0.10	0.15	0.10	...
W2	W209, W210	T72302	0.85-1.50(e)	0.10-0.40	0.10-0.40	0.15	0.20	0.10	0.15	0.15-0.35	...
W5	...	T72305	1.05-1.15	0.10-0.40	0.10-0.40	0.40-0.60	0.20	0.10	0.15	0.10	...

(a) Maximum unless a range is specified. All steels except group W contain 0.25 max Cu, 0.03 max P and 0.025 max S; group W steels contain 0.20 max Cu, 0.03 max P and 0.025 max S. Where specified, sulfur may be increased to 0.06 to 0.15% to improve machinability of group H, M and T steels. (b) Available in several carbon ranges. (c) Contains free graphite in the microstructure. (d) Optional. (e) Specified carbon ranges are designated by suffix numbers

Properties and Characteristics

Density and thermal expansion of selected tool steels

Type	Density Mg/m³	Density lb/in.³	Coefficient of linear thermal expansion μm/m · K from 20 °C to 100 °C	200 °C	425 °C	540 °C	650 °C	μin./in. °F from 68 °F to 200 °F	400 °F	800 °F	1000 °F	1200 °F
W1	7.84	0.282	10.4	11.0	13.1	13.8(a)	14.2(b)	5.76	6.13	7.28	7.64(a)	7.90(b)
W2	7.85	0.283	...	...	14.4	14.8	14.9	...	...	8.0	8.2	8.3
S1	7.88	0.255	12.4	12.6	13.5	13.9	14.2	6.9	7.0	7.5	7.7	7.9
S2	7.79	0.281	10.9	11.9	13.5	14.0	14.2	6.0	6.6	7.5	7.8	7.9
S5	7.76	0.280	...	...	12.6	13.3	13.7	...	...	7.0	7.4	7.6
S6	7.75	0.279	...	...	12.6	13.3	...	...	...	7.0	7.4	...
S7	7.76	0.280	...	12.6	13.3	13.7(a)	13.3	...	7.0	7.4	7.6(a)	7.4
O1	7.85	0.283	...	10.6(c)	12.8	14.0(d)	14.4(d)	...	5.9(c)	7.1	7.8(d)	8.0(d)
O2	7.66	0.277	11.2	12.6	13.9	14.6	15.1	6.2	7.0	7.7	8.1	8.4
O7	7.80	0.282	...	...	...	...	...	...	...	...	...	...
A2	7.86	0.284	10.7	10.6(c)	12.9	14.0	14.2	5.96	5.91(c)	7.2	7.8	7.9
A6	7.84	0.283	11.5	12.4	13.5	13.9	14.2	6.4	6.9	7.5	7.7	7.9
A7	7.66	0.277	...	...	12.4	12.9	13.5	...	...	6.9	7.2	7.5
A8	7.87	0.284	...	...	12.0	12.4	12.6	...	...	6.7	6.9	7.0
A9	7.78	0.281	...	...	12.0	12.4	12.6	...	...	6.7	6.9	7.0
D2	7.70	0.278	10.4	10.3	11.9	12.2	12.2	5.8	5.7	6.6	6.8	6.8
D3	7.70	0.278	12.0	11.7	12.9	13.1	13.5	6.7	6.5	7.2	7.3	7.5
D4	7.70	0.278	...	...	12.4	...	...	...	...	6.9	...	...
D5	...	...	...	...	...	12.0	...	...	...	...	6.7	...
H10	7.81	0.281	...	...	12.2	13.3	13.7	...	...	6.8	7.4	7.6
H11	7.75	0.280	11.9	12.4	12.8	12.9	13.3	6.6	6.9	7.1	7.2	7.4
H13	7.76	0.280	10.4	11.5	12.2	12.4	13.1	5.8	6.4	6.8	6.9	7.3
H14	7.89	0.285	11.0	...	...	...	...	6.1	...	...	...	...
H19	7.98	0.288	11.0	11.0	12.0	12.4	12.9	6.1	6.1	6.7	6.9	7.2
H21	8.28	0.299	12.4	12.6	12.9	13.5	13.9	6.9	7.0	7.2	7.5	7.7
H22	8.36	0.302	11.0	...	11.5	12.0	12.4	6.1	...	6.4	6.7	6.9
H26	8.67	0.313	...	...	...	12.4	...	...	...	...	6.9	...
H42	8.15	0.295	...	...	...	11.9	...	...	...	...	6.6	...
T1	8.67	0.313	...	9.7	11.2	11.7	11.9	...	5.4	6.2	6.5	6.6
T2	8.67	0.313	...	...	...	...	...	...	...	...	...	...
T4	8.68	0.313	...	...	...	11.9	...	...	...	...	6.6	...
T5	8.75	0.316	11.2	...	...	11.5	...	6.2	...	...	6.4	...
T6	8.89	0.321	...	...	...	...	...	...	...	...	...	...
T8	8.43	0.305	...	...	...	...	...	...	...	...	...	...
T15	8.19	0.296	...	9.9	11.0	11.5	...	...	5.5(c)	6.1	6.4	...
M1	7.89	0.285	...	10.6(c)	11.3	12.0	12.4	...	5.9(c)	6.3	6.7	6.9
M2	8.16	0.295	10.1	9.4(c)	11.2	11.9	12.2	5.6	5.2(c)	6.2	6.6	6.8
M3, class 1	8.15	0.295	...	...	11.5	12.0	12.2	...	...	6.4	6.7	6.8
M3, class 2	8.16	0.295	...	...	11.5	12.0	12.8	...	...	6.4	6.7	7.1
M4	7.97	0.288	...	9.5(c)	11.2	12.0	12.2	...	5.3(c)	6.2	6.7	6.8
M7	7.95	0.287	...	9.5(c)	11.5	12.2	12.4	...	5.3(c)	6.4	6.8	6.9
M10	7.88	0.255	...	...	11.0	11.9	12.4	...	...	6.1	6.6	6.9
M30	8.01	0.289	...	...	11.2	11.7	12.2	...	...	6.2	6.5	6.8
M33	8.03	0.290	...	...	11.0	11.7	12.0	...	...	6.1	6.5	6.7
M36	8.18	0.296	...	...	...	...	...	...	...	...	...	...
M41	8.17	0.295	...	9.7	10.4	11.2	...	...	5.4	5.8	6.2	...
M42	7.98	0.288	...	...	...	...	...	...	...	...	...	...
M46	7.83	0.283	...	...	...	...	...	...	...	...	...	...
M47	7.96	0.288	10.6	11.0	11.9	...	12.6	5.9	6.1	6.6	...	7.0
L2	7.86	0.284	...	...	14.4	14.6	14.8	...	...	8.0	8.1	8.2
L6	7.86	0.284	11.3	12.6	12.6	13.5	13.7	6.3	7.0	7.0	7.5	7.6
P2	7.86	0.284	...	...	13.7	...	...	...	...	7.6	...	...
P5	7.80	0.282	...	...	...	...	...	...	...	...	...	...
P6	7.85	0.284	...	...	...	...	...	...	...	...	...	...
P20	7.85	0.284	...	...	12.8	13.7	14.2	...	...	7.1	7.6	7.9

(a) From 20 °C to 500 °C (68 °F to 930 °F). (b) From 20 °C to 600 °C (68 °F to 1110 °F). (c) From 20 °C to 260 °C (68 °F to 500 °F). (d) From 38 °C (100 °F)

Thermal conductivity of selected tool steels

Type	Temperature °C	°F	Thermal conductivity W/m · K	Btu/ft · h · °F	Type	Temperature °C	°F	Thermal conductivity W/m · K	Btu/ft · h · °F
W1	100	200	48.3	27.9	H21 (continued)				
	260	500	41.5	24.0		400	750	29.8	17.2
	400	750	38.1	22.0		540	1000	29.4	17.0
	540	1000	34.6	20.0		675	1250	29.1	16.8
	675	1250	29.4	17.0	T1	100	20	19.9	11.5
	815	1500	24.2	14.0		260	500	21.6	12.5
H11	100	200	42.2	24.4		400	750	23.2	13.4
	260	500	36.3	21.0		540	1000	24.7	14.3
	400	750	33.4	19.3					
	540	1000	31.5	18.2	T15	100	200	20.9	12.1
	675	1250	30.1	17.4		200	500	24.1	13.9
	815	1500	28.6	16.5		400	750	25.4	14.7
						540	1000	26.3	15.2
H13	215	420	28.6	16.5					
	350	660	28.4	16.4	M2	100	200	21.3	12.3
	475	890	28.4	16.4		200	500	23.5	13.6
	605	1120	28.7	16.6		400	750	25.6	14.8
						540	1000	27.0	15.6
H21	100	200	27.0	15.6		675	1250	28.9	16.7
	260	500	29.8	17.2					

Nominal room-temperature mechanical properties of group L and group S tool steels

Type	Condition	Tensile strength MPa	ksi	0.2% yield strength MPa	ksi	Elongation(a), %	Reduction in area, %	Hardness, HRC	Impact energy J	ft · lb
L2	Annealed	710	103	510	74	25	50	96 HRB	...	...
	Oil quenched from 855 °C (1575 °F) and single tempered at:									
	205 °C (400 °F)	2000	290	1790	260	5	15	54	28(b)	21(b)
	315 °C (600 °F)	1790	260	1655	240	10	30	52	19(b)	14(b)
	425 °C (800 °F)	1550	225	1380	200	12	35	47	26(b)	19(b)
	540 °C (1000 °F)	1275	185	1170	170	15	45	41	39(b)	29(b)
	650 °C (1200 °F)	930	135	760	110	25	55	30	125(b)	92(b)
L6	Annealed	655	95	380	55	25	55	93 HRB	...	...
	Oil quenched from 845 °C (1550 °F) and single tempered at:									
	315 °C (600 °F)	2000	290	1790	260	4	9	54	12(b)	9(b)
	425 °C (800 °F)	1585	230	1380	200	8	20	46	18(b)	13(b)
	540 °C (1000 °F)	1345	195	1100	160	12	30	42	23(b)	17(b)
	650 °C (1200 °F)	965	140	830	120	20	48	32	81(b)	60(b)
S1	Annealed	690	100	415	60	24	52	96 HRB	...	...
	Oil quenched from 930 °C (1700 °F) and single tempered at:									
	205 °C (400 °F)	2070	300	1895	275	...	...	57.5	249(c)	184(c)
	315 °C (600 °F)	2030	294	1860	270	4	12	54	233(c)	172(c)
	425 °C (800 °F)	1790	260	1690	245	5	17	50.5	203(c)	150(c)
	540 °C (1000 °F)	1680	244	1525	221	9	23	47.5	230(c)	170(c)
	650 °C (1200 °F)	1345	195	1240	180	12	37	42	...	...
S5	Annealed	725	105	440	64	25	50	96 HRB	...	...
	Oil quenched from 870 °C (1600 °F) and single tempered at:									
	205 °C (400 °F)	2345	340	1930	280	5	20	59	206(c)	152(c)
	315 °C (600 °F)	2240	325	1860	270	7	24	58	232(c)	171(c)
	425 °C (800 °F)	1895	275	1690	245	9	28	52	243(c)	179(c)
	540 °C (1000 °F)	1520	220	1380	200	10	30	48	188(c)	139(c)
	650 °C (1200 °F)	1035	150	1170	170	15	40	37	...	...
S7	Annealed	640	93	380	55	25	55	95 HRB	...	...
	Fan cooled from 940 °C (1725 °F) and single tempered at:									
	205 °C (400 °F)	2170	315	1450	210	7	20	58	244(c)	180(c)
	315 °C (600 °F)	1965	285	1585	230	9	25	55	309(c)	228(c)
	425 °C (800 °F)	1895	275	1410	205	10	29	53	243(c)	179(c)
	540 °C (1000 °F)	1820	264	1380	200	10	33	51	324(c)	239(c)
	650 °C (1200 °F)	1240	180	1035	150	14	45	39	358(c)	264(c)

(a) In 50 mm or 2 in. (b) Charpy V-notch. (c) Charpy unnotched

Processing and service characteristics of tool steels

AISI designation	Resistance to decarburization	Hardening and tempering			Approximate hardness(b), HRC	Machinability	Fabrication and service		
		Hardening response	Amount of distortion(a)	Resistance to cracking			Toughness	Resistance to softening	Resistance to wear
Molybdenum high speed steels									
M1	Low	Deep	A or S, low; O, medium	Medium	60-65	Medium	Low	Very high	Very high
M2	Medium	Deep	A or S, low; O, medium	Medium	60-65	Medium	Low	Very high	Very high
M3 (class 1 and class 2)	Medium	Deep	A or S, low; O, medium	Medium	61-66	Medium	Low	Very high	Very high
M4	Medium	Deep	A or S, low; O, medium	Medium	61-66	Low to medium	Low	Very high	Highest
M6	Low	Deep	A or S, low; O, medium	Medium	61-66	Medium	Low	Highest	Very high
M7	Low	Deep	A or S, low; O, medium	Medium	61-66	Medium	Low	Very high	Very high
M10	Low	Deep	A or S, low; O, medium	Medium	60-65	Medium	Low	Very high	Very high
M30	Low	Deep	A or S, low; O, medium	Medium	60-65	Medium	Low	Highest	Very high
M33	Low	Deep	A or S, low; O, medium	Medium	60-65	Medium	Low	Highest	Very high
M34	Low	Deep	A or S, low; O, medium	Medium	60-65	Medium	Low	Highest	Very high
M36	Low	Deep	A or S, low; O, medium	Medium	60-65	Medium	Low	Highest	Very high
M41	Low	Deep	A or S, low; O, medium	Medium	65-70	Medium	Low	Highest	Very high
M42	Low	Deep	A or S, low; O, medium	Medium	65-70	Medium	Low	Highest	Very high
M43	Low	Deep	A or S, low; O, medium	Medium	65-70	Medium	Low	Highest	Very high
M44	Low	Deep	A or S, low; O, medium	Medium	62-70	Medium	Low	Highest	Very high
M46	Low	Deep	A or S, low; O, medium	Medium	67-69	Medium	Low	Highest	Very high
M47	Low	Deep	A or S, low; O, medium	Medium	65-70	Medium	Low	Highest	Very high
Tungsten high speed steels									
T1	High	Deep	A or S, low; O, medium	High	60-65	Medium	Low	Very high	Very high
T2	High	Deep	A or S, low; O, medium	High	61-66	Medium	Low	Very high	Very high
T4	Medium	Deep	A or S, low; O, medium	Medium	62-66	Medium	Low	Highest	Very high
T5	Low	Deep	A or S, low; O, medium	Medium	60-65	Medium	Low	Highest	Very high
T6	Low	Deep	A or S, low; O, medium	Medium	60-65	Low to medium	Low	Highest	Very high
T8	Medium	Deep	A or S, low; O, medium	Medium	60-65	Medium	Low	Highest	Very high
T15	Medium	Deep	A or S, low; O, medium	Medium	63-68	Low to medium	Low	Highest	Highest
Chromium hot work steels									
H10	Medium	Deep	Very low	Highest	39-56	Medium to high	High	High	Medium
H11	Medium	Deep	Very low	Highest	38-54	Medium to high	Very high	High	Medium
H12	Medium	Deep	Very low	Highest	38-55	Medium to high	Very high	High	Medium
H13	Medium	Deep	Very low	Highest	38-53	Medium to high	Very high	High	Medium
H14	Medium	Deep	Low	Highest	40-47	Medium	High	High	Medium
H19	Medium	Deep	A, low; O, medium	High	40-57	Medium	High	High	Medium to high
Tungsten hot work steels									
H21	Medium	Deep	A, low; O, medium	High	36-54	Medium	High	High	Medium to high
H22	Medium	Deep	A, low; O, medium	High	39-52	Medium	High	High	Medium to high
H23	Medium	Deep	Medium	High	34-47	Medium	Medium	Very high	Medium to high
H24	Medium	Deep	A, low; O, medium	High	45-55	Medium	Medium	Very high	High
H25	Medium	Deep	A, low; O, medium	High	35-44	Medium	High	Very high	Medium
H26	Medium	Deep	A or S, low; O, medium	High	43-58	Medium	Medium	Very high	High
Molybdenum hot work steels									
H42	Medium	Deep	A or S, low; O, medium	Medium	50-60	Medium	Medium	Very high	High

(a) A, air cool; B, brine quench; O, oil quench; S, salt bath quench; W, water quench. (b) After tempering in temperature range normally recommended for this steel. (c) Carburized case hardness. (d) After aging at 510 to 550 °C (950 to 1025 °F). (e) Toughness decreases with increasing carbon content and depth of hardening

(continued)

Processing and service characteristics of tool steels(continued)

AISI designation	Hardening and tempering					Fabrication and service			
	Resistance to decarburization	Hardening response	Amount of distortion(a)	Resistance to cracking	Approximate hardness(b), HRC	Machinability	Toughness	Resistance to softening	Resistance to wear
Air-hardening medium-alloy cold work steels									
A2	Medium	Deep	Lowest	Highest	57-62	Medium	Medium	High	High
A3	Medium	Deep	Lowest	Highest	57-65	Medium	Medium	High	Very high
A4	Medium to high	Deep	Lowest	Highest	54-62	Low to medium	Medium	Medium	Medium to high
A6	Medium to high	Deep	Lowest	Highest	54-60	Low to medium	Medium	Medium	Medium to high
A7	Medium	Deep	Lowest	Highest	57-67	Low	Low	High	Highest
A8	Medium	Deep	Lowest	Highest	50-60	Medium	High	High	Medium to high
A9	Medium	Deep	Lowest	Highest	35-56	Medium	High	High	Medium to high
A10	Medium to high	Deep	Lowest	Highest	55-62	Medium to high	Medium	Medium	High
High-carbon, high-chromium cold work steels									
D2	Medium	Deep	Lowest	Highest	54-61	Low	Low	High	High to very high
D3	Medium	Deep	Very low	High	54-61	Low	Low	High	Very high
D4	Medium	Deep	Lowest	Highest	54-61	Low	Low	High	Very high
D5	Medium	Deep	Lowest	Highest	54-61	Low	Low	High	High to very high
D7	Medium	Deep	Lowest	Highest	58-65	Low	Low	High	Highest
Oil-hardening cold work steels									
O1	High	Medium	Very high	Very high	57-62	High	Medium	Low	Medium
O2	High	Medium	Very low	Very high	57-62	High	Medium	Low	Medium
O6	High	Medium	Very low	Very high	58-63	Highest	Medium	Low	Medium
O7	High	Medium	W, high; O, very low	W, low; O, very high	58-64	High	Medium	Low	Medium
Shock-resisting steels									
S1	Medium	Medium	Medium	High	40-58	Medium	Very high	Medium	Low to medium
S2	Low	Medium	High	Low	50-60	Medium to high	Highest	Low	Low to medium
S5	Low	Medium	Medium	High	50-60	Medium to high	Highest	Low	Low to medium
S6	Low	Medium	Medium	High	54-56	Medium	Very high	Low	Low to medium
S7	Medium	Deep	A, lowest; O, low	A, highest; O, high	45-57	Medium	Very high	High	Low to medium
Low-alloy special-purpose steels									
L2	High	Medium	W, low; O, medium	W, high; O, medium	45-63	High	Very high(c)	Low	Low to medium
L6	High	Medium	Low	High	45-62	Medium	Very high	Low	Medium
Low-carbon mold steels									
P2	High	Medium	Low	High	58-64(c)	Medium to high	High	Low	Medium
P3	High	Medium	Low	High	58-64(c)	Medium	High	Low	Medium
P4	High	High	Very low	High	58-64(c)	Low to medium	High	Medium	High
P5	High	...	W, high; O, low	High	58-64(c)	Medium	High	Low	Medium
P6	High	...	A, very low; O, low	High	58-61(c)	Medium to high	High	Low	Low to medium
P20	High	Medium	Low	High	28-37	Medium	Medium	Medium	Medium
P21	High	Deep	Lowest	Highest	30-40(d)	Medium	Medium	Medium	Medium
Water-hardening steels									
W1	Highest	Shallow	High	Medium	50-64	Highest	High(e)	Low	Low to medium
W2	Highest	Shallow	High	Medium	50-64	Highest	High(e)	Low	Low to medium
W5	Highest	Shallow	High	Medium	50-64	Highest	High(e)	Low	Low to medium

(a) A, air cool; B, brine quench; O, oil quench; S, salt bath quench; W, water quench. (b) After tempering in temperature range normally recommended for this steel. (c) Carburized case hardness. (d) After aging at 510 to 550 °C (950 to 1025 °F). (e) Toughness decreases with increasing carbon content and depth of hardening

Properties of refractory metal carbides

Carbide	Hardness, HV	Crystal system	Melting point		Theoretical density, Mg/m³	Modulus of elasticity	
			°C	°F		GPa	10⁶ psi
TiC	3200	Cubic	3065 ± 15	5550 ± 30	4.92	448	65
VC	2950	Cubic	2730 ± 75	4950 ± 150	5.48	434	63
HfC	2700	Cubic	3925 ± 50	7100 ± 100	12.67	…	…
ZrC	2600	Cubic	3440 ± 20	6225 ± 40	6.56	474	68.8
NbC	2400	Cubic	3500 ± 75	6330 ± 135	7.82	~290	~42
Cr_3C_2	2280	Orthorhombic	~1900	~3440	6.68	386	56
WC	2080	Hexagonal	~2800	~5030	15.8	669	97
Mo_2C	1950	Hexagonal	2490-2520	4510-4570	9.12	227	33
TaC	1790	Cubic	3915 ± 50	7080 ± 100	14.50	276	40

Typical applications of cobalt-bonded cemented carbides

Grade	Grain size	Application
Straight grades		
97WC-3Co	Medium	Machining of cast iron, nonferrous metals and nonmetallic materials; excellent abrasion resistance and low shock resistance; the most wear resistant of the straight WC-Co grades; maintains a sharp cutting edge and makes long finishing cuts to close tolerances possible; also used for fine wire dies and small nozzles
94WC-6Co	Fine	Machining nonferrous and high-temperature alloys
94WC-6Co	Medium	General-purpose machining of work materials other than steel; also used for small and medium size compacting dies, coating dies, burnishing rings and nozzles
94WC-6Co	Coarse	Machining of cast iron, nonferrous metals and nonmetallic materials; also used for small wire-drawing dies, compacting dies, small drawing dies and caps and rings. The hardest grade used in mining applications where impact is encountered, as in rotary percussive bits
90WC-10Co	Fine	Machining steel and milling high-temperature metals (including titanium and its alloys) at low feeds and speeds: face mills, end mills, form tools, cutoff tools and screw-machine tools
90WC-10Co	Coarse	Primarily used for mining roller bits and percussive drilling bits
84WC-16Co	Fine	Primarily used for mining and metal-forming components
84WC-16Co	Coarse	Metal-forming and mining components: medium and large dies where great toughness is required, blanking dies for punch presses, and large mandrels
75WC-25Co	Medium	Metal-forming components for heavy impact applications, such as heading dies, cold extrusion dies, and punches and dies for blanking heavy stock
Complex grades		
71-74.5WC-10-12.5TiC-11-12.0TaC-4.5Co	Medium	Finishing, semifinishing and light roughing operations on plain carbon and alloy steels and alloy cast irons
72-73WC-7-8TiC-11.5-12TaC-8-8.5Co	Medium	Tough, wear-resistant grade for heavy-duty roughing cuts. Successfully withstands high temperatures encountered in heavy-duty machining, interrupted turning, scale cuts and milling of plain carbon and alloy steels and alloy cast irons
64TiC-28WC-2TaC-2Cr₃C₂-4Co	Medium	High-speed finishing of steels and cast irons
57WC-27TaC-16Co	Coarse	Cutting hot flash formed in the manufacture of welded tubing; also used to make dies for hot extrusion of aluminum wirebar and tubing

Properties of representative cobalt-bonded cemented carbides

Nominal composition	Grain size	Hardness, HRA	Density		Transverse strength		Compressive strength		Proportional limit, compression		Modulus of elasticity	
			Mg/m³	lb/in.³	MPa	ksi	MPa	ksi	MPa	ksi	GPa	10⁶ psi
97WC-3Co	Medium	92.5-93.2	15.3	0.55	1590	230	5860	850	2410	350	641	93
94WC-6Co	Fine	92.5-93.1	15.0	0.54	1790	260	5930	860	2550	370	614	89
	Medium	91.7-92.2	15.0	0.54	2000	290	5450	790	1930	280	648	94
	Coarse	90.5-91.5	15.0	0.54	2210	320	5170	750	1450	210	641	93
90WC-10Co	Fine	90.7-91.3	14.6	0.53	3100	450	5170	750	1590	230	620	90
	Coarse	87.4-88.2	14.5	0.52	2760	400	4000	580	1170	170	552	80
84WC-16Co	Fine	89	13.9	0.50	3380	490	4070	590	970	140	524	76
	Coarse	86.0-87.5	13.9	0.50	2900	420	3860	560	700	100	524	76
75WC-25Co	Medium	83-85	13.0	0.47	2550	370	3100	450	410	60	483	70
71WC-12.5TiC-12TaC-4.5Co	Medium	92.1-92.8	12.0	0.43	1380	200	5790	840	1170	170	565	82
72WC-8TiC-11.5TaC-8.5Co	Medium	90.7-91.5	12.6	0.45	1720	250	5170	750	1720	250	558	81
64TiC-28WC-2TaC-2Cr₂C₃-4.0Co	Medium	94.5-95.2	6.6	0.24	690	100	4340	630	...	...	...	...
57WC-27TaC-16Co	Coarse	84.0-86.0	13.7	0.49	2690	390	3720	540	1170	170	441	64

Nominal composition	Tensile strength		Impact strength		Relative abrasion resistance(a)	Coefficient of linear thermal expansion				Thermal conductivity, W/m · K	Electrical conductivity, % LACS
						μm/m · °C		μin./in/ · °F			
	MPa	ksi	J	in. · lb		at 200 °C	at 1000 °C	at 400 °F	at 1800 °F		
97WC-3Co	...	...	1.13	10	100	4.0	...	2.2	...	121	5.3
94WC-6Co	...	...	1.02	9	100	4.3	5.9	2.4	3.3	...	...
	1450	210	1.36	12	58	4.3	5.4	2.4	3.0	100	7.8
	1520	220	1.36	12	25	4.3	5.6	2.4	3.1	121	10.0
90WC-10Co	...	...	1.69	15	22	...	...	...	...	...	...
	1340	195	2.03	18	7	5.2	...	2.9	...	112	11.4
84WC-16Co	...	...	3.05	27	5	...	...	...	...	...	...
	1860	270	2.83	25	5	5.8	7.0	3.2	3.9	88	9.2
75WC-25Co	1380	200	3.05	27	3	6.3	...	3.5	...	71	9.8
71WC-12.5TiC-12TaC-4.5Co	...	...	0.79	7	11	5.2	6.5	2.9	3.6	35	4.3
72WC-8TiC-11.5TaC-8.5Co	...	...	0.90	8	13	5.8	6.8	3.2	3.8	50	5.2
64TiC-28WC-2TaC-2Cr₂C₃-4.0Co	...	...	...	...	8	...	...	...	...	...	...
57WC-27TaC-16Co	...	...	2.03	18	3	5.9	7.7	3.3	4.3	...	...

(a) Based on a value of 100 for the most abrasion-resistant grade

Test methods for determining properties of cemented carbides

Property	ASTM/ANSI	CCPA	ISO	SAE
Abrasive wear resistance	B611	P112	...	...
Apparent grain size	B390	M203	...	J439
Apparent porosity	B276	M201	4505	J439
Axial load fatigue	...	...	...	...
Coefficient of sliding friction	...	P111	...	...
Coercive force	...	...	3326	...
Compressive strength	E9	P104	4506	...
Density	B311	P101	3369	J439
Diametral compression testing	B485	P115	...	...
Electrical resistivity	B421	P107	...	...
Fracture toughness	...	...	...	...
Hardness, HRA	B294	P103	3738	J439
Hardness, HV	E92	...	3878	...
Linear thermal expansion	B95	P108	...	...
Magnetic permeability	A342	P109	...	...
Metallographic preparation of samples	...	...	...	...
Microstructure	B657	M202	4499	...
Poisson's ratio	E132	P105	...	...
Powder sampling and testing	...	...	4884	...
Sampling and testing	...	...	4889	...
Tensile testing	B437	P113	...	...
Thermal shock resistance	...	P110	...	...
Transverse rupture strength	B406	P102	...	...
Young's modulus	E111	P106	3327	...
			3312	...

Compositions and selected properties of three principal grades of steel-bonded titanium carbide

Characteristic	Grade C	Grade CM	Grade SK
Composition:			
Titanium carbide	45 vol%	45 vol%	40 vol%
Steel matrix	0.6C-3Cr-3Mo	0.85C-10Cr-3Mo	0.40C-5Cr-4Mo-0.50Ni
Hardness:			
Annealed, HRC	40	45	37
Hardened(a), HRC	70	69	65
Max service temperature: °C	200	540	540
°F	400	1000	1000
Density: Mg/m^3	6.60	6.45	6.80
$lb/in.^3$	0.238	0.232	0.245
Tranverse strength: MPa	2070	2140	2070
ksi	300	310	300
Modulus of elasticity: GPa	305	305	270
10^6 psi	44	44	39
Thermal expansion: $\mu m/m \cdot °C$	7.83(b)	8.3(c)	9.47(c)
$\mu in./in. \cdot °F$	4.35	4.6	5.26
Electrical conductivity, % IACS	3.2	2.8	3.0

(a) Grade C: austenitized at 950 °C (1750 °F), oil quenched and tempered 1 h at 190 °C (375 °F). Grade CM: austenitized at 1100 °C (2000 °F), oil quenched (gas quenched if heat treated in vacuum), and double tempered at 525 °C (975 °F). Grade SK: austenitized at 1000 °C (1850 °F), oil quenched (gas quenched if heat treated in vacuum), and double tempered at 525 °C. (b) At 21 to 200 °C (70 to 400 °F). (c) At 21 to 540 °C (70 to 1000 °F)

Typical properties of ceramic tool materials

Property	Group A-1 General	Group A-1 Example(a)	Group A-2	Group A-3
Hardness: HRA	93 to 94	93 to 94	93 to 94	93 to 94
Density:				
Mg/m^3	3.96 to 3.98	4.1	4.0	4.24
$lb/in.^3$	0.142 to 0.143	0.148	0.144	0.153
Transverse strength:				
MPa	480 to 690	620	640	760
ksi	70 to 100	90	92.5	110
Compressive strength:				
MPa	3790 to 4480	2140(b)	4140	3930 to 4070
ksi	550 to 650	310(b)	600	570-590
Modulus of elasticity:				
GPa	390	400	390	...
10^6 psi	57	58	57	...
Impact strength:				
J	...	0.23	...	...
in. · lb	...	2	...	...
Thermal expansion(c): µin./in. · °F	...	3.4(d)	4.0	4.3
µm/m · °C				
Thermal conductivity:	...	6.1(d)	7.2	7.7
At room temperature				
W/m · K	29	...	...	17 to 21
Btu/ft · h · °F	17	...	...	10 to 12
At 100 °C (212 °F):				
W/m · K	22	...	29	...
Btu/ft · h · °F	13	...	17	...
At 450 °C (850 °F):				
W/m · K	11	...	...	...
Btu/ft · h · °F	6.5	...	...	...
At 600 °C (1100 °F):				
W/m · K	...	...	...	14.7
Btu/ft · h · °F	...	...	...	8.4

(a) $89Al_2O_3$-$11TiO$, cold pressed and sintered. (b) Proportional limit. (c) At 21 to 200 °C (70 to 400 °F). (d) 8.3 µm/m· °C (4.6 µin./in.· °F) at 21 to 980 °C (70 to 1800 °F)

Typical elevated-temperature hardness and strength of groups A-2 and A-3 ceramic tool materials

Temperature °C	Temperature °F	Hardness, HV Group A-2	Hardness, HV Group A-3	Transverse rupture strength Group A-2 MPa	Group A-2 ksi	Group A-3 MPa	Group A-3 ksi
RT	RT	2100	2400	690	100	735	107
480	900	2000	2000	690	100	715	104
650	1200	1950	1850	...	...	...	...
815	1500	1850	1700	...	...	...	...
980	1800	1700	1500	700	101	700	102
1200	2200	1400	1400	610	90	690	100

Properties and uses of nickel-bonded titanium carbides

Property	High TiC, plus Mo$_2$C; low nickel	High TiC, plus Mo$_2$C; low nickel	TiC, plus Mo$_2$C; intermediate nickel	Lower TiC, plus Mo$_2$C; high nickel
Grain size	Fine	...	Fine	Fine
Hardness: HRA	93.3	93.0	91.7	90.5
HV	1970	1890	1600	1440
Density: Mg/m^3	5.50	5.63	5.71	5.82
lb/in.^{3}	0.198	0.203	0.206	0.210
Transverse strength: MPa	1170	1380	1720	1890
ksi	170	200	250	275
Compressive strength: MPa	3585	3450	3270	2960
ksi	520	500	475	430
Tensile strength: MPa	970	1100	1170	1240
ksi	140	160	170	180
Modulus of elasticity: GPa	462	448	414	379
10^6 psi	67	65	60	55
Impact strength: J	0.79	0.90	1.02	1.24
in.·lb	7	8	9	11
Thermal expansion(a): μm/m · °C	7.5	7.8	8.4	9.1
μin./in. · °F	4.2	4.3	4.7	5.1
Thermal conductivity(b): W/m · K	16.7	16.7	16.7	16.7
Btu/ft · h · °F	9.6	9.6	9.6	9.6
Typical classification for machining use: C-grade	C-8	C-7	C-6	C-6
ISO	P01	P10	P20	P30

(a) At 21 to 650 °C (70 to 1200 °F). (b) At 100 to 300 °C (200 to 575 °F).

Selected properties of cubic boron nitride

Crystal structure	Zinc blende ($F\bar{4}3m$)
Density	3.48 Mg/m^3 (0.125 lb/in.^{3})
Hardness:	
At 20 °C (70 °F)	4000 HV
At 1000 °C (1800 °F)	~4000 HV
Temperature for dislocation mobility	1300-1400 °C (2400-2550 °F)
Melting point (triple point)	3500 K
Thermal conductivity (theoretical)	13 W/m · K (7.5 Btu/ft · h · °F)
Thermal stability: Limit of oxidation resistance in air	~1300 °C (~2400 °F)
Metastable reversion temperature	~1500 °C (~2700 °F)
Thermal expansion, 21 to 500 °C (70 to 900 °F)	4.8 μm/m · °C (2.7 μin./in. · °F)

Typical properties of cast Tantung G

Property	Chill cast	Refractory mold cast
Melting temperature:		
°C	1150 to 1200	
°F	2100 to 2200	
Casting temperature:		
°C	1370	
°F	2500	
Density:		
Mg/m^3	8.3	8.3
lb/in.^{3}	0.30	0.30
Thermal expansion:		
μm/m · °C	4.2	4.2
μin./in. · °F	2.3	2.3
Thermal conductivity:		
W/m · K	26.8	26.8
Btu/ft · h · °F	15.5	15.5
Hardness: HRC	60 to 63	53 to 58
Transverse strength:		
MPa	2240	1030 to 1200
ksi	325	150 to 175
Modulus of elasticity:		
GPa	265	...
10^6 psi	41	...
Tensile strength:		
MPa	585 to 620	450
ksi	85 to 90	65
Compressive strength:		
MPa	2760	2930
ksi	400	425
Impact strength:		
J	6.1	6.1
ft · lb	4.5	4.5

Mutual indentation hot hardness of cast Tantung G

Temperature		Hardness,	Equivalent hardness(b)		
°C	°F	HB(a)	HRA	HRC	HRB
RT	RT	654	81.3	60.1	...
425	800	479	75.7	49.8	...
650	1200	479	75.7	49.8	...
870	1600	267	63.8	27.1	104
980	1800	114	...	...	66.7

(a) 3000 kg load, applied for 30 s. (b) Converted values

SAE J1072 system for classification of superhard tool materials

Basic classification

Material classification	Designation
Material compound	
Nitride	1
Carbide	2
Oxide	3
Other	9(a)
Binder material	
None	0
Nickel	1
Iron	2
Cobalt	3
Other	9(a)
Base metal	
None	0
Niobium	1
Tungsten	2
Titanium	3
Tantalum	4
Chromium	5
Aluminum	6
Other	9(a)

Suffixes

Material property	Identifier(b)
Binder metal quantity (wt% to nearest 0.1%)	A
Base metal quantity (wt% to nearest 0.1%)	B
Hardness(c) (HRA to nearest 0.1)	C
Specific gravity(c) (to nearest 0.1)	D
Grain size(c) (maximum amount of each type)	E
Apparent porosity(c) (the first digit indicates the amount of type A, the second the amount of type B, and the third the amount of type C porosity)	F
Transverse rupture strength(d) (minimum, in ksi)	G
Other properties (written description required)	Z

(a) Material in this category shall be described by suffix Z. (b) Complete description consists of the letter identifier followed by one to three digits that express a quantitative value for the specific property. (c) Determined according to procedures outlined in SAE J439. (d) Determined according to procedures outlined in ASTM B406

C-grade classification system for cemented carbides

C grade	Application category
Machining of cast iron, nonferrous and nonmetallic materials	
C-1	Roughing
C-2	General-purpose machining
C-3	Finishing
C-4	Precision finishing
Machining of carbon and alloy steels	
C-5	Roughing
C-6	General-purpose machining
C-7	Finishing
C-8	Precision finishing
Wear-surface applications	
C-9	No shock
C-10	Light shock
C-11	Heavy shock
Impact applications	
C-12	Light impact
C-13	Medium impact
C-14	Heavy impact
Miscellaneous applications	
C-15	Hot weld-flash removal, light cuts
C15A	Hot weld-flash removal, heavy cuts
C-16	Rock bits
C-17	Cold header dies
C-18	Wear at elevated temperatures and/or resistance to chemicals
C-19	Radioactive shielding, counterbalances and kinetic-energy devices

Heat Treating

Normalizing and annealing temperatures of tool steels

Type	Normalizing treatment/ temperature(a) °C	Normalizing treatment/ temperature(a) °F	Annealing(b) Temperature °C	Annealing(b) Temperature °F	Rate of cooling, max °C/h	Rate of cooling, max °F/h	Hardness, HB
Molybdenum high speed steels							
M1, M10	Do not normalize		815 to 870	1500 to 1600	22	40	207 to 235
M2	Do not normalize		870 to 900	1600 to 1650	22	40	212 to 241
M3, M4	Do not normalize		870 to 900	1600 to 1650	22	40	223 to 255
M6	Do not normalize		870	1600	22	40	248 to 277
M7	Do not normalize		815 to 870	1500 to 1600	22	40	217 to 255
M30, M33, M34, M36, M41, M42, M46, M47	Do not normalize		870 to 900	1600 to 1650	22	40	235 to 269
M43	Do not normalize		870 to 900	1600 to 1650	22	40	248 to 269
M44	Do not normalize		870 to 900	1600 to 1650	22	40	248 to 293
Tungsten high speed steels							
T1	Do not normalize		870 to 900	1600 to 1650	22	40	217 to 255
T2	Do not normalize		870 to 900	1600 to 1650	22	40	223 to 255
T4	Do not normalize		870 to 900	1600 to 1650	22	40	229 to 269
T5	Do not normalize		870 to 900	1600 to 1650	22	40	235 to 277
T6	Do not normalize		870 to 900	1600 to 1650	22	40	248 to 293
T8	Do not normalize		870 to 900	1600 to 1650	22	40	229 to 255
T15	Do not normalize		870 to 900	1600 to 1650	22	40	241 to 277
Chromium hot work steels							
H10, H11, H12, H13	Do not normalize		845 to 900	1550 to 1650	22	40	192 to 229
H14	Do not normalize		870 to 900	1600 to 1650	22	40	207 to 235
H19	Do not normalize		870 to 900	1600 to 1650	22	40	207 to 241
Tungsten hot work steels							
H21, H22, H25	Do not normalize		870 to 900	1600 to 1650	22	40	207 to 235
H23	Do not normalize		870 to 900	1600 to 1650	22	40	212 to 255
H24, H26	Do not normalize		870 to 900	1600 to 1650	22	40	217 to 241
Molybdenum hot works steels							
H41, H43	Do not normalize		815 to 870	1500 to 1600	22	40	207 to 235
H42	Do not normalize		845 to 900	1550 to 1650	22	40	207 to 235
High-carbon high-chromium cold work steels							
D2, D3, D4	Do not normalize		870 to 900	1600 to 1650	22	40	217 to 255
D5	Do not normalize		870 to 900	1600 to 1650	22	40	223 to 255
D7	Do not normalize		870 to 900	1600 to 1650	22	40	235 to 262
Medium-alloy air-hardening cold work steels							
A2	Do not normalize		845 to 870	1550 to 1600	22	40	201 to 229
A3	Do not normalize		845 to 870	1550 to 1600	22	40	207 to 229
A4	Do not normalize		740 to 760	1360 to 1400	14	25	200 to 241
A6	Do not normalize		730 to 745	1350 to 1375	14	25	217 to 248
A7	Do not normalize		870 to 900	1600 to 1650	14	25	235 to 262
A8	Do not normalize		845 to 870	1550 to 1600	22	40	192 to 223
A9	Do not normalize		845 to 870	1550 to 1600	14	25	212 to 248
A10	790	1450	765 to 795	1410 to 1460	8	15	235 to 269

(a) Time held at temperature varies from 15 min for small sections to 1 h for large sizes. Cooling is done in still air. Normalizing should not be confused with low-temperature annealing. (b) The upper limit of ranges should be used for large sections and the lower limit for smaller sections. Time held at temperature varies from 1 h for light sections to 4 h for heavy sections and large furnace charges of high-alloy steel. (c) For 0.25 Si type, 183 to 207 HB; for 1.00 Si type, 207 to 229 HB. (d) Temperature varies with carbon content: 0.60 to 0.75 C, 815 °C (1500 °F); 0.75 to 0.90 C, 790 °C (1450 °F); 0.90 to 1.10 C, 870 °C (1600 °F); 1.10 to 1.40 C, 870 to 925 °C (1600 to 1700 °F). (e) Temperature varies with carbon content: 0.60 to 0.90 C, 740 to 790 °C (1360 to 1450 °F); 0.90 to 1.40 C, 760 to 790 °C (1400 to 1450 °F)

(continued)

Normalizing and annealing temperatures of tool steels (continued)

| Type | Normalizing treatment/ temperature(a) | | Annealing(b) | | | | |
| | °C | °F | Temperature | | Rate of cooling, max | | Hardness, HB |
			°C	°F	°C/h	°F/h	
Oil-hardening cold work steels							
O1	870	1600	760 to 790	1400 to 1450	22	40	183 to 212
O2	845	1550	745 to 775	1375 to 1425	22	40	183 to 212
O6	870	1600	765 to 790	1410 to 1450	11	20	183 to 217
O7	900	1650	790 to 815	1450 to 1500	22	40	192 to 217
Shock-resisting steels							
S1	Do not normalize		790 to 815	1450 to 1500	22	40	183 to 229(c)
S2	Do not normalize		760 to 790	1400 to 1450	22	40	192 to 217
S5	Do not normalize		775 to 800	1425 to 1475	14	25	192 to 229
S7	Do not normalize		815 to 845	1500 to 1550	14	25	187 to 223
Mold steels							
P2	Not required		730 to 815	1350 to 1500	22	40	103 to 123
P3	Not required		730 to 815	1350 to 1500	22	40	109 to 137
P4	Do not normalize		870 to 900	1600 to 1650	14	25	116 to 128
P5	Not required		845 to 870	1550 to 1600	22	40	105 to 116
P6	Not required		845	1550	8	15	183 to 217
P20	900	1650	760 to 790	1400 to 1450	22	40	149 to 179
P21	900	1650	Do not anneal				
Low-alloy special-purpose steels							
L2	871 to 900	1600 to 1650	760 to 790	1400 to 1450	22	40	163 to 197
L3	900	1650	790 to 815	1450 to 1500	22	40	174 to 201
L6	870	1600	760 to 790	1400 to 1450	22	40	183 to 212
Carbon-tungsten special-purpose steels							
F1	900	1650	760 to 800	1400 to 1475	22	40	183 to 207
F2	900	1650	790 to 815	1450 to 1500	22	40	207 to 235
Water-hardening steels							
W1, W2	790 to 925(d)	1450 to 1700(d)	740 to 790(e)	1360 to 1450(e)	22	40	156 to 201
W5	870 to 925	1600 to 1700	760 to 790	1400 to 1450	22	40	163 to 201

(a) Time held at temperature varies from 15 min for small sections to 1 h for large sizes. Cooling is done in still air. Normalizing should not be confused with low-temperature annealing. (b) The upper limit of ranges should be used for large sections and the lower limit for smaller sections. Time held at temperature varies from 1 h for light sections to 4 h for heavy sections and large furnace charges of high-alloy steel. (c) For 0.25 Si type, 183 to 207 HB; for 1.00 Si type, 207 to 229 HB. (d) Temperature varies with carbon content: 0.60 to 0.75 C, 815 °C (1500 °F); 0.75 to 0.90 C, 790 °C (1450 °F); 0.90 to 1.10 C, 870 °C (1600 °F); 1.10 to 1.40 C, 870 to 925 °C (1600 to 1700 °F). (e) Temperature varies with carbon content: 0.60 to 0.90 C, 740 to 790 °C (1360 to 1450 °F); 0.90 to 1.40 C, 760 to 790 °C (1400 to 1450 °F)

Hardening and tempering of tool steels

Type	Rate of heating	Hardening					Time at temperature, min.	Quenching medium(a)	Tempering temperature	
		Preheat temperature		Hardening temperature						
		°C	°F	°C	°F				°C	°F
Molybdenum high speed steels										
M1, M7, M10	Rapidly from preheat	730 to 845	1350 to 1550	1175 to 1220	2150 to 2225(b)	2 to 5	O, A or S		540 to 595(c)	1000 to 1100(c)
M2	Rapidly from preheat	730 to 845	1350 to 1550	1190 to 1230	2175 to 2250(b)	2 to 5	O, A or S		540 to 595(c)	1000 to 1100(c)
M3, M4, M30, M33, M34	Rapidly from preheat	730 to 845	1350 to 1550	1205 to 1230(b)	2200 to 2250(b)	2 to 5	O, A or S		540 to 595(c)	1000 to 1100(c)
M6	Rapidly from preheat	790	1450	1175 to 1205(b)	2150 to 2200(b)	2 to 5	O, A or S		540 to 595(c)	1000 to 1100(c)
M36	Rapidly from preheat	730 to 845	1350 to 1550	1220 to 1245(b)	2225 to 2275(b)	2 to 5	O, A or S		540 to 595(c)	1000 to 1100(c)
M41	Rapidly from preheat	730 to 845	1350 to 1550	1190 to 1215(b)	2175 to 2220(b)	2 to 5	O, A or S		540 to 595(d)	1000 to 1100(d)
M42	Rapidly from preheat	730 to 845	1350 to 1550	1190 to 1210(b)	2175 to 2210(b)	2 to 5	O, A or S		510 to 595(d)	950 to 1100(d)
M43	Rapidly from preheat	730 to 845	1350 to 1550	1190 to 1215(b)	2175 to 2220(b)	2 to 5	O, A or S		510 to 595(d)	950 to 1100(d)
M44	Rapidly from preheat	730 to 845	1350 to 1550	1200 to 1225(b)	2190 to 2240(b)	2 to 5	O, A or S		540 to 625(d)	1000 to 1160(d)
M46	Rapidly from preheat	730 to 845	1350 to 1550	1190 to 1220(b)	2175 to 2225(b)	2 to 5	O, A or S		525 to 565(d)	975 to 1050(d)
M47	Rapidly from preheat	730 to 845	1350 to 1550	1180 to 1205(b)	2150 to 2200(b)	2 to 5	O, A or S		525 to 595(d)	975 to 1100(d)
Tungsten high speed steels										
T1, T2, T4, T8	Rapidly from preheat	815 to 870	1500 to 1600	1260 to 1300(b)	2300 to 2375(b)	2 to 5	O, A or S		540 to 595(c)	1000 to 1100(c)
T5, T6	Rapidly from preheat	815 to 870	1500 to 1600	1275 to 1300(b)	2325 to 2375(b)	2 to 5	O, A or S		540 to 595(c)	1000 to 1100(c)
T15	Rapidly from preheat	815 to 870	1500 to 1600	1205 to 1260(b)	2200 to 2300(b)	2 to 5	O, A or S		540 to 650(d)	1000 to 1200(d)
Chromium hot work steels										
H10	Moderately from preheat	815	1500	1010 to 1040	1850 to 1900	15 to 40(e)	A		540 to 650	1000 to 1200
H11, H12	Moderately from preheat	815	1500	995 to 1025	1825 to 1875	15 to 40(e)	A		540 to 650	1000 to 1200
H13	Moderately from preheat	815	1500	995 to 1040	1825 to 1900	15 to 40(e)	A		540 to 650	1000 to 1200
H14	Moderately from preheat	815	1500	1010 to 1065	1850 to 1950	15 to 40(e)	A		540 to 650	1000 to 1200
H19	Moderately from preheat	815	1500	1095 to 1205	2000 to 2200	2 to 5	A or O		540 to 705	1000 to 1300
Molybdenum hot work steels										
H41, H43	Rapidly from preheat	730 to 845	1350 to 1550	1095 to 1190	2000 to 2175	2 to 5	O, A or S		565 to 650	1050 to 1200
H42	Rapidly from preheat	730 to 845	1350 to 1550	1120 to 1220	2050 to 2225	2 to 5	O, A or S		565 to 650	1050 to 1200
Tungsten hot work steels										
H21, H22	Rapidly from preheat	815	1500	1095 to 1205	2000 to 2200	2 to 5	A or O		595 to 675	1100 to 1250
H23	Rapidly from preheat	845	1550	1205 to 1260	2200 to 2300	2 to 5	O		650 to 815	1200 to 1500
H24	Rapidly from preheat	815	1500	1095 to 1230	2000 to 2250	2 to 5	O		565 to 650	1050 to 1200
H25	Rapidly from preheat	815	1500	1150 to 1260	2100 to 2300	2 to 5	A or O		565 to 675	1050 to 1250
H26	Rapidly from preheat	870	1600	1175 to 1260	2150 to 2300	2 to 5	O, A or S		565 to 675	1050 to 1250
Medium-alloy air-hardening cold work steels										
A2	Slowly	790	1450	925 to 980	1700 to 1800	20 to 45	A		175 to 540	350 to 1000
A3	Slowly	790	1450	955 to 980	1750 to 1800	25 to 60	A		175 to 540	350 to 1000
A4	Slowly	675	1250	815 to 870	1500 to 1600	20 to 45	A		175 to 425	350 to 800
A6	Slowly	650	1200	830 to 870	1525 to 1600	20 to 45	A		150 to 425	300 to 800
A7	Very slowly	815	1500	955 to 980	1750 to 1800	30 to 60	A		150 to 540	300 to 1000
A8	Slowly	790	1450	980 to 1010	1800 to 1850	20 to 45	A		175 to 595	350 to 1100
A9	Slowly	790	1450	980 to 1025	1800 to 1875	20 to 45	A		510 to 620	950 to 1150
A10	Slowly	650	1200	790 to 815	1450 to 1500	30 to 60	A		175 to 425	350 to 800

(a) O, oil quench; A, air cool; S, salt bath quench; W, water quench; B, brine quench. (b) When the high-temperature heating is carried out in a salt bath, the range of temperatures should be about 15 °C (25 °F) lower. (c) Double tempering recommended for not less than 1 h at temperature each time. (d) Triple tempering recommended for not less than 1 h at temperature each time. (e) Times apply to open-furnace heat treatment. For pack hardening, a common rule is to heat 30 min/in. of cross section of the pack. (f) Preferable for large tools to minimize decarburization. (g) Carburizing temperature. (h) After carburizing. (j) Carburized case hardness. (k) P21 is a precipitation-hardening steel having a thermal treatment which involves solution treating and aging rather than hardening and tempering. (m) Recommended for large tools and tools with intricate sections

(continued)

Hardening and tempering of tool steels (continued)

Type	Rate of heating	Hardening				Time at temperature, min.	Quenching medium(a)	Tempering temperature	
		Preheat temperature		Hardening temperature					
		°C	°F	°C	°F			°C	°F
Oil-hardening cold work steels									
O1	Slowly	650	1200	790 to 815	1450 to 1500	10 to 30	O	175 to 260	350 to 500
O2	Slowly	650	1200	760 to 800	1400 to 1475	5 to 20	O	175 to 260	350 to 500
O6	Slowly	...	...	790 to 815	1450 to 1500	10 to 30	O	175 to 315	350 to 600
O7	Slowly	650	1200	790 to 830	W:1450 to 1525	10 to 30	O or W	175 to 290	350 to 550
				845 to 885	O:1550 to 1625				
Shock-resisting steels									
S1	Slowly	...	...	900 to 955	1650 to 1750	15 to 45	O	205 to 650	400 to 1200
S2	Slowly	650(f)	1200(f)	845 to 900	1550 to 1650	5 to 20	B or W	175 to 425	350 to 800
S5	Slowly	760	1400	870 to 925	1600 to 1700	5 to 20	O	175 to 425	350 to 800
S7	Slowly	650 to 705	1200 to 1300	925 to 955	1700 to 1750	15 to 45	A or O	205 to 620	400 to 1150
Mold steels									
P2	...	900 to 925(g)	1650 to 1700(g)	830 to 845(h)	1525 to 1550(h)	15	O	175 to 260	350 to 500
P3	...	900 to 925(g)	1650 to 1700(g)	800 to 830(h)	1475 to 1525(h)	15	O	175 to 260	350 to 500
P4	...	970 to 995(g)	1775 to 1825(g)	970 to 995(h)	1775 to 1825(h)	15	A	175 to 480	350 to 900
P5	...	900 to 925(g)	1650 to 1700(g)	845 to 870(h)	1550 to 1600(h)	15	O or W	175 to 260	350 to 500
P6	...	900 to 925(g)	1650 to 1700(g)	790 to 815(h)	1450 to 1500(h)	15	A or O	175 to 230	350 to 450
P20	...	870 to 900(h)	1600 to 1650(h)	815 to 870	1500 to 1600	15	O	480 to 595(j)	900 to 1100(j)
P21(k)	Slowly	Do not preheat		705 to 730	1300 to 1350	60 to 180	A or O	510 to 550	950 to 1025
Low-alloy special-purpose steels									
L2	Slowly	...	...	W:790 to 845	W:1450 to 1550	10 to 30	O or W	175 to 540	350 to 1000
				O: 845 to 925	O:1550 to 1700				
L3	Slowly	...	...	W: 775 to 815	W: 1425 to 1500	10 to 30	O or W	175 to 315	350 to 600
				O: 815 to 870	O: 1500 to 1600				
L6	Slowly	...	...	790 to 845	1450 to 1550	10 to 30	O	175 to 540	350 to 1000
Carbon-tungsten special-purpose steels									
F1, F2	Slowly	650	1200	790 to 870	1450 to 1600	15	W or B	175 to 260	350 to 500
Water-hardening steels									
W1, W2, W3	Slowly	565 to 650(m)	1050 to 1200(m)	760 to 815	1400 to 1550	10 to 30	B or W	175 to 345	350 to 650
High-carbon, high-chromium cold work steels									
D1, D5	Very slowly	815	1500	980 to 1025	1800 to 1875	15 to 45	A	205 to 540	400 to 1000
D3	Very slowly	815	1500	925 to 980	1700 to 1800	15 to 45	O	205 to 540	400 to 1000
D4	Very slowly	815	1500	970 to 1010	1775 to 1850	15 to 45	A	205 to 540	400 to 1000
D7	Very slowly	815	1500	1010 to 1065	1850 to 1950	30 to 60	A	150 to 540	300 to 1000

(a) O, oil quench; A, air cool; S, salt bath quench; W, water quench; B, brine quench. (b) When the high-temperature heating is carried out in a salt bath, the range of temperatures should be about 15 °C (25 °F) lower. (c) Double tempering recommended for not less than 1 h at temperature each time. (d) Triple tempering recommended for not less than 1 h at temperature each time. (e) Times apply to open-furnace heat treatment. For pack hardening, a common rule is to heat 30 min/in. of cross section of the pack. (f) Preferable for large tools to minimize decarburization. (g) Carburizing temperature. (h) After carburizing. (j) Carburized case hardness. (k) P21 is a precipitation-hardening steel having a thermal treatment which involves solution treating and aging rather than hardening and tempering. (m) Recommended for large tools and tools with intricate sections

Microconstituents in four tool steels after hardening

Steel	Hardening treatment	As-quenched hardness, HRC	Martensite, vol%	Retained austenite, vol%	Undissolved carbides, vol%
W1	790 °C (1450 °F), 30 min; WQ	67.0	88.5	9	2.5
L3	840 °C (1550 °F), 30 min; OQ	66.5	90	7	3.0
M2	1225 °C (2235 °F), 6 min; OQ	64	71.5	20	8.5
D2	1040 °C (1900 °F), 30 min; AC	62	45	40	15

Typical dimensional changes in hardening and tempering

Tool steel	Hardening treatment		Quenching medium	Total change in linear dimensions, % after quenching	Total change in linear dimensions, %, after tempering at:										
	°C	°F			°C 150 / °F 300	205 / 400	260 / 500	315 / 600	370 / 700	425 / 800	480 / 900	510 / 950	540 / 1000	565 / 1050	595 °C / 1100 °F
O1	816	1500	Oil	0.22	0.17	0.16	0.18	...	...	...	...	...	...	...	...
O1	788	1450	Oil	0.18	0.09	0.12	0.13	...	...	...	...	...	...	...	...
O6	788	1450	Oil	0.12	0.07	0.10	0.14	0.10	0.00	−0.05	−0.06	...	−0.07	...	...
A2	954	1750	Air	0.09	0.06	0.06	0.08	0.07	...	0.05	0.04	...	0.06	...	...
A10	788	1450	Air	0.04	0.00	0.00	0.08	0.08	0.01	0.01	0.02	...	0.01	...	0.02
D2	1010	1850	Air	0.06	0.03	0.03	0.02	0.00	...	−0.01	−0.02	...	0.06	...	...
D3	954	1750	Oil	0.07	0.04	0.02	0.01	−0.02	...	...	...	...	...	...	...
D4	1038	1900	Air	0.07	0.03	0.01	−0.01	−0.03	...	−0.04	−0.03	...	0.05	...	...
D5	1010	1850	Air	0.07	0.03	0.02	0.01	0.00	...	0.3	0.03	...	0.05	...	...
H11	1010	1850	Air	0.11	0.06	0.07	0.08	0.08	...	0.3	0.01	...	0.12	...	...
H13	1010	1850	Air	−0.01	...	...	...	...	...	...	0.00	...	0.06	...	...
M2	1210	2210	Oil	−0.02	...	...	...	...	...	...	...	−0.06	0.10	0.14	0.16
M41	1210	2210	Oil	−0.16	...	...	...	...	...	...	...	−0.17	0.08	0.21	0.23

Machining

Standard machining allowances for hot rolled square and flat bars

| Specified width | | Machining allowance(a) | | | |
| | | Top and bottom surfaces | | Edges | |
mm	in.	mm	in.	mm	in.
Specified thickness, up to 12.7 mm (1/2 in.)					
0 to 12.7	0 to 1/2	0.64	0.025	0.64	0.025
>12.7 to 25.4	>1/2 to 1	0.64	0.025	0.89	0.035
>25.4 to 50.8	>1 to 2	0.76	0.030	1.02	0.040
>50.8 to 76.2	>2 to 3	0.89	0.035	1.27	0.050
>76.2 to 101.6	>3 to 4	1.02	0.040	1.65	0.065
>101.6 to 127.0	>4 to 5	1.14	0.045	2.03	0.080
>127.0 to 152.4	>5 to 6	1.27	0.050	2.41	0.095
>152.4 to 177.8	>6 to 7	1.40	0.055	2.67	0.105
>177.8 to 203.2	>7 to 8	1.52	0.060	3.05	0.120
>203.2 to 228.6	>8 to 9	1.52	0.060	3.30	0.130
>228.6 to 304.8	>9 to 12	1.52	0.060	3.56	0.140
Specified thickness, >12.7 to 25.4 mm (>1/2 to 1 in.)					
>12.7 to 25.4	>1/2 to 1	1.14	0.045	1.14	0.045
>25.4 to 50.8	>1 to 2	1.14	0.045	1.27	0.050
>50.8 to 76.2	>2 to 3	1.27	0.050	1.52	0.060
>76.2 to 101.6	>3 to 4	1.40	0.055	1.90	0.075
>101.6 to 127.0	>4 to 5	1.52	0.060	2.41	0.095
>127.0 to 152.4	>5 to 6	1.65	0.065	2.92	0.115
>152.4 to 177.8	>6 to 7	1.78	0.070	3.30	0.130
>177.8 to 203.2	>7 to 8	1.90	0.075	3.81	0.150
>203.2 to 228.6	>8 to 9	1.90	0.075	3.94	0.155
>228.6 to 304.8	>9 to 12	1.90	0.075	3.94	0.155
Specified thickness, >25.4 to 50.8 mm (>1 to 2 in.)					
>25.4 to 50.8	>1 to 2	1.65	0.065	1.65	0.065
>50.8 to 76.2	>2 to 3	1.65	0.065	1.78	0.070
>76.2 to 101.6	>3 to 4	1.78	0.070	2.16	0.085
>101.6 to 127.0	>4 to 5	1.78	0.070	2.67	0.105
>127.0 to 152.4	>5 to 6	1.90	0.075	3.18	0.125
>152.4 to 177.8	>6 to 7	2.03	0.080	3.68	0.145
>177.8 to 203.2	>7 to 8	2.03	0.080	4.19	0.165
>203.2 to 228.6	>8 to 9	2.41	0.095	4.32	0.170
>228.6 to 304.8	>9 to 12	2.54	0.100	4.32	0.170
Specified thickness, >50.8 to 76.2 mm (>2 to 3 in.)					
>50.8 to 76.2	>2 to 3	2.16	0.085	2.16	0.085
>76.2 to 101.6	>3 to 4	2.16	0.085	2.54	0.100
>101.6 to 127.0	>4 to 5	2.16	0.085	3.05	0.120
>127.0 to 152.4	>5 to 6	2.16	0.085	3.43	0.135
>152.4 to 177.8	>6 to 7	2.29	0.090	3.94	0.155
>177.8 to 203.2	>7 to 8	2.54	0.100	4.32	0.170
>203.2 to 228.6	>8 to 9	2.54	0.100	4.83	0.190
>228.6 to 304.8	>9 to 12	2.54	0.100	4.83	0.190
Specified thickness, >76.2 to 101.6 mm (>3 to 4 in.)					
>76.2 to 101.6	>3 to 4	2.92	0.115	2.92	0.115
>101.6 to 127.0	>4 to 5	2.92	0.115	3.18	0.125
>127.0 to 152.4	>5 to 6	2.92	0.115	3.56	0.140
>152.4 to 177.8	>6 to 7	2.92	0.115	4.32	0.170
>177.8 to 203.2	>7 to 8	3.18	0.125	4.83	0.190
>203.2 to 228.6	>8 to 9	3.18	0.125	4.83	0.190
>228.6 to 304.8	>9 to 12	3.18	0.125	4.83	0.190

(a) Minimum allowance per side for machining prior to heat treatment. Maximum decarburization limit, 80% of machining allowance

ISO R513 classification of carbides according to use for machining

Designation(a)	Material to be machined	Use and working conditions
P 01	Steel, steel castings	Finish turning and boring; high cutting speeds, small chip section, accuracy of dimensions and fine finish, vibration-free operation
P 10	Steel, steel castings	Turning, copying, threading and milling; high cutting speeds, small or medium chip sections
P 20	Steel, steel castings Malleable cast iron with long chips	Turning, copying, milling, medium cutting speeds and chip sections; planing with small chip sections
P 30	Steel, steel castings Malleable cast iron with long chips	Turning, milling, planing, medium or low cutting speeds, medium or large chip sections, and machining in unfavorable conditions(b)
P 40	Steel Steel castings with sand inclusion and cavities	Turning, planing, slotting, low cutting speeds, large chip sections with the possibility of large cutting angles for machining in unfavorable conditions(b) and work on automatic machines
P 50	Steel Steel castings of medium or low tensile strength, with sand inclusion and cavities	For operations demanding very tough carbide: turning, planing, slotting, low cutting speeds, large chip sections, with the possibility of large cutting angles for machining in unfavorable conditions(b) and work on automatic machines
M 10	Steel, steel castings, manganese steel Grey cast iron, alloy cast iron	Turning, medium or high cutting. Small or medium chip sections
M 20	Steel, steel castings, austenitic or manganese steel, grey cast iron	Turning, milling. Medium cutting speeds and chip sections
M 30	Steel, steel castings, austenitic steel, grey cast iron, high temperature resistant alloys	Turning, milling, planing. Medium cutting speeds, medium or large chip sections
M 40	Mild free cutting steel, low tensile steel Nonferrous metals and light alloys	Turning, parting off, particularly on automatic machines
K 01	Very hard grey cast iron, chilled castings of over 85 scleroscope hardness, high silicon aluminum alloys, hardened steel, highly abrasive plastics, hard card-board, ceramics	Turning, finish turning, boring, milling, scraping
K 10	Grey cast iron over 220 HB malleable cast iron with short chips, hardened steel, silicon aluminum alloys, copper alloys, plastics, glass, hard rubber, hard card-board, porcelain, stone	Turning, milling, drilling, boring, broaching, scraping
K 20	Grey cast iron up to 220 HB nonferrous metals: cop-per, brass, aluminum	Turning, milling, planing, boring, broaching, demanding very tough carbide
K 30	Low hardness grey cast iron, low tensile steel, com-pressed wood	Turning, milling, planing, slotting, for machining in unfavorable conditions(b) and with the possibility of large cutting angles
K 40	Soft wood or hard work Nonferrous metals	Turning, milling, planing, slotting, for machining in unfavorable conditions(b) and with the possibility of large cutting angles

(a) In each letter category, low designation numbers are for high speeds and light feeds, higher numbers for slower speeds and/or heavier feeds. Also, increasing designation numbers imply increasing toughness and decreasing wear resistance of the cemented carbide materials. (b) Unfavorable conditions include: shapes that are awkward to machine; material having a casting or forging skin; material having variable hardness; and machining that involves variable depth of cut, interrupted cut or moderate to severe vibrations

Approximate machinability ratings for annealed tool steels

Type	Machinability rating
O6	125
W1, W2, W5	100(a)
A10	90
P2, P3, P4, P5, P6	75 to 90
P20, P21	65 to 80
L2, L6	65 to 75
S1, S2, S5, S6, S7	60 to 70
H10, H11, H13, H14, H19	60 to 70(b)
O1, O2, O7	45 to 60
A2, A3, A4, A6, A8, A9	45 to 60
H21, H22, H24, H25, H26, H42	45 to 55(b)
T1	40 to 50
M2	40 to 50
T4	35 to 40
M3 (class 1)	35 to 40
D2, D3, D4, D5, D7, A7	30 to 40
T15	25 to 30
M15	25 to 30

(a) Equivalent to approximately 30% of the machinability of B1112. (b) For hardness range 150 to 200 HB

Wrought Stainless Steels

Family Relationships

Family relationships for standard austenitic stainless steels

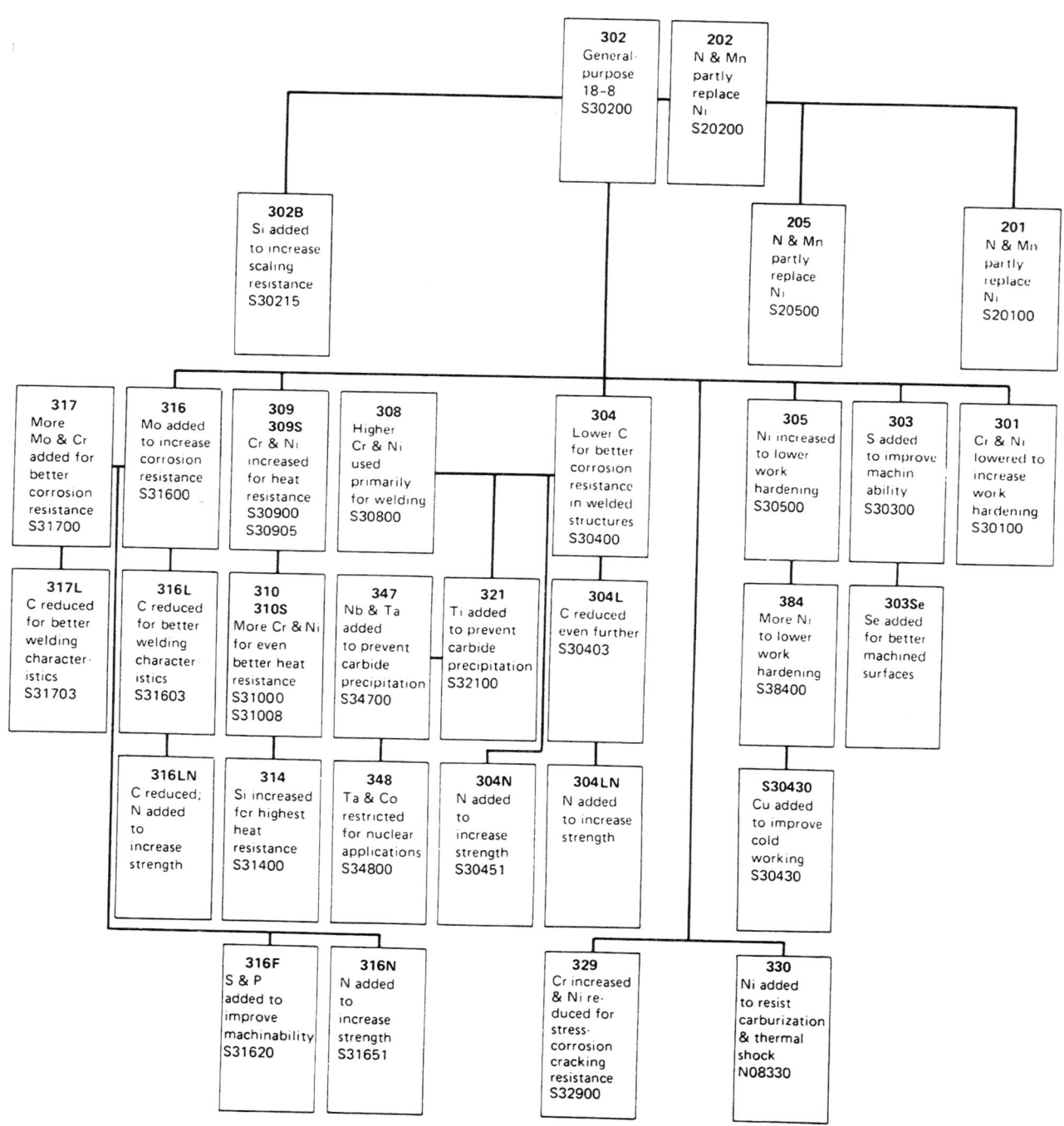

Family relationships for standard ferritic stainless steels

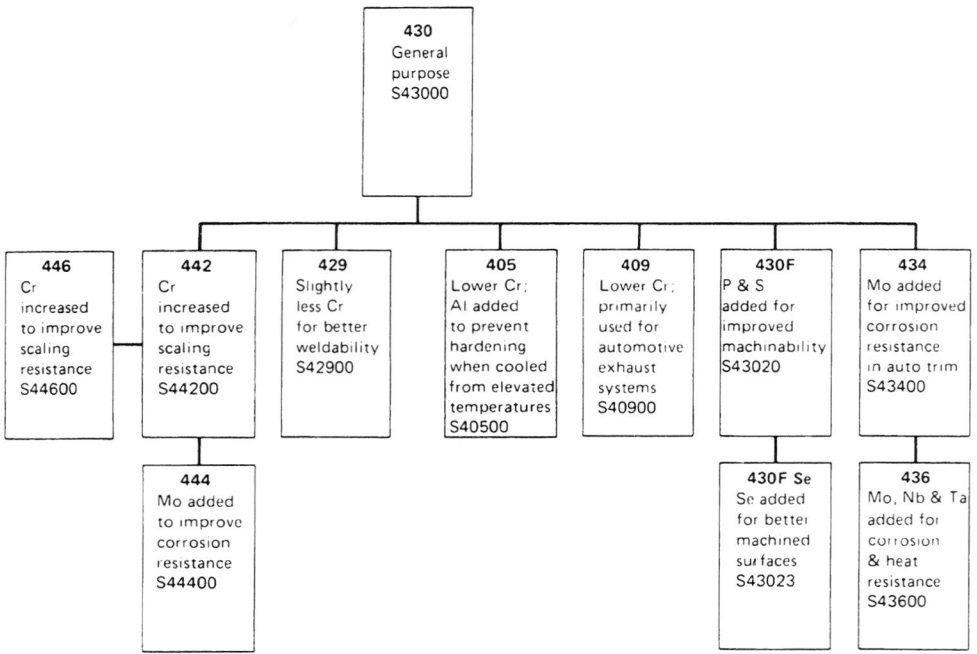

Family relationships for standard martensitic stainless steels

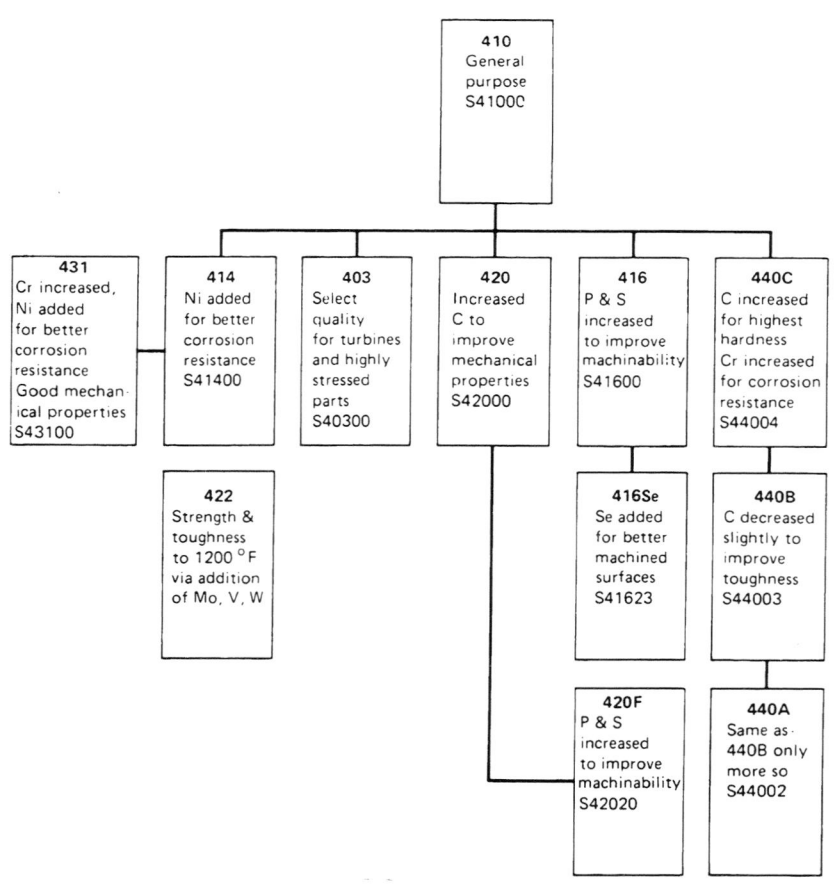

A guide for the selection of stainless steels

Type	Composition or alloy content	Microstructure	Mechanical properties		Elongation in 50 mm, %	Physical characteristics	Applications
			Strength, MPa				
			Tensile	Yield			
Austenitic	15 to 27% Cr, 8 to 35% Ni, 0 to 6% Mo, Cu, N	Austenite	490 to 860	205 to 575	30 to 60	Non heat-treatable-nonmagnetic	Most widely used in general applications
Ferritic	11 to 30% Cr, 0 to 4% Ni, 0 to 4% Mo	Ferrite	415 to 650	275 to 550	10 to 25	Non heat-treatable; magnetic; good resistance to chloride stress-corrosion-cracking	Parts requiring combination of good general corrosion resistance with good stress corrosion resistance, seawater applications
Martensitic	11 to 18% Cr, 0 to 6% Ni, 0 to 2% Mo	Martensite	480 to 1000	275 to 860	14 to 30	Hardenable by heat treatment; high strength	High-strength parts, pumps, valves and paper machinery
Duplex	18 to 27% Cr, 4 to 7% Ni; 2 to 4% Mo, Cu, N	Austenite and ferrite	680 to 900	410 to 900	10 to 48	Non heat-treatable	Shell-and-tube heat exchangers, wastewater treatment and cooling coils
Precipitation hardening	12 to 28% Cr, 4 to 24% Ni, 1 to 5% Mo, Al, Ti, Co	Austenite and martensite	895 to 1100	276 to 1000	10 to 35	Hardenable by heat treatment; very high strength	Parts requiring high strength, corrosion- and/or high-temperature resistance

Characteristics and typical applications for standard stainless steels

Type (UNS)	Characteristics and typical applications	Type (UNS)	Characteristics and typical applications
Austenitic		(S30430)	Lower work-hardening rate than type 305. Severe cold-heading applications
201 (S20100)	High work-hardening rate; low-nickel equivalent of type 301. Flatware, automobile wheel covers, trim		
202 (S20200)	General-purpose low-nickel equivalent of type 302. Kitchen equipment, hub caps, milk handling	304N (S30451)	Higher nitrogen than type 304 to increase strength with minimum effect on ductility and corrosion resistance. More resistant to increased magnetic permeability. Type 304 applications requiring higher strength
205 (S20500)	Lower work-hardening rate than type 202; used for spinning and special drawing operations. Nonmagnetic and cryogenic parts	305 (S30500)	Low work-hardening rate; used for spin forming, severe drawing, cold heading, and forming. Coffee urn tops, mixing bowls, reflectors
301 (S30100)	High work-hardening rate; used for structural applications where high strength plus high ductility is required. Railroad cars, trailer bodies, aircraft structurals, fasteners, automobile wheel covers, trim, pole line hardware	308 (S30800)	Higher-alloy steel having high corrosion and heat resistance. Welding filler metals to compensate for alloy loss in welding, industrial furnaces
302 (S30200)	General-purpose austenitic stainless steel. Trim, food handling equipment, aircraft cowlings, antennas, springs, cookware, building exteriors, hospital tanks, household appliances, jewelry, oil refining equipment, signs	309 (S30900)	High temperature strength and scale resistance. Aircraft heaters, heat treating equipment, annealing covers, furnace parts, heat exchangers, heat treating trays, oven linings, pump parts
302B (S30215)	More resistant to scale than type 302. Furnace parts: still liners, heating elements, annealing covers, burner sections	309S (S30908)	Low-carbon modification of type 309. Welded constructions: assemblies subject to moist corrosion conditions
303 (S30300)	Free-machining modification of type 302. For heavier cuts. Screw machine products, shafts, valves, bolts, bushings, nuts	310 (S31000)	Higher elevated temperature strength and scale resistance than type 309. Heat exchangers, furnace parts, combustion chambers, welding filler metals, gas turbine parts, incinerators, recuperators, rolls for roller hearth furnaces
303Se (S30323)	Free-machining modification of type 302. For lighter cuts: used where hot working or cold heading may be involved. Aircraft fittings, bolts, nuts, rivets, screws, studs	310S (S31008)	Low-carbon modification of type 310. Welded construction, jet engine rings
304 (S30400)	Low-carbon modification of type 302 for restriction of carbide precipitation during welding. Chemical and food processing equipment, brewing equipment, cryogenic vessels, gutters, downspouts, flashings	314 (S31400)	More resistant to scale than type 310. Severe cold heading or forming applications. Annealing and carburizing boxes, heat treating fixtures, radiant tubes
304L (S30403)	Extra-low-carbon modification of type 304 for further restriction of carbide precipitation during welding. Coal hopper linings, tanks for liquid fertilizer and tomato paste	316 (S31600)	Higher corrosion resistance than types 302 and 304; high creep strength. Chemical and pulp handling equipment, photographic equipment, brandy vats, fertilizer parts, ketchup cooking kettles, yeast tubs

(continued)

Characteristics and typical applications for standard stainless steels (continued)

Type (UNS)	Characteristics and typical applications	Type (UNS)	Characteristics and typical applications
316L (S31603)	Extra-low-carbon modification of type 316. Welded construction where intergranular carbide precipitation must be avoided. Type 316 applications requiring extensive welding	442 (S44200)	High-chromium steel, principally for parts which must resist high service temperatures without scaling. Furnace parts, nozzles, combustion chambers
316F (S31620)	Higher phosphorus and sulfur than type 316 to improve machining and nonseizing characteristics. Automatic screw machine parts	446 (S44600)	High-resistance to corrosion and scaling at high temperatures especially for intermittent service; often used in sulfur-bearing atmosphere. Annealing boxes, combustion chambers, glass molds, heaters, pyrometer tubes, recuperators, stirring rods, valves
316N (S31651)	Higher nitrogen than type 316 to increase strength with minimum effect on ductility and corrosion resistance. Type 316 applications requiring extra strength	**Martensitic**	
317 (S31700)	Higher corrosion and creep resistance than type 316. Dyeing and ink manufacturing equipment	403 (S40300)	"Turbine quality" grade. Steam turbine blading and other highly stressed parts including jet engine rings
317L (S31703)	Extra-low-carbon modification of type 317 for restriction of carbide precipitation during welding. Welded assemblies	410 (S41000)	General-purpose heat treatable type. Machine parts, pump shafts, belts, bushings, coal chutes, cutlery, finishing tackle, hardware, jet engine parts, mining machinery, rifle barrels, screws, valves
321 (S32100)	Stabilized for weldments subject to severe corrosive conditions, and for service from 149 to 871 °C (300 to 1600 °F). Aircraft exhaust manifolds, boiler shells, process equipment, expansion joints, cabin heaters, fire walls, flexible couplings, pressure vessels	414 (S41400)	High hardenability steel. Springs, tempered rules, machine parts, bolts, mining machinery, scissors, ships' belts, spindles, valve seats
329 (S32900)	Austenitic-ferritic type with general corrosion resistance similar to type 316 but with better resistance to stress-corrosion cracking; capable of age hardening. Valves, valve fittings, piping, pump parts	416 (S41600)	Free-machining modification of type 410; for heavier cuts. Aircraft fittings, bolts, nuts, fire extinguisher inserts, rivets, screws
330 (N08330)	Good resistance to carburization and to heat and thermal shock. Heat treating fixtures	416Se (S41623)	Free-machining modification of type 410; for lighter cuts. Machined parts requiring hot working or cold heading
347 (S34700)	Similar to type 321 with higher creep strength. Airplane exhaust stacks, welded tank cars for chemicals, jet engine parts	420 (S42000)	Higher carbon modification of type 410. Cutlery, surgical instruments, valves, wear-resisting parts, glass molds, hand tools, vegetable choppers
348 (S34800)	Similar to type 321; low retentivity. Tubes and pipes for radioactive systems, nuclear energy uses	420F (S42020)	Free-machining modification of type 420. Applications similar to those for type 420 requiring better machinability
384 (S38400)	Suitable for severe cold-heading or cold-forming; lower cold-work hardening rate than type 304. Bolts, rivets, screws, instrument parts	422 (S42200)	High strength and toughness at service temperatures up to 649 °C (1200 °F). Steam turbine blades, fasteners
Ferritic		431 (S43100)	Special-purpose hardenable steel used where particularly high mechanical properties are required. Aircraft fittings, beater bars, paper machinery, belts
405 (S40500)	Nonhardenable grade for assemblies where air-hardening types such as 410 or 403 are objectionable. Annealing boxes, quenching racks, oxidation-resistant partitions	440A (S44002)	Hardenable to higher hardness than type 420 with good corrosion resistance. Cutlery, bearings, surgical tools
409 (S40900)	General-purpose construction stainless. Automotive exhaust systems, transformer and capacitor cases, dry fertilizer spreaders, tanks for agricultural sprays	440B (S44003)	Cutlery grade. Cutlery, valve parts, instrument bearings
429 (S42900)	Improved weldability as compared to type 430. Nitric acid and nitrogen fixation-equipment	440C (S44004)	Yields highest hardnesses of hardenable stainless steels. Balls, bearings, races, nozzles, balls and seats for oil well pumps, valve parts
430 (S43000)	General-purpose nonhardenable chromium type. Decorative trim, nitric acid tanks, annealing baskets, combustion chambers, dishwashers, heaters, mufflers, range hoods, recuperators, restaurant equipment	**Heat resisting**	
		501 (S50100)	Heat resistance; good mechanical properties at moderately elevated temperatures. Heat exchangers, petroleum refining equipment
430F (S43020)	Free-machining modification of type 430, for heavier cuts. Screw machine parts	502 (S50200)	More ductility and less strength than type 501. Heat exchangers, petroleum refining equipment, gaskets
430FSe (S43023)	Free-machining modification of type 430, for lighter cuts. Machined parts requiring light cold heading or forming	**Precipitation-hardening**	
434 (S43400)	Modification of type 430 designed to resist atmospheric corrosion in the presence of winter road-conditioning and dust-laying compounds. Automotive trim and fasteners	PH13-8 Mo (S13800)	Martensitic (maraging) stainless that can be hardened by a single low-temperature heat treatment. Forged aircraft parts
436 (S43600)	Similar to types 430 and 434. Used where low "roping" or "ridging" required. General corrosion and heat-resistant applications such as automobile trim	15-5 PH (S15500)	Martensitic (maraging) stainless with high strength, hardness, and corrosion resistance. Gears, cams, cutlery, shafting, aircraft parts
		17-4 PH (S17400)	Similar to S15500, but with slightly higher chromium content. Gears, springs, cutlery, fasteners, aircraft and turbine parts
		17-7 PH (S17700)	Semiaustenitic stainless. Can be cold drawn and then hardened by a low-temperature heat treatment. Springs, knives, pressure vessels

Composition Limits

Compositions of standard stainless steels

Type	UNS number	C	Mn	Si	Cr	Ni(b)	P	S	Others
Austenitic									
201	S20100	0.15	5.5-7.5	1.00	16.0-18.0	3.5-5.5	0.06	0.03	0.25 N
202	S20200	0.15	7.5-10.0	1.00	17.0-19.0	4.0-6.0	0.06	0.03	0.25 N
205	S20500	0.12-0.25	14.0-15.5	1.00	16.5-18.0	1.0-1.75	0.06	0.03	0.32-0.40 N
301	S30100	0.15	2.00	1.00	16.0-18.0	6.0-8.0	0.045	0.03	…
302	S30200	0.15	2.00	1.00	17.0-19.0	8.0-10.0	0.045	0.03	…
302B	S30215	0.15	2.00	2.0-3.0	17.0-19.0	8.0-10.0	0.045	0.03	…
303	S30300	0.15	2.00	1.00	17.0-19.0	8.0-10.0	0.20	0.15 min	0.6 Mo(c)
303Se	S30323	0.15	2.00	1.00	17.0-19.0	8.0-10.0	0.20	0.06	0.15 Se min
304	S30400	0.08	2.00	1.00	18.0-20.0	8.0-10.5	0.045	0.03	…
304H	S30409	0.04-0.10	2.00	1.00	18.0-20.0	8.0-10.5	0.045	0.03	…
304L	S30403	0.03	2.00	1.00	18.0-20.0	8.0-12.0	0.045	0.03	…
304LN	…	0.03	2.00	1.00	18.0-20.0	8.0-10.5	0.045	0.03	0.10-0.15 N
S30430	S30430	0.08	2.00	1.00	17.0-19.0	8.0-10.0	0.045	0.03	3.0-4.0 Cu
304N	S30451	0.08	2.00	1.00	18.0-20.0	8.0-10.5	0.045	0.03	0.10-0.16 N
305	S30500	0.12	2.00	1.00	17.0-19.0	10.5-13.0	0.045	0.03	…
308	S30800	0.08	2.00	1.00	19.0-21.0	10.0-12.0	0.045	0.03	…
309	S30900	0.20	2.00	1.00	22.0-24.0	12.0-15.0	0.045	0.03	…
309S	S30908	0.08	2.00	1.00	22.0-24.0	12.0-15.0	0.045	0.03	…
310	S31000	0.25	2.00	1.50	24.0-26.0	19.0-22.0	0.045	0.03	…
310S	S31008	0.08	2.00	1.50	24.0-26.0	19.0-22.0	0.045	0.03	…
314	S31400	0.25	2.00	1.5-3.0	23.0-26.0	19.0-22.0	0.045	0.03	…
316	S31600	0.08	2.00	1.00	16.0-18.0	10.0-14.0	0.045	0.03	2.0-3.0 Mo
316F	S31620	0.08	2.00	1.00	16.0-18.0	10.0-14.0	0.20	0.10 min	1.75-2.5 Mo
316H	S31609	0.04-0.10	2.00	1.00	16.0-18.0	10.0-14.0	0.045	0.03	2.0-3.0 Mo
316L	S31603	0.03	2.00	1.00	16.0-18.0	10.0-14.0	0.045	0.03	2.0-3.0 Mo
316LN	…	0.03	2.00	1.00	16.0-18.0	10.0-14.0	0.045	0.03	2.0-3.0 Mo; 0.10-0.30 N
316N	S31651	0.08	2.00	1.00	16.0-18.0	10.0-14.0	0.045	0.03	2.0-3.0 Mo; 0.10-0.16 N
317	S31700	0.08	2.00	1.00	18.0-20.0	11.0-15.0	0.045	0.03	3.0-4.0 Mo
317L	S31703	0.03	2.00	1.00	18.0-20.0	11.0-15.0	0.045	0.03	3.0-4.0 Mo
321	S32100	0.08	2.00	1.00	17.0-19.0	9.0-12.0	0.045	0.03	5 × %C Ti min
321H	S32109	0.04-0.10	2.00	1.00	17.0-19.0	9.0-12.0	0.045	0.03	5 × %C Ti min
329	S32900	0.10	2.00	1.00	25.0-30.0	3.0-6.0	0.045	0.03	1.0-2.0 Mo
330	N08330	0.08	2.00	0.75-1.5	17.0-20.0	34.0-37.0	0.04	0.03	…
347	S34700	0.08	2.00	1.00	17.0-19.0	9.0-13.0	0.045	0.03	10 × %C Nb + Ta(d) min
347H	S34709	0.04-0.10	2.00	1.00	17.0-19.0	9.0-13.0	0.045	0.03	10 × %C Nb + Ta min
348	S34800	0.08	2.00	1.00	17.0-19.0	9.0-13.0	0.045	0.03	0.2 Cu; 10 × %C Nb + Ta(d) min
348H	S34809	0.04-0.10	2.00	1.00	17.0-19.0	9.0-13.0	0.045	0.03	0.2 Cu; 10 × %C Nb + Ta(d) min
384	S38400	0.08	2.00	1.00	15.0-17.0	17.0-19.0	0.045	0.03	…
Ferritic									
405	S40500	0.08	1.00	1.00	11.5-14.5	…	0.04	0.03	0.10-0.30 Al
409	S40900	0.08	1.00	1.00	10.5-11.75	…	0.045	0.045	6 × %C Ti(e) min
429	S42900	0.12	1.00	1.00	14.0-16.0	…	0.04	0.03	…
430	S43000	0.12	1.00	1.00	16.0-18.0	…	0.04	0.03	…
430F	S43020	0.12	1.25	1.00	16.0-18.0	…	0.06	0.15 min	0.6 Mo(c)
430FSe	S43023	0.12	1.25	1.00	16.0-18.0	…	0.06	0.06	0.15 Se min
434	S43400	0.12	1.00	1.00	16.0-18.0	…	0.04	0.03	0.75-1.25 Mo
436	S43600	0.12	1.00	1.00	16.0-18.0	…	0.04	0.03	0.75-1.25 Mo; 5 × %C Nb + Ta(f) min
442	S44200	0.20	1.00	1.00	18.0-23.0	…	0.04	0.03	…
446	S44600	0.20	1.50	1.00	23.0-27.0	…	0.04	0.03	0.25 N
Martensitic									
403	S40300	0.15	1.00	0.50	11.5-13.0	…	0.04	0.03	…
410	S41000	0.15	1.00	1.00	11.5-13.0	…	0.04	0.03	…

(a) Single values are maximum values unless otherwise indicated. (b) For some tubemaking processes, the nickel content of certain austenitic types must be slightly higher than shown. (c) Optional. (d) 0.10% Ta max. (e) 0.75% maximum. (f) 0.70% maximum

(continued)

Compositions of standard stainless steels (continued)

Type	UNS number	C	Mn	Si	Cr	Ni(b)	P	S	Others
414	S41400	0.15	1.00	1.00	11.5-13.5	1.25-2.50	0.04	0.03	...
416	S41600	0.15	1.25	1.00	12.0-14.0	...	0.04	0.03	0.6 Mo(c)
416Se	S41623	0.15	1.25	1.00	12.0-14.0	...	0.06	0.06	0.15 Se min
420	S42000	0.15 min	1.00	1.00	12.0-14.0	...	0.04	0.03	...
420F	S42020	0.15 min	1.25	1.00	12.0-14.0	...	0.06	0.15 min	0.6 Mo(c)
422	S42200	0.20-0.25	1.00	0.75	11.0-13.0	0.5-1.0	0.025	0.025	0.75-1.25 Mo; 0.75-1.25 W; 0.15-0.3 V
431	S43100	0.20	1.00	1.00	15.0-17.0	1.25-2.50	0.04	0.03	...
440A	S44002	0.60-0.75	1.00	1.00	16.0-18.0	...	0.04	0.03	0.75 Mo
440B	S44003	0.75-0.95	1.00	1.00	16.0-18.0	...	0.04	0.03	0.75 Mo
440C	S44004	0.95-1.20	1.00	1.00	16.0-18.0	...	0.04	0.03	0.75 Mo
501	S50100	0.10 min	1.00	1.00	4.0-6.0	...	0.04	0.03	0.40-0.65 Mo
501A	S50300	0.15	0.30-0.60	0.50-1.00	6.0-8.0	...	0.03	0.03	0.45-0.65 Mo
501B	S50400	0.15	0.30-0.60	0.50-1.00	8.0-10.0	...	0.03	0.03	0.9-1.1 Mo
502	S50200	0.10	1.00	1.00	4.0-6.0	...	0.04	0.03	0.40-0.65 Mo
503	S50300	0.15	1.00	1.00	6.0-8.0	...	0.04	0.04	0.45-0.65 Mo
504	S50400	0.15	1.00	1.00	8.0-10.0	...	0.04	0.04	0.9-1.1 Mo
Precipitation-hardening									
PH 13-8Mo	S13800	0.05	0.10	0.10	12.25-13.25	7.5-8.5	0.01	0.008	2.0-2.5 Mo; 0.90-1.35 Al; 0.01 N
15-5 PH	S15500	0.07	1.00	1.00	14.0-15.5	3.5-5.5	0.04	0.03	2.5-4.5 Cu; 0.15-0.45 Nb + Ta
17-4 PH	S17400	0.07	1.00	1.00	15.5-17.5	3.0-5.0	0.04	0.03	3.0-5.0 Cu; 0.15-0.45 Nb + Ta
17-7 PH	S17700	0.09	1.00	1.00	16.0-18.0	6.5-7.75	0.04	0.03	0.75-1.5 Al

(a) Single values are maximum values unless otherwise indicated. (b) For some tubemaking processes, the nickel content of certain austenitic types must be slightly higher than shown. (c) Optional. (d) 0.10% Ta max. (e) 0.75% maximum. (f) 0.70% maximum

Compositions of nonstandard stainless steels

Designation(a)	UNS number	C	Mn	Si	Cr	Ni	P	S	Others
Austenitic									
Type 216 (XM-17)	S21600	0.08	7.5-9.0	1.00	17.5-22.0	5.0-7.0	0.045	0.03	2.0-3.0 Mo; 0.25-0.50 N
Type 304HN	S30452	0.04-0.10	2.00	1.00	18.0-20.0	8.0-10.5	0.045	0.03	0.10-0.16 N
Type 308	S30800	0.08	2.00	1.00	19.0-21.0	10.0-12.0	0.045	0.03	...
Type 308L	...	0.03	2.00	1.00	19.0-21.0	10.0-12.0	0.045	0.03	...
Type 309S	S30908	0.08	2.00	1.00	22.0-24.0	12.0-15.0	0.045	0.03	...
Type 309S Cb	S30940	0.08	2.00	1.00	22.0-24.0	12.0-15.0	0.045	0.03	8 × %C Nb min
Type 309 Cb + Ta	...	0.08	2.00	1.00	22.0-24.0	12.0-15.0	0.045	0.03	8 × %C Nb + Ta min
Type 312	...	0.15	2.00	1.00	30.0 nom	9.0 nom	0.045	0.03	...
Type 317LM	...	0.03	2.00	1.00	18.0-20.0	12.0-16.0	0.045	0.03	4.0-5.0 Mo
Type 330HC	...	0.40	1.50	1.25	19.0 nom	35.0 nom	...	...	...
Type 332	...	0.04	1.00	0.50	21.5 nom	32.0 nom	0.045	0.03	...
Type 385	...	0.08	2.00	1.00	11.5-13.5	14.0-16.0	0.045	0.03	...
904L	N08904	0.02	2.00	1.00	19.0-23.0	23.0-28.0	0.045	0.035	4.0-5.0 Mo; 1.0-2.0 Cu
18-18-2 (XM-15)	S38100	0.08	2.00	1.5-2.5	17.0-19.0	17.5-18.5	0.03	0.03	0.08-0.18 N
18-18 Plus	S28200	0.15	17.0-19.0	1.00	17.5-19.5	...	0.045	0.03	0.5-1.5 Mo; 0.5-1.5 Cu; 0.4-0.6 N
20Cb-3	N08020	0.07	2.00	1.00	19.0-21.0	32.0-38.0	0.045	0.035	2.0-3.0 Mo; 3.0-4.0 Cu; 8 × %C Nb min(c)
AL-6X	N08366	0.03	2.00	0.75	20.0-22.0	23.5-25.5	0.030	0.003	6.0-7.0 Mo
303 Plus X (XM-5)	...	0.15	2.5-4.5	1.00	17.0-19.0	7.0-10.0	0.20	0.25 min	0.6 Mo
HNM(d)	...	0.30	3.5	0.5	18.5	9.5	0.25	...	...
Crutemp 25(d)	...	0.05	1.5	0.4	25.0	25.0	...	...	...

(a) XM designations in this column are ASTM designations for the listed alloy. Type numbers in parentheses are obsolete AISI designations. (b) Single values are maximum values unless otherwise indicated. (c) 1.00% maximum. (d) Nominal composition; composition limits not available. (e) 0.50% maximum. (f) 0.75% maximum. (g) 0.80 maximum

(continued)

Compositions of nonstandard stainless steels (continued)

Designation(a)	UNS number	Composition(b), %							
		C	Mn	Si	Cr	Ni	P	S	Others
JS-700	N08700	0.04	2.00	1.00	19.0-23.0	24.0-26.0	0.04	0.03	4.3-5.0 Mo; 0.5 Cu; 8 × %C Nb min(e); 0.005 Pb; 0.035 Sn
JS-777	...	0.04	2.00	1.00	19.0-23.0	24.0-26.0	0.045	0.035	4.0-5.0 Mo; 1.9-2.5 Cu
Nitronic 32(d)	S24100	0.10	12.0	0.5	18.0	1.6	...	...	0.35 N
Nitronic 33(d)	S24000	0.06	13.0	0.5	18.0	3.0	...	...	0.30 N
Nitronic 40 (21-6-9) (XM-10)	S21900	0.08	8.0-10.0	1.00	18.0-20.0	5.0-7.0	0.06	0.03	0.15-0.40 N
Nitronic 50 (22-13-5) (XM-19)	S20910	0.06	4.0-6.0	1.00	20.5-23.5	11.5-13.5	0.04	0.03	1.5-3.0 Mo; 0.2-0.3 N; 0.1-0.3 Nb; 0.1-0.3 V
Nitronic 60	S21800	0.10	7.0-9.0	3.5-4.5	16.0-18.0	8.0-9.0	0.04	0.03	...
Tenelon (XM-31)	S21400	0.12	14.5-16.0	0.3-1.0	17.0-18.5	0.75	0.045	0.03	0.35 N
Cryogenic Tenelon (XM-14)	S21460	0.12	14.0-16.0	1.00	17.0-19.0	5.0-6.0	0.06	0.03	0.35-0.50 N
Ferritic									
Type 404	...	0.05	1.00	0.50	11.0-12.5	1.25-2.00	0.03	0.03	...
Type 430Ti	S43036	0.10	1.00	1.00	16.0-19.5	0.75	0.04	0.03	5 × %C Ti min(f)
Type 444 (18-2)	S44400	0.025	1.00	1.00	17.5-19.5	1.00	0.04	0.03	1.75-2.5 Mo; 0.035 N max; 0.2 + 4(%C + %N) (Ti + Nb) min
18SR(d)	...	0.04	0.3	1.00	18.0	...	...	...	2.0 Al; 0.4 Ti
18-2 FM	S18200	0.08	2.50	1.00	17.5-19.5	...	0.04	0.15 min	...
E Brite 26-1 (XM-27)	S44625	0.01	0.40	0.40	25.0-27.5	0.50	0.02	0.02	0.75-1.5 Mo; 0.015 N; 0.2 Cu; 0.5 Ni + Cu
26-1 Ti (XM-33)	S44626	0.06	0.75	0.75	25.0-27.0	0.50	0.04	0.02	0.75-1.5 Mo; 0.04 N; 0.2 Cu; 0.2-1.0 Ti(g)
29-4	S44700	0.010	0.30	0.20	28.0-30.0	0.15	0.025	0.02	3.5-4.2 Mo
29-4-2	S44800	0.010	0.30	0.20	28.0-30.0	2.0-2.5	0.025	0.02	3.5-4.2 Mo
Monit	S44635	0.25	1.00	0.75	24.5-26.0	3.5-4.5	0.04	0.03	3.5-4.5 Mo; 0.3-0.6 (Ti + Nb)
Sea-cure/Sc-1	S44660	0.025	1.00	0.75	25.0-27.0	1.5-3.5	0.04	0.03	2.5-3.5 Mo; 0.2 + 4 (%C + %N) (Ti + Nb) min
Martensitic									
Type 410Cb (XM-30)	S41040	0.18	1.00	1.00	11.5-13.5	...	0.04	0.03	0.05-0.30 Nb
Type 410S	S41008	0.08	1.00	1.00	11.5-13.5	0.60	0.04	0.03	...
Type 414L	...	0.06	0.50	0.15	12.5-13.0	2.5-3.0	0.04	0.03	0.5 Mo; 0.03 Al
416 Plus X (XM-6)	S41610	0.15	1.5-2.5	1.00	12.0-14.0	...	0.06	0.15 min	0.6 Mo
Precipitation-hardening									
AM-350 (Type 633)	S35000	0.07-0.11	0.5-1.25	0.50	16.0-17.0	4.0-5.0	0.04	0.03	2.5-3.25 Mo; 0.07-0.13 N
AM-355 (Type 634)	S35500	0.10-0.15	0.5-1.25	0.50	15.0-16.0	4.0-5.0	0.04	0.03	2.5-3.25 Mo
AM-363(d)	...	0.04	0.15	0.05	11.0	4.0	...	...	0.25 Ti
Custom 450 (XM-25)	S45000	0.05	1.00	1.00	14.0-16.0	5.0-7.0	0.03	0.03	1.25-1.75 Cu; 0.5-1.0 Mo; 8× %C Nb min
Custom 455 (XM-16)	S45500	0.05	0.50	0.50	11.0-12.5	7.5-9.5	0.04	0.03	0.5 Mo; 1.5-2.5 Cu; 0.8-1.4 Ti; 0.1-0.5 Nb
PH 15-7 Mo (Type 632)	S15700	0.09	1.00	1.00	14.0-16.0	6.5-7.75	0.04	0.03	2.0-3.0 Mo; 0.75-1.5 Al
Stainless W (Type 635)	S17600	0.08	1.00	1.00	16.0-17.5	6.0-7.5	0.04	0.03	0.4 Al; 0.4-1.2 Ti
17-10 P(d)	...	0.07	0.75	0.5	17.0	10.5	0.28	...	...

(a) XM designations in this column are ASTM designations for the listed alloy. Type numbers in parentheses are obsolete AISI designations. (b) Single values are maximum values unless otherwise indicated. (c) 1.00% maximum. (d) Nominal composition; composition limits not available. (e) 0.50% maximum. (f) 0.75% maximum. (g) 0.80 maximum

Physical Properties

Typical physical properties of wrought stainless steels, annealed condition

Type	UNS designation	Density Mg/m³	Density lb/in.³	Elastic modulus GPa	Elastic modulus 10⁶ psi	Mean coefficient of thermal expansion μm/m · °C 0 °C to: 100 / 32 °F to: 212	μm/m · °C 315 600	μm/m · °C 538 1000	μin./in. · °F 100 212	μin./in. · °F 315 600	μin./in. · °F 538 1000
201	S20100	7.8	0.28	197	28.6	15.7	17.5	18.4	8.7	9.7	10.2
202	S20200	7.8	0.28	...	...	17.5	18.4	19.2	9.7	10.2	10.7
205	S20500	7.8	0.28	197	28.6	...	17.9	19.1	...	9.9	10.6
301	S30100	8.0	0.29	193	27.8	17.0	17.2	18.2	9.4	9.6	10.1
302	S30200	8.0	0.29	193	27.8	17.2	17.8	18.4	9.6	9.9	10.2
302B	S30215	8.0	0.29	193	27.8	16.2	18.0	19.4	9.0	10.0	10.8
303	S30300	8.0	0.29	193	27.8	17.2	17.8	18.4	9.6	9.9	10.2
304	S30400	8.0	0.29	193	27.8	17.2	17.8	18.4	9.6	9.9	10.2
304L	S30403	8.0	0.29	...	...	...	...	...	...	...	...
S30430	S30430	8.0	0.29	193	27.8	17.2	17.8	...	9.6	9.9	...
304N	S30451	8.0	0.29	196	28.4	...	...	...	...	...	...
305	S30500	8.0	0.29	193	27.8	17.2	17.8	18.4	9.6	9.9	10.2
308	S30800	8.0	0.29	193	27.8	17.2	17.8	18.4	9.6	9.9	10.2
309	S30900	8.0	0.29	200	29.0	15.0	16.6	17.2	8.3	9.2	9.6
310	S31000	8.0	0.29	200	29.0	15.9	16.2	17.0	8.8	9.0	9.4
314	S31400	7.8	0.28	200	29.0	...	15.1	...	...	8.4	...
316	S31600	8.0	0.29	193	27.8	15.9	16.2	17.5	8.8	9.0	9.7
316L	S31603	8.0	0.29	...	...	...	...	...	...	...	...
316N	S31651	8.0	0.29	196	28.4	...	...	...	...	...	...
317	S31700	8.0	0.29	193	27.8	15.9	16.2	17.5	8.8	9.0	9.7
317L	S31703	8.0	0.29	200	29.0	16.5	...	18.1	9.2	...	10.1
321	S32100	8.0	0.29	193	27.8	16.6	17.2	18.6	9.2	9.6	10.3
329	S32900	7.8	0.28	...	...	...	...	...	...	...	...
330	N08330	8.0	0.29	196	28.4	14.4	16.0	16.7	8.0	8.9	9.3
347	S34700	8.0	0.29	193	27.8	16.6	17.2	18.6	9.2	9.6	10.3
384	S38400	8.0	0.29	193	27.8	17.2	17.8	18.4	9.6	9.9	10.2
405	S40500	7.8	0.28	200	29.0	10.8	11.6	12.1	6.0	6.4	6.7
409	S40900	7.8	0.28	...	...	11.7	...	...	6.5	...	...
410	S41000	7.8	0.28	200	29.0	9.9	11.4	11.6	5.5	6.3	6.4
414	S41400	7.8	0.28	200	29.0	10.4	11.0	12.1	5.8	6.1	6.7
416	S41600	7.8	0.28	200	29.0	9.9	11.0	11.6	5.5	6.1	6.4
420	S42000	7.8	0.28	200	29.0	10.3	10.8	11.7	5.7	6.0	6.5
422	S42200	7.8	0.28	...	...	11.2	11.4	11.9	6.2	6.3	6.6
429	S42900	7.8	0.28	200	29.0	10.3	...	...	5.7	...	...
430	S43000	7.8	0.28	200	29.0	10.4	11.0	11.4	5.8	6.1	6.3
430F	S43020	7.8	0.28	200	29.0	10.4	11.0	11.4	5.8	6.1	6.3
431	S43100	7.8	0.28	200	29.0	10.2	12.1	...	5.7	6.7	...
434	S43400	7.8	0.28	200	29.0	10.4	11.0	11.4	5.8	6.1	6.3
436	S43600	7.8	0.28	200	29.0	9.3	...	...	5.2	...	...
440A	S44002	7.8	0.28	200	29.0	10.2	...	...	5.7	...	...
440C	S44004	7.8	0.28	200	29.0	10.2	...	...	5.7	...	...
444	S44400	7.8	0.28	200	29.0	10.0	10.6	11.4	5.6	5.9	6.3
446	S44600	7.5	0.27	200	29.0	10.4	10.8	11.2	5.8	6.0	6.2
PH 13-8 Mo	S13800	7.8	0.28	203	29.4	10.6	11.2	11.9	5.9	6.2	6.6
15-5 PH	S15500	7.8	0.28	196	28.4	10.8	11.4	...	6.0	6.3	...
17-4 PH	S17400	7.8	0.28	196	28.4	10.8	11.6	...	6.0	6.4	...
17-7 PH	S17700	7.8	0.28	204	29.5	11.0	11.6	...	6.1	6.4	...

(a) At 0 to 100 °C (32 to 212 °F). (b) Approximate values

(continued)

Typical physical properties of wrought stainless steels, annealed condition (continued)

Type	Thermal conductivity W/m·K °C: 100 / °F: 212	W/m·K 500 / 932	Btu/h·ft·°F 100 / 212	Btu/h·ft·°F 500 / 932	Specific heat(a) J/kg·K	Specific heat(a) Btu/lb·°F	Electrical resistivity, nΩ·m	Magnetic permeability(b)	Melting range °C	Melting range °F
201	16.2	21.5	9.4	12.4	500	0.12	690	1.02	1400-450	2550-2650
202	16.2	21.6	9.4	12.5	500	0.12	690	1.02	1400-1450	2550-2650
205	...	...	...	...	500	0.12	...	...	...	...
301	16.2	21.5	9.4	12.4	500	0.12	720	1.02	1400-1420	2550-2590
302	16.2	21.5	9.4	12.4	500	0.12	720	1.02	1400-1420	2550-2590
302B	15.9	21.6	9.2	12.5	500	0.12	720	1.02	1375-1400	2500-2550
303	16.2	21.5	9.4	12.4	500	0.12	720	1.02	1400-1420	2550-2590
304	16.2	21.5	9.4	12.4	500	0.12	720	1.02	1400-1450	2550-2650
304L	...	...	...	...	...	...	...	1.02	1400-1450	2550-2650
S30430	11.2	21.5	6.5	12.4	500	0.12	720	1.02	1400-1450	2550-2650
304N	...	...	...	...	500	0.12	720	1.02	1400-1450	2550-2650
305	16.2	21.5	9.4	12.4	500	0.12	720	1.02	1400-1450	2550-2650
308	15.2	21.6	8.8	12.5	500	0.12	720	...	1400-1420	2550-2590
309	15.6	18.7	9.0	10.8	500	0.12	780	1.02	1400-1450	2550-2650
310	14.2	18.7	8.2	10.8	500	0.12	780	1.02	1400-1450	2550-2650
314	17.5	20.9	10.1	12.1	500	0.12	770	1.02	...	...
316	16.2	21.5	9.4	12.4	500	0.12	740	1.02	1375-1400	2500-2550
316L	...	...	...	...	...	...	...	1.02	1375-1400	2500-2550
316N	...	...	...	...	500	0.12	740	1.02	1375-1400	2500-2550
317	16.2	21.5	9.4	12.4	500	0.12	740	1.02	1375-1400	2500-2550
317L	14.4	...	8.3	...	500	0.12	790	...	1375-1400	2500-2550
321	16.1	22.2	9.3	12.8	500	0.12	720	1.02	1400-1425	2550-2600
329	...	...	...	...	460	0.11	750	...	...	...
330	...	...	...	...	460	0.11	1020	1.02	1400-1425	2550-2600
347	16.1	22.2	9.3	12.8	500	0.12	730	1.02	1400-1425	2550-2600
384	16.2	21.5	9.4	12.4	500	0.12	790	1.02	1400-1450	2550-2650
405	27.0	...	15.6	...	460	0.11	600	...	1480-1530	2700-2790
409	...	...	...	...	...	...	...	...	1480-1530	2700-2790
410	24.9	28.7	14.4	16.6	460	0.11	570	700-1000	1480-1530	2700-2790
414	24.9	28.7	14.4	16.6	460	0.11	700	...	1425-1480	2600-2700
416	24.9	28.7	14.4	16.6	460	0.11	570	700-1000	1480-1530	2700-2790
420	24.9	...	14.4	...	460	0.11	550	...	1450-1510	2650-2750
422	23.9	27.3	13.8	15.8	460	0.11	...	...	1470-1480	2675-2700
429	25.6	...	14.8	...	460	0.11	590	...	1450-1510	2650-2750
430	26.1	26.3	15.1	15.2	460	0.11	600	600-1100	1425-1510	2600-2750
430F	26.1	26.3	15.1	15.2	460	0.11	600	...	1425-1510	2600-2750
431	20.2	...	11.7	...	460	0.11	720	...	...	...
434	...	26.3	...	15.2	460	0.11	600	600-1100	1425-1510	2600-2750
436	23.9	26.0	13.8	15.0	460	0.11	600	600-1100	1425-1510	2600-2750
440A	24.2	...	14.0	...	460	0.11	600	...	1370-1480	2500-2700
440C	24.2	...	14.0	...	460	0.11	600	...	1370-1480	2500-2700
444	26.8	...	15.5	...	420	0.10	620	...	...	...
446	20.9	24.4	12.1	14.1	500	0.12	670	400-700	1425-1510	2600-2750
PH 13-8 Mo	14.0	22.0	8.1	12.7	460	0.11	1020	...	1400-1440	2560-2625
15-5 PH	17.8	23.0	10.3	13.1	420	0.10	770	95	1400-1440	2560-2625
17-4 PH	18.3	23.0	10.6	13.1	460	0.11	800	95	1400-1440	2560-2625
17-7 PH	16.4	21.8	9.5	12.6	460	0.11	830	...	1400-1440	2560-2625

(a) At 0 to 100 °C (32 to 212 °F). (b) Approximate values

Resistance of standard types of stainless steel to various classes of environments

| Type | Mild atmospheric and fresh water | Atmospheric | | Salt water | Chemical | | |
		Industrial	Marine		Mild	Oxidizing	Reducing
Austenitic							
201	x	x	x		x	x	
202	x	x	x		x	x	
205	x	x	x		x	x	
301	x	x	x		x	x	
302	x	x	x		x	x	
302B	x	x	x		x	x	
303	x	x			x		
303Se	x	x			x		
304	x	x	x		x	x	
304H	x	x	x		x	x	
304L	x	x	x		x	x	
304N	x	x	x		x	x	
S30430	x	x	x		x	x	
305	x	x	x		x	x	
308	x	x	x		x	x	
309	x	x	x		x	x	
309S	x	x	x		x	x	
310	x	x	x		x	x	
310S	x	x	x		x	x	
314	x	x	x		x	x	
316	x	x	x	x	x	x	x
316F	x	x	x	x	x	x	x
316H	x	x	x	x	x	x	x
316L	x	x	x	x	x	x	x
316N	x	x	x	x	x	x	x
317	x	x	x	x	x	x	x
317L	x	x	x	x	x	x	x
321	x	x	x		x	x	
321H	x	x	x		x	x	
329	x	x	x	x	x	x	x
330	x	x	x	x	x	x	x
347	x	x	x		x	x	
347H	x	x	x		x	x	
348	x	x	x		x	x	
348H	x	x	x		x	x	
384	x	x	x		x	x	
Ferritic							
405	x				x		
409	x				x		
429	x	x			x	x	
430	x	x			x	x	
430F	x	x			x		
430FSe	x	x			x		
434	x	x	x		x	x	
436	x	x	x		x	x	
442	x	x			x	x	
446	x	x	x		x	x	
Martensitic							
403	x				x		
410	x				x		
414	x				x		
416	x						
416Se	x						
420	x						
420F	x						
422	x						
431	x	x	x		x		
440A	x				x		
440B	x						
440C	x						
501							
502							
503							
504							
Precipitation-hardening							
PH 13-8 Mo	x	x			x	x	
15-5 PH	x	x	x		x	x	
17-4 PH	x	x	x		x	x	
17-7 PH	x	x	x		x	x	

Note: An "x" notation above indicates that the specific type is considered resistant to the corrosive environment

Aqueous environments capable of causing stress-corrosion cracking in stainless steels

Material	Cl^-, acid	Cl^-, neutral	Cl^-, oxidizing	Br^-	I^-	OH^-	F^-	$S^=$	$S^= \cdot Cl^-$	$S_4O_6^=$	$SO_3^=$	$SO_4^=$	$CrO_4^=$	NO_3^-	NH_3	Ultra-pure $H_2O + O_2$	Seawater
405		1				1		4					1	5	1	1	X
18-2	4	1	4			3	1						1		1	1	X
26-1	4	1	4			4		2	1						1		X
26-1S	4	1	4			4		2				5	1	1	1		X
29-4	1	1	1			4		2				5	1	1	1		X
430		1				4		1					1	1	1		4
434		1						1	3				1	1	1		X
431	5	5	5			4	2	5				2	1	1	1		X
410	5	2		5	5	2	5	5	5				1	2	1	2	X
CA-6NM		5						5	2								X
440A, B, C	5	5	5					5									X
PH 15-7 Mo	5	4					1	5				5	1	1	1		X
17-4 PH	5	4					1	5	5			5	1	1	1		X
PH 13-8 Mo	5	4						5									X
17-7 PH	5	5					1	5									X
Custom 450	4	1	4			4	2	2	2			2	1	1	1		X
Custom 455	4	1	4			4	2	2				2	1	1	1		X
202	5	5	5		1	4	1	2		3	3	1	1	1	1		X
216	5	5	5		1	4	1	2		3	3	1	1	1	1		X
216L	5	5	5		1	4	1	2		3,4	3,4	1	1	1	1		X
Nitronic 50	5	4	5		1	4	1	2		3	3	4	1	1	1		X
Nitronic 60	5	5	5		1	4	1	2		3	3	1	1	1	1	1	4
304	5	5	5		1	4		2	5	3	3	4	1	1	1	3	X
304L	5	5	5		1	4	1	2	5	3,4	3,4	4	1	1	1	3,4	X
309S	5	5	5		1	4	1	2		3	3	1	1	1	1	3	X
310S	5	5	5		1	4	1	2		3	3	1	1	1	1	3	X
316	5	5	5	4	1	4	1	2	5	3	3	1	1	1	1	3	X
316L	5	5	5			4		2		3,4	3,4	1	1	1	1	3,4	X
317	5	5	5		1	4	1	2		3	3	1	1	1	1	3	X
317L	5	5	5		1	4	1	2		3,4	3,4	1	1	1	1	3	X
321	5	5	5		1	4	1	2	5	1	1	1	1	1	1	1	X
347	5	5	5		1	4	1	2	5	1	1	4	1	1	1	1	X
329	4	5	5		2	4	2					1	1	1	1		X
3RE60	4	1	5			3	4		4			1	1	1	1	3	4
18-18-2	4	4	5		1	4	1					1	1	1	1		X

1, resistant; 2, resistant unless cold-worked or hardened; 3, resistant unless sensitized; 4, resistant except at high temperatures and concentrations; 5, nonresistant; X, not recommended for this environment

Mechanical Properties

Minimum mechanical properties of austenitic stainless steels

Product form(a)	Condition	Tensile strength MPa	Tensile strength ksi	0.2% yield strength MPa	0.2% yield strength ksi	Elongation, %	Reduction in area, %	Hardness, HRB	ASTM specification
Type 301 (UNS S30100)									
B, W, P, Sh, St	Annealed	515	75	205	30	40	...	88 max	A167
Sh, St	1/4 hard	860	125	515	75	25	...	...	A177
	1/2 hard	1030	150	760	110	18	...	...	A177
	3/4 hard	1210	175	930	135	12	...	...	A177
	Full hard	1280	185	965	140	9	...	...	A177
Type 302 (UNS S30200)									
B	Hot finished and annealed	515	75	205	30	40	50	...	A276
	Cold finished and annealed(b)	620	90	310	45	30	40	...	A276
	Cold finished and annealed(c)	515	75	205	30	30	40	...	A276
W	Annealed	515	75	205	30	...	...	...	A580
	Cold finished	620	90	310	45	...	...	...	A580
P, Sh, St	Annealed	515	75	205	30	40	...	88 max	A167
	High tensile, grade B	585	85	310	45	40	...	...	A666
	High tensile, grade C	860	125	515	75	...	...	...	A666
	High tensile, grade D	1030	150	760	110	...	...	...	A666
B, W	High tensile(d)	2240	325	...	...	...	...	...	A313
Type 302B (UNS S30215)									
B	Hot finished and annealed	515	75	205	30	40	50	...	A276
	Cold finished and annealed(b)	620	90	310	45	30	40	...	A276
	Cold finished and annealed(c)	515	75	205	30	30	40	...	A276
W	Annealed	515	75	205	30	...	...	...	A580
	Cold finished	620	90	310	45	...	...	...	A580
P, Sh, St	Annealed	515	75	205	30	...	...	...	A167
Type 302Cu (UNS S30430)									
B	Annealed	450 to 585	65 to 85	...	...	...	...	...	A493
	Lightly drafted	485 to 620	70 to 90	...	...	...	...	...	A493
W(e)	Annealed	485 to 620	70 to 90	...	...	...	...	...	A493
	Lightly drafted	485 to 620	70 to 90	...	...	...	...	...	A493
W(f)	Annealed	485 to 690	70 to 100	...	...	...	...	...	A493
	Lightly drafted	520 to 725	75 to 105	...	...	...	...	...	A493
Types 303 (UNS S30300) and 303Se (UNS S30323)									
B	Annealed	585(g)	85(g)	240(g)	35(g)	50(g)	55(g)	...	A581
W	Annealed	585 to 860	85 to 125	...	...	...	...	...	A581
	Cold worked	790 to 1000	115 to 145	...	...	...	...	...	A581
Type 304 (UNS S30400)									
B	Hot finished and annealed	515	75	205	30	40	50	...	A276
	Cold finished and annealed(b)	620	90	310	45	30	40	...	A276
	Cold finished and annealed(c)	515	75	205	30	30	40	...	A276
W	Annealed	515	75	205	30	...	...	...	A580
	Cold finished	620	90	310	45	...	...	...	A580
P, Sh, St	Annealed	515	75	205	30	40	...	88 max	A167
Sh, St	High tensile, grade B	550	80	310	45	...	...	...	A666
	High tensile, grade C	860	125	515	75	...	...	...	A666
	High tensile, grade D	1030	150	690	110	...	...	...	A666
B, W	High tensile(d)	2240	325	...	...	...	...	...	A313
Type 304L (UNS S30403)									
B	Hot finished and annealed	480	70	170	25	40	50	...	A276
	Cold finished and annealed(b)	620	90	310	45	30	40	...	A276
	Cold finished and annealed(c)	480	70	170	25	30	40	...	A276
W	Annealed	480	70	170	25	...	...	...	A580

(a) B, bar; W, wire; P, plate; Sh, sheet; St, strip. (b) Up to 13 mm (0.5 in.) thick. (c) Over 13 mm (0.5 in.) thick. (d) Depending on size and amount of cold reduction. (e) 4 mm (0.156 in.) in diameter and over. (f) Under 4 mm (0.156 in.) in diameter. (g) Values given are typical. (h) Not a basis for acceptance or rejection

(continued)

Minimum mechanical properties of austenitic stainless steels (continued)

Product form(a)	Condition	Tensile strength		0.2% yield strength		Elonga-tion, %	Reduction in area, %	Hardness, HRB	ASTM specifi-cation
		MPa	ksi	MPa	ksi				
Type 304L (UNS S30403) (continued)									
P, Sh, St	Cold finished	620	90	310	45	...	...	...	A580
	Annealed	480	70	170	25	40	...	88 max	A167
Types 304N (UNS S30451) and 316N (UNS S31651)									
B	Annealed	550	80	240	35	30	...	...	A276
Type 304LN									
B	Annealed	515	75	205	30	...	...	...	...
Type 305 (UNS S30500)									
B	Hot finished and annealed	515	75	205	30	40	50	...	A276
	Cold finished and annealed(b)	260	90	310	45	30	40	...	A276
	Cold finished and annealed(c)	515	75	205	30	30	40	...	A276
W	Annealed	515	75	205	30	...	...	...	A580
	Cold finished	620	90	310	45	...	...	...	A580
P, Sh, St	Annealed	480	70	170	25	40	...	88 max	A167
B, W	High tensile(d)	1690	245	...	...	...	...	...	...
Types 308 (UNS S30800), 321 (UNS S32100), 347 (UNS34700) and 348 (UNS S34800)									
B	Hot finished and annealed	515	75	205	30	40	50	...	A276
	Cold finished and annealed(b)	620	90	310	45	30	40	...	A276
	Cold finished and annealed(c)	515	75	205	30	30	40	...	A276
W	Annealed	515	75	205	30	...	...	...	A580
	Cold finished	620	90	310	45	...	...	...	A580
P, Sh, St	Annealed	515	75	205	30	40	...	88 max	A167
Type 308L									
B	Annealed	550(g)	80(g)	207(g)	30(g)	60(g)	70(g)	...	...
Types 309 (UNS S30900), 309S (UNS S30908), 310 (UNS S31000) and 310S (UNS S31008)									
B	Hot finished and annealed	515	75	205	30	40	50	...	A276
	Cold finished and annealed(b)	620	90	310	45	30	40	...	A276
	Cold finished and annealed(c)	515	75	205	30	30	40	...	A276
W	Annealed	515	75	205	30	...	...	...	A580
	Cold finished	620	90	310	45	...	...	...	A580
P, Sh, St	Annealed	515	75	205	30	40	...	95 max	A167
Type 312									
Weld metal	...	655	95	...	...	20	...	...	MIL-E-19933
Type 314 (UNS S31400)									
B	Hot finished and annealed	515	75	205	30	40	50	...	A276
	Cold finished and annealed(b)	620	90	310	45	30	40	...	A276
	Cold finished and annealed(c)	515	75	205	30	30	40	...	A276
W	Annealed	515	75	205	30	...	...	...	A580
	Cold finished	620	90	310	45	...	...	...	A580
Type 316 (UNS S31600)									
B	Hot finished and annealed	515	75	205	30	40	50	...	A276
	Cold finished and annealed(b)	620	90	310	45	30	40	...	A276
	Cold finished and annealed(c)	515	75	205	30	30	40	...	A276
W	Annealed	515	75	205	30	...	...	...	...
	Cold finished	620	90	310	45	...	...	...	A580
P, Sh, St	Annealed	515	75	205	30	40	...	95 max	A580
B, W	High tensile(d)	1690	245	...	...	...	...	...	A167
Type 316F (UNS S31620)									
B	Annealed	585(g)	85(g)	240(g)	35(g)	40(g)	55(g)	...	...

(a) B, bar; W, wire; P, plate; Sh, sheet; St, strip. (b) Up to 13 mm (0.5 in.) thick. (c) Over 13 mm (0.5 in.) thick. (d) Depending on size and amount of cold reduction. (e) 4 mm (0.156 in.) in diameter and over. (f) Under 4 mm (0.156 in.) in diameter. (g) Values given are typical. (h) Not a basis for acceptance or rejection

(continued)

Minimum mechanical properties of austenitic stainless steels (continued)

Product form(a)	Condition	Tensile strength MPa	ksi	0.2% yield strength MPa	ksi	Elonga- tion, %	Reduction in area, %	Hardness, HRB	ASTM specifi- cation
Type 316L (UNS S31603)									
B	Hot finished and annealed	480	70	170	25	40	50	...	A276
	Cold finished and annealed(b)	620	90	310	45	30	40	...	A276
	Cold finished and annealed(c)	480	70	170	25	30	40	...	A276
W	Annealed	480	70	170	25	...	...	...	A580
	Cold finished	620	90	310	45	...	...	...	A580
Type 316LN									
B	Annealed	515(g)	75(g)	205(g)	30(g)	60(g)	70(g)	...	...
Type 317 (UNS S31700)									
B	Hot finished and annealed	515	75	205	30	40	50	...	A276
	Cold finished and annealed(b)	620	90	310	45	30	40	...	A276
	Cold finished and annealed(c)	515	75	205	30	30	40	...	A276
W	Annealed	515	75	205	30	...	...	...	A580
	Cold finished	620	90	310	45	...	...	...	A580
P, Sh, St	Annealed	515	75	205	30	35	...	95 max	A167
Type 317L (UNS S31703)									
B	Annealed	585(g)	85(g)	240(g)	35(g)	55(g)	65(g)	85 max(g)	...
P, Sh, St	Annealed	515	75	205	30	35	...	95 max	A167
Type 317LM									
B, P, Sh, St	Annealed	515	75	205	30	35	50	95 max	...
Type 329 (UNS S32900)									
B	Annealed	724(g)	105(g)	550(g)	80(g)	25(g)	50(g)	...	...
Type 330 (UNS N08330)									
B	Annealed	480	70	210	30	30	...	...	B511
P, Sh, St	Annealed	480	70	210	30	30	...	75 to 85(h)	B536
Type 330HC									
B, W, St	Annealed	585(g)	85(g)	290(g)	42(g)	45(g)	65(g)	...	...
Type 332									
B, W, Sh, St	Annealed	550(g)	80(g)	240(g)	35(g)	45(g)	70(g)	...	...
Types 384 (UNS S38400) and 385 (UNS38500)									
B	Annealed	415 to 550	60 to 80	...	...	...	...	...	A493
	Lightly drafted	450 to 585	65 to 85	...	...	...	...	...	A493
W(e)	Annealed	450 to 585	65 to 85	...	...	...	...	...	A493
	Lightly drafted	485 to 620	70 to 90	...	...	...	...	...	A493
W(f)	Annealed	450 to 655	65 to 95	...	...	...	...	...	A493
	Lightly drafted	485 to 690	70 to 100	...	...	...	...	...	A493
904L (UNS N08904)									
B, P, Sh, St	Annealed	490	71	220	31	35	...	95 max	B625
AL-6X (UNS N08366)									
Sh, St	Annealed	515	75	205	30	30	...	...	B676
18-18-2 (UNS S38100)									
P, Sh, St	Annealed	515	75	205	30	40	...	96 max	A167
Crutemp 25									
P, Sh, St	Annealed	615(g)	89(g)	275(g)	40(g)	40(g)	...	...	...
JS-700 (UNS N08700)									
P, Sh, St	Annealed	550	80	205	30	30	40	...	B599
JS-777									
B, P, Sh, St	Annealed	550	80	240	35	30	40	95 max	...
20Cb-3 (UNS N08020)									
B	Annealed	585	85	240	35	30	50	...	B473
Shapes	Cold finished and annealed	585	85	240	35	15	50	...	B473
W	Annealed	620 to 825	90 to 120	...	...	...	...	...	B473
P, Sh, St	Annealed	585	85	275	40	30	...	95 max	B463

(a) B, bar; W, wire; P, plate; Sh, sheet; St, strip. (b) Up to 13 mm (0.5 in.) thick. (c) Over 13 mm (0.5 in.) thick. (d) Depending on size and amount of cold reduction. (e) 4 mm (0.156 in.) in diameter and over. (f) Under 4 mm (0.156 in.) in diameter. (g) Values given are typical. (h) Not a basis for acceptance or rejection

Minimum mechanical properties of high-nitrogen austenitic stainless steels

| Product form(a) | Condition | Tensile strength | | 0.2% yield strength | | Elonga-tion, % | Reduction in area, % | Hardness, HRB | ASTM specifi-cation |
		MPa	ksi	MPa	ksi				
Type 201 (UNS S20100)									
B	Annealed	515	75	275	40	40	45	...	A276
W, P, Sh, St	Annealed	655	95	310	45	40	...	...	A276, A412
Sh, St	1/4 hard	860	125	515	75	20	...	...	A412
	1/2 hard	1030	150	760	110	10	...	...	A412
	3/4 hard	1210	175	930	135	7	...	...	A412
	Full hard	1280	185	965	140	5	...	...	A412
Type 202 (UNS S20200)									
B	Annealed	515	75	275	40	40	...	...	A276
W, P, Sh, St	Annealed	655	95	310	45	40	...	...	A412
Sh, St	1/4 hard	860	125	515	75	12	...	...	A412
	1/2 hard	1030	150	760	110	10	...	...	A666
Type 205 (UNS S20500)									
P	Annealed(b)	830(b)	120(b)	475(b)	69(b)	58(b)	62(b)	98 max(b)	...
Type 216 (UNS S21600)									
Sh, St	Annealed	690	100	415	60	40	...	100 max	A240
P	Annealed	620	90	345	50	40	...	100 max	A240
Type 304N (UNS S30451)									
B	Annealed	550	80	240	35	30	...	...	A276
P, Sh, St	Annealed	550	80	240	35	30	...	88 max	A240
Type 304HN (UNS S30452)									
B	Annealed	620	90	345	50	30	50	...	...
Sh, St	Annealed	620	90	345	50	30	...	100 max	A240
P	Annealed	585	85	275	40	30	...	100 max	A240
Type 316N (UNS S31651)									
B	Annealed	550	80	240	35	30	...	...	A276
P, Sh, St	Annealed	550	80	240	35	30	...	95 max	A240
Nitronic 32 (UNS S24100)									
B	Annealed	690	100	380	55	30	50	...	A276
W	Annealed	690	100	380	55	30	50	...	A580
Nitronic 33 (UNS S24000)									
B	Annealed	690	100	380	55	30	50	...	A276
W	Annealed	690	100	380	55	30	50	...	A580
Sh, St	Annealed	690	100	415	60	40	...	...	A412
P	Annealed	690	100	380	55	40	...	...	A412
Nitronic 40 (UNS S21900)									
B	Annealed	550	80	345	50	45	...	...	A276
W	Annealed	620	90	345	50	45	60	...	A580
Sh, St	Annealed	690	100	415	60	40	...	...	A412
	10% cold rolled	895	130	795	115	15	...	...	A412
P	Annealed	620	90	345	50	45	...	...	A412
Nitronic 50 (UNS S20910)									
B	Annealed	690	100	380	55	35	55	...	A276
W	Annealed	690	100	380	55	35	55	...	A580
Sh, St	Annealed	825	120	515	75	30	...	...	A412
P	Annealed	690	100	380	55	35	...	...	A412
Nitronic 60 (UNS S21800)									
B	Annealed	655	95	345	50	35	55	...	A276
W	Annealed	655	95	345	50	35	55	...	A580
18-18 Plus (UNS S28200)									
B	Annealed	825(b)	120(b)	450(b)	65(b)	60(b)	70(b)	95 min(b)	...
	Annealed	760	110	415	60	35	55	...	A276
W	Annealed	760 to 930	110 to 135	...	...	...	...	...	A493
Tenelon (UNS S21400)									
Sh	Annealed	860	125	485	70	40	...	...	A240
St	Annealed	725	105	380	55	40	...	...	A240

(a) B, bar; W, wire; P, plate; Sh, sheet; St, strip. (b) Typical values

Minimum mechanical properties of ferritic stainless steels

Product form(a)	Condition	Tensile strength		0.2% yield strength		Elonga-tion, %	Reduction in area, %	Hardness, HRB	ASTM specifi-cation
		MPa	ksi	MPa	ksi				
Type 405 (UNS S40500)									
W	Annealed	480	70	275	40	20	45	...	A580
	Annealed, cold finished	480	70	275	40	16	45	...	A580
P, Sh, St	Annealed	415	60	170	25	20	...	88 max	A176
Type 409 (UNS S40900)									
B	Annealed	450(b)	65(b)	240(b)	35(b)	25(b)	...	75 max(b)	...
P, Sh, St	Annealed	415	60	205	30	22(c)	...	80 max	A176
Type 429 (UNS S42900)									
B	Annealed	490(b)	71(b)	310(b)	45(b)	30(b)	65(b)	...	...
P, Sh, St	Annealed	450	65	205	30	22(c)	...	88 max	A176
Type 430 (UNS S43000)									
B	Annealed, hot finished	480	70	275	40	20	45	...	A276
	Annealed, cold finished	480	70	275	40	16	45	...	A276
W	Annealed	480	70	275	40	20	45	...	A580
	Annealed, cold finished	480	70	275	40	16	45	...	A580
P, Sh, St	Annealed	450	65	205	30	22(c)	...	88 max	A176
Type 430F (UNS S43020)									
W	Annealed	585 to 860	85 to 125	...	...	...	...	...	A581
Type 430Ti (UNS S43036)									
B	Annealed	515(b)	75(b)	310(b)	45(b)	30(b)	65(b)	...	...
Type 434 (UNS S43400)									
W	Annealed	545(b)	79(b)	415(b)	60(b)	33(b)	78(b)	90 max(b)	...
Sh	Annealed	530(b)	77(b)	365(b)	53(b)	23(b)	...	83 max(b)	...
Type 436 (UNS S43600)									
Sh, St	Annealed	530(b)	77(b)	365(b)	53(b)	23(b)	...	83 max(b)	...
Type 442 (UNS S44200)									
B	Annealed	550(b)	80(b)	310(b)	45(b)	20(b)	40(b)	90 max(b)	...
P, Sh, St	Annealed	515	75	275	40	20	...	95 max	A176
Type 444 (UNS S44400)									
P, Sh, St	Annealed	415	60	275	40	20	...	95 max	A176
Type 446 (UNS S44600)									
B	Annealed, hot finished	480	70	275	40	20	45	...	A276
	Annealed, cold finished	480	70	275	40	16	45	...	A276
W	Annealed	480	70	275	40	20	45	...	A580
	Annealed, cold finished	480	70	275	40	16	45	...	A580
P, Sh, St	Annealed	515	75	275	40	20	...	95 max	A176
18 SR									
Sh, St	Annealed	620(b)	90(b)	450(b)	65(b)	25(b)	...	90 min(b)	...
E-Brite 26-1 (UNS S44625)									
B	Annealed, hot finished	450	65	275	40	20	45	...	A276
	Annealed, cold finished	450	65	275	40	16	45	...	A276
P, Sh, St	Annealed	450	65	275	40	22(c)	...	90 max	A176
26-1 Ti (UNS S44626)									
P, Sh, St	Annealed	470	68	310	45	20	...	95 max	A176
Monit (UNS S44635)									
B, P, Sh, St	Annealed	650	94	550	80	20	...	100 max	A176
Sea-cure/SC-1 (UNS S44600)									
B, P, Sh, St	Annealed	550	80	380	55	20	...	100 max	A176
29-4 (UNS S44700)									
B, P, Sh, St	Annealed	550	80	415	60	20	...	98 max	A176
29-4-2 (UNS S44800)									
B, P, Sh, St	Annealed	550	80	415	60	20	...	98 max	A176

(a) B, bar; W, wire; P, plate; Sh, sheet; St, strip. (b) Typical values. (c) 20% reduction for 1.3 mm (0.050 in.) in thickness and under

Minimum mechanical properties of martensitic stainless steels

Product form(a)	Condition	Tensile strength MPa	ksi	0.2% yield strength MPa	ksi	Elonga- tion, %	Reduction in area, %	Rockwell hardness	ASTM specifi- cation
Type 403 (UNS S40300)									
B	Annealed, hot finished	485	70	275	40	20	45	...	A276
	Annealed, cold finished	485	70	275	40	16	45	...	A276
	Intermediate temper, hot finished	690	100	550	80	15	45	...	A276
	Intermediate temper, cold finished	690	100	550	80	12	40	...	A276
	Hard temper, hot finished	825	120	620	90	12	40	...	A276
	Hard temper, cold finished	825	120	620	90	12	40	...	A276
W	Annealed	485	70	275	40	20	45	...	A580
	Annealed, cold finished	485	70	275	40	16	45	...	A580
	Intermediate temper, cold finished	690	100	550	80	12	40	...	A580
	Hard temper, cold finished	825	120	620	90	12	40	...	A580
P, Sh, St	Annealed	485	70	205	30	25(b)	...	88 HRB max	A176
Type 404 (UNS S40400)									
B	Tempered 260 °C (500 °F)	1120(c)	162(c)	910(c)	132(c)	15(c)	50(c)	35 HRC(c)	...
	Tempered 595 °C (1100 °F)	745(c)	108(c)	655(c)	95(c)	23(c)	70(c)	20 HRC(c)	...
Type 410 (UNS S41000)									
B	Annealed, hot finished	485	70	275	40	20	45	...	A276
	Annealed, cold finished	485	70	275	40	16	45	...	A276
	Intermediate temper, hot finished	690	100	550	80	15	45	...	A276
	Intermediate temper, cold finished	690	100	550	80	12	40	...	A276
	Hard temper, hot finished	825	120	620	90	12	40	...	A276
	Hard temper, cold finished	825	120	620	90	12	40	...	A276
W	Annealed	485	70	275	40	20	45	...	A580
	Annealed, cold finished	485	70	275	40	16	45	...	A580
	Intermediate temper, cold finished	690	100	550	80	12	40	...	A580
	Hard temper, cold finished	825	120	620	90	12	40	...	A580
P, Sh, St	Annealed	450	65	205	30	22(b)	...	95 HRB max	A176
Type 410S (UNS S41008)									
P, Sh, St	Annealed	415	60	205	30	22	...	95 HRB max	A176
Type 410Cb (UNS S41040)									
B	Annealed, hot finished	485	70	275	40	13	45	...	A276
	Annealed, cold finished	485	70	275	40	12	35	...	A276
	Intermediate temper, hot finished	860	125	690	100	13	45	...	A276
	Intermediate temper, cold finished	860	125	690	100	12	35	...	A276
Type 414 (UNS S41400)									
B	Intermediate temper, hot finished	795	115	620	90	15	45	...	A276
	Intermediate temper, cold finished	795	115	620	90	15	45	...	A276
W	Annealed, cold finished	1030 max	150 max	...	...	...	...	...	A580
Type 414L									
B	Annealed	795(c)	115(c)	550(c)	80(c)	20(c)	60(c)	...	...
Types 416 (UNS S41600) and 416Se (UNS S41623)									
W	Annealed	585 to 860	85 to 125	...	...	...	...	...	A581
	Intermediate temper	795 to 1000	115 to 145	...	...	...	...	...	A581
	Hard temper	965 to 1210	140 to 175	...	...	...	...	...	A581
Type 416 Plus X									
B	Annealed	515	75	275	40	30	60	...	...
Type 418 (UNS S41800)									
B	Tempered 260 °C (500 °F)	1450(c)	210(c)	1210(c)	175(c)	18(c)	52(c)	...	...
	Tempered 650 °C (1200 °F)	930(c)	135(c)	725(c)	105(c)	20(c)	60(c)	...	...
Type 420 (UNS S42000)									
B	Tempered 205 °C (400 °F)	1720	250	1480(c)	215(c)	8(c)	25(c)	52 HRC(c)	...
W	Annealed, cold finished	860 max	125 max	...	...	...	...	...	A580
Type 422 (UNS S42200)									
B	Intermediate and hard tempers(d)	965	140	760	110	13	30	...	A565

(a) B, bar; W, wire; P, plate; Sh, sheet; St, strip. (b) 20% elongation for 1.3 mm (0.050 in.) and under in thickness. (c) Typical values. (d) Heat treated for high-temperature service

(continued)

Minimum mechanical properties of martensitic stainless steels (continued)

Product form(a)	Condition	Tensile strength		0.2% yield strength		Elonga- tion, %	Reduction in area, %	Rockwell hardness	ASTM specifi- cation
		MPa	ksi	MPa	ksi				
Type 431 (UNS S43100)									
B	Tempered 260 °C (500 °F)	1370(c)	198(c)	1030(c)	149(c)	16(c)	55(c)	...	...
	Tempered 595 °C (1100 °F)	965(c)	140(c)	795(c)	115(c)	19(c)	57(c)	...	...
W	Annealed, cold finished	965 max	140 max	...	...	...	...	...	A580
Type 440A (UNS S44002)									
B	Annealed	725(c)	105(c)	415(c)	60(c)	20(c)	...	95 HRB(c)	...
	Tempered 315 °C (600 °F)	1790(c)	260(c)	1650(c)	240(c)	5(c)	20(c)	51 HRC(c)	...
W	Annealed, cold finished	965 max	140 max	...	...	...	...	...	A580
Type 440B (UNS S44003)									
B	Annealed	740(c)	107(c)	425(c)	62(c)	18(c)	...	96 HRB(c)	...
	Tempered 315 °C (600 °F)	1930(c)	280(c)	1860(c)	270(c)	3(c)	15(c)	55 HRC(c)	...
W	Annealed, cold finished	965 max	140 max	...	...	...	...	...	A580
Type 440C (UNS S44004)									
B	Annealed	760(c)	110(c)	450(c)	65(c)	14(c)	...	97 HRB(c)	...
	Tempered 315 °C (600 °F)	1970(c)	285(c)	1900(c)	275(c)	2(c)	10(c)	57 HRC(c)	...
W	Annealed, cold finished	965 max	140 max	...	...	...	...	...	A580
Type 501 (UNS S50100)									
B, P	Annealed	485(c)	70(c)	205(c)	30(c)	28(c)	65(c)	...	...
	Tempered 540 °C (1000 °F)	1210(c)	175(c)	965(c)	140(c)	15(c)	50(c)	...	...
Type 502 (UNS S50200)									
B, P	Annealed	485(c)	70(c)	205(c)	30(c)	30(c)	70(c)	...	...

(a) B, bar; W, wire; P, plate; Sh, sheet; St, strip. (b) 20% elongation for 1.3 mm (0.050 in.) and under in thickness. (c) Typical values. (d) Heat treated for high-temperature service

Minimum mechanical properties of precipitation-hardening stainless steels

Product form(a)	Condition	Tensile strength		Yield strength		Elongation, %	Reduction in area(b), %	Hardness(c), HRC	
		MPa	ksi	MPa	ksi			min	max
PH 13-8 Mo (UNS S13800)									
B, P, Sh, St	H950	1520	220	1410	205	6-10(c)	45	45	...
	H1000	1380	200	1310	190	6-10(c)	45	43	...
15-5 PH (UNS S15500) and									
17-4 PH (UNS S17400)									
B, P, Sh, St	H900	1310	190	1170	170	10(d)	35(e)	40	48
	H925	1170	170	1070	155	10(d)	38(e)	38	47(d)
	H1025	1070	155	1000	145	12(d)	45(e)	35(d)	42(d)
	H1075	1000	145	860	125	13(d)	45(e)	32(d)	38(d)
	H1100	965	140	795	115	14(d)	45(e)	31(d)	38(d)
	H1150	930	135	725	105	16(d)	50(e)	28(d)	36(d)
	H1150M	795	115	515	75	18(d)	55(e)	24(d)	34(d)
17-7 PH (UNS S17700)									
B	RH950	1275	185	1030	150	6	10	41	...
	TH1050	1170	170	965	140	6	25	38	...
P, Sh, St	RH950	1450	210	1310	190	1-6(d)	...	41(d)	44(d)
	TH1050	1240	180	1030	150	3-7(d)	...	38	...
	Cold rolled (condition C)	1380	200	1210	175	1	...	41	...
	CH900	1650	240	1590	230	1	...	46	...
Custom 450 (UNS S45000)									
B	Annealed	860	125	655	95	10	40	...	33
P, Sh, St	Annealed	895	130	620	90	4	...	25	...
B, P, Sh, St	H900	1240	180	1170	170	10	40	40	...
	H1000	1100	160	1030	150	12	45	36	...
	H1150	860	125	515	75	15	50	26	...
Custom 455 (UNS S45500)									
B	H900	1620	235	1520	220	8	30	47	...
	H950	1520	220	1410	205	10	40	44	...
P, Sh, St	H950	1530	222	1410	205	Up to 4	...	44	...
AM-350 (UNS S35000)									
P, Sh, St	H850	1275	185	1030	150	2-8	...	42	...
	H1000	1140	165	1000	145	2-8	...	36	...
AM-355 (UNS S35500)									
B	Equalize plus overtemper 540 °C (1000 °F)	1170	170	1070	155	12	25	39	...
	H850	1310	190	1140	165	10	...	37	...
	H1000	1170	170	1030	150	12	...	28(f)	...
AM-363									
St	Annealed	850(f)	123(f)	730(f)	106(f)	12(f)	...	...	...
Stainless W (UNS S17600)									
B, P, Sh, St	H950	1310	190	1170	170	8	25	39	...
	H1000	1240	180	1100	160	8	30	37	...
	H1050	1170	170	1070	150	10	40	35	...
PH 15-7 Mo (UNS S15700)									
B	RH950	1380	200	1210	175	7	25	...	...
	TH1050	1240	180	1100	160	8	25	...	...
P, Sh, St	RH950	1550	225	1380	200	1-5(d)	...	43(d)	46(d)
	TH1050	1310	190	1170	170	2-5(d)	...	38(d)	40(d)
	Cold rolled	1380	200	1210	175	1	...	41	...
	Cold rolled and aged	1650	240	1590	230	1	...	46	...
17-10 P									
B	Annealed	615(f)	89(f)	255(f)	37(f)	70(f)	76(f)	82(f)(g)	...
	Aged	945(f)	137(f)	605(f)	88(f)	25(f)	39(f)	30(f)	...
HNM									
B, W, F	Aged at 705 °C (1300 °F)	825	120	550	80	18	30	...	...

(a) B, bar; P, plate; Sh, sheet; St, strip; W, wire; F, forgings. (b) Values are for bar products. (c) Where minimum value is also given, maximum value applies only to flat-rolled products. Both max and min values may vary with thickness for flat-rolled products. (d) Value varies with thickness for flat-rolled products. (e) Value generally lower for flat-rolled products and varies with thickness. (f) Values are typical. (g) Rockwell B hardness

Nominal mechanical properties of stainless steels at low and elevated temperatures(a)

AISI type (UNS)	Temperature, °F	Nominal properties of annealed material at low temperature					Mechanical properties at elevated temperatures					Scaling temperature	
		Tensile strength, ksi	Yield strength, ksi	Elongation, %	Reduction in area, %	Izod impact energy, ft·lb	Creep strength — Load for 1% elongation in 10 000 h, ksi					Max continuous service in air, °F	Max intermittent service in air, °F
							1000 °F	1100 °F	1200 °F	1300 °F	1500 °F		
Austenitic													
201	+70	...	...	...	...	110-20	...	...	...	...	...	1550	1450
(S20100)	−300					38-70							
	+70	100	55	55		110-120							
	−100	145	95	38		...							
202	−300	200	150	15	...	42-120	...	...	...	...	...	1550	1450
(S20200)	−423	220	170	5	...	...							
205	...	...	...	...	...	200(e)	...	...	...	...	...	...	...
(S20500)													
	+70	105	40	60	70	100							
	+32	155	43	53	64	110							
301	−40	180	48	42	63	110	19	12.5	8	4.5	1.8	1650	1500
(S30100)	−80	195	50	40	62	110							
	−320	275	75	30	57	110							
	−70	94	37	68	78	110							
	+32	122	40	65	76	110							
302	−40	145	48	60	73	110	20	12.5	7.5	4.3	1.5	1650	1500
(S30200)	−80	161	50	57	70	110							
	−320	219	68	46	70	110							
	−423	250	125	41	55	...							
302B	+70	(b)	(b)	(b)	(b)	90	...	...	7	4.5	1	1750	1600
(S30215)													
303													
(S30300)	+70	100	40	67	67	85							
	+32	114	40	61	65	90							
	−40	145	40	45	62	100	16.5	11.5	6.5	3.5	0.7	1650	1400
	−80	162	40	40	60	106							
303Se	−320	235	37	35	52	125							
(S30323)	−452	267	...	30	37	...							
304													
(S30400)	+70	95	35	65	71	110							
	+32	130	34	55	68	110							
	−40	155	34	47	64	110	20	12	7.5	4	1.5	1650	1550
	−80	170	34	39	63	110							
	−320	221	39	40	55	110							
304L	−423	243	50	40	50	110							
(S30403)													
(S30430)	...	...	...	...	...	240(c)	...	...	...	...	...	...	...
305	−70	...	...	...	...	110	19	12.5	8	4.5	2	1650	...
(S30500)													
308	+70	...	...	...	...	110	...	...	...	...	...	1700	1550
(S30800)													
309													
(S30900)													
	+70	...	...	...	...	110	16.5	12.5	10	6	3	1950	1850
309S													
(S30906)													
310	+70	86	37	55	70	110							
(S31000)	+32	85	32	64	75	110							
	−40	95	39	57	75	110	33	23	15	10	3	2050	1900
	−80	100	40	55	75	110							
	−320	152	74	54	64	85							
	−423	176	106	56	61	...							

(a) Single values are maximums, except as noted. (b) Not applicable. Silicon added to type 302 for oxidation resistance. (c) Charpy V-notch. (d) Not applicable. Silicon added to type 310 for carburization resistance. (e) Same as type 316. (f) Approximately same as type 410 in annealed condition. (g) Same as type 410. (h) Solution-treated condition

(continued)

Nominal mechanical properties of stainless steels at low and elevated temperatures(a) (continued)

AISI type (UNS)	Nominal properties of annealed material at low temperature						Mechanical properties at elevated temperatures					Scaling temperature	
							Creep strength — Load for 1% elongation in 10 000 h, ksi					Max continuous service in air, °F	Max intermittent service in air, °F
	Temperature, °F	Tensile strength, ksi	Yield strength, ksi	Elongation, %	Reduction in area, %	Izod impact energy, ft·lb	1000 °F	1100 °F	1200 °F	1300 °F	1500 °F		
310S (S31008)													
314 (S31400)	(d)	(d)	(d)	(d)	(d)	(d)	20	13	7.5	5	2.5	…	…
316 (S31600)													
	+70	85	37	65	76	110							
	+32	90	39	60	75	110							
	−40	104	41	59	75	110	25	17.4	11.6	7.5	2.4	1650	1550
	−80	118	44	57	73	110							
	−320	185	75	59	76	…							
	−423	210	84	52	60	…							
316L (S31603)													
317 (S31700)	(e)	(e)	(e)	(e)	(e)	(e)	23	16.8	11.2	6.9	2.0	1700	1600
	+70	89	37	62	76	110							
	+32	99	38	58	73	110							
321 (S32100)	−40	117	44	58	70	115	18	17	9	5	1.5	1650	1550
	−80	130	45	57	68	117							
	−320	208	64	44	57	110							
	−423	238	92	35	…	…							
329 (S32900)	…	…	…	…	…	40(c)	…	…	…	…	…	…	…
330 (N08330)	…	…	…	…	…	240(c)	…	…	…	…	…	…	…
347 (S34700)	+70	93	38	55	69	110							
	−32	105	42	62	72	110							
	−40	117	44	63	71	117	32	23	16	10	2	1650	1550
348 (S34800)	−80	130	45	57	70	110							
	−320	200	47	43	65	95							
	−423	228	65	39	53	60							
Ferritic													
405 (S40500)	+70	(f)	(f)	(f)	(f)	20-35	8.4	…	…	…	…	1400	1450
	+70	65	38	37	73	35							
	−32	69	40	37	72	20							
430 (S43000)	−40	76	41	36	72	10							
	−80	81	44	36	70	8	8.5	4.7	2.6	1.4	…	1550	1650
	−320	90	87	2	4	2							
430F (S43020)													
	+70					5-50							
	−100	…	…	…	…	4	8.5	4.6	1.9	1.3	…	1500	1600
	−300					1							
430FSe (S43023)													
442 (S44200)	+70	…	…	…	…	5-15	8.5	5	1.6	1	0.6	1800	1900
446 (S44600)	+70	…	…	…	…	2-10	6.4	2.9	1.4	0.6	0.4	1950	2050
Martensitic													
403 (S40300)	(g)	(g)	(g)	(g)	(g)	(g)	11	4.5	2	1.4	…	1300	1450
	+70	110	87	21	68	85							
	+32	115	89	24	69	40							

(a) Single values are maximums, except as noted. (b) Not applicable. Silicon added to type 302 for oxidation resistance. (c) Charpy V-notch. (d) Not applicable. Silicon added to type 310 for carburization resistance. (e) Same as type 316. (f) Approximately same as type 410 in annealed condition. (g) Same as type 410. (h) Solution-treated condition

(continued)

Nominal mechanical properties of stainless steels at low and elevated temperatures(a) (continued)

AISI type (UNS)	Nominal properties of annealed material at low temperature						Mechanical properties at elevated temperatures					Scaling temperature	
							Creep strength Load for 1% elongation in 10 000 h, ksi						
	Temperature, °F	Tensile strength, ksi	Yield strength, ksi	Elongation, %	Reduction in area, %	Izod impact energy, ft · lb	1000 °F	1100 °F	1200 °F	1300 °F	1500 °F	Max continuous service in air, °F	Max intermittent service in air, °F
410 (S41000)	−40	122	90	23	64	25							
	−80	128	94	22	60	25	11.5	4.3	2	1.5	...	1300	1450
	−320	158	148	10	11	5							
	...	...	...	...	...	...							
414 (S41400)	+70	...	...	...	...	40-80	...	...	...	...	...	1300	1450
416 (S41600)													
	+70					20-64							
	−100	...	...	...	...	50	11	4.6	2	1.2	...	1250	1400
416Se (S41623)	−300					3							
	+70					10							
420 (S42000)	+32					10							
	−40	...	...	...	...	8	9.2	4.2	2	1	...	1200	1400
	−80					7							
	+70					50							
431 (S43100)	+32					50							
	−40	...	...	...	...	30	6.8	3.5	...	...	...	1500	1600
	−80					17							
440A (S44002)	...	...	...	...	...	...	...	...	...	...	...	1400	1500
440B (S44003)	...	...	...	...	...	...	...	...	...	...	...	1400	1500
440C (S44004)	...	...	...	...	...	...	...	...	...	...	...	1400	1500
Precipitation hardening(h)													
PH 13-8 Mo (S13800)	+70	...	...	...	...	60(c)	...	...	...	...	...	...	...
15-5 PH (S15500)	+70	...	...	...	...	30(c)	...	...	...	...	...	...	...
17-4PH (S17400)	+70	...	...	...	...	30(c)	...	...	...	...	...	...	...

(a) Single values are maximums, except as noted. (b) Not applicable. Silicon added to type 302 for oxidation resistance. (c) Charpy V-notch. (d) Not applicable. Silicon added to type 310 for carburization resistance. (e) Same as type 316. (f) Approximately same as type 410 in annealed condition. (g) Same as type 410. (h) Solution-treated condition

Critical stress intensities for some hardenable stainless steels

Material	Heat treat condition	Critical stress intensity, ksi $\sqrt{\text{in.}}$		
		Air	Seacoast	3.5% NaCl
15-5 PH	H900	71.8	35.9	>32.3
	H1150	75.7	>72.0	>72.0
AM-355	SCT850	140	>18.3	5.5
	SCT1000	70.0	>35.0	28.0
431	(a)	75.7	>37.8	43.0
	(b)	79.2	>39.6	11.9
PH 13-8 Mo	H950	62.6	>31.3	>45.9
	H1050	87.8	>43.9	>65.8
PH 15-7 Mo	RH950	30.6	15.3	10.1
	RH1050	40.7	20.1	12.1
17-7 PH	RH1050	47.0	11.7	9.4

(a) Heat treated to a tensile strength of 125 ksi. (b) Heat treated to a tensile strength of 200 ksi

Heat Treating

Recommended thermal treatments for stainless steels(a)

AISI type (UNS)	Initial forging temperature, °F	Annealing temperature, °F	Stress-relief annealing temperature, °F(b)
Austenitic			
201 (S20100)	2100 to 2250	1850 to 2050	...
202 (S20200)	2100 to 2250	1850 to 2050	...
205 (S20500)	2250	1950	...
301 (S30100)	2100 to 2300	1850 to 2050	400 to 750
302 (S30200)	2100 to 2300	1850 to 2050	400 to 750
302B (S30215)	2050 to 2250	1850 to 2050	...
303 (S30300)	2100 to 2350	1850 to 2050	400 to 750
303Se (S30323)	2100 to 2350	1850 to 2050	400 to 750
304 (S30400)	2100 to 2300	1850 to 2050	400 to 750
304L (S30403)	2100 to 2300	1850 to 2050	400 to 750
(S30430)	2100 to 2300	1850 to 2050	...
304N (S30451)	2100 to 2300	1850 to 2050	...
305 (S30500)	2100 to 2300	1850 to 2050	...
308 (S30800)	2100 to 2300	1850 to 2050	...
309 (S30900)	2050 to 2250	1900 to 2050	...
309S (S30908)	2050 to 2250	1900 to 2050	...
310 (S31000)	2000 to 2250	1900 to 2100	400 to 750
310S (S31008)	2000 to 2250	1900 to 2100	400 to 750
314 (S31400)	1900 to 2050	2100	...
316 (S31600)	2100 to 2300	1850 to 2050	400 to 750
316L (S31603)	2100 to 2300	1850 to 2050	400 to 750
316F (S31620)	2200	2000	...
316N (S31651)	2100 to 2300	1850 to 2050	...
317 (S31700)	2100 to 2300	1850 to 2050	...
317L (S31703)	2250	1900 to 2000	...
321 (S32100)	2100 to 2300	1750 to 2050	400 to 750(c)
329 (S32900)	2000	1750 to 1800	M1350
330 (N08330)	2100 to 2150	1950 to 2150	...
347 (S34700)	2100 to 2300	1850 to 2050	400 to 750(h)
348 (S34800)	2100 to 2300	1850 to 2050	400 to 750(h)
384 (S38400)	2100 to 2250	1900 to 2100	...
Ferritic			
405 (S40500)	1950 to 2050	1350 to 1500(d)	...

AISI type (UNS)	Initial forging temperature, °F	Annealing temperature, °F	Stress-relief annealing temperature, °F(b)
Ferritic (continued)			
409 (S40900)	...	1625	...
429 (S42900)	1900 to 2050	1450 to 1550	...
430 (S43000)	1900 to 2050	1400 to 1500(d)	...
430F (S43020)			
430FSe (S43023)	1950 to 2100	1250 to 1400(d)	...
434 (S43400)	1900 to 2050	1450 to 1550	...
436 (S43600)	1900 to 2050	1450 to 1550	...
442 (S44200)	1600 to 2100	1300	...
446 (S44600)	1950 to 2050	1450 to 1600	...
Martensitic			
403 (S40300)	2000 to 2200(e)	1500 to 1650(f) / 1200 to 1400(d)	H1700 to 1850 / T 400 to 1400 (g)
410 (S41000)	2000 to 2200(e)	1500 to 1650(f) / 1200 to 1400(d)	H1700 to 1850 / T 400 to 1400(g)
414 (S41400)	2100 to 2200	... / 1200 to 1300(d)	H1800 to 1900 / T 400 to 1300(g)
416 (S41600)			
416Se (S41623)	2100 to 2300(e)	1500 to 1650(f) / 1200 to 1400(d)	H1700 to 1850 / T 400 to 1400(g)
420 (S42000)	2000 to 2200(h)	1550 to 1650(f) / 1350 to 1450(d)	H1800 to 1900 / T 300 to 700
420F (S42020)	2050 to 2250	1550 to 1650(e)	H1800 to 1900 / T 300 to 700
422 (S42200)	2100	1350 to 1450	H1900
431 (S43100)	2100 to 2250(h)	... / 1150 to 1225(d)	H1800 to 1900 / T 400 to 1200(g)
440A (S44002)	1900 to 2200(h)	1550 to 1650(f) / 1350 to 1450(d)	H1850 to 1950 / T 300 to 800
440B (S44003)	1900 to 2150(h)	1550 to 1650(f) / 1350 to 1450(d)	H1850 to 1950 / T 300 to 800
440C (S44004)	1900 to 2100(h)	1550 to 1650(f) / 1350 to 1450(d)	H1850 to 1950 / T 300 to 800
Heat resisting			
501 (S50100)	2100 to 2200(h)	1525 to 1600(f) / 1325 to 1375(d)	H1600 to 1700 / T 400 to 1400
502 (S50200)	2100 to 2200	1525 to 1600(f) / 1325 to 1375(d)	...
Precipitation hardening			
PH 13-8 Mo (S13800)	2150	...	H950 to 1150
15-5 PH (S15500)	2150	...	H900 to 1150
17-4 PH (S17400)	2150	...	H900 to 1150
17-7 PH (S17700)	2150	...	H900 to 1050

(a) Single values are maximums, except as noted. (b) Followed by rapid cooling, H is hardening temperature, T is tempering temperature. (c) Stabilizing temperature, 1550 to 1650 °F. (d) Low anneal. (e) Retarded cool. (f) Full anneal, followed by slow cooling. (g) Tempering between 800 and 1100 °F is not recommended because of resulting low and erratic impact properties and reduced corrosion resistance. Time at temperature and temperatures may vary depending on part size. (h) Retarded cool and anneal

Procedures for hardening and tempering wrought martensitic stainless steels to specific strength and hardness levels

Type	Austenitizing(a) Temperature(b) °C	°F	Quenching medium(c)	Tempering temperature(d) °C min	max	°F min	max	Tensile strength MPa	ksi	Hardness, HRC
403, 410	925-1010	1700-1850	Air or oil	565	605	1050	1125	760-965	110-140	25-31
				205	370	400	700	1105-1515	160-220	38-47
414	925-1050	1700-1925	Air or oil	595	650	1100	1200	760-965	110-140	25-31
				230	370	450	700	1105-1515	160-220	38-49
416, 416Se	925-1010	1700-1850	Oil	565	605	1050	1125	760-965	110-140	25-31
				230	370	450	700	1105-1515	160-220	35-45
420	985-1065	1800-1950	Air or oil(e)	205	370	400	700	1550-1930	225-280	48-56
431	985-1065	1800-1950	Air or oil(e)	565	605	1050	1125	860-1035	125-150	26-34
				230	370	450	700	1210-1515	175-220	40-47
440A	1010-1065	1850-1950	Air or oil(e)	150	370	300	700	...	...	49-57
440B	1010-1065	1850-1950	Air or oil(e)	150	370	300	700	...	...	53-59
440C, 440F	1010-1065	1850-1950	Air or oil(e)	...	160	...	325	...	...	60 min
				...	190	...	375	...	...	58 min
				...	230	...	450	...	...	57 min
				...	355	...	675	...	...	52-56

(a) Preheating to a temperature within the process annealing range is recommended for thin-gage parts, heavy sections, previously hardened parts, parts with extreme variations in section or with sharp re-entrant angles, and parts that have been straightened or heavily ground or machined, to avoid cracking and minimize distortion, particularly for types 420, 431, and 440A, B, C and F. (b) Usual time at temperature ranges from 30 to 90 min. The low side of the austenitizing range is recommended for all types subsequently tempered to 25 to 31 HRC; generally, however, corrosion resistance is enhanced by quenching from the upper limit of the austenitizing range. (c) Where air or oil is indicated, oil quenching should be used for parts more than 6.4 mm (1/4 in.) thick; martempering baths at 150 to 400 °C (300 to 750 °F) may be substituted for an oil quench. (d) Generally, the low end of the tempering range of 150 to 370 °C (300 to 700 °F) is recommended for maximum hardness, the middle for maximum toughness, and the high end for maximum yield strength. Tempering in the range of 370 to 565 °C (700 to 1050 °F) is not recommended, because it results in low and erratic impact properties and poor resistance to corrosion and stress corrosion. (e) For minimum retained austenite and maximum dimensional stability, a subzero treatment −75 °C ± 10 °C (−100 °F ± 20 °F) is recommended; this should incorporate continuous cooling from the austenitizing temperature to the cold transformation temperature

Annealing temperatures and procedures for wrought martensitic stainless steels

Type	Process (subcritical) annealing Temperature(a), °C	Hardness	Full annealing Temperature(b)(c), °C	Hardness	Isothermal annealing(c) Procedure(d)	Hardness
403, 410	650-760	86-92 HRB	830-885	75-85 HRB	Heat to 830 to 885 °C; hold 6 h at 705 °C	85 HRB
414	650-730	99 HRB-24 HRC	Not recommended		Not recommended	
416, 416Se	650-760	86-92 HRB	830-885	75-85 HRB	Heat to 830 to 885 °C; hold 2 h at 720 °C	85 HRB
420	675-760	94-97 HRB	830-885	86-95 HRB	Heat to 830 to 885 °C; hold 2 h at 705 °C	95 HRB
431	620-705	99 HRB-30 HRC	Not recommended		Not recommended	
440A	675-760	90 HRB-22 HRC	845-900	94-98 HRB	Heat to 845 to 900 °C; hold 4 h at 690 °C	98 HRB
440B	675-760	98 HRB-23 HRC	845-900	95 HRB-20 HRC	Same as 440A	20 HRC
440C, 440F	675-760	98 HRB-23 HRC	845-900	98 HRB-25 HRC	Same as 440A	25 HRC

(a) Air cool from temperature; maximum softness is obtained by heating to temperature at high end of range. (b) Soak thoroughly at temperature within range indicated; furnace cool to 790 °C; continue cooling at 15 to 25 °C/h to 595 °C; air cool to room temperature. (c) Recommended for applications in which full advantage may be taken of the rapid cooling to the transformation temperature and from it to room temperature. (d) Preheating to a temperature within the process annealing range is recommended for thin-gage parts, heavy sections, previously hardened parts, parts with extreme variations in section or with sharp re-entrant angles, and parts that have been straightened or heavily ground or machined to avoid cracking and minimize distortion, particularly for types 420 and 431, and 440A, B, C and F

Recommended annealing temperatures for austenitic stainless steels

UNS no.	Designation	Temperature(a)	
		°C	°F
Conventional grades			
S30100, S30200, S30215	301, 302, 302B	1010 to 1120	1850 to 2050
S30300, S30323	303, 303Se	1010 to 1120	1850 to 2050
S30400, S30500, S30800	304, 305, 308	1010 to 1120	1850 to 2050
S30900, S30908	309, 309S	1040 to 1120	1900 to 2050
S31000, S31008	310, 310S	1040 to 1065	1900 to 1950
S31600	316	1040 to 1120	1900 to 2050
S31700	317	1065 to 1120	1950 to 2050
Stabilized grades			
S32100	321	955 to 1065	1750 to 1950
S34700, S34800	347, 348	980 to 1065	1800 to 1950
N08020	Carpenter 20Cb-3	925 to 955	1700 to 1750
Low-carbon grades			
S30403	304L, 304LN	1010 to 1120	1850 to 2050
S31603, S31703	316L, 316LN, 317L	1040 to 1110	1900 to 2025
High-nitrogen grades			
S20100, S20200	201, 202	1010 to 1120	1850 to 2050
S30451	304N	1010 to 1120	1850 to 2050
S31651	316N	1010 to 1120	1850 to 2050
S24100	Nitronic 32, Carpenter 18Cr-2Ni-12Mn	1010 to 1065	1850 to 1950
S24000	Nitronic 33		
S21904	Nitronic 40, Carpenter 21Cr-6Ni-9Mn	1040 to 1095	1900 to 2000
		980 to 1175	1800 to 2150
S20910	Nitronic 50, Carpenter 22Cr-13Ni-5Mn	1065 to 1120	1950 to 2050
S21800	Nitronic 60	1040 to 1095	1900 to 2000
S28200	Carpenter 18-18 PLUS	1040 to 1095	1900 to 2000
Highly alloyed grades			
...	317LM, 317LX, 317L PLUS, 317MO, 7L4	1120 to 1150	2050 to 2100
...	JS700, JS777	1065 to 1150	1950 to 2100
N08904	904L, AL-4X, 2RK65	1075 to 1125	1965 to 2055
N08028	Sanicro 28	...	...
N08366	AL-6X	1205 to 1230	2200 to 2250
S31254	254 SMO	1150 to 1205	2100 to 2200

(a) Temperatures given are for annealing a composite structure. Time at temperature and method of cooling depend on thickness. Light sections may be held at temperature for 3 to 5 min. per 2.5 mm (0.10 in.) of thickness, followed by rapid air cooling. Thicker sections are water quenched. For many of these grades, a postweld heat treatment is not necessary. For proprietary alloys, alloy producers may be consulted for details. Although cooling from the annealing temperature must be rapid, it must also be consistent with limitations of distortion

Recommended annealing treatments for ferritic stainless steels

UNS no.	Designation	Treatment temperature	
		°C	°F
Conventional ferritic grades			
S40500	405	650 to 815	1200 to 1500
S40900	409	870 to 900	1600 to 1650
S43000	430	705 to 790	1300 to 1450
S43020	430F	705 to 790	1300 to 1450
S43400	434	705 to 790	1300 to 1450
S44600	446	760 to 830	1400 to 1525
Low-interstitial ferritic grades			
S43035	439	870 to 925	1600 to 1700
S44400	444	955 to 1010	1750 to 1850
S44626	E-BRITE	760 to 955	1400 to 1750
S44660	SEA-CURE, SC-1	1010 to 1065	1850 to 1950
...	AL 29-4C	1010 to 1065	1850 to 1950
S44800	AL29-4-2	1010 to 1065	1850 to 1950
S44635	MONIT	1010 to 1065	1850 to 1950

Note: Postweld heat treating of the low interstitial ferritic stainless steels is generally unnecessary and frequently undesirable. Any annealing of these grades should be followed by water quenching or very rapid cooling

Machining

Machinability of wrought stainless steels

Operation or type of cutting machine	Machining speeds, ft/min.(a), for type							
	403(b), 405, 410(b): 180 to 240 HB	416(b): 180 to 240 HB	420, 420F(c): 180 to 230 HB	430: 170 to 230 HB	414(b), 430F: 170 to 230 HB	440B, 431: 230 to 280 HB	440A, 440C, 440F(c): 200 to 265 HB	446: 170 to 230 HB
Automatic screw machine(d)	90 to 100	120 to 150	80 to 100	90 to 100	120 to 150	80 to 100	60 to 80	80 to 100
Heavy-duty single or multiple spindle(d)	80 to 100	110 to 130	60 to 80	80 to 100	110 to 130	70 to 90	50 to 70	60 to 80
Turret lathe(d)	80 to 100	100 to 130	60 to 80	80 to 100	110 to 130	70 to 90	50 to 70	60 to 80
Automatic screw machine(e)	110 to 140	120 to 150	90 to 120	110 to 140	120 to 150	100 to 140	60 to 100	100 to 140
Milling(f)	40 to 60	50 to 80	30 to 50	40 to 60	50 to 80	40 to 60	30 to 50	40 to 60
Reaming(f)								
Smooth finish	15 to 40	15 to 40	15 to 40	15 to 40	15 to 40	15 to 40	15 to 40	15 to 40
Work sizing	40 to 120	40 to 120	40 to 120	40 to 120	40 to 120	40 to 120	40 to 120	40 to 120
Threading(g)	10 to 25	10 to 25	10 to 25	10 to 25	10 to 25	10 to 25	10 to 25	10 to 25
Tapping(g)	10 to 25	10 to 25	10 to 25	10 to 25	10 to 25	10 to 25	10 to 25	10 to 25
Drill press(g)	40 to 80	60 to 90	30 to 50	40 to 80	60 to 90	40 to 60	30 to 50	40 to 60
Single-point turning:								
Carbide tooling:								
Roughing	150 to 200	150 to 200	100 to 150	150 to 200	150 to 200	140 to 180	100 to 150	140 to 180
Finishing	200 to 400	200 to 400	150 to 250	200 to 400	200 to 400	150 to 350	150 to 200	150 to 350
High-cobalt or cast alloy tooling:								
Roughing	100 to 130	100 to 150	80 to 100	100 to 130	100 to 150	90 to 120	60 to 80	100 to 130
Finishing	100 to 150	150 to 200	100 to 150	100 to 150	150 to 200	90 to 140	80 to 100	100 to 150
High speed steel tooling:								
Roughing	80 to 100	80 to 100	60 to 80	80 to 100	80 to 100	60 to 80	40 to 60	60 to 90
Finishing	80 to 130	100 to 150	80 to 120	80 to 130	100 to 150	80 to 100	60 to 80	90 to 120

Operation or type of cutting machine	Machining speeds, ft/min.(a), for type						
	301, 302, 304, 304L: 150 to 250 HB	303: 150 to 240 HB	309, 309S, 310, 310S, 316, 316L: 150 to 240 HB	321, 347: 150 to 240 HB	347F: 150 to 240 HB	17-4 PH: 300 to 360 HB	17-7 PH: 150 to 240 HB
Automatic screw machine(d)	70 to 90	100 to 130	60 to 80	70 to 90	90 to 110	60 to 80	60 to 80
Heavy-duty single or multiple spindle(d)	60 to 80	90 to 120	60 to 80	60 to 80	80 to 100	50 to 70	50 to 70
Turret lathe(d)	60 to 80	90 to 120	60 to 80	60 to 80	80 to 100	50 to 70	50 to 70
Automatic screw machine(e)	80 to 120	110 to 130	80 to 120	80 to 120	100 to 120	80 to 120	80 to 120
Milling(f)	40 to 60	40 to 60	30 to 50	40 to 60	40 to 60	40 to 60	40 to 60
Reaming(f)							
Smooth finish	15 to 40	15 to 40	15 to 40	15 to 40	15 to 40	15 to 40	15 to 40
Work sizing	40 to 80	40 to 120	40 to 80	40 to 80	40 to 80	40 to 80	40 to 80
Threading(g)	10 to 25	10 to 25	10 to 25	10 to 25	10 to 25	10 to 25	10 to 25
Tapping(g)	10 to 25	10 to 25	10 to 25	10 to 25	10 to 25	10 to 25	10 to 25
Drill press(g)	30 to 50	50 to 80	30 to 50	30 to 50	30 to 50	40 to 60	40 to 60
Single-point turning:							
Carbide tooling:							
Roughing	130 to 180	150 to 250	130 to 180	130 to 180	150 to 250	130 to 180	130 to 180
Finishing	150 to 300	200 to 400	150 to 300	150 to 300	200 to 400	150 to 300	150 to 300
High-cobalt or cast alloy tooling:							
Roughing	100 to 130	100 to 150	100 to 130	100 to 130	100 to 140	100 to 130	100 to 130
Finishing	100 to 150	150 to 200	100 to 150	100 to 150	140 to 190	100 to 150	100 to 150
High speed steel tooling:							
Roughing	60 to 90	70 to 90	60 to 90	60 to 90	60 to 90	60 to 90	60 to 90
Finishing	100 to 120	100 to 140	100 to 120	100 to 120	100 to 130	100 to 120	100 to 120

(a) To obtain equivalent values in m/min., multiply listed values by 0.3. (b) Harder stock in the 260 to 320 HB range may be machined by reducing these speeds approximately 20%. (c) When using an automatic screw machine, cutting speeds may be increased about 10% over those shown. (d) Based on tungsten or molybdenum high speed steel tooling. Rates may be increased 15 to 30% with high-cobalt or cast alloys. (e) Based on the use of tools made of cemented carbide or cast cobalt-chromium-tungsten alloy. (f) Based on tungsten or molybdenum high speed steel tooling. Greatly increased speeds can be used with carbide tooling. (g) Based on tungsten or molybdenum high speed steel tooling

Heat- and Corrosion-Resistant Alloys

Nominal Compositions

Nominal compositions of wrought iron-base heat-resistant alloys

Designation	UNS number	C	Cr	Ni	Mo	N	Nb	Ti	Others
Ferritic stainless steels									
405	S40500	0.15 max	13.0	...	...	...	...	...	0.2 Al
406	...	0.15 max	13.0	...	...	...	...	...	4.0 Al
409	S40900	0.08 max	11.0	0.5	...	...	...	...	...
430	S43000	0.12 max	16.0	...	...	...	...	6 × C min	...
434	S43400	0.12 max	17.0	...	1.0	...	...	...	...
439	S43027	0.07 max	18.25	...	...	...	...	...	...
18 SR	...	0.05	18.0	0.5	...	...	...	0.2 + 4 (C + N)	...
18Cr-2Mo	...	...	18.0	...	2.0	...	...	0.40	2.0 Al
446	S44600	0.20 max	25.0	...	...	0.25	...	...	...
E-Brite 26-1	S44627	0.01 max	26.0	...	1.0	0.015 max	0.1	...	...
26-1Ti	...	0.04	26.0	...	1.0	...	...	10 × C min	...
29Cr-4Mo	...	0.01 max	29.0	...	4.0	0.02 max	...	...	...
Quenched and tempered martensitic stainless steels									
403	S40300	0.15 max	12.0	...	...	...	...	...	...
410	S41000	0.15 max	12.5	...	...	...	...	...	...
416	S41600	0.15 max	13.0	...	0.6(a)	...	...	...	0.15 S min
422	S42200	0.20	12.5	0.75	1.0	...	...	...	1.0 W, 0.22 V
H-46	...	0.12	10.75	0.50	0.85	0.07	0.30	...	0.20 V
Moly Ascoloy	...	0.14	12.0	2.4	1.80	0.05	...	...	0.35 V
Greek Ascoloy	...	0.15	13.0	2.0	...	...	...	...	3.0 W
Jethete M-152	...	0.12	12.0	2.5	1.7	...	...	...	0.30 V
Almar 363	...	0.05	11.5	4.5	...	...	...	10 × C min	...
431	S43100	0.20 max	16.0	2.0	...	...	...	...	...
Precipitation-hardening martensitic stainless steels									
Custom 450	...	0.05 max	15.5	6.0	0.75	...	8 × C min	...	1.5 Cu
Custom 455	...	0.03	11.75	8.5	...	...	0.30	1.2	2.25 Cu
15-5 PH	S15500	0.07	15.0	4.5	...	...	0.30	...	3.5 Cu
17-4 PH	S17400	0.04	16.5	4.25	...	...	0.25	...	3.6 Cu
PH 13-8 Mo	S13800	0.05	12.5	8.0	2.25	...	...	...	1.1 Al
Precipitation-hardening semiaustenitic stainless steels									
AM-350	S35000	0.10	16.5	4.25	2.75	0.10	...	...	...
AM-355	S35500	0.13	15.5	4.25	2.75	0.10	...	...	...
17-7 PH	S17700	0.07	17.0	7.0	...	...	...	...	1.15 Al
PH 15-7 Mo	S15700	0.07	15.0	7.0	2.25	...	...	...	1.15 Al
Austenitic stainless steels									
304	S30400	0.08 max	19.0	10.0	...	...	...	...	...
304L	S30403	0.03 max	19.0	10.0	...	...	...	...	...
304N	S30451	0.08 max	19.0	9.25	...	0.13	...	...	...
309	S30900	0.20 max	23.0	13.0	...	...	...	...	...
310	S31000	0.25 max	25.0	20.0	...	...	...	...	...
316	S31600	0.08 max	17.0	12.0	2.5	...	...	...	...
316L	S31603	0.03 max	17.0	12.0	2.5	...	...	...	...
316N	S31651	0.08 max	17.0	12.0	2.5	0.13	...	...	...
317	S31700	0.08 max	19.0	13.0	3.5	...	...	...	...
321	S32100	0.08 max	18.0	10.0	...	...	...	5 × C min	...
347	S34700	0.08 max	18.0	11.0	...	...	10 × C min	...	...
19-9 DL	K63198	0.30	19.0	9.0	1.25	...	0.4	0.3	1.25 W
19-9 DX	K63199	0.30	19.2	9.0	1.5	...	...	0.55	1.2 W
17-14-CuMo	...	0.12	16.0	14.0	2.5	...	0.4	0.3	3.0 Cu
202	S20200	0.09	18.0	5.0	...	0.10	...	...	8.0 Mn
216	S21600	0.05	20.0	6.0	2.5	0.35	...	...	8.5 Mn
21-6-9	S21900	0.04 max	20.25	6.5	...	0.30	...	...	9.0 Mn
Nitronic 32	...	0.10	18.0	1.6	...	0.34	...	...	12.0 Mn
Nitronic 33	...	0.08 max	18.0	3.0	...	0.30	...	...	13.0 Mn
Nitronic 50	...	0.06 max	21.0	12.0	2.0	0.30	0.20	...	5.0 Mn
Nitronic 60	...	0.10 max	17.0	8.5	2.0	...	...	...	8.0 Mn, 0.20 V, 4.0 Si
Carpenter 18-18 Plus	...	0.10	18.0	<0.50	1.0	0.50	...	...	16.0 Mn, 0.40 Si, 1.0 Cu

(a) Optional

Nominal compositions of wrought superalloys

Alloy	UNS number	Composition, %										
		Cr	Ni	Co	Mo	W	Nb	Ti	Al	Fe	C	Other
Iron-base solid-solution alloys												
16-25-6	...	16.0	25.0	...	6.00	...	...	...	...	50.7	0.06	1.35 Mn; 0.70 Si; 0.15 N
17-14CuMo	...	16.0	14.0	...	2.50	...	0.4	0.3	...	62.4	0.12	0.75 Mn; 0.50 Si; 3.0 Cu
19-9DL	K63198	19.0	9.0	...	1.25	1.25	0.4	0.3	...	66.8	0.30	1.10 Mn; 0.60 Si
Carpenter 20Cb-3	N08020	20.0	34.0	...	2.50	...	1.0 max	...	...	42.4	0.07 max	3.5 Cu
Incoloy 800	N08800	21.0	32.5	...	...	...	...	0.38	0.38	45.7	0.05	...
Incoloy 801	N08801	20.5	32.0	...	...	...	...	1.13	...	46.3	0.05	...
Incoloy 802	...	21.0	32.5	...	...	...	...	0.75	0.58	44.8	0.35	...
N-155	R30155	21.0	20.0	20.0	3.00	2.5	1.0	...	...	32.2	0.15	0.15 N; 0.02 La; 0.02 Zr
RA330	N08330	19.0	36.0	...	...	...	...	...	...	45.1	0.05	...
Cobalt-base solid-solution alloys												
Haynes 25 (L-605)	R30605	20.0	10.0	50.0	...	15.0	...	...	...	3.0	0.10	1.5 Mn
Haynes 188	R30188	22.0	22.0	37.0	...	14.5	...	...	...	3.0 max	0.10	0.90 La
S-816	R30816	20.0	20.0	42.0	4.0	4.0	4.0	...	...	4.0	0.38	...
Stellite 6B	...	30.0	1.0	61.5	...	4.5	...	...	...	1.0	1.0	...
UMCo-50	...	28.0	...	49.0	...	...	...	...	...	21.0	0.12 max	...
Nickel-base solid-solution alloys												
Hastelloy B	N10001	1.0 max	63.0	2.5 max	28.0	...	...	...	...	5.0	0.05 max	...
Hastelloy B-2	N10665	1.0 max	69.0	1.0 max	18.0	...	...	...	...	2.0 max	0.02 max	...
Hastelloy C	N10002	16.5	56.0	...	17.0	4.5	...	...	...	6.0	0.15 max	0.03 V
Hastelloy C4	N06455	16.0	63.0	2.0 max	15.5	...	...	0.7 max	...	3.0 max	0.015 max	...
Hastelloy C-276	N10276	15.5	59.0	...	16.0	3.7	...	...	...	5.0	0.02 max	...
Hastelloy N	N10003	7.0	72.0	...	16.0	...	...	0.5 max	...	5.0 max	0.06	...
Hastelloy S	...	15.5	67.0	...	15.5	...	...	...	0.2	1.0	0.02 max	0.02 La
Hastelloy W	N10004	5.0	61.0	2.5 max	24.5	...	...	...	...	5.5	0.12 max	0.06 V
Hastelloy X	N06002	22.0	49.0	1.5 max	9.0	0.6	...	...	2.0	15.8	0.15	...
Inconel 600	N06600	15.5	76.0	...	...	...	...	...	...	8.0	0.08	0.25 Cu max
Inconel 601	N06601	23.0	60.5	...	...	...	...	...	1.35	14.1	0.05	0.5 Cu max
Inconel 604	...	16.0	74.0	...	...	...	2.25	...	1.0	7.5	0.02	0.03 Cu max
Inconel 617	...	22.0	55.0	12.5	9.0	...	...	...	1.0	5.0	0.07	...
Inconel 625	N06625	21.5	61.0	...	9.0	...	3.6	0.2	0.2	2.5	0.05	...
NA-224	...	27.0	48.0	...	...	6.0	...	...	...	18.5	0.50	...
Nimonic 75	...	19.5	75.0	...	...	...	...	0.4	...	2.5	0.12	0.25 Cu max
RA-333	N06333	25.0	45.0	3.0	3.0	3.0	...	...	0.15	18.0	0.05	...
Iron-base precipitation-hardening alloys												
A-286	K66286	15.0	26.0	...	1.25	...	...	2.0	0.2	55.2	0.04	0.005 B; 0.3 V
Discaloy	K66220	14.0	26.0	...	3.0	...	...	1.7	0.25	55.0	0.06	...
Haynes 556	...	22.0	21.0	20.0	3.0	2.5	0.1	...	0.3	29.0	0.10	0.50 Ta; 0.02 La; 0.002 Zr
Incoloy 903	...	0.1 max	38.0	15.0	0.1	...	3.0	1.4	0.7	41.0	0.04	...
Pyromet CTX-1	...	0.1 max	37.7	16.0	0.1	...	3.0	1.7	1.0	39.0	0.03	...
V-57	...	14.8	27.0	...	1.25	...	...	3.0	0.25	48.6	0.08 max	0.01 B; 0.05 V max
W-545	K66545	13.5	26.0	...	1.5	...	...	2.85	0.2	55.8	0.08	0.05 B

(continued)

Nominal compositions of wrought superalloys (continued)

Alloy	UNS number	Composition, %										
		Cr	Ni	Co	Mo	W	Nb	Ti	Al	Fe	C	Other
Cobalt-base precipitation-hardening alloys												
AR-213	...	19.0	0.5 max	65.0	...	4.5	...	...	3.5	0.05 max	0.17	6.5 Ta; 0.15 Zr; 0.1Y
MP-35N	R30035	20.0	35.0	35.0	10.0	...	...	...	...	...	...	...
MP-159	...	19.0	25.0	36.0	7.0	...	0.6	3.0	0.2	9.0	...	...
Nickel-base precipitation-hardening alloys												
Astroloy	...	15.0	56.5	15.0	5.25	...	...	3.5	4.4	<0.3	0.06	0.03 B; 0.06 Zr
D-979	N09979	15.0	45.0	...	4.0	4.0	...	3.0	1.0	27.0	0.05	0.01 B
IN 100	N13100	10.0	60.0	15.0	3.0	...	...	4.7	5.5	<0.6	0.15	1.0 V; 0.06 Zr; 0.015 B
IN 102	N06102	15.0	67.0	...	2.9	3.0	2.9	0.5	0.5	7.0	0.06	0.005 B; 0.02 Mg; 0.03 Zr
Incoloy 901	N09901	12.5	42.5	...	6.0	...	...	2.7	...	36.2	0.10 max	...
Inconel 706	N09706	16.0	41.5	...	...	...	...	1.75	0.2	37.5	0.03	2.9 (Nb+Ta); 0.15 Cu max
Inconel 718	N07718	19.0	52.5	...	3.0	...	5.1	0.9	0.5	18.5	0.08 max	0.15 Cu max
Inconel 751	...	15.5	72.5	...	...	...	1.0	2.3	1.2	7.0	0.05	0.25 Cu max
Inconel X750	N07750	15.5	73.0	...	...	...	1.0	2.5	0.7	7.0	0.04	0.25 Cu max
M252	N07252	19.0	56.5	10.0	10.0	...	...	2.6	1.0	<0.75	0.15	0.005 B
Nimonic 80A	N07080	19.5	73.0	1.0	...	...	...	2.25	1.4	1.5	0.05	0.10 Cu max
Nimonic 90	N07090	19.5	55.5	18.0	...	...	...	2.4	1.4	1.5	0.06	...
Nimonic 95	...	19.5	53.5	18.0	...	...	...	2.9	2.0	...	0.15 max	+B; +Zr
Nimonic 100	...	11.0	56.0	20.0	5.0	...	...	1.5	5.0	5.0 max	0.30 max	+B; +Zr
Nimonic 105	...	15.0	54.0	20.0	5.0	...	...	1.2	4.7	2.0 max	0.08	0.005 B
Nimonic 115	...	15.0	55.0	15.0	4.0	...	...	4.0	5.0	...	0.20	0.04 Zr
Nimonic 263	...	20.0	51.0	20.0	5.9	...	...	2.1	0.45	1.0	0.06	...
Pyromet 860	...	13.0	44.0	4.0	6.0	...	...	3.0	1.0	0.7 max	0.05	0.01 B
Refractory 26	...	18.0	38.0	20.0	3.2	...	...	2.6	0.2	28.9	0.03	0.015 B
René 41	N07041	19.0	55.0	11.0	10.0	...	...	3.1	1.5	16.0	0.09	0.01 B
René 95	...	14.0	61.0	8.0	3.5	3.5	3.5	2.5	3.5	<0.3	0.16	0.01 B; 0.05 Zr
René 100	...	9.5	61.0	15.0	3.0	...	...	4.2	5.5	<0.3	0.16	0.015 B; 0.06 Zr; 1.0 V
Udimet 500	N07500	19.0	48.0	19.0	4.0	...	...	3.0	3.0	1.0 max	0.08	0.005 B
Udimet 520	...	19.0	57.0	12.0	6.0	1.0	...	3.0	2.0	4.0 max	0.08	0.005 B
Udimet 630	...	17.0	50.0	...	3.0	3.0	6.5	1.0	0.7	18.0	0.04	0.004 B
Udimet 700	...	15.09	53.0	18.5	5.0	...	...	3.4	4.3	<1.0	0.07	0.03 B
Udimet 710	...	18.0	55.0	14.8	3.0	1.5	...	5.0	2.5	...	0.07	0.01 B
Unitemp AF2-1DA	...	12.0	59.0	10.0	3.0	6.0	...	3.0	4.6	<0.5	0.35	1.5 Ta; 0.015 B; 0.1 Zr
Waspaloy	N07001	19.5	57.0	13.5	4.3	...	...	3.0	1.4	2.0 max	0.07	0.006 B; 0.09 Zr

Compositions of ACI heat-resistant casting alloys

ACI designation	UNS number	ASTM specifications(a)	Composition, %(b)			
			C	Cr	Ni	Si (max)
HA	...	A217	0.20 max	8 to 10	...	1.00
HC	J92605	A297, A608	0.50 max	26 to 30	4 max	2.00
HD	J93005	A297, A608	0.50 max	26 to 30	4 to 7	2.00
HE	J93403	A297, A608	0.20 to 0.50	26 to 30	8 to 11	2.00
HF	J92603	A297, A608	0.20 to 0.40	19 to 23	9 to 12	2.00
HH	J93503	A297, A608	0.20 to 0.50	24 to 28	11 to 14	2.00
HI	J94003	A297, A567, A608	0.20 to 0.50	26 to 30	14 to 18	2.00
HK	J94224	A297, A351, A567, A608	0.20 to 0.60	24 to 28	18 to 22	2.00
HL	J94604	A297, A608	0.20 to 0.60	28 to 32	18 to 22	2.00
HN	J94213	A297, A608	0.20 to 0.50	19 to 32	23 to 27	2.00
HP	...	A297	0.35 to 0.75	24 to 28	33 to 37	2.00
HP-50WZ(c)	...	...	0.45 to 0.55	24 to 28	33 to 37	2.50
HT	J94605	A297, A351, A567, A608	0.35 to 0.75	13 to 17	33 to 37	2.50
HU	...	A297, A608	0.35 to 0.75	17 to 21	37 to 41	2.50
HW	...	A297, A608	0.35 to 0.75	10 to 14	58 to 62	2.50
HX	...	A297, A608	0.35 to 0.75	15 to 19	64 to 68	2.50

(a) ASTM designations are same as ACI designations. (b) Rem Fe in all compositions. Manganese content: 0.35 to 0.65% for HA, 1% for HC, 1.5% for HD and 2% for the other alloys. Phosphorus and sulfur contents: 0.04% max for all but HP-50WZ. Molybdenum is intentionally added only to HA, which has 0.90 to 1.20% Mo; maximum for other alloys is set at 0.5% Mo. HH also contains 0.2% max N. (c) Also contains 4 to 6% W, 0.1 to 1.0% Zr, and 0.035% max S and P

Compositions of nickel-base heat-resistant casting alloys

Alloy designation	Nominal composition, %											
	C	Ni	Cr	Co	Mo	Fe	Al	B	Ti	W	Zr	Others
B-1900	0.1	64	8	10	6	...	6	0.015	1	...	0.10	4 Ta(a)
Hastelloy X	0.1	50	21	1	9	18	...	...	...	1	...	...
IN-100	0.18	60.5	10	15	3	...	5.5	0.01	5	...	0.06	1 V
IN-738X	0.17	61.5	16	8.5	1.75	...	3.4	0.01	3.4	2.6	0.1	1.75 Ta, 0.9 Nb
IN-792	0.2	60	13	9	2.0	...	3.2	0.02	4.2	4	0.1	4 Ta
Inconel 713C	0.12	74	12.5	...	4.2	...	6	0.012	0.8	...	0.1	2 Nb
Inconel 713LC	0.05	75	12	...	4.5	...	6	0.01	0.6	...	0.1	2 Nb
Inconel 718	0.04	53	19	...	3	18	0.5	...	0.9	...	...	0.1 Cu, 5 Nb
Inconel X-750	0.04	73	15	...	...	7	0.7	...	2.5	...	...	0.25 Cu, 0.9 Nb
M-252	0.15	56	20	10	10	...	1	0.005	2.6	...	...	...
MAR-M 200	0.15	59	9	10	...	1	5	0.015	2	12.5	0.05	1 Nb(b)
MAR-M 246	0.15	60	9	10	2.5	...	5.5	0.015	1.5	10	0.05	1.5 Ta
MAR-M 247	0.15	59	8.25	10	0.7	0.5	5.5	0.015	1	10	0.05	1.5 Hf, 3 Ta
NX 188 (DS)	0.04	74	...	...	18	...	8	...	...	...	...	...
René 77	0.07	58	15	15	4.2	...	4.3	0.015	3.3	...	0.04	...
René 80	0.17	60	14	9.5	4	...	3	0.015	5	4	0.03	...
René 100	0.18	61	9.5	15	3	...	5.5	0.015	4.2	...	0.06	1 V
TRW-NASA VIA	0.13	61	6	7.5	2	...	5.5	0.02	1	6	0.13	0.4 Hf, 0.5 Nb, 0.5 Re, 9 Ta
Udimet 500	0.1	53	18	17	4	2	3	...	3	...	...	...
Udimet 700	0.1	53.5	15	18.5	5.25	...	4.25	0.03	3.5	...	...	...
Udimet 710	0.13	55	18	15	3	...	2.5	...	5	1.5	0.08	...
Waspaloy	0.07	57.5	19.5	13.5	4.2	1	1.2	0.005	3	...	0.09	...
WAZ-20 (DS)	0.20	72	...	...	...	...	6.5	...	...	20	1.5	...

(a) B-1900 + Hf also contains 1.5% Hf. (b) MAR-M 200 + Hf also contains 1.5% Hf

Compositions of cobalt-base heat-resistant casting alloys

Alloy designation	Nominal composition, %										
	C	Co	Cr	Ni	Al	B	Fe	Ta	W	Zr	Others
AiResist 13	0.45	62	21	...	3.4	...	...	2	11	...	0.1 Y
AiResist 213	0.20	64	20	0.5	3.5	...	0.5	6.5	4.5	...	0.1 Y
AiResist 215	0.35	63	19	0.5	4.3	...	0.5	7.5	4.5	0.1	0.1 Y
Haynes 21	0.25	64	27	3	...	...	1	...	...	...	5 Mo
Haynes 25; L-605	0.1	54	20	10	...	...	1	...	15	...	...
J-1650	0.20	36	19	27	...	0.02	...	2	12	...	3.8 Ti
MAR-M 302	0.85	58	21.5	...	...	0.005	...	9	10	0.2	...
MAR-M 322	1.0	60.5	21.5	...	...	...	0.5	4.5	9	2	0.75 Ti
MAR-M 509	0.6	54.5	23.5	10	...	...	...	3.5	7	0.5	0.2 Ti
MAR-M 918	0.05	52	20	20	...	...	...	7.5	...	0.1	...
NASA Co-W-Re	0.40	67.5	3	...	...	...	...	...	25	1	2 Re, 1 Ti
S-816	0.4	42	20	20	...	...	4	...	4	...	4 Mo, 4 Nb, 1.2 Mn, 0.4 Si
V-36	0.27	42	25	20	...	...	3	...	2	...	4 Mo, 2 Nb, 1 Mn, 0.4 Si
WI-52	0.45	63.5	21	...	...	...	2	...	11	...	2 Nb + Ta
X-40	0.50	57.5	22	10	...	...	1.5	...	7.5	...	0.5 Mn, 0.5 Si

Standard designations and composition ranges for corrosion-resistant steel castings

ACI type(a)	Wrought alloy type(b)	Composition, %(c)					
		C	Mn	Si	Cr	Ni	Others(d)
CA-6NM	...	0.06	1.00	1.00	11.5 to 14.0	3.5 to 4.5	0.40 to 1.0 Mo
CA-15	410	0.15	1.00	1.50	11.5 to 14.0	1.0	0.5 Mo max(e)
CA-40	420	0.40	1.00	1.50	11.5 to 14.0	1.0	0.5 Mo max(e)
CB-7Cu-1	...	0.07	0.70	1.00	15.5 to 17.7	3.6 to 4.6	2.5 to 3.2 Cu; 0.20 to 0.35 Nb; 0.05 N max
CB-7Cu-2	...	0.07	0.70	1.00	14.0 to 15.5	4.5 to 5.5	2.5 to 3.2 Cu; 0.20 to 0.35 Nb; 0.05 N max
CB-30	431	0.30	1.00	1.50	18.0 to 22.0	2.0	...
CC-50	446	0.50	1.00	1.50	26.0 to 30.0	4.0	...
CD-4MCu	...	0.04	1.00	1.00	25.0 to 26.5	4.75 to 6.0	1.75 to 2.25 Mo; 2.75 to 3.25 Cu
CE-30	312	0.30	1.50	2.00	26.0 to 30.0	8.0 to 11.0	...
CF-3(f)	304L	0.03	1.50	2.00	17.0 to 21.0	8.0 to 12.0	...
CF-3M(f)	316L	0.03	1.50	2.00	17.0 to 21.0	8.0 to 12.0	2.0 to 3.0 Mo
CF-8(f)	304	0.08	1.50	2.00	18.0 to 21.0	8.0 to 11.0	...
CF-8C	347	0.08	1.50	2.00	18.0 to 21.0	9.0 to 12.0	Nb(g)
CF-8M	316	0.08	1.50	2.00	18.0 to 21.0	9.0 to 12.0	2.0 to 3.0 Mo
CF-12M	316	0.12	1.50	2.00	18.0 to 21.0	9.0 to 12.0	2.0 to 3.0 Mo
CF-16F	303	0.16	1.50	2.00	18.0 to 21.0	9.0 to 12.0	1.5 Mo max; 0.20 to 0.35 Se
CF-20	302	0.20	1.50	2.00	18.0 to 21.0	8.0 to 11.0	...
CG-8M	317	0.08	1.50	1.50	18.0 to 21.0	9.0 to 13.0	3.0 to 4.0 Mo
CH-20	309	0.20	1.50	2.00	22.0 to 26.0	12.0 to 15.0	...
CK-20	310	0.20	2.00	2.00	23.0 to 27.0	19.0 to 22.0	...
CN-7M	...	0.07	1.50	1.50	19.0 to 22.0	27.5 to 30.5	2.0 to 3.0 Mo; 3.0 to 4.0 Cu
CN-7MS	...	0.07	1.50	3.50(h)	18.0 to 20.0	22.0 to 25.0	2.5 to 3.0 Mo; 1.5 to 2.0 Cu

(a) Most of these standard grades are covered by ASTM A743 and A744. (b) Type numbers of wrought alloys are listed only for nominal identification of corresponding wrought and cast grades. Composition ranges of cast alloys are not the same as for corresponding wrought alloys; cast alloy designations should be used for castings only. (c) Maximum unless a range is specified. (d) Phosphorus content is 0.04% max except in CF-16F, which has 0.17% max P; sulfur content is 0.04% max in all grades. (e) Molybdenum not intentionally added. (f) CF-3A, CF-3MA and CF-8A have the same composition ranges as CF-3, CF-3M and CF-8, respectively, but have balanced compositions so that ferrite contents are at levels that permit higher mechanical-property specifications than those for related grades. They are covered by ASTM A351. (g) Nb, 8 × %C min (1.0% max); or Nb + Ta, 9 × %C (1.1% max). (h) For CN-7MS, silicon ranges from 2.50 to 3.50%

Physical properties of selected superalloys

Alloy	Density, Mg/m³	Melting temperature Liquidus °C	Liquidus °F	Solidus °C	Solidus °F	Specific heat(a) J/kg	Specific heat(a) Btu/lb	Electrical conductivity, % IACS	Electric resistivity, nΩ·m	Magnetic permeability	Curie temperature °C	Curie temperature °F
Iron-base alloys												
Carpenter 20-Cb3	8.055	1425	2600	1370	2500	...	...	...	1040	...	...	...
Haynes 556	8.23	...	...	...	...	472	0.113	...	970	...	...	...
Incoloy 800	7.94	1385	2525	1355	2475	502	0.117	1.7	989	1.0092	...	...
Incoloy 801	7.94	1385	2525	1355	2475	452	0.105	1.7	1012	...	...	...
Incoloy 825	8.14	1400	2500	1370	2500	...	...	1.5	1127	1.005	<-196	<-520
Cobalt-base alloys												
Haynes 25 (L-605)	9.13	1410	2570	1329	2425	374	0.090	...	890	<1.00	...	...
Haynes 188	9.13	1398	2550	1302-1330	2375-2425	423(b)	0.101(b)	...	...	1.01	...	...
Stellite 6B	8.38	1354	2470	1265	2310	421	0.101	...	910	<1.2	...	...
UMCo 50	8.05	1395	2540	1380	2515	...	...	...	825	...	...	...
Nickel-base alloys												
Hastelloy B-2	9.21	...	...	...	...	389(b)	0.093(b)	...	1380(b)	...	...	...
Hastelloy C-4	8.64	...	...	...	...	426(b)	0.102(b)	...	1250	...	...	...
Hastelloy C-276	8.90	1371	2500	1323	2415	427	0.102	...	1330	...	...	...
Hastelloy N	8.93	...	...	...	...	419(b)	0.100(b)	...	1200(b)	...	...	...
Hastelloy S	8.76	1380	2516	1335	2435	427(b)	0.102(b)	...	...	...	...	...
Hastelloy W	9.03	1315	2400	...	...	...	...	...	...	...	...	...
Hastelloy X	8.23	1290	2350	1250	2280	486	0.116	...	1180	<1.002(c)	...	...
Inconel 600	8.42	1415	2575	1354	2470	444	0.103	1.7	1030	1.010	-124	-192
Inconel 617	...	...	...	1333	2430	...	...	...	...	...	...	...
Inconel 625	8.44	1350	2460	1290	2350	410	0.095	1.3	1290	1.006	-196	-320
Inconel 671	7.86	1350	2460	1305	2385	456	0.106	2.0	869	...	...	...
Inconel 690	8.03	1375	2510	1345	2450	...	...	1.5	148	...	...	...
Inconel X750	8.25	1425	2600	1393	2540	431	0.103	...	1215	1.0020	-143	-225
Nimonic 75	...	...	...	1380	2515	...	...	...	...	...	...	...
Nimonic 80A	...	...	...	1360	2480	...	...	...	...	...	...	...
Nimonic 90	...	...	...	1310	2390	...	...	...	...	...	...	...
Nimonic 100	...	...	...	...	2256	...	...	...	...	...	...	...
Nimonic 105	...	...	...	1290	2350	...	...	...	...	...	...	...
René 41	8.25	1371	2500	1232	2250	452	0.108	...	1308	1.002	...	...
Udimet 500	8.14	1345	2450	1260	2300	...	...	...	1203	...	...	...
Udimet 700	7.92	1345	2450	1216	2220	...	...	...	...	...	...	...
Waspaloy	8.20	1355	2475	1339	2425	523(c)	0.125(c)	...	1240	...	...	...

(a) At room temperature. (b) At 100 °C (212 °F). (c) At 93 °C (200 °F)

Typical mechanical properties of cobalt-base and nickel-base superalloys

Temperature		Tensile strength		Yield strength		Elongation, %
°C	°F	MPa	ksi	MPa	ksi	
Cobalt-base alloys						
Haynes 25 (L-605) sheet						
21	70	1010	146	460	67	64
540	1000	800	116	250	36	59
650	1200	710	103	240	35	35
760	1400	455	66	260	38	12
870	1600	325	47	240	35	30
Haynes 188, sheet						
21	70	960	139	485	70	56
540	1000	740	107	305	44	70
650	1200	710	103	305	44	61
760	1400	635	92	290	42	43
870	1600	420	61	260	38	73
S-816, bar						
21	70	965	140	385	56	30
540	1000	840	122	310	45	27
650	1200	765	111	305	44	25
760	1400	650	94	285	41	21
870	1600	360	52	240	35	16
Nickel-base alloys						
Astroloy, bar						
21	70	1410	205	1050	152	16
540	1000	1240	180	965	140	16
650	1200	1310	190	965	140	18
760	1400	1160	168	910	132	21
870	1600	770	112	690	100	25
D-979, bar						
21	70	1410	204	1010	146	15
540	1000	1300	188	925	134	15
650	1200	1100	160	890	129	21
760	1400	7	104	655	95	17
870	1600	345	50	305	44	18
Hastelloy X, sheet						
21	70	785	114	360	52	43
540	1000	650	94	290	42	45
650	1200	570	83	275	40	37
760	1400	435	63	260	38	37
870	1600	255	37	180	26	50
IN-102, bar						
21	70	960	139	505	73	47
540	1000	825	120	400	58	48
650	1200	710	103	400	58	64
760	1400	440	64	385	56	110
870	1600	215	31	200	29	110
Inconel 600, bar						
21	70	620	90	250	36	47
540	1000	580	84	195	28	47
650	1200	450	65	180	26	39
760	1400	185	27	115	17	46
870	1600	105	15	62	9	80
Inconel 601, sheet						
21	70	740	107	340	49	45
540	1000	725	105	150	22	38
650	1200	525	76	180	26	45
760	1400	290	42	200	29	73
870	1600	160	23	140	20	92
Inconel 625, bar						
21	70	855	124	490	71	50
540	1000	745	108	405	59	50
650	1200	710	103	420	61	35
760	1400	505	73	420	61	42
870	1600	285	41	475	40	125
Inconel 706, bar						
21	70	1300	188	980	142	19
540	1000	1120	163	895	130	19
650	1200	1010	147	825	120	21
760	1400	690	100	675	98	32

(continued)

Typical mechanical properties of cobalt-base and nickel-base superalloys (continued)

Temperature		Tensile strength		Yield strength		Elongation,
°C	°F	MPa	ksi	MPa	ksi	%

Nickel-base alloys (continued)

Inconel 718, bar						
21	70	1430	208	1190	172	21
540	1000	1280	185	1060	154	18
650	1200	1230	178	1020	148	19
760	1400	950	138	740	107	25
870	1600	340	49	330	48	88
Inconel 718, sheet						
21	70	1280	185	1050	153	22
540	1000	1140	166	945	137	26
650	1200	1030	150	870	126	15
760	1400	675	98	625	91	8
Inconel X-750, bar						
21	70	1120	162	635	92	24
540	1000	965	140	580	84	22
650	1200	825	120	565	82	9
760	1400	485	70	455	66	9
870	1600	235	34	165	24	47
M-252, bar						
21	70	1240	180	840	122	16
540	1000	1230	178	765	111	15
650	1200	1160	168	745	108	11
760	1400	945	137	715	104	10
870	1600	510	74	485	70	18
Nimonic 75, bar						
21	70	750	109	...	...	41
540	1000	635	92	...	...	41
650	1200	538	78	...	...	42
760	1400	290	42	...	...	70
870	1600	145	21	...	...	68
Nimonic 80A, bar						
21	70	1240	179	620	90	24
540	1000	1100	160	530	77	24
650	1200	1000	145	550	80	18
760	1400	760	110	505	73	20
870	1600	400	58	260	38	34
Nimonic 90, bar						
21	70	1240	180	805	117	23
540	1000	1100	160	725	105	23
650	1200	1030	150	685	99	20
760	1400	825	120	540	78	10
870	1600	430	62	260	38	16
Nimonic 105, bar						
21	70	1140	166	815	118	12
540	1000	1100	160	775	112	18
650	1200	1080	156	800	116	24
760	1400	965	140	655	95	22
870	1600	605	88	365	53	25
Nimonic 115, bar						
21	70	1240	180	860	125	25
540	1000	1090	158	795	115	26
650	1200	1120	163	815	118	25
760	1400	1080	157	800	116	22
870	1600	825	120	550	80	18
Pyromet 860, bar						
21	70	1300	188	835	121	22
540	1000	1250	182	840	122	15
650	1200	1110	161	850	123	17
760	1400	910	132	835	121	18
René 41, bar						
21	70	1420	206	1060	154	14
540	1000	1400	203	1010	147	14
650	1200	1340	194	1000	145	14
760	1400	1100	160	940	136	11
870	1600	620	90	550	80	19

(continued)

Typical mechanical properties of cobalt-base and nickel-base superalloys (continued)

Temperature		Tensile strength		Yield strength		Elongation, %
°C	°F	MPa	ksi	MPa	ksi	

Nickel-base alloys (continued)

Temperature		Tensile strength		Yield strength		Elongation, %
René 95, bar						
21	70	1620	235	1310	190	15
540	1000	1540	224	1250	182	12
650	1200	1460	212	1220	177	14
760	1400	1170	170	1100	160	15
Udimet 500, bar						
21	70	1310	190	840	122	32
540	1000	1240	180	795	115	28
650	1200	1210	176	760	110	28
760	1400	1040	151	730	106	39
870	1600	640	93	495	72	20
Udimet 520, bar						
21	70	1310	190	860	125	21
540	1000	1240	180	825	120	20
650	1200	1170	170	795	115	17
760	1400	725	105	725	105	15
870	1600	515	75	515	75	20
Udimet 700, bar						
21	70	1410	204	965	140	17
540	1000	1280	185	895	130	16
650	1200	1240	180	855	124	16
760	1400	1030	150	825	120	20
870	1600	690	100	635	92	27
Udimet 710, bar						
21	70	1190	172	910	132	7
540	1000	1150	167	850	123	10
650	1200	1290	187	860	125	15
760	1400	1020	148	815	118	25
870	1600	705	102	635	92	29
Unitemp AF2-1DA, bar						
21	70	1290	187	1050	152	10
540	1000	1340	194	1080	157	13
650	1200	1360	197	1080	157	13
760	1400	1150	167	1010	146	8
870	1600	830	120	715	104	8
Waspaloy, bar						
21	70	1280	185	795	115	25
540	1000	1170	170	725	105	23
650	1200	1120	162	690	100	34
760	1400	795	115	675	98	28
870	1600	525	76	515	75	35

Mechanical properties of representative refractory metal alloys

Common designation	Alloying additions, %	Product form	Condition(a)	Low-temperature ductility(b)	Typical high-temperature strength		
					Temperature, °F	Tensile, ksi	10-h rupture, ksi
Niobium alloys							
Unalloyed	None	All	RX	A	2000	10	5.4
Nb-1Zr	1 Zr	All	RX	A	2000	23	14
C103 (KbI-3)	10 Hf, 1 Ti, 0.7 Zr	All	RX	A	2000	27	...
SCb291	10 Ta, 10 W	Bar, sheet	RX	A	2000	32	9
C129	10 W, 10 Hf, 0.1 Y	Sheet	RX	A	2400	26	15
FS85	28 Ta, 11 W, 0.8 Zr	Sheet	RX	A	2400	23	12
SU31	17 W, 3.5 Hf, 0.12 C, 0.03 Si	Bar, sheet	SP	C	2400	40	22
Molybdenum alloys							
Unalloyed	None	All	SRA	B-C	1800	52	25
Doped Mo	K, Si; ppm levels	Wire, sheet	CW	B	3000	30	...
Low C Mo	None	All	SRA	B	1800	50	24
TZM(c)	0.5 Ti, 0.08 Zr, 0.015 C	All	SRA	B-C	2400	45	23
TZC	1.0 Ti, 0.14 Zr, 0.02 to 0.08 C	All	SRA	B-C	2400	55	28
Mo-5Re	5 Re	All	SRA	B	3000	2	1
Mo-30W	30 W	All	SRA	B-C	2000	50	20
Tantalum alloys							
Unalloyed	None	All	RX	A	2400	8.5	2.5
FS61	7.5 W (P/M)	Wire, sheet	CW	A	75	165	...
FS63	2.5 W, 0.15 Nb	All	RX	A	200	46	...
TA-10W	10 W	All	RX	A	2400	50	20
KBI-40	40 Nb	All	RX	A	500	42	...
Tungsten alloys							
Unalloyed	None	Bar, sheet, wire	SRA	D	3000	25	6.8
Doped	K, Si, Al; ppm levels	Wire	CW	C	3000	94	...
W-1 ThO₂	1 ThO₂	Bar, sheet, wire	SRA	D	3000	37	...
W-2 ThO₂	2 ThO₂	Bar, sheet, wire	SRA	D	3000	30	18
W-3 ThO₂	3 ThO₂	Bar, wire	SRA	D	3000	30	18
W-4 ThO₂	4 ThO₂	Bar	SRA	D	3000	30	18
W-15 Mo	15 Mo	Bar, wire	SRA	D	3000	36	12
W-50 Mo	50 Mo	Bar, wire	SRA	D	3000	20	12
W-3 Re	3 Re	Wire	CW	C	...	...	...
W-25 Re	25 Re	Bar, sheet, wire	SRA	B	3000	33	10

(a) Cw, cold worked; RX, recrystallized; SRA, stres-relief annealed; SP, special thermal processing. (b) A, excellent cryogenic ductility; B, excellent room-temperature ductility; C, may have marginal ductility at room temperature; D, normally brittle at room temperature. (c) Available in both powder metallurgy and arc cast forms

Typical rupture strengths of selected superalloys

Temperature °C	Temperature °F	For stress rupture at: 100 h MPa	ksi	1000 h MPa	ksi
Incoloy 800					
650	1200	220	32	145	21
760	1400	115	17	69	10
870	1600	45	6.5	33	4.8
Incoloy 801					
650	1200	250	36	...	...
730	1350	145	21	...	...
815	1500	62	9	...	...
Incoloy 802					
650	1200	240	35	170	24
760	1400	145	21	105	15
870	1600	97	14	62	9
Inconel 600					
815	1500	55	8	39	5.6
870	1600	37	5.3	24	3.5
Inconel 601(a)					
540	1000	...	...	400	58
870	1600	48	7	30	4.3
980	1800	23	3.4	14	2.1
Inconel 617(b)					
815	1500	140	20	97	14
925	1700	62	9	...	5.5
980	1800	41	6	...	3.5
Inconel 625(a)					
650	1200	440	64	370	54
815	1500	130	19	93	13.5
870	1600	72	10.5	48	7
Inconel 718(c)					
540	1000	...	...	951	138
595	1100	860	125	760	110
650	1200	690	100	585	85
Inconel 751(d)					
815	1500	200	29	125	18
870	1600	120	17	69	10
Inconel X-750(e)					
540	1000	...	...	827	120
870	1600	83	12	45	6.5
925	1700	58	8.4	21	3.1
N-155, bar(f)					
650	1200	360	52	295	43
730	1350	195	28	150	22
870	1600	97	14	66	9.5
N-155(g)					
650	1200	380	55	290	42
N-155, sheet(f)					
980	1800	39	5.6	20	2.9
Nimonic 75(h)					
815	1500	38	5.5	24	3.5
870	1600	23	3.4	15	2.2
925	1700	14	2.1	10	1.5
980	1800	...	...	7.6	1.1
Nimonic 80A(j)					
540	1000	...	...	825	120
815	1500	185	27	115	17
870	1600	105	15	...	...
Nimonic 90(j)					
815	1500	240	35	155	22.5
870	1600	150	22	69	10
925	1700	69	10	...	...
Nimonic 105(k)					
815	1500	325	47	225	32
870	1600	210	30.2	135	19
Nimonic 115(m)					
815	1500	425	62	315	46
870	1600	315	46	205	30
925	1700	205	30	130	18.5
Nimonic 263(n)					
815	1500	170	24.5	105	15
870	1600	93	13.5	46	6.7
925	1700	45	6.5	...	...

(a) Solution treat 1150 °C (2100 °F). (b) Solution treat 1175 °C (2150 °F). (c) Heat treat to 980 °C (1800 °F) plus 720 °C (1325 °F) hold for 8 h, furnace cool to 620 °C (1150 °F) hold for 8 h. (d) 730 °C (1350 °F) hold for 2 h. (e) Heat treat to 1150 °C (2100 °F) plus 840 °C (1550 °F) hold for 24 h, plus 705 °C (1300 °F) hold for 20 h. (f) Solution treated and aged. (g) Stress-relieved forging. (h) Heat treat to 1050 °C (1920 °F) hold for 1 h. (j) Heat treat to 1080 °C (1975 °F) hold for 8 h, plus 700 °C (1290 °F) hold for 16 h. (k) Heat treat to 1150 °C (2100 °F) hold for 4 h, plus 1050 °C (1920 °F) hold for 16 h, plus 850 °C (1560 °F) hold for 16 h. (m) Heat treat to 1190 °C (2175 °F) hold for 1.5 h, plus 1100 °C (2010 °F) hold for 6 h. (n) Heat treat to 1150 °C (2100 °F) hold for 2 h, water quench, plus 800 °C (1475 °F) hold for 8 h

Typical room-temperature properties of ACI heat-resistant casting alloys

Alloy	Condition	Tensile strength MPa	ksi	Yield strength MPa	ksi	Elongation, %	Hardness, HB
HC	As cast	760	110	515	75	19	223
	Aged(a)	790	115	550	80	18	...
HD	As cast	585	85	330	48	16	90
HE	As cast	655	95	310	45	20	200
	Aged(a)	620	90	380	55	10	270
HF	As cast	635	92	310	45	38	165
	Aged(a)	690	100	345	50	25	190
HH, type 1	As cast	585	85	345	50	25	185
	Aged(a)	595	86	380	55	11	200
HH, type 2	As cast	550	80	275	40	15	180
	Aged(a)	635	92	310	45	8	200
HI	As cast	550	80	310	45	12	180
	Aged(a)	620	90	450	65	6	200
HK	As cast	515	75	345	50	17	170
	Aged(a)	585	85	345	50	10	190
HL	As cast	565	82	360	52	19	192
HN	As cast	470	68	260	38	13	160
HP	As cast	490	71	275	40	11	170
HT	As cast	485	70	275	40	10	180
	Aged(b)	515	75	310	45	5	200
HU	As cast	485	70	275	40	9	170
	Aged(c)	505	73	295	43	5	190
HW	As cast	470	68	250	36	4	185
	Aged(d)	580	84	360	52	4	205
HX	As cast	450	65	250	36	9	176
	Aged(c)	505	73	305	44	9	185

(a) 24 h at 760 °C (1400 °F), furnace cool. (b) 24 h at 760 °C (1400 °F), air cool. (c) 48 h at 980 °C (1800 °F), air cool. (d) 48 h at 980 °C (1800 °F), furnace cool

Approximate rates of corrosion for ACI heat-resistant casting alloys in air and in flue gas

Alloy	Oxidation rate in air, mils/yr(a), at: 870 °C (1600 °F)	980 °C (1800 °F)	1090 °C (2000 °F)	Corrosion rate, mils/yr(a)(b), in flue gas with sulfur content of: 0.12 g/m³ Oxidizing	Reducing	2.3 g/m³ Oxidizing	Reducing
HB	25–	250–	500–	100+	500–	250–	500
HC	10	50	50	25–	25+	25	25–
HD	10–	50–	50–	25–	25–	25–	25–
HE	5–	25–	35–	25–	25–	25–	25–
HF	5–	50+	100	50+	100+	50+	250–
HH	5–	25–	50	25–	25	25	25–
HI	5–	10+	35–	25–	25–	25–	25–
HK	10–	10–	35–	25–	25–	25–	25–
HL	10+	25–	35	25–	25–	25–	25–
HN	5	10+	50–	25–	25–	25	25
HP	25–	25	50	25–	25–	25–	25–
HT	5–	10+	50	25	25–	25	100
HU	5–	10–	35–	25–	25–	25–	25
HW	5–	10–	35	25	25–	50–	250
HX	5–	10–	35–	25–	25–	25–	25–

(a) Data based on 100-h tests. To convert to µm/yr multiply by 25. (b) At 980 °C (1800 °F)

Results of in-plant corrosion testing of CF-8, CF-8M, and CN-7M alloys

Type and composition of corroding solution	Temperature of solution		Alloy	Metal loss on surface		Surface condition by visual examination	Remarks
	°C	°F		µm/yr	mils/yr		
Neutralizer after formation of ammonium sulfate: ammonium sulfate plus small excess of sulfuric acid, ammonia vapor and steam	100	212	CF-8 CF-8M CN-7M	665 28 18	26.2 1.1 0.7	Very heavy etch(a) Light tarnish(b) Bright	CF-8M was installed for low-corrosion-tolerance equipment in this service and performed satisfactorily
Settling tank after neutralizer: ammonium sulfate plus excess of sulfuric acid	50	122	CF-8 CF-8M CN-7M	385 10 2.5	15.2 0.4 0.1	Very heavy etch(a) Slight tarnish Bright(b)	CF-8 in service showed excessive corrosion rate plus heavy concentration-cell attack
Ammonium sulfate processing solution: ammonium sulfate at pH of 8.0	50	122	CF-8 CF-8M CN-7M	685 175 5	27.0 6.8 2.0	Heavy etch Moderate etch Light etch	CF-8M had too high a corrosion rate in service for good valve life, although suitable for equipment of greater corrosion tolerance. CN-7M was installed in this service
99 to 100% fuming nitric acid	20	68	CF-8 CN-7M CF-8M	245 79 345	9.6 3.1 13.5	Moderate etch Light etch Moderate etch	CF-8 was satisfactory except for low-tolerance equipment such as valves. CN-7M valves performed satisfactorily in service
Saturated solution of sodium chloride plus 15% sodium sulfate; pH, 4.5	60	140	CF-8M CF-8	2.5 240	0.1 9.5	Bright Concentration-cell corrosion at various small areas of specimen	CF-8M was installed for valves in service

(a) Concentration-cell attack under insulating washer. (b) Slight concentration-cell attack under insulating washer

Heat Treating

Typical solution treating and aging cycles for heat-resisting casting alloys

| | Solution treating | | | | Aging | | | |
| | Temperature(a) | | Time, | Cooling | Temperature(b) | | Time, | Cooling |
Alloy	°C	°F	h	procedure	°C	°F	h	procedure
A-286	1095	2000	2	Rapid cool	720	1325	16	Air cool
B-1900	As cast		...	...	...	...	...	...
FSX-414	1150	2100	4	Rapid cool	980	1800	4	Air cool
Hastelloy B	1175	2150	2	Rapid cool	(c)	(c)	...	...
Hastelloy C	1220	2225	1	Rapid cool	(c)	(c)	...	...
HS-31 (X-40)	As cast		...	...	...	...	...	...
IN-100	As cast		...	...	...	...	...	...
IN-713C	As cast		...	...	...	...	...	...
IN-738	1120	2050	2	Air cool	845	1550	24	Air cool
IN-792	1120	2050	2	Air cool	845	1550	24	Air cool
IN-939	1160	2120	4	Air cool	850	1560	16	Air cool
Inconel 718	1095	2000	1	Air cool	620	1150	10	Air cool
MAR-M 200	...	...	...	...	870	1600	50	Air cool
MAR-M 200 DS	1230	2250	4	Air cool	870	1600	32	Air cool
MAR-M 246	...	...	...	...	845	1550	50	Air cool
MAR-M 247	...	...	...	...	870	1600	16	Air cool
MAR-M 302	As cast		...	...	...	...	...	...
MAR-M 509	As cast		...	...	...	...	...	...
René 41	1095	2000	½	Rapid cool	900	1650	4	Air cool
René 80	1220	2225	2	Air cool	1095	2000	4	Air cool
					1055	1925	4	Air cool
					845	1550	16	Air cool
Udimet 700	1150	2100	2	Air cool	760	1400	16	Air cool

(a) Furnace temperature tolerance of ± 15 °C (± 25 °F) is satisfactory. (b) Furnace temperature tolerance of ±10 °C (± 15 °F) is recommended. (c) Aging occurs in service at elevated temperature. Use a vacuum or protective atmosphere for heat treating at temperatures above 1040 °C (1900 °F) and subsequent cooling

Prepared atmospheres suitable for annealing of nickels and nickel alloys

Atmospheres 2 through 7 can be used for bright annealing of nickel, modified nickels, and nickel-copper alloys; atmosphere 4 or atmosphere 7 must be used for bright annealing of nickel alloys that contain chromium or molybdenum, or both

| Atmosphere | Air-to-gas ratio(a) | Composition, vol% | | | | | | Dew point (approx) | |
		H_2	CO	CO_2	CH_4	O_2	N_2	°C	°F
1—Completely burned fuel, lean atmosphere	10 to 1	0.5	0.5	10.0	0.0	0.0	89.0	Saturated(b)	
2—Partially burned fuel, medium-rich atmosphere	6 to 1	15.0	10.0	5.0	1.0	0.0	69.0	Saturated(b)	
3—Reacted fuel, rich atmosphere	3 to 1	38.0	19.0	1.0	2.0	0.0	40.0	+20	+70
4—Dissociated ammonia (complete dissociation)	No air	75.0	0.0	0.0	0.0	0.0	25.0	−55 to −75	−70 to −100
5—Dissociated ammonia, partially burned	1.25 to 1(c)	15.0	0.0	0.0	0.0	0.0	85.0	Saturated(b)	
6—Dissociated ammonia, completely burned	1.8 to 1(c)	1.0	0.0	0.0	0.0	0.0	99.0	Saturated(b)	
7—Electrolytic hydrogen, dried(d)	No air	100.0	0.0	0.0	0.0	0.0	0.0	−55 to −75	−70 to −100

(a) Based on use of natural gas containing nearly 100% methane and rated at 37 MJ/m³ (1000 Btu/ft³). For high-hydrogen manufactured gas (20 MJ/m³, or 550 Btu/ft³), ratios are about 50% of values listed. For manufactured gas with lower hydrogen and high carbon monoxide contents (17 MJ/m³, or 450 Btu/ft³), ratios are about 40% of values listed. For propane, ratios are about twice those listed. For butane, multiply listed values by three. (b) When atmosphere is cooled by tap-water heat exchangers, dew point will be about 6 to 8 °C (10 to 15 °F) above the temperature of the tap water. Dew point may be reduced to about 5 °C (40 °F) by refrigeration equipment, and to −55 °C (−70 °F) or lower by activated-absorption equipment. (c) Ratio of air to dissociated ammonia. (d) Dried to a dew point of −55 to −75 °C (−70 to −100 °F) by alumina plus molecular sieve

Typical solution treating and aging cycles for wrought heat-resisting alloys

Alloy	Solution treating Temperature °C	°F	Time, h	Cooling procedure	Aging Temperature °C	°F	Time, h	Cooling procedure
Iron-base alloys								
A-286	980	1800	1	Oil quench	720	1325	16	Air cool
Discaloy	1010	1850	2	Oil quench	730	1350	20	Air cool
					650	1200	20	Air cool
N-155	1175	2150	1	Water quench	815	1500	4	Air cool
Nickel-base alloys								
Astroloy	1175	2150	4	Air cool	845	1550	24	Air cool
	1080	1975	4	Air cool	760	1400	16	Air cool
Hastelloy B	1175	2150	½	(a)	(b)	(b)	…	…
Hastelloy B-2	1065	1950	½	Rapid quench	…	…	…	…
Hastelloy C-4	1065	1950	½	Rapid quench	…	…	…	…
Hastelloy C-276	1120	2050	½	Rapid quench	…	…	…	…
Hastelloy N	1175	2150	½	Rapid quench	…	…	…	…
Hastelloy S	1065	1950	½	Rapid quench	…	…	…	…
Hastelloy C	1220	2225	1	(a)	(b)	(b)	…	…
Hastelloy W	1175	2150	1	(a)	(b)	(b)	…	…
Hastelloy X	1175	2150	1	(a)	…	…	…	…
Inconel 901	1095	2000	2	Water quench	790	1450	2	Air cool
					720	1325	24	Air cool
Inconel 600	1120	2050	2	Air cool	…	…	…	…
Inconel 601	1150	2100	1	Air cool	…	…	…	…
Inconel 617	1175	2150	2	(a)	…	…	…	…
Inconel 625	1150	2100	2	(a)	…	…	…	…
Inconel 706	925 to 1010	1700 to 1850	…	…	845	1550	3	Air cool
					720	1325	8	Furnace cool
					620	1150	8	Air cool
	925 to 1010	1700 to 1850	…	…	730	1350	8	Furnace cool
					620	1150	8	Air cool
Inconel 718	980	1800	1	Air cool	720	1325	8	Furnace cool
					620	1150	8	Air cool
Inconel X-750 (AMS 5667)	855	1625	24	Air cool	705	1300	20	Air cool
Inconel X-750 (AMS 5668)	1150	2100	2	Air cool	845	1550	24	Air cool
					705	1300	20	Air cool
Nimonic 80A	1080	1975	8	Air cool	705	1300	16	Air cool
Nimonic 90	1080	1975	8	Air cool	705	1300	16	Air cool
René 41	1065	1950	½	Air cool	760	1400	16	Air cool
Udimet 500	1080	1975	4	Air cool	845	1550	24	Air cool
					760	1400	16	Air cool
Udimet 700	1175	2150	4	Air cool	845	1550	24	Air cool
	1080	1975	4	Air cool	760	1400	16	Air cool
Waspaloy	1080	1975	4	Air cool	845	1550	24	Air cool
					760	1400	16	Air cool
Cobalt-base alloys								
Haynes 25; L-605	1230	2250	1	Rapid air cool	(b)	(b)	…	…
Haynes 188	1175	2150	½	Rapid air cool	…	…	…	…
Haynes 556	1175	2150	½	Rapid air cool	…	…	…	…
S-816	1175	2150	1	(a)	760	1400	12	Air cool
Stellite 6B	1230	2250	1	Air cool	…	…	…	…

Note: Alternate treatments may be used to improve specific properties. (a) To provide an adequate quench after solution treating, it is necessary to cool below about 540 °C (1000 °F) rapidly enough to prevent precipitation in the intermediate temperature range. For sheet metal parts of most alloys, rapid air cooling will suffice. Oil or water quenching is frequently required for heavier sections that are not subject to cracking. (b) Aging occurs in service at elevated temperatures

Typical stress relieving and annealing cycles for wrought heat-resisting alloys

Alloy	Stress relieving Temperature °C	°F	Holding time per inch of section, h	Annealing(a) Temperature °C	°F	Holding time per inch of section, h
Iron-base and iron-nickel-chromium alloys						
RA-330	900	1650	1(b)	1110(c)	2025(c)	¼(d)
19-9 DL	675(e)	1250(e)	4	980	1800	1
A-286	(f)	(f)	…	980	1800	1
Discaloy	(f)	(f)	…	1035	1900	1
Nickel-base alloys						
Astroloy	(f)	(f)	…	1135	2075	4
Hastelloy B	(f)	(f)	…	1175	2150	1
Hastelloy C	(f)	(f)	…	1215	2225	1
Hastelloy W	(f)	(f)	…	1175	2150	1
Hastelloy X	(f)	(f)	…	1175	2150	1
Incoloy 800	870	1600	1½	980	1800	¼
Incoloy 800H	…	…	…	1175	2150	…
Incoloy 825	…	…	…	980	1800	…
Incoloy 901	(f)	(f)	…	1095	2000	2
Inconel 600	900	1650	1	1010	1850	¼(d)
Inconel 601	…	…	…	980	1800	…
Inconel 625	870	1600	1	980	1800	1
Inconel 690	…	…	…	1040	1900	½
Inconel 718	(f)	(f)	…	955	1750	1
Inconel X750	880(g)	1625(g)	…	1035	1900	½
Nimonic 80A	(f)	(f)	…	1080	1975	2
Nimonic 90	(f)	(f)	…	1080	1975	2
René 41	(f)	(f)	…	1080	1975	2
Udimet 500	(f)	(f)	…	1080	1975	4
Udimet 700	(f)	(f)	…	1135	2075	4
Waspaloy	(f)	(f)	…	1010	1850	4
Cobalt-chromium-nickel-base alloys						
L-605 (HS-25)	(h)	(h)	…	1230	2250	1
N-155 (HS-95)	(h)	(h)	…	1175	2150	…
S-816	(h)	(h)	…	1205	2200	1
Refractory metals(j)						
Ta-10W	1205(k)	2200(k)	1	1425(k)	2600(k)	1
FS-80	1095(k)	2000(k)	1	1315(k)	2400(k)	1
FS-82	1095(k)	2000(k)	1	1315(k)	2400(k)	1
Mo-0.5 Ti	1095(m)	2000(m)	½	1315(m)(n)	2400(m)(n)	1
TZM	1205(m)	2200(m)	1	1425(m)(n)(p)	2600(m)(n)(p)	1

(a) Minimum hardness is achieved by cooling rapidly from the annealing temperature, to prevent precipitation of hardening phases. Water quenching is preferred, and is usually necessary for heavy sections; air cooling is preferred for heavy sections of Waspaloy, Udimet 500, Udimet 700 and Inconel X-750, because water quenching causes cracking. However, for complex shapes subject to excessive distortion, oil quenching is often adequate and more practical. Rapid air cooling usually is adequate for parts formed from strip or sheet. Rapid cooling from the annealing or solution treating temperature does not suppress the aging reaction of some alloys, such as Astroloy; these alloys become harder and stronger. (b) Time given is minimum; some plants use as long as 3 h per in. (c) Nominal temperature; 1035 to 1175 °C (1900 to 2150 °F) is commonly used. (d) Short time is required for prevention of grain coarsening. (e) Nominal temperature; 650 to 705 °C (1200 to 1300 °F) is permissible. (f) Full annealing is recommended, because intermediate temperatures cause aging. (g) Used only for stress equalizing of warm worked grades. (h) Full annealing is recommended if further fabrication is performed; otherwise, material can be stress relieved at approximately 55 °C (100 °F) below annealing temperature. (j) Annealing temperatures depend on prior plastic deformation, degree of cold work, alloy content and interstitial purity. Annealing temperatures given are those most frequently used for cold worked sheet or plate; in many instances, more precise determination of the recrystallization temperature is necessary for a specific application. (k) Heat and cool in vacuum or inert-gas atmosphere. (m) Heat and cool in hydrogen or vacuum. (n) Seldom used as finished product in annealed condition, because recrystallization raises the ductile-brittle transition temperature, resulting in brittleness at low temperatures. (p) For vacuum-arc-cast material with a minimum of 50% cold work

Machining

Machining data for ACI heat-resistant casting alloys

ACI designation	Typical hardness, HB	Rough turning(a) Speed, ft/min.(b)	Feed, ipr(c)	Finishing Speed, ft/min.(b)	Feed, ipr(c)	Drilling speed(d), ft/min.(b)
HA	220	40 to 50	0.010 to 0.030	80 to 100	0.005 to 0.010	35 to 70
HC	220	40 to 50	0.025 to 0.035	80 to 100	0.010 to 0.015	40 to 60
HD	190	40 to 50	0.025 to 0.035	80 to 100	0.010 to 0.015	40 to 60
HE	270	30 to 40	0.020 to 0.025	60 to 80	0.005 to 0.010	30 to 60
HF	190	25 to 35	0.020 to 0.025	50 to 70	0.005 to 0.010	20 to 40
HH	200	25 to 35	0.015 to 0.020	50 to 70	0.005 to 0.010	20 to 40
HI	200	25 to 35	0.015 to 0.020	50 to 70	0.005 to 0.010	20 to 40
HK	190	25 to 35	0.020 to 0.025	50 to 70	0.005 to 0.010	20 to 40
HL	190	30 to 40	0.020 to 0.025	60 to 80	0.005 to 0.010	30 to 60
HN	160	35 to 45	0.020 to 0.025	70 to 90	0.005 to 0.010	40 to 60
HP	…	35 to 45	0.020 to 0.025	70 to 90	0.005 to 0.010	40 to 60
HT	200	40 to 45	0.025 to 0.035	80 to 90	0.005 to 0.010	40 to 60
HU	190	40 to 45	0.025 to 0.035	80 to 90	0.010 to 0.015	40 to 60
HW	200	40 to 45	0.025 to 0.035	80 to 90	0.010 to 0.015	40 to 60
HX	185	40 to 45	0.025 to 0.035	80 to 90	0.010 to 0.015	40 to 60

(a) Single-point high speed steel tools usually are ground to 4 to 10° side and back rake, 4 to 7° side relief, 7 to 10° end relief, 8 to 15° end cutting-edge angle, 10 to 15° side cutting-edge angle, and 1/32- to 1/8-in. nose radius. (b) To convert to m/s, multiply by 0.005. (c) To convert to mm/rev, multiply by 25. (d) Recommended drilling feeds are as follows: for drill diameters up to 1/8 in., 0.001 to 0.002 ipr; 1/8 to 1/4 in., 0.002 to 0.004 ipr; 1/4 to 1/2 in., 0.004 to 0.007 ipr; 1/2 to 1 in., 0.007 to 0.015 ipr; over 1 in., 0.015 to 0.025 ipr. Tapping speeds recommended for HA, HC, HD, HE and HL are 10 to 25 sfm; for HF, HH, HI, and HK, 10 to 20 sfm; and for HN, HT, HU, HW and HX, 5 to 15 sfm

Speeds and feeds for machining corrosion-resistant steel castings

Operation	Approximate feed, ipr(a)	Machining speed, ft/min.(b) CF-20 CF-8	CF-16F	CE-30 CF-8M CH-20 CK-20	CF-8C	CN-7M	CA-6NM CA-15	CA-40	CB-30	CC-50
Broaching	0.001 to 0.005	8 to 15	10 to 20	8 to 15	8 to 15	8 to 15	10 to 20	8 to 15	10 to 20	10 to 20
Tapping	0.003 to 0.007	10 to 20	15 to 30	10 to 25	10 to 25	12 to 20	10 to 25	10 to 20	10 to 25	10 to 25
Threading	0.003 to 0.008	10 to 20	10 to 25	10 to 25	10 to 25	10 to 20	10 to 20	10 to 20	10 to 25	10 to 25
Reaming	0.003 to 0.008	20 to 60	30 to 100	40 to 80	40 to 80	20 to 60	20 to 60	20 to 60	40 to 120	40 to 120
Drilling	0.003 to 0.007	15 to 40	35 to 85	30 to 50	30 to 50	30 to 60	35 to 75	30 to 60	40 to 60	40 to 60
Turret lathe	0.003 to 0.008	60 to 90	90 to 130	60 to 80	60 to 90	60 to 80	80 to 110	60 to 100	70 to 100	60 to 100
Milling	0.003 to 0.008	35 to 65	75 to 110	30 to 50	40 to 60	35 to 70	70 to 105	35 to 70	40 to 60	40 to 60
Turning, boring	0.003 to 0.008	40 to 85	85 to 120	60 to 120	60 to 120	60 to 80	80 to 115	40 to 80	60 to 100	60 to 120
Screw machine	0.003 to 0.008	60 to 90	90 to 130	60 to 80	60 to 90	60 to 80	80 to 110	60 to 100	70 to 100	60 to 100
Hack sawing		Use a coarse-tooth blade (not over 10 teeth per in.) at about 50 strokes per minute with positive pressure								

(a) For feeds in mm/rev, multiply listed values by 25. (b) For speeds in m/min, multiply listed values by 0.3

Aluminum and Aluminum Alloys

Wrought Aluminum

Temper designations for aluminum alloys

F **As fabricated.** Applies to products of shaping processes in which no special control over thermal conditions or strain-hardening is employed.

H **Strain-hardened (wrought products only).** Applies to products which have their strength increased by strain-hardening, with or without supplementary thermal treatments to produce some reduction in strength.

H1 **Strain-hardened only.** Applies to products which are strain-hardened to obtain the desired strength without supplementary thermal treatment.

H111 Applies to products which are strain-hardened less than the amount required for a controlled H11 temper.

H112 Applies to products which acquire some temper from shaping processes not having special control over the amount of strain-hardening or thermal treatment, but for which there are mechanical property limits.

H2 **Strain-hardened and partially annealed.** Applies to products which are strain-hardened more than the desired final amount, and then reduced in strength to the desired level by partial annealing. For alloys that age-soften at room temperature, the H2 tempers have the same minimum tensile strength as the corresponding H3 tempers. For other alloys, the H2 tempers have the same minimum tensile strength as the corresponding H1 tempers and slightly higher elongation. The number following this designation indicates the degree of strain-hardening remaining after the product has been partially annealed.

H3 **Strain-hardened and stabilized.** Applies to products which are strain-hardened and whose mechanical properties are stabilized by a low temperature thermal treatment which results in slightly lowered tensile strength and improved ductility. This designation is applicable only to those alloys which, unless stabilized, gradually age-soften at room temperature. The number following this designation indicates the degree of strain-hardening before the stabilization treatment.

H311 Applies to products which are strain-hardened less than the amount required for a controlled H31 temper.

H321 Applies to products which are strain-hardened less than the amount required for a controlled H32 temper.

H323 Applies to products which are specially fabri-
H343 cated to have acceptable resistance to stress corrosion cracking.

O **Annealed.** Applies to wrought products which are annealed to obtain the lowest strength temper, and to cast products which are annealed to improve ductility and dimensional stability.

T **Thermally treated to produce stable tempers other than F, O or H.** Applies to products which are thermally treated, with or without supplementary strain-hardening, to produce stable tempers.

T1 **Cooled from an elevated temperature shaping process and naturally aged to a substantially stable condition.** Applies to products which are not cold worked after cooling from an elevated temperature shaping process, or in which the effect of cold work in flattening or straightening may not be recognized in mechanical property limits.

T2 **Cooled from an elevated temperature shaping process, cold worked, and naturally aged to a substantially stable condition.** Applies to products which are cold worked to improve strength after cooling from an elevated temperature shaping process, or in which the effect of cold work in flattening or straightening is recognized in mechanical property limits.

T3 **Solution heat-treated, cold worked, and naturally aged to a substantially stable condition.** Applies to products which are cold worked to improve strength after solution heat-treatment, or in which the effect of cold work in flattening or straightening is recognized in mechanical property limits.

T4 **Solution heat-treated and naturally aged to a substantially stable condition.** Applies to products which are not cold worked after solution heat-treatment, or in which the effect of cold work in flattening or straightening may not be recognized in mechanical property limits. (T42 indicates material is solution heat-treated from the O or F temper to demonstrate response to heat-treatment, and naturally aged to a substantially stable condition.)

T5 **Cooled from an elevated temperature shaping process and then artificially aged.** Applies to products which are not cold worked after cooling from an elevated temperature shaping process, or in which the effect of cold work in flattening or straightening may not be recognized in mechanical property limits.

T51 **Stress relieved by stretching.** Applies to the following products when stretched the indicated amounts after solution heat-treatment or cooled from an elevated temperature shaping process.

Plate—1½ to 3% permanent set
Rod, bar, shapes, extruded tube—1 to 3% permanent set
Drawn—tube 1/2 to 3% permanent set

Applies directly to plate and rolled or cold-finished rod and bar, which receive no further straightening after stretching. Applies to extruded rod, bar, shapes, tubing, and to drawn tubing when designated as follows:

T510 Products that receive no further straightening after stretching.

T511 Products that may receive minor straightening after stretching to comply with standard tolerances.

T52 **Stress-relieved by compressing.** Applies to products which are stress-relieved by compressing after solution heat-treatment, or cooled from an elevated temperature shaping process to produce a permanent set of 1 to 5%.

T54 **Stress-relieved by combined stretching and compressing.** Applies to die forgings which are stress-relieved by restriking cold in the finish die.

T6 **Solution heat-treated and then artificially aged.** Applies to products which are not cold worked after solution heat-treatment, or in which the effect of cold work in flattening or straightening may not be recognized in mechanical property limits. (T62 indicates material is solution heat-treated from the O or F temper to demonstrate response to heat-treatment, and artificially aged.)

T7 **Solution heat-treated and stabilized.** Applies to products which are stabilized after solution heat-treatment to carry them beyond the point of maximum strength to provide control of some special characteristic.

T8 **Solution heat-treated, cold worked, and artificially aged.** Applies to products which are cold worked to improve strength, or in which the effect of cold work in flattening or straightening is recognized in mechanical property limits.

T9 **Solution heat-treated, artificially aged, and cold worked.** Applies to products which are cold worked to improve strength.

T10 **Cooled from an elevated temperature shaping process, cold worked, and artificially aged.** Applies to products which are cold worked to improve strength, or in which the effect of cold work in flattening or straightening is recognized in mechanical property limits.

W **Solution heat-treated.** An unstable temper applicable only to alloys which spontaneously age at room temperature after solution heat-treatment.

Source: Aluminum Association

Refining bauxite to alumina; smelting alumina to aluminum

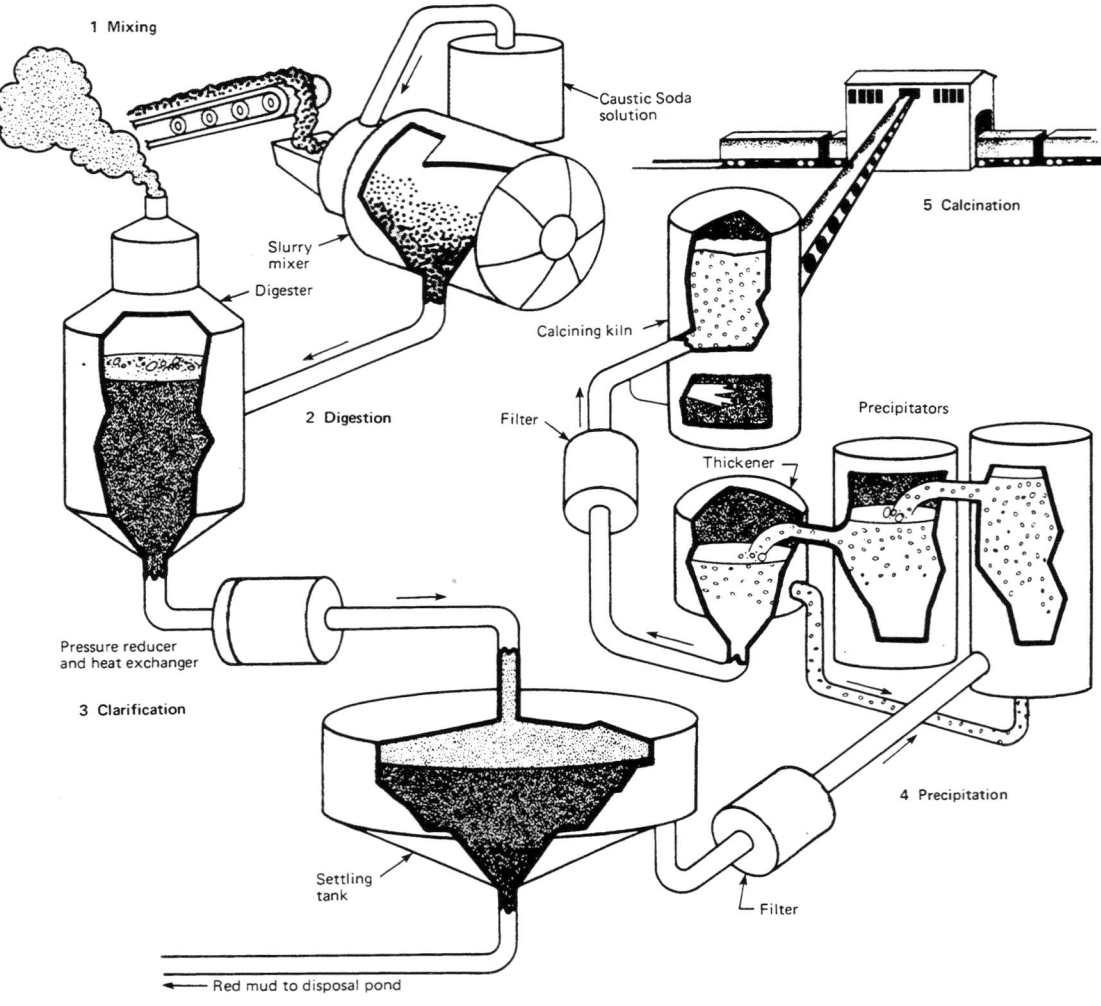

1. Mixing: crushed and mixed with caustic soda, bauxite (aluminum ore) is pumped into huge digesters
2. Digestion: under high pressure and heat, the caustic soda dissolves the alumina, or aluminum oxide, in the bauxite to form sodium aluminate
3. Clarification: while the sodium aluminate remains in solution, iron oxides and other solid impurities drop to the bottom of the settling tank, where, as red mud, they are pumped to a disposal pond
4. Precipitation: after the liquid sodium aluminate is further cooled, it is agitated and seeded with aluminum hydroxide crystals. These form larger crystals, which gradually settle out of the solution. Seed crystals and sodium aluminate remaining in solution are recirculated
5. Calcination: the aluminum hydroxide crystals are roasted at more than 980 °C (1800 °F) to remove the water. A fine white powder, alumina, remains—half aluminum and half oxygen—ready for transport to a smelter

(continued)

Refining bauxite to alumina; smelting alumina to aluminum (continued)

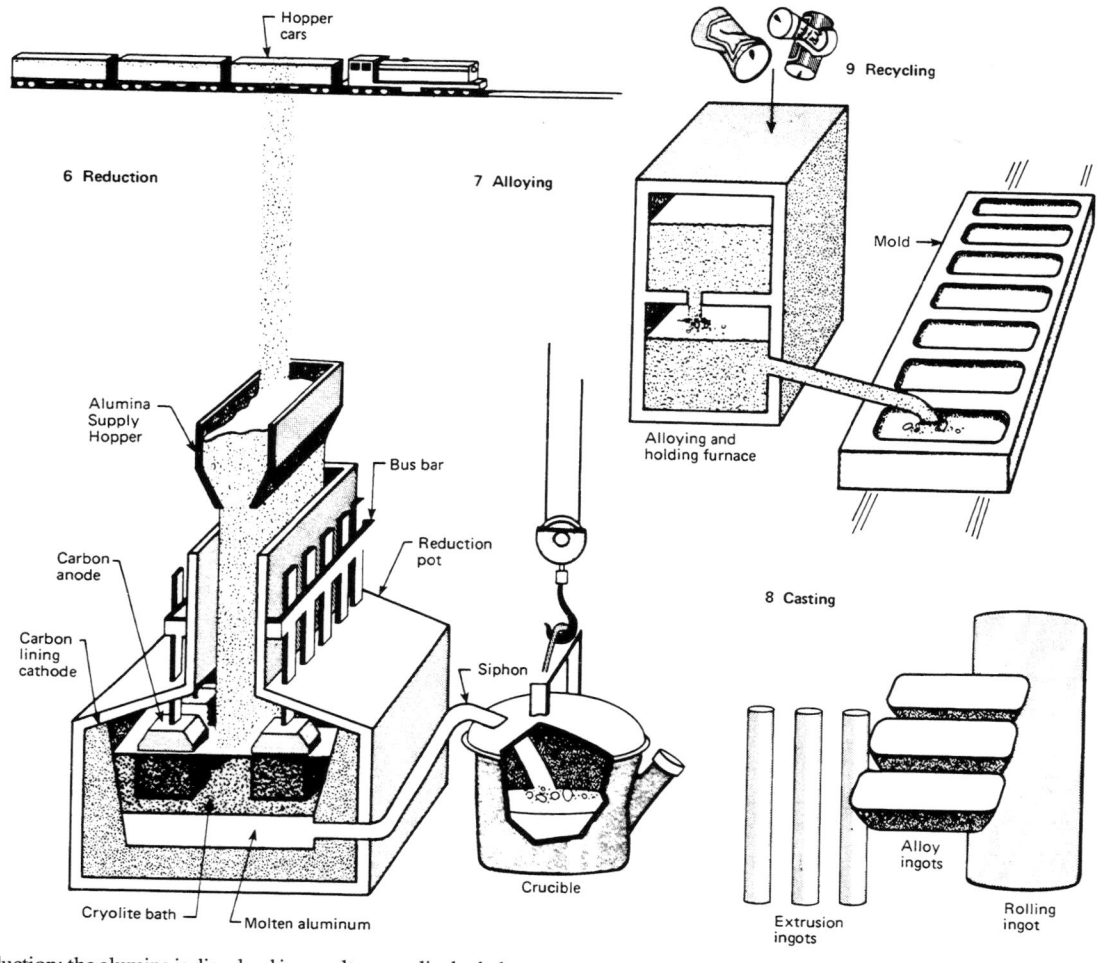

6. Reduction: the alumina is dissolved in a molten cryolite bath that acts as an electrolyte, in which a powerful electric current wrests aluminum from the oxygen. Molten metal settles to the bottom of the pot
7. Alloying: carried by crucible to a furnace, aluminum is alloyed with small amounts of other metals. Copper adds strength; magnesium imparts additional marine-corrosion resistance
8. Casting: molten aluminum is cast into various shapes, from 180-kN (20-ton) rolling ingots for sheet metal to 20-N (4-lb) alloy ingots for further casting
9. Recycling: nearly indestructible, aluminum can be remelted over and over. Depending on energy used to collect and transport cans and scrap, recycling saves up to 95% of the energy used to make aluminum from bauxite

Source: National Geographic Society

Approximate correlation between Brinell and Rockwell hardness of wrought aluminum alloys

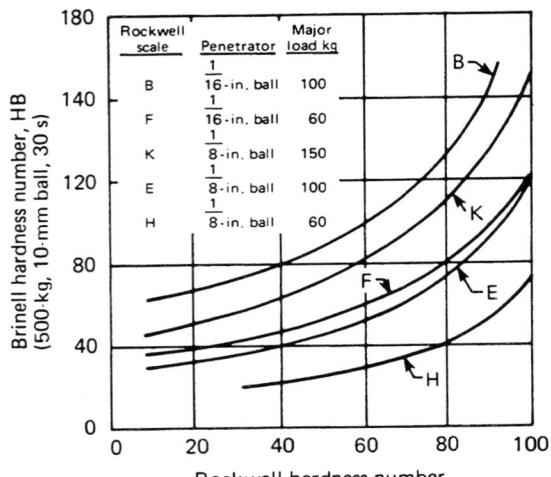

Product forms and nominal compositions of common wrought aluminum alloys

AA number	Product(a)	Al	Si	Cu	Mn	Mg	Cr	Zn	Others
					Composition, %				
1050	DT	99.50 min	...	...	...	...	...	...	...
1060	S, P, ET, DT	99.60 min	...	...	...	...	...	...	...
1100	S, P, F, E, ES, ET, C, DT, FG	99.00 min	...	0.12	...	...	...	...	...
1145	S, P, F	99.45 min	...	...	...	...	...	...	...
1199	F	99.99 min	...	...	...	...	...	...	...
1350	S, P, E, ES, ET, C	99.50 min	...	...	...	...	...	...	...
2011	E, ES, ET, C, DT	93.7	...	5.5	...	...	...	...	0.4 Bi; 0.4 Pb
2014	S, P, E, ES, ET, C, DT, FG	93.5	0.8	4.4	0.8	0.5	...	...	...
2024	S, P, E, ES, ET, C, DT	93.5	...	4.4	0.6	1.5	...	...	...
2036	S	96.7	...	2.6	0.25	0.45	...	...	...
2048	S, P	94.8	...	3.3	0.4	1.5	...	...	...
2124	P	93.5	...	4.4	0.6	1.5	...	...	...
2218	FG	92.5	...	4.0	...	1.5	...	...	2.0 Ni
2219	S, P, E, ES, ET, C, FG	93.0	...	6.3	0.3	...	...	...	0.06 Ti; 0.10 V; 0.18 Zr
2319	C	93.0	...	6.3	0.3	...	...	...	0.18 Zn; 0.15 Ti; 0.10 V
2618	FG	93.7	0.18	2.3	...	1.6	...	...	1.1 Fe; 1.0 Ni; 0.07 Ti
3003	S, P, F, E, ES, ET, C, DT, FG	98.6	...	0.12	1.2	...	...	...	...
3004	S, P, ET, DT	97.8	...	...	1.2	1.0	...	...	...
3105	S	99.0	...	...	0.55	0.50	...	...	...
4032	FG	85.0	12.2	0.9	...	1.0	...	...	0.9 Ni
4043	C	94.8	5.2	...	...	...	...	...	...
5005	S, P, C	99.2	...	...	...	0.8	...	...	...
5050	S, P, C, DT	98.6	...	...	...	1.4	...	...	...
5052	S, P, F, C, DT	97.2	...	...	...	2.5	0.25	...	...
5056	F, C	95.0	...	...	0.12	5.0	0.12	...	...
5083	S, P, E, ES, ET, FG	94.7	...	...	0.7	4.4	0.15	...	...
5086	S, P, E, ES, ET, DT	95.4	...	...	0.4	4.0	0.15	...	...
5154	S, P, E, ES, ET, C, DT	96.2	...	...	...	3.5	0.25	...	...
5182	S	95.2	...	...	0.35	4.5	...	...	...
5252	S	97.5	...	...	...	2.5	...	...	...
5254	S, P	96.2	...	...	...	3.5	0.25	...	...
5356	C	94.6	...	...	0.12	5.0	0.12	...	0.13 Ti
5454	S, P, E, ES, ET	96.3	...	...	0.8	2.7	0.12	...	...
5456	S, P, E, ES, ET, DT, FG	93.9	...	...	0.8	5.1	0.12	...	...
5457	S	98.7	...	...	0.3	1.0	...	...	...
5652	S, P	97.2	...	...	...	2.5	0.25	...	...
5657	S	99.2	...	...	...	0.8	...	...	...
6005	E, ES, ET	98.7	0.8	...	...	0.5	...	...	...
6009	S	97.7	0.8	0.35	0.5	0.6	...	...	...
6010	S	97.3	1.0	0.35	0.5	0.8	...	...	...
6061	S, P, E, ES, ET, C, DT, FG	97.9	0.6	0.28	...	1.0	0.2	...	...
6063	E, ES, ET, DT	98.9	0.4	...	...	0.7	...	...	...
6066	E, ES, ET, DT, FG	95.7	1.4	1.0	0.8	1.1	...	...	...
6070	E, ES, ET	96.8	1.4	0.28	0.7	0.8	...	...	...
6101	E, ES, ET	98.9	0.5	...	...	0.6	...	...	...
6151	FG	98.2	0.9	...	...	0.6	0.25	...	...
6201	C	98.5	0.7	...	...	0.8	...	...	...
6205	E, ES, ET	98.4	0.8	...	0.1	0.5	0.1	...	0.1 Zr
6262	E, ES, ET, C, DT	96.8	0.6	0.28	...	1.0	0.09	...	0.6 Bi; 0.6 Pb
6351	E, ES	97.8	1.0	...	0.6	0.6	...	...	...
6463	E, ES	98.9	0.4	...	...	0.7	...	...	...
7005	E, ES	93.3	...	...	0.45	1.4	0.13	4.5	0.04 Ti; 0.14 Zr
7049	P, E, ES, FG	88.2	...	1.5	...	2.5	0.15	7.6	...
7050	P, E, ES, FG	89.0	...	2.3	...	2.3	...	6.2	0.12 Zr
7072	S, F	99.0	...	...	...	...	...	1.0	...
7075	S, P, E, ES, ET, C, DT, FG	90.0	...	1.6	...	2.5	0.23	5.6	...
7175	S, P, FG	90.0	...	1.6	...	2.5	0.23	5.6	...
7178	S, P, E, ES, C	88.1	...	2.0	...	2.7	0.26	6.8	...
7475	S, P, FG	90.3	1.5	...	...	2.3	0.22	5.7	...

(a) S, sheet; P, plate; F, foil; E, extruded rod, bar and wire; ES, extruded shapes; ET, extruded tubes; C, cold finished rod, bar and wire; DT, drawn tube; FG, forgings

Typical physical properties of wrought aluminum alloys

Alloy	Temper	Electrical conductivity(a)		Electrical resistivity(b)		Thermal conductivity(c)	
		Volume	Weight	$n\,\Omega \cdot m$	ohm/cir mil/ft	W/m · K	Btu/ ft · η · °F
1050	O	61	190	28	17	231	133
1060	O	62	204	28	17	234	135
	H18	61	201	28	17	234	135
1100	O	59	194	29	18	222	128
	H18	57	187	30	18	218	126
1145	O	61	202	28	17	230	133
	H18	60	198	29	18	227	131
1199	O	65	215	27	16	243	140
1350	O	62	204	28	17	234	135
	H1x(d)	61	201	28	17	230	133
2011	T3, T4	39	123	44	27	152	88
	T8	45	142	38	23	173	100
2014	O	50	159	34	21	192	111
	T3, T4	34	108	51	31	134	77
	T6	40	127	43	26	155	89
2024	O	50	160	34	21	190	110
	T3, T4	30	96	57	35	120	69
	T6	37	119	46	28	145	84
	T8	39	125	44	27	152	88
2036	O	52	169	33	20	198	114
	T4	41	135	42	25	159	92
2048	T851	42	137	40	24	159	92
2124	O	50	161	35	21	191	110
	T851	39	126	44	27	152	88
2218	T61	38	121	45	27	148	86
	T72	40	128	43	26	155	90
2219	O	44	138	39	24	170	98
	T31, T37, T351	28	88	62	37	116	67
	T62, T81, T87, T851	30	95	57	35	130	75
2319	O	44	139	39	24	170	98
2618	T61	37	120	47	28	146	84
3003	O	47	154	37	22	180	104
	H12	42	138	41	25	162	94
	H14	41	134	42	25	159	92
	H18	40	130	43	26	155	90
3004	O (all)	42	137	41	25	162	94
3105	O (all)	45	148	38	23	173	100
4032	O	40	132	43	26	155	90
	T6	36	120	48	29	141	82
4043	O	42	140	41	25	163	94
5005	O, H38	52	172	33	20	205	118
5050	O, H38	50	165	34	21	191	110
5052	O, H38	35	116	49	30	137	79
5056	O	29	98	59	36	120	69
	H38	27	91	64	38	112	65
5083	All	29	98	60	36	120	69
5086	All	31	104	56	33	127	73
5154	All	32	108	54	32	127	73
5182	All	31	105	56	33	123	71
5252	All	35	117	49	30	138	80
5254	All	32	107	54	32	127	73
5356	O	29	98	59	36	116	67
5454	All	34	113	51	31	134	77
5456	All	29	98	60	36	116	67
5457	All	46	153	38	23	177	102
5652	All	35	116	49	30	137	79
5657	All	54	179	32	19	…	…
6005	T5	49	162	35	21	167	97
6009	O	54	184	32	19	205	118
	T4	44	150	39	24	172	99
	T6	47	160	37	22	180	104
6010	O	53	175	33	20	202	117
	T4	39	129	44	27	151	87
	T6	44	146	39	24	180	104

(a) % IACS at 20 °C (68 °F). (b) At 20 °C (68 °F). (c) At 25 °C (77 °F). (d) All H1x-type tempers

(continued)

Typical physical properties of wrought aluminum alloys (continued)

Alloy	Temper	Electrical conductivity(a)		Electrical resistivity(b)		Thermal conductivity(c)	
		Volume	Weight	n Ω · m	ohm/cir mil/ft	W/m · K	Btu/ ft · η · °F
6061	O	47	155	37	22	180	104
	T4	40	132	43	26	154	89
	T6	43	142	40	24	167	97
6063	O	58	191	30	18	218	126
	T1	50	165	35	21	193	112
	T5	55	181	32	19	209	121
	T6	53	175	33	20	201	116
6066	O	40	132	43	26	147	85
	T6	37	122	47	28	147	85
6070	T6	44	145	39	24	172	99
6101	T6	57	188	30	18	218	138
	T8	54	178	32	19	218	138
6151	O	54	178	32	19	205	118
	T4	42	138	41	25	163	94
	T6	45	148	38	23	175	101
6201	T81	54	179	32	19	205	118
6205	T1	45	149	37	22	172	99
	T5	49	162	35	21	188	109
6262	T9	44	145	39	24	172	99
6351	T6	46	152	38	23	176	102
6463	T1	50	165	34	21	192	111
	T5	55	181	31	19	209	121
	T6	53	175	33	20	201	116
7005	O	43	138	40	24	166	96
	T53	38	122	45	27	148	86
	T6	35	113	49	30	137	79
	T63	38	122	45	27	148	86
7049	T73	38	120	43	27	154	89
7050	O	47	148	37	22	180	104
	T73	40	127	43	26	157	91
	T76	40	125	44	26	154	89
7072	O	60	197	29	17	227	131
7075	T6	33	105	52	31	130	75
	T73	40	128	43	26	155	90
	T76	38	123	45	27	150	87
7175	O	46	147	38	23	177	102
	T66	36	115	48	29	142	82
	T73	40	128	43	26	155	90
7475	O	46	147	38	23	177	102
	T6	36	115	48	29	142	82
	T7351	40	128	43	26	155	90
	T76	42	134	41	25	163	94

(a) % IACS at 20 °C (68 °F). (b) At 20 °C (68 °F). (c) At 25 °C (77 °F). (d) All H1x-type tempers

Comparative corrosion and fabrication characteristics and typical applications of wrought aluminum alloys

Alloy	Temper	Resistance to corrosion General(a)	Stress-corrosion cracking(b)	Workability (cold)(c)	Machinability(c)	Weldability(d) Gas	Arc	Resistance spot and seam	Brazeability(d)	Solderability(e)	Typical applications
1050	O	A	A	A	E	A	A	B	A	A	Chemical equipment, railroad tank cars
	H12	A	A	A	E	A	A	A	A	A	
	H14	A	A	A	D	A	A	A	A	A	
	H16	A	A	B	D	A	A	A	A	A	
	H18	A	A	B	D	A	A	A	A	A	
1060	O	A	A	A	E	A	A	B	A	A	Chemical equipment, railroad tank cars
	H12	A	A	A	E	A	A	A	A	A	
	H14	A	A	A	D	A	A	A	A	A	
	H16	A	A	B	D	A	A	A	A	A	
	H18	A	A	B	D	A	A	A	A	A	
1100	O	A	A	A	E	A	A	B	A	A	Sheet-metal work, spun hollowware, fin stock
	H12	A	A	A	E	A	A	A	A	A	
	H14	A	A	A	D	A	A	A	A	A	
	H16	A	A	B	D	A	A	A	A	A	
	H16	A	A	C	D	A	A	A	A	A	
1145	O	A	A	A	E	A	A	B	A	A	Foil, fin stock
	H12	A	A	A	E	A	A	A	A	A	
	H14	A	A	A	D	A	A	A	A	A	
	H16	A	A	B	D	A	A	A	A	A	
	H18	A	A	B	D	A	A	A	A	A	
1199	O	A	A	A	E	A	A	B	A	A	Electrolytic capacitor foil, chemical equipment, railroad tank cars
	H12	A	A	A	E	A	A	A	A	A	
	H14	A	A	A	D	A	A	A	A	A	
	H16	A	A	B	D	A	A	A	A	A	
	H18	A	A	B	D	A	A	A	A	A	
1350	O	A	A	A	E	A	A	B	A	A	Electrical conductors
	H12, H111	A	A	A	E	A	A	A	A	A	
	H14, H24	A	A	A	D	A	A	A	A	A	
	H16, H26	A	A	B	D	A	A	A	A	A	
	H18	A	A	B	D	A	A	A	A	A	
2011	T3	D(f)	D	C	A	D	D	D	D	C	Screw-machine products
	T4, T451	D(f)	D	B	A	D	D	D	D	C	
	T8	D	B	D	A	D	D	D	D	C	
2014	O	…	…	…	D	D	D	B	D	C	Truck frames, aircraft structures
	T3, T4, T451	D(f)	C	C	B	D	B	B	D	C	
	T6, T651, T6510, T6511	D	C	D	B	D	B	B	D	C	

(a) Ratings A through E are relative ratings in decreasing order of merit, based on exposures to sodium chloride solution by intermittent spraying or immersion. Alloys with A and B ratings can be used in industrial and seacoast atmospheres without protection. Alloys with C, D and E ratings generally should be protected at least on faying surfaces. (b) Stress-corrosion cracking ratings are based on service experience and on laboratory tests of specimens exposed to the 3.5% sodium chloride alternate immersion test. A = No known instance of failure in service or in laboratory tests. B = No known instance of failure in service; limited failures in laboratory tests of short transverse specimens. C = Service failures with sustained tension stress acting in short transverse direction relative to grain structure; limited failures in laboratory tests of long transverse specimens. D = Limited service failures with sustained longitudinal or long transverse stress. (c) Ratings A through D for workability (cold), and A through E for machinability, are relative ratings in decreasing order of merit. (d) Ratings A through D for weldability and brazeability are relative ratings defined as follows: A, generally weldable by all commercial procedures and methods; B, weldable with special techniques or for specific applications which justify preliminary trials or testing to develop welding procedure and weld performance; C, limited weldability because of crack sensitivity or loss in resistance to corrosion and mechanical properties; D, no commonly used welding methods have been developed. (e) Ratings A through D and NA for solderability are relative ratings defined as follows: A, excellent; B, good; C, fair; D, poor; NA, not applicable. (f) In relatively thick sections the rating would be E. (g) This rating may be different for material held at elevated temperature for long periods

(continued)

Comparative corrosion and fabrication characteristics and typical applications of wrought aluminum alloys (continued)

Alloy	Temper	Resistance to corrosion			Machinability(c)	Weldability(d)				Solderability(e)	Typical applications
		General(a)	Stress-corrosion cracking(b)	Workability (cold)(c)		Gas	Arc	Resistance spot and seam	Brazeability(d)		
2024	O	...	...	...	D	D	D	D	D	C	
	T4, T3, T351, T3510, T3511	D(f)	C	C	B	C	B	B	D	C	Truck wheels, screw-machine products, aircraft structures
	T361	D(f)	C	D	B	D	C	B	D	C	
	T6	D	B	C	B	D	C	B	D	C	
	T861, T81, T851, T8510, T8511	D	B	D	B	D	C	B	D	C	
	T72	...	...	...	B	...	...	...	...	...	
2036	T4	C	...	B	C	...	B	B	D	...	Auto-body panel sheet
2124	T851	D	B	D	B	D	C	B	D	...	Military supersonic aircraft
2218	T61	D	C	...	...	...	...	C	...	C	Jet engine impellers and rings
	T72	D	C	...	B	D	C	B	D	C	
2219	O	...	...	...	...	D	A	B	D	NA	Structural uses at high temperatures (to 315 °C or 600 °F) high-strength weldments
	T31, T351, T3510, T3511	D(f)	C	C	B	A	A	A	D	NA	
	T37	D(f)	C	D	B	A	A	A	D	NA	
	T81, T851, T8510, T8511	D	B	D	B	A	A	A	D	NA	
	T87	D	B	D	B	A	A	A	D	NA	
2618	T61	D	C	...	B	D	C	B	D	NA	Aircraft engines
3003	O	A	A	A	E	A	A	B	A	A	Cooking utensils, chemical equipment, pressure vessels, sheet-metal work, builders' hardware, storage tanks
	H12	A	A	A	E	A	A	A	A	A	
	H14	A	A	B	D	A	A	A	A	A	
	H16	A	A	C	D	A	A	A	A	A	
	H18	A	A	C	D	A	A	A	A	A	
	H25	A	A	B	D	A	A	A	A	A	
3004	O	A	A	A	D	B	A	B	B	B	Sheet-metal work, storage tanks
	H32	A	A	B	D	B	A	A	B	B	
	H34	A	A	B	C	B	A	A	B	B	
	H36	A	A	C	C	B	A	A	B	B	
	H38	A	A	C	C	B	A	A	B	B	
3105	O	A	A	A	E	B	A	B	B	B	Residential siding, mobile homes, rain-carrying goods, sheet-metal work
	H12	A	A	B	E	B	A	A	B	B	
	H14	A	A	B	D	B	A	A	B	B	
	H16	A	A	C	D	B	A	A	B	B	
	H18	A	A	C	D	B	A	A	B	B	
	H25	A	A	B	D	B	A	A	B	B	
4032	T6	C	B	...	B	D	B	C	D	NA	Pistons
4043		B	A	NA	C	NA	NA	NA	NA	NA	Welding electrode

(a) Ratings A through E are relative ratings in decreasing order of merit, based on exposures to sodium chloride solution by intermittent spraying or immersion. Alloys with A and B ratings can be used in industrial and seacoast atmospheres without protection. Alloys with C, D and E ratings generally should be protected at least on faying surfaces. (b) Stress-corrosion cracking ratings are based on service experience and on laboratory tests of specimens exposed to the 3.5% sodium chloride alternate immersion test. A = No known instance of failure in service or in laboratory tests. B = No known instance of failure in service; limited failures in laboratory tests of short transverse specimens. C = Service failures with sustained tension stress acting in short transverse direction relative to grain structure; limited failures in laboratory tests of long transverse specimens. D = Limited service failures with sustained longitudinal or long transverse stress. (c) Ratings A through D for workability (cold), and A through E for machinability, are relative ratings in decreasing order of merit. (d) Ratings A through D for weldability and brazeability are relative ratings defined as follows: A, generally weldable by all commercial procedures and methods; B, weldable with special techniques or for specific applications which justify preliminary trials or testing to develop welding procedure and weld performance; C, limited weldability because of crack sensitivity or loss in resistance to corrosion and mechanical properties; D, no commonly used welding methods have been developed. (e) Ratings A through D and NA for solderability are relative ratings defined as follows: A, excellent; B, good; C, fair; D, poor; NA, not applicable. (f) In relatively thick sections the rating would be E. (g) This rating may be different for material held at elevated temperature for long periods

(continued)

Comparative corrosion and fabrication characteristics and typical applications of wrought aluminum alloys (continued)

Alloy	Temper	Resistance to corrosion General(a)	Stress-corrosion cracking(b)	Workability (cold)(c)	Machinability(c)	Weldability(d) Gas	Arc	Resistance spot and seam	Brazeability(d)	Solderability(e)	Typical applications
5005	O	A	A	A	E	A	A	B	B	B	Appliances, utensils, architectural, electrical conductors
	H12	A	A	A	E	A	A	A	B	B	
	H14	A	A	B	D	A	A	A	B	B	
	H16	A	A	C	D	A	A	A	B	B	
	H18	A	A	C	D	A	A	A	B	B	
	H32	A	A	B	E	A	A	A	B	B	
	H34	A	A	C	D	A	A	A	B	B	
	H36	A	A	C	D	A	A	A	B	B	
	H38	A	A	…	D	A	A	A	B	B	
5050	O	A	A	A	E	A	A	B	B	C	Builders' hardware, refrigerator trim, coiled tubes
	H32	A	A	A	D	A	A	A	B	C	
	H34	A	A	B	D	A	A	A	B	C	
	H36	A	A	C	C	A	A	A	B	C	
	H38	A	A	C	C	A	A	A	B	C	
5052	O	A	A	A	D	A	A	B	C	D	Sheet-metal work, hydraulic tube, appliances
	H32	A	A	B	D	A	A	A	C	D	
	H34	A	A	B	C	A	A	A	C	D	
	H36	A	A	C	C	A	A	A	C	D	
	H38	A	A	C	C	A	A	A	C	D	
5056	O	A(g)	B(g)	A	D	C	A	B	D	D	Cable sheathing, rivets for magnesium, screen wire, zippers
	H111	A(g)	B(g)	A	D	C	A	A	D	D	
	H12, H32	A(g)	B(g)	B	D	C	A	A	D	D	
	H14, H34	A(g)	B(g)	B	C	C	A	A	D	D	
	H18, H38	A(g)	C(g)	C	C	C	A	A	D	D	
	H192	B(g)	D(g)	D	B	C	A	A	D	D	
	H392	B(g)	D(g)	D	B	C	A	B	D	D	
5083	O	A(g)	B(g)	B	D	C	A	B	D	D	Unfired, welded pressure vessels, marine, auto aircraft cryogenics, TV towers, drilling rigs, transportation equipment, missile components
	H321, H116	A(g)	B(g)	C	D	C	A	A	D	D	
	H323	A(g)	B(g)	C	D	C	A	A	D	D	
	H343	A(g)	B(g)	C	C	C	A	A	D	D	
	H111	A(g)	B(g)	C	D	C	A	A	D	D	
5086	O	A(g)	A(g)	A	D	C	A	B	D	D	
	H32, H116	A(g)	A(g)	B	D	C	A	A	D	D	
	H34	A(g)	B(g)	B	C	C	A	A	D	D	
	H36	A(g)	B(g)	C	C	C	A	A	D	D	
	H38	A(g)	B(g)	C	C	C	A	A	D	D	
	H111	A(g)	A(g)	B	D	C	A	A	D	D	
5154	O	A(g)	A(g)	A	D	C	A	B	D	D	Welded structures, storage tanks, pressure vessels, salt-water service
	H32	A(g)	A(g)	B	D	C	A	A	D	D	
	H34	A(g)	A(g)	B	C	C	A	A	D	D	
	H36	A(g)	A(g)	C	C	C	A	A	D	D	
	H38	A(g)	A(g)	C	C	C	A	A	D	D	

(a) Ratings A through E are relative ratings in decreasing order of merit, based on exposures to sodium chloride solution by intermittent spraying or immersion. Alloys with A and B ratings can be used in industrial and seacoast atmospheres without protection. Alloys with C, D and E ratings generally should be protected at least on faying surfaces. (b) Stress-corrosion cracking ratings are based on service experience and on laboratory tests of specimens exposed to the 3.5% sodium chloride alternate immersion test. A = No known instance of failure in service or in laboratory tests. B = No known instance of failure in service; limited failures in laboratory tests of short transverse specimens. C = Service failures with sustained tension stress acting in short transverse direction relative to grain structure; limited failures in laboratory tests of long transverse specimens. D = Limited service failures with sustained longitudinal or long transverse stress. (c) Ratings A through D for workability (cold), and A through E for machinability, are relative ratings in decreasing order of merit. (d) Ratings A through D for weldability and brazeability are relative ratings defined as follows: A, generally weldable by all commercial procedures and methods; B, weldable with special techniques or for specific applications which justify preliminary trials or testing to develop welding procedure and weld performance; C, limited weldability because of crack sensitivity or loss in resistance to corrosion and mechanical properties; D, no commonly used welding methods have been developed. (e) Ratings A through D and NA for solderability are relative ratings defined as follows: A, excellent; B, good; C, fair; D, poor; NA, not applicable. (f) In relatively thick sections the rating would be E. (g) This rating may be different for material held at elevated temperature for long periods

(continued)

Comparative corrosion and fabrication characteristics and typical applications of wrought aluminum alloys (continued)

Alloy	Temper	Resistance to corrosion General(a)	Stress-corrosion cracking(b)	Workability (cold)(c)	Machinability(c)	Weldability(d) Gas	Arc	Resistance spot and seam	Brazeability(d)	Solderability(e)	Typical applications
5182	O	A	A(g)	A	D	C	A	B	D	D	
	H19	A	A(g)	D	B	C	A	A	D	D	Automobile body sheet, can ends
5252	H24	A	A	B	D	A	A	A	D	D	Automotive and appliance trim
	H25	A	A	B	C	A	A	A	C	D	
	H28	A	A	C	C	A	A	A	C	D	
5254	O	A(g)	A(g)	A	D	C	A	B	D	D	
	H32	A(g)	A(g)	B	D	C	A	A	D	D	Hydrogen peroxide and chemical storage vessels
	H34	A(g)	A(g)	B	C	C	A	A	D	D	
	H36	A(g)	A(g)	C	C	C	A	A	D	D	
	H38	A(g)	A(g)	C	C	C	A	A	D	D	
5356		A	A	NA	B	NA	NA	NA	NA	NA	Welding electrode
5454	O	A	A	A	D	C	A	B	D	NA	Welded structures, pressure vessels, marine service
	H32	A	A	B	D	C	A	A	D	NA	
	H34	A	A	B	C	C	A	A	D	NA	
	H111	A	A	B	D	C	A	A	D	NA	
5456	O	A(g)	B(g)	B	D	C	A	B	D	NA	High-strength welded structures, storage tanks, pressure vessels, marine applications
	H111	A(g)	B(g)	C	D	C	A	A	D	NA	
	H321, H115	A(g)	B(g)	C	D	C	A	A	D	NA	
	H323	A(g)	B(g)	C	D	C	A	A	D	NA	
	H343	A(g)	B(g)	C	C	C	A	A	D	NA	
5457	O	A	A	A	E	A	A	B	B	B	
5652	O	A	A	A	D	A	A	B	C	D	
	H32	A	A	B	D	A	A	A	C	D	Hydrogen peroxide and chemical storage vessels
	H34	A	A	B	C	A	A	A	C	D	
	H36	A	A	C	C	A	A	A	C	D	
	H38	A	A	C	C	A	A	A	C	D	
5657	H241	A	A	A	D	A	A	A	B	NA	Anodized auto and appliance trim
	H25	A	A	B	D	A	A	A	B	NA	
	H26	A	A	B	D	A	A	A	B	NA	
	H28	A	A	C	D	A	A	A	B	NA	
6005	T5	B	A	C	C	A	A	A	A	NA	Heavy-duty structures requiring good corrosion resistance applications, truck and marine, railroad cars, furniture, pipelines
6009	T4	A	A	A	C	A	A	A	A	B	Automobile body sheet
6010	T4	A	A	B	C	A	A	A	A	B	Automobile body sheet
6061	O	B	A	A	D	A	A	B	A	B	Heavy-duty structures requiring good corrosion resistance, truck and marine, railroad cars, furniture, pipelines
	T4, T451, T4510, T4511	B	B	B	C	A	A	A	A	B	
	T6, T651, T652, T6510, T6511	B	A	C	C	A	A	A	A	B	

(a) Ratings A through E are relative ratings in decreasing order of merit, based on exposures to sodium chloride solution by intermittent spraying or immersion. Alloys with A and B ratings can be used in industrial and seacoast atmospheres without protection. Alloys with C, D and E ratings generally should be protected at least on faying surfaces. (b) Stress-corrosion cracking ratings are based on service experience and on laboratory tests of specimens exposed to the 3.5% sodium chloride alternate immersion test. A = No known instance of failure in service or in laboratory tests. B = No known instance of failure in service; limited failures in laboratory tests of short transverse specimens. C = Service failures with sustained tension stress acting in short transverse direction relative to grain structure; limited failures in laboratory tests of long transverse specimens. D = Limited service failures with sustained longitudinal or long transverse stress. (c) Ratings A through D for workability (cold), and A through E for machinability, are relative ratings in decreasing order of merit. (d) Ratings A through D for weldability and brazeability are relative ratings defined as follows: A, generally weldable by all commercial procedures and methods; B, weldable with special techniques or for specific applications which justify preliminary trials or testing to develop welding procedure and weld performance; C, limited weldability because of crack sensitivity or loss in resistance to corrosion and mechanical properties; D, no commonly used welding methods have been developed. (e) Ratings A through D and NA for solderability are relative ratings defined as follows: A, excellent; B, good; C, fair; D, poor; NA, not applicable. (f) In relatively thick sections the rating would be E. (g) This rating may be different for material held at elevated temperature for long periods

(continued)

Comparative corrosion and fabrication characteristics and typical applications of wrought aluminum alloys (continued)

Alloy	Temper	Resistance to corrosion General(a)	Stress-corrosion cracking(b)	Workability (cold)(c)	Machinability(c)	Weldability(d) Gas	Arc	Resistance spot and seam	Brazeability(d)	Solderability(e)	Typical applications
6063	T1	A	A	B	D	A	A	A	A	B	Pipe railing, furniture, architectural extrusions
	T4	A	A	B	D	A	A	A	A	B	
	T5, T52	A	A	B	C	A	A	A	A	B	
	T6	A	A	C	C	A	A	A	A	B	
	T83, T831, T832	A	A	C	C	A	A	A	A	B	
6066	O	C	A	B	D	D	B	B	D	NA	Forgings and extrusions for welded structures
	T4, T4510, T4511	C	B	C	C	D	B	B	D	NA	
	T6, T6510, T6511	C	B	C	B	D	B	B	D	NA	
6070	T4, T4511	B	B	B	C	A	A	A	B	NA	Heavy-duty welded structures, pipelines
	T6	B	B	C	C	A	A	A	B	NA	
6101	T6, T63	A	A	C	C	A	A	A	A	NA	High-strength bus conductors
	T61, T64	A	A	B	D	A	A	A	A	B	
6151	T6, T652	…	…	…	…	…	…	…	…		Moderate-strength, intricate forgings for machine and auto parts
6201	T81	A	A	…	C	A	A	A	A	NA	High-strength electric conductor wire
6262	T6, T651, T6510, T6511	B	A	C	B	A	A	A	A	NA	Screw-machine products
	T9	B	A	D	B	A	A	A	A		
6351	T5, T6	B	A	C	C	A	A	A	A	B	Heavy-duty structures requiring good corrosion resistance, truck and tractor extrusions
6463	T1	A	A	B	D	A	A	A	A	…	Extruded architectural and trim sections
	T5	A	A	B	C	A	A	A	A	NA	
	T6	A	A	C	C	A	A	A	A	…	
7005	T53, T63	B	B	C	A	B	B	B	B	B	Heavy-duty structures requiring good corrosion resistance, trucks, trailers, dump bodies
7049	T73, T7351, T7352	C	B	D	B	D	C	B	D	D	Aircraft and other structures
	T76, T7651	C	B	D	B	D	C	B	D	D	
7050	T736, T73651, T73652	C	B	D	B	D	C	B	D	D	Aircraft and other structures
	T76, T761	C	B	D	B	D	C	B	D	D	
7072		A	A	A	D	A	A	A	A	A	Fin stock, cladding alloy
7075	O	…	…	…	D	D	C	B	D	D	Aircraft and other structures
	T6, T651, T652, T6510, T6511	C(f)	C	D	B	D	C	B	D	D	
	T73, T7351	C	B	D	B	D	C	B	D	D	

(a) Ratings A through E are relative ratings in decreasing order of merit, based on exposures to sodium chloride solution by intermittent spraying or immersion. Alloys with A and B ratings can be used in industrial and seacoast atmospheres without protection. Alloys with C, D and E ratings generally should be protected at least on faying surfaces. (b) Stress-corrosion cracking ratings are based on service experience and on laboratory tests of specimens exposed to the 3.5% sodium chloride alternate immersion test. A = No known instance of failure in service or in laboratory tests. B = No known instance of failure in service; limited failures in laboratory tests of short transverse specimens. C = Service failures with sustained tension stress acting in short transverse direction relative to grain structure; limited failures in laboratory tests of long transverse specimens. D = Limited service failures with sustained longitudinal or long transverse stress. (c) Ratings A through D for workability (cold), and A through E for machinability, are relative ratings in decreasing order of merit. (d) Ratings A through D for weldability and brazeability are relative ratings defined as follows: A, generally weldable by all commercial procedures and methods; B, weldable with special techniques or for specific applications which justify preliminary trials or testing to develop welding procedure and weld performance; C, limited weldability because of crack sensitivity or loss in resistance to corrosion and mechanical properties; D, no commonly used welding methods have been developed. (e) Ratings A through D and NA for solderability are relative ratings defined as follows: A, excellent; B, good; C, fair; D, poor; NA, not applicable. (f) In relatively thick sections the rating would be E. (g) This rating may be different for material held at elevated temperature for long periods

(continued)

Comparative corrosion and fabrication characteristics and typical applications of wrought aluminum alloys (continued)

| Alloy | Temper | Resistance to corrosion | | Workability (cold)(c) | Machinability(c) | Weldability(d) | | | Brazeability(d) | Solderability(e) | Typical applications |
		General(a)	Stress-corrosion cracking(b)			Gas	Arc	Resistance spot and seam			
7175	T736, T73652	C	B	D	B	D	C	B	D	D	Aircraft and other structures, forgings
7178	O	...	...	...	...	D	C	B	D	D	Aircraft and other structures
	T6, T651, T6510, T6511	C(f)	C	D	B	D	C	B	D	D	
7475	T6, T651	C	C	D	B	D	C	B	D	D	Aircraft and other structures
	T73, T7351, T7352	C	B	D	B	D	C	B	D	D	
	T76, T7651	C	B	D	B	D	C	B	D	D	

(a) Ratings A through E are relative ratings in decreasing order of merit, based on exposures to sodium chloride solution by intermittent spraying or immersion. Alloys with A and B ratings can be used in industrial and seacoast atmospheres without protection. Alloys with C, D and E ratings generally should be protected at least on faying surfaces. (b) Stress-corrosion cracking ratings are based on service experience and on laboratory tests of specimens exposed to the 3.5% sodium chloride alternate immersion test. A = No known instance of failure in service or in laboratory tests. B = No known instance of failure in service; limited failures in laboratory tests of short transverse specimens. C = Service failures with sustained tension stress acting in short transverse direction relative to grain structure; limited failures in laboratory tests of long transverse specimens. D = Limited service failures with sustained longitudinal or long transverse stress. (c) Ratings A through D for workability (cold), and A through E for machinability, are relative ratings in decreasing order of merit. (d) Ratings A through D for weldability and brazeability are relative ratings defined as follows: A, generally weldable by all commercial procedures and methods; B, weldable with special techniques or for specific applications which justify preliminary trials or testing to develop welding procedure and weld performance; C, limited weldability because of crack sensitivity or loss in resistance to corrosion and mechanical properties; D, no commonly used welding methods have been developed. (e) Ratings A through D and NA for solderability are relative ratings defined as follows: A, excellent; B, good; C, fair; D, poor; NA, not applicable. (f) In relatively thick sections the rating would be E. (g) This rating may be different for material held at elevated temperature for long periods

Typical mechanical properties of wrought aluminum alloys

Alloy	Temper	Tensile strength MPa	ksi	Yield strength MPa	ksi	Elongation(a), % (b)	(c)	Hardness(d)	Shear strength MPa	ksi	Fatigue strength(e) MPa	ksi
1050	O	76	11	28	4	...	...	...	62	9	...	...
	H14	110	16	105	15	...	...	...	69	10	...	...
	H16	130	19	125	18	...	...	...	76	11	...	...
	H18	160	23	145	21	...	...	...	83	12	...	...
1060	O	69	10	28	4	43	...	19	48	7	21	3
	H12	83	12	76	11	16	...	23	55	8	28	4
	H14	97	14	90	13	12	...	26	62	9	34	5
	H16	110	16	105	15	8	...	30	69	10	45	6.5
	H18	130	19	125	18	6	...	35	76	11	45	6.5
1100	O	90	13	34	5	35	45	23	62	9	34	5
	H12	110	16	105	15	12	25	28	69	10	41	6
	H14	125	18	115	17	9	20	32	76	11	48	7
	H16	145	21	140	20	6	17	38	83	12	62	9
	H18	165	24	150	22	5	15	44	90	13	62	9
1350	O	83	12	28	4	23(f)	...	...	55	8	...	...
	H12	97	14	83	12	...	...	...	62	9	...	...
	H14	110	16	97	14	...	...	...	69	10	...	...
	H16	125	18	110	16	...	...	...	76	11	...	...
	H19	185	27	165	24	1.5(f)	...	...	105	15	48	7
2011	T3	380	55	295	43	...	15	95	220	32	125	18
	T8	405	59	310	45	...	15	100	240	35	125	18
2014	O	185	27	97	14	...	18	45	125	18	90	13
	T4	425	62	290(g)	42(g)	...	20	105	260	38	140	20
	T6(h)	485	70	415	60	...	13	135	290	42	125	18
Alclad 2014	O	170	25	69	10	21	...	...	125	18	...	...
	T3	435	63	275	40	20	...	...	255	37	...	...
	T4	420	61	255	37	22	...	...	255	37	...	...
	T6	470	68	415	60	10	...	...	285	41	...	...
2024	O	185	27	76	11	20	22	47	125	18	90	13
	T3	485	70	345	50	18	...	120	285	41	140	20
	T4 T351	470	68	325	47	20	19	120	285	41	140	20
	T361	495	72	395	57	13	...	130	290	42	125	18
Alclad 2024	O	180	26	76	11	20	...	...	125	18	...	...
	T	450	65	310	45	18	...	...	275	40	...	...
	T4, T351	440	64	290	42	19	...	...	275	40	...	...
	T361	460	67	365	53	11	...	...	285	41	...	...
	T81,T851	450	65	415	60	6	...	...	275	40	...	...
	T861	485	70	455	66	6	...	...	290	42	...	...
2036	T4	340	49	195	28	24	...	...	...	...	125(j)	18(j)
2048		455	66	415	60	8.3	...	...	...	...	220(k)	32(k)
2124	T851	490	71	440	64	9.4	...	...	...	...	...	...
2218	T61	405	59	305	44	...	13	115	...	...	...	...
	T71	345	50	275	40	...	11	105	...	...	...	...
	T72	330	48	255	37	...	11	95	205	30	...	...
2219	O	170	25	76	11	18	...	...	...	...	...	...
	T42	360	52	185	27	20	...	...	...	...	...	...
	T31, T351	360	52	250	36	17	...	...	...	...	...	...
	T37	395	57	315	46	11	...	...	...	...	...	...
	T62	415	60	290	42	10	...	...	...	...	105	15
	T81, T851	455	66	350	51	10	...	...	...	...	105	15
	T87	475	69	395	57	10	...	...	...	...	105	15
2618	All	440	64	370	54	10	10	...	260	38	125	18
3003 and Alclad 3003(m)	O	110	16	42	6	30	40	28	76	11	48	7
	H12	130	19	125	18	10	20	35	83	12	55	8
	H14	150	22	145	21	8	16	40	97	14	62	9
	H16	180	26	170	25	5	14	47	105	15	69	10
	H18	200	29	185	27	4	10	55	110	16	69	10
3004 and Alclad 3004(m)	O	180	26	69	10	20	25	45	110	16	97	14
	H32	215	31	170	25	10	17	52	115	17	105	15
	H34	240	35	200	29	9	12	63	125	18	105	15
	H36	260	38	230	33	5	9	70	140	20	110	16
	H38	285	41	250	36	5	6	77	145	21	110	16

(a) In 50 mm or 2 in. (b) Specimen 1.6 mm (1/16 in.) thick. (c) Specimen 12.5 mm (0.50 in.) in diameter. (d) 500-kg load; 10-mm ball. (e) In 5×10^8 cycles; R. R. Moore-type test. (f) In 254 mm or 10 in. (g) Die forgings are about 20% lower in yield strength. (h) Extruded products more than 19 mm (3/4 in.) thick are 15 to 20% higher in strength. (j) In 10^7 cycles using flexural-type testing of sheet specimens. (k) In 10^7 cycles; axially loaded specimens tested at $R = 0.1$. (m) No shear-strength or fatigue-strength values for Alclad. (n) Properties for this temper are those of container end stock 0.25 to 0.38 mm (0.010 to 0.015 in.) thick. (p) HR15T. (q) Specimen 6.35 mm (0.25 in.) thick

(continued)

Typical mechanical properties of wrought aluminum alloys (continued)

Alloy	Temper	Tensile strength MPa	Tensile strength ksi	Yield strength MPa	Yield strength ksi	Elongation(a), % (b)	Elongation(a), % (c)	Hardness(d)	Shear strength MPa	Shear strength ksi	Fatigue strength(e) MPa	Fatigue strength(e) ksi
3105	O	115	17	55	8	24	...	...	83	12	...	...
	H12	150	22	130	19	7	...	...	97	14	...	...
	H14	170	25	150	22	5	...	...	105	15	...	...
	H16	195	28	170	25	4	...	...	110	16	...	...
	H18	215	31	195	28	3	...	...	115	17	...	...
	H25	180	26	160	23	8	...	...	105	15	...	...
4032	T6	380	55	315	46	...	9	120	260	38	110	16
4043	O	145	21	69	10	22	...	...	...	...	...	...
	H18	285	41	270	39	0.5	...	...	...	...	...	...
5005	O	125	18	41	6	25	...	28	76	11	...	...
	H12	140	20	130	19	10	...	...	97	14	...	...
	H14	160	23	150	22	6	...	...	97	14	...	...
	H16	180	26	170	25	5	...	...	105	15	...	...
	H18	200	29	195	28	4	...	...	110	16	...	...
	H32	140	20	115	17	11	...	36	97	14	...	...
	H34	160	23	140	20	8	...	41	97	14	...	...
	H36	180	26	165	24	6	...	46	105	15	...	...
	H38	200	29	185	27	5	...	51	110	16	...	...
5050	O	145	21	55	8	24	...	36	105	15	83	12
	H32	170	25	145	21	9	...	46	115	17	90	13
	H34	195	28	165	24	8	...	53	125	18	90	13
	H36	205	30	180	26	7	...	58	130	19	97	14
	H38	220	32	200	29	6	...	63	140	20	97	14
5052	O	195	28	90	13	25	27	47	125	18	110	16
	H32	230	33	195	28	12	16	60	140	20	115	17
	H34	260	38	215	31	10	12	68	145	21	125	18
	H36	275	40	240	35	8	9	73	160	23	130	19
	H38	290	42	255	37	7	7	77	165	24	140	20
5056	O	290	42	150	22	...	35	65	180	26	140	20
	H18	435	63	405	59	...	10	105	235	34	150	22
	H38	415	60	345	50	...	15	100	220	32	150	22
5083	O	290	42	145	21	...	22	...	170	25	...	...
	H112	305	44	195	28	...	16	...	...	...	...	...
	H113	315	46	230	33	...	16	...	...	...	...	...
	H321	315	46	230	33	...	16	...	...	...	160	23
	H323, H32	325	47	250	36	...	10	...	...	...	...	...
	H343, H34	345	50	285	41	...	9	...	...	...	...	...
5086	O	260	38	115	17	22	...	...	160	23	...	...
	H32, H116, H117	290	42	205	30	12	...	...	...	...	...	...
	H34	325	47	255	37	10	...	...	185	27	...	...
	H112	270	39	130	19	14	...	...	...	...	...	...
5154	O	240	35	115	17	27	...	58	150	22	115	17
	H32	270	39	205	30	15	...	67	150	22	125	18
	H34	290	42	230	33	13	...	73	165	24	130	19
	H36	310	45	250	36	12	...	78	180	26	140	20
	H38	330	48	270	39	10	...	80	195	28	145	21
	H112	240	35	115	17	25	...	63	...	...	115	17
5182	O	275	40	140	19	25	...	58	150	22	140	20
	H32	315	46	235	34	12	...	...	...	...	...	...
	H34	340	49	285	41	10	...	...	...	...	...	...
	H19(n)	420	61	395	57	4	...	...	...	...	...	...
5252	H25	235	34	170	25	11	...	68	145	21	...	...
	H28, H38	285	41	240	35	5	...	75	160	23	...	...
5254	O	240	35	115	17	27	...	58	150	22	115	17
	H32	270	39	205	30	15	...	67	150	22	125	18
	H34	290	42	230	33	13	...	73	165	24	130	19
	H36	310	45	250	36	12	...	78	180	26	140	20
	H38	330	48	270	39	10	...	80	195	28	145	21
	H112	240	35	115	17	25	...	63	...	...	115	17

(a) In 50 mm or 2 in. (b) Specimen 1.6 mm (1/16 in.) thick. (c) Specimen 12.5 mm (0.50 in.) in diameter. (d) 500-kg load; 10-mm ball. (e) In 5×10^8 cycles; R. R. Moore-type test. (f) In 254 mm or 10 in. (g) Die forgings are about 20% lower in yield strength. (h) Extruded products more than 19 mm (3/4 in.) thick are 15 to 20% higher in strength. (j) In 10^7 cycles using flexural-type testing of sheet specimens. (k) In 10^7 cycles; axially loaded specimens tested at $R = 0.1$. (m) No shear-strength or fatigue-strength values for Alclad. (n) Properties for this temper are those of container end stock 0.25 to 0.38 mm (0.010 to 0.015 in.) thick. (p) HR15T. (q) Specimen 6.35 mm (0.25 in.) thick

(continued)

Typical mechanical properties of wrought aluminum alloys (continued)

Alloy	Temper	Tensile strength MPa	ksi	Yield strength MPa	ksi	Elongation(a), % (b)	(c)	Hardness(d)	Shear strength MPa	ksi	Fatigue strength(e) MPa	ksi
5454	O	250	36	115	17	22	...	62	160	23	...	...
	H32	275	40	205	30	10	...	73	165	24	...	...
	H34	305	44	240	35	10	...	81	180	26	...	...
	H36	340	49	275	40	8	...	...	...	...	...	...
	H38	370	54	310	45	8	...	...	...	...	...	...
	H111	260	38	180	26	14	...	70	160	23	...	...
	H112	250	36	125	18	18	...	62	160	23	...	...
	H311	260	38	180	26	18	...	70	160	23	...	...
5456	O	310	45	160	23	...	24	...	...	...	...	...
	H111	325	47	230	33	...	18	...	...	...	...	...
	H112	310	45	165	24	...	22	...	...	...	...	...
	H321, H116	350	51	255	27	...	16	90	205	30	...	...
5457	O	130	19	48	7	22	...	32	83	12	...	...
	H25	180	26	160	23	12	...	48	110	16	...	...
	H28, H38	205	30	185	27	6	...	55	125	18	...	...
5652	O	195	28	90	13	25	30	47	125	18	110	16
	H32	230	33	195	28	12	18	60	140	20	115	17
	H34	260	38	215	31	10	14	68	145	21	125	18
	H36	275	40	240	35	8	10	73	160	23	130	19
	H38	290	42	255	34	7	8	77	165	24	140	20
5657	H25	160	23	140	20	12	...	40	97	14	...	...
	H28, H38	195	28	165	24	7	...	50	105	15	...	...
6005	T1	170	25	105	15	16	...	...	...	...	97	14
	T5	260	38	240	35	8	10	95	205	30	97	14
6009	T4	235	34	130	19	24	...	70(p)	150	22	115	17
	T6	345	50	325	47	12	...	...	...	...	...	...
6010	T4	255	37	170	25	24	...	76(p)	...	...	115	17
6061	O	125	18	55	8	25	30	30	83	12	62	9
	T4, T451	240	35	145	21	22	25	65	165	24	97	14
	T6, T651	310	45	275	40	12	17	95	205	30	97	14
Alclad 6061	O	115	17	48	7	25	...	...	76	11	...	...
	T4, T451	230	33	130	19	22	...	...	150	22	...	...
	T6, T651	290	42	255	37	12	...	...	185	27	...	...
6063	O	90	13	48	7	...	...	25	69	10	55	8
	T1	150	22	90	13	20	...	42	97	14	62	9
	T4	170	25	90	13	22	...	...	...	...	...	...
	T5	185	27	145	21	12	...	60	115	17	69	10
	T6	240	35	215	31	12	...	73	150	22	69	10
	T83	255	37	240	35	9	...	82	150	22	...	...
	T831	205	30	185	27	10	...	70	125	18	...	...
	T832	290	42	270	39	12	...	95	185	27	...	...
6066	O	150	22	83	12	...	18	43	97	14	...	...
	T4, T451	360	52	205	30	...	18	90	200	29	...	...
	T6, T651	395	57	360	52	...	12	120	235	34	110	16
6070	O	145	21	69	10	20	...	35	97	14	62	9
	T4	315	46	170	25	20	...	90	205	30	90	13
	T6	380	55	350	51	10	...	120	235	34	97	14
6101	H111	97	14	76	11	...	...	...	...	...	...	...
6151	T6	220	32	195	28	15(q)	...	71	140	20	...	...
6201	T6	330	48	300	43	17	...	90	...	...	...	...
	T81	330	48	310	45	6(f)	...	...	...	...	...	...
6205	T1	260	38	140	20	19	...	65	...	...	...	...,
	T5	310	45	290	42	11	...	95	205	30	105	15
6262	T9	400	58	380	55	...	10	120	240	35	90	13
6351	T4	250	36	150	22	20	...	...	...	...	...	...
	T6	310	45	285	41	14	...	95	200	29	90	13
6463	T1	150	22	90	13	20	...	42	97	14	69	10
	T5	185	27	145	21	12	...	60	115	17	69	10
	T6	240	35	215	31	12	...	74	150	22	69	10
7005	O	193	28	83	12	20	...	...	117	17	...	...
	T53	393	57	345	50	15	...	...	221	32	140	20
	T6, T63, T6351	372	54	315	46	12	...	...	214	31	125	18
7049	T73	...	...	...	...	...	...	135	...	...	295(k)	43(k)
7050	T736	515	75	455	66	11	15	...	...	...	240(k)	35(k)

(a) In 50 mm or 2 in. (b) Specimen 1.6 mm (1/16 in.) thick. (c) Specimen 12.5 mm (0.50 in.) in diameter. (d) 500-kg load; 10-mm ball. (e) In 5×10^8 cycles; R. R. Moore-type test. (f) In 254 mm or 10 in. (g) Die forgings are about 20% lower in yield strength. (h) Extruded products more than 19 mm (3/4 in.) thick are 15 to 20% higher in strength. (j) In 10^7 cycles using flexural-type testing of sheet specimens. (k) In 10^7 cycles; axially loaded specimens tested at $R = 0.1$. (m) No shear-strength or fatigue-strength values for Alclad. (n) Properties for this temper are those of container end stock 0.25 to 0.38 mm (0.010 to 0.015 in.) thick. (p) HR15T. (q) Specimen 6.35 mm (0.25 in.) thick

(continued)

Typical mechanical properties of wrought aluminum alloys (continued)

Alloy	Temper	Tensile strength		Yield strength		Elongation(a), %		Hardness(d)	Shear strength		Fatigue strength(e)	
		MPa	ksi	MPa	ksi	(b)	(c)		MPa	ksi	MPa	ksi
7072	O	…	…	…	…	…	…	20	55	8	…	…
	H12	…	…	…	…	…	…	28	62	9	…	…
	H14	…	…	…	…	…	…	32	69	10	…	…
7075	O	230	38	105	15	17	16	60	150	22	…	…
	T6, T651	570	83	505	73	11	11	150	330	48	160	23
	T73	505	73	435	63	13	…	…	…	…	…	…
Alclad 7075	O	220	32	95	14	17	…	…	150	22	…	…
	T6, T651	525	76	460	67	11	…	…	315	46	…	…
7175	T66	595	86	525	76	11	…	150	325	47	160	23
	T736	525	76	455	66	14	…	145	290	42	160	23
7475	T61	525	76	460	67	12	…	…	…	…	…	…
	T651	…	…	…	…	…	…	…	295	43	…	…
	T7351	…	…	…	…	…	…	…	270	9	220	32
	T7651	…	…	…	…	…	…	…	270	39	…	…

(a) In 50 mm or 2 in. (b) Specimen 1.6 mm (1/16 in.) thick. (c) Specimen 12.5 mm (0.50 in.) in diameter. (d) 500-kg load; 10-mm ball. (e) In 5×10^8 cycles; R. R. Moore-type test. (f) In 254 mm or 10 in. (g) Die forgings are about 20% lower in yield strength. (h) Extruded products more than 19 mm (3/4 in.) thick are 15 to 20% higher in strength. (j) In 10^7 cycles using flexural-type testing of sheet specimens. (k) In 10^7 cycles; axially loaded specimens tested at $R = 0.1$. (m) No shear-strength or fatigue-strength values for Alclad. (n) Properties for this temper are those of container end stock 0.25 to 0.38 mm (0.010 to 0.015 in.) thick. (p) HR15T. (q) Specimen 6.35 mm (0.25 in.) thick

Cast Aluminum

Designations and nominal compositions of common aluminum alloys used for casting

	Alloys			Composition, %				
AA number	Former AA designation	Former ASTM number	Product(a)	Cu	Mg	Mn	Si	Others
201.0	...	...	S	4.6	0.35	0.35	...	0.7 Ag, 0.25 Ti
206.0	...	...	S or P	4.6	0.25	0.35	0.10(b)	0.22 Ti, 0.15 Fe(b)
A206.0	...	...	S or P	4.6	0.25	0.35	0.05(b)	0.22 Ti, 0.10 Fe(b)
208.0	108	CS43A	S	4.0	...	...	3.0	...
242.0	142	CN42A	S or P	4.0	1.5	...	...	2.0 Ni
295.0	195	C4A	S	4.5	...	...	0.8	...
296.0	B295.0, B195	...	P	4.5	...	...	2.5	...
308.0	A108	SC64A	S or P	4.5	...	...	5.5	...
319.0	319, Allcast	SC64D	S or P	3.5	...	...	6.0	...
336.0	A332.0, A132	SN122A	P	1.0	1.0	...	12.0	2.5 Ni
354.0	354	SC92A	P	1.8	0.50	...	9.0	...
355.0	355	SC51A	S or P	1.2	0.50	0.50(b)	5.0	0.6 Fe(b), 0.35Zn(b)
C355.0	C355	SC51B	S or P	1.2	0.50	0.10(b)	5.0	0.20 Fe(b), 0.10Zn(b)
356.0	356	SG70A	S or P	0.25(b)	0.32	0.35(b)	7.0	0.6 Fe(b), 0.35 Zn(b)
A356.0	A356	SG70B	S or P	0.20(b)	0.35	0.10(b)	7.0	0.20 Fe(b), 0.10 Zn(b)
357.0	357	...	S or P	...	0.50	...	7.0	...
A357.0	A357	...	S or P	...	0.6	...	7.0	0.15 Ti, 0.005 Be
359.0	359	SG91A	S or P	...	0.6	...	9.0	...
360.0	360	SG100B	D	...	0.50	...	9.5	2.0 Fe(b)
A360.0	A360	SG100A	D	...	0.50	...	9.5	1.3 Fe(b)
380.0	380	SC84B	D	3.5	...	...	8.5	2.0 Fe(b)
A380.0	A380	SC84A	D	3.5	...	...	8.5	1.3 Fe(b)
383.0	...	SC102A	D	2.5	...	...	10.5	...
384.0	384	SC114A	D	3.8	...	...	11.2	3.0 Zn(b)
A384.0	384	SC114A	D	3.8	...	...	11.2	1.0 Zn(b)
390.0	390	...	D	4.5	0.6	...	17.0	1.3 Zn(b)
A390.0	A390	...	S or P	4.5	0.6	...	17.0	0.5 Zn(b)
413.0	13	S12B	D	...	...	...	12.0	2.0 Fe(b)
A413.0	A13	S12A	D	...	...	...	12.0	1.3 Fe(b)
4430	43	S5B	S	0.6(b)	...	...	5.2	...
A443.0	43	...	S	0.30(b)	...	...	5.2	...
B443.0	43	S5A	S or P	0.15(b)	...	...	5.2	...
C443.0	A43	S5C	D	0.6(b)	...	...	5.2	2.0 Fe(b)
514.0	214	G4A	S	...	4.0	...	...	...
518.0	218	G8A	D	...	8.0	...	...	...
520.0	220	G10A	S	...	10.0	...	...	...
535.0	Almag 35	GM70B	S	...	6.8	0.18	...	0.18 Ti
A535.0	A218	...	S	...	7.0	0.18	...	...
B535.0	B218	...	S	...	7.0	...	...	0.18 Ti
712.0	D712.0, D612, 40E	ZG61A	S or P	...	0.6	...	...	5.8 Zn, 0.5 Cr, 0.20 Ti
713.0	613, Tenzaloy	ZC81A, B	S or P	0.7	0.35	...	...	7.5 Zn, 0.7 Cu
771.0	Precedent 71A	ZG71B	S	...	0.9	...	...	7.0 Zn, 0.13 Cr, 0.15 Ti
850.0	750	...	S or P	1.0	...	...	...	6.2 Sn, 1.0 Ni

(a) S, sand casting; P, permanent mold casting; D, die casting. (b) Maximum

Characteristics of common aluminum alloys used in sand and permanent mold castings(a)

Alloy	Type of mold(b)	Fluidity	Resistance to hot cracking	Pressure tightness	Heat treatment	Strength at elevated temperatures	General corrosion resistance	Machining	Polishing	Anodizing appearance	Weldability
208.0	S	2	2	2	Optional	3	4	3	3	3	2
213.0	P	2	3	3	No	3	4	2	2	3	3
222.0	S or P	3	3	3	Yes	1	4	1	2	3	3
242.0	S or P	3	4	4	Yes	1	4	2	2	3	4
295.0	S	3	4	4	Yes	3	4	2	2	3	3
296.0	P	3	4	3	Yes	2	4	3	2	3	3
308.0	P	2	2	2	No	3	3	3	3	4	2
319.0	S or P	2	2	2	Optional	3	3	3	4	4	2
328.0	S	1	1	2	Optional	2	3	3	3	4	1
332.0	P	1	2	2	Yes	1	3	4	4	4	2
333.0	P	1	2	2	Yes	2	3	3	3	4	3
336.0	P	1	2	2	Yes	1	3	4	4	4	3
354.0	P	1	1	1	Yes	2	3	4	4	4	3
355.0	S or P	1	1	1	Yes	2	3	3	3	4	1
C355.0	S or P	1	1	1	Yes	2	3	3	3	4	1
356.0	S or P	1	1	1	Yes	3	2	3	4	4	1
A356.0	S or P	1	1	1	Yes	3	2	3	4	4	1
357.0	S or P	1	1	1	Yes	3	2	3	4	4	1
A357.0	S or P	1	1	1	Yes	2	2	3	4	4	1
359.0	S or P	1	2	2	Yes	2	2	4	4	4	1
B443.0	S or P	1	1	1	No	4	2	5	4	4	1
512.0	S	3	3	4	No	3	1	2	2	2	3
513.0	P	4	4	4	No	3	1	1	1	1	3
514.0	S	4	4	5	No	3	1	1	1	1	3
520.0	S	4	4	5	Yes	5	1	1	1	1	4
535.0	S	5	4	5	Optional	3	1	1	1	1	4
705.0	S or P	4	4	4	No	4	2	1	2	2	4
707.0	S or P	4	4	4	No	4	2	1	2	2	4
710.0	S	4	5	4	No	4	4	1	2	2	4
711.0	S	3	5	4	No	4	3	1	2	2	4
713.0	S or P	3	4	4	No	4	3	1	1	1	4
771.0	S	3	4	4	Yes	4	3	1	1	1	4
850.0	S or P	4	5	5	Yes	5	4	1	3	...	5
851.0	S or P	4	5	5	Yes	5	4	1	3	...	5
852.0	S or P	4	5	5	Yes	5	4	1	3	...	5

(a) Characteristics are comparatively rated from 1 to 5; 1 is the highest or best possible rating. (b) S = sand; P = permanent

Characteristics of aluminum die casting alloys(a)

Alloy	Approximate melting temperature, °C	Resistance to Hot cracking	Resistance to Die soldering	Corrosion	Die filling capacity	Machining	Polishing	Electroplating	Anodized surface Appearance	Anodized surface Protection	Elevated temperature strength	Pressure tightness
360.0	557-596	1	2	2	3	3	3	2	3	3	1	2
A360.0	557-596	1	2	2	3	3	3	2	3	3	1	2
380.0	538-593	2	1	4	2	3	3	1	3	4	3	2
A380.0	583-593	2	1	4	2	3	3	1	3	4	3	2
383.0	516-582	1	2	3	1	2	3	1	3	4	2	2
384.0	516-582	2	2	5	1	3	3	2	4	5	2	2
413.0	574-582	1	1	2	1	4	5	3	5	3	3	1
A413.0	574-582	1	1	2	1	4	5	3	5	3	3	1
C443.0	574-632	3	4	2	4	5	4	2	2	2	5	3
518.0	535-621	5	5	1	5	1	1	5	1	1	4	5

(a) Relative rating of die casting alloys from 1 to 5; 1 is the highest or best possible rating. A rating of 5 in one or more categories does not rule an alloy out of commercial use if other attributes are favorable; however, ratings of 5 may present manufacturing difficulties

Factors affecting selection of casting process for aluminum alloys

| | Casting process | | |
Factor	Sand casting	Permanent mold casting	Die casting
Cost of equipment	Lowest cost if only a few items required	Less than die casting	Highest
Casting rate	Lowest rate	11 kg/h (25 lb/h) common; higher rates possible	4.5 kg/h (10 lb/h) common; 45 kg/h (100 lb/h) possible
Size of casting	Largest of any casting method	Limited by size of machine	Limited by size of machine
External and internal shape	Best suited for complex shapes where coring required	Simple sand cores can be used, but more difficult to insert than in sand castings	Cores must be able to be pulled because they are metal; undercuts can be formed only by collapsing cores or loose pieces
Minimum wall thickness	3.0 to 5.0 mm (0.125 to 0.200 in.) required; 4.0mm (0.150 in.) normal	3.0 to 5.0 mm (0.125 to 0.200 in.) required; 3.5 mm (0.140 in.) normal	1.0 to 2.5 mm (0.100 to 0.40 in.); depends on casting size
Type of cores	Complex baked sand cores can be used	Reusable cores can be made of steel, or nonreusable baked cores can be used	Steel cores; must be simple and straight so they can be pulled
Tolerance obtainable	Poorest; best linear tolerance is 300 mm/m (300 mils/in.)	Best linear tolerance is 10 mm/m (10 mils/in.)	Best linear tolerance is 4 mm/m (4 mils/in.)
Surface finish	6.5 to 12.5 μm (250 to 500 μin.)	4.0 to 10 μm (150 to 400 μin.)	1.5 μm (50 μin.); best finish of the three casting processes
Gas porosity	Lowest porosity possible with good technique	Best pressure tightness; low porosity possible with good technique	Porosity may be present
Cooling rate	0.1 to 0.5 °C/s (0.2 to 0.9 °F/s)	0.3 to 1.0 °C/s (0.5 to 1.8 °F/s)	50 to 500 °C/s (90 to 900 °F/s)
Grain size	Coarse	Fine	Very fine on surface
Strength	Lowest	Excellent	Highest, usually used in the "as cast" condition
Fatigue properties	Good	Good	Excellent
Wear resistance	Good	Good	Excellent
Over-all quality	Depends on foundry technique	Highest quality	Tolerance and repeatability very good
Remarks	Very versatile as to size, shape, internal configurations	…	Excellent for fast production rates

Typical mechanical properties of several aluminum sheet alloys

| Alloy and temper | Thickness | | Tensile-test values(a) | | | | | | | Cup test values | | Bend diameter | |
| | | | Ultimate | | Yield | | | | | Olsen, | Swift, | Longi- | Trans- |
	mm	in.	MPa	ksi	MPa	ksi	$\bar{E}$, %	$\bar{n}$	$\bar{r}$	O_d	LDR	tudinal	verse
2036-T4	0.99	0.039	349	50.6	203	29.5	23	0.226	0.75	0.306	2.10	1t	2t
3003-O	0.89	0.035	107	15.5	52	7.6	33	0.235	0.66	0.300	2.04	0	0
5052-O	0.86	0.034	211	30.6	97	14.0	22	0.282	0.62	0.316	2.08	0	0
5056-O	0.86	0.034	283	41.0	147	21.3	27	0.279	0.79	0.325	2.16	0	0
5086-O	1.63	0.064	284	41.2	136	19.7	24	0.359	0.68	0.365	2.06	2t	2t
5086-H32	1.53	0.064	303	43.9	203	29.5	16	0.291	0.71	0.340	2.01	2t	2t
5182-O	1.02	0.040	280	40.6	132	19.2	26	0.340	0.78	0.340	2.09	0	1/2t
5252-H25	0.79	0.031	247	35.8	192	27.8	14	0.140	1.04	0.216	1.98	…	…
6009-T4	1.32	0.052	261	37.8	150	21.7	24	0.264	0.68	0.330	2.04	0	1/2t
6010-T4	0.81	0.032	296	42.9	170	24.6	27	0.260	0.61	0.320	2.04	1/2t	1/2t
6061-T4	1.02	0.040	301	43.6	193	28.0	23	0.194	0.71	0.290	2.08	2t	2t
6151-T4	1.02	0.040	279	40.5	180	26.1	21	0.195	0.62	0.305	2.07	2t	2t
7021-O	2.36	0.093	195	28.3	132	19.1	20	0.202	0.990	0.345	2.15	2t	2t
7021-O	4.75	0.187	215	31.2	140	20.3	20	0.176	0.372	…	…	2t	2t
7029-O	2.56	0.101	273	39.6	123	17.9	22	0.298	0.611	0.293	2.15	2t	2t
7029-O	5.08	0.200	228	33.0	110	15.9	25	0.264	0.544	0.377	…	2t	2t
7029-W	2.54	0.100	222	32.2	79	11.4	26	0.428	…	0.387	2.30	2t	2t
7146-O	4.72	0.186	174	25.2	133	19.3	21	0.122	0.680	0.370	…	2t	2t
7146-O	2.54	0.100	169	24.5	137	19.8	19	0.123	0.642	0.347	2.26	2t	2t

(a) Values shown are planar averages of tensile strength ($\bar{T}$), yield strength ($\bar{Y}$), elongation ($\bar{E}$), strain-hardening exponent ($\bar{n}$) and plastic strain ratio ($\bar{r}$)

Typical tensile properties for separately cast test bars of common aluminum casting alloys

Alloy	Product(a)	Temper	Tensile strength		Yield strength(b)		Elongation(c), %
			MPa	ksi	MPa	ksi	
201.0	S	T4	365	53	215	31	20
	S	T6	485	70	435	63	7
	S	T7	460	67	415	60	4.5
206.0, A206.0	S	T7	435	63	345	50	11.7
208.0	S	F	145	21	97	14	2.5
242.0	S	T21	185	27	125	18	1.0
	S	T571	220	32	205	30	0.5
	S	T77	205	30	160	23	2.0
	P	T571	275	40	235	34	1.0
	P	T61	325	47	290	42	0.5
295.0	S	T4	220	32	110	16	8.5
	S	T6	250	36	165	24	5.0
	S	T62	285	41	220	32	2.0
296.0	P	T4	255	37	130	19	9.0
	P	T6	275	40	180	26	5.0
	P	T7	270	39	140	20	4.5
308.0	P	F	195	28	110	16	2.0
319.0	S	F	185	27	125	18	2.0
	S	T6	250	36	165	24	2.0
	P	F	235	34	130	19	2.5
	P	T6	280	40	185	27	3.0
336.0	P	T551	250	36	195	28	0.5
	P	T65	325	47	295	43	0.5
354.0	P	T61	380	55	285	41	6.0
355.0	S	T51	195	28	160	23	1.5
	S	T6	240	35	175	25	3.0
	S	T61	270	39	240	35	1.0
	S	T7	265	38	250	36	0.5
	S	T71	175	35	200	29	1.5
	P	T51	210	30	165	24	2.0
	P	T6	290	42	190	27	4.0
	P	T62	310	45	280	40	1.5
	P	T7	280	40	210	30	2.0
	P	T71	250	36	215	31	3.0
356.0	S	T51	175	25	140	20	2.0
	S	T6	230	33	165	24	3.5
	S	T7	235	34	210	30	2.0
	S	T71	195	28	145	21	3.5
	P	T6	265	38	185	27	5.0
	P	T7	220	32	165	24	6.0
357.0, A357.0	S	T62	360	52	290	42	8.0
359.0	P	T61	330	48	255	37	6.0
		T62	345	50	290	42	5.5
360.0	D	F	325	47	170	25	3.0
A360.0	D	F	320	46	165	24	5.0
380.0	D	F	330	48	165	24	3.0
383.0	D	F	310	45	150	22	3.5
384.0, A384.0	D	F	330	48	165	24	2.5
390.0	D	F	280	41	240	35	1.0
	D	T5	300	43	260	38	1.0
A390.0	S	F, T5	180	26	180	26	<1.0
	S	T6	280	40	280	40	<1.0
	S	T7	250	36	250	36	<1.0
	P	F, T5	200	29	200	29	1.0
	P	T6	310	45	310	45	<1.0
	P	T7	260	38	260	38	<1.0
413.0	D	F	300	43	140	21	2.5
A413.0	D	F	290	42	130	19	3.5
443.0	S	F	130	19	55	8	8.0
B443.0	P	F	159	23	62	9	10.0
C443.0	D	F	228	33	110	16	9.0
514.0	S	F	170	25	85	12	9.0
518.0	D	F	310	45	190	28	5.0 to 8.0
520.0	S	T4	330	48	180	26	16
535.0	S	F	275	40	140	20	13
712.0	S	F	240	35	170	25	5.0
713.0	S	T5	210	30	150	22	3.0
	P	T5	220	32	150	22	4.0
771.0	S	T6	345	50	275	40	9.0
850.0	P	T5	160	23	75	11	10.0

(a) S = sand casting, P = permanent mold casting, D = die casting. (b) 0.2% offset. (c) 12.7-mm (1/2-in.) diam specimen

Extraction of aluminum from bauxite ore

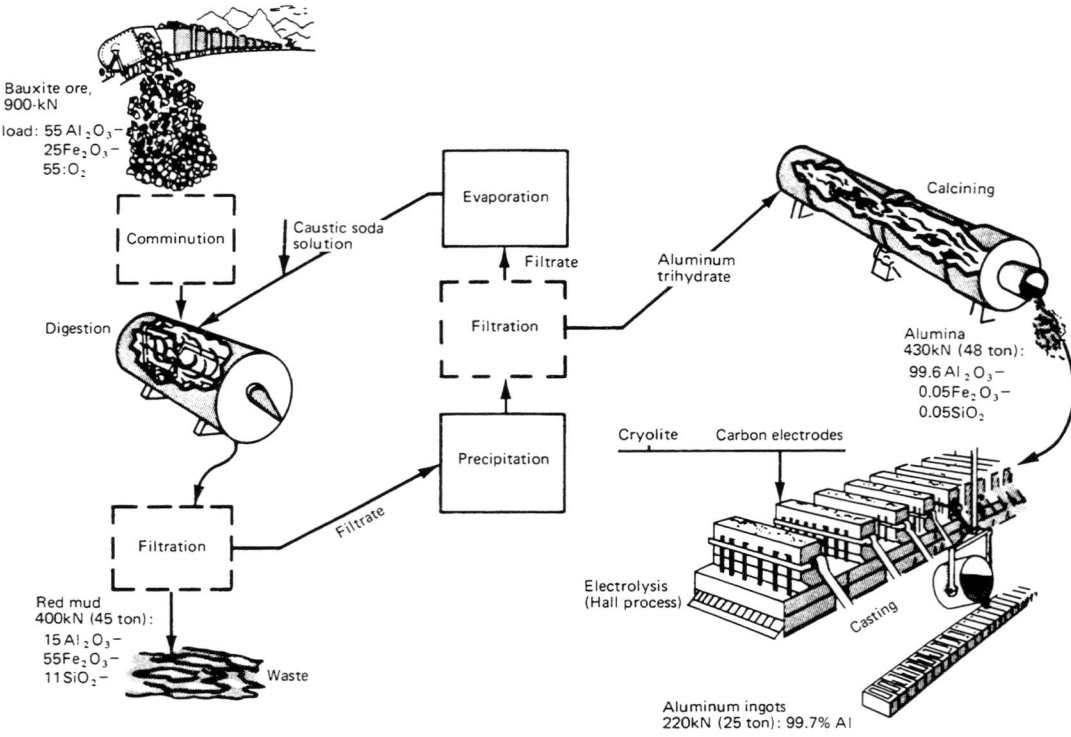

Bauxite ore,
900-kN

load: 55 Al$_2$O$_3$ –
25Fe$_2$O$_3$ –
55:O$_2$

Comminution

Caustic soda
solution

Evaporation

Filtrate

Digestion

Filtration

Aluminum
trihydrate

Calcining

Alumina
430kN (48 ton):
99.6 Al$_2$O$_3$ –
0.05Fe$_2$O$_3$ –
0.05SiO$_2$

Precipitation

Cryolite   Carbon electrodes

Filtration

Filtrate

Red mud
400kN (45 ton):
15 Al$_2$O$_3$ –
55Fe$_2$O$_3$ –
11SiO$_2$ –

Waste

Electrolysis
(Hall process)

Casting

Aluminum ingots
220kN (25 ton): 99.7% Al

Heat Treating

Typical solution and precipitation heat treatments for aluminum alloy mill products

These times and temperatures are typical for various forms, sizes, and methods of manufacture and may not exactly describe optimum treatments for specific items

Alloy	Product form	Solution heat treatment(a) Metal temperature(b) °C	°F	Temper designation	Precipitation heat treatment Metal temperature(b) °C	°F	Time(c), h	Temper designation
2011	Rolled or cold finished rod and bar	525	975	T3(d)	160	320	14	T8(d)
				T4	...	...	...	...
				T451(e)	...	...	...	...
2014(f)	Flat sheet	500	935	T3(d)	160	320	18	T6
				T42	160	320	18	T62
	Coiled sheet	500	935	T4	160	320	18	T6
				T42	160	320	18	T62
	Plate	500	935	T42	160	320	18	T62
				T451(e)	160	320	18	T651(e)
	Rolled or cold finished wire, rod and bar	500	935	T4	160(g)	320(g)	18	T6
				T42	160(g)	320(g)	18	T62
				T451(e)	160(g)	320(g)	18	T651(e)
	Extruded rod, bar, shapes and tube	500	935	T4	160(g)	320(g)	18	T6
				T42	160(g)	320(g)	18	T62
				T4510(e)	160(g)	320(g)	18	T6510(e)
				T4511(e)	160(g)	320(g)	18	T6511(e)
	Drawn tube	500	935	T4	160(g)	320(g)	18	T6
				T42	160(g)	320(g)	18	T62
	Die forgings	500(h)	935(h)	T4	170	340	10	T6
	Hand forgings and rolled rings	500(h)	935(h)	T4	170	340	10	T6
				T452(j)	170	340	10	T652(j)
2017	Rolled or cold finished wire, rod and bar	500	935	T4	...	...	...	...
				T42	...	...	...	...
				T451(e)	...	...	...	...
2018	Die forgings	510(k)	950(k)	T4	170	340	10	T61

(a) Material should be quenched from the solution-treating temperature as rapidly as possible and with minimum delay after removal from the furnace. When material is quenched by total immersion in water, unless otherwise indicated, the water should be at room temperature, and should be suitably cooled so that it remains below 38 °C (100 °F) during the quenching cycle. Use of high-velocity, high-volume jets of cold water also is effective for some materials. (b) The nominal temperatures listed should be attained as rapidly as possible and maintained within ± 6 °C (±10 °F) of nominal during the time at temperature. (c) Approximate time at temperature. The specific time will depend on the time required for the load to reach temperature. The times shown are based on rapid heating, with soak time measured from the time the load reaches a temperature within 6 °C (10 °F) of the applicable temperature. (d) Cold working subsequent to solution heat treatment and prior to any precipitation heat treatment is necessary to attain the specified properties for this temper. (e) Stress relieved by stretching to produce a specified amount of permanent set subsequent to solution heat treatment and prior to any precipitation heat treatment. (f) These heat treatments also apply to alclad sheet and plate in these alloys. (g) An alternative treatment of 8 h at 177 °C (350 °F) also may be used. (h) Solution heat treatment is followed by quenching in water 60 to 82 °C (140 to 180 °F). (j) Stress relieved by 1 to 5% cold reduction subsequent to solution heat treatment and prior to precipitation heat treatment. (k) Solution heat treatment is followed by quenching in water at 100 °C (212 °F). (m) Solution heat treatment is followed by quenching in room-temperature air blast. (n) By suitable control of extrusion temperature, product may be quenched directly from extrusion press to provide specified properties for this temper. Some products may be adequately quenched in room-temperature air blast. (p) See U.S. Patent 4 082 578. (q) Applicable to tread plate only. (r) An alternative treatment of 8 h at 171 °C (340 °F) also may be used. (s) Cold working subsequent to precipitation heat treatment is necessary to attain the specified properties for this temper. (t) An alternative treatment of 3 h at 182 °C (360 °F) also may be used. (u) An alternative treatment of 6 h at 182 °C (360 °F) also may be used. (v) No solution heat treatment; 72 h at room temperature following press quench, followed by two-stage precipitation heat treatment comprised of 8 h at 107 °C (225 °F) plus 16 h at 149 °C (300 °F). (w) Aging practice varies with product, size, nature of equipment, loading procedures and furnace-control capabilities. The optimum practice for a specific item can be ascertained only by actual trial treatment of the item under specific conditions. Typical procedures involve a two-stage treatment comprised of 3 to 30 h at 121 °C (250 °F) followed by 15 to 18 h at 163 °C (325 °F) for extrusions. An alternative two-stage treatment of 8 h at 99 °C (210 °F) followed by 24 to 28 h at 163 °C (325 °F) also may be used. (x) Aging of aluminum alloys 7050, 7075, 7175 and 7475 from any temper to the T73 or T76 temper series requires closer-than-normal controls on aging variables such as time, temperature, heatup rate, etc., for any given item. In addition, when material in a T6-type temper is reaged to a T73- or T76-type temper, the specific condition of the T6 material (such as property levels and other effects of processing variables) is extremely important and will affect the capability of the reaged material to conform to the requirements specified for the applicable T73- or T76-type temper. (y) Two-stage treatment comprised of 6 to 8 h at 107 °C (225 °F) followed by: 24 to 30 h at 163 °C (325 °F) for sheet and plate; 8 to 10 h at 177 °C (350 °F) for rolled or cold finished rod and bar; 6 to 8 h at 177 °C (350 °F) for extrusions and tube; 8 to 10 h at 177 °C (350 °F) for forgings in the T73 temper and 6 to 8 h at 177 °C (350 °F) for forgings in the T7352 temper. (z) An alternative two-stage treatment of 4 h at 96 °C (205 °F) followed by 8 h at 157 °C (315 °F) also may be used. (aa) For sheet, plate, tube and extrusions, an alternative two-stage treatment comprised of 6 to 8 h at 107 °C (225 °F) followed by 14 to 18 h at 168 °C (335 °F) may be used, provided that a heatup rate of approximately 14 °C/h (25 °F/h) is employed. For rolled or cold finished rod and bar, the alternative treatment is 10 h at 177 °C (350 °F). (bb) An alternative three-stage treatment comprised of 5 h at 99 °C (210 °F), 4 h at 121 °C (250 °F) and then 4 h at 149 °C (300 °F) also may be used. (cc) 7175-T736 and -T73652 heat treatments are directed to specific results, may vary from supplier to supplier and are either proprietary or patented. (dd) Must be preceded by soak at 466 to 477 °C (870 to 890 °F). See U.S. Patent 3 791 880

(continued)

Typical solution and precipitation heat treatments for aluminum alloy mill products (continued)

These times and temperatures are typical for various forms, sizes, and methods of manufacture and may not exactly describe optimum treatments for specific items

Alloy	Product form	Solution heat treatment(a) Metal temperature(b) °C	°F	Temper designation	Precipitation heat treatment Metal temperature(b) °C	°F	Time(c), h	Temper designation
2024(f)	Flat sheet	495	920	T3(d)	190	375	12	T81(d)
				T361(d)	190	375	8	T861(d)
				T42	190	375	9	T62
					190	375	16	T72
	Coiled sheet	495	920	T4	…	…	…	…
				T42	190	375	9	T62
					190	375	16	T72
	Plate	495	920	T351(e)	190	375	12	T851(e)
				T361(d)	190	375	8	T861(d)
				T42	190	375	9	T62
	Rolled or cold finished wire, rod and bar	495	920	T4	190	375	12	T6
				T351(e)	190	375	12	T851(e)
				T36(d)	190	375	8	T86(d)
				T42	190	375	16	T62
	Extruded rod, bar, shapes and tube	495	920	T3	190	375	12	T81
				T3510(e)	190	375	12	T8510(e)
				T3511(e)	190	375	12	T8511(e)
				T42	190	375	16	T62
	Drawn tube	495	920	T3(d)	…	…	…	…
				T42	…	…	…	…
2025	Die forgings	515	960	T4	170	340	10	T6
2036	Sheet	500	930	T4	…	…	…	…
2117	Rolled and cold finished wire and rod	500	935	T4	…	…	…	…
				T42	…	…	…	…
2218	Die forgings	510(k)	950(k)	T4	170	340	10	T61
		510(m)	950(m)	T41	240	460	6	T72
2219(f)	Flat sheet	535	995	T31(d)	175	350	18	T81(d)
				T37(d)	165	325	24	T87(d)
				T42	190	375	36	T62
	Plate	535	995	T31(d)	175	350	18	T81(d)
				T37(d)	175	350	18	T87(d)
				T351(e)	175	350	18	T851(e)
				T42	190	375	36	T62
	Rolled or cold finished wire, rod and bar	535	995	T351(e)	190	375	18	T851(e)

(a) Material should be quenched from the solution-treating temperature as rapidly as possible and with minimum delay after removal from the furnace. When material is quenched by total immersion in water, unless otherwise indicated, the water should be at room temperature, and should be suitably cooled so that it remains below 38 °C (100 °F) during the quenching cycle. Use of high-velocity, high-volume jets of cold water also is effective for some materials. (b) The nominal temperatures listed should be attained as rapidly as possible and maintained within ± 6 °C (±10 °F) of nominal during the time at temperature. (c) Approximate time at temperature. The specific time will depend on the time required for the load to reach temperature. The times shown are based on rapid heating, with soak time measured from the time the load reaches a temperature within 6 °C (10 °F) of the applicable temperature. (d) Cold working subsequent to solution heat treatment and prior to any precipitation heat treatment is necessary to attain the specified properties for this temper. (e) Stress relieved by stretching to produce a specified amount of permanent set subsequent to solution heat treatment and prior to any precipitation heat treatment. (f) These heat treatments also apply to alclad sheet and plate in these alloys. (g) An alternative treatment of 8 h at 177 °C (350 °F) also may be used. (h) Solution heat treatment is followed by quenching in water 60 to 82 °C (140 to 180 °F). (j) Stress relieved by 1 to 5% cold reduction subsequent to solution heat treatment and prior to precipitation heat treatment. (k) Solution heat treatment is followed by quenching in water at 100 °C (212 °F). (m) Solution heat treatment is followed by quenching in room-temperature air blast. (n) By suitable control of extrusion temperature, product may be quenched directly from extrusion press to provide specified properties for this temper. Some products may be adequately quenched in room-temperature air blast. (p) See U.S. Patent 4 082 578. (q) Applicable to tread plate only. (r) An alternative treatment of 8 h at 171 °C (340 °F) also may be used. (s) Cold working subsequent to precipitation heat treatment is necessary to attain the specified properties for this temper. (t) An alternative treatment of 3 h at 182 °C (360 °F) also may be used. (u) An alternative treatment of 6 h at 182 °C (360 °F) also may be used. (v) No solution heat treatment; 72 h at room temperature following press quench, followed by two-stage precipitation heat treatment comprised of 8 h at 107 °C (225 °F) plus 16 h at 149 °C (300 °F). (w) Aging practice varies with product, size, nature of equipment, loading procedures and furnace-control capabilities. The optimum practice for a specific item can be ascertained only by actual trial treatment of the item under specific conditions. Typical procedures involve a two-stage treatment comprised of 3 to 30 h at 121 °C (250 °F) followed by 15 to 18 h at 163 °C (325 °F) for extrusions. An alternative two-stage treatment of 8 h at 99 °C (210 °F) followed by 24 to 28 h at 163 °C (325 °F) also may be used. (x) Aging of aluminum alloys 7050, 7075, 7175 and 7475 from any temper to the T73 or T76 temper series requires closer-than-normal controls on aging variables such as time, temperature, heatup rate, etc., for any given item. In addition, when material in a T6-type temper is reaged to a T73- or T76-type temper, the specific condition of the T6 material (such as property levels and other effects of processing variables) is extremely important and will affect the capability of the reaged material to conform to the requirements specified for the applicable T73- or T76-type temper. (y) Two-stage treatment comprised of 6 to 8 h at 107 °C (225 °F) followed by: 24 to 30 h at 163 °C (325 °F) for sheet and plate; 8 to 10 h at 177 °C (350 °F) for rolled or cold finished rod and bar; 6 to 8 h at 177 °C (350 °F) for extrusions and tube; 8 to 10 h at 177 °C (350 °F) for forgings in the T73 temper and 6 to 8 h at 177 °C (350 °F) for forgings in the T7352 temper. (z) An alternative two-stage treatment comprised of 4 h at 96 °C (205 °F) followed by 8 h at 157 °C (315 °F) also may be used. (aa) For sheet, plate, tube and extrusions, an alternative two-stage treatment comprised of 6 to 8 h at 107 °C (225 °F) followed by 14 to 18 h at 168 °C (335 °F) may be used, provided that a heatup rate of approximately 14 °C/h (25 °F/h) is employed. For rolled or cold finished rod and bar, the alternative treatment is 10 h at 177 °C (350 °F). (bb) An alternative three-stage treatment comprised of 5 h at 99 °C (210 °F), 4 h at 121 °C (250 °F) and then 4 h at 149 °C (300 °F) also may be used. (cc) 7175-T736 and -T73652 heat treatments are directed to specific results, may vary from supplier to supplier and are either proprietary or patented. (dd) Must be preceded by soak at 466 to 477 °C (870 to 890 °F). See U.S. Patent 3 791 880

(continued)

Typical solution and precipitation heat treatments for aluminum alloy mill products(continued)

These times and temperatures are typical for various forms, sizes, and methods of manufacture and may not exactly describe optimum treatments for specific items

Alloy	Product form	Solution heat treatment(a)		Temper designation	Precipitation heat treatment		Time(c), h	Temper designation
		Metal temperature(b)			Metal temperature(b)			
		°C	°F		°C	°F		
2219(f) (continued)	Extruded rod, bar, shapes and tube	535	995	T31(d)	190	375	18	T81(d)
				T3510(e)	190	375	18	T8510(e)
				T3511(e)	190	375	18	T8511(e)
				T42	190	375	36	T62
	Die forgings and rolled rings	535	995	T4	190	375	26	T6
	Hand forgings	535	995	T4	190	375	26	T6
				T352(j)	175	350	18	T852(j)
2618	Forgings and rolled rings	530	985	T4	200	390	20	T61
4032	Die forgings	510(h)	950(h)	T4	170	340	10	T6
6005	Extruded rod, bar, shapes and tube	530(n)	985(n)	T1	175	350	8	T5
6009(p)	Coiled sheet	555	1030	T4	175	350	8	T6
6010(p)	Coiled sheet	565	1050	T4	175	350	8	T6
6053	Die forgings	520	970	T4	170	340	10	T6
6061(f)	Sheet	530	985	T4	160	320	18	T6
				T42	160	320	18	T62
	Plate	530	985	T4(q)	160	320	18	T6(q)
				T42	160	320	18	T62
				T451(e)	160	320	18	T651(e)
	Rolled or cold finished wire, rod and bar	530	985	T4	160(r)	320(r)	18	T6
					160(r)	320(r)	18	T89(d)
					160(r)	320(r)	18	T93(s)
					160(r)	320(r)	18	T913(s)
					160(r)	320(r)	18	T94(s)
				T42	160(r)	320(r)	18	T62
				T451(e)	160(r)	320(r)	18	T651(e)
	Extruded rod, bar, shapes and tube	530(n)	985(n)	T4	175	350	8	T6
				T4510(e)	175	350	8	T6510(e)
				T4511(e)	175	350	8	T6511(e)
		530	985	T42	175	350	8	T62
	Drawn tube	530	985	T4	160(r)	320(r)	18	T6
				T42	160(r)	320(r)	18	T62

(a) Material should be quenched from the solution-treating temperature as rapidly as possible and with minimum delay after removal from the furnace. When material is quenched by total immersion in water, unless otherwise indicated, the water should be at room temperature, and should be suitably cooled so that it remains below 38 °C (100 °F) during the quenching cycle. Use of high-velocity, high-volume jets of cold water also is effective for some materials. (b) The nominal temperatures listed should be attained as rapidly as possible and maintained within ± 6 °C (±10 °F) of nominal during the time at temperature. (c) Approximate time at temperature. The specific time will depend on the time required for the load to reach temperature. The times shown are based on rapid heating, with soak time measured from the time the load reaches a temperature within 6 °C (10 °F) of the applicable temperature. (d) Cold working subsequent to solution heat treatment and prior to any precipitation heat treatment is necessary to attain the specified properties for this temper. (e) Stress relieved by stretching to produce a specified amount of permanent set subsequent to solution heat treatment and prior to any precipitation heat treatment. (f) These heat treatments also apply to alclad sheet and plate in these alloys. (g) An alternative treatment of 8 h at 177 °C (350 °F) also may be used. (h) Solution heat treatment is followed by quenching in water 60 to 82 °C (140 to 180 °F). (j) Stress relieved by 1 to 5% cold reduction subsequent to solution heat treatment and prior to precipitation heat treatment. (k) Solution heat treatment is followed by quenching in water at 100 °C (212 °F). (m) Solution heat treatment is followed by quenching in room-temperature air blast. (n) By suitable control of extrusion temperature, product may be quenched directly from extrusion press to provide specified properties for this temper. Some products may be adequately quenched in room-temperature air blast. (p) See U.S. Patent 4 082 578. (q) Applicable to tread plate only. (r) An alternative treatment of 8 h at 171 °C (340 °F) also may be used. (s) Cold working subsequent to precipitation heat treatment is necessary to attain the specified properties for this temper. (t) An alternative treatment of 3 h at 182 °C (360 °F) also may be used. (u) An alternative treatment of 6 h at 182 °C (360 °F) also may be used. (v) No solution heat treatment; 72 h at room temperature following press quench, followed by two-stage precipitation heat treatment comprised of 8 h at 107 °C (225 °F) plus 16 h at 149 °C (300 °F). (w) Aging practice varies with product, size, nature of equipment, loading procedures and furnace-control capabilities. The optimum practice for a specific item can be ascertained only by actual trial treatment of the item under specific conditions. Typical procedures involve a two-stage treatment comprised of 3 to 30 h at 121 °C (250 °F) followed by 15 to 18 h at 163 °C (325 °F) for extrusions. An alternative two-stage treatment of 8 h at 99 °C (210 °F) followed by 24 to 28 h at 163 °C (325 °F) also may be used. (x) Aging of aluminum alloys 7050, 7075, 7175 and 7475 from any temper to the T73 or T76 temper series requires closer-than-normal controls on aging variables such as time, temperature, heatup rate, etc., for any given item. In addition, when material in a T6-type temper is reaged to a T73- or T76-type temper, the specific condition of the T6 material (such as property levels and other effects of processing variables) is extremely important and will affect the capability of the reaged material to conform to the requirements specified for the applicable T73- or T76-type temper. (y) Two-stage treatment comprised of 6 to 8 h at 107 °C (225 °F) followed by: 24 to 30 h at 163 °C (325 °F) for sheet and plate; 8 to 10 h at 177 °C (350 °F) for rolled or cold finished rod and bar; 6 to 8 h at 177 °C (350 °F) for extrusions and tube; 8 to 10 h at 177 °C (350 °F) for forgings in the T73 temper and 6 to 8 h at 177 °C (350 °F) for forgings in the T7352 temper. (z) An alternative two-stage treatment comprised of 4 h at 96 °C (205 °F) followed by 8 h at 157 °C (315 °F) also may be used. (aa) For sheet, plate, tube and extrusions, an alternative two-stage treatment comprised of 6 to 8 h at 107 °C (225 °F) followed by 14 to 18 h at 168 °C (335 °F) may be used, provided that a heatup rate of approximately 14 °C/h (25 °F/h) is employed. For rolled or cold finished rod and bar, the alternative treatment is 10 h at 177 °C (350 °F). (bb) An alternative three-stage treatment comprised of 5 h at 99 °C (210 °F), 4 h at 121 °C (250 °F) and then 4 h at 149 °C (300 °F) also may be used. (cc) 7175-T736 and -T73652 heat treatments are directed to specific results, may vary from supplier to supplier and are either proprietary or patented. (dd) Must be preceded by soak at 466 to 477 °C (870 to 890 °F). See U.S. Patent 3 791 880

(continued)

Typical solution and precipitation heat treatments for aluminum alloy mill products (continued)

These times and temperatures are typical for various forms, sizes, and methods of manufacture and may not exactly describe optimum treatments for specific items

Alloy	Product form	Solution heat treatment(a)			Precipitation heat treatment			
		Metal temperature(b)		Temper designation	Metal temperature(b)		Time(c), h	Temper designation
		°C	°F		°C	°F		
6061(f) (continued)	Die and hand forgings	530	985	T4	175	350	8	T6
	Rolled rings	530	985	T4	175	350	8	T6
				T452(j)	175	350	8	T652(j)
6063	Extruded rod, bar, shapes and tube	(n)	(n)	T1	205(t)	400(t)	1	T5
		520(n)	970(n)	T4	175(u)	350(u)	8	T6
		520	970	T42	175(u)	350(u)	8	T62
	Drawn tube	520	970	T4	175	350	8	T6
					175	350	8	T83(d)(n)
					175	350	8	T831(d)(n)
					175	350	8	T832(d)(n)
				T42	175	350	8	T62
6066	Extruded rod, bar, shapes and tube	530	990	T4	175	350	8	T6
				T42	175	350	8	T62
				T4510(e)	175	350	8	T6510(e)
				T4511(e)	175	350	8	T6511(e)
	Drawn tube	530	990	T4	175	350	8	T6
				T42	175	350	8	T62
	Die forgings	530	990	T4	175	350	8	T6
6070	Extruded rod, bar, shapes and tube	545(n)	1015(n)	T4	160	320	18	T6
				T42	160	320	18	T62
6151	Die forgings	515	960	T4	170	340	10	T6
	Rolled rings	515	960	T4	170	340	10	T6
				T452(j)	170	340	10	T652(j)
6262	Rolled or cold finished wire, rod and bar	540	1000	T4	170	340	8	T6
					170	340	12	T9(s)
				T451	170	340	8	T651(e)
				T42	170	340	8	T62
	Extruded rod, bar, shapes and tube	540(n)	1000(n)	T4	175	350	12	T6
				T4510(e)	175	350	12	T6510(e)
				T4511(e)	175	350	12	T6511(e)
		540	1000	T42	175	350	12	T62

(a) Material should be quenched from the solution-treating temperature as rapidly as possible and with minimum delay after removal from the furnace. When material is quenched by total immersion in water, unless otherwise indicated, the water should be at room temperature, and should be suitably cooled so that it remains below 38 °C (100 °F) during the quenching cycle. Use of high-velocity, high-volume jets of cold water also is effective for some materials. (b) The nominal temperatures listed should be attained as rapidly as possible and maintained within ± 6 °C (±10 °F) of nominal during the time at temperature. (c) Approximate time at temperature. The specific time will depend on the time required for the load to reach temperature. The times shown are based on rapid heating, with soak time measured from the time the load reaches a temperature within 6 °C (10 °F) of the applicable temperature. (d) Cold working subsequent to solution heat treatment and prior to any precipitation heat treatment is necessary to attain the specified properties for this temper. (e) Stress relieved by stretching to produce a specified amount of permanent set subsequent to solution heat treatment and prior to any precipitation heat treatment. (f) These heat treatments also apply to alclad sheet and plate in these alloys. (g) An alternative treatment of 8 h at 177 °C (350 °F) also may be used. (h) Solution heat treatment is followed by quenching in water 60 to 82 °C (140 to 180 °F). (j) Stress relieved by 1 to 5% cold reduction subsequent to solution heat treatment and prior to precipitation heat treatment. (k) Solution heat treatment is followed by quenching in water at 100 °C (212 °F). (m) Solution heat treatment is followed by quenching in room-temperature air blast. (n) By suitable control of extrusion temperature, product may be quenched directly from extrusion press to provide specified properties for this temper. Some products may be adequately quenched in room-temperature air blast. (p) See U.S. Patent 4 082 578. (q) Applicable to tread plate only. (r) An alternative treatment of 8 h at 171 °C (340 °F) also may be used. (s) Cold working subsequent to precipitation heat treatment is necessary to attain the specified properties for this temper. (t) An alternative treatment of 3 h at 182 °C (360 °F) also may be used. (u) An alternative treatment of 6 h at 182 °C (360 °F) also may be used. (v) No solution heat treatment; 72 h at room temperature following press quench. (w) Aging practice varies with product, size, nature of equipment, loading procedures and furnace-control capabilities. The optimum practice for a specific item can be ascertained only by actual trial treatment of the item under specific conditions. Typical procedures involve a two-stage treatment comprised of 3 to 30 h at 121 °C (250 °F) followed by 15 to 18 h at 163 °C (325 °F) for extrusions. An alternative two-stage treatment of 8 h at 99 °C (210 °F) followed by 24 to 28 h at 163 °C (325 °F) also may be used. (x) Aging of aluminum alloys 7050, 7075, 7175 and 7475 from any temper to the T73 or T76 temper series requires closer-than-normal controls on aging variables such as time, temperature, heatup rate, etc., for any given item. In addition, when material in a T6-type temper is reaged to a T73- or T76-type temper, the specific condition of the T6 material (such as property levels and other effects of processing variables) is extremely important and will affect the capability of the reaged material to conform to the requirements specified for the applicable T73- or T76-type temper. (y) Two-stage treatment comprised of 6 to 8 h at 107 °C (225 °F) followed by: 24 to 30 h at 163 °C (325 °F) for sheet and plate; 8 to 10 h at 177 °C (350 °F) for rolled or cold finished rod and bar; 6 to 8 h at 177 °C (350 °F) for extrusions and tube; 8 to 10 h at 177 °C (350 °F) for forgings in the T73 temper and 6 to 8 h at 177 °C (350 °F) for forgings in the T7352 temper. (z) An alternative two-stage treatment comprised of 4 h at 96 °C (205 °F) followed by 8 h at 157 °C (315 °F) also may be used. (aa) For sheet, plate, tube and extrusions, an alternative two-stage treatment comprised of 6 to 8 h at 107 °C (225 °F) followed by 14 to 18 h at 168 °C (335 °F) may be used, provided that a heatup rate of approximately 14 °C/h (25 °F/h) is employed. For rolled or cold finished rod and bar, the alternative treatment is 10 h at 177 °C (350 °F). (bb) An alternative three-stage treatment comprised of 5 h at 99 °C (210 °F), 4 h at 121 °C (250 °F) and then 4 h at 149 °C (300 °F) also may be used. (cc) 7175-T736 and -T73652 heat treatments are directed to specific results, may vary from supplier to supplier and are either proprietary or patented. (dd) Must be preceded by soak at 466 to 477 °C (870 to 890 °F). See U.S. Patent 3 791 880

(continued)

Typical solution and precipitation heat treatments for aluminum alloy mill products(continued)

These times and temperatures are typical for various forms, sizes, and methods of manufacture and may not exactly describe optimum treatments for specific items

Alloy	Product form	Solution heat treatment(a) Metal temperature(b) °C	°F	Temper designation	Precipitation heat treatment Metal temperature(b) °C	°F	Time(c), h	Temper designation
6262 (continued)	Drawn tube	540	1000	T4	170	340	8	T6
					170	340	8	T9(s)
				T42	170	340	8	T62
6463	Extruded rod, bar, shapes and tube	(n)	(n)	T1	205(t)	400(t)	1	T5
		520(n)	970(n)	T4	175(u)	350(u)	8	T6
		520	970	T42	175(u)	350(u)	8	T62
6951	Sheet	530	985	T4	160	320	18	T6
				T42	160	320	18	T62
7001	Extruded rod, shapes and tube	465	870	W	120	250	24	T6
					120	250	24	T62
				W510(e)	120	250	24	T6510(e)
				W511(e)	120	250	24	T6511(e)
7005	Extruded rod, bar and shapes	...	...	...	...	...	...	T53(v)
7050	Plate	475	890	W51(e)	(w)	(w)	(w)	T7651(x)
					(y)	(y)	(y)	T73651(x)
	Extrusions	475	890	W510(e)	(w)	(w)	(w)	T76510(x)
				W511(e)	(w)	(w)	(w)	T76511(x)
	Die and hand forgings	475	890	W	(y)	(y)	(y)	T736(x)
				W52(e)	(y)	(y)	(y)	T73652(x)
7075(f)	Sheet	480	900	W	120(z)	250(z)	24	T6
					120(z)	250(z)	24	T62
					(w)	(w)	(w)	T76(x)
					(y)(aa)	(y)(aa)	(y)(aa)	T73(x)
	Plate	480	900	W	120(z)	250(z)	24	T62
				W51(e)	(y)(aa)	(y)(aa)	(y)(aa)	T7351(e)(x)
					120(z)	250(z)	24	T651(e)
					(w)	(w)	(w)	T7651(x)
	Rolled or cold finished wire, rod and bar	490	915	W	120	250	24	T6
					120	250	24	T62
					(y)(aa)	(y)(aa)	(y)(aa)	T73(x)
				W51(e)	120	250	24	T651(e)
					(y)(aa)	(y)(aa)	(y)(aa)	T7351(e)(x)

(a) Material should be quenched from the solution-treating temperature as rapidly as possible and with minimum delay after removal from the furnace. When material is quenched by total immersion in water, unless otherwise indicated, the water should be at room temperature, and should be suitably cooled so that it remains below 38 °C (100 °F) during the quenching cycle. Use of high-velocity, high-volume jets of cold water also is effective for some materials. (b) The nominal temperatures listed should be attained as rapidly as possible and maintained within ± 6 °C (±10 °F) of nominal during the time at temperature. (c) Approximate time at temperature. The specific time will depend on the time required for the load to reach temperature. The times shown are based on rapid heating, with soak time measured from the time the load reaches a temperature within 6 °C (10 °F) of the applicable temperature. (d) Cold working subsequent to solution heat treatment and prior to any precipitation heat treatment is necessary to attain the specified properties for this temper. (e) Stress relieved by stretching to produce a specified amount of permanent set subsequent to solution heat treatment and prior to any precipitation heat treatment. (f) These heat treatments also apply to alclad sheet and plate in these alloys. (g) An alternative treatment of 8 h at 177 °C (350 °F) also may be used. (h) Solution heat treatment is followed by quenching in water 60 to 82 °C (140 to 180 °F). (j) Stress relieved by 1 to 5% cold reduction subsequent to solution heat treatment and prior to precipitation heat treatment. (k) Solution heat treatment is followed by quenching in water at 100 °C (212 °F). (m) Solution heat treatment is followed by quenching in room-temperature air blast. (n) By suitable control of extrusion temperature, product may be quenched directly from extrusion press to provide specified properties for this temper. Some products may be adequately quenched in room-temperature air blast. (p) See U.S. Patent 4 082 578. (q) Applicable to tread plate only. (r) An alternative treatment of 8 h at 171 °C (340 °F) also may be used. (s) Cold working subsequent to precipitation heat treatment is necessary to attain the specified properties for this temper. (t) An alternative treatment of 3 h at 182 °C (360 °F) also may be used. (u) An alternative treatment of 6 h at 182 °C (360 °F) also may be used. (v) No solution heat treatment; 72 h at room temperature following press quench, followed by two-stage precipitation heat treatment comprised of 8 h at 107 °C (225 °F) plus 16 h at 149 °C (300 °F). (w) Aging practice varies with product, size, nature of equipment, loading procedures and furnace-control capabilities. The optimum practice for a specific item can be ascertained only by actual trial treatment of the item under specific conditions. Typical procedures involve a two-stage treatment comprised of 3 to 30 h at 121 °C (250 °F) followed by 15 to 18 h at 163 °C (325 °F). An alternative two-stage treatment comprised of 8 h at 99 °C (210 °F) followed by 24 to 28 h at 163 °C (325 °F) also may be used. (x) Aging of aluminum alloys 7050, 7075, 7175 and 7475 from any temper to the T73 or T76 temper series requires closer-than-normal controls on aging variables such as time, temperature, heatup rate, etc., for any given item. In addition, when material in a T6-type temper is reaged to a T73- or T76-type temper, the specific condition of the T6 material (such as property levels and other effects of processing variables) is extremely important and will affect the capability of the reaged material to conform to the requirements specified for the applicable T73- or T76-type temper. (y) Two-stage treatment comprised of 6 to 8 h at 107 °C (225 °F) followed by: 24 to 30 h at 163 °C (325 °F) for sheet and plate; 8 to 10 h at 177 °C (350 °F) for rolled or cold finished rod and bar; 6 to 8 h at 177 °C (350 °F) for extrusions and tube; 8 to 10 h at 177 °C (350 °F) for forgings in the T73 temper and 6 to 8 h at 177 °C (350 °F) for forgings in the T7352 temper. (z) An alternative two-stage treatment comprised of 4 h at 96 °C (205 °F) followed by 8 h at 157 °C (315 °F) also may be used. (aa) For sheet, plate, tube and extrusions, an alternative two-stage treatment comprised of 6 to 8 h at 107 °C (225 °F) followed by 14 to 18 h at 168 °C (335 °F) may be used, provided that a heatup rate of approximately 14 °C/h (25 °F/h) is employed. For rolled or cold finished rod and bar, the alternative treatment is 10 h at 177 °C (350 °F). (bb) An alternative three-stage treatment comprised of 5 h at 99 °C (210 °F), 4 h at 121 °C (250 °F) and then 4 h at 149 °C (300 °F) also may be used. (cc) 7175-T736 and -T73652 heat treatments are directed to specific results, may vary from supplier to supplier and are either proprietary or patented. (dd) Must be preceded by soak at 466 to 477 °C (870 to 890 °F). See U.S. Patent 3 791 880

Typical solution and precipitation heat treatments for aluminum alloy mill products (continued)

These times and temperatures are typical for various forms, sizes, and methods of manufacture and may not exactly describe optimum treatments for specific items

Alloy	Product form	Solution heat treatment(a) Metal temperature(b) °C	°F	Temper designation	Precipitation heat treatment Metal temperature(b) °C	°F	Time(c), h	Temper designation
7075(f) (continued)	Extruded rod, bar, shapes and tube	465	870	W	120(bb)	250(bb)	24	T6
					120(bb)	250(bb)	24	T62
					(y)(aa)	(y)(aa)	(y)(aa)	T73(x)
					(w)	(w)	(w)	T76(x)
				W510(e)	120(bb)	250(bb)	24	T6510(e)
					(y)(aa)	(y)(aa)	(y)(aa)	T73510(e)(x)
					(w)	(w)	(w)	T76510(x)
				W511(e)	120(bb)	250(bb)	24	T6511(e)
					(y)(aa)	(y)(aa)	(y)(aa)	T73511(e)(x)
					(w)	(w)	(w)	T76511(x)
	Drawn tube	465	870	W	120	250	24	T6
					120	250	24	T62
					(y)(aa)	(y)(aa)	(y)(aa)	T73(x)
	Die forgings	470(h)	880(h)	W	120	250	24	T6
					(y)	(y)	(y)	T73(x)
				W52(j)	(y)	(y)	(y)	T7352(j)(x)
	Hand forgings	470(h)	880(h)	W	120	250	24	T6
					(y)	(y)	(y)	T73(x)
				W52(j)	120	250	24	T652(j)
					(y)	(y)	(y)	T7352(j)(x)
	Rolled rings	470	880	W	120	250	24	T6
7175	Die forgings	(cc)	(cc)	W	(cc)	(cc)	(cc)	T66(cc)
		(cc)	(cc)	W	(cc)	(cc)	(cc)	T736(x)(cc)
		(cc)	(cc)	W52(j)	(cc)	(cc)	(cc)	T73652(j)(x)(cc)
	Hand forgings	(cc)	(cc)	W	(cc)	(cc)	(cc)	T736(x)(cc)
		(cc)	(cc)	W52(j)	(cc)	(cc)	(cc)	T7365(j)(x)(cc)
7475	Sheet	515(dd)	960(dd)	W	120	250	3	
					plus 155	315	3	T61(dd)
					(w)	(w)	(w)	T761(x)(dd)
	Plate	510(dd)	950(dd)	W51(e)	120	250	24	T651(dd)
					(w)	(w)	(w)	T7651(x)(dd)
					(y)	(y)	(y)	T7351(x)(dd)
Alclad 7475	Sheet	495	920	W	120	250	3	
					plus 155	315	3	T61(dd)
					(w)	(w)	(w)	T761(x)(dd)

(a) Material should be quenched from the solution-treating temperature as rapidly as possible and with minimum delay after removal from the furnace. When material is quenched by total immersion in water, unless otherwise indicated, the water should be at room temperature, and should be suitably cooled so that it remains below 38 °C (100 °F) during the quenching cycle. Use of high-velocity, high-volume jets of cold water also is effective for some materials. (b) The nominal temperatures listed should be attained as rapidly as possible and maintained within ± 6 °C (±10 °F) of nominal during the time at temperature. (c) Approximate time at temperature. The specific time will depend on the time required for the load to reach temperature. The times shown are based on rapid heating, with soak time measured from the time the load reaches a temperature within 6 °C (10 °F) of the applicable temperature. (d) Cold working subsequent to solution heat treatment and prior to any precipitation heat treatment is necessary to attain the specified properties for this temper. (e) Stress relieved by stretching to produce a specified amount of permanent set subsequent to solution heat treatment and prior to any precipitation heat treatment. (f) These heat treatments also apply to alclad sheet and plate in these alloys. (g) An alternative treatment of 8 h at 177 °C (350 °F) also may be used. (h) Solution heat treatment is followed by quenching in water 60 to 82 °C (140 to 180 °F). (j) Stress relieved by 1 to 5% cold reduction subsequent to solution heat treatment and prior to precipitation heat treatment. (k) Solution heat treatment is followed by quenching in water at 100 °C (212 °F). (m) Solution heat treatment is followed by quenching in room-temperature air blast. (n) By suitable control of extrusion temperature, product may be quenched directly from extrusion press to provide specified properties for this temper. Some products may be adequately quenched in room-temperature air blast. (p) See U.S. Patent 4 082 578. (q) Applicable to tread plate only. (r) An alternative treatment of 8 h at 171 °C (340 °F) also may be used. (s) Cold working subsequent to precipitation heat treatment is necessary to attain the specified properties for this temper. (t) An alternative treatment of 3 h at 182 °C (360 °F) also may be used. (u) An alternative treatment of 6 h at 182 °C (360 °F) also may be used. (v) No solution heat treatment; 72 h at room temperature following press quench, followed by two-stage precipitation heat treatment comprised of 8 h at 107 °C (225 °F) plus 16 h at 149 °C (300 °F). (w) Aging practice varies with product, size, nature of equipment, loading procedures and furnace-control capabilities. The optimum practice for a specific item can be ascertained only by actual trial treatment of the item under specific conditions. Typical procedures involve a two-stage treatment comprised of 3 to 30 h at 121 °C (250 °F) followed by 15 to 18 h at 163 °C (325 °F) for extrusions. An alternative two-stage treatment of 8 h at 99 °C (210 °F) followed by 24 to 28 h at 163 °C (325 °F) also may be used. (x) Aging of aluminum alloys 7050, 7075, 7175 and 7475 from any temper to the T73 or T76 temper series requires closer-than-normal controls on aging variables such as time, temperature, heatup rate, etc., for any given item. In addition, when material in a T6-type temper is reaged to a T73- or T76-type temper, the specific condition of the T6 material (such as property levels and other effects of processing variables) is extremely important and will affect the capability of the reaged material to conform to the requirements specified for the applicable T73- or T76-type temper. (y) Two-stage treatment comprised of 6 to 8 h at 107 °C (225 °F) followed by: 24 to 30 h at 163 °C (325 °F) for sheet and plate; 8 to 10 h at 177 °C (350 °F) for rolled or cold finished rod and bar; 6 to 8 h at 177 °C (350 °F) for extrusions and tube; 8 to 10 h at 177 °C (350 °F) for forgings in the T73 temper and 6 to 8 h at 177 °C (350 °F) for forgings in the T7352 temper. (z) An alternative two-stage treatment comprised of 4 h at 96 °C (205 °F) followed by 8 h at 157 °C (315 °F) also may be used. (aa) For sheet, plate, tube and extrusions, an alternative two-stage treatment comprised of 6 to 8 h at 107 °C (225 °F) followed by 14 to 18 h at 168 °C (335 °F) may be used, provided that a heatup rate of approximately 14 °C/h (25 °F/h) is employed. For rolled or cold finished rod and bar, the alternative treatment is 10 h at 177 °C (350 °F). (bb) An alternative three-stage treatment comprised of 5 h at 99 °C (210 °F), 4 h at 121 °C (250 °F) and then 4 h at 149 °C (300 °F) also may be used. (cc) 7175-T736 and -T73652 heat treatments are directed to specific results, may vary from supplier to supplier and are either proprietary or patented. (dd) Must be preceded by soak at 466 to 477 °C (870 to 890 °F). See U.S. Patent 3 791 880

Typical heat treatments for aluminum alloy sand and permanent mold castings

Alloy	Temper	Type of casting(a)	Solution heat treatment(b) Temperature(c) °C	°F	Time, h	Aging treatment Temperature(c) °C	°F	Time, h
201.0	T6	S	510 to 515; 525 to 530	950 to 960; 980 to 990	2 / 14 to 20	155	310	20
	T7	S	510 to 515; 525 to 530	950 to 960; 980 to 990	2 / 14 to 20	190	370	5
204.0	T4	S or P	520	970	10	...	...	...
208.0	T55	S	...	...	...	155	310	16
222.0	O(d)	S	...	...	...	315	600	3
	T61	S	510	950	12	155	310	11
	T551	P	...	...	...	170	340	16 to 22
	T65	...	510	950	4 to 12	170	340	7 to 9
242.0	O(e)	S	...	...	...	345	650	3
	T571	S	...	...	...	205	400	8
		P	...	...	...	165 to 170	330 to 340	22 to 26
	T77	S	515	960	5(f)	330 to 355	625 to 675	2 (min)
	T61	S or P	515	960	4 to 12(f)	205 to 230	400 to 450	3 to 5
295.0	T4	S	515	960	12	...	...	...
	T6	S	515	960	12	155	310	3 to 6
	T62	S	515	960	12	155	310	12 to 24
	T7	S	515	960	12	260	500	4 to 6
296.0	T4	P	510	950	8	...	...	...
	T6	P	510	950	8	155	310	1 to 8
	T7	P	510	950	8	260	500	4 to 6
319.0	T5	S	...	...	...	205	400	8
	T6	S	505	940	12	155	310	2 to 5
		P	505	940	4 to 12	155	310	2 to 5
328.0	T6	S	515	960	12	155	310	2 to 5
332.0	T5	P	...	...	...	205	400	7 to 9
333.0	T5	P	...	...	...	205	400	7 to 9
	T6	P	505	940	6 to 12	155	310	2 to 5
	T7	P	505	940	6 to 12	260	500	4 to 6
336.0	T551	P	...	...	...	205	400	7 to 9
	T65	P	515	960	8	205	400	7 to 9
354.0	...	(g)	525 to 535	980 to 995	10 to 12	(h)	(h)	(h)
355.0	T51	S or P	...	...	...	225	440	7 to 9
	T6	S	525	980	12	155	310	3 to 5
		P	525	980	4 to 12	155	310	2 to 5
	T62	P	525	980	4 to 12	170	340	14 to 18
	T7	S	525	980	12	225	440	3 to 5
		P	525	980	4 to 12	225	440	3 to 9
	T71	S	525	980	12	245	475	4 to 6
		P	525	980	4 to 12	245	475	3 to 6
C355.0	T6	S	525	980	12	155	310	3 to 5
	T61	P	525	980	6 to 12	20 / 155	68 / 310	8 (min) / 10 to 12
356.0	T51	S or P	...	...	...	225	440	7 to 9
	T6	S	540	1000	12	155	310	3 to 5
		P	540	1000	4 to 12	155	310	2 to 5
	T7	S	540	1000	12	205	400	3 to 5
		P	540	1000	4 to 12	225	440	7 to 9
	T71	S	540	1000	10 to 12	245	475	3
		P	540	1000	4 to 12	245	475	3 to 6
A356.0	T6	S	540	1000	12	155	310	3 to 5
	T61	P	540	1000	6 to 12	20 / 155	68 / 310	8 (min) / 6 to 12
357.0	T6	P	540	1000	8	175	350	6
	T61	S	540	1000	10 to 12	155	310	10 to 12

(a) S, sand; P, permanent mold. (b) Unless otherwise indicated, solution treating is followed by quenching in water at 65 to 100 °C (150 to 212 °F). (c) Except where ranges are given, listed temperatures are ±6 °C or ±10 °F. (d) Stress relieve for dimensional stability as follows: hold 5 h at 413 ± 14 °C (775 ± 25 °F); furnace cool to 345 °C (650 °F) over a period of 2 h or more; furnace cool to 230 °C (450 °F) over a period of not more than ½ h; furnace cool to 120 °C (250 °F) over a period of approximately 2 h; cool to room temperature in still air outside the furnace. (e) No quench required; cool in still air outside furnace. (f) Air-blast quench from solution-treating temperature. (g) Casting process varies (sand, permanent mold or composite) depending on desired mechanical properties. (h) Solution heat treat as indicated, then artificially age by heating uniformly at the temperature and for the time necessary to develop the desired mechanical properties. (j) Quench in water at 65 to 100 °C (150 to 212 °F) for 10 to 20 s only. (k) Cool to room temperature in still air outside furnace

Typical heat treatments for aluminum alloy sand and permanent mold castings (continued)

Alloy	Temper	Type of casting(a)	Solution heat treatment(b) Temperature(c) °C	°F	Time, h	Aging treatment Temperature(c) °C	°F	Time, h
A357.0	...	(g)	540	1000	8 to 12	(h)	(h)	(h)
359.0	...	(g)	540	1000	10 to 14	(h)	(h)	(h)
A444.0	T4	P	540	1000	8 to 12	...	...	...
520.0	T4	S	430	810	18(j)	...	...	...
535.0	T5(d)	S	400	750	5	...	...	...
705.0	T5	S	...	...	...	20	68	21 days
						100	210	8
		P	...	...	...	20	68	21 days
						100	210	10
						155	310	3 to 5
707.0	T5	S	...	...	...	20	68	21 days
		P	...	...	...	100	210	8
	T7	S	530	990	8 to 16	175	350	4 to 10
	T7	P	530	990	4 to 8	175	350	4 to 10
710.0	T5	S	...	...	...	20	68	21 days
711.0	T1	P	...	...	...	20	68	21 days
712.0	T5	S	...	...	...	20	68	21 days
						155	315	6 to 8
713.0	T5	S or P	...	...	...	20	68	21 days
						120	250	16
771.0	T53(d)	S	415(k)	775(k)	5(k)	180(k)	360(k)	4(k)
	T5	S	...	...	...	180(k)	355(k)	3 to 5(k)
	T51	S	...	...	...	205	405	6
	T52	S	...	...	...	(d)	(d)	(d)
	T6	S	590(k)	1090(k)	6(k)	130	265	3
	T71	S	590(e)	1090(e)	6(e)	140	285	15
850.0	T5	S or P	...	...	...	220	430	7 to 9
851.0	T5	S or P	...	...	...	220	430	7 to 9
	T6	P	480	900	6	220	430	4
852.0	T5	S or P	...	...	...	220	430	7 to 9

(a) S, sand; P, permanent mold. (b) Unless otherwise indicated, solution treating is followed by quenching in water at 65 to 100 °C (150 to 212 °F). (c) Except where ranges are given, listed temperatures are ±6 °C or ±10 °F. (d) Stress relieve for dimensional stability as follows: hold 5 h at 413 ± 14 °C (775 ± 25 °F); furnace cool to 345 °C (650 °F) over a period of 2 h or more; furnace cool to 230 °C (450 °F) over a period of not more than ½ h; furnace cool to 120 °C (250 °F) over a period of approximately 2 h; cool to room temperature in still air outside the furnace. (e) No quench required; cool in still air outside furnace. (f) Air-blast quench from solution-treating temperature. (g) Casting process varies (sand, permanent mold or composite) depending on desired mechanical properties. (h) Solution heat treat as indicated, then artificially age by heating uniformly at the temperature and for the time necessary to develop the desired mechanical properties. (j) Quench in water at 65 to 100 °C (150 to 212 °F) for 10 to 20 s only. (k) Cool to room temperature in still air outside furnace

Effects of annealing treatments on ductility of 7075-O sheet

Annealing treatment	Elongation in tension(a), % in 50 mm or 2 in., for thickness of:			Bend angle(b), degrees, for thickness of:		Elongation in bending(c), % in 50 mm on 2 in., for thickness of:	
	0.5 mm (0.020 in.)	1.6 mm (0.064 in.)	2.6 mm (0.102 in.)	1.6 mm (0.064 in.)	2.6 mm (0.102 in.)	1.6 mm (0.064 in.)	2.6 mm (0.102 in.)
Treatment 1(d)	12	12	12	82	73	48	50
Treatment 2(e)	14	14	14	91	76	58	57
Treatment 3(f)	16	16	…	92.5	84	56	60

(a) Uniform elongation of gridded tension specimens. (b) Bend angle at first fracture. (c) Elongation in bend test for 1.3-mm (0.05-in.) gage spanning fracture. (d) Soak 2 h at 415 ± 14 °C (775 ± 25 °F); furnace cool to 260 °C (500 °F) at 30 °C/h (50 °F/h); air cool. (e) Soak 2 h at 425 °C (800 °F), air cool; soak 2 h at 230 °C (450 °F), air cool. (f) Soak 1 at 425 °C (800 °F); furnace cool to 230 °C (450 °F) at 30 °C/h (50 °F/h); soak 6 h at 230 °C (450 °F), air cool

Typical full annealing treatments for some common wrought aluminum alloys

These treatments, which anneal the material to the "O" temper, are typical for various sizes and methods of manufacture and may not exactly describe optimum treatments for specific items

Alloy	Metal temperature		Approximate time at temperature, h	Alloy	Metal temperature		Approximate time at temperature, h
	°C	°F			°C	°F	
1060	345	650	(a)	5457	345	650	(a)
1100	345	650	(a)	5652	345	650	(a)
1350	345	650	(a)	6005	415(b)	775(b)	2-3
2014	415(b)	775(b)	2-3	6009	415(b)	775(b)	2-3
2017	415(b)	775(b)	2-3	6010	415(b)	775(b)	2-3
2024	415(b)	775(b)	2-3	6053	415(b)	775(b)	2-3
2036	385(b)	725(b)	2-3	6061	415(b)	775(b)	2-3
2117	415(b)	775(b)	2-3	6063	415(b)	775(b)	2-3
2124	415(b)	775(b)	2-3	6066	415(b)	775(b)	2-3
2219	415(b)	775(b)	2-3	7001	415(c)	775(c)	2-3
3003	415	775	(a)	7005	345(d)	650(d)	2-3
3004	345	650	(a)	7049	415(c)	775(c)	2-3
3105	345	650	(a)	7050	415(c)	775(c)	2-3
5005	345	650	(a)	7075	415(c)	775(c)	2-3
5050	345	650	(a)	7079	415(c)	775(c)	2-3
5052	345	650	(a)	7178	415(c)	775(c)	2-3
5056	345	650	(a)	7475	415(c)	775(c)	2-3
5083	345	650	(a)	**Brazing sheet**			
5086	345	650	(a)	No. 11 and 12	345	650	(a)
5154	345	650	(a)	No. 21 and 22	345	650	(a)
5182	345	650	(a)	No. 23 and 24	345	650	(a)
5254	345	650	(a)				
5454	345	650	(a)				
5456	345	650	(a)				

(a) Time in the furnace need not be longer than necessary to bring all parts of the load to annealing temperature. Cooling rate is unimportant. (b) These treatments are intended to remove the effects of solution treatment and include cooling at a rate of about 30 °C/h (50 °F/h) from the annealing temperature to 260 °C (500 °F). Rate of subsequent cooling is unimportant. Treatment at 345 °C (650 °F), followed by uncontrolled cooling, may be used to remove the effects of cold work, or to partly remove the effects of heat treatment. (c) These treatments are intended to remove the effects of solution treatment and include cooling at an uncontrolled rate to 205 °C (400 °F) or less, followed by reheating to 230 °C (450 °F) for 4 h. Treatment at 345 °C (650 °F), followed by uncontrolled cooling, may be used to remove the effects of cold work, or to partly remove the effects of heat treatment. (d) Cooling rate to 205 °C (400 °F) or below is less than or equal to 30 °C/h (50 °F/h)

Typical acceptable hardness values for wrought aluminum alloys

Acceptable hardness does not guarantee acceptable properties; acceptance should be based on acceptable hardness plus written evidence of compliance with specified heat treating procedures. Hardness values higher than the listed maximums are acceptable provided that the material is positively identified as the correct alloy

Alloy and temper	Product form(a)	Hardness			
		HRB	HRE	HRH	HR15T
2014-T3, -T4, -T42	All	65 to 70	87 to 95	...	...
2014-T6, -T62, -T65	Sheet(b)	80 to 90	103 to 110	...	...
	All others	81 to 90	104 to 110	...	...
2014-T61	All	...	100 to 109	...	...
2024-T3	Not clad(c)	69 to 83	97 to 106	111 to 118	82.5 to 87.5
	Clad, through 1.60 mm (0.063 in.)	52 to 71	91 to 100	109 to 116	80 to 84.5
	Clad, over 1.60 mm (0.063 in.)	52 to 71	93 to 102	109 to 116	...
2024-T36	All	76 to 90	100 to 110	...	85 to 90
2024-T4, -T42(d)	Not clad	69 to 83	97 to 106	111 to 118	82.5 to 87.5
	Clad, through 1.60 mm (0.063 in.)	52 to 71	91 to 100	109 to 116	80 to 84.5
	Clad, over 1.60 mm (0.063 in.)	52 to 71	93 to 102	109 to 116	...
2024-T6, -T62	All	74.5 to 83.5	99 to 106	...	84 to 88
2024-T81	Not clad	74.5 to 83.5	99 to 106	...	84 to 88
	Clad	...	99 to 106	...	...
2024-T86	All	83 to 90	105 to 110	...	87.5 to 90
6053-T6	All	...	79 to 87	...	74.5 to 78.5
6061-T4(d)	Sheet	...	60 to 75	88 to 100	64 to 75
	Extrusions; bar	...	70 to 81	82 to 103	67 to 78
6061-T6	Not clad, 0.41 mm (0.016 in.)	...	...	...	75 to 84
	Not clad, 0.51 mm (0.020 in.) and over	47 to 72	85 to 97	...	78 to 84
	Clad	...	84 to 96	...	...
6063-T5	All	...	55 to 70	89 to 97	62.5 to 70
6063-T6	All	...	70 to 85	...	...
6151-T6	All	...	91 to 102	...	...
7075-T6, T65	Not clad(e)	85 to 94	106 to 114	...	87.5 to 92
	Clad: Through 0.91 mm (0.036 in.)	...	102 to 110	...	86 to 90
	Over 0.91 through 1.27 mm (over 0.036 through 0.050 in.)	78 to 90	104 to 110	...	...
	Over 1.27 through 1.57 mm (over 0.050 in. through 0.062 in.)	76 to 90	104 to 110	...	...
	Over 1.57 through 1.78 mm (over 0.062 through 0.070 in.)	76 to 90	102 to 110	...	...
	Over 1.78 mm (0.070 in.)	73 to 90	102 to 110	...	...
7079-T6, -T65	All(e)	81 to 93	104 to 114	...	87.5 to 92
7178-T6	Not clad(f)	85 min	105 min	...	88 min
	Clad: Through 0.91 mm (0.036 in.)	...	102 min	...	86 min
	Over 0.91 through 1.57 mm (over 0.036 through 0.062 in.)	85 min	...	...	...
	Over 1.57 mm (0.062 in.)	88 min	...	...	...

(a) Minimum hardness values shown for clad products are valid for thicknesses up to and including 2.31 mm (0.091 in.); for heavier-gage material, cladding should be locally removed for hardness testing or test should be performed on edge of sheet. (b) 126 to 158 HB (10-mm bal, 500-kg load). (c) 100 to 130 HB (10-mm ball, 500-kg load). (d) Alloys 2024-T4, 2024-T42 and 6061-T4 should not be rejected for low hardness until they have remained at room temperature for at least three days following solution treatment. (e) 136 to 164 HB (10-mm ball, 500-kg load). (f) 136 HB min (10-mm ball, 500-kg load)

Machining

Machinability ratings of aluminum alloys

Alloy(a)	Temper(a)	Rating(b)	Alloy(a)	Temper(a)	Rating(b)
Casting alloys			3003(c)	O, H112, H12	E
208.0	F	C		H14 to H18	D
242.0	T21, T571, T61, T77	B	3004(c)	O, H112, H32	D
295.0	T4, T6, T7, T62	B		H34 to H38	C
308.0	F	C	5005	O, H112, H12, H32	E
319.0	F	C		H14 to H18	D
	T5, T6, T7	C		H34 to H38	D
354.0	T61, T62	C	5050	O, H112, H32	D
355.0	T51, T6, T61, T62, T7, T71	C		H34 to H38	C
C355.0	T61	B	5052	O, H112, H32	D
356.0	T51, T6, T7, T71	C		H34 to H38	C
A356.0, A357.0	T61	C	5056	O	D
357.0	T6	C		H18, H38	C
359.0	T61, T62	C	5083	O, H111, H116, H321, H323	D
360.0	F	C		H131, H343	C
A360.0	F	C	5086, 5154	O, H111, H116, H32	D
380.0, A380.0	F, T5	B		H34 to H38	C
390.0, A390.0	T5	E	5182	O	D
413.0, 443.0	F	E		H19	C
514.0	F	A	5454, 5456	O, H112, H311	D
518.0, B535.0	F	B		H343	C
520.0	T4	A	5457	O	D
712.0	F	A		H25, H28, H38	C
713.0	F	A	5657	O	E
850.0	T5	A		H25, H28, H38	D
Wrought alloys			6061(c)	O	D
1100	O, H112, H12	E		T4, T6	C
	H14 to H18	D	6009, 6010	T4	D
1350	O, H111, H112, H12	E	6063	O, T2, T4	D
	H14 to H19	D		T5, T6, T8	C
2011	T3, T4, T6, T8	A	6262	T4, T9	B
2014(c), 2124	O	C	6463	O, T1	D
	T3, T4, T6	B		T4, T5, T6	C
2017	O	C	7049, 7050	T73, T76	B
	T4	B		T736, T76	B
2024(c)	T3, T4, T6, T8	B	7075(c)	T6, T73, T76	B
2036	T4	C	7175	T736	B
2219	T3, T6, T8	B	7178(c)	T6	B
2618	T61	B	7475	T6, T73, T76	B

(a) Alloys and tempers are those commonly used. Alloy modifications designated by other second digits and temper variations designated by added numerals will have the same ratings.
(b) A, B, C, D and E are relative ratings in increasing order of chip length and decreasing order of quality of finish and are defined as: A, free cutting, very small broken chips and excellent finish; B, curled or easily broken chips and good-to-excellent finish; C, continuous chips and good finish; D, continuous chips and satisfactory finish; E, optimum tool design and machine settings required to obtain satisfactory control of chip and finish. (c) Includes clad alloys and tempers

Joining

Weldability of aluminum alloys by the gas metal-arc and gas tungsten-arc processes

Readily weldable

Wrought alloys

    Unalloyed aluminum, 1060, 1100, 1350, 2219

    3003, 3004, 3105

    5005, 5050, 5052, 5056, 5083, 5086, 5154, 5252, 5254, 5454, 5456,
      5457, 5652, 5657

    6061, 6063, 6070, 6101, 6201, 6262, 6463

    7005

Casting alloys

    328.0, 355.0, C355.0, 356.0, A356.0,
      357.0, A357.0, 359.0
      443.0, A443.0, B443.0

Weldable in most applications(a)

Wrought alloys: 2014, 4032, 6066

Casting alloys

    208.0, 308.0, 319.0, 332.0

    413.0, 712.0

Limited weldability(b)

Wrought alloys: 2024, 2218, 2618

Casting alloys

    213.0, 222.0, 295.0, 296.0

    333.0, 336.0, 354.0

    512.0, 513.0, 514.0

    Die casting alloys

Welding not recommended

Wrought alloys: 2011, 7075, 7178

Casting alloys

    242.0, 520.0, 535.0

    705.0, 707.0, 710.0, 711.0, 713.0, 771.0

(a) May require special techniques for some applications. (b) Require special techniques

Minimum expected properties at room temperature for butt-welded aluminum alloys(a)(b)

Alloy and temper	Filler wire	Product forms	Thickness range, mm	Tensile strength				Compressive yield strength(c)		Shear strength				Bearing strength			
				Ultimate(b)		Yield(c)				Ultimate		Yield		Ultimate		Yield	
				MPa	ksi	MPa	ksi	MPa	ksi	MPa	ksi	MPa	ksi	MPa	ksi	MPa	ksi
1100-H12, H14	1100	All	All	76	11	31	4.5	31	4.5	55	8	17	2.5	160	23	55	8
3003-H12, H14, H16, H18	1100	All	All	97	14	48	7	48	7	69	10	28	4	205	30	83	12
Alclad 3003-H12, H14, H16, H18	1100	All	All	90	13	41	6	41	6	69	10	24	3.5	205	30	76	11
3004-H32, H34, H36, H38	4043	All	All	150	22	76	11	76	11	97	14	45	6.5	315	46	140	20
Alclad 3004-H32, H34, H36, H38, H14, H16	4043	All	All	145	21	76	11	76	11	90	13	45	6.5	305	44	130	19
3003-H25	1100	Sheet	All	115	17	62	9	62	9	83	12	34	5	250	36	105	15
5005-H12, H14, H32, H34	4043	All	All	97	14	48	7	48	7	62	9	28	4	195	28	69	10
5050-H32, H34	4043	All	All	125	18	55	8	55	8	83	12	30	4.5	250	36	83	12
5052-H32, H34	5654	All	All	170	25	90	13	90	13	110	16	52	7.5	345	50	130	19
5083-H111	5183	Extrusions	All	270	39	145	21	140	20	160	23	83	12	540	78	220	32
H321	5183	Sheet and plate	4.7-38.1	275	40	165	24	165	24	165	24	97	14	550	80	250	36
H321	5183	Plate	38.1-76.2	270	39	160	23	165	23	160	24	90	13	540	78	235	34
H323, H343	5183	Sheet	All	275	40	165	24	165	24	165	24	97	14	550	80	250	36
5086-H111	5356	Extrusions	All	240	35	125	18	115	17	145	21	69	10	485	70	195	28
H112	5356	Plate	6.4-12.7	240	35	115	17	115	17	145	21	66	9.5	485	70	195	28
H112	5356	Plate	12.7-25.4	240	35	110	16	110	16	145	21	62	9	485	70	195	28
H112	5356	Plate	25.4-50.8	240	35	97	14	97	14	145	21	55	8	485	70	195	28
H32, H34	5356	Sheet and plate	All	240	35	130	19	130	19	145	21	76	11	485	70	195	28
5154-H38	5654	Sheet	All	205	30	105	15	105	15	130	19	59	8.5	415	60	160	23
5454-H111	5554	Extrusions	All	215	31	110	16	105	15	130	19	66	9.5	425	62	165	24
H112	5554	Extrusions	All	215	31	83	12	83	12	130	19	49	7	425	62	165	24
H32, H34	5554	Sheet and plate	All	215	31	110	16	110	16	130	19	66	9.5	425	62	165	24
5456-H111	5556	Extrusions	All	285	41	165	24	150	22	165	24	97	14	565	82	260	38
H112	5556	Extrusions	All	285	41	130	19	130	19	165	24	76	11	565	82	260	38
H321	5556	Sheet and plate	4.7-38.1	290	42	180	26	165	24	170	25	105	15	580	84	260	38
H321	5556	Plate	38.1-76.2	285	41	165	24	160	23	170	25	97	14	565	82	260	38
H323, H343	5556	Sheet	All	290	42	180	26	180	26	170	25	105	15	580	84	260	38
6061-T6, T651, T6510, T6511(d)	4043	All	Over 9.5	165	24	140	20	140	20	105	15	83	12	345	50	205	30
6061-T6, T651, T6510, T6511(e)	4043	All	Over 9.5	165	24	105	15	105	15	105	15	62	9	345	50	205	30
6063-T5, T6	4043	All	All	115	17	76	11	76	11	76	11	45	6.5	235	34	150	22
6351-T5(d)	(d)	Extrusions	Over 9.5	165	24	140	20	140	20	105	15	83	12	345	50	205	30
T5(e)	(e)	Extrusions	Over 9.5	165	24	105	15	105	15	105	15	62	9	345	50	205	30

(a) Gas tungsten-arc or gas metal-arc welding with no postweld heat treatment. (b) Ultimate tensile values are ASME weld-qualification-test values. (c) 0.2% offset in 250-mm (10-in.) gage length across a butt weld. (d) Values are for welding with 5183, 5356 or 5556 filler wire, regardless of thickness. Values also apply to thicknesses less than 9.5 mm (0.375 in.) when welding is done with 4043, 5554 or 5654 filler wire. (e) Values are for welding with 4043, 5554 or 5654 filler wire

Copper and Copper Alloys

Wrought Copper Alloys

Standard color controlled wrought copper alloys

UNS number	Common name	Color description
C11000	Electrolytic tough pitch copper	Soft pink
C21000	Gilding, 95%	Red brown
C22000	Commercial bronze, 90%	Bronze gold
C23000	Red brass, 85%	Tan gold
C26000	Cartridge brass, 70%	Green gold
C28000	Muntz metal, 60%	Light brown gold
C61200	Aluminum bronze	Brown gold
C65500	High-silicon bronze, A	Lavender-brown
C70600	Copper nickel, 10%	Soft lavender
C74500	Nickel silver, 65-10	Gray white
C75200	Nickel silver, 65-18	Silver

Classification of copper and copper alloys

Family	Principal alloying element	Solid solubility, at.%(a)	UNS numbers(b)
Coppers, high copper alloys	(c)	...	C10000
Brasses	Zn	37	C20000, C30000, C40000, C66400 to C69800
Phosphor bronzes	Sn	9	C50000
Aluminum bronzes	Al	19	C60600 to C64200
Silicon bronzes	Si	8	C64700 to C66100
Copper nickels, nickel silvers	Ni	100	C70000

(a) At 20 °C (68 °F). (b) Wrought alloys. (c) Various elements having less than 8 at.% solid solubility at 20 °C (68 °F)

ASTM B 601 temper designation codes for copper and copper alloys

Temper designation	Temper name or material condition	Temper designation	Temper name or material condition
Cold-worked tempers(a)		**As-manufactured tempers (continued)**	
H00	1/8 hard	M04	As-pressure die cast
H01	1/4 hard	M05	As-permanent mold cast
H02	1/2 hard	M06	As-investment cast
H03	3/4 hard	M07	As-continuous cast
H04	Hard	M10	As-hot forged and air cooled
H06	Extra hard	M11	As-forged and quenched
H08	Spring	M20	As-hot rolled
H10	Extra spring	M30	As-hot extruded
H12	Special spring	M40	As-hot pierced
H13	Ultra spring	M45	As-hot pierced and rerolled
H14	Super spring	**Annealed tempers(d)**	
Cold-worked tempers(b)		O10	Cast and annealed (homogenized)
H50	Extruded and drawn	O11	As-cast and precipitation heat treated
H52	Pierced and drawn	O20	Hot forged and annealed
H55	Light drawn; light cold rolled	O25	Hot rolled and annealed
H58	Drawn general purpose	O30	Hot extruded and annealed
H60	Cold heading; forming	O31	Extruded and precipitation heat treated
H63	Rivet	O40	Hot pierced and annealed
H64	Screw	O50	Light annealed
H66	Bolt	O60	Soft annealed
H70	Bending	O61	Annealed
H80	Hard drawn	O65	Drawing annealed
H85	Medium-hard-drawn electrical wire	O68	Deep-drawing annealed
H86	Hard-drawn electrical wire	O70	Dead-soft annealed
H90	As-finned	O80	Annealed to temper, 1/8 hard
Cold-worked and stress-relieved tempers		O81	Annealed to temper, 1/4 hard
HR01	H01 and stress relieved	O82	Annealed to temper, 1/2 hard
HR02	H02 and stress relieved	**Annealed tempers(e)**	
HR04	H04 and stress relieved	OS005	Average grain size, 0.005 mm
HR08	H08 and stress relieved	OS010	Average grain size, 0.010 mm
HR10	H10 and stress relieved	OS015	Average grain size, 0.015 mm
HR20	As-finned	OS025	Average grain size, 0.025 mm
HR50	Drawn and stress relieved	OS035	Average grain size, 0.035 mm
Cold-rolled and order-strengthened tempers(c)		OS050	Average grain size, 0.050 mm
HT04	H04 and order heat treated	OS060	Average grain size, 0.060 mm
HT08	H08 and order heat treated	OS070	Average grain size, 0.070 mm
As-manufactured tempers		OS100	Average grain size, 0.100 mm
M01	As-sand cast	OS120	Average grain size, 0.120 mm
M02	As-centrifugal cast	OS150	Average grain size, 0.150 mm
M03	As-plaster cast	OS200	Average grain size, 0.200 mm

(a) Cold-worked tempers to meet standard requirements based on cold rolling or cold drawing. (b) Cold-worked tempers to meet standard requirements based on temper names applicable to specific products. (c) Tempers produced by controlled amounts of cold work followed by a thermal treatment to produce order strengthening. (d) Annealed to meet specific mechanical property requirements. (e) Annealed to meet prescribed nominal average grain size. (f) Tempers of fully finished tubing that has been drawn or annealed to produce specified mechanical properties or that has been annealed to produce a prescribed nominal average grain size are commonly identified by the appropriate H, O, or OS temper designation

(continued)

ASTM B 601 temper designation codes for copper and copper alloys (continued)

Temper designation	Temper name or material condition	Temper designation	Temper name or material condition
Solution-treated temper		**Precipitation-hardened, cold-worked, and thermal-stress-relieved tempers**	
TB00	Solution heat treated	TR01	TL01 and stress relieved
Solution-treated and cold-worked tempers		TR02	TL02 and stress relieved
TD00	TB00 cold worked to $\frac{1}{8}$ hard	TR04	TL04 and stress relieved
TD01	TB00 cold worked to $\frac{1}{4}$ hard	**Solution-treated and spinodal-heat-treated temper**	
TD02	TB00 cold worked to $\frac{1}{2}$ hard	TX00	Spinodal hardened
TD03	TB00 cold worked to $\frac{3}{4}$ hard	**Tempers of welded tubing(f)**	
TD04	TB00 cold worked to full hard	WH00	Welded and drawn to $\frac{1}{8}$ hard
Solution-treated and precipitation-hardened temper		WH01	Welded and drawn to $\frac{1}{4}$ hard
TF00	TB00 and precipitation hardened	WH02	Welded and drawn to $\frac{1}{2}$ hard
Cold-worked and precipitation-hardened tempers		WH03	Welded and drawn to $\frac{3}{4}$ hard
TH01	TD01 and precipitation hardened	WH04	Welded and drawn to full hard
TH02	TD02 and precipitation hardened	WH06	Welded and drawn to extra hard
TH03	TD03 and precipitation hardened	WM00	As welded from H00 ($\frac{1}{8}$-hard) strip
TH04	TD04 and precipitation hardened	WM01	As welded from H01 ($\frac{1}{4}$-hard) strip
Precipitation-hardened and cold-worked tempers		WM02	As welded from H02 ($\frac{1}{2}$-hard) strip
TL00	TF00 cold worked to $\frac{1}{8}$ hard	WM03	As welded from H03 ($\frac{3}{4}$-hard) strip
TL01	TF00 cold worked to $\frac{1}{4}$ hard	WM04	As welded from H04 (full-hard) strip
TL02	TF00 cold worked to $\frac{1}{2}$ hard	WM06	As welded from H06 (extra-hard) strip
TL04	TF00 cold worked to full hard	WM08	As welded from H08 (spring) strip
TL08	TF00 cold worked to spring	WM10	As welded from H10 (extra-spring) strip
TL10	TF00 cold worked to extra spring	WM15	WM50 and stress relieved
Mill-hardened tempers		WM20	WM00 and stress relieved
TM00	AM	WM21	WM01 and stress relieved
TM01	$\frac{1}{4}$ HM	WM22	WM02 and stress relieved
TM02	$\frac{1}{2}$ HM	WM50	As welded from annealed strip
TM04	HM	WO50	Welded and light annealed
TM06	XHM	WR00	WM00; drawn and stress relieved
TM08	XHMS	WR01	WM01; drawn and stress relieved
Quench-hardened tempers		WR02	WM02; drawn and stress relieved
TQ00	Quench hardened	WR03	WM03; drawn and stress relieved
TQ50	Quench hardened and temper annealed	WR04	WM04; drawn and stress relieved
TQ55	Quench hardened and temper annealed, cold drawn and stress relieved	WR06	WM06; drawn and stress relieved
TQ75	Interrupted quench hardened		

(a) Cold-worked tempers to meet standard requirements based on cold rolling or cold drawing. (b) Cold-worked tempers to meet standard requirements based on temper names applicable to specific products. (c) Tempers produced by controlled amounts of cold work followed by a thermal treatment to produce order strengthening. (d) Annealed to meet specific mechanical property requirements. (e) Annealed to meet prescribed nominal average grain size. (f) Tempers of fully finished tubing that has been drawn or annealed to produce specified mechanical properties or that has been annealed to produce a prescribed nominal average grain size are commonly identified by the appropriate H, O, or OS temper designation

Extraction of copper from a low-grade ore

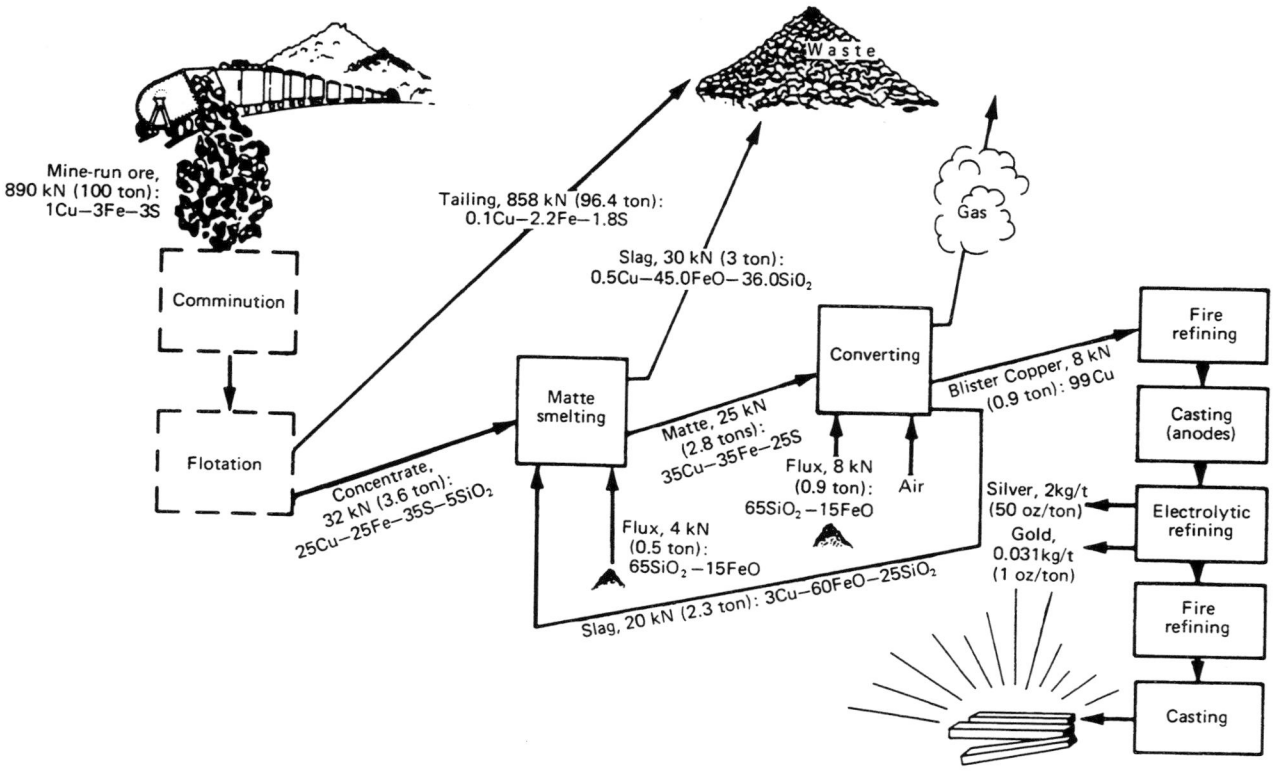

Mine-run ore, 890 kN (100 ton): 1Cu−3Fe−3S

Commination

Flotation

Tailing, 858 kN (96.4 ton): 0.1Cu−2.2Fe−1.8S

Slag, 30 kN (3 ton): 0.5Cu−45.0FeO−36.0SiO₂

Matte smelting

Converting

Concentrate, 32 kN (3.6 ton): 25Cu−25Fe−35S−5SiO₂

Matte, 25 kN (2.8 tons): 35Cu−35Fe−25S

Flux, 4 kN (0.5 ton): 65SiO₂−15FeO

Flux, 8 kN (0.9 ton): 65SiO₂−15FeO

Air

Gas

Slag, 20 kN (2.3 ton): 3Cu−60FeO−25SiO₂

Blister Copper, 8 kN (0.9 ton): 99 Cu

Fire refining

Casting (anodes)

Silver, 2kg/t (50 oz/ton)

Gold, 0.031kg/t (1 oz/ton)

Electrolytic refining

Fire refining

Casting

Wire bar copper, 8 kN (0.9 ton): 99.95 Cu

Properties and applications of wrought coppers and copper alloys

UNS number and name	Nominal composition, %	Commercial forms(a)	Mechanical properties(b)				Elongation in 2 in., %	Corrosion resistance (c)	Machinability rating (d)	Fabricating characteristics and typical applications
			Tensile strength		Yield strength					
			MPa	ksi	MPa	ksi				
C10100 Oxygen-free electronic	99.99 Cu	F, R, W, T, P, S	221-455	32-66	69-365	10-53	55(4)	G-E	20	Excellent hot and cold workability; good forgeability. Fabricated by coining, coppersmithing, drawing and upsetting, hot forging and pressing, spinning, swaging, stamping. **Uses:** busbars, bus conductors, waveguides, hollow conductors, lead-in wires and anodes for vacuum tubes, vacuum seals, transistor components, glass to metal seals, coaxial cables and tubes, klystrons, microwave tubes, rectifiers
C10200 Oxygen-free copper	99.95 Cu	F, R, W, T, P, S	221-455	32-66	69-365	10-53	55(4)	G-E	20	Fabricating characteristics same as C10100. **Uses:** busbars, waveguides
C10300 Oxygen-free, extra-low phosphorus	99.95 Cu, 0.003 P	F, R, T, P, S	221-379	32-55	69-345	10-50	50(6)	G-E	20	Fabricating characteristics same as C10100. **Uses:** busbars, electrical conductors, tubular bus and applications requiring good conductivity and welding or brazing properties
C10400, C10500, C10700 Oxygen-free, silver-bearing	99.95 Cu(e)	F, R, W, S	221-455	32-66	69-365	10-53	55(4)	G-E	20	Fabricating characteristics same as C10100. **Uses:** auto gaskets, radiators, busbars, conductivity wire, contacts, radio parts, winding, switches, terminals, commutator segments; chemical process equipment, printing rolls, clad metals, printed circuit foil
C10800 Oxygen-free, low phosphorus	99.95 Cu, 0.009 P	F, R, T, P	221-379	32-55	69-345	10-50	50(4)	G-E	20	Fabricating characteristics same as C10100. **Uses:** refrigerators, air conditioners, gas and heater lines, oil burner tubes, plumbing pipe and tube, brewery tubes, condenser and heat exchanger tubes, dairy and distiller tubes, pulp and paper lines, tanks, air, gasoline, hydraulic and oil lines
CS11000 Electrolytic tough pitch copper	99.90 Cu, 0.04 O	F, R, W, T, P, S	221-455	32-66	69-365	10-53	55(4)	G-E	20	Fabricating characteristics same as C10100. **Uses:** downspouts, gutters, roofing, gaskets, auto radiators, busbars, nails, printing rolls, rivets, radio parts
C11100 Electrolytic tough pitch, anneal resistant	99.90 Cu, 0.04 O, 0.01 Cd	W	455	66	...	2.5	(60)	G-E	20	Fabricating characteristics same as C10100. **Uses:** electrical power transmission where resistance to softening under overloads is desired
C11300, C11400, C11500, C11600 Silver-bearing tough pitch copper	99.90 Cu, 0.04 O, Ag(f)	F, R, W, T, S	221-455	32-66	69-365	10-53	55(4)	G-E	20	Fabricating characteristics same as C10100. **Uses:** gaskets, radiators, busbars, windings, switches, chemical process equipment, clad metals, printed circuit foil
C12000, C12100	99.9 Cu(g)	F, T, P	221-393	32-57	69-365	10-53	55(4)	G-E	20	Fabricating characteristics same as C10100. **Uses:** busbars, electrical conductors, tubular bus, and applications requiring welding or brazing

(a) F, flat products; R, rod; W, wire; T, tube; P, pipe; S, shapes. (b) Softest to hardest commercial forms. The strength of the standard copper alloys depends on the temper (annealed grain size or degree of cold work) and the section thickness of the mill product. Ranges cover standard tempers for each alloy. Elongation in 2-in. gage length, unless another is specified (in inches in parentheses). (c) E, excellent; G, good; F, fair. (d) Based on 100% for C360000. (e) C10400, 8 oz/ton Ag; C10500, 10 oz/ton; C10700, 25 oz/ton. (f) C11300, 8 oz/ton Ag; C11400, 10 oz/ton; C11500, 16 oz/ton; C11600, 25 oz/ton. (g) C12000, 0.008 P; C12100, 0.008 P and 4 oz/ton Ag. (h) C12700, 8 oz/ton Ag; C12800, 10 oz/ton; C12900, 16 oz/ton; C13000, 25 oz/ton. (i) 8.30 oz/ton Ag. (j) C18200, 0.9 Cr; C18400, 0.8 Cr; C18500, 0.7 Cr. (k) Rod, 61.0 Cu min. Source: Copper Development Assn. Inc., New York

(continued)

Properties and applications of wrought coppers and copper alloys (continued)

UNS number and name	Nominal composition, %	Commercial forms(a)	Mechanical properties(b) Tensile strength		Yield strength		Elongation in 2 in., %	Corrosion resistance (c)	Machinability rating (d)	Fabricating characteristics and typical applications
			MPa	ksi	MPa	ksi				
C12200 Phosphorus deoxidized copper, high residual phosphorus	99.90 Cu, 0.02 P	F, R, T, P	221-379	32-55	69-345	10-50	45(8)	G-E	20	Fabricating characteristics same as C10100. **Uses:** gas and heater lines; oil burner tubing; plumbing pipe and tubing; condenser, evaporator, heat exchanger, dairy, and distiller tubing; steam and water lines; air, gasoline, and hydraulic lines
C12500, C12700, C12800, C12900, C13000 Fire-refined tough pitch with silver	99.88 Cu(h)	F, R, W, S	221-462	32-67	69-365	10-53	55(4)	G-E	20	Fabricating characteristics same as C10100. Uses: same as C11000
C14200 Phosphorus deoxidized, arsenical	99.68 Cu, 0.3 As, 0.02 P	F, R, T	221-379	32-55	69-345	10-50	45(8)	G-E	20	Fabricating characteristics same as C10100. **Uses:** plates for locomotive fireboxes, staybolts, heat exchanger and condenser tubes
C19200	98.97 Cu, 1.0 Fe, 0.03 P	F, T	255-531	37-77	76-510	11-74	40	G-E	20	Excellent hot and cold workability. **Uses:** automotive hydraulic brake lines, flexible hose, electrical terminals, fuse clips, gaskets, gift hollow ware, applications requiring resistance to softening and stress corrosion, air conditioning and heat exchanger tubing
C14300	99.9 Cu, 0.1 Cd	F	221-400	32-58	76-386	11-56	42(1)	G-E	20	Fabricating characteristics same as C10100. **Uses:** anneal resistant electrical applications requiring thermal softening and embrittlement resistance, lead frames, contacts, terminals, solder-coated and solder-fabricated parts, furnace-brazed assemblies and welded components, cable wrap
C14310	99.8 Cu, 0.2 Cd	F	221-400	32-58	76-386	11-56	42(1)	G-E	20	Same as C14300
C14500 Phosphorus deoxidized, tellurium bearing	99.5 Cu, 0.50 Te, 0.008 P	F, R, W, T	221-386	32-56	69-352	10-51	50(3)	G-E	85	Fabricating characteristics same as C10100. **Uses:** forgings and screw machine products, and parts requiring high conductivity, extensive machining, corrosion resistance, copper color, or a combination of these; electrical connectors, motor and switch parts, plumbing fittings, soldering coppers, welding torch tips, transistor bases and furnace-brazed articles
C14700 Sulfur bearing	99.6 Cu, 0.40 S	R, W	221-393	32-57	69-379	10-55	52(8)	G-E	85	Fabricating characteristics same as C1010. **Uses:** screw machine products and parts requiring high conductivity, extensive machining, corrosion resistance, copper color, or a combination of these; electrical connectors, motor and switch components, plumbing fittings, cold headed and machined parts, cold forgings, furnace brazed articles, screws, soldering coppers, rivets and welding torch tips

(a) F, flat products; R, rod; W, wire; T, tube; P, pipe; S, shapes. (b) Softest to hardest commercial forms. The strength of the standard copper alloys depends on the temper (annealed grain size or degree of cold work) and the section thickness of the mill product. Ranges cover standard tempers for each alloy. Elongation in 2-in. gage length, unless another is specified (in inches in parentheses). (c) E, excellent; G, good; F, fair. (d) Based on 100% for C360000. (e) C10400, 8 oz/ton Ag; C10500, 10 oz/ton; C10700, 25 oz/ton. (f) C11300, 8 oz/ton Ag; C11400, 10 oz/ton; C11500, 16 oz/ton; C11600, 25 oz/ton. (g) C12000, 0.008 P; C12100, 0.008 P and 4 oz/ton Ag. (h) C12700, 8 oz/ton Ag; C12800, 10 oz/ton; C12900, 16 oz/ton; C13000, 25 oz/ton. (i) 8.30 oz/ton Ag. (j) C18200, 0.9 Cr; C18400, 0.8 Cr; C18500, 0.7 Cr. (k) Rod, 61.0 Cu min. Source: Copper Development Assn. Inc., New York

(continued)

Properties and applications of wrought coppers and copper alloys (continued)

UNS number and name	Nominal composition, %	Commercial forms(a)	Mechanical properties(b) Tensile strength		Yield strength		Elongation in 2 in., %	Corrosion resistance (c)	Machinability rating (d)	Fabricating characteristics and typical applications
			MPa	ksi	MPa	ksi				
C15000 Zirconium copper	99.8 Cu, 0.15 Zr	R, W	200-524	29-76	41-496	6-72	54(1.5)	G-E	20	Fabricating characteristics same as C10100. **Uses:** switches, high-temperature circuit breakers; commutators, stud bases for power transmitters, rectifiers, soldering welding tips
C15500	99.75 Cu, 0.06 P, 0.11 Mg, Ag(i)	F	276-552	40-80	124-496	18-72	40(3)	G-E	20	Fabricating characteristics same as C10100. **Uses:** high-conductivity light-duty springs, electrical contacts, fittings, clamps, connectors, diaphragms, electronic components, resistance welding electrodes
C15710	99.8 Cu, 0.2 Al_2O_3	R, W	324-724	47-105	268-689	39-100	20(10)	...	...	Excellent cold workability. Fabricated by extrusion, drawing, rolling, impacting, heading, swaging, bending, machining, blanking, roll threading. **Uses:** electrical connectors, light-duty current-carrying springs, inorganic insulated wire, thermocouple wire, lead wire, resistance welding electrodes for aluminum, heat sinks
C15720	99.6 Cu, 0.4 Al_2O_3	F, R	462-614	67-89	365-586	53-85	20(3.5)	...	...	Excellent cold workability. Fabricated by extrusion, drawing, rolling, impacting, heading, swaging, machining, blanking. **Uses:** relay and switch springs, lead frames, contact supports, heat sinks, circuit breaker parts, rotor bars, resistance welding electrodes and wheels, connectors, high-strength high-temperature parts
C15735	99.3 Cu, 0.7 Al_2O_3	R	483-586	70-85	414-565	60-82	16(10)	...	...	Excellent cold workability. Fabricated by extrusion, drawing, heading, impacting, machining. **Uses:** resistance welding electrodes, circuit breakers, feed-through conductors, heat sinks, motor parts, high-strength high-temperature parts
C15760	98.9 Cu, 1.1 Al_2O_3	F, R	483-648	70-94	386-552	56-80	20(8)	...	...	Excellent cold workability. Fabricated by extrusion and drawing. **Uses:** resistance welding electrodes, circuit breakers, electrical connectors, wire feed contact tips, plasma spray nozzles, high-strength high-temperature parts
C16200 Cadmium copper	99.0 Cu, 1.0 Cd	F, R, W	241-689	35-100	48-476	7-69	57(1)	G-E	20	Excellent cold workability; good hot formability. **Uses:** trolley wire, heating pad, electric-blanket elements, spring contacts, railbands, high-strength transmission lines, connectors, cable wrap, switch gear components and wave-guide cavities

(a) F, flat products; R, rod; W, wire; T, tube; P, pipe; S, shapes. (b) Softest to hardest commercial forms. The strength of the standard copper alloys depends on the temper (annealed grain size or degree of cold work) and the section thickness of the mill product. Ranges cover standard tempers for each alloy. Elongation in 2-in. gage length, unless another is specified (in inches in parentheses). (c) E, excellent; G, good; F, fair. (d) Based on 100% for C360000. (e) C10400, 8 oz/ton Ag; C10500, 10 oz/ton; C10700, 25 oz/ton. (f) C11300, 8 oz/ton Ag; C11400, 10 oz/ton; C11500, 16 oz/ton; C11600, 25 oz/ton. (g) C12000, 0.008 P; C12100, 0.008 P and 4 oz/ton Ag. (h) C12700, 8 oz/ton Ag; C12800, 10 oz/ton; C12900, 16 oz/ton; C13000, 25 oz/ton. (i) 8.30 oz/ton Ag. (j) C18200, 0.9 Cr; C18400, 0.8 Cr; C18500, 0.7 Cr. (k) Rod, 61.0 Cu min. Source: Copper Development Assn. Inc., New York

(continued)

Properties and applications of wrought coppers and copper alloys (continued)

UNS number and name	Nominal composition, %	Commercial forms(a)	Mechanical properties(b) Tensile strength MPa	ksi	Yield strength MPa	ksi	Elongation in 2 in., %	Corrosion resistance (c)	Machinability rating (d)	Fabricating characteristics and typical applications
C16500	98.6 Cu, 0.8 Cd, 0.6 Sn	F, R, W	276-655	40-95	97-490	14-71	53(1.5)	G-E	20	Fabricating characteristics same as C16200. **Uses:** electrical springs and contacts, trolley wire, clips, flat cable, resistance welding electrodes
C17000 Beryllium copper	99.5 Cu, 1.7 Be, 0.20 Co	F, R	483-1310	70-190	221-1172	32-170	45(3)	G-E	20	Fabricating characteristics same as C16200. Commonly fabricated by blanking, forming and bending, turning, drilling, tapping. **Uses:** bellows, bourdon tubing, diaphragms, fuse clips, fasteners, lock-washers, springs, switch parts, roll pins, valves, welding equipment
C17200 Beryllium copper	99.5 Cu, 1.9 Be, 0.20 Co	F, R, W, T, P, S	469-1462	68-212	172-1344	25-195	48(1)	G-E	20	Similar to C17000, particularly for its nonsparking characteristics
C17300 Beryllium copper	99.5 Cu, 1.9 Be, 0.40 Pb	R	469-1479	68-200	172-1255	25-182	48(3)	G-E	50	Combines superior machinability with good fabricating characteristics of C17200
C17500 Copper-cobalt-beryllium alloy	99.5 Cu, 2.5 Co, 0.6 Be	F, R	310-793	45-115	172-758	25-110	28(5)	G-E	...	Fabricating characteristics same as C16200. **Uses:** fuse clips, fasteners, springs, switch and relay parts, electrical conductors, welding equipment
C18200, C18400, C18500 Chromium copper	99.5 Cu(j)	F, W, R, S, T	234-593	34-86	97-531	14-77	40(5)	G-E	20	Excellent cold workability; good hot workability. **Uses:** resistance welding electrodes, seam welding wheels, switch gear, electrode holder jaws, cable connectors, current carrying arms and shafts, circuit breaker parts, molds, spot welding tips, flash welding electrodes, electrical and thermal conductors requiring strength, switch contacts
C18700 Leaded copper	99.0 Cu, 1.0 Pb	R	221-379	32-55	69-345	10-50	45(8)	G-E	85	Good cold workability; poor hot formability. **Uses:** connectors, motor and switch parts, screw machine parts requiring high conductivity
C18900	98.75 Cu, 0.75 Sn, 0.3 Si, 0.20 Mn	R, W	262-655	38-95	62-359	9-52	48(14)	G-E	20	Fabricating characteristics same as C10100. **Uses:** welding rod and wire for inert gas tungsten arc and metal arc welding and oxyacetylene welding of copper
C19000 Copper-nickel-phosphorus alloy	98.7 Cu, 1.1 Ni, 0.25 P	F, R, W	262-793	38-115	138-552	20-80	50	G-E	30	Fabricating characteristics same as C10100. **Uses:** springs, clips, electrical connectors, power tube and electron tube components, high-strength electrical conductors, bolts, nails, screws, cotter pins, and parts requiring some combination of high-strength, high-electrical or thermal conductivity, high resistance to fatigue and creep, and good workability

(a) F, flat products; R, rod; W, wire; T, tube; P, pipe; S, shapes. (b) Softest to hardest commercial forms. The strength of the standard copper alloys depends on the temper (annealed grain size or degree of cold work) and the section thickness of the mill product. Ranges cover standard tempers for each alloy. Elongation in 2-in. gage length, unless another is specified (in inches in parentheses). (c) E, excellent; G, good; F, fair. (d) Based on 100% for C360000. (e) C10400, 8 oz/ton Ag; C10500, 10 oz/ton; C10700, 25 oz/ton. (f) C11300, 8 oz/ton Ag; C11400, 10 oz/ton; C11500, 16 oz/ton; C11600, 25 oz/ton. (g) C12000, 0.008 P; C12100, 0.008 P and 4 oz/ton Ag. (h) C12700, 8 oz/ton Ag; C12800, 10 oz/ton; C12900, 16 oz/ton; C13000, 25 oz/ton. (i) 8.30 oz/ton Ag. (j) C18200, 0.9 Cr; C18400, 0.8 Cr; C18500, 0.7 Cr. (k) Rod, 61.0 Cu min. Source: Copper Development Assn. Inc., New York

(continued)

Properties and applications of wrought coppers and copper alloys (continued)

UNS number and name	Nominal composition, %	Commercial forms(a)	Mechanical properties(b)				Elongation in 2 in., %	Corrosion resistance (c)	Machinability rating (d)	Fabricating characteristics and typical applications
			Tensile strength		Yield strength					
			MPa	ksi	MPa	ksi				
C19100 Copper-nickel-phosphorus-tellurium alloy	98.15 Cu, 1.1 Ni, 0.50 Te, 0.25 P	R, F	248-717	36-104	69-634	10-92	27(6)	G-E	75	Good hot and cold workability. **Uses:** forgings and screw machine parts requiring high strength, hardenability, extensive machining, corrosion resistance, copper color, good conductivity, or a combination of these; bolts, bushings, electrical connectors, gears, marine hardware, nuts, pinions, tie rods, turnbuckle barrels, welding torch tips
C19400	97.5 Cu, 2.4 Fe, 0.13 Zn, 0.03 P	F	310-524	45-76	165-503	24-73	32	G-E	20	Excellent hot and cold workability. **Uses:** circuit breaker components, contact springs, electrical clamps, electrical springs, electrical terminals, flexible hose, fuse clips, gaskets, gift hollow ware, plug contacts, rivets, and welded condenser tubes
C19500	97.0 Cu, 1.5 Fe, 0.6 Sn, 0.10 P, 0.80 Co	F	552-669	80-97	448-655	65-95	15	G-E	20	Excellent hot and cold workability. **Uses:** electrical springs, sockets, terminals, connectors, clips and other current carrying parts having strength
C21000 Gilding, 95%	95.0 Cu, 5.0 Zn	F, W	234-441	34-64	69-400	10-58	45(4)	G-E	20	Excellent cold workability, good hot workability for blanking, coining, drawing, piercing and punching, shearing, spinning, squeezing and swaging, stamping. **Uses:** coins, medals, bullet jackets, fuse caps, primers, plaques, jewelry base for gold plate
C22000 Commercial bronze, 90%	90.0 Cu, 10.0 Zn	F, R, W, T	255-496	37-72	69-427	10-62	50(3)	G-E	20	Fabricating characteristics same as C21000, plus heading and upsetting, roll threading and knurling, hot forging and pressing. **Uses:** etching bronze, grillwork, screen cloth, weatherstripping, lipstick cases, compacts, marine hardware, screws, rivets
C22600 Jewelry bronze, 87.5%	87.5 Cu, 12.5 Zn	F, W	269-669	39-97	76-427	11-62	46(3)	G-E	30	Fabricating characteristics same as C21000, plus heading and upsetting, roll threading and knurling. **Uses:** angles, channels, chain, fasteners, costume jewelry, lipstick cases, compacts, base for gold plate
C23000 Red brass, 85%	85.0 Cu, 15.0 Zn	F, W, T, P	269-724	39-105	69-434	10-63	55(3)	G-E	30	Excellent cold workability; good hot formability. **Uses:** weatherstripping, conduit, sockets, fasteners, fire extinguishers, condenser and heat exchanger tubing, plumbing pipe, radiator cores
C24000 Low brass, 80%	80.0 Cu, 20.0 Zn	F, W	290-862	42-125	83-448	12-65	55(3)	F-E	30	Excellent cold workability. Fabricating characteristics same as C23000. **Uses:** battery caps, bellows, musical instruments, clock dials, pump lines, flexible hose

(a) F, flat products; R, rod; W, wire; T, tube; P, pipe; S, shapes. (b) Softest to hardest commercial forms. The strength of the standard copper alloys depends on the temper (annealed grain size or degree of cold work) and the section thickness of the mill product. Ranges cover standard tempers for each alloy. Elongation in 2-in. gage length, unless another is specified (in inches in parentheses). (c) E, excellent; G, good; F, fair. (d) Based on 100% for C360000. (e) C10400, 8 oz/ton Ag; C10500, 10 oz/ton; C10700, 25 oz/ton. (f) C11300, 8 oz/ton Ag; C11400, 10 oz/ton; C11500, 16 oz/ton; C11600, 25 oz/ton. (g) C12000, 0.008 P; C12100, 0.008 P and 4 oz/ton Ag. (h) C12700, 8 oz/ton Ag; C12800, 10 oz/ton; C12900, 16 oz/ton; C13000, 25 oz/ton. (i) 8.30 oz/ton Ag. (j) C18200, 0.9 Cr; C18400, 0.8 Cr; C18500, 0.7 Cr. (k) Rod, 61.0 Cu min. Source: Copper Development Assn. Inc., New York

(continued)

Properties and applications of wrought coppers and copper alloys (continued)

UNS number and name	Nominal composi-tion, %	Com-mercial forms(a)	Mechanical properties(b) Tensile strength MPa	ksi	Yield strength MPa	ksi	Elon-gation in 2 in., %	Corro-sion resis-tance (c)	Machin-ability rating (d)	Fabricating characteristics and typical applications
C26000 Cartridge brass, 70%	70.0 Cu, 30.0 Zn	F, R, W, T	303-896	44-130	76-448	11-65	66(3)	F-E	30	Excellent cold workability. Fabricating characteristics same as C23000, except for coining, roll threading, and knurling. Uses: radiator cores and tanks, flashlight shells, lamp fixtures, fasteners, locks, hinges, ammunition components, plumbing accessories, pins, rivets
C26800, C27000 Yellow brass	65.0 Cu, 35.0 Zn	F, R, W	317-883	46-128	97-427	14-62	65(3)	F-E	30	Excellent cold workability. Fabricating characteristics same as C23000. Uses: same as C26000 except not used for ammunition
C28000 Muntz metal	60.0 Cu, 40.0 Zn	F, R, T	372-510	54-74	145-379	21-55	52(10)	F-E	40	Excellent hot formability and forgeability for blanking, forming and bending, hot forging and pressing, hot heading and upsetting, shearing. Uses: architectural, large nuts and bolts, brazing rod, condenser plates, heat exchanger and condenser tubing, hot forgings
C31400 Leaded commercial bronze	89.0 Cu, 1.75 Pb, 9.25 Zn	F, R	255-414	37-60	83-379	12-55	45(10)	G-E	80	Excellent machinability. Uses: screws, machine parts, pickling crates
C31600 Leaded commercial bronze, nickel-bear-ing	89.0 Cu, 1.9 Pb, 1.0 Ni, 8.1 Zn	F, R	255-462	37-67	83-407	12-59	45(12)	G-E	80	Good cold workability; poor hot formability. Uses: electrical connectors, fasteners, hardware, nuts, screws, screw machine parts
C33000 Low-leaded brass tube	66.0 Cu, 0.5 Pb, 33.5 Zn	T	324-517	47-75	103-414	15-60	60(7)	F-E	60	Combines good machinability and excellent cold workability. Fabricated by forming and bending, machining, piercing and punching. Uses: pump and power cylinders and liners, ammunition primers, plumbing accessories
C33200 High-leaded brass tube	66.0 Cu, 1.6 Pb, 32.4 Zn	T	359-517	52-75	138-414	20-60	50(7)	F-E	80	Excellent machinability. Fabricated by piercing, punching, and machining. Uses: general-purpose screw machine parts
C33500 Low-leaded brass	65.0 Cu, 0.5 Pb, 34.5 Zn	F	317-510	46-74	97-414	14-60	65(8)	F-E	60	Similar to C33200. Commonly fabricated by blanking, drawing, machining, piercing and punching, stamping. Uses: butts, hinges, watch backs
C34000 Medium-leaded brass	65.0 Cu, 1.0 Pb, 34.0 Zn	F, R, W, S	324-607	47-88	103-414	15-60	60(7)	F-E	70	Similar to C33200. Fabricated by blanking, heading and upsetting, machining, piercing and punching, roll threading and knurling, stamping. Uses: butts, gears, nuts, rivets, screws, dials, engravings, instrument plates
C34200 High-leaded brass	64.5 Cu, 2.0 Pb, 33.5 Zn	F, R	338-586	49-85	117-427	17-62	52(5)	F-E	90	Combines excellent machinability with moderate cold workability. Uses: clock plates and nuts, clock and watch backs, gears, wheels and channel plate

(a) F, flat products; R, rod; W, wire; T, tube; P, pipe; S, shapes. (b) Softest to hardest commercial forms. The strength of the standard copper alloys depends on the temper (annealed grain size or degree of cold work) and the section thickness of the mill product. Ranges cover standard tempers for each alloy. Elongation in 2-in. gage length, unless another is specified (in inches in parentheses). (c) E, excellent; G, good; F, fair. (d) Based on 100% for C360000. (e) C10400, 8 oz/ton Ag; C10500, 10 oz/ton; C10700, 25 oz/ton. (f) C11300, 8 oz/ton Ag; C11400, 10 oz/ton; C11500, 16 oz/ton; C11600, 25 oz/ton. (g) C12000, 0.008 P; C12100, 0.008 P and 4 oz/ton Ag. (h) C12700, 8 oz/ton Ag; C12800, 10 oz/ton; C12900, 16 oz/ton; C13000, 25 oz/ton. (i) 8.30 oz/ton Ag. (j) C18200, 0.9 Cr; C18400, 0.8 Cr; C18500, 0.7 Cr. (k) Rod, 61.0 Cu min. Source: Copper Development Assn. Inc., New York

(continued)

Properties and applications of wrought coppers and copper alloys (continued)

UNS number and name	Nominal composition, %	Commercial forms(a)	Mechanical properties(b) Tensile strength MPa	ksi	Yield strength MPa	ksi	Elongation in 2 in., %	Corrosion resistance (c)	Machinability rating (d)	Fabricating characteristics and typical applications
C34900	62.2 Cu, 0.35 Pb, 37.45 Zn	R, W	365-469	53-68	110-379	16-55	72(18)	F-E	50	Good cold workability, fair hot workability for bending and forming, heading and upsetting, machining, roll threading and knurling. **Uses:** building hardware, rivets and nuts, plumbing goods, and parts requiring moderate cold working combined with some machining
C35000 Medium-leaded brass	62.5 Cu, 1.1 Pb, 36.4 Zn	F, R	310-655	45-95	90-483	13-70	66(1)	F-E	70	Fair cold workability; poor hot formability. **Uses:** bearing cages, book dies, clock plates, engraving plates, gears, hinges, hose couplings, keys, lock parts, lock tumblers, meter parts, nuts, sink strainers, strike plates, templates, type characters, washers, wear plates
C35300 High-leaded brass	62.0 Cu, 1.8 Pb, 36.2 Zn	F, R	338-586	49-85	117-427	17-62	52(5)	F-E	90	Similar to C34200
C35600 Extra-high-leaded brass	63.0 Cu, 2.5 Pb, 34.5 Zn	F	338-510	49-74	117-414	17-60	50(7)	F-E	100	Excellent machinability. Fabricated by blanking, machining, piercing and punching, stamping. **Uses:** same as C34200 and C35300.
C36000 Free-cutting brass	61.5 Cu, 3.0 Pb, 35.5 Zn	F, R, S	338-469	49-68	124-310	18-45	53(18)	F-E	100	Excellent machinability. Fabricated by machining, roll threading and knurling. **Uses:** gears, pinions, automatic high-speed screw machine parts
C36500 to C36800 Leaded Muntz metal	60.0 Cu(k), 0.6 Pb, 39.4 Zn	F	372 (As hot rolled)	54	138	20	45	F-E	60	Combines good machinability with excellent hot formability. **Uses:** condenser tube plates
C37000 Free-cutting Muntz metal	60.0 Cu, 1.0 Pb, 39.0 Zn	T	372-552	54-80	138-414	20-60	40(6)	F-E	70	Fabricating characteristics similar to C36500 to C36800. **Uses:** automatic screw machine parts
C37700 Forging brass	59.0 Cu, 2.0 Pb, 39.0 Zn	R, S	359 (As extruded)	52	138	20	45	F-E	80	Excellent hot workability. Fabricated by heading and upsetting, hot forging and pressing, hot heading and upsetting, machining. **Uses:** forgings and pressings of all kinds
C38500 Architectural bronze	57.0 Cu, 3.0 Pb, 40.0 Zn	R, S	414 (As extruded)	60	138	20	30	F-E	90	Excellent machinability and hot workability. Fabricated by hot forging and pressing, forming, bending and machining. **Uses:** architectural extrusions, store fronts, thresholds, trim, butts, hinges, lock bodies and forgings
C40500	95 Cu, 1 Sn, 4 Zn	F	269-538	39-78	83-483	12-70	49(3)	G-E	20	Excellent cold workability. Fabricated by blanking, forming and drawing. **Uses:** meter clips, terminals, fuse clips, contact and relay springs, washers
C40800	95 Cu, 2Sn, 3Zn	F	290-545	42-79	90-517	13-75	43(3)	G-E	20	Excellent cold workability. Fabricated by blanking, stamping and shearing. **Uses:** electrical connectors

(a) F, flat products; R, rod; W, wire; T, tube; P, pipe; S, shapes. (b) Softest to hardest commercial forms. The strength of the standard copper alloys depends on the temper (annealed grain size or degree of cold work) and the section thickness of the mill product. Ranges cover standard tempers for each alloy. Elongation in 2-in. gage length, unless another is specified (in inches in parentheses). (c) E, excellent; G, good; F, fair. (d) Based on 100% for C360000. (e) C10400, 8 oz/ton Ag; C10500, 10 oz/ton; C10700, 25 oz/ton. (f) C11300, 8 oz/ton Ag; C11400, 10 oz/ton; C11500, 16 oz/ton; C11600, 25 oz/ton. (g) C12000, 0.008 P; C12100, 0.008 P and 4 oz/ton Ag. (h) C12700, 8 oz/ton Ag; C12800, 10 oz/ton; C12900, 16 oz/ton; C13000, 25 oz/ton. (i) 8.30 oz/ton Ag. (j) C18200, 0.9 Cr; C18400, 0.8 Cr; C18500, 0.7 Cr. (k) Rod, 61.0 Cu min. Source: Copper Development Assn. Inc., New York

(continued)

Properties and applications of wrought coppers and copper alloys (continued)

UNS number and name	Nominal composition, %	Commercial forms(a)	Mechanical properties(b) Tensile strength MPa	ksi	Yield strength MPa	ksi	Elongation in 2 in., %	Corrosion resistance (c)	Machinability rating (d)	Fabricating characteristics and typical applications
C41100	91 Cu, 0.5 Sn, 8.5 Zn	F, W	269-731	39-106	76-496	11-72	13	G-E	20	Excellent cold workability, good hot formability. Fabricated by blanking, forming and drawing. **Uses:** bushings, bearing sleeves, thrust washers, terminals, connectors, flexible metal hose, electrical conductors
C41300	90.0 Cu, 1.0 Sn, 9.0 Zn	F, R, W	283-724	41-105	83-565	12-82	45	G-E	2	Excellent cold workability; good hot formability. **Uses:** plater bar for jewelry products, flat springs for electrical switchgear
C41500	91Cu, 1.8 Sn, 7.2 Zn	F	317-558	46-81	117-517	17-75	44	G-E	30	Excellent cold workability. Fabricated by blanking, drawing, bending, forming, shearing and stamping. **Uses:** spring applications for electrical switches
C42200	87.5 Cu, 1.1Sn, 11.4 Zn	F	296-607	43-88	103-517	15-75	46	G-E	30	Excellent cold workability; good hot formability. Fabricated by blanking, piercing, forming and drawing. **Uses:** sash chains, fuse clips, terminals, spring washers, contact springs, electrical connectors
C42500	88.5 Cu, 2.0 Sn, 9.5 Zn	F	310-634	45-92	124-524	18-76	49	G-E	30	Excellent cold workability. Fabricated by blanking, piercing, forming and drawing. **Uses:** electrical switches, springs, terminals, connectors, fuse clips, pen clips, weather stripping
C43000	87.0Cu, 2.2 Sn, 10.8 Zn	F	317-648	46-94	124-503	18-73	55(3)	G-E	30	Excellent cold workability; good hot formability. Fabricated by blanking, coining, drawing, forming, bending, heading, and upsetting. **Uses:** Same as C42500
C43400	85.0 Cu, 0.7 Sn, 14.3 Zn	F	310-607	45-88	103-517	15-75	49(3)	G-E	30	Excellent cold workability. Fabricated by blanking, drawing, bonding, forming, stamping and shearing. **Uses:** electrical switch parts, blades, relay springs, contacts
C43500	81.0 Cu, 0.9 Sn, 18.1 Zn	F, T	317-552	46-80	110-469	16-68	46(7)	G-E	30	Excellent cold workability for fabrication by forming and bending. **Uses:** bourdon tubing and musical instruments
C44300, C44400, C44500 Inhibited admiralty	71.0 Cu, 28.0 Zn, 1.0 Sn	F, W., T	331-379	48-55	124-152	18-22	65(0)	G-E	30	Excellent cold workability for forming and bending. **Uses:** condenser, evaporator and heat exchanger tubing, condenser tubing plates, distiller tubing, ferrules
C46400 to C46700 Naval brass	60.0 Cu, 39.25 Zn, 0.75 Sn	F, R, T, S	379-607	55-88	172-455	25-66	50(17)	F-E	30	Excellent hot workability and hot forgeability. Fabricated by blanking, drawing, bending, heading and upsetting, hot forging, pressing. **Uses:** aircraft turnbuckle barrels, balls, bolts, marine hardware, nuts, propeller shafts, rivets, valve stems, condenser plates, welding rod

(a) F, flat products; R, rod; W, wire; T, tube; P, pipe; S, shapes. (b) Softest to hardest commercial forms. The strength of the standard copper alloys depends on the temper (annealed grain size or degree of cold work) and the section thickness of the mill product. Ranges cover standard tempers for each alloy. Elongation in 2-in. gage length, unless another is specified (in inches in parentheses). (c) E, excellent; G, good; F, fair. (d) Based on 100% for C360000. (e) C10400, 8 oz/ton Ag; C10500, 10 oz/ton; C10700, 25 oz/ton. (f) C11300, 8 oz/ton Ag; C11400, 10 oz/ton; C11500, 16 oz/ton; C11600, 25 oz/ton. (g) C12000, 0.008 P; C12100, 0.008 P and 4 oz/ton Ag. (h) C12700, 8 oz/ton Ag; C12800, 10 oz/ton; C12900, 16 oz/ton; C13000, 25 oz/ton. (i) 8.30 oz/ton Ag. (j) C18200, 0.9 Cr; C18400, 0.8 Cr; C18500, 0.7 Cr. (k) Rod, 61.0 Cu min. Source: Copper Development Assn. Inc., New York

(continued)

Properties and applications of wrought coppers and copper alloys (continued)

UNS number and name	Nominal composition, %	Commercial forms(a)	Mechanical properties(b)				Elongation in 2 in., %	Corrosion resistance (c)	Machinability rating (d)	Fabricating characteristics and typical applications
			Tensile strength		Yield strength					
			MPa	ksi	MPa	ksi				
C48200 Naval brass, medium-leaded	60.5 Cu, 0.7 Pb, 0.8 Sn, 38.0 Zn	F, R, S	386-517	56-75	172-365	25-53	43(15)	F-E	50	Good hot workability for hot forging, pressing, and machining operations. **Uses:** marine hardware, screw machine products, valve stems
C48500 Leaded naval brass	60.0 Cu, 1.75 Pb, 37.5 Zn, 0.75 Sn	F, R, S	379-531	55-77	172-365	25-53	40(15)	F-E	70	Combines excellent hot forgeability and machinability. Fabricated by hot forging and pressing, machining. **Uses:** marine hardware, screw machine parts, valve stems
C50500 Phosphor bronze, 1.25% E	98.75 Cu, 1.25 Sn, trace P	F, W	276-545	40-79	97-345	14-50	48(4)	G-E	20	Excellent cold workability; good hot formability. Fabricated by blanking, bending, heading and upsetting, shearing and swaging. **Uses:** electrical contacts, flexible hose, pole-line hardware
C51000 Phosphor bronze, 5% A	95.0 Cu, 5.0 Sn, trace P	F, R, W, T	324-965	47-140	131-552	19-80	64	G-E	20	Excellent cold workability. Fabricated by blanking, drawing, bending, heading and upsetting, roll threading and knurling, shearing, stamping. **Uses:** bellows, bourdon tubing, clutch discs, cotter pins, diaphragms, fasteners, lock washers, wire brushes, chemical hardware, textile machinery, welding rod
C51100	95.6 Cu, 4.2 Sn, 0.2 P	F	317-710	46-103	345-552	50-80	48	G-E	20	Excellent cold workability. **Uses:** bridge bearing plates, locator bars, fuse clips, sleeve bushings, springs, switch parts, truss wire, wire brushes, chemical hardware, perforated sheets, textile machinery, welding rod
C52100 Phosphor bronze, 8% C	92.0 Cu, 8.0 Sn, trace P	F, R, W	379-965	55-140	165-552	24-80	70	G-E	20	Good cold workability for blanking, drawing, forming and bending, shearing, stamping. **Uses:** generally for more severe service conditions than C51000
C52400 Phosphor bronze, 10% D	90.0 Cu, 10.0 Sn, trace P	F, R, W	455-147	66-147	193	28 (Annealed)	70(3)	G-E	20	Good cold workability for blanking, forming and bending, shearing. **Uses:** heavy bars and plates for severe compression, bridge and expansion plates and fittings, articles requiring good spring qualities, resiliency, fatigue resistance, good wear and corrosion resistance
C54400 Free-cutting phosphor bronze	88.0 Cu, 4.0 Pb, 4.0 Zn, 4.0 Sn	F, R	303-517	44-75	131-434	19-63	50(15)	G-E	80	Excellent machinability; good cold workability. Fabricated by blanking, drawing, bending, machining, shearing, stamping. **Uses:** bearings, bushings, gears, pinions, shafts, thrust washers, valve parts
C60800 Aluminum bronze, 5%	95.0 Cu, 5.0 Al	T	414	60	186	27	55	G-E	20	Good cold workability; fair hot formability. **Uses:** condenser, evaporator and heat exchanger tubes, distiller tubes, ferrules
C61000	92.0 Cu, 8.0 Al	R, W	483-552	70-80	207-379	30-55	65(25)	G-E	20	Good hot and cold workability. **Uses:** bolts, pump parts, shafts, tie rods, overlay on steel for wearing surfaces

(a) F, flat products; R, rod; W, wire; T, tube; P, pipe; S, shapes. (b) Softest to hardest commercial forms. The strength of the standard copper alloys depends on the temper (annealed grain size or degree of cold work) and the section thickness of the mill product. Ranges cover standard tempers for each alloy. Elongation in 2-in. gage length, unless another is specified (in inches in parentheses). (c) E, excellent; G, good; F, fair. (d) Based on 100% for C360000. (e) C10400, 8 oz/ton Ag; C10500, 10 oz/ton; C10700, 25 oz/ton. (f) C11300, 8 oz/ton Ag; C11400, 10 oz/ton; C11500, 16 oz/ton; C11600, 25 oz/ton. (g) C12000, 0.008 P; C12100, 0.008 P and 4 oz/ton Ag. (h) C12700, 8 oz/ton Ag; C12800, 10 oz/ton; C12900, 16 oz/ton; C13000, 25 oz/ton. (i) 8.30 oz/ton Ag. (j) C18200, 0.9 Cr; C18400, 0.8 Cr; C18500, 0.7 Cr. (k) Rod, 61.0 Cu min. Source: Copper Development Assn. Inc., New York

(continued)

Properties and applications of wrought coppers and copper alloys (continued)

UNS number and name	Nominal composition, %	Commercial forms(a)	Mechanical properties(b) Tensile strength		Yield strength		Elongation in 2 in., %	Corrosion resistance (c)	Machinability rating (d)	Fabricating characteristics and typical applications
			MPa	ksi	MPa	ksi				
C61300	92.65 Cu, 0.35 Sn, 7.0 Al	F, R, T, P, S	483-586	70-85	207-400	30-58	42(35)	G-E	30	Good hot and cold formability. **Uses:** nuts, bolts, stringers and threaded members, corrosion-resistant vessels and tanks, structural components, machine parts, condenser tube and piping systems, marine protective sheathing and fastening, munitions mixing troughs and blending chambers
C61400 Aluminum bronze, D	91.0 Cu, 7.0 Al, 2.0 Fe	F, R, W, T, P, S	524-614	76-89	228-414	33-60	45(32)	G-E	20	Similar to C61300
C61500	90.0 Cu, 8.0 Al, 2.0 Ni	F	483-1000	70-145	152-965	22-140	55(1)	G-E	30	Good hot and cold workability. Fabricating characteristics similar to C52100. **Uses:** hardware, decorative metal trim, interior furnishings and other articles requiring high tarnish resistance
C61800	89.0 Cu, 1.0 Fe, 10.0 Al	R	552-586	80-85	269-293	39-42.5	28(23)	G-E	40	Fabricated by hot forging and hot pressing. **Uses:** bushings, bearings, corrosion-resistant applications, welding rods
C61900	86.5 Cu, 4.0 Fe, 9.5 Al	F	634-1048	92-152	338-1000	49-145	30(1)	G-E	...	Excellent hot formability for fabricating by blanking, forming, bending, shearing, and stamping. **Uses:** springs, contacts, and switch components
C62300	87.0 Cu, 3.0 Fe, 10.0 Al	F, R	517-676	75-98	241-359	35-52	35(22)	G-E	50	Good hot and cold formability. Fabricated by bending, hot forging, hot pressing, forming, and welding. **Uses:** bearings, bushings, valve guides, gears, valve seats, nuts, bolts, pump rods, work gears, and cams
C62400	86.0 Cu, 3.0 Fe, 11.0 Al	F, R	621-724	90-105	276-359	40-52	18(14)	G-E	50	Excellent hot formability for fabrication by hot forging and hot bending. **Uses:** bushings, gears, cams, wear strips, nuts, drift pins, tie rods
C62500	82.7 Cu, 4.3 Fe, 13.0 Al	F, R	689	100	379 (as extruded)		55	1	G-E	20	Excellent hot formability for fabrication by hot forging and machining. **Uses:** guide bushings, wear strips, cams, dies, forming rolls
C63000	82.0 Cu, 3.0 Fe, 10.0 Al, 5.0 Ni	F, R	621-814	90-118	345-517	50-75	20(15)	G-E	30	Good hot formability. Fabricated by hot forming and forging. **Uses:** nuts, bolts, valve seats, plunger tips, marine shafts, valve guides, aircraft parts, pump shafts, structural members
C63200	82.0 Cu, 4.0 Fe, 9.0 Al, 5.0 Ni	F, R	621-724	90-105	310-365	45-53	25(20)	G-E	30	Good hot formability. Fabricated by hot forming and welding. Uses: nuts, bolts, structural pump parts, shafting requiring corrosion resistance
C63600	95.5 Cu, 3.5 Al, 1.0 Si	R, W	414-579	60-84	...	...	64(29)	G-E	40	Excellent cold workability; fair hot formability. Fabricated by cold heading. **Uses:** components for pole-line hardware, cold-headed nuts for wire and cable connectors, bolts and screw products

(a) F, flat products; R, rod; W, wire; T, tube; P, pipe; S, shapes. (b) Softest to hardest commercial forms. The strength of the standard copper alloys depends on the temper (annealed grain size or degree of cold work) and the section thickness of the mill product. Ranges cover standard tempers for each alloy. Elongation in 2-in. gage length, unless another is specified (in inches in parentheses). (c) E, excellent; G, good; F, fair. (d) Based on 100% for C360000. (e) C10400, 8 oz/ton Ag; C10500, 10 oz/ton; C10700, 25 oz/ton. (f) C11300, 8 oz/ton Ag; C11400, 10 oz/ton; C11500, 16 oz/ton; C11600, 25 oz/ton. (g) C12000, 0.008 P; C12100, 0.008 P and 4 oz/ton Ag. (h) C12700, 8 oz/ton Ag; C12800, 10 oz/ton; C12900, 16 oz/ton; C13000, 25 oz/ton. (i) 8.30 oz/ton Ag. (j) C18200, 0.9 Cr; C18400, 0.8 Cr; C18500, 0.7 Cr. (k) Rod, 61.0 Cu min. Source: Copper Development Assn. Inc., New York

(continued)

Properties and applications of wrought coppers and copper alloys (continued)

UNS number and name	Nominal composi- tion, %	Com- mercial forms(a)	Mechanical properties(b)				Elon- gation in 2 in., %	Corro- sion resis- tance (c)	Machin- ability rating (d)	Fabricating characteristics and typical applications
			Tensile strength		Yield strength					
			MPa	ksi	MPa	ksi				
C63800	99.5 Cu, 2.8 Al, 1.8 Si, 0.40 Co	F	565-896	82-130	372-786	54-114	36(4)	G-E	...	Excellent cold workability and hot formability. Uses: springs, switch parts, contacts, relay springs, glass sealing and porcelain enameling
C64200	91.2 Cu, 7.0 Al	F, R	517-703	75-102	241-469	35-68	32(22)	G-E	60	Excellent hot formability. Fabricated by hot forming, forging, machining. Uses: valve stems, gears, marine hardware, pole-line hardware, bolts, nuts, valve bodies and components
C65100 Low-silicon bronze, B	98.5 Cu, 1.5 Si	R, W, T	276-655	40-95	103-476	15-69	55(11)	G-E	30	Excellent hot and cold workability. Fabricated by forming and bending, heading and upsetting, hot forging and pressing, roll threading and knurling, squeezing and swaging. Uses: hydraulic pressure lines, anchor screws, bolts, cable clamps, cap screws, machine screws, marine hardware, nuts, pole-line hardware, rivets, U-bolts, electrical conduits, heat exchanger tubing, welding rod
C65500 High-silicon bronze, A	97.0 Cu, 3.0 Si	F, R, W, T	386-1000	56-145	145-483	21-70	63(3)	G-E	30	Excellent hot and cold workability. Fabricated by blanking, drawing, forming and bending, heading and upsetting, hot forging and pressing, roll threading and knurling, shearing, squeezing and swaging. Uses: similar to C65100 including propeller shafts
C66700 Manganese brass	70.0 Cu, 28.8 Zn, 1.2 Mn	F, W	315-689	45.8-100	83-638	12-92.5	60	G-E	30	Excellent cold formability. Fabricated by blanking, bending, forming, stamping, welding. Uses: brass products resistance welded by spot, seam, and butt welding
C67400	58.5 Cu, 36.5 Zn, 1.2 Al, 2.8 Mn, 1.0 Sn	F, R	483-634	70-92	234-379	34-55	28(20)	F-E	25	Excellent hot formability. Fabricated by hot forging and pressing, machining. Uses: bushings, gears, connecting rods, shafts, wear plates
C67500 Manganese bronze, A	58.5 Cu, 1.4 Fe, 39.0 Zn, 1.0 Sn, 0.1 Mn	R, S	448-579	65-84	207-414	30-60	33(19)	F-E	30	Excellent hot workability. Fabricated by hot forging and pressing, hot heading and upsetting. Uses: clutch discs, pump rods, shafting, balls, valve stems and bodies
C68700 Aluminum brass, arsenical	77.5 Cu, 20.5 Zn, 2.0 Al, 0.1 As	T	414	60	186	27	55	G-E	30	Excellent cold workability for forming and bending. Uses: condenser, evaporator and heat exchanger tubing, condenser tubing plates, distiller tubing, ferrules
C68800	73.5 Cu, 22.7 Zn, 3.4 Al, 0.40 Co	F	565-889	82-129	379-786	55-114	36	G-E	...	Excellent hot and cold formability. Fabricated by blanking, drawing, forming and bending, shearing and stamping. Uses: springs, switches, contacts, relays, drawn parts
C69000	73.3 Cu, 3.4 Al, 0.6 Ni, 22.7 Zn	F	496-896	72-130	345-807	50-117	40	G-E	...	Fabricating characteristics same as C68800. Uses: wiring devices, relays, switches, springs, high-strength shells

(a) F, flat products; R, rod; W, wire; T, tube; P, pipe; S, shapes. (b) Softest to hardest commercial forms. The strength of the standard copper alloys depends on the temper (annealed grain size or degree of cold work) and the section thickness of the mill product. Ranges cover standard tempers for each alloy. Elongation in 2-in. gage length, unless another is specified (in inches in parentheses). (c) E, excellent; G, good; F, fair. (d) Based on 100% for C36000. (e) C10400, 8 oz/ton Ag; C10500, 10 oz/ton; C10700, 25 oz/ton. (f) C11300, 8 oz/ton Ag; C11400, 10 oz/ton; C11500, 16 oz/ton; C11600, 25 oz/ton. (g) C12000, 0.008 P; C12100, 0.008 P and 4 oz/ton Ag. (h) C12700, 8 oz/ton Ag; C12800, 10 oz/ton; C12900, 16 oz/ton; C13000, 25 oz/ton. (i) 8.30 oz/ton Ag. (j) C18200, 0.9 Cr; C18400, 0.8 Cr; C18500, 0.7 Cr. (k) Rod, 61.0 Cu min. Source: Copper Development Assn. Inc., New York

(continued)

Properties and applications of wrought coppers and copper alloys (continued)

UNS number and name	Nominal composition, %	Commercial forms(a)	Mechanical properties(b) Tensile strength		Yield strength		Elongation in 2 in., %	Corrosion resistance (c)	Machinability rating (d)	Fabricating characteristics and typical applications
			MPa	ksi	MPa	ksi				
C69400 Silicon red brass	81.5 Cu, 14.5 Zn, 4.0 Si	R	552-689	80-100	276-393	40-57	25(20)	G-E	30	Excellent hot formability for fabrication by forging, screw machine operations. **Uses:** valve stems where corrosion resistance and high strength are critical
C70400	92.4 Cu, 1.5 Fe, 5.5 Ni, 0.6 Mn	F, T	262-531	38-77	276-524	40-76	46	G-E	20	Excellent cold workability; good hot formability. Fabricated by forming, bending and welding. **Uses:** condensers, evaporators, heat exchangers, ferrules, salt water piping, lithium bromide absorption tubing, shipboard condenser intake systems
C70600 Copper nickel, 10%	88.7 Cu, 1.3 Fe, 10.0 Ni	F, T	303-414	44-60	110-393	16-57	42(10)	E	20	Good hot and cold workability. Fabricated by forming and bending, welding. **Uses:** condensers, condenser plates, distiller tubing, evaporator and heat exchanger tubing, ferrules, salt water piping
C71000 Copper nickel, 20%	79.0 Cu, 21.0 Ni	F, W, T	338-655	49-95	90-586	13-85	40(3)	E	20	Good hot and cold formability. Fabricated by blanking, forming and bending, welding. **Uses:** communication relays, condensers, condenser plates, electrical springs, evaporator and heat exchanger tubes, ferrules, resistors
C71500 Copper nickel, 30%	70.0 Cu, 30.0 Ni	F, R, T	372-517	54-75	138-483	20-70	45(15)	E	20	Similar to C70600
C71700	67.8 Cu, 0.7 Fe, 31.0 Ni, 0.5 Be	F, R, W	483-1379	70-200	207-1241	30-180	40(4)	G-E	20	Good hot and cold formability. **Uses:** high-strength constructional parts for sea water corrosion resistance, hydrophone cases, mooring cable wire, springs, retainer rings, bolts, screws, pins for ocean telephone cable applications
C72500	88.2 Cu, 9.5 Ni, 2.3 Sn	F, R, W, T	379-827	55-120	152-745	22-108	35(1)	E	20	Excellent cold and hot formability. Fabricated by blanking, brazing, coining, drawing, etching, forming and bending, heading and upsetting, roll threading and knurling, shearing, spinning, squeezing, stamping and swaging. **Uses:** relay and switch springs, connectors, brazing alloy, lead frames, control and sensing bellows
C73500	72.0 Cu, 10.0 Zn, 18.0 Ni	F, R, W, T	345-758	50-110	103-579	15-84	37(1)	E	20	Fabricating characteristics same as C74500. **Uses:** hollowware, medallions, jewelry, base for silver plate, cosmetic cases, musical instruments, name plates, contacts

(a) F, flat products; R, rod; W, wire; T, tube; P, pipe; S, shapes. (b) Softest to hardest commercial forms. The strength of the standard copper alloys depends on the temper (annealed grain size or degree of cold work) and the section thickness of the mill product. Ranges cover standard tempers for each alloy. Elongation in 2-in. gage length, unless another is specified (in inches in parentheses). (c) E, excellent; G, good; F, fair. (d) Based on 100% for C360000. (e) C10400, 8 oz/ton Ag; C10500, 10 oz/ton; C10700, 25 oz/ton. (f) C11300, 8 oz/ton Ag; C11400, 10 oz/ton; C11500, 16 oz/ton; C11600, 25 oz/ton. (g) C12000, 0.008 P; C12100, 0.008 P and 4 oz/ton Ag. (h) C12700, 8 oz/ton Ag; C12800, 10 oz/ton; C12900, 16 oz/ton; C13000, 25 oz/ton. (i) 8.30 oz/ton Ag. (j) C18200, 0.9 Cr; C18400, 0.8 Cr; C18500, 0.7 Cr. (k) Rod, 61.0 Cu min. Source: Copper Development Assn. Inc., New York

(continued)

Properties and applications of wrought coppers and copper alloys (continued)

UNS number and name	Nominal composition, %	Commercial forms(a)	Mechanical properties(b)				Elongation in 2 in., %	Corrosion resistance (c)	Machinability rating (d)	Fabricating characteristics and typical applications
			Tensile strength		Yield strength					
			MPa	ksi	MPa	ksi				
C74500 Nickel silver, 65-10	65.0 Cu, 25.0 Zn, 10.0 Ni	F, W	338-896	49-130	124-524	18-76	50(1)	E	20	Excellent cold workability. Fabricated by blanking, drawing, etching, forming and bending, heading and upsetting, roll threading and knurling, shearing, spinning, squeezing and swaging. **Uses:** rivets, screws, slide fasteners, optical parts, etching stock, hollow ware, nameplates, platers' bars
C75200 Nickel silver, 65-18	65.0 Cu, 17.0 Zn, 18.0 Ni	F, R, W	386-710	56-103	172-621	25-90	45(3)	E	20	Fabricating characteristics similar to C74500. **Uses:** rivets, screws, table flatware, truss wire, zippers, bows, camera parts, core bars, temples, base for silver plate, costume jewelry, etching stock, hollow ware, nameplates, radio dials
C75400 Nickel silver, 65-15	65.0 Cu, 20.0 Zn, 15.0 Ni	F	365-634	53-92	124-545	18-79	43	E	20	Fabricating characteristics similar to C74500. **Uses:** camera parts, optical equipment, etching stock, jewelry
C75700 Nickel silver, 65-12	65.0 Cu, 23.0 Zn, 12.0 Ni	F, W	359-641	52-93	124-545	18-79	48	E	20	Fabricating characteristics similar to C74500. **Uses:** slide fasteners, camera parts, optical parts, etching stock, nameplates
C76200	59.0 Cu, 29.0 Zn, 12.0 Ni	F, T	393-841	57-122	145-758	21-110	50(1)	G-E	...	Fabricating characteristics same as C77000. **Uses:** electrical terminals, contact springs, release brackets, ornamental bits and spurs, optical parts, surgical instruments, electrical contacts
C77000 Nickel silver, 55-18	55.0 Cu, 27.0 Zn, 18.0 Ni	F, R, W	414-1000	60-145	186-621	27-90	40	E	30	Good cold workability. Fabricated by blanking, forming and bending, and shearing. **Uses:** optical goods, springs and resistance wire
C72200	82.0 Cu, 16.0 Ni, 0.5 Cr, 0.8 Fe, 0.5 Mn	F, T	317-483	46-70	124-455	18-66	46(6)	G-E	...	Good hot and cold workability. Fabricated by forming, bending and welding. **Uses:** condenser and heat exchanger tubing, salt water piping
C78200 Leaded nickel silver, 65-8-2	65.0 Cu, 2.0 Pb, 25.0 Zn, 8.0 Ni	F	365-627	53-91	159-524	23-76	40(3)	E	60	Good cold formability. Fabricated by blanking, milling and drilling. **Uses:** key blanks, watch plates, watch parts

(a) F, flat products; R, rod; W, wire; T, tube; P, pipe; S, shapes. (b) Softest to hardest commercial forms. The strength of the standard copper alloys depends on the temper (annealed grain size or degree of cold work) and the section thickness of the mill product. Ranges cover standard tempers for each alloy. Elongation in 2-in. gage length, unless another is specified (in inches in parentheses). (c) E, excellent; G, good; F, fair. (d) Based on 100% for C360000. (e) C10400, 8 oz/ton Ag; C10500, 10 oz/ton; C10700, 25 oz/ton. (f) C11300, 8 oz/ton Ag; C11400, 10 oz/ton; C11500, 16 oz/ton; C11600, 25 oz/ton. (g) C12000, 0.008 P; C12100, 0.008 P and 4 oz/ton Ag. (h) C12700, 8 oz/ton Ag; C12800, 10 oz/ton; C12900, 16 oz/ton; C13000, 25 oz/ton. (i) 8.30 oz/ton Ag. (j) C18200, 0.9 Cr; C18400, 0.8 Cr; C18500, 0.7 Cr. (k) Rod, 61.0 Cu min. Source: Copper Development Assn. Inc., New York

Temper designations for wrought copper and brass based on cold reduction

Nominal temper designation	Increase in B&S gage numbers	Rolled sheet			Drawn wire		
		Reduction in thickness and area, %	True strain		Reduction in diameter, %	Reduction in area, %	True strain
1/4 hard	1	10.9	0.116		10.9	20.7	0.232
1/2 hard	2	20.7	0.232		20.7	37.1	0.463
3/4 hard	3	29.4	0.347		29.4	50.1	0.694
Hard	4	37.1	0.463		37.1	60.5	0.926
Extra hard	6	50.1	0.696		50.1	75.1	1.39
Spring	8	60.5	0.928		60.5	84.4	1.86
Extra spring	10	68.6	1.16		68.6	90.2	2.32
Special spring	12	75.1	1.39		75.1	93.8	2.78
Super spring	14	80.3	1.62		80.3	96.1	3.25

Copper tube alloys and typical applications

UNS number	Alloy type	ASTM specifications	Typical uses
C10200	Oxygen-free copper	B68, B75, B88, B111, B188, B280, B359, B372, B395, B447	Bus tube, conductors, wave guides
C12200	Phosphorus deoxidized copper	B68, B75, B88, B111, B280, B306, B359, B360, B395, B447, B543	Water tubes; condenser, evaporator and heat exchanger tubes; air conditioning and refrigeration, gas, heater and oil burner lines; plumbing pipe and steam tubes; brewery and distillery tubes; gasoline, hydraulic and oil lines; rotating bands
C19200	Copper	B111, B359, B395, B469	Automotive hydraulic brake lines; flexible hose
C23000	Red brass, 85%	B111, B135, B359, B395, B543	Condenser and heat exchanger tubes, flexible hose; plumbing pipe; pump lines
C26000	Cartridge brass, 70%	B135	Plumbing brass goods
C33000	Low-leaded brass (tube)	B135	Pump and power cylinders and liners; plumbing brass goods
C36000	Free-cutting brass		Screw machine parts; plumbing goods
C43500	Tin brass		Bourdon tubes; musical instruments
C44300	Inhibited admiralty metal	B111, B359, B395	Condenser, evaporator and heat exchanger tubes; distiller tubes
C44400			
C44500			
C46400	Naval brass		Marine hardware, nuts
C46500			
C46600			
C46700			
C60800	Aluminum bronze, 5%	B111, B359, B395	Condenser, evaporator and heat exchanger tubes; distiller tubes
C65100	Silicon bronze B	B315	Heater exchanger tubes; electrical conduits
C65500	Silicon bronze A	B315	Chemical equipment, heat exchanger tubes; piston rings
C68700	Arsenical aluminum brass	B111, B359, B395	Condenser, evaporator and heat exchanger tubes; distiller tubes
C70600	Copper nickel, 10%	B111, B359, B395, B466, B467, B543, B552	Condenser, evaporator and heat exchanger tubes; salt water piping; distiller tubes
C71500	Copper nickel, 30%	B111, B359, B395, B446, B467, B543, B552	Condenser, evaporator and heat exchanger tubes; distiller tubes; salt water piping

456

Typical mechanical properties for copper alloy tube(a)

Temper	Tensile strength MPa	ksi	Yield strength(b) MPa	ksi	Elongation(c), %
C10200					
OS050	220	32	69	10	45
OS025	235	34	76	11	45
H55	275	40	220	32	25
H80	380	55	345	50	8
C12200					
OS050	220	32	69	10	45
OS025	235	34	76	11	45
H55	275	40	220	32	25
H80	380	55	345	50	8
C19200					
H55(d)	290	42	205(e)	30(e)	35
C23000					
OS050	275	40	83	12	55
OS015	305	44	125	18	45
H55	345	50	275	40	30
H80	485	70	400	58	8
C26000					
OS050	325	47	105	15	65
OS025	360	52	140	20	55
H80	540	78	440	64	8
C33000					
OS050	325	47	105	15	60
OS025	360	52	140	20	50
H80	515	75	415	60	7
C43500					
OS035	315	46	110	16	46
H80	515	75	415	60	10
C44300, C44400, C44500					
OS025	365	53	150	22	65
C46400, C46500, C46600, C46700(f)					
H80	605	88	455	66	18
C60800					
OS025	415	60	185	27	55
C65100					
OS015	310	45	140	20	55
H80	450	65	275	40	20
C65500					
OS050	395	57	…	…	70
H80	640	93	…	…	22
C68700					
OS025	415	60	185	27	55
C70600					
OS025	305	44	110	16	42
H55	415	60	395	57	10
C71500					
OS025	415	60	170	25	45

(a) Tube size: 25 mm (1 in.) OD by 1.65 mm (0.065 in.) wall. (b) 0.5% extension under load. (c) In 50 mm or 2 in. (d) Tube size: 4.8 mm (0.1875 in.) OD by 0.76 mm (0.030 in.) wall. (e) 0.2% offset. (f) Tube size: 9.5 mm (0.375 in.) OD by 2.5 mm (0.097 in.) wall

Characteristics of solid round copper wire: ASTM B1, B3, B258

Conductor size, AWG	Conductor diameter, mils	Conductor area, circular mils	Net weight, lb/1000 ft	Soft (annealed) wire		Hard drawn wire		
				Minimum elongation(a), %	Nominal resistance, Ω/1000 ft	Nominal breaking strength, lb	Nominal tensile strength, ksi	Nominal resistance, Ω/1000 ft
4/0	460.0	211 600	640.5	35	0.0491	8143	49.0	0.05044
3/0	409.6	167 800	507.8	35	0.06180	6720	51.0	0.06361
2/0	364.8	133 100	402.8	35	0.07791	5519	52.8	0.08019
1.0	324.9	105 600	319.5	35	0.09821	4518	54.5	0.1011
1	289.3	83 690	253.3	30	0.1239	3888	56.1	0.1289
2	257.6	66 360	200.9	30	0.1563	3002	57.6	0.1625
3	229.4	52 620	159.3	30	0.1971	2439	59.0	0.2050
4	204.3	41 740	126.3	30	0.2485	1970	60.1	0.2584
5	181.9	33 090	100.2	30	0.3134	1590	61.2	0.3259
6	162.0	26 240	79.44	30	0.3952	1280	62.1	0.4110
7	144.3	20 820	63.03	30	0.4981	1030	63.1	0.5180
8	128.5	16 510	49.98	30	0.6281	826.1	63.7	0.6532
9	114.4	13 090	39.62	30	0.7923	660.9	64.3	0.8239
10	101.9	10 380	31.43	25	0.9991	529.3	64.9	1.039
11	90.7	8 230	24.9	25	1.26	423	65.4	1.31
12	80.8	6 530	19.8	25	1.59	337	65.7	1.65
13	72.0	5 180	15.7	25	2.00	268	65.9	2.08
14	64.1	4 110	12.4	25	2.52	214	66.2	2.62
15	57.1	3 260	9.87	25	3.18	170	66.4	3.31
16	50.8	2 580	7.81	25	4.02	135	66.6	4.18
17	45.3	2 050	6.21	25	5.06	108	66.8	5.26
18	40.3	1 620	4.92	25	6.40	85.5	67.0	6.66
19	35.9	1 290	3.90	25	8.04	68.0	67.2	8.36
20	32.0	1 020	3.10	25	10.2	54.2	67.4	10.6
21	28.5	812	2.46	25	12.8	43.2	67.7	13.3
22	25.3	640	1.94	25	16.2	34.1	67.9	16.9
23	22.6	511	1.55	25	20.3	27.3	68.1	21.1
24	20.1	404	1.22	20	25.7	21.7	68.3	26.7
25	17.9	320	0.970	20	32.4	17.3	68.6	33.7
26	15.9	253	0.765	20	41.0	13.7	68.8	42.6
27	14.2	202	0.610	20	51.4	10.9	69.0	53.4
28	12.6	159	0.481	20	65.2	8.64	69.3	67.8
29	11.3	128	0.387	20	81.0	6.96	69.4	84.3
30	10.0	100	0.303	15	104.0	5.47	69.7	108.0
31	8.9	79.2	0.240	15	131.0	4.35	69.9	136.0
32	8.0	64.0	0.194	15	162.0	3.53	70.2	169.0
33	7.1	50.4	0.153	15	206.0	2.79	70.4	214.0
34	6.3	39.7	0.120	15	261.0	2.20	70.6	272.0
35	5.6	31.4	0.0949	15	330.0	1.75	70.9	343.0
36	5.0	25.0	0.0757	15	415.0	1.40	71.1	431.0
37	4.5	20.2	0.0613	15	513.0	1.13	71.3	534.0
38	4.0	16.0	0.0484	15	648.0	0.898	71.5	674.0
39	3.5	12.2	0.0371	15	850.0	0.691	71.8	884.0
40	3.1	9.61	0.0291	15	1079.0	0.543	72.0	1122.0
41	2.8	7.84	0.0237	15	1323.0	0.443	72.0	1376.0
42	2.5	6.25	0.0189	15	1659.0	0.353	72.0	1726.0
43	2.2	4.48	0.0147	15	2143.0	0.274	72.0	2228.0
44	2.0	4.00	0.0121	15	2593.0	0.226	72.0	2696.0

(a) In 10 in.

Copper Casting Alloys

Nominal compositions of principal copper casting alloys

UNS number	Common name	Previous ASTM designation	Composition,%						
			Cu	Sn	Pb	Zn	Fe	Al	Others
ASTM B22									
C86300	Manganese bronze	B22-E	63	...	...	25	3	6	3 Mn
C90500	Tin bronze	B22-D	88	10	...	2	...	...	...
C91100	Tin bronze	B22-B	84	16	...	...	...	...	...
C91300	Tin bronze	B22-A	81	19	...	...	...	...	...
C93700	High-lead tin bronze	B22-C	80	10	10	...	...	...	...
ASTM B61									
C92200	Valve bronze	...	88	6	1.5	4	...	...	1 Ni max
ASTM B62									
C83600	Leaded red brass	...	85	5	5	5	...	...	...
ASTM B66									
C93800	High-lead tin bronze	...	78	7	15	...	...	...	...
C94300	High-lead tin bronze	...	70	5	25	...	...	...	...
C94400	Leaded phosphor bronze	...	81	8	11	...	...	...	0.35 P
C94500	High-lead tin bronze	...	73	7	19	1	...	...	...
ASTM B67									
C94100	High-lead tin bronze	...	70	5.5	18.5	3 max	...	...	...
ASTM B148									
C95200	Aluminum bronze	B148-9A	88	...	...	...	3	9	...
C95300	Aluminum bronze	B148-9B	89	...	...	...	1	10	...
C95400	Aluminum bronze	B148-9C	85	...	...	...	4	11	4 Ni
C95500	Nickel-aluminum bronze	B148-9D	81	...	...	...	4	11	
C95800	Nickel-aluminum bronze	...	81.3	...	...	...	4	9	4.5 Ni 1.2 Mn
ASTM B176 (Die casting alloys)									
C85800	Yellow brass	Z30A	58	1	1	40	...	...	...
C87800	Silicon brass	ZS144A	82	...	...	14	...	...	4 Si
C87900	Silicon brass	ZS331A	65	...	...	33	...	...	1 Si
ASTM B584									
C83600	Leaded red brass	B145-4A	85	5	5	5	...	...	...
C83800	Leaded red brass	B145-4B	83	4	6	7	...	...	...
C84400	Leaded semi-red brass	B145-5A	81	3	7	9	...	...	...
C84800	Leaded semi-red brass	B145-5B	76	3	6	15	...	...	...
C85200	Leaded yellow brass	B146-6A	72	1	3	24	...	...	...
C85400	Leaded yellow brass	B146-6B	67	1	3	29	...	...	...
C85700	Leaded naval brass	B146-6C	63	1	1	34.7	0.03	...	...
C86200	High-strength manganese bronze	B147-8B	64	...	...	26	3	4	3 Mn
C86300	High-strength manganese bronze	B147-8C	63	...	...	25	3	6	3 Mn
C86400	Leaded manganese bronze	B147-7A or B132-A	59	...	1	38	1	0.5	0.5 Mn
C86500	Manganese bronze	B147-8A	58	...	...	39	1	1	1 Mn
C86700	Leaded manganese bronze	B132-B	58	1	1	34	2	2	2Mn
C87200	Silicon bronze	B198-12A	Several nominal compositions available						
C87400	Silicon brass	B198-13A	82	...	0.5	14	...	...	3.5 Si
C87500	Silicon brass	B198-13B	82	...	...	14	...	...	4Si
C87600	Silicon brass	B198-13C	89	...	...	6	...	...	5 Si
C90300	Modified G bronze	B143-1B	88	8	...	4	...	...	...
C90500	G bronze	B143-1A	88	10	...	2	...	...	...
C92200	Steam bronze(a)	B143-2A	88	6	1.5	4.5	...	...	...
C92300	Leaded tin bronze	B143-2B	87	8	1	4	...	...	...
C93200	High-lead tin bronze	B144-3B	83	7	7	3	...	...	...
C93500	High-lead tin bronze	B144-3C	85	5	9	1	...	...	...
C93700	High-lead tin bronze	B144-3A	80	10	10	...	...	...	...
C93800	High-lead tin bronze	B144-3D	78	7	15	...	...	...	...
C94300	High-lead tin bronze	B144-3E	70	5	25	...	...	...	...
C94700	Nickel-tin bronze	B292-A	88	5	...	2	...	...	5 Ni
C94800	Leaded nickel-tin bronze	B292-B	87	5	1	2	...	...	5 Ni
C94900	Leaded nickel-tin bronze	...	80	5	5	5	...	...	5 Ni
C97300	Leaded nickel silver	B149-10A	56	2	10	20	...	...	12 Ni
C97600	Leaded nickel silver	B149-11A	64	4	4	8	...	...	20 Ni
C97800	Leaded nickel silver	B149-11B	66	5	2	2	...	...	25 Ni

(a) Also known as valve bronze or Navy M bronze

Foundry properties for principal copper alloys for sand casting

UNS number	Common name	Shrinkage allowance, %	Approximate liquidus temperature		Castability rating(a)	Fluidity rating(a)
			°C	°F		
C83600	Leaded red brass	5.7	1010	1850	2	6
C84400	Leaded semi-red brass	2.0	980	1795	2	6
C84800	Leaded semi-red brass	1.4	955	1750	2	6
C85400	Leaded yellow brass	1.5 to 1.8	940	1725	4	4
C85800	Yellow brass	2.0	925	1700	4	4
C86300	Manganese bronze	2.3	920	1690	6	2
C86500	Manganese bronze	1.9	880	1615	6	2
C87200	Silicon bronze	1.8 to 2.0	...	...	8	3
C87500	Silicon brass	1.9	915	1680	7	1
C90300	Tin bronze	1.5 to 1.8	980	1795	3	6
C92200	Leaded tin bronze	1.5	990	1810	3	6
C93700	High-lead tin bronze	2.0	930	1705	1	6
C94300	High-lead tin bronze	1.5	925	1700	1	6
C95300	Aluminum bronze	1.6	1045	1910	8	5
C95800	Aluminum bronze	1.6	1060	1940	8	5
C97600	Nickel silver	2.0	1145	2090	5	7
C97800	Nickel silver	1.6	1180	2160	5	7

(a) Relative rating for casting in sand molds. The alloys are ranked from 1 to 8 in both over-all castability and fluidity; 1 is the highest or best possible rating

Composition and typical properties of heat treated copper casting alloys of high strength and conductivity

UNS number	Nominal composition	Tensile strength		Yield strength		Elongation, %	Hardness	Electrical conductivity, % IACS
		MPa	ksi	MPa	ksi			
C81400	99Cu-0.8Cr-0.06Be	365	53	250	36	11	69 HRB	70
C81500	99Cu-1Cr	350	51	275	40	17	105 HB	85
C81800	97Cu-1.5Co-1Ag-0.4Be	705	102	515	75	8	96 HRB	48
C82000	97Cu-2.5Co-0.5Be	660	96	515	75	6	96 HRB	48
C82200	98Cu-1.5Ni-0.5Be	655	95	515	75	7	96 HRB	48
C82500	97Cu-2Be-0.5Co-0.3Si	1105	160	1035	150	1	43 HRC	20
C82800	96.6Cu-2.6Be-0.5Co 0.3Si	1140	165	1070	155	1	46 HRC	18

Typical properties of copper casting alloys

UNS number	Tensile strength		Yield strength(a)		Compressive yield strength(b)		Elongation, %	Hardness, HB(c)	Electrical conductivity, % IACS
	MPa	ksi	MPa	ksi	MPa	ksi			
ASTM B22									
C86300	820	119	570(d)	82(d)	490	71	18	177	9.05
C90500	275-345	40-50	140-160	20-23	...	...	24-43	75-85	10.5-11.5
C93700	270	39	125	18	125	18	30	67	10.0
ASTM B61									
C92200	280	41	110	16	105	15	45	64	14.5
ASTM B62									
C83600	240	35	105	15	100	14	32	62	15.0
ASTM B66									
C94300	160-205	23-30	75-105	11-15	80-95	12-14	7-16	42-55	...
ASTM B147									
C86200	625-670	91-97	315-345	45-50	345	50	19-25	170-195(e)	7-8
C86300	820	119	570(d)	82(d)	490	71	18	177	9.0
C86400	415-540	60-78	170-275	25-40	140-180	20-26	15-30	80-95	20-24
C86500	490	71	180	26	165	24	40	98	20.5
ASTM B148									
C95200	480-600	70-87	170-205	25-30	185-215	27-31	22-38	110-140(e)	12-14
C95300(f)	480-585	70-85	205-240	30-35	110-140	16-20	20-35	110-160(e)	12-15
C95300(g)	550-655	80-95	275-380	40-55	240-275	35-45	12-16	160-225(e)	13.8
C95400(f)	515-655	75-95	205-285	30-41	...	...	12-20	150-185(e)	13-15
C95400(g)	620-690	90-100	310-360	45-52	...	...	6-15	190-235(e)	...
C95500(f)	620-725	90-105	275-345	40-50	...	...	7-20	175-210(e)	8-9.5
C95500(g)	760-855	110-124	415-550	60-80	...	...	5-12	215-260(e)	...
ASTM B176									
C85800(h)	380	55	205(d)	30(d)	...	...	15	...	22
C87800(h)	620	90	205(d)	30(d)	...	...	25	...	...
C87900(h)	400	58	205(d)	30(d)	...	...	15	...	...
ASTM B584									
C83600	243	35	105	15	100	14	32	62	15
C83800	205-260	30-38	85-115	12-17	76-83	11-12	15-27	50-60	...
C84400	200-270	29-39	90-115	13-17	...	...	18-30	50-60	18
C84800	260	38	105	15	85	12	37	59	16.5
C85200	240-275	35-40	85-95	12-14	55-70	8-10	25-40	40-55	15-22
C85400	205-260	30-38	75-105	11-15	62	9	20-35	40-60	18-25
C85700	275-310	40-45	95-140	14-20	...	...	15-25	50-75	20-26
C86200	625-670	91-97	315-345	46-50	345	50	19-25	170-195(e)	7-8
C86300	820	119	573	82(d)	490	71	18	177	9.0
C86400	415-540	60-78	170-275	25-40	...	20-26	15-30	80-95	20-24
C86500	490	71	180	26	140-180	24	40	98	20.5
C86700	550	80	220	32	165	...	15	...	...
C87200	380-450	55-65	150-205	22-30	105-150	15-22	25-55	85-120	4.5-6.4
C87400	345-485	50-70	145-225	21-33	...	...	20-50	70-130	...
C87500	470	68	207	30	185	27	17	115	6.0
C87600	414 min	60 min	207 min	30 min	...	...	16 min	...	...
C90300	275-345	40-50	125-150	18-22	...	...	25-50	60-75	12-13
C90500	275-345	40-50	140-160	20-23	...	...	24-43	75-85	10.5-11.5
C92200	280	41	110	16	105	15	45	64	14.5
C92300	225-295	33-43	110-165	16-24	62-76	9-11	18-30	60-75	10-12
C93200	205-260	30-38	115-145	17-21	...	...	12-20	55-65	...
C93500	195-240	28-35	83-105	12-15	90	13	20-35	55-65	15
C93700	270	39	125	18	125	18	30	67	10.0
C93800	170-225	25-33	95-140	14-20	90-110	13-16	10-18	50-60	...
C94300	160-205	23-30	76-105	11-15	83-97	12-14	7-16	42-55	...
C94700(f)	310	45	140	20	...	...	25	...	...
C94700(j)	515	75	345	50	...	...	5	...	...
C94800(f)	275	40	140	20	...	...	20	...	...
C94900	262 min	38 min	97 min	15 min	...	...	15 min	...	...
C97300	205-275	30-40	105-140	15-20	...	...	10-25	50-60	5.7
C97600	325	47	180	26	168	24	22	85	4.8
C97800	345-450	50-65	180-275	26-40	...	...	15-25	120-150	4-5

(a) At 0.5% extension under load. (b) At permanent set of 0.1%. (c) 500 kg load; 10 mm diam ball. (d) At 0.2% offset. (e) 3000 kg load. (f) MO1 temper. (g) TQ00 temper. (h) MO4 temper. (j) TF00 temper

Corrosion ratings of cast copper metals in various media

Corrosive medium	Copper	Tin bronze	Leaded tin bronze	High-leaded tin bronze	Leaded red brass	Leaded semi-red brass	Leaded yellow brass	Leaded high-strength yellow brass	High-strength yellow brass	Aluminum bronze	Leaded nickel brass	Leaded nickel bronze	Silicon bronze	Silicon brass
Acetate solvents	B	A	A	A	A	A	B	A	A	A	A	A	A	B
Acetic acid														
20%	A	C	B	C	B	C	C	C	C	A	C	A	A	B
50%	A	C	B	C	B	C	C	C	C	A	C	B	A	B
Glacial	A	A	A	C	A	C	C	C	C	C	B	B	A	B
Acetone	A	A	A	A	A	A	A	A	A	A	A	A	A	A
Acetylene(a)	C	C	C	C	C	C	C	C	C	C	C	C	C	C
Alcohols(b)	A	A	A	A	A	A	A	A	A	A	A	A	A	C
Aluminum chloride	C	C	C	C	C	C	C	C	C	B	C	C	C	A
Aluminum sulfate	B	B	B	B	B	C	C	C	C	C	C	C	C	C
Ammonia, moist gas	C	C	C	C	C	C	C	C	C	A	C	C	C	A
Ammonia, moisture-free	A	A	A	A	A	A	A	A	A	A	A	A	A	C
Ammonium chloride	C	C	C	C	C	C	C	C	C	C	C	C	C	C
Ammonium hydroxide	C	C	C	C	C	C	C	C	C	C	C	C	C	C
Ammonium nitrate	C	C	C	C	C	C	C	C	C	C	C	C	C	C
Ammonium sulfate	B	B	B	B	B	C	C	C	C	A	C	C	A	C
Aniline and aniline dyes	C	C	C	C	C	C	C	C	C	B	C	C	C	A
Asphalt	A	A	A	A	A	A	A	A	A	A	A	A	A	C
Barium chloride	A	A	A	A	A	C	C	C	C	A	A	A	A	A
Barium sulfide	C	C	C	C	C	C	C	C	B	C	C	C	A	C
Beer(b)	A	A	B	B	B	C	B	C	A	A	A	A	A	B
Beet sugar syrup	A	A	B	B	B	A	A	A	B	A	C	C	B	B
Benzine	A	A	A	A	A	A	A	A	A	A	A	A	B	B
Benzol	A	A	A	A	A	A	A	A	A	A	A	A	A	A
Boric acid	A	A	A	A	A	A	A	A	A	A	A	A	A	A
Butane	A	A	A	A	A	A	A	B	A	A	A	A	A	A
Calcium bisulfite	B	A	B	B	B	C	C	A	C	A	A	A	A	B
Calcium chloride (acid)	B	B	B	B	B	B	C	C	C	A	B	B	A	B
Calcium chloride (alkaline)	C	C	C	C	C	C	C	C	C	A	C	C	A	C
Calcium hydroxide	C	C	C	C	C	C	C	C	C	A	C	C	C	B
Calcium hypochlorite	C	C	B	B	B	C	C	C	C	B	C	C	C	C
Cane sugar syrups	A	A	C	C	B	A	A	A	A	B	A	A	A	C
Carbonated beverages(b)	A	C	C	C	B	C	C	C	C	A	A	A	A	B
Carbon dioxide, dry	A	A	A	A	A	A	A	A	A	A	A	A	A	C
Carbon dioxide, moist(b)	B	B	B	B	B	C	C	C	C	A	C	C	A	A
Carbon tetrachloride, dry	A	A	A	A	A	A	A	A	A	A	A	A	A	B
Carbon tetrachloride, moist	B	B	B	B	B	B	B	B	B	B	B	B	A	A
Chlorine, dry	A	A	A	A	A	A	A	A	A	A	A	A	A	A
Chlorine, moist	C	C	B	B	B	C	C	C	C	C	C	B	C	C
Chromic acid	C	C	C	C	C	C	C	C	C	C	C	C	C	C
Citric acid	A	A	A	A	A	A	A	A	A	A	A	A	A	A
Copper sulfate	B	A	A	A	A	C	C	C	C	B	B	B	A	A
Cottonseed oil(b)	A	A	B	B	B	A	A	A	B	A	A	A	A	A
Creosote	B	B	B	B	B	B	C	C	C	A	C	B	B	B
Ethers	A	A	A	A	A	A	A	A	A	A	A	A	B	A

Note: A, recommended; B, acceptable; C, not recommended. (a) Acetylene forms an explosive compound with copper when moist or when certain impurities are present and the gas is under pressure. Alloys containing less than 65% Cu are satisfactory under this use. When gas is not under pressure other copper alloys are satisfactory. (b) Copper and copper alloys resist corrosion by most food products. Traces of copper may be dissolved and affect taste or color. In such cases, copper metals are often tin coated

(continued)

Corrosion ratings of cast copper metals in various media (continued)

Corrosive medium	Copper	Tin bronze	Leaded tin bronze	High-leaded tin bronze	Leaded red brass	Leaded semi-red brass	Leaded yellow brass	Leaded high-strength yellow brass	High-strength yellow brass	Aluminum bronze	Leaded nickel brass	Leaded nickel bronze	Silicon bronze	Silicon brass
Ethylene glycol	A	A	A	A	A	A	A	A	A	A	A	A	A	A
Ferric chloride, sulfate	C	C	C	C	C	C	C	C	C	C	C	C	C	C
Ferrous chloride, sulfate	C	C	C	C	C	C	C	C	C	C	C	C	C	C
Formaldehyde	A	A	A	A	A	A	A	A	A	A	A	A	A	A
Formic acid	A	A	A	A	B	B	B	B	B	B	B	B	B	B
Freon	A	A	A	A	A	A	A	A	A	A	A	A	C	B
Fuel oil	A	A	A	A	A	A	A	A	A	A	A	A	A	A
Furfural	A	A	A	A	A	A	A	A	A	A	A	A	A	A
Gasoline	A	A	A	A	A	A	A	A	A	A	A	A	A	A
Gelatin(b)	A	A	A	A	A	A	A	A	A	A	A	A	A	A
Glucose	A	A	A	A	A	A	A	A	A	A	A	A	A	A
Glue	A	A	A	A	A	A	A	A	A	A	A	A	A	A
Glycerin	A	A	A	A	A	A	A	A	A	A	A	A	A	A
Hydrochloric or muriatic acid	C	C	C	C	C	C	C	C	C	B	C	C	C	C
Hydrofluoric acid	B	B	B	B	B	B	B	B	B	A	B	B	B	B
Hydrofluosilicic acid	B	B	B	B	B	B	C	C	C	B	C	B	B	C
Hydrogen	A	A	A	A	A	A	A	A	A	A	A	A	A	A
Hydrogen peroxide	C	C	C	C	C	C	C	C	C	B	C	C	C	C
Hydrogen sulfide, dry	C	C	C	C	C	C	C	C	C	B	C	C	B	C
Hydrogen sulfide, moist	C	C	C	C	C	C	C	C	C	B	C	C	C	C
Lacquers	A	A	A	A	A	A	A	A	A	A	A	A	A	A
Lacquer thinners	A	A	A	A	A	A	A	A	A	A	A	A	A	A
Lactic acid	A	A	A	A	A	C	C	C	C	A	C	C	A	C
Linseed oil	A	A	A	A	A	A	A	A	A	A	A	A	A	A
Liquors														
Black liquor	B	B	B	B	B	C	C	C	C	B	C	C	B	B
Green liquor	C	C	C	C	C	C	C	C	C	B	C	C	C	B
White liquor	C	C	C	C	C	C	C	C	C	A	C	C	A	B
Magnesium chloride	A	A	A	A	A	A	A	A	A	A	A	A	A	B
Magnesium hydroxide	B	B	B	B	B	B	B	B	B	A	B	B	B	B
Magnesium sulfate	A	A	A	A	A	B	B	B	B	A	B	B	A	B
Mercury, mercury salts	C	C	C	C	C	C	C	C	C	C	C	C	C	C
Milk(b)	A	A	A	A	A	A	A	A	A	A	A	A	A	A
Molasses(b)	A	A	A	A	A	A	A	A	A	A	A	A	A	A
Natural gas	A	A	A	A	A	A	A	A	A	A	A	A	A	A
Nickel chloride	C	C	C	C	C	C	C	B	B	B	B	B	C	C
Nickel sulfate	A	A	A	A	A	A	A	A	A	A	A	A	A	A
Nitric acid	C	C	C	C	C	C	C	C	C	C	C	C	C	C
Oleic acid	B	B	B	B	B	B	B	B	B	A	B	B	B	B
Oxalic acid	A	A	A	A	A	A	A	A	A	A	A	A	A	A
Phosphoric acid	A	A	A	A	A	A	A	A	C	C	C	C	A	A
Picric acid	C	C	C	C	C	C	C	C	C	C	C	C	C	C
Potassium chloride	A	A	A	A	A	A	A	A	A	A	A	A	A	A
Potassium cyanide	C	C	C	C	C	C	C	C	C	A	C	C	C	C
Potassium hydroxide	C	C	C	C	C	C	C	C	C	A	C	C	C	C

Note: A, recommended; B, acceptable; C, not recommended. (a) Acetylene forms an explosive compound with copper when moist or when certain impurities are present and the gas is under pressure. Alloys containing less than 65% Cu are satisfactory under this use. When gas is not under pressure other copper alloys are satisfactory. (b) Copper and copper alloys resist corrosion by most food products. Traces of copper may be dissolved and affect taste or color. In such cases, copper metals are often tin coated

(continued)

Corrosion ratings of cast copper metals in various media (continued)

Corrosive medium	Copper	Tin bronze	Leaded tin bronze	High-leaded tin bronze	Leaded red brass	Leaded semi-red brass	Leaded yellow brass	Leaded high-strength yellow brass	High-strength yellow brass	Aluminum bronze	Leaded nickel brass	Leaded nickel bronze	Silicon bronze	Silicon brass
Potassium sulfate	A	A	A	A	A	C	C	C	C	A	C	C	A	C
Propane gas	A	A	A	A	A	A	A	A	A	A	A	A	A	A
Sea water	A	A	A	A	A	C	C	C	C	A	C	C	B	B
Soap solutions	A	A	A	A	B	C	C	C	C	A	C	C	A	C
Sodium bicarbonate	A	A	A	A	A	C	C	C	A	A	A	A	A	B
Sodium bisulfate	C	C	C	C	C	C	C	C	C	A	C	C	A	B
Sodium carbonate	C	A	A	A	A	C	C	C	C	B	C	C	C	C
Sodium chloride	A	A	A	A	A	B	C	C	C	A	A	A	A	A
Sodium cyanide	C	C	C	C	C	C	C	C	C	B	C	C	C	C
Sodium hydroxide	C	C	C	C	C	C	C	C	C	C	C	C	C	C
Sodium hypochlorite	C	C	C	C	C	C	C	C	C	C	C	C	C	C
Sodium nitrate	B	B	B	B	B	B	B	B	B	A	B	B	A	A
Sodium peroxide	B	B	B	B	B	B	B	B	B	B	B	B	B	B
Sodium phosphate	A	A	A	A	A	A	A	A	A	A	A	A	A	A
Sodium sulfate, silicate	A	A	B	B	B	B	C	C	C	B	C	C	C	B
Sodium sulfide, thiosulfate	C	C	C	C	C	C	C	C	C	B	C	C	C	C
Stearic acid	A	A	A	A	A	A	A	A	A	A	A	A	A	A
Sulfur, solid	C	C	C	C	C	C	C	C	C	A	C	C	A	A
Sulfur chloride	C	C	C	C	C	C	C	C	C	A	C	C	C	C
Sulfur dioxide, dry	A	A	A	A	A	A	A	A	A	C	A	A	C	C
Sulfur dioxide, moist	A	A	A	B	B	C	C	C	C	A	A	A	A	A
Sulfur trioxide, dry	A	A	A	A	A	A	A	A	A	A	A	A	A	B
Sulfuric acid														
78% or less	B	B	B	B	B	C	C	C	C	A	C	C	B	B
78% to 90%	C	C	C	C	C	C	C	C	C	B	C	C	C	C
90% to 95%	C	C	C	C	C	C	C	C	C	B	C	C	C	C
Fuming	C	C	C	C	A	A	A	A	C	A	A	A	A	A
Tannic acid	A	A	A	A	A	A	A	A	A	A	A	A	A	A
Tartaric acid	B	A	A	A	A	A	A	A	A	A	A	A	A	A
Toluene	B	B	A	A	A	B	B	B	B	B	B	B	B	B
Trichlorethylene, dry	A	A	A	A	A	A	A	A	A	A	A	A	A	A
Trichlorethylene, moist	A	A	A	A	A	A	A	A	A	A	A	A	A	A
Turpentine	A	A	A	A	A	A	A	A	A	A	A	A	A	A
Varnish	A	A	A	A	A	A	A	A	A	A	A	A	A	A
Vinegar	A	A	B	B	B	C	C	C	C	B	C	C	C	B
Water, acid mine	C	C	C	C	C	C	C	C	C	C	C	C	C	C
Water, condensate	A	A	A	A	A	A	A	A	A	A	A	A	A	A
Water, potable	A	A	A	A	B	A	B	B	B	A	B	B	A	A
Whiskey(b)	A	A	C	C	C	C	C	C	B	A	A	A	A	A
Zinc chloride	C	C	C	C	C	C	C	C	C	B	C	C	C	C
Zinc sulfate	A	A	A	A	A	C	C	C	C	B	A	B	B	C

Note: A, recommended; B, acceptable; C, not recommended. (a) Acetylene forms an explosive compound with copper when moist or when certain impurities are present and the gas is under pressure. Alloys containing less than 65% Cu are satisfactory under this use. When gas is not under pressure other copper alloys are satisfactory. (b) Copper and copper alloys resist corrosion by most food products. Traces of copper may be dissolved and affect taste or color. In such cases, copper metals are often tin coated

Properties and applications of cast coppers and copper alloys

UNS designation	Nominal composition, %(a)	Tensile strength MPa	Tensile strength ksi	Yield strength MPa	Yield strength ksi	Elongation in 2 in., %	Hardness Rockwell	Hardness Brinell 500 kg	Hardness Brinell 3000 kg	Machinability rating(c)	Casting types(d)	Typical applications
C80100	99.95 Cu + Ag min, 0.05 others max	172	25	62	9	40	…	44	…	10	C, T, I, M, P, S	Electrical and thermal conductors; corrosion and oxidation resistant applications
C80300	99.95 Cu + Ag min, 0.034 Ag min, 0.05 others max	172	25	62	9	40	…	44	…	10	C, T, I, M, P, S	Electrical and thermal conductors; corrosion and oxidation resistant applications
C80500	99.75 Cu + Ag min, 0.034 Ag min, 0.02 B max, 0.23 others max	172	25	62	9	40	…	44	…	10	C, T, I, M, P, S	Electrical and thermal conductors; corrosion and oxidation resistant applications
C80700	99.75 Cu + Ag min, 0.02 B max, 0.23 others max	172	25	62	9	40	…	44	…	10	C, T, I, M, P, S	Electrical and thermal conductors; corrosion and oxidation resistant applications
C80900	99.70 Cu + Ag min, 0.034 Ag min, 0.30 others max	172	25	62	9	40	…	44	…	10	C, T, I, M, P, S	Electrical and thermal conductors; corrosion and oxidation resistant applications
C81100	99.70 Cu + Ag min, 0.30 others max	172	25	62	9	40	…	44	…	10	C, T, I, M, P, S	Electrical and thermal conductors; corrosion and oxidation resistant applications
High-copper alloys												
C81300	98.5 Cu min, 0.06 Be, 0.80 Co, 0.40 others max	(365)	(53)	(248)	(36)	(11)	…	(39)	…	20	C, T, I, M, P, S	Higher hardness electrical and thermal conductors
C81400	98.5 Cu min, 0.06 Be, 0.80 Cr, 0.40 others max	(365)	(53)	(248)	(36)	(11)	(B 69)	…	…	20	C, T, I, M, P, S	Higher hardness electrical and thermal conductors
C81500	98.0 Cu min, 1.0 Cr, 0.50 others max	(352)	(51)	(276)	(40)	(17)	…	(105)	…	20	C, T, I, M, P, S	Electrical and/or thermal conductors used as structural members where strength and hardness greater than that of C80100-81100 are required
C81700	94.25 Cu min, 1.0 Ag, 0.4 Be, 0.9 Co, 0.9 Ni	(634)	(92)	(469)	(68)	(8)	…	…	(217)	30	C, T, I, M, P, S	Electrical and/or thermal conductors used as structural members where strength and hardness greater than that of C80100-81100 are required. Also used in place of C81500 where electrical and/or thermal conductivities can be sacrificed for hardness and strength
C81800	95.6 Cu min, 1.0 Ag, 0.4 Be, 1.6 Co	345 (703)	50 (102)	172 (517)	25 (75)	20 (8)	B 55 (B 96)	…	…	20	C, T, I, M, P, S	Resistance welding electrodes, dies

(a) Nominal composition, unless otherwise noted. For seldom-used alloys, only compositions are available. (b) Values for C82700, 84200, 96200, 96300 are minimum, not typical. As-cast values are for sand casting except C93900, continuous cast; and C85800, 87800, 87900, die cast. Heat treated values, in parentheses, indicate that the alloy responds to heat treatment. If heat treated values are not shown, the copper or copper alloy does not respond. (c) Free cutting brass = 100. (d) C, centrifugal; T, continuous; D, die; I, investment; M, permanent mold; P, plaster; S, sand. (Note: C82000, 82400, 82500, 82600, 82800 are also pressure cast.) (e) As-heat treated value for C94700, 20; for C94800, 40. Source: Copper Development

(continued)

Properties and applications of cast coppers and copper alloys (continued)

UNS designation	Nominal composition, %(a)	Tensile strength MPa	Tensile ksi	Yield strength MPa	Yield ksi	Elongation in 2 in., %	Hardness Rockwell	Brinell 500 kg	Brinell 3000 kg	Machinability rating(c)	Casting types(d)	Typical applications
C82000	96.8 Cu, 0.6 Be, 2.6 Co	345 (689)	50 (100)	138 (517)	20 (75)	20 (8)	B 55 (B 95)	...	(195)	20	C, T, I, M, P, S	Current carrying parts, contact and switch blades, bushings and bearings, soldering iron and resistance welding tips
C82100	97.7 Cu, 0.5 Be, 0.9 Co, 0.9 Ni	(634)	(92)	(469)	(68)	(8)	...	...	(217)	30	C, T, I, M, P, S	Electrical and/or thermal conductors used as structural members where strength and hardness greater than that of C80100-81100 are required. Also used in place of C81500 where electrical and/or thermal conductivities can be sacrificed for hardness and strength
C82200	96.5 Cu min, 0.6 Be, 1.5 Ni	393 (655)	57 (95)	207 (517)	30 (75)	20 (8)	B 60 (B 96)	...	...	20	C, T, I, M, P, S	Clutch rings, brake drums, seam welder electrodes, projection welding dies, spot welding tips, beam welder shapes, bushings, water-cooled holders
C82400	96.4 Cu min, 1.70 Be, 0.25 Co	496 (1034)	72 (150)	255 (965)	37 (140)	20 (1)	B 78 (C 38)	...	...	20	C, I, M, P, S	Safety tools, molds for plastic parts, cams, bushings, bearings, valves, pump parts, gears
C82500	97.2 Cu, 2.0 Be, 0.5 Co, 0.25 Si	552 (1103)	80 (160)	310	45	20 (1)	B 82 (C 40)	...	...	20	C, I, M, P, S	Safety tools, molds for plastic parts, cams, bushings, bearings, valves, pump parts
C82600	95.2 Cu min, 2.3 Be, 0.5 Co, 0.25 Si	565 (1138)	82 (165)	324 (1069)	47 (155)	20 (1)	B 83 (C 43)	...	...	20	C, I, M, P, S	Bearings and molds for plastic parts
C82700	96.3 Cu, 2.45 Be, 1.25 Ni	(1069)	(155)	(896)	(130)	(0)	(C 39)	...	...	20	C, I, M, P, S	Bearings and molds for plastic parts
C82800	96.6 Cu, 2.6 Be, 0.5 Co, 0.25 Si	669 (1138)	97 (165)	379 (1000)	55 (145)	20 (1)	B 85 (C 45)	...	...	10	C, I, M, P, S	Molds for plastic parts, cams, bushings, bearings, valves, pump parts, sleeves
Red brasses and leaded red brasses												
C83300	93 Cu, 1.5 Sn, 1.5 Pb, 4 Zn	221	32	69	10	35	...	35	...	35	S	Terminal ends for electrical cables
C83400	90 Cu, 10 Zn	241	35	69	10	30	F 50	...	...	60	C, S	Moderate strength, moderate conductivity castings; rotating bands
C83600	85 Cu, 5 Sn, 5 Pb, 5 Zn	255	37	117	17	30	...	60	...	84	C, T, I, S	Valves, flanges, pipe fittings, plumbing goods, pump castings, water pump impellers and housings, ornamental fixtures, small gears

(a) Nominal composition, unless otherwise noted. For seldom-used alloys, only compositions are available. (b) Values for C82700, 84200, 96200, 96300, are minimum, not typical. As-cast values are for sand casting except C93900, continuous cast; and C85800, 87800, 87900, die cast. Heat treated values, in parentheses, indicate that the alloy responds to heat treatment. If heat treated values are not shown, the copper or copper alloy does not respond. (c) Free cutting brass = 100. (d) C, continuous; T, centrifugal; I, investment; D, die; I, investment; M, permanent mold; P, plaster; S, sand. (Note: C82000, 82400, 82500, 82600, 82800, are also pressure cast.). (e) As-heat treated value for C94700, 20; for C94800, 40. Source: Copper Development Assn. Inc., New York

(continued)

Properties and applications of cast coppers and copper alloys (continued)

UNS designa-tion	Nominal composition, %(a)	Typical mechanical properties, as cast (heat treated)(b)									Machin-ability rating(c)	Casting types(d)	Typical applications
		Tensile strength		Yield strength		Elon-gation in 2 in., %	Hardness						
							Rockwell	Brinell					
		MPa	ksi	MPa	ksi			500 kg	3000 kg				
C83800	83 Cu, 4 Sn, 6 Pb, 7 Zn	241	35	110	16	25	...	60	...	90	C, T, S	Low-pressure valves and fittings, plumbing supplies and fittings, general hardware, air-gas-water fittings, pump components, railroad catenary fittings
Semi-red brasses and leaded semi-red brasses													
C84200	80 Cu, 5 Sn, 2.5 Pb, 12.5 Zn	193	28	103	15	27	...	60	...	80	C, T, S	Pipe fittings, elbows, T's, couplings, bushings, locknuts, plugs, unions
C84400	81 Cu, 3 Sn, 7 Pb, 9 Zn	234	34	103	15	26	...	55	...	90	C, T, S	General hardware, ornamental castings, plumbing supplies and fixtures, low-pressure valves and fittings
C84500	78 Cu, 3 Sn, 7 Pb, 12 Zn	241	35	97	14	28	...	55	...	90	C, T, S	Plumbing fixtures, cocks, faucets, stops, waste, air and gas fittings, low-pressure valve fittings
C84800	76 Cu, 3Sn, 6 Pb, 15 Zn	248	36	97	14	30	...	55	...	90	C, S	Plumbing fixtures, cocks, faucets, stops, waste, air, and gas, general hardware, and low-pressure valve fittings
Yellow brasses and leaded yellow brasses													
C85200	72 Cu, 1 Sn, 3 Pb, 24 Zn	262	38	90	13	35	...	45	...	80	C, T	Plumbing fittings and fixtures, ferrules, valves, hardware, ornamental brass, chandeliers, and irons
C85400	67 Cu, 1 Sn, 3 Pb, 29 Zn	234	34	83	12	35	...	50	...	80	C, T, M, P, S	General purpose yellow casting alloy not subject to high internal pressure. Furniture hardware, ornamental castings, radiator fittings, ship trimmings, cocks, battery clamps, valves and fittings
C85500	61 Cu, 0.8 Al, bal Zn	414	60	159	23	40	B 55	85	...	80	C, S	Ornamental castings
C85700	63 Cu, 1 Sn, 1 Pb, 34.7 Zn, 0.3 Al	345	50	124	18	40	...	75	...	80	C, M, P, S	Bushings, hardware fittings, ornamental castings
C85800	58 Cu, 1 Sn, 1 Pb, 40 Zn	379	55	207	30	15	B 55	...	...	80	D	General purpose die casting alloy having moderate strength
Manganese and leaded manganese bronze alloys													
C86100	67 Cu, 21 Zn, 3 Fe, 5 Al, 4 Mn	655	95	345	50	20	...	...	180	30	C, I, P, S	Marine castings, gears, gun mounts, bushings and bearings, marine racing propellers
C86200	64 Cu, 26 Zn, 3 Fe, 4 Al, 3 Mn	655	95	331	48	20	...	...	180	30	C, T, D, I, P, S	Marine castings, gears, gun mounts, bushings and bearings

(a) Nominal composition, unless otherwise noted. For seldom-used alloys, only compositions are available. (b) Values for C82700, 84200, 96200, 96300, are minimum, not typical. As-cast values are for sand casting except C93900, continuous cast; and C85800, 87800, 87900, die cast. Heat treated values, in parentheses, indicate that the alloy responds to heat treatment. If heat treated values are not shown, the copper or copper alloy does not respond. (d) C, cast; T, continuous; D, die; I, investment; M, permanent mold; P, plaster; S, sand. (Note: C82000, 82400, 82500, 82600, 82800 are also pressure cast.). (e) As-heat treated value for C94700, 20; for C94800, 40. Source: Copper Development Assn. Inc., New York

(continued)

Properties and applications of cast coppers and copper alloys (continued)

UNS designation	Nominal composition, %(a)	Tensile strength (MPa)	Tensile strength (ksi)	Yield strength (MPa)	Yield strength (ksi)	Elongation in 2 in., %	Hardness Rockwell	Hardness Brinell 500 kg	Hardness Brinell 3000 kg	Machinability rating(c)	Casting types(d)	Typical applications
C86300	63 Cu, 25 Zn, 3 Fe, 6 Al, 3 Mn	793	115	572	83	15	...	...	225	8	C, I, P, S	Extra-heavy duty, high-strength alloy. Large valve stems, gears, cams, slow-speed heavy-load bearings, screwdown nuts, hydraulic cylinder parts
C86400	59 Cu, 1 Pb, 40 Zn	448	65	172	25	20	...	90	105	65	C, D, M, P, S	Free-machining manganese bronze. Valve stems, marine fittings, lever arms, brackets, light-duty gears
C86500	58 Cu, 0.5 Sn, 39.5 Zn, 1 Fe, 1 Al	490	71	193	28	30	...	100	130	26	C, I, P, S	Machinery parts requiring strength and toughness, lever arms, valve stems, gears
C86700	58 Cu, 1 Pb, 41 Zn	586	85	290	42	20	B 80	...	155	55	C, S	High strength, free-machining manganese bronze. Valve stems
C86800	55 Cu, 37 Zn, 3 Ni, 2 Fe, 3 Mn	565	82	262	38	22	...	...	80	30	S	Marine fittings, marine propellers
Silicon bronzes and silicon brasses												
C87200	89 Cu min, 4 Si	379	55	172	25	30	...	85	...	40	C, I, M, P, S	Bearings, bells, impellers, pump and valve components, marine fittings, corrosion-resistant castings
C87400	83 Cu, 14 Zn, 3 Si	379	55	165	24	30	...	70	100	50	C, D, I, M, P, S	Bearings, gears, impellers, rocker arms, valve stems, clamps
C87500	82 Cu, 14 Zn, 4 Si	462	67	207	30	21	...	115	134	50	C, D, I, M, P, S	Bearings, gears, impellers, rocker arms, valve stems, small boat propellers
C87600	90 Cu, 5.5 Zn, 4.5 Si	455	66	221	32	20	B 76	110	135	40	S	Valve stems
C87800	82 Cu, 14 Zn, 4 Si	586	85	345	50	25	B 85	...	...	40	D	High-strength, thin-wall die castings; brush holders, lever arms, brackets, clamps, hexagonal nuts
C87900	65 Cu, 34 Zn, 1 Si	483	70	241	35	25	B 70	...	...	80	D	General purpose die casting alloy having moderate strength
Tin bronzes												
C90200	93 Cu, 7 Sn	262	38	110	16	30	...	70	...	20	C, S	Bearings and bushings
C90300	88 Cu, 8 Sn, 4 Zn	310	45	145	21	30	...	70	...	30	C, T, I, P, S	Bearings, bushings, pump impellers, piston rings, valve components, seal rings, steam fittings, gears
C90500	88 Cu, 10 Sn, 2 Zn	310	45	152	22	25	...	75	...	30	C, T, I, S	Bearings, bushings, pump impellers, piston rings, valve components, steam fittings, gears
C90700	89 Cu, 11 Sn	303 (379)	44 (55)	152 (207)	22 (30)	20 (16)	...	80 (102)	...	20	C, T, I, M, S	Gears, bearings, bushings
C90800	87 Cu, 12 Sn											
C90900	87 Cu, 13 Sn	276	40	138	20	15	...	90	...	20	C, S	Bearings and bushings
C91000	85 Cu, 14 Sn, 1 Zn	221	32	172	25	2	...	105	...	20	C, T, I, S	Piston rings and bearings

Typical mechanical properties, as cast (heat treated)(b)

(a) Nominal composition, unless otherwise noted. For seldom-used alloys, only compositions are available. (b) Values for C82700, 84200, 96200, 96300, are minimum, not typical. As-cast values are for sand casting except C93900, continuous cast; and C85800, 87800, 87900, die cast. Heat treated values, in parentheses, indicate that the alloy responds to heat treatment. If heat treated values are not shown, the copper or copper alloy does not respond. (c) Free cutting brass = 100. (d) C, continuous cast; T, continuous; D, die; I, investment; M, permanent mold; P, plaster; S, sand. (Note: C82000, 82400, 82500, 82600, 82800 are also pressure cast.). (e) As-heat treated value for C94700, 20; for C94800, 40. Source: Copper Development Assn. Inc., New York

(continued)

Properties and applications of cast coppers and copper alloys (continued)

UNS designation	Nominal composition, %(a)	Tensile strength		Yield strength		Elongation in 2 in., %	Hardness			Machinability rating(c)	Casting types(d)	Typical applications
							Rockwell	Brinell				
		MPa	ksi	MPa	ksi			500 kg	3000 kg			
C91100	84 Cu, 16 Sn	241	35	172	25	2	...	...	135	10	S	Piston rings, bearings, bushings, bridge plates
C91300	81 Cu, 19 Sn	241	35	207	30	0.5	...	...	170	10	C, T, M, S	Piston rings, bearings, bushings, bridge plates, bells
C91600	88 Cu, 10.5 Sn, 1.5 Ni	303 (414)	44 (60)	152 (221)	22 (32)	16 (16)	...	85 (106)	...	20	C, T, M, S	Gears
C91700	86.5 Cu, 12 Sn, 1.5 Ni	303 (414)	44 (60)	152 (221)	22 (32)	16 (16)	...	85 (106)	...	20	C, T, I, M, S	Gears
Leaded tin bronzes												
C92200	88 Cu, 6 Sn, 1.5 Pb, 4.5 Zn	276	40	138	20	30	...	65	...	42	C, T, I, M, P, S	Valves, fittings, and pressure-containing parts for use up to 550 °F
C92300	87 Cu, 8 Sn, 4 Zn	276	40	138	20	25	...	70	...	42	C, T, S	Valves, pipe fittings and high-pressure steam castings. Superior machinability to C90300
C92400	88 Cu, 10 Sn, 2 Pb, 2 Zn											
C92500	87 Cu, 11 Sn, 1 Pb, 1 Ni	303	44	138	20	20	...	80	...	30	C, T, M, S	Gears, automotive synchronizer rings
C92600	87 Cu, 10 Sn, 1 Pb, 2 Zn	303	44	138	20	30	F 78	70	...	40	C, T, S	Bearings, bushings, pump impellers, piston rings, valve components, steam fittings, and gears. Superior machinability to C90500
C92700	88 Cu, 10 Sn, 2 Pb	290	42	145	21	20	...	77	...	45	C, T, S	Bearings, bushings, pump impellers, piston rings, valve components, steam fittings, and gears. Superior machinability to C90500
C92800	79 Cu, 16 Sn, 5 Pb	276	40	207	30	1	B 80	...	...	70	C, S	Piston rings
C92900	84 Cu, 10 Sn, 2.5 Pb, 3.5 Ni	324 (324)	47 (47)	179 (179)	26 (26)	20 (20)	...	80 (80)	...	40	C, T, M, S	Gears, wear plates, guides, cams, parts requiring machinability superior to that of C91600 or 91700
High-leaded tin bronzes												
C93200	83 Cu, 7 Sn, 7 Pb, 3 Zn	241	35	124	18	20	...	65	...	70	C, T, M, S	General-utility bearings and bushings
C93400	84 Cu, 8 Sn, 8 Pb	221	32	110	16	20	...	60	...	70	C, T, S	Bearings and bushings
C93500	85 Cu, 5 Sn, 9 Pb	221	32	110	16	20	...	60	,	70	C, T, S	Small bearings and bushings, bronze backing for babbit-lined automotive bearings
C93700	80 Cu, 10 Sn, 10 Pb	241	35	124	18	20	...	60	...	80	C, T, M, S	Bearings for high speed and heavy pressures, pumps, impellers, corrosion-resistant applications, pressure tight castings
C93800	78 Cu, 7 Sn, 15 Pb	207	30	110	16	18	...	55	...	80	C, T, M, S	Bearings for general service and moderate pressures, pump impellers and bodies for use in acid mine water

(a) Nominal composition, unless otherwise noted. For seldom-used alloys, only compositions are available. (b) Values for C82700, 84200, 96200, 96300, are minimum, not typical. As-cast values are for sand casting except C93900, continuous cast; and C85800, 87800, 87900, die cast. Heat treated values, in parentheses, indicate that the alloy responds to heat treatment. If heat treated values are not shown, the copper or copper alloy does not respond. (c) Free cutting brass = 100. (d) C, centrifugal; T, continuous; D, die; I, investment; M, permanent mold; P, plaster; S, sand. (Note: C82000, 82400, 82500, 82600, 82800 are also pressure cast.) (e) As-heat treated value for C94700, 20; for C94800, 40. Source: Copper Development Assn. Inc., New York

(continued)

Properties and applications of cast coppers and copper alloys (continued)

UNS designation	Nominal composition, %(a)	Typical mechanical properties, as cast (heat treated)(b)									Machinability rating(c)	Casting types(d)	Typical applications
		Tensile strength		Yield strength		Elongation in 2 in., %	Hardness						
		MPa	ksi	MPa	ksi		Rockwell	Brinell 500 kg	Brinell 3000 kg				
C93900	79 Cu, 6 Sn, 15 Pb	221	32	152	22	7	...	63	...		80	T	Continuous castings only. Bearings for general service, pump bodies and impellers for mine waters
C94000	70.5 Cu, 13.0 Sn, 15.0 Pb, 0.50 Zn, 0.75 Ni, 0.25 Fe, 0.05 P, 0.35 Sb												
C94100	70.0 Cu, 5.5 Sn, 18.5 Pb, 3.0 Zn, 1.0 others max												
C94300	70 Cu, 5 Sn, 25 Pb	186	27	90	13	15	...	48	...		80	C, S	High-speed bearings for light loads
C94400	81 Cu, 8 Sn, 11 Pb	221	32	110	16	18	...	55	...		80	C, T, S	General-utility alloy for bushings and bearings
C94500	73 Cu, 7 Sn, 20 Pb	172	25	83	12	12	...	50	...		80	C, S	Locomotive wearing parts, high-speed low-load bearings
Nickel-tin bronzes													
C94700	88 Cu, 5 Sn, 2 Zn, 5 Ni	345 (586)	50 (85)	159 (414)	23 (60)	35 (10)	...	85	(180)		30 (e)	C, T, I, M, S	Valve stems and bodies, bearings, wear guides, shift forks, feeding mechanisms, circuit breaker parts, gears, piston cylinders, nozzles
C94800	87 Cu, 5 Sn, 5 Ni	310 (414)	45 (60)	159 (207)	23 (30)	35 (8)	...	80 (120)	...		50 (e)	M, S	Structural castings, gear components, motion translation devices, machinery parts, bearings
C94900	80 Cu, 5Sn, 5 Pb, 5 Zn, 5 Ni												
Aluminum bronzes													
C95200	88 Cu, 3 Fe, 9 Al	552	80	186	27	35	...	...	125		50	C, T, M, P, S	Acid-resisting pumps, bearings, gears, valve seats, guides, plungers, pump rods, bushings
C95300	89 Cu, 1 Fe, 10 Al	517 (586)	75 (85)	186 (290)	27 (42)	25 (15)	...	...	140 (174)		55	C, T, M, P, S	Pickling baskets, nuts, gears, steel mill slippers, marine equipment, welding jaws
C95400	85 Cu, 4 Fe, 11 Al	586 (724)	85 (105)	241 (372)	35 (54)	18 (8)	...	...	170 (195)		60	C, T, M, P, S	Bearings, gears, worms, bushings, valve seats and guides, pickling hooks
C95410	85 Cu, 4 Fe, 11 Al, 2 Ni												
C95500	81 Cu, 4 Ni, 4 Fe, 11 Al	689 (827)	100 (120)	303 (469)	44 (68)	12 (10)	...	...	192 (230)		50	C, T, M, P, S	Valve guides and seats in aircraft engines, corrosion-resistant parts, bushings, gears, worms, pickling hooks and baskets, agitators
C95600	91 Cu, 7 Al, 2 Si	517	75	234	34	18	...	...	140		60	C, T, M, P, S	Cable connectors, terminals, valve stems, marine hardware, gears, worms, pole-line hardware

(a) Nominal composition, unless otherwise noted. For seldom-used alloys, only compositions are available. (b) Values for C82700, 84200, 96200, 96300, are minimum, not typical. As-cast values are for sand casting except C93900, continuous cast; and C85800, 87800, 87900, die cast. Heat treated values, in parentheses, indicate that the alloy responds to heat treatment. If heat treated values are not shown, the copper or copper alloy does not respond. (c) Free cutting brass = 100. (d) C, continuous; T, centrifugal; T, continuous; D, die; I, investment; M, permanent mold; P, plaster; S, sand. (Note: C82000, 82400, 82500, 82600, 82800 are also pressure cast.). (e) As-heat treated value for C94700, 20; for C94800, 40. Source: Copper Development Assn. Inc., New York

(continued)

470

Properties and applications of cast coppers and copper alloys (continued)

UNS designation	Nominal composition, %(a)	Tensile strength MPa	ksi	Yield strength MPa	ksi	Elongation in 2 in., %	Hardness Rockwell	Brinell 500 kg	Brinell 3000 kg	Machinability rating(c)	Casting types(d)	Typical applications
C95700	75 Cu, 2 Ni, 3 Fe, 8 Al, 12 Mn	655	95	310	45	26	...	...	180	50	C, T, M, P, S	Propellers, impellers, stator clamp segments, safety tools, welding rods, valves, pump casings
C95800	81 Cu, 5 Ni, 4 Fe, 9 Al, 1 Mn	655	95	262	38	25	...	...	159	50	C, T, M, P, S	Propeller hubs, blades, and other parts in contact with salt water
Copper nickels												
C96200	88.6 Cu, 10 Ni, 1.4 Fe	310	45	172	25	20	...	...	...	10	C, S	Components of items being used for sea water corrosion resistance
C96300	79.3 Cu, 20 Ni, 0.7 Fe	517	75	379	55	10	...	150	...	15	C, S	Centrifugally cast tailshaft sleeves
C96400	69.1 Cu, 30 Ni, 0.9 Fe	469	68	255	37	28	...	...	140	20	C, T, S	Valves, pump bodies, flanges, elbows used for sea-water corrosion resistance
C96600	68.5 Cu, 30 Ni, 1 Fe, 0.5 Be	(758)	(110)	(482)	(70)	(7)	...	...	(230)	20	C, T, I, M, S	High-strength constructional parts for sea-water corrosion resistance
C96700	67.6 Cu, 30 Ni, 0.9 Fe, 1.15 Be, 0.15 Zr, 0.15 Ti	(1207)	(175)	(552)	(80)	(10)	C26	...	...	40	I, M, S	Corrosion-resistant molds for plastics, high-strength constructional parts for sea-water use
Nickel silvers												
C97300	56 Cu, 2 Sn, 10 Pb, 12 Ni, 20 Zn	241	35	117	17	20	...	55	...	70	I, M, S	Hardware fittings, valves and valve trim, statuary, ornamental castings
C97400	59 Cu, 3 Sn, 5 Pb, 17 Ni, 16 Zn	262	38	117	17	20	...	70	...	60	C, I, S	Valves, hardware, fittings, ornamental castings
C97600	64 Cu, 4 Sn, 4 Pb, 20 Ni, 8 Zn	310	45	165	24	20	...	80	...	70	C, I, S	Marine castings, sanitary fittings, ornamental hardware, valves, pumps
C97800	66 Cu, 5 Sn, 2 Pb, 25 Ni, 2 Zn	379	55	207	30	15	...	...	130	60	I, M, S	Ornamental and sanitary castings, valves and valve seats, musical instrument components
Leaded coppers												
C98200	76.0 Cu, 24.0 Pb											
C98400	70.5 Cu, 28.5 Pb, 1.5 Ag											
C98600	65.0 Cu, 35.0 Pb, 1.5 Ag											
C98800	59.5 Cu, 40.0 Pb, 5.5 Ag											
Special alloys												
C99300	71.8 Cu, 15 Ni, 0.7 Fe, 11 Al, 1.5 Co	655	95	379	55	2	...	200	20	20	T, S	Glass making molds, plate glass rolls, marine hardware

(a) Nominal composition, unless otherwise noted. For seldom-used alloys, only compositions are available. (b) Values for C82700, 84200, 96200, 96300, are minimum, not typical. As-cast values are for sand casting except C93900, continuous cast; and C85800, 87800, 87900, die cast. Heat treated values, in parentheses, indicate that the alloy responds to heat treatment. If heat treated values are not shown, the copper or copper alloy does not respond. (c) Free cutting brass = 100. (d) C, cast; T, continuous; D, die; I, investment; M, permanent mold; P, plaster; S, sand. (Note: C82000, 82400, 82500, 82600, 82800 are also pressure cast.). (e) As-heat treated value for C94700, 20; for C94800, 40. Source: Copper Development Assn. Inc., New York

(continued)

Properties and applications of cast coppers and copper alloys (continued)

UNS designation	Nominal composition, %(a)	Tensile strength		Yield strength		Elongation in 2 in., %	Hardness			Machinability rating(c)	Casting types(d)	Typical applications
		MPa	ksi	MPa	ksi		Rockwell	Brinell 500 kg	Brinell 3000 kg			
C99400	90.4 Cu, 2.2 Ni, 2.0 Fe, 1.2 Al, 1.2 Si, 3.0 Zn	455 (545)	66 (79)	234 (372)	34 (54)	25	...	...	125 (170)	50	C, T, I, S	Valve stems, marine and other uses requiring resistance to dezincification and dealuminification, propeller wheels, electrical parts, mining equipment gears
C99500	87.9 Cu, 4.5 Ni, 4.0 Fe, 1.2 Al, 1.2 Si, 1.2 Zn	483	70	276	40	12	...	145	50	50	C, T, S	Same as C99400, but where higher yield strength is required
C99600	58 Cu, 2 Al, 40 Mn	558 (558)	81 (81)	248 (303)	36 (44)	34 (27)	B 72	...	130	...	C, T, M, S	Damping alloy to reduce noise and vibration
C99700	56.5 Cu, 1 Al, 1.5 Pb, 12 Mn, 5 Ni, 24 Zn	379	55	172	25	25	...	...	110	80	C, D, I, M, P, S	
C99750	58 Cu, 1 Al, 1 Pb, 20 Mn, 20 Zn	448 (517)	65 (75)	221 (276)	32 (40)	30 (20)	B 77 (B 82)	110 (119)	...	...	D, I, M, P, S	

(a) Nominal composition, unless otherwise noted. For seldom-used alloys, only compositions are available. (b) Values for C82700, 84200, 96200, 96300, are minimum, not typical. As-cast values are for sand casting except C93900, continuous cast; and C85800, 87800, 87900, die cast. Heat treated values, in parentheses, indicate that the alloy responds to heat treatment. If heat treated values are not shown, the copper or copper alloy does not respond. (c) Free cutting brass = 100. (d) C, centrifugal; T, continuous; D, die; I, investment; M, permanent mold; P, plaster; S, sand. (Note: C82000, 82400, 82500, 82600, 82800 are also pressure cast.). (e) As-heat treated value for C94700, 20; for C94800, 40. Source: Copper Development Assn. Inc., New York

471

Heat Treating

Typical stress-relieving temperatures for 19 wrought copper alloys

Alloy	Common name	Stress-relieving temperature(a)	
		°C	°F
C21000	Gilding metal	190	375
C22000	Commercial bronze	205	400
C23000	Red brass	230	450
C24000	Low brass	260	500
C26000	Cartridge brass	260	500
C27000	Yellow brass	260	500
C28000	Muntz metal	205	400
C36000	Free-cutting brass	245	475
C44300, C44400, C44500	Inhibited admiralty	290	550
C51000, C52100	Phosphor bronze	205	400
C61300, C61400	Aluminum bronze	345	650
C65500	High-silicon bronze	345	650
C70600, C71500	Copper nickel	260	500
C75200	Nickel silver	260	500

(a) Time at temperature, 1 h

Typical heat treatments and resulting properties for several low-temperature-hardening alloys

Alloy	Solution-treating temperature(a)		Aging treatment			Hardness	Electrical conductivity, % IACS
			Temperature		Time, h		
	°C	°F	°C	°F			
Precipitation-hardening alloys							
C15000	980	1795	500 to 550	930 to 1025	3	30 HRB	87 to 95
C17000, C17200, C17300	760 to 800	1400 to 1475	300 to 350	575 to 660	1 to 3	35 to 44 HRC	22
C17500, C17600	900 to 950	1650 to 1740	455 to 490	850 to 915	1 to 4	95 to 98 HRB	48
C18000(b), C81540	900 to 930	1650 to 1705	425 to 540	800 to 1000	2 to 3	92 to 96 HRB	42 to 48
C18200, C18400, C18500, C81500	980 to 1000	1795 to 1830	425 to 500	800 to 930	2 to 4	68 HRB	80
C94700	775 to 800	1425 to 1475	305 to 325	580 to 620	5	180 HB	15
C99400	885	1625	482	900	1	170 HB	17
Spinodal-hardening alloys							
C71900	900 to 950	1650 to 1740	425 to 760	800 to 1400	1 to 2	86 HRC	4
C72800	815 to 845	1500 to 1550	350 to 360	660 to 680	4	32 HRC	...

(a) Solution treating is followed by water quenching. (b) Alloy C18000 (81540) must be double aged—typically, 3 h at 540 °C (1000 °F) followed by 3 h at 425 °C (800 °F) (U.S. Patent No. 4 191 601)—to develop the higher levels of electrical conductivity and hardness

Recommended precipitation-hardening schedules and resulting properties for solution-treated copper-beryllium castings

Alloy	Solution treatment			Aging treatment			Tensile strength		Yield strength(a)		Elongation(b), %	Hardness	Electrical conductivity, % IACS
	Temperature		Time, min	Temperature		Time, min							
	°C	°F		°C	°F		MPa	ksi	MPa	ksi			
C81300	980-1010	1800-1850	60	480	900	120	365	53	250	36	11	89 HB(c)	60
C81700	900-925	1650-1700	60	455	850	180	635	92	470	68	8	217 HB(d)	48
C81800	900-925	1650-1700	60	480	900	180	705	102	515	75	8	92 HRB	45
C82000	900-925	1650-1700	180	480	900	180	690	100	515	75	8	195 HB(d)	45
C82100	900-925	1650-1700	60	455	850	180	635	92	470	68	8	217 HB(d)	48
C82200	900-925	1650-1700	60	445-455	835-850	120	655	95	515	75	8	96 HRB	45
C82400	785-850	1450-1560	60	345	650	180	1035	150	965	140	1	34 HRC	25
C82500	785-800	1450-1475	60	345	650	180	1105	160	795	115	1	40 HRC	20
C82600	785-800	1450-1475	60	345	650	180	1105	160	1035	150	1	40 HRC	19
C82700	785-800	1450-1475	180	345	650	180	1070	155	895	130	0	39 HRC	20
C82800	785-800	1450-1475	60	345	650	180	1140	165	1000	145	1	42 HRC	18

(a) At 0.2% extension under load. (b) In 50 mm or 2 in. (c) 500-kg load. (d) 3000-kg load

Annealing temperatures for cold-worked coppers and copper alloys

Alloy	Common name	Annealing temperature	
		°C	°F
Wrought coppers			
C10200	Oxygen-free copper	425 to 650	800 to 1200
C11000	Electrolytic tough pitch copper	250 to 650	500 to 1200
C11300, C11400, C11500, C11600	Silver-bearing tough pitch copper	400 to 475	750 to 900
C12000	Phosphorus-deoxidized copper, low residual phosphorus	325 to 650	600 to 1200
C12200	Phosphorus-deoxidized copper, high residual phosphorus	375 to 650	700 to 1200
C14500	Phosphorus-deoxidized, tellurium-bearing copper	425 to 650	800 to 1200
Wrought copper alloys			
C17000, C17200, C17500	Beryllium copper	775 to 925(a)	1425 to 1700(a)
C21000	Gilding metal	425 to 800	800 to 1450
C22000	Commercial bronze	425 to 800	800 to 1450
C22600	Jewelry bronze	425 to 750	800 to 1400
C23000	Red brass	425 to 725	800 to 1350
C24000	Low brass	425 to 700	800 to 1300
C26000	Cartridge brass	425 to 750	800 to 1400
C26800, C27000, C27400	Yellow brass	425 to 700	800 to 1300
C28000	Muntz metal	425 to 600	800 to 1100
C31400	Leaded commercial bronze	425 to 650	800 to 1200
C33000, C33500	Low-leaded brass	425 to 650	800 to 1200
C33200, C34200, C35300	High-leaded brass	425 to 650	800 to 1200
C34000, C35000	Medium-leaded brass	425 to 650	800 to 1200
C35600	Extra-high-leaded brass	425 to 650	800 to 1200
C36000	Free-cutting brass	425 to 600	800 to 1100
C36500, C36600, C36700, C36800	Leaded Muntz metal	425 to 600	800 to 1100
C37000	Free-cutting Muntz metal	425 to 650	800 to 1200
C37700	Forging brass	425 to 600	800 to 1100
C38500	Architectural bronze	425 to 600	800 to 1100
C44300, C44400, C44500	Inhibited admiralty	425 to 600	800 to 1100
C46200	Naval brass	425 to 600	800 to 1100
C48200, C48500	Leaded naval brass	425 to 600	800 to 1100
C50500	Phosphor bronze	475 to 650	900 to 1200
C51000, C52100, C54200	Phosphor bronze	475 to 675	900 to 1250
C53200, C53400, C54400	Free-cutting phosphor bronze	475 to 675	900 to 1250
C60600, C60800	Aluminum bronze	550 to 650	1000 to 1200
C61000	Aluminum bronze	600 to 675	1100 to 1250
C61300, C61400	Aluminum bronze	750 to 875	1400 to 1600
C61800, C61900, C62400	Aluminum bronze	600 to 650(b)	1100 to 1200(b)
C63000	Aluminum bronze	650 to 700(c)	1200 to 1300(c)
C63200	Aluminum bronze	675 to 725(c)	1250 to 1350(c)
C64200	Aluminum bronze	Above 650	Above 1200
C65100	Low-silicon bronze	475 to 675	900 to 1250
C65500	High-silicon bronze	475 to 700	900 to 1300
C67000, C67500	Manganese bronze	425 to 600	800 to 1100
C68700	Aluminum brass	425 to 600	800 to 1100
C70600	Copper nickel, 10%	600 to 825	1100 to 1500
C71500	Copper nickel, 30%	650 to 825	1200 to 1500
C75200, C75700, C77000	Nickel silver	600 to 825	1100 to 1500

(a) Solution-treating temperature. (b) Cool rapidly (cooling method important in determining result of annealing). (c) Air cool (cooling method important in determining result of annealing)

Effects of special precipitation-hardening treatments on mechanical properties and electrical conductivity of Cu-1.9Be strip

Initial condition	Aging treatment Time, min	Temperature °C	°F	Tensile strength MPa	ksi	Yield strength(a) MPa	ksi	Elongation(b), %	Electrical conductivity, % IACS	Fatigue strength(c) MPa	ksi	Modulus of elasticity GPa	10^6 psi
Alloy C17200													
Annealed	None	...	...	465	67.5	250	36	49	18.0	205	30	115	16.5
	5	370	700	855	124	695	101	18	19.5	...	...	120	17.5
	15	370	700	1195	173	1055	153	10	22.0	...	...	125	18.0
	30	370	700	1260	182.5	1060	153.5	6	23.0	...	...	125	18.0
	60	370	700	1240	180	1055	153	5	25.5	255	37	130	18.5
	120	370	700	1195	173.5	1040	151	6	26.0	...	...	130	18.5
	240	370	700	1150	167	980	142	6	26.5	...	...	130	19.0
¼ hard	None	...	...	570	82.5	485	70.5	21	17.0	220	32	115	17.0
	5	370	700	1115	162	945	137	9	18.5	...	...	125	18.0
	15	370	700	1250	181	1115	162	6	20.5	...	...	130	18.5
	30	370	700	1290	187	1125	163.5	4	23.5	290	42	130	18.5
	60	370	700	1230	178.5	1060	154	3	25.5	...	...	130	18.5
	120	370	700	1185	172	1000	145	4	26.5	...	...	130	19.0
	240	370	700	1155	167.5	970	141	6	27.0	...	...	130	19.0
½ hard	None	...	...	605	87.5	555	80.5	17	16.0	230	33	115	17.0
	3	370	700	1010	146.5	885	128	11	18.0	230	33	125	18.0
	5	370	700	1280	186	1110	161	3	21.0	295	43	125	18.0
	15	370	700	1310	190	1175	170.5	2	23.0	305	44	130	18.5
	30	370	700	1325	192.5	1180	171	2	24.5	305	44	130	18.5
	60	370	700	1280	185.5	1105	160	2	25.0	295	43	130	18.5
	120	370	700	1200	174	1040	150.5	3	26.0	275	40	130	18.5
	240	370	700	1185	172	1035	150	3	27.0	275	40	130	19.0
	420	370	700	1010	146.5	860	125	10	27.0	200	29	130	19.0
Hard	None	...	...	730	106	690	100	5	15.0	270	39	120	17.5
	5	370	700	1300	188.5	1125	163	3	18.0	...	...	125	18.0
	15	370	700	1360	197	1195	173	2	21.0	...	...	130	18.5
	30	370	700	1310	190	1170	170	1	24.5	315	46	130	19.0
	60	370	700	1295	188	1105	160	1	26.5	...	...	130	19.0
	120	370	700	1240	180	1090	158	2	27.5	...	...	130	19.0
	240	370	700	1215	176	1055	153	2	27.5	...	...	130	19.0
Alloy C17500													
Annealed	None	...	...	350	51	170	25	30	25	...	...	110	16.3
	120	425	800	805	117	625	91	14	44	...	...	135	19.3
	120	455	850	835	121	675	98	14	48	...	...	140	20.0
	120	480	900	805	116.5	625	91	14	48	215	31	140	20.0
	120	510	950	795	115	600	87	16	48.5	...	...	140	20.0
Hard	None	...	...	440	63.5	425	61.5	2	27.8	...	...	125	18.3
	120	425	800	985	142.5	860	125	11	44.0	...	...	140	20.0
	120	455	850	915	133	800	116	13	45.0	...	...	140	20.0
	120	480	900	850	123	760	110.5	13	47.5	250	36	140	20.0
	120	510	950	800	116	705	102	12	49.0	...	...	140	20.0

(a) At 0.2% offset. (b) In 50 mm or 2 in. (c) 10^7 cycles

Properties and precipitation treatments usually specified for copper-beryllium alloys

Initial condition	Standard aging treatment Time, h	Temperature °C	Temperature °F	Tensile strength MPa	Tensile strength ksi	Yield strength(a) MPa	Yield strength(a) ksi	Elongation(b), %	Hardness(c)	Electrical conductivity, % IACS
C17200										
Flat products:										
Annealed	None	...	...	415-540	60-78	195-380	28-55	35-60	45-78 HRB	17-19
¼ hard	None	...	...	515-605	75-88	415-550	60-80	10-40	68-90 HRB	16-18
½ hard	None	...	...	585-690	85-100	515-655	75-95	10-25	88-96 HRB	15-17
Hard	None	...	...	690-825	100-120	620-770	90-112	2-8	96-102 HRB	15-17
Annealed(d)	3	315	600	1140-1345	165-195	965-1205	140-175	4-10	35-40 HRC	22-25
Annealed	½	370	700	1105-1310	160-190	895-1205	130-175	3-10	34-40 HRC	22-25
¼ hard(d)	2	315	600	1205-1415	175-205	1035-1275	150-185	3-6	37-42 HRC	22-25
¼ hard	⅓	370	700	1170-1380	170-200	965-1275	140-185	2-6	36-42 HRC	22-25
½ hard(d)	2	315	600	1275-1485	185-215	1105-1345	160-195	2-5	39-44 HRC	22-25
½ hard	¼	370	700	1240-1450	180-210	1070-1345	155-195	2-5	38-44 HRC	22-25
Hard(d)	2	315	600	1310-1575	190-220	1140-1415	165-205	1-4	40-45 HRC	22-25
Hard	¼	370	700	1275-1480	185-215	1105-1415	160-205	1-4	39-45 HRC	22-25
Rod, bar and plate:										
Annealed	None	...	...	415-585	60-85	185-205	20-30	35-60	45-85 HRB	17-19
Hard	None	...	...	585-895	85-130	515-725	75-105	10-20	88-103 HRB	15-17
Annealed(d)	3	315	600	1140-1345	165-200	1000-1205	145-175	3-10	36-41 HRC	22-25
Hard(d)	2	315	600	1205-1550	175-225	1035-1380	150-200	2-5	39-45 HRC	22-25
Wire(e):										
Annealed	None	...	...	450-590	65-85	185-240	20-35	35-55	...	17-19
¼ hard	None	...	...	620-795	90-115	485-655	70-95	10-35	...	15-17
½ hard	None	...	...	760-930	110-135	620-760	90-110	4-10	...	15-17
¾ hard	None	...	...	895-1070	130-155	760-930	110-135	2-8	...	15-17
Annealed(d)	3	315	600	1140-1310	165-190	1000-1205	145-175	3-8	...	22-25
Annealed	½	370	700	1105-1310	160-190	930-1205	135-175	3-8	...	22-25
¼ hard(d)	2	315	600	1205-1415	175-205	1105-1310	160-190	2-5	...	22-25
¼ hard	¼	370	700	1170-1415	170-205	1035-1310	150-190	2-5	...	22-25
½ hard(d)	1½	315	600	1310-1480	190-215	1205-1380	175-200	1-3	...	22-25
½ hard	¼	370	700	1275-1480	185-215	1170-1380	170-200	1-3	...	22-25
¾ hard(d)	1	315	600	1345-1585	195-230	1245-1415	180-205	1-3	...	22-25
¾ hard	¼	370	700	1310-1585	190-230	1205-1415	175-205	1-3	...	22-25
C17000										
Flat products:										
Annealed	None	...	...	415-540	60-78	170-365	25-55	35-60	45-78 HRB	17-19
¼ hard	None	...	...	515-605	75-88	310-515	45-75	10-40	68-90 HRB	16-18
½ hard	None	...	...	585-690	85-100	450-620	65-90	10-25	88-96 HRB	15-17
Hard	None	...	...	690-825	100-120	550-760	80-110	2-8	96-102 HRB	15-17
Annealed	3	315	600	1035-1240	150-180	895-1105	130-165	4-10	33-39 HRC	22-25
Annealed(d)	3	345	650	1105-1275	160-185	860-1140	125-165	4-10	34-40 HRC	22-25
¼ hard	2	315	600	1105-1310	160-190	860-1140	135-170	3-6	34-40 HRC	22-25
¼ hard(d)	3	330	625	1170-1345	170-195	895-1170	130-170	3-6	36-41 HRC	22-25
½ hard	2	315	600	1170-1380	170-200	895-1170	145-175	2-5	36-41 HRC	22-25
½ hard(d)	2	330	625	1240-1380	180-200	965-1240	140-180	2-5	38-42 HRC	22-25
Hard	2	315	600	1240-1450	180-210	965-1240	155-180	2-5	38-42 HRC	22-25
Hard(d)	2	330	625	1275-1415	185-205	1070-1345	155-195	2-5	39-43 HRC	22-25
Rod and bar:										
Annealed	None	...	...	415-585	60-85	185-205	20-30	35-60	45-85 HRB	17-19
Hard	None	...	...	585-895	85-130	515-725	75-105	10-20	88-103 HRB	15-17
Annealed	3	315	600	1035-1240	150-180	860-1070	125-155	4-10	32-39 HRC	22-25
Annealed(d)	3	345	650	1105-1275	160-185	930-1140	135-165	4-10	34-40 HRC	22-25
Hard	2	315	600	1140-1380	165-200	930-1140	135-165	2-5	36-41 HRC	22-25
Hard(d)	2	345	650	1205-1415	175-205	965-1170	140-170	2-5	38-42 HRC	22-25
C17500, C17510										
Rod, bar, plate and flat products:										
Annealed	None	...	...	240-380	35-55	185-205	20-30	20-35	20-43 HRB	25-30
Hard	None	...	...	515-585	75-85	380-550	55-80	3-10	78-88 HRB	20-30
Annealed	3	480	900	690-760	100-120	550-690	80-100	10-20	92-100 HRB	45-60
Annealed(d)	3	455	850	725-825	105-120	550-725	80-105	8-12	93-100 HRB	45-52
Hard	2	480	900	760-860	110-130	690-825	100-120	8-15	95-103 HRB	45-60
Hard(d)	2	455	850	795-930	115-135	725-860	105-125	5-8	97-104 HRB	45-52

(a) At 0.2% offset. (b) In 50 mm or 2 in. (c) Rockwell B and C hardness values are accurate only if metal is at least 1 mm (0.040 in.) thick. (d) Heat treatment that provides optimum strength. (e) For wire diameters greater than 1.3 mm (0.050 in.)

Typical effects of heat treatment and cold work on properties of copper-1% chromium alloys

Condition	Ultimate tensile strength		Yield strength(a)		Elongation(b), %	Hardness	Electrical conductivity, % IACS
	MPa	ksi	MPa	ksi			
Alloy C18200							
Solution treated	240	35	105	15	42	50 HRF	35 to 42
Solution treated and aged	350	51	275	40	15	90 HB(c)	75 to 82
Solution treated and drawn 40%	415	60	310	45	15	65 HRB	40
Solution treated hard drawn and aged	435	63	385	56	18	68 to 75 HRB	80
Solution treated, aged and drawn 30%	480	70	425	62	18	75 to 80 HRB	80
Alloy C81500							
Cast, solution treated and aged	350	51	275	40	17	105 HB(c)	75 to 80

(a) At 0.5% extension under load. (b) In 50 mm or 2 in. (c) 500-kg load

Effect of heat treatment and cold work on properties of copper-zirconium alloy C15000

Solution treating temperature(a)		Amount of cold work, %	Aging		Time, h	Tensile strength		Yield strength		Elongation(b), %	Hardness, HRB	Electrical conductivity, % IACS
°C	°F		Temperature									
			°C	°F		MPa	ksi	MPa	ksi			
900	1650	20	475	885	1	310	45	260	38	25	48	85 min
		80	425	795	1	425	62	380	55	12	64	85 min
980	1795	None	...	...	...	200	29	41(c)	6(c)	54	...	64
		20	...	...	...	270	39	250(c)	36(c)	26	37	64
		80	...	...	...	440	64	420(c)	61(c)	19	73	64
		None	500	930	3	205	30	90	13	51	...	87
		None	550	1025	3	205	30	90	13	49	...	95
		20	400	750	3	330	48	260	38	31	50	80
		20	450	840	3	330	48	275	40	28	57	92
		85	400	750	3	495	72	440	64	24	79	85
		85	450	840	3	470	68	425	62	23	74	91

(a) Hold 30 min, water quench. (b) In 50 mm or 2 in. (c) 0.5% extension under load

Typical heat treating schedules and resulting properties for precipitation-hardened miscellaneous alloys

Alloy	Solution treatment			Tempering treatment			Tensile strength		Yield strength(a)		Elongation(b), %	Hardness, HB(c)
	Temperature		Time, min	Temperature		Time, min						
	°C	°F		°C	°F		MPa	ksi	MPa	ksi		
C94700	775-800	1425-1475	120	305-325	580-620	300	585	85	415	60	10	180
C94800	...	...	...	305-325	580-620	360-1000	415	60	205	30	8	120
C96600	995	1825	60	510	950	180	760	110	485	70	7	230
C99400	885	1625	60	480	900	60	545	79	370	54	...	170
C99500	885	1625	60	480	900	60	595	86	425	62	8	196

(a) At 0.2% extension under load for C96600; at 0.5% extension under load for all other alloys. (b) In 50 mm or 2 in. (c) 3000-kg load

Typical strengths and recommended aging times for various spinodal alloys

UNS no.	Nominal composition	Solution-treated and cold-worked temper	Aging time, h(a)	Tensile strength MPa	Tensile strength ksi	Yield strength(b) MPa	Yield strength(b) ksi	Elongation,%
C72600	Cu-4Ni-4Sn	TD 02(½H)	1½	635 to 690	92 to 100	495 to 570	72 to 83	12
		TD 06(XH)	1½	690 to 725	100 to 105	565 to 620	82 to 90	9
		TD 08(S)	1½	705 to 795	102 to 115	565 to 655	82 to 95	7
C72700	Cu-9Ni-6Sn	TD 04(H)	1½	860 to 1035	125 to 150	760 to 895	110 to 130	8
		TD 14(SS)	1½	1055 to 1145	153 to 166	930 to 985	135 to 143	...
C72800	Cu-10Ni-8Sn-0.2Nb	TB 00 cast and solution treated	4 to 6	830 to 965	120 to 140	550 to 690	80 to 100	3
		TB 00 hot work and solution treated	3 to 5	965 to 1070	140 to 155	690 to 825	100 to 120	6 to 14
		TD 01(¼H)	3	1140 to 1240	165 to 180	895 to 930	130 to 135	7
		TD 04(H)	3	1205 to 1380	175 to 200	930 to 1000	135 to 145	7
		TD 06(XH)	3	1205 to 1380	175 to 200	965 to 1035	140 to 150	5
		TD 08(S)	3	1240 to 1380	180 to 200	1000 to 1070	145 to 155	4
		TD 14(SS)	1½	1240 to 1380	180 to 200	1070 to 1140	155 to 165	2.5
C72900	Cu-15Ni-8Sn	TD 14(SS)	1½	1140 to 1380	165 to 200	1035 to 1170	150 to 170	3
C71900	Cu-30Ni-3Cr	Hot extruded	1½	550	80	345	50	25

(a) At 350 °C (660 °F), except C71900, at 405 °C (760 °F). (b) 0.05% offset yield; C72800, 0.01% offset yield; C71900, 0.20% offset yield

Typical heat treatments and resulting properties for complex (alpha-beta) aluminum bronzes

Alloy	Typical condition(a)	Tensile strength MPa	Tensile strength ksi	Yield strength(b) MPa	Yield strength(b) ksi	Elongation(c), %	Hardness, HB
C62400	As forged or extruded	620 to 690	90 to 100	240 to 260	35 to 38	14 to 16	163 to 183
	Solution treated at 870 °C (1600 °F) and quenched, tempered 2 h at 620 °C (1150 °F)	675 to 725	98 to 105	345 to 385	50 to 56	8 to 14	187 to 202
C63000	As forged or extruded	730	106	365	53	13	187
	Solution treated at 855 °C (1575 °F) and quenched, tempered 2 h at 650 °C (1200 °F)	760	110	425	62	13	212
C95300	As cast	495 to 530	72 to 77	185 to 205	27 to 30	27 to 30	137 to 140
	Solution treated at 855 °C (1575 °F) and quenched, tempered 2 h at 620 °C (1150 °F)	585	85	290	42	14 to 16	159 to 179
C95400	As cast	585 to 690	85 to 100	240 to 260	35 to 38	14 to 18	156 to 179
	Solution treated at 870 °C (1600 °F) and quenched, tempered 2 h at 620 °C (1150 °F)	655 to 725	95 to 105	330 to 370	48 to 54	8 to 14	187 to 202
C95500	As cast	640 to 710	93 to 103	290 to 310	42 to 45	10 to 14	183 to 192
	Solution treated at 855 °C (1575 °F) and quenched, tempered 2 h at 650 °C (1200 °F)	775 to 800	112 to 116	440 to 470	64 to 68	10 to 14	217 to 234

(a) As-cast condition is typical for moderate sections shaken out at temperatures above 540 °C (1000 °F) and fan cooled; or mold cooled, annealed at 620 °C (1150 °F) and fan (rapid) cooled. (b) At 0.5% extension under load. (c) In 50 mm or 2 in.

Machining

Machinability ratings of several copper casting alloys

UNS number	Common name	Machinability rating (a), %
Free-cutting alloys		
C83600	Leaded red brass	90
C83800	Leaded red brass	90
C84400	Leaded semi-red brass	90
C84800	Leaded semi-red brass	90
C94300	High-leaded tin bronze	90
C85200	Leaded yellow brass	80
C85400	Leaded yellow brass	80
C93700	High-leaded tin bronze	80
C93800	High-leaded tin bronze	80
C93200	High-leaded tin bronze	70
C93500	High-leaded tin bronze	70
C97300	Leaded nickel brass	70
Moderately machinable alloys		
C86400	Leaded high-strength manganese bronze	60
C92200	Leaded tin bronze	60
C92300	Leaded tin bronze	60
C90300	Tin bronze	50
C90500	Tin bronze	50
C95600	Silicon-aluminum bronze	50
C95300	Aluminum bronze	35
C86500	High-strength manganese bronze	30
Hard-to-machine alloys		
C86300	High-strength manganese bronze	20
C95200	9% aluminum bronze	20
C95400	11% aluminum bronze	20
C95500	Nickel-aluminum bronze	20

(a) Machinability rating expressed as a percentage of the machinability of C36000, free-cutting brass. The rating is based on relative speed for equivalent tool life. For instance, a material having a rating of 50 should be machined at about half the speed that would be used to make a similar cut in C36000

Joining

Relative solderability of various types of copper metals

Type of copper metal	Solderability and remarks
Coppers(a)	Excellent. Need only rosin or other noncorrosive flux
Copper-tin alloys	Good. Easily soldered with activated rosin and intermediate fluxes
Copper-zinc alloys	Good. Easily soldered with activated rosin and intermediate fluxes
Copper-nickel alloys	Good. Easily soldered with intermediate and corrosive type fluxes
Chromium copper and beryllium copper	Good. Require intermediate and corrosive type fluxes
Copper-silicon alloys	Fair. Silicon produces refractory oxides that require use of corrosive fluxes
Copper-aluminum alloys	Difficult. May be soldered with help of very corrosive fluxes
High-strength manganese bronze	Not recommended. Should be plated to ensure consistent solderability

(a) Includes tough-pitch, oxygen-free, phosphorized, arsenical, silver-bearing, leaded, tellurium, and selenium coppers

Comparative behavior of copper metals in resistance welding

Alloy type	Common name	Conductivity, % IACS	Rating
Cu-Si	Silicon bronzes	7 to 12	Excellent
Cu-Ni	Copper nickels	4 to 10	Excellent
Cu-Ni-Zn	Nickel silvers	5 to 9	Good
Cu-Al	Aluminum bronzes	7 to 18	Good
Cu-Sn	Phosphor bronzes	10 to 22	Fair
Cu-Zn-Mn	Manganese brass	15 to 16	Fair
Cu-Zn-Si	Silicon red brass	15 to 16	Fair
Cu-Zn (high Zn)	Yellow brasses	22 to 28	Fair
Cu-Zn-Mn-Fe	Manganese bronzes	22 to 28	Fair
Cu-Zn (low Zn)	Red brasses	32 to 43	Poor
Cu	High-copper alloys	50 to 65	Poor

Joining-process selection guide for copper alloys based on service requirements

| Service requirement | Fusion welding | | | | | | | Resistance welding | | | | Solid-state welding | | | | Brazing | Soldering | Adhesive bonding | Mechanical fastening |
| | Arc welding | | | | Other welding processes | | | | | | | | | | | | | | |
	Gas metal-arc	Gas tungsten-arc	Plasma-arc	Shielded metal-arc	Electron beam	Laser beam	Oxyfuel gas	Resistance spot	Resistance seam	Projection	Flash	Diffusion	Explosive	Ultrasonic	Friction	Brazing	Soldering	Adhesive bonding	Mechanical fastening
Primary structural	B	A	E	B	A	E	C	A	A	D	A	A	C	C	D	B	D	C	B
Elevated temperature	A	A	E	B	C	E	C	A	A	D	A	A	C	C	D	A	C	A	A
Ambient temperature	B	A	E	C	A	E	D	A	A	D	A	A	C	C	D	B	D	D	B
Cryogenic	B	A	E	D	A	E	D	D	B	D	C	A	C	C	D	B	D	C	B
Vacuum	A	A	E	B	A	E	C	B	A	D	B	A	C	C	D	A	D	C	C
Atmospheric pressure	B	A	E	B	C	E	D	D	C	D	C	A	C	C	D	C	D	D	C
High pressure	A	A	E	D	C	E	B	A	A	C	A	A	C	C	C	C	D	A	C
Secondary structural	A	A	C	A	A	E	B	A	A	A	A	A	C	B	B	A	B	A	A
Noncritical	C	C	C	B	C	E	C	D	D	D	C	A	B	A	B	B	B	A	A
Dissimilar metal joining																			

Note: A, most satisfactory; B, satisfactory; C, restricted use; D, prohibited use; E, experimental

Joining-process selection guide for copper alloys based on joint configuration

| Joint configuration | Fusion welding | | | | | | | Resistance welding | | | | Solid-state welding | | | | Brazing | Soldering | Adhesive bonding | Mechanical fastening |
| | Arc welding | | | | Other welding processes | | | | | | | | | | | | | | |
	Gas metal-arc	Gas tungsten-arc	Plasma-arc	Shielded metal-arc	Electron beam	Laser beam	Oxyfuel gas	Resistance spot	Resistance seam	Projection	Flash	Diffusion	Explosive	Ultrasonic	Friction	Brazing	Soldering	Adhesive bonding	Mechanical fastening
Butt	A	A	A	A	A	A,E	B	D	D	D	A	C	D	D	B	D	D	D	D
Tee	A	A	B	A	C	C,E	B	D	D	D	C	B	D	D	D	B	B	C	D
Edge	A	A	B	B	A	A,E	B	D	D	D	D	C	D	D	D	C	D	D	D
Corner	A	A	B	B	A	A,E	B	D	D	D	D	C	D	D	B	B	B	D	D
Flange	A	A	B	B	A	A,E	B	D	D	C	D	C	B,E	D	D	C	C	C	B
Scarf	D	D	D	D	C	D	D	B	B	D	D	C	C	B	D	A	B	B	B
Strap butt (splice)	C	C	C	C	C	B,E	D			C	D				D		B	A	A
Lap																			
Shear load	B	B	B	B	A	A,E	D	A	A	C	D	A	B	B	D	A	B	A	A
Tensile load	B	B	B	B	A	A,E	D	B	B	C	D	A	B	B	D	D	D	C	A

Note: A, most satisfactory; B, satisfactory; C, restricted use; D, prohibited use; E, experimental

Joining-process selection guide for copper alloys based on thickness of parts being joined

| Metal thickness | | Fusion welding — Arc welding | | | | Fusion welding — Other welding processes | | | Resistance welding | | | | Solid-state welding | | | | Brazing | Soldering | Adhesive bonding | Mechanical fastening |
mm	in.	Gas metal-arc	Gas tungsten-arc	Plasma-arc	Shielded metal-arc	Electron beam	Laser beam	Oxyfuel gas	Resistance spot	Resistance seam	Projection	Flash	Diffusion	Explosive	Ultrasonic	Friction				
0.025–0.25	0.001–0.010	D	B	A	D	A	B,E	D	C	C	D	D	A	D	A	D	B	C	B	C
0.25–0.50	0.010–0.020	D	A	B	D	A	C,E	D	A	A	B	D	A	D	A	D	B	B	A	B
0.50–1.25	0.020–0.050	C	A	B	D	A	D	B	A	A	A	D	A	D	C	B	B	A	A	A
1.25–2.50	0.050–0.100	A	A	A	B	A	D	B	A	A	A	A	A	D	D	B	B	C	A	A
2.50–3.75	0.100–0.150	A	A	A	A	A	D	B	C	C	C	A	A	C	D	B	B	D	A	A
3.75–6.25	0.150–0.250	A	A	A	A	A	D	C	D	D	C	A	A	C	D	B	C	D	B	A
6.25–12.50	0.250–0.500	A	A	C	A	A	D	D	D	D	D	A	A	C	D	B	C	D	C	A
12.50–25.0	0.500–1.00	A	A	D	A	A	D	D	D	D	D	A	A	C	D	C	D	D	D	A
25.0–62.5	1.00–2.50	B	B	D	A	A	D	D	C	C	C	C	A	C	D	C	D	D	D	A
Over 62.5	Over 2.50	C	B	D	A	A	D	D	C	D	D	C	A	A	D	A	…	D	D	D	A
Thick to thin		A	A	C	B	A	B, E	D	C	C	C	C	A	A	A	…	B	A	A	A

Note: A, most satisfactory; B, satisfactory; C, restricted use; D, prohibited use; E, experimental

Joining-process selection guide for copper alloys based on alloy composition

| | Fusion welding | | | | | | | | | Resistance welding | | | | | | | |
| | Arc welding | | | | | Other welding processes | | | | | | | | | | | |
	Gas metal-arc	Gas tungsten-arc	Submerged arc	Shielded metal-arc	Stud	Electron beam	Electroslag	Laser beam	Oxyfuel gas	Spot	Seam	Projection	Flash butt	Brazing	Soldering	Adhesive bonding	Mechanical fastening
Coppers																	
Oxygen free	B	B	D	D	D	B, E	D	E	C	D	D	D	B	A	A	C	A
Deoxidized	A	A	D	D	D	B, E	D	E	B	D	D	D	B	A	A	C	A
Tough pitch	C	C	D	D	D	B, E	D	E	D	D	D	D	B	B	A	C	A
High copper alloys																	
Cadmium	B	B	D	D	D	…	D	…	B	D	D	D	B	A	A	C	A
Beryllium	B	B	D	B	D	…	D	…	D	B	C	B	C	B	B	C	A
Chromium	B	B	D	D	D	…	D	…	D	D	D	D	C	B	B	C	A
Leaded	D	D	D	D	D	…	D	…	D	D	D	D	C	B	A	C	A
Oxide dispersion-strengthened	D	D	D	D	D	…	D	…	D	C	C	C	C	B	A	C	A
Brasses																	
Red	B	B	D	D	D	…	D	…	B	C	D	C	B	A	A	C	A
Yellow	C	C	D	D	D	…	D	…	B	B	D	B	B	B	A	C	A
Leaded	D	D	D	D	D	…	D	…	D	D	D	D	C	B	A	C	A
Tin	B	B	D	C	D	…	D	…	B	B	C	B	B	A	A	C	A
Bronzes																	
Phosphor	B	B	D	C	D	…	D	…	C	B	C	B	A	A	A	C	A
Leaded phosphor	D	D	D	D	D	…	D	…	D	D	D	D	C	B	C	C	A
Aluminum	A	A	D	B	D	…	D	…	D	B	B	B	B	C	C	C	A
Silicon	A	A	D	C	D	…	D	…	B	A	A	A	A	A	A	C	A
Copper nickels																	
10% Ni	A	A	D	B	D	…	D	…	C	B	B	B	A	A	A	C	A
30% Ni	C	C	D	C	D	…	D	…	B	A	A	A	A	A	A	C	A
Nickel silvers	C	C	D	C	D	…	D	…	B	B	C	B	B	A	A	C	A

Note: A, most satisfactory; B, satisfactory; C, restricted use; D, prohibited use; E, experimental

Comparison of soldering with other bonding methods for electrical applications

Factor	Metallurgical bonding			Mechanical bonding			Adhesive bonding conductive cement
	Soldering	Brazing	Welding	Crimping	Screwing	Wrapping	
Temperature limit of joint, °C (°F)	73-460 (100-800)	460-900 (800-1600)	Conductor melting temperature	No limit except wire	No limit except wire	No limit except wire	100-180 (160-300)
Heating effect on assembly	Small	Large	Small (quick)	None	None	None	Cures at ambient to 120 °C (250 °F)
Ease of rework and rebonding	Simple	Simple	Not practical	Not practical	Simple	Simple	Not practical
Process economy							
Equipment cost	Low	Medium	High	Low	Low	Low	Low
Ease of automation	Easiest	More difficult	More difficult	More difficult	More difficult	More difficult	More difficult
Extra hardware	No	No	No	Yes	Yes	No	No
Joint stable under vibration	Yes	Yes	Yes	Yes	No	Yes	Yes
Joint stable in oxidizing environment	Yes	Yes	Yes	No	No	Yes	Yes

Representative list of solders used for joining copper metals

ASTM classification	Nominal composition	Solidus		Liquidus		Melting range		Density		Typical applications
		°C	°F	°C	°F	°C	°F	Mg/m³	lb/in.³	
Tin-lead(a)										
5A	95Pb-5Sn	308	586	312	594	4	8	11.3	0.408	Coating and joining
10B(b)	90Pb-10Sn	268	514	301	573	33	59	10.8	0.389	Coating and joining
15B(b)	85Pb-15Sn	225	437	290	553	65	116	10.5	0.379	Coating and joining
20A	80Pb-20Sn	183	361	280	535	97	174	10.2	0.368	Coating and joining
25A	75Pb-25Sn	183	361	267	511	84	150	9.99	0.361	Machine and torch soldering
30A	70Pb-30Sn	183	361	255	491	72	130	9.69	0.350	Machine and torch soldering
35A	65Pb-35Sn	183	361	247	477	64	116	9.69	0.350	General purpose; wiping
40A	60Pb-40Sn	183	361	235	455	52	94	9.27	0.335	Wiping; auto radiators
45A	55Pb-45Sn	183	361	228	441	45	80	8.97	0.324	Auto radiators; roofing seams
50A	50Pb-50Sn	183	361	217	421	34	60	8.83	0.319	General purpose; most widely used on copper
60A	40Pb-60Sn	183	361	190	374	7	13	8.64	0.312	"Fine solder"; general purpose, especially where low soldering temperature is essential
63A	37Pb-63Sn	183	361	183	361	0	0	8.40	0.303	"Eutectic solder"; lowest-melting lead-tin solder
70A	30Pb-70Sn	183	361	192	378	9	17	8.32	0.301	…
Tin-lead-antimony and tin-antimony										
20C	79Pb-20Sn-1.0Sb	184	363	270	517	86	154	10.2	0.367	Machine soldering and coating(c)
25C	73.7Pb-25Sn-1.3 Sb	185	364	262	504	77	140	9.94	0.359	Torch and machine soldering(c)
30C	68.4Pb-30Sn-1.6Sb	185	364	250	482	65	118	9.63	0.348	Torch and machine soldering(c)
35C	63.2Pb-35Sn-1.8Sb	186	365	243	470	57	105	9.44	0.341	Wiping(c)
40C	58Pb-40Sn-2.0Sb	186	365	231	448	45	83	9.22	0.333	General purpose(c)
95TA	95Sn-5Sb	232	450	240	464	8	14	7.80	0.260	Copper joints in electrical, plumbing and heating systems
Lead-silver, lead-tin-silver and tin-silver										
2.5S	97.5Pb-2.5Ag	304	579	579	579	275	0	11.3	0.409	Torch soldering of copper and brass
5.5S	94.5Pb-5.5Ag	304	579	343	689	39	110	…	…	Torch soldering of copper and brass
1.5S	97.5Pb-1.0Sn-1.5Ag	313	588	313	588	0	0	11.3	0.409	Torch soldering of copper and brass
	97Pb-2.5Sn-0.5Ag	303	577	310	590	7	13	…	…	Torch soldering of copper and brass
	94.5 Pb-5Sn-0.5Ag	294	561	301	574	7	13	…	…	Torch soldering of copper and brass
	36Pb-62Sn-2Ag	180	354	190	372	10	18	…	…	…
96TS	96Sn-4Ag	221	430	221	430	0	0	10.4	0.375	Delicate instruments; electrical conductors for use at high temperature

(a) Most class A solders also available as antimonial class B solders (0.20 to 0.50% Sb). (b) This alloy has virtually no strnegth at temperatures above 183 °C (361 °F). (c) Not recommended for soldering to galvanized iron

Lead, Tin, and Zinc

Lead

Composition limits of pig leads (ASTM B29)

Element	Composition(a), % Corroding lead(b)	Common lead(c)	Chemical lead(d)	Acid-copper lead(e)
Silver, max	0.0015	0.005	0.020	0.002
Silver, min	...	...	0.002	...
Copper, max	0.0015	0.0015	0.080	0.080
Copper, min	...	...	0.040	0.040
Silver + copper, max	0.0025	...	...	...
Arsenic + antimony + tin, max.	0.002	0.002	0.002	0.002
Zinc, max	0.001	0.001	0.001	0.001
Iron, max	0.002	0.002	0.002	0.002
Bismuth, max	0.050	0.050(f)	0.005	0.025
Lead(g), min	99.94	99.94	99.90	99.90

(a) By agreement between the purchaser and the supplier, analyses may be required and limits established for elements (or compounds) not specified here. (b) "Corroding lead" is a designation used in the trade to describe lead that has been refined to a high degree of purity. (c) "Common lead" is fully refined desilverized lead. (d) "Chemical lead" designates the undesilverized lead produced from southeastern Missouri ores. (e) "Copper-bearing lead" is made by adding copper to fully refined lead. (f) By agreement between the purchaser and the supplier, bismuth levels of up to 0.150% may be allowed. (g) By difference

Tin and Tin Alloys

Designations, chemical compositions, and applications of commercially pure tins

| Grade designation | | | Composition, % max(a) | | | | | | | | | | | | General |
ASTM B339	Designation	Class	Sn	Sb	As	Bi	Cd	Cu	Fe	Pb	Ni + Co	S	Zn	applications
AAA	Electrolytic	Extra-high purity	99.98	0.008	0.0005	0.001	0.001	0.002	0.005	0.010	0.005	0.002	0.001	Analytical standards, research
AA	Electrolytic	High purity	99.95	0.02	0.01	0.01	0.01	0.02	0.01	0.02	0.01	0.01	0.005	Research, pharmaceuticals, fine chemicals
A(b)	A, Straits	High purity; commercial	99.80	0.04	0.05	0.015	0.001	0.04	0.015	0.05	0.01	0.01	0.005	Tinplate, foil, collapsible tubes, block tin products, pewter
B(c)	B	General purpose	99.80	...	0.05	...	...	...	...	...	...	...	...	Less exacting, general purpose
C	C	Intermediate grade	99.65	...	...	...	...	...	...	...	...	...	...	General purpose alloys
D	D	Lower intermediate grade	99.50	...	...	...	...	...	...	...	...	...	...	General purpose alloys
E	E	Common	99.00	(d)	...	...	...	(d)	...	(d)	...	...	...	Cast bronze, bearing metal, general purpose alloys, lead base alloys

(a) The maximum impurity limits, which are from ASTM Standard Classification B339, are not specification limits, but simply guides to the maximum impurity contents commonly found in the various brands of tin that fall into these grades. (b) ASTM Grade A includes about 80 to 90% of the refined tin produced. (c) Grade B is intended for those uses where the specific impurity limitations of Grade A are not critical. (d) Limits of these impurities may be specified for some uses

Tensile properties of commercially pure tin

| Temperature | | Yield strength | | Elongation in 25 mm or 1 in., % | Reduction in area, % |
°C	°F	MPa	ksi		
Strained at 0.2 mm/m · min. (0.0002 in./in. · min.)					
−200	−328	36.2	5.25	6	6
−160	−256	90.3	13.10	15	10
−120	−184	87.6	12.71	60	97
−80	−112	38.9	5.64	89	100
−40	−40	20.1	2.92	86	100
0	32	12.5	1.81	64	100
23	73	11.0	1.60	57	100
Strained at 0.4 mm/m · min. (0.0004 in./in. · min.)					
15	59	14.5	2.10	75	...
50	122	12.4	1.80	85	...
100	212	11.0	1.60	55	...
150	302	7.6	1.10	55	...
200	392	4.5	0.65	45	...

Note: It is uncertain if the inconsistencies among these data are due to differences in purity or the difference in straining rate

Zinc and Zinc Alloys

Grades and compositions of slab zinc (ASTM B6)

Grade	Composition % max(a)			
	Pb	Fe	Cd	Zn, min (by difference)
Special high grade(b)	0.003	0.003	0.003	99.990
High grade	0.03	0.02	0.02	99.90
Prime western(c)	1.4	0.05	0.20	98.0

(a) When specified for use in manufacture of rolled zinc or brass, aluminum is held to 0.005% max. (b) Tin in special high grade zinc is held to 0.001% max. (c) Aluminum in prime western zinc is held to 0.05% max

Designations of zinc die casting alloys

Alloy	UNS number	SAE number	Government specification	ASTM specification
AG40A	Z33520	903	QQ-Z-363	B240 B86
AC41A	Z35530	925	QQ-Z-363	B240 B86
Alloy 7	…	…	…	…
ILZRO 16	…	…	…	…

Classification of wrought zinc alloys

Composition, %(a)							Characteristics	Typical uses
Pb	Fe	Cd	Cu	Mg	Al	Other		
0.05 to 0.10	0.012	0.005	0.001	…	…	…	High ductility with low hardness and stiffness. Very little work hardening possible	Drawn battery cans, eyelets, fuse links, and a variety of articles drawn, formed and spun. Address plates
0.05 to 0.10	0.012	0.06	0.005	…	…	…	High ductility with low hardness. Can be work hardened slightly	Drawn battery cans, eyelets and grommets. Extruded battery cans. Address plates, laundry tags
0.15 to 0.35	0.017	0.15 to 0.30	0.005	…	…	…	High hardness and stiffness. Uniform etching quality. Can be work hardened	Soldered battery cans, photoengraver's plate, lithographer sheet, boiler and ship plates, weather-strips
0.05 to 0.10	0.012	0.005	0.85 to 1.25	…	…	…	High hardness and stiffness. Good ductility. Good creep resistance. Work hardens easily	Weatherstrips and drawn and formed articles requiring stiffness
0.05 to 0.10	0.015	0.005	0.85 to 1.25	0.007 to 0.02	…	…	High stiffness and creep resistance. Can be severely work hardened	Flat or formed articles requiring high stiffness and strength
0.005 to 0.10	0.012	0.05	0.50 to 1.50	…	…	0.12 to 1.50 Ti(b)	Outstanding creep resistance. Can be severely work hardened. Lowest thermal expansivity with the grain. Very high resistance to grain growth during annealing	Corrugated roofing, leaders and gutters, and other uses requiring maximum creep resistance
0.15 to 0.35	0.014 to 0.025	0.15 to 0.30	0.005	0.005 to 0.025	…	…	High hardness. Can be baked without severe softening. Good etching characteristics	Photoengraver's sheet
…	…	…	…	0 to 0.025	0.25 to 0.60	…	High hardness. Can be baked without severe softening. Good etching characteristics	Photoengraver's sheet
0.007	0.10	0.007	0 to 3.5	0.02 to 0.10	3.5 to 4.5	0.005 Sn	High strength and hardness	Shearing and forming dies. Extruded rod, tube and moldings
0.005 to 0.10	0.012	0.05	0.50 to 1.5	…	…	0.12 to 0.50 Ti	Good creep resistance	Corrugated roofing, leaders and gutters, and formed articles requiring maximum creep resistance

(a) Maximum unless a range is specified. (b) U.S. Patent 2472402

Composition of zinc die casting alloys

Alloy	Form	Composition, % max								
		Cu	Al	Mg	Pb	Cd	Sn	Fe	Others	Zn
AG40A	Ingot	0.10	3.9 to 4.3	0.025 to 0.05	0.004	0.003	0.002	0.075	(a)	rem
AC41A	Die castings	0.25(b)	3.5 to 4.3	0.020 to 0.05(c)	0.005	0.004	0.003	0.100	(a)	rem
Alloy 7	Ingot	0.75 to 1.25	3.9 to 4.3	0.03 to 0.06	0.004	0.003	0.002	0.075	(a)	rem
ILZRO 16	Die castings	0.75 to 1.25	3.5 to 4.3	0.03 to 0.08(c)	0.005	0.004	0.003	0.100	(a)	rem
	Die castings	0.25	3.5 to 4.3	0.010 to 0.02	0.0020	0.0020	0.0010	0.050	...	rem
	Die castings	1.0 to 1.5	0.01 to 0.04	...	...	...	...	...	(d)	rem

(a) May contain nickel, chromium, silicon and manganese in amount of 0.02, 0.02, 0.035 and 0.5%, respectively. No harmful effects have ever been noted due to the presence of these elements in these concentrations; therefore, analyses are not required for these elements. (b) For the majority of commercial applications, a copper content in the range of 0.25 to 0.75% will not adversely affect the serviceability of die castings and should not serve as a basis for rejection. (c) Magnesium content may be as low as 0.015% provided that lead, cadmium and tin contents do not exceed 0.003, 0.003 and 0.001%, respectively. (d) 0.15 to 0.25 Ti, 0.10 to 0.20 Cr, 0.30 to 0.40 Ti + Cr

Average properties of zinc die casting alloys

Properties	ASTM AG-40A (SAE 903)	ASTM AC41A (SAE 925)	Alloy 7	ILZRO 16
Mechanical properties				
Charpy impact strength, ¼-by-¼-in. bar:				
As cast, J(ft · lb)	58 (43)	65 (48)	...	...
After aging indoors 10 yr, J(ft · lb)	56 (41)	54 (40)	54 (40)	...
Tensile strength:				
As cast, MPa (ksi)	285 (41)	330 (47.6)	285 (41)	230 to 235 (33 to 34)
After aging indoors 10 yr, MPa (ksi)	240 (35)	270 (39.3)	...	...
Elongation, % in 50 mm or 2 in.:				
As cast	10	7	14	5
After aging indoors 10 yr	16	13	...	...
Expansion, after aging indoors 10 yr at room temperature, μm/m	80	70	...	...
Other properties and constants of as-cast alloys				
Brinell hardness (HB)	82	91	76	75 to 77
Compressive strength, MPa (ksi)	415 (60)	600 (87)	...	...
Electrical conductivity, % IACS	27.5	26.5	...	...
Liquidus temperature, °C (°F)	387 (728)	386 (727)	...	417 (785)
Solidus temperature, °C (°F)	381 (717)	380 (716)	...	415 (780)
Modulus of rupture, MPa (ksi)	655 (95)	725 (105)	...	...
Shear strength, MPa (ksi)	215 (31)	260 (38)	...	...
Specific heat, J/kg (Btu/lb)	420 (0.10)	420 (0.10)	...	...
Thermal conductivity, W/m · K (Btu/ft · h · °F)(a)	113 (65.3)	109 (62.9)	...	...
Thermal expansion, μm/m · K (μin./in · °F)	27.4 (15.2)	27.4 (15.2)	...	...
Transverse deflection, mm (in.)	6.9 (0.27)	4.1 (0.16)	...	...
Density, Mg/m³ (lb/in.³)	6.6 (0.238)	6.7 (0.242)	...	...

(a) At 18 °C (64 °F)

A comparison between zinc and bronze bearing alloys

	Zinc alloys				Bronze alloys		
	ZA-12		ZA-27		SAE 660 (CAD 93200)	SAE 64 (CDA 93700)	SAE 40 (CDA 83600)
	Sand cast	Permanent mold	Sand cast	Sand cast(a)			
Nominal composition	11 Al, 0.75 Cu, 0.02 Mg, bal Zn		27 Al, 2.2 Cu, 0.015 Mg, bal Zn		83 Cu, 7 Pb, 7 Sn, 3 Zn	80 Cu, 10 Sn, 10 Pb	85 Cu, 5 Pb, 5 Sn, 5 Zn
Tensile strength, ksi	40 to 45	45 to 50	58 to 64	45 to 47	35	35	37
0.2% offset yield strength, ksi	30	31	53	37	18(b)	18(b)	17(b)
Elongation in 2 in., %	1 to 3	1 to 3	3 to 6	8 to 11	20	20	30
Hardness, HB	105 to 120	105 to 125	110 to 120	90 to 100	65	60	60
Young's modulus, 10^6 psi	12	12	10.9	11.5	14.5	11	13.5
Density, lb/in.^{3}	0.218	0.218	0.181	0.181	0.322	0.320	0.318
Melting range, °F	710 to 810	710 to 810	708 to 903	708 to 903	1570 to 1790	1403 to 1705	1570 to 1850
Electrical conductivity, % IACS	28.3	28.3	29.7	29.7	12	10	15
Thermal conductivity, Btu · ft/h · ft · °F	67		72.5		33.6	27.1	41.6

Note: Zinc-alloy bearings have a higher load-bearing capability than bronze-alloy bearings. However, bronze-alloy bearings can be used at higher temperatures (above 300 °F) and in more corrosive environments than zinc-alloy bearings. (a) Heat treated 3 h at 610 °F and slow furnace cooled. (b) 0.5% offset. Source: Systems Design Reference Manual, *Machine Design*, June, 1990, p 196

Load and wear data for zinc bearing alloys

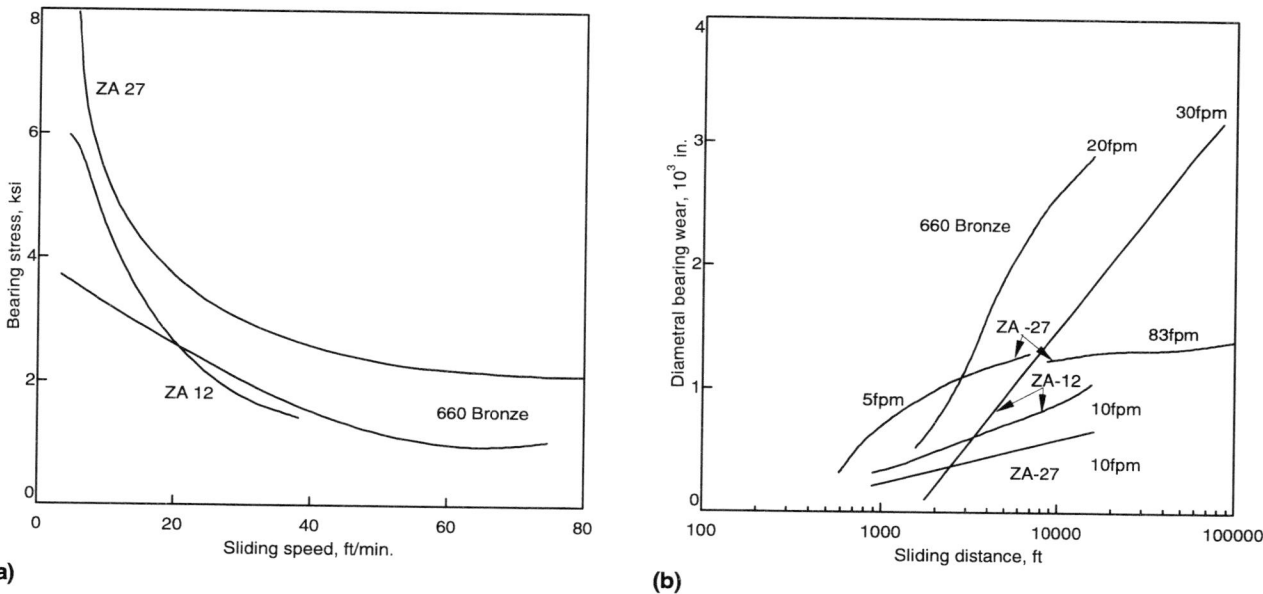

(a)

(b)

(a) Operating limits. Load capacity is the highest bearing pressure reached when the coefficient of friction remains stable and the bearing temperature does not exceed 140 °C. Friction coefficients for ZA-27 range from 0.03 to 0.07; a coefficient of 0.1 is characterized of bronze at maximum load capacity. (b) Bearing wear for the three materials was evaluated at 1,000-psi stress. Wear rates were measured during a 24-h break-in period at 2, 4, 8, and 24 h. Running times were converted to sliding distances to permit comparisons of bearings at different speeds. Source: Systems Design Reference Manual, *Machine Design*, June, 1990, p 196

Magnesium and Magnesium Alloys

Compositions and Properties

Nominal compositions and typical room-temperature mechanical properties of magnesium alloys

Alloy	Al	Mn(a)	Th	Zn	Zr	Other	Tensile strength MPa	ksi	Yield strength Tensile MPa	ksi	Compressive MPa	ksi	Bearing MPa	ksi	Elongation in 50 mm or 2 in., %	Shear strength MPa	ksi	Hardness, HRB(b)
Sand and permanent mold castings																		
AM100A-T61	10.0	0.1	...	...	...	...	275	40	150	22	150	22	...	...	1	...	...	69
AZ63A-T6	6.0	0.15	...	3.0	...	...	275	40	130	19	130	19	360	52	5	145	21	73
AZ81A-T4	7.6	0.13	...	0.7	...	...	275	40	83	12	83	12	305	44	15	125	18	55
AZ91C-T6	8.7	0.13	...	0.7	...	...	275	40	195	21	145	21	360	52	6	145	21	66
AZ92A-T6	9.0	0.10	...	2.0	...	...	275	40	150	22	150	22	450	65	3	150	22	84
EZ33A-T5	...	...	...	2.7	0.6	3.3 RE	160	23	110	16	110	16	275	40	2	145	21	50
HK31A-T6	...	...	3.3	...	0.7	...	220	32	105	15	105	15	275	40	8	145	21	55
HZ32A-T5	...	...	3.3	2.1	0.7	...	185	27	90	13	90	13	255	37	4	140	20	57
K1A-F	...	...	...	...	0.7	...	180	26	55	8	...	...	125	18	1	55	8	...
QE22A-T6	...	...	...	...	0.7	2.5 Ag, 2.1 Di	260	38	195	28	195	28	...	...	3	...	...	80
QH21A-T6	...	...	60	...	0.7	2.5 Ag, 1.0 Di	275	40	205	30	...	...	...	...	4	...	...	...
ZE41A-T5	...	...	...	4.2	0.7	1.2 RE	205	30	140	20	140	20	350	51	3.5	160	23	62
ZE63A-T6	...	...	...	5.8	0.7	2.6 RE	300	44	190	28	195	28	...	...	10	...	...	60-85
ZH62A-T5	...	...	1.8	5.7	0.7	...	240	35	170	25	170	25	340	49	4	165	24	70
ZK51A-T5	...	...	...	4.6	0.7	...	205	30	165	24	165	24	325	47	3.5	160	23	65
ZK61A-T5	...	...	...	6.0	0.7	...	310	45	185	27	185	27	...	...	...	170	25	68
ZK61A-T6	...	...	...	6.0	0.7	...	310	45	195	28	195	28	...	...	10	180	26	70
Die castings																		
AM60A-F	6.0	0.13	...	...	...	...	205	30	115	17	115	17	...	...	6	...	...	...
AS41A-F(c)	4.3	0.35	...	...	...	1.0 Si	220	32	150	22	150	22	...	...	4	...	...	...
AZ91A and B-F(d)	9.0	0.13	...	0.7	...	...	230	33	150	22	165	24	...	...	3	140	20	63
Extruded bars and shapes																		
AZ10A-F	1.2	0.2	...	0.4	...	...	240	35	145	21	69	10	...	...	10	...	...	...
AZ21X1-F	1.8	0.02	...	1.2	...	...	...	...	...	...	...	...	...	...	...	...	...	...
AZ31 B and C-F(e)	3.0	...	...	1.0	...	...	260	38	200	29	97	14	230	33	15	130	19	49
AZ61A-F	6.5	...	...	1.0	...	...	310	45	230	33	130	19	285	41	16	140	20	60
AZ80A-T5	8.5	...	...	0.5	...	...	380	55	275	40	240	35	...	...	7	165	24	82
HM31A-F	...	1.2	3.0	...	...	...	290	42	230	33	185	27	345	50	10	150	22	...
M1A-F	...	1.2	...	...	...	...	255	37	180	26	83	12	195	28	12	125	18	44
ZK21A-F	...	...	...	2.3	0.45(a)	...	260	38	195	28	135	20	...	...	4	...	...	...
ZK40A-T5	...	...	...	4.0	0.45(a)	...	276	40	255	37	140	20	...	...	4	...	...	...
ZK60A-T5	...	...	...	5.5	0.45(a)	...	365	53	305	44	250	36	405	59	11	180	26	88
Sheet and plate																		
AZ31B-H24	3.0	...	...	1.0	...	...	290	42	220	32	180	26	325	47	15	160	23	73
HK31A-H24	...	...	3.0	...	0.6	...	255	33	200	29	160	23	285	41	9	140	20	68
HM21A-T8	...	0.6	2.0	...	...	...	235	34	170	25	130	19	270	39	11	125	18	...
PE(f)	3.3	...	...	0.7	...	...	...	...	...	...	...	...	...	...	...	...	...	...

(a) Minimum. (b) 500-kg load, 10-mm ball. (c) For battery applications. (d) A and B are identical except that 0.30% Cu max residual is allowable in AZ91B. (e) Properties of B and C are identical, but AZ31C has 0.15% Mn min, 0.1% Cu max and 0.03% Ni max. (f) Photoengraving grade

Some low-temperature tensile properties of various magnesium alloys

Alloy	Thickness		Tensile strength		Yield strength		Elongation, %
	mm	in.	MPa	ksi	MPa	ksi	
Transverse tests of plate alloys at 24 °C (75 °F)							
HK31A-H24	6.35	0.250	240	35.2	180	25.9	21.0
HK31A-O	6.35	0.250	200	29.0	125	18.0	30.5
HM21A-T8	6.35	0.250	240	35.0	170	24.8	13.7
Longitudinal tests of sheet and plate alloys at 24 °C (75 °F)							
HK31A-H24	1.63	0.064	250	36.3	200	29.0	7.5
HK31A-H24	6.35	0.250	240	34.5	190	27.3	14.2
Welded(a)	6.35	0.250	200	28.8	150	21.7	2.4
HK31A-O	1.63	0.064	205	29.7	125	17.9	27.5
HK31A-O	6.35	0.250	200	28.9	120	17.7	29.7
Welded(a)	6.35	0.250	160	23.4	120	17.3	3.2
HM21A-T5	(b)	(b)	210	30.4	105	15.5	8.0
HM21A-T8	1.63	0.064	220	32.2	160	23.1	7.2
HM21A-T8	6.35	0.250	235	32.4	175	25.1	5.6
Welded(a)	6.35	0.250	195	28.6	130	18.6	2.7
Longitudinal tests of sheet and plate alloys at –54 °C (–65 °F)							
HK31A-H24	1.63	0.064	300	43.3	220	32.0	5.0
	6.35	0.250	280	40.8	230	33.4	9.0
HK31A-O	1.63	0.064	275	39.9	150	21.4	20.7
	6.35	0.250	265	38.3	150	21.5	18.0
HM21A-T5	(b)	(b)	270	39.5	110	15.8	9.3
HM21A-T8	1.63	0.064	275	39.6	175	25.6	6.2
	6.35	0.250	265	38.4	205	29.7	4.7
Longitudinal tests of sheet and plate alloys at –72 °C (–98 °F)							
HK31A-H24	1.63	0.064	295	42.7	210	30.6	4.2
HK31A-H24	6.35	0.250	290	42.4	235	33.8	11.5
Welded(a)	6.35	0.250	195	28.1	165	23.6	0.5
HK31A-O	1.63	0.064	285	41.3	145	21.1	17.5
HK31A-O	6.35	0.250	275	40.2	150	21.9	20.2
Welded(a)	6.35	0.250	205	29.4	145	21.0	2.2
HM21A-T5	(b)	(b)	275	40.0	110	16.3	8.3
HM21A-T8	1.63	0.064	280	40.8	150	22.1	17.5
	6.35	0.250	275	40.1	215	31.3	5.0
	6.35	0.250	200	29.2	120	17.5	1.5
Longitudinal tests of sheet and plate alloys at –196 °C (–320 °F)							
HK31A-H24	1.63	0.064	370	54.0	225	33.0	6.2
HK31A-H24	6.35	0.250	365	52.9	240	34.7	8.0
Welded(a)	6.35	0.250	230	33.7	180	25.9	1.5
HK31A-O	1.63	0.064	330	47.9	170	24.3	12.7
HK31A-O	6.35	0.250	325	47.2	170	24.7	12.5
Welded(a)	6.35	0.250	205	29.7	150	21.6	2.2
HM21A-T5	...	(b)	320	46.6	125	18.1	8.0
HM21A-T8	1.63	0.064	330	47.6	170	24.9	4.0
HM21A-T8	6.35	0.250	325	47.3	210	30.6	4.2
Welded(a)	6.35	0.250	330	33.1	145	20.9	1.5

Note: Values for wrought alloys are averages of two to four tests at room temperature (2-in. gage length). Values of duplicate tests at low temperatures are also averages (1-in. gage length). Values for cast alloys are averages of two to four tests on separately cast bars. (a) Welding rod was EZ33A; weld bead intact. (b) Specimen machined from a forging

Alloy	Tensile strength		Yield strength		Elongation, %	Charpy impact			
						(a)		(b)	
	MPa	ksi	MPa	ksi		J	ft · lb	J	ft · lb
Cast alloys at 24 °C (75 °F)									
AZ91C-T6	290	41.8	130	19.2	6.3	7.96	5.87	1.36	1.00
AZ92A-T6	290	41.8	160	23.4	4.0	7.62	5.62	0.68	0.50
EZ33A-T5	190	27.5	115	16.9	7.6	7.46	5.50	0.84	0.62
HK31A-T6	225	32.7	110	16.3	9.5	16.61	12.25	3.80	2.81
ZH62A-T5	275	39.9	190	27.9	5.7	15.02	11.08	1.02	0.75
Cast alloys at –78 °C (–109 °F)									
AZ91C-T6	305	44.3	150	21.6	5.1	6.26	4.62	1.36	1.00
AZ92A-T6	295	42.7	170	24.6	2.3	6.44	4.75	0.76	0.56
EZ33A-T5	190	27.6	125	18.0	3.1	4.83	3.56	0.68	0.50
HK31A-T6	300	43.3	120	17.5	8.6	16.43	12.12	3.21	2.37
ZH62A-T5	330	47.6	200	29.2	2.7	18.99	14.00	1.02	0.75
Cast alloys at –196 °C (–321 °F)									
AZ91C-T6	310	44.9	180	26.0	1.7	4.06	3.00	1.02	0.75
AZ92A-T6	320	46.5	195	28.5	0.8	4.57	3.37	0.68	0.50
EZ33A-T5	200	29.0	140	20.3	2.2	5.00	3.69	0.68	0.50
HK31A-T6	330	48.1	135	19.6	6.1	13.72	10.12	3.05	2.25
ZH62A-T5	320	46.6	235	34.1	1.0	8.56	6.31	1.02	0.75

Note: Values for wrought alloys are averages of two to four tests at room temperature (2-in. gage length). Values of duplicate tests at low temperatures are also averages (1-in. gage length). Values for cast alloys are averages of two to four tests on separately cast bars. (a) Unnotched specimens. (b) Notched specimens

Effect of elevated temperature on values of creep stress and elastic modulus for magnesium alloys

Alloy	Creep stress(a) at 205 °C (400 °F)		315 °C (600 °F)		Elastic modulus at 205 °C (400 °F)		315 °C (600 °F)	
	MPa	ksi	MPa	ksi	GPa	10⁶ psi	GPa	10⁶ psi
Castings								
AZ92A-T6	3.4	0.5	...	...	31	4.5	21	3.0
EZ33A-T5	38	5.5	6.9	1.0	40	5.8	38	5.5
HK31A-T6	64	9.3	14	2.0	40	5.8	39	5.6
HZ32A-T5	52	7.5	22	3.2	40	5.8	39	5.6
ZH62A-T5	17	2.5	...	...	40	5.8	38	5.5
Extrusions								
ZK60A-T5	7	1.0(b)	...	...	...	...	...	...
HM31A-F	83	12.0	41	6.0	40	5.8	38	5.5
Sheet								
AZ31B-H24	7	1.0(b)	...	...	30	4.3	17	2.5
HK31A-T6	69	10.0	17	2.5	40	5.8	25	3.6
HM21A-T8	76	11.0	34	5.0	40	5.8	34	5.0

(a) Stress to produce 0.2% total extension in 1000 h for cast alloys and 100 h for wrought alloys. (b) Tested at 150 °C (300 °F)

Effect of elevated temperature on tensile strength of magnesium alloys

Alloy	Tested at exposure temperature — Exposed 10 min at 20 °C (70 °F)		150 °C (300 °F)		315 °C (600 °F)		Exposed 1000 h at 205 °C (400 °F)		315 °C (600 °F)		Tested at room temperature — Exposed at 1000 h at 205 °C (400 °F)		315 °C (600 °F)	
	MPa	ksi	MPa	ksi	MPa	ksi	MPa	ksi	MPa	ksi	MPa	ksi	MPa	ksi
Castings														
AZ63A-T6	275	40	165	24	55	8	110	16	...	...	255	37	...	...
AZ92A-T6	275	40	195	28	55	8	115	17	...	...	270	39	...	...
EZ33A-T5	160	23	145	21	83	12	130	19	76	11	170	25	180	26
HK31A-T6	215	31	195	28	125	18	180	26	62	9	240	35	180	26
HZ32A-T5	200	29	145	21	83	12	115	17	76	11	220	32	235	34
ZH62A-T5	290	42	195	28	69	10	...	...	...	...	...	...	...	...
QH21A-T6	275	40	235	34	97	14	...	...	...	...	...	...	...	...
Extrusions														
AZ80A-T5	380	55	235	34	69	10	...	...	...	...	...	...	...	...
ZK60A-T5	365	53	180	26	41	6	...	...	...	...	315	46	315	46
HM31A-F	275	40	195(a)	28(a)	115	17	...	...	...	...	...	...	...	...
Sheet														
AZ31B-H24	285	41	145	21	48	7	90	13	62(a)	9(a)	255	37	260	38
HK31A-T6	255	37	180	26	115	17	...	...	55	8	255	37	215	31
HM21A-T8	235	34	140	20	97	14	...	...	...	...	...	...	...	...

(a) Tested at 260 °C (500 °F)

Principal advanced aerospace magnesium alloys and applications

Alloy	Producer	Processing route	Application
AZ91E	Dow, AMAX	Casting	Corrosion resistance
WE54	Magnesium Elektron	Casting	High-temperature strength(a), corrosion resistance
EA55B	Allied-Signal	Rapid solidification	High room-temperature strength(b)
EA65	Magnesium Elektron	Chill casting	Corrosion resistance

(a) <300 °C (<570 °F). (b) <150 °C (<300°F). Source: W.E. Frazier, *et al.*, "Advanced Lightweight Alloys for Aerospace Application," *Journal of Metals*, Vol 41, No. 5, p 22-26, 1989; and *Journal of Metals*, Vol 41, No. 7, p 58, 1989

Nominal composition of advanced aerospace magnesium alloys

Alloy	Composition, wt.%					
	Al	Zn	Mn	Zr	Rare Earth	Other
AZ91C(a)	8.7	0.7	0.2		...	...
AZ91E	8.7	0.7	>0.15		...	Fe <0.005 Ni<0.001 Cu<0.03
WE54	...	...	...	>0.3	5.25 Y	3.5 other RE
ES55	5.0	5.0	...	...	4.9 Nd	...
ES65	5.0	5.0	...	...	5.9 Y	

(a) AZ91C is a currently used alloy shown for comparison. Source: W.E. Frazier, *et al.*, "Advanced Lightweight Alloys for Aerospace Application," *Journal of Metals*, Vol 41, No. 5, p 22-26, 1989; and *Journal of Metals*, Vol 41, No. 7, p 58, 1989

Typical ambient temperature properties of advanced aerospace magnesium alloys

Alloy	Density, strength, g/cm^3	Yield strength, Elongation, Corrosion MPa	Ultimate tensile Product MPa	%	rate(a), mpy	form
AZ91C-T6(b)	1.81	145	275	6	1000	Casting
AZ91E-T6	1.81	145	275	6	8	Casting
WE54-T6	1.90	225	280	4	4	Casting
EA55B-T6	1.94	167	271	8	20	Casting
EA55RS-T4	1.94	370	425	14	8	Extrusion
EA65B-T6	1.92	130	235	7	20	Casting
EA65RS-T4	1.92	460	515	5	8	Extrusion

(a) 20 days in a neutral salt fog. (b) AZ91C is a currently used alloy shown for comparison. Source: W.E. Frazier, *et al.*, "Advanced Lightweight Alloys for Aerospace Application," *Journal of Metals*, Vol 41, No. 5, p 22-26, 1989; and *Journal of Metals*, Vol 41, No. 7, p 58, 1989

Heat Treating

Annealing temperatures for wrought magnesium alloys

Alloy	Original temper	Annealing temperature(a)	
		°C	°F
AZ31B	F, H10, H11, H23, H24, H26	345	650
AZ31C	F	345	650
AZ61A	F	345	650
AZ80A	F, T5, T6	385	725
HK31A	H24	400	750
HM21A	T5, T8, T81	455	850
HM31A	T5	455	850
ZK60A	F, T5, T6	290	550

(a) Time at temperature, 1 h or more

Recommended stress-relieving treatments for wrought magnesium alloys

	Sheet						Extrusions and forgings		
	Annealed			Hard rolled					
	Temperature		Time,	Temperature		Time,	Temperature		Time,
Alloy	°C	°F	min.	°C	°F	min.	°C	°F	min.
AZ31B	345	650	120	150	300	60	...	...	...
AZ31B-F	...	...	...	...	...	...	260	500	15
AZ61A	345	650	120	205	400	60	...	...	...
AZ61A-F	...	...	...	...	...	...	260	500	15
AZ80A-F	...	...	...	...	...	...	260	500	15
AZ80A-T5	...	...	...	...	...	...	205	400	60
HK31A	345	650	60	290	550	30	...	...	...
HM21A-T5	...	...	...	...	...	...	370	700	30
HM21A-T8	...	...	...	370	700	30	...	...	...
HM21A-T81	...	...	...	400	750	30	...	...	...
HM31A-T5	...	...	...	...	...	...	425	800	60
ZK60A-F	230	450	180	...	...	...	260	500	15
ZK60A-T5	...	...	...	...	...	...	150	300	60

Note: Stress relieving after welding, to prevent stress-corrosion cracking, is necessary only for alloys that contain more than 1.5% aluminum

Heat treatments commonly applied to magnesium alloys

Alloy	Heat treatment(a)
Casting alloys	
AM100A	T4, T5, T6, T61(b)
AZ63A	T4, T5, T6
AZ81A	T4
AZ91C	T4, T6
AZ92A	T4, T6
EZ33A	T5
HK31A	T6
HZ32A	T5
QE22A	T6
QH21A	T6
ZE41A	T5
ZE63A	T6(c)
ZH62A	T5
ZK51A	T5
ZK61A	T4, T6
Wrought alloys	
AZ80A	T5
HM21A	T5, T8, T81(d)
HM31A	T5
ZK60A	T5

(a) Indicated by temper designations. (b) Same as T6 except aged for longer time to increase yield strength. (c) Thermal treatment must include hydriding. (d) Mill modification of T8 to improve mechanical properties

Recommended solution-treating and aging schedules for magnesium alloy castings
For castings up to 51 mm (2 in.) in section thickness; heavier sections may require longer times at temperature

Alloy	Final temper	Aging(a) Temperature °C, ±6(b)	°F, ±10(b)	Time, h	Solution treating(c) Temperature °C, ±6(b)	°F, ±10(b)	Time, h	Maximum temperature °C	°F	Aging after solution treating Temperature °C, ±6(b)	°F, ±10(b)	Time, h
Magnesium-aluminum-zinc alloys(d)												
AM100A	T5	232	450	5	...	...	...	...	...	...	...	...
	T4	...	...	...	424(e)	795(e)	16-24(e)	432	810	...	...	...
	T6	...	...	...	424(e)	795(e)	16-24(e)	432	810	232	450	5
	T61	...	...	...	424(e)	795(e)	16-24(e)	432	810	218	425	25
AZ63A	T5	260(f)	500(f)	4(f)	...	...	...	...	...	...	...	...
	T4	...	...	...	385	725	10-14	391	735	...	...	...
	T6	...	...	...	385	725	10-14	391	735	218(f)	425(f)	5(f)
AZ81A	T4	...	...	...	413(e)	775(e)	16-24(e)	418	785	...	...	...
AZ91C	T5	168(g)	335(g)	16(g)	...	...	...	...	...	...	...	...
	T4	...	...	...	413(e)	775(e)	16-24(e)	418	785	...	...	...
	T6	...	...	...	413(e)	775(e)	16-24(e)	418	785	168(h)	335(h)	16(h)
AZ92A	T5	260	500	4	...	...	...	...	...	...	...	...
	T4	...	...	...	407(j)	765(j)	16-24(j)	413	775	...	...	...
	T6	...	...	...	407(j)	765(j)	16-24(j)	413	775	218	425	5
Magnesium-zirconium alloys												
EZ33A	T5	216(k)	420(k)	5(k)	...	...	...	...	...	...	...	...
HK31A(m)	T6	...	...	...	566	1050	2	571	1060	204	400	16
HZ32A	T5	316	600	16	...	...	...	...	...	...	...	...
QE22A(n)	T6	...	...	...	527	980	4-8	538	1000	204	400	8
QH21A(n)	T6	...	...	...	527	980	4-8	538	1000	204	400	8
ZE41A	T5	329(p)	625(p)	2(p)	...	...	...	...	...	...	...	...
ZE63A(q)	T6	...	...	...	480	895	10-72	491	915	141	285	48
ZH62A	T5	329	625	2	...	...	...	...	...	...	...	...
		plus: 177	350	16	...	...	...	...	...	...	...	...
ZK51A	T5	177(r)	350(r)	12(r)	...	...	...	...	...	...	...	...
ZK61A	T5	149	300	48	...	...	...	...	...	...	...	...
	T6	...	...	...	499(s)	930(s)	2(s)	502	935	129	265	48

(a) Aging of castings to the T5 temper is done from the as-cast condition. (b) Except where quoted differently. (c) After solution treatment and before subsequent aging, castings are cooled to room temperature by fast fan cooling, except where otherwise indicated. Use carbon dioxide or sulfur dioxide atmosphere above 400 °C (750 °F). (d) For solution treating, Mg-Al-Zn alloys are loaded into the furnace at 260 °C (500 °F) and brought to temperature over a 2-h period at a uniform rate of temperature increase. (e) Alternative treatment, to prevent germination (excessive grain growth): 6 h at 413 ± 6 °C (775 ± 10 °F), 2 h at 352 ± 6 °C (665 ± 10 °F), 10 h at 413 ± 6 °C (775 ± 10 °F). (f) Alternative treatment: 5 h at 232 ± 6 °C (450 ± 10 °F). (g) Alternative treatment: 4 h at 216 ± 6 °C (420 ± 10 °F). (h) Alternative treatment: 5 to 6 h at 216 ± 6 °C (420 ± 10 °F). (j) Alternative treatment, to prevent germination (excessive grain growth): 6 h at 407 ± 6 °C (765 ± 10 °F), 2 h at 352 ± 6 °C (665 ± 10 °F), 10 h at 407 ± 6 °C (765 ± 10 °F). (k) Alternative treatment, which can be used where maximum resistance to creep at elevated temperature is not of prime importance: 2 h at 343 ± 6 °C (650 ± 10 °F). (m) Alloy HK31A castings must be loaded into the furnace already at temperature and brought back to temperature as quickly as possible. (n) Quench from solution-treating medium. (p) This treatment is adequate for development of satisfactory properties; it may be followed by 16 h at 177 ± 6 °C (350 ± 10 °F) to provide very slight improvements in mechanical properties. (q) Alloy ZE63A must be solution treated in a special hydrogen atmosphere, because its mechanical properties are developed through hydriding of some of its alloying elements. Hydriding time depends on section thickness; as a guide, 6.4 mm (¼-in) sections require approximately 10 h, and 19-mm (¾-in.) sections require about 72 h. Following solution treatment, ZE63A should be quenched in oil, water spray or air blast. (r) Alternative treatment: 8 h at 218 ± 6 °C (425 ± 10 °F). (s) Alternative treatment: 10 h at 482 ± 6 °C (900 ± 10 °F)

Forming

Recommended minimum radii for 90° bends in magnesium sheet(a)

Alloy and temper	Forming temperature							
	20 °C (70 °F)	95 °C (200 °F)	150 °C (300 °F)	205 °C (400 °F)	260 °C (500 °F)	315 °C (600 °F)	370 °C (700 °F)	425 °C (800 °F)
AZ31B-O	5.5t	5.5t	4t	3t	2t	...	...	...
AZ31B-H24	8t	8t	6t	3t	2t	...	...	...
HK31A-O	6t	6t	6t	5t	4t	3t	2t	1t
HK31A-H24	13t	13t	13t	9t	8t	5t	3t	...
HM21A-T8	9t	9t	9t	9t	9t	8t	6t	4t

(a) Numerical values of bend radii are given as multiples of sheet thickness t

Joining

Weldability of magnesium alloys

Alloy	Thickness		Welding rod	Joint efficiency, %	Joint ductility(a)
	mm	in.			
AZ31B-O	1.63	0.064	AZ61A, AZ92A	97	12.0
AZ31B-H24	1.63	0.064	AZ61A, AZ92A	88	10.0
ZE10A-O	1.63	0.064	AZ61A, AZ92A	94	7.0
ZE10A-H24	1.63	0.064	AZ61A, AZ92A	87	3.0
M1A-F	3.17	0.125	M1A	55	2.0
AZ31N-F	3.17	0.125	AZ61A, AZ92A	92	12.0
AZ61A-F	3.17	0.125	AZ61A, AZ92A	89	8.0
AZ80A-F	3.17	0.125	AZ61A, AZ92A	86	4.0
AZ63A-F	12.70	0.5	AZ63A	83	2.5
AZ63A-T4	12.70	0.5	AZ63A	70	5.0
AZ63A-T6	12.70	0.6	AZ63A	75	2.0
AZ92A-F	12.70	0.5	AZ92A	100	2.5
AZ92A-T4	12.70	0.5	AZ92A	70	4.0
AZ92A-T6	12.70	0.5	AZ92A	75	2.0
AZ91C-F	12.70	0.5	AZ92A	100	2.5
AZ91C-T4	12.70	0.5	AZ92A	78	4.0
AZ91C-T6	12.70	0.5	AZ92A	75	2.0
AZ81A-F	12.70	0.5	AZ92A	100	2.5
AZ81A-T4	12.70	0.5	AZ92A	85	8.0
EK41A-T5	12.70	0.5	EK41A	100	1.0
EK41A-T6	12.70	0.5	EK41A	93	6.2
EZ33A-T5	12.70	0.5	EZ33A	100	1.1
HK31A-T6	12.70	0.5	HK31A	100	9.5
HK31A-H24	...	...	EZ33A	83	1.0
HZ32A-T5	12.70	0.5	HZ32A	93	3.8
HM21A-T8	1.63	0.064	EZ33A	88	1.5
	...	...	HM31A	74	1.5
HM31A-F	15.88	0.625	EZ33A	71	1.8
			HM31A	58	2.5

(a) Percentage elongation across the weld over a 50-mm (2-in.) gage length from tension tests

Titanium and Titanium Alloys

Compositions and Properties

Commercial and semicommercial grades and alloys of titanium

Designation	Tensile strength (min)		0.2% yield strength (min)		Impurity limits, wt % max					Nominal composition, wt %				
	MPa	ksi	MPa	ksi	N	C	H	Fe	O	Al	Sn	Zr	Mo	Other
Unalloyed grades														
ASTM Grade 1	240	35	170	25	0.03	0.10	0.015	0.20	0.18	...	...	...	...	...
ASTM Grade 2	340	50	280	40	0.03	0.10	0.015	0.30	0.25	...	...	...	...	...
ASTM Grade 3	450	65	380	55	0.05	0.10	0.015	0.30	0.35	...	...	...	...	...
ASTM Grade 4	550	80	480	70	0.05	0.10	0.015	0.50	0.40	...	...	...	...	...
ASTM Grade 7	340	50	280	40	0.03	0.10	0.015	0.30	0.25	...	...	...	...	0.2Pd
Alpha and near-alpha alloys														
Ti Code 12	480	70	380	55	0.03	0.10	0.015	0.30	0.25	...	...	...	0.3	0.8Ni
Ti-5Al-2.5Sn	790	115	760	110	0.05	0.08	0.02	0.50	0.20	5	2.5	...	...	...
Ti-5Al-2.5Sn-ELI	690	100	620	90	0.07	0.08	0.0125	0.25	0.12	5	2.5	...	...	...
Ti-8Al-1Mo-1V	900	130	830	120	0.05	0.08	0.015	0.30	0.12	8	...	...	1	1V
Ti-6Al-2Sn-4Zr-2Mo	900	130	830	120	0.05	0.05	0.0125	0.25	0.15	6	2	4	2	...
Ti-6Al-2Nb-1Ta-0.8Mo	790	115	690	100	0.02	0.03	0.0125	0.12	0.10	6	...	...	1	2Nb, 1Ta
Ti-2.25Al-11Sn-5Zr-1Mo	1000	145	900	130	0.04	0.04	0.008	0.12	0.17	2.25	11.0	5.0	1.0	0.2Si
Ti-5Al-5Sn-2Zr-2Mo(a)	900	130	830	120	0.03	0.05	0.0125	0.15	0.13	5	5	2	2	0.25Si
Alpha-beta alloys														
Ti-6Al-4V(b)	900	130	830	120	0.05	0.10	0.0125	0.30	0.20	6.0	...	...	...	4.0V
Ti-6Al-4V-ELI(b)	830	120	760	110	0.05	0.08	0.0125	0.25	0.13	6.0	...	...	...	4.0V
Ti-6Al-6V-2Sn(b)	1030	150	970	140	0.04	0.05	0.015	1.0	0.20	6.0	2.0	...	...	0.75Cu, 6.0V
Ti-8Mn(b)	860	125	760	110	0.05	0.08	0.015	0.50	0.20	...	...	...	...	8.0Mn
Ti-7Al-4Mo(b)	1030	150	970	140	0.05	0.10	0.013	0.30	0.20	7.0	...	...	4.0	...
Ti-6Al-2Sn-4Zr-6Mo(c)	1170	170	1100	160	0.04	0.04	0.0125	0.15	0.15	6.0	2.0	4.0	6.0	...
Ti-5Al-2Sn-2Zr-4Mo-4Cr(a)(c)	1125	163	1055	153	0.04	0.05	0.0125	0.30	0.13	5.0	2.0	2.0	4.0	4.0Cr
Ti-6Al-2Sn-2Zr-2Mo-2Cr(a)(b)	1030	150	970	140	0.03	0.05	0.0125	0.25	0.14	5.7	2.0	2.0	2.0	2.0Cr, 0.25Si
Ti-10V-2Fe-3Al(a)(c)	1170	170	1100	160	0.05	0.05	0.015	2.5	0.16	3.0	...	...	...	10.0 V
Ti-3Al-2.5V(d)	620	90	520	75	0.015	0.05	0.015	0.30	0.12	3.0	...	...	...	2.5V
Beta alloys														
Ti-13V-11Cr-3Al(c)	1170	170	1100	160	0.05	0.05	0.025	0.35	0.17	3.0	...	...	...	11.0Cr, 13.0V
Ti-8Mo-8V-2Fe-3Al(a)(c)	1170	170	1100	160	0.05	0.05	0.015	2.5	0.17	3.0	...	...	8.0	8.0V
Ti-3Al-8V-6Cr-4Mo-4Zr(a)(b)	900	130	830	120	0.03	0.05	0.020	0.25	0.12	3.0	...	4.0	4.0	6.0Cr, 8.0V
Ti-11.5Mo-6Zr-4.5Sn(b)	690	100	620	90	0.05	0.10	0.020	0.35	0.18	...	4.5	6.0	11.5	...

(a) Semicommercial alloy; mechanical properties and composition limits subject to negotiation with suppliers. (b) Mechanical properties given for annealed condition; may be solution treated and aged to increase strength. (c) Mechanical properties given for solution treated and aged condition; alloy not normally applied in annealed condition. Properties may be sensitive to section size and processing. (d) Primarily a tubing alloy; may be cold drawn to increase strength

AMS specifications for titanium and titanium alloys

AMS no.	Mill form	Condition	Alloy	Similar military specification
4900	Plate, sheet, strip	Annealed	Unalloyed; 55-ksi YS	MIL-T-9046
4901	Plate, sheet, strip	Annealed	Unalloyed; 70-ksi YS	MIL-T-9046
4902	Plate, sheet, strip	Annealed	Unalloyed; 40-ksi YS	MIL-T-9046
4906	Sheet, strip; continuously rolled	Annealed	Ti-6Al-4V	...
4907	Plate, sheet, strip	Annealed	Ti-6Al-4V-ELI	MIL-T-9046
4908	Sheet, strip	Annealed	Ti-8Mn; 110-ksi YS	MIL-T-9046
4909	Plate, sheet, strip	Annealed	Ti-5Al-2.5Sn-ELI	MIL-T-9046
4910	Plate, sheet, strip	Annealed	Ti-5Al-2.5Sn	MIL-T-9046
4911	Plate, sheet, strip	Annealed	Ti-6Al-4V	MIL-T-9046
4915	Plate, sheet, strip	Single annealed	Ti-8Al-1Mo-1V	MIL-T-9046
4916	Plate, sheet, strip	Duplex annealed	Ti-8Al-1Mo-1V	MIL-T-9046
4917	Plate, sheet, strip	Solution treated	Ti-13V-11Cr-3Al	MIL-T-9046
4918	Plate, sheet, strip	Annealed	Ti-6Al-6V-2Sn	MIL-T-9046
4921	Bar, forgings, rings	Annealed	Unalloyed; 70-ksi YS	MIL-T-9047
4924	Bar, forgings, rings	Annealed	Ti-5Al-2.5Sn-ELI; 90-ksi YS	MIL-T-9047
4926	Bar, rings	Annealed	Ti-5Al-2.5Sn; 110-ksi YS	MIL-T-9047
4928	Bar, forgings	Annealed	Ti-6Al-4V; 120-ksi YS	MIL-T-9047
4930	Bar, forgings, rings	Annealed	Ti-6Al-4V-ELI	MIL-T-9047
4935	Extrusions	Annealed	Ti-6Al-4V	...
4936	Extrusions	...	Ti-6Al-6V-2Sn	...
4941	Tubing, welded	Annealed	Unalloyed; 40-ksi YS	...
4942	Tubing, seamless	Annealed	Unalloyed; 40-ksi YS	...
4943	Tubing, seamless	Annealed	Ti-3Al-2.5V	...
4944	Tubing, seamless hydraulic	Cold worked and stress relieved	Ti-3Al-2.5V	...
4951	Wire, welding	...		...
4953	Wire, welding	Annealed	Ti-5Al-2.5Sn	...
4954	Wire, welding	...	Ti-6Al-4V	...
4955	Wire, welding	...	Ti-8Al-1Mo-1V	...
4956	Wire, welding	...	Ti-6Al-4V-ELI	...
4965	Bar, forgings, rings	Precipitation heat treated	Ti-6Al-4V	...
4966	Forgings	Annealed	Ti-5Al-2.5Sn; 110-ksi YS	MIL-F-83142
4967	Bar, forgings	Annealed	Ti-6Al-4V	MIL-T-9047
4970	Bar, forgings	Precipitation heat treated	Ti-7Al-4Mo	MIL-T-9047
4971	Bar, forgings, rings	Annealed	Ti-6Al-6V-2Sn	MIL-T-9047, MIL-F-83142
4972	Bar, rings	Solution treated and stabilized	Ti-8Al-1Mo-1V	...
4973	Forgings	Solution treated and stabilized	Ti-8Al-1Mo-1V	...
4974	Bar, forgings	Precipitation heat treated	Ti-11Sn-5Zr-2.3Al-1Mo-0.21Si	...
4975	Bar, rings	Precipitation heat treated	Ti-6Al-2Sn-4Zr-2Mo	MIL-T-9047
4976	Forgings	Precipitation heat treated	Ti-6Al-2Sn-4Zr-2Mo	...
4977	Bar, wire	Solution treated	Ti-11.5Mo-6Zr-4.5Sn	MIL-T-9047
4978	Bar, forgings, rings	Annealed	Ti-6Al-6V-2Sn; 140-ksi YS	MIL-T-9047, MIL-F-83142
4979	Bar, forgings, rings	Precipitation heat treated	Ti-6Al-6V-2Sn	MIL-T-9047, MIL-F-83142
4980	Bar, wire	Solution treated at 745 °C (1375 °F)	Ti-11.5Mo-6Zr-4.5Sn	...
4981	Bar, forgings	Precipitation heat treated	Ti-6Al-2Sn-4Zr-6Mo	MIL-T-9047

ASTM specifications for titanium and titanium alloys

Specification(a)	Grade	Alloy	0.2% yield strength(b) MPa	ksi	Similar AMS specification(a)
Plate, sheet and strip					
B265	1	Unalloyed	170	25	...
	2	Unalloyed	280	40	4902
	3	Unalloyed	380	55	4900
	4	Unalloyed	480	70	4901
	5	Ti-6Al-4V	830	120	4911
	6	Ti-5Al-2.5Sn	790	115	4910
	7	Ti-0.2Pd	280	40	...
	10	Ti-4.5Sn-11.5Mo-6Zr	620	90	4977
	11	Ti-0.2Pd	170	25	...
Seamless and welded pipe					
B337	1	Unalloyed	170	25	...
	2	Unalloyed	280	40	4941(c), 4942(d)
	3	Unalloyed	380	55	...
	7	Ti-0.2Pd	280	40	...
	9	Ti-3Al-2.5V	480	70	4943
	9	Ti-3Al-2.5V	720(e)	105(e)	4943
	10	Ti-11.5Mo-6Zr-4.5Sn	620(f)	90(f)	4977, 4980
	11	Ti-0.2Pd	170	25	...
Seamless and welded tube for condensers and heat exchangers					
B338	1	Unalloyed	170	25	...
	2	Unalloyed	280	40	...
	3	Unalloyed	380	55	...
	7	Ti-0.2Pd	280	40	...
	9	Ti-3Al-2.5V	720	105	4943, 4944
	10	Ti-11.5Mo-6Zr-4.5Sn	620	90	4977, 4980
	11	Ti-0.2Pd	170	25	...
Bar and billet					
B348	1	Unalloyed	170	25	...
	2	Unalloyed	280	40	...
	3	Unalloyed	380	55	...
	4	Unalloyed	480	70	4921(g)
	5	Ti-6Al-4V	830	120	4928(g)
	6	Ti-5Al-2.5Sn	790	115	4926(g)
	7	Ti-0.2Pd	280	40	...
	10	Ti-4.5Sn-11.5Mo-6Zr	620	90	4977, 4980
	11	Ti-0.2Pd	170	25	...
Castings					
B367	C-1	Unalloyed	170	25	...
	C-2	Unalloyed	280	40	...
	C-3	Unalloyed	380	55	...
	C-4	Unalloyed	480	70	...
	C-5	Ti-6Al-4V	830	120	...
	C-6	Ti-5Al-2.5Sn	720	105	...
	C-7A	Ti-0.2Pd	170	25	...
	C-7B	Ti-0.2Pd	280	40	...
	C-8A	Ti-0.2Pd	380	55	...
	C-8B	Ti-0.2Pd	480	70	...
Forgings					
B381	F1	Unalloyed	170	25	...
	F2	Unalloyed	280	40	...
	F3	Unalloyed	380	55	...
	F4	Unalloyed	480	70	4921
	F5	Ti-6Al-4V	830	120	4928
	F6	Ti-5Al-2.5Sn	790	115	4966
	F7	Ti-0.2Pd	280	40	...
	F11	Ti-0.2Pd	170	25	...

(a) Interstitial and impurity levels, and mechanical property requirements, may show minor differences compared with ASTM specifications. (b) Minimum in annealed condition. (c) Welded tubing. (d) Seamless tubing. (e) Cold worked and stress relieved. (f) Solution treated. (g) AMS specifications cover bar and forgings but not billet

Specifications and applications of wrought titanium alloys

Alloy	Condition	ASTM no.	Military	Aerospace material specifications				Applications and characteristics
				Bars and forgings	Sheet and plate	Tubing	Wire	
Commercially pure								
99.5 Ti	Annealed	B265 (Gr. 1)	...	...	...	...	...	Airframes; chemical, desalination, and marine parts; plate-type heat exchangers; cold spun or pressed parts; platinized anodes; high formability
		B348 (Gr. 1)	...	...	...	...	...	
		B381 (Gr. 1)	...	...	...	...	...	
99.2 Ti	Annealed	B265 (Gr. 2)	MIL-T-9046	...	4902	4941	4951	Airframes; aircraft engines; marine and chemical parts; heat exchangers; condenser and evaporator tubing; high formability
		B348 (Gr. 2)	...	...	...	4942	...	
		B381 (Gr. 2)	...	...	...	...	...	
99.1 Ti	Annealed	B265 (Gr. 3)	MIL-T-9046	...	4900	...	...	Chemical, marine, airframe, and aircraft engine parts which require formability strength, weldability, and corrosion resistance
		B348 (Gr. 3)	...	...	...	...	...	
		B381 (Gr. 3)	...	...	...	...	...	
99.0 Ti	Annealed	B265 (Gr. 4)	MIL-T-9046	4921	4901	...	...	Chemical, marine, airframe, and aircraft engine parts; surgical implants; high speed fans; gas compressors; good formability and corrosion resistance, high strength
		B348 (Gr. 4)	MIL-T-9047	...	...	...	...	
		B381 (Gr. 4)	...	...	...	...	...	
99.2Ti-0.2Pd	Annealed	B265 (Gr. 7)	...	...	...	...	...	Good corrosion resistance for chemical industry applications where media is mildly reducing or varies between oxidizing and reducing
		B348 (Gr. 7)	...	...	...	...	...	
		B381 (Gr. 7)	...	...	...	...	...	
Ti-0.8Ni-0.3Mo	Annealed	...	...	...	...	...	...	Same as 0.2 Pd alloy (above)
Alpha alloys								
Ti-5Al-2.5Sn	Annealed	B265 (Gr. 6)	MIL-T-9046	4926	4910	...	4953	Weldable alloy for forgings and sheet metal parts such as aircraft engine compressor blades and ducting; steam turbine blades; good oxidation resistance and strength at 315 to 595 °C (600 to 1100 °F); good stability at elevated temperatures
		B348 (Gr. 6)	MIL-T-9047	4966	...	...	...	
		B381 (Gr. 6)	...	...	...	...	...	
Ti-5Al-2.5Sn (low O₂)	Annealed	...	MIL-T-9046	4924	4909	...	...	Special grade for high-pressure cryogenic vessels operating down to –255 °C (–423 °F)
		...	MIL-T-9047	...	...	...	...	
Near alpha alloys								
Ti-8Al-1Mo-1V	Duplex annealed	...	MIL-T-9046	4972	4915	...	4955	Airframe and jet engine parts requiring high strength to 455 °C (850 °F); good creep and toughness properties; good weldability
		...	MIL-T-9047	4973	4916	...	...	
Ti-11Sn-1Mo-2.25Al 5.0Zr-1Mo-0.2Si		...	MIL-T-9047	4974	...	...	...	Airframes; blades, discs, wheels, spacers, and fasteners for turbine engines
Ti-6Al-2Sn-4Zr-2Mo		...	MIL-T-9046	4975	...	...	...	Parts and cases for jet-engine compressors; airframe skin components
Ti-5Al-5Sn-2Zr-2Mo 0.25Si	975 °C (1785 °F) (½ h), AC + 595 °C (1100 °F) (2 h), AC	...	...	...	...	...	...	Jet engine parts; high creep strength to 540 °C (1000 °F)
Ti-6Al-2Nb-1Ta-1Mo	As rolled (1 in. plate)	...	MIL-T-9046	...	...	...	...	High toughness; moderate strength; good resistance to seawater and hot-salt stress corrosion; good weldability
Ti-6Al-2Sn-1.5Zr-1Mo 0.35Bi-0.1Si	Beta forge + duplex anneal	...	...	...	...	...	...	Jet engine discs and blades requiring extra creep resistance and stability
Alpha-beta alloys								Aircraft sheet components, structural sections, and skins, good formability, moderate strength
Ti-8Mn	Annealed	...	MIL-T-9046	...	4908	...	...	...
Ti-3Al-2.5V	Annealed	...	...	...	...	4943	...	Aircraft hydraulic tubing, foil; combines strength, weldability, and formability
						4944	4954	
Ti-6Al-4V	Annealed	B265 (Gr. 5)	MIL-T-9046	4928	4911	...	...	Rocket motor cases; blades and discs for aircraft turbines and compressors; structural forgings and fasteners; pressure vessels; gas and chemical pumps; cryogenic parts; ordnance equipment; marine components; steam-turbine blades
		B348 (Gr. 5)	MIL-T-9047	...	4906	...	...	
		B381 (Gr. 5)	...	4965	...	...	...	
	Solution + age							
Ti-6Al-4V (low O₂)	Annealed	...	MIL-T-9046	4930	4907	...	4956	High pressure cryogenic vessels operating down to –195 °C (–320 °F)
		...	MIL-T-9047	...	...	...	...	
Ti-6Al-6V-2Sn	Annealed	...	MIL-T-9046	4971	4918	...	...	Rocket motor cases; ordnance components; structural aircraft parts and landing gears; responds well to heat treatments; good hardenability
	Solution + age	...	MIL-T-9047	4978	...	...	...	
		...	...	4979	...	...	...	

(continued)

Specifications and applications of wrought titanium alloys (continued)

Alloy	Condition	ASTM no.	Military	Aerospace material specifications				Applications and characteristics
				Bars and forgings	Sheet and plate	Tubing	Wire	
Ti-7Al-4Mo	Solution	...	MIL-T-9047	4970	...	...	...	Airframes and jet engine parts for operation at up to 425 °C (800 °F); missile forgings; ordnance equipment
Ti-6Al-2Sn-4Zr-6Mo	Solution + age	...	...	4981	...	...	...	Components for advanced jet engines
Ti-6Al-2Sn-2Zr-2Mo 2Cr-2Cr-0.25Si	Solution + age	...	...	...	...	...	...	Strength, fracture toughness in heavy sections; landing gear wheels
Ti-10V-2Fe-3Al	Solution + age	...	...	...	...	...	...	Heavy airframe structural components requiring toughness at high strengths
Beta alloys	Solution + age							High strength fasteners
13V, 11Cr, 3Al	Solution + age	...	MIL-T-9046	...	4917	...	...	High strength fasteners, aerospace components, honeycomb panels; good formability, heat treatable
	Solution + age	...	MIL-T-9047	...	...	...	...	
8Mo, 8V, 2Fe, 3Al	Solution + age	...	...	...	...	...	...	High-strength, tough airframe sheet, plate, fasteners, and forged components
3Al, 8V, 6Cr, 4Mo, 4Zr	Solution + age	...	...	...	...	...	...	High strength fasteners, torsion bars, aerospace components
	Annealed							Parts requiring formability and corrosion resistance
11.5Mo, 6Zr, 4.5Sn	Solution + age	...	MIL-T-9047	4977	...	...	4980	High-strength fasteners, high-strength aircraft sheet parts

Titanium and titanium alloy product characteristics and typical applications

Alloy	Guaranteed minimum room-temperature tensile strength		Processing characteristics					Typical applications
	Ultimate, ksi	Yield, ksi	Resistance to cracking during forging	Sheet-forming rating	Weld-ability rating	Heat treatable to high strength?	Harden-ability, section depth, in.	
Unalloyed	50 65 80	40 55 70	Excellent	Excellent	Excellent	No	Not harden-able	Hydraulic control valve, gyro-wheel structure, fittings, attach brackets, welded-duct halves, complex tube shapes, heat-pump channel, skin-stringer structures
Ti-5Al-2.5Sn	120	115	Fair to good	Fair	Excellent	No	Not harden-able	Transmission and gear housing, jet-engine-compressor case assembly and stator housing, droop leading edge in boundary-layer control system and duct structure
Ti-5Al-1Mo-1V	130 to 135	120 to 125	Fair	Fair	Good	No	Not harden-able	Jet-engine compressor blades, discs and housings, gyro-scope gimbal housing, inner skin and frame for jet-engine nozzle assembly, experimental sheet-stringer structures, bulkhead forgings
Ti-6Al-4V	130 Age hardenable to 180	120 170	Good	Good	Fair to good	Yes	1	Jet-engine compressor blades, discs, etc., landing-gear wheels and structures, fasteners, brackets, fittings, pressure bottles, primary and secondary sheet-stringer structures, frames, fire-walls, stiffeners, gussets, and ducts
Ti-6Al-6V-2Sn	150 Age hardenable to 180	140 170	Good	…	Poor	Yes	2	Fasteners and air-intake control track, experimental structural forgings
Ti-13V-11Cr-3Al	125 to 130 Age hardenable to 175	120 to 125 165	Fair	Excellent to fair	Fair to poor(a)	Yes	7	Structural forgings, primary and secondary sheet-stringer structures, skins, frames, brackets, fittings, fasteners, tension-torsion rotor straps and specialty uses
Ti-2.25Al-11Sn-5Zr-1Mo-0.2Si	145 Age hardenable to 180	130 160	Fair to good	…	…	Yes	2	Jet-engine compressor blades, discs, wheels, and spacers, airframe, fasteners
Ti-6Al-2Sn-4Zr-2Mo	130	120	Good	Good	Fair to good	No	Slightly harden-able	Jet-engine compressor blades, discs, wheels, and spacers, compressor case assemblies, airframe skin components
Ti-4Al-3Mo-1V	125 Age hardenable to 180	115 155	Good	Good	Fair to good	Yes	…	Airframe components

(a) Welds are generally not heat treated because of embrittling reactions. Source: *Titanium Alloys Handbook*, MCIC-HB-02, R.A. Wood and R.J. Favor, Metals and Ceramics Information Center, Columbus, OH, 1972

Physical properties of wrought titanium alloys

Alloy	Coefficient of linear thermal expansion, μm/m · K (μin./in./°F)							Modulus of elasticity		Modulus of rigidity		Poisson's ratio	Density	
	20-100 °C (70-212 °F)	20-205 °C (70-400 °F)	20-315 °C (70-600 °F)	20-425 °C (70-800 °F)	20-540 °C (70-1000 °F)	20-650 °C (70-1200 °F)	20-815 °C (70-1500 °F)	GPa	10^6 psi	GPa	10^6 psi		Mg/m^3	lb/in.^{3}
Commercially pure														
99.5 Ti	8.6(4.8)	...	9.2(5.1)	...	9.7(5.4)	10.1(5.6)	10.1(5.6)	102.7	14.9	38.6	5.6	0.34	4.51	0.163
99.2 Ti	8.6(4.8)	...	9.2(5.1)	...	9.7(5.4)	10.1(5.6)	10.1(5.6)	102.7	14.9	38.6	5.6	0.34	4.51	0.163
99.1 Ti	8.6(4.8)	...	9.2(5.1)	...	9.7(5.4)	10.1(5.6)	10.1(5.6)	103.4	15.0	38.6	5.6	0.34	4.51	0.163
99.0 Ti	8.6(4.8)	...	9.2(5.1)	...	9.7(5.4)	10.1(5.6)	10.1(5.6)	104.1	15.1	38.6	5.6	0.34	4.51	0.163
99.2 Ti-0.2Pd	8.6(4.8)	...	9.2(5.1)	...	9.7(5.4)	10.1(5.6)	10.1(5.6)	102.7	14.9	38.6	5.6	0.34	4.51	0.163
Ti-0.8Ni-0.3Mo	...	...	...	...	...	...	...	...	...	102.7	14.9	...	4.54	0.164
Alpha alloys														
Ti-5Al-2.5Sn	9.4(5.2)	...	9.5(5.3)	...	9.5(5.3)	9.7(5.4)	10.1(5.6)	110.3	16.0	...	...	...	4.48	0.162
Ti-5Al-2.5Sn (low O$_2$)	9.4(5.2)	...	9.5(5.3)	...	9.7(5.4)	9.9(5.5)	10.1(5.6)	110.3	16.0	...	...	...	4.48	0.162
Near alpha alloys														
Ti-8Al-1Mo-1V	8.5(4.7)	...	9.0(5.0)	...	10.1(5.6)	10.3(5.7)	...	124.1	18.0	46.9	6.8	0.32	4.37	0.158
Ti-11Sn-1Mo-2.25Al-5.0Zr-1Mo-0.2Si	8.5(4.7)	...	9.2(5.1)	...	9.4(5.2)	...	...	113.8	16.5	...	...	...	4.82	0.174
Ti-6Al-2Sn-4Zr-2Mo	7.7(4.3)	...	8.1(4.5)	...	8.1(4.5)	...	...	113.8	16.5	...	...	...	4.54	0.164
Ti-5Al-5Sn-2Zr-2Mo-0.25Si	...	...	...	...	...	...	10.3(5.7)	113.8	16.5	...	...	0.326	4.51	0.163
Ti-6Al-2Nb-1Ta-1Mo	...	...	...	...	...	9.0(5.0)	...	113.8	17.5	...	...	...	4.48	0.162
Ti-6Al-2Sn-1.5Zr-1Mo-0.35Bi-0.1Si	...	...	...	...	...	...	...	...	...	...	...	...	...	...
Alpha-beta alloys														
Ti-8Mn	8.6(4.8)	9.2(5.1)	9.7(5.4)	10.3(5.7)	10.8(6.0)	11.7(6.5)	12.6(7.0)	113.1	16.4	48.3	7.0	...	4.73	0.171
Ti-3Al-2.5V	9.5(5.3)	9.9(5.5)	9.9(5.5)	...	9.9(5.5)	...	...	106.9	15.5	...	...	...	4.48	0.162
Ti-6Al-4V	8.6(4.8)	9.0(5.0)	9.2(5.1)	9.4(5.2)	9.5(5.3)	9.7(5.4)	...	113.8	16.5	42.1	6.1	0.342	4.43	0.160
Ti-6Al-4V (low O$_2$)	8.6(4.8)	9.0(5.0)	9.2(5.1)	9.4(5.2)	9.5(5.3)	9.7(5.4)	...	113.8	16.5	42.1	6.1	0.342	4.43	0.160
Ti-6Al-6V-2Sn	9.0(5.0)	...	9.4(5.2)	...	9.5(5.3)	...	...	110.3	16.0	...	...	...	4.54	0.164
Ti-7Al-4Mo	9.0(5.0)	9.2(5.1)	9.4(5.2)	9.7(5.4)	10.1(5.6)	10.4(5.8)	11.2(6.2)	113.8	16.5	44.8	6.5	...	4.48	0.162
Ti-6Al-2Sn-4Zr-6Mo	9.0(5.0)	9.2(5.1)	9.4(5.2)	9.5(5.3)	9.5(5.3)	...	...	113.8	16.5	...	...	...	4.65	0.168
Ti-6Al-2Sn-2Zr-2Mo-2Cr-0.25Si	...	...	9.2(5.1)	...	...	...	...	122.0	17.7	46.2	6.7	0.327	4.57	0.165
Ti-10V-2Fe-3Al	...	...	...	...	...	...	...	111.7	16.2	...	...	...	4.65	0.168
Beta alloys														
Ti-13V-11Cr-3Al	9.4(5.2)	...	10.1(5.6)	10.6(5.9)	...	...	...	101.4	14.7	42.7	6.2	0.304	4.84	0.175
Ti-8Mo-8V-2Fe-3Al	...	...	...	...	...	...	...	106.9	15.5	...	...	...	4.84	0.175
Ti-3Al-8V-6Cr-4Mo-4Zr	...	...	...	9.68(5.38) (to 900 °F)	...	...	...	105.5	15.3	...	...	...	4.82	0.174
Ti-11.5Mo-6Zr-4.5Sn	...	...	...	...	...	...	...	103.4	15.0	...	...	...	...	...

Mechanical properties of wrought titanium alloys

Alloy	Condition	Room temperature properties Tensile strength MPa	ksi	Yield strength MPa	ksi	Elongation, %	Reduction in area, %	Test temperature °C	°F	Extreme-temperature properties Tensile strength MPa	ksi	Yield strength MPa	ksi	Elongation, %	Reduction in area, %	Charpy impact strength J	ft·lb	Hardness, HRC
Commercially pure																		
99.5 Ti	Annealed	331	48	241	35	30	55	315	600	152	22	97	14	32	80	...	...	120(a)
99.2 Ti	Annealed	434	63	345	50	28	50	315	600	193	28	117	17	35	75	...	...	200(a)
99.1 Ti	Annealed	517	75	448	65	25	45	315	600	234	34	138	20	34	75	43	32	225(a)
99.0 Ti	Annealed	662	96	586	85	20	40	315	600	310	45	172	25	25	70	38	28	265(a)
99.2Ti-0.2Pd	Annealed	434	63	345	50	28	50	315	600	186	27	110	16	37	75	20	15	200(a)
Ti-0.8Ni-0.3Mo	Annealed	517	75	448	65	25	42	205	400	345	50	248	36	37	...	43	32	...
								315	600	324	47	207	30	32	...	...	...	...
Alpha alloys																		
Ti-5Al-2.5Sn	Annealed	862	125	807	117	16	40	315	600	565	82	448	65	18	45	26	19	36
Ti-5Al-2.5Sn (low O₂)	Annealed	807	117	745	108	16	...	−195	−320	1241	180	1158	168	16	...	27	20	35
								−255	−423	1579	229	1420	206	15	...	...	...	...
Near alpha alloys																		
Ti-8Al-1Mo-1V	Duplex annealed	1000	145	951	138	15	28	315	600	793	115	621	90	20	38	32	24	35
								425	800	738	107	565	82	20	44	...	...	...
Ti-11Sn-1Mo-2.25Al-5.0Zr-1Mo-0.25Si	Duplex annealed	1103	160	993	144	15	35	540	1000	621	90	517	75	25	55	...	...	36
								315	600	896	130	758	110	20	44	...	...	...
Ti-6Al-2Sn-4Zr-2Mo	Duplex annealed	979	142	896	130	15	35	425	800	827	120	676	98	22	48	...	...	32
								540	1000	758	110	586	85	24	50	...	...	...
								315	600	772	112	586	85	16	42	...	...	...
Ti-5Al-5Sn-2Zr-2Mo-0.25Si	975 °C (1785 °F) (½h), AC + 595 °C (1100 °F) (2 h), AC	1048	152	965	140	13	...	425	800	703	102	489	71	21	55	...	...	...
								540	1000	648	94	565	82	26	60	...	...	...
								315	600	793	115	...	...	15	...	...	...	...
Ti-6Al-2Nb-1Ta-1Mo	As rolled 2.5 cm (1 in.) plate	855	124	758	110	13	34	425	800	779	113	531	77	17	...	31	23	30
								540	1000	689	100	503	73	19	...	...	...	...
								315	600	586	85	462	67	20	...	...	...	...
Ti-6Al-2Sn-1.5Zr-1Mo-0.35Bi-0.1Si	Beta forge + duplex anneal	1014	147	945	137	11	...	425	800	517	75	414	60	20	...	...	...	...
								540	1000	483	70	379	55	20	...	...	...	...
								480	900	724	105	586	85	15	...	...	...	...
Alpha-beta alloys																		
Ti-8Mn	Annealed	945	137	862	125	15	32	315	600	717	104	565	82	18	35	...	...	...
Ti-3Al-2.5V	Annealed	689	100	586	85	20	...	315	600	483	70	345	50	25	30	...	...	...
Ti-6Al-4V	Annealed	993	144	924	134	14	30	315	600	724	105	655	95	14	35	19	14	36
								425	800	669	97	572	83	18	40	...	...	...
								540	1000	531	77	427	62	35	50	...	...	...
	Solution + age	1172	170	1103	160	10	25	315	600	862	125	703	102	10	28	...	...	41
								425	800	800	116	621	90	12	35	...	...	...
								540	1000	655	95	483	70	22	45	...	...	...
Ti-6Al-4V (low O₂)	Annealed	896	130	827	120	15	35	160	320	1517	220	1413	205	14	42	24	18	35
Ti-6Al-6V-2Sn	Annealed	1069	155	1000	145	14	30	315	600	931	135	807	117	18	28	18	13	38
	Solution + age	1276	185	1172	170	10	20	315	600	979	142	896	130	12	50	...	...	42
Ti-7Al-4Mo	Solution + age	1103	160	1034	150	16	22	315	600	976	127	745	108	18	55	18	13	38
Ti-6Al-2Sn-4Zr-6Mo	Solution + age	1269	184	1172	170	10	23	315	600	848	123	717	104	20	55	...	...	42
								425	800	1020	148	841	122	18	67	...	...	...
Ti-6Al-2Sn-2Zr-2Mo-2Cr-0.25Si	Solution + age	1276	185	1138	165	11	33	315	600	951	138	758	110	19	70	...	...	...
								540	1000	848	123	655	95	19	27	...	...	...
Ti-10V-2Fe-3Al	Solution + age	1276	185	1200	174	10	19	205	400	1117	162	1048	152	13	33	...	...	...
								315	600	1103	160	979	142	13	42	...	...	...
Beta alloys																		
Ti-13V-11Cr-3Al	Solution + age	1220	177	1172	170	8	...	315	600	883	128	793	115	19	...	...	...	...
								425	800	1103	160	827	120	12	...	...	...	...
Ti-8Mo-8V-2Fe-3Al	Solution + age	1276	185	1207	175	8	...	315	600	1131	164	979	142	15	...	11	8	40
Ti-3Al-8V-6Cr-4Mo-4Zr	Solution + age	1310	190	1241	180	8	...	315	600	1034	150	896	130	20	...	...	...	40
	Solution + age	1448	210	1379	200	7	...	425	800	938	136	758	110	17	...	10	7.5	42
Ti-11.5Mo-6Zr-4.5Sn	Annealed	883	128	834	121	15	...	315	600	724	105	655	95	22	...	...	...	...
	Solution + age	1386	201	1317	191	11	...	315	600	903	131	848	123	16	...	...	...	...

(a) Hardness, HB

Approximate threshold for stress-corrosion cracking of titanium alloys in hot salt

Alloy	Condition	100-h threshold stress, ksi								
		550	600	650	700	750	800	850	900	950
Ti-4Al-3Mo-1V	Aged	...	95	...	25	...	25	...	...	...
	Annealed	84	78	...	28	...	15-49	...	...	...
Ti-2.5Al-1Mo-10Sn-5Zr	Aged	...	...	...	70	...	40	...	35	...
Ti-5Al-5Sn-5Zr-1Mo-1V		69	...	...	...	...	35	...	...	...
Ti-6Al-4V	Aged	...	95	65	25	30	12	15	...	...
	Annealed	50	50	...	22	...	18-24	...	...	...
Ti-5Al-2.75Cr-1.25Fe	Aged	...	80	...	25	...	18	...	...	...
	Annealed	...	45	...	...	...	15	...	...	...
Ti-8Al-1Mo-1V	Aged	...	...	...	...	25	...	20	...	15
	Annealed	25	55	...	23	...	18	...	...	...
Ti-5Al-2.5Sn	Annealed	28	30	...	15	...	10-20	...	...	...
Ti-7Al-12Zr		...	...	...	...	...	<5	...	...	...
Ti-5Al-5Sn-5Zr		...	...	...	...	...	<5	...	<5	...

Source: *Titanium Alloys Handbook*, MCIC-HB-02, R.A. Wood and R.J. Favor, Metals and Ceramics Information, Columbus, OH, 1972

Heat Treating

Recommended stress-relief treatments for titanium and titanium alloys
Parts can be cooled from stress relief by either air cooling or slow cooling

Alloy	Temperature °C	°F	Time, h
Commercially pure Ti			
(all grades)	480 to 595	900 to 1100	¼ to 4
Alpha or near-alpha alloys			
Ti-5Al-2.5Sn	540 to 650	1000 to 1200	¼ to 4
Ti-8Al-1Mo-1V	595 to 705	1100 to 1300	¼ to 4
Ti-6Al-2Sn-4Zr-2Mo	595 to 705	1100 to 1300	¼ to 4
Ti-6Al-2Nb-1Ta-0.8Mo	595 to 650	1100 to 1200	¼ to 2
Ti-0.3Mo-0.8Ni (Ti Code 12)	480 to 595	900 to 1100	¼ to 4
Alpha-beta alloys			
Ti-6Al-4V	480 to 650	900 to 1200	1 to 4
Ti-6Al-6V-2Sn (Cu + Fe)	480 to 650	900 to 1200	1 to 4
Ti-3Al-2.5V	540 to 650	1000 to 1200	½ to 2
Ti-6Al-2Sn-4Zr-6Mo	595 to 705	1100 to 1300	¼ to 4
Ti-5Al-2Sn-4Mo-2Zr-4Cr (Ti-17)	480 to 650	900 to 1200	1 to 4
Ti-7Al-4Mo	480 to 705	900 to 1300	1 to 8
Ti-6Al-2Sn-2Zr-2Mo-2Cr-0.25 Si	480 to 650	900 to 1200	1 to 4
Ti-8Mn	480 to 595	900 to 1100	¼ to 2
Beta or near-beta alloys			
Ti-13V-11Cr-3Al	705 to 730	1300 to 1350	1/12 to ¼
Ti-11.5Mo-6Zr-4.5Sn (Beta III)	720 to 730	1325 to 1350	1/12 to ¼
Ti-3Al-8V-6Cr-4Zr-4Mo (Beta C)	705 to 760	1300 to 1400	1/6 to ½
Ti-10V-2Fe-3Al	675 to 705	1250 to 1300	½ to 2
Ti-15V-3Al-3Cr-3Sn	790 to 815	1450 to 1500	1/12 to ¼

Recommended annealing treatments for titanium and titanium alloys

Alloy	Temperature °C	°F	Time, h	Cooling method
Commercially pure Ti (all grades)	650 to 760	1200 to 1400	1/10 to 2	Air
Alpha or near-alpha alloys				
Ti-5Al-2.5Sn	720 to 845	1325 to 1550	1/6 to 4	Air
Ti-8Al-1Mo-1V	790(a)	1450(a)	1 to 8	Air or furnace
Ti-6Al-2Sn-4Zr-2Mo	900(b)	1650(b)	½ to 1	Air
Ti-6Al-2Nb-1Ta-0.8Mo	790 to 900	1450 to 1650	1 to 4	Air
Alpha-beta alloys				
Ti-6Al-4V	705 to 790	1300 to 1450	1 to 4	Air or furnace
Ti-6Al-6V-2Sn (Cu + Fe)	705 to 815	1300 to 1500	¾ to 4	Air or furnace
Ti-3Al-2.5V	650 to 760	1200 to 1400	½ to 2	Air
Ti-6Al-2Sn-4Zr-6Mo	(c)	(c)	…	…
Ti-5Al-2Sn-4Mo-2Zr-4Cr (Ti-17)	(c)	(c)	…	…
Ti-7Al-4Mo	705 to 790	1300 to 1450	1 to 8	Air
Ti-6Al-2Sn-2Zr-2Mo-2Cr-0.25Si	705 to 815	1300 to 1500	1 to 2	Air
Ti-8Mn	650 to 760	1200 to 1400	½ to 1	(d)
Beta or near-beta alloys				
Ti-13V-11Cr-3Al	705 to 790	1300 to 1450	1/6 to 1	Air or water
Ti-11.5Mo-6Zr-4.5Sn (Beta III)	690 to 760	1275 to 1400	1/6 to 1	Air or water
Ti-3Al-8V-6Cr-4Zr-4Mo (Beta C)	790 to 815	1450 to 1500	¼ to 1	Air or water
Ti-10V-2Fe-3Al	(c)	(c)	…	…
Ti-15V-3Al-3Cr-3Sn	790 to 815	1450 to 1500	1/12 to ¼	Air

(a) For sheet and plate, follow by ¼ h at 790 °C (1450 °F), then air cool. (b) For sheet, follow by ¼ h at 790 °C (1450 °F), then air cool (plus 2 h at 595 °C or 1100 °F, then air cool, in certain applications). For plate, follow by 8 h at 595 °C (1100 °F), then air cool. (c) Not normally supplied or used in annealed condition. (d) Furnace or slow cool to 540 °C (1000 °F), then air cool

Recommended solution treating and aging (stabilizing) treatments for titanium alloys

Alloy	Solution temperature		Solution time, h	Cooling rate	Aging temperature		Aging time, h
	°C	°F			°C	°F	
Alpha or near-alpha alloys							
Ti-8Al-1Mo-1V	980 to 1010(a)	1800 to 1850(a)	1	Oil or water	565 to 595	1050 to 1100	...
Ti-6Al-2Sn-4Zr-2Mo	955 to 980	1750 to 1800	1	Air	595	1100	8
Alpha-beta alloys							
Ti-6Al-4V	955 to 970(b)(c)	1750 to 1775(b)(c)	1	Water	480 to 595	900 to 1100	4 to 8
	955 to 970	1750 to 1775	1	Water	705 to 760	1300 to 1400	2 to 4
Ti-6Al-6V-2Sn (Cu + Fe)	885 to 910	1625 to 1675	1	Water	480 to 595	900 to 1100	4 to 8
Ti-6Al-2Sn-4Zr-6Mo	845 to 890	1550 to 1650	1	Air	580 to 605	1075 to 1125	4 to 8
Ti-5Al-2Sn-2Zr-4Mo-4Cr	845 to 870	1550 to 1600	1	Air	580 to 605	1075 to 1125	4 to 8
Ti-6Al-2Sn-2Zr-2Mo-2Cr-0.25Si	870 to 925	1600 to 1700	1	Water	480 to 595	900 to 1100	4 to 8
Beta or near-beta alloys							
Ti-13V-11Cr-3Al	775 to 800	1425 to 1475	$\frac{1}{4}$ to 1	Air or Water	425 to 480	800 to 900	4 to 100
Ti-11.5Mo-6Zr-4.5Sn (Beta III)	690 to 790	1275 to 1450	$\frac{1}{8}$ to 1	Air or water	480 to 595	900 to 1100	8 to 32
Ti-3Al-8V-6Cr-4Mo-4Zr (Beta C)	815 to 925	1500 to 1700	1	Water	455 to 540	850 to 1000	8 to 24
Ti-10V-2Fe-3Al	760 to 780	1400 to 1435	1	Water	495 to 525	925 to 975	8
Ti-15V-3Al-3Cr-3Sn	790 to 815	1450 to 1500	$\frac{1}{4}$	Air	510 to 595	950 to 1100	8 to 24

(a) For certain products, use solution temperature of 890 °C (1650 °F) for 1 h, then air cool or faster. (b) For thin plate or sheet, solution temperature can be used down to 890 °C (1650 °F) for 6 to 30 min., then water quench. (c) This treatment is used to develop maximum tensile properties in this alloy

Minimum metal removal after thermal exposure of titanium alloys

Heat treating temperature		Time at temperature, h	Minimum stock removal per surface(a)	
°C	°F		mm	in.
480 to 593	900 to 1100	Up to 12	0.005	0.0002
594 to 648	1101 to 1200	Up to 4	0.008	0.0003
		4 to 12	0.015	0.0006
649 to 704	1201 to 1300	Up to 1	0.013	0.0005
		1 to 8	0.020	0.0008
		8 to 12	0.025	0.0010
705 to 760	1301 to 1400	Up to 1	0.025	0.0010
		1 to 4	0.036	0.0014
		4 to 8	0.038	0.0015
		8 to 12	0.043	0.0017
761 to 787	1401 to 1450	Up to 1	0.030	0.0012
		1 to 2	0.038	0.0015
		2 to 4	0.046	0.0018
		4 to 8	0.051	0.0020
		8 to 12	0.056	0.0022
788 to 815	1451 to 1500	Up to $\frac{1}{2}$	0.036	0.0014
		$\frac{1}{2}$ to 1	0.041	0.0016
		1 to 2	0.051	0.0020
816 to 871	1501 to 1600	Up to $\frac{1}{2}$	0.058	0.0023
		$\frac{1}{2}$ to 1	0.066	0.0026
		1 to 2	0.076	0.0030
872 to 898	1601 to 1650	Up to $\frac{1}{2}$	0.058	0.0023
		$\frac{1}{2}$ to 1	0.081	0.0032
		1 to 2	0.089	0.0035
899 to 926	1651 to 1700	Up to $\frac{1}{2}$	0.086	0.0034
		$\frac{1}{2}$ to 1	0.091	0.0036
		1 to 2	0.107	0.0042
927 to 954	1701 to 1750	Up to $\frac{1}{2}$	0.097	0.0038
		$\frac{1}{2}$ to 1	0.107	0.0042
		1 to 2	0.122	0.0048

(a) Values shown are typical; actual values may vary with alloy type

Variation of tensile properties of Ti-6Al-4V bar stock with solution-treating temperature

Solution treating temperature		Room-temperature tensile properties(a)				Elongation in 4D,
		Tensile strength		Yield strength(b)		
°C	°F	MPa	ksi	MPa	ksi	%
845	1550	1025	149	980	142	18
870	1600	1060	154	985	143	17
900	1650	1095	159	995	144	16
925	1700	1110	161	1000	145	16
940	1725	1140	165	1055	153	16

(a) Properties determined on 13-mm ($\frac{1}{2}$-in.) bar after solution treating, quenching and aging. Aging treatment: 8 h at 480 °C (900 °F), air cool. (b) 0.2% offset

Relation of tensile strength to size of solution treated and aged titanium alloys

Alloy	Tensile strength of square bar in section size of:											
	13 mm ($\frac{1}{2}$ in.)		25 mm (1 in.)		50 mm (2 in.)		75 mm (3 in.)		100 mm (4 in.)		150 mm (6 in.)	
	MPa	ksi	MPa	ksi	MPa	ksi	MPa	ksi	MPa	ksi	MPa	ksi
Ti-6Al-4V	1105	160	1070	155	1000	145	930	135	...	...	...	...
Ti-6Al-6V-2Sn (Cu + Fe)	1205	175	1205	175	1070	155	1035	150	...	...	...	...
Ti-6Al-2Sn-4Zr-6Mo	1170	170	1170	170	1170	170	1140	165	1105	160	...	...
Ti-5Al-2Sn-2Zr-4Mo-4Cr (Ti-17)	1170	170	1170	170	1170	170	1105	160	1105	160	1105	160
Ti-10V-2Fe-3Al	1240	180	1240	180	1240	180	1240	180	1170	170	1170	170
Ti-13V-11Cr-3Al	1310	190	1310	190	1310	190	1310	190	1310	190	1310	190
Ti-11.5Mo-6Zr-4.5Sn (Beta III)	1310	190	1310	190	1310	190	1310	190	1310	190	...	...
Ti-3Al-8V-6Cr-4Zr-4Mo (Beta C)	1310	190	1310	190	1240	180	1240	180	1170	170	1170	170

Beta transformation temperatures of titanium alloys

Alloy	Beta transus	
	°C, ± 15	°F, ± 25
Commercially pure Ti, 0.25 max O_2	910	1675
Commercially pure Ti, 0.40 max O_2	945	1735
Alpha and near-alpha alloys		
Ti-5Al-2.5Sn	1050	1925
Ti-8Al-1Mo-1V	1040	1900
Ti-6Al-2Sn-4Zr-2Mo	995	1820
Ti-6Al-2Cb-1Ta-0.8Mo	1015	1860
Ti-0.3Mo-0.8Ni (Ti Code 12)	880	1615
Alpha-beta alloys		
Ti-6Al-4V	1000(a)	1830(b)
Ti-6Al-6V-2Sn (Cu + Fe)	945	1735
Ti-3Al-2.5V	935	1715
Ti-6Al-2Sn-4Zr-6Mo	940	1720
Ti-5Al-2Sn-2Zr-4Mo-4Cr (Ti-17)	900	1650
Ti-7Al-4Mo	1000	1840
Ti-6Al-2Sn-2Zr-2Mo-2Cr-0.25Si	970	1780
Ti-8Mn	800(c)	1475(d)
Beta or near-beta alloys		
Ti-13V-11Cr-3Al	720	1330
Ti-11.5Mo-6Zr-4.5Sn (Beta III)	760	1400
Ti-3Al-8V-6Cr-4Zr-4Mo (Beta C)	795	1460
Ti-10V-2Fe-3Al	805	1480
Ti-15V-3Al-3Cr-3Sn	760	1400

(a) ± 20. (b) ± 30. (c) ± 35. (d) ± 50

Joining

Typical tensile, bend, and hardness data for as-welded titanium and several titanium alloys

Material condition	Tensile strength		Yield strength		Elongation, %	Minimum bend radius(a)	Hardness	
	MPa	ksi	MPa	ksi			Knoop	HRC
Ti Grade 1								
Unwelded sheet	315	46	215	31	50.4	0.7t	140	63.5 (b)
Single-bead weld	345	50	255	37	37.5	1.0t	140	55.8(b)
Multiple-bead weld	365	53	270	39	37.7	...	...	...
Transverse weld	325	47(c)	...	...	...	...	...	...
Ti Grade 2								
Unwelded sheet	460	67	325	47	26.2	2.9t	165	80.6(b)
Single-bead weld	505	73	380	55	18.3	2.9t	175	83.1(b)
Multiple-bead weld	510	74	385	56	13.3	...	...	...
Transverse weld	475	69(c)	...	...	...	...	...	...
Ti Grade 3								
Unwelded sheet	545	79	395	57	25.9	1.9t	175	94.4(b)
Single-bead sheet	605	88	475	69	15.5	4.7t	220	92.4(b)
Multiple-bead weld	615	89	480	70	14.7	...	...	...
Transverse weld	560	81(c)	...	...	...	...	...	...
Ti Grade 4								
Unwelded sheet	660	96	530	77	22.3	3.2t	215	23.4
Single-bead weld	695	101	580	84	16.4	5.6t	240	21.2
Multiple-bead weld	710	103	585	85	16.0	...	...	...
Transverse weld	660	96(c)	...	...	...	...	...	...
Ti-5Al-2.5Sn-ELI								
Unwelded sheet	850	123	805	117	15.7	3.8t	265	33.2
Single-bead weld	920	133	770	112	9.8	5.9t	310	28.0
Multiple-bead weld	935	136	820	119	7.5	...	...	...
Transverse weld	850	123(c)	...	...	...	...	...	...
Ti-6Al-2Nb-1Ta-1Mo								
Unwelded sheet	895	130	855	124	9.7	2.8t	275	29.6
Single-bead weld	930	135	800	116	5.9	7.7t	30	27.7
Multiple-bead weld	945	137	815	118	5.7	...	...	...
Transverse weld	890	129(c)	...	...	...	...	...	...
Ti-3Al-2.5V								
Unwelded sheet	705	102	670	97	15.2	4.0t	230	23.6
Single-bead weld	705	102	600	87	12.7	5.4t	250	19.6
Multiple-bead weld	745	108	625	91	11.2	...	...	...
Transverse weld	710	103(c)	...	...	...	...	...	...
Ti-6Al-4V								
Unwelded sheet	1000	145	945	137	11.0	2.6t	320	32.2
Single-bead weld	1060	154	920	133	3.5	10.5t	350	35.9
Multiple-bead weld	1090	158	945	137	3.2	...	...	...
Transverse weld	1015	147(c)	...	...	...	...	...	...
Ti-8Al-1Mo-1V								
Unwelded sheet	1060	154	1020	148	15.0	2.9t	325	36.0
Single-bead weld	1085	157	930	135	5.5	7.0t	345	35.2
Multiple-bead weld	1115	162	960	139	3.2	...	...	...
Transverse weld	1060	154(c)	...	...	...	...	...	...
Ti-6Al-6V-2Sn								
Unwelded sheet	1060	154	1005	146	9.8	2.8t	350	34.0
Single-bead weld	1295	188	1255	182	0.3	25.6t	420	46.8
Multiple-bead weld	1280	186	...	...	0.1	...	...	...
Transverse weld	1103	160(c)	...	...	...	...	...	...
Ti-13V-11Cr-3Al								
Unwelded sheet	965	140	910	132	13.9	2.7t	300	30.6
Single-bead weld	950	138	925	134	11.6	2.7t	320	30.1
Multiple-bead weld	925	134	875	127	9.1	...	...	...
Transverse weld	950	138(c)	...	...	...	...	...	...

(a) Sheet thickness, t. (b) Hardness, HRB. (c) Fracture occurred in base metal

Powder Metals

Ferrous Materials

Typical density designations and ranges of ferrous P/M materials

| MPIF density suffix | Designation | | Density, g/cm^3 |
	ASTM type(a)	SAE type	
N	I	1(b)	Less than 6.0
P	II	2	6.0 to 6.4
R	III	3	6.4 to 6.8
S	IV	4	6.8 to 7.2
T	V(c)	5(c)	7.2 to 7.6
U	...	...	7.6 to 8.0

Note: Density of pure iron is 7.87 g/cm^3. (a) ASTM B 426 only; different density ranges used in ASTM B 310 and B 484. (b) Density range of 5.6 to 6.0 g/cm^3 is specified. (c) Minimum density of 7.2 g/cm^3 is specified

Applications of ferrous P/M materials

Material and specification designation	MPIF density suffix(a)	Condition	Application
P/M iron F-0000-N through T: 0.3 C max	N	As sintered	Structural (lightly loaded gears); magnetic (motor pole pieces); self-lubricating bearings; structural wear-resisting (small levers and cams) as carbonitrided
	P	As sintered	
	R	As sintered	
	S	As sintered	
	T	As sintered	
P/M steel F-0005-N through T: 0.3 to 0.6 C	N	As sintered	Structural (moderately loaded gears, levers, cams); structural (moderately loaded gears, levers, and cams requiring wear resistance) as heat treated
	P	As sintered	
	R	As sintered	
	S	Heat treated	
		As sintered	
		Heat treated	
P/M steel F-0008-N through T: 0.6 to 1.0 C	N	As sintered	Structural (moderately loaded gears, levers, cams); structural (moderately loaded gears, levers, and cams requiring wear resistance) as heat treated
		Heat treated	
	P	As sintered	
		Heat treated	
	R	As sintered	
		Heat treated	
	S	As sintered	
		Heat treated	
P/M copper iron FC-0200-P through S: 1.5 to 3.9 Cu, 0.3 C max	P	As sintered	Bearings or mechanical components
	R	As sintered	Mechanical components
	S	As sintered	
P/M copper steel FC-0205-P through S: 1.5 to 3.9 Cu, 0.3 to 0.6 C		As sintered	
	P	As sintered	Bearings or mechanical components
	R	Heat treated	Mechanical components
		As sintered	
	S	Heat treated	
P/M copper steel FC-0208-N through S: 1.5 to 3.9 Cu, 0.6 to 1.0 C	N	As sintered	Bearings or mechanical components
		Heat treated	Mechanical components
	P	As sintered	
		Heat treated	
	R	As sintered	
		Heat treated	
	S	As sintered	
		Heat treated	

(a) See table on density designations. (b) Generally heat treated for wear resistance rather than strength

(continued)

Applications of ferrous P/M materials (continued)

Material and specification designation	MPIF density suffix(a)	Condition	Application
P/M copper steel FC-0505-N through S: 4.0 to 6.0 Cu, 0.3 to 0.6 C	N	As sintered	
		Heat treated(b)	
	P	As sintered	
		Heat treated(b)	
	R	As sintered	
		Heat treated(b)	
P/M copper steel FC-0508-N through R: 4.0 to 6.0 Cu, 0.6 to 1.0 C	N	As sintered	Mechanical components
		Heat treated(b)	
	P	As sintered	
		Heat treated(b)	
	R	As sintered	
		Heat treated(b)	
P/M copper steel FC-0808-N: 6 to 11 Cu, 0.6 to 1.0 C	N	As sintered	Mechanical components
P/M copper iron FC-1000-N: 9.5 to 10.5 Cu, 0.3 C max	N	As sintered	Bearings or mechanical components
P/M iron-nickel FN-0200-R through T: 1 to 3 Ni, 2.5 Cu max, 0.3 C max	R	As sintered	Mechanical components (can be case hardened)
	S	As sintered	
	T	As sintered	
	R	As sintered	Structural (couplings) as sintered; structural, wear resisting (oil pump gears and heavily loaded support brackets) as heat treated; structural, wear and impact resisting (oil pump gears to 3000 psi and heavily loaded transmission gears)
		Heat treated	
	S	Sintered and sized	
		Heat treated	
	T	Sintered and sized	
		Heat treated	
P/M nickel steel FN-0208-R through T: 1 to 3 Ni, 0.6 to 0.9 C, 2.5 Cu max	R	As sintered	Mechanical components
		Heat treated	
	S	As sintered	
		Heat treated	
	T	As sintered	
		Heat treated	
P/M iron-nickel FN-0400-R through T: 3 to 5.5 Ni, 0.3 to 0.6 C, 2.0 Cu max	R	As sintered	Mechanical components (can be case hardened)
P/M nickel steel FN-0405-R through T: 3 to 5 Ni, 0.3 to 0.6 C, 2.0 Cu max	R	As sintered	Mechanical components
		Heat treated	
	S	As sintered	
		Heat treated	
	T	As sintered	
		Heat treated	
P/M nickel steel FN-0408-R through T: 3.0 to 5.0 Ni, 0.6 to 0.9 C, 2.0 Cu max	R	As sintered	Structural, wear resisting, high stress (planetary differential and transmission gears up to 6 hp) as heat treated; structural, wear resisting, high stress and requiring welded assembly (welded assembly of pinion and sprocket) as carbonitrided
	S	As sintered	
	T	As sintered	
P/M iron-nickel FN-0700-R through T: 6 to 8 Ni, 0.3 C max, 2.0 Cu max	R	As sintered	Mechanical components (can be case hardened)
	S	As sintered	
	T	As sintered	
P/M nickel steel FN-0705-R through T: 6 to 8 Ni, 0.3 to 0.6 C, 2.0 Cu max	R	As sintered	Mechanical components
		Heat treated	
	S	As sintered	
		Heat treated	
	T	As sintered	
		Heat treated	

(a) See table on density designations. (b) Generally heat treated for wear resistance rather than strength

(continued)

Applications of ferrous P/M materials (continued)

Material and specification designation	MPIF density suffix(a)	Condition	Application
P/M nickel steel FN-0708-R through T: 6 to 8 Ni, 0.6 to 0.9 C, 2.0 Cu max	R	As sintered	Mechanical components
	S	As sintered	
	T	As sintered	
P/M infiltrated steel FX-1005-T: 8 to 14.9 Cu, 0.3 to 0.6 C	T	As sintered	Mechanical components (special shapes)
		Heat treated	
FX-1008-T: 8 to 14.9 Cu, 0.6 to 1.0 C	T	As sintered	Mechanical components (special shapes)
		Heat treated	
FX-2000-T: 15 to 25 Cu, 10.3 C max	T	As sintered	Mechanical components
FX-2005-T: 15 to 25 Cu, 0.3 to 0.6 C	T	As sintered	Mechanical components
		Heat treated	
FX-2008-T: 15 to 25 Cu, 0.6 to 1.0 C	T	As sintered	Mechanical components
		Heat treated	
P/M austenitic stainless steel	P	As sintered	Type 303, mechanical components requiring secondary machining; type 316, structural, corrosion resisting, nonmagnetic (small gears, levers, cams, and other parts for exposure to salt water and specific industrial acids); type 410, structural, corrosion resisting (small gears, levers, cams, and other parts where applications require heat treating for wear resistance)
SS-303-P	R	As sintered	
SS-303-R	P	As sintered	
SS-316-P	R	As sintered	
SS-316-R	N	As sintered	
SS-410-N	P	As sintered	
SS-410-P			

(a) See table on density designations. (b) Generally heat treated for wear resistance rather than strength

Effects of density on elastic modulus, Poisson's ratio, and coefficient of thermal expansion of ferrous P/M materials

MPIF density suffix(a)	Density, g/cm^3	Elastic modulus		Poisson's ratio	Coefficient of thermal expansion, 10^{-6} / K
		GPa	10^6 psi		
N	5.6 to 6.0	72	10.5	0.18	8.1
P	6.0 to 6.4	90	13	0.20	8.7
R	6.4 to 6.8	110	16	0.21	9.2
S	6.8 to 7.2	130	19	0.23	9.8
T	7.2 to 7.6	160	23	0.26	10.4
Theoretical	7.86	205	30	0.28	11–12

(a) See table on density designations

Typical mechanical properties of ferrous P/M materials

Designation	MPIF density suffix(a)	Condition(b)	Tensile strength MPa	ksi	Yield strength MPa	ksi	Elongation in 25 mm (1 in.), %	Fatigue strength MPa	ksi	Impact energy(c) J	ft·lb	Apparent hardness	Elastic modulus GPa	10⁶ psi
F-0000	N	AS	110	16	75	11	2.0	40	6(d)	4.1	3.0	10 HRH	70	10.5
	P	AS	130	19	95	14	2.5	50	7(d)	6.1	4.5	70 HRH	90	13
	R	AS	165	24	110	16	5	60	9(d)	13	9.5	80 HRH	110	16
	S	AS	205	30	150	22	9	80	11(d)	20	15	15 HRB	130	19
	T	AS	275	40	180	26	15	105	15(d)	34	25	30 HRB	160	23
F-0005	N	AS	125	18	105	15	1.0	45	7(d)	3.4	2.5	5 HRB	70	10.5
	P	AS	170	25	140	20	1.5	65	10(d)	4.7	3.5	20 HRB	90	13
	R	AS	220	32	160	23	2.5	85	12(d)	6.8	5.0	45 HRB	110	16
		HT	415	60	395	57	0.5	155	23(d)	...	...	100 HRB	110	16
	S	AS	295	43	195	28	3.5	110	16(d)	12	9.0	60 HRB	130	19
		HT	550	80	515	75	0.5	210	30(d)	...	...	25 HRC	130	19
F-0008	N	AS	200	29	170	25	0.5	75	11(d)	2.7	2.0	35 HRB	70	10.5
		HT	290	42	...	...	< 0.5	110	16(d)	...	...	90 HRB	70	10.5
	P	AS	240	35	205	30	1.0	90	13(d)	4.1	3.0	50 HRB	90	13
		HT	400	58	...	...	< 0.5	150	22(d)	...	...	100 HRB	90	13
	R	AS	290	42	250	36	1.5	110	14(d)	4.7	3.5	65 HRB	110	16
		HT	510	74	...	...	< 0.5	195	28(d)	...	...	25 HRC	110	16
	S	AS	395	57	275	40	2.5	150	22(d)	9.5	7.0	75 HRB	130	19
		HT	650	94	625	91	< 0.5	245	36(d)	...	...	30 HRC	130	19
FC-0200	P	AS	160	23	115	17	2.5	60	9(d)	7.5	5.5	80 HRH	90	13
	R	AS	205	30	145	21	4	80	11(d)	9.5	7.0	15 HRB	110	16
	S	AS	255	37	160	23	7	95	14(d)	23	17	30 HRB	130	19
FC-0205	P	AS	275	40	235	34	1.0	105	15(d)	4.7	3.5	35 HRB	90	13
	R	AS	345	50	260	38	1.5	130	19(d)	7.5	5.5	60 HRB	110	16
		HT	585	85	560	81	< 0.5	220	31(d)	...	...	30 HRC	110	16
	S	AS	425	62	310	45	3.0	160	24(d)	13	9.5	75 HRB	130	19
		HT	690	100	655	95	< 0.5	260	38(d)	...	...	35 HRC	130	19
FC-0208	N	AS	225	33	205	30	< 0.5	85	13(d)	3.4	2.5	45 HRB	70	10.5
		HT	295	43	...	...	< 0.5	110	16(d)	...	...	95 HRB	70	10.5
	P	AS	310	45	280	41	< 0.5	115	17(d)	4.1	3.0	50 HRB	90	13
		HT	380	55	...	...	< 0.5	145	21(d)	...	...	25 HRC	90	13
	R	AS	415	60	330	48	1.0	155	23(d)	6.8	5.0	70 HRB	110	16
		HT	550	80	...	...	< 0.5	210	30(d)	...	...	35 HRC	110	16
	S	AS	550	80	395	57	1.5	210	30(d)	11	8.0	80 HRB	130	19
		HT	690	100	655	95	< 0.5	260	38(d)	...	...	40 HRC	130	19
FC-0505	N	AS	240	35	205	30	0.5	90	13(d)	4.1	3.0	50 HRB	70	10.5
	P	AS	345	50	290	42	1.0	130	19(d)	6.1	4.5	60 HRB	90	13
	R	AS	455	66	380	55	1.5	170	25(d)	6.8	5.0	75 HRB	116	16
FC-0508	N	AS	330	48	295	43	< 0.5	125	18(d)	4.1	3.0	60 HRB	70	10.5
	P	AS	425	62	395	57	1.0	160	24(d)	4.7	3.5	65 HRB	90	13
		HT	480	70	480	70	< 0.5	185	27(d)	...	...	30 HRC	90	13
	R	AS	515	75	480	70	1.0	195	29(d)	6.1	4.5	85 HRB	116	16
FC-0808	N	AS	250	36	...	...	< 0.5	...	...	...	...	55 HRB	...	...
FC-1000	N	AS	205	30	...	...	0.5	...	...	...	...	70 HRF	...	...
FN-0200	R	AS	195	28	125	18	4	75	11	19	14	38 HRB	115	17
	S	AS	260	38	170	25	7	105	15	43	32	42 HRB	145	21
	T	AS	310	45	205	30	11	125	18	68	50	51 HRB	160	23
FN-0205	R	AS	255	37	160	23	3.0	105	15	14	10	50 HRB	115	17
		HT	565	82	450	65	0.5	225	33	8.1	6	32 HRC	115	17
	S	SS	345	50	215	31	3.5	140	20	24	18	70 HRB	145	21
		HT	760	110	605	88	1.0	305	44	22	16	42 HRC	145	21
	T	SS	420	61	255	37	4.5	165	24	43	32	85 HRB	160	23
		HT	925	134	725	105	2.0	370	54	38	28	46 HRC	160	23
FN-0208	R	AS	330	48	205	30	2.0	130	19	11	8	62 HRB	115	17
		HT	690	100	650	94	0.5	275	40	8.1	6	34 HRC	115	17
	S	AS	450	65	280	41	3.0	180	26	19	14	79 HRB	145	21
		HT	930	135	880	128	0.5	370	54	16	12	45 HRC	145	21
	T	AS	545	79	345	50	3.5	220	32	30	22	87 HRB	160	23
		HT	1105	160	1070	155	0.5	415	60	24	18	47 HRC	160	23
FN-0400	R	AS	250	36	150	22	5	95	14	22	16	40 HRB	115	17
	S	AS	340	49	205	30	6	140	20	47	35	60 HRB	145	21
	T	AS	400	58	250	36	6.5	160	23	68	50	67 HRB	160	23
FN-0405	R	AS	310	45	180	26	3.0	125	18	14	10	63 HRB	115	17
		HT	770	112	650	94	0.5	310	45	8.1	6	27 HRC	115	17

(a) See table on density designations. (b) AS, as sintered; SS, sintered and sized; HT, heat treated, typically austenitized at 870 °C (1600 °F), oil quenched and tempered 1 h at 200 °C (400 °F). (c) Unnotched Charpy test. (d) Estimated as 38% of tensile strength. (e) X indicates infiltrated steel

(continued)

Typical mechanical properties of ferrous P/M materials (continued)

Designation	MPIF density suffix(a)	Condition(b)	Tensile strength MPa	ksi	Yield strength MPa	ksi	Elongation in 25 mm (1 in.), %	Fatigue strength MPa	ksi	Impact energy(c) J	ft·lb	Apparent hardness	Elastic modulus GPa	10^6 psi
	S	AS	425	62	240	35	4.5	165	24	20	15	72 HRB	145	21
		HT	1060	154	880	128	1.0	415	60	14	10	39 HRC	145	21
	T	AS	510	74	295	43	6.0	205	30	41	30	80 HRB	160	23
		HT	1240	180	1060	154	1.5	450	65	19	14	44 HRC	160	23
FN-0408	R	AS	395	57	290	42	1.5	160	23	8.1	6	72 HRB	115	17
	S	AS	530	77	390	57	3.0	215	31	14	10	88 HRB	145	21
	T	AS	640	93	470	68	4.5	255	37	22	16	95 HRB	160	23
FN-0700	R	AS	560	52	205	30	2.5	145	21	16	12	60 HRB	115	17
	S	AS	490	71	275	40	4	195	28	28	21	72 HRB	145	21
	T	AS	585	85	330	48	6	240	34	35	26	83 HRB	160	23
FN-0705	R	AS	370	54	240	35	2.0	150	22	12	9	69 HRB	115	17
		HT	705	102	550	80	0.5	280	41	11	8	24 HRC	115	17
	S	AS	525	76	330	48	3.5	205	30	23	17	83 HRB	145	21
		HT	965	140	760	110	1.0	385	56	20	15	38 HRC	145	21
	T	AS	620	90	390	57	5.0	250	36	33	24	90 HRB	160	23
		HT	1160	168	895	130	1.5	500	65	27	20	40 HRC	160	23
FN-0708	R	AS	395	57	280	41	1.5	160	23	8	6	75 HRB	115	17
	S	AS	550	80	380	55	2.5	220	32	16	12	88 HRB	145	21
	T	AS	655	95	455	66	3.0	260	38	22	16	96 HRB	160	23
FX-1005(e)	T	AS	570	83	440	64	4.0	...	...	19	14	75 HRB	135	20
		HT	830	120	740	107	1.0	...	...	9.5	7.0	35 HRC	135	20
FX-1008(e)	T	AS	620	90	515	75	2.5	...	...	16	12	80 HRB	135	20
		HT	895	130	725	105	60.5	...	...	9.5	7.0	40 HRC	135	20
FX-2000(e)	T	AS	450	65	...	...	1.0	...	...	20	15	60 HRB	...	...
FX-2005(e)	T	AS	515	75	345	50	1.5	...	...	12.9	9.5	75 HRB	125	18
		HT	790	115	655	95	< 0.5	...	...	8.1	6.0	30 HRC	125	18
FX-2008(e)	T	AS	585	85	515	75	1.0	...	...	14	10	80 HRB	125	18
		HT	860	125	740	107	< 0.5	...	...	6.8	5.0	42 HRC	125	18

(a) See table on density designations. (b) AS, as sintered; SS, sintered and sized; HT, heat treated, typically austenitized at 870 °C (1600 °F), oil quenched and tempered 1 h at 200 °C (400 °F). (c) Unnotched Charpy test. (d) Estimated as 38% of tensile strength. (e) X indicates infiltrated steel

Effects of steam treating on density and apparent hardness of ferrous P/M materials

MPIF designation	MPIF density suffix(a)	Density, g/cm^3 Sintered	Steam treated	Apparent hardness, HRB Sintered	Steam treated
F-0000	N	5.8	6.2	7(b)	75
	P	6.2	6.4	32(b)	61
	R	6.5	6.6	45(b)	51
F-0008	M	5.8	6.1	44	100
	P	6.2	6.4	58	98
	R	6.5	6.6	60	97
FC-0700	N	5.7	6.0	14	73
	P	6.35	6.5	49	78
	R	6.6	6.6	58	77
FC-0708	N	5.7	6.0	52	97
	P	6.3	6.4	72	94
	R	6.6	6.6	79	93

(a) See table on density designations. (b) Hardness, HRF

Iron

Green density and green strength for various types of iron powders

Powder	Apparent density, g/cm^3	Compaction pressure MPa	Compaction pressure tsi	Green density, g/cm^3	Green strength MPa	Green strength psi
Sponge(a)	2.4	410	30	6.2	14	2100
		550	40	6.6	22	3200
		690	50	6.8	28	4100
Atomized sponge(b)	2.5	410	30	6.55	13	1900
		550	40	6.8	19	2700
		690	50	7.0	...	...
Reduced(a)	2.5	410	30	6.5	16	2300
		550	40	6.7	21	3000
		690	50	6.9	24	3500
Sponge(a)	2.6	410	30	6.6	19	2700
		550	40	6.8	25	3600
		690	50	7.0	27	3900
Electrolytic(c)	2.6	410	30	6.3	32	4600
		550	40	6.7	43	6200
		690	50	6.95	54	7800

(a) Powders contained 1% zinc stearate blended in. (b) Powder contained 0.75% zinc stearate blended in. (c) Isostatically pressed

Mechanical properties of electrolytic iron powder compacts hot pressed at 140 MPa (20 ksi)

Temperature °C	Temperature °F	Dwell time at temperature and pressure, s	Tensile strength MPa	Tensile strength ksi	Elongation in 25 mm (1 in.), %	Hardness, HB
500	930	50	180	26.2	0	50
		150	176	25.5	0	51
		450	274	39.8	1	63
600	1110	50	254	36.9	0.5	62
		150	281	40.8	1	77
		450	336	48.8	2	80
700	1290	50	330	47.8	1	90
		150	395	57.3	12	95
		450	397	57.5	27	100
780	1435	50	373	54.1	22	101
		150	361	52.4	32	93
		450	365	52.9	37	96

Effect of hydrogen chloride on iron sintered in hydrogen

Temperature °C	Temperature °F	Time, min	Atmosphere, % hydrogen chloride	Density, g/cm^3	Strength MPa	Strength ksi	Elongation %
950	1740	30	0	6.20	131	19	6
		30	1	6.30	159	23	10
		120	0	6.30	138	20	6
		120	1	6.30	159	23	10
1375	2505	30	0	7.00	193	28	11
		30	1	7.20	234	34	20
		120	0	7.50	234	34	17
		120	1	7.80	283	41	25

Low-Alloy Steels

Typical mechanical properties of P/M forged low-alloy steels
All materials are in the hardened and tempered condition unless otherwise indicated

Material	Processing	Ultimate tensile strength MPa	ksi	0.2% yield strength MPa	ksi	Elongation in 25 mm (1 in.), %	Reduction in area, %	Charpy V-notch impact energy J	ft·lb	Hardness	Fracture toughness (K_{Ic}) MPa √m	ksi √in.	Density, % of theoretical
Fe-2MCM-0.67C(a)(b)	...	960	139.3	590	86	...	12	...	...	98 HRB	...	...	...
Fe-2MCM-0.67C(a)	...	1900	275.6	1500	218	...	4.5	...	...	49 HRC	...	...	100
4120	Sintered at 1315 °C (2400 °F), re-pressed	701	101.7	616	89.4	14	46	38	28	20-25 HRC	...	...	100
1520	Sintered at 1315 °C (2400 °F), re-pressed	936	135.7	...	...	9	13	39	29	20-25 HRC	...	...	100
4130	Gas atomized, −65 mesh	1586	230	1303	189	5	3	10	7.5	46 HRC	49	45	100
4640	Gas atomized, −65 mesh	...	...	...	...	...	...	7	5	55 HRC	36	33	100
	Water atomized	...	...	...	...	...	...	7	5	42 HRC	37	34	100
	Sintered at 1200 °C (2190 °F)	1040	150.8	1000	145	20	40	36	26	310-350 HV	...	...	99
Fe-2Ni-0.35C	Mixed elemental powders	938	136	600	87	13	44	...	13	31 HRC	...	...	99
Fe-0.55Ni-0.32Mo-0.47Mn-0.23Cr-0.30C	Sintered at 1200 °C (2190 °F)	1020	147.9	970	141	17	37	46	34	...	...	...	...
Fe-3Cu-0.5C-0.3S	...	873	127	...	...	6.5	...	...	...	274 HV	...	...	99
Fe-9Cu-0.34Mn-0.43Ni-0.65Mo-0.31C	...	1675	245	1410	205	13	31	19	14	49 HRC	...	...	99
Fe-0.35Mn-0.57Mo-1.95Ni-0.5C	...	1200	174	1120	162	10	19	30	22	475 HV	...	...	99
4630 modified	Sintered at 1205 °C (2200 °F)	148	215	1331	193	6	10	8	6	42 HRC	...	...	98

(a) MCM is a master alloy containing 20% Mn, 20% Cr, 20% Mo, and 7% C. (b) As-sintered condition

Tool Steels

Mechanical properties of tool steels made from cold compacted and vacuum sintered water-atomized powders

Grade	Hardness, HRC	Ultimate tensile strength(a)		Elongation(b), %	Impact strength(c)	
		MPa	ksi		J	ft · lb
M2	62 to 65	750 to 2000	109 to 290	12 to 14	9 to 12	7 to 9
M35	63 to 66	770 to 2000	112 to 290	6 to 9	8 to 11	6 to 8
T15	64 to 67	770 to 2000	112 to 290	3 to 6	8 to 11	6 to 8

(a) Values depend on heat treatment. Lowest values are for fully annealed condition. (b) Fully annealed. (c) Triple temper. Izod unnotched impact bar (ASTM E 23)

Comparison of mechanical properties and compositions of gas-atomized and hot isostatically pressed P/M and I/M tool steels

Alloy	Hardness, HRC	Bend fracture stress		Bend deflection at fracture		Charpy V-notch impact strength(a)		Fracture toughness		0.2% compressive yield strength	
		MPa	ksi	mm	in.	J	ft · lb	MPa √m	ksi √in.	MPa	ksi
Ingot metallurgy alloys											
M2	64 to 65	3819	554	...	...	23	17	...	...	...	...
M4	64 to 65	3585	520	...	...	16	12	...	...	...	...
M42	67 to 68	2565	372	...	...	7	5	...	...	...	...
T15	66 to 67	2151	312	...	...	5	4	...	...	...	...
D2	62	2068	300	...	...	23	17	...	...	...	...
P/M alloys											
M2 CPM	64 to 65	4991	724	...	...	41	30	...	...	...	...
M4 CPM	64 to 65	5377	780	...	...	43	32	...	...	...	...
M42 CPM	67 to 68	4005	581	...	...	16	12	...	...	...	...
T15 CPM	66 to 67	4674	678	...	...	...	14	...	...	...	...
CPM Rex 20	67 to 68	4005	581	...	...	19	12.5	...	...	...	...
CPM Rex 25 (M61)	66 to 67	4323	627	...	...	15	11	...	...	...	...
CPM Rex 76 (M48)	69	4088	593	...	...	14	10	...	...	...	...
CPM 10V	63	4240	615	...	...	23	17	...	...	...	...
ASP 23	66	4800	696	...	...	...	...	13	12	3500	508
ASP 23	62	...	...	...	...	...	...	19	17	2800	406
ASP 30	66	5100	740	2.1	0.083	...	...	...	...	3600	522
ASP 60	67	4600	667	...	...	...	...	...	...	...	...

AISI designation	Composition, wt%							
	C	Mn	Si	Cr	V	W	Mo	Co
M2	0.85	0.30	0.30	4.0	2.0	6.0	5.0	...
M4	1.30	0.30	0.30	4.0	4.0	5.5	4.5	...
M42	1.10	0.30	0.30	3.75	1.15	1.5	9.5	8.0
CPM Rex 20	1.30	0.30	0.30	3.75	2.0	6.25	10.5	...
T15	1.55	0.30	0.30	4.0	5.0	12.25	...	5.0
CPM Rex 25	1.80	0.30	0.35	4.0	5.0	12.5	6.5	...
CPM Rex 76	1.50	0.30	0.30	3.75	3.1	10.0	5.25	9.0
CPM 10V	2.45	0.50	0.90	5.25	9.75	...	1.30	...
D2	1.55	0.35	0.45	11.5	0.9	...	0.8	...
ASP 23	1.3	...	...	4.2	3.1	6.4	5.0	...
ASP 30	1.3	...	...	4.2	3.1	6.4	5.0	8.5
ASP 60	2.3	...	...	4.0	6.5	6.5	7.0	10.5

(a) 12.7-mm (1/2-in.) radius notch

Stainless Steels

Commercial P/M grades of stainless steel powder

Alloy	Cr	Ni	Si	Mo	Cu	Sn	Mn	C	S	P	Fe	Oxygen content, ppm	Sieve analysis, %(a) +100 (>150 μm)	−325 (<44 μm)	Apparent density, g/cm³	Flow rate, s/50 g
Austenitic																
303	17-18	12-13	0.6-0.8	...	...	...	0.3	0.03	0.1-0.3	0.03	rem	...	3	40-60	3.0-3.2	24-28
304L	18-19	10-12	0.7-0.9	...	...	...	0.3	0.03	0.03	0.03	rem	1000-2000	1-4	30-45	2.5-2.8	28-32
304LSC	18-20	10-12	0.8-1.0	...	2(b)	1(b)	0.3	0.03	0.03	0.03	rem	...	3	30-45	2.7-2.9	26-30
316L	16.5-17.5	13-14	0.7-0.9	2-2.5	...	...	0.3	0.03	0.03	0.03	rem	1000-2000	1-4	35-45	2.6-3.0	24-32
Martensitic																
410L	12-13	...	0.7-0.9	...	...	...	0.1-0.5	0.05(a)	0.03	0.03	rem	1500-2500	3	30-45	2.6-2.9	26-30
Ferritic																
430L	16-17	...	0.7-0.9	...	...	...	0.3	0.03	0.03	0.03	rem	...	3	30-45	2.5-2.9	26-32
434L	16-18	...	0.7-0.9	0.5-1.5	...	...	0.3	0.03	0.03	0.03	rem	...	3	30-45	2.5-2.9	26-32

(a) Maximum unless a range is specified. (b) Typical

Typical mechanical properties of medium-density P/M stainless steels

MPIF designation	MPIF density suffix(a)	Composition, %(b) Cr	Ni	Mo	Si	Tensile strength MPa	ksi	0.2% yield strength MPa	ksi	Elongation in 25 mm (1 in.), %	Density, g/cm³
SS-303	P	17	12	...	0.7	241	35	220	32	1	6.2
	R	17	12	...	0.7	358	52	324	47	2	6.6
SS-316	P	16	13	2	0.7	262	38	220	32	2	6.2
	R	16	13	2	0.7	372	54	275	40	4	6.6
SS-410	N	12	...	...	0.8	289	42	283	41	< 1	5.8
	P	12	...	...	0.8	379	55	372	54	< 1	6.2

Note: All materials sintered in dissociated ammonia. (a) See table on density designations. (b) Remainder Fe

Mechanical properties of type 316L(a)

Property	Sintering atmosphere Dissociated ammonia	Hydrogen
Yield strength (0.2% offset), MPa (ksi)	274 (39)	183 (26)
Ultimate tensile strength, MPa (ksi)	365 (53)	288 (41.8)
Elongation in 25 mm (1 in.), %	7.0	10.9
Apparent hardness, HRB	67	47

(a) Pressed to 6.85/cm³ and sintered for 30 min. at 1120 °C (2050 °F)

Typical mechanical properties of fully dense stainless steel

Property	P/M material	Wrought material
Extruded 0.3- by 15.5-mm (0.1- by 0.61-in.) 317LM tube(a)		
Ultimate tensile strength, MPa (ksi)	693 (100)	693 (100)
0.2% yield strength, MPa (ksi)	324 (47)	353 (51)
Reduction in area, %	71	73
Elongation in 25 mm (1 in.), %	47	50
Hot isostatically pressed 316		
Ultimate tensile strength, MPa (ksi)	579 (84)	...
0.2% yield strength, MPa (ksi)	288 (42)	...
Elongation in 25 mm (1 in.), %	58	...

(a) Gas-atomized powder, canned, cold isostatically pressed, and extruded

Properties of sintered type 410L stainless steel

Processing treatment	Graphite added, %	Sintering atmosphere(a)	Tempering temperature		Tensile strength		Apparent hardness, HRB
			°C	°F	MPa	ksi	
As sintered and cooled in water-jacketed zone of furnace	0	Dissociated ammonia	...	...	724	105	102
	0.10	Dissociated ammonia	205	400	683	99	103
	0	Hydrogen	...	...	393	57	68
	0.10	Hydrogen	175	350	710	103	95
Reheated in dissociated ammonia and oil quenched from 950 °C (1750 °F)	0	Dissociated ammonia	205	400	627	91	106
	0.10	Dissociated ammonia	220	430	703	102	102
	0	Hydrogen	...	...	752	109	106
	0.10	Hydrogen	220	430	717	104	105
Reheated in hydrogen and oil quenched from 950 °C (1750 °F)	0	Dissociated ammonia	205	400	731	106	104
	0.10	Dissociated ammonia	205	400	745	108	104
	0	Hydrogen	205	400	641	93	95
	0.10	Hydrogen	220	430	800	116	101

(a) Sintered for 30 min at 1120 °C (2050 °F)

Typical mechanical properties of nearly dense P/M stainless steel
Based on high-temperature sintering

Alloy	Condition	Ultimate tensile strength		0.2% yield strength		Elongation in 25 mm (1 in.) %	Hardness	Impact strength		Density, g/cm³	Theoretical density, g/cm³
		MPa	ksi	MPa	ksi			J	ft · lb		
Ultimet 04, 304	Sintered	593	86	248	36	36	80 HRB	10.8(a)	8(a)	7.8	7.9
Ultimet 16, 316	Sintered	687	99.6	308	44.7	26	94 HRB	8.1(a)	6(a)	7.7	7.8
	Solution treated and quenched	684	99.3	329	47.7	45	90 HRB	5.4(a)	40(a)	7.7	7.8
Ultimet 40C, 440C	Sintered	...	...	...	...	...	20 to 30 HRC	2.7(b)	2(b)	7.6	7.7
	Hardened and tempered	...	...	...	...	...	50 to 60 HRC	2.7(b)	2(b)	7.6	7.7

(a) Charpy V-notch. (b) Unnotched

Nonferrous Materials

Aluminum and Aluminum Alloys

Thermal treatments for aluminum P/M alloys

Temper	Description
O	Annealed 1 h at 413 °C (775 °F), furnace cooled at maximum rate of about 30 °C (50 °F) per hour to 260 °C (500 °F) or below
T1	Cooled from sintering temperature to 425 °C (800 °F) (601AB and 602AB) or 260 °C (500 °F) (201AB) in nitrogen, air cooled to room temperature
T4	Heat treated 30 min at 520 °C (970 °F) (601AB and 602AB) or 505 °C (940 °F) (201AB) in air, cold water quenched, and aged minimum of 4 days at room temperature
T6	Heat treated 30 min at 520 °C (970 °F) (601AB and 602AB) or 505 °C (940 °F) (201AB) in air, cold water quenched, and aged 18h at 160 °C (320 °F)

Applications of aluminum P/M alloys

Material, manufacturer, and nominal composition	Application
601AB (Alcoa): 0.25 Cu, 0.6 Si, 1.0 Mg, 1.5 lubricant, bal 1202(a)	Similar to wrought 6061; strength, ductility, corrosion resistance
201AB (Alcoa): 4.4 Cu, 0.8 Si, 0.5 Mg, 1.5 lubricant, bal 1202(a)	Similar to wrought 2014 but without manganese. Good strength properties
202AB (Alcoa): 4.0 Cu, 1.5 lubricant, bal 1202(a)	Good ductility. Suitable for cold formed parts
602AB (Aloca): 0.4 Si, 0.6 Mg, 1.5 lubricant, bal 1202(a)	Good electrical conductivity (from 42.0 to 48.5% IACS, depending on treatment), ductility, and finishability
601AC (Alcoa): 0.25 Cu, 0.6 Si, 1.0 Mg, bal 1202(a)	Same as 601AB, without lubricant; for isostatic compacting
201AC (Alcoa): 4.4 Cu, 0.8 Si, 0.5 Mg, bal 1202(a)	Same as 201AB, without lubricant; for isostatic compacting
22 (Alcan)(b): 2.0 Cu, 1.0 Mg, 0.3 Si, bal Al	Has good mechanical properties in sintered or heat treated forms
24 (Alcan)(b) (2014): 4.4 Cu, 0.5 Mg, 0.9 Si, 0.4 Mn, bal Al	Properties resemble those of its wrought counterpart, 2014. Good mechanical properties
67 (Alcan)(b): 0.5 Cu, bal Al	High electrical conductivity (48% IACS) and ductility. Similar to wrought 1100
68 (Alcan)(b): 0.6 Mg, 0.4 Si, bal Al	Good surface finish; high ductility and conductivity (42% IACS). Similar to wrought 6101
69 (Alcan)(b) (6061): 0.25 Cu, 1.0 Mg, 0.61 Si, 0.10 Cr	Properties are similar to those of 6061. Good strength, corrosion resistance, ductility, and conductivity (40% IACS)
76 (Alcan)(b) (7075): 1.6 Cu, 2.5 Mg, 0.20 Cr, 5.6 Zn	Properties are similar to those of 7075. High strength and hardness
91 (Alcan)(b): 26.3 tribaloy	Excellent wear resistance
4040 (Alcan): 1.0 Cu, 1.0 Si, bal Al −150 + 325 mesh	High porosity parts for controlling contamination, pressure, sound, catalytic reactions, etc.

(a) 1202 composition: 99.4% Al, 0.3% Al_2O_3, 0.15% Fe, 0.07% Si, balance other metallics. (b) Grade number includes suffix: FF, premix with 1.5% lubricant; NL, premix without lubricant

Typical properties of nitrogen-sintered aluminum P/M alloys

Alloy	Compacting pressure		Green density		Green strength		Sintered density		Temper	Tensile strength(a)		Yield strength(a)		Elongation,	Hardness
	MPa	tsi	%	g/cm³	MPa	psi	%	g/cm³		MPa	ksi	MPa	ksi	%	
601AB	96	7	85	2.29	3.1	450	91.1	2.45	T1	110	16	48	7	6	55-60 HRH
									T4	141	20.5	96	14	5	80-85HRH
									T6	183	26.5	176	25.5	1	70-75 HRE
	165	12	90	2.42	6.55	950	93.7	2.52	T1	139	20.1	88	12.7	5	60-65 HRH
									T4	172	24.9	114	16.6	5	80-85 HRH
									T6	232	33.6	224	32.5	2	75-80 HRE
	345	25	95	2.55	10.4	1500	96.0	2.58	T1	145	21	94	13.7	6	65-70 HRH
									T4	176	25.6	117	17	6	85-90 HRH
									T6	238	34.5	230	33.4	2	80-85 HRE
602AB	165	12	90	2.42	6.55	950	93.0	2.55	T1	121	17.5	59	8.5	9	55-60 HRH
									T4	121	17.5	62	9	7	65-70 HRH
									T6	179	26	169	24.5	2	55-60 HRE
	345	25	95	2.55	10.4	1500	96.0	2.58	T1	131	19	62	9	9	55-60 HRH
									T4	134	19.5	65	9.5	10	70-75 HRH
									T6	186	27	172	25	3	65-70 HRE
201AB	110	8	85	2.36	4.2	600	91.0	2.53	T1	169	24.5	145	24	2	60-65 HRE
									T4	210	30.5	179	26	3	70-75 HRE
									T6	248	36	248	36	0	80-85 HRE
	180	13	90	2.50	8.3	1200	92.9	2.58	T1	201	29.2	170	24.6	3	70-75 HRE
									T4	245	35.6	205	29.8	3.5	75-80 HRE
									T6	323	46.8	322	46.7	0.5	85-90 HRE
	413	30	95	2.64	13.8	2000	97.0	2.70	T1	209	30.3	181	26.2	3	70-75 HRE
									T4	262	38	214	31	5	80-85 HRE
									T6	332	48.1	327	47.5	2	90-95 HRE
202AB: Compacts	180	13	90	2.49	5.4	780	92.4	2.56	T1	160	23.2	75	10.9	10	55-60 HRH
									T4	194	28.2	119	17.2	8	70-75 HRH
									T6	227	33	147	21.3	7.3	45-50 HRE
Cold formed parts (19% strain)	180	13	90	2.49	5.4	780	92.4	2.56	T2	238	33.9	216	31.4	2.3	80 HRE
									T4	236	34.3	148	21.5	8	70 HRE
									T6	274	39.8	173	25.1	8.7	85 HRE
									T8	280	40.6	250	36.2	3	87 HRE

(a) Tensile properties determined using powder metal flat tension bar (MPIF standard 10-63), sintered 15 min at 620 °C (1150 °F) in nitrogen

Properties of aluminum P/M alloys sintered in dissociated ammonia

Alloy	Green density		Temper	Tensile strength		Yield strength		Elongation in 25 mm (1 in.), %
	%	g/cm³		MPa	ksi	MPa	ksi	
601AB(a)	90	2.42	T1	93	13.5	76	11.0	2.5
			T4	108	15.7	88	12.7	3.5
			T6	159	23.1	...	...	1.0
	95	2.55	T1	121	17.6	87	12.6	3.5
			T4	146	21.2	99	14.3	5.0
			T6	207	30.1	205	29.7	1.5
201AB(b)	90	2.50	T1	161	23.3	141	20.5	2.0
			T4	198	28.8	163	23.7	2.5
			T6	247	35.8	...	...	0.5
	95	2.64	T1	174	25.2	152	22.0	2.0
			T4	221	32.0	180	26.1	3.0
			T6	288	41.8	287	41.6	1.0

(a) Sinter 10 to 30 min at 620 °C (1150 °F) at a dew point of –40 to –50 °C (–40 to –60 °F). (b) Sinter 10 to 30 min at 595 °C (1100 °F) at a dew point of –40 to –50 °C (–40 to –60 °F)

Properties of aluminum P/M alloys sintered in vacuum

Alloy	Green density		Temper	Tensile strength		Yield strength		Elongation in 25 mm (1 in.), %
	%	g/cm³		MPa	ksi	MPa	ksi	
601AB(a)	90	2.42	T1	112	16.3	68	9.9	4.5
			T4	140	20.3	91	13.2	4.0
			T6	223	32.3	211	30.6	2.0
	95	2.55	T1	131	19.0	80	11.6	5.0
			T4	161	23.4	99	14.4	7.0
			T6	230	33.3	219	31.8	2.0
201AB(b)	90	2.50	T1	185	26.8	143	20.7	4.0
			T4	241	35.0	187	27.1	5.5
			T6	296	43.0	287	41.6	2.0
	95	2.64	T1	184	26.7	146	21.2	4.0
			T4	250	36.3	185	26.9	6.5
			T6	312	45.3	290	42.0	2.0

(a) Sinter 10 to 30 min at 605 °C (1125 °F) at 0.0013 to 27 Pa (0.01 to 200 µm Hg) vacuum level. (b) Sinter 10 to 30 min at 580 °C (1075 °F) at 0.0013 to 27 Pa (0.01 to 200 µm Hg) vacuum level

Effect of vacuum level on tensile strengths of aluminum P/M alloys

Alloy	Density, %	Tensile strength(a)		Tensile strength(b)	
		MPa	ksi	MPa	ksi
601AB	85	176	25.5	208	30.2
	90	223	32.4	225	32.6
	95	227	32.9	245	35.5
201AB	85	186	27.0	263	38.2
	90	303	43.9	304	44.1

Note: All specimens heat treated to T6 temper after sintering. (a) 27 Pa (200 µm Hg). (b) 1.3 mPa (0.01 µm Hg)

Beryllium

Mechanical properties at ambient and elevated temperatures of vacuum hot pressed and vacuum hot pressed and extruded beryllium

Property	Temperature	Vacuum hot pressed powder	Powder vacuum hot pressed and extruded at 425 °C (800 °F), 2.25 to 1 reduction		Powder vacuum hot pressed and extruded at 1050 °C (1920 °F), 12 to 1 reduction	
			Longitudinal	Transverse	Longitudinal	Transverse
Tensile strength, MPa (ksi)	25 °C (75 °F)	225 to 350 (33 to 51)	440 (64)	260 to 315 (38 to 46)	565 to 620 (82 to 90)	345 to 435 (50 to 63)
	300 °C (570 °F)	160 to 240 (23 to 35)	330 (48)	235 to 250 (34 to 36)	340 (49)	295 to 310 (43 to 45)
	500 °C (930 °F)	150 to 170 (22 to 25)	250 (36)	205 to 240 (30 to 35)	240 (35)	240 (35)
	700 °C (1290 °F)	95 (14)	115 (17)	90 to 110 (13 to 16)	90 (13)	90 (13)
Tensile yield strength, MPa (ksi)	25 °C (75 °F)	220 (32)	...	...	310 (45)	...
Modulus of elasticity, GPa (10^6 psi)	25 °C (75 °F)	305 (44)	...	...	285 (41)	...
Elongation in 50 mm, % (2 in., %)	25 °C (75 °F)	1 to 3	4	1	11 to 17	1
	300 °C (570 °F)	12 to 30	13	2 to 4	23	2
	500 °C (930 °F)	23 to 40	14	3 to 11	15	3 to 8
	700 °C (1290 °F)	10 to 14	15	11 to 13	7	4 to 6
Contraction, %	25 °C (75 °F)	1 to 4	1	1 to 4	17	1
	300 °C (570 °F)	15 to 35	5	2 to 25	28	1 to 2
	500 °C (930 °F)	40 to 53	33	14 to 29	24	3 to 12
	700 °C (1290 °F)	10 to 13	10	14 to 17	5	3
Compressive yield strength, MPa (ksi)	25 °C (75 °F)	170 (25)	...	...	260 (38)	...
Unnotched Charpy impact strength, J (ft·lb)	25 °C (75 °F)	1.1 (0.8)	...	...	5.6 (4.1)	...
Tensile impact strength, J (ft·lb)	25 °C (75 °F)	1.9 (1.4)	...	...	6.1 (4.5)	...

Copper and Copper Alloys

Properties of commercial grades of copper powder produced by the copper oxide process

						Physical properties							Compacted properties		
	Chemical properties, %					Apparent density, g/cm^3	Hall flow rate, s/50 g	Tyler sieve analysis, %					Green density, g/cm^3	Green strength, MPa (psi), at:	
Copper	Tin	Graphite	Lubricant	Hydrogen loss	Acid insolubles			+100	+150	+200	+325	−325		165 MPa (12 tsi)	6.30 g/cm^3
99.53	...	...	...	0.23	0.04	2.99	23	0.3	11.1	26.7	24.1	37.8	6.04	6 (890)	...
99.64	...	...	...	0.24	0.03	2.78	24	...	0.6	8.7	34.1	56.6	5.95	7.8 (1140)(a)	...
99.62	...	...	...	0.26	0.03	2.71	27	...	0.3	5.7	32.2	61.8	5.95	9.3 (1350)(a)	...
99.36	...	...	...	0.39	0.12	1.56	...	0.1	1.0	4.9	12.8	81.2	5.79	21.4 (3100)(a)	...
99.25	...	...	...	0.30	0.02	2.63	30	0.08	7.0	13.3	16.0	63.7	...	...	8.3 (1200)(a)
90	10	...	0.75	...	...	3.23	30.6	0.0	1.4	9.0	32.6	57.0	6.32	...	3.80 (550)
88.5	10	0.5	0.80	...	...	3.25	12(b)	...	...	...	...	...	...	...	3.6 (525)

(a) Measured with die wall lubricant only. (b) Carney flow

Properties and applications of sintered copper alloy P/M parts

Material and specification designation	MPIF density suffix	Density, g/cm³	Tensile strength		Compressive yield strength(a)		Apparent hardness, HB	Young's modulus		Impact strength(b)		Elongation, %	Application
			MPa	ksi	MPa	ksi		GPa	10⁶ psi	J	ft·lb		
P/M bronze(c): 86.3 to 90.5 Cu, 9.5 to 10.5 Sn, 1.75 C max(d), 1 Fe max; MPIF CT-0010 (N, R, S, T); ASTM B438 (N, R), B255 (S); SAE 840 (N), 841 (R), 842 (S); MIL-B-5687-C (R, S)	N	5.8	55	8	48	7	...	...	...	...	...	1	Bearings or mechanical components resistant to atmospheric corrosion. Sleeve bearings, flange bearings, thrust washers, load-carrying bearing plates
	R	6.6	197	14	76	11	...	...	...	...	...	1	
	S	7.0	124	18	120.7	17.5	...	...	...	...	...	2	
P/M brass: 88.0 to 91.0 Cu, 8.3 to 12.0 Zn, 0 to 0.3 Fe; MPIF CZ-0010 (T, U)	T	7.4	138	20	62	9	57	...	...	...	...	13	Mechanical components requiring corrosion resistance and a pleasing appearance
	U	7.8	186	27	69	10	70	...	...	...	...	10	
P/M brass (leaded): 88.0 to 91.0 Cu, 1.0 to 2.0 Pb, bal Zn; MPIF CZP-0010 (T, U)	T	7.4	124	18	48	7	46	...	...	...	...	14	Same as MPIF CZ-0010; free-machining quality
	U	7.8	175.8	25.5	55	8	60	...	...	...	...	20	
P/M brass (leaded): 77.0 to 80.0 Cu, 1.0 to 2.0 Pb, 0.3 Fe max, 0.1 Sn max, bal Zn; MPIF CZP-0218 (T, U, W); ASTM B282 (T, U, W); SAE 890 (T), 891 (U); MIL-B-12128 C (T, U)	T	7.4	165	24	83	12	55	82.7	12.0	13.6	10.0	13	Mechanical components resistant to atmospheric corrosion. Ordnance components, builders' hardware, lock parts, housing, nuts, gears
	U	7.8	193	28	97	14	68	89.6	13.0	20.3	15.0	19	
	W	8.2	221	32	110	16	75	96.5	14.0	28.5	21.0	23	
P/M brass: 68.5 to 71.5 Cu, 27.8 to 31.5 Zn, 0 to 0.3 Fe; MPIF CZ-0030 (T, U)	T	7.4	214	31	90	13	76	...	...	...	...	20	Mechanical components requiring corrosion resistance and a pleasing appearance
	U	7.8	255	37	103	15	85	...	...	...	...	26	
P/M brass (leaded): 68.5 to 71.5 Cu, 1.0 to 2.0 Pb, bal Zn; MPIF CZP-0030 (T, U)	T	7.4	193	28	76	11	65	...	...	...	...	22	Same as MPIF CZ-0030; free-machining quality
	U	7.8	234	34	90	13	76	...	...	...	...	27	
P/M nickel silver: 62.5 to 65.5 Cu, 16.5 to 19.5 Ni, bal Zn; MPIF CZN-1818 (U, W); ASTM B458 (U, W)	U	7.8	207	30	110	16	75	96.5	14.0	13.6	10.0	10	Mechanical components, corrosion resisting. Gears, levers, chuck jaws, electrical components, parts for marine exposure
	W	8.2	255	37	124	18	85	96.5	14.0	17.6	13.0	12	
P/M nickel silver (leaded): 62.5 to 65.5 Cu, 16.5 to 19.5 Ni, 1.0 to 1.8 Pb, bal Zn; MPIF CZNP-1618 (U, W); ASTM B458 (U, W)	U	7.8	207	30	110	16	75	89.6	13.0	12.2	9.0	10	Same as MPIF CZN-1818; free-machining quality
	W	8.2	241	35	117	17	85	96.5	14.0	16.2	12.0	12	

(a) 1% offset. (b) Unnotched Charpy. (c) At 300×, microstructure should be substantially alpha brass without visible free tin. (d) Commonly graphite; up to 1.7% of another type of solid lubricant can be added as authorized by the producer

Mechanical properties of unalloyed hot pressed copper powder compacts

Temperature		Pressure(a)		Yield strength		Tensile strength		Elongation,	Compressive strength		Compression,	Increase in area,
°C	°F	MPa	ksi	MPa	ksi	MPa	ksi	%	MPa	ksi	%	%
500	930	350	50	...	...	...	...	...	2220	322	55	245
500	930	700	100	...	...	...	...	...	2600	377	70	270
610	1130	330	48	...	...	230	33.5	...	...	...	...	...
715	1320	330	48	...	...	210	30.5	...	...	...	...	...
800	1470	70	10	186	27	186	27	4	...	...	...	...
810	1490	330	48	...	...	203	29.5	...	...	...	...	...
940	1740	70	10	76	11	207	30	60	...	...	...	...

(a) Dwell time: 1 min.

Typical mechanical properties of copper-base P/M materials

MPIF designation	Sn	Zn	Ni	Pb	MPIF density suffix	Density, g/cm³	Tensile strength MPa	ksi	0.2% yield strength MPa	ksi	Elongation in 25 mm (1 in.), %	0.1% compressive yield strength MPa	ksi	Apparent hardness, HRH
Bronzes														
CT-0010	10	...	...	...	N	5.6 to 6.0	55	8	...	...	1	48	7	...
					R	6.4 to 6.8	96	14	...	...	1	76	11	...
					S	6.8 to 7.2	124	18	...	...	2.5	121	17.5	...
Brasses														
CZ-0010	...	10	...	...	T	7.2 to 7.6	138	20	62	9	13	...	...	57
					U	7.6 to 8.0	186	27	69	10	18	...	...	70
CZ-0030	...	30	...	...	T	7.2 to 7.6	214	31	89	13	20	...	...	76
					U	7.6 to 8.0	255	37	103	15	26	...	...	85
CZP-0210	...	10	...	2	T	7.2 to 7.6	124	18	48	7	14	...	...	46
					U	7.6 to 8.0	176	25.5	55	8	20	...	...	60
CZP-0220	...	20	...	2	T	7.2 to 7.6	165	24	76	11	13	...	...	55
					U	7.6 to 8.0	193	28	89	13	19	96	14	68
					W	8.0 to 8.4	221	32	103	15	23	110	16	75
CZP-0230	...	30	...	2	T	7.2 to 7.6	193	28	76	11	22	...	...	65
					U	7.6 to 8.0	234	34	89	13	27	...	...	76
Nickel silvers														
CZN-1818	...	18	18	...	U	7.6 to 8.0	206	30	...	...	10	110	16	75
					W	8.0 to 8.4	255	37	...	...	12	124	18	85
CZNP-1818	...	18	18	1.5	U	7.6 to 8.0	206	30	...	...	10	110	16	75
					W	8.0 to 8.4	241	35	...	...	12	117	17	85

(a) Remainder Cu

Typical mechanical properties of brass and nickel silver P/M compacts pressed at 414 MPa (30 tsi)

Nominal composition	Sintered density, g/cm³	Tensile strength MPa	ksi	Elongation in 25 mm (1 in.), %	Hardness, HRH
Brass					
90Cu-10Zn	8.1	207	30	20	77
85Cu-15Zn	8.2	217	31.5	20	82
70Cu-30Zn	8.1	262	38	21	87
88.5Cu-10Zn-1.5Pb	8.4	207	30	25	76
80Cu-18.5Zn-1.5Pb	8.2	238	34.5	31	82
68.5Cu-30Zn-1.5Pb	7.7	239	34.6	29	71
Nickel silver					
64Cu-18Ni-18Zn	7.9	234	34	12	83(a)
64Cu-18Ni-16.5Zn-1.5Pb	7.8	193	28	11	84(a)

(a) Hardness, HRB

Cobalt- and Nickel-Base Alloys

Typical mechanical properties of hydrometallurgical powder rolled cobalt strip

Strip	Strip thickness		Strip density,	Ultimate tensile strength		Yield strength		Elongation,
	mm	in.	%	MPa	ksi	MPa	ksi	%
Green strip	2.1	0.084	86	22	3	22	3	0
Sintered strip	2.1	0.084	86	201	29	195	28	5
Hot rolled strip	1.2	0.048	100	758	110	413	60	15
Cold rolled strip	0.9	0.036	100	1103	160	1100	159	1
Annealed strip	0.9	0.036	100	793	115	345	50	20

Note: Compacting roll diameter, 254 mm (10 in.). Roll speed, 6.0 rpm. Roll gap (green strip), 1.5 mm (0.06 in.)

Typical mechanical properties of hydrometallurgical powder rolled nickel strip

Strip	Strip thickness		Strip density,	Ultimate tensile strength		Yield strength		Elongation,
	mm	in.	%	MPa	ksi	MPa	ksi	%
Green strip	4.0	0.158	79	4	0.6	4	0.6	0
Sintered strip	4.1	0.161	79	138	20	136	19	0
Hot rolled strip	2.1	0.084	100	358	52	165	24	38
Cold rolled strip	1.3	0.052	100	579	84	572	83	5
Annealed strip	1.3	0.052	100	362	53	83	12	48

Note: Compacting roll diameter, 560 mm (22 in.). Roll speed, 2.2 rpm. Roll gap (green strip), 3.5 mm (0.140 in.)

Nominal compositions of nickel-base P/M superalloys

Alloy	Composition, wt%												
	C	Cr	Mo	Fe	Co	Al	Ti	B	Nb	V	Hf	W	Zr
IN-100	0.1	10.0	3.5	1.0	14.0	4.5	5.5	0.01	...	1.0	...	...	0.05
René 95	0.1	14.0	3.5	...	8.0	3.5	2.5	0.01	3.5	...	...	3.6	0.05
Astroloy	0.05	15.0	5.0	...	18.0	4.0	3.5	0.03	...	...	...	...	...
MERL 76	0.025	12.5	3.0	...	18.5	5.0	4.3	0.02	1.4	...	0.4	...	0.06
AF-115	0.05	10.5	2.8	...	15.0	3.8	3.9	0.02	1.8	...	0.8	5.9	0.05
Udimet 100	0.03	14.72	4.90	0.62	17.72	3.86	3.53	0.026	...	...	...	0.04	0.03

Typical mechanical properties of P/M superalloys

Property	René 95(a)	Low-carbon Astroloy(b)	Low-carbon Astroloy(c)	Low-carbon Astroloy(d)	IN-1000(e)	MERL 76(f)	Udimet 700(g)	Udimet 700(a)
0.2% yield strength at 210 °C (410 °F), MPa (ksi)	1257 (182)	973 (141)	928 (135)	994 (143)	1095 (159)	1188 (172)	860 (125)	1115 (162)
Ultimate tensile strength at 210 °C (410 °F), MPa (ksi)	1671 (242)	1376 (200)	1338 (194)	1359 (197)	1594 (231)	1674 (243)	1355 (197)	1515 (219)
Elongation, %	20	22	26	28	26	21	25	18.5
Reduction in area, %	20.3	23	28	32	27	22	27	18.5
Creep at 595 °C (1110 °F) at 1034 MPa (150 ksi), 100 h/% strain	0.15	...	...	...	...	...	...	...
Stress rupture at 650 °C (1200 °F) at 1034 MPa (150 ksi), service life (hours)/% elongation	29.5/5.4	...	...	...	...	...	...	...
	28.4/4.7	...	...	...	...	...	...	...
Strain-controlled low-cycle fatigue at 535 °C (1000 °F), strain/cycles to failure	0.78/26 948	...	...	...	...	...	...	...
	0.66/94 447	...	...	...	...	...	...	...
Stress rupture at 621 MPa (90 ksi) at 730 °C (1350 °F) at 151 h, % elongation/% reduction in area	...	14/17	17/22	16/21	...	...	...	...
0.2% yield strength at 705 °C (1300 °F), MPa (ksi)	...	...	...	...	1044 (151)	...	...	...
Ultimate tensile strength at 705 °C (1300 °F), MPa (ksi)	...	...	...	...	1265 (184)	...	...	...
% elongation/% reduction in area at 705 °C (1300 °F)	...	...	...	...	19/23	...	...	...
Stress rupture at 730 °C (1350 °F) at 655 MPa (95 ksi), hours to failure/% elongation	...	...	...	...	35.6/16.6	...	...	...
	...	...	...	...	25.5/11.0	...	...	...
	...	...	...	...	37.0/14.5	...	...	...
Creep at 705 °C (1300 °F) at 551 MPa (80 ksi), time for 0.1%/time for 0.2%	...	...	...	...	140.5/193.5	...	...	...
	...	...	...	...	100.0/142.0	...	...	...
	...	...	...	...	91.0/125.0	...	...	...
0.2% yield strength at 620 °C (1150 °F), MPa (ksi)	...	...	...	...	...	1136 (165)	...	...
Ultimate tensile strength at 620 °C (1150 °F), MPa (ksi)	...	...	...	...	...	1492 (216)	...	...
% elongation/% reduction in area at 620 °C (1150 °F)	...	...	...	...	...	18.5/17	...	...

(a) Hot isostatically pressed and hardened and tempered. (b) Produced by rapid omnidirectional compaction. Consolidated at 811 MPa (58.8 tsi); 0.5-s dwell in composite of copper-nickel fluid dies. Preheated to 1075 °C (1970 °F); held at temperature 1 h. Powder was electrodynamically degassed prior to vacuum filling. Post-consolidation solution treated at 1165 °C (2125 °F) for 4 h, fan air cooled. (c) Hot isostatically pressed at 1150 °C (2100 °F) at 104 MPa (15 ksi) for 4 h. Hot loaded at 925 °C (1700 °F). Hot unloaded at 980 °C (1800 °F). Post-consolidation solution treated at 1120 °C (2050 °F) for 2 h, fan air cooled. (d) Hot isostatically pressed and forged. Forging conditions: open die side upset at 1095 °C (2000 °F). Average reduction is 52%. Aging heat treatment for all processes: 650 °C (1200 °F) for 24 h, air cooled plus holding at 760 °C (1400 °F) for 8 h, air cooled. All tensile specimens tested normal to the forging direction. (e) Hot isostatically pressed and gatorized billet. (f) Hot isostatically pressed and gatorized. (g) As hot isostatically pressed

Comparison of mechanical properties of several P/M superalloys

Material	Condition	Test temperature °C	Test temperature °F	0.2% offset yield strength MPa	0.2% offset yield strength ksi	Ultimate tensile strength MPa	Ultimate tensile strength ksi	Reduction in area, %	Total elongation, %
René 95	Hot isostatically pressed(a)	23	74	1214	176	1636	237	15	16
	Hot isostatically pressed and forged	23	74	1179	171	1629	236	23	18
	Cast and wrought	23	74	1144	166	1434	208	12	10
	Minimum hot isostatically pressed	650	1202	1120	162	1514	220	17	16
	Hot isostatically pressed and forged	650	1202	1122	163	1480	215	14	13
	Cast and wrought	650	1202	1055	5	1282	186	10	8
Astroloy	Hot isostatically pressed	23	74	936	136	1379	200	31	27
	Hot isostatically pressed and forged	23	74	1055	153	1517	220	23	27
	Hot isostatically pressed	650	1202	881	128	1234	179	36	31
	Hot isostatically pressed and forged	650	1202	975	141	1261	183	25	38
IN-100	Hot isostatically pressed	650	1202	1286	187	942	137	...	21
	Hot isostatically pressed and forged	650	1202	1200	174	1000	145	...	8
	Hot isostatically pressed and extruded	650	1202	1350	196	1000	145	...	18

(a) 1120 °C (2050 °F) at 103 MPa (15 ksi) for 3 h, solution treated at 1150 °C (2100 °F) 1 h, hot salt quench to 535 °C (1000 °F), aged at 870 °C (1600 °F) 1 h, then 650 °C (1200 °F) 24 h; air cooled

Titanium and Titanium Alloys

Typical tensile properties of hot isostatically pressed Ti-6Al-4V rotating electrode processed powder

Orientation	Tensile strength MPa	Tensile strength ksi	0.2% offset yield strength MPa	0.2% offset yield strength ksi	Elongation (4D), %	Reduction in area, %
L	938.4	136.1	850.8	123.4	20.0	37.0
	936.3	135.8	868.1	125.9	18.0	37.4
T	950.8	137.9	863.3	125.2	18.0	40.2
	936.3	135.8	848.8	123.1	18.0	35.6
S	962.9	135.3	843.3	122.3	23.0	42.2
	941.9	136.6	848.8	123.1	20.0	39.1
AMS 4928-H	896.4(a)	130(a)	827.4(a)	120(a)	10(a)	25(a)

Note: Consolidated material made by hot isostatic pressing at 950 °C (1750 °F) for 10 h at 100 MPa (ksi). Vacuum annealed for 10 h at 700 °C (1300 °F). Hydrogen after vacuum annealing equals 0.0057%. (a) Minimum

Mechanical properties of hot pressed copper alloy powder compacts

Material	Temperature °C	Temperature °F	Pressure(a) MPa	Pressure(a) ksi	Yield strength MPa	Yield strength ksi	Tensile strength MPa	Tensile strength ksi	Elongation %	Compressive strength MPa	Compressive strength ksi	Compression %	Increase in area %	Hardness HB
Elemental powder mixture														
90Cu-10Zn	900	1650	60	9	117	17	210	31	22	...	...	...	...	...
85Cu-15Zn	500	930	700	100	...	...	...	...	...	1450	210	52	128	103
80Cu-20Zn	900	1650	60	9	124	18	255	37	34	...	...	...	...	...
75Cu-25Zn	500	930	700	100	...	...	...	...	...	2150	312	46	112	105
70Cu-30Zn	800	1470	60	9	152	22	262	38	16	...	...	...	...	111
65Cu-35Zn	500	930	700	100	...	...	...	...	...	2300	333	43	108	...
55Cu-45Zn	500	930	700	100	...	...	...	...	...	1460	212	33	100	160
50Cu-50Zn	775	1425	60	9	145	21	145	21	0	...	...	...	...	...
95Cu-5Sn	500	930	700	100	...	...	...	...	...	2200	319	55	103	114
95Cu-5Sn	700	1290	60	9	179	26	240	35	9	...	...	...	...	110
95Cu-5Sn	800	1470	60	9	165	24	310	45	47	...	...	...	...	114
93Cu-7Sn	800	1470	60	9	165	24	325	47	75	...	...	...	...	114
91Cu-9Sn	800	1470	60	9	207	30	290	42	17	...	...	...	...	...
90Cu-10Sn	500	930	700	100	...	...	...	...	...	2300	333	53	98	130
85Cu-15Sn	500	930	700	100	...	...	...	...	...	1360	197	33	37	165
80Cu-20Sn	500	930	700	100	...	...	...	...	...	880	128	17	23	211
86Cu-10.5Zn-3.5Sn	900	1650	60	9	124	18	262	38	53	...	...	...	...	...
83Cu-10.5Zn-2.5Sn-4Ni	900	1650	60	9	124	18	270	39	32	...	...	...	...	...
89Cu-5.5Sn-4.5Ni-1Si	900	1650	60	9	220	32	310	45	13	...	...	...	...	...
88.5 Cu-5.5Sn-5Ni-1Si	900	1650	60	9	303	44	358	52	5	...	...	...	...	...
Prealloyed powder														
85Cu-15Zn	500	930	700	100	...	...	...	...	...	2635	382	66	193	115
75Cu-25Zn	500	930	700	100	...	...	...	...	...	3000	435	69	192	129
70Cu-30Zn	900	1650	60	9	117	17	206	30	21	...	...	...	...	...
65Cu-35Zn	500	930	700	100	...	...	...	...	...	3075	446	67	200	122
55Cu-45Zn	500	930	700	100	...	...	...	...	...	2425	352	54	173	125
95Cu-5Sn	500	930	700	100	...	...	...	...	...	2950	428	66	210	111
90Cu-10Sn	500	930	700	100	...	...	...	...	...	3215	466	59	197	133
85Cu-15Sn	500	930	700	100	...	...	...	...	...	2500	362	44	66	153
80Cu-20Sn	500	930	700	100	...	...	...	...	...	1035	150	25	35	217
83Cu-12Zn-4Sn-1Fe	870	1600	60	9	172	25	296	43	46	...	...	...	...	...

(a) Dwell time: 5 min. for powder mixtures, 1 min. for prealloyed powder

Typical mechanical properties and compositions of nickel- and cobalt-base P/M alloys

Alloy designation	Room temperature			540 °C (1000 °F)			650 °C (1200 °F)			760 °C (1400 °F)		
	Ultimate tensile strength, MPa (ksi)	Elongation, %	Hardness, HRC	Ultimate tensile strength, MPa (ksi)	Elongation, %	Hardness, HRC	Ultimate tensile strength, MPa (ksi)	Elongation, %	Hardness, HRC	Ultimate tensile strength, MPa (ksi)	Elongation, %	Hardness, HRC
Stellite 3	863 (125)	<1	54	725 (105)	<1	40	690 (100)	<1	39	621 (90)	1	28
Stellite 6	897 (130)	<1	40	828 (120)	1	37	766 (111)	1	30	518 (75)	10	15
Stellite 19	1035 (150)	<1	49	...	<1	...	...	...	...	...	...	...
Stellite 31	828 (120)	4	...	676 (98)	14	...	614 (89)	13	46	518 (75)	<1	34
Stellite 190	621 (90)	<1	58	518 (75)	<1	54	518 (75)	<1	43	573 (83)	0.1	31
Star J Metal	523 (76)	0.1	56	539 (78)	0.1	52	569 (82)	0.1	...	656 (95)	0.5	...
Stellite 98 M2	794 (115)	0.3	58	725 (105)	0.3	...	690 (100)	0.5	37	483 (70)	1	25
Haynes 208	690 (100)	<1	44	552 (80)	<1	41	552 (80)	<1	...	428 (62)	7	82(a)
Haynes N-6	656 (95)	2	30	545 (79)	3	25	545 (79)	4	20	504 (73)	<1	27
Haynes 711	559 (81)	<1	50	490 (71)	<1	43	490 (71)	<1	43	...	...	...

	Composition, %											
	Ni	Si(b)	Fe	Mn(b)	Cr	Mo	W	C	V	B(b)	Co	Other(total)(b)
Stellite 3	3	1	3	1	31	...	12.5	2.4	...	1	rem	1
Stellite 6	3	1.5	3	1	29	1.5	4.5	1.2	...	1	rem	2
Stellite 19	3	1	3	1	31	...	10.5	1.9	...	1	rem	2
Stellite 31	10.5	1	2	1	25.5	...	7.5	0.5	...	...	rem	2
Stellite 190	3	1	5	1	26	1	1.4	3.1	...	1	rem	2
Star J Metal	3	1	3	1	32.5	...	17.5	2.5	...	...	rem	2
Stellite 98 M2	3.5	1	5	1	30	0.8	18.5	2	4.2	1	rem	2
Haynes 208	rem	1	12.5	0.75	26	10	10	2.6	...	1	10	2
Haynes N-6	rem	1.5	3	1	29	5.5	2	1.1	...	0.6	3	...
Haynes 711	rem	...	2.3	...	27	7	3	2.7	...	1	12	2

(a) Hardness, HRB. (b) Maximum

Tensile properties of forgings made from electric spark-activated hot pressed Ti-6Al-4V alloy powder preforms

Billet material	Forging	Condition(a)	Yield strength		Tensile strength		Elongation in 25 mm (1 in.), %	Reduction in area, %
			MPa	ksi	MPa	ksi		
Hydrided and dehydrided powder, 97.6 to 98.0% dense	Upset	A	1006 to 1020	146 to 148	1028 to 1042	149 to 151	11 to 16	24 to 46
		STA	1192 to 1206	173 to 175	1254 to 1268	182 to 184	10 to 12	27 to 42
	Step, 55% reduced	A	944 to 958	137 to 139	980 to 986	142 to 143	14 to 16	35 to 40
		STA	1130 to 1138	164 to 165	1220 to 1228	177 to 178	8 to 10	20 to 15
	Step, 95% reduced	A	930 to 938	135 to 136	972 to 980	141 to 142	15 to 17	29 to 35
Rotating electrode powder, 97.6 to 98.5% dense	Upset	A	952 to 986	138 to 143	980 to 1028	142 to 149	12 to 15	43 to 47
		STA	1124 to 1138	163 to 165	1186 to 1200	172 to 174	12 to 14	39 to 46
	Step, 55% reduced	A	924 to 930	134 to 135	958 to 972	139 to 141	15 to 17	38 to 42
		STA	1152 to 1158	167 to 168	1214 to 1220	176 to 177	7 to 9	11 to 15
	Step, 95% reduced	A	966 to 972	140 to 141	1006 to 1014	146 to 147	16 to 18	48 to 52
Rotating electrode powder, 99.5 to 99.7% dense	Upset	A	952 to 992	138 to 144	1000 to 1028	145 to 149	10 to 14	37 to 40
		STA	1110 to 1130	161 to 164	1166 to 1200	169 to 174	10 to 12	32 to 38
	Step, 55% reduced	A	882 to 890	128 to 129	938 to 958	136 to 137	9 to 11	15 to 18
		STA	1192 to 1200	173 to 174	1254 to 1262	182 to 183	5 to 7	11 to 14
	Step, 95% reduced	A	992 to 1000	144 to 145	1028 to 1042	149 to 151	17 to 19	40 to 44

(a) A, annealed: heated in air at 705 °C (1300 °F) for 2 h, air cooled. STA, solution treated and aged: heated in air at 955 °C (1750 °F) for 1 h, water quenched; aged at 540 °C (1000 °F) for 8 h, air cooled

Mechanical properties of P/M and wrought titanium and alloys

Alloy	Processing	Density, %	Ultimate tensile strength, MPa (ksi)	Yield strength, MPa (ksi)	Elongation, %	Reduction in area, %	Elastic modulus, GPa (10^6 psi)	Fatigue limit, notched MPa (ksi)	Fracture toughness, MPa$\sqrt{m}$ (ksi$\sqrt{in.}$)
Wrought commercial purity titanium grade II	...	100	345 (50)	344 (50)	5	35	103 (14.9)	...	...
	...	95.5	414 (60)	324 (47)	15	14	103 (15)	...	...
Sponge commercial purity P/M titanium(a)	...	94	427 (62)	338 (49)	15	23	...	...	...
	Forged	100	455 (66)	365 (53)	23	30	...	...	...
Wrought Ti-6Al-4V (AMS 4298)	...	100	896 (130)	827 (120)	10	20	114 (16.5)	427 (62)	55 (50)(b)
P/M Ti-6Al-4V	Blended elemental alloy, cold pressed	95.5	876 (127)	786 (114)	8	14	117 (17)	193 (28)	45 (40)(b)
	Blended elemental alloy, forged preforms or vacuum hot pressed	98+	919 (133)	839 (121.6)	10.9	19.0	...	262 (38)	56 (51)(b)
		99 min	937 (136)	862 (125)	12 to 18	15 to 40	116 (16.8)	414 (60)	61 (56)(b)
	Hot isostatically pressed prealloy	100	947 (137.4)	868 (125.9)	18.8	43.2	117 (17)	414 (60)	...
	Solution treated and aged	99	1103 (160)	1013 (147)	4.9	7.6	...	...	...
	Rapid omnidirectional compacted(c)	100	1014 (147)	944 (137)	18.4	40.9	...	...	...
Plasma rotating electrode processed Ti-6Al-6V-2Sn	Hot isostatically pressed	100	1053 (152.7)	1008 (146.3)	18	36.5	110 (16)	448 (65)	...
Plasma rotating electrode processed Ti-6Al-4V	Hot isostatically pressed	100	951 (138)	910 (132)	15	39	...	414 (60)(d)	83 (76)(e)
P/M Ti-6Al-4V(a)	Forged	94	827 (120)	738 (107)	5	8	...	...	...
		100	920 (133.5)	841 (122)	11.5	25	...	...	...
P/M Ti-6Al-4V(f)	Hot isostatically pressed	100	917 (133)	827 (120)	13	26	...	...	...
P/M Ti-6Al-6V-2Sn	...	99	1067 (155)	977 (142)	10	14	...	...	...

(a) 0.12% oxygen. (b) K_{Ic}. (c) Consolidated at 811 MPa (58.8 tsi), 0.5-s dwell in low-carbon steel fluid dies. Preheat temperature was 940 °C (1725 °F), held at temperature ¾ h. Powder was vacuum filled into fluid dies following cold static outgassing for 24 h. (d) K_t = 3. (e) K_{Ic}. (f) 0.2% oxygen

Properties of electric spark-activated hot pressed elemental titanium powder compacts

Material	Billet analysis (center), %				Density, g/cm³	Density, % of theoretical	Billet properties						
	Oxygen	Hydrogen	Magnesium	Sodium			Yield strength		Tensile strength		Elongation in 12.7 mm (0.5 in.), %	Reduction in area, %		
							MPa	ksi	MPa	ksi				
Titanium sponge, magnesium-reduced, graded	0.187	0.052	0.212	...	4.47	99.2	470 to 520	68 to 75	600 to 680	87 to 99	4 to 6	4 to 6		
Titanium sponge, sodium-reduced, high-purity	0.120	...	0.008	1.156	4.47	99.2	310 to 390	45 to 56	430 to 520	63 to 75	14 to 17	27 to 29		
Titanium sponge, sodium-reduced, extra-low-interstitials	0.070	...	...	0.097	4.47	99.2	205 to 235	30 to 34	290 to 345	42 to 50	41 to 47	55 to 66		
Titanium powder, electrolytic, acicular, porous	0.196	0.053	...	0.066	4.44	98.5	335 to 365	49 to 53	455 to 510	66 to 74	10 to 14	20 to 25		

Properties of electric spark-activated hot pressed Ti-6Al-4V alloy powder compacts

Powder	Billet condition	Billet analysis range, %			Billet properties						
		Oxygen	Hydrogen	Nitrogen	Yield strength		Tensile strength		Elongation in 12.7 mm (0.5 in.), %	Reduction in area, %	
					MPa	ksi	MPa	ksi			
Hydrided and dehydrided, angular, porous	As hot pressed	0.121 to 0.128	0.0056 to 0.0091	...	855 to 882	124 to 128	980 to 1000	142 to 145	6 to 7	7 to 12	
	Annealed(a)	...	...	...	896 to 910	130 to 132	966 to 980	140 to 142	11 to 13	18 to 23	
	Solution treated and aged(b)	...	...	...	952 to 986	138 to 143	1124 to 1186	163 to 172	4 to 5	5 to 7	
Rotating electrode, spherical, solid	As hot pressed	0.120 to 0.150	0.0035 to 0.0075	0.004 to 0.081	828 to 882	120 to 128	958 to 1006	139 to 146	9 to 11	11 to 21	
	Annealed(a)	...	...	...	868 to 802	126 to 128	930 to 945	135 to 137	12 to 14	20 to 25	
	Solution treated and aged(b)	...	...	...	1055 to 1068	153 to 155	1172 to 1192	170 to 173	5 to 7	8 to 10	

(a) Heated in air at 705 °C (1300 °F) for 2 h, air cooled. (b) Heated in air at 955 °C (1750 °F) for 1 h, water quenched; aged at 540 °C (1000 °F) for 4 h, air cooled

General Engineering Data

Conversion Factors and Tables

Metric Conversion

Units for converting from the English to the metric (SI) system

The Système Internationale d'Unités (SI) is built upon seven base units and two supplementary units. Derived units are related to base and supplementary units by formulas in the right-hand column of the table. Symbols for units with specific names are given in parentheses. The information is adapted from the *Standard for Metric Practice*, ASTM E 380

Quantity	Unit	Formula
Base SI units		
Length	metre (m)	...
Mass	kilogram (kg)	...
Time	second (s)	...
Electric current	ampere (A)	...
Thermodynamic temperature	kelvin (K)	...
Amount of substance	mole (mol)	...
Luminous intensity	candela (cd)	...
Supplementary SI units		
Plane angle	radian (rad)	...
Solid angle	steradian (sr)	...
Derived SI units with special names		
Frequency (of a periodic phenomenon)	hertz (Hz)	$1/s$
Force	newton (N)	$kg{\cdot}m/s^2$
Pressure, stress	pascal (Pa)	N/m^2
Energy, work, quantity of heat	joule (J)	$N{\cdot}m$
Powder, radiant flux	watt (W)	J/s
Quantity of electricity, electric charge	coulomb (C)	$A{\cdot}s$
Electric potential, potential difference, electromotive force	volt (V)	W/A
Electric capacitance	farad (F)	C/V
Electric resistance	ohm (Ω)	V/A
Conductance	siemens (S)	A/V
Magnetic flux	weber (Wb)	$V{\cdot}s$
Magnetic flux density	tesla (T)	Wb/m^2
Inductance	henry (H)	Wb/A
Celsius temperature	degree Celsius (°C)	K
Luminous flux	lumen (lm)	$cd{\cdot}sr$
Illuminance	lux (lx)	lm/m^2
Activity (of a radionuclide)	becquerel (Bq)	$1/s$
Absorbed dose	gray (Gy)	J/kg
Dose equivalent	sievert (Sv)	J/kg

Quantity	Unit
Some common derived units of SI	
Absorbed dose rate	gray per second (Gy/s)
Acceleration	metre per second squared (m/s^2)
Angular acceleration	radian per second squared (rad/s^2)
Angular velocity	radian per second (rad/s)
Area	square metre (m^2)
Concentration (of amount)	mole per cubic metre (mol/m^3)
Current density	ampere per square metre (A/m^2)
Density, mass	kilogram per cubic metre (kg/m^3)
Electric charge density	coulomb per cubic metre (C/m^3)
Electric field strength	volt per metre (V/m)
Electric flux density	coulomb per square metre (C/m^2)
Energy density	joule per cubic metre (J/m^3)
Entropy	joule per kelvin (J/K)
Exposure (X and gamma rays)	coulomb per kilogram (C/kg)
Heat capacity	joule per kelvin (J/K)
Heat flux density, irradiance	watt per square metre (W/m^2)
Luminance	candela per square metre (cd/m^2)
Magnetic field strength	ampere per metre (A/m)
Molar energy	joule per mole (J/mol)
Molar entropy	joule per mole kelvin (J/[mol·K])
Molar heat capacity	joule per mole kelvin (J/[mol·K])

(continued)

Units for converting from the English to the metric (SI) system (continued)

The Système Internationale d'Unités (SI) is built upon seven base units and two supplementary units. Derived units are related to base and supplementary units by formulas in the right-hand column of the table. Symbols for units with specific names are given in parentheses. The information is adapted from the *Standard for Metric Practice*, ASTM E 380

Quantity	Unit
Moment of force[A]	newton metre (N·m)
Permeability (magnetic)	henry per metre (H/m)
Permittivity	farad per metre (F/m)
Power density	watt per square metre (W/m^2)
Radiance	watt per square metre steradian (W/[m^2·sr])
Radiant intensity	watt per steradian (W/sr)
Specific heat capacity	joule per kilogram kelvin (J/[kg·K])
Specific energy	joule per kilogram (J/kg)
Specific entropy	joule per kilogram kelvin (J/[kg·K])
Specific volume	cubic metre per kilogram (m^3/kg)
Surface tension	newton per metre (N/m)
Thermal conductivity	watt per metre kelvin (W/[m·K])
Velocity	metre per second (m/s)
Viscosity, dynamic	pascal second (Pa·s)
Viscosity, kinematic	square metre per second (m^2/s)
Volume	cubic metre (m^3)
Wave number	1 per metre (1/m)

Multiplication factor	Prefix	Symbol
SI prefixes		
1 000 000 000 000 000 000 = 10^{18}	exa	E
1 000 000 000 000 000 = 10^{15}	peta	P
1 000 000 000 000 = 10^{12}	tera	T
1 000 000 000 = 10^9	giga	G
1 000 000 = 10^6	mega	M
1 000 = 10^3	kilo	k
100 = 10^2	hecto(a)	h
10 = 10^1	deka(a)	da
0.1 = 10^{-1}	deci(a)	d
0.01 = 10^{-2}	centi(a)	c
0.001 = 10^{-3}	milli	m
0.000 001 = 10^{-6}	micro	μ
0.000 000 001 = 10^{-9}	nano	n
0.000 000 000 001 = 10^{-12}	pico	p
0.000 000 000 000 001 = 10^{-15}	femto	f
0.000 000 000 000 000 001 = 10^{-18}	atto	a

(a) To be avoided where practical

Metric conversion factors

To convert from	To	Multiply by	To convert from	To	Multiply by
angstrom	m	$1.0000 \cdot 10^{-10}$(a)	hp(e)	W	7.4570×10^2
atm	Pa	1.0133×10^5	hp(f)	W	7.4600×10^2
Btu(b)	J	1.054×10^3	in.	m	2.5400×10^{-2}
Btu(b)/ft$^2 \cdot$h	W/m^2	3.1525	in.^{2}	m^2	6.4516×10^{-4}
Btu(b)/ft$^2 \cdot$h$\cdot$°F	W/m$^2 \cdot$K	5.6745	in.^{3}	m^3	1.6387×10^{-5}
Btu(b)$\cdot$ft/h$\cdot$ft$^2 \cdot$°F	W/m$\cdot$K	1.7296	in. of Hg(g)	Pa	3.3864×10^3
Btu(b)/ft$^2 \cdot$s	W/m^2	1.135×10^4	in. of water(c)	Pa	2.4908×10^2
Btu(b)$\cdot$in./ft$^2 \cdot$h$\cdot$°F	W/m$\cdot$K	1.4413×10^{-1}	K	°C	$t_{°C} = t_K - 273.15$
Btu(b)$\cdot$in./s$\cdot$ft$^2 \cdot$°F	W/m$\cdot$K	5.1887×10^2	kgf	N	9.80665(a)
Btu(b)/lbm$\cdot$°F	J/kg$\cdot$K	4.1840×10^3	kgf/mm^2	Pa	9.80665×10^6(a)
cal(b)	J	4.1840(a)	ksi	MPa	6.8948
cal(b)/cm$\cdot$s$\cdot$°C	W/m$\cdot$K	4.1840×10^2(a)	ksi	Pa	6.8948×10^6
cal(b)/g	J/kg	4.1840×10^3(a)	ksi$\sqrt{\text{in.}}$	MPa$\sqrt{m}$	1.089
cal(b)/g$\cdot$°C	J/kg$\cdot$K	4.1840×10^3(a)	lb(h)	kg	4.5359×10^{-1}
circ mil	m^2	5.0671×10^{-10}	lb/in.^{3}	kg/m^3	2.7680×10^4
°C	K	$t_K = t_{°C} + 273.15$	lbf	N	4.4482
degree	rad	1.7453×10^{-2}	lbf$\cdot$in.	N$\cdot$m	1.1298×10^{-1}
dyne/cm^2	Pa	1.0000×10^{-1}(a)	lbf$\cdot$ft	N$\cdot$m	1.3558
°F	°C	$t_{°C} = (t_{°F} + 32)/1.8$	MPa$\sqrt{m}$	MNm$^{-3/2}$	1.0000(a)
°F	K	$t_K = (t_{°F} + 459.67)/1.8$	μin.	m	2.5400×10^{-8}(a)
ft	m	3.0480×10^{-1}	mil	m	2.5400×10^{-5}(a)
ft^2	m^2	9.2903×10^{-2}	N/m^2	Pa	1.0000(a)
ft^3	m^3	2.8317×10^{-2}	oersted	A/m	79.578
ft of water(c)	Pa	2.9890×10^3	oz/ft^2	kg/m^2	3.0515×10^{-1}
ft^2/h (thermal diffusivity)	m^2/s	2.58064×10^{-5}(a)	psi	Pa	6.8948×10^3
ft$\cdot$lbf	J	1.3558	°R	K	$t_K = t_{°R}/1.8$
ft$\cdot$lbf/s	W	1.3558	ton(j)	kg	9.0718×10^2
ft/s	m/s	3.0480×10^{-1}	ton(k)	kg	1.0160×10^3
gauss	T	1.0000×10^{-4}(a)	ton/in.^{2}	Pa	1.3786×10^4
gallon(d)	m^3	3.7854×10^{-3}	tonne	kg	1.0000×10^3(a)
g/cm^3	kg/m^3	1.0000×10^3(a)	torr	Pa	1.3332×10^2
g/cm^3	Mg/m^3	1.0000(a)	Ω/circ mil$\cdot$ft	Ω$\cdot$m	1.6624×10^{-9}

(a) Exactly. (b) Thermochemical. (c) At 4 °C (39.2 °F). (d) U.S. liquid. (e) Mechanical (1 hp = 550 ft·lbf/s). (f) Electrical. (g) At 0 °C (32 °F). (h) Avoirdupois. (j) Short; equal to 2000 lbm. (k) Long; 2240 lbm.

Temperature conversions

The general arrangement of this table was devised by Sauveur and Boylston more than 40 years ago. The middle column of figures (in bold-faced type) contains the reading (°F or °C) to be converted. If converting from degrees Fahrenheit to degrees Centigrade, read the Centigrade equivalent in the column head "°C." If converting from Centigrade to Fahrenheit, read the Fahrenheit equivalent in the column headed "°F." $°C = \frac{5}{9}(°F - 32)$

°F		°C	°F		°C	°F		°C	°F		°C	°F		°C
...	**-458**	-272.22	...	**-324**	-197.78	-310.0	**-190**	-123.33	-68.8	**-56**	-48.89	172.4	**78**	25.56
...	**-456**	-271.11	...	**-322**	-196.67	-306.4	**-188**	-122.22	-65.2	**-54**	-47.78	176.0	**80**	26.67
...	**-454**	-270.00	...	**-320**	-195.56	-302.8	**-186**	-121.11	-61.6	**-52**	-46.67	179.6	**82**	27.78
...	**-452**	-268.89	...	**-318**	-194.44	-299.2	**-184**	-120.00	-58.0	**-50**	-45.56	183.2	**84**	28.89
...	**-450**	-267.78	...	**-316**	-193.33	-295.6	**-182**	-118.89	-54.4	**-48**	-44.44	186.8	**86**	30.00
...	**-448**	-266.67	...	**-314**	-192.22	-292.0	**-180**	-117.78	-50.8	**-46**	-43.33	190.4	**88**	31.11
...	**-446**	-265.56	...	**-312**	-191.11	-288.4	**-178**	-116.67	-47.2	**-44**	-42.22	194.0	**90**	32.22
...	**-444**	-264.44	...	**-310**	-190.00	-284.8	**-176**	-115.56	-43.6	**-42**	-41.11	197.6	**92**	33.33
...	**-442**	-263.33	...	**-308**	-188.89	-281.2	**-174**	-114.44	-40.0	**-40**	-40.00	201.2	**94**	34.44
...	**-440**	-262.22	...	**-306**	-187.78	-277.6	**-172**	-113.33	-36.4	**-38**	-38.89	204.8	**96**	35.56
...	**-438**	-261.11	...	**-304**	-186.67	-274.0	**-170**	-112.22	-32.8	**-36**	-37.78	208.4	**98**	36.67
...	**-436**	-260.00	...	**-302**	-185.56	-270.4	**-168**	-111.11	-29.2	**-34**	-36.67	212.0	**100**	37.78
...	**-434**	-258.89	...	**-300**	-184.44	-266.8	**-166**	-110.00	-25.6	**-32**	-35.56	215.6	**102**	38.89
...	**-432**	-257.78	...	**-298**	-183.33	-263.2	**-164**	-108.89	-22.0	**-30**	-34.44	219.2	**104**	40.00
...	**-430**	-256.67	...	**-296**	-182.22	-259.6	**-162**	-107.78	-18.4	**-28**	-33.33	222.8	**106**	41.11
...	**-428**	-255.56	...	**-294**	-181.11	-256.0	**-160**	-106.67	-14.8	**-26**	-32.22	226.4	**108**	42.22
...	**-426**	-254.44	...	**-292**	-180.00	-252.4	**-158**	-105.56	-11.2	**-24**	-31.11	230.0	**110**	43.33
...	**-424**	-253.33	...	**-290**	-178.89	-248.8	**-156**	-104.44	-7.6	**-22**	-30.00	233.6	**112**	44.44
...	**-422**	-252.22	...	**-288**	-177.78	-245.2	**-154**	-103.33	-4.0	**-20**	-28.89	237.2	**114**	45.56
...	**-420**	-251.11	...	**-286**	-176.67	-241.6	**-152**	-102.22	-0.4	**-18**	-27.78	240.8	**116**	46.67
...	**-418**	-250.00	...	**-284**	-175.56	-238.0	**-150**	-101.11	3.2	**-16**	-26.67	244.4	**118**	47.78
...	**-416**	-248.89	...	**-282**	-174.44	-234.4	**-148**	-100.00	6.8	**-14**	-25.56	248.0	**120**	48.89
...	**-414**	-247.78	...	**-280**	-173.33	-230.8	**-146**	-98.89	10.4	**-12**	-24.44	251.6	**122**	50.00
...	**-412**	-246.67	...	**-278**	-172.22	-227.2	**-144**	-97.78	14.0	**-10**	-23.33	255.2	**124**	51.11
...	**-410**	-245.56	...	**-276**	-171.11	-223.6	**-142**	-96.67	17.6	**-8**	-22.22	258.8	**126**	52.22
...	**-408**	-244.44	...	**-274**	-170.00	-220.0	**-140**	-95.56	21.2	**-6**	-21.11	262.4	**128**	53.33
...	**-406**	-243.33	-457.6	**-272**	-168.89	-216.4	**-138**	-94.44	24.8	**-4**	-20.00	266.0	**130**	54.44
...	**-404**	-242.22	-454.0	**-270**	-167.78	-212.8	**-136**	-93.33	28.4	**-2**	-18.89	269.6	**132**	55.56
...	**-402**	-241.11	-450.4	**-268**	-166.67	-209.2	**-134**	-92.22	32.0	**±0**	-17.78	273.2	**134**	56.67
...	**-400**	-240.00	-446.8	**-266**	-165.56	-205.6	**-132**	-91.11	35.6	**2**	-16.67	276.8	**136**	57.78
...	**-398**	-238.89	-443.2	**-264**	-164.44	-202.0	**-130**	-90.00	39.2	**4**	-15.56	280.4	**138**	58.89
...	**-396**	-237.78	-439.6	**-262**	-163.33	-198.4	**-128**	-88.89	42.8	**6**	-14.44	284.0	**140**	60.00
...	**-394**	-236.67	-436.0	**-260**	-162.22	-194.8	**-126**	-87.78	46.4	**8**	-13.33	287.6	**142**	61.11
...	**-392**	-235.56	-432.4	**-258**	-161.11	-191.2	**-124**	-86.67	50.0	**10**	-12.22	291.2	**144**	62.22
...	**-390**	-234.44	-428.8	**-256**	-160.00	-187.6	**-122**	-85.56	53.6	**12**	-11.11	294.8	**146**	63.33
...	**-388**	-233.33	-425.2	**-254**	-158.89	-184.0	**-120**	-84.44	57.2	**14**	-10.00	298.4	**148**	64.44
...	**-386**	-232.22	-421.6	**-252**	-157.78	-180.4	**-118**	-83.33	60.8	**16**	-8.89	302.0	**150**	65.56
...	**-384**	-231.11	-418.0	**-250**	-156.67	-176.8	**-116**	-82.22	64.4	**18**	-7.78	305.6	**152**	66.67
...	**-382**	-230.00	-414.4	**-248**	-155.56	-173.2	**-114**	-81.11	68.0	**20**	-6.67	309.2	**154**	67.78
...	**-380**	-228.89	-410.8	**-246**	-154.44	-169.6	**-112**	-80.00	71.6	**22**	-5.56	312.8	**156**	68.89
...	**-378**	-227.78	-407.2	**-244**	-153.33	-166.0	**-110**	-78.89	75.2	**24**	-4.44	316.4	**158**	70.00
...	**-376**	-226.67	-403.6	**-242**	-152.22	-162.4	**-108**	-77.78	78.8	**26**	-3.33	320.0	**160**	71.11
...	**-374**	-225.56	-400.0	**-240**	-151.11	-158.8	**-106**	-76.67	82.4	**28**	-2.22	323.6	**162**	72.22
...	**-372**	-224.44	-396.4	**-238**	-150.00	-155.2	**-104**	-75.56	86.0	**30**	-1.11	327.2	**164**	73.33
...	**-370**	-223.33	-392.8	**-236**	-148.89	-151.6	**-102**	-74.44	89.6	**32**	±0.00	330.8	**166**	74.44
...	**-368**	-222.22	-389.2	**-234**	-147.78	-148.0	**-100**	-73.33	93.2	**34**	1.11	334.4	**168**	75.56
...	**-366**	-221.11	-385.6	**-232**	-146.67	-144.4	**-98**	-72.22	96.8	**36**	2.22	338.0	**170**	76.67
...	**-364**	-220.00	-382.0	**-230**	-145.56	-140.8	**-96**	-71.11	100.4	**38**	3.33	341.6	**172**	77.78
...	**-362**	-218.89	-378.4	**-228**	-144.44	-137.2	**-94**	-70.00	104.0	**40**	4.44	345.2	**174**	78.89
...	**-360**	-217.78	-374.8	**-226**	-143.33	-133.6	**-92**	-68.89	107.6	**42**	5.56	348.8	**176**	80.00
...	**-358**	-216.67	-371.2	**-224**	-142.22	-130.0	**-90**	-67.78	111.2	**44**	6.67	352.4	**178**	81.11
...	**-356**	-215.56	-367.6	**-222**	-141.11	-126.4	**-88**	-66.67	114.8	**46**	7.78	356.0	**180**	82.22
...	**-354**	-214.44	-364.0	**-220**	-140.00	-122.8	**-86**	-65.56	118.4	**48**	8.89	359.6	**182**	83.33
...	**-352**	-213.33	-360.4	**-218**	-138.89	-119.2	**-84**	-64.44	122.0	**50**	10.00	363.2	**184**	84.44
...	**-350**	-212.22	-356.8	**-216**	-137.78	-115.6	**-82**	-63.33	125.6	**52**	11.11	366.8	**186**	85.56
...	**-348**	-211.11	-353.2	**-214**	-136.67	-112.0	**-80**	-62.22	129.2	**54**	12.12	370.4	**188**	86.67
...	**-346**	-210.00	-349.6	**-212**	-135.56	-108.4	**-78**	-61.11	132.8	**56**	13.33	374.0	**190**	87.78
...	**-344**	-208.89	-346.0	**-210**	-134.44	-104.8	**-76**	-60.00	136.4	**58**	14.44	377.6	**192**	88.89
...	**-342**	-207.78	-342.4	**-208**	-133.33	-101.2	**-74**	-58.89	140.0	**60**	15.56	381.2	**194**	90.00
...	**-340**	-206.67	-338.8	**-206**	-132.22	-97.6	**-72**	-57.78	143.6	**62**	16.67	384.8	**196**	91.11
...	**-338**	-205.56	-335.2	**-204**	-131.11	-94.0	**-70**	-56.67	147.2	**64**	17.78	388.4	**198**	92.22
...	**-336**	-204.44	-331.6	**-202**	-130.00	-90.4	**-68**	-55.56	150.8	**66**	18.89	392.0	**200**	93.33
...	**-334**	-203.33	-328.0	**-200**	-128.89	-86.8	**-66**	-54.44	154.4	**68**	20.00	395.6	**202**	94.44
...	**-332**	-202.22	-324.4	**-198**	-127.78	-83.2	**-64**	-53.33	158.0	**70**	21.11	399.2	**204**	95.56
...	**-330**	-201.11	-320.8	**-196**	-126.67	-79.6	**-62**	-52.22	161.6	**72**	22.22	402.8	**206**	96.67
...	**-328**	-200.00	-317.2	**-194**	-125.56	-76.0	**-60**	-51.11	165.2	**74**	23.33	406.4	**208**	97.78
...	**-326**	-198.89	-313.6	**-192**	-124.44	-72.4	**-58**	-50.00	168.8	**76**	24.44	410.0	**210**	98.89

(continued)

Temperature conversions(continued)

The general arrangement of this table was devised by Sauveur and Boylston more than 40 years ago. The middle column of figures (in bold-faced type) contains the reading (°F or °C) to be converted. If converting from degrees Fahrenheit to degrees Centigrade, read the Centigrade equivalent in the column head "°C." If converting from Centigrade to Fahrenheit, read the Fahrenheit equivalent in the column headed "°F." °C = 5/9 (°F − 32)

°F		°C	°F		°C	°F		°C	°F		°C	°F		°C	°F		°C
413.6	**212**	100.00	654.8	**346**	174.44	896.0	**480**	248.89	1598.0	**870**	465.56	2804.0	**1540**	837.78			
417.2	**214**	101.11	658.4	**348**	175.56	899.6	**482**	250.00	1616.0	**880**	471.11	2822.0	**1550**	843.33			
420.8	**216**	102.22	662.0	**350**	176.67	903.2	**484**	251.11	1634.0	**890**	476.67	2840.0	**1560**	848.89			
424.4	**218**	103.33	665.6	**352**	177.78	906.8	**486**	252.22	1652.0	**900**	482.22	2858.0	**1570**	854.44			
428.0	**220**	104.44	669.2	**354**	178.89	910.4	**488**	253.33	1670.0	**910**	487.78	2876.0	**1580**	860.00			
431.6	**222**	105.56	672.8	**356**	180.00	914.0	**490**	254.44	1688.0	**920**	493.33	2894.0	**1590**	865.56			
435.2	**224**	106.67	676.4	**358**	181.11	917.6	**492**	255.56	1706.0	**930**	498.89	2912.0	**1600**	871.11			
438.8	**226**	107.78	680.0	**360**	182.22	921.2	**494**	256.67	1724.0	**940**	504.44	2930.0	**1610**	876.67			
442.4	**228**	108.89	683.6	**362**	183.33	924.8	**496**	257.78	1742.0	**950**	510.00	2948.0	**1620**	882.22			
446.0	**230**	110.00	687.2	**364**	184.44	928.4	**498**	258.89	1760.0	**960**	515.56	2966.0	**1630**	887.78			
449.6	**232**	111.11	690.8	**366**	185.56	932.0	**500**	260.00	1778.0	**970**	521.11	2984.0	**1640**	893.33			
453.2	**234**	112.22	694.4	**368**	186.67	935.6	**502**	261.11	1796.0	**980**	526.67	3002.0	**1650**	898.89			
456.8	**236**	113.33	698.0	**370**	187.78	939.2	**504**	262.22	1814.0	**990**	532.22	3020.0	**1660**	904.44			
460.4	**238**	114.44	701.6	**372**	188.89	942.8	**506**	263.33	1832.0	**1000**	537.78	3038.0	**1670**	910.00			
464.0	**240**	115.56	705.2	**374**	190.00	946.4	**508**	264.44	1850.0	**1010**	543.33	3056.0	**1680**	915.56			
467.6	**242**	116.67	708.8	**376**	191.11	950.0	**510**	265.56	1868.0	**1020**	548.89	3074.0	**1690**	921.11			
471.2	**244**	117.78	712.4	**378**	192.22	953.6	**512**	266.67	1886.0	**1030**	554.44	3092.0	**1700**	926.67			
474.8	**246**	118.89	716.0	**380**	193.33	957.2	**514**	267.78	1904.0	**1040**	560.00	3110.0	**1710**	932.22			
478.4	**248**	120.00	719.6	**382**	194.44	960.8	**516**	268.89	1922.0	**1050**	565.56	3128.0	**1720**	937.78			
482.0	**250**	121.11	723.2	**384**	195.56	964.4	**518**	270.00	1940.0	**1060**	571.11	3146.0	**1730**	943.33			
485.6	**252**	122.22	726.8	**386**	196.67	968.0	**520**	271.11	1958.0	**1070**	576.67	3164.0	**1740**	948.89			
489.2	**254**	123.33	730.4	**388**	197.78	971.6	**522**	272.22	1976.0	**1080**	582.22	3182.0	**1750**	954.44			
492.8	**256**	124.44	734.0	**390**	198.89	975.2	**524**	273.33	1994.0	**1090**	587.78	3200.0	**1760**	960.00			
496.4	**258**	125.56	737.6	**392**	200.00	978.8	**526**	274.44	2012.0	**1100**	593.33	3218.0	**1770**	965.56			
500.0	**260**	126.67	741.2	**394**	201.11	982.4	**528**	275.56	2030.0	**1110**	598.89	3236.0	**1780**	971.11			
503.6	**262**	127.78	744.8	**396**	202.22	986.0	**530**	276.67	2048.0	**1120**	604.44	3254.0	**1790**	976.67			
507.2	**264**	128.89	748.4	**398**	203.33	989.6	**532**	277.78	2066.0	**1130**	610.00	3272.0	**1800**	982.22			
510.8	**266**	130.00	752.0	**400**	204.44	993.2	**534**	278.89	2084.0	**1140**	615.56	3290.0	**1810**	987.78			
514.4	**268**	131.11	755.6	**402**	205.56	996.8	**536**	280.00	2102.0	**1150**	621.11	3308.0	**1820**	993.33			
518.0	**270**	132.22	759.2	**404**	206.67	1000.4	**538**	281.11	2120.0	**1160**	626.67	3326.0	**1830**	998.89			
521.6	**272**	133.33	762.8	**406**	207.78	1004.0	**540**	282.22	2138.0	**1170**	632.22	3344.0	**1840**	1004.4			
525.2	**274**	134.44	766.4	**408**	208.89	1007.6	**542**	283.33	2156.0	**1180**	637.78	3362.0	**1850**	1010.0			
528.8	**276**	135.56	770.0	**410**	210.00	1011.2	**544**	284.44	2174.0	**1190**	643.33	3380.0	**1860**	1015.6			
532.4	**278**	136.67	773.6	**412**	211.11	1014.8	**546**	285.56	2192.0	**1200**	648.89	3398.0	**1870**	1021.1			
536.0	**280**	137.78	777.2	**414**	212.22	1018.4	**548**	286.67	2210.0	**1210**	654.44	3416.0	**1880**	1026.7			
539.6	**282**	138.89	780.8	**416**	213.33	1022.0	**550**	287.78	2228.0	**1220**	660.00	3434.0	**1890**	1032.2			
543.2	**284**	140.00	784.4	**418**	214.44	1040.0	**560**	293.33	2246.0	**1230**	665.56	3452.0	**1900**	1037.8			
546.8	**286**	141.11	788.0	**420**	215.56	1058.0	**570**	298.89	2264.0	**1240**	671.11	3470.0	**1910**	1043.3			
550.4	**288**	142.22	791.6	**422**	216.67	1076.0	**580**	304.44	2282.0	**1250**	676.67	3488.0	**1920**	1048.9			
554.0	**290**	143.33	795.2	**424**	217.78	1094.0	**590**	310.00	2300.0	**1260**	682.22	3506.0	**1930**	1054.4			
557.6	**292**	144.44	798.8	**426**	218.89	1112.0	**600**	315.56	2318.0	**1270**	687.78	3524.0	**1940**	1060.0			
561.2	**294**	145.56	802.4	**428**	220.00	1130.0	**610**	321.11	2336.0	**1280**	693.33	3542.0	**1950**	1065.6			
564.8	**296**	146.67	806.0	**430**	221.11	1148.0	**620**	326.67	2354.0	**1290**	698.89	3560.0	**1960**	1071.1			
568.4	**298**	147.78	809.6	**432**	222.22	1166.0	**630**	332.22	2372.0	**1300**	704.44	3578.0	**1970**	1076.7			
572.0	**300**	148.89	813.2	**434**	223.33	1184.0	**640**	337.78	2390.0	**1310**	710.00	3596.0	**1980**	1082.2			
575.6	**302**	150.00	816.8	**436**	224.44	1202.0	**650**	343.33	2408.0	**1320**	715.56	3614.0	**1990**	1087.8			
579.2	**304**	151.11	820.4	**438**	225.56	1220.0	**660**	348.89	2426.0	**1330**	721.11	3632.0	**2000**	1093.3			
582.8	**306**	152.22	824.0	**440**	226.67	1238.0	**670**	354.44	2444.0	**1340**	726.67	3650.0	**2010**	1098.9			
586.4	**308**	153.33	827.6	**442**	227.78	1256.0	**680**	360.00	2462.0	**1350**	732.22	3668.0	**2020**	1104.4			
590.0	**310**	154.44	831.2	**444**	228.89	1274.0	**690**	365.56	2480.0	**1360**	737.78	3686.0	**2030**	1110.0			
593.6	**312**	155.56	834.8	**446**	230.00	1292.0	**700**	371.11	2498.0	**1370**	743.33	3704.0	**2040**	1115.6			
597.2	**314**	156.67	838.4	**448**	231.11	1310.0	**710**	376.67	2516.0	**1380**	748.89	3722.0	**2050**	1121.1			
600.8	**316**	157.78	842.0	**450**	232.22	1328.0	**720**	382.22	2534.0	**1390**	754.44	3740.0	**2060**	1126.7			
604.4	**318**	158.89	845.6	**452**	233.33	1346.0	**730**	387.78	2552.0	**1400**	760.00	3758.0	**2070**	1132.2			
608.0	**320**	160.00	849.2	**454**	234.44	1364.0	**740**	393.33	2570.0	**1410**	765.56	3776.0	**2080**	1137.8			
611.6	**322**	161.11	852.8	**456**	235.56	1382.0	**750**	398.89	2588.0	**1420**	771.11	3794.0	**2090**	1143.3			
615.2	**324**	162.22	856.4	**458**	236.67	1400.0	**760**	404.44	2606.0	**1430**	776.67	3812.0	**2100**	1148.9			
618.8	**326**	163.33	860.0	**460**	237.78	1418.0	**770**	410.00	2624.0	**1440**	782.22	3830.0	**2110**	1154.4			
622.4	**328**	164.44	863.6	**462**	238.89	1436.0	**780**	415.56	2642.0	**1450**	787.78	3848.0	**2120**	1160.0			
626.0	**330**	165.56	867.2	**464**	240.00	1454.0	**790**	421.11	2660.0	**1460**	793.33	3866.0	**2130**	1165.6			
629.6	**332**	166.67	870.8	**466**	241.11	1472.0	**800**	426.67	2678.0	**1470**	798.89	3884.0	**2140**	1171.1			
633.2	**334**	167.78	874.4	**468**	242.22	1490.0	**810**	432.22	2696.0	**1480**	804.44	3902.0	**2150**	1176.7			
636.8	**336**	168.89	878.0	**470**	243.33	1508.0	**820**	437.76	2714.0	**1490**	810.00	3920.0	**2160**	1182.2			
640.4	**338**	170.00	881.6	**472**	244.44	1526.0	**830**	443.33	2732.0	**1500**	815.56	3938.0	**2170**	1187.8			
644.0	**340**	171.11	885.2	**474**	245.56	1544.0	**840**	448.89	2750.0	**1510**	821.11	3956.0	**2180**	1193.3			
647.6	**342**	172.22	888.8	**476**	246.67	1562.0	**850**	454.44	2768.0	**1520**	826.67	3974.0	**2190**	1198.9			
651.2	**344**	173.33	892.4	**478**	247.78	1580.0	**860**	460.00	2786.0	**1530**	832.22	3992.0	**2200**	1204.4			

(continued)

Temperature conversions (continued)

The general arrangement of this table was devised by Sauveur and Boylston more than 40 years ago. The middle column of figures (in bold-faced type) contains the reading (°F or °C) to be converted. If converting from degrees Fahrenheit to degrees Centigrade, read the Centigrade equivalent in the column head "°C." If converting from Centigrade to Fahrenheit, read the Fahrenheit equivalent in the column headed "°F." °C = $\frac{5}{9}$ (°F – 32)

°F		°C	°F		°C	°F		°C	°F		°C	°F		°C
4010.0	**2210**	1210.0	4514.0	**2490**	1365.6	5018.0	**2770**	1521.1	5522.0	**3050**	1676.7	7682.0	**4250**	2343.3
4028.0	**2220**	1215.6	4532.0	**2500**	1371.1	5036.0	**2780**	1526.7	5540.0	**3060**	1682.2	7772.0	**4300**	2371.1
4046.0	**2230**	1221.1	4550.0	**2510**	1376.7	5054.0	**2790**	1532.2	5558.0	**3070**	1687.8	7862.0	**4350**	2398.8
4064.0	**2240**	1226.7	4568.0	**2520**	1382.2	5072.0	**2800**	1537.8	5576.0	**3080**	1693.3	7952.0	**4400**	2426.6
4082.0	**2250**	1232.2	4586.0	**2530**	1387.8	5090.0	**2810**	1543.3	5594.0	**3090**	1698.9	8042.0	**4450**	2454.4
4100.0	**2260**	1237.8	4604.0	**2540**	1393.3	5108.0	**2820**	1548.9	5612.0	**3100**	1704.4	8132.0	**4500**	2482.2
4118.0	**2270**	1243.3	4622.0	**2550**	1398.9	5126.0	**2830**	1554.4	5702.0	**3150**	1732.2	8222.0	**4550**	2510.0
4136.0	**2280**	1248.9	4640.0	**2560**	1404.4	5144.0	**2840**	1560.0	5792.0	**3200**	1760.0	8312.0	**4600**	2537.7
4154.0	**2290**	1254.4	4658.0	**2570**	1410.0	5162.0	**2850**	1565.6	5882.0	**3250**	1787.7	8402.0	**4650**	2565.5
4172.0	**2300**	1260.0	4676.0	**2580**	1415.6	5180.0	**2860**	1571.1	5972.0	**3300**	1815.5	8492.0	**4700**	2593.3
4190.0	**2310**	1265.6	4694.0	**2590**	1421.1	5198.0	**2870**	1576.7	6062.0	**3350**	1843.3	8582.0	**4750**	2621.1
4208.0	**2320**	1271.1	4712.0	**2600**	1426.7	5216.0	**2880**	1582.2	6152.0	**3400**	1871.1	8672.0	**4800**	2648.8
4226.0	**2330**	1276.7	4730.0	**2610**	1432.2	5234.0	**2890**	1587.8	6242.0	**3450**	1898.8	8762.0	**4850**	2676.6
4244.0	**2340**	1282.2	4748.0	**2620**	1437.8	5252.0	**2900**	1593.3	6332.0	**3500**	1926.6	8852.0	**4900**	2704.4
4262.0	**2350**	1287.8	4766.0	**2630**	1443.3	5270.0	**2910**	1598.9	6422.0	**3550**	1954.4	8942.0	**4950**	2732.2
4280.0	**2360**	1293.3	4784.0	**2640**	1448.9	5288.0	**2920**	1604.4	6512.0	**3600**	1982.2	9032.0	**5000**	2760.0
4298.0	**2370**	1298.9	4802.0	**2650**	1454.4	5306.0	**2930**	1610.0	6602.0	**3650**	2010.0	9122.0	**5050**	2787.7
4316.0	**2380**	1304.4	4820.0	**2660**	1460.0	5324.0	**2940**	1615.6	6692.0	**3700**	2037.7	9212.0	**5100**	2815.5
4334.0	**2390**	1310.0	4838.0	**2670**	1465.6	5342.0	**2950**	1621.1	6782.0	**3750**	2065.5	9302.0	**5150**	2843.3
4352.0	**2400**	1315.6	4856.0	**2680**	1471.1	5360.0	**2960**	1626.7	6872.0	**3800**	2093.3	9392.0	**5200**	2871.1
4370.0	**2410**	1321.1	4874.0	**2690**	1476.7	5378.0	**2970**	1632.2	6962.0	**3850**	2121.1	9482.0	**5250**	2898.8
4388.0	**2420**	1326.7	4892.0	**2700**	1482.2	5396.0	**2980**	1637.8	7052.0	**3900**	2148.8	9572.0	**5300**	2926.6
4406.0	**2430**	1332.2	4910.0	**2710**	1487.8	5414.0	**2990**	1643.3	7142.0	**3950**	2176.6	9662.0	**5350**	2954.4
4424.0	**2440**	1337.8	4928.0	**2720**	1493.3	5432.0	**3000**	1648.9	7232.0	**4000**	2204.4	9752.0	**5400**	2982.2
4442.0	**2450**	1343.3	4946.0	**2730**	1498.9	5450.0	**3010**	1654.4	7322.0	**4050**	2232.2	9842.0	**5450**	3010.0
4460.0	**2460**	1348.9	4964.0	**2740**	1504.4	5468.0	**3020**	1660.0	7412.0	**4100**	2260.0	9932.0	**5500**	3037.7
4478.0	**2470**	1354.4	4982.0	**2750**	1510.0	5486.0	**3030**	1665.6	7502.0	**4150**	2287.7	10022.0	**5550**	3065.5
4496.0	**2480**	1360.0	5000.0	**2760**	1515.6	5504.0	**3040**	1671.1	7592.0	**4200**	2315.5	10112.0	**5600**	3093.3

Metric stress or pressure conversions

The middle column of figures (in bold-faced type) contains the reading (in MPa or ksi) to be converted. If converting from ksi to MPa, read the MPa equivalent in the column headed "MPa." If converting from MPa to ksi, read the ksi equivalent in the column headed "ksi." 1 ksi = 6.894757 MPa. 1 psi = 6.894757 kPa

ksi		MPa	ksi		MPa	ksi		MPa	ksi		MPa
0.14504	1	6.895	9.8626	68	468.84	65.267	450	3102.6	179.85	1240	...
0.29008	2	13.790	10.008	69	475.74	66.717	460	3171.6	182.75	1260	...
0.43511	3	20.684	10.153	70	482.63	68.168	470	3240.5	185.65	1280	...
0.58015	4	27.579	10.298	71	489.53	69.618	480	3309.5	188.55	1300	...
0.72519	5	34.474	10.443	72	496.42	71.068	490	3378.4	191.45	1320	...
0.87023	6	41.369	10.588	73	503.32	72.519	500	3447.4	194.35	1340	...
1.0153	7	48.263	10.733	74	510.21	73.969	510	...	197.25	1360	...
1.1603	8	55.158	10.878	75	517.11	75.420	520	...	200.15	1380	...
1.3053	9	62.053	11.023	76	524.00	76.870	530	...	203.05	1400	...
1.4504	10	68.948	11.168	77	530.90	78.320	540	...	205.95	1420	...
1.5954	11	75.842	11.313	78	537.79	79.771	550	...	208.85	1440	...
1.7405	12	82.737	11.458	79	544.69	81.221	560	...	211.76	1460	...
1.8855	13	89.632	11.603	80	551.58	82.672	570	...	214.66	1480	...
2.0305	14	96.527	11.748	81	558.48	84.122	580	...	217.56	1500	...
2.1756	15	103.42	11.893	82	565.37	85.572	590	...	220.46	1520	...
2.3206	16	110.32	12.038	83	572.26	87.023	600	...	223.36	1540	...
2.4656	17	117.21	12.183	84	579.16	88.473	610	...	226.26	1560	...
2.6107	18	124.11	12.328	85	586.05	89.923	620	...	229.16	1580	...
2.7557	19	131.00	12.473	86	592.95	91.374	630	...	232.06	1600	...
2.9008	20	137.90	12.618	87	599.84	92.824	640	...	234.96	1620	...
3.0458	21	144.79	12.763	88	606.74	94.275	650	...	237.86	1640	...
3.1908	22	151.68	12.909	89	613.63	95.725	660	...	240.76	1660	...
3.3359	23	158.58	13.053	90	620.53	97.175	670	...	243.66	1680	...
3.4809	24	165.47	13.198	91	627.42	98.626	680	...	246.56	1700	...
3.6259	25	172.37	13.343	92	634.32	100.08	690	...	249.46	1720	...
3.7710	26	179.26	13.489	93	641.21	101.53	700	...	252.37	1740	...
3.9160	27	186.16	13.634	94	648.11	102.98	710	...	255.27	1760	...
4.0611	28	193.05	13.779	95	655.00	104.43	720	...	258.17	1780	...
4.2061	29	199.95	13.924	96	661.90	105.88	730	...	261.07	1800	...
4.3511	30	206.84	14.069	97	668.79	107.33	740	...	263.97	1820	...
4.4962	31	213.74	14.214	98	675.69	108.78	750	...	266.87	1840	...
4.6412	32	220.63	14.359	99	682.58	110.23	760	...	269.77	1860	...
4.7862	33	227.53	14.504	100	689.48	111.68	770	...	272.67	1880	...
4.9313	34	234.42	15.954	110	758.42	113.13	780	...	275.57	1900	...
5.0763	35	241.32	17.405	120	827.37	114.58	790	...	278.47	1920	...
5.2214	36	248.21	18.855	130	896.32	116.03	800	...	281.37	1940	...
5.3664	37	255.11	20.305	140	965.27	117.48	810	...	284.27	1960	...
5.5114	38	262.00	21.756	150	1034.2	118.93	820	...	287.17	1980	...
5.6565	39	268.90	23.206	160	1103.2	120.38	830	...	290.08	2000	...
5.8015	40	275.79	24.656	170	1172.1	121.83	840	...	292.98	2020	...
5.9465	41	282.69	26.107	180	1241.1	123.28	850	...	295.88	2040	...
6.0916	42	289.58	27.557	190	1310.0	124.73	860	...	298.78	2060	...
6.2366	43	296.47	29.008	200	1379.0	126.18	870	...	301.68	2080	...
6.3817	44	303.37	30.458	210	1447.9	127.63	880	...	304.58	2100	...
6.5267	45	310.26	31.908	220	1516.8	129.08	890	...	307.48	2120	...
6.6717	46	317.16	33.359	230	1585.8	130.53	900	...	310.38	2140	...
6.8168	47	324.05	34.809	240	1654.7	131.98	910	...	313.28	2160	...
6.9618	48	330.95	36.259	250	1723.7	133.43	920	...	316.18	2180	...
7.1068	49	337.84	37.710	260	1792.6	134.89	930	...	319.08	2200	...
7.2519	50	344.74	39.160	270	1861.6	136.34	940	...	321.98	2220	...
7.3969	51	351.63	40.611	280	1930.5	137.79	950	...	324.88	2240	...
7.5420	52	358.53	42.061	290	1999.5	139.24	960	...	327.79	2260	...
7.6870	53	365.42	43.511	300	2068.4	140.69	970	...	330.69	2280	...
7.8320	54	372.32	44.962	310	2137.4	142.14	980	...	333.59	2300	...
7.9771	55	379.21	46.412	320	2206.3	143.59	990	...	336.49	2320	...
8.1221	56	386.11	47.862	330	2275.3	145.04	1000	...	339.39	2340	...
8.2672	57	393.00	49.313	340	2344.2	147.94	1020	...	342.29	2360	...
8.4122	58	399.90	50.763	350	2413.2	150.84	1040	...	345.19	2380	...
8.5572	59	406.79	52.214	360	2482.1	153.74	1060	...	348.09	2400	...
8.7023	60	413.69	53.664	370	2551.1	156.64	1080	...	350.99	2420	...
8.8473	61	420.58	55.114	380	2620.0	159.54	1100	...	353.89	2440	...
8.9923	62	427.47	56.565	390	2689.0	162.44	1120	...	356.79	2460	...
9.1374	63	434.37	58.015	400	2757.9	165.34	1140	...	359.69	2480	...
9.2824	64	441.26	59.465	410	2826.9	168.24	1160	...	362.59	2500	...
9.4275	65	448.16	60.916	420	2895.8	171.14	1180	...			
9.5725	66	455.05	62.366	430	2964.7	174.05	1200	...			
9.7175	67	461.95	63.817	440	3033.7	176.95	1220	...			

Metric stress-intensity conversions

The middle column of figures (in bold-faced type) contains the reading (in MPa√m or ksi√in.) to be converted. If converting from ksi√in. to MPa√m, read the MPa√m equivalent in the column headed "MPa√m." If converting from MPa√m to ksi√in., read the ksi√in. equivalent in the column headed "ksi√in.." 1 ksi√in. = 1.098845 MPa√m

ksi√in.		MPa√m	ksi√in.		MPa√m	ksi√in.		MPa√m	ksi√in.		MPa√m	ksi√in.		MPa√m
0.91005	1	1.0988	37.312	41	45.051	73.714	81	89.003	110.12	121	132.95	146.52	161	176.91
1.8201	2	2.1976	38.222	42	46.150	74.624	82	90.102	111.03	122	134.05	147.43	162	178.01
2.7301	3	3.2964	39.132	43	47.248	75.534	83	91.200	111.94	123	135.15	148.34	163	179.10
3.6402	4	4.3952	40.042	44	48.347	76.444	84	92.300	112.85	124	136.25	149.25	164	180.20
4.5502	5	5.4940	40.952	45	49.446	77.354	85	93.398	113.76	125	137.35	150.16	165	181.30
5.4603	6	6.5928	41.862	46	50.545	78.264	86	94.497	114.67	126	138.45	151.07	166	182.40
6.3703	7	7.6916	42.772	47	51.644	79.174	87	95.596	115.58	127	139.55	151.98	167	183.50
7.2804	8	8.7904	43.682	48	52.742	80.084	88	96.694	116.49	128	140.65	152.89	168	184.60
8.1904	9	9.8892	44.592	49	53.841	80.994	89	97.793	117.40	129	141.75	153.80	169	185.70
9.1005	10	10.988	45.502	50	54.940	81.904	90	98.892	118.31	130	142.84	154.71	170	186.80
10.011	11	12.087	46.412	51	56.039	82.814	91	99.991	119.22	131	143.94	155.62	171	187.90
10.921	12	13.186	47.322	52	57.138	83.724	92	101.09	120.13	132	145.04	156.53	172	189.00
11.831	13	14.284	48.232	53	58.236	84.634	93	102.19	121.04	133	146.14	157.44	173	190.10
12.741	14	15.383	49.143	54	59.335	85.544	94	103.29	121.95	134	147.24	158.35	174	191.19
13.651	15	16.482	50.053	55	60.434	86.454	95	104.39	122.86	135	148.34	159.26	175	192.29
14.561	16	17.581	50.963	56	61.533	87.364	96	105.48	123.77	136	149.44	160.17	176	193.39
15.471	17	18.680	51.873	57	62.632	88.275	97	106.58	124.68	137	150.54	161.08	177	194.49
16.381	18	19.778	52.783	58	63.730	89.185	98	107.68	125.59	138	151.63	161.99	178	195.59
17.291	19	20.877	53.693	59	64.829	90.095	99	108.78	126.50	139	152.73	162.90	179	196.69
18.201	20	21.976	54.603	60	65.928	91.005	100	109.88	127.41	140	153.83	163.81	180	197.78
19.111	21	23.075	55.513	61	67.027	91.915	101	110.98	128.32	141	154.93	164.72	181	198.88
20.021	22	24.174	56.423	62	68.126	92.825	102	112.08	129.23	142	156.03	165.63	182	199.98
20.931	23	25.272	57.333	63	69.224	93.735	103	113.18	130.14	143	157.13	166.54	183	201.08
21.841	24	26.371	58.243	64	70.323	94.645	104	114.28	131.05	144	158.23	167.45	184	202.18
22.751	25	27.470	59.153	65	71.422	95.555	105	115.37	131.96	145	159.33	168.36	185	203.28
23.661	26	28.569	60.063	66	72.521	96.465	106	116.47	132.87	146	160.42	169.27	186	204.38
24.571	27	29.668	60.973	67	73.620	97.375	107	117.57	133.78	147	161.52	170.18	187	205.48
25.481	28	30.766	61.883	68	74.718	98.285	108	118.67	134.69	148	162.62	171.09	188	206.57
26.391	29	31.865	62.793	69	75.817	99.195	109	119.77	135.60	149	163.72	172.00	189	207.67
27.301	30	32.964	63.703	70	76.916	100.11	110	120.87	136.51	150	164.82	172.91	190	208.77
28.211	31	34.063	64.613	71	78.015	101.02	111	121.97	137.42	151	165.92	173.82	191	209.87
29.121	32	35.162	65.523	72	79.114	101.93	112	123.07	138.33	152	167.02	174.73	192	210.97
30.032	33	36.260	66.433	73	80.212	102.84	113	124.16	139.24	153	168.12	175.64	193	212.07
30.942	34	37.359	67.343	74	81.311	103.75	114	125.26	140.15	154	169.22	176.55	194	213.17
31.852	35	38.458	68.253	75	82.410	104.66	115	126.36	141.06	155	170.31	177.46	195	214.27
32.762	36	39.557	69.164	76	83.509	105.57	116	127.46	141.97	156	171.41	178.37	196	215.36
33.672	37	40.656	70.074	77	84.608	106.48	117	128.56	142.88	157	172.51	179.28	197	216.46
34.582	38	41.754	70.984	78	85.706	107.39	118	129.66	143.79	158	173.61	180.19	198	217.56
35.492	39	42.853	71.893	79	86.805	108.30	119	130.76	144.70	159	174.71	181.10	199	218.66
36.402	40	43.952	72.804	80	87.904	109.21	120	131.86	145.61	160	175.81	182.01	200	219.76

Metric energy conversions

The middle column of figures (in bold-faced type) contains the reading (in J or ft · lb) to be converted. If converting from ft · lb to J, read the J equivalent in the column headed "J." If converting from J to ft · lb, read the equivalent in the column headed "ft · lb." 1 ft · lb = 1.355818 J

ft · lb		J	ft · lb		J	ft · lb		J	ft · lb		J
0.7376	1	1.3558	28.7649	39	52.8769	56.7923	77	104.3980	129.0734	175	237.2681
1.4751	2	2.7116	29.5025	40	54.2327	57.5298	78	105.7538	132.7612	180	244.0472
2.2127	3	4.0675	30.2400	41	55.5885	58.2674	79	107.1096	136.4490	185	250.8263
2.9502	4	5.4233	30.9776	42	56.9444	59.0050	80	108.4654	140.1368	190	257.6054
3.6878	5	6.7791	31.7152	43	58.3002	59.7425	81	109.8212	143.8246	195	264.3845
4.4254	6	8.1349	32.4527	44	59.6560	60.4801	82	111.1771	147.5124	200	271.1636
5.1629	7	9.4907	33.1903	45	61.0118	61.2177	83	112.5329	154.8880	210	284.7218
5.9005	8	10.8465	33.9279	46	62.3676	61.9552	84	113.8887	162.2637	220	298.2799
6.6381	9	12.2024	34.6654	47	63.7234	62.6928	85	115.2445	169.6393	230	311.8381
7.3756	10	13.5582	35.4030	48	65.0793	63.4303	86	116.6003	177.0149	240	325.3963
8.1132	11	14.9140	36.1405	49	66.4351	64.1679	87	117.9562	184.3905	250	338.9545
8.8507	12	16.2698	36.8781	50	67.7909	64.9055	88	119.3120	191.7661	260	352.5126
9.5883	13	17.6256	37.6157	51	69.1467	65.6430	89	120.6678	199.1418	270	366.0708
10.3259	14	18.9815	38.3532	52	70.5025	66.3806	90	122.0236	206.5174	280	379.6290
11.0634	15	20.3373	39.0908	53	71.8583	67.1182	91	123.3794	213.8930	290	393.1872
11.8010	16	21.6931	39.8284	54	73.2142	67.8557	92	124.7452	221.2686	300	406.7454
12.5386	17	23.0489	40.5659	55	74.5700	68.5933	93	126.0911	228.6442	310	420.3036
13.2761	18	24.4047	41.3035	56	75.9258	69.3308	94	127.4469	236.0199	320	433.8617
14.0137	19	25.7605	42.0410	57	77.2816	70.0684	95	128.8027	243.3955	330	447.4199
14.7512	20	27.1164	42.7786	58	78.6374	70.8060	96	130.1585	250.7711	340	460.9781
15.4888	21	28.4722	43.5162	59	79.9933	71.5435	97	131.5143	258.1467	350	474.5363
16.2264	22	29.8280	44.2537	60	81.3491	72.2811	98	132.8702	265.5224	360	488.0944
16.9639	23	31.1838	44.9913	61	82.7049	73.0186	99	134.2260	272.8980	370	501.6526
17.7015	24	32.5396	45.7288	62	84.0607	73.7562	100	135.5818	280.2736	380	515.2108
18.4390	25	33.8954	46.4664	63	85.4165	77.4440	105	142.3609	287.6492	390	528.7690
19.1766	26	35.2513	47.2040	64	86.7723	81.1318	110	149.1400	295.0248	400	542.3272
19.9142	27	36.6071	47.9415	65	88.1282	84.8196	115	155.9191	302.4005	410	555.8854
20.6517	28	37.9629	48.6791	66	89.4840	88.5075	120	162.6982	309.7761	420	569.4435
21.3893	29	39.3187	49.4167	67	90.8398	92.1953	125	169.4772	317.1517	430	583.0017
22.1269	30	40.6745	50.1542	68	92.1956	95.8831	130	176.2563	324.5273	440	596.5599
22.8644	31	42.0304	50.8918	69	93.5514	99.5709	135	183.0354	331.9029	450	610.1181
23.6020	32	43.3862	51.6293	70	94.9073	103.2587	140	189.8145	339.2786	460	623.6762
24.3395	33	44.7420	52.3669	71	96.2631	106.9465	145	196.5936	346.6542	470	637.2344
25.0771	34	46.0978	53.1045	72	97.6189	110.6343	150	203.3727	354.0298	480	650.7926
25.8147	35	47.4536	53.8420	73	98.9747	114.3221	155	210.1518	361.4054	490	664.3508
26.5522	36	48.8094	54.5796	74	100.3305	118.0099	160	216.9308	368.7811	500	677.9090
27.2898	37	50.1653	55.3172	75	101.6863	121.6977	165	223.7099			
28.0274	38	51.5211	56.0547	76	103.0422	125.3856	170	230.4890			

Metric length and weight conversion factors

Unit	Inches to millimeters	Millimeters to inches	Pounds to kilograms	Kilograms to pounds
1	25.400 1	0.039 371	0.453 59	2.204 62
2	50.800 1	0.078 742	0.907 19	4.409 24
3	76.200 2	0.118 112	1.360 78	6.613 86
4	101.600 2	0.157 483	1.814 37	8.818 49
5	127.000 3	0.196 854	2.267 96	11.023 11
6	152.400 3	0.236 225	2.721 56	13.227 73
7	177.800 4	0.275 596	3.175 15	15.432 35
8	203.200 4	0.314 966	3.628 74	17.636 97
9	228.600 5	0.354 337	4.082 33	19.841 59
10	254.000 6	0.393 708	4.355 92	22.046 22

Conversion of inches to millimeters

Inches	Millimeters	Inches	Millimeters	Inches	Millimeters	Inches	Millimeters
0.001	0.025	0.200	5.08	0.480	12.19	0.760	19.30
0.002	0.051	0.210	5.33	0.490	12.45	0.770	19.56
0.003	0.076	0.220	5.59	0.500	12.70	0.780	19.81
0.004	0.102	0.230	5.84	0.510	12.95	0.790	20.07
0.005	0.127	0.240	6.10	0.520	13.21	0.800	20.32
0.006	0.152	0.250	6.35	0.530	13.46	0.810	20.57
0.007	0.178	0.260	6.60	0.540	13.72	0.820	20.83
0.008	0.203	0.270	6.86	0.550	13.97	0.830	21.08
0.009	0.229	0.280	7.11	0.560	14.22	0.840	21.34
0.010	0.254	0.290	7.37	0.570	14.48	0.850	21.59
0.020	0.508	0.300	7.62	0.580	14.73	0.860	21.84
0.030	0.762	0.310	7.87	0.590	14.99	0.870	22.10
0.040	1.016	0.320	8.13	0.600	15.24	0.880	22.35
0.050	1.270	0.330	8.38	0.610	15.49	0.890	22.61
0.060	1.524	0.340	8.64	0.620	15.75	0.900	22.86
0.070	1.778	0.350	8.89	0.630	16.00	0.910	23.11
0.080	2.032	0.360	9.14	0.640	16.26	0.920	23.37
0.090	2.286	0.370	9.40	0.650	16.51	0.930	23.62
0.100	2.540	0.380	9.65	0.660	16.76	0.940	23.88
0.110	2.794	0.390	9.91	0.670	17.02	0.950	24.13
0.120	3.048	0.400	10.16	0.680	17.17	0.960	24.38
0.130	3.302	0.410	10.41	0.690	17.53	0.970	24.64
0.140	3.56	0.420	10.67	0.700	17.78	0.980	24.89
0.150	3.81	0.430	10.92	0.710	18.03	0.990	25.15
0.160	4.06	0.440	11.18	0.720	18.29	1.000	25.40
0.170	4.32	0.450	11.43	0.730	18.54		
0.180	4.57	0.460	11.68	0.740	18.80		
0.190	4.83	0.470	11.94	0.750	19.05		

Conversion of millimeters to inches

Millimeters	Inches	Millimeters	Inches	Millimeters	Inches	Millimeters	Inches
0.01	0.0004	0.26	0.0102	0.51	0.0201	0.76	0.0299
0.02	0.0008	0.27	0.0106	0.52	0.0205	0.77	0.0303
0.03	0.0012	0.28	0.0110	0.53	0.0209	0.78	0.0307
0.04	0.0016	0.29	0.0114	0.54	0.0213	0.79	0.0311
0.05	0.0020	0.30	0.0118	0.55	0.0217	0.80	0.0315
0.06	0.0024	0.31	0.0122	0.56	0.0220	0.81	0.0319
0.07	0.0028	0.32	0.0126	0.57	0.0224	0.82	0.0323
0.08	0.0031	0.33	0.0130	0.58	0.0228	0.83	0.0327
0.09	0.0035	0.34	0.0134	0.59	0.0232	0.84	0.0331
0.10	0.0039	0.35	0.0138	0.60	0.0236	0.85	0.0335
0.11	0.0043	0.36	0.0142	0.61	0.0240	0.86	0.0339
0.12	0.0047	0.37	0.0146	0.62	0.0244	0.87	0.0343
0.13	0.0051	0.38	0.0150	0.63	0.0248	0.88	0.0346
0.14	0.0055	0.39	0.0154	0.64	0.0252	0.89	0.0350
0.15	0.0059	0.40	0.0157	0.65	0.0256	0.90	0.0354
0.16	0.0063	0.41	0.0161	0.66	0.0260	0.91	0.0358
0.17	0.0067	0.42	0.0165	0.67	0.0264	0.92	0.0362
0.18	0.0071	0.43	0.0169	0.68	0.0268	0.93	0.0366
0.19	0.0075	0.44	0.0173	0.69	0.0272	0.94	0.0370
0.20	0.0079	0.45	0.0177	0.70	0.0276	0.95	0.0374
0.21	0.0083	0.46	0.0181	0.71	0.0280	0.96	0.0378
0.22	0.0087	0.47	0.0185	0.72	0.0283	0.97	0.0382
0.23	0.0091	0.48	0.0189	0.73	0.0287	0.98	0.0386
0.24	0.0094	0.49	0.0193	0.74	0.0291	0.99	0.0390
0.25	0.0098	0.50	0.0197	0.75	0.0295	1.00	0.0394

Conversion factors and measurements

Equivalents

1 gram = 15 432 grains
1 meter = 39.371 inches or 3.28083 feet
1 millimeter = 0.03937 inch or 1/25 in. approx

$\left.\begin{array}{l}1\text{ metric ton}\\1000\text{ kilograms}\end{array}\right\}$ = 2204.6 pounds or 0.9842 ton or 2240 pounds

$\left.\begin{array}{l}1.016\text{ metric ton}\\1016\text{ kilograms}\end{array}\right\}$ = 1 ton or 2240 pounds

1 kilogram per sq centimeter = 14.2234 lb per sq in.
1 kilogram per sq millimeter = 1422.32 lb per sq in.

1000 lb per sq in. = $\left\{\begin{array}{l}0.70308\text{ kilograms per sq mm}\\70.308\text{ kilograms per sq cm}\end{array}\right.$

Linear measure

12 inches = 1 foot
3 feet = 1 yard = 36 inches
5½ yards = 1 rod or pole = 16½ feet
40 rods = 1 furlong = 220 yards = 660 feet = ⅛ mile
8 furlongs = 1 statute mile = 1760 yards = 5280 feet
3 miles = 1 league = 5280 yards = 15 840 feet

Square measure

144 square inches = 1 square foot
9 square feet = 1 square yard = 1296 square inches
30¼ square yards = 1 square rod = 272¼ square feet
160 square rods = 1 acre = 4840 square yards
640 acres = 1 square mile = 3 097 600 square yards

Cubic measure

1728 cubic inches = 1 cubic foot
27 cubic feet = 1 cubic yard
144 cubic inches = 1 board foot
128 cubic feet = 1 cord

Liquid measure

4 gills = 1 pint
2 pints = 1 quart = 8 gills
4 quarts = 1 gallon = 8 pints = 32 gills
31½ gallons = 1 barrel = 126 quarts
2 barrels = 1 hogshead = 63 gallons = 252 quarts

Nautical measure

6 feet = 1 fathom
100 fathoms = 1 cable's length (ordinary) = 608 ft (Br.) = 607.61 ft (U.S.)
120 fathoms = 1 cable's length (U.S. Navy)
10 cable's lengths = 1 nautical mile = 6080 ft (Br.) = 6076.1033 ft (U.S.)
1 nautical mile = 1.1508 statute miles
3 nautical miles = 1 league (marine)
60 nautical miles = 1 degree (of a terrestial great circle)

Avoirdupois weight

$27^{11}/_{32}$ grains = 1 dram
16 drams = 1 ounce = 437½ grains
16 ounces = 1 pound = 256 drams = 7000 grains
100 pounds = 1 hundredweight = 1600 ounces
20 hundredweight = 1 ton = 2000 pounds
112 pounds = 1 long hundredweight
20 long hundredweight = 1 long ton = 2240 pounds

Troy weight

24 grains = 1 pennyweight
20 pennyweights = 1 ounce = 480 grains
12 ounces = 1 pound = 240 pennyweights = 5760 grains

Apothecaries' weight

20 grains = 1 scruple
3 scruples = 1 dram = 60 grains
8 drams = 1 ounce = 24 scruples = 430 grains
12 ounces = 1 pound = 96 drams = 283 scruples = 5760 grains

Dry measure

2 pints = 1 quart
8 quarts = 1 peck = 16 pints
4 pecks = 1 bushel = 32 quarts = 64 pints
105 quarts = 1 barrel (for fruits, vegetables, and other dry commodities) = 7056 cubic inches

Circular measure

60 seconds (″) = 1 minute (′)
60 minutes = 1 degree (°)
90 degrees = 1 quadrant
4 quadrants = 1 circle of circumference

Roman numerals

1	I	8	VIII
2	II	9	IX
3	III	10	X
4	IV	50	L
5	V	100	C
6	VI	500	D
7	VII	1000	M

The chief symbols are I = 1; V = 5; X = 10; L = 50; C = 100; D = 500; and M = 1000. Note that IV = 4, means 1 short of 5; IX = 9, means 1 short of 10; XL = 40, means 10 short of 50; and XC = 90, means 10 short of 100. Any symbol following one of equal or greater value adds its value—II = 2. Any symbol preceding one of greater value subtracts its value—IV = 4. When a symbol stands between two of greater value, its value is subtracted from the second and the remainder is added to the first—XIV = 14; LIX = 59. Of two equivalent ways of representing a number, that in which the symbol of larger denomination preceded is preferred—XIV instead of VIX for 14.

Numerical data

1 cubic foot of water at 4 °C (weight)	62.43 lb
1 foot of water at 4 °C (pressure)	0.4335 lb/in.^{3}
Velocity of light in vacuum, c	186,280 mi/s = 2.998×10^{10} cm/s
Velocity of sound in dry air at 20 °C, 76 cm Hg	1127 ft/s
Degree of longitude at equator	69.173 miles
Acceleration due to gravity at sea-level, 40′ Latitude, g	32.1578 ft/s^2
$\sqrt{2g}$	8.020
Base of natural logs ε	2.718
1 radian	180° ÷ π = 57.3
360 degrees	2π radians
π	3.1416
Sine 1′	0.00029089
Arc 1°	0.01745 radian
Side of square	0.707 × (diagonal of square)

Mathematical symbols

× or ·	Multiplied by
÷ or :	Divided by
+	Positive. Plus. Add
−	Negative. Minus. Subtract
±	Positive or negative. Plus or minus
∓	Negative or positive. Minus or plus
= or ::	Equals
??	Identity
≡	Approximately equal to
≠	Not equal to
>	Greater than
>>	Much greater than
<	Less than
<<	Much less than
≥	Greater than or equal to
≤	Less than or equal to
∴	Therefore
∠	Angle
Δ	Increment. Decrement
⊥	Perpendicular to
‖	Parallel to
$|n|$	Absolute value of n
	a, b, c used for known quantities
	x, y, z used for unknown quantities

Mathematical constants

$\pi = 3.14$	$\sqrt{\pi} = 1.77$
$2\pi = 6.28$	$\sqrt{\dfrac{\pi}{2}} = 1.25$
$(2\pi)^2 = 39.5$	$\sqrt{2} = 1.41$
$4\pi = 12.6$	$\sqrt{3} = 1.73$
$\pi^2 = 9.87$	$\dfrac{1}{\sqrt{2}} = 0.707$
$\dfrac{\pi}{2} = 1.57$	$\dfrac{1}{\sqrt{3}} = 0.577$
$\dfrac{1}{\pi} = 0.318$	$\log \pi = 0.497$
$\dfrac{1}{2\pi} = 0.159$	$\log \dfrac{\pi}{2} = 0.196$
$\dfrac{1}{\pi^2} = 0.101$	$\log \pi^2 = 0.994$
$\dfrac{1}{\sqrt{\pi}} = 0.564$	$\log \sqrt{\pi} = 0.248$

Greek alphabet

Capital	Small	Name	Commonly used to designate
A	α	Alpha	Angles, coefficients, attenuation constant, absorption factor, area
B	β	Beta	Angles, coefficients, phase constant
Γ	γ	Gamma	Complex propagation constant (cap), specific gravity, angles, electrical conductivity, propagation constant
Δ	δ	Delta	Increment or decrement (cap or small), determinant (cap), permittivity (cap), density, angles
E	ε	Epsilon	Dielectric constant, permittivity, base of natural logarithms, electric intensity
Z	ζ	Zeta	Coordinates, coefficients
H	η	Eta	Intrinsic impedance, efficiency, surface charge density, hysteresis, coordinates
Θ	θ	Theta	Angular phase displacement, time constant, reluctance, angles
I	ι	Iota	Unit vector
K	κ	Kappa	Susceptibility, coupling coefficient
Λ	λ	Lambda	Permeance (cap), wavelength, attenuation constant
M	μ	Mu	Permeability, amplification factor, prefix micro
N	ν	Nu	Reluctivity, frequency
Ξ	ξ	Xi	Coordinates
O	ο	Omicron	
Π	π	Pi	3.1416
P	ρ	Rho	Resistivity, volume charge density, coordinates
Σ	σ	Sigma	Summation (cap), surface charge density, complex propagation constant, electrical conductivity, leakage coefficient
T	τ	Tau	Time constant, volume resistivity, time-phase displacement, transmission factor, density
Y	υ	Upsilon	
Φ	φ	Phi	Scalar potential (cap), magnetic flux, angles
X	χ	Chi	Electric susceptibility, angles
Ψ	ψ	Psi	Dielectric flux, phase difference, coordinates, angles
Ω	ω	Omega	Resistance in ohms (cap), solid angle (cap), angular velocity

Note: Use small letter except where capital (cap) is specified

Miscellaneous Electrical Information

Fusing currents of wires

The current I in amperes at which a wire will melt can be calculated from $I = Kd^2$ where d is the wire diameter in inches and K is a constant that depends on the metal concerned; a wide variety of factors influence the rate of heat loss and these figures must be considered as approximations

AWG B & S gage	Wire diameter (d), in.	Fusing current, A, for wire type:				
		Copper, K = 10 244	Aluminum, K = 7585	German silver, K = 5230	Iron, K = 3148	Tin K = 1642
40	0.0031	1.77	1.31	0.90	0.54	0.28
38	0.0039	2.50	1.85	1.27	0.77	0.40
36	0.0050	3.62	2.68	1.85	1.11	0.58
34	0.0063	5.12	3.79	2.61	1.57	0.82
32	0.0079	7.19	5.32	3.67	2.21	1.15
30	0.0100	10.2	7.58	5.23	3.15	1.64
28	0.0126	14.4	10.7	7.39	4.45	2.32
26	0.0159	20.5	15.2	10.5	6.31	3.29
24	0.0201	29.2	21.6	14.9	8.97	4.68
22	0.0253	41.2	30.5	21.0	12.7	6.61
20	0.0319	58.4	43.2	29.8	17.9	9.36
19	0.0359	69.7	51.6	35.5	21.4	11.2
18	0.0403	82.9	61.4	42.3	25.5	13.3
17	0.0452	98.4	72.9	50.2	30.2	15.8
16	0.0508	117.0	86.8	59.9	36.0	18.8
15	0.0571	140.0	103.0	71.4	43.0	22.4
14	0.0641	166.0	123.0	84.9	51.1	26.6
13	0.0719	197.0	146.0	101.0	60.7	31.7
12	0.0808	235.0	174.0	120.0	72.3	37.7
11	0.0907	280.0	207.0	143.0	86.0	44.9
10	0.1019	333.0	247.0	170.0	102.0	53.4
9	0.1144	396.0	298.0	202.0	122.0	63.5
8	0.1285	472.0	349.0	241.0	145.0	75.6
7	0.1443	561.0	416.0	287.0	173.0	90.0
6	0.1620	668.0	495.0	341.0	205.0	107.0

Electrical formulae

$$\text{Resistance} = \frac{\text{voltage}}{\text{ampere turns}} \times \text{turns} \left(\Omega = \frac{V}{IT} \times T\text{'s} \right)$$

$$\text{Ampere turns} = \frac{\text{voltage}}{\text{resistance}} \times \text{turns} \left(IT = \frac{V}{\Omega} \times T\text{'s} \right)$$

$$\text{Effective turns} = \frac{\text{total resistance}}{\text{resistance of inductive coil}} \times \text{turns of inductive coil}$$

$$\text{Amperes} = \frac{\text{ampere turns}}{\text{turns of inductive coil}} \left(I = \frac{IT}{\Omega} \right)$$

Ohm's law for direct current

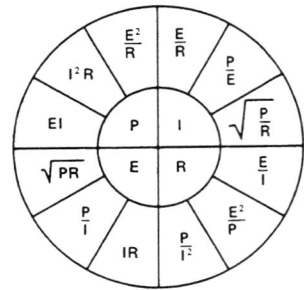

P = power in watts
I = current in amperes
E = electromotive force in volts
R = resistance in ohms

Two resistances in parallel combination:

$$Req = \frac{R_1 \quad R_2}{R_1 + R_2}$$

Any number of resistances in parallel combination:

$$\frac{1}{Req} = \frac{1}{r_1} + \frac{1}{r_2} + \cdots \frac{1}{r_n}$$

For calculating capacitances in series combinations, substitute C for R in the above formulas

Ohm's law for alternating current

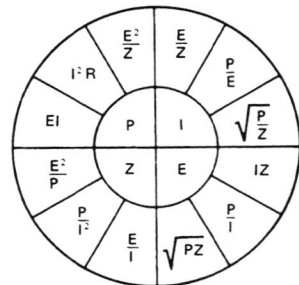

$$f = \frac{1}{2\pi\sqrt{LC}} = \frac{1}{2\pi CXc} = \frac{XL}{2\pi L}$$

$$XL = 2\pi fL$$

$$Xc = \frac{1}{2\pi fC}$$

$$L = \frac{XL}{2\pi f} = \frac{1}{(2\pi f)^2 C}$$

$$C = \frac{1}{2\pi fXc} = \frac{1}{(2\pi f)^2 L}$$

$$Z = \sqrt{R^2 + X^2} = \sqrt{R^2 + (XL - Xc)^2}$$

$Z = R$ when $XL = Xc$
Z = impedance in ohms
XL = inductive reactance in ohms
Xc = capacitive reactance in ohms
L = inductance in henrys
C = capacitance in farads
f = frequency in cycles per second
$2\pi f \approx 377$ for 60 cps

Miscellaneous Factors and Tables

Physical constants

Name and symbol	Value and units
Velocity of light, c	2.997902×10^{10} cm/s
Planck constant, h	6.62377×10^{-27} erg s/molecule
Avogadro constant, N	6.02380×10^{23} molecule mol
Faraday constant, F	96 493.1 C/equivalent
Absolute temperature of ice point, T (0 °C)	273.15 K
Pressure-volume product for 1 mol of gas at 0 °C and zero pressure (PV) $P = 0$ $\quad\quad T = 0$ °C	2271.16 J/mol
Gas constant $P = 0$ $R = \underline{(PV)T = 0 \text{ °C}}$ $\quad\quad T$ (0 °C)	8.31469 J/mol°
	1.98726 cal/mol°
Boltzmann constant $k = R/N$	1.38031×10^{16} erg/molecule°
	11.96171 Jcm/mol
Constant relating wave number and energy $Z = Nhc$	2.858917 cal cm/mole
Standard atmosphere, atm	1 013 250 dynes/cm^2
Thermochemical calorie	4.1840 J (exact)
	4.18331 J (international)

Miscellaneous conversion factors

To convert from	To	Multiply by	To convert from	To	Multiply by
Atmospheres	cm of Hg at 0 °C	76	Inches	mils	1000
Atmospheres	gm/cm^2	1033.3	Inches of Hg at 32 °C	atmospheres	0.033421
Atmospheres	inches of Hg at 32 °F	29.921	Inches of Hg at 32 °C	feet of water at 39.1 °F	1.13299
Atmospheres	lb/in.^{2}	14.696	Inches of water at 39.2 °F	inches of Hg	0.073554
Btu	calories (gram)	252	Kilograms	ounces (avoir.)	35.274
Calories, gram	Btu	3.968×10^{-3}	Kilograms	pounds (avoir.)	2.2046
Centimeters	angstrom units	1×10^{8}	Liters	cubic feet	0.035316
Centimeters	feet	0.032808	Liters	gallons (U.S.)	0.2642
Cubic cm	cubic inches	0.061023	Liters	ounces (U.S. fluid)	33.8143
Cubic cm	gallons (U.S.)	2.6417×10^{-4}	Liters	pints (U.S. liquid)	2.11336
Cubic cm	ounces (U.S. fluid)	0.033814	Liters	quarts (U.S. liquid)	1.05668
Cubic cm	pints (U.S. fluid)	0.0021134	Meters	angstrom units	1×10^{10}
Cubic ft	cubic cm	28317	Meters	feet (U.S.)	3.28083
Cubic ft	cubic meters	0.02832	Meters	inches (U.S.)	39.3700
Cubic ft	gallons (U.S.)	7.481	Microns	angstrom units	1×10^{4}
Cubic ft	liters	28.316	Microns	inches	3.937×10^{-5}
Cubic in. (U.S.)	cubic cm	16.3872	Microns	millimeters	0.001
Cubic in. (U.S.)	liters	0.016387	Microns	mils	0.03937
Cubic yd (U.S.)	cubic meters	0.7646	Millimeters	inches (U.S.)	0.03937
Cubic yd of sand	pounds	2700	Millimeters	microns	1000
Feet (U.S.)	centimeters	30.48	Millimeters	mils	39.37
Feet (U.S.)	meters	0.3048	Ounces (avoir.)	grams	28.3495
Gallons (U.S.)	cubic centimeters	3785.4	Ounces (U.S. fluid)	cubic cm	29.5737
Gallons (U.S.)	cubic feet	0.13368	Ounces (U.S. fluid)	cubic in.	1.8047
Gallons (U.S.)	cubic inches	231	Ounces (U.S. fluid)	liters	0.02957
Gallons (U.S.)	liters	3.7854	Pints (U.S. liquid)	cubic cm	473.179
Grams	ounces (avoir.)	0.03527	Pints (U.S. liquid)	liters	0.473168
Grams	pounds (avoir.)	0.002205	Pounds (avoir.)	grams	453.5924
Horsepower	Btu (mean)/min	42.418	Square cm	square in.	0.1550
Horsepower	calories, log (mean)/min	10.688	Square in. (U.S.)	square cm	6.5416
Inches (U.S.)	angstrom units	2.5400×10^{8}	Years (leap)	hours	8784

Load conversion table, tsi to psi

tsi	psi	tsi	psi	tsi	psi	tsi	psi	tsi	psi	tsi	psi
10.0	22 400	22.5	50 400	35.0	78 400	47.5	106 400	70	156 800	95	212 800
10.5	23 520	23.0	51 520	35.5	79 520	48.0	107 520	71	159 040	96	215 040
11.0	24 640	23.5	52 640	36.0	80 640	48.5	108 640	72	161 280	97	217 280
11.5	25 760	24.0	53 760	36.5	81 760	49.0	109 760	73	163 520	98	219 520
12.0	26 880	24.5	54 880	37.0	82 880	49.5	110 880	74	165 760	99	221 760
12.5	28 000	25.0	56 000	37.5	84 000	50	112 000	75	168 000	100	224 000
13.0	29 120	25.5	57 120	38.0	85 120	51	114 240	76	170 240	101	226 240
13.5	30 240	26.0	58 240	38.5	86 240	52	116 480	77	172 480	102	228 480
14.0	31 360	26.5	59 360	39.0	87 360	53	118 720	78	174 720	103	230 720
14.5	32 480	27.0	60 480	39.5	88 480	54	120 960	79	176 960	104	232 960
15.0	33 600	27.5	61 600	40.0	89 600	55	123 200	80	179 200	105	235 200
15.5	34 720	28.0	62 720	40.5	90 720	56	125 440	81	181 440	106	237 440
16.0	35 840	28.5	63 840	41.0	91 840	57	127 680	82	183 680	107	239 680
16.5	36 960	29.0	64 960	41.5	92 960	58	129 920	83	185 920	108	241 920
17.0	38 080	29.5	66 080	42.0	94 080	59	132 160	84	188 160	109	244 160
17.5	39 200	30.0	67 200	42.5	95 200	60	134 400	85	190 400	110	246 400
18.0	40 320	30.5	68 320	43.0	96 320	61	136 640	86	192 640	111	248 640
18.5	41 440	31.0	69 440	43.5	97 440	62	138 880	87	194 880	112	250 880
19.0	42 560	31.5	70 560	44.0	98 560	63	141 120	88	197 120	113	253 120
19.5	43 680	32.0	71 680	44.5	99 680	64	143 360	89	199 360	114	255 360
20.0	44 800	32.5	72 800	45.0	100 800	65	145 600	90	201 600	115	257 600
20.5	45 920	33.0	73 920	45.5	101 920	66	147 840	91	203 840	116	259 840
21.0	47 040	33.5	75 040	46.0	103 040	67	150 080	92	206 080	117	262 080
21.5	48 160	34.0	76 160	46.5	104 160	68	152 320	93	208 320	118	264 320
22.0	49 280	34.5	77 280	47.0	105 280	69	154 560	94	210 560	119	266 560

Load conversion table, kg/mm² to psi

kg/mm²	psi	kg/mm²	psi	kg/mm²	psi	kg/mm²	psi	kg/mm²	psi
10	14 223	40	56 894	70	99 564	100	142 234	130	184 904
11	15 646	41	58 316	71	100 986	101	143 656	131	186 327
12	17 068	42	59 738	72	102 408	102	145 079	132	187 749
13	18 490	43	61 161	73	103 831	103	146 501	133	189 171
14	19 913	44	62 583	74	105 253	104	147 923	134	190 594
15	21 335	45	64 005	75	106 675	105	149 346	135	192 016
16	22 757	46	65 428	76	108 098	106	150 768	136	193 438
17	24 180	47	66 850	77	109 520	107	152 190	137	194 861
18	25 602	48	68 272	78	110 943	108	153 613	138	196 283
19	27 024	49	69 695	79	112 365	109	155 035	139	197 705
20	28 447	50	71 117	80	113 787	110	156 457	140	199 128
21	29 869	51	72 539	81	115 210	111	157 880	141	200 550
22	31 291	52	73 962	82	116 632	112	159 302	142	201 972
23	32 714	53	75 384	83	118 054	113	160 724	143	203 395
24	34 136	54	76 806	84	119 477	114	162 147	144	204 817
25	35 558	55	78 229	85	120 899	115	163 569	145	206 239
26	36 981	56	79 651	86	122 321	116	164 991	146	207 662
27	38 403	57	81 073	87	123 744	117	166 414	147	209 084
28	39 826	58	82 496	88	125 166	118	167 836	148	210 506
29	41 248	59	83 918	89	126 588	119	169 258	149	211 929
30	42 670	60	85 340	90	128 011	120	170 681	150	213 351
31	44 093	61	86 763	91	129 433	121	172 103	151	214 773
32	45 515	62	88 185	92	130 855	122	173 525	152	216 196
33	46 937	63	89 607	93	132 278	123	174 948	153	217 618
34	48 360	64	91 030	94	133 700	124	176 370	154	219 040
35	49 782	65	92 452	95	135 122	125	177 792	155	220 463
36	51 204	66	93 874	96	136 545	126	179 215	156	221 885
37	52 627	67	95 297	97	137 967	127	180 637	157	223 307
38	54 049	68	96 719	98	139 389	128	182 059	158	224 730
39	55 471	69	98 141	99	140 812	129	183 482	159	226 152

Approximate hourly production

Time to make one piece, s	Gross production per hour, pieces	Gross time per 1000 pieces, h	Time to make one piece, s	Gross production per hour, pieces	Gross time per 1000 pieces, h	Time to make one piece, s	Gross production per hour, pieces	Gross time per 1000 pieces, h
0.5	7200	0.14	25	144	6.95	50	72	13.9
1	3600	0.28	26	138	7.22	52	69	14.5
2	1800	0.55	27	133	7.50	54	66	15.0
3	1200	0.83	28	128	7.78	56	64	15.6
4	900	1.11	29	124	8.06	58	62	16.1
5	720	1.39	30	120	8.33	60	60	16.7
6	600	1.67	31	116	8.62	62	58	17.2
7	514	1.94	32	112	8.90	64	56	17.8
8	450	2.22	33	109	9.17	66	54	18.4
9	400	2.50	34	106	9.45	68	53	18.9
10	360	2.78	35	103	9.73	70	51	19.5
11	327	3.05	36	100	10.00	72	50	20.0
12	300	3.33	37	97	10.30	74	49	20.6
13	276	3.62	38	95	10.56	76	47	21.1
14	257	3.89	39	92	10.83	78	46	21.7
15	240	4.17	40	90	11.11	80	45	22.2
16	225	4.44	41	88	11.39	82	44	22.8
17	212	4.72	42	86	11.67	84	43	23.3
18	200	5.00	43	84	11.94	86	42	23.9
19	189	5.28	44	82	12.22	88	41	24.5
20	180	5.56	45	80	12.50	90	40	25.0
21	171	5.83	46	78	12.78	92	39	25.5
22	164	6.12	47	77	13.05	94	38	26.1
23	156	6.40	48	75	13.34	96	37	26.7
24	150	6.67	49	73	13.61	100	36	27.8

Decimal and metric equivalents of fractions of an inch

Fraction of an inch	Equivalents in.	Equivalents mm	Fraction of an inch	Equivalents in.	Equivalents mm	Fraction of an inch	Equivalents in.	Equivalents mm
1/64	0.015625	0.39687	23/64	0.359375	9.12801	45/64	0.703125	17.85915
1/32	0.03125	0.79374	3/8	0.375	9.52491	23/32	0.71875	18.25608
3/64	0.046875	1.19061	25/64	0.390625	9.92175	47/64	0.734375	18.65289
1/16	0.0625	1.58748	13/32	0.40625	10.31865	3/4	0.75	19.04982
5/64	0.078125	1.98435	27/64	0.421875	10.71549	49/64	0.765625	19.44663
3/32	0.09375	2.38123	7/16	0.4375	11.11240	25/32	0.78125	19.84356
7/64	0.109375	2.77809	29/64	0.453125	11.50923	51/64	0.796875	20.24037
1/8	0.125	3.17497	15/32	0.46875	11.90614	13/16	0.8125	20.63731
9/64	0.140625	3.57183	31/64	0.484375	12.30297	53/64	0.828125	21.03411
5/32	0.15625	3.96871	1/2	0.5	12.69988	27/32	0.84375	21.43105
11/64	0.171875	4.36557	33/64	0.515625	13.09671	55/64	0.859275	21.82785
3/16	0.1875	4.76245	17/32	0.53125	13.49362	7/8	0.875	22.22479
13/64	0.203125	5.15931	35/64	0.546875	13.89045	57/64	0.890625	22.62159
7/32	0.21875	5.55620	9/16	0.5625	14.28737	29/32	0.90625	23.01853
15/64	0.234375	5.95305	37/64	0.578125	14.68419	59/64	0.921875	23.41533
1/4	0.25	6.34994	19/32	0.59375	15.08111	15/16	0.9375	23.81228
17/64	0.265625	6.74679	39/64	0.609375	15.47793	61/64	0.953125	24.20907
9/32	0.28125	7.14368	5/8	0.625	15.87485	31/32	0.96875	24.60602
19/64	0.296875	7.54053	41/64	0.640625	16.27167	63/64	0.984375	25.00281
5/16	0.3125	7.93743	21/32	0.65625	16.66859	1	1.0	25.4
21/64	0.328125	8.33427	43/64	0.671875	17.06541			
11/32	0.34375	8.73117	11/16	0.6875	17.46234			

Comparison of standard gages(a)

Gage No.	Thickness, in., for manufacturer:					
	A(a)	B(b)	C(c)	D(d)	E(e)	F(f)
0000000	...	...	0.4900	0.5000	0.500	...
000000	...	0.580000	0.4615	0.4687	0.464	...
00000	...	0.516500	0.4305	0.4375	0.432	...
0000	0.454	0.460000	0.3938	0.4062	0.400	...
000	0.425	0.409642	0.3625	0.3750	0.372	...
00	0.380	0.364796	0.3310	0.3437	0.348	...
0	0.340	0.324861	0.3065	0.3125	0.324	...
1	0.300	0.289297	0.2830	0.2812	0.300	...
2	0.284	0.257627	0.2625	0.2656	0.276	...
3	0.259	0.229423	0.2437	0.2500	0.252	0.2391
4	0.238	0.204307	0.2253	0.2344	0.232	0.2242
5	0.220	0.181940	0.2070	0.2187	0.212	0.2092
6	0.203	0.162023	0.1920	0.2031	0.192	0.1943
7	0.180	0.144285	0.1770	0.1875	0.176	0.1793
8	0.165	0.128490	0.1620	0.1719	0.160	0.1644
9	0.148	0.114423	0.1483	0.1562	0.144	0.1495
10	0.134	0.101897	0.1350	0.1406	0.128	0.1345
11	0.120	0.090742	0.1205	0.1250	0.116	0.1196
12	0.109	0.080808	0.1055	0.1094	0.104	0.1046
13	0.095	0.071962	0.0915	0.0937	0.092	0.0897
14	0.083	0.064084	0.0800	0.0781	0.080	0.0747
15	0.072	0.057068	0.0720	0.0703	0.072	0.0673
16	0.065	0.050821	0.0625	0.0625	0.064	0.0598
17	0.058	0.045257	0.0540	0.0562	0.056	0.0538
18	0.049	0.040303	0.0475	0.0500	0.048	0.0478
19	0.042	0.035890	0.0410	0.0437	0.040	0.0418
20	0.035	0.031961	0.0348	0.0375	0.036	0.0359
21	0.032	0.028462	0.03175	0.0344	0.032	0.0329
22	0.028	0.025346	0.0286	0.0312	0.028	0.0299
23	0.025	0.022572	0.0258	0.0281	0.024	0.0269
24	0.022	0.020101	0.0230	0.0250	0.022	0.0239
25	0.020	0.017900	0.0204	0.0219	0.020	0.0209
26	0.018	0.015941	0.0181	0.0187	0.018	0.0179
27	0.016	0.014195	0.0173	0.0172	0.0164	0.0164
28	0.014	0.012641	0.0162	0.0156	0.0148	0.0149
29	0.013	0.011257	0.0150	0.0141	0.0136	0.0135
30	0.012	0.010025	0.0140	0.0125	0.0124	0.0120
31	0.010	0.008928	0.0132	0.0109	0.0116	0.0105
32	0.009	0.007950	0.0128	0.0102	0.0108	0.0097
33	0.008	0.007080	0.0118	0.0094	0.0100	0.0090
34	0.007	0.006305	0.0104	0.0086	0.0092	0.0082
35	0.005	0.005615	0.0095	0.0078	0.0084	0.0075
36	0.004	0.005000	0.0090	0.0070	0.0076	0.0067
37	...	0.004453	0.0085	0.0066	0.0068	0.0064
38	...	0.003965	0.0080	0.0062	0.0060	0.0060
39	...	0.003531	0.0075	...	0.0052	...
40	...	0.003144	0.0070	...	0.0048	...

(a) Birmingham Wire (BWG) and Stubs' Iron Wire. Birmingham Wire gages used principally for strips, bands, hoops, and wire. (b) American Wire (AWG) and Brown and Sharpe; gages used principally for nonferrous sheets, rod, and wire. (c) U.S. Steel Wire; American Steel and Wire; Washburn and Moen; and Steel Wire. U.S. Steel Wire gages used principally for steel wire, except music wire. (d) U.S. Standard (old); gages used principally for stainless steel sheets. (e) British Imperial Standard Wire (SWG); gages used principally for English legal standard wire gage. (f) Manufacturers' Standard; gages used principally for uncoated steel sheets

Lengths, Areas, Volumes, Weights

Mensuration: Lengths, Areas, Volumes

In the figures and equations that follow, **a, b, c, d, s** denote lengths, **A** denotes area, **V** denotes volume.

Right triangle

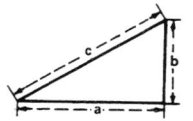

$A = \frac{1}{2}\,ab$ $a = \sqrt{c^2 - b^2}$
$c = \sqrt{a^2 + b^2}$ $b = \sqrt{c^2 - a^2}$

Equilateral triangle

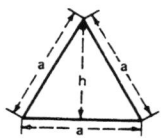

$A = \frac{1}{2}\,ah = \frac{1}{4}\,a^2\sqrt{3}$
$h = \frac{1}{2}\,a\sqrt{3}$

Square

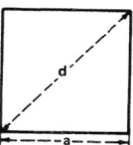

$A = a^2$
$d = a\sqrt{2}$

Oblique triangle

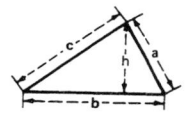

$A = \frac{1}{2}\,bh$

Rectangle

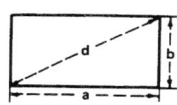

$A = ab$
$d = \sqrt{a^2 + b^2}$

Trapezoid

One pair of opposite sides parallel

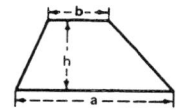

$A = \frac{1}{2}\,h\,(a + b)$

Parallelogram

Opposite sides parallel

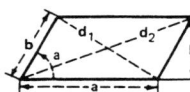

$A = ah = ab \sin \alpha$

$d_1 = \sqrt{a^2 + b^2 - 2\,ab \cos \alpha}$

$d_2 = \sqrt{a^2 + b^2 + 2\,ab \cos \alpha}$

Isosceles Trapezoid

Nonparallel sides equal

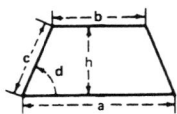

$A = \frac{1}{2}\,h\,(a + b) =$

$\frac{1}{2}\,c \sin \alpha\,(a + b) =$

$c \sin \alpha\,(a - c \cos \alpha) =$

$c \sin \alpha\,(b + c \cos \alpha)$

Cube

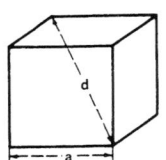

$V = a^3$

$d = a\sqrt{3}$

Total surface $= 6\,a^2$

Ellipsoid

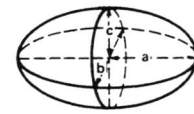

$V = \frac{4}{3}\,\pi abc$

Circle

$C = $ circumference

$\alpha = $ central angle in radians

$C = \pi D = 2\,\pi R$

$c = R\alpha = \frac{1}{2}\,D\alpha = D \cos^{-1} \frac{d}{R} = D \tan^{-1} \frac{l}{2\,d}$

$l = 2\sqrt{R^2 - d^2} = 2\,R \sin \frac{\alpha}{2} = 2\,d \tan \frac{\alpha}{2} = 2\,d \tan \frac{c}{D}$

$d = \frac{1}{2}\sqrt{4\,R^2 - l^2} = \frac{1}{2}\sqrt{D^2 - l^2} = R \cos \frac{\alpha}{2} = \frac{1}{2}\,l \cot \frac{\alpha}{2} = \frac{1}{2}\,l \cot \frac{c}{D}$

$h = R - d$

$\alpha = \frac{c}{R} = \frac{2\,c}{D} = 2 \cos^{-1} \frac{d}{R} = 2 \tan^{-1} \frac{l}{2\,d} = 2 \sin^{-1} \frac{l}{D}$

$A_{(circle)} = \pi R^2 = \frac{1}{4}\,\pi D^2 = \frac{1}{2}\,RC = \frac{1}{4}\,DC$

$A_{(sector)} = \frac{1}{2}\,Rc = \frac{1}{2}\,R^2\alpha = \frac{1}{8}\,D^2\alpha$

$A_{(segment)} = A_{(sector)} - A_{(triangle)} = \frac{1}{2}\,R^2(\alpha - \sin \alpha) = \frac{1}{2}\,R\left(c - R \sin \frac{c}{R}\right)$

$\quad = R^2 \sin^{-1} \frac{l}{2\,R} - \frac{1}{4}\,l\sqrt{4\,R^2 - l^2} = R^2 \cos^{-1} \frac{d}{R} - d\sqrt{R^2 - d^2}$

$\quad = R^2 \cos^{-1} \frac{R - h}{R} - (R - h)\sqrt{2\,Rh - h^2}$

Prism or cylinder

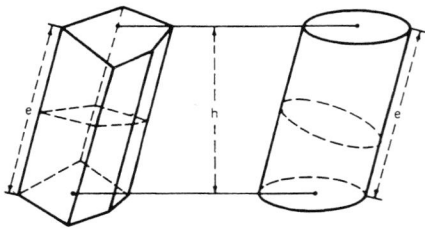

$V = $ area of base $\times$ altitude

Lateral area $= $ perimeter of right section $\times$ lateral edge

Rectangular parallelopiped

$V = abc$

$d = \sqrt{a^2 + b^2 + c^2}$

Total surface $= 2(ab + bc + ca)$

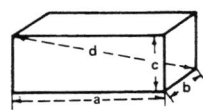

Regular polygon of n sides

All sides equal

All angles equal

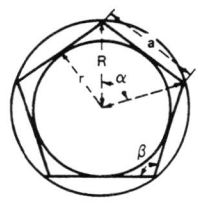

$$\beta = \frac{n-2}{n} 180° = \frac{n-2}{n} \pi \text{ radians}$$

$$\alpha = \frac{360°}{n} = \frac{2\pi}{n} \text{ radians}$$

n	a	r	R	A	
3	$2r\sqrt{3} = R\sqrt{3}$	$\frac{1}{6}a\sqrt{3}$	$\frac{1}{3}a\sqrt{3}$	$\frac{1}{4}a^2\sqrt{3}$	$= 3r^2\sqrt{3}$
					$= \frac{3}{4}R^2\sqrt{3}$
4	$2r = R\sqrt{2}$	$\frac{1}{2}a$	$\frac{1}{2}a\sqrt{2}$	a^2	$= 4r^2 = 2R^2$
6	$\frac{2}{3}r\sqrt{3} = R$	$\frac{1}{2}a\sqrt{3}$	a	$\frac{3}{2}a^2\sqrt{3}$	$= 2r^2\sqrt{3}$
					$= \frac{3}{2}R^2\sqrt{3}$
8	$2r(\sqrt{2}-1)$	$\frac{1}{2}a(\sqrt{2}+1)$	$\frac{1}{2}a\sqrt{4+2\sqrt{2}}$	$2a^2(\sqrt{2}+1)$	$= 8r^2(\sqrt{2}-1)$
	$= R\sqrt{2-\sqrt{2}}$				$= 2R^2\sqrt{2}$
n	$2r\tan\frac{\alpha}{2}$	$\frac{a}{2}\cot\frac{\alpha}{2}$	$\frac{a}{2}\csc\frac{\alpha}{2}$	$\frac{na^2}{4}\cot\frac{\alpha}{2}$	$= nr^2\tan\frac{\alpha}{2}$
	$= 2R\sin\frac{\alpha}{2}$				$= \frac{nR^2}{2}\sin\alpha$

Area by approximation

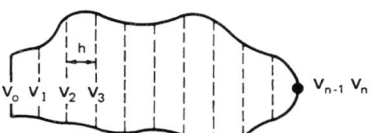

Let $y_0, y_1, y_2, \ldots, y_n$ be the measured lengths of a series of equidistant parallel chords, and let **h** be their distance apart, then the area enclosed by any boundary is given approximately by one of the following rules.

Trapezoidal rule:

$$A_T = h[\tfrac{1}{2}(y_0 + y_n) + y_1 + y_2 + \cdots + y_{n-1}]$$

Durand's rule:

$$A_D = h[0.4(y_0 + y_n) + 1.1(y_1 + y_{n-1}) + y_2 + y_3 + \cdots + y_{n-2}]$$

Simpson's rule:

$$A_s = \tfrac{1}{3}h[(y_0 + y_n) + 4(y_1 + y_3 + \cdots + y_{n-1}) + 2(y_2 + y_4 + \cdots + y_{n-2})]$$

where n is even

The larger the value of n, the greater is the accuracy of approximation. In general, for the same number of chords, A_s gives the most accurate, A_T, the least accurate approximation.

Trapezium

No sides parallel

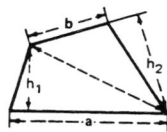

$$A = \tfrac{1}{2}(ah_1 + bh_2) =$$
sum of areas of two triangles

Cycloid

r = radius of generating circle

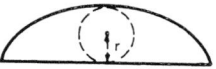

$$A = 3\pi r^2$$
Length of arc (s) = 8r

Catenary

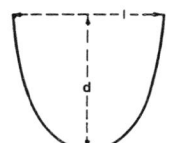

Length of arc (s) =
$$l\left[1 + \tfrac{2}{3}\left(\frac{2d}{l}\right)^2\right] \text{approx}$$
if **d** is small in comparison with **l**

Torus

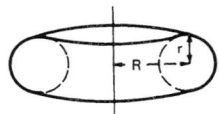

$$V = 2\pi^2 R r^2$$
Surface (S) = $4\pi^2 R r$

Ellipse

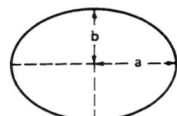

$A = \pi ab$

Perimeter $(s) =$

$$\pi(a + b)\left[1 + \tfrac{1}{4}\left(\frac{a - b}{a + b}\right)^2 + \frac{1}{64}\left(\frac{a - b}{a + b}\right)^4 + \frac{1}{256}\left(\frac{a - b}{a + b}\right)^6 + \cdots\right].$$

$$\cong \pi \ \frac{a + b}{4}\left[3(1 + \lambda) + \frac{1}{1 - \lambda}\right] \qquad \lambda = \left[\frac{a - b}{2(a + b)}\right]^2$$

Pyramid or cone

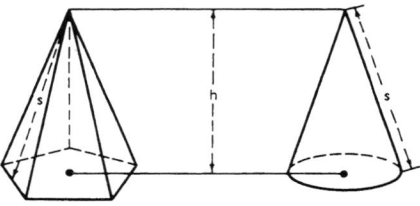

$V = \tfrac{1}{3}$ (area of base) $\times$ (altitude)
Lateral area of regular figure =
$\tfrac{1}{2}$ (perimeter of base) $\times$ (slant height)

Frustum of pyramid or cone

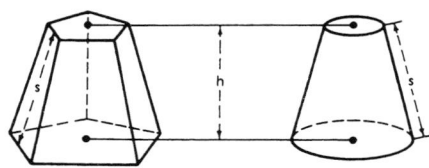

$$V = \tfrac{1}{3}\left(A_1 + A_2 + \sqrt{A_1 \times A_2}\right) h$$

where A_1 and A_2 are areas of bases, and h is altitude.
Lateral area of regular figure =
$\tfrac{1}{2}$ (sum of perimeters of bases) $\times$ (slant height)

Prismatoid

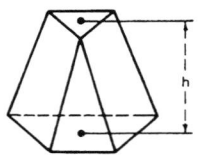

Bases are in parallel planes, lateral faces are triangles or trapezoids

$$V = \tfrac{1}{6}\left(A_1 + A_2 + 4 A_m\right) h$$

where A_1, A_2 are areas of bases, A_m is area of midsection, and h is altitude

Sphere

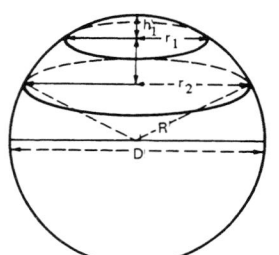

$A_{(sphere)} = 4 \pi R^2 = \pi D^2$

$A_{(zone)} = 2 \pi Rh = \pi Dh$

$V_{(sphere)} = \tfrac{4}{5} \pi R^3 = \tfrac{1}{6} \pi D^3$

$V_{(spherical\ sector)} = \tfrac{2}{3} \pi R^2 h = \tfrac{1}{6} \pi D^2 h$

$V_{(spherical\ segment\ of\ one\ base)} =$
$\tfrac{1}{6} \pi h_1 (3 r_1^2 + h_1^2) = \tfrac{1}{3} \pi h_1^2 (3 R - h_1)$

$V_{(spherical\ segment\ of\ two\ bases)} =$
$\tfrac{1}{6} \pi h (3 r_1^2 + 3 r_2^2 + h^2)$

Parabola

$$A = \tfrac{2}{3}\,\mathbf{l}\mathbf{d}$$

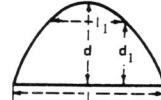

Length of arc ($\mathbf{s}$) =

$$\tfrac{1}{2}\sqrt{16\,\mathbf{d}^2 + \mathbf{l}^2} + \frac{\mathbf{l}^2}{8\,\mathbf{d}}\,\ln\left(\frac{4\,\mathbf{d} + \sqrt{16\,\mathbf{d}^2 + \mathbf{l}^2}}{\mathbf{l}}\right)$$

$$= \mathbf{l}\left[1 + \tfrac{2}{3}\left(\frac{2\,\mathbf{d}}{\mathbf{l}}\right)^2 - \tfrac{2}{5}\left(\frac{2\,\mathbf{d}}{\mathbf{l}}\right)^4 + \cdots\right]$$

Height of segment ($\mathbf{d}_1$) = $\dfrac{\mathbf{d}}{\mathbf{l}^2}(\mathbf{l}^2 - \mathbf{l}_1^2)$.

Width of segment ($\mathbf{l}_1$) = $\mathbf{l}\sqrt{\dfrac{\mathbf{d} - \mathbf{d}_1}{\mathbf{d}}}$

Solid (V) or surface (S) of revolution

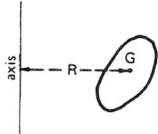

Generated by revolving any plane area ($\mathbf{A}$) or arc ($\mathbf{s}$) about an axis in its plane, and not crossing the area or arc

$$\mathbf{V} = 2\,\pi\mathbf{R}\mathbf{A}; \quad \mathbf{S} = 2\,\pi\mathbf{R}\mathbf{s}$$

where $\mathbf{R}$ = distance of center of gravity ($\mathbf{G}$) of area or arc from axis

Solid angle

At any point (P) subtended by any surface (S), the solid angle (Ψ) is equal to the portion (A) of the surface of a sphere of unit radius which is cut out by a conical surface with vertex at P and the perimeter of S for base. The unit solid angle (Ψ) is called a steradian. The total solid angle about a point = $4\,\pi$ steradians

Paraboloidal segment

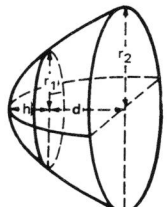

$$\mathbf{V}_{(\text{segment of one base})} = \tfrac{1}{2}\,\pi\mathbf{r}_1^2\mathbf{h}$$

$$\mathbf{V}_{(\text{segment of two bases})} = \tfrac{1}{2}\,\pi\mathbf{d}\,(\mathbf{r}_1^2 + \mathbf{r}_2^2)$$

Weight Formulas and Conversions

Weight formulas

Steel weights are based on 0.2833 lb/in.³. Aluminum weights are based on 0.0979 lb/in.³, which applies to 1100 alloy.

Rounds

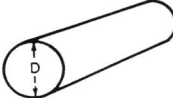

Steel:
Pounds per lineal foot = 2.67036 × D²
Aluminum:
Pounds per lineal foot = 0.9227 × D²
D = size in inches

Squares

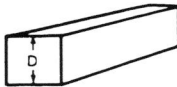

Steel:
Pounds per lineal foot = 3.4 × D²

Aluminum:
Pounds per lineal foot = 1.1748 × D²
D = size in inches

Hexagons

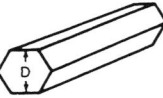

Steel:
Pounds per lineal foot = 2.9446 × D²

Aluminum:
Pounds per lineal foot = 1.0175 × D²
D = size in inches

Tubing

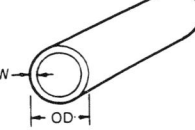

Steel:
Pounds per lineal foot =
10.68 × (OD − W) × W
Aluminum:
Pounds per lineal foot =
3.6904 × (OD − W) × W
OD = outside diameter
to 3 decimal places
W = wall thickness
to 3 decimal places

Flats

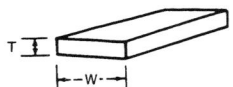

Steel:
Pounds per lineal foot = 3.4 × T × W
Aluminum:
Pounds per lineal foot = 1.1748 × T × W
T = thickness in inches
W = width in inches

Octagons

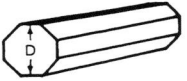

Steel:
Pounds per lineal foot = 2.8166 × D²

Aluminum:
Pounds per lineal foot = 0.9733 × D²
D = size in inches

Circles

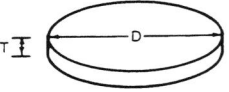

Steel:
Weight of circle in pounds =
0.22253 × T × D²
Aluminum:
Weight of circle in pounds = 0.0769 × T × D²
D = diameter in inches
T = thickness in inches

Rings

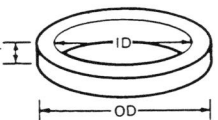

Steel:
Weight of ring in pounds =
0.22253 × T × (OD² − ID²)
Aluminum:
Weight of ring in pounds =
0.07690 × T × (OD² − ID²)
OD = outside diameter in inches
ID = inside diameter in inches
T = thickness in inches

Source: Earl M. Jorgensen Co.

Weight conversion factors

To obtain weight of	Density (lb/in.^{3})	Multiply weight of steel by	To obtain weight of	Density (lb/in.^{3})	Multiply weight of steel by
Aluminum	0.098	0.3462	Lead	0.410	1.448
1100 Aluminum	0.098	0.3462	Silver	0.379	1.339
2011 Aluminum	0.102	0.3604	Molybdenum	0.369	1.303
2014 Aluminum	0.101	0.3568	Copper	0.324	1.144
2017 Aluminum	0.101	0.3568	Nickel	0.322	1.137
2024 Aluminum	0.100	0.3533	Niobium	0.310	1.095
3003 Aluminum	0.099	0.3498	Brass	0.307	1.084
5005 Aluminum	0.098	0.3462	Monel	0.307	1.084
5052 Aluminum	0.097	0.3427	**Stainless steels**		
5056 Aluminum	0.095	0.3356	300 series	0.286	1.010
5083 Aluminum	0.096	0.3392	400 series	0.283	1.000
5086 Aluminum	0.096	0.3392	Carbon, alloy steels	0.283	1.000
6061 Aluminum	0.098	0.3462	Tin	0.264	0.932
6063 Aluminum	0.098	0.3462	Zinc	0.258	0.911
7075 Aluminum	0.101	0.3568	Cast iron	0.258	0.911
7178 Aluminum	0.102	0.3604	Zirconium	0.230	0.812
Gold	0.698	2.466	Titanium	0.163	0.575
Tungsten	0.697	2.462	Beryllium	0.067	0.236
Tantalum	0.600	2.120	Magnesium	0.065	0.229

Theoretical weights of carbon-steel bars(a)

Thickness or diameter, in.	Round lb/in.	Round lb/ft	Square lb/in.	Square lb/ft	Hexagon lb/in.	Hexagon lb/ft
1/32	0.0002	0.0026	0.0003	0.0033	0.0002	0.0028
1/16	0.0009	0.0104	0.0011	0.0133	0.0010	0.0115
3/32	0.0020	0.0235	0.0025	0.0299	0.0022	0.0259
1/8	0.0035	0.0417	0.0044	0.0531	0.0038	0.0460
5/32	0.0054	0.0652	0.0069	0.0830	0.0060	0.0719
3/16	0.0078	0.0939	0.0100	0.1195	0.0086	0.1035
7/32	0.0106	0.1278	0.0136	0.1627	0.0117	0.1409
1/4	0.0139	0.1669	0.0177	0.2125	0.0153	0.1840
9/32	0.0176	0.2112	0.0224	0.2689	0.0194	0.2329
5/16	0.0217	0.2608	0.0277	0.3320	0.0240	0.2875
11/32	0.0263	0.3155	0.0335	0.4018	0.0290	0.3479
3/8	0.0313	0.3755	0.0398	0.4781	0.0345	0.4141
13/32	0.0367	0.4407	0.0468	0.5611	0.0405	0.4860
7/16	0.0426	0.5111	0.0542	0.6508	0.0470	0.5636
15/32	0.0489	0.5867	0.0623	0.7471	0.0538	0.6470
1/2	0.0556	0.6676	0.0708	0.8500	0.0613	0.7361
17/32	0.0628	0.7536	0.0800	0.9596	0.0693	0.8310
9/16	0.0704	0.8449	0.0896	1.076	0.0776	0.9317
19/32	0.0785	0.9414	0.0999	1.199	0.0865	1.038
5/8	0.0869	1.043	0.1107	1.328	0.0958	1.150
21/32	0.0958	1.150	0.1220	1.464	0.1057	1.268
11/16	0.1052	1.262	0.1339	1.607	0.1160	1.392
23/32	0.1150	1.380	0.1464	1.756	0.1268	1.521
3/4	0.1252	1.502	0.1594	1.913	0.1380	1.656
25/32	0.1358	1.630	0.1729	2.075	0.1498	1.797
13/16	0.1469	1.763	0.1870	2.245	0.1620	1.944
27/32	0.1584	1.901	0.2017	2.421	0.1747	2.096
7/8	0.1704	2.044	0.2169	2.603	0.1879	2.254
29/32	0.1828	2.193	0.2327	2.792	0.2015	2.418
15/16	0.1956	2.347	0.2490	2.988	0.2157	2.588
31/32	0.2088	2.506	0.2659	3.191	0.2303	2.763
1	0.2225	2.670	0.2833	3.400	0.2454	2.944
1-1/16	0.2512	3.015	0.3199	3.838	0.2770	3.324
1-1/8	0.2816	3.380	0.3586	4.303	0.3106	3.727
1-3/16	0.3138	3.766	0.3995	4.795	0.3460	4.152
1-1/4	0.3477	4.172	0.4427	5.313	0.3834	4.601
1-5/16	0.3833	4.600	0.4881	5.857	0.4227	5.072
1-3/8	0.4207	5.049	0.5357	6.428	0.4639	5.567

(a) Theoretical weight per cubic inch = 0.2833 lb. Theoretical weight per cubic foot = 489.6 lb. For flats, to determine the theoretical weight in pounds per linear foot, multiply the width in inches times the thickness in inches times 3.4. Source: American Iron and Steel Institute

(continued)

Theoretical weights of carbon-steel bars(a) (continued)

Thickness or diameter, in.	Round		Square		Hexagon	
	lb/in.	lb/ft	lb/in.	lb/ft	lb/in.	lb/ft
1-7/16	0.4598	5.518	0.5855	7.026	0.5070	6.085
1-1/2	0.5007	6.008	0.6375	7.650	0.5521	6.625
1-9/16	0.5433	6.519	0.6917	8.301	0.5991	7.189
1-5/8	0.5876	7.051	0.7482	8.978	0.6479	7.775
1-11/16	0.6337	7.604	0.8068	9.682	0.6988	8.385
1-3/4	0.6815	8.178	0.8677	10.41	0.7515	9.018
1-13/16	0.7310	8.773	0.9308	11.17	0.8060	9.67
1-7/8	0.7823	9.388	0.9961	11.95	0.8626	10.35
1-15/16	0.8354	10.02	1.064	12.76	0.9211	11.05
2	0.8901	10.68	1.133	13.60	0.9815	11.78
2-1/16	0.9466	11.36	1.205	14.46	1.044	12.53
2-1/8	1.055	12.06	1.279	15.35	1.108	13.30
2-3/16	1.065	12.78	1.356	16.27	1.174	14.09
2-1/4	1.127	13.52	1.434	17.21	1.242	14.91
2-5/16	1.190	14.28	1.515	18.18	1.312	15.75
2-3/8	1.255	15.06	1.598	19.18	1.384	16.61
2-7/16	1.322	15.87	1.683	20.20	1.458	17.49
2-1/2	1.391	16.69	1.771	21.25	1.534	18.40
2-5/8	1.533	18.40	1.952	23.43	1.691	20.29
2-3/4	1.683	20.19	2.143	25.71	1.856	22.27
2-7/8	1.839	22.07	2.342	28.10	2.028	24.34
3	2.003	24.03	2.550	30.60	2.208	26.50
3-1/8	2.173	26.08	2.767	33.20	2.396	28.75
3-1/4	2.350	28.21	2.993	35.91	2.592	31.10
3-3/8	2.535	30.42	3.227	38.73	2.795	33.54
3-1/2	2.726	32.71	3.471	41.65	3.006	36.07
3-5/8	2.924	35.09	3.723	44.68	3.224	38.69
3-3/4	3.129	37.55	3.984	47.81	3.451	41.41
3-7/8	3.341	40.10	4.254	51.05	3.684	44.21
4	3.560	42.73	4.533	54.40	3.926	47.11
4-1/8	3.786	45.44	4.821	57.85	4.175	50.10
4-1/4	4.019	48.23	5.118	61.41	4.432	53.18
4-3/8	4.259	51.11	5.423	65.08	4.700	56.36
4-1/2	4.506	54.07	5.738	68.85	4.970	59.63
4-5/8	4.760	57.12	6.061	72.73	5.248	62.98
4-3/4	5.021	60.25	6.393	76.71	5.536	66.44
4-7/8	5.289	63.46	6.734	80.80	5.831	69.98
5	5.563	66.76	7.083	85.00	6.134	73.61
5-1/8	5.845	70.14	7.442	89.30	6.445	77.34
5-1/4	6.133	73.60	7.809	93.71	6.763	81.16
5-3/8	6.429	77.15	8.186	98.23	7.089	85.07
5-1/2	6.732	80.78	8.571	102.85	7.422	89.07
5-5/8	7.041	84.49	8.965	107.58	7.763	93.16
5-3/4	7.357	88.29	9.368	112.41	8.112	97.35
5-7/8	7.681	92.17	9.779	117.35	8.470	101.63
6	8.011	96.13	10.200	122.40	8.833	106.00

(a) Theoretical weight per cubic inch = 0.2833 lb. Theoretical weight per cubic foot = 489.6 lb. For flats, to determine the theoretical weight in pounds per linear foot, multiply the width in inches times the thickness in inches times 3.4. Source: American Iron and Steel Institute

Diameters, weights and thicknesses of large-diameter steel pipe

Outside diameter in.	Weight		Thickness		Outside diameter in.	Weight		Thickness	
	lb/ft	kg/m	in.	mm		lb/ft	kg/m	in.	mm
12¾	25.22	37.57	0.188	4.78		78.93	117.57	0.312	7.92
	27.20	40.51	0.203	5.16		86.91	129.45	0.344	8.74
	28.23	42.05	0.210	5.33		94.62	140.94	0.375	9.52
	29.31	43.66	0.219	5.56		102.31	152.39	0.406	10.31
	33.38	49.72	0.250	6.35		110.22	164.17	0.438	11.13
	37.42	55.74	0.281	7.14		117.86	175.55	0.469	11.91
	41.45	61.74	0.312	7.92		125.49	186.92	0.500	12.70
	45.58	67.89	0.344	8.74		140.68	209.54	0.562	14.27
14	27.73	41.30	0.188	4.78		156.03	232.41	0.625	15.88
	29.91	44.55	0.203	5.16		171.29	255.14	0.688	17.48
	30.93	46.07	0.210	5.33		186.23	277.39	0.750	19.05
	32.23	48.01	0.219	5.56	26	68.75	102.40	0.250	6.35
	36.71	54.68	0.250	6.35		77.18	114.96	0.281	7.14
	41.17	61.32	0.281	7.14		85.60	127.50	0.312	7.92
	45.61	67.94	0.312	7.92		94.26	140.40	0.344	8.74
	50.17	74.73	0.344	8.74		102.63	152.87	0.375	9.52
16	31.35	47.29	0.188	4.78		110.98	165.30	0.406	10.31
	34.25	51.02	0.203	5.16		119.57	178.10	0.438	11.13
	35.38	52.70	0.210	5.33		127.88	190.48	0.469	11.91
	36.91	54.98	0.219	5.56		136.17	202.83	0.500	12.70
	42.05	62.63	0.250	6.35		152.68	227.42	0.562	14.27
	47.17	70.26	0.281	7.14		169.38	252.29	0.625	15.88
	52.27	77.86	0.312	7.92		185.99	277.03	0.688	17.48
	57.52	85.68	0.344	8.74		202.25	301.25	0.750	19.05
	62.58	93.21	0.375	9.52	28	74.09	110.36	0.250	6.35
	67.62	100.72	0.406	10.31		83.19	123.91	0.281	7.14
18	35.76	53.26	0.188	4.78		92.26	137.42	0.312	7.92
	38.55	57.42	0.203	5.16		101.61	151.35	0.344	8.74
	39.86	59.38	0.210	5.33		110.64	164.80	0.375	9.52
	41.59	61.95	0.219	5.56		119.65	178.22	0.406	10.31
	47.39	70.59	0.250	6.35		128.93	192.04	0.438	11.13
	53.18	79.21	0.281	7.14		137.90	205.40	0.469	11.91
	58.94	87.79	0.312	7.92		146.85	218.73	0.500	12.70
	64.87	96.62	0.344	8.74		164.69	245.31	0.562	14.27
	70.59	105.14	0.375	9.52		182.73	272.18	0.625	15.88
	76.29	113.63	0.406	10.31		200.68	298.91	0.688	17.48
	82.15	122.36	0.438	11.13		218.27	325.11	0.750	19.05
	87.81	130.79	0.469	11.91		235.78	351.19	0.812	20.62
	93.45	139.19	0.500	12.70	30	79.43	118.31	0.250	6.35
	104.67	155.91	0.562	14.27		89.19	132.85	0.281	7.14
	115.98	172.75	0.625	15.88		98.93	147.36	0.312	7.92
20	46.27	68.92	0.219	5.56		108.95	162.28	0.344	8.74
	52.73	78.54	0.250	6.35		118.65	176.73	0.375	9.52
	59.18	88.15	0.281	7.14		128.32	191.13	0.406	10.31
	65.60	97.71	0.312	7.92		138.29	205.98	0.438	11.13
	72.21	107.56	0.344	8.74		147.92	220.33	0.469	11.91
	78.60	117.07	0.375	9.52		157.53	234.64	0.500	12.70
	84.96	126.55	0.406	10.31		176.69	263.18	0.562	14.27
	91.51	136.30	0.438	11.13		196.08	292.06	0.625	15.88
	97.83	145.72	0.469	11.91		215.38	320.81	0.688	17.48
	104.13	155.10	0.500	12.70		234.29	348.97	0.750	19.05
	116.67	173.78	0.562	14.27		253.12	377.02	0.812	20.62
	129.33	192.64	0.625	15.88		272.17	405.40	0.875	22.22
22	50.94	75.88	0.219	5.56		291.14	433.65	0.938	23.83
	58.07	86.50	0.250	6.35		309.72	461.33	1.000	25.40
	65.18	97.09	0.281	7.14	32	84.77	126.26	0.250	6.35
	72.27	107.65	0.312	7.92		95.19	141.79	0.281	7.14
	79.56	118.50	0.344	8.74		105.59	157.28	0.312	7.92
	86.61	129.01	0.375	9.52		116.30	173.23	0.344	8.74
	93.63	139.46	0.406	10.31		126.66	188.66	0.375	9.52
	100.86	150.23	0.438	11.13		136.99	204.05	0.406	10.31
	107.85	160.64	0.469	11.91		147.64	219.91	0.438	11.13
	114.81	171.01	0.500	12.70		157.94	235.25	0.469	11.91
	128.67	191.65	0.562	14.27		168.21	250.55	0.500	12.70
	142.68	212.52	0.625	15.88		188.70	281.07	0.562	14.27
	156.60	233.26	0.688	17.48		209.43	311.95	0.625	15.88
24	63.41	94.45	0.250	6.35		230.08	342.70	0.688	17.48
	71.18	106.02	0.281	7.14		250.31	372.84	0.750	19.05

Source: Italsider (IRI-Finsider Group)

(continued)

Diameters, weights and thicknesses of large-diameter steel pipe (continued)

Outside diameter in.	Weight lb/ft	kg/m	Thickness in.	mm	Outside diameter in.	Weight lb/ft	kg/m	Thickness in.	mm
	270.47	402.87	0.812	20.62	42	153.04	227.95	0.344	8.74
	290.86	433.24	0.875	22.22		166.71	248.31	0.375	9.52
	311.17	463.49	0.938	23.83		180.35	268.63	0.406	10.31
	331.08	493.14	1.000	25.40		194.42	289.59	0.438	11.13
34	90.11	134.22	0.250	6.35		208.03	309.86	0.469	11.91
	101.19	150.72	0.281	7.14		221.61	330.09	0.500	12.70
	112.25	167.20	0.312	7.92		248.72	370.47	0.562	14.27
	123.65	184.18	0.344	8.74		276.18	411.37	0.625	15.88
	134.67	200.59	0.375	9.52		303.55	452.14	0.688	17.48
	145.67	216.98	0.406	10.31		330.41	492.15	0.750	19.05
	157.00	233.85	0.438	11.13		357.19	532.03	0.812	20.62
	167.95	250.16	0.469	11.91		384.31	572.43	0.875	22.22
	178.89	266.46	0.500	12.70		411.35	612.71	0.938	23.83
	200.70	298.94	0.562	14.27		437.88	652.22	1.000	25.40
	222.78	331.83	0.625	15.88	44	160.39	238.90	0.344	8.74
	244.77	364.58	0.688	17.48		174.72	260.25	0.375	9.52
	266.33	396.70	0.750	19.05		189.03	281.56	0.406	10.31
	287.81	428.69	0.812	20.62		203.78	303.53	0.438	11.13
	309.55	461.07	0.875	22.22		218.04	324.77	0.469	11.91
	331.21	493.34	0.938	23.83		232.29	346.00	0.500	12.70
	357.44	524.96	1.000	25.40		260.72	388.34	0.562	14.27
36	95.45	142.17	0.250	6.35		289.53	430.87	0.625	15.88
	107.20	159.67	0.281	7.14		318.25	474.03	0.688	17.48
	118.92	177.13	0.312	7.92		346.43	516.01	0.750	19.05
	131.00	195.12	0.344	8.74		374.53	557.86	0.812	20.62
	142.68	212.52	0.375	9.52		403.00	600.27	0.875	22.22
	154.34	229.89	0.406	10.31		431.39	642.56	0.938	23.83
	166.35	247.78	0.438	11.13		459.24	684.04	1.000	25.40
	177.97	265.09	0.469	11.91	46	167.74	249.85	0.344	8.74
	189.57	282.36	0.500	12.70		182.63	272.18	0.375	9.52
	212.70	316.82	0.562	14.27		197.70	294.47	0.406	10.31
	236.13	351.72	0.625	15.88		213.13	317.46	0.438	11.13
	259.47	396.48	0.688	17.48		228.06	339.70	0.469	11.91
	282.35	420.56	0.750	19.05		242.97	361.90	0.500	12.70
	305.16	454.54	0.812	20.62		272.73	406.23	0.562	14.27
	328.24	488.91	0.875	22.22		302.88	451.14	0.625	15.88
	351.25	523.19	0.938	23.83		332.95	495.93	0.688	17.48
	373.80	556.78	1.000	25.40		362.45	539.87	0.750	19.05
38	125.58	187.05	0.312	7.92		391.88	583.71	0.812	20.62
	138.35	206.07	0.344	8.74		421.69	628.11	0.875	22.22
	150.69	224.45	0.375	9.52		451.42	671.79	0.938	23.83
	163.01	242.80	0.406	10.31		480.60	715.85	1.000	25.40
	175.71	261.72	0.438	11.13	48	175.08	260.78	0.344	8.74
	187.99	280.01	0.469	11.91		190.74	284.11	0.375	9.52
	200.25	298.27	0.500	12.70		206.37	307.39	0.406	10.31
	224.71	324.71	0.562	14.27		222.49	331.40	0.438	11.13
	249.48	371.60	0.625	15.88		238.08	354.62	0.469	11.91
	274.16	408.36	0.688	17.48		253.65	377.81	0.500	12.70
	298.37	444.42	0.750	19.05		284.73	424.11	0.562	14.27
	322.50	480.36	0.812	20.62		316.23	471.02	0.625	15.88
	346.93	516.75	0.875	22.22		347.64	517.81	0.688	17.48
	371.28	553.02	0.938	23.83		378.47	563.73	0.750	19.05
	395.16	588.59	1.000	25.40		409.22	609.53	0.812	20.62
40	132.25	196.99	0.312	7.92		440.38	655.95	0.875	22.22
	145.69	217.01	0.344	8.74		471.46	702.24	0.938	23.83
	158.70	236.38	0.375	9.52		501.96	747.67	1.000	25.40
	171.68	255.72	0.406	10.31	50	198.56	295.76	0.375	9.52
	185.06	275.65	0.438	11.13		214.85	320.02	0.406	10.31
	198.01	294.94	0.469	11.91		231.63	345.02	0.438	11.13
	210.93	314.18	0.500	12.70		246.53	367.21	0.469	11.91
	236.71	352.58	0.562	14.27		264.09	393.37	0.500	12.70
	262.83	391.49	0.625	15.88		296.46	441.58	0.562	14.27
	288.86	430.26	0.688	17.48		329.27	490.46	0.625	15.88
	314.39	468.28	0.750	19.05		362.00	539.21	0.688	17.48
	339.84	506.19	0.812	20.62		394.14	587.07	0.750	19.05
	365.62	544.59	0.875	22.22		426.17	634.79	0.812	20.62
	391.32	582.87	0.938	23.83		458.65	683.17	0.875	22.22
	416.52	620.41	1.000	25.40		491.05	731.42	0.938	23.83

Source: Italsider (IRI-Finsider Group)

(continued)

Diameters, weights and thicknesses of large-diameter steel pipe (continued)

Outside diameter in.	Weight lb/ft	kg/m	Thickness in.	mm
	522.85	778.78	1.000	25.40
52	206.57	307.69	0.375	9.52
	223.52	332.93	0.406	10.31
	240.98	358.95	0.438	11.13
	257.88	384.11	0.469	11.91
	274.76	409.26	0.500	12.70
	308.46	459.45	0.562	14.27
	342.70	510.44	0.625	15.88
	376.70	561.09	0.688	17.48
	410.14	610.90	0.750	19.05
	443.51	660.61	0.812	20.62
	477.39	711.08	0.875	22.22
	511.06	761.23	0.938	23.83
	544.18	810.56	1.000	25.40
54	232.31	346.03	0.406	10.31
	250.46	373.07	0.438	11.13
	267.88	399.02	0.469	11.91
	285.43	425.15	0.500	12.70
	320.44	477.30	0.562	14.27
	355.95	530.20	0.625	15.88
	391.44	583.05	0.688	17.48
	426.15	634.75	0.750	19.05
	460.82	686.40	0.812	20.62
	496.19	739.07	0.875	22.22
	531.08	791.05	0.938	23.83
	565.52	842.35	1.000	25.40
56	240.84	358.73	0.406	10.31
	259.60	386.80	0.438	11.13
	277.95	414.02	0.469	11.91
	296.09	441.04	0.500	12.70
	333.03	496.06	0.562	14.27
	369.29	550.06	0.625	15.88
	406.18	605.02	0.688	17.48
	442.14	658.58	0.750	19.05
	478.16	712.23	0.812	20.62
	514.70	766.60	0.875	22.22
	551.09	820.86	0.938	23.83
	586.86	874.14	1.000	25.40
58	249.50	371.63	0.406	10.31
	268.59	400.07	0.438	11.13
	287.90	428.83	0.469	11.91
	306.83	457.03	0.500	12.70
	344.43	513.04	0.562	14.27
	382.73	570.08	0.625	15.88
	420.73	626.69	0.688	17.48
	458.16	682.43	0.750	19.05
	495.49	738.04	0.812	20.62
	533.35	794.42	0.875	22.22
	570.07	850.07	0.938	23.83
	608.20	905.92	1.000	25.40
60	258.17	384.54	0.406	10.31
	278.36	414.63	0.438	11.13
	297.91	443.74	0.469	11.91
	317.44	472.83	0.500	12.70
	356.49	531.00	0.562	14.27
	396.10	590.00	0.625	15.88
	435.41	648.55	0.688	17.48
	474.15	706.26	0.750	19.05
	512.95	764.05	0.812	20.62
	552.02	822.24	0.875	22.22
	591.14	880.51	0.938	23.83
	629.55	937.72	1.000	25.40
62	266.83	397.45	0.406	10.31
	287.71	428.55	0.438	11.13
	307.92	458.65	0.469	11.91
	328.31	489.02	0.500	12.70
	368.62	549.07	0.562	14.27
	409.30	609.66	0.625	15.88

Outside diameter in.	Weight lb/ft	kg/m	Thickness in.	mm
	450.09	670.42	0.688	17.48
	490.16	730.10	0.750	19.05
	530.15	789.66	0.812	20.62
	570.69	850.05	0.875	22.22
	611.15	910.32	0.938	23.83
	650.88	969.50	1.000	25.40
64	338.77	504.61	0.500	12.70
	380.68	567.03	0.562	14.27
	422.64	629.53	0.625	15.88
	464.78	692.30	0.688	17.48
	506.23	754.04	0.750	19.05
	547.47	815.47	0.812	20.62
	589.49	878.06	0.875	22.22
	631.17	940.13	0.938	23.83
	672.22	1001.28	1.000	25.40
66	349.45	520.51	0.500	12.70
	392.42	584.51	0.562	14.27
	436.41	650.04	0.625	15.88
	479.46	714.16	0.688	17.48
	522.18	777.75	0.750	19.05
	564.80	841.28	0.812	20.62
	608.03	905.67	0.875	22.22
	651.26	970.05	0.938	23.83
	693.56	1033.07	1.000	25.40
68	404.40	602.36	0.562	14.27
	449.85	670.06	0.625	15.88
	494.14	736.03	0.688	17.48
	538.18	801.62	0.750	19.05
	582.14	867.10	0.812	20.62
	626.72	933.50	0.875	22.22
	671.20	999.76	0.938	23.83
	715.04	1065.06	1.000	25.40
70	416.40	620.23	0.562	14.27
	463.26	690.03	0.625	15.88
	508.89	758.00	0.688	17.48
	554.18	825.46	0.750	19.05
	599.53	893.00	0.812	20.62
	645.38	961.30	0.875	22.22
	691.21	1029.57	0.938	23.83
	736.52	1097.05	1.000	25.40
72	428.39	638.09	0.562	14.27
	475.99	709.00	0.625	15.88
	523.66	790.00	0.688	17.48
	570.19	849.30	0.750	19.05
	616.79	918.71	0.812	20.62
	664.06	989.12	0.875	22.22
	711.23	1059.39	0.938	23.83
	757.58	1128.43	1.000	25.40
74	440.41	656.00	0.562	14.27
	489.42	729.00	0.625	15.88
	538.19	801.64	0.688	17.48
	586.19	873.14	0.750	19.05
	634.11	944.52	0.812	20.62
	682.77	1017.00	0.875	22.22
	731.25	1089.21	0.938	23.83
	778.93	1160.22	1.000	25.40
76	452.51	674.02	0.562	14.27
	502.66	748.72	0.625	15.88
	552.87	823.51	0.688	17.48
	602.21	897.00	0.750	19.05
	651.44	970.33	0.812	20.62
	701.60	1045.04	0.875	22.22
	751.27	1119.02	0.938	23.83
	800.26	1192.00	1.000	25.40
78	464.63	692.08	0.562	14.27
	516.01	768.60	0.625	15.88
	567.55	845.38	0.688	17.48
	618.20	920.82	0.750	19.05

Source: Italsider (IRI-Finsider Group)

(continued)

Diameters, weights and thicknesses of large-diameter steel pipe (continued)

Outside diameter in.	Weight		Thickness		Outside diameter in.	Weight		Thickness	
	lb/ft	kg/m	in.	mm		lb/ft	kg/m	in.	mm
	668.78	996.15	0.812	20.62		655.65	976.60	0.688	17.48
	720.08	1072.56	0.875	22.22		714.37	1064.06	0.750	19.05
	771.39	1149.00	0.938	23.83		772.75	1151.01	0.812	20.62
	821.78	1224.05	1.000	25.40		832.12	1239.44	0.875	22.22
80	476.36	709.54	0.562	14.27		891.60	1328.03	0.938	23.83
	529.34	788.46	0.625	15.88		949.32	1414.02	1.000	25.40
	575.52	857.25	0.688	17.48	92	548.32	816.73	0.562	14.27
	634.48	945.06	0.750	19.05		609.37	907.66	0.625	15.88
	686.17	1022.06	0.812	20.62		670.33	998.47	0.688	17.48
	738.75	1100.37	0.875	22.22		730.24	1087.70	0.750	19.05
	791.54	1179.01	0.938	23.83		790.22	1177.03	0.812	20.62
	843.29	1256.08	1.000	25.40		850.78	1267.25	0.875	22.22
82	488.35	727.41	0.562	14.27		911.41	1357.55	0.938	23.83
	542.68	808.33	0.625	15.88		970.99	1446.30	1.000	25.40
	596.92	889.12	0.688	17.48	94	560.32	834.60	0.562	14.27
	650.21	968.50	0.750	19.05		622.71	927.53	0.625	15.88
	703.63	1048.06	0.812	20.62		685.01	1020.33	0.688	17.48
	757.31	1128.02	0.875	22.22		746.25	1111.54	0.750	19.05
	811.32	1208.47	0.938	23.83		807.67	1203.03	0.812	20.62
	864.29	1287.36	1.000	25.40		869.46	1295.06	0.875	22.22
84	500.35	745.27	0.562	14.27		931.43	1387.37	0.938	23.83
	556.01	828.19	0.625	15.88		992.34	1478.09	1.000	25.40
	611.61	911.00	0.688	17.48	96	572.31	852.46	0.562	14.27
	666.23	992.35	0.750	19.05		636.05	947.40	0.625	15.88
	720.77	1073.60	0.812	20.62		699.70	1042.20	0.688	17.48
	776.09	1156.00	0.875	22.22		762.25	1135.38	0.750	19.05
	831.34	1238.29	0.938	23.83		824.73	1228.44	0.812	20.62
	885.63	1319.15	1.000	25.40		888.27	1323.08	0.875	22.22
86	512.35	763.14	0.562	14.27		951.44	1417.18	0.938	23.83
	569.35	848.06	0.625	15.88		1013.81	1510.07	1.000	25.40
	626.42	933.06	0.688	17.48	98	584.30	870.32	0.562	14.27
	682.24	1016.20	0.750	19.05		649.38	967.26	0.625	15.88
	738.10	1099.40	0.812	20.62		714.38	1064.07	0.688	17.48
	794.91	1184.02	0.875	22.22		778.26	1159.22	0.750	19.05
	851.36	1268.10	0.938	23.83		842.06	1254.25	0.812	20.62
	907.01	1351.00	1.000	25.40		906.81	1350.70	0.875	22.22
88	524.30	781.00	0.562	14.27		971.46	1447.00	0.938	23.83
	582.76	868.03	0.625	15.88		1035.28	1542.06	1.000	25.40
	640.97	954.73	0.688	17.48	100	596.30	888.19	0.562	14.27
	698.24	1040.03	0.750	19.05		662.72	987.13	0.625	15.88
	755.42	1125.20	0.812	20.62		729.13	1086.04	0.688	17.48
	813.71	1212.03	0.875	22.22		794.26	1183.06	0.750	19.05
	871.43	1298.00	0.938	23.83		859.39	1280.06	0.812	20.62
	928.31	1382.72	1.000	25.40		925.47	1378.50	0.875	22.22
90	536.46	799.07	0.562	14.27		991.62	1477.02	0.938	23.83
	596.03	887.80	0.625	15.88		1056.36	1573.45	1.000	25.40

Source: Italsider (IRI-Finsider Group)

Standards Organizations

Argentina

Instituto Argentino de Recionalizacion de Materiales (IRAM)
Chile 1192
- 1098 - Buenos Aires
Argentina

Tel: 54 1 383 3751
Fax: 54 1 383 8463

IRAM is the Argentine standardization institute and the only organization in Argentina authorized by the government to deal, both in national and international ambits, with all the affairs related to standardization and quality control certification.

IRAM is a private society, and a member of ISO (International Organization for Standardization) and COPANT (Pan American Standards Commission).

Australia

Standards Australia (SAA)
Standards House
80-86 Arthur Street
P.O. Box 458
North Sydney, NSW 2059
Australia

Tel: 61 2 963 41 11
Fax: 61 2 959 38 96
Telex: 26514 astan

The Standards Association of Australia (SAA) issues standards used primarily by firms doing business in Australia and the southwest Pacific area. Australian standards appear as 1- to 4-digit numerical codes preceded by the upper case letters AS. Designations may also appear with the standard, and these should be separated by a space.

Example: AS 1446; AS 1565 80 A; 1867 1050

Austria

Osterreichisches Normungeinstitut (ON)
Heinestrasse 38, Postlach 130
A-1021, Wien
Austria

Tel: 43 1 26 75 35
Fax: 43 1 26 75 52
Telex: 11 59 60 onorm a

The Austrian Standards Institute (ON), founded in 1920, creates and publishes Austrian standards. The organization also recommends foreign standards for use in Austria.

Example: ONORM M3430; ONORM M3429; ONORM M3421 AI99.98

Belgium

Institut Belge de Normalisation (IBN)
Avenue de la Brabanconne 29
B-1040 Bruxelles
Belgium

Tel: 32 2 734 92 05
Fax: 32 2 733 42 64
Telex: 2 38 77 benor b

Created in 1946, the IBN consists of approximately 600 members, both individuals and businesses. The designations are prefixed with the letters NBN. Belgian designations are different for nonalloyed and alloyed steel. For nonalloyed steel (including carbon steel), the conventional designation usually consists of a letter, a number code, and a possible variable third part. There are three different criteria for classification: mechanical characteristics, technological characteristics, and chemical composition.

For alloyed steels, the designation system varies for heavily alloyed (above 5%) and slightly alloyed steel.

Example: NBN D 02-002

Brazil

Associacao Brasileira de Normas Tecnicas (ABNT)
Av. 13 de Maio, 13-27o andar
Caixa Postal 1680
CEP 20003-Rio de Janeiro-RJ
Brazil

Tel: 55 21 210 31 22
Fax: 55 21 532 21 43
        55 21 240 82 49
Telex: 213 43 33 ABNT BR

The Brazilian Association of Technical Standards (ABNT) issues national standards. These designations now begin with upper case letters NBR and are followed by a 4-digit numerical code. Projects are coded as Committee: Subcommittee. Working Group- Sequential number.

Example: NBR 5000. Other examples are ABNT 1040; NB 82/79

Bulgaria

Committee for Standardization and Metrology at the Council of Ministers
21 6th September Street
1000 Sofia
Bulgaria

Tel: 359 2 85 91
Fax: 359 2 80 1402
Telex: 2 25 70 dks bg

Bulgarian standards are issued by the Committee for Standardization and Metrology. These designations begin with the upper case letters BDS and are followed by the standard's numerical code.

Example: BDS 7938; BDS 6751

Canada

Canadian Standards Association (CSA)
178 Rexdale Blvd.
Rexdale, Ontario M9W 1R3
Canada

Tel: (416) 747-4000
Fax: (416) 747-4149

The Canadian Standards Association (CSA) is a membership organization that issues standards used primarily in Canada but also in commerce between Canada and the U.S. The Association provides more than 1500 standards in 8 major programs.

All Canadian standards are preceded by the upper case letters CSA. The standard or designation then follows.

Examples: CSA GR20; CSA SG121; CSA S5; CSA GH.1.7.3; CSA HA.4.1100

China

China State Bureau of Technical Supervision
P.O. Box 8010
Beijing
Peoples Republic of China

Tel: 86 1 2025835
Fax: 86 1 2031010
Telex: 22 29 28 csbts cn

Czechoslovakia

Federalni urad pro normalizaci a mereni (CSN)
Vaclavske Namesti 19
113 47 Praha 1
Czechoslovakia

Tel: 42 2 2365706
Fax: 42 2 2365706
Telex: 12 1 948 FUNM C

The Czechoslovakian Office for Standards and Measurements (CSN) is a government agency concerned with standardization, metrology, testing, certification, and accreditation. The CSN was founded in 1952 and is a member of the ISO and IEC, an affiliate of CEN and CENELEC.

Czechoslovakian standards are arranged according to classes and subgroups by a six-digit reference number. All standards are preceded by CSN.

Example: CSN 01 0010

Denmark

Dansk Standard (DS)
Bauegardsvej 73
DK-2900-Hellerup
Denmark

Tel: 45 39 77 01 01
Fax: 45 39 77 02 02
Telex: 11 92 03 DS STAND

The Danish Standards Association (DS) was founded in 1926 and is involved in the standardization of all fields except tele-communications.

DS is accredited to certify quality assurance systems according to ISO 9000 series. The organization is composed of 110 members.

Example: DS/EN 10025, DS/ISO 3798, DS/IEC 141-1, DS 13080-1.

Egypt

Egyptian Organization for Standardization and Quality Control (EOS)
2 Latin America Street
Garden City
Cairo
Egypt

Tel: 20 2 354 97 20
Telex: 9 32 96 eos un

The Egyptian Organization for Standardization was established in 1957 to be responsible for elaborating standard specifica-tions for raw materials, products, technical operations, apparatuses, machines, measurement units, terminology, definitions, unified symbols classification. In 1979 the organization name was changed to include Quality Control.

Europe

European Committee for Iron and Steel Standardization (ECISS)
C/O European Committee for Standardization (Comite Europeen de Normalisation) (CEN)
36 rue de Stassart
B-1050 Brussels
Belgium

Tel: 32 2 5196811
Fax: 32 2 5196819
Telex: 172210097

The European Committee for Iron and Steel Standardization (ECISS) issues standards for European steel. This effort is being carried out in close cooperation with the European Committee for Standardization (CEN) which produces European Stan-dards (EN).

Europe

Association Europeenne des Constructeurs de Materiel Aerospatial (AECMA)
88 Bd Malesherbes
F-75008 Paris
France

Tel: 33 1 4563 82 85
Fax: 33 1 4225 15 48
Telex: 642701 F AECMA

AECMA. Within the European Association of Manufacturers of Aerospace Material (AECMA) standardization is carried out by the Standardization Committee (CN). The subcommittees responsible for this work are represented by the aerospace industry, the processing industries, public bodies and authorities, and commerce and science. The number of members is not limited. AECMA standards and designation begin with the prefix AECMA. A standard's numerical code is preceded with the lower and upper case letters prEN. Designations are alphanumeric.

Example: AECMA prEN2002-03; AECMA prEN2389; AECMA Co-P 92-HT; AECMA A1-P13 PI-T3

Europe

Commission of the European Communities (CEC or CCE)
2100 M Street NW
7th Floor
Washington DC 20037
USA

Tel: 202 862 9500
Fax: 202 429 1766
Telex: 64215 EURCOM UW

Distribution of Euronorm standards is now handled by the national standards institutions of the Member States of the European Community. Standards issued are prefaced by the letters EURONORM. These are followed by a numerical code which simply numbers the standards chronologically followed by an indication of the year when issued or last updated.

Example: EURONORM 137-87

Finland

Finnish Electrotechnical Standards Assoc. (SESKO)
P.O. Box 134
SF-00211 Helsinki
Finland

Tel: 358 0 696 31
Fax: 358 0 677 059
Telex: 122877

The Finnish Electrotechnical Standards Association (SESKO) is a private organization established by common agreement in 1965. SESKO is a member of the Finnish Standards Association (SFS) which is a central coordinating body, also private by constitution.

The SESKO is composed of 22 member bodies representing professional associations and governmental institutions.

The SESKO forms the Finnish National Committee of the International Electrotechnical Commission (IEC), the European Committee for Electrotechnical Standardization (CENELEC), the CENELEC Electronic Components Committee (CECC) and the Nordic standardization co-operation (NOREK).

Finland

Suomen Standardisoimisliitto r.y. (SFS)
Finnish Standards Association
Maistraatin portti 2
SF-00240, Helsinki
Finland

Tel: 358 0 1499 331

The Finnish Standards Association (SFS) consists of approximately 37 organizations. It was founded in 1924 and develops about 300 standards per year. Finnish standards and designations were preceded by the letters SFS.

Most of the Finnish national steel standards will disappear. As a member of the European Standards Committee CEN all European steel standards will be implemented.

France

Delegation Generale pour L'Armement (AIR)
Centre de Documentation de l'Armement
26, Boulevard Victor
00460 - Armees
France

Tel: 33 1 4552 45 24
Fax: 33 1 4552 45 74

The French Ministry of Defense issues AIR standards. The prefix AIR in upper case letters appears with these designations.

Example: AIR 9165-001; AIR 9165-211

France

Association Francaise de Normalisation (AFNOR)
Tour Europe - Cedex 7
92049 Paris La Defense
France

Tel: 33 1 42 91 55 55
Fax: 33 1 42 91 56 56
Telex: 611974 AFNOR F

The Association Francaise de Normalisation is a non-profit organization founded in 1926. Of its nearly 15,600 standards, more than 1,000 relate to metallurgy and are used widely in Europe, Africa, Asia, the Middle East, and the Caribbean. AFNOR standards usually begin with the letters NF.

Examples: NF A 35-550 for steel grade XC 38 (special unalloyed steel for heat treatment). NF A 35-557 for steel grade 35 N CD 16 (a steel in which no alloying element exceeds a proportion of 5% by weight). NF F 80-107 for steel grade Z 120 M 12 (a steel in which the carbon concentration lies between 1.05 and 1.35% and the manganese concentration lies between 11 and 14%). NF A 32-058 for cast steel grade Z 120 M 12 M.

Germany

Deutsches Institut fur Normung e.V. (DIN)
Burggrafenstrasse 6-10, Postfach 1107
D-1007 Berlin 30
Germany

Tel: 49 30 26 01 1
Fax: 49 30 260 12 31
Telex: 184 273 din d

The German Institute for Normalization (DIN) standards are developed by a non-profit organization of approximately 130 standards committees with representatives from all technical circles. More than 20,000 standards have been created. Membership is voluntary and open to both German and foreign companies.

All German standards are preceded by the upper case letters DIN and followed by a numerical or alphanumerical code. An upper case letter sometimes precedes this code. German designations are reported in one of two methods. One method uses a descriptive code number with chemical symbols and numbers in the designation; the second, known as the Werkstoff number, uses numbers only with a decimal point after the first digit. (The latter method was devised to be more compatible with computerization.)

With one exception, German standards for steel and cast-iron are preceded by the prefix DIN. Standards for heat-resisting steels are prefixed with the letters SEW (Stahl-Eisen-Werkstoffblatter, Steel-Iron Material Sheets).

Examples: DIN E17440 X5CrNi1810 and DIN 17442 G-X20CrMo13 (standard and designation); DIN 17745 1.4120 (standard and Werkstoff number).

Hungary

Magyar Szabvarnyugyi Hivatal (MSZH)
Postafiok 24
1450 Budapest 9
Hungary

Tel: 36 1 118 30 11
Fax: 36 1 118 51 25
Telex: 225723 norm h

Hungarian standards are developed by the Hungarian Office for Standardization (MSZH) which was founded in 1921. The agency is also a member of the ISO.

Traditional steel grade designations have been adopted from other standards for Hungarian use. In some cases new designation methods have been developed, but for cast steel alloys Germany's DIN designation system has been adopted.

Example: MSZ 1300; MSZ NI 499; MSZ 5744; MSZ KGST 483

India

Bureau of Indian Standards (BIS)
Manak Bhavan
6 Bahadur Shah Zafar Marg
New Delhi 110002
India

Tel: 91 11 331 79 91
Fax: 91 11 331 40 62
Telex: 316 58 70 bis in

The Bureau of Indian Standards (BIS) is responsible for issuing national standards. Indian standards begin with the prefix IS and are followed by a numerical code.

Example: IS:3930; IS:5517

Indonesia

Dewan Standardisasi Nasional (DSN)
(Standardization Council of Indonesia)
Sasana Widya Sarwono Lantai 5
Jl. Jenderal Gatot Subroto No. 10
Jakarta 12710
Indonesia

Tel: 62 21 520 6574
Fax: 62 21 520 6574
Telex: 62 875 PDII IA

International

International Organization for Standardization (ISO)
1, rue de Varembe
Case postale 56
CH-1211 Geneve 20
Switzerland

Tel: 41 22 749 0111
Fax: 41 22 7333430
Telex: 412205 iso ch

ISO (the International Organization for Standardization) is a worldwide federation of national standards bodies, at present comprising 90 members, one in each country. The object of ISO is to promote the development of standardization and related activities in the world with a view to facilitating international exchange of goods and services, and to developing co-operation in the sphere of intellectual, scientific, technological and economic activity. The results of ISO technical work are published as International Standards.

These standards and designations are prefixed with the letters ISO. Standards appear as a numerical code, and the designations as an alphanumeric code relating to the composition of the metal or alloy.

Example: ISO 3522; ISO AlMn1Cu; ISO AlZn6MgCu

Israel

Standards Institution of Israel (SII)
42 Chaim Levanon Street
Tel Aviv 69977
Israel

Tel: 972 3 64 65 154
Fax: 972 3 64 19 683
Telex: 35508 SIIT IL

The Standards Institution of Israel (SII) is Israel's official body for the preparation and publication of Israel Standards, of which there are over 1600. In its more than 65 years of operation, SII has established itself as a factor in promoting the quality of consumer and industrial goods, and building products. SII's laboratories are responsible for testing a product's compliance with Israel Standards, and its eligibility to receive the Institute's Certification Mark. The Quality and Certification Division prepares Quality Assurance manuals, provides follow-up supervision of quality control systems in those factories which received the Institute's Certification Mark and certifies companies according to ISO 9000.

Italy

Ente Nazionale Italiano di Unificazione (UNI)
Via Battistotti Sassi, 11
20133 Milano
Italy

Tel: 39 2 700241
Fax: 39 2 70 105 992
       39 2 70 106 106
Telex: 312481 UNI I

UNI, The Italian National Standards Body, was founded in 1921, and is a member of both ISO (International Organisation of Standardisation) and CEN (European Committee for Standardisation).

Italian standards are preceded by the upper case letters UNI and followed by an alphanumeric code.

When UNI takes over an international standard the upper case letter UNI is followed by ISO or EN and by an alphanumeric code.

Example: UNI 3159, UNI ISO 9000, UNI EN 29000

Japan

Japanese Industrial Standards Committee (JISC)
C/O Standards Department
Agency of Industrial Science & Technology
Ministry of International Trade & Industry
1-3-1, Kasumigasoki,
Chiyoda-ku, Tokyo 100
Japan

Tel: 81 3 3501 9295/6
Fax: 81 3 3580 1418
Telex: 02 42 42 45 jsatyo j

The Japanese Industrial Standards Committee (JISC) issues standards that cover industrial or mineral products with the exception of those regulated by their own special standards organizations. The standards are divided into 17 divisions and are used both by commercial and government organizations involved in design engineering, quality assurance, research and development, construction, testing and maintenance.

JIS standards begin with the upper case letters JIS and are followed by an upper case letter which designates the standard's division. This is then followed by a space and a series of digits.

Example: JIS G 3311; JIS S 20CK

Japan

Japanese Standards Association (JSA)
1-24-4, Akasaka
Minato-ku
Tokyo 107
Japan
    Tel: 81 3 3583 8001
    Fax: 81 3 3580 1418

Korea

Korean Bureau of Standards,
Industrial Advancement Administration (KBS)
2, Chungang-dong Kwachon-city
Kyonggi-Do 427-010
Republic of Korea

    Tel: +82 2 503 79 38
    Fax: +82 2 503 79 41
    Telex: 2 84 56 fincen k

The Korean Bureau of Standards is responsible for industrial standardization. It has several divisions including Standards Planning, International Standards (established March 1991 for cooperation with international organizations), and Metals and Materials.

As of 1991 there were 829 industrial standards dealing with metals. The codes used for KS standards consist of KS (Korean Industrial Standards), D (metals division) and a 4-digit number.

Mexico

Direccion General de Normas (DGN)
Calle Puente de Tecamachalco No 6
Lomas de Tecamachalco
Seccion Fuentes
Naucalpan de Juarez
53 950 Mexico

    Tel: 52 5 520 84 94
    Fax: 52 5 540 51 53
    Telex: 177 58 40 imceme

The General Directorate of Standards (DGN) issues national standards for the country. Mexican standards begin with the upper case letters NOM (Normas Oficiales Mexicanas). The code that follows consists of an upper case letter which denotes the standard's classification, followed by a hyphen and a number.

Example: NOM C-189

Netherlands

Nederlands Normalisatie-Instituut (NNI)
Kalfjeslaan 2
Postbus 5059
2600 GB Delft
The Netherlands

Tel: 31 15 69 03 90
Fax: 31 15 69 01 90
Telex: 3 81 44 nni nl

The Netherlands Normalization Institute (NNI) is composed of approximately 3000 individual firms and companies, and 200 various organizations. This association helps prepare Dutch standards and cooperates in the development of international standardization.

Dutch standards are prefixed by the letters NEN and are followed by a numerical code.

Example: NEN 213; NEN 3077

New Zealand

Standards Association of New Zealand (SANZ)
Private Bag
Wellington
New Zealand

Tel: 64 4 3842108
Fax: 64 4 384398

SANZ is the national standards body for New Zealand, is a member of the International Standards Organization (ISO) and International Electrotechnical Commission (IEC). It publishes New Zealand Standards (NZS) and joint Australian/New Zealand Standards.

Norway

Norwegian Standards Association (NSF)
P.O. Box 7072 Homansbyen
N-0306 Oslo
Norway

Tel: 47 2 46 60 94
Fax: 47 2 46 44 57
Telex: 1 90 50 nsf n

The Norwegian Standards Association (NSF) is the principal organization for standardization in Norway. The standardizing bodies affiliated with NSF are as follows: The Norwegian Council for Building Standardization - NBR, Norwegian Engineering Industries Standardization Centre - NVS, The Norwegian Electrotechnical Committee - NEK, General Standardization - Norsk Allmenstandardisering NAS, Norwegian Telecommunication Regulatory Authority.

The Norwegian Standards Association (NSF) is the national member of ISO and CEN and the body responsible for the approval and publishing of all Norwegian Standards (Norsk Standard - NS).

Example: NS 824; NS 6097; NS 17570

Pan American

Pan American Standards Commission (COPANT)
Avenida Andres Bello-Torre Fondo Comun
Piso 11
Caracas 1050
Venezuela

Tel: 58 2 5742941
Fax: 58 2 5741312 and
        58 2 922791
Telex: 24235 MINFO VC and
        28735 CNBCA VC

The Pan American Standards Commission (COPANT) is comprised of national standards bodies of 18 countries (from the United States and many Latin American countries). COPANT is the Regional Standards Organization for America for the iron and steel industry and other areas such as foods and agricultural products, plastics, quality assurance, etc. For its designations the acronym COPANT in upper case letters precedes the numeric code and the year of its approbation.

Example: COPANT 1590-1992

Poland

Polish Committee for Standardization
Measures and Quality Control (PKNiM)
el. Elektorania 2
00-139 Warszawa
Poland

Tel: 48 22 20 02 41
Fax: 48 22 20 83 78
Telex: 813642

A national agency, the Polish Committee for Standardization issues standards for that country. They are prefixed with the upper case letters PN. The designations or standards may appear in a number of ways.

Example: PN-79/H-88026

Portugal

Instituto Postugues da Qualidade (IPQ)
Rua Jose Estevao 83-A
1199 Lisboa codex
Portugal

Tel: 351 1 52 39 78
        351 1 52 37 35
        351 1 52 37 59
Fax: 351 1 53 00 33
Telex: 13042 QUALIT P

The Portuguese Institute for Quality (IPQ) (cx-DGQ) is the national body that manages and develops the National Quality Management System (SNGQ), the legal framework for Quality matters in Portugal. IPQ is the organization responsible for certification, standardization, and metrology activities in Portugal.

Example: DGQ R 001 FC 10; DGQ NP 968 C 720

Romania

Romanian Institute for Standards
13 Rue Jean-Louie Calderon
70201 Bucharest 2
Romania

Tel: 40 0 11 14 40
Fax: 40 0 12 08 23
Telex: 1 13 12 irs ro

Russia

Gosudarstvennyi Komitet Standartow (GOST)
USSR State Committee for Standards
Lenisky, Prospekt 9
Moskva 117049
Russia

Tel: 7 095 236 40 44
Fax: 7 095 236 82 09
Telex: 411378 gost su

State standards for Russia number more than 23,000 and cover most areas of commerce, industry, agriculture and public health. The standards are defined within groups, i.e. mining minerals, petroleum products, metals and metallic products, etc. The standards are prefaced with the upper case letters GOST and are followed by a numerical code.

Example: GOST 13819; GOST 5.1491; GOST 22974.9

Saudi Arabia

Saudi Arabian Standards Organization (SASO)
P.O. Box 3437
Riyadh 11471
Saudi Arabia

Tel: 996 1 479 30 46
Fax: 966 1 479 30 63
Telex: 40 16 10 saso sj

The Kingdom of Saudi Arabia established a national standards organization by Royal Decree No. M/10 dated 03/03/1392 (April 16, 1972). SASO is the Saudi organization responsible for all of the activities related to standards and measurements, including the formulation, adoption, publication and distribution of national standards for all commodities and products as well as metrology, symbols, definitions of commodities and products, methods of sampling and testing and any other assignment approved by the Board of Directors. Participating in and cooperating with the Arab, regional and international standards organizations, including ISO and IEC.

South Africa

South African Bureau of Standards (SABS)
Private Bag X191
Pretoria
0001 Republic of South Africa

Tel: 27 12 428 79 11
Fax: 27 12 344 15 68
Telex: 32 13 08 sa

The South African Bureau of Standards (SABS) was officially established by the South African government in 1945, although work in the area of standardization by other organizations began in the early 1900's. The number of the standard is preceded by the letters SABS and followed by the numeric and alphanumeric material type or grade designation.

Example: SABS 407 Type 1; SABS 1431 Grade 300WA; SABS 1465-2 Grade W4

Spain

Asociacion Espanola de Normalizacion y Certificacion (AENOR)
Calle Fernandez de la Hoz, 52
28010 Madrid
Spain

Tel: 34 1 410 48 51
Fax: 34 1 410 49 76
Telex: 4 65 45 unor e

The Spanish Association for Standardization and Certification (AENOR) is an independent organization of a private nature, set up to carry out Standardization and Certification activities, as a tool to improve the quality and competitiveness of products and services. AENOR is designated as a recognized body to develop Standardization and Certification (S+C) activities in Spain.

The designations begin with the letters UNE, representing the Spanish words *une normal Espanola*.

Sweden

Standardiseringskommissionen i Sverige (SIS)
Box 3295
103 66 Stockholm
Sweden

Tel: 46 8 613 52 00
Fax: 46 8 11 70 35
Telex: 17453 sis s

The Swedish Standards Institution (SIS) was founded in 1922 and has a membership of approximately 29 organizations. SIS is a member of the ISO and European Committee for Standardization. SIS is the central body with overall responsibility for standardization in Sweden. SIS is responsible for approval of Swedish standards. SIS is also responsible for publishing, marketing and selling of Swedish standards and for information, promotion and publicity of Swedish standardization.

There are nine independent standardizing bodies affiliated to SIS. Each of them is responsible for drafting standards within its field of activity. Within the field of steel and iron industry there are three standardizing bodies: Materialnormcentralen, MNC (Materials Standards Institution); Sveriges Mekanstandardisering, SMS (Swedish Mechanical Standards Institution); and Allmanna Standardiseringsgruppen, SIS-STG (General Standards Group) (Corrosion etc).

All standards begin with the prefix SS or, if the standard was written prior to 1978, SIS.

Example: SS 11 21 19; SIS 14 01 00

Switzerland

Swiss Association for Standardization (SNV)
Kirchenweg 4
CH-8032 Zurich
Switzerland

Tel: 41 1 384 47 47
Fax: 41 1 384 47 74
Telex: 755931 SNV CH

Thailand

Thai Industrial Standards Institute (TISI)
Ministry of Industry
Rama VI Street
Bangkok 10400
Thailand

Tel: 66 2 2461174
Fax: 66 2 2464327 or 66 2 2478741
Telex: 84375 MINIDUS TH

Turkey

Turk Standardlari Enstitusu (TSE)
Necatibey Caddesi 112
Bakanliklar, Ankara
Turkey

Tel: 90 4 417 83 30
Fax: 90 4 425 43 99
Telex: 42 047 TSE-TR

Founded in 1960, the Turkish Standards Institution (TSE) is a government agency dedicated to the preparation and publication of standards. It is also a member of the ISO. The prefix for Turkish standards are the letters TS. These are followed by a code number, or, in the case of a designation, an alphanumeric code.

Example: TS 2276; TS Mg-Al6Zn1

United Kingdom

British Standards Institution (BSI)
2 Park Street
London W1A 2BS
England

Tel: 44 01 629 9000
Fax: 44 01 629 0506
Telex: 269933 BSILON G

The British Standards Institution (BSI) develops and publishes standards that are used extensively by exporters and importers. They are used both in government and industry by those who are involved in engineering, designing, production, testing and construction. The letters BS precede the standard's numerical code and may also include the alloy's designation.

Example: BS 3100; BS EN 10083-1

USA

Aerospace Materials Specifications (AMS)
Metric Aerospace Materials Specifications (MAM)
SAE International
400 Commonwealth Drive
Warrendale, PA 15096-0001
USA

Tel: 412 776 4841
Fax: 412 776 0002/5760
Telex: 866 355

Aerospace Materials Specifications (AMS) and Metric Aerospace Materials Specifications (MAM) are published by SAE International. AMS and MAM designations pertain to materials intended for aerospace applications; the specifications typically include mechanical property requirements significantly more severe than those for nonaerospace applications. Processing requirements are common in AMS steels. These specifications are generally used for procurement purposes.

Example: AMS 5356, AMS 5598B, MAM 5598

USA

American Iron and Steel Institute (AISI)
1101 17th Street NW
13th Floor
Washington, DC 20036
USA

Tel: 202 452 7100
Fax: 202 463 6573

The American Iron and Steel Institute (AISI) is not a material specification writing body, although many times steels are referred to as AISI Standard steels. The steels are actually part of a designation system that refers only to the chemical composition ranges and limits of the different steels. AISI designations are reported in the same manner as the SAE steel designations, except AISI is placed in front of the code.

The most widely used system for designating carbon and alloy steels in the United States is that of the American Iron and Steel Institute (AISI) and the Society of Automotive Engineers (SAE). Although they are two separate systems, they are nearly identical and are carefully coordinated by the two groups. In this joint system, a particular designation implies the same limits and ranges of chemical composition for both an AISI steel and the corresponding SAE steel. The differences in listings occur as a result of differences in determining eligibility for listing. AISI uses production tonnage as the basis for including a steel. SAE includes a steel if it is used in significant quantity by at least two users or if it has unique engineering characteristics. The fact that a particular steel is listed by AISI or SAE implies only that it has been produced in appreciable quantity. It does not imply that other grades are unavailable, nor does it imply that any particular steel producer makes all of the listed grades.

AISI designations and standard practices are not specifications. The SAE designations are published in the annual SAE handbook under various SAE standards. These standards are comprised entirely of listings of SAE designations and the limits and ranges of chemical composition defined by these designations. Either designation contains only a portion of the information necessary to describe properly a steel product for procurement purposes.

Example: Refer to the example under SAE steel designations.

USA

American National Standards Institute (ANSI)
11 W. 42nd Street
13th Floor
New York, NY 10036
USA

Tel: 212 642 4900
Fax: 212 398 0023

The American National Standards Institute (ANSI) standards are used widely throughout industry. They cover a tremendous variety of items, from architectural products to consumer goods to nuclear safety standards. The Institute is the coordinator of the United States voluntary standards system and assists participants in the voluntary system to reach agreement on standards needs and priorities; arranging for competent organizations to undertake standards development work; providing fair and effective procedures for standards development; and resolving conflicts and preventing duplication of efforts.

An ANSI standard begins with the prefix ANSI. This group is followed by an alphanumeric code which begins with an upper case letter that is followed by 1 to 3 digits. These groups are then followed by additional digits that are separated by decimal points. ANSI standards can also have a standards developer's acronym in the title.

Example: ANSI H35.2; ANSI A156.2; ANSI B18.2.3.6M; ANSI/ASME NQA-2-1989; ANSI/EIA 534-1988; ANSI/NFPA 170-1991

USA

American Petroleum Institute (API)
Publications and Distribution Section
1220 L Street NW
Washington, DC 20005
USA

Tel: 202 682 8000
Fax: 202 962 4776

The American Petroleum Institute (API) fosters the development of standards, codes, and safe practices within the petroleum industries. These standards and codes are used by persons involved in the engineering, production, transportation, handling, and use of petroleum products. The API Standards appear with the letters API before the specifications.

Example: API Spec 5AC; API Spec 5L

USA

American Society for Testing and Materials (ASTM)
1916 Race Street
Philadelphia, PA 19103
USA

Tel: 215 299 5400
Fax: 215 977 9679

The American Society for Testing and Materials (ASTM), founded in 1898, is a scientific and technical organization formed for the development of standards on characteristics and performance of materials, products, systems, and services. The organization issues the most widely used—in the United States—standard specifications for steel products, many of which are complete and generally adequate for procurement purposes. These frequently apply to specific products, which are generally oriented toward the performance of the fabricated end product.

ASTM is the world's largest source of voluntary "consensus" standards. That is, its documents represent a consensus drawn from producers, specifiers, fabricators, and users of steel mill products. In many cases, the dimensions, tolerances, limits and restrictions in the ASTM specifications are the same as corresponding items of the standard practices in the AISI Steel Products Manuals.

Many of the ASTM specifications have been adopted by the American Society of Mechanical Engineers (ASME) with little or no modification; ASME uses the prefix "S" along with the ASTM designation for these specifications. For example, ASME SA-213 and ASTM A213 are identical.

All ASTM standards begin with the prefix ASTM, followed by the actual standard code number.

Example: ASTM A311; ASTM A372 Class V Type B; ASTM A723 Grade 1 Class 1; ASTM A336 Grade F31

USA

American Society of Mechanical Engineers (ASME)
Codes and Standards Department
345 East 47th Street
New York, NY 10017
USA

Tel: 212 605 3333
Fax: 212 605 8750

The American Society of Mechanical Engineers (ASME) standards are used by personnel in research, testing, and design of power-producing machines such as internal combustion engines, steam and gas turbines, and jet and rocket engines. They are also used for the design and development of power-using machines such as refrigeration and air-conditioning equipment, elevators, machine tools, printing presses and steel-rolling mills. ASME committees report to five different boards covering the following areas respectively; safety, nuclear, pressure technology, dimensional standards and performance test codes. The upper case letters ASME appear at the left of the specification followed by an alphanumeric code. When referenced by other ASME standards, standards dated prior to 1989 which carry ANSI in their designation should be shown as the current ones (i.e. ASME followed by the alphanumeric designation given by ANSI) followed by the title and the ANSI designation in parentheses.

Example: ASME B16.14-1991; ASME SA194

USA

American Welding Society (AWS)
550 N.W. Lejeune Road
P.O. Box 351040
Miami, FL 33135
USA

Tel: 305 443 9353 or
       800 443 9353
Fax: 305 443 7559
Telex: 519245

The American Welding Society (AWS) standards are used to support welding design, fabrication, testing, quality assurance and other related joining functions found in shipbuilding (design/construction), heavy construction, and a wide variety of other industries.

These standards always begin with the upper case letters AWS.

Example: AWS A5.24; AWS C5.7; AWS B4.0

USA

Defense Printing Service
700 Robbins Avenue
Bldg. 4, Section D
Attn: Standardization Document Order Desk
Philadelphia, PA 19111-5094
USA

Tel: 215 697 2179 or
       215 697 2667
Fax: 215 697 2978

The Department of Defense Single Stock Point (DODSSP) was created to centralize control and distribution, and provide access to extensive technical information within the collection of Military and Federal Specifications and Standards and related documents produced or adopted by the DoD. The DODSSP mission was assumed by the Defense Printing Service in October 1990.

Military specifications (MIL) are issued by the United States Department of Defense (DOD) to define materials, products, or services used only or predominantly by military entities. Military standards provide procedures for design, manufacturing, and testing, rather than giving a particular material description.

All military specifications begin with the upper case letters MIL. The actual specification that follows begins with an upper case code letter that represents the first letter of the title for the item, followed immediately by a hyphen and then the serial number or digits.

Federal (QQ) specifications and standards are similar to the military, except they are issued by the General Services Administration (GSA) and are primarily for use by federal agencies. Their use, however, is now acceptable to the United States military establishment when there are no separate MIL specifications available.

Federal specifications begin with the upper case letters QQ followed by the code numbers and letters.

Example: MIL S-862 (military); FED QQ-S-763 (federal)

Both military and federal standards and specifications can be obtained through this address or by faxing the request on company letterhead.

USA

National Center for Standards and Certification Information
National Institute of Standards & Technology (NIST)
TRF Building, Room A163
Gaithersburg, MD 20899
USA

Tel: 301 974 4040 ext. 4038, 4036
Fax: 301 926 1559

The National Institute of Standards & Technology (NIST) is the centralized reference repository within the United States for the national standards of the world. Although the organization issues no standards or specifications, it attempts to maintain the most current copies of those standards issued by other organizations both inside and outside of the United States.

Two publications relating to standards, Directory of International and Regional Organizations Conducting Standards-Related Activities and Standards Activities of Organizations in the United States, are available from NIST.

USA

SAE, International (SAE)
400 Commonwealth Drive
Warrendale, PA 15096-0001
USA

Tel: 412 776 4841
Fax: 412 776 5760
Telex: 866 355 SAE IN WNDE

The Society of Automotive Engineers (SAE) standards are used primarily by designers, manufacturers, and maintenance personnel in the automotive and aerospace industries. These standards are also a useful and effective series for the metals, plastics, rubber, chemical, and fastener industries in their standardization efforts. Automotive SAE standards begin with the upper case letters SAE. Immediately after this prefix the letter J appears, and it is followed by a numerical code. Prefixes for other standards vary, such as AS, AIR, and ARP with a numerical code.

Example: SAE J450; SAE J993b; ARP 111

USA

Steel Founders' Society of America (SFSA)
Cast Metals Federation Bldg.
455 State Street
Des Plaines, IL 60016
USA

Tel: 708 299 9160
Fax: 708 299 3105

The Alloy Casting Institute (ACI) established a standards-designating system for cast high-alloy stainless steels. Originally a separate organization, it was absorbed by the Steel Founders' Society of America in 1970. Designations consist of an alpha-numeric code.

Example: CF-8M; CF-3MA; CB-7Cu-1; CA-6NM-B

Venezuela

Comision Venezolana de Normas Industriales (COVENIN)
Avda. Andres Bello
Edf. Torre Fondo Comun
Piso 11
Caracas 1050
Venezuela

Tel: +58 2 575 22 98
Fax: +58 2 574 13 12
Telex: 2 42 35 minfo vc

COVENIN, created in 1958, is the Organization in charge of directing, planning and coordinating all the standardization activities in Venezuela. A series of measures of legal order culminated with the law on "Technical Standards and Quality Control" on December 31, 1979. This law represents the consolidation, from the legal view point, of the standardization and quality control process in Venezuela.

COVENIN coordinates the elaboration of the technical standards through committees and sub-committees dependent on COVENIN, which include: Construction, Petroleum and Derivatives, Automotive, Ferrous Materials, Non-Ferrous Materials, Dentistry, Electricity and Electronics, Chemistry, Metrology, Documentation, Containers and Packing, Maintenance, Mechanics, Non-Destructive Testing, and Quality. COVENIN has approved a total of 2822 technical standards; 300 of them having been declared compulsory by the National Government.

Yugoslavia

Jugoslovenski zavodza Standardizeciju (SZS)
Sonodana Penozica Krcunabr. 35
Post. Pregr. 933
11000 Beograd
Yugoslavia

Tel: 38 11 64 40 66
Fax: 38 11 235 1036
Telex: 1 20 89 jus yu

The Yugoslavian Standardization Institute (SZS) was founded in 1946 and is concerned with the adoption and application of standards, technical norms for product quality and services, and regulations covered by legislation. Yugoslavian standards begin with the prefix JUS, which is followed by an alphanumeric code. The first letter of the code denotes the section under which the standard is classified. Most standards relating to metallurgy are in section C.

Example: JUS C.AO.003; JUS C.K6.150; JUS C.T3.005

Colleges and Universities in the United States and Canada with Metallurgy / Materials Science Faculties

University of Alabama

Dept. of Metallurgical and Materials Engineering
Tuscaloosa AL 35487-0202
TEL: (205) 348-1740
FAX: (205) 348-8573

B.S., M.S. Met. Eng., Ph.D. Met. Eng., Mat. Sc.

University of Alabama at Birmingham

Dept. of Materials Science and Engineering
UAB Station
Birmingham AL 35294
TEL: (205) 934-8450
FAX: (205) 934-8485

B.S., M.S., Ph.D.

University of Alberta

Mining, Metallurgical and Petroleum Engineering
Edmonton ALB T6G 2G6 Canada
TEL: (403) 492-3337
FAX: (403) 492-7219

B.S., M., M.S., Ph.D. Met. Eng., B.S., M.S., Ph.D. Mnl. Eng.,
M. Eng., M., M.S., Ph.D. Met.

Arizona State University

Dept. of Chemical, Bio and Materials Science Engineering
Tempe AZ 85287
TEL: (602) 965-3313
FAX: (602) 965-8296

B.S., M.S., Ph.D.

University of Arizona

Dept. of Materials Science and Engineering
Tucson AZ 85721
TEL: (602) 621-6070
FAX: (602) 621-8159

B.S., M.S., Ph.D. Mat. Sc., Eng.

Auburn University

Dept. of Mechanical Engineering
Auburn AL 36849-5351
TEL: (205) 844-3326
FAX: (205) 844-3307

B.S., M.S., Ph.D. Mat. Eng.

University of British Columbia

Dept. of Metals and Materials Engineering
309-6350 Stores Road
Vancouver BC V6T 1W5 Canada
TEL: (604) 228-2676
FAX: (604) 228-3619

B.A. Sc., M.A. Sc., M., Ph.D. Met. Mat. Eng., M. Sc. Mat.,
Met., M. Eng.

Brown University

Division of Engineering
Providence RI 02912
TEL: (401) 863-2276
FAX: (401) 863-1157

Undergraduate: Mat. Eng. Concentration; Graduate: Mat. Sc.

University of California, Berkeley

Dept. of Materials Science and Mineral Engineering
Berkeley CA 94720
TEL: (415) 642-3801
FAX: (415) 643-8426

B.S., M.S., Ph.D., D. Eng.

University of California, Davis

Dept. of Mechanical, Aeronautical and Materials Engineering
Davis CA 95616-5294
TEL: (916) 752-0580
FAX: (916) 752-4158

B.S., M.S., Ph.D., D. Eng.

University of California, Los Angeles
Materials Science and Engineering Dept.
Los Angeles CA 90024
TEL: (213) 825-5534
FAX: (213) 206-7353

M.S., Ph.D. Phys. Met., Cer., Sc. of Mat.

University of California, San Diego
Materials Science Program
La Jolla CA 92093-0411
TEL: (619) 534-7715
FAX: (619) 634-7080

M.S., Ph.D. Mat. Sc.

California Institute of Technology
Dept. of Materials Science MS 138-78
Keck Laboratory
Pasadena CA 91125
TEL: (818) 356-4411
FAX: (818) 795-1547

B.S. Eng., Appl. Sc., M.S., Ph.D. Mat. Sc.

California Polytechnic State University
Materials Engineering
San Luis Obispo CA 93407
TEL: (805) 756-2568
FAX: (605) 756-6503

B.S. Mat. Eng.

Carnegie Mellon University
Dept. of Metallurgical Engineering and Materials Science
Pittsburgh PA 15213
TEL: (412) 268-2700
FAX: (412) 268-6421

B.S., M.S., M.E., Ph.D. Met. Eng., Mat. Sc.

Case Western Reserve University
Dept. of Materials Science and Engineering
Cleveland OH 44108
TEL: (216) 368-4230
FAX: (216) 368-3209

B.S., M.S., Ph.D.

University of Cincinnati
Dept. of Materials Science and Engineering
Cincinnati OH 45221
TEL: (513) 556-3096
FAX: (513) 556-2569

B.S. Mat. Eng., M.S., Ph.D. Mat. Sc., Met. Eng.

The Cleveland State University
Dept. of Chemical Engineering
Cleveland OH 44115
TEL: (216) 687-2569
FAX: (216) 687-9366

M.S., D. Eng.

Colorado School of Mines
Dept. of Metallurgical and Materials Engineering
Golden CO 80401
TEL: (303) 273-3770
FAX: (303) 273-3795

B.S., M.S., M.E., Ph.D. Met. Eng.

Columbia University
Division of Metallurgy and Materials Engineering
Henry Krumb School of Mines
New York NY 10027
TEL: (212) 854-8008
    (212) 854-2905
FAX: (212) 749-0397

B.S., M.S., Ph.D., Sc.D.

University of Connecticut
Dept. of Metallurgy
Storrs CT 06268
TEL: (203) 486-4620
FAX: (203) 486-4745

M.S., Ph.D. Met.

Cornell University
Dept. of Materials Science and Engineering
Bard Hall
Ithaca NY 14853-1501
TEL: (607) 255-9617
FAX: (607) 225-2365

B.S., M. Eng. (Mat.), M.S., Ph.D.

Dartmouth College
Thayer School of Engineering
Hanover NH 03755
TEL: (603) 646-2230
FAX: (603) 646-3856

A.B., B.E., M.S., M.E., Ph.D., D.E.

University of Dayton
Materials Engineering Graduate Program
School of Engineering
Dayton OH 45469
TEL: (513) 229-2627
FAX: (513) 229-3433

M.S., D.E., Ph.D. Mat. Eng.

University of Delaware
Materials Science Program
College of Engineering
Newark DE 19716
TEL: (302) 451-2062
FAX: (302) 451-1048

M.S., Ph.D. Mat. Sc.

Drexel University
Dept. of Materials Engineering
Philadelphia PA 19104
TEL: (215) 895-2323
FAX: (215) 895-4929

B.S., M.S., Ph.D. Mat. Eng.

Duke University
Dept. of Mechanical Engineering and Material Science
Durham NC 27706
TEL: (919) 684-2832
FAX: (919) 684-4860

Ecole Polytechnique
Dept. of Metallurgical Engineering
C.P. 6079, Succursale A
Montreal, QUE H3C 3A7 Canada
TEL: (514) 340-4787
FAX: (514) 340-4468

B.A. Sc., M.A. Sc., M. Eng., Ph.D.

University of Florida
Dept. of Material Science and Engineering
Gainesville FL 32611
TEL: (904) 392-1454
FAX: (904) 392-6359

B.S., M.S., Ph.D.

Georgia Institute of Technology
School of Materials Engineering
Atlanta GA 30332-0245
TEL: (404) 894-2850
FAX: (404) 853-9140

B.S. Mat. Eng., B.S., M.S., Ph.D. Met., Cer. Eng.

The Hartford Graduate Center
School of Engineering and Science
Hartford CT 06120
TEL: (203) 548-2450
FAX: (203) 548-2473

M.S. Met.

Harvard University
Division of Applied Sciences
Cambridge MA 02138
TEL: (617) 495-2833
FAX: (617) 495-9837

Ph.D.

University of Idaho
Dept. of Metallurgy and Mining Engineering
Moscow ID 83843
TEL: (208) 885-6376
FAX: (208) 885-6911

B.S., M.S., Ph.D.

University of Illinois at Chicago
Dept. of Civil Engineering, Mechanics and Metallurgy
P.O. Box 4348
Chicago IL 60680
TEL: (312) 996-3428
FAX: (312) 996-7149

B.S., M.S., Ph.D.

University of Illinois, Urbana-Champaign
Dept. of Materials Science and Engineering
Urbana IL 61801
TEL: (217) 333-1441
FAX: (217) 333-2736

B.S., M.S., Ph.D. Cer. Eng., Met. Eng., Mat. Sc. Eng., M.S., Ph.D. Cer. Sc.

Illinois Institute of Technology
Dept. of Metallurgical and Materials Engineering
Chicago IL 60616
TEL: (312) 567-3050
FAX: (312) 567-8875

B.S. Met. Eng., M.S. Mfg. Eng., M.S., Ph.D. Met., Mat. Eng.

Iowa State University of Science and Technology
Dept. of Materials Science and Engineering
Ames IA 50011
TEL: (515) 294-1214

B.S., M.S., Ph.D. Cer. Eng., B.S. Met. Eng., M.S., Ph.D. Met.

The Johns Hopkins University
Materials Science and Engineering
Baltimore MD 21218
TEL: (301) 338-6145
FAX: (301) 338-5293

B.S., M.S.E., Ph.D.

University of Kentucky
Dept. of Materials Science and Engineering
Lexington KY 40506-0046
TEL: (606) 257-8884
FAX: (606) 258-1929

B.S., M.S., Ph.D.

Laval University
Dept. of Mining and Metallurgy
Quebec G1K 7P4 Canada
TEL: (418) 656-2160
FAX: (418) 616-5343

B. Sc., M.S., D.Sc.

Lehigh University
Dept. of Materials Science and Engineering
Whitaker Laboratory No. 5
Bethlehem PA 18015
TEL: (215) 758-4220
FAX: (215) 758-4244

B.S., M.S., Ph.D. Mat. Sc. Eng.

Louisiana State University
Dept. of Mechanical Engineering
Materials Engineering Program
Baton Rouge LA 70803-6413
TEL: (504) 388-5802
FAX: (504) 388-5990

B.S., M.S., Ph.D., D. Eng. Sc.

Marquette University
Dept. of Mechanical Engineering and Industrial Engineering
Milwaukee WI 53233
TEL: (414) 288-7259
FAX: (414) 288-7082

B.S. Mech. Eng., Elec. Eng., M.S., Ph.D. Met., Mat. Sc., Elec. Eng.

University of Maryland
Engineering Materials Program
Dept. of Materials and Nuclear Engineering
College Park MD 20742
TEL: (301) 405-5211
FAX: (301) 314-9467

M.S., Ph.D.

University of Massachusetts
Dept. of Mechanical Engineering
Amherst MA 01002
TEL: (413) 545-2505
FAX: (413) 545-0724

M.S., Ph.D. Mech. Eng., Mat.

Massachusetts Institute of Technology
Dept. of Materials Science and Engineering
Cambridge MA 02139
TEL: (617) 253-3300
FAX: (617) 258-6886

S.B. Mat. Sc. Eng., Mat. Eng., Met. Eng., S.M. Sc.D., Ph.D. Cer., E. Mat., Mat. Eng., Mat. Sc., Met., Polymerics

McGill University
Dept. of Mining Metallurgical Engineering
Montreal QUE Canada
TEL: (514) 398-4350
FAX: (514) 398-4492

B.E., M.E., Ph.D. Met. Eng., Ph.D. Met.

McMaster University
Dept. of Materials Science and Engineering
Hamilton ONT Canada
TEL: (416) 525-4140 ext. 4295/4293
FAX: (416) 528-9295

M.E., M.S. Met., M.S., Ph.D. Mat. Sc.

University of Michigan
Dept. of Materials Science and Engineering
Ann Arbor MI 48109-2136
TEL: (313) 763-4970
FAX: (313) 763-4788

B.S., M.S., Ph.D. Mat. Sc., Eng.

Michigan State University
Dept. of Metallurgy, Mechanics and Material Science
East Lansing MI 48824-1226
TEL: (517) 355-5141
FAX: (517) 353-9842

B.S., M.S., Ph.D. Mat. Sc., Mech., M.S., Ph.D. Met.

Michigan Technological University
Dept. of Metallurgical and Materials Engineering
Houghton MI 49931
TEL: (906) 487-2630
FAX: (906) 487-2934

B.S., M.S., Ph.D. Met. Eng.

University of Minnesota
Dept. of Chemical Engineering and Materials Science
Minneapolis MN 55455
TEL: (612) 625-1313
FAX: (612) 626-7246

B.S., M.S., Ph.D. Mat. Sc., Eng.

University of Minnesota
Dept. of Civil and Mineral Engineering
Mineral Resources Research Center
Minneapolis MN 55455
TEL: (612) 625-3344
FAX: (612) 625-1882

B.S., M.S., Ph.D., M. Min. Eng., M.S., Ph.D. Min. Eng.

University of Missouri, Rolla
Metallurgical Engineering
Rolla MO 65401
TEL: (314) 341-4711
FAX: (314) 341-2071

B.S., M.S., Ph.D. Met. Eng.

Montana College of Mineral Science and Technology
Dept. of Metallurgy and Mineral Processing Engineering
Butte MT 59701
TEL: (406) 496-4341
FAX: (406) 496-4133

B.S., M.S. Met. Eng., M.S. Met., B.S., M.S. Min. Proc. Eng.

University of Nebraska
Dept. of Mechanical Engineering
Metallurgical Engineering Program
255 WSEC Lincoln NE 68588-0656
TEL: (402) 472-2375
FAX: (402) 472-1465

M.S. Met. Eng., Ph.D. Chem., Mat. Eng., Eng.

University of Nevada, Reno
Dept. of Chemical and Metallurgical Engineering
Reno NV 89557-0047
TEL: (702) 784-4307
FAX: (702) 784-1766

B.S., M.S., Ph.D. Met. Eng.

New Mexico Institute of Mining and Technology
Dept. of Metallurgy and Materials Engineering
Socorro NM 87801
TEL: (505) 835-5229
FAX: (505) 835-5626

B.S. Met. Eng., B.S., M. Sc., Ph.D. Mat. Eng.

State University of New York
Dept. of Material Science and Engineering
Stony Brook LI NY 11794
TEL: (516) 632-8484
FAX: (516) 632-8052

B.S. Eng. Sc., M.S., Ph.D. Mat. Sc.

State University of New York, Buffalo
Materials Engineering Program
School of Engineering and Applied Sciences
Buffalo NY 14260
TEL: (716) 636-2520
FAX: (716) 636-2495

B.S., M.S., Ph.D.

North Carolina State University
Dept. of Materials Science and Engineering
Raleigh NC 27695-7907
TEL: (919) 737-2377
FAX: (919) 737-7724

B.S., M.S., Ph.D. Mat. Eng.

Northwestern University
Dept. of Materials Science and Engineering
Evanston IL 60208
TEL: (708) 491-3537
     (708) 491-3587
FAX: (708) 491-4133

B.S., M.S., Ph.D. Mat. Sc., Eng.

University of Notre Dame
Center for Materials Science and Engineering
Dept. of Electrical Engineering
Notre Dame IN 46556
TEL: (219) 239-5330
FAX: (219) 239-8007

B.S., M.S., Ph.D. Mat. Sc., Eng.

Technical University of Nova Scotia
Mining and Metallurgical Engineering Department
P.O. Box 1000, Halifax
Nova Scotia B3J 2X4 Canada
TEL: (902) 420-7674
FAX: (902) 425-1037

B. Eng., M.A. Sc.

The Ohio State University
Dept. of Materials Science and Engineering
Columbus OH 43210
TEL: (614) 292-2553
FAX: (614) 292-1537

B.S., M.S., Ph.D. Met. Eng., B.S., M.S., Ph.D. Cer. Eng.

University of Oklahoma
School of Chemical Engineering and Material Science
Norman OK 73019
TEL: (405) 325-5811
FAX: (405) 325-5813

B.S., M.S., Ph.D. Chem. Eng., M.S., Ph.D. Met. Eng.

Oregon Graduate Institute of Science and Technology
Beaverton OR 97006-1999
TEL: (503) 690-1170
FAX: (503) 690-1029

M.S., Ph.D. Mat. Sc., Eng.

The Pennsylvania State University
Dept. of Material Science and Engineering
University Park PA 16802
TEL: (814) 865-0497
     (814) 865-0498
FAX: (814) 865-2917

B.S., M.S., Ph.D. Cer. Sc., Met., Fuel Sc., Polymer Sc.

University of Pennsylvania
Dept. of Materials Science and Engineering
Philadelphia PA 19104-6272
TEL: (215) 898-8337
FAX: (215) 898-8296

B.S., M.S., Ph.D. Mat. Sc., Eng.

University of Pittsburgh
Materials Science and Engineering Dept.
Pittsburgh PA
TEL: (412) 624-9720
FAX: (412) 624-1108

B.S., M.S., Ph.D. Met. Eng., M.S., Ph.D. Mat. Eng.

Polytechnic University
Dept. of Metallurgy and Materials Science
Brooklyn NY 11201
TEL: (718) 260-3250
FAX: (718) 260-3136

B.S., M.S., Ph.D. Met. Eng., M.S. Mat. Sc. Eng., Ph.D. Mat. Sc.

Purdue University
School of Materials Engineering
West Lafayette IN 47907
TEL: (317) 494-4100
FAX: (317) 494-1204

B.S., M.S. Met. Eng., Eng., M.S., Ph.D.

Queen's University
Dept. of Metallurgical Engineering
Kingston ONT K7L 3N6 Canada
TEL: (613) 545-2754
FAX: (613) 545-5610

B. Sc., M. Sc., Ph.D. Met. Eng.

Rensselaer Polytechnic Institute
Materials Engineering Dept.
Troy NY 12180-3590
TEL: (518) 276-6372
FAX: (518) 276-8554

B.S. Mat. Eng., M.E., M.S. Mat., D. Eng., Ph.D.

Rice University
Dept. of Mechanical Engineering and Material Science
P.O. Box 1892
Houston TX 77251
TEL: (713) 527-4993
FAX: (713) 285-5136

M., M.S., Ph.D. Mat. Sc.

University of Rochester
College of Engineering and Applied Science
Rochester NY 14627
TEL: (716) 275-4151
FAX: (716) 256-2509

M.S., Ph.D. Mat. Sc.

Rutgers University
College of Engineering
Dept. of Mechanics and Materials Science
P.O. Box 909
Piscataway NJ 08854
TEL: (908) 932-2245
FAX: (908) 932-5977

M.S, Ph.D.

San Jose State University
Dept. of Materials Engineering
San Jose CA 95192
TEL: (408) 924-4050

B.S., M.S. Mat. Eng.

South Dakota School of Mines and Technology
Dept. of Metallurgical Engineering
Rapid City SD 57701-3995
TEL: (605) 394-2341
FAX: (605) 394-6131

B. Sc., M. Sc. Met. Eng., Ph.D. Mat. Sc., Eng.

University of Southern California
Dept. of Materials Science and Engineering
University Park, Los Angeles CA 90089-0241
TEL: (213) 740-4339
FAX: (213) 740-7797

M.S. Mat. Eng., M.S., Ph.D. Mat. Sc.

Stanford University
Material Science and Engineering
Stanford CA 94305-2205
TEL: (415) 723-2534
FAX: (415) 725-4034

B.S., M.S., Eng. (MS&E), Ph.D.

Stevens Institute of Technology
Dept. of Materials Science and Engineering
Hoboken NJ 07030
TEL: (201) 420-5270
FAX: (201) 963-3017

B.E., B.S., M.S., Ph.D.

Stevens Institute of Technology
Dept. of Chemistry and Chemical Engineering
Hoboken NJ 07030
TEL: (201) 420-5546
FAX: (201) 420-1606

B.E., B.S., M.E., M.S. Chem. Eng., Ph.D.

Syracuse University
Solid State Science and Technology
Syracuse NY 13244-1200
TEL: (315) 443-2359
FAX: (315) 443-4070

M.S., Ph.D. Solid State Sc. and Tech.

The University of Tennessee
Dept. of Materials Science and Engineering
Knoxville TN 37966-2200
TEL: (615) 974-5336
FAX: (615) 974-2669

B.S. Mat. Sc., Eng., M.S. Ph.D. Met. Eng., M.S.,
Ph.D. Polymer Eng.

Texas A & M University
Dept. of Mechanical Engineering
College Station TX 77643
FAX: (409) 845-3081

B.S. Mech. Eng., Mechs. Mat., Ph.D., D. Eng.

The University of Texas, Austin
Materials Science and Engineering Program
Austin TX 76712
TEL: (512) 471-1504
FAX: (512) 471-8727

M.S., Ph.D.

The University of Texas, El Paso
Dept. of Metallurgical and Materials Engineering
El Paso TX 79968
TEL: (915) 747-5468
FAX: (915) 747-5616

B.S., M.S. Met. Eng.

University of Toronto
Dept. of Metallurgy and Materials Science
Toronto ONT M5S 1A4 Canada
TEL: (416) 978-3013
       (416) 978-4429
FAX: (416) 978-4155

B.A. Sc. Mat., Met. Eng., Min. Eng., Geo. Eng., Mat. Sc.,
M. Eng., B.A. Sc., M.A. Sc., Ph.D. Met., Mat. Sc.

University of Utah
Dept. of Metallurgical Engineering
Salt Lake City UT 94112
TEL: (801) 581-6386
FAX: (801) 581-5560

B.S., M.S., M.E., Ph.D. Met.

University of Utah
Dept. of Material Science and Engineering
College of Engineering
304 EMRO Building
Salt Lake City UT 84112
TEL: (801) 581-6863
FAX: (801) 581-4816

B.S., M.S., Ph.D. Mat. Sc. Eng.

Vanderbilt University
Mechanical and Materials Engineering
Box 1593, Station B
Nashville TN 37225
TEL: (615) 343-6868
FAX: (615) 322-7062

M.S., Ph.D.

University of Virginia
Dept. of Materials Science and Engineering
Charlottesville VA 22901
TEL: (804) 924-3264
FAX: (804) 982-2794

M.M.S., M.S., Ph.D. Mat. Sc.

Virginia Polytechnic Institute and State University
Dept. of Materials Engineering
Blacksburg VA 24061
TEL: (703) 231-6640
FAX: (703) 231-8919

B.S., M.S. Mat. Eng., Cer. Eng., Ph.D. Mat. Eng. Sc.

University of Washington
Dept. of Materials Science and Engineering
Seattle WA 98195
TEL: (206) 543-2500
FAX: (206) 543-3100

B.S. Met. Eng., Cer. Eng., M.S. Eng., M.S., Ph.D.

Washington State University
Dept. of Mechanical and Materials Engineering
Pullman WA 99164-2920
TEL: (509) 335-8521

B.S., M.S. Mat. Sc. and Eng., Ph.D. Eng. Sc.

University of Waterloo
Dept. of Mechanical Engineering
Waterloo ONT N2L 3G1 Canada
TEL: (519) 885-1211

M.A. Sc., Ph.D. Mech. Eng., Mat. Eng.

Wayne State University
Dept. of Chemical and Metallurgical Engineering
Detroit MI 48202
TEL: (313) 577-3800
FAX: (313) 577-3881

B.S., M.S., Ph.D. Mat. Eng.

The University of Western Ontario
Dept. of Materials Engineering
Faculty of Engineering Science
London ONT N6A 5B9 Canada
TEL: (519) 661-3757
FAX: (519) 661-3808

B.E. Sc., M.E. Sc., Ph.D.

University of Windsor
Dept. of Mechanical Engineering
Engineering Materials Group
Windsor ONT N9B 3P4 Canada
TEL: (519) 253-4232
FAX: (519) 973-7062

B.A. Sc. Mech. Eng., B.A. Sc., M.A. Sc., Ph.D. Eng. Mat.

University of Wisconsin, Madison
Dept. of Materials Science and Engineering
1509 University Avenue
Madison WI 53706
TEL: (608) 262-1478
FAX: (608) 262-8353

B.S., M.S., Ph.D. Met. Eng., M.S., Ph.D. Mat. Sc.

University of Wisconsin, Milwaukee
Materials Dept.
P.O. Box 784
Milwaukee WI 53201
TEL: (414) 229-5181
FAX: (414) 229-6958

B.S., M.S., M.E., Ph.D. Eng.

Worcester Polytechnic Institute
Dept. of Mechanical Engineering
100 Institute Road
Worcester MA 01609
TEL: (508) 831-5236
FAX: (508) 831-5483

M.S., Ph.D. Mech. Eng., Mat. Eng.

Wright State University
Materials Science and Engineering Program
Dept. of Mechanical and Materials Engineering
Dayton OH 45435
TEL: (513) 873-2476
FAX: (513) 873-3301

B.S., M.S., Mat. Sc. Eng.

Yale University
Council of Engineering
New Haven CT 06520
TEL: (203) 432-4346
FAX: (203) 432-2797

B.S., M.S., Ph.D.

Youngstown State University
Dept. of Materials Engineering
Youngstown OH 44555
TEL: (216) 742-1735
FAX: (216) 742-1998

B.E. Mat. Eng., M.S. Mat. Sc.

Schools in the United States with Faculties of Ceramics

Alfred University
New York State College of Ceramics
Division of Engineering and Science
Alfred NY 14802
TEL: (607) 871-2448
FAX: (607) 871-3469

B.S., M.S. Cer. Eng., Cer. Sc., Glass Sc., Ph.D. Cer.

University of California, Berkeley
Dept. of Materials Science and Mineral Engineering
Berkeley CA 94720
TEL: (415) 642-3801
FAX: (415) 643-8426

B.S., M.S., Ph.D., D. Eng.

University of California, Los Angeles
Dept. of Materials Science and Engineering
Los Angeles CA 90024
TEL: (213) 825-5473
FAX: (213) 206-7353

M.S., Ph.D. Phys. Met., Cer., Sc. of Mat.

Clemson University
Dept. of Ceramic Engineering
Clemson SC 29634-0907
TEL: (803) 656-3093
    (803) 656-2698

B.S., M.S., Ph.D. Cer. Eng.

University of Florida
Ceramics Division
Gainesville FL 32611
TEL: (904) 392-1454
FAX: (904) 392-6359

B.S., M.S., Ph.D.

University of Illinois
Dept. of Material Science and Engineering
Ceramics Division
Urbana IL 61801
TEL: (217) 333-1770
FAX: (217) 244-6917

B.S., M.S., Ph.D. Cer. Eng., M.S., Ph.D. Cer. Sc.

Iowa State University of Science and Technology
Dept. of Materials Science and Engineering
Ames IA 50011
TEL: (515) 294-1214
FAX: (515) 294-9273

B.S., M.S., Ph.D. Cer. Eng.

Lehigh University
Dept. of Materials Science and Engineering
Whitaker Laboratory No. 5
Bethlehem PA 18015
TEL: (215) 758-4227
FAX: (215) 758-4244

B.S., M.S., Ph.D. Mat. Sc. Eng.

Massachusetts Institute of Technology
Ceramics Program
Dept. of Materials Science and Engineering
Cambridge MA 02139
TEL: (617) 253-3300
FAX: (617) 258-6886

S.M., Sc.D., Ph.D. Cer.

The University of Missouri, Rolla
School of Mines and Metallurgy
Dept. of Ceramic Engineering
Rolla MO 65401
TEL: (314) 341-4401
FAX: (314) 341-4192

B.S., M.S., Ph.D. Cer. Eng.

State University of New York, Buffalo
Materials Engineering Program
School of Engineering and Applied Sciences
Buffalo NY 14260
TEL: (716) 636-2520
FAX: (716) 636-2495

B.S., M.S., Ph.D.

The Ohio State University
Dept. of Materials Science and Engineering
Columbus OH 43210
TEL: (614) 292-2553
FAX: (614) 292-1537

B.S., M.S., Ph.D. Cer. Eng.

The Pennyslvania State University
Materials Science and Engineering Department
Ceramic Science and Engineering Program
University Park PA 16802
TEL: (814) 865-4992
FAX: (814) 865-2917

B.S., M.S., Ph.D. Cer. Sc.

Rutgers, The State University
Department of Ceramics
Piscataway NJ 08854
TEL: (201) 932-5700
FAX: (201) 932-3258

B.S., M.S., Ph.D. Cer., Cer. Eng.

University of Utah
Dept. of Materials Science and Engineering
304 EMRO Building
Salt Lake City UT 84112
TEL: (801) 581-6863
FAX: (801) 581-4816

B.S., M.S., Ph.D. Mat. Sc. Eng.

Virginia Polytechnic Institute and State University
Dept. Materials Engineering
Blacksburg VA 24061
TEL: (703) 231-6640
FAX: (703) 231-8919

B.S., M.S. Cer. Eng.

University of Washington
Dept. of Materials Science and Engineering
Seattle WA 98195
TEL: (206) 543-2600
FAX: (206) 543-3100

B.S., M.S., Cer. Eng., M.S. Eng., Ph.D.

Schools in the United States with Faculties of Polymer Science

University of Akron
College of Polymer Science and Engineering
Akron OH 44325
TEL: (216) 972-7500

B.S., M.S., Ph.D.

University of Akron
Dept. of Polymer Science
Whitby Hall
Akron OH 44325
TEL: (216) 972-7542
FAX: (216) 972-5121

B.S., M.S., Ph.D.

University of Akron
Institute of Polymer Science
Auburn Science and Engineering Center
Akron OH 44325
TEL: (216) 972-5110

B.S., M.S., Ph.D.

Case Western Reserve University
Case Institute of Technology
Dept. of Macromolecular Science
University Circle
Cleveland OH 44106
TEL: (216) 368-4172
FAX: (216) 368-4202

B.S., M.S., Ph.D.

University of Connecticut
Polymer Science Program
Institute of Materials Science
Storrs CT 06288
TEL: (203) 486-3582
FAX: (203) 486-4745

M.S., Ph.D.

University of Florida
Dept. of Material Science and Engineering
Gainesville FL 32611
TEL: (904) 392-1454
FAX: (904) 392-6359

B.S., M.S., Ph.D.

University of Illinois, Urbana-Champaign
Polymer Group
Urbana IL 61801
TEL: (217) 333-1440
FAX: (217) 333-2736

B.S., M.S., Ph.D. Mat. Sc. Eng. (Polymer concentration),
B.S. Polymer minor in College of Engineering

Lehigh University
Center for Polymer Science and Engineering Program
Mountaintop Campus Building A
Bethlehem PA 18015
TEL: (215) 758-3590
FAX: (215) 758-5423

M.S., Ph.D. Polymer Sc. Eng.

University of Lowell
Plastics Engineering Dept.
Lowell MA 01854
TEL: (508) 934-3420
FAX: (508) 452-1445

B.S., M.S., D. Eng. (Plast. Eng.), Ph.D. Chem. (Polymer Sc./Plast. Eng.)

University of Massachusetts
Dept. of Polymer Science and Technology
Amherst MA 01003
TEL: (413) 545-0433
FAX: (413) 545-0082

Massachusetts Institute of Technology
Program in Polymer Science and Technology (PPST)
Room 13-5034, 77 Massachusetts Avenue
Cambridge MA 02139
TEL: (617) 258-6175
FAX: (617) 258-7874

S.M., Sc.D., Ph.D. Polymerics

University of Michigan
Macromolecular Science and Engineering Program
Ann Arbor MI 48109
TEL: (313) 763-2316
FAX: (313) 747-0036

M.S., Ph.D.

University of Southern Mississippi
Dept. of Polymer Science
Hattlesburg MS 39406-0076
TEL: (601) 266-4868
FAX: (601) 266-5829

B.S., M.S., Ph.D. Polymer Sc.

State University of New York, Buffalo
Materials Engineering Program
School of Engineering and Applied Sciences
Buffalo NY 14260
TEL: (716) 636-2520

B.S., M.S., Ph.D.

North Dakota State University
Polymers and Coatings Dept.
Fargo ND 58105
TEL: (701) 237-7633
FAX: (701) 237-8831

M.S., Ph.D.

The Pennsylvania State University
Materials Science and Engineering Dept.
Polymer Science Program
University Park PA 16802
TEL: (814) 865-1288
FAX: (814) 865-2917

B.S., M.S., Ph.D. Polymer Sc.

Stevens Institute of Technology
Dept. of Chemistry and Chemical Engineering
Hoboken NY 07030
TEL: (201) 420-5546
FAX: (201) 420-1606

B.E., B.S., M.E., M.S. Chem. Eng., Ph.D.

University of Tennessee
Polymer Science and Engineering Group
TEL: (615) 974-5336
FAX: (615) 974-2669

M.S., Ph.D. Polymer Eng.

University of Utah
Dept. of Materials Science and Engineering
304 EMRO Building
Salt Lake City UT 84112
TEL: (801) 581-6863
FAX: (801) 581-4816

B.S., M.S., Ph.D. Mat. Sc. Eng.